Quantum Mechanics through Problems

This book contains more than 300 problems in quantum mechanics, with accompanying solutions, covering topics that are commonly taught in first-year graduate physics programs. Special care is given to each problem's formulation, with detailed and extensive solutions provided to support understanding. The problems span a range of difficulty, from basic exercises to more challenging applications and extensions of the standard material. Students are required to think critically and incorporate physics and mathematical techniques learned previously or concurrently to solve the more challenging problems. Each chapter begins by framing the particular topic being examined with a short theory section, which sets the context for and motivates the problems that follow. This text is well suited for self-study or as a useful supplement to existing quantum mechanics textbooks for upper-undergraduate and graduate students and their instructors.

Rocco Schiavilla is Professor and Eminent Scholar in the Physics Department of Old Dominion University and Senior Staff Scientist in the Theory Center of the Thomas Jefferson National Accelerator Facility. His research interests are in nuclear theory with over 170 publications in this area. He is a Fellow of the American Physical Society and a recipient of the Excellence in Graduate Teaching Award at Old Dominion University, among other awards.

Quantum Mechanics through Problems

With Complete Solutions

ROCCO SCHIAVILLA

Old Dominion University and Theory Center, Jefferson Laboratory

CAMBRIDGE
UNIVERSITY PRESS

Shaftesbury Road, Cambridge CB2 8EA, United Kingdom

One Liberty Plaza, 20th Floor, New York, NY 10006, USA

477 Williamstown Road, Port Melbourne, VIC 3207, Australia

314–321, 3rd Floor, Plot 3, Splendor Forum, Jasola District Centre, New Delhi – 110025, India

103 Penang Road, #05–06/07, Visioncrest Commercial, Singapore 238467

Cambridge University Press is part of Cambridge University Press & Assessment,
a department of the University of Cambridge.

We share the University's mission to contribute to society through the pursuit of
education, learning and research at the highest international levels of excellence.

www.cambridge.org
Information on this title: www.cambridge.org/9781009473651

DOI: 10.1017/9781009473637

When citing this work, please include a reference to the DOI 10.1017/9781009473637

First published 2025

A catalogue record for this publication is available from the British Library.

A Cataloging-in-Publication data record for this book is available from the Library of Congress

ISBN 978-1-009-47365-1 Hardback
ISBN 978-1-009-47362-0 Paperback

To Lynn Louise

Contents

Preface

This book grew out of handwritten class notes, problems, and solutions for the two-semester graduate quantum mechanics course, which I have been teaching in the Physics Department at Old Dominion University over the past three decades or so. Undeniably, the start of the covid pandemic in March 2020, and the ensuing forced isolation, provided the impetus for typesetting the notes and problems into LaTex. While, for a short time, I was tempted by the idea of writing a new graduate quantum mechanics textbook, I quickly realized that such a textbook – or, at least, the one I had in mind – would not add a sufficiently novel perspective on the subject to justify its publication. In contrast, I thought that the collection of problems and solutions had pedagogical value and could be useful not just to students but also to instructors, as a resource for additional problems and as a reference for derivations and/or topics not usually covered in class.

A glance at the table of Contents reveals that the problems cover core material, which is standard fare for the two-semester quantum mechanics sequence. Regretfully, some important topics – for example, entanglement and path integrals – have been deliberately left out. In my many years of teaching graduate quantum mechanics at Old Dominion University, I have never had the time to cover those topics satisfactorily. The reason is mostly the varied backgrounds of an incoming class, which has made it pedagogically necessary to spend the first several lectures reviewing some more introductory material, corresponding roughly to the first few chapters of the present book. I suspect that this state of affairs is not uncommon in other physics departments. In order to keep up the interest of the more advanced students, the weekly homework assignments have usually included one or two more challenging problems, aimed at those students. The solutions of these problems were then discussed, for the benefit of *all* students, during the following homework follow-up sessions.

Each chapter begins by framing the particular topic being examined with a concise theory section that sets the context for, and motivates, the problems that follow. Special care has been devoted to the formulation of these problems and to their solutions. They span a range of difficulties, from basic exercises to more challenging applications and extensions of the material covered in the chapter. Examples of the latter type include: Problem 12 in Chapter 3 on the derivation of Ehrenfest's relations for a charged particle in an electromagnetic field; Problem 15 in Chapter 5 on the phase-shift method for describing scattering in one dimension; Problem 11 in Chapter 8 on a model of a one-dimensional crystal with nearest-neighbor harmonic interactions; Problem 14 in Chapter 11 on the algebraic derivation of the hydrogen-atom spectrum; Problem 18 in Chapter 14 on the derivation of the leading-order correction to the hydrogen atom ground-state energy in an electric field; Problems 24 and 25 in Chapter 15 on the high-energy eikonal approximation for phase shifts and scattering amplitudes; Problem 21 in Chapter 17 on irreducible tensor operators and time reversal; Problem 17 in Chapter 18 on the theory of line widths; Problem 10 in Chapter 19 on the derivation of the Chandrasekhar limit for white dwarf stars.

The aim of these more challenging problems is to encourage students to think critically and to incorporate physics and mathematical techniques learned previously or concurrently into their work in quantum mechanics. My hope is that the extensive and detailed solutions accompanying the problems will make the present text well suited for self-study and a useful supplement to existing quantum mechanics textbooks for upper-undergraduate and graduate students.

Finally, a word about notation: I have used the standard Dirac notation in terms of kets, bras, etc. that has been adopted by the books on quantum mechanics of the last several decades. However, when I felt that the Dirac notation was ambiguous, in particular for the inner product $\langle \psi | \phi \rangle$, as is the case, for example, in the discussion of antiunitary operators and time reversal symmetry in Chapter 16, I used the notation (ψ, ϕ), prevalent in the mathematical literature, to denote this product. Also, an operator in the state space of a system is generally denoted by a caret on the letter representing it. The emphasis here is on the word "generally;" when I felt that it was clear from the context that the discussion involved operators, I avoided including the caret in order to keep the notation simple. I hope the reader will not be confused, or too bothered, by this slight lack of consistency.

Acknowledgments

I owe much to the many first-year graduate students to whom I have had the pleasure to teach quantum mechanics at Old Dominion University and who have had to endure, through no fault of their own, the weekly homework assignments which form the body of the present book. Their questions and comments and, yes, facial expressions, particularly during the homework follow-up sessions, have motivated me to constantly try to improve and clarify each problem's formulation and solution and, whenever appropriate, expand its scope. It is for the reader to judge whether these efforts have been successful.

I am especially indebted to Vince Higgs and Stephanie Windows at Cambridge University Press. Vince's gently persistent nudges have been decisive in helping me to see and define the scope and focus of the book, while Stephanie's patient responses to my repeated inquiries have been helpful in navigating the practicalities of preparing the manuscript for publication. I am also indebted to the four seasoned and anonymous reviewers, who gave me valuable criticism and advice at an earlier stage, and to Susan Parkinson for her careful editing of the final version of the manuscript.

Finally, I would like to recognize a special person in my life, my wife Lynn Louise. To her go my deepest thanks for the patience and understanding she has extended to me, undeservedly, during the completion of this project and, truly, over the past many years together, when professional commitments and obligations have intruded into, and sometimes disrupted, our personal lives. *Quos amor verus tenuit tenebit.*

1 The Failure of Classical Physics

At the beginning of the twentieth century, classical physics – Newtonian mechanics and electromagnetic theory (Maxwell's equations, wave phenomena, and optics) – could not explain a number of experimental facts, including the observed black-body radiation spectrum, the photoelectric effect, the stability of atoms and the associated spectral lines, the heat capacities of solids, and several others. The following problems are intended to illustrate the failure of classical physics to explain these phenomena and how this failure pointed to the need for a radically new treatment.

1.1 Problems

Problem 1 Black-Body Radiation Spectrum

Consider an enclosure of volume V whose walls are kept at temperature T, and define as $u(v, T)\, dv$ the energy density (energy per unit volume) of electromagnetic radiation in the frequency interval v to $v + dv$. In the mid nineteenth century (at the height of classical physics!), Gustav Kirchoff was able to show, on the basis of purely thermodynamic arguments, that the distribution $u(v, T)$ has a universal character. He also calculated the energy per unit time of radiation of given frequency that strikes a small area A of the enclosure walls. He introduced polar coordinates r, θ, and ϕ, where r is the distance from a point P in the enclosure to the area A, the polar angle θ is measured from the normal to A and the azimuthal angle ϕ is measured around the normal to A. The area subtends a solid angle at P given by $A \cos \theta / r^2$ and the fraction of radiation energy from P that is directed to A is given by $A \cos \theta / (4\pi r^2)$. The total energy in the frequency interval v to $v + dv$ that strikes the area A at time t is then obtained as

$$\int_0^{2\pi} d\phi \int_0^{\pi/2} d\theta \sin\theta \int_0^{ct} dr\, r^2\, \frac{A \cos \theta}{4\pi r^2}\, u(v, T)\, dv = \frac{ctA}{4}\, u(v, T)\, dv\ ,$$

where the integration is restricted to a hemisphere of radius ct, c being the speed of light, with θ varying in the range 0 to $\pi/2$. Given the finite velocity c of propagation, only radiation within this hemisphere will reach the area A in the time t. Denoting by $f(v, T)$ the fraction of this energy that is absorbed by the enclosure walls, we have that the total absorbed energy per unit time and area is

$$E(v, T) = \frac{c}{4}\, f(v, T)\, u(v, T)\, dv\ .$$

In a situation of equilibrium, $E(v, T)$ must equal the energy per unit time and area emitted by the enclosure walls in the same frequency interval. The fraction $f(v, T)$ of absorbed radiation can be at the most equal to unity. Indeed, a material for which $f(v, T) = 1$ is called black, and hence the name "black-body radiation" used to describe the present phenomenon.

Electromagnetic radiation in an enclosure can be described in terms of an infinite set of uncoupled harmonic oscillators. The equipartition theorem of classical statistical mechanics then leads to a prediction for $u(\nu, T)$ that is in contradiction with the experimental data and produces nonsensical results in the limit of high frequency ν, leading to the so-called ultraviolet catastrophe. The goal of the present problem is to see how this comes about.

1. Write down Maxwell's equations for the electric and magnetic fields, $\mathbf{E}(\mathbf{r}, t)$ and $\mathbf{B}(\mathbf{r}, t)$, in the absence of charge and current distributions (use CGS units).

2. Consider the enclosure to be a cubical box of side $V^{1/3}$, and impose periodic boundary conditions on the fields, namely

$$\mathbf{E}(x + V^{1/3}, y, z, t) = \mathbf{E}(x, y, z, t) , \qquad \mathbf{B}(x + V^{1/3}, y, z, t) = \mathbf{B}(x, y, z, t) ,$$

and similarly for y and z. Given that the $\mathbf{E}$ and $\mathbf{B}$ components are periodic functions, they can be expanded in Fourier series of the form

$$E_i(\mathbf{r}, t) = \sum_{\mathbf{k}} \widetilde{E}_i(\mathbf{k}, t)\, e^{i\mathbf{k}\cdot\mathbf{r}} , \qquad B_i(\mathbf{r}, t) = \sum_{\mathbf{k}} \widetilde{B}_i(\mathbf{k}, t)\, e^{i\mathbf{k}\cdot\mathbf{r}} ,$$

where the quantities carrying tildes are coefficients and the wave number $\mathbf{k} = (k_x, k_y, k_z)$ is given by

$$\mathbf{k} = \frac{2\pi}{V^{1/3}}\, \mathbf{n} , \qquad \mathbf{n} = (n_x, n_y, n_z) , \qquad n_i = 0, \pm 1, \pm 2, \ldots$$

Insert these expansions into Maxwell's equations and show that $\widetilde{\mathbf{E}}(\mathbf{k}, t)$ satisfies

$$\frac{\partial^2 \widetilde{\mathbf{E}}(\mathbf{k}, t)}{\partial t^2} = -\omega_k^2\, \widetilde{\mathbf{E}}(\mathbf{k}, t) , \qquad \omega_k = c|\mathbf{k}| = c\frac{2\pi}{V^{1/3}}\, |\mathbf{n}| .$$

How many independent directions of $\widetilde{\mathbf{E}}$ are there? Further, show how to obtain $\widetilde{\mathbf{B}}(\mathbf{k}, t)$.

3. In the limit of large $V^{1/3}$, the wave numbers $\mathbf{k}$ are densely distributed. Show that in this limit the number of independent harmonic oscillators (normal modes) in $d\mathbf{k}$ is as follows:

$$\text{number of modes in } d\mathbf{k} = \rho(\mathbf{k})\, d\mathbf{k} = 2\, \frac{V}{(2\pi)^3}\, d\mathbf{k} .$$

Recalling that in classical statistical mechanics the average energy of a harmonic oscillator kept at temperature T is simply $k_B T$, where k_B is Boltzmann's constant, and that the frequency ν is related to the wave number $|\mathbf{k}|$ via $c|\mathbf{k}|/(2\pi)$, show that the energy density of radiation with frequencies between ν and $\nu + d\nu$ is given by the Rayleigh–Jeans law

$$u(\nu, T)\, d\nu = 8\pi \frac{k_B T}{c^3}\, \nu^2\, d\nu .$$

For a fixed temperature the above prediction is in agreement with the data only for small values of the frequency; it fails spectacularly at larger values. Furthermore, the total energy density of the radiation, obtained by integrating over ν, is found to be infinite – the aforementioned ultraviolet catastrophe; see Fig. 1.1.

4. Following Einstein, suppose that the radiation energy of frequency ν is quantized in integer multiples of $h\nu$, where h is Planck's constant. Calculate the average energy of radiation of frequency ν, given that the probability that there are n quanta is

$$p_n = \frac{e^{-nh\nu/k_B T}}{\sum_{n=0}^{\infty} e^{-nh\nu/k_B T}} .$$

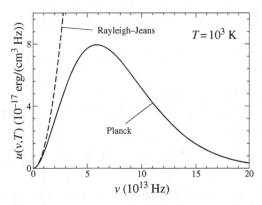

Fig. 1.1 The black-body radiation energy density per unit frequency: classical (Rayleigh–Jeans) versus quantum (Planck) description. The experimental data are in agreement with the quantum description.

Obtain the energy density in this case – the correct black-body radiation formula first derived by Planck. Show that the total energy density, that is, the energy density integrated over frequency ν, is proportional to T^4.

Solution

Part 1

In the absence of charge and current distributions, Maxwell's equations read (in the CGS system of units)

$$\boldsymbol{\nabla} \cdot \mathbf{E}(\mathbf{r}, t) = 0\,, \qquad \boldsymbol{\nabla} \times \mathbf{E}(\mathbf{r}, t) = -\frac{\partial \mathbf{B}(\mathbf{r}, t)}{c\,\partial t}\,,$$

$$\boldsymbol{\nabla} \cdot \mathbf{B}(\mathbf{r}, t) = 0\,, \qquad \boldsymbol{\nabla} \times \mathbf{B}(\mathbf{r}, t) = \frac{\partial \mathbf{E}(\mathbf{r}, t)}{c\,\partial t}\,,$$

where c is the speed of light.

Part 2

By inserting Fourier expansions into the set of Maxwell's equations, we obtain

$$\sum_{\mathbf{k}} i\,\mathbf{k} \cdot \widetilde{\mathbf{E}}(\mathbf{k}, t)\,e^{i\mathbf{k}\cdot\mathbf{r}} = 0\,, \qquad \sum_{\mathbf{k}} i\,\mathbf{k} \times \widetilde{\mathbf{E}}(\mathbf{k}, t)\,e^{i\mathbf{k}\cdot\mathbf{r}} = -\sum_{\mathbf{k}} \frac{\partial \widetilde{\mathbf{B}}(\mathbf{k}, t)}{c\,\partial t}\,e^{i\mathbf{k}\cdot\mathbf{r}}\,,$$

$$\sum_{\mathbf{k}} i\,\mathbf{k} \cdot \widetilde{\mathbf{B}}(\mathbf{k}, t)\,e^{i\mathbf{k}\cdot\mathbf{r}} = 0\,, \qquad \sum_{\mathbf{k}} i\,\mathbf{k} \times \widetilde{\mathbf{B}}(\mathbf{k}, t)\,e^{i\mathbf{k}\cdot\mathbf{r}} = \sum_{\mathbf{k}} \frac{\partial \widetilde{\mathbf{E}}(\mathbf{k}, t)}{c\,\partial t}\,e^{i\mathbf{k}\cdot\mathbf{r}}\,,$$

from which we deduce that the vectors $\widetilde{\mathbf{E}}$ and $\widetilde{\mathbf{B}}$ are perpendicular to the wave number $\mathbf{k}$,

$$\mathbf{k} \cdot \widetilde{\mathbf{E}}(\mathbf{k}, t) = 0\,, \qquad \mathbf{k} \cdot \widetilde{\mathbf{B}}(\mathbf{k}, t) = 0\,,$$

and satisfy the differential equations

$$\frac{\partial \widetilde{\mathbf{B}}(\mathbf{k}, t)}{c\,\partial t} = -i\,\mathbf{k} \times \widetilde{\mathbf{E}}(\mathbf{k}, t)\,, \qquad \frac{\partial \widetilde{\mathbf{E}}(\mathbf{k}, t)}{c\,\partial t} = i\,\mathbf{k} \times \widetilde{\mathbf{B}}(\mathbf{k}, t)\,.$$

By taking the partial derivative $\partial/(c\,\partial t)$ of both sides of the second equation above, we find

$$\frac{\partial^2 \widetilde{\mathbf{E}}(\mathbf{k},t)}{c^2\,\partial t^2} = i\,\mathbf{k} \times \frac{\partial \widetilde{\mathbf{B}}(\mathbf{k},t)}{c\,\partial t} = i\,\mathbf{k} \times \left[-i\,\mathbf{k} \times \widetilde{\mathbf{E}}(\mathbf{k},t) \right] = \mathbf{k}\,\underbrace{\left[\mathbf{k} \cdot \widetilde{\mathbf{E}}(\mathbf{k},t) \right]}_{\text{vanishes}} - k^2\,\widetilde{\mathbf{E}}(\mathbf{k},t) \, ,$$

where we have used the cross product property $\mathbf{A} \times (\mathbf{B} \times \mathbf{C}) = \mathbf{B}\,(\mathbf{A} \cdot \mathbf{C}) - \mathbf{C}\,(\mathbf{A} \cdot \mathbf{B})$. The equation above reduces to

$$\frac{\partial^2 \widetilde{\mathbf{E}}(\mathbf{k},t)}{\partial t^2} = -\omega_k^2\,\widetilde{\mathbf{E}}(\mathbf{k},t) \, , \qquad \omega_k = c|\mathbf{k}| = c\,\frac{2\pi}{V^{1/3}}\,|\mathbf{n}| \, ,$$

which shows that Maxwell's equations in the absence of sources are equivalent to an infinite set of uncoupled harmonic oscillators. The initial condition $\widetilde{\mathbf{E}}(\mathbf{k},t_0)$ and that on $\partial \widetilde{\mathbf{E}}(\mathbf{k},t)/\partial t$ at time t_0, which follows from $ic\,\mathbf{k}\times\widetilde{\mathbf{B}}(\mathbf{k},t_0)$, determine $\widetilde{\mathbf{E}}(\mathbf{k},t)$. For each wave number $\mathbf{k}$ and angular frequency ω_k there are two harmonic oscillators, corresponding to the two possible independent directions of the vector $\widetilde{\mathbf{E}}(\mathbf{k},t)$ in the plane perpendicular to $\mathbf{k}$, that is, the two independent polarizations of the electric field. Once the electric field $\widetilde{\mathbf{E}}(\mathbf{k},t)$ has been determined, the magnetic field $\widetilde{\mathbf{B}}(\mathbf{k},t)$ follows from direct integration of the equation obtained above,

$$\frac{\partial \widetilde{\mathbf{B}}(\mathbf{k},t)}{c\,\partial t} = -i\,\mathbf{k} \times \widetilde{\mathbf{E}}(\mathbf{k},t) \, .$$

Part 3

In the limit of large $V^{1/3}$, we can describe the distribution of modes with a function $\rho(\mathbf{k})$. Since there is a single wave number in each cell centered at $\mathbf{k}$ and with volume $(2\pi)^3/V$, given that the allowed $\mathbf{k}$ values are close to each other, the function $\rho(\mathbf{k})$ must satisfy the condition

$$\rho(\mathbf{k})\,\frac{(2\pi)^3}{V} = 2 \implies \rho(\mathbf{k}) = 2\,\frac{V}{(2\pi)^3} \, ,$$

where the factor 2 accounts for the two independent polarizations associated with each given $\mathbf{k}$. We then obtain

$$\text{number of modes in } d\mathbf{k} = \rho(\mathbf{k})\,d\mathbf{k} = 2\,\frac{V}{(2\pi)^3}\,d\mathbf{k} \, ,$$

and the energy density of radiation between ν and $\nu + d\nu$ is given by

$$u(\nu,T)\,d\nu = \frac{1}{V}\,k_B T \times (\text{number of modes in } d\nu) \, .$$

Recalling the relation between $|\mathbf{k}|$ and ν, it follows that

$$\text{number of modes in } d\nu = 2\,\frac{V}{(2\pi)^3}\,4\pi k^2\,dk = 8\pi\,\frac{V}{c^3}\,\nu^2\,d\nu \, ,$$

yielding the Rayleigh–Jeans law for the energy density of radiation:

$$u(\nu,T)\,d\nu = 8\pi\,\frac{k_B T}{c^3}\,\nu^2\,d\nu \, .$$

Part 4

Setting $\gamma = h\nu/(k_B T)$ for the time being, the average energy is now given by

$$\langle \text{energy} \rangle = h\nu \, \frac{\sum_{n=0}^{\infty} n e^{-n\gamma}}{\sum_{n=0}^{\infty} e^{-n\gamma}} = -h\nu \, \frac{d}{d\gamma} \ln\left(\sum_{n=0}^{\infty} e^{-n\gamma} \right) = -h\nu \, \frac{d}{d\gamma} \ln\left(\frac{1}{1-e^{-\gamma}} \right) \, ,$$

where the last step follows from summing the geometric series (here $e^{-\gamma} < 1$):

$$\sum_{n=0}^{\infty} (e^{-\gamma})^n = \frac{1}{1 - e^{-\gamma}} \, .$$

After carrying out the derivative in γ, we find

$$\langle \text{energy} \rangle = \frac{h\nu}{e^{h\nu/k_B T} - 1} \, ,$$

and therefore

$$u(\nu, T) \, d\nu = \frac{1}{V} \langle \text{energy} \rangle \times (\text{number of modes in } d\nu) = \frac{8\pi h}{c^3} \frac{\nu^3}{e^{h\nu/k_B T} - 1} \, d\nu \, ,$$

the correct black-body radiation formula first derived by Planck. Comparison with observation gives $k_B \approx 1.38 \times 10^{-16}$ erg/K and $h \approx 6.63 \times 10^{-27}$ erg sec. The formula above reproduces the Rayleigh–Jeans law at small $h\nu/(k_B T)$ but predicts an exponential fall-off at large frequency, in agreement with experimental data; see Fig. 1.1. In particular, the total energy density is now finite and is proportional to T^4, since

$$\int_0^{\infty} u(\nu, T) \, d\nu = \frac{8\pi h}{c^3} \left(\frac{k_B T}{h} \right)^4 \overbrace{\int_0^{\infty} \frac{x^3}{e^x - 1} \, dx}^{\pi^4/15} = \frac{8\pi^5}{15} \frac{k_B^4}{h^3 c^3} T^4 \, ,$$

where in the integral we have introduced the non-dimensional variable $x = h\nu/(k_B T)$. Thus, the successful explanation of the black-body energy spectrum sketched above suggests that light of frequency ν consists of quanta of energy $h\nu$.

Problem 2 Compton Scattering for an Electron in Motion in the Lab Frame

Consider the scattering of monochromatic photons by free electrons (Compton scattering).

1. Assuming that the electron is initially at rest and using energy and momentum conservation, show that the shift in the photon wavelength $\Delta\lambda = \lambda_f^{\gamma} - \lambda_i^{\gamma}$ is given by

$$\Delta\lambda = 2\lambda_e \sin^2 \theta_0/2 \qquad \text{with} \qquad \lambda_e = \frac{h}{mc} \, ,$$

where m is the electron mass, c is the speed of light, and θ_0 is the angle between the momentum of the scattered photon and the momentum of the incident photon. Determine (under the same assumption) the magnitude and direction of the recoil momentum of the electron as a function of the incident-photon energy E_i^{γ} and scattering angle θ_0.

2. Assume that the electron has an initial momentum $\mathbf{p}_i$ parallel to the incident-photon momentum $\mathbf{p}_i^{\gamma}$. Using energy and momentum conservation, show that the wavelength shift is given by

$$\Delta\lambda = 2\lambda_i^{\gamma} \, \frac{p_i^{\gamma} + p_i}{E_i/c - p_i} \, \sin^2 \theta/2 \, ,$$

where λ_i^γ is the wavelength of the incident photon, θ is the angle of the scattered photon, and $E_i = c\sqrt{p_i^2 + (mc)^2}$ is the initial energy of the electron.

3. Show that the result in part 2 can be derived from the expressions in part 1 when the electron is initially at rest, by a suitable Lorentz transformation.

Solution

Part 1

In the rest frame of the electron, let $\mathbf{p}_i^{\gamma\prime}$ and $\mathbf{p}_f^{\gamma\prime}$ be the incident- and scattered-photon momenta. Energy and momentum conservation give

$$\mathbf{p}_i^{\gamma\prime} = \mathbf{p}_f^{\gamma\prime} + \mathbf{p}_f', \qquad E_i^{\gamma\prime} + mc^2 = E_f^{\gamma\prime} + E_f',$$

where $\mathbf{p}_f'$ and E_f' denote the scattered-electron momentum and energy. Exploiting momentum conservation, we have

$$E_i^{\gamma\prime} + mc^2 - E_f^{\gamma\prime} = c\sqrt{(mc)^2 + (\mathbf{p}_i^{\gamma\prime} - \mathbf{p}_f^{\gamma\prime})^2}.$$

Squaring both sides and recalling that the photon is massless (and hence $E^\gamma = c|\mathbf{p}^\gamma|$), we arrive at

$$2mc^2 (E_i^{\gamma\prime} - E_f^{\gamma\prime}) - 2E_i^{\gamma\prime} E_f^{\gamma\prime} = -2E_i^{\gamma\prime} E_f^{\gamma\prime} \cos\theta_0,$$

yielding

$$E_f^{\gamma\prime} = \frac{E_i^{\gamma\prime}}{1 + (E_i^{\gamma\prime}/mc^2)(1 - \cos\theta_0)} = \frac{E_i^{\gamma\prime}}{1 + 2(E_i^{\gamma\prime}/mc^2)\sin^2(\theta_0/2)},$$

where θ_0 is the photon scattering angle (in the electron rest frame). Recalling that for a photon

$$E^\gamma = \hbar\omega^\gamma = \hbar c|\mathbf{k}^\gamma| = \frac{2\pi\hbar c}{\lambda^\gamma} = \frac{hc}{\lambda^\gamma},$$

the relationship between the initial and final photon energies can be cast in terms of the initial and final wavelengths as

$$\frac{hc}{\lambda_f^{\gamma\prime}} = \frac{hc}{\lambda_i^{\gamma\prime}} \frac{1}{1 + 2(hc/(\lambda_i^{\gamma\prime} mc^2))\sin^2(\theta_0/2)} \implies \lambda_f^{\gamma\prime} = \lambda_i^{\gamma\prime} + 2\underbrace{\frac{h}{mc}}_{\lambda_e}\sin^2(\theta_0/2),$$

the required expression for the wavelength shift $\Delta\lambda' = \lambda_f^{\gamma\prime} - \lambda_i^{\gamma\prime}$ is

$$\Delta\lambda' = 2\lambda_e \sin^2(\theta_0/2),$$

where λ_e is the Compton wavelength of the electron. In order to determine the electron scattering angle in the initial electron rest frame, we use momentum conservation to obtain

$$p_i^{\gamma\prime} = p_f^{\gamma\prime} \cos\theta_0 + p_f' \cos\theta_0^e, \qquad 0 = p_f^{\gamma\prime} \sin\theta_0 + p_f' \sin\theta_0^e,$$

where θ_0^e is the electron scattering angle. We have

$$p_f' \sin\theta_0^e = -p_f^{\gamma\prime} \sin\theta_0, \qquad p_f' \cos\theta_0^e = p_i^{\gamma\prime} - p_f^{\gamma\prime} \cos\theta_0,$$

yielding

$$\tan \theta_0^e = -\frac{p_f^{\gamma'} \sin \theta_0}{p_i^{\gamma'} - p_f^{\gamma'} \cos \theta_0} = -\frac{\sin \theta_0}{E_i^{\gamma'}/E_f^{\gamma'} - \cos \theta_0} \ .$$

Using

$$E_i^{\gamma'}/E_f^{\gamma'} = 1 + (E_i^{\gamma'}/mc^2)(1 - \cos \theta_0) \ ,$$

we arrive at

$$\tan \theta_0^e = -\frac{\sin \theta_0}{1 + (E_i^{\gamma'}/mc^2)(1 - \cos \theta_0) - \cos \theta_0} = -\frac{\sin \theta_0}{1 - \cos \theta_0} \frac{1}{1 + E_i^{\gamma'}/mc^2} = -\frac{\cot(\theta_0/2)}{1 + E_i^{\gamma'}/mc^2} \ .$$

To determine the magnitude of the electron's final momentum, we square both sides of the energy-conservation relation to find

$$\underbrace{c^2(p_f'^2 + m^2 c^2)}_{E_f'^2} = \underbrace{E_i^{\gamma'2} + E_f^{\gamma'2} + m^2 c^4 + 2mc^2(E_i^{\gamma'} - E_f^{\gamma'}) - 2E_i^{\gamma'} E_f^{\gamma'}}_{(E_i^{\gamma'} + mc^2 - E_f^{\gamma'})^2} \ ,$$

which reduces to

$$c^2 p_f'^2 = (E_i^{\gamma'} - E_f^{\gamma'})(E_i^{\gamma'} - E_f^{\gamma'} + 2mc^2) \ .$$

Inserting into $p_f'^2$ the expression for the difference between the initial and final photon energies,

$$E_i^{\gamma'} - E_f^{\gamma'} = E_i^{\gamma'} \frac{2(E_i^{\gamma'}/mc^2) \sin^2(\theta_0/2)}{1 + 2(E_i^{\gamma'}/mc^2) \sin^2(\theta_0/2)} \ ,$$

yields the required relation between the magnitude of the electron's final momentum and the photon's initial energy and final scattering angle.

Part 2

We call the frame in which the electron has initial momentum $\mathbf{p}_i$ the lab frame. By assumption $\mathbf{p}_i$ is parallel to $\mathbf{p}_i^{\gamma}$. We will denote the momenta and energies of the electron and photons by unprimed symbols, so $\mathbf{p}_i^{\gamma}$, E_i^{γ} and $\mathbf{p}_f^{\gamma}$, E_f^{γ} are the initial and final photon momenta and energies and $\mathbf{p}_i$, E_f and $\mathbf{p}_f$, E_f are the initial and final electron momenta and energies, respectively. Energy and momentum conservation in this frame read

$$\mathbf{p}_i^{\gamma} + \mathbf{p}_i = \mathbf{p}_f^{\gamma} + \mathbf{p}_f \ , \qquad E_i^{\gamma} + E_i = E_f^{\gamma} + E_f \ .$$

These relations imply

$$E_i + E_i^{\gamma} - E_f^{\gamma} = c\sqrt{(\mathbf{p}_i + \mathbf{p}_i^{\gamma} - \mathbf{p}_f^{\gamma})^2 + (mc)^2} \ ,$$

and squaring both sides yields

$$E_i(E_i^{\gamma} - E_f^{\gamma}) - E_i^{\gamma} E_f^{\gamma} = cp_i E_i^{\gamma} - cp_i E_f^{\gamma} \cos \theta - E_i^{\gamma} E_f^{\gamma} \cos \theta \ ,$$

where we have used the fact that $\mathbf{p}_i$ and $\mathbf{p}_i^{\gamma}$ are parallel. Rearranging terms, we find

$$E_i^{\gamma}(E_i - cp_i) = E_f^{\gamma}(E_i + E_i^{\gamma}) - E_f^{\gamma}(E_i^{\gamma} + cp_i) \cos \theta \ ,$$

which can be further simplified by using the identity $\cos\theta = 1 - 2\sin^2(\theta/2)$ to obtain

$$E_i^\gamma(E_i - cp_i) = \underbrace{E_f^\gamma(E_i + E_i^\gamma) - E_f^\gamma(E_i^\gamma + cp_i)}_{E_f^\gamma(E_i - cp_i)} + 2E_f^\gamma(E_i^\gamma + cp_i)\sin^2(\theta/2) \, ,$$

or, after dividing both sides by $E_i - cp_i$,

$$E_i^\gamma = E_f^\gamma + 2E_f^\gamma \frac{E_i^\gamma + cp_i}{E_i - cp_i}\sin^2(\theta/2) \, .$$

In terms of photon wavelengths, the above expression is written as

$$\frac{hc}{\lambda_i^\gamma} = \frac{hc}{\lambda_f^\gamma} + 2\frac{hc}{\lambda_f^\gamma}\frac{p_i^\gamma + p_i}{E_i/c - p_i}\sin^2(\theta/2) \implies \lambda_f^\gamma = \lambda_i^\gamma + 2\lambda_i^\gamma\frac{p_i^\gamma + p_i}{E_i/c - p_i}\sin^2(\theta/2) \, ,$$

resulting in a wavelength shift given by

$$\Delta\lambda = 2\lambda_i^\gamma\frac{p_i^\gamma + p_i}{E_i/c - p_i}\sin^2(\theta/2) \, .$$

Part 3

Another (instructive) way to solve the problem in Part 2 is to work in the electron's initial rest frame, and then transform back to the lab frame, in which the electron has initial momentum $\mathbf{p}_i$ (parallel to the initial photon momentum $\mathbf{p}_i^\gamma$) and energy E_i. The energy of the scattered photon in the laboratory frame follows from the Lorentz transformation relation:

$$E_f^\gamma = \gamma(E_f^{\gamma\,\prime} + \beta cp_{f,x}^{\gamma\,\prime}) = \gamma E_f^{\gamma\,\prime}(1 + \beta\cos\theta_0),$$

where $\beta\hat{\mathbf{x}}$ with $\beta = cp_i/E_i$ is the velocity of the electron in the lab frame, and $\gamma = 1/\sqrt{1 - \beta^2}$. By substituting for $E_f^{\gamma\,\prime}$ the expression found in part 1, we obtain

$$E_f^\gamma = \gamma\frac{E_i^{\gamma\,\prime}(1 + \beta\cos\theta_0)}{1 + (E_i^{\gamma\,\prime}/mc^2)(1 - \cos\theta_0)} \, .$$

We need to express the photon energy $E_i^{\gamma\,\prime}$ and scattered photon angle θ_0 in terms of the corresponding lab frame quantities. The energy of the initial photon in the electron's rest frame is related to its energy in the lab frame by a Lorentz transformation:

$$E_i^{\gamma\,\prime} = \gamma(E_i^\gamma - \beta cp_{i,x}^\gamma) = \gamma E_i^\gamma(1 - \beta),$$

where the momentum of the initial photon is along the $\hat{\mathbf{x}}$-direction, and hence $p_{i,x}^\gamma = p_i^\gamma$. The Lorentz transformation also gives

$$p_{f,x}^{\gamma\,\prime} = \gamma(p_{f,x}^\gamma - \beta E_f^\gamma/c) \, , \qquad p_{f,y}^{\gamma\,\prime} = p_{f,y}^\gamma \, ,$$

which implies that

$$\tan\theta_0 = \frac{p_{f,y}^{\gamma\,\prime}}{p_{f,x}^{\gamma\,\prime}} = \frac{p_{f,y}^\gamma}{\gamma(p_{f,x}^\gamma - \beta E_f^\gamma/c)} = \frac{\sin\theta}{\gamma(\cos\theta - \beta)} \, ,$$

where θ is the angle of the scattered photon in the lab frame. Making use of the identity $\cos \theta_0 = 1/\sqrt{1 + \tan^2 \theta_0}$, we find

$$\cos \theta_0 = \frac{1}{\sqrt{1 + \sin^2 \theta/[\gamma^2(\cos \theta - \beta)^2]}} = \frac{\gamma(\cos \theta - \beta)}{\sqrt{\gamma^2(\cos \theta - \beta)^2 + \sin^2 \theta}}$$

$$= \frac{\cos \theta - \beta}{\sqrt{\cos^2 \theta + \beta^2 - 2\beta \cos \theta + (1 - \beta^2) \sin^2 \theta}} = \frac{\cos \theta - \beta}{1 - \beta \cos \theta},$$

from which we have the following relations:

$$1 - \cos \theta_0 = (1 + \beta) \frac{1 - \cos \theta}{1 - \beta \cos \theta}, \qquad 1 + \beta \cos \theta_0 = \frac{1 - \beta^2}{1 - \beta \cos \theta} = \frac{1/\gamma^2}{1 - \beta \cos \theta}.$$

We insert all these relations into the expression for E_f^γ found above to obtain

$$E_f^\gamma = \frac{\gamma^2 E_i^\gamma(1 - \beta)/[\gamma^2(1 - \beta \cos \theta)]}{1 + \gamma(E_i^\gamma/mc^2) \underbrace{(1 - \beta)(1 + \beta)}_{1/\gamma^2}(1 - \cos \theta)/(1 - \beta \cos \theta)} = E_i^\gamma \frac{(1 - \beta)/(1 - \beta \cos \theta)}{1 + (E_i^\gamma/\gamma mc^2)(1 - \cos \theta)/(1 - \beta \cos \theta)}.$$

Recalling that $\beta = cp_i/E_i$ and $E_i = \gamma mc^2$, we finally arrive at

$$E_f^\gamma = E_i^\gamma \frac{1 - \beta}{1 - \beta \cos \theta + (E_i^\gamma/E_i)(1 - \cos \theta)} = E_i^\gamma \frac{E_i - cp_i}{E_i - cp_i \cos \theta + E_i^\gamma(1 - \cos \theta)},$$

which can be simplified by expressing $\cos \theta$ as $1 - 2\sin^2(\theta/2)$:

$$E_f^\gamma = E_i^\gamma \frac{E_i - cp_i}{E_i - cp_i + 2\sin^2(\theta/2)(E_i^\gamma + cp_i)} = \frac{E_i^\gamma}{1 + 2\sin^2(\theta/2)(E_i^\gamma + cp_i)/(E_i - cp_i)}.$$

In terms of wavelengths this gives

$$\lambda_f^\gamma = \lambda_i^\gamma \left[1 + 2\sin^2(\theta/2) \frac{p_i^\gamma + p_i}{E_i/c - p_i} \right],$$

and the wavelength shift is, of course, identical to that found in part 2 above.

Problem 3 The Thomson Model of the Atom and Rutherford's Experiment

After the discovery of the electron by Thomson in 1897, it was believed that "atoms were like puddings, with negatively charged electrons stuck in like raisins in a smooth background of positive charge" (S. Weinberg). This picture was drastically changed by experiments performed by Rutherford and collaborators, who scattered α particles (^{4}He nuclei, which, as we now know, consist of two protons and two neutrons bound together by the nuclear force, having electric charge $2e$) off a thin foil of gold. Rutherford and collaborators observed α particles scattered at large backward angles. This was totally unexpected, since electrons are much lighter than α particles.

1. Consider a particle of mass M and velocity v hitting a particle of mass m at rest and continuing along the same line with velocity v'. Show that, for a given v, energy and momentum conservation lead to two possible solutions for v'. If a certain condition is satisfied, one of these solutions corresponds to the case in which particle M inverts its direction of motion. What is this condition?

2. Suppose the α particles (which were in fact emitted by a radium source in Rutherford's experiment) have velocity $v \approx 2.1 \times 10^9$ cm/sec, and that the target particles (much heavier than the α particles) have each charge Ze. If the α particles and target particles interact via the Coulomb repulsion, what is the distance of closest approach? Show that this distance is of the order $3Z \times 10^{-14}$ cm, and therefore (even for $Z \approx 100$) it is much smaller than atomic radii.

Solution

Part 1

Energy and momentum conservation require

$$\frac{Mv^2}{2} = \frac{Mv'^2}{2} + \frac{mu^2}{2} , \qquad Mv = Mv' + mu ,$$

where, as assumed in the problem, the particle of mass M proceeds after the collision along the same trajectory as it followed before the collision. Replacing u with $M(v-v')/m$ in the energy-conservation relation leads to an equation for v'/v:

$$\left(1 + \frac{M}{m}\right) \left(\frac{v'}{v}\right)^2 - 2\frac{M}{m}\frac{v'}{v} - 1 + \frac{M}{m} = 0 ,$$

which has the solutions

$$v' = v , \qquad v' = -\frac{m - M}{m + M} v .$$

The first solution ($v' = v$) says that the particle continues along its trajectory undisturbed, which is unphysical. However, the second solution says that if $m > M$ the particle inverts its trajectory, since in that case $v' < 0$.

Part 2

At the distance of closest approach, the kinetic energy of the α particle must have been converted into potential energy (we are neglecting here the recoil energy of the target particle, that is, we are assuming $m \gg M$),

$$\frac{Mv^2}{2} = \frac{(Ze)(2e)}{r_0} \implies r_0 = \frac{4Ze^2}{Mv^2} .$$

We have

$$e^2 \approx 2.3 \times 10^{-19} \text{ g cm}^3/\text{sec}^2 , \qquad v \approx 2.1 \times 10^9 \text{ cm/sec} , \qquad M \approx 6.6 \times 10^{-24} \text{ g} ,$$

where we have expressed e^2 as $\alpha \hbar c$ and α is the (non-dimensional) fine-structure constant having the approximate value $\approx 1/137$ and have used $\hbar \approx 1.05 \times 10^{-27}$ g cm^2/sec and $c = 3.00 \times 10^{10}$ cm/sec. For the mass of the ^{4}He nucleus we have used $Mc^2 = 2(m_p + m_n)c^2 - 28.3$ MeV ≈ 3727 MeV. Here 28.3 MeV is the nuclear binding energy of ^{4}He, and MeV/c^2 $\approx 1.78 \times 10^{-27}$ g. We obtain

$$r_0 \approx 3Z \times 10^{-14} \text{ cm} ,$$

which is much smaller than the size of the atom, of the order of 10^{-8} cm.

Problem 4   The Stability Problem for the Rutherford Model of the Atom

Consider the Rutherford model of the atom: an electron of electric charge $-e$ orbiting a point-like nucleus (much heavier, and hence effectively at rest) of electric charge Ze in a circular orbit of radius R. Knowing that the electron radiates energy away at a rate $dE(t)/dt$, given by

$$\frac{dE(t)}{dt} = -\frac{2}{3}\frac{e^2\,|\mathbf{a}(t)|^2}{c^3}\,,$$

where $\mathbf{a}(t)$ is the electron's acceleration and c is the speed of light, show that it will take a time

$$\tau = \frac{m^2c^3}{Ze^4}\frac{R^3}{4}\,,$$

for the electron to spiral into the nucleus. Assume that τ is much larger than the revolution period. By taking $Z=1$ and $R \approx 10^{-8}$ cm, as is appropriate for the hydrogen atom, justify this assumption *a posteriori* by comparing τ with the revolution period.

Solution

The electron energy is given by

$$E = \frac{mv^2}{2} - \frac{Ze^2}{r}\,.$$

It is assumed that the electron is in a circular orbit and that the energy lost by emission of radiation per revolution is tiny relative to E. The electron is subject to a centripetal acceleration whose magnitude is given by $a = v^2/r$, where r is the radius of the orbit, so that

$$ma = \frac{Ze^2}{r^2} \implies v^2 = \frac{Ze^2}{mr}\,,$$

and hence

$$E = -\frac{Ze^2}{2r}\,.$$

The radius does change with time, albeit very slowly, corresponding to the loss of energy:

$$\frac{dE}{dt} = \frac{d}{dt}\left(-\frac{Ze^2}{2r}\right) = \frac{Ze^2}{2r^2}\frac{dr}{dt} = -\frac{2}{3}\frac{e^2a^2}{c^3} = -\frac{2e^2}{3c^3}\left(\frac{Ze^2}{mr^2}\right)^2\,,$$

which leads to

$$\frac{dr}{dt} = -\frac{4}{3}\frac{Ze^4}{m^2c^3r^2} \qquad \text{or} \qquad dt = -\frac{3}{4}\frac{m^2c^3}{Ze^4}r^2\,dr\,.$$

At time $t=0$ the electron is in an orbit of radius $r(0)=R$, while at time τ the electron has "fallen into the nucleus", and so $r(\tau)=0$; therefore, we find by integrating above the differential equation,

$$\int_0^\tau dt = -\frac{3}{4}\frac{m^2c^3}{Ze^4}\int_R^0 dr\,r^2 \implies \tau = \frac{m^2c^3}{Ze^4}\frac{R^3}{4}\,.$$

We take

$$m \approx 9.1\times10^{-28}\,\text{g}\,, \qquad c = 3.0\times10^{10}\,\text{cm/sec}\,, \qquad e^2 \approx 2.3\times10^{-19}\,\text{g cm}^3/\text{sec}^2\,,$$

and, for the hydrogen atom, $Z = 1$ and $R \approx 10^{-8}$ cm, and so we find $\tau \approx 10^{-10}$ sec, while the revolution period is

$$T = 2\pi \frac{R}{v} = 2\pi R \left(\frac{mR}{Ze^2} \right)^{1/2} \approx 4 \times 10^{-16} \text{ sec} ,$$

giving $T \ll \tau$ as assumed.

Problem 5 Bohr's Calculation of the Energy Spectrum of the Hydrogen Atom

In order to solve the stability problem, Niels Bohr proposed in 1913 that the atom can exist only in certain states having energies $E_1 < E_2 < \cdots$, that is, atomic energies are quantized. To obtain these energies, Bohr assumed that the angular momentum of an electron of mass m and electric charge $-e$ in a *stable circular orbit* of radius r around a nucleus of electric charge Ze is an integer multiple n of the Planck constant $\hbar = h/(2\pi)$. Following Bohr, calculate the energies E_n.

Solution

The magnitude of the angular momentum of the electron in a circular orbit of radius r is given by mvr, where v is the magnitude of the velocity, and according to Bohr's hypothesis

$$mvr = n\hbar .$$

The attractive Coulomb force acting on the electron is responsible for its centripetal acceleration, which reads (in magnitude)

$$\frac{v^2}{r} = \frac{Ze^2}{mr^2} \implies r = \frac{Ze^2}{mv^2} .$$

When combined with the quantization condition, this leads to

$$v = \frac{Ze^2}{n\hbar} ,$$

and hence to the energy levels, given by (after substituting for r and v the expressions above)

$$E = \frac{mv^2}{2} - \frac{Ze^2}{r} = -\frac{mv^2}{2} \implies E_n = -\frac{Z^2 e^4 m}{2n^2 \hbar^2} = -\frac{(Z\alpha)^2 mc^2}{2n^2} ,$$

which turns out to give the (correct!) result obtained in Schrödinger's wave mechanics.

Problem 6 The Bohr–Sommerfeld Quantization Rule and the Harmonic Oscillator Energy Spectrum

In the old quantum theory, one assumes that the particles follow the laws of classical mechanics but one postulates further that, of all the possible solutions of the equations of motion, one must retain only those which satisfy certain *ad hoc* quantization rules. One therefore selects a discontinuous family of motions; these are, by hypothesis, the only motions which are realized in nature. The discontinuous sequence of energy values thus obtained constitutes the spectrum of quantized energy levels.

For a one-dimensional periodic motion, the quantization rule, known as the Bohr–Sommerfeld quantization rule, is

$$\oint_E dq\, p = nh \qquad n = 1, 2, \ldots ,$$

where h is Planck's constant – recall that $\hbar = h/(2\pi)$ – and the symbol $\oint_E$ means that one must integrate over a complete period of the motion corresponding to the energy E. Here q and p are the position and momentum variables, respectively. The integral is known as the action integral. Apply this rule to the case of the one-dimensional harmonic oscillator, for which

$$E = \frac{p^2}{2m} + \frac{m\omega^2}{2} q^2 .$$

Calculate the energy, period, and amplitude of the quantized trajectories.

Solution

For a fixed energy E, the momentum p is given by

$$p = \sqrt{2mE \left(1 - \frac{m\omega^2}{2E} q^2 \right)} ,$$

and it vanishes at the endpoints $\pm q_0$, where

$$q_0 = \sqrt{\frac{2E}{m\omega^2}} .$$

The Bohr–Sommerfeld rule requires that

$$2 \int_{-q_0}^{q_0} dq \sqrt{2mE \left(1 - \frac{q^2}{q_0^2} \right)} = nh ,$$

where the factor 2 in front of the integral accounts for the fact that over a full period the particle goes from $-q_0$ to q_0 and back to $-q_0$. Substituting $x = q/q_0$, the left-hand side can written as follows:

$$\text{l.h.s.} = 2 \sqrt{2mE}\, q_0 \underbrace{\int_{-1}^{1} dx \sqrt{1 - x^2}}_{\pi/2} \implies \text{l.h.s.} = 2\pi \frac{E}{\omega} ,$$

after substituting for q_0. The Bohr–Sommerfeld gives

$$2\pi \frac{E}{\omega} = nh \implies E_n = n\hbar\omega \qquad \text{with } n = 1, 2, \ldots$$

Note that the exact quantum result is $E_n = (n + 1/2)\hbar\omega$ with $n = 0, 1, 2, \ldots$ The period of the harmonic oscillator is $2\pi/\omega$, while its amplitude A_n is simply given by q_0 and is therefore quantized,

$$A_n = \sqrt{\frac{2\hbar}{m\omega}} \sqrt{n} .$$

Problem 7 An Application of the Bohr–Sommerfeld Quantization Rule
to the Hydrogen Atom

Quantize the *circular* electronic orbits of the hydrogen atom by applying the Bohr–Sommerfeld rule introduced in the previous problem. Determine the energy, period, and radius of the quantized orbits. Calculate specifically the numerical values of the energy, period, and radius of the lowest orbit. Use $mc^2 \approx 0.51 \times 10^6$ eV and $\hbar c/e^2 \approx 137$.

Solution

The energy of a hydrogen-like atom is given by

$$E = \frac{p^2}{2m} - \frac{Ze^2}{r} \,,$$

where m and $-e$ are the electron mass and charge; we assume that the nucleus of charge Ze is fixed at the origin (we will neglect reduced-mass corrections). In a circular orbit the centripetal acceleration of the electron is provided by the attractive Coulomb force, and hence

$$m\frac{v^2}{r} = \frac{Ze^2}{r^2} \implies \frac{p^2}{m} = \frac{Ze^2}{r} \,,$$

where r is the radius of the circular orbit; the magnitude p of the electron momentum is constant (but not its direction, of course). The momentum corresponding to a given energy for such an orbit is then given by

$$E = \frac{p^2}{2m} - \frac{p^2}{m} = -\frac{p^2}{2m} \implies p = \sqrt{2m|E|} \,,$$

and the action integral for such an orbit follows from

$$\oint_E dq\,p = \oint_E d\mathbf{s} \cdot \mathbf{p} = p \int_0^{2\pi} r\,d\theta = 2\pi r p = 2\pi r \sqrt{2m|E|} \,.$$

For a given E, the radius of the circular orbit is given by

$$E = \frac{p^2}{2m} - \frac{Ze^2}{r} = \frac{Ze^2}{2r} - \frac{Ze^2}{r} = -\frac{Ze^2}{2r} \implies r = \frac{Ze^2}{2|E|} \,.$$

Inserting this expression into the action–integral result and imposing the Bohr–Sommerfeld rule yields

$$2\pi \frac{Ze^2}{2|E|} \sqrt{2m|E|} = nh \implies E_n = -\frac{m}{2n^2} \frac{Z^2 e^4}{\hbar^2} = -\frac{(Z\alpha)^2}{2n^2} mc^2 \,,$$

where we have introduced the fine-structure constant $\alpha = e^2/(\hbar c) \approx 1/137$ and $mc^2 \approx 0.51$ MeV is the electron rest mass. The result above for E_n turns out to agree with that obtained in Schrödinger's wave mechanics. The radii of these circular orbits are quantized,

$$r_n = \frac{Ze^2}{2|E_n|} = \frac{e^2}{Z\alpha^2 mc^2} n^2 = \frac{1}{Z\alpha} \frac{\hbar}{mc} n^2 = \frac{a_0}{Z} n^2 \,,$$

where we have expressed e^2 in terms of the fine-structure constant as $\alpha\hbar c$ and have introduced the Bohr radius a_0:

$$a_0 = \frac{\hbar}{\alpha mc} \approx 0.53 \times 10^{-8} \text{ cm} \,.$$

The periods of the orbits also turn out to be quantized,

$$T_n = 2\pi \frac{r_n}{p_n/m} \implies T_n = 2\pi \frac{n^2 a_0}{Z} \sqrt{\frac{m}{2|E_n|}} = 2\pi \frac{n^3}{Z^2 \alpha} \frac{a_0}{c} \,.$$

For the hydrogen atom having $Z = 1$ and for the most bound orbit, corresponding to $n = 1$, we find

$$E_1 = -13.6 \text{ eV}, \qquad r_1 = 0.53 \times 10^{-8} \text{ cm}, \qquad T_1 = 1.5 \times 10^{-16} \text{ sec} \,.$$

Problem 8 Heat Capacity of Solids

In addition to the failures in explaining the black-body radiation spectrum, the photoelectric effect, and the stability of atoms and spectral lines, classical physics could not explain the heat capacity of a solid.

1. Assume that a solid of volume V with N atoms (or molecules) can be modeled as a set of $3N$ independent one-dimensional harmonic oscillators of frequency ν_0 (that is, all oscillators have the same frequency ν_0). Making use of the equipartition theorem, calculate the total average energy E of the solid at temperature T and derive the Dulong–Petit law for the heat capacity (at constant volume),

$$c_V = \left(\frac{\partial E}{\partial T}\right)_V = 3N k_B \,,$$

where k_B is Boltzmann's constant.

2. The observed heat capacity of a solid is not in fact a constant independent of T but rather vanishes as T^3 at low temperature and only approaches the classical prediction (the Dulong–Petit law) at high temperatures. Einstein (1907) proposed that the energy of each harmonic oscillator is quantized and that its average energy at temperature T is given as follows:

$$\text{average energy} = \frac{h\nu_0}{e^{h\nu_0/k_B T} - 1} \,.$$

Show that the heat capacity is now found to be

$$c_V = 3N k_B \frac{x_0^2 \, e^{x_0}}{(e^{x_0} - 1)^2} \,, \qquad x_0 = \frac{h\nu_0}{k_B T} \,.$$

Does c_V vanish as $T \longrightarrow 0$? What happens at high temperature?

3. Einstein's theory predicts that c_V vanishes exponentially at low temperature, a result that is at variance with experimental observations. A more realistic model of a solid is that it consists of $3N$ independent harmonic oscillators (normal modes) with a distribution in frequency $g(\nu)$ given by

$$g(\nu) = 4\pi \frac{V}{v_S^3} \nu^2 \,,$$

where v_S is the sound velocity in the solid, such that

$$\text{number of modes} = \int_0^{\nu_{\max}} d\nu \, g(\nu) = 3N \,.$$

This condition fixes $\nu_{\max}$ as a function of the density N/V. Introduce the parameter T_D (the Debye temperature), defined as

$$T_D = \frac{h\nu_{\max}}{k_B} \,,$$

and show that the total average energy at temperature T is given by

$$E = 9Nh\nu_{\max}\left(\frac{T}{T_D}\right)^4 \int_0^{T_D/T} dx\,\frac{x^3}{e^x - 1}\,,$$

and that the constant-volume heat capacity is proportional to T^3 for low T, in agreement with experiment. Show that it also satisfies the Dulong–Petit law at high temperatures.

Solution

Part 1

Each oscillator contributes $k_B T$ to the total average energy E of the solid held at temperature T, and, since there are $3N$ oscillators, this total energy is simply $3Nk_B T$, which immediately yields the Dulong–Petit law given in the text of the problem.

Part 2

Einstein's theory gives for the total average energy

$$E = 3N\frac{h\nu_0}{e^{h\nu_0/k_B T} - 1} = 3Nk_B T\underbrace{\frac{x_0}{e^{x_0} - 1}}_{f(x_0)}\,,$$

and the heat capacity at constant volume is then as follows:

$$c_V = \left(\frac{\partial E}{\partial T}\right)_V = 3Nk_B f(x_0) + 3Nk_B T\frac{\partial f(x_0)}{\partial x_0}\frac{\partial x_0}{\partial T}$$

$$= 3Nk_B\frac{x_0}{e^{x_0} - 1} - 3Nk_B T\left[\frac{1}{e^{x_0} - 1} - \frac{x_0\,e^{x_0}}{(e^{x_0} - 1)^2}\right]\frac{x_0}{T} = 3Nk_B\frac{x_0^2\,e^{x_0}}{(e^{x_0} - 1)^2}\,.$$

In the limit of low T, x_0 becomes large and $c_V \approx 3Nk_B x_0^2\,e^{-x_0}$ vanishes exponentially. In the high-T limit, we have $x_0 \longrightarrow 0$, and hence

$$c_V = 3Nk_B\frac{x_0^2(1 + x_0 + \cdots)}{(x_0 + x_0^2/2 + \cdots)^2} = 3Nk_B\frac{1 + x_0 + \cdots}{(1 + x_0/2 + \cdots)^2} \approx 3Nk_B\,,$$

and the Dulong–Petit law is reproduced in this limit up to corrections proportional to x_0^2.

Part 3

We have

$$\int_0^{\nu_{\max}} d\nu\,g(\nu) = 3N \qquad \text{or} \qquad \frac{4\pi}{3}\frac{V}{v_S^3}\nu_{\max}^3 = 3N \implies \nu_{\max} = v_S\left(\frac{9}{4\pi}\rho\right)^{1/3}\,.$$

The total average energy follows from

$$E = \int_0^{\nu_{\max}} d\nu\,g(\nu)\,\frac{h\nu}{e^{h\nu/k_B T} - 1} = 4\pi V\frac{h}{v_S^3}\int_0^{\nu_{\max}} d\nu\,\frac{\nu^3}{e^{h\nu/k_B T} - 1}\,,$$

which, after introducing the integration variable $x = h\nu/(k_B T)$, can also be expressed as

$$E = 4\pi V \frac{h}{v_S^3} \left(\frac{k_B T}{h}\right)^4 \int_0^{T_D/T} dx \, \frac{x^3}{e^x - 1} = 4\pi V \frac{h}{v_S^3} v_{max}^4 \left(\frac{T}{T_D}\right)^4 \int_0^{T_D/T} dx \, \frac{x^3}{e^x - 1}$$

or, using the relation $(v_{max}/v_S)^3 = (9/4\pi)\rho$,

$$E = 9Nh\nu_{max} \left(\frac{T}{T_D}\right)^4 \int_0^{T_D/T} dx \, \frac{x^3}{e^x - 1} \; .$$

In the limit of low T, the ratio $T_D/T \longrightarrow \infty$, and hence we have

$$E \approx 9Nh\nu_{max} \left(\frac{T}{T_D}\right)^4 \underbrace{\int_0^\infty dx \, \frac{x^3}{e^x - 1}}_{T \text{ independent}} \implies c_V \propto T^3 \; .$$

By contrast, at high T we have $T_D/T \longrightarrow 0$, and the integral can be approximated as

$$\int_0^{T_D/T} dx \, \frac{x^3}{e^x - 1} \approx \int_0^{T_D/T} dx \, x^2 = \frac{1}{3}\left(\frac{T}{T_D}\right)^3 \; ,$$

where we have expanded the integrand for small x, since $x \ll 1$ in the range $0 \leq x \leq T_D/T$. Thus, we find for E,

$$E \approx 3Nh\nu_{max} \left(\frac{T}{T_D}\right)^4 \left(\frac{T_D}{T}\right)^3 = 3N \frac{h\nu_{max}}{T_D} T = 3Nk_B T \implies c_V = 3Nk_B \; ,$$

in agreement with the Dulong–Petit law.

2 Wave–Particle Duality and Wave Mechanics

Light exhibits both wave behavior (interference and diffraction phenomena) and particle behavior (the photoelectric and Compton effects). The proposal that particles too could exhibit wave behavior and that this wave–particle duality was in fact a property of matter particles was made by de Broglie in 1923. He postulated that a particle of momentum $\mathbf{p}$ and energy E has a plane wave associated with it of wave number $\mathbf{k} = \mathbf{p}/\hbar$ and angular frequency $\omega = E/\hbar$. The de Broglie hypothesis found experimental confirmation (Davisson and Germer, 1927) when diffractive phenomena were observed in the scattering of a beam of electrons by a crystal. When X-rays are scattered by a crystal, intense peaks of reflected light are observed. These peaks depend on the wavelength of the incident light and on the incidence angle (Bragg reflection). The condition for Bragg reflection, which results from constructive interference of the waves reflected by different crystal planes, is

$$2d \sin \theta = n\lambda \, , \tag{2.1}$$

where λ and θ are respectively the wavelength and angle of the incident wave, d is the separation between two adjacent crystal planes, and n is an integer (see Fig. 2.1). Davisson and Germer showed that electrons scattered by a single crystal of nickel produced a diffraction pattern similar to that observed in X-ray scattering by crystals.

2.1 Wave Packets

Following de Broglie, we represent the wave associated with a particle as a superposition (a wave packet). In one dimension it is written as

$$\Psi(x, t) = \int \frac{dk'}{(2\pi)^{1/2}} \, f(k') \, \mathrm{e}^{i\varphi(k')} \, , \qquad \varphi(k') = k'x - \omega_{k'} t \, , \tag{2.2}$$

and its generalization to three dimensions should be obvious. The generally complex function $f(k')$ is such that its magnitude $|f(k')|$ takes on appreciable values only in a small region surrounding $k = p/\hbar$; the factor $1/(2\pi)^{d/2}$, with d the number of dimensions, is introduced for convenience. The condition $\varphi(k') = \varphi_0$, a constant, defines a plane of constant phase, which moves according to

$$x_{\mathrm{ph}}(t) = (\omega_{k'} t + \varphi_0)/k' \, , \qquad v_{\mathrm{ph}} = \dot{x}_{\mathrm{ph}} = \frac{\omega_{k'}}{k'} = \frac{E_{p'}}{p'} = \frac{p'}{2m} \, , \tag{2.3}$$

where v_{ph} is the phase velocity. The phase velocity depends on the wave number. Thus, different wave-number components in $\Psi(x, t)$ move at different phase velocities, in contrast with electromagnetic wave components propagating in vacuum, for which the phase velocity is independent of the wave number and equal to c. As a consequence, in matter waves there is a tendency for the original coherence to be lost, and for the wave packet to become distorted in space as time goes on.

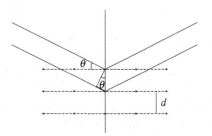

Fig. 2.1 Bragg reflection: d is the separation between adjacent crystal planes, θ is the incidence angle, and the path difference is $2d \sin \theta$.

Since the function $f(k') = |f(k')|\, e^{i\delta(k')}$ is appreciable only over a narrow range around k, the wave packet $\Psi(x,t)$, given by

$$\Psi(x,t) = \int \frac{dk'}{(2\pi)^{1/2}}\, |f(k')|\, e^{i[\varphi(k')+\delta(k')]} , \tag{2.4}$$

will be negligible in this range if the phase factor $e^{i[\varphi(k')+\delta(k')]}$ oscillates rapidly, since positive and negative contributions in $\Psi(x,t)$ will tend to cancel out. By contrast, if the phase factor hardly oscillates around $k' \approx k$ then $\Psi(x,t)$ will be large. Indeed, the condition that $\varphi(k') + \delta(k')$ changes slowly around k implies that

$$\frac{d}{dk'}\left[\varphi(k')+\delta(k')\right]\Big|_{k'=k} \approx 0 , \tag{2.5}$$

and this in turn defines the "center" of the wave packet,

$$x_0(t) \approx t\, \frac{d}{dk'}\, \omega_{k'}\Big|_{k'=k} - \frac{d}{dk'}\, \delta(k')\Big|_{k'=k} . \tag{2.6}$$

It moves with a velocity, called the group velocity, given by

$$v_{\mathrm{gr}} = \dot{x}_0(t) = \frac{d}{dk'}\, \omega_{k'}\Big|_{k'=k} = \frac{\hbar k}{m} = \frac{p}{m} , \tag{2.7}$$

i.e., with the particle's velocity $v = p/m$. By expanding $\varphi(k') + \delta(k')$ around k up to linear terms in $k' - k$, we have

$$\varphi(k') + \delta(k') \approx \varphi(k) + \delta(k) + (k'-k)\, \frac{d}{dk'}\left[\varphi(k')+\delta(k')\right]\Big|_{k'=k} , \tag{2.8}$$

and by noting that

$$\frac{d}{dk'}\left[\varphi(k')+\delta(k')\right]\Big|_{k'=k} = x - t\, \frac{d}{dk'}\, \omega_{k'}\Big|_{k'=k} + \frac{d}{dk'}\, \delta(k')\Big|_{k'=k} = x - x_0(t) , \tag{2.9}$$

the wave packet can be written approximately as

$$\Psi(x,t) \approx e^{i[\varphi(k)+\delta(k)]} \int \frac{dk'}{(2\pi)^{1/2}}\, |f(k')|\, e^{i(k'-k)[x-x_0(t)]} . \tag{2.10}$$

It follows that $|\Psi(x,t)| \approx |\Psi(x-x_0(t),0)|$; that is, the wave packet that was concentrated at $t=0$ in a range Δx centered at x, is approximately concentrated at t in about the same range Δx but centered at $x + x_0(t)$.

2.2 Wave Mechanics

A free particle is represented by the wave packet

$$\Psi(\mathbf{r}, t) = \int \frac{d\mathbf{k}'}{(2\pi)^{3/2}} f(\mathbf{k}') \, e^{i(\mathbf{k}' \cdot \mathbf{r} - \omega_{k'} t)} \, , \tag{2.11}$$

and use of the de Broglie condition $\hbar \omega_k = \hbar^2 k^2 / 2m$ shows that it satisfies a partial differential equation, known as the Schrödinger equation,

$$i\hbar \frac{\partial}{\partial t} \Psi(\mathbf{r}, t) = -\frac{\hbar^2}{2m} \nabla^2 \Psi(\mathbf{r}, t) \, . \tag{2.12}$$

Since this equation is first order in time, its solution is uniquely specified by the initial condition $\Psi(\mathbf{r}, t_0)$ at some reference time t_0. The physical meaning of the wave packet (or wave function, hereafter) is that $|\Psi(\mathbf{r}, t)|^2 \, d\mathbf{r}$ gives the probability of finding the particle in a volume element $d\mathbf{r}$ centered at $\mathbf{r}$. It is natural to define a probability density (probability per unit volume)

$$\rho(\mathbf{r}, t) = |\Psi(\mathbf{r}, t)|^2 \qquad \text{(probability density)} \, , \tag{2.13}$$

which obviously must satisfy

$$\int d\mathbf{r} \, \rho(\mathbf{r}, t) = 1 \, , \tag{2.14}$$

the integration being over all space (i.e., the particle is to be found somewhere!). In particular, $\Psi(\mathbf{r}, t)$ must be a square–integrable function (note that the set of square-integrable functions forms a linear vector space).

The requirement in Eq. (2.14) is independent of time. To verify this fact, consider

$$\frac{\partial}{\partial t} \rho(\mathbf{r}, t) = \left[\frac{\partial}{\partial t} \Psi^*(\mathbf{r}, t) \right] \Psi(\mathbf{r}, t) + \Psi^*(\mathbf{r}, t) \frac{\partial}{\partial t} \Psi(\mathbf{r}, t) \, . \tag{2.15}$$

From the Schrödinger equation, we have

$$\frac{\partial}{\partial t} \Psi(\mathbf{r}, t) = -\frac{\hbar}{2mi} \nabla^2 \Psi(\mathbf{r}, t) \, , \qquad \frac{\partial}{\partial t} \Psi^*(\mathbf{r}, t) = \frac{\hbar}{2mi} \nabla^2 \Psi^*(\mathbf{r}, t) \, , \tag{2.16}$$

yielding

$$\frac{\partial}{\partial t} \rho(\mathbf{r}, t) = \frac{\hbar}{2mi} \left[\Psi(\mathbf{r}, t) \nabla^2 \Psi^*(\mathbf{r}, t) - \Psi^*(\mathbf{r}, t) \nabla^2 \Psi(\mathbf{r}, t) \right] = -\frac{\hbar}{2mi} \nabla \cdot \left[\Psi^*(\mathbf{r}, t) \nabla \Psi(\mathbf{r}, t) - \Psi(\mathbf{r}, t) \nabla \Psi^*(\mathbf{r}, t) \right] \, . \tag{2.17}$$

We are then led to define a probability current density $\mathbf{j}(\mathbf{r}, t)$ as

$$\mathbf{j}(\mathbf{r}, t) = \frac{\hbar}{2mi} \left[\Psi^*(\mathbf{r}, t) \nabla \Psi(\mathbf{r}, t) - \Psi(\mathbf{r}, t) \nabla \Psi^*(\mathbf{r}, t) \right] \qquad \text{(probability current density)} \, , \tag{2.18}$$

so as to obtain a local conservation law for the probability,

$$\frac{\partial}{\partial t} \rho(\mathbf{r}, t) + \nabla \cdot \mathbf{j}(\mathbf{r}, t) = 0 \, , \tag{2.19}$$

which is formally identical to the local conservation law satisfied, for example, by the electromagnetic charge and current densities. However, in the present case, rather than a flow of charge (or matter as in hydrodynamics), we are dealing with the flow of a "probability fluid."

Having established the local conservation law, we immediately obtain by Gauss' theorem

$$\frac{d}{dt} \int d\mathbf{r} \, \rho(\mathbf{r}, t) = \int d\mathbf{r} \, \frac{\partial}{\partial t} \rho(\mathbf{r}, t) = - \int d\mathbf{r} \, \boldsymbol{\nabla} \cdot \mathbf{j}(\mathbf{r}, t) = - \int dS \, \hat{\mathbf{n}} \cdot \mathbf{j}(\mathbf{r}, t) = 0 \,, \qquad (2.20)$$

where $\hat{\mathbf{n}}$ is the outwardly directed unit vector perpendicular to the surface element dS. Since the volume integration is over all space, the surface is obviously at ∞, and therefore the surface integral vanishes, since $\Psi(\mathbf{r}, t)$ and hence $\mathbf{j}(\mathbf{r}, t)$ vanish as $|\mathbf{r}| \longrightarrow \infty$ (recall that the wave function is square integrable). We obtain

$$\frac{d}{dt} \int d\mathbf{r} \, \rho(\mathbf{r}, t) = 0 \implies \int d\mathbf{r} \, \rho(\mathbf{r}, t) = \text{constant} \,. \qquad (2.21)$$

Given a wave function $\Psi(\mathbf{r}, t)$ we can calculate the average position of the particle at time t as

$$\langle \mathbf{r}(t) \rangle = \int d\mathbf{r} \, \Psi^*(\mathbf{r}, t) \, \mathbf{r} \, \Psi(\mathbf{r}, t) \,, \qquad (2.22)$$

where we have assumed $\Psi(\mathbf{r}, t)$ to be normalized:

$$\int d\mathbf{r} \, |\Psi(\mathbf{r}, t)|^2 = 1 \,. \qquad (2.23)$$

We note that the average position is time dependent, since $\Psi(\mathbf{r}, t)$ changes in time. The meaning of this average is the following: if we make a large number of position measurements with the particle always in the *same state*, represented by the wave function $\Psi(\mathbf{r}, t)$, then the average of these position measurements is $\langle \mathbf{r}(t) \rangle$. Similarly, the average of any function of position $f(\mathbf{r})$ follows:

$$\langle f(t) \rangle = \int d\mathbf{r} \, \Psi^*(\mathbf{r}, t) \, f(\mathbf{r}) \, \Psi(\mathbf{r}, t) \,. \qquad (2.24)$$

In order to evaluate the average momentum of the particle at time t, we need first to consider the distribution of momenta $\mathbf{p} = \hbar \, \mathbf{k}$ associated with the wave function $\Psi(\mathbf{r}, t)$. To obtain this distribution, we evaluate the Fourier transform

$$\begin{aligned}
\widetilde{\Psi}(\mathbf{p}, t) &= \int \frac{d\mathbf{r}}{(2\pi\hbar)^{3/2}} \, e^{-i\mathbf{p}\cdot\mathbf{r}/\hbar} \, \Psi(\mathbf{r}, t) = \int \frac{d\mathbf{r}}{(2\pi\hbar)^{3/2}} \, e^{-i\mathbf{p}\cdot\mathbf{r}/\hbar} \left[\int \frac{d\mathbf{p}'}{(2\pi\hbar)^{3/2}} \, f(\mathbf{p}') \, e^{i(\mathbf{p}'\cdot\mathbf{r} - E_{p'} t)/\hbar} \right] \\
&= \int d\mathbf{p}' \, f(\mathbf{p}') \, e^{-iE_{p'} t/\hbar} \, \frac{1}{(2\pi\hbar)^3} \int d\mathbf{r} \, e^{i(\mathbf{p}'-\mathbf{p})\cdot\mathbf{r}/\hbar} = \int d\mathbf{p}' \, f(\mathbf{p}') \, e^{-iE_{p'} t/\hbar} \, \delta(\mathbf{p}' - \mathbf{p}) = f(\mathbf{p}) \, e^{-iE_p t/\hbar} \,,
\end{aligned}$$
$$(2.25)$$

where we have used the δ-function property

$$\int d\mathbf{r} \, e^{i(\mathbf{p}'-\mathbf{p})\cdot\mathbf{r}/\hbar} = (2\pi\hbar)^3 \, \delta(\mathbf{p}' - \mathbf{p}) \,, \qquad (2.26)$$

and $E_p = p^2/(2m)$. We interpret $f(\mathbf{p}) \, e^{-iE_p t/\hbar}$ as the momentum-space wave function $\widetilde{\Psi}(\mathbf{p}, t)$ of the (free) particle. The associated momentum-space probability density is then

$$|\widetilde{\Psi}(\mathbf{p}, t)|^2 = |f(\mathbf{p})|^2 \,, \qquad (2.27)$$

and is independent of time. Note that this density is normalized if the wave function $\Psi(\mathbf{r}, t)$ is normalized (Parseval's identity). The average momentum of the particle at time t is obtained as

$$\langle \mathbf{p}(t) \rangle = \int d\mathbf{p} \, \widetilde{\Psi}^*(\mathbf{p}, t) \, \mathbf{p} \, \widetilde{\Psi}(\mathbf{p}, t) \,, \qquad (2.28)$$

and is constant for a free particle,

$$\langle \mathbf{p}(t) \rangle = \int d\mathbf{p} \, \mathbf{p} \, |f(\mathbf{p})|^2 \,. \tag{2.29}$$

The average value of the particle's momentum can also be calculated from the coordinate-space wave function $\Psi(\mathbf{r}, t)$, by noting that

$$\mathbf{p} \, \widetilde{\Psi}(\mathbf{p}, t) = \int \frac{d\mathbf{r}}{(2\pi\hbar)^{3/2}} \left(i\hbar \boldsymbol{\nabla} \, e^{-i\mathbf{p}\cdot\mathbf{r}/\hbar} \right) \Psi(\mathbf{r}, t) = \int \frac{d\mathbf{r}}{(2\pi\hbar)^{3/2}} \, e^{-i\mathbf{p}\cdot\mathbf{r}/\hbar} \left[-i\hbar \boldsymbol{\nabla}\Psi(\mathbf{r}, t) \right] \,, \tag{2.30}$$

after an integration by parts, and hence

$$\langle \mathbf{p}(t) \rangle = \int d\mathbf{p} \underbrace{\left[\int \frac{d\mathbf{r}'}{(2\pi\hbar)^{3/2}} \, e^{i\mathbf{p}\cdot\mathbf{r}'/\hbar} \, \Psi^*(\mathbf{r}', t) \right]}_{\Psi^*(\mathbf{p},t)} \underbrace{\left[\int \frac{d\mathbf{r}}{(2\pi\hbar)^{3/2}} \, e^{-i\mathbf{p}\cdot\mathbf{r}/\hbar} \left[-i\hbar\boldsymbol{\nabla}\Psi(\mathbf{r}, t) \right] \right]}_{\mathbf{p}\,\Psi(\mathbf{p},t)}$$

$$= \int d\mathbf{r}' \, d\mathbf{r} \, \Psi^*(\mathbf{r}', t) \left[-i\hbar\boldsymbol{\nabla}\Psi(\mathbf{r}, t) \right] \underbrace{\frac{1}{(2\pi\hbar)^3} \int d\mathbf{p} \, e^{i\mathbf{p}\cdot(\mathbf{r}'-\mathbf{r})/\hbar}}_{\delta(\mathbf{r}-\mathbf{r}')} = \int d\mathbf{r} \, \Psi^*(\mathbf{r}, t) \left[-i\hbar\boldsymbol{\nabla}\Psi(\mathbf{r}, t) \right] \,. \tag{2.31}$$

Therefore, in coordinate space we have $\mathbf{p} \longrightarrow -i\hbar\boldsymbol{\nabla}$, and so in this space the momentum is a differential operator such that $\mathbf{p}\Psi(\mathbf{r}, t) = -i\hbar\boldsymbol{\nabla}\Psi(\mathbf{r}, t)$. By contrast, in momentum space the action of $\mathbf{p}$ is to multiply the (momentum-space) wave function by the vector $\mathbf{p}$: $\mathbf{p}\,\widetilde{\Psi}(\mathbf{p}, t)$.

While it is obvious from the calculation in momentum space that the average value of the particle's momentum is real (as it must be to have physical meaning), this can also be verified directly in coordinate space, since

$$\langle \mathbf{p}(t) \rangle = \int d\mathbf{r} \, \Psi^*(\mathbf{r}, t) \left[-i\hbar\boldsymbol{\nabla}\Psi(\mathbf{r}, t) \right] = \int d\mathbf{r} \left[i\hbar\boldsymbol{\nabla}\Psi^*(\mathbf{r}, t) \right] \Psi(\mathbf{r}, t)$$

$$= \int d\mathbf{r} \left[-i\hbar\boldsymbol{\nabla}\Psi(\mathbf{r}, t) \right]^* \Psi(\mathbf{r}, t) = \langle \mathbf{p}(t) \rangle^* \,. \tag{2.32}$$

This property follows from the fact that, when viewed as a differential operator $-i\hbar\boldsymbol{\nabla}$ acting on square-integrable wave functions, the momentum operator is hermitian, namely, it satisfies

$$\int d\mathbf{r} \, \Psi_1^*(\mathbf{r}, t) \left[-i\hbar\boldsymbol{\nabla}\Psi_2(\mathbf{r}, t) \right] = \int d\mathbf{r} \left[-i\hbar\boldsymbol{\nabla}\Psi_1(\mathbf{r}, t) \right]^* \Psi_2(\mathbf{r}, t) \,, \tag{2.33}$$

where $\Psi_1(\mathbf{r}, t)$ and $\Psi_2(\mathbf{r}, t)$ are any two (square-integrable) wave functions.

Of course, we can also calculate the average value of the particle's position from the momentum-space wave function, since inversion of the Fourier transform in Eq. (2.25) gives

$$\Psi(\mathbf{r}, t) = \int \frac{d\mathbf{p}}{(2\pi\hbar)^{3/2}} \, e^{i\mathbf{p}\cdot\mathbf{r}/\hbar} \, \widetilde{\Psi}(\mathbf{p}, t) \,, \tag{2.34}$$

so that

$$\mathbf{r}\,\Psi(\mathbf{r}, t) = \int \frac{d\mathbf{p}}{(2\pi\hbar)^{3/2}} \left(-i\hbar\boldsymbol{\nabla}_p \, e^{i\mathbf{p}\cdot\mathbf{r}/\hbar} \right) \widetilde{\Psi}(\mathbf{p}, t) = \int \frac{d\mathbf{p}}{(2\pi\hbar)^{3/2}} \, e^{i\mathbf{p}\cdot\mathbf{r}/\hbar} \left[i\hbar\boldsymbol{\nabla}_p \, \widetilde{\Psi}(\mathbf{p}, t) \right] \,, \tag{2.35}$$

where $\boldsymbol{\nabla}_p$ is the gradient with respect to the momentum variables. The remaining steps are the same as previously, yielding

$$\langle \mathbf{r}(t) \rangle = \int d\mathbf{p} \, \widetilde{\Psi}^*(\mathbf{p}, t) \left[i\hbar\boldsymbol{\nabla}_p \, \widetilde{\Psi}(\mathbf{p}, t) \right] \,. \tag{2.36}$$

Therefore the descriptions based on coordinate-space and momentum-space wave functions, respectively $\Psi(\mathbf{r}, t)$ and $\widetilde{\Psi}(\mathbf{p}, t)$, are completely equivalent. The position and momentum operators are given in these spaces by

$$
\begin{array}{ccc}
 & \text{position} & \text{momentum} \\
\text{coordinate space} & \mathbf{r} & -i\hbar\boldsymbol{\nabla} \\
\text{momentum space} & i\hbar\boldsymbol{\nabla}_p & \mathbf{p}
\end{array}
\qquad (2.37)
$$

The $\mathbf{r}$ and $\mathbf{p}$ operators acting on either coordinate-space or momentum-space wave functions satisfy the following relations (known as "commutation relations")

$$
r_i p_j - p_j r_i \equiv \left[r_i, p_j \right] = i\hbar\, \delta_{ij}\,, \qquad (2.38)
$$

where i and j denote the cartesian components. They are independent of the adopted representation (coordinate or momentum space).

2.3 Problems

Problem 1 Group and Phase Velocities of the Matter Wave for a Relativistic Particle

In relativistic mechanics, the velocity of a particle is given by

$$
\mathbf{v} = \frac{c^2\,\mathbf{p}}{E}\,,
$$

where c is the speed of light, and $\mathbf{p}$ and $E = c\sqrt{p^2 + (mc)^2}$ (with $p = |\mathbf{p}|$) are the momentum and energy, respectively. Show that the group velocity $\mathbf{v}_{\text{gr}}$ of the de Broglie wave associated with such a particle is $\mathbf{v}$. Calculate the phase velocity $\mathbf{v}_{\text{ph}}$ of this wave and show that it is larger than c and that $v_{\text{gr}}\, v_{\text{ph}} = c^2$.

Solution

The wave packet describing a free relativistic particle is given by

$$
\Psi(\mathbf{r}, t) = \int \frac{d\mathbf{k}'}{(2\pi)^{3/2}}\, f(\mathbf{k}')\, e^{i(\mathbf{k}'\cdot\mathbf{r} - \omega_{k'} t)}\,,
$$

where $|f(\mathbf{k})|$ vanishes unless $\mathbf{k}'$ is close to $\mathbf{k}$, and

$$
\mathbf{p}' = \hbar\,\mathbf{k}'\,, \qquad \omega_{k'} = \frac{E'}{\hbar} = \frac{c}{\hbar}\sqrt{\hbar^2 k'^2 + m^2 c^2}\,.
$$

The group velocity is given by

$$
\mathbf{v}_{\text{gr}} = \boldsymbol{\nabla}'\omega_{k'}\big|_{\mathbf{k}} = \frac{c}{\hbar}\,\frac{\hbar^2\mathbf{k}}{\sqrt{\hbar^2 k^2 + m^2 c^2}} = \frac{c^2\,\mathbf{p}}{E}\,.
$$

The phase velocity follows from

$$
k'\,\hat{\mathbf{n}}\cdot\mathbf{r} - \omega_{k'} t = \text{constant} \implies \mathbf{v}'_{\text{ph}} = \frac{\omega_{k'}}{k'}\,\hat{\mathbf{n}} = \frac{E'}{p'}\,\hat{\mathbf{n}} = c\,\frac{\sqrt{p'^2 + (mc)^2}}{p'}\,\hat{\mathbf{n}}\,,
$$

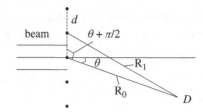

Fig. 2.2   A beam of neutrons incident on a linear chain of nuclei.

where $\mathbf{k}' = k'\,\hat{\mathbf{n}}$. This shows that $v'_{\text{ph}} > c$. For $\mathbf{k}' = \mathbf{k}$ we obtain

$$v_{\text{gr}}\, v_{\text{ph}} = c^2\, \frac{p}{E}\, \frac{E}{p} = c^2\,.$$

Problem 2   Diffraction of Neutrons by a Linear Chain of Nuclei

A beam of neutrons having energy E is incident on a linear chain of atomic nuclei. The chain is perpendicular to the beam, as indicated in Fig. 2.2, and the nuclei are located along it at regular intervals separated by a distance d, which is much larger than the radius of a nucleus. A neutron detector is positioned far from the chain and in a direction making an angle θ with the direction of the incident beam. Intensity peaks are observed at the detector (just as in Bragg reflection): under what condition do they occur?

Solution

One expects to see peaks in the counting rate of diffracted neutron waves when the length difference in the paths from two contiguous nuclei is a multiple of the wavelength associated with a neutron of energy E. To determine this length difference, we refer to Fig. 2.2. By the law of cosines, we have

$$R_1^2 = d^2 + R_0^2 - 2dR_0\,\cos(\pi/2 + \theta) = d^2 + R_0^2 + 2dR_0\,\sin\theta\,,$$

and the length difference between the two paths is $|R_1 - R_0|$ and we are given that $d \ll R_0$. We find that

$$|R_1 - R_0| = R_0\left(\sqrt{1 + 2d\,\sin\theta/R_0 + d^2/R_0^2} - 1\right) \approx R_0\,(1 + d\,\sin\theta/R_0 - 1) = d\,\sin\theta\,.$$

Constructive interference occurs if

$$d\sin\theta = n\lambda \qquad n = 1, 2, \dots\,, \qquad \lambda = \frac{h}{\sqrt{2ME}}\,.$$

As the energy of the neutrons changes, peaks will be seen in the counting rate when the wavelength λ matches this condition.

Problem 3   Bragg Reflection with Index of Refraction

Davisson and Germer, while carrying out their experiment on the reflection of electrons from a nickel crystal, actually observed a slight deviation from the Bragg reflection formula, as if there were an index of refraction $\mu > 1$ in the medium. In this problem the reader can explain this deviation.

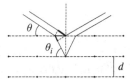

Fig. 2.3   Bragg reflection with index of refraction.

1. Begin by considering the reflection of two light rays by two contiguous crystal planes, as illustrated in Fig. 2.3. Let μ be the index of refraction of the light in the crystal (the index of refraction is 1 in vacuum), and let θ and θ_i be the angles in vacuum and in the medium, respectively, as specified in the figure. Snell's law of optics then states that $\cos\theta = \mu\cos\theta_i$. Knowing that the optical path length is defined as the geometric length times the index of refraction, show that there is no difference in the optical path lengths of s (extra-bold line in vacuum) and s_i (bold line in the medium).

2. Show that the difference in optical path lengths between the two reflected light rays is given as follows:

$$\text{difference in optical path length} = 2\mu d\sin\theta_i \,,$$

and that intensity peaks in the reflected light occur when

$$2d\sqrt{\mu^2 - \cos^2\theta} = n\lambda \,,$$

where λ is the wavelength of the incident light and n is an integer (Bragg reflection taking into account the index of refraction).

3. Now, consider the Davisson–Germer experiment. Assume that an electron in the medium sees, approximately, a uniform attractive potential $-|V|$. If the energy E of the electron is conserved, show that its wavelengths inside and outside the medium, respectively λ_i and λ, are related to each other via

$$\frac{\lambda}{\lambda_i} = \sqrt{\frac{E + |V|}{E}} > 1 \,,$$

and, knowing that $\mu = \lambda/\lambda_i$, conclude that intensity peaks in the waves associated with the reflected electrons occur when

$$2d\sqrt{\frac{E + |V|}{E} - \cos^2\theta} = n\lambda \,.$$

Solution

Part 1

Referring to Fig. 2.3, we see that the segment s (extra-bold line) is the adjacent side of a right-angled triangle with hypothenuse, say, b; the length of this side is $b\cos\theta$. The segment s_i (bold line) is the adjacent side of another right triangle sharing the same hypothenuse, and its length is $b\cos\theta_i$. The optical path lengths in vacuum and in medium are, respectively, $b \cos \theta$ and $\mu b \cos \theta_i$. By Snell's law we conclude that these optical lengths are the same.

Part 2

The difference in the optical path lengths of the two light rays is given by 2μ times $d \sin \theta_i$, and the condition for constructive interference becomes

$$2\mu d \sin \theta_i = n \lambda \,,$$

which by Snell's law can also be expressed as

$$2\mu d \sqrt{1 - \cos^2 \theta_i} = 2\mu d \sqrt{1 - \cos^2 \theta / \mu^2} = 2d \sqrt{\mu^2 - \cos^2 \theta} = n\lambda \,.$$

Part 3

An electron of energy E (which is assumed to be conserved) has momentum $p_i = \sqrt{2m(E + |V|)}$ in the crystal and momentum $p = \sqrt{2mE}$ in vacuum, The wavelengths of the associated waves in the medium and in vacuum are $\lambda_i = h/p_i$ and $\lambda = h/p$, respectively, and hence

$$\frac{\lambda}{\lambda_i} = \frac{p_i}{p} = \sqrt{\frac{E + |V|}{E}} \,,$$

and the condition for Bragg reflection of the electron waves is

$$2d \sqrt{\frac{E + |V|}{E} - \cos^2 \theta} = n\lambda \,.$$

Problem 4 Shape of a Wave Packet in Three Dimensions as a Function of Time

Consider the following wave packet in three dimensions:

$$\Psi(\mathbf{r}, t) = \int \frac{d\mathbf{k}'}{(2\pi)^{3/2}} \, f(\mathbf{k}') \, e^{i(\mathbf{k}' \cdot \mathbf{r} - \omega' t)} \,,$$

and assume that ω' is some function of $\mathbf{k}'$.

1. Suppose $f(\mathbf{k}')$ is $\neq 0$ only for $\mathbf{k}'$ in a small region around $\mathbf{k}$. By expanding ω' in a Taylor series around $\mathbf{k}' = \mathbf{k}$ up to linear terms in $\mathbf{k}' - \mathbf{k}$, show that

$$\Psi(\mathbf{r}, t) \approx e^{-i(\omega - \mathbf{k} \cdot \boldsymbol{\nabla}_{\mathbf{k}} \, \omega) t} \, \Psi(\mathbf{r} - \boldsymbol{\nabla}_{\mathbf{k}} \, \omega t, 0)$$

   where we have defined $\omega = \omega'(\mathbf{k})$ and $\boldsymbol{\nabla}_{\mathbf{k}} \, \omega = \boldsymbol{\nabla}_{\mathbf{k}'} \, \omega'|_{\mathbf{k}'=\mathbf{k}}$. Consider the shape of the wave packet $|\Psi(\mathbf{r}, t)|$. What does the relation above imply about the change of this shape with time?
2. From the relation obtained in part 1, deduce that the wave packet travels with velocity – the so-called group velocity $\mathbf{v}_{\mathrm{gr}}$ – given by $\boldsymbol{\nabla}_{\mathbf{k}} \, \omega$. Identifying this velocity as the velocity $\hbar\mathbf{k}/m$ of the particle, where m is its mass, shows that

$$\omega = \frac{\hbar \mathbf{k}^2}{2m} + \text{constant} \,.$$

Solution

Part 1

To linear terms in $\mathbf{k}' - \mathbf{k}$, we have

$$\omega(\mathbf{k}') = \omega(\mathbf{k}) + (\mathbf{k}' - \mathbf{k}) \cdot \nabla_{\mathbf{k}'} \omega'(\mathbf{k}')|_{\mathbf{k}'=\mathbf{k}} + \cdots = \omega + (\mathbf{k}' - \mathbf{k}) \cdot \nabla_{\mathbf{k}} \omega + \cdots ,$$

in the notation adopted in the text of the problem. Inserting this expansion into the expression for the wave packet, we find

$$\Psi(\mathbf{r}, t) \approx \int \frac{d\mathbf{k}'}{(2\pi)^3} f(\mathbf{k}') \, e^{i\mathbf{k}' \cdot \mathbf{r}} \, e^{-i[\omega + (\mathbf{k}'-\mathbf{k}) \cdot \nabla_{\mathbf{k}} \omega]t} = e^{-i(\omega - \mathbf{k} \cdot \nabla_{\mathbf{k}} \omega)t} \int \frac{d\mathbf{k}'}{(2\pi)^3} f(\mathbf{k}') \, e^{i\mathbf{k}' \cdot (\mathbf{r} - \nabla_{\mathbf{k}} \omega \, t)} .$$

At $t = 0$ the wave packet is given by

$$\Psi(\mathbf{r}, 0) = \int \frac{d\mathbf{k}'}{(2\pi)^3} f(\mathbf{k}') \, e^{i\mathbf{k}' \cdot \mathbf{r}} ,$$

and we see that the magnitude of $\Psi(\mathbf{r}, t)$ is the same as that of $\Psi(\mathbf{r} - \mathbf{r}_0(t), 0)$ with $\mathbf{r}_0(t) = \nabla_{\mathbf{k}} \omega t$, that is,

$$|\Psi(\mathbf{r}, t)| \approx \left| \int \frac{d\mathbf{k}'}{(2\pi)^3} f(\mathbf{k}') \, e^{i\mathbf{k}' \cdot (\mathbf{r} - \nabla_{\mathbf{k}} \omega \, t)} \right| = |\Psi(\mathbf{r} - \mathbf{r}_0(t), 0)| .$$

Thus, the shape of the wave packet is unchanged as time increases (of course, this is true only under the approximation that higher-order corrections in the expansion of ω' can be neglected). In particular, if the wave packet was localized at a position $\mathbf{r}'$ at $t = 0$, it will be localized at the position $\mathbf{r}' + \mathbf{r}_0(t)$ at time t.

Part 2

The wave packet travels with the group velocity, given by $\dot{\mathbf{r}}_0(t) = \nabla_{\mathbf{k}} \omega$, which we identify with the particle's velocity $\hbar \mathbf{k}/m$. It follows that

$$\nabla_{\mathbf{k}} \omega = \frac{\hbar \mathbf{k}}{m} \quad \Longrightarrow \quad \omega = \frac{\hbar k^2}{2m} + \text{constant} .$$

Problem 5 The Green's Function for a Free Particle

1. Show that the wave function at time t for a free particle of mass m can be written as

$$\Psi(\mathbf{r}, t) = \int d\mathbf{r}_0 \, G(\mathbf{r} - \mathbf{r}_0, t - t_0) \, \Psi(\mathbf{r}_0, t_0) ,$$

where $\Psi(\mathbf{r}, t_0)$ is the wave function at the initial time t_0 and the function $G(\mathbf{r} - \mathbf{r}_0, t - t_0)$, known as the free-particle Green's function, is given by

$$G(\mathbf{r} - \mathbf{r}_0, t - t_0) = \int \frac{d\mathbf{p}}{(2\pi\hbar)^3} \, e^{i[\mathbf{p} \cdot (\mathbf{r} - \mathbf{r}_0) - E_p(t - t_0)]/\hbar} , \qquad E_p = \frac{p^2}{2m} .$$

2. Obtain an explicit expression for the Green's function.

   Hint: Use the following integral:

$$\int_{-\infty}^{\infty} dx \, e^{-\alpha^2 (x - \beta)^2} = \frac{\sqrt{\pi}}{\alpha} ,$$

   where α and β are generally complex numbers with $-\pi/4 < \arg \alpha < \pi/4$ for convergence.

Solution

Part 1

Using the general expression for the free-particle wave function, we find for its Fourier transform

$$
\int \frac{d\mathbf{r}}{(2\pi\hbar)^{3/2}}\, e^{-i\mathbf{q}\cdot\mathbf{r}/\hbar}\, \Psi(\mathbf{r},t) = \int \frac{d\mathbf{r}}{(2\pi\hbar)^{3/2}}\, e^{-i\mathbf{q}\cdot\mathbf{r}/\hbar} \underbrace{\int \frac{d\mathbf{p}}{(2\pi\hbar)^{3/2}}\, f(\mathbf{p})\, e^{i(\mathbf{p}\cdot\mathbf{r}-E_p t)/\hbar}}_{\Psi(\mathbf{r},t)}
$$

$$
= \int d\mathbf{p}\, f(\mathbf{p})\, e^{-iE_p t/\hbar}\, \underbrace{\frac{1}{(2\pi\hbar)^3} \int d\mathbf{r}\, e^{i(\mathbf{p}-\mathbf{q})\cdot\mathbf{r}/\hbar}}_{\delta(\mathbf{p}-\mathbf{q})} = f(\mathbf{q})\, e^{-iE_q t/\hbar}\,.
$$

Therefore at t_0 we have

$$
\int \frac{d\mathbf{r}}{(2\pi\hbar)^{3/2}}\, e^{-i\mathbf{q}\cdot\mathbf{r}/\hbar}\, \Psi(\mathbf{r},t_0) = f(\mathbf{q})\, e^{-iE_q t_0/\hbar}\,.
$$

Now in the wave packet $\Psi(\mathbf{r},t)$ we replace $f(\mathbf{p})$ with the expression above to obtain

$$
\Psi(\mathbf{r},t) = \int \frac{d\mathbf{p}}{(2\pi\hbar)^{3/2}}\, e^{i(\mathbf{p}\cdot\mathbf{r}-E_p t)/\hbar}\, \underbrace{e^{iE_p t_0/\hbar} \int \frac{d\mathbf{r}_0}{(2\pi\hbar)^{3/2}}\, e^{-i\mathbf{p}\cdot\mathbf{r}_0/\hbar}\, \Psi(\mathbf{r}_0,t_0)}_{f(\mathbf{p})}
$$

$$
= \int d\mathbf{r}_0 \underbrace{\left[\frac{1}{(2\pi\hbar)^3} \int d\mathbf{p}\, e^{i[\mathbf{p}\cdot(\mathbf{r}-\mathbf{r}_0)-E_p(t-t_0)]/\hbar} \right]}_{G(\mathbf{r}-\mathbf{r}_0,t-t_0)} \Psi(\mathbf{r}_0,t_0)\,.
$$

Part 2

We are left with the evaluation of the Fourier transform

$$
G(\mathbf{r},t) = \frac{1}{(2\pi\hbar)^3} \int d\mathbf{p}\, e^{i[\mathbf{p}\cdot\mathbf{r}-tp^2/(2m)]/\hbar} = \frac{1}{(2\pi\hbar)^3} \int d\mathbf{p}\, e^{-ia\{[\mathbf{p}-\mathbf{r}/(2a\hbar)]^2-r^2/(4a^2\hbar^2)\}}\,.
$$

where we have completed the square in the exponent and have defined

$$
a = \frac{t}{2m\hbar}\,.
$$

We shift the integration in $\mathbf{p}$, namely replacing $\mathbf{p} - \mathbf{r}/(2a\hbar)$ by $\mathbf{p}$, and then use the result for Gaussian integrals to find

$$
G(\mathbf{r},t) = \frac{e^{ir^2/(4a\hbar^2)}}{(2\pi\hbar)^3} \underbrace{\left(\int_{-\infty}^{\infty} dp_x\, e^{-iap_x^2} \right) \left(\int_{-\infty}^{\infty} dp_y\, e^{-iap_y^2} \right) \left(\int_{-\infty}^{\infty} dp_z\, e^{-iap_z^2} \right)}_{\int d\mathbf{p}\, e^{-iap^2}}
$$

$$
= \frac{e^{ir^2/(4a\hbar^2)}}{(2\pi\hbar)^3} \left(\frac{\pi}{ia} \right)^{3/2} = \left(\frac{m}{2\pi i\hbar t} \right)^{3/2} \exp\left(i\frac{mr^2}{2\hbar t} \right)\,.
$$

Problem 6 Dominant Contribution to the Free-Particle Wave Packet

The wave packet at time t for a free particle of mass m can be written as

$$
\Psi(\mathbf{r},t) = \int d\mathbf{r}_0\, G(\mathbf{r}-\mathbf{r}_0, t-t_0)\, \Psi(\mathbf{r}_0,t_0)\,,
$$

where $\Psi(\mathbf{r}, t_0)$ is the wave function at the initial time t_0 and the function $G(\mathbf{r}, t)$, known as the free-particle Green's function, reads

$$G(\mathbf{r}, t) = \left(\frac{m}{2\pi i\hbar t}\right)^{3/2} \exp\left(i\frac{m r^2}{2\hbar t}\right) .$$

Deduce from this that the main contribution to $\Psi(\mathbf{r}, t)$ comes from a region surrounding $\mathbf{r}$ whose radius is of order $\sqrt{2\hbar(t - t_0)/m}$.

Solution

It is convenient to make the shift in integration variable $\mathbf{r} - \mathbf{r}_0 \longrightarrow \mathbf{r}_0$ to obtain

$$\Psi(\mathbf{r}, t) = \int d\mathbf{r}_0 \, G(\mathbf{r}_0, t - t_0) \, \Psi(\mathbf{r} - \mathbf{r}_0, t_0) ,$$

where the Green's function,

$$G(\mathbf{r}_0, t) = \left(\frac{1}{i\pi a_0^2}\right)^{3/2} e^{i r_0^2/a_0^2}, \qquad a_0^2 = \frac{2\hbar(t - t_0)}{m} ,$$

is a rapidly oscillating function of $\mathbf{r}_0^2$. As a consequence, we expect the dominant contribution to $\Psi(\mathbf{r}, t)$ to come from the region $|\mathbf{r}_0| \lesssim a_0$. We expand $\Psi(\mathbf{r} - \mathbf{r}_0, t_0)$ in a Taylor series,

$$\Psi(\mathbf{r} - \mathbf{r}_0, t_0) = \Psi(\mathbf{r}, t_0) - \mathbf{r}_0 \cdot \boldsymbol{\nabla}\Psi(\mathbf{r}, t_0) + \frac{1}{2}\sum_{ij} r_{0,i}\, r_{0,j} \frac{\partial}{\partial r_i}\frac{\partial}{\partial r_j}\Psi(\mathbf{r}, t_0) + \cdots ,$$

which, when inserted into the integral for $\Psi(\mathbf{r}, t)$, leads to

$$\Psi(\mathbf{r}, t) \approx \Psi(\mathbf{r}, t_0) + \frac{1}{2}\sum_{ij}\left[\int d\mathbf{r}_0 \, r_{0,i}\, r_{0,j}\, G(\mathbf{r}_0, t - t_0)\right]\frac{\partial}{\partial r_i}\frac{\partial}{\partial r_j}\Psi(\mathbf{r}, t_0) .$$

Note that the term linear in $\mathbf{r}_0$ vanishes (the integrand is odd under $\mathbf{r}_0 \longrightarrow -\mathbf{r}_0$), and we have used[1]

$$\int d\mathbf{r}_0 \, G(\mathbf{r}_0, t - t_0) = 1 .$$

Because of the spherical symmetry of $G(\mathbf{r}_0, t - t_0)$, the only non-vanishing terms in the integral are those having $i = j$; they each have the same value, so $r_{0,i}\, r_{0,i} \longrightarrow \mathbf{r}_0^2/3$. Thus, we find

$$\Psi(\mathbf{r}, t) \approx \Psi(\mathbf{r}, t_0) + \frac{1}{6}\left[\int d\mathbf{r}_0 \, \mathbf{r}_0^2 \, G(\mathbf{r}_0, t - t_0)\right]\sum_{ij}\delta_{ij}\frac{\partial}{\partial r_i}\frac{\partial}{\partial r_j}\Psi(\mathbf{r}, t_0) ,$$

or

$$\Psi(\mathbf{r}, t) \approx \Psi(\mathbf{r}, t_0) + i\frac{a_0^2}{4}\boldsymbol{\nabla}^2\Psi(\mathbf{r}, t_0) ,$$

since

$$\int d\mathbf{r}_0 \, \mathbf{r}_0^2 \, G(\mathbf{r}_0, t - t_0) = i\frac{3a_0^2}{2} .$$

---

[1] In general, we have, for a Gaussian integral,

$$\int_{-\infty}^{\infty} dx \, e^{-\alpha^2(x-\beta)^2} = \frac{\sqrt{\pi}}{\alpha} ,$$

with α and β complex numbers and $-\pi/4 < \arg\alpha < \pi/4$ for convergence.

Problem 7 Time Evolution of a Gaussian Free-Particle Wave Packet

A one-dimensional wave packet is given by

$$\Psi(x, t) = \int_{-\infty}^{\infty} \frac{dk}{\sqrt{2\pi}} \, f(k) \, e^{i(kx - \omega_k t)} \, ,$$

describing a free particle of energy $\hbar\omega_k = \hbar^2 k^2/(2m)$, and with a (real) profile function given by a Gaussian centered at k_0,

$$f(k) = \frac{\sqrt{a}}{(2\pi)^{1/4}} \, e^{-a^2 (k - k_0)^2/4} \, , \qquad a > 0 \, .$$

In the following the integral

$$\int_{-\infty}^{\infty} dx \, e^{-\alpha^2 (x - \beta)^2} = \frac{\sqrt{\pi}}{\alpha}$$

will be needed, where α and β are generally complex numbers with $-\pi/4 < \arg\alpha < \pi/4$ for convergence.

1. Evaluate the integral at time $t = 0$, and obtain $\Psi(x, 0)$.
2. Evaluate the integral at a generic time t, and obtain $\Psi(x, t)$ as

$$\Psi(x, t) = \left(\frac{2a^2}{\pi} \right)^{1/4} \frac{e^{-i\theta}}{\left(a^4 + 4\hbar^2 t^2/m^2 \right)^{1/4}} \, e^{i(k_0 x - \omega_{k_0} t)} \exp\left[-\frac{(x - \hbar k_0 t/m)^2}{a^2 + 2i\hbar t/m} \right] \, ,$$

where the phase factor is given by

$$\tan(2\,\theta) = 2\hbar t/(ma^2) \, .$$

You may want to either (i) use the Green's function or (ii) expand the phase $e^{i(kx - \omega_k t)}$ in a Taylor series around k_0.

3. Calculate $|\Psi(x, t)|$ and compare it with $|\Psi(x, 0)|$. Discuss the result. In particular, does the shape of $|\Psi(x, t)|$ change with time?

Solution

Part 1

By using the result given for the Gaussian integral, we find

$$\Psi(x, 0) = \int_{-\infty}^{\infty} \frac{dk}{\sqrt{2\pi}} f(k) \, e^{ikx} = \frac{\sqrt{a}}{(2\pi)^{3/4}} \int_{-\infty}^{\infty} dk \, e^{-a^2 (k - k_0)^2/4} \, e^{ikx} = \frac{\sqrt{a}}{(2\pi)^{3/4}} \, e^{ik_0 x} \int_{-\infty}^{\infty} dk \, e^{-a^2 k^2/4} \, e^{ikx}$$

$$= \frac{\sqrt{a}}{(2\pi)^{3/4}} \, e^{ik_0 x} \int_{-\infty}^{\infty} dk \, e^{-[a^2 (k - 2ix/a^2)^2/4 + x^2/a^2]} = \frac{\sqrt{a}}{(2\pi)^{3/4}} \, e^{ik_0 x} \frac{2\sqrt{\pi}}{a} \, e^{-x^2/a^2} = \left(\frac{2}{\pi a^2} \right)^{1/4} e^{ik_0 x} \, e^{-x^2/a^2} \, .$$

Part 2

We use the result from Problem 5 to write

$$\Psi(x, t) = \int_{-\infty}^{\infty} dx_0 \, G(x - x_0, t) \, \Psi(x_0, 0) \, ,$$

where $\Psi(x,0)$ is the wave function at the initial time $t = 0$,

$$\Psi(x,0) = \left(\frac{2}{\pi a^2}\right)^{1/4} e^{ik_0 x} e^{-x^2/a^2} ,$$

obtained above, and the Green's function $G(x,t)$ in one dimension reads

$$G(x,t) = \left(\frac{m}{2\pi i\hbar t}\right)^{1/2} \exp\left(i\frac{mx^2}{2\hbar t}\right) .$$

We therefore find that

$$\Psi(x,t) = \left(\frac{2}{\pi a^2}\right)^{1/4} \left(\frac{m}{2\pi i\hbar t}\right)^{1/2} \int_{-\infty}^{\infty} dx_0 \exp\left[i\frac{m(x-x_0)^2}{2\hbar t} + ik_0 x_0 - \frac{x_0^2}{a^2}\right] .$$

We now complete the square in order to evaluate the Gaussian integral:

$$[\cdots] = -\underbrace{\left(\frac{1}{a^2} + \frac{m}{2i\hbar t}\right)}_{A} x_0^2 + i\underbrace{\frac{m}{\hbar t}\left(\frac{\hbar k_0 t}{m} - x\right)}_{B} x_0 + i\underbrace{\frac{mx^2}{2\hbar t}}_{C}$$

$$= -Ax_0^2 + iBx_0 + C = -A\left(x_0 - i\frac{B}{2A}\right)^2 - \frac{B^2}{4A} + C ,$$

and

$$\Psi(x,t) = \left(\frac{2}{\pi a^2}\right)^{1/4} \left(\frac{m}{2\pi i\hbar t}\right)^{1/2} \left(\frac{\pi}{A}\right)^{1/2} \exp\left(-\frac{B^2}{4A} + C\right) .$$

We need to manipulate the above expression. To this end, we define

$$\lambda = 2\hbar t/m , \qquad v_0 = \hbar k_0/m ,$$

and consider first the exponent:

$$(\cdots) = -\frac{(x-v_0 t)^2}{(\lambda/a)^2 - i\lambda} + i\frac{x^2}{\lambda} = \frac{-x^2 - (v_0 t)^2 + 2xv_0 t + x^2 + i\lambda x^2/a^2}{(\lambda/a)^2 - i\lambda}$$

$$= \frac{-(v_0 t)^2 + 2xv_0 t + i\lambda x^2/a^2}{(\lambda/a)^2 - i\lambda} = \frac{-x^2 + i(a^2/\lambda)2xv_0 t - i(a^2/\lambda)(v_0 t)^2}{a^2 + i\lambda}$$

$$= \frac{-(x-v_0 t)^2 + (v_0 t)^2(1 - ia^2/\lambda) - 2xv_0 t(1 - ia^2/\lambda)}{a^2 + i\lambda}$$

$$= -\frac{(x-v_0 t)^2}{a^2 + i\lambda} + \left[(v_0 t)^2 - 2xv_0 t\right]\frac{1 - ia^2/\lambda}{a^2 + i\lambda}$$

$$= -\frac{(x-v_0 t)^2}{a^2 + i\lambda} - \frac{i}{\lambda}\left[(v_0 t)^2 - 2xv_0 t\right]$$

$$= -\frac{(x-v_0 t)^2}{a^2 + 2i\hbar t/m} + ik_0 x - i\frac{\hbar k_0^2 t}{2m} .$$

Next, we consider

$$\left(\frac{2}{\pi a^2}\right)^{1/4} \left(\frac{m}{2\pi i\hbar t}\right)^{1/2} \left(\frac{\pi}{A}\right)^{1/2} = \left(\frac{2}{\pi a^2}\right)^{1/4} \left(\frac{1}{i\lambda}\frac{1}{1/a^2 - i/\lambda}\right)^{1/2}$$

$$= \left(\frac{2}{\pi a^2}\right)^{1/4} \frac{a}{(a^2 + i\lambda)^{1/2}} = \left(\frac{2a^2}{\pi}\right)^{1/4} \frac{1}{\sqrt{a^2 + i\lambda}} ,$$

and, since

$$a^2 + i\lambda = \sqrt{a^4 + \lambda^2}\, e^{2i\theta}\,, \qquad \tan(2\,\theta) = \lambda/a^2 = \frac{2\hbar t}{ma^2}\,,$$

we finally arrive at

$$\left(\frac{2}{\pi a^2}\right)^{1/4} \left(\frac{m}{2\pi i\hbar t}\right)^{1/2} \left(\frac{\pi}{A}\right)^{1/2} = \left(\frac{2a^2}{\pi}\right)^{1/4} \frac{1}{[a^4 + (2\hbar t/m)^2]^{1/4}}\, e^{-i\theta}\,.$$

Collecting results, we obtain

$$\Psi(x,t) = \left(\frac{2a^2}{\pi}\right)^{1/4} \frac{e^{-i\theta}}{(a^4 + 4\hbar^2 t^2/m^2)^{1/4}} \underbrace{e^{ik_0 x}\, e^{-i\hbar k_0^2 t/(2m)}}_{\text{plane wave}} \exp\left[-\frac{(x - \hbar k_0 t/m)^2}{a^2 + 2i\hbar t/m}\right]\,,$$

where the plane wave represents the particle of momentum $p_0 = \hbar k_0$ and corresponding energy $E_0 = \hbar^2 k_0^2/(2m)$, namely

$$\text{plane wave} = e^{i(p_0 x - E_0 t)/\hbar}\,.$$

Part 3

We easily find

$$|\Psi(x,0)| = \left(\frac{2}{\pi a^2}\right)^{1/4} e^{-x^2/a^2}$$

and

$$|\Psi(x,t)| = \left(\frac{2}{\pi a^2}\right)^{1/4} \frac{1}{\left(1 + \Delta_t^2\right)^{1/4}} \exp\left[-\frac{(x - x_t)^2}{a^2(1 + \Delta_t^2)}\right]\,,$$

where we have defined

$$x_t = \frac{\hbar k_0 t}{m}\,, \qquad \Delta_t = \frac{2\hbar t}{ma^2}\,.$$

In obtaining the expression for $|\Psi(x,t)|$, we have used the identity

$$\exp\left[-\frac{(x - \hbar k_0 t/m)^2}{a^2 + 2i\hbar t/m}\right] = \exp\left[-\frac{(x - x_t)^2}{a^2(1 + i\Delta_t)}\right] = \exp\left[-\frac{(x - x_t)^2(1 - i\Delta_t)}{a^2(1 + i\Delta_t)(1 - i\Delta_t)}\right]$$

$$= \exp\left[-\frac{(x - x_t)^2}{a^2(1 + \Delta_t^2)}\right] \underbrace{\exp\left[i\,\Delta_t\, \frac{(x - x_t)^2}{a^2(1 + \Delta_t^2)}\right]}_{\text{phase factor}}\,,$$

and the phase factor drops out when taking the magnitude. As expected, the "center" of the wave packet is at x_t and travels with the velocity of the particle (this is the group velocity associated with the wave packet). As time progresses, the Gaussian is "squashed": its maximum at $x = x_t$ is reduced by the factor $\left(1 + \Delta_t^2\right)^{1/4}$ and its width is broadened so as to preserve the normalization square, which does not change with time (by the conservation of total probability). Indeed, we have

$$\int_{-\infty}^{\infty} dx\, |\Psi(x,0)|^2 = \left(\frac{2}{\pi a^2}\right)^{1/2} \int_{-\infty}^{\infty} dx\, e^{-2x^2/a^2} = 1\,,$$

and

$$\int_{-\infty}^{\infty} dx \, |\Psi(x,t)|^2 = \left(\frac{2}{\pi a^2}\right)^{1/2} \frac{1}{\left(1 + \Delta_t^2\right)^{1/2}} \int_{-\infty}^{\infty} dx \exp\left[-2\frac{(x - x_t)^2}{a^2(1 + \Delta_t^2)}\right] = \left(\frac{2}{\pi a^2}\right)^{1/2} \int_{-\infty}^{\infty} dy \, e^{-2y^2/a^2} = 1 \, ,$$

where we have made the replacements

$$y = \frac{x - x_t}{(1 + \Delta_t^2)^{1/2}} \, , \qquad dx = (1 + \Delta_t^2)^{1/2} \, dy \, .$$

Problem 8   Probability Density and Probability Current Density as Expectation Values of Operators

Show that the probability density and probability current density at position $\mathbf{r}_0$ can be expressed as expectation values of the operators $\rho(\mathbf{r}_0)$ and $\mathbf{j}(\mathbf{r}_0)$ defined as

$$\rho(\mathbf{r}_0) = \delta(\mathbf{r} - \mathbf{r}_0) \, , \qquad \mathbf{j}(\mathbf{r}_0) = \frac{1}{2m} \left[\mathbf{p} \, \delta(\mathbf{r} - \mathbf{r}_0) + \delta(\mathbf{r} - \mathbf{r}_0)\,\mathbf{p}\right] \, ,$$

where $\mathbf{r}$ and $\mathbf{p}$ are the position and momentum operators. Derive expressions for these densities in both coordinate and momentum space.

Solution

Assuming that the (coordinate-space) wave function $\psi(\mathbf{r})$ is normalized, we have

$$\langle \rho(\mathbf{r}_0) \rangle = \int d\mathbf{r} \, \psi^*(\mathbf{r}) \, \delta(\mathbf{r} - \mathbf{r}_0) \, \psi(\mathbf{r}) = \int d\mathbf{r} \, |\psi(\mathbf{r})|^2 \delta(\mathbf{r} - \mathbf{r}_0) = |\psi(\mathbf{r}_0)|^2 \, ,$$

that is, the probability density at $\mathbf{r}_0$. Similarly, for the probability current density we find

$$\langle \mathbf{j}(\mathbf{r}_0) \rangle = -\frac{i\hbar}{2m} \underbrace{\int d\mathbf{r} \, \psi^*(\mathbf{r}) \left[\boldsymbol{\nabla} \, \delta(\mathbf{r} - \mathbf{r}_0)\psi(\mathbf{r})\right]}_{\text{term 1}} -\frac{i\hbar}{2m} \underbrace{\int d\mathbf{r} \, \psi^*(\mathbf{r}) \, \delta(\mathbf{r} - \mathbf{r}_0) \left[\boldsymbol{\nabla}\psi(\mathbf{r})\right]}_{\text{term 2}} \, .$$

Integrating by parts, term 1 yields (recall that the wave function $\psi(\mathbf{r})$ is assumed to be square integrable and hence vanishing as $|\mathbf{r}| \longrightarrow \infty$)

$$\text{term 1} = \underbrace{\int d\mathbf{r} \, \boldsymbol{\nabla} \left[\psi^*(\mathbf{r}) \, \delta(\mathbf{r} - \mathbf{r}_0)\psi(\mathbf{r})\right]}_{\text{vanishes}} - \int d\mathbf{r} \left[\boldsymbol{\nabla} \, \psi^*(\mathbf{r})\right] \delta(\mathbf{r} - \mathbf{r}_0)\psi(\mathbf{r}) \, ,$$

and therefore, combining and rearranging terms 1 and 2,

$$\langle \mathbf{j}(\mathbf{r}_0) \rangle = -\frac{i\hbar}{2m} \int d\mathbf{r} \, \delta(\mathbf{r} - \mathbf{r}_0) \left[\psi^*(\mathbf{r}) \, \boldsymbol{\nabla}\psi(\mathbf{r}) - \psi(\mathbf{r}) \, \boldsymbol{\nabla}\psi^*(\mathbf{r})\right] = -\frac{i\hbar}{2m} \left[\psi^*(\mathbf{r}_0) \, \boldsymbol{\nabla}\psi(\mathbf{r})|_{\mathbf{r}=\mathbf{r}_0} - \text{c.c.}\right] \, .$$

where c.c. is the complex conjugate. The (normalized) coordinate- and momentum-space wave functions are related to each other by a Fourier transform,

$$\widetilde{\psi}(\mathbf{p}) = \int \frac{d\mathbf{r}}{(2\pi\hbar)^{3/2}} \, e^{-i\mathbf{p}\cdot\mathbf{r}/\hbar} \, \psi(\mathbf{r}) \, , \qquad \psi(\mathbf{r}) = \int \frac{d\mathbf{p}}{(2\pi\hbar)^{3/2}} \, e^{i\mathbf{p}\cdot\mathbf{r}/\hbar} \, \widetilde{\psi}(\mathbf{p}) \, .$$

The probability density at $\mathbf{r}_0$ in momentum space is given by

$$\langle \rho(\mathbf{r}_0) \rangle = \underbrace{\left[\int \frac{d\mathbf{p}}{(2\pi\hbar)^{3/2}} \, e^{i\mathbf{p}\cdot\mathbf{r}_0/\hbar} \, \widetilde{\psi}(\mathbf{p}) \right]^*}_{\psi^*(\mathbf{r}_0)} \underbrace{\left[\int \frac{d\mathbf{p}'}{(2\pi\hbar)^{3/2}} \, e^{i\mathbf{p}'\cdot\mathbf{r}_0/\hbar} \, \widetilde{\psi}(\mathbf{p}') \right]}_{\psi(\mathbf{r}_0)} = \int \frac{d\mathbf{p}\,d\mathbf{p}'}{(2\pi\hbar)^3} \, e^{-i(\mathbf{p}-\mathbf{p}')\cdot\mathbf{r}_0/\hbar} \, \widetilde{\psi}^*(\mathbf{p}) \, \widetilde{\psi}(\mathbf{p}') \,.$$

Similarly, we find

$$\langle \mathbf{j}(\mathbf{r}_0) \rangle = -\frac{i\hbar}{2m} \underbrace{\left[\int \frac{d\mathbf{p}}{(2\pi\hbar)^{3/2}} \, e^{-i\mathbf{p}\cdot\mathbf{r}_0/\hbar} \, \widetilde{\psi}^*(\mathbf{p}) \right]}_{\psi^*(\mathbf{r}_0)} \underbrace{\left[\mathbf{\nabla} \int \frac{d\mathbf{p}'}{(2\pi\hbar)^{3/2}} \, e^{i\mathbf{p}'\cdot\mathbf{r}/\hbar} \, \widetilde{\psi}^*(\mathbf{p}) \right]}_{\mathbf{\nabla}\psi(\mathbf{r})} \Bigg|_{\mathbf{r}=\mathbf{r}_0} + \text{c.c.}$$

$$= \int \frac{d\mathbf{p}\,d\mathbf{p}'}{(2\pi\hbar)^3} \, e^{-i(\mathbf{p}-\mathbf{p}')\cdot\mathbf{r}_0/\hbar} \, \frac{\mathbf{p}'}{2m} \, \widetilde{\psi}^*(\mathbf{p}) \, \widetilde{\psi}(\mathbf{p}') + \int \frac{d\mathbf{p}\,d\mathbf{p}'}{(2\pi\hbar)^3} \, e^{i(\mathbf{p}-\mathbf{p}')\cdot\mathbf{r}_0/\hbar} \, \frac{\mathbf{p}'}{2m} \, \widetilde{\psi}(\mathbf{p}) \, \widetilde{\psi}^*(\mathbf{p}')$$

$$= \int \frac{d\mathbf{p}\,d\mathbf{p}'}{(2\pi\hbar)^3} \, e^{-i(\mathbf{p}-\mathbf{p}')\cdot\mathbf{r}_0/\hbar} \left(\frac{\mathbf{p}}{2m} + \frac{\mathbf{p}'}{2m} \right) \widetilde{\psi}^*(\mathbf{p}) \, \widetilde{\psi}(\mathbf{p}') \,,$$

where in the second integral of the second line we made the exchange $\mathbf{p} \rightleftharpoons \mathbf{p}'$ (the Jacobian of the transformation is unity).

Problem 9   Wigner Distribution

Suppose that a particle is described by the normalized wave function $\psi(\mathbf{r})$. The Wigner distribution is defined as

$$W(\mathbf{r}, \mathbf{p}) = \frac{1}{(2\pi\hbar)^3} \int d\mathbf{x} \, e^{i\mathbf{p}\cdot\mathbf{x}/\hbar} \, \psi^*(\mathbf{r} + \mathbf{x}/2) \, \psi(\mathbf{r} - \mathbf{x}/2) \,.$$

1. Show that $W(\mathbf{r}, \mathbf{p})$ is a real function of the (real) variables $\mathbf{r}$ and $\mathbf{p}$ (note that these are not to be interpreted as the position and momentum operator).
2. Show that

$$\int d\mathbf{p} \, W(\mathbf{r}, \mathbf{p}) = |\psi(\mathbf{r})|^2 \,,$$

and that the expectation value of a function of the operator $\mathbf{r}$ is given by

$$\langle f(\mathbf{r}) \rangle = \int d\mathbf{r} \int d\mathbf{p} \, f(\mathbf{r}) \, W(\mathbf{r}, \mathbf{p}) \,.$$

Solution

Part 1

We have

$$W^*(\mathbf{r}, \mathbf{p}) = \frac{1}{(2\pi\hbar)^3} \int d\mathbf{x} \, e^{-i\mathbf{p}\cdot\mathbf{x}/\hbar} \, \psi(\mathbf{r} + \mathbf{x}/2) \, \psi^*(\mathbf{r} - \mathbf{x}/2) \,,$$

and making the exchange $\mathbf{x} \longrightarrow -\mathbf{x}$ yields

$$W^*(\mathbf{r}, \mathbf{p}) = \frac{1}{(2\pi\hbar)^3} \int d\mathbf{x} \, e^{i\mathbf{p}\cdot\mathbf{x}/\hbar} \, \psi(\mathbf{r} - \mathbf{x}/2) \, \psi^*(\mathbf{r} + \mathbf{x}/2) = W(\mathbf{r}, \mathbf{p}) \,;$$

thus the Wigner distribution function is real.

Part 2

We find that

$$\int d\mathbf{p}\, W(\mathbf{r}, \mathbf{p}) = \int d\mathbf{x}\, \psi^*(\mathbf{r} + \mathbf{x}/2)\, \psi(\mathbf{r} - \mathbf{x}/2) \underbrace{\int \frac{d\mathbf{p}}{(2\pi\hbar)^3}\, e^{i\mathbf{p}\cdot\mathbf{x}/\hbar}}_{\delta(\mathbf{x})},$$

and the $\mathbf{p}$-integral can be viewed as the inverse Fourier transform of $\delta(\mathbf{x})$; indeed,

$$\widetilde{\delta}(\mathbf{p}) = \int \frac{d\mathbf{x}}{(2\pi\hbar)^{3/2}}\, e^{-i\mathbf{p}\cdot\mathbf{x}/\hbar}\, \delta(\mathbf{x}) = \frac{1}{(2\pi\hbar)^{3/2}} \quad \text{and} \quad \delta(\mathbf{x}) = \int \frac{d\mathbf{p}}{(2\pi\hbar)^{3/2}}\, e^{i\mathbf{p}\cdot\mathbf{x}/\hbar}\, \widetilde{\delta}(\mathbf{p}) = \int \frac{d\mathbf{p}}{(2\pi\hbar)^3}\, e^{i\mathbf{p}\cdot\mathbf{x}/\hbar}.$$

Thus, exploiting the δ-function, we have

$$\int d\mathbf{p}\, W(\mathbf{r}, \mathbf{p}) = \int d\mathbf{x}\, \psi^*(\mathbf{r} + \mathbf{x}/2)\, \psi(\mathbf{r} - \mathbf{x}/2)\, \delta(\mathbf{x}) = |\psi(\mathbf{r})|^2.$$

This also implies that the average of $f(\mathbf{r})$ follows as

$$\int d\mathbf{r} \int d\mathbf{p}\, f(\mathbf{r})\, W(\mathbf{r}, \mathbf{p}) = \int d\mathbf{r}\, f(\mathbf{r})\, |\psi(\mathbf{r})|^2 = \langle f(\mathbf{r}) \rangle.$$

3

Schrödinger Equation; Uncertainty Relations

In this section we introduce the Schrödinger equation for a particle under the influence of a potential and Heisenberg's uncertainty relations for the position and momentum operators (as a matter of fact, for any two hermitian operators). Heisenberg's relations put limitations on our ability to measure simultaneously the position and momentum of a particle (or the observables associated with any two non-commuting hermitian operators).

3.1 Schrödinger Equation for a Particle in a Potential

In a great leap of faith, which however, turns out to be justified *a posteriori* by the success of the theory in describing correctly all quantum phenomena, we write the Schrödinger equation for a particle in a potential $V(\mathbf{r})$, a real function of $\mathbf{r}$, as

$$i\hbar\frac{\partial}{\partial t}\Psi(\mathbf{r},t) = \underbrace{\left[-\frac{\hbar^2}{2m}\nabla^2 + V(\mathbf{r})\right]}_{H(\mathbf{r},-i\hbar\nabla)}\Psi(\mathbf{r},t)\,, \tag{3.1}$$

where the right-hand side is the classical Hamiltonian $H(\mathbf{r},\mathbf{p}) = \mathbf{p}^2/(2m) + V(\mathbf{r})$, with $\mathbf{r}$ and $\mathbf{p}$ replaced by the corresponding operators. The resulting Hamiltonian operator is hermitian, which ensures that its average (or expectation) value on $\Psi(\mathbf{r},t)$ is real. The interpretation of the wave function is the same as in the free-particle case, and

$$\rho(\mathbf{r},t) = |\Psi(\mathbf{r},t)|^2\,, \qquad \mathbf{j}(\mathbf{r},t) = \frac{\hbar}{2mi}\left[\Psi^*(\mathbf{r},t)\nabla\Psi(\mathbf{r},t) - \Psi(\mathbf{r},t)\nabla\Psi^*(\mathbf{r},t)\right] \tag{3.2}$$

represent, respectively, the probability density and probability current density associated with $\Psi(\mathbf{r},t)$. They satisfy the same local conservation law as that obtained in the free-particle case.

The time dependence of the wave function can be made explicit by setting $\Psi(\mathbf{r},t) = \psi(\mathbf{r})\,\phi(t)$ and solving the Schrödinger equation by the separation of variables, to obtain $\Psi(\mathbf{r},t) = e^{-iE(t-t_0)/\hbar}\,\psi(\mathbf{r})$, where $\psi(\mathbf{r})$ satisfies the time-independent Schrödinger equation

$$H(\mathbf{r},-i\hbar\nabla)\,\psi(\mathbf{r}) = E\psi(\mathbf{r})\,. \tag{3.3}$$

This partial differential equation can be viewed as an eigenvalue problem for the hermitian operator $H(\mathbf{r},-i\hbar\nabla)$. For its solution, we need to impose suitable boundary conditions on $\psi(\mathbf{r})$. Since the wave function must be square integrable for its probabilistic interpretation to make sense, we must have[1]

---

[1] These boundary conditions are appropriate for bound states. Later, we will expand the class of wave functions by including those that are only bounded as $|\mathbf{r}| \longrightarrow \infty$ (typically, they oscillate at large $|\mathbf{r}|$) and for which $\int d\mathbf{r}\,|\psi(\mathbf{r})|^2$ is not defined. They will be interpreted as describing scattering states.

$$\lim_{|\mathbf{r}|\to\infty} \psi(\mathbf{r}) = 0 \,. \tag{3.4}$$

Furthermore, $\psi(\mathbf{r})$ and its derivatives $\boldsymbol{\nabla}\psi(\mathbf{r})$ are continuous, unless the potential $V(\mathbf{r})$ has an infinite singularity as is the case for a δ-function potential or an infinite barrier potential. In such cases, only $\psi(\mathbf{r})$ is continuous; its derivatives are not. Once these boundary conditions have been imposed, solutions are found only for specific values of E,

$$H(\mathbf{r}, -i\hbar\boldsymbol{\nabla})\, \psi_n(\mathbf{r}) = E_n \psi_n(\mathbf{r}) \,, \qquad n = 1, 2, \ldots \tag{3.5}$$

The solutions $\psi_n(\mathbf{r})$ represent the eigenfunctions of the differential operator $H(\mathbf{r}, -i\hbar\boldsymbol{\nabla})$, and the constants E_n represent the associated eigenvalues.[2] The solutions satisfy a number of properties: (i) since H is hermitian, the eigenvalues E_n are real; (ii) the eigenfunctions corresponding to different eigenvalues are orthogonal, in the sense that

$$\int d\mathbf{r}\, \psi_m^*(\mathbf{r})\, \psi_n(\mathbf{r}) = 0 \qquad \text{if } E_m \neq E_n \,; \tag{3.6}$$

and (iii) the eigenfunctions $\psi_n(\mathbf{r})$ form a *complete* set (a basis),[3] and so any square-integrable function $\phi(\mathbf{r})$ can be expanded as

$$\phi(\mathbf{r}) = \sum_n c_n \psi_n(\mathbf{r}) \,, \qquad c_n = \int d\mathbf{r}\, \psi_n^*(\mathbf{r})\, \phi(\mathbf{r}) \,. \tag{3.7}$$

Once an initial condition is specified,

$$\Psi(\mathbf{r}, t = t_0) = \phi(\mathbf{r}) \,, \tag{3.8}$$

a unique solution $\Psi(\mathbf{r}, t)$ can be constructed by linear superposition:

$$\Psi(\mathbf{r}, t) = \sum_n c_n \underbrace{\psi_n(\mathbf{r})\, e^{-iE_n(t-t_0)/\hbar}}_{\Psi_n(\mathbf{r},t)} \,, \tag{3.9}$$

with the c_n as given in Eq. (3.7). Note that $\Psi(\mathbf{r}, t)$ satisfies the same boundary conditions as $\psi_n(\mathbf{r})$.

The above treatment is in coordinate space. Because of the equivalence of the descriptions based on coordinate- and momentum-space representations, however, we can write the Schrödinger equation in momentum space as

$$i\hbar\frac{\partial}{\partial t}\widetilde{\Psi}(\mathbf{p}, t) = \left[\frac{\mathbf{p}^2}{2m} + V(i\hbar\boldsymbol{\nabla}_p)\right]\widetilde{\Psi}(\mathbf{p}, t) \,, \tag{3.10}$$

where $\boldsymbol{\nabla}_p$ denotes the gradient with respect to the $\mathbf{p}$ variables and $V(i\hbar\boldsymbol{\nabla}_p)$ is the differential operator resulting from the replacement $\mathbf{r} \longrightarrow i\hbar\boldsymbol{\nabla}_p$. The coordinate- and momentum-space wave functions are related to each other by a Fourier transform, again as in the free-particle case.

---

[2] A particular eigenvalue may have two or more independent eigenfunctions corresponding to it, in which case it is said to be degenerate.

[3] Mathematically, *completeness* requires that for any square-integrable $\phi(\mathbf{r})$

$$\lim_{N\to\infty} \int d\mathbf{r}\, \left|\phi(\mathbf{r}) - \phi_N(\mathbf{r})\right|^2 = 0 \,, \qquad \phi_N(\mathbf{r}) = \sum_{n=1}^{N} c_n \psi_n(\mathbf{r}) \,.$$

3.2 Heisenberg's Uncertainty Relations

Along with the average value, it is useful to define the variance of $\mathbf{r}$ (or $\mathbf{p}$) as

$$(\Delta r_x)^2 = \int d\mathbf{r}\, \Psi^*(\mathbf{r}, t)\, (r_x - \langle r_x \rangle)^2\, \Psi(\mathbf{r}, t)\,, \tag{3.11}$$

and similarly for the other components (it is understood that the average and variance depend generally on time).[4] Note that $(\Delta r_x)^2 \geq 0$ gives a measure of the spread of r_x values around the average $\langle r_x \rangle$. Similarly, we define

$$(\Delta p_x)^2 = \int d\mathbf{r}\, \Psi^*(\mathbf{r}, t)\, (p_x - \langle p_x \rangle)^2\, \Psi(\mathbf{r}, t)\,. \tag{3.12}$$

If we make a large number of measurements of r_x and p_x with the particle always in the *same state*, described by the wave function $\Psi(\mathbf{r}, t)$, the variances $(\Delta r_x)^2$ and $(\Delta p_x)^2$ give the spreads in values (fluctuations or uncertainties) around $\langle r_x \rangle$ and $\langle p_x \rangle$. In this context, however, it is worthwhile pointing out that, since the expectation value of any power p of the Hamiltonian on its eigenfunctions Ψ_n is given by $\langle H^p \rangle_n = E_n^p$, the variance of the Hamiltonian on these states vanishes, and so the energy value is sharp in Ψ_n; there are no energy fluctuations.

The uncertainties $(\Delta r_x)^2$ and $(\Delta p_x)^2$ depend on the state of the particle, and not on the instrument and/or method of measurement, which in fact will introduce additional uncertainties. It can be shown that

$$\Delta r_\alpha\, \Delta p_\alpha \geq \underbrace{\frac{1}{2}\left| \int d\mathbf{r}\, \Psi^*(\mathbf{r}, t)\, [\, r_\alpha\,,\, p_\alpha\,]\, \Psi(\mathbf{r}, t) \right|}_{\hbar/2} \qquad \text{(Heisenberg's uncertainty relation)}\,. \tag{3.13}$$

The above relation limits the precision with which a given component α of the position and momentum of a particle can be measured simultaneously. However, it puts no *a priori* limitations on the precision with which we can measure different components of the particle's position and momentum, say the x and y components, since in that case the commutator $[\, r_x\,,\, p_y\,]$ vanishes.

Generally, uncertainty relations hold for any two hermitian operators (corresponding to observables) A and B, that is,

$$\Delta A\, \Delta B \geq \frac{1}{2}\left| \int d\mathbf{r}\, \Psi^*(\mathbf{r}, t)\, [\, A\,,\, B\,]\, \Psi(\mathbf{r}, t) \right|\,, \tag{3.14}$$

where

$$(\Delta A)^2 = \int d\mathbf{r}\, \Psi^*(\mathbf{r}, t)\, (A - \langle A \rangle)^2\, \Psi(\mathbf{r}, t)\,, \qquad \langle A \rangle = \int d\mathbf{r}\, \Psi^*(\mathbf{r}, t)\, A\, \Psi(\mathbf{r}, t)\,, \tag{3.15}$$

and similarly for B; note again that, if A and B commute, there is no restriction on the precision with which A and B can be measured simultaneously.

---

[4] The square root of the variance is by definition the root-mean-square deviation.

3.3 Problems

Problem 1 Average Values of Position and Momentum for a Wave Function
$$e^{-i\langle p\rangle x/\hbar}\,\psi(x+\langle x\rangle)$$

Suppose that a system described by the normalized wave function $\psi(x)$ has average values of the position and momentum operators given by, respectively, $\langle x\rangle$ and $\langle p\rangle$. Now, consider the system being described by the wave function

$$\overline{\psi}(x) = e^{-i\langle p\rangle x/\hbar}\,\psi(x+\langle x\rangle)\,.$$

Show that $\overline{\psi}(x)$ is normalized and calculate the expectation values of x and p in this case.

Solution

We have

$$\int_{-\infty}^{\infty} dx\,|\overline{\psi}(x)|^2 = \int_{-\infty}^{\infty} dx\,|\psi(x+\langle x\rangle)|^2 = \int_{-\infty}^{\infty} dx'\,|\psi(x')|^2 = 1\,,$$

where we have shifted the integration variable $x+\langle x\rangle$ to x'. Similarly, we find

$$\langle x\rangle|_{\overline{\psi}} = \int_{-\infty}^{\infty} dx\,x\,|\overline{\psi}(x)|^2 = \int_{-\infty}^{\infty} dx\,x\,|\psi(x+\langle x\rangle)|^2 = \int_{-\infty}^{\infty} dx'\,(x'-\langle x\rangle)|\psi(x')|^2 = \langle x\rangle - \langle x\rangle = 0\,.$$

Using

$$-i\hbar\frac{d}{dx}\overline{\psi}(x) = e^{-i\langle p\rangle x/\hbar}\left[-\langle p\rangle - i\hbar\frac{d}{dx}\right]\psi(x+\langle x\rangle)\,,$$

we also find

$$\langle p\rangle|_{\overline{\psi}} = \int_{-\infty}^{\infty} dx\,\overline{\psi}^{\,*}(x)\left[-i\hbar\frac{d}{dx}\overline{\psi}(x)\right] = \int_{-\infty}^{\infty} dx\,\psi^*(x+\langle x\rangle)\left[-\langle p\rangle - i\hbar\frac{d}{dx}\right]\psi(x+\langle x\rangle)$$

$$= -\langle p\rangle + \int_{-\infty}^{\infty} dx\,\psi^*(x+\langle x\rangle)\left[-i\hbar\frac{d}{dx}\psi(x+\langle x\rangle)\right]$$

$$= -\langle p\rangle + \int_{-\infty}^{\infty} dx'\,\psi^*(x')\left[-i\hbar\frac{d}{dx'}\psi(x')\right] = -\langle p\rangle + \langle p\rangle = 0\,.$$

Problem 2 Probability and Current Densities for a Particle in a Potential

The time-dependent Schrödinger equation in coordinate space for a particle under the influence of a potential $V(\mathbf{r})$ is given by

$$i\hbar\frac{\partial}{\partial t}\Psi(\mathbf{r},t) = \left[-\frac{\hbar^2}{2m}\nabla^2 + V(\mathbf{r})\right]\Psi(\mathbf{r},t)\,.$$

Verify that there exists a probability current given by

$$\mathbf{j}(\mathbf{r}, t) = \frac{\hbar}{2mi} \left[\Psi^*(\mathbf{r}, t) \, \mathbf{\nabla}\Psi(\mathbf{r}, t) - \Psi(\mathbf{r}, t) \, \mathbf{\nabla}\Psi^*(\mathbf{r}, t) \right] ,$$

such that

$$\frac{\partial}{\partial t} \rho(\mathbf{r}, t) + \mathbf{\nabla} \cdot \mathbf{j}(\mathbf{r}, t) = 0 ,$$

where $\rho(\mathbf{r}, t) = |\Psi(\mathbf{r}, t)|^2$.

Solution

Consider (the dependences on $\mathbf{r}$ and t are suppressed for brevity)

$$\frac{\partial \rho}{\partial t} = \left(\frac{\partial}{\partial t} \Psi^* \right) \Psi + \Psi^* \frac{\partial}{\partial t} \Psi .$$

From the Schrödinger equation, we have

$$\frac{\partial \Psi}{\partial t} = \left(-\frac{\hbar}{2im} \nabla^2 + \frac{V}{i\hbar} \right) \Psi , \qquad \frac{\partial \Psi^*}{\partial t} = \left(\frac{\hbar}{2im} \nabla^2 - \frac{V}{i\hbar} \right) \Psi^* ,$$

where we have used the fact that V is real. Inserting these equations into the expression for ρ, we obtain

$$\frac{\partial \rho}{\partial t} = \left[\frac{\hbar}{2im} \nabla^2\Psi^* - \frac{V}{i\hbar} \Psi^* \right] \Psi + \Psi^* \left[-\frac{\hbar}{2im} \nabla^2\Psi + \frac{V}{i\hbar} \Psi \right]$$

$$= \frac{\hbar}{2im} \left[(\nabla^2\Psi^*)\Psi - \Psi^*\nabla^2\Psi \right] = -\frac{\hbar}{2im} \mathbf{\nabla} \cdot (\Psi^*\mathbf{\nabla}\Psi - \Psi\mathbf{\nabla}\Psi^*) ,$$

which leads to the definition of the probability current density $\mathbf{j}$ as

$$\mathbf{j} = \frac{\hbar}{2mi} (\Psi^*\mathbf{\nabla}\Psi - \Psi\mathbf{\nabla}\Psi^*) ,$$

and the local conservation law for the probability fluid,

$$\frac{\partial \rho}{\partial t} + \mathbf{\nabla} \cdot \mathbf{j} = 0 .$$

Problem 3 Probability and Current Densities for a Charged Particle in an Electromagnetic Field

The time-dependent Schrödinger equation for a charged particle in an electromagnetic field reads

$$i\hbar \frac{\partial}{\partial t} \Psi(\mathbf{r}, t) = \left\{ \frac{1}{2m} \left[-i\hbar\mathbf{\nabla} - \frac{q}{c} \mathbf{A}(\mathbf{r}, t) \right]^2 + qU(\mathbf{r}, t) \right\} \Psi(\mathbf{r}, t),$$

where $U(\mathbf{r}, t)$ and $\mathbf{A}(\mathbf{r}, t)$ are the (real) scalar potential and vector potential, respectively; c is the speed of light; and q is the charge of the particle. Show that the probability density $\rho(\mathbf{r}, t)$ and probability current density $\mathbf{j}(\mathbf{r}, t)$ are given in this case by

$$\rho(\mathbf{r}, t) = |\Psi(\mathbf{r}, t)|^2,$$

$$\mathbf{j}(\mathbf{r}, t) = \frac{\hbar}{2mi} \left[\Psi^*(\mathbf{r}, t)\mathbf{\nabla}\Psi(\mathbf{r}, t) - \Psi(\mathbf{r}, t)\mathbf{\nabla}\Psi^*(\mathbf{r}, t) \right] - \frac{q}{mc} \mathbf{A}(\mathbf{r}, t)|\Psi(\mathbf{r}, t)|^2 ,$$

with

$$\frac{\partial}{\partial t}\rho(\mathbf{r}, t) + \mathbf{\nabla} \cdot \mathbf{j}(\mathbf{r}, t) = 0 .$$

Solution

Suppressing the dependence on $\mathbf{r}$ and t for brevity, the Schrödinger equation and its complex conjugate are given by

$$i\hbar\frac{\partial}{\partial t}\Psi = \left[\frac{1}{2m}\left(-i\hbar\mathbf{\nabla} - \frac{q}{c}\mathbf{A}\right)^2 + qU\right]\Psi ,$$

$$-i\hbar\frac{\partial}{\partial t}\Psi^* = \left[\frac{1}{2m}\left(i\hbar\mathbf{\nabla} - \frac{q}{c}\mathbf{A}\right)^2 + qU\right]\Psi^* .$$

Multiply the first equation by Ψ^* and the second by Ψ, and then subtract the second from the first to obtain

$$i\hbar(\Psi^* \dot{\Psi} + \Psi \dot{\Psi}^*) = \frac{1}{2m}\left[\Psi^*\left(-i\hbar\mathbf{\nabla} - \frac{q}{c}\mathbf{A}\right)^2\Psi - \Psi\left(i\hbar\mathbf{\nabla} - \frac{q}{c}\mathbf{A}\right)^2\Psi^*\right] .$$

This expression can be simplified by expanding the squares,

$$i\hbar\frac{\partial\rho}{\partial t} = \frac{1}{2m}\left[(i\hbar)^2(\Psi^*\mathbf{\nabla}^2\Psi - \Psi\mathbf{\nabla}^2\Psi^*) + i\hbar\frac{q}{c}(\Psi^*\mathbf{\nabla}\cdot\mathbf{A}\Psi + \Psi\mathbf{\nabla}\cdot\mathbf{A}\Psi^* + \Psi^*\mathbf{A}\cdot\mathbf{\nabla}\Psi + \Psi\mathbf{A}\cdot\mathbf{\nabla}\Psi^*)\right]$$

$$= \frac{1}{2m}\left[(i\hbar)^2\mathbf{\nabla}\cdot(\Psi^*\mathbf{\nabla}\Psi - \Psi\mathbf{\nabla}\Psi^*) + i\hbar\frac{q}{c}\mathbf{\nabla}\cdot\underbrace{(\Psi^*\mathbf{A}\Psi + \Psi\mathbf{A}\Psi^*)}_{2\mathbf{A}|\Psi|^2}\right] ,$$

where it is important to keep track of the ordering, since for example $\mathbf{\nabla}\cdot\mathbf{A}\Psi$ is not the same $\mathbf{A}\cdot\mathbf{\nabla}\Psi$ (in the first case the $\mathbf{\nabla}$ operator acts on both $\mathbf{A}$ *and* Ψ). We have also used the identity $\mathbf{\nabla}\cdot\mathbf{F}g = g\mathbf{\nabla}\cdot\mathbf{F} + \mathbf{F}\cdot\mathbf{\nabla}g$, where g and $\mathbf{F}$ are generic scalar and vector functions, respectively. Dividing both sides by $i\hbar$, we arrive at

$$\frac{\partial\rho}{\partial t} = \frac{1}{2m}\mathbf{\nabla}\cdot\left[i\hbar(\Psi^*\mathbf{\nabla}\Psi - \Psi\mathbf{\nabla}\Psi^*) + \frac{2q}{c}\mathbf{A}|\Psi|^2\right] ,$$

and hence identify $\mathbf{j}$:

$$\mathbf{j} = -\frac{i\hbar}{2m}(\Psi^*\mathbf{\nabla}\Psi - \Psi\mathbf{\nabla}\Psi^*) - \frac{q}{mc}\mathbf{A}|\Psi|^2 .$$

The local conservation law $\partial\rho/\partial t + \mathbf{\nabla}\cdot\mathbf{j} = 0$ follows.

Problem 4 Gram–Schmidt Orthogonalization

Consider the case where there are two independent eigenfunctions $\psi_1(\mathbf{r})$ and $\psi_2(\mathbf{r})$ corresponding to the same eigenvalue E of the Hamiltonian (so E has degeneracy 2). In general, $\psi_1(\mathbf{r})$ and $\psi_2(\mathbf{r})$ are not orthogonal to each other, albeit each is assumed to be normalized.

1. Define the linear combinations

$$\psi_1'(\mathbf{r}) = \psi_1(\mathbf{r}) , \qquad \psi_2'(\mathbf{r}) = \psi_2(\mathbf{r}) - \lambda\psi_1(\mathbf{r}) ,$$

and determine λ such that $\psi_1'(\mathbf{r})$ and $\psi_2'(\mathbf{r})$ are orthogonal. Show that $|\lambda| \le 1$.

2. Construct the eigenfunctions

$$\overline{\psi}_1'(\mathbf{r}) = \psi_1'(\mathbf{r})\,, \qquad \overline{\psi}_2'(\mathbf{r}) = \frac{1}{\sqrt{1 - |\lambda|^2}}\,\psi_2'(\mathbf{r})\,,$$

and show that they are normalized. Explain why we must have $|\lambda| < 1$.

3. Consider the case in which the eigenvalue E is n-fold degenerate with independent but not mutually orthogonal eigenfunctions $\psi_1(\mathbf{r})$, ..., $\psi_n(\mathbf{r})$ (each is assumed to be normalized). Generalize the above procedure and construct an orthonormal set.

Solution

Part 1

We require

$$\int d\mathbf{r}\,\psi_1'^*(\mathbf{r})\,\psi_2'(\mathbf{r}) = 0 \implies \lambda = \int d\mathbf{r}\,\psi_1^*(\mathbf{r})\,\psi_2(\mathbf{r})\,,$$

where we have used the fact that $\psi_1(\mathbf{r})$ is normalized. This amounts to subtracting from $\psi_2(\mathbf{r})$ its projection along $\psi_1(\mathbf{r})$. From Schwarz's inequality (see the next problem for a derivation) it follows that

$$|\lambda| = \left| \int d\mathbf{r}\,\psi_1^*(\mathbf{r})\,\psi_2(\mathbf{r}) \right| \leq \left[\int d\mathbf{r}\,|\psi_1(\mathbf{r})|^2 \right]^{1/2} \left[\int d\mathbf{r}\,|\psi_2(\mathbf{r})|^2 \right]^{1/2} = 1\,.$$

Part 2

While $\psi_1'(\mathbf{r})$ is obviously normalized, the wave function $\psi_2'(\mathbf{r})$ is not. We have

$$\int d\mathbf{r}\,\psi_2'^*(\mathbf{r})\,\psi_2'(\mathbf{r}) = \int d\mathbf{r}\,\left[\psi_2^*(\mathbf{r}) - \lambda^*\psi_1^*(\mathbf{r}) \right]\left[\psi_2(\mathbf{r}) - \lambda\psi_1(\mathbf{r}) \right]$$

$$= 1 - \lambda^*\underbrace{\int d\mathbf{r}\,\psi_1^*(\mathbf{r})\,\psi_2(\mathbf{r})}_{\lambda} - \lambda\underbrace{\int \psi_2^*(\mathbf{r})\,\psi_1(\mathbf{r})}_{\lambda^*} + |\lambda|^2 = 1 - |\lambda|^2\,,$$

and so the normalized wave function is given by

$$\overline{\psi}_2'(\mathbf{r}) = \frac{1}{\sqrt{1 - |\lambda|^2}}\,\psi_2'(\mathbf{r})\,.$$

We cannot have $|\lambda| = 1$ in this case, otherwise $\psi_2(\mathbf{r})$ would differ from $\psi_1(\mathbf{r})$ only by a phase factor and therefore could not be an independent solution of the Schrödinger equation corresponding to the same eigenvalue E.

Part 3

We choose any two eigenfunctions, say $\psi_1(\mathbf{r})$ and $\psi_2(\mathbf{r})$, from the set and make them orthonormal, proceeding as in part 2. We choose another eigenfunction, say $\psi_3(\mathbf{r})$, from the remaining $n-2$ and construct

$$\psi_3'(\mathbf{r}) = \psi_3(\mathbf{r}) - \lambda_1\overline{\psi}_1'(\mathbf{r}) - \lambda_2\overline{\psi}_2'(\mathbf{r})\,,$$

where

$$\lambda_i = \int d\mathbf{r}\, \overline{\psi}_i^{\prime *}(\mathbf{r})\, \psi_3(\mathbf{r}) \qquad \text{with } i = 1, 2 \,.$$

It is easily seen that $\psi_3'(\mathbf{r})$ is orthogonal to both $\overline{\psi}_1'(\mathbf{r})$ and $\overline{\psi}_2'(\mathbf{r})$. However, it is not normalized and a straightforward calculation shows that

$$\int d\mathbf{r}\, \psi_3^{\prime *}(\mathbf{r})\, \psi_3'(\mathbf{r}) = 1 - |\lambda_1|^2 - |\lambda_2|^2 \,.$$

We therefore define

$$\overline{\psi}_3'(\mathbf{r}) = \frac{1}{\sqrt{1 - |\lambda_1|^2 - |\lambda_2|^2}}\, \psi_3'(\mathbf{r}) \,,$$

and $\overline{\psi}_1'(\mathbf{r})$, $\overline{\psi}_2'(\mathbf{r})$, $\overline{\psi}_3'(\mathbf{r})$ form an orthonormal set. We now choose another eigenfunction, say $\psi_4(\mathbf{r})$, subtract from it its projections along the orthonormal set, and normalize it to obtain $\overline{\psi}_4'(\mathbf{r})$; and so on.

Problem 5   Heisenberg's Uncertainty Relations: A Derivation

Using Schwarz's inequality, which holds for any two square-integrable functions $f(\mathbf{r})$ and $g(\mathbf{r})$,

$$\left[\int d\mathbf{r}\, |f(\mathbf{r})|^2\right]^{1/2} \left[\int d\mathbf{r}\, |g(\mathbf{r})|^2\right]^{1/2} \geq \left|\int d\mathbf{r}\, g^*(\mathbf{r})\, f(\mathbf{r})\right| \,,$$

and the hermiticity of r_x and p_x to express their variances as

$$(\Delta r_x)^2 = \int d\mathbf{r}\, \underbrace{[(r_x - \langle r_x \rangle)\, \Psi(\mathbf{r}, t)]^*}_{f^*}\, \underbrace{[(r_x - \langle r_x \rangle)\, \Psi(\mathbf{r}, t)]}_{f} \,,$$

$$(\Delta p_x)^2 = \int d\mathbf{r}\, \underbrace{[(p_x - \langle p_x \rangle)\, \Psi(\mathbf{r}, t)]^*}_{g^*}\, \underbrace{[(p_x - \langle p_x \rangle)\, \Psi(\mathbf{r}, t)]}_{g} \,,$$

obtain Heisenberg's uncertainty relations.

Solution

We first derive Schwarz's equality. For any complex number λ we have

$$\int d\mathbf{r}\, |f(\mathbf{r}) - \lambda\, g(\mathbf{r})|^2 = \int d\mathbf{r}\, |f(\mathbf{r})|^2 + |\lambda|^2 \int d\mathbf{r}\, |g(\mathbf{r})|^2 - \lambda^* \int d\mathbf{r}\, g^*(\mathbf{r})\, f(\mathbf{r}) - \lambda \int d\mathbf{r}\, g(\mathbf{r})\, f^*(\mathbf{r}) \geq 0 \,.$$

In particular, it is valid for the specific complex number

$$\lambda = \frac{\int d\mathbf{r}\, g^*(\mathbf{r})\, f(\mathbf{r})}{\int d\mathbf{r}\, |g(\mathbf{r})|^2} \,,$$

with

$$\lambda^* = \frac{\int d\mathbf{r}\, g(\mathbf{r})\, f^*(\mathbf{r})}{\int d\mathbf{r}\, |g(\mathbf{r})|^2} \,, \qquad |\lambda|^2 = \frac{\left|\int d\mathbf{r}\, g^*(\mathbf{r})\, f(\mathbf{r})\right|^2}{\left(\int d\mathbf{r}\, |g(\mathbf{r})|^2\right)^2} \,,$$

which, when inserted into the above relation, leads to Schwarz's inequality after multiplying both sides by $\int d\mathbf{r}\, |g(\mathbf{r})|^2$.

Now, it follows that

$$(\Delta r_x)^2 \, (\Delta p_x)^2 \geq \left| \int d\mathbf{r} \, \underbrace{[(p_x - \langle p_x \rangle)\Psi(\mathbf{r}, t)]^*}_{g^*} \underbrace{[(r_x - \langle r_x \rangle)\Psi(\mathbf{r}, t)]}_{f} \right|^2 .$$

We can write the above right-hand side as

$$\text{r.h.s.} = \frac{1}{2}\left| \int d\mathbf{r} \, \underbrace{[(p_x - \langle p_x \rangle)\Psi(\mathbf{r}, t)]^*}_{g^*} \underbrace{[(r_x - \langle r_x \rangle)\Psi(\mathbf{r}, t)]}_{f} \right|^2 + \frac{1}{2}\left| \int d\mathbf{r} \, \underbrace{[(r_x - \langle r_x \rangle)\Psi(\mathbf{r}, t)]^*}_{f^*} \underbrace{[(p_x - \langle p_x \rangle)\Psi(\mathbf{r}, t)]}_{g} \right|^2$$

$$= \frac{1}{2}\left| \underbrace{\int d\mathbf{r}\Psi^*(\mathbf{r}, t) \, [(p_x - \langle p_x \rangle) \, (r_x - \langle r_x \rangle)\Psi(\mathbf{r}, t)]}_{\text{complex number } a} \right|^2 + \frac{1}{2}\left| \underbrace{\int d\mathbf{r}\Psi^*(\mathbf{r}, t) \, [(r_x - \langle r_x \rangle) \, (p_x - \langle p_x \rangle)\Psi(\mathbf{r}, t)]}_{\text{complex number } b} \right|^2$$

$$= \frac{1}{2}\left(|a|^2 + |b|^2 \right) .$$

For any two complex numbers we have

$$|a|^2 + |b|^2 = \frac{1}{2}\left(|a+b|^2 + |a-b|^2 \right) \geq \frac{1}{2}\,|a-b|^2 ,$$

and therefore

$$(\Delta r_x)^2 \, (\Delta p_x)^2 \geq \frac{1}{4}\left| \int d\mathbf{r}\Psi^*(\mathbf{r}, t) \, \underbrace{[(p_x - \langle p_x \rangle)(r_x - \langle r_x \rangle) - (r_x - \langle r_x \rangle)(p_x - \langle p_x \rangle)]}_{\text{expand products}} \Psi(\mathbf{r}, t) \right|^2$$

$$= \frac{1}{4}\left| \int d\mathbf{r} \, \Psi^*(\mathbf{r}, t) \, \underbrace{(p_x r_x - r_x p_x)}_{\text{commutator}} \Psi(\mathbf{r}, t) \right|^2 = \frac{1}{4}\left| \int d\mathbf{r} \, \Psi^*(\mathbf{r}, t) \, \underbrace{[p_x \, , \, r_x]}_{-i\hbar} \Psi(\mathbf{r}, t) \right|^2 ,$$

or (recall that the wave function is normalized) $\Delta r_x \, \Delta p_x \geq \hbar/2$.

Problem 6 Heisenberg's Uncertainty Relations: An Alternative Derivation

Let $\psi(\mathbf{r})$ be a normalized (square-integrable) wave function, and consider

$$\varphi_\alpha(\mathbf{r}; \lambda) = (r_\alpha' + i\lambda \, p_\alpha')\psi(\mathbf{r}) ,$$

where λ is real, and, for $\alpha = x, y, z$,

$$r_\alpha' = r_\alpha - \langle r_\alpha \rangle , \qquad p_\alpha' = p_\alpha - \langle p_\alpha \rangle ,$$

$\langle r_\alpha \rangle$ and $\langle p_\alpha \rangle$ being the expectation values of the operators r_α and $p_\alpha = -i\hbar\nabla_\alpha$. Define the second-order polynomial

$$P_\alpha(\lambda) = \int d\mathbf{r} \, |\varphi_\alpha(\mathbf{r}; \lambda)|^2 ,$$

and, observing that $P_\alpha(\lambda) \geq 0$ for any (real) λ, deduce Heisenberg's uncertainty relations $\Delta r_\alpha \, \Delta p_\alpha \geq \hbar/2$.

Solution

We have

$$P_\alpha(\lambda) = \int d\mathbf{r} \, |\varphi_\alpha(\mathbf{r}; \lambda)|^2 \geq 0 ,$$

or, by inserting the expression for $\varphi_\alpha(\mathbf{r}; \lambda)$,

$$0 \leq \int d\mathbf{r} \, [(r'_\alpha + i\lambda p'_\alpha)\psi(\mathbf{r})]^* [(r'_\alpha + i\lambda p'_\alpha)\psi(\mathbf{r})]$$

$$= \int d\mathbf{r} [[r'_\alpha \psi(\mathbf{r})]^* [r'_\alpha \psi(\mathbf{r})] + \lambda^2 [p'_\alpha \psi(\mathbf{r})]^* [p'_\alpha \psi(\mathbf{r})] - i\lambda [p'_\alpha \psi(\mathbf{r})]^* [r'_\alpha \psi(\mathbf{r})] + i\lambda [r'_\alpha \psi(\mathbf{r})]^* [p'_\alpha \psi(\mathbf{r})]] \, .$$

Since r'_α and p'_α are hermitian, the above expression can be written as

$$(\Delta r_\alpha)^2 + \lambda^2 \, (\Delta p_\alpha)^2 - i\lambda \int d\mathbf{r} \, \psi^*(\mathbf{r}) \left[(p'_\alpha r'_\alpha - r'_\alpha p'_\alpha)\psi(\mathbf{r}) \right] \geq 0 \, ,$$

where we have introduced the variance

$$(\Delta r_\alpha)^2 = \int d\mathbf{r} \, \psi^*(\mathbf{r})[r'^2_\alpha \, \psi(\mathbf{r})] = \int d\mathbf{r} \, \psi^*(\mathbf{r})[(r_\alpha - \langle r_\alpha \rangle)^2 \, \psi(\mathbf{r})] \, ,$$

and similarly for $(\Delta p_\alpha)^2$. We note that

$$p'_\alpha r'_\alpha - r'_\alpha p'_\alpha = (p_\alpha - \langle p_\alpha \rangle)(r_\alpha - \langle r_\alpha \rangle) - (r_\alpha - \langle r_\alpha \rangle)(p_\alpha - \langle p_\alpha \rangle) = p_\alpha r_\alpha - r_\alpha p_\alpha = -i\hbar \, ,$$

and, by exploiting the fact that $\psi(\mathbf{r})$ is normalized, the above inequality reads

$$\underbrace{(\Delta r_\alpha)^2 + \lambda^2 (\Delta p_\alpha)^2 - \lambda\hbar}_{P_\alpha(\lambda)} \geq 0 \, .$$

The above polynomial as a function of λ has a minimum when

$$P'_\alpha(\lambda) = 0 \implies \lambda_{\min} = \frac{\hbar}{2(\Delta p_\alpha)^2} \, ,$$

and so

$$P_\alpha(\lambda_{\min}) = (\Delta r_\alpha)^2 + \frac{\hbar^2}{4(\Delta p_\alpha)^2} - \frac{\hbar^2}{2(\Delta p_\alpha)^2} = (\Delta r_\alpha)^2 - \frac{\hbar^2}{4(\Delta p_\alpha)^2} \, .$$

Since $P_\alpha(\lambda) \geq 0$ for any λ, we must have $P_\alpha(\lambda_{\min}) \geq 0$, and therefore

$$(\Delta r_\alpha)^2 - \frac{\hbar^2}{4(\Delta p_\alpha)^2} \geq 0 \implies \Delta r_\alpha \, \Delta p_\alpha \geq \frac{\hbar}{2} \, .$$

Problem 7 Schrödinger Equation in the p-Representation

In the coordinate representation, the Schrödinger equation for a particle under the influence of a potential $V(\mathbf{r})$ is given by

$$i\hbar \frac{\partial}{\partial t} \Psi(\mathbf{r}, t) = \left[-\frac{\hbar^2}{2m} \nabla^2 + V(\mathbf{r}) \right] \Psi(\mathbf{r}, t) \, .$$

1. Show that the Schrödinger equation in the momentum representation can be written as

$$\left(i\hbar \frac{\partial}{\partial t} - \frac{\mathbf{p}^2}{2m} \right) \widetilde{\Psi}(\mathbf{p}, t) = \frac{1}{(2\pi\hbar)^{3/2}} \int d\mathbf{p}' \, \widetilde{V}(\mathbf{p} - \mathbf{p}') \, \widetilde{\Psi}(\mathbf{p}', t) \, ,$$

where

$$\widetilde{\Psi}(\mathbf{p}, t) = \int \frac{d\mathbf{r}}{(2\pi\hbar)^{3/2}} \, e^{-i\mathbf{p}\cdot\mathbf{r}/\hbar} \, \Psi(\mathbf{r}, t) \, , \qquad \widetilde{V}(\mathbf{p}) = \int \frac{d\mathbf{r}}{(2\pi\hbar)^{3/2}} \, e^{-i\mathbf{p}\cdot\mathbf{r}/\hbar} \, V(\mathbf{r}) \, .$$

Assume that $V(\mathbf{r})$ vanishes as $|\mathbf{r}| \longrightarrow \infty$ fast enough for its Fourier transform to exist. Note that in the momentum-space Schrödinger equation, as given above, the term with the potential reduces to a convolution product involving the Fourier transforms $\widetilde{V}(\mathbf{p})$ and $\widetilde{\Psi}(\mathbf{p}, t)$.

2. Show that the momentum-space Schrödinger equation can also be written as

$$i\hbar \frac{\partial}{\partial t} \widetilde{\Psi}(\mathbf{p}, t) = \left[\frac{\mathbf{p}^2}{2m} + V(i\hbar \boldsymbol{\nabla}_p)\right] \widetilde{\Psi}(\mathbf{p}, t) .$$

Solution

Part 1

Introduce the Fourier transform given by

$$\Psi(\mathbf{r}, t) = \int \frac{d\mathbf{q}}{(2\pi\hbar)^{3/2}} e^{i\mathbf{q}\cdot\mathbf{r}/\hbar} \widetilde{\Psi}(\mathbf{q}, t)$$

into the Schrödinger equation to obtain

$$i\hbar \frac{\partial}{\partial t} \int \frac{d\mathbf{q}}{(2\pi\hbar)^{3/2}} e^{i\mathbf{q}\cdot\mathbf{r}/\hbar} \widetilde{\Psi}(\mathbf{q}, t) = \left[-\frac{\hbar^2}{2m}\boldsymbol{\nabla}^2 + V(\mathbf{r})\right] \int \frac{d\mathbf{q}}{(2\pi\hbar)^{3/2}} e^{i\mathbf{q}\cdot\mathbf{r}/\hbar} \widetilde{\Psi}(\mathbf{q}, t) ,$$

or more explicitly

$$\int \frac{d\mathbf{q}}{(2\pi\hbar)^{3/2}} e^{i\mathbf{q}\cdot\mathbf{r}/\hbar} \, i\hbar \frac{\partial}{\partial t}\widetilde{\Psi}(\mathbf{q}, t) = \int \frac{d\mathbf{q}}{(2\pi\hbar)^{3/2}} e^{i\mathbf{q}\cdot\mathbf{r}/\hbar} \frac{\mathbf{q}^2}{2m}\widetilde{\Psi}(\mathbf{q}, t) + V(\mathbf{r}) \int \frac{d\mathbf{q}}{(2\pi\hbar)^{3/2}} e^{i\mathbf{q}\cdot\mathbf{r}/\hbar}\widetilde{\Psi}(\mathbf{q}, t) .$$

Multiply both sides by $e^{-i\mathbf{p}\cdot\mathbf{r}/\hbar}$ and integrate over $d\mathbf{r}/(2\pi\hbar)^{3/2}$ to find

$$\int d\mathbf{q} \, i\hbar \frac{\partial}{\partial t}\widetilde{\Psi}(\mathbf{q}, t) \underbrace{\int \frac{d\mathbf{r}}{(2\pi\hbar)^3} e^{-i(\mathbf{p}-\mathbf{q})\cdot\mathbf{r}/\hbar}}_{\delta(\mathbf{p}-\mathbf{q})} = \int d\mathbf{q} \, \frac{\mathbf{q}^2}{2m}\widetilde{\Psi}(\mathbf{q}, t) \underbrace{\int \frac{d\mathbf{r}}{(2\pi\hbar)^3} e^{-i(\mathbf{p}-\mathbf{q})\cdot\mathbf{r}/\hbar}}_{\delta(\mathbf{p}-\mathbf{q})}$$

$$+ \int d\mathbf{q}\left[\underbrace{\int \frac{d\mathbf{r}}{(2\pi\hbar)^3} e^{-i(\mathbf{p}-\mathbf{q})\cdot\mathbf{r}/\hbar} V(\mathbf{r})}_{\widetilde{V}(\mathbf{p}-\mathbf{q})/(2\pi\hbar)^{3/2}}\right]\widetilde{\Psi}(\mathbf{q}, t) ,$$

where we have defined

$$\widetilde{V}(\mathbf{p}) = \int \frac{d\mathbf{r}}{(2\pi\hbar)^{3/2}} e^{-i\mathbf{p}\cdot\mathbf{r}/\hbar} V(\mathbf{r}) .$$

Exploiting the δ-function property, we finally arrive at

$$i\hbar \frac{\partial}{\partial t}\widetilde{\Psi}(\mathbf{p}, t) = \frac{\mathbf{p}^2}{2m}\widetilde{\Psi}(\mathbf{p}, t) + \frac{1}{(2\pi\hbar)^{3/2}} \int d\mathbf{q} \, \widetilde{V}(\mathbf{p}-\mathbf{q})\,\widetilde{\Psi}(\mathbf{q}, t) .$$

We could have obtained this result by observing that the Fourier transform of a convolution product is proportional to the product of the individual Fourier transforms, namely

$$\int \frac{d\mathbf{p}}{(2\pi\hbar)^{3/2}} e^{i\mathbf{p}\cdot\mathbf{r}/\hbar}\left[\int d\mathbf{q} \, \widetilde{V}(\mathbf{p}-\mathbf{q})\,\widetilde{\Psi}(\mathbf{q}, t)\right] = (2\pi\hbar)^{3/2} V(\mathbf{r})\,\Psi(\mathbf{r}, t) .$$

Part 2

In momentum space the position and momentum operators are given, respectively, by $i\hbar\boldsymbol{\nabla}_p$ and $\mathbf{p}$, and the wave function is given by $\widetilde{\Psi}(\mathbf{p}, t)$. Thus, it follows that

$$i\hbar\frac{\partial}{\partial t}\widetilde{\Psi}(\mathbf{p}, t) = \frac{\mathbf{p}^2}{2m}\widetilde{\Psi}(\mathbf{p}, t) + V(i\hbar\boldsymbol{\nabla}_p)\widetilde{\Psi}(\mathbf{p}, t) .$$

This can also be seen in the following way. As shown in part 1, we have (illustrating the argument in one dimension for simplicity)

$$\int\frac{dq_x}{(2\pi\hbar)^{1/2}}e^{iq_xr_x/\hbar}\,i\hbar\frac{\partial}{\partial t}\widetilde{\Psi}(q_x, t) = \int\frac{dq_x}{(2\pi\hbar)^{1/2}}e^{iq_xr_x/\hbar}\frac{q_x^2}{2m}\widetilde{\Psi}(q_x, t) + V(r_x)\int\frac{dq_x}{(2\pi\hbar)^{1/2}}e^{iq_xr_x/\hbar}\widetilde{\Psi}(q_x, t) .$$

Expand $V(r_x)$ in a Taylor series around $r_x = 0$,

$$V(r_x) = \sum_{n=0}^{\infty}\frac{V^n(0)}{n!}r_x^n \qquad \text{with} \quad V^{(n)}(0) = \frac{d^n}{dr_x^n}V(r_x)\bigg|_{r_x=0} ,$$

and consider

$$V(r_x)\int\frac{dq_x}{(2\pi\hbar)^{1/2}}e^{iq_xr_x/\hbar}\widetilde{\Psi}(q_x, t) = \sum_{n=0}^{\infty}\frac{V^n(0)}{n!}\int\frac{dq_x}{(2\pi\hbar)^{1/2}}\underbrace{\left[\left(-i\hbar\frac{d}{dq_x}\right)^n e^{iq_xr_x/\hbar}\right]}_{r_x^n\,e^{iq_xr_x/\hbar}}\widetilde{\Psi}(q_x, t)$$

$$= \sum_{n=0}^{\infty}\frac{V^n(0)}{n!}\underbrace{\int\frac{dq_x}{(2\pi\hbar)^{1/2}}e^{iq_xr_x/\hbar}\left(i\hbar\frac{d}{dq_x}\right)^n\widetilde{\Psi}(q_x, t)}_{\text{integrating by parts } n \text{ times}}$$

$$= \int\frac{dq_x}{(2\pi\hbar)^{1/2}}e^{iq_xr_x/\hbar}\underbrace{\left[\sum_{n=0}^{\infty}\frac{V^n(0)}{n!}\left(i\hbar\frac{d}{dq_x}\right)^n\right]}_{V(i\hbar\nabla_{q_x})}\widetilde{\Psi}(q_x, t) ,$$

where we have assumed that $\widetilde{\Psi}(q_x, t)$ and its derivatives vanish as $|q_x| \longrightarrow \infty$. We then have

$$\int\frac{dq_x}{(2\pi\hbar)^{1/2}}e^{iq_xr_x/\hbar}\left[i\hbar\frac{\partial}{\partial t} - \frac{q_x^2}{2m} - V(i\hbar\nabla_{q_x})\right]\widetilde{\Psi}(q_x, t) = 0 ,$$

yielding the required result.

Problem 8   Averages and Variances of Position and Momentum Operators for Gaussian Wave Packet

As we showed in a previous problem, given the following Gaussian wave packet $\Psi(x, 0)$ at time $t = 0$,

$$\Psi(x, 0) = \left(\frac{2}{\pi a^2}\right)^{1/4}e^{ip_0x/\hbar}\,e^{-x^2/a^2} ,$$

the wave packet at time t reads

$$\Psi(x, t) = \left(\frac{2a^2}{\pi}\right)^{1/4}\frac{e^{-i\theta}}{(a^4 + 4\hbar^2t^2/m^2)^{1/4}}\,e^{i(p_0x - E_{p_0}t)/\hbar}\exp\left[-\frac{(x - p_0t/m)^2}{a^2 + 2i\hbar t/m}\right] ,$$

where we define

$$\tan(2\theta) = 2\hbar t/(ma^2) \ .$$

The corresponding momentum-space wave packet is given by

$$\widetilde{\Psi}(p,t) = \frac{\sqrt{a/\hbar}}{(2\pi)^{1/4}} \, e^{-a^2(p-p_0)^2/(2\hbar)^2} \, e^{-iE_p t/\hbar} \ .$$

Calculate $\langle x \rangle$ and $\langle p \rangle$, and the variances $(\Delta x)^2$ and $(\Delta p)^2$ at both $t=0$ and a generic time t. Is Heisenberg's uncertainty relation satisfied?

Solution

In r-space we have

$$|\Psi(x,0)|^2 = \left(\frac{2}{\pi a^2}\right)^{1/2} e^{-2x^2/a^2}$$

and

$$|\Psi(x,t)|^2 = \left(\frac{2}{\pi a^2}\right)^{1/2} \frac{1}{\left(1+\Delta_t^2\right)^{1/2}} \exp\left[-2\frac{(x-x_t)^2}{a^2(1+\Delta_t^2)}\right] \ ,$$

where we have defined

$$x_t = \frac{p_0 t}{m} \ , \qquad \Delta_t = \frac{2\hbar t}{ma^2} \ .$$

By contrast, in p-space we have

$$|\widetilde{\Psi}(p,t)|^2 = \left(\frac{a^2}{2\pi\hbar^2}\right)^{1/2} e^{-2a^2(p-p_0)^2/(2\hbar)^2} \ .$$

We can now easily calculate $\langle x(t) \rangle$ as follows:

$$\langle x(0) \rangle = \int_{-\infty}^{\infty} dx \, x \, |\Psi(x,0)|^2 = \left(\frac{2}{\pi a^2}\right)^{1/2} \int_{-\infty}^{\infty} dx \, x \, e^{-2x^2/a^2} = 0 \ ,$$

$$\langle x(t) \rangle = \int_{-\infty}^{\infty} dx \, x \, |\Psi(x,t)|^2 = \left(\frac{2}{\pi a^2}\right)^{1/2} \frac{1}{\left(1+\Delta_t^2\right)^{1/2}} \int_{-\infty}^{\infty} dx \, x \exp\left[-2\frac{(x-x_t)^2}{a^2(1+\Delta_t^2)}\right]$$

$$= \left(\frac{2}{\pi a^2}\right)^{1/2} \int_{-\infty}^{\infty} dy \left[\left(1+\Delta_t^2\right)^{1/2} y + x_t\right] e^{-2y^2/a^2} = x_t \ ,$$

where in the last line we have shifted the integration variable,

$$y = \frac{x-x_t}{\left(1+\Delta_t^2\right)^{1/2}} \implies dx = \left(1+\Delta_t^2\right)^{1/2} dy \ ,$$

and then used the fact that the first term in the integrand is an odd function of y and therefore its integral vanishes. In p-space we have

$$\langle p(t) \rangle = \int_{-\infty}^{\infty} dp \, p \, |\widetilde{\Psi}(p,t)|^2 = \left(\frac{a^2}{2\pi\hbar^2}\right)^{1/2} \int_{-\infty}^{\infty} dp \, p \, e^{-2a^2(p-p_0)^2/(2\hbar)^2} = p_0 \ ,$$

which is independent of time. In order to calculate the variance, we need (setting $\alpha = 2/a^2$)

$$\langle x^2(0) \rangle = \left(\frac{2}{\pi a^2}\right)^{1/2} \int_{-\infty}^{\infty} dx\, x^2\, e^{-2x^2/a^2} = -\left(\frac{\alpha}{\pi}\right)^{1/2} \frac{d}{d\alpha} \int_{-\infty}^{\infty} dx\, e^{-\alpha x^2} = \frac{1}{2\alpha} = \frac{a^2}{4} \,,$$

$$\langle x^2(t) \rangle = \left(\frac{2}{\pi a^2}\right)^{1/2} \frac{1}{\left(1 + \Delta_t^2\right)^{1/2}} \int_{-\infty}^{\infty} dx\, x^2 \exp\left[-2\frac{(x - x_t)^2}{a^2(1 + \Delta_t^2)}\right]$$

$$= \left(\frac{\alpha}{\pi}\right)^{1/2} \int_{-\infty}^{\infty} dy \left[\left(1 + \Delta_t^2\right)^{1/2} y + x_t\right]^2 e^{-\alpha y^2}$$

$$= \left(\frac{\alpha}{\pi}\right)^{1/2} \left(1 + \Delta_t^2\right) \int_{-\infty}^{\infty} dy\, y^2\, e^{-\alpha y^2} + x_t^2 = \frac{a^2}{4}\left(1 + \Delta_t^2\right) + x_t^2 \,,$$

and therefore

$$(\Delta x_0)^2 = \langle x^2(0) \rangle - \langle x(0) \rangle^2 = \frac{a^2}{4} \,,$$

$$(\Delta x_t)^2 = \langle x^2(t) \rangle - \langle x(t) \rangle^2 = \frac{a^2}{4}\left(1 + \Delta_t^2\right) \,.$$

Similarly, we have

$$\langle p^2(t) \rangle = \left(\frac{a^2}{2\pi\hbar^2}\right)^{1/2} \int_{-\infty}^{\infty} dp\, p^2 e^{-2a^2(p-p_0)^2/(2\hbar)^2}$$

$$= \left(\frac{a^2}{2\pi\hbar^2}\right)^{1/2} \int_{-\infty}^{\infty} dq\, (q + p_0)^2 e^{-2a^2 q^2/(2\hbar)^2} = \frac{\hbar^2}{a^2} + p_0^2$$

and

$$(\Delta p)^2 = \langle p^2 \rangle - \langle p \rangle^2 = \frac{\hbar^2}{a^2} \,.$$

We finally arrive at

$$\Delta x_0\, \Delta p = \frac{\hbar}{2} \,, \qquad \Delta x_t\, \Delta p = \frac{\hbar}{2}\left(1 + \Delta_t^2\right)^{1/2} \,,$$

and $|\Psi(x, t)|$ is "minimal" at $t = 0$ but spreads out as time increases. Heisenberg's uncertainty relation is obeyed.

Problem 9 Formulation of the Schrödinger Equation for $\Psi(\mathbf{r}, t) = e^{iS(\mathbf{r},t)/\hbar}$

Consider writing the wave function as

$$\Psi(\mathbf{r}, t) = e^{iS(\mathbf{r},t)/\hbar} \,,$$

where $\Psi(\mathbf{r}, t)$ satisfies the Schrödinger equation in the presence of a potential $V(\mathbf{r})$ and $S(\mathbf{r}, t)$ is generally a complex function.

1. When $\Psi(\mathbf{r}, t) \neq 0$, show that $S(\mathbf{r}, t)$ satisfies the equation

$$\frac{\partial}{\partial t} S(\mathbf{r}, t) + \frac{[\boldsymbol{\nabla} S(\mathbf{r}, t)]^2}{2m} - \frac{i\hbar}{2m}\, \boldsymbol{\nabla}^2 S(\mathbf{r}, t) + V(\mathbf{r}) = 0 \,.$$

It is interesting to note that in the limit $\hbar \longrightarrow 0$ this reduces to the Hamilton–Jacobi equation of classical mechanics.

2. Express the probability density $\rho(\mathbf{r}, t)$ and probability current density $\mathbf{j}(\mathbf{r}, t)$ in terms of the real and imaginary parts of $S(\mathbf{r}, t)$, denoted respectively as $S_R(\mathbf{r}, t)$ and $S_I(\mathbf{r}, t)$ with $S(\mathbf{r}, t) = S_R(\mathbf{r}, t) + iS_I(\mathbf{r}, t)$. In particular, show that the wave function can be written as

$$\Psi(\mathbf{r}, t) = \sqrt{\rho(\mathbf{r}, t)}\, e^{iS_R(\mathbf{r}, t)/\hbar} \ .$$

3. By taking the real and imaginary parts of the equation for $S(\mathbf{r}, t)$ derived in part 1, show that this equation reduces to a set of coupled partial differential equations for $\rho(\mathbf{r}, t)$ and $S_R(\mathbf{r}, t)$, given by

$$\frac{\partial}{\partial t}\rho(\mathbf{r}, t) + \boldsymbol{\nabla} \cdot \left[\rho(\mathbf{r}, t)\, \frac{\boldsymbol{\nabla} S_R(\mathbf{r}, t)}{m}\right] = 0 \ ,$$

$$\frac{\partial}{\partial t} S_R(\mathbf{r}, t) + \frac{[\boldsymbol{\nabla} S_R(\mathbf{r}, t)]^2}{2m} - \frac{\hbar^2}{2m} \frac{\boldsymbol{\nabla}^2 \sqrt{\rho(\mathbf{r}, t)}}{\sqrt{\rho(\mathbf{r}, t)}} + V(\mathbf{r}) = 0 \ .$$

Show that the first equation is just the continuity equation for the "probability fluid."

Solution

Part 1

Note that

$$\frac{\partial \Psi}{\partial t} = \frac{\partial\, e^{iS/\hbar}}{\partial t} = \frac{i}{\hbar}\left(\frac{\partial S}{\partial t}\right) e^{iS/\hbar}$$

and

$$\boldsymbol{\nabla}^2 \Psi = \boldsymbol{\nabla} \cdot \left(\boldsymbol{\nabla} e^{iS/\hbar}\right) = \boldsymbol{\nabla} \cdot \left[\frac{i}{\hbar}\,(\boldsymbol{\nabla} S)\, e^{iS/\hbar}\right] = \frac{i}{\hbar}\left(\boldsymbol{\nabla}^2 S\right) e^{iS/\hbar} - \frac{1}{\hbar^2}\,(\boldsymbol{\nabla} S)^2\, e^{iS/\hbar} \ .$$

Inserting these relations into the Schrödinger equation yields

$$-\left(\frac{\partial S}{\partial t}\right) e^{iS/\hbar} = -\frac{\hbar^2}{2m}\left[\frac{i}{\hbar}\,\boldsymbol{\nabla}^2 S - \frac{1}{\hbar^2}\,(\boldsymbol{\nabla} S)^2\right] e^{iS/\hbar} + V e^{iS/\hbar} \ ,$$

which implies (reinstating the $\mathbf{r}$ and t dependence)

$$\frac{\partial}{\partial t} S(\mathbf{r}, t) + \frac{[\boldsymbol{\nabla} S(\mathbf{r}, t)]^2}{2m} - \frac{i\hbar}{2m}\,\boldsymbol{\nabla}^2 S(\mathbf{r}, t) + V(\mathbf{r}) = 0 \ .$$

Part 2

Given $S = S_R + iS_I$, we have for the wave function

$$\Psi = e^{-S_I/\hbar}\, \underbrace{e^{iS_R/\hbar}}_{\text{phase}} \ ,$$

which immediately leads to

$$\rho = \Psi^* \Psi = e^{-2\,S_I/\hbar} \ .$$

Using

$$\boldsymbol{\nabla} \Psi = \frac{i}{\hbar}\,(\boldsymbol{\nabla} S)\, e^{iS/\hbar} \ , \qquad \boldsymbol{\nabla} \Psi^* = -\frac{i}{\hbar}\,(\boldsymbol{\nabla} S^*)\, e^{-iS^*/\hbar}$$

we obtain for the probability current

$$\mathbf{j} = \frac{\hbar}{2mi}\left(\Psi^*\boldsymbol{\nabla}\Psi - \Psi\boldsymbol{\nabla}\Psi^*\right) = \frac{1}{2m}\left(e^{-iS^*/\hbar}\,e^{iS/\hbar}\boldsymbol{\nabla}S + e^{iS/\hbar}\,e^{-iS^*/\hbar}\boldsymbol{\nabla}S^*\right)$$

$$= \frac{1}{2m}\,e^{i(S-S^*)/\hbar}\,\boldsymbol{\nabla}\,(S+S^*) = \frac{1}{m}\,e^{-2\,S_I/\hbar}\,\boldsymbol{\nabla}S_R = \frac{\rho}{m}\,\boldsymbol{\nabla}S_R\,,$$

since $S + S^* = 2S_R$ and $S - S^* = 2iS_I$. We also find that

$$e^{-S_I/\hbar} = \sqrt{\rho} \implies \Psi = \sqrt{\rho}\,e^{iS_R/\hbar}\,.$$

Part 3

Start from

$$\frac{\partial S}{\partial t} + \frac{(\boldsymbol{\nabla}S)^2}{2m} - \frac{i\hbar}{2m}\,\boldsymbol{\nabla}^2 S + V = 0$$

and substitute $S = S_R + iS_I$, to obtain

$$\frac{\partial S_R}{\partial t} + i\,\frac{\partial S_I}{\partial t} + \frac{(\boldsymbol{\nabla}S_R + i\,\boldsymbol{\nabla}S_I)^2}{2m} - \frac{i\hbar}{2m}\,(\boldsymbol{\nabla}^2 S_R + i\,\boldsymbol{\nabla}^2 S_I) + V = 0\,.$$

In order for this equation to be satisfied, both its real and imaginary parts must vanish, yielding for the imaginary part

$$\frac{\partial S_I}{\partial t} + \frac{1}{m}\,(\boldsymbol{\nabla}S_R)\cdot(\boldsymbol{\nabla}S_I) - \frac{\hbar}{2m}\,\boldsymbol{\nabla}^2 S_R = 0$$

and for the real part

$$\frac{\partial S_R}{\partial t} + \frac{1}{2m}\left[(\boldsymbol{\nabla}S_R)^2 - (\boldsymbol{\nabla}S_I)^2\right] + \frac{\hbar}{2m}\,\boldsymbol{\nabla}^2 S_I + V = 0\,.$$

Using

$$S_I = -\frac{\hbar}{2}\,\ln\rho$$

gives

$$\frac{\partial S_I}{\partial t} = -\frac{\hbar}{2\rho}\,\frac{\partial\rho}{\partial t}\,,\qquad \boldsymbol{\nabla}S_I = -\frac{\hbar}{2\rho}\,\boldsymbol{\nabla}\rho\,,\qquad \boldsymbol{\nabla}^2 S_I = \boldsymbol{\nabla}\cdot\left(-\frac{\hbar}{2\rho}\,\boldsymbol{\nabla}\rho\right) = \frac{\hbar}{2}\left[\frac{1}{\rho^2}(\boldsymbol{\nabla}\rho)^2 - \frac{1}{\rho}\,\boldsymbol{\nabla}^2\rho\right]\,,$$

and the equation for S_I can now be expressed in term of ρ and S_R as

$$-\frac{\hbar}{2\rho}\,\frac{\partial\rho}{\partial t} - \frac{\hbar}{2m\rho}\,(\boldsymbol{\nabla}\rho)\cdot(\boldsymbol{\nabla}S_R) - \frac{\hbar}{2m}\,\boldsymbol{\nabla}^2 S_R = 0 \implies \frac{\partial\rho}{\partial t} + \frac{1}{m}\,(\boldsymbol{\nabla}\rho)\cdot(\boldsymbol{\nabla}S_R) + \frac{1}{m}\,\rho\boldsymbol{\nabla}^2 S_R = 0\,,$$

which can be further manipulated to yield

$$\frac{\partial\rho}{\partial t} + \frac{1}{m}\,\boldsymbol{\nabla}\cdot\left(\rho\,\boldsymbol{\nabla}S_R\right) = 0\,.$$

Inserting the expression for the probability current density obtained above, this relation reduces to

$$\frac{\partial\rho}{\partial t} + \boldsymbol{\nabla}\cdot\mathbf{j} = 0\,,$$

that is, the conservation of the probability current.

The equation for S_R can be written as

$$\frac{\partial S_R}{\partial t} + \frac{1}{2m} (\boldsymbol{\nabla} S_R)^2 - \frac{\hbar^2}{8m\rho^2} (\boldsymbol{\nabla}\rho)^2 + \frac{\hbar^2}{4m} \left[\frac{1}{\rho^2}(\boldsymbol{\nabla}\rho)^2 - \frac{1}{\rho} \boldsymbol{\nabla}^2\rho \right] + V = 0 \,,$$

or, combining terms,

$$\frac{\partial S_R}{\partial t} + \frac{1}{2m} (\boldsymbol{\nabla} S_R)^2 + \frac{\hbar^2}{4m} \left[\frac{1}{2\rho^2}(\boldsymbol{\nabla}\rho)^2 - \frac{1}{\rho} \boldsymbol{\nabla}^2\rho \right] + V = 0 \,.$$

Note that

$$\frac{1}{\sqrt{\rho}} \boldsymbol{\nabla}^2 \sqrt{\rho} = \frac{1}{\sqrt{\rho}} \boldsymbol{\nabla} \cdot \underbrace{\left(\frac{1}{2\sqrt{\rho}} \boldsymbol{\nabla}\rho \right)}_{\boldsymbol{\nabla}\sqrt{\rho}} = \frac{1}{\sqrt{\rho}} \left[\frac{1}{2\sqrt{\rho}} \boldsymbol{\nabla}^2\rho - \frac{1}{4\,\rho\sqrt{\rho}}(\boldsymbol{\nabla}\rho)^2 \right] \,,$$

which, when compared with the above equation, yields

$$\frac{\partial S_R}{\partial t} + \frac{1}{2m} (\boldsymbol{\nabla} S_R)^2 - \frac{\hbar^2}{2m\sqrt{\rho}} \boldsymbol{\nabla}^2 \sqrt{\rho} + V = 0 \,,$$

proving the result given in the text of the problem.

Problem 10 Time Evolution of the Averages of the Position and Momentum Operators

Consider a particle in a potential $V(\mathbf{r})$ with associated wave function satisfying the time-dependent Schrödinger equation

$$i\hbar\frac{\partial}{\partial t} \Psi(\mathbf{r}, t) = \left[-\frac{\hbar^2}{2m} \boldsymbol{\nabla}^2 + V(\mathbf{r}) \right] \Psi(\mathbf{r}, t) \,.$$

1. Show that

$$\frac{d}{dt}\langle \mathbf{r}(t) \rangle = \int d\mathbf{r} \; \mathbf{j}(\mathbf{r}, t) \,,$$

   where $\langle \mathbf{r}(t) \rangle$ is the average position of the particle and $\mathbf{j}(\mathbf{r}, t)$ is the probability current density. Using the definition of $\mathbf{j}(\mathbf{r}, t)$, show that the above equation can also be written as

$$m \frac{d}{dt}\langle \mathbf{r}(t) \rangle = \langle \mathbf{p}(t) \rangle \,.$$

2. Show that

$$\frac{d}{dt}\langle \mathbf{p}(t) \rangle = -\langle \boldsymbol{\nabla} V \rangle = - \int d\mathbf{r} \, \Psi^*(\mathbf{r}, t) \, [\boldsymbol{\nabla} V(\mathbf{r})] \, \Psi(\mathbf{r}, t) = \int d\mathbf{r} \, \Psi^*(\mathbf{r}, t) \, \mathbf{F}(\mathbf{r}) \, \Psi(\mathbf{r}, t) \,,$$

   where we have introduced the force $\mathbf{F}(\mathbf{r})$.

3. The above equation looks like Newton's second law but for average values. As a matter of fact, if $\langle \mathbf{F} \rangle = \mathbf{F}(\langle \mathbf{r} \rangle)$ then $\langle \mathbf{r}(t) \rangle$ changes in time as the position of a classical particle under the action of the force $\mathbf{F}(\mathbf{r})$. Under what condition(s) can this happen? Obtain $\langle \mathbf{r}(t) \rangle$ and $\langle \mathbf{p}(t) \rangle$ for a particle in a harmonic potential

$$V(\mathbf{r}) = \frac{m\omega^2}{2} \mathbf{r}^2 \,.$$

Solution

Part 1

Start from

$$\langle \mathbf{r}(t) \rangle = \int d\mathbf{r} \, \Psi^*(\mathbf{r}, t) \, \mathbf{r} \, \Psi(\mathbf{r}, t) = \int d\mathbf{r} \, \mathbf{r} \, \rho(\mathbf{r}, t) \,,$$

and therefore

$$\frac{d}{dt}\langle \mathbf{r}(t) \rangle = \int d\mathbf{r} \, \mathbf{r} \, \frac{\partial}{\partial t} \rho(\mathbf{r}, t) = - \int d\mathbf{r} \, \mathbf{r} \, \boldsymbol{\nabla} \cdot \mathbf{j}(\mathbf{r}, t) \,.$$

Consider the x-component (suppressing the $\mathbf{r}$ and t dependence for the time being)

$$\frac{d}{dt}\langle r_x \rangle = - \int d\mathbf{r} \, r_x \, \boldsymbol{\nabla} \cdot \mathbf{j} = - \int d\mathbf{r} \, [\boldsymbol{\nabla} \cdot (r_x \mathbf{j}) - \mathbf{j} \cdot \boldsymbol{\nabla} r_x] \,,$$

where in the last step we have used the identity

$$\boldsymbol{\nabla} \cdot (f \mathbf{g}) = (\boldsymbol{\nabla} f) \cdot \mathbf{g} + f \boldsymbol{\nabla} \cdot \mathbf{g} \implies f \boldsymbol{\nabla} \cdot \mathbf{g} = \boldsymbol{\nabla} \cdot (f \mathbf{g}) - (\boldsymbol{\nabla} f) \cdot \mathbf{g} \,,$$

with $f = r_x$ and $\mathbf{g} = \mathbf{j}$. Now, the first term can be written as a surface integral with the surface at ∞ (Gauss's theorem), and hence vanishes. Noting that the vector $\boldsymbol{\nabla} r_x$ has components $\boldsymbol{\nabla} r_x = (1, 0, 0)$, we conclude that

$$\frac{d}{dt}\langle r_x(t) \rangle = \int d\mathbf{r} \, j_x(\mathbf{r}, t) \,.$$

A similar result holds for r_y and r_z, which proves that

$$\frac{d}{dt}\langle \mathbf{r}(t) \rangle = \int d\mathbf{r} \, \mathbf{j}(\mathbf{r}, t) \,.$$

Inserting the expression for $\mathbf{j}$, we find that

$$\frac{d}{dt}\langle \mathbf{r}(t) \rangle = \int d\mathbf{r} \, \frac{\hbar}{2mi} [\Psi^* \boldsymbol{\nabla} \Psi - (\boldsymbol{\nabla} \Psi^*) \Psi] = \frac{1}{2m} \int d\mathbf{r} \, [\Psi^*(\mathbf{p} \Psi) + (\mathbf{p} \Psi)^* \Psi] = \frac{1}{m} \int d\mathbf{r} \, \Psi^* \mathbf{p} \, \Psi \,,$$

where we have introduced the momentum operator $\mathbf{p} = -i\hbar \boldsymbol{\nabla}$ and in the last step we have used the fact that $\mathbf{p}$ is hermitian; hence,

$$\frac{d}{dt}\langle \mathbf{r}(t) \rangle = \frac{1}{m} \, \langle \mathbf{p}(t) \rangle \,.$$

Part 2

Consider

$$\frac{d}{dt}\langle p_x \rangle = \int d\mathbf{r} \left[\frac{\partial \Psi^*}{\partial t} p_x \Psi + \Psi^* p_x \frac{\partial \Psi}{\partial t} \right]$$

$$= -\frac{\hbar^2}{2m} \int d\mathbf{r} \left[(\boldsymbol{\nabla}^2 \Psi^*) \frac{\partial \Psi}{\partial x} - \Psi^* \boldsymbol{\nabla}^2 \frac{\partial \Psi}{\partial x} \right] + \int d\mathbf{r} \left[V \Psi^* \frac{\partial \Psi}{\partial x} - \Psi^* \frac{\partial (V\Psi)}{\partial x} \right] \,,$$

where we have used the Schrödinger equation, and its complex conjugate,

$$\frac{\partial \Psi}{\partial t} = \left(-\frac{\hbar}{2im} \boldsymbol{\nabla}^2 + \frac{V}{i\hbar} \right) \Psi \,, \qquad \frac{\partial \Psi^*}{\partial t} = \left(\frac{\hbar}{2im} \boldsymbol{\nabla}^2 - \frac{V}{i\hbar} \right) \Psi^* \,,$$

and have replaced p_x by $-i\hbar\partial/\partial x$. The integrand in the first term can be written as

$$(\boldsymbol{\nabla}^2\Psi^*)\frac{\partial\Psi}{\partial x} - \Psi^*\boldsymbol{\nabla}^2\frac{\partial\Psi}{\partial x} = \boldsymbol{\nabla}\cdot\left[(\boldsymbol{\nabla}\Psi^*)\frac{\partial\Psi}{\partial x} - \Psi^*\boldsymbol{\nabla}\frac{\partial\Psi}{\partial x}\right],$$

and therefore the first term reduces to a vanishing surface integral (the surface is at ∞, where Ψ vanishes). In the second term, we observe that

$$\Psi^*\frac{\partial(V\Psi)}{\partial x} = \Psi^*\left(\frac{\partial V}{\partial x}\right)\Psi + \Psi^*V\frac{\partial\Psi}{\partial x},$$

which yields

$$\frac{d}{dt}\langle p_x\rangle = -\int d\mathbf{r}\,\Psi^*\left(\frac{\partial V}{\partial x}\right)\Psi.$$

Identical manipulations for the other components lead to

$$\frac{d}{dt}\langle\mathbf{p}(t)\rangle = -\int d\mathbf{r}\,\Psi^*(\mathbf{r},t)\,[\boldsymbol{\nabla}V(\mathbf{r})]\,\Psi(\mathbf{r},t) = -\langle\boldsymbol{\nabla}V\rangle.$$

Part 3

If the force is linear in $\mathbf{r}$, as is the case for a particle in a harmonic potential, then $\langle\mathbf{F}\rangle$ is proportional to $\langle\mathbf{r}(t)\rangle$ and the latter will change in time as for a classical particle under the action of the force $\mathbf{F}(\mathbf{r})$. This will also occur, albeit approximately, if the magnitude $|\Psi(\mathbf{r},t)|$ of the wave function is sharply peaked at $\mathbf{r}_0(t)$, in which case $\langle\mathbf{r}(t)\rangle \approx \mathbf{r}_0(t)$ and $\langle\mathbf{F}\rangle \approx \mathbf{F}(\mathbf{r}_0)$.

For the harmonic potential given in the problem, the force is $\mathbf{F} = -m\,\omega^2\mathbf{r}$, and we have

$$\frac{d}{dt}\langle\mathbf{r}(t)\rangle = \frac{1}{m}\langle\mathbf{p}(t)\rangle, \qquad \frac{d}{dt}\langle\mathbf{p}(t)\rangle = -m\,\omega^2\langle\mathbf{r}(t)\rangle.$$

One way to solve this set of coupled first-order differential equations is to introduce an auxiliary function $\xi(t)$, defined as

$$\xi(t) = \langle\mathbf{p}(t)\rangle + im\omega\,\langle\mathbf{r}(t)\rangle,$$

from which we obtain (note that the averages are real)

$$\langle\mathbf{p}(t)\rangle = \mathrm{Re}[\xi(t)], \qquad \langle\mathbf{r}(t)\rangle = \frac{1}{m\,\omega}\mathrm{Im}[\xi(t)].$$

The system of differential equations is easily seen (by taking the real and imaginary parts of both sides) to be equivalent to $\dot{\xi}(t) = i\omega\,\xi(t)$, which has the solution (the reference time is $t=0$)

$$\xi(t) = \xi(0)e^{i\omega t} = [\langle\mathbf{p}(0)\rangle + im\omega\,\langle\mathbf{r}(0)\rangle]\,[\cos(\omega t) + i\sin(\omega t)].$$

By taking the real and imaginary parts of both sides, we obtain

$$\langle\mathbf{p}(t)\rangle = \langle\mathbf{p}(0)\rangle\cos(\omega t) - m\omega\langle\mathbf{r}(0)\rangle\sin(\omega t),$$

$$\langle\mathbf{r}(t)\rangle = \langle\mathbf{r}(0)\rangle\cos(\omega t) + \frac{\langle\mathbf{p}(0)\rangle}{m\omega}\sin(\omega t),$$

and so in the harmonic oscillator the average values of the position and momentum operators oscillate in time with angular frequency ω.

Problem 11 Time Evolution of the Average of a Generic Time-Dependent Operator

Suppose that the wave function $\Psi(\mathbf{r}, t)$ satisfies the time-dependent Schrödinger equation

$$i\hbar \frac{\partial}{\partial t} \Psi(\mathbf{r}, t) = H(\mathbf{p}, \mathbf{r}) \, \Psi(\mathbf{r}, t) \,, \qquad H(\mathbf{p}, \mathbf{r}) = \frac{\mathbf{p}^2}{2m} + V(\mathbf{r}) \,,$$

and $\mathbf{r}$ and $\mathbf{p} = -i\hbar\nabla$ are the position and momentum operators. Consider the expectation value $\langle O \rangle$ of a generic operator $O(\mathbf{p}, \mathbf{r}, t)$, where

$$\langle O(t) \rangle \equiv \int d\mathbf{r} \, \Psi^*(\mathbf{r}, t) \, O(\mathbf{p}, \mathbf{r}, t) \Psi(\mathbf{r}, t) \,.$$

Show that

$$\frac{d}{dt} \langle O(t) \rangle = \frac{i}{\hbar} \langle [\, H(\mathbf{p}, \mathbf{r}) \,, \, O(\mathbf{p}, \mathbf{r}, t) \,] \rangle + \left\langle \frac{\partial O(\mathbf{p}, \mathbf{r}, t)}{\partial t} \right\rangle \,,$$

where $[H, O]$ denotes the commutator of H and O.

Solution

Suppressing the dependence on the arguments for brevity, it follows from the Schrödinger equation that

$$\frac{\partial \Psi}{\partial t} = -\frac{i}{\hbar} H\Psi \,, \qquad \frac{\partial \Psi^*}{\partial t} = \frac{i}{\hbar} (H\Psi)^* \,.$$

Thus, we have

$$\frac{d}{dt} \langle O \rangle = \int d\mathbf{r} \left(\frac{\partial \Psi^*}{\partial t} \right) O\Psi + \int d\mathbf{r} \, \Psi^* O \left(\frac{\partial \Psi}{\partial t} \right) + \int d\mathbf{r} \, \Psi^* \left[\frac{\partial}{\partial t} O \right] \Psi$$

$$= \frac{i}{\hbar} \int d\mathbf{r} \, (H\Psi)^* \, O\Psi - \frac{i}{\hbar} \int d\mathbf{r} \, \Psi^* OH\Psi + \left\langle \frac{\partial O}{\partial t} \right\rangle$$

$$= \frac{i}{\hbar} \int d\mathbf{r} \, \Psi^* \, (HO - OH) \, \Psi + \left\langle \frac{\partial O}{\partial t} \right\rangle = \frac{i}{\hbar} \langle [H, O] \rangle + \left\langle \frac{\partial O}{\partial t} \right\rangle \,.$$

This result is known as Ehrenfest's theorem.

Problem 12 Charged Particle in an Electromagnetic Field: Lorentz Force

Consider a particle of mass m and charge q in an electromagnetic field described by the scalar potential $U(\mathbf{r}, t)$ and vector potential $\mathbf{A}(\mathbf{r}, t)$ such that

$$\mathbf{E}(\mathbf{r}, t) = -\nabla U(\mathbf{r}, t) - \frac{\partial}{c \, \partial t} \mathbf{A}(\mathbf{r}, t) \,, \qquad \mathbf{B}(\mathbf{r}, t) = \nabla \times \mathbf{A}(\mathbf{r}, t) \,,$$

where $\mathbf{E}(\mathbf{r}, t)$ and $\mathbf{B}(\mathbf{r}, t)$ are the electric and magnetic fields, and c is the speed of light.

1. Knowing that the Hamiltonian is given by

$$H(\mathbf{p}, \mathbf{r}) = \frac{1}{2m} \left[\mathbf{p} - \frac{q}{c} \mathbf{A}(\mathbf{r}, t) \right]^2 + q \, U(\mathbf{r}, t) \,,$$

show that Hamilton's equations of classical mechanics,

$$\frac{d\mathbf{r}}{dt} = \frac{\partial H}{\partial \mathbf{p}} \,, \qquad \frac{d\mathbf{p}}{dt} = -\frac{\partial H}{\partial \mathbf{r}} \,,$$

lead to Newton's second law for a particle under the influence of the Lorentz force,

$$m \frac{d^2 \mathbf{r}}{dt^2} = q \left[\mathbf{E}(\mathbf{r}, t) + \frac{\mathbf{v}}{c} \times \mathbf{B}(\mathbf{r}, t) \right] , \qquad \mathbf{v} = \frac{d\mathbf{r}}{dt} .$$

Hint: Recall that the total time derivative of a function $f[\mathbf{r}(t), t]$ is given by (using the chain rule of differentiation)

$$\frac{d}{dt} f[\mathbf{r}(t), t] = \frac{\partial f(\mathbf{r}, t)}{\partial r_x} \frac{dr_x}{dt} + \frac{\partial f(\mathbf{r}, t)}{\partial r_y} \frac{dr_y}{dt} + \frac{\partial f(\mathbf{r}, t)}{\partial r_z} \frac{dr_z}{dt} + \frac{\partial f(\mathbf{r}, t)}{\partial t} = \mathbf{v} \cdot \boldsymbol{\nabla} f + \frac{\partial f}{\partial t} .$$

2. Now, switch to quantum mechanics, and interpret $\mathbf{r}$ and $\mathbf{p}$ – the position and conjugate momentum of classical mechanics – as operators, with $\mathbf{p} = -i\hbar \boldsymbol{\nabla}$. Note that the scalar and vector potential $U(\mathbf{r}, t)$ and $\mathbf{A}(\mathbf{r}, t)$ are also operators, since they are functions of the particle's position operator $\mathbf{r}$. Using the result of the previous problem, show that

$$\frac{d}{dt} \langle r_\alpha \rangle = \frac{1}{m} \langle p_\alpha \rangle - \frac{q}{mc} \langle A_\alpha \rangle$$

and

$$\frac{d}{dt} \langle p_\alpha \rangle = -q \left\langle \frac{\partial U}{\partial r_\alpha} \right\rangle + \frac{q}{2c} \left\langle \mathbf{v} \cdot \frac{\partial \mathbf{A}}{\partial r_\alpha} + \frac{\partial \mathbf{A}}{\partial r_\alpha} \cdot \mathbf{v} \right\rangle ,$$

where the velocity operator is defined as

$$\mathbf{v} = \frac{1}{m} \left[\mathbf{p} - \frac{q}{c} \mathbf{A}(\mathbf{r}, t) \right] .$$

By taking the time derivative of the first equation, show that

$$m \frac{d^2 \langle \mathbf{r} \rangle}{dt^2} = q \langle \mathbf{E} \rangle + \frac{q}{2c} \langle \mathbf{v} \times \mathbf{B} - \mathbf{B} \times \mathbf{v} \rangle ,$$

where $\mathbf{E}(\mathbf{r}, t)$ and $\mathbf{B}(\mathbf{r}, t)$ are the electric and magnetic field operators as defined above; this is the quantum mechanical analogue of Newton's second law for a charged particle in an electromagnetic field.

Hint: Several commutators need to be evaluated. The following identity, valid for any three generic operators A, B, and C, is easily seen to hold,

$$[AB, C] = A[B, C] + [A, C]B ,$$

and may be useful in the solution of this part of the problem. Also note that

$$[A, B] = -[B, A] , \qquad [\mathbf{p}, F(\mathbf{r}, t)] = -i\hbar \boldsymbol{\nabla} F(\mathbf{r}, t) ,$$

where F is an operator function of the position operator (and possibly of time).

Solution

Part 1

Consider a component α; we find, suppressing the dependence of the external fields on $\mathbf{r}$ and t for brevity,

$$\frac{dr_\alpha}{dt} = \frac{\partial H}{\partial p_\alpha} = \underbrace{\frac{1}{m} \left(p_\alpha - \frac{q}{c} A_\alpha \right)}_{v_\alpha} , \qquad \frac{dp_\alpha}{dt} = -\frac{\partial H}{\partial r_\alpha} = \underbrace{\frac{q}{mc} \sum_\beta \left(p_\beta - \frac{q}{c} A_\beta \right) \frac{\partial A_\beta}{\partial r_\alpha}}_{(q/c)\mathbf{v} \cdot (\partial \mathbf{A}/\partial r_\alpha)} - q \frac{\partial U}{\partial r_\alpha} ,$$

where we observe that the particle's velocity $\mathbf{v}$ is *not* simply given by $\mathbf{p}/m$ in the presence of electromagnetic fields. We need to evaluate the acceleration, and so from the first equation we obtain

$$m \frac{d^2 r_\alpha}{dt^2} = \frac{dp_\alpha}{dt} - \frac{q}{c} \frac{dA_\alpha}{dt} .$$

The total time derivative of $A_\alpha[\mathbf{r}(t), t]$ is given by

$$\frac{dA_\alpha}{dt} = \sum_\beta \left(\frac{\partial A_\alpha}{\partial r_\beta} \right) \frac{dr_\beta}{dt} + \frac{\partial A_\alpha}{\partial t} = \mathbf{v} \cdot \boldsymbol{\nabla} A_\alpha + \frac{\partial A_\alpha}{\partial t} .$$

Newton's second law can be written as

$$m \frac{d^2 r_\alpha}{dt^2} = \frac{q}{c} \mathbf{v} \cdot \frac{\partial \mathbf{A}}{\partial r_\alpha} - q \frac{\partial U}{\partial r_\alpha} - \frac{q}{c} \left(\mathbf{v} \cdot \boldsymbol{\nabla} A_\alpha + \frac{\partial A_\alpha}{\partial t} \right)$$

$$= q \underbrace{\left(-\frac{\partial U}{\partial r_\alpha} - \frac{1}{c} \frac{\partial A_\alpha}{\partial t} \right)}_{\text{term 1}} + \frac{q}{c} \underbrace{\left(\mathbf{v} \cdot \frac{\partial \mathbf{A}}{\partial r_\alpha} - \mathbf{v} \cdot \boldsymbol{\nabla} A_\alpha \right)}_{\text{term 2}} ,$$

and term 1 is just q times the α component of the electric field. Expanding the scalar products in term 2 yields

$$\text{term 2} = \frac{q}{c} \left(v_x \frac{\partial A_x}{\partial r_\alpha} + v_y \frac{\partial A_y}{\partial r_\alpha} + v_z \frac{\partial A_z}{\partial r_\alpha} - v_x \frac{\partial A_\alpha}{\partial r_x} - v_y \frac{\partial A_\alpha}{\partial r_y} - v_z \frac{\partial A_\alpha}{\partial r_z} \right) ,$$

where $\alpha = x, y, z$. Consider the x component, for example: the term $v_x(\partial A_x / \partial r_x)$ drops out, and so

$$\text{term 2} = \frac{q}{c} \left(v_y \frac{\partial A_y}{\partial r_x} + v_z \frac{\partial A_z}{\partial r_x} - v_y \frac{\partial A_x}{\partial r_y} - v_z \frac{\partial A_x}{\partial r_z} \right)$$

$$= \frac{q}{c} \left[v_y \left(\frac{\partial A_y}{\partial r_x} - \frac{\partial A_x}{\partial r_y} \right) - v_z \left(\frac{\partial A_x}{\partial r_z} - \frac{\partial A_z}{\partial r_x} \right) \right]$$

$$= \frac{q}{c} \left[v_y (\boldsymbol{\nabla} \times \mathbf{A})_z - v_z (\boldsymbol{\nabla} \times \mathbf{A})_y \right] = \frac{q}{c} \left[\mathbf{v} \times (\boldsymbol{\nabla} \times \mathbf{A}) \right]_x = \frac{q}{c} (\mathbf{v} \times \mathbf{B})_x .$$

The other components can be treated similarly. Thus, we arrive at

$$m \frac{d^2 r_\alpha}{dt^2} = q \left[E_\alpha + \frac{1}{c} (\mathbf{v} \times \mathbf{B})_\alpha \right] ,$$

that is, Newton's second law for a particle under the action of the Lorentz force.

Part 2

Ehrenfest's theorem states that

$$\frac{d}{dt} \langle r_\alpha \rangle = \frac{i}{\hbar} \langle [H, r_\alpha] \rangle , \qquad \frac{d}{dt} \langle p_\alpha \rangle = \frac{i}{\hbar} \langle [H, p_\alpha] \rangle ,$$

since these operators have no explicit time dependence. We begin by working out the first of these commutators by letting it act on a generic wave function $\psi(\mathbf{r})$:

$$[H, r_\alpha] \psi = \frac{1}{2m} \left[\underbrace{\sum_\beta \left(p_\beta - \frac{q}{c} A_\beta \right)^2 r_\alpha \psi - r_\alpha \left(\mathbf{p} - \frac{q}{c} \mathbf{A} \right)^2 \psi}_{[\mathbf{p} - (q/c)\mathbf{A}]^2} \right] + q \underbrace{\left(U r_\alpha \psi - r_\alpha U \psi \right)}_{\text{vanishes}} .$$

Now, consider the term (note that r_α and A_β commute)

$$
\begin{aligned}
\sum_\beta \left(p_\beta - \frac{q}{c} A_\beta \right)^2 r_\alpha \psi &= \sum_\beta \left(p_\beta - \frac{q}{c} A_\beta \right) \left(-i\hbar \frac{\partial}{\partial r_\beta} r_\alpha \psi - \frac{q}{c} r_\alpha A_\beta \psi \right) \\
&= \sum_\beta \left(p_\beta - \frac{q}{c} A_\beta \right) \left[-i\hbar \delta_{\alpha\beta} \psi + r_\alpha \underbrace{\left(p_\beta - \frac{q}{c} A_\beta \right) \psi}_{\psi'_\beta} \right] \\
&= -i\hbar \left(p_\alpha - \frac{q}{c} A_\alpha \right) \psi - i\hbar \sum_\beta \delta_{\alpha\beta} \psi'_\beta + r_\alpha \sum_\beta \left(p_\beta - \frac{q}{c} A_\beta \right) \psi'_\beta \\
&= -2 i\hbar \left(p_\alpha - \frac{q}{c} A_\alpha \right) \psi + r_\alpha \left(\mathbf{p} - \frac{q}{c} \mathbf{A} \right)^2 \psi ,
\end{aligned}
$$

and the last term in the last line exactly cancels the identical term in $[H, r_\alpha] \psi$. Since this is true for any ψ, we conclude that

$$
[H, r_\alpha] = -\frac{i\hbar}{m} \left(p_\alpha - \frac{q}{c} A_\alpha \right) \implies \frac{d}{dt} \langle r_\alpha \rangle = \frac{1}{m} \langle p_\alpha \rangle - \frac{q}{mc} \langle A_\alpha \rangle .
$$

The above method is correct but cumbersome. There is a simpler way of working out the commutator. It is based on the following identity, valid for any three operators A, B, and C,

$$
[AB, C] = ABC - CAB = ABC \underbrace{-ACB + ACB}_{\text{cancel out}} - CAB = A[B, C] + [A, C]B ,
$$

which also implies that

$$
[C, AB] = A[C, B] + [C, A]B ,
$$

since for any A and B we have $[A, B] = -[B, A]$. Going back to the case under consideration, we obtain (since taking the commutator is a linear operation and so the commutator of a sum is the sum of the individual commutators)

$$
\begin{aligned}
[H, r_\alpha] &= \frac{1}{2m} \sum_\beta \left[\left(p_\beta - \frac{q}{c} A_\beta \right)^2 , r_\alpha \right] + q \underbrace{[U, r_\alpha]}_{\text{vanishes}} \\
&= \frac{1}{2m} \sum_\beta \left\{ \left(p_\beta - \frac{q}{c} A_\beta \right) \left[p_\beta - \frac{q}{c} A_\beta , r_\alpha \right] + \left[p_\beta - \frac{q}{c} A_\beta , r_\alpha \right] \left(p_\beta - \frac{q}{c} A_\beta \right) \right\} \\
&= \frac{1}{2m} \sum_\beta \left\{ \left(p_\beta - \frac{q}{c} A_\beta \right) \underbrace{[p_\beta , r_\alpha]}_{-i\hbar\delta_{\alpha\beta}} + [p_\beta , r_\alpha] \left(p_\beta - \frac{q}{c} A_\beta \right) \right\} = -\frac{i\hbar}{m} \left(p_\alpha - \frac{q}{c} A_\alpha \right) ,
\end{aligned}
$$

the result obtained earlier (again, note that A_β and r_α commute, since the former is only a function of the position operator).

We now turn to the commutator of p_α with H,

$$
[H, p_\alpha] = \frac{1}{2m} \sum_\beta \left[\left(p_\beta - \frac{q}{c} A_\beta \right)^2 , p_\alpha \right] + q [U, p_\alpha]
$$

and first make the following observations: (i) components of the momentum operator commute, and (ii) the commutator $[F, p_\alpha]$, where $F(\mathbf{r})$ is a function of the position operator, reduces to[5]

$$[F, p_\alpha] = i\hbar \frac{\partial F}{\partial r_\alpha} ,$$

as can easily be verified by acting with $[F, p_\alpha]$ on a generic wave function $\psi(\mathbf{r})$. On account of these observations, we have

$$[H, p_\alpha] = \frac{1}{2m} \sum_\beta \left\{ \left(p_\beta - \frac{q}{c} A_\beta \right) \left[p_\beta - \frac{q}{c} A_\beta, p_\alpha \right] + \left[p_\beta - \frac{q}{c} A_\beta, p_\alpha \right] \left(p_\beta - \frac{q}{c} A_\beta \right) \right\} + i\hbar q \frac{\partial U}{\partial r_\alpha}$$

$$= -\frac{q}{2mc} \sum_\beta \left\{ \left(p_\beta - \frac{q}{c} A_\beta \right) \underbrace{\left[A_\beta, p_\alpha \right]}_{i\hbar \partial A_\beta / \partial r_\alpha} + \left[A_\beta, p_\alpha \right] \left(p_\beta - \frac{q}{c} A_\beta \right) \right\} + i\hbar q \frac{\partial U}{\partial r_\alpha}$$

$$= -\frac{i\hbar q}{2mc} \sum_\beta \left[\left(p_\beta - \frac{q}{c} A_\beta \right) \frac{\partial A_\beta}{\partial r_\alpha} + \frac{\partial A_\beta}{\partial r_\alpha} \left(p_\beta - \frac{q}{c} A_\beta \right) \right] + i\hbar q \frac{\partial U}{\partial r_\alpha}$$

$$= -\frac{i\hbar q}{2c} \left(\mathbf{v} \cdot \frac{\partial \mathbf{A}}{\partial r_\alpha} + \frac{\partial \mathbf{A}}{\partial r_\alpha} \cdot \mathbf{v} \right) + i\hbar q \frac{\partial U}{\partial r_\alpha} ,$$

and hence

$$\frac{d}{dt} \langle p_\alpha \rangle = -q \left\langle \frac{\partial U}{\partial r_\alpha} \right\rangle + \frac{q}{2c} \left\langle \mathbf{v} \cdot \frac{\partial \mathbf{A}}{\partial r_\alpha} + \frac{\partial \mathbf{A}}{\partial r_\alpha} \cdot \mathbf{v} \right\rangle .$$

Taking the time derivative of the relation for $m \, d\langle r_\alpha \rangle / dt$, we find

$$m \frac{d^2}{dt^2} \langle r_\alpha \rangle = \frac{d}{dt} \langle p_\alpha \rangle - \frac{q}{c} \frac{d}{dt} \langle A_\alpha \rangle$$

and

$$\frac{d}{dt} \langle A_\alpha \rangle = \frac{i}{\hbar} \langle [H, A_\alpha] \rangle + \left\langle \frac{\partial A_\alpha}{\partial t} \right\rangle ,$$

where the second term on the right-hand side accounts for the explicit time dependence of A_α; the commutator in the first term is given by

$$[H, A_\alpha] = \frac{1}{2m} \sum_\beta \left\{ \left(p_\beta - \frac{q}{c} A_\beta \right) \left[p_\beta - \frac{q}{c} A_\beta, A_\alpha \right] + \left[p_\beta - \frac{q}{c} A_\beta, A_\alpha \right] \left(p_\beta - \frac{q}{c} A_\beta \right) \right\}$$

$$= \frac{1}{2mc} \sum_\beta \left\{ \left(p_\beta - \frac{q}{c} A_\beta \right) \underbrace{\left[p_\beta, A_\alpha \right]}_{-i\hbar \partial A_\alpha / \partial r_\beta} + \left[p_\beta, A_\alpha \right] \left(p_\beta - \frac{q}{c} A_\beta \right) \right\}$$

$$= -\frac{i\hbar}{2m} \sum_\beta \left[\left(p_\beta - \frac{q}{c} A_\beta \right) \frac{\partial A_\alpha}{\partial r_\beta} + \frac{\partial A_\alpha}{\partial r_\beta} \left(p_\beta - \frac{q}{c} A_\beta \right) \right] ,$$

$$= -\frac{i\hbar}{2} \left(\mathbf{v} \cdot \boldsymbol{\nabla} A_\alpha + \boldsymbol{\nabla} A_\alpha \cdot \mathbf{v} \right) ,$$

[5] The above relation has a counterpart in classical mechanics, where one defines the Poisson brackets of any two functions $B(q_i, p_i)$ and $C(q_i, p_i)$ of the generalized conjugate variable q_i and p_i for a system with n degrees of freedom as

$$\{B, C\} = \sum_{k=1}^{n} \left[\left(\frac{\partial B}{\partial q_k} \right) \frac{\partial C}{\partial p_k} - \left(\frac{\partial B}{\partial p_k} \right) \frac{\partial C}{\partial q_k} \right] .$$

In particular, note that, for the fundamental Poisson brackets, we have $\{q_i, p_j\} = \delta_{ij}$. The transition from classical to quantum mechanics is implemented via the replacement $q_i, p_i \longrightarrow q_i^{op}, p_i^{op}$ and by requiring that $[q_i^{op}, p_j^{op}] = i\hbar \{q_i, p_j\}$. Indeed, we have the general correspondence $[B^{op}, C^{op}] \longrightarrow i\hbar \{B, C\}$.

where, since A_α and U are functions of the position operator, the commutators satisfy

$$[U, A_\alpha] = [A_\beta, A_\alpha] = 0 .$$

Collecting results, we arrive at

$$m \frac{d^2}{dt^2} \langle r_\alpha \rangle = -q \left\langle \frac{\partial U}{\partial r_\alpha} \right\rangle + \frac{q}{2c} \left\langle \mathbf{v} \cdot \frac{\partial \mathbf{A}}{\partial r_\alpha} + \frac{\partial \mathbf{A}}{\partial r_\alpha} \cdot \mathbf{v} \right\rangle - \frac{q}{2c} \left\langle \mathbf{v} \cdot \boldsymbol{\nabla} A_\alpha + \boldsymbol{\nabla} A_\alpha \cdot \mathbf{v} \right\rangle - \frac{q}{c} \left\langle \frac{\partial A_\alpha}{\partial t} \right\rangle ,$$

which, introducing the electric and magnetic field operators and using the identity obtained in part 1,[6] can be expressed as

$$m \frac{d^2}{dt^2} \langle r_\alpha \rangle = q \langle E_\alpha \rangle + \frac{q}{2c} \langle (\mathbf{v} \times \mathbf{B})_\alpha - (\mathbf{B} \times \mathbf{v})_\alpha \rangle ,$$

where it should be noted that the components of $\mathbf{v}$ and $\mathbf{B}$ do not commute.

---

[6] Note, however, that the combination with inverted ordering

$$\frac{\partial A_x}{\partial r_\alpha} v_x + \frac{\partial A_y}{\partial r_\alpha} v_y + \frac{\partial A_z}{\partial r_\alpha} v_z - \frac{\partial A_\alpha}{\partial r_x} v_x - \frac{\partial A_\alpha}{\partial r_y} v_y - \frac{\partial A_\alpha}{\partial r_z} v_z$$

reduces for $\alpha = x$ to

$$\frac{\partial A_y}{\partial r_x} v_y + \frac{\partial A_z}{\partial r_x} v_z - \frac{\partial A_x}{\partial r_y} v_y - \frac{\partial A_x}{\partial r_z} v_z = (\boldsymbol{\nabla} \times \mathbf{A})_z v_y - (\boldsymbol{\nabla} \times \mathbf{A})_y v_z = B_z v_y - B_y v_z = -(\mathbf{B} \times \mathbf{v})_x;$$

hence the minus sign in the second term. Incidentally, this minus sign ensures that the operator $\mathbf{v} \times \mathbf{B} - \mathbf{B} \times \mathbf{v}$ is hermitian.

4

The One-Dimensional Schrödinger Equation; Bound States

In this chapter we review the one-dimensional Schrödinger equation. This is not just an academic exercise, as many problems of physical interest, for example the problem of two particles interacting via a central potential (as in the hydrogen atom), ultimately involve the solution of a one-dimensional Schrödinger equation.

The time-dependent wave function is given by

$$\Psi(x, t) = e^{-iE(t-t_0)/\hbar} \, \psi(x) \,, \tag{4.1}$$

where $\psi(x)$ satisfies the time-independent Schrödinger equation

$$\left[-\frac{\hbar^2}{2m} \frac{d^2}{dx^2} + V(x) \right] \psi(x) = E\psi(x) \,. \tag{4.2}$$

The latter can be more conveniently written as

$$\psi''(x) = [v(x) - \epsilon] \, \psi(x) \qquad \text{with} \qquad v(x) = \frac{2m}{\hbar^2} \, V(x) \quad \text{and} \quad \epsilon = \frac{2m}{\hbar^2} \, E \,. \tag{4.3}$$

Note that ϵ and $v(x)$ have dimensions of $(\text{length})^{-2}$, since $\hbar^2/m$ has dimensions of energy $\times (\text{length})^2$. Here, we are interested in solutions that either vanish or are bounded in the asymptotic regions $|x| \longrightarrow \infty$; specifically, either $\lim_{|x|\to\infty} \psi(x) = 0$ or $\left| \psi(x) \right| \leq M$, where M is a finite nonzero constant. The former solutions exist only for discrete values of ϵ (the *eigenenergy* or, simply, the *energy*), are square integrable, and correspond to the bound states of the particle. Their number can be finite or infinite, depending on the potential. The latter solutions exist for a continuum of ϵ values, are not square integrable, and represent the scattering states of the particle.

4.1 Nature of the Energy Spectrum and General Properties of the Eigenfunctions

Here, we recall some results regarding the nature of the energy spectrum for a generic potential $v(x)$; in particular, under what conditions discrete and/or continuous energies ϵ are obtained. We also briefly review some of the properties satisfied by the associated solutions. For definiteness, we assume that $v(x)$ is piecewise continuous, has an absolute minimum v_0, and is such that

$$v_{\pm} = \lim_{x\to\pm\infty} v(x) \,, \qquad v_- \geq v_+ \,, \tag{4.4}$$

where the constants $v_{\pm}$ can be zero (several potentials of physical interest satisfy these requirements). No other assumptions are made on $v(x)$.

We are looking for continuous solutions $\psi(x)$ that are either vanishing or bounded as $|x| \longrightarrow \infty$. The first derivative of $\psi(x)$ can be expressed as

$$\psi'(x) = \psi'(x_0) + \int_{x_0}^{x} dx' \left[v(x') - \epsilon \right] \psi(x') \,, \tag{4.5}$$

which shows that $\psi'(x)$ is also continuous, if the integral on the right-hand side is finite. This is certainly the case for a potential $v(x)$ with finite discontinuities (piecewise continuous) but not for a singular $v(x)$. For example, in the case of an infinite barrier $v(x > a) = \infty$, we require only $\psi(x \geq a) = 0$ and impose no condition on $\psi'(a)$; in the case of a δ-function potential $v(x) = v_0\, \delta(x-a)$, we require $\psi(x)$ to be continuous at a but its first derivative to have a jump discontinuity, given by

$$\psi'(a^+) - \psi'(a^-) = v_0 \psi(a) \,, \tag{4.6}$$

which can be easily derived from Eq. (4.5).

For a given ϵ, there are two types of x-regions that can occur:

$$\text{region I,} \quad v(x) - \epsilon < 0 \,; \qquad \text{region II,} \quad v(x) - \epsilon > 0 \,. \tag{4.7}$$

The nature of the energy spectrum, that is, whether it is discrete or continuous, depends on whether, respectively, $v(x) - \epsilon$ as $x \longrightarrow -\infty$ and as $x \longrightarrow \infty$ are both of type II or not. It is only the asymptotic behavior that is important. There are no acceptable solutions for $\epsilon < v_0$, the absolute minimum of $v(x)$. The remaining three possibilities are: (1) $v_0 < \epsilon < v_+$; (2) $v_+ \leq \epsilon \leq v_-$; and (3) $\epsilon > v_-$. We examine them in turn.

(1) $v_0 < \epsilon < v_+$. In the asymptotic regions $x \longrightarrow \pm\infty$, the Schrödinger equation is obtained as

$$\psi''_{\pm}(x) = \kappa_{\pm}^2\, \psi_{\pm}(x) \,, \qquad \kappa_{\pm} = \sqrt{v_{\pm} - \epsilon} \,. \tag{4.8}$$

In each of these regions, the solution consists of a linear combination of increasing and decreasing exponentials. However, in order to satisfy the boundary conditions that $\psi_{\pm}(x)$ vanishes as $|x| \longrightarrow \infty$, we must require that

$$\psi_{\pm}(x) = \alpha_{\pm}\, e^{\mp \kappa_{\pm} x} \qquad \text{as} \quad x \longrightarrow \pm\infty \,, \tag{4.9}$$

where $\alpha_{\pm}$ are constants. Since $\psi(x)$ and its derivative are continuous, $\psi_-(x)$ and $\psi_+(x)$ must match: given a point x_0, we must have

$$\psi_-(x_0; \epsilon) = \psi_+(x_0; \epsilon) \,, \qquad \psi'_-(x_0; \epsilon) = \psi'_+(x_0; \epsilon) \,, \tag{4.10}$$

where we have emphasized the implicit dependence on ϵ of the solution. This matching can only occur for certain values of ϵ (or possibly not at all). The total number of allowed energies ϵ depends on $v(x)$, and can be finite (possibly zero) or infinite.

Suppose that $\epsilon_0 < \epsilon_1 < \cdots < \epsilon_n < \cdots$ are the discrete eigenenergies, and $\psi_0(x), \psi_1(x), \ldots, \psi_n(x), \ldots$ the corresponding eigenfunctions. These eigenenergies are non-degenerate. Further, $\psi_0(x)$, the ground-state wave function, has no nodes (zeros), $\psi_1(x)$, the first excited-state wave function, has one node, $\ldots$, $\psi_n(x)$, the nth excited-state wave function, has n nodes, and so on.

(2) $v_+ \leq \epsilon < v_-$. In the regions $x \longrightarrow \pm\infty$ the Schrödinger equations read

$$\psi''_-(x) = \kappa^2_- \, \psi_-(x) \, , \qquad \kappa_- = \sqrt{v_- - \epsilon} \, , \tag{4.11}$$

and

$$\psi''_+(x) = -k^2_+ \, \psi_+(x) \, , \qquad k_+ = \sqrt{\epsilon - v_+} \, , \tag{4.12}$$

with

$$\psi_-(x) = \alpha_- \, e^{\kappa_- x} \qquad x \longrightarrow -\infty \, , \qquad \psi_+(x) = \alpha_+ \, e^{ik_+ x} + \beta_+ \, e^{-ik_+ x} \qquad x \longrightarrow \infty \, ; \tag{4.13}$$

namely, the solution vanishes for $x \longrightarrow -\infty$ and oscillates (that is, it is bounded) for $x \longrightarrow \infty$. The matching of $\psi_-(x)$ and $\psi_+(x)$ can now be satisfied by any value of ϵ in the range $v_+ \leq \epsilon < v_-$, since there are three available constants but only two conditions. These continuous eigenenergies are non-degenerate.

(3) $\epsilon \geq v_-$. In the regions $x \longrightarrow \pm\infty$ the Schrödinger equation is given by

$$\psi''_\pm(x) = -k^2_\pm \, \psi_\pm(x) \, , \qquad k_\pm = \sqrt{\epsilon - v_\pm} \, , \tag{4.14}$$

with solutions

$$\psi_\pm(x) = \alpha_\pm \, e^{ik_\pm x} + \beta_\pm \, e^{-ik_\pm x} \, , \tag{4.15}$$

and $\psi_\pm(x)$ oscillate as $|x| \longrightarrow \infty$. Since the Schrödinger equation is a linear second-order differential equation, it admits two independent solutions,[1] both of which are oscillating as $|x| \longrightarrow \infty$. We conclude that each (continuous) eigenenergy $\epsilon \geq v_-$ is doubly degenerate.

4.2 Parity

Consider a potential $v(x)$ that is invariant under the parity transformation $x \longmapsto -x$, that is,

$$v(x) \longmapsto v(-x) = v(x) \, . \tag{4.16}$$

As a consequence, the Schrödinger equation is also invariant under such transformation,

$$\frac{d^2}{dx^2} \, \psi(x) = [v(x) - \epsilon]\psi(x) \longmapsto \frac{d^2}{dx^2} \, \psi(-x) = [v(x) - \epsilon]\psi(-x) \, , \tag{4.17}$$

and therefore $\psi(x)$ and $\psi(-x)$ are both solutions corresponding to the *same* ϵ. We can construct two additional solutions by forming the linear combinations

$$\psi_e(x) = \psi(x) + \psi(-x) \, , \qquad \psi_o(x) = \psi(x) - \psi(-x), \tag{4.18}$$

where $\psi_e(x)$ and $\psi_o(x)$ are, respectively, even and odd under parity, that is, $\psi_e(-x) = \psi_e(x)$ and $\psi_o(-x) = -\psi_o(x)$. Two cases can then occur.

---

[1] Linear independence of two functions $f_1(x)$ and $f_2(x)$ means that $\lambda f_1(x) + \mu f_2(x) = 0$ is satisfied for any x in $-\infty < x < \infty$ iff $\lambda = \mu = 0$. Equivalently, the two functions are linearly dependent iff the Wronskian $W[f_1(x), f_2(x)] = 0$ for any x in $-\infty < x < \infty$ (just take the derivative of the above relation, and view the resulting equations as a linear system in the unknowns λ and μ).

1. ϵ is non-degenerate: the four wave functions $\psi(x)$, $\psi(-x)$, $\psi_e(x)$, and $\psi_o(x)$ are multiples of each other. Thus, $\psi(x)$ is necessarily proportional either to $\psi_e(x)$ with $\psi_o(x) = 0$ or to $\psi_o(x)$ with $\psi_e(x) = 0$. We conclude that the ground-state wave function, being nodeless, must be even; the first excited-state wave function, having one node, must be odd; and so on.

2. ϵ is doubly degenerate: any eigenfunction can be written as a linear combination

$$\alpha \psi_1(x) + \beta \psi_2(x) , \tag{4.19}$$

where $\psi_1(x)$ and $\psi_2(x)$ are two independent eigenfunctions corresponding to the same ϵ. Suppose one of them, say $\psi_1(x)$, has no definite parity. We can construct $\psi_e(x)$ and $\psi_o(x)$; neither of these is zero, since $\psi_1(x)$ has no definite parity by assumption. But $\psi_e(x)$ and $\psi_o(x)$ are linearly independent, and therefore $\psi_1(x)$ and $\psi_2(x)$ can be expressed as linear combinations of $\psi_e(x)$ and $\psi_o(x)$. In other words, the eigenfunctions of the degenerate spectrum can be written as a superposition of $\psi_e(x)$ and $\psi_o(x)$.

4.3 Problems

Problem 1   No Acceptable Solutions for Energies Less Than the Minimum of *v(x)*

Suppose v_0 is the absolute minimum of the potential $v(x)$, that is, $v(x) \geq v_0$ for any x in $-\infty < x < \infty$. Show that the Schrödinger equation admits no vanishing or bounded solution for $\epsilon < v_0$.

Hint: Consider the the sign of the ratio $\psi''(x)/\psi(x)$.

Solution

Given $\epsilon < v_0$, the Schrödinger equation implies that

$$\psi''(x)/\psi(x) = v(x) - \epsilon > 0 \quad \text{in} \quad -\infty < x < \infty ,$$

and the curvature of $\psi(x)$ is either *up* or *down* depending on whether $\psi(x) > 0$ or $\psi(x) < 0$, respectively, as illustrated in Fig. 4.1. Furthermore, possible zeros of $\psi(x)$ are inflection points. Consider a point x_0 such that, without loss of generality, $\psi(x_0) > 0$ (if $\psi(x_0) < 0$, multiply $\psi(x)$ by -1; this is permissible, since the Schrödinger equation is homogeneous). Draw a tangent at x_0.

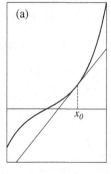 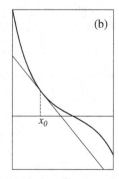

Fig. 4.1   (a) $\psi'(x_0) > 0$; (b) $\psi'(x_0) < 0$.

If $\psi'(x_0) > 0$, then $\psi(x)$ is always above the tangent for $x > x_0$, and hence $\psi(x)$ diverges as $x \longrightarrow \infty$. If $\psi'(x_0) < 0$, then $\psi(x)$ is always above the tangent for $x < x_0$, and again $\psi(x)$ diverges as $x \longrightarrow -\infty$. In either cases, $\psi(x)$ is neither bounded nor vanishing as $|x| \longrightarrow \infty$, and cannot be a solution.

Problem 2 Degeneracy of the Energy Eigenvalues Corresponding to a Potential $v(x)$

Show that the discrete eigenenergies corresponding to a potential $v(x)$ are non-degenerate. Does this hold for the continuum solutions that vanish at either $-\infty$ or ∞? Justify your answer.

Solution

Suppose that there are two wave functions $\psi_1(x)$ and $\psi_2(x)$ corresponding to the *same* ϵ; then

$$\psi_1''(x) = [v(x) - \epsilon]\psi_1(x) , \qquad \psi_2''(x) = [v(x) - \epsilon]\psi_2(x) .$$

Multiply both sides of the first Schrödinger equation by $\psi_2(x)$ and both sides of the second by $\psi_1(x)$ and subtract one from the other, to obtain

$$\psi_1(x)\psi_2'' - \psi_2(x)\psi_1''(x) = 0 \qquad \text{or} \qquad \frac{d}{dx} \underbrace{\left[\psi_1(x)\psi_2' - \psi_2(x)\psi_1'(x) \right]}_{W[\psi_1(x),\psi_2(x)]} = 0 ,$$

and the quantity in square brackets, that is, the Wronskian $W[\psi_1(x), \psi_2(x)]$, is a constant in $-\infty < x < \infty$. Given that ϵ belongs to the discrete spectrum, $\psi_1(x)$ and $\psi_2(x)$ vanish as $|x| \longrightarrow \infty$. Therefore, we evaluate the Wronskian at, say, $x \longrightarrow \infty$ to find that this constant is in fact zero, which yields

$$\psi_2(x)\psi_1'(x) = \psi_1(x)\psi_2'(x) \qquad \text{or} \qquad \psi_1'(x)/\psi_1(x) = \psi_2'(x)/\psi_2(x) ,$$

and therefore

$$\frac{d}{dx} \left[\ln|\psi_1(x)| - \ln|\psi_2(x)| \right] = 0 \implies \ln\left| \frac{\psi_1(x)}{\psi_2(x)} \right| = \text{constant} .$$

We conclude that $\psi_1(x)$ and $\psi_2(x)$ are proportional to each other and, as such, are not independent solutions of the Schrödinger equation.

The same argument holds for the continuous energies corresponding to solutions that vanish, say, at $x = -\infty$: the Wronskian is zero in this limit, implying that $\psi_1(x)$ and $\psi_2(x)$ are proportional to each other.

Problem 3 Nodes of Excited-State Wave Functions

Suppose that $\epsilon_n > \epsilon_m$. Show that $\psi_n(x)$ has at least one node between two consecutive nodes of $\psi_m(x)$.

Solution

The Schrödinger equations satisfied by these two wave functions are

$$\psi_m''(x) = [v(x) - \epsilon_m]\psi_m(x) , \qquad \psi_n''(x) = [v(x) - \epsilon_n]\psi_n(x) .$$

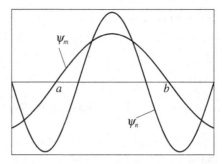

Fig. 4.2 The wave functions $\psi_m(x)$ and $\psi_n(x)$.

Multiply both sides of the first Schrödinger equation by $\psi_n(x)$ and both sides of the second by $\psi_m(x)$ and subtract one from the other, to obtain

$$\underbrace{\psi_n(x)\psi_m''(x) - \psi_m(x)\psi_n''(x)}_{[\psi_n(x)\psi_m'(x) - \psi_m(x)\psi_n'(x)]'} = (\epsilon_n - \epsilon_m)\psi_m(x)\psi_n(x) .$$

Now, let a and b be two consecutive nodes of $\psi_m(x)$, so that $\psi_m(a) = \psi_m(b) = 0$. Integrate the above identity between a and b; the left-hand side is

$$\int_a^b dx \frac{d}{dx}\left[\psi_n(x)\psi_m'(x) - \psi_m(x)\psi_n'(x)\right] = \psi_n(x)\psi_m'(x) - \psi_m(x)\psi_n'(x)\Big|_a^b = \psi_n(b)\psi_m'(b) - \psi_n(a)\psi_m'(a) ,$$

and therefore

$$\psi_n(b)\psi_m'(b) - \psi_n(a)\psi_m'(a) = (\epsilon_n - \epsilon_m)\int_a^b dx\, \psi_m(x)\psi_n(x) .$$

Assume $\psi_m(x)$ is positive in the interval $[a, b]$, see Fig. 4.2; then $\psi_m'(a) > 0$ and $\psi_m'(b) < 0$. If $\psi_n(x)$ does *not* have a node in $[a, b]$ then take it to be negative in that interval. The above identity cannot be satisfied since the left-hand side is positive and the right-hand side is negative. The identical conclusion is reached if $\psi_n(x)$ is taken to be nodeless and positive in $[a, b]$. Therefore $\psi_n(x)$ must change sign at least once in $[a, b]$.

Problem 4   Coordinate- and Momentum-Space Bound-State Wave Functions in a δ-Function Potential

Consider the problem of a particle in an attractive δ-function potential given by

$$v(x) = -v_0\,\delta(x) \qquad v_0 > 0 .$$

1. Obtain the energy and wave function of the bound state.
2. Calculate the probability $dP(p)$ that a measurement of the momentum in this bound state will give a result between p and $p + dp$. For what value of p is this probability largest?

Solution

Part 1

The Schrödinger equation reads

$$\psi''(x) = -[v_0\,\delta(x) + \epsilon]\psi(x)\,.$$

A bound-state wave function vanishes as $|x| \longrightarrow \infty$ and is continuous at $x = 0$. What condition needs to be imposed on the wave function's first derivative? To obtain this condition, integrate both sides of the Schrödinger equation over a small interval containing the origin,

$$\int_{-\eta}^{\eta} dx\,\psi''(x) = \psi'(\eta) - \psi'(-\eta) = -\int_{-\eta}^{\eta} dx\,[v_0\,\delta(x) + \epsilon]\psi(x) = -v_0\psi(0) - \epsilon\int_{-\eta}^{\eta} dx\,\psi(x)\,.$$

Taking the limit $\eta \longrightarrow 0$ yields

$$\psi'(0^+) - \psi'(0^-) = -v_0\psi(0)\,,$$

and the integral of $\psi(x)$ vanishes, since $\psi(x)$ is continuous. We conclude that the first derivative of $\psi(x)$ must have a finite discontinuity at the location of the δ-function.

To solve the Schrödinger equation, we split the real axis into two regions: $x < a$ (region $<$) and $x > a$ (region $>$). In these regions, we have (recall that $\epsilon = -|\epsilon|$ is negative)

$$\psi''_<(x) = |\epsilon|\,\psi_<(x)\,, \qquad\qquad \psi''_>(x) = |\epsilon|\,\psi_>(x)\,,$$

and, since $\psi_<(x) \longrightarrow 0$ as $x \longrightarrow -\infty$ and $\psi_>(x) \longrightarrow 0$ as $x \longrightarrow \infty$ in order to satisfy the boundary conditions at ∞, we find

$$\psi_<(x) = \alpha\,e^{\sqrt{|\epsilon|}x}\,, \qquad\qquad \psi_>(x) = \beta\,e^{-\sqrt{|\epsilon|}x}\,.$$

We now impose the condition of continuity of $\psi(x)$ at $x = 0$ and the condition on its first derivative, to obtain

$$\alpha = \beta\,, \qquad -\beta\sqrt{|\epsilon|} - \alpha\sqrt{|\epsilon|} = -v_0\beta\,.$$

The linear system for the unknown α and β has a non-trivial solution if the determinant of the coefficient matrix vanishes, namely

$$\det\begin{bmatrix} 1 & -1 \\ \sqrt{|\epsilon|} & \sqrt{|\epsilon|} - v_0 \end{bmatrix} = 0 \qquad\text{or}\qquad 2\sqrt{|\epsilon|} - v_0 = 0 \implies \epsilon = -v_0^2/4\,,$$

and we find a single energy. The corresponding normalized eigenfunction follows:

$$\psi_<(x) = \sqrt{\frac{v_0}{2}}\,e^{v_0x/2}\,, \qquad\qquad \psi_>(x) = \sqrt{\frac{v_0}{2}}\,e^{-v_0x/2}\,.$$

The larger the value of v_0 the more the wave function is localized at the origin.

Part 2

Recalling that $v_0 = 2mV_0/\hbar^2$, the wave function in momentum space follows from the Fourier transform,

$$\widetilde{\psi}(p) = \frac{1}{\sqrt{2\pi\hbar}} \int_{-\infty}^{\infty} dx\, e^{-ipx/\hbar}\, \psi(x) = \frac{1}{\sqrt{2\pi\hbar}} \sqrt{\frac{mV_0}{\hbar^2}} \int_{-\infty}^{\infty} dx\, [\cos(px/\hbar) - i\,\sin(px/\hbar)] e^{-mV_0|x|/\hbar^2}$$

$$= \sqrt{\frac{2mV_0}{\pi\hbar^3}} \int_0^{\infty} dx\, \cos(px/\hbar)\, e^{-mV_0x/\hbar^2} \, .$$

The integral can easily be found (integrate by parts twice):

$$\widetilde{\psi}(p) = \sqrt{\frac{2p_0^3}{\pi}} \frac{1}{p^2 + p_0^2} \, , \qquad p_0 = \frac{mV_0}{\hbar} \, .$$

The probability of measuring a momentum p between p and $p + dp$ is given by $|\widetilde{\psi}(p)|^2 dp$, and the momentum-space probability density is simply

$$P(p) = |\widetilde{\psi}(p)|^2 = \frac{2p_0^3}{\pi} \frac{1}{(p^2 + p_0^2)^2} \, ;$$

$P(p)$ is largest when $p = 0$.

Problem 5 The Schrödinger Equation in Momentum Space for an Attractive δ-Function Potential

This problem deals with the solution of the the Schrödinger equation in momentum space.

1. Solve directly in momentum space for the bound-state energy and wave function in an attractive δ-function potential, given in coordinate space by $v(x) = -v_0\,\delta(x)$ with $v_0 > 0$.
2. From the momentum-space wave function thus obtained, calculate the average kinetic energy of the particle. Repeat the calculation of this average kinetic energy but in coordinate space, using the wave function you derived in the Problem 4. Note that the first derivative of the coordinate-space bound-state wave function is discontinuous at $x = 0$.

Solution

Part 1

We start from

$$\left(i\hbar\frac{\partial}{\partial t} - \frac{p^2}{2m} \right) \widetilde{\Psi}(p, t) = \frac{1}{(2\pi\hbar)^{1/2}} \int_{-\infty}^{\infty} dp'\, \widetilde{V}(p - p')\, \widetilde{\Psi}(p', t) \, ,$$

where

$$\widetilde{\Psi}(p, t) = \int \frac{dx}{(2\pi\hbar)^{1/2}} e^{-ipx/\hbar}\, \Psi(x, t) \, , \qquad \widetilde{V}(p) = \int \frac{dx}{(2\pi\hbar)^{1/2}} e^{-ipx/\hbar}\, V(x) \, .$$

Using the separation of variables $\widetilde{\Psi}(p, t) = \widetilde{\psi}(p)\phi(t)$, we easily find

$$\widetilde{\Psi}(p, t) = \widetilde{\psi}(p)\, e^{-iE(t-t_0)/\hbar} \, ,$$

with

$$\left(E - \frac{p^2}{2m}\right)\widetilde{\psi}(p) = \frac{1}{(2\pi\hbar)^{1/2}}\int_{-\infty}^{\infty} dp' \, \widetilde{V}(p - p') \, \widetilde{\psi}(p') \,.$$

The momentum-space potential follows from

$$\widetilde{V}(p) = -V_0 \int \frac{dx}{(2\pi\hbar)^{1/2}} \, e^{-ipx/\hbar} \, \delta(x) = -\frac{V_0}{(2\pi\hbar)^{1/2}} \,,$$

and therefore the Schrödinger equation can be written as

$$\left(E - \frac{p^2}{2m}\right)\widetilde{\psi}(p) = -\frac{V_0}{2\pi\hbar}\underbrace{\int_{-\infty}^{\infty} dp' \, \widetilde{\psi}(p')}_{\text{constant } C} \implies \widetilde{\psi}(p) = -\frac{V_0}{2\pi\hbar}\frac{C}{E - p^2/(2m)} \,.$$

The requirement that

$$C = \int_{-\infty}^{\infty} dp \, \widetilde{\psi}(p) = -\frac{V_0}{2\pi\hbar} C \int_{-\infty}^{\infty} dp \, \frac{1}{E - p^2/(2m)} \implies -\frac{V_0}{2\pi\hbar}\int_{-\infty}^{\infty} dp \, \frac{1}{E - p^2/(2m)} = 1$$

determines the eigenvalue E, which must be negative (corresponding to a bound state). Carrying out the integral gives

$$1 = \frac{V_0}{2\pi\hbar}\int_{-\infty}^{\infty} dp \, \frac{1}{|E| + p^2/(2m)} = \frac{V_0}{2\pi\hbar}\pi\sqrt{\frac{2m}{|E|}} \implies E = -\frac{mV_0^2}{2\hbar^2} \,,$$

in agreement with the result found in coordinate space (see Problem 4). After normalizing $\widetilde{\psi}(p)$, we obtain the momentum-space wave function found in Problem 4, namely

$$\widetilde{\psi}(p) = \sqrt{\frac{2p_0^3}{\pi}}\frac{1}{p^2 + p_0^2} \,, \qquad p_0 = \frac{mV_0}{\hbar} \,.$$

Part 2

The expectation value of the kinetic energy is given in momentum space by

$$T = \int_{-\infty}^{\infty} dp \, \frac{p^2}{2m} \, |\widetilde{\psi}(p)|^2 = \frac{p_0^2}{2m}\frac{2p_0}{\pi}\int_{-\infty}^{\infty} dp \, \frac{p^2}{(p^2 + p_0^2)^2} = \frac{p_0^2}{2m}\frac{p_0}{\pi}\int_{-\infty}^{\infty} dp \, p \left(-\frac{d}{dp}\right)\frac{1}{p^2 + p_0^2} \,,$$

which can be integrated by parts to obtain

$$T = \frac{p_0^2}{2m}\frac{p_0}{\pi}\int_{-\infty}^{\infty} dp \, \frac{1}{p^2 + p_0^2} = \frac{p_0^2}{2m} \,.$$

The coordinate-space expectation value follows from

$$T = -\frac{\hbar^2}{2m}\int_{-\infty}^{\infty} dx \, \psi^*(x)\psi''(x) = \frac{\hbar^2}{2m}\int_{-\infty}^{\infty} dx \, \psi'^*(x)\psi'(x) \,,$$

where in the last step we have integrated by parts and used the fact that $\psi(x) \longrightarrow 0$ as $|x| \longrightarrow \infty$. Using

$$\psi(x) = \sqrt{\frac{mV_0}{\hbar^2}}\, e^{-mV_0|x|/\hbar^2} = \sqrt{\frac{p_0}{\hbar}}\, e^{-p_0|x|/\hbar} \,,$$

we find

$$x > 0, \quad \psi'(x) = -\frac{p_0}{\hbar}\psi(x); \qquad x < 0, \quad \psi'(x) = \frac{p_0}{\hbar}\psi(x),$$

and therefore

$$T = \frac{\hbar^2}{2m}\int_{-\infty}^{\infty} dx\,|\psi'(x)|^2 = \frac{p_0^2}{2m}\int_{-\infty}^{\infty} dx\,|\psi(x)|^2 = \frac{p_0^2}{2m},$$

since the wave function is normalized.

We could have also calculated directly the second derivative of the wave function. To this end, we first note that

$$\psi'(x) = -\frac{p_0}{\hbar}\psi(x)[\theta(x) - \theta(-x)],$$

where the step function $\theta(x) = 0$ if $x < 0$ and $\theta(x) = 1$ if $x > 0$. The second derivative is then obtained as

$$\psi''(x) = -\frac{p_0}{\hbar}\frac{d}{dx}\left\{\psi(x)\left[\theta(x) - \theta(-x)\right]\right\} = \frac{p_0^2}{\hbar^2}\underbrace{[\theta(x) - \theta(-x)]^2}_{=1}\psi(x) - \frac{p_0}{\hbar}\psi(x)\left[\delta(x) + \delta(-x)\right],$$

where we have used the fact the derivative of the step function is a δ-function:

$$\frac{d}{dx}\theta(x) = \delta(x), \qquad \frac{d}{dx}\theta(-x) = -\delta(-x).$$

Therefore, we find that

$$T = -\frac{\hbar^2}{2m}\int_{-\infty}^{\infty} dx\,\psi^*(x)\left[\frac{p_0^2}{\hbar^2}\psi(x) - \frac{p_0}{\hbar}\psi(x)[\delta(x) + \delta(-x)]\right] = -\frac{p_0^2}{2m} + \frac{\hbar p_0}{m}|\psi(0)|^2 = \frac{p_0^2}{2m},$$

as before.

Problem 6 Inequality for Ground-State Energies Corresponding to Potentials $v(x) \leq \bar{v}(x)$

Let $\psi_0(x)$ and $\bar{\psi}_0(x)$ be the ground-state wave functions for a particle in potentials $v(x)$ and $\bar{v}(x)$, respectively. Assume $v(x) \leq \bar{v}(x)$ for $-\infty < x < \infty$. Show that the corresponding ground-state energies ϵ_0 and $\bar{\epsilon}_0$ satisfy $\epsilon_0 \leq \bar{\epsilon}_0$. What can you conclude for energies other than the ground-state energy?

Solution

We have

$$\psi_0''(x) = [v(x) - \epsilon_0]\psi_0(x), \qquad \bar{\psi}_0''(x) = [\bar{v}(x) - \bar{\epsilon}_0]\bar{\psi}_0(x).$$

Multiplying the first equation by $\bar{\psi}_0(x)$ and the second equation by $\psi_0(x)$ and subtracting one from the other, we obtain

$$\underbrace{\bar{\psi}_0(x)\psi_0''(x) - \psi_0(x)\bar{\psi}_0''(x)}_{W'(x)} = [v(x) - \bar{v}(x) - (\epsilon_0 - \bar{\epsilon}_0)]\bar{\psi}_0(x)\psi_0(x).$$

The left-hand side of the above relation consists of the derivative of the Wronskian $W(x) = \overline{\psi}_0(x)\psi'_0(x) - \psi_0(x)\overline{\psi}'_0(x)$, and hence integrating both sides yields

$$\underbrace{\int_{-\infty}^{\infty} dx\, W'(x)}_{W(\infty) - W(-\infty) = 0} = \int_{-\infty}^{\infty} dx\, [v(x) - \overline{v}(x) - (\epsilon_0 - \overline{\epsilon}_0)]\overline{\psi}_0(x)\psi_0(x)\,,$$

where the left-hand side vanishes given that $\psi_0(x)$ and $\overline{\psi}_0(x)$ are bound-state wave functions. We arrive at

$$\epsilon_0 - \overline{\epsilon}_0 = \frac{\int_{-\infty}^{\infty} dx\, [v(x) - \overline{v}(x)]\overline{\psi}_0(x)\psi_0(x)}{\int_{-\infty}^{\infty} dx\, \overline{\psi}_0(x)\psi_0(x)}\,.$$

Note that, since $\psi_0(x)$ and $\overline{\psi}_0(x)$ are ground-state wave functions, they are nodeless and can be made to have the same sign (if they do not originally have the same sign then just multiply one of the Schrödinger equations by -1). Thus, we have

$$\int_{-\infty}^{\infty} dx\, \overline{\psi}_0(x)\psi_0(x) > 0\,, \qquad \int_{-\infty}^{\infty} dx\, [v(x) - \overline{v}(x)]\overline{\psi}_0(x)\psi_0(x) \le 0\,,$$

where the second inequality follows from $\psi_0(x)\overline{\psi}_0(x) > 0$ and the assumption $v(x) \le \overline{v}(x)$. We conclude that $\epsilon_0 \le \overline{\epsilon}_0$. Note that the above proof fails for energies other than ground-state energies, the reason being that the presence of nodes in the corresponding wave functions $\psi_n(x)$ and $\overline{\psi}_n(x)$ does not ensure that the overlap integral of these wave functions is positive definite.

Problem 7 Particle in an Asymmetric One-Dimensional Potential Well

Consider the problem of a particle of mass m in a potential given by

$$v(x) = v_1, \qquad x < 0\,; \qquad v(x) = 0, \qquad 0 < x < a\,; \qquad v(x) = v_2, \qquad x > a\,,$$

where $v_1 > v_2 > 0$.

1. Discuss the nature of the energy spectrum.
2. Solve graphically for the energies of the bound states (if any).
3. Consider the case $v_1 = v_2 = v$ and show that there is at least one bound state for any values of v and a. How many bound states are there for a given v and a?

Hint: In part 2 you should find the equations

$$\cot\beta = \kappa_1/k\,, \qquad \cot(ka + \beta) = -\kappa_2/k\,,$$

where β is a constant and the wave numbers are defined as

$$\kappa_1 = \sqrt{v_1 - \epsilon}\,, \qquad \kappa_2 = \sqrt{v_2 - \epsilon}\,, \qquad k = \sqrt{\epsilon}\,.$$

The above equations have the solutions (verify this!)

$$\beta = n_1\pi + \sin^{-1}(k/K_1)\,, \qquad ka + \beta = n_2\pi - \sin^{-1}(k/K_2)\,,$$

where n_1 and n_2 are integers, and we have defined the wave numbers

$$K_1 = \sqrt{v_1}\,, \qquad K_2 = \sqrt{v_2}\,.$$

After eliminating β, you should be able to carry out a graphical solution of the eigenvalue equation.

Solution

Part 1

There are no valid solutions for $\epsilon < 0$. For $0 < \epsilon < v_2$ only bound states are possible, while for $\epsilon > v_2$ only continuum states occur, which are either non-degenerate if $v_2 < \epsilon < v_1$ or doubly degenerate if $\epsilon > v_1$.

Part 2

Assume $0 < \epsilon < v_2$. There are three regions: region I for $x < 0$; region II for $0 < x < a$; region III for $x > a$. We have

$$\psi_I''(x) = \kappa_1^2\,\psi_I(x)\,, \qquad\qquad \kappa_1^2 = v_1 - \epsilon\,,$$
$$\psi_{II}''(x) = -k^2\,\psi_{II}(x)\,, \qquad\qquad k^2 = \epsilon\,,$$
$$\psi_{III}''(x) = \kappa_2^2\,\psi_{III}(x)\,, \qquad\qquad \kappa_2^2 = v_2 - \epsilon\,,$$

with corresponding (normalizable) solutions

$$\psi_I(x) = \alpha_1 e^{\kappa_1 x}\,, \qquad \psi_{II}(x) = \alpha\,\sin(kx + \beta)\,, \qquad \psi_{III}(x) = \alpha_2 e^{-\kappa_2 x}\,.$$

We now impose the continuity of the wave function and its first derivative at $x = 0$ and $x = a$, to obtain

$$x = 0:\ \alpha_1 = \alpha\,\sin\beta\,,\ \alpha_1\kappa_1 = \alpha k\,\cos\beta$$
$$x = a:\ \alpha_2\,e^{-\kappa_2 a} = \alpha\,\sin(ka + \beta)\,,\ -\alpha_2\kappa_2\,e^{-\kappa_2 a} = \alpha k\,\cos(ka + \beta)\,,$$

which imply

$$\cot\beta = \kappa_1/k\,, \qquad \cot(ka + \beta) = -\kappa_2/k\,.$$

To solve these equations, it is convenient to re-express them as

$$\sin\beta = \frac{1}{\sqrt{1 + (\kappa_1/k)^2}} = \frac{k}{K_1}\,, \qquad \sin(ka + \beta) = -\frac{1}{\sqrt{1 + (\kappa_2/k)^2}} = -\frac{k}{K_2}\,,$$

where we have used $\sin x = 1/\sqrt{1 + \cot^2 x}$ and have defined the constants

$$K_1 = \sqrt{v_1}\,, \qquad K_2 = \sqrt{v_2}\,.$$

In particular, we observe that k/K_1 and k/K_2 are both < 1 here. In the intervals $-\pi/2 \le \beta \le \pi/2$ and $-\pi/2 \le ka + \beta \le \pi/2$, these equations have unique solutions, given by, respectively, $\sin^{-1}(k/K_1)$ and $-\sin^{-1}(k/K_2)$. However, the cotangent has period π, and hence the complete sets of solutions are given by

$$\beta = n_1\pi + \sin^{-1}(k/K_1)\,, \qquad ka + \beta = n_2\pi - \sin^{-1}(k/K_2)\,,$$

where n_1 and n_2 are integers. Eliminating β, we finally arrive at

$$ka = n\pi - \sin^{-1}(k/K_1) - \sin^{-1}(k/K_2)\,,$$

where n must be a positive integer, $n = 1, 2, \ldots$, for the equation to have a solution (the left-hand side is positive). Note that the values of $\sin^{-1}$ lie in the interval $[0, \pi/2]$. The above equation can be solved graphically; see Fig. 4.3. Since for bound states $0 \leq \epsilon \leq v_2$, the wave number k must lie in the range $0 \leq k \leq K_2$. We define

$$f(k) = ak, \qquad g(k) = n\pi - \sin^{-1}(k/K_1) - \sin^{-1}(k/K_2) ,$$

and $f(k)$ is a straight line with slope a while, for a fixed n, $g(k)$ is monotonically decreasing function of k with

$$g(0) = n\pi , \qquad g(K_2) = n\pi - \sin^{-1}(K_2/K_1) - \sin^{-1}(1) = (n - 1/2)\pi - \sin^{-1}(K_2/K_1) .$$

We see that if a is too small, there is no intercept and therefore no solution. For fixed a, there is in general a finite number of bound states.

Part 3

When $K_1 = K_2 = K$, as is apparent from Fig. 4.3 there will be at least one solution for any K and a. Note that in this case $g(K) = (n - 1)\pi$ and so $g(K)$ vanishes for $n = 1$. In the limit $Ka \ll 1$, we must have $n = 1$ and

$$ak = \pi - 2\sin^{-1}(k/K) \implies \sin^{-1}(k/K) = \pi/2 - ak/2 \implies k/K = \sin(\pi/2 - ak/2) .$$

Using $\sin(\pi/2 - ak/2) = \cos(ak/2)$, in the limit $aK \ll 1$ we can approximate as follows:

$$\cos(ak/2) = \cos(\lambda k/K) = 1 - \frac{\lambda^2}{2}\frac{k^2}{K^2} + \cdots \qquad \lambda = aK/2 \ll 1 ,$$

and the equation for k reduces to

$$\frac{k}{K} \approx 1 - \frac{\lambda^2}{2}\frac{k^2}{K^2} ,$$

which can easily be solved, yielding

$$\frac{k}{K} \approx 1 - \frac{\lambda^2}{2} \implies \frac{\epsilon}{v} \approx \left(1 - \frac{\lambda^2}{2}\right)^2 \approx 1 - \lambda^2 .$$

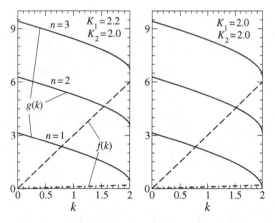

Fig. 4.3 Graphical solution of $f(k) = g(k)$ (arbitrary units). Only three branches of $g(k)$ are shown, corresponding to $n = 1, 2, 3$. The left-hand panel is for the case $K_1 > K_2$; if $a < \pi/2 - \sin^{-1}(K_2/K_1)$ (dash-dotted curve close to the k-axis) there is no intercept with the solid curve, and hence no solution; the right-hand panel is for the case $K_1 = K_2$, there is always at least one solution.

The bound-state energy in the limit $\lambda \ll 1$ is

$$\epsilon_0 = v\left(1 - \frac{a^2 v_0}{4}\right) \, .$$

It is also easy to see from Fig. 4.3 that the number N of eigenvalues, for given v and a, follows from the relation

$$(N-1)\pi < Ka < N\pi \, .$$

Problem 8 Particle in a Potential $v(x) = \infty$ for $x < 0$ and $v(x) = -v_0\,\delta(x-a)$ for $x > 0$

Consider a particle in a one-dimensional potential $v(x)$ such that $v(x) = \infty$ for $x < 0$ and

$$v(x) = -v_0\,\delta(x-a) \qquad \text{for } x > 0 \, ,$$

where v_0 is a positive constant. Determine whether this potential admits any bound states.

Solution

The Schrödinger equation is given by

$$\psi''(x) = -[v_0\,\delta(x-a) + \epsilon]\psi(x) \, ,$$

and bound states, if they exist, occur for $\epsilon < 0$. The wave functions in regions $0 < x < a$ and $x > a$ are denoted, respectively, as $\psi_<(x)$ and $\psi_>(x)$. The boundary conditions are

$$\psi_<(0) = 0 \, , \qquad \psi_<(a) = \psi_>(a) \, , \qquad \psi_>'(a) - \psi_<'(a) = -v_0\,\psi_>(a) \, , \qquad \lim_{x\to\infty}\psi_>(x) = 0 \, .$$

In the region $0 < x < a$ we have

$$\psi_<''(x) = |\epsilon|\,\psi_<(x) \implies \psi_<(x) = \alpha_< \sinh(\sqrt{|\epsilon|}\,x) + \beta_< \cosh(\sqrt{|\epsilon|}\,x) \, ,$$

and the boundary condition at the origin requires $\beta_< = 0$. In the region $x > a$ we have

$$\psi_>''(x) = |\epsilon|\,\psi_>(x) \implies \psi_>(x) = \alpha_> \, e^{-\sqrt{|\epsilon|}\,x} + \beta_> \, e^{\sqrt{|\epsilon|}\,x} \, ,$$

and the boundary condition at ∞ imposes $\beta_> = 0$. The remaining boundary conditions yield

$$\alpha_< \sinh(\sqrt{|\epsilon|}\,a) = \alpha_> \, e^{-\sqrt{|\epsilon|}\,a} \, ,$$

$$-\alpha_> \sqrt{|\epsilon|}\,e^{-\sqrt{|\epsilon|}\,a} - \alpha_< \sqrt{|\epsilon|}\,\cosh(\sqrt{|\epsilon|}\,a) = -v_0\,\alpha_> \, e^{-\sqrt{|\epsilon|}\,a} \, ,$$

and non-trivial solutions for $\alpha_<$ and $\alpha_>$ occur if

$$\det\begin{bmatrix} \sinh(\sqrt{|\epsilon|}\,a) & -e^{-\sqrt{|\epsilon|}\,a} \\ \sqrt{|\epsilon|}\,\cosh(\sqrt{|\epsilon|}\,a) & \left(\sqrt{|\epsilon|} - v_0\right)e^{-\sqrt{|\epsilon|}\,a} \end{bmatrix} = 0 \, ,$$

or

$$\left(\sqrt{|\epsilon|} - v_0\right)\sinh\left(\sqrt{|\epsilon|}\,a\right) + \sqrt{|\epsilon|}\,\cosh\left(\sqrt{|\epsilon|}\,a\right) = 0 \implies \tanh(\sqrt{|\epsilon|}\,a) = \frac{\sqrt{|\epsilon|}}{v_0 - \sqrt{|\epsilon|}} \, .$$

Introduce the non-dimensional variable $z = \sqrt{|\epsilon|}\,a$, to obtain

$$\tanh(z) = \frac{z}{z_0 - z} \qquad \text{with } z_0 = av_0 > 0 \, .$$

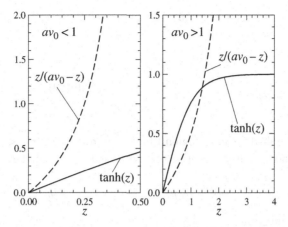

Fig. 4.4 Graphical solution of eigenvalue equation: the case $av_0 < 1$, with no solution (left-hand panel) and the case $av_0 > 1$, with one solution (right-hand panel).

Since $0 \leq \tanh(z) < 1$ for $z \geq 0$, solutions can occur for $0 \leq z/(z_0 - z) < 1$, which requires $0 \leq z \leq z_0/2$. Figure 4.4 shows that in fact there is a single solution if the slope of $z/(z_0 - z)$ is less than that of $\tanh(z)$ at $z = 0$, yielding

$$\frac{z}{z_0 - z} = \frac{z}{z_0}\frac{1}{1 - z/z_0} = \sum_{n=0}^{\infty}\left(\frac{z}{z_0}\right)^{n+1} \implies \frac{d}{dz}\frac{z}{z_0 - z}\Big|_{z=0} = \frac{1}{z_0} < 1 \,,$$

and therefore the condition for bound states is that $av_0 > 1$.

Problem 9 Particle in a Potential $v(x) = -v_0\left[\delta(x) + \delta(x - a)\right]$: The Ionized H_2^+ Molecule

A particle in one dimension is subject to a potential consisting of two attractive δ-functions, placed at the origin and at $x = a$ (this could be a model for the H_2^+ molecule)

$$v(x) = -v_0\,\delta(x) - v_0\,\delta(x - a)\,, \qquad\qquad v_0 > 0\,.$$

1. Bound states can occur for energy $\epsilon < 0$. Set $\rho = \sqrt{|\epsilon|}$ and show that the possible energies are given by

$$e^{-\rho a} = \pm\left(1 - \frac{2\,\rho}{v_0}\right)\,.$$

   Provide a graphical solution for this equation.
2. Denote by $\epsilon_0 = -v_0^2/4$ the energy of a particle bound in a single (attractive) δ-function potential of strength v_0. Show that the ground-state energy ϵ_S of this system is less than ϵ_0. Further, show that, if a is larger than a certain value a_0, to be determined, there exists an excited state of energy $\epsilon_A > \epsilon_0$. How do ϵ_S and ϵ_A vary as function of a? What happens in the limits $a \longrightarrow 0$ and $a \longrightarrow \infty$? Explain why the ground- and excited-state wave functions are, respectively, even and odd with respect to a reflection about the point $x = a/2$. Sketch these wave functions.
3. Justify in what sense the above system provides a rough model for the ionized H_2^+ molecule, in which the two protons, assumed to be fixed in space, are separated by a distance a. Include in the total energy of the molecule the contribution of the Coulomb repulsion between the protons, and discuss the variation of the total energy with respect to a. Argue why, under certain conditions,

one can predict the existence of bound states of the H_2^+ molecule, thus illustrating a mechanism of chemical bonding.

Solution

Part 1

We define three regions: region I for $x < 0$; region II for $0 < x < a$; and region III for $x > a$. The wave function must be continuous at $x = 0$ and $x = a$, its first derivatives satisfying the conditions

$$\psi_I'(0) - \psi_{II}'(0) = v_0 \psi_I(0) , \qquad \psi_{II}'(a) - \psi_{III}'(a) = v_0 \psi_{III}(a) .$$

The solutions are

$$\psi_I(x) = A\,e^{\rho x} , \qquad \psi_{II}(x) = B\,e^{-\rho x} + C e^{\rho x} , \qquad \psi_{III}(x) = D\,e^{-\rho x} .$$

Imposing the boundary conditions leads to the homogenous linear system,

$$A = B + C$$
$$A\rho - (-B\rho + C\rho) = v_0 A$$
$$B\,e^{-\rho a} + C\,e^{\rho a} = D\,e^{-\rho a}$$
$$-B\rho\,e^{-\rho a} + C\rho\,e^{\rho a} + D\rho\,e^{-\rho a} = v_0 D\,e^{-\rho a} ,$$

which we rewrite as

$$A - B - C = 0$$
$$\left(1 - \frac{v_0}{\rho}\right)A + B - C = 0$$
$$B + e^{2\rho a} C - D = 0$$
$$-B + e^{2\rho a} C + \left(1 - \frac{v_0}{\rho}\right)D = 0 .$$

The condition for having non-trivial solutions is that the determinant of the coefficients vanishes,

$$\det \begin{bmatrix} 1 & -1 & -1 & 0 \\ 1 - v_0/\rho & 1 & -1 & 0 \\ 0 & 1 & e^{2\rho a} & -1 \\ 0 & -1 & e^{2\rho a} & 1 - v_0/\rho \end{bmatrix} = 0 ,$$

which is easily evaluated to give

$$e^{2\rho a} - 1 + 2(1 - v_0/\rho)(e^{2\rho a} + 1) + (1 - v_0/\rho)^2(e^{2\rho a} - 1) = 0 ,$$

or

$$e^{2\rho a}(2 - v_0/\rho)^2 = (v_0/\rho)^2 \implies e^{\rho a} = \pm \frac{v_0/\rho}{2 - v_0/\rho} \implies e^{-\rho a} = \pm(1 - 2\rho/v_0) .$$

A graphical solution is provided in Fig. 4.5

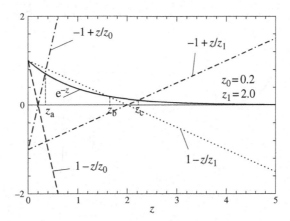

Fig. 4.5 Graphical solution of the eigenvalue equation; here, $z = \rho a$. Two cases are shown: in one case $av_0/2 = z_0 = 0.2$ (dashed and long-dash-dotted lines) and there is only one solution z_a, while in the other case $av_0/2 = z_1 = 2.0$ (dotted and short-dashed-dotted lines) and there are two solutions, z_b and z_c.

Part 2

We introduce the variables $z = a\rho$ and $z_0 = av_0/2$. As should be clear from Fig. 4.5, there is always at least one bound state, given by the intercept of e^{-z} and $-(1 - z/z_0)$. The intercept z_a or z_c in the figure is such that $\rho_0 > v_0/2$, which leads to

$$|\epsilon_S| > \frac{v_0^2}{4} \implies \epsilon_S < \epsilon_0 \equiv -\frac{v_0^2}{4} ,$$

where ϵ_0 is the ground-state energy of a particle in a single δ-function potential; indeed, a numerical solution gives

$$\epsilon_S(a) \approx \epsilon_0 \left(1 + e^{-av_0}\right)^2 .$$

By contrast, there is no intercept of e^{-z} with $1 - z/z_0$ unless the slope of $1 - z/z_0$ is larger than the slope of e^{-z} at $z = 0$, yielding the condition $-1/z_0 > -1$ or $z_0 > 1$ and hence $a > 2/v_0 \equiv a_0$. Thus, if $a < a_0$, there is no excited state, but if $a > a_0$ then there exists an excited state with energy $\epsilon_A > \epsilon_0$. In the limit $a \longrightarrow 0$, we have $v(x) = -2v_0 \, \delta(x)$, so that $\epsilon_S \longrightarrow 4\epsilon_0$ and there is no excited state. In the limit $a \longrightarrow \infty$, we have $\epsilon_S = \epsilon_A = \epsilon_0$. The potential can be written as

$$v(x) = -v_0 \, \delta[(x - a/2) + a/2] - v_0 \, \delta[(x - a/2) - a/2] ,$$

which shows that $v(x)$ is even under a reflection $(x - a/2) \longrightarrow -(x - a/2)$. The (non-degenerate) bound-state wave functions corresponding to this $v(x)$ are either even or odd with respect to reflections about $a/2$. Since the ground-state wave function is nodeless, it must be even whereas the excited-state wave function, when it exists, must be odd. A sketch of the even and odd wave functions is provided in the left-hand panel of Fig. 4.6.

Part 3

In the ionized molecule H_2^+, the electron feels the Coulomb attraction due to the two protons:

$$v(x) = -\frac{2me^2}{\hbar^2} \left(\frac{1}{|x|} - \frac{1}{|x - a|} \right) .$$

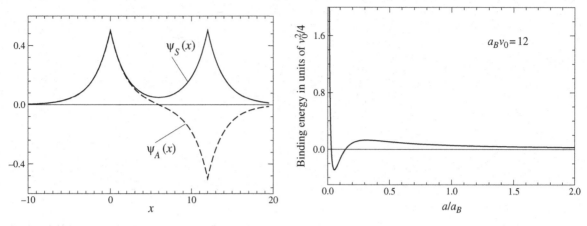

Fig. 4.6 Left-hand panel: the ground- and excited-state wave functions, respectively $\psi_S(x)$ and $\psi_A(x)$, roughly corresponding to potential parameters $a = 12$ and $v_0 = 1$ (arbitrary units). Right-hand panel: binding energy of H_2^+ in units of $v_0^2/4$ as a function of a/a_B, where a_B is the Bohr radius.

This potential can be roughly approximated by two δ-functions, positioned at $x = 0$ and $x = a$. The total energy includes, in addition to the electronic binding energy, the Coulomb repulsive energy between the two protons, and so

$$\epsilon_S^T(a) = \epsilon_S(a) + \frac{2me^2}{a\hbar^2} \approx \epsilon_0 \left(1 + e^{-av_0}\right)^2 + \frac{2}{a_B a} \,,$$

where a_B is the Bohr radius. The function $\epsilon_S^T(a)$ diverges as $a \longrightarrow 0$, and approaches $\epsilon_0 = -v_0^2/4$ as the separation between the two protons becomes larger and larger. We define the H_2^+ binding energy in units of $|\epsilon_0|$ as

$$B(H_2^+; a) = \frac{\epsilon_S^T(a) - \epsilon_0}{|\epsilon_0|} = -e^{-av_0} \left(2 + e^{-av_0}\right) + \frac{8}{v_0^2 a_B a} \,,$$

where ϵ_0 is the energy of a isolated "hydrogen atom," which is obtained when one of the protons is at ∞. For appropriate choices of the parameters v_0 and a, the function $B(H_2^+; a)$ has a minimum, as shown in the right-hand panel of Fig. 4.6, and the a_{min} at which it occurs is the equilibrium separation between the two protons in the molecule.

Problem 10 Particle in a Potential $v(x) = -v_0\,\delta(x)$ for $|x| < a/2$ and $v(x) = \infty$ for $|x| > a/2$

Consider the one-dimensional potential

$$\begin{aligned} v(x) &= -v_0\,\delta(x) && |x| < a/2 \\ &= \infty && |x| > a/2 \,, \end{aligned}$$

where $v_0 > 0$.

1. Assume the particle's energy $\epsilon < 0$. Show that there is a bound state if a certain condition is satisfied.

2. Assume the particle's energy $\epsilon > 0$. Find the bound-state energies in this case (a graphical solution of the relevant condition suffices).

3. Compare the bound-state energies found in part 2 with those obtained in an infinitely deep well for which $v(x) = 0$ if $|x| < a/2$ and $v(x) = \infty$ if $|x| > a/2$. Comment on the results.

Solution

We denote $-a/2 < x < 0$ as region L and $0 < x < a/2$ as region R. The boundary conditions are

$$\psi_L(-a/2) = 0 = \psi_R(a/2)\,, \qquad \psi_L(0) = \psi_R(0)\,, \qquad \psi_R'(0) - \psi_L'(0) = -v_0\,\psi_R(0)\,.$$

The potential is even under parity. We consider the cases $\epsilon < 0$ and $\epsilon > 0$ separately.

Part 1

For $\epsilon < 0$ the Schrödinger equation reads

$$\psi_L''(x) = \kappa^2 \psi_L(x)\,, \qquad \psi_R''(x) = \kappa^2 \psi_R(x)\,, \qquad \kappa^2 = |\epsilon|\,.$$

The solutions in each region can be written as combinations of hyperbolic sine and cosine functions. However, the boundary conditions at $x = \pm a/2$ lead to

$$\psi_L(x) = A_L\,\sinh[\kappa(x + a/2)]\,, \qquad \psi_R(x) = A_R\,\sinh[\kappa(x - a/2)]\,.$$

The boundary conditions at the origin give

$$A_L\,\sinh(\kappa a/2) = -A_R\,\sinh(\kappa a/2) \implies A_L = -A_R$$

and

$$A_R\kappa\,\cosh(\kappa a/2) - A_L\kappa\,\cosh(\kappa a/2) = v_0 A_R\,\sinh(\kappa a/2) \implies \tanh(\kappa a/2) = \frac{2\kappa}{v_0}\,.$$

Define $x = \kappa a/2$; the above condition requires that

$$\tanh x = \frac{4x}{av_0}\,,$$

and, recalling that $\tanh x \approx x$ for small x, we see that there is always a solution if

$$\frac{4}{av_0} < 1 \implies v_0 > \frac{4}{a}\,.$$

The corresponding solution is found as the intercept x_0 of $\tanh x$ with $(4/av_0)x$, giving the energy

$$\epsilon_0 = -\frac{4x_0^2}{a^2}\,.$$

Part 2

For $\epsilon > 0$ the Schrödinger equation reads

$$\psi_L''(x) = -k^2 \psi_L(x)\,, \qquad \psi_R''(x) = -k^2 \psi_R(x)\,, \qquad k^2 = \epsilon\,.$$

The solutions in each region can be written as combinations of sine and cosine functions. However, the boundary conditions at $x = \pm a/2$ lead to

$$\psi_L(x) = A_L \sin[k(x + a/2)] , \qquad \psi_R(x) = A_R \sin[k(x - a/2)] .$$

The boundary conditions at the origin give

$$A_L \sin(ka/2) = -A_R \sin(ka/2)$$

and

$$A_R k \cos(ka/2) - A_L k \cos(ka/2) = v_0 A_R \sin(ka/2) .$$

These can be satisfied by either

$$A_L = A_R \qquad \text{and} \qquad \frac{ka}{2} = n\pi \qquad n = 1, 2, \ldots ,$$

which lead to the same ϵ_n and $\psi_n(x)$ with n even for the infinite well (see part 3), or

$$A_L = -A_R \qquad \text{and} \qquad \frac{ka}{2} \neq n\pi \qquad n = 1, 2, \ldots$$

which require that

$$\cot(ka/2) = \frac{v_0}{2k} ,$$

or, in terms of $x = ka/2$,

$$\cot x = \frac{\alpha}{x} , \qquad \alpha = \frac{a v_0}{4} .$$

Note that $\cot x$ is periodic with period π, and a graphical solution of the equations shows that there is an infinite number of intercepts. If $\alpha < 1$, one solution is close to $x = 0$; if $\alpha > 1$, this solution is "pushed" into negative energy.

Part 3

In the absence of the δ-function potential, the (bound-state) energies would be

$$\epsilon_p = \frac{\pi^2}{a^2} p^2 , \qquad p = 1, 2, \ldots ,$$

with corresponding (normalized) eigenfunctions given by

$$\psi_p(x) = \sqrt{\frac{2}{a}} \cos\left(\frac{p\pi}{a} x\right) \qquad p = 1, 3, \ldots$$

$$= \sqrt{\frac{2}{a}} \sin\left(\frac{p\pi}{a} x\right) \qquad p = 2, 4, \ldots ,$$

Even when the δ-function is present, we note that the (odd-under-parity) wave functions $\psi_p(x)$ with p even satisfy the boundary condition at the origin and $|x| = a/2$. Therefore, ϵ_p and $\psi_p(x)$ with p even must be the eigenfunctions found in part 2 (up to phase factors). Indeed, we have, for $A_L = A_R = A$ and $ka/2 = n\pi$ with $n = 1, 2, \ldots$,

$$\psi_L(x) = A \sin\left(\frac{2n\pi}{a} x + n\pi\right) , \qquad \psi_R(x) = A \sin\left(\frac{2n\pi}{a} x - n\pi\right) ,$$

which reduce to $\psi_p(x)$ with $2n \longrightarrow p$ even, since

$$\sin\left(\frac{2n\pi}{a}x \pm n\pi\right) \longrightarrow (-)^{p/2}\sin\left(\frac{p\pi}{a}x\right) .$$

Problem 11 Particle in a Potential $v(x) = v_0\,\theta(a - |x|) - w_0\,\delta(x)$

Consider the one-dimensional potential given by

$$\begin{aligned}
v(x) &= v_0 - w_0\,\delta(x) & |x| \le a \\
&= 0 & |x| > a,
\end{aligned}$$

consisting of a repulsive square well superimposed on an attractive δ-function potential. Note that the potential is even under parity.

1. Explain why there cannot be bound states of odd parity.
2. Determine under what condition there is a single bound state (of even parity), and explain why its energy is higher than the bound-state energy for the attractive δ-function potential alone. Show that in the limit $v_0 = 0$, that is, when there is no repulsive well potential, the bound-state energy is that of a particle in the potential $-w_0\,\delta(x)$.
3. Provide a graphical solution of the eigenvalue equation.

Solution

Part 1

The Schrödinger equation reads

$$\psi'' = [v_0\,\theta(|x| - a) - w_0\,\delta(x) - \epsilon]\,\psi(x) ,$$

where we have introduced the step function $\theta(z) = 1$ if $z > 0$ and $\theta(z) = 0$ if $z < 0$. Bound states, if they exist, occur when $\epsilon < 0$. These bound states are necessarily even or odd under parity, since the corresponding energies are non-degenerate. However, odd bound states are not possible given that their wave functions must vanish at $x = 0$ and are therefore unaffected by the presence of the attractive δ-function potential. This fact will become obvious below.

Part 2

In regions 1 and 2, defined by, respectively, $x > a$ and $0 < x < a$, we have

$$\psi_1'' = |\epsilon|\psi_1(x) , \qquad \psi_2'' = (v_0 + |\epsilon|)\psi_2(x) ,$$

with solutions

$$\psi_1(x) = \alpha\,e^{-\kappa_1 x} , \qquad \psi_2(x) = \beta e^{-\kappa_2 x} + \gamma e^{\kappa_2 x} ,$$

where we define

$$\kappa_1 = \sqrt{|\epsilon|} , \qquad \kappa_2 = \sqrt{v_0 + |\epsilon|} .$$

The even solutions in regions 3 $(-a < x < 0)$ and 4 $(x < -a)$ are obtained as $\psi_3(x) = \psi_2(-x)$ and $\psi_4(x) = \psi_1(-x)$, while the odd solutions follow from $\psi_3(x) = -\psi_2(-x)$ and $\psi_4(x) = -\psi_1(-x)$. The boundary conditions at $x = a$ require $\psi_1(a) = \psi_2(a)$ and $\psi_1'(a) = \psi_2'(a)$, so that

$$\alpha \, e^{-\kappa_1 a} = \beta \, e^{-\kappa_2 a} + \gamma \, e^{\kappa_2 a} \, , \qquad -\alpha \kappa_1 \, e^{-\kappa_1 a} = -\beta \kappa_2 \, e^{-\kappa_2 a} + \gamma \kappa_2 e^{\kappa_2 a} \, .$$

We solve this (inhomogeneous) linear system with respect to β and γ as functions of α to find

$$\beta = \frac{e^{-(\kappa_1 - \kappa_2)a}}{2} \left(1 + \frac{\kappa_1}{\kappa_2} \right) \alpha \, , \qquad \gamma = \frac{e^{-(\kappa_1 + \kappa_2)a}}{2} \left(1 - \frac{\kappa_1}{\kappa_2} \right) \alpha \, .$$

We will treat the even and odd solutions separately. The even solution is necessarily continuous at $x = 0$, while its first derivative must satisfy

$$\psi'_{e,2}(0) - \psi'_{e,3}(0) = -w_0 \psi_{e,2}(0) \, ,$$

which yields

$$-2\kappa_2(\beta - \gamma) = -w_0 \left(\beta + \gamma \right) \, ,$$

where

$$\beta + \gamma = \alpha \, e^{-\kappa_1 a} \left[\cosh(\kappa_2 a) + \frac{\kappa_1}{\kappa_2} \sinh(\kappa_2 a) \right] \, , \qquad \beta - \gamma = \alpha \, e^{-\kappa_1 a} \left[\sinh(\kappa_2 a) + \frac{\kappa_1}{\kappa_2} \cosh(\kappa_2 a) \right] \, .$$

Assuming $\alpha \neq 0$ (otherwise we have a null solution), the jump discontinuity of the first derivative requires that

$$\sinh(\kappa_2 a) + \frac{\kappa_1}{\kappa_2} \cosh(\kappa_2 a) = \frac{w_0}{2\kappa_2} \left[\cosh(\kappa_2 a) + \frac{\kappa_1}{\kappa_2} \sinh(\kappa_2 a) \right] \, ,$$

which can also be written as

$$\frac{1}{\kappa_2} \frac{\kappa_2 + \kappa_1 \tanh(\kappa_2 a)}{\kappa_1 + \kappa_2 \tanh(\kappa_2 a)} = \frac{2}{w_0} \, .$$

Before examining whether the above equation has any solutions at all, we turn to the odd wave functions. In this case, the condition on the first derivative gives

$$\underbrace{\psi'_{o,2}(0) - \psi'_{o,3}(0)}_{\text{vanishes}} = -w_0 \psi_{o,2}(0) \implies \psi_{o,2}(0) = 0 \, ,$$

or, inserting the expression for $\psi_{o,2}(0) = \beta + \gamma$,

$$\alpha \, e^{-\kappa_1 a} \underbrace{\left[\cosh(\kappa_2 a) + \frac{\kappa_1}{\kappa_2} \sinh(\kappa_2 a) \right]}_{> 0 \text{ for any } \kappa_1, \kappa_2 > 0} = 0 \implies \alpha = 0 \, .$$

Thus, we conclude that there are no odd solutions, as anticipated.

Going back to the eigenvalue equation, we first note that the ground state, if it exists, should have energy higher than $\epsilon_0 = -w_0^2/4$, which is the ground-state energy corresponding to the δ-function potential $-w_0 \, \delta(x)$. The latter is less than or equal to the present potential $v(x) = v_0 \theta(|x| - a) - w_0 \, \delta(x)$ (see Problem 4.6). Note also that in the limit $v_0 \longrightarrow 0$, that is, when there is no repulsive well potential, we have $\kappa_2 \longrightarrow \kappa_1$ and the eigenvalue equation reduces to $1/\kappa_1 = 2/w_0$ or $\epsilon = \epsilon_0$.

We set

$$\kappa_2 \, a = x \implies |\epsilon| = (x/a)^2 - v_0 \, , \qquad \kappa_1 a = a\sqrt{|\epsilon|} = \sqrt{x^2 - a^2 v_0}$$

and the eigenvalue equation becomes

$$f(x) \equiv \frac{1}{x} \frac{x + \sqrt{x^2 - a^2 v_0} \, \tanh x}{\sqrt{x^2 - a^2 v_0} + x \, \tanh x} = \frac{2}{a w_0} \, , \qquad x \geq x_0 \equiv a \sqrt{v_0} \, .$$

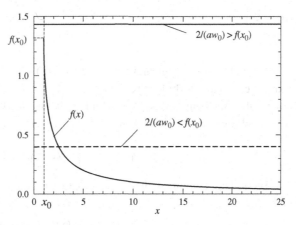

Fig. 4.7 Graphical solution of $f(x) = 2/(aw_0)$ with $x_0 = a\sqrt{v_0}$. If $2/(aw_0) > f(x_0)$ there are no solutions; if $2/(aw_0) < f(x_0)$ there is one solution.

When the particle is only just bound we have $\epsilon \longrightarrow 0^-$, and in this limit $x \longrightarrow x_0$, so that the eigenvalue equation reduces as follows:

$$f(x) \longrightarrow f(x_0) = \frac{1}{x_0 \tanh x_0} = \frac{2}{aw_0} \implies w_0^\star = 2\sqrt{v_0} \tanh x_0 \ ;$$

for $w_0 = w_0^\star$ there is a bound state with energy 0^-, and hence for $w_0 < w_0^\star$ there are no bound states. The other interesting limit is $x \longrightarrow \infty$ so that $|\epsilon| \gg v_0$, in which case $f(x) \approx 1/x$ and hence

$$\frac{1}{a\sqrt{|\epsilon| + v_0}} \approx \frac{1}{a\sqrt{|\epsilon|}} = \frac{2}{aw_0} \implies \epsilon = -\frac{w_0^2}{4} = \epsilon_0 \ ,$$

and the bound-state energy reduces to that of the δ-function potential. Thus, the above condition also says that $w_0 \gg 2\sqrt{v_0}$, namely the repulsive barrier has no effect and can be ignored.

Part 3

We provide a graphical solution of the eigenvalue equation; see Fig. 4.7. Note that the bound-state energy approaches the limit ϵ_0 as $2/(aw_0) \longrightarrow 0$.

Problem 12   The WKB Approximation for Bound-State Solutions of the Schrödinger Equation

This problem deals with the derivation of the WKB (Wentzel, Kramers, and Brillouin) method for solving approximately the one-dimensional Schrödinger equation

$$\psi''(x) = [v(x) - \epsilon]\psi(x) \ .$$

1. Write the wave function $\psi(x)$ as

$$\psi(x) = e^{iu(x)}$$

and obtain a differential equation satisfied by $u(x)$,

$$[u'(x)]^2 = \epsilon - v(x) + iu''(x) \ .$$

Hereafter define

$$k(x) = \sqrt{\epsilon - v(x)} \quad \text{for } \epsilon > v(x) , \qquad\qquad \kappa(x) = \sqrt{v(x) - \epsilon} \quad \text{for } \epsilon < v(x) ,$$

where the conditions ϵ larger and smaller than $v(x)$ characterize, respectively, the classically allowed and classically forbidden regions.

2. Assume for the time being that $\epsilon > v(x)$ and, by neglecting the term proportional to $u''(x)$ on the r.h.s. of the differential equation for $u(x)$, obtain the zeroth-order approximation

$$u_0(x) = \pm \int^x dx' \, k(x') + C ,$$

where C is constant. Obtain a better approximation $u_1(x)$ for $u(x)$ by substituting $u''(x)$ by $u_0''(x)$ in the r.h.s., that is,

$$[u_1'(x)]^2 = [k(x)]^2 + i u_0''(x) ,$$

and show that

$$u_1(x) = \pm \int^x dx' \, k(x') + \frac{i}{2} \ln k(x) + C .$$

Obtain the corresponding (independent) solutions as (up to constants)

$$\psi(x) \propto \frac{1}{\sqrt{k(x)}} \exp\left[\pm i \int^x dx' \, k(x')\right] \quad \text{for } \epsilon > v(x) .$$

In the classically forbidden region, show that

$$\psi(x) \propto \frac{1}{\sqrt{\kappa(x)}} \exp\left[\pm \int^x dx' \, \kappa(x')\right] \quad \text{for } \epsilon < v(x) ,$$

which also follows from the replacement $k(x) = -i\kappa(x)$ in the solutions for the classically allowed region.

3. The approximation outlined above is valid if $u_1(x)$ is close to $u_0(x)$. Show that this requires, in the classically allowed region,

$$|k'(x)| \ll [k(x)]^2 ,$$

with a similar relation in the classically forbidden region. By introducing the x-dependent wavelength (the "local wavelength") for the de Broglie wave associated with the particle,

$$\lambda(x) = \frac{2\pi}{k(x)} ,$$

show that the above condition can also be written in a number of equivalent ways:

$$|\lambda'(x)| \ll 2\pi , \qquad \lambda(x) \, |p'(x)| \ll 2\pi \, |p(x)| , \qquad \lambda(x) \, |V'(x)| \ll 4\pi \, \frac{[p(x)]^2}{2m} ,$$

where $p(x)$ is the local momentum of the particle and $V(x) = (\hbar^2/2m)v(x)$ is the potential. What does each of these equivalent conditions require? Explain. What happens close to a classical inversion point? Recall that a classical inversion point is defined by the condition $\epsilon = v(x)$.

4. The WKB approximation as formulated so far fails near a classical inversion point. Suppose that $x = a$ is such a classical inversion point and that $x > a$ is the classically forbidden region and $x < a$ is the classically allowed region. Then, away from a we have the general solutions

$$\psi(x) \approx \frac{A_-}{\sqrt{\kappa(x)}} \exp\left[-\int_a^x dx' \, \kappa(x')\right] + \frac{A_+}{\sqrt{\kappa(x)}} \exp\left[\int_a^x dx' \, \kappa(x')\right] \quad \text{for } x \gg a$$

and

$$\psi(x) \approx \frac{B_-}{\sqrt{k(x)}} \exp\left[-i \int_a^x dx'\, k(x')\right] + \frac{B_+}{\sqrt{k(x)}} \exp\left[i \int_a^x dx'\, k(x')\right] \qquad \text{for } x \ll a .$$

Since $\psi(x)$ must represent the same state, we need to relate the constants $A_\pm$ to $B_\pm$. We cannot impose the continuity of $\psi(x)$ and $\psi'(x)$ at $x=a$, since the above approximation breaks down there. Proceed in the following way. Expand $v(x) - \epsilon$ near the inversion point so that

$$v(x) - \epsilon \approx g(x - a) \qquad\qquad \text{with } g > 0 ;$$

g must be larger than zero since the classically forbidden region is $x > a$. Consider the resulting Schrödinger equation in the region $x \approx a$. Show that it can be written as

$$\frac{d^2 \psi(z)}{dz^2} - z\psi(z) = 0 ,$$

by introducing the auxiliary variable

$$z = g^{1/3}(x - a) .$$

5. Noting that for this linear approximation of $v(x) - \epsilon$ the square of the wavenumber $k^2(z)$ or $\kappa^2(z)$ is given by $g^{2/3}|z|$, show that the condition for the validity of the WKB approximation $|k'(x)| \ll k^2(x)$ established in part 3 is equivalent to

$$|z|^{3/2} \gg \frac{1}{2} .$$

6. The solutions of the Schrödinger equation $\psi''(z) = z\psi(z)$ are known to be the Airy functions $\mathrm{Ai}(z)$ and $\mathrm{Bi}(z)$. They behave asymptotically as follows:

$$\mathrm{Ai}(z) \simeq \frac{1}{2\sqrt{\pi}} \frac{e^{-\xi}}{z^{1/4}} \qquad \text{and} \qquad \mathrm{Bi}(z) \simeq \frac{1}{\sqrt{\pi}} \frac{e^{\xi}}{z^{1/4}} \qquad \text{for } z \longrightarrow \infty$$

and

$$\mathrm{Ai}(z) \simeq \frac{1}{\sqrt{\pi}} \frac{\cos(\xi - \pi/4)}{|z|^{1/4}} \qquad \text{and} \qquad \mathrm{Bi}(z) \simeq -\frac{1}{\sqrt{\pi}} \frac{\sin(\xi - \pi/4)}{|z|^{1/4}} \qquad \text{for } z \longrightarrow -\infty$$

where

$$\xi = \frac{2}{3} |z|^{3/2} .$$

The next step involves matching the Airy functions to the WKB solutions obtained in part 4. To this end, first show that, for $v(x) - \epsilon = g(x - a)$,

$$\int_a^x dx'\, \kappa(x') = \xi \qquad \text{and} \qquad \int_x^a dx'\, k(x') = \xi .$$

Next, show that

$$\psi(x \gg a) \approx \frac{\sqrt{\pi}}{g^{1/6}} \left[2A_- \,\mathrm{Ai}(z) + A_+ \,\mathrm{Bi}(z) \right] .$$

Last, obtain the constants $B_\pm$ in terms of $A_\pm$ by matching the above wave function expressed in terms in Airy functions to the asymptotic form for $\psi(x \ll a)$ in part 4. Show that

$$B_- = e^{-i\pi/4}\left(A_- + \frac{i}{2} A_+ \right) , \qquad B_+ = e^{i\pi/4}\left(A_- - \frac{i}{2} A_+ \right) .$$

These are connection formulae for the WKB solutions: given the solution for $x \gg a$,

$$\psi(x) \approx \frac{A_-}{\sqrt{\kappa(x)}} \exp\left[-\int_a^x dx' \, \kappa(x')\right] + \frac{A_+}{\sqrt{\kappa(x)}} \exp\left[\int_a^x dx' \, \kappa(x')\right] ,$$

one can connect it to the solution for $x \ll a$,

$$\psi(x) \approx \frac{B_-}{\sqrt{k(x)}} \exp\left[i\int_x^a dx' \, k(x')\right] + \frac{B_+}{\sqrt{k(x)}} \exp\left[-i\int_x^a dx' \, k(x')\right] ,$$

via the above $B_\pm$ formulae. Vice versa, by expressing $A_\pm$ in terms of $B_\pm$,

$$A_- = \frac{1}{2}\left(e^{-i\pi/4} B_+ + e^{i\pi/4} B_-\right) , \qquad A_+ = i\left(e^{-i\pi/4} B_+ - e^{i\pi/4} B_-\right) ,$$

they connect the solution for $x \ll a$ to that for $x \gg a$.

By going through steps 4–6 it is easy to see that, when the classically forbidden region is to the left of the inversion point $x = b$ rather than to the right, the WKB solutions are

$$\psi(x) \approx \frac{A_-}{\sqrt{\kappa(x)}} \exp\left[-\int_x^b dx' \, \kappa(x')\right] + \frac{A_+}{\sqrt{\kappa(x)}} \exp\left[\int_x^b dx' \, \kappa(x')\right] \qquad \text{for } x \ll b$$

and

$$\psi(x) \approx \frac{B_-}{\sqrt{k(x)}} \exp\left[i\int_b^x dx' \, k(x')\right] + \frac{B_+}{\sqrt{k(x)}} \exp\left[-i\int_b^x dx' \, k(x')\right] \qquad \text{for } x \gg b .$$

with $B_\pm$ in terms of $A_\pm$ as obtained above.

7. Consider a generic potential $v(x)$, and assume that for any ϵ in a given range there are only two inversion points b and a with $a > b$ and that the classically forbidden regions are $x < b$ and $x > a$. Then for $x < b$ the WKB solution corresponding to a bound state must be

$$\psi(x < b) \approx \frac{A_-}{\sqrt{\kappa(x)}} \exp\left[-\int_x^b dx' \, \kappa(x')\right] .$$

By matching this solution to that in $b < x < a$, and then the latter to the solution for $x > a$, show that one arrives at the WKB quantization condition for the energy:

$$\int_b^a dx \, k(x) = \int_b^a dx \, \sqrt{\epsilon - v(x)} = (n + 1/2)\pi \qquad n = 0, 1, 2, \ldots$$

Solution

Part 1

We have

$$\psi'(x) = iu'(x)\, e^{iu(x)} , \qquad \psi''(x) = iu''(x)e^{iu(x)} + [iu'(x)]^2 e^{iu(x)} ,$$

and hence

$$\left[i u''(x) - u'^2(x)\right] e^{iu(x)} = [v(x) - \epsilon]\, e^{iu(x)} \implies u'^2(x) = \epsilon - v(x) + iu''(x) .$$

Part 2

The zeroth-order approximation $u_0(x)$ results from neglecting $u''(x)$ on the r.h.s. of the differential equation, which yields

$$u_0'^2(x) = k^2(x) , \qquad k(x) = \sqrt{\epsilon - v(x)} \qquad \text{for } \epsilon > v(x) ,$$

with solution

$$u_0(x) = \pm \int^x dx' \, k(x') + C .$$

We obtain a better approximation $u_1(x)$ on replacing $u''(x)$ by $u_0''(x)$:

$$u_1'^2(x) = k^2(x) + i u_0''(x) .$$

Note that

$$u_0'(x) = \pm k(x) , \qquad u_0''(x) = \pm k'(x) ,$$

and hence, assuming that $|k'(x)|/k^2(x) \ll 1$,

$$u_1'(x) = \pm \sqrt{k^2(x) \pm i k'(x)} = \pm k(x) \left[1 \pm i \frac{k'(x)}{k^2(x)} \right]^{1/2} \approx \pm k(x) \left[1 \pm i \frac{k'(x)}{2k^2(x)} \right]$$

or

$$u_1'(x) \approx \pm k(x) + \frac{i}{2} \frac{k'(x)}{k(x)} = \pm k(x) + \frac{i}{2} \frac{d}{dx} \ln k(x) ,$$

which can immediately be integrated to give

$$u_1(x) = \pm \int^x dx' \, k(x') + i \ln \sqrt{k(x)} + C .$$

The corresponding wave function reads

$$\psi(x) \approx e^{i u_1(x)} = e^{\pm i \int^x dx' \, k(x') - \ln \sqrt{k(x)} + iC} = \frac{e^{iC}}{\sqrt{k(x)}} \, e^{\pm i \int^x dx' \, k(x')}$$

or

$$\psi(x) \propto \frac{1}{\sqrt{k(x)}} \exp \left[\pm i \int^x dx' \, k(x') \right] \qquad \text{for } \epsilon > v(x) .$$

An identical derivation in the classically forbidden region leads to

$$\psi(x) \propto \frac{1}{\sqrt{\kappa(x)}} \exp \left[\pm \int^x dx' \, \kappa(x') \right] \qquad \text{for } \epsilon < v(x) .$$

Part 3

For the approximation in part 2 to be valid, we require that $|k'(x)|/k^2(x) \ll 1$, which can be expressed as

$$k(x) = \frac{2\pi}{\lambda(x)} \implies k'(x) = -\frac{2\pi}{\lambda^2(x)} \lambda'(x) ,$$

and hence

$$\left| \frac{2\pi}{\lambda^2(x)} \lambda'(x) \right| \ll \frac{4\pi^2}{\lambda^2(x)} \implies |\lambda'(x)| \ll 2\pi ,$$

that is, the rate of the change of the local wavelength should be small. Using $k(x) = p(x)/\hbar$, where $p(x)$ is the local momentum, this requirement can also be expressed as

$$\frac{|p'(x)|}{\hbar} \ll \frac{|p(x)|}{\hbar} |k(x)| \implies \lambda(x) |p'(x)| \ll 2\pi |p(x)| \,,$$

which says that the rate of change of the local momentum over a length $\lambda(x)$ should be small compared with the local momentum. Last, using

$$p'(x) = \hbar k'(x) = -\frac{\hbar v'(x)}{2\sqrt{\epsilon - v(x)}} = -\frac{m}{\hbar} \frac{V'(x)}{k(x)} = -\frac{m}{2\pi\hbar} \lambda(x) V'(x)$$

and

$$|p'(x)| \ll \frac{p^2(x)}{\hbar} \,,$$

we arrive at a different formulation for the validity of the approximation, namely

$$\lambda(x) |V'(x)| \ll 4\pi \frac{p^2(x)}{2m} \,,$$

that is, the rate of change of the potential over a length $\lambda(x)$ should be much smaller the local kinetic energy. Clearly, all these conditions become invalid in the proximity of a classical inversion point, where $v(x) = \epsilon$ and therefore $|p(x)| = 0$ and $|\lambda'(x)| \longrightarrow \infty$.

Part 4

Close to the inversion point $x = a$ we expand $v(x)$ in a Taylor series and keep only the term linear in x,

$$v(x) = v(a) + v'(a)(x - a) + \cdots = \epsilon + g(x - a) + \cdots \,,$$

where g – the derivative of $v(x)$ at the inversion point – is positive, since $v(x)$ is an increasing function of x (since $v(x) > \epsilon$ for $x > a$). Thus

$$\psi''(x) = g(x - a)\psi(x) \,.$$

In terms of the variable z, we have

$$\frac{d}{dx} = \frac{dz}{dx} \frac{d}{dz} = g^{1/3} \frac{d}{dz} \,, \qquad \frac{d^2}{dx^2} = g^{2/3} \frac{d^2}{dz^2}$$

and therefore

$$g^{2/3} \frac{d^2}{dz^2} \psi(z) = g \frac{z}{g^{1/3}} \psi(z) \implies \frac{d^2\psi(z)}{dz^2} - z\psi(z) = 0 \,.$$

Part 5

Using $|x - a| = |z|/g^{1/3}$, we have

$$k^2(x) \text{ or } \kappa^2(x) = g|x - a| = g^{2/3}|z| \implies k(z) \text{ or } \kappa(z) = g^{1/3} \sqrt{|z|} \,.$$

and

$$\frac{d}{dx}[k(x) \text{ or } \kappa(x)] = \frac{dz}{dx} \frac{d}{dz}[k(z) \text{ or } \kappa(z)] = g^{1/3} \frac{g^{1/3}}{2\sqrt{|z|}} = \frac{g^{2/3}}{2\sqrt{|z|}} \,,$$

yielding

$$|k'(x)| \ll k^2(x) \text{ or } |\kappa'(x)| \ll \kappa^2(x) \implies \frac{g^{2/3}}{2\sqrt{|z|}} \ll g^{2/3}|z| \text{ or } |z|^{3/2} \gg \frac{1}{2}.$$

Part 6

For the linear expansion of $v(x)$, we have

$$\kappa(x) = \sqrt{v(x) - \epsilon} = g^{1/2}\sqrt{x-a}, \qquad k(x) = \sqrt{\epsilon - v(x)} = g^{1/2}\sqrt{a-x},$$

and

$$\int_a^x dx'\,\kappa(x') = \frac{2}{3}g^{1/2}(x'-a)^{3/2}\Big|_a^x = \xi, \qquad \int_x^a dx'\,k(x') = -\frac{2}{3}g^{1/2}(a-x')^{3/2}\Big|_x^a = \xi.$$

Then, for the linear expansion of the potential, the WKB solution is a linear combination of the type

$$\psi(x) \approx \frac{A_-}{g^{1/6}\,z^{1/4}}\,e^{-\xi} + \frac{A_+}{g^{1/6}\,z^{1/4}}\,e^{\xi} \qquad \text{for } x \gg a,$$

and

$$\psi(x) \approx \frac{B_-}{g^{1/6}\,|z|^{1/4}}\,e^{i\xi} + \frac{B_+}{g^{1/6}\,|z|^{1/4}}\,e^{-i\xi} \qquad \text{for } x \ll a,$$

where we have used $\int_a^x dx'\cdots = -\int_x^a dx'\cdots$ in the solution for $x \ll a$. The first set matches the (exact) solution in terms of Airy functions for $x \gg a$:

$$\psi(x) \approx \frac{\sqrt{\pi}}{g^{1/6}}\,[\,2A_-\,\text{Ai}(z) + A_+\,\text{Bi}(z)\,] \qquad \text{for } x \gg a.$$

In the region $x \ll a$ the above wave function reads

$$\psi(x) = \frac{1}{g^{1/6}}\left[2A_-\frac{\cos(\xi - \pi/4)}{|z|^{1/4}} - A_+\frac{\sin(\xi - \pi/4)}{|z|^{1/4}}\right] \qquad \text{for } x \ll a,$$

which can also be written as

$$\psi(x) = \frac{1}{g^{1/6}\,|z|^{1/4}}\left[2A_-\frac{e^{i(\xi-\pi/4)} + e^{-i(\xi-\pi/4)}}{2} - A_+\frac{e^{i(\xi-\pi/4)} - e^{-i(\xi-\pi/4)}}{2i}\right]$$

$$= e^{-i\pi/4}\left(A_- + \frac{i}{2}A_+\right)\frac{e^{i\xi}}{g^{1/6}\,|z|^{1/4}} + e^{i\pi/4}\left(A_- - \frac{i}{2}A_+\right)\frac{e^{-i\xi}}{g^{1/6}\,|z|^{1/4}}.$$

The latter expression matches the WKB solution for $x \ll a$ if

$$B_- = e^{-i\pi/4}\left(A_- + \frac{i}{2}A_+\right), \qquad B_+ = e^{i\pi/4}\left(A_- - \frac{i}{2}A_+\right).$$

These are the connection formulae: they connect the WKB solution in the classically forbidden region $(x > a)$ to that in the classically allowed region $(x < a)$. By inverting the above relations, obtaining

$$A_- = \frac{1}{2}\left(e^{-i\pi/4}B_+ + e^{i\pi/4}B_-\right), \qquad A_+ = i\left(e^{-i\pi/4}B_+ - e^{i\pi/4}B_-\right),$$

we can connect the WKB solution in the classically allowed region $(x < a)$ to that in the classically forbidden region $(x > a)$.

Part 7

Since the wave function for a bound state vanishes as $x \longrightarrow -\infty$, for $x < b$ the WKB solution must be

$$\psi(x < b) = \frac{A_-}{\sqrt{\kappa(x)}} \, e^{-\int_x^b dx' \, \kappa(x')} \, .$$

In the classically allowed region $b < x < a$, we have

$$\psi(b < x < a) = \frac{B_-}{\sqrt{k(x)}} \, e^{i \int_b^x dx' \, k(x')} + \frac{B_+}{\sqrt{k(x)}} \, e^{-i \int_b^x dx' \, k(x')}$$

and, using the connection formulae for the $B_\pm$ with $A_+ = 0$, we find

$$\psi(b < x < a) = \frac{A_-}{\sqrt{k(x)}} \, e^{i[\int_b^x dx' \, k(x') - \pi/4]} + \frac{A_-}{\sqrt{k(x)}} \, e^{-i[\int_b^x dx' \, k(x') - (\pi/4)]}$$

$$= \underbrace{A_- \, e^{i[\int_b^a dx' \, k(x') - \pi/4]}}_{B_+'} \, \frac{e^{-i \int_x^a dx' \, k(x')}}{\sqrt{k(x)}} + \underbrace{A_- \, e^{-i[\int_b^a dx' \, k(x') - \pi/4]}}_{B_-'} \, \frac{e^{i \int_x^a dx' \, k(x')}}{\sqrt{k(x)}} \, ,$$

where in the second line we have used

$$\int_b^x dx' \, k(x') = \int_b^a dx' \, k(x') - \int_x^a dx' \, k(x') \, .$$

This solution must now be matched to that in the classically forbidden region for $x > a$,

$$\psi(x > a) = \frac{A_-'}{\sqrt{\kappa(x)}} \, e^{-\int_a^x dx' \, \kappa(x')} + \frac{A_+'}{\sqrt{\kappa(x)}} \, e^{\int_a^x dx' \, \kappa(x')} \, .$$

The connection formulae yield

$$A_-' = \frac{1}{2} \left(e^{-i\pi/4} B_+' + e^{i\pi/4} B_-' \right) \, , \qquad A_+' = i \left(e^{-i\pi/4} B_+' - e^{i\pi/4} B_-' \right) \, .$$

However, the WKB wave function must vanish as $x \longrightarrow \infty$, which requires

$$A_+' = 0 \implies e^{-i\pi/4} B_+' - e^{i\pi/4} B_-' = 0 \, .$$

Substituting for $B_\pm'$, found earlier, we arrive at

$$e^{-i\pi/4} A_- \, e^{i[\int_b^a dx' \, k(x') - \pi/4]} - e^{i\pi/4} A_- \, e^{-i[\int_b^a dx' \, k(x') - \pi/4]} = 0$$

or

$$\sin \left[\int_b^a dx' \, k(x') - \pi/2 \right] = 0 \implies \int_b^a dx \, k(x) = (n + 1/2)\pi, \qquad n = 0, 1, 2, \dots ,$$

that is, the WKB quantization condition, which looks similar to the Bohr–Sommerfeld rule.

Problem 13   Bound States for an Infinite Barrier at $x = 0$ and a Finite Barrier for $a < x < b$

Consider a particle in a square well potential consisting of an infinite barrier at $x = 0$ and

$$v(x) = -v_0 \theta(a - x) + w_0 [\theta(x - a) - \theta(x - b)] \qquad \text{for } x > 0 \, ,$$

with $v_0, w_0 > 0$ and $b > a$.

1. For bound states to exist, we must have $-v_0 < \epsilon < 0$. Show that these energies result from

$$\tan(k_1 a) = -\frac{k_1}{\kappa_2} \frac{1 + (\kappa_3/\kappa_2) \tanh[\kappa_2(b-a)]}{\tanh[\kappa_2(b-a)] + \kappa_3/\kappa_2} ,$$

   where

$$k_1 = \sqrt{v_0 - |\epsilon|} , \qquad \kappa_2 = \sqrt{w_0 + |\epsilon|} , \qquad \kappa_3 = \sqrt{|\epsilon|} .$$

2. Introduce the variable $x = k_1 a$, and show that the above relation can be written as

$$\tan(x) = g(x) , \qquad 0 < x < x_0 \equiv a\sqrt{v_0} .$$

   Provide a graphical solution, and verify that there is one solution if the potential parameters are such that $a\sqrt{v_0} = 3$, $a\sqrt{w_0} = 1$, and $\Delta = b/a - 1 = 0.5$. Generally, what condition must be satisfied for at least one bound state to exist?

3. Consider the limit $w_0 = 0$. Show that the solutions of the eigenvalue equation agree with those of Problem 5. Now consider the limit $w_0 \longrightarrow \infty$. What do you expect? Do the solutions of the eigenvalue equation conform to your expectations?

Solution

Part 1

Bound states are possible for $-v_0 < \epsilon < 0$. The Schrödinger equation reads

$$\psi_1''(x) = -(v_0 - |\epsilon|)\psi_1(x) \qquad 0 < x < a , \qquad \psi_2''(x) = (w_0 + |\epsilon|)\psi_2(x) \qquad a < x < b ,$$

and

$$\psi_3''(x) = |\epsilon|\psi_3(x) \qquad x > b .$$

We define

$$k_1 = \sqrt{v_0 - |\epsilon|} , \qquad \kappa_2 = \sqrt{w_0 + |\epsilon|} , \qquad \kappa_3 = \sqrt{|\epsilon|} .$$

After imposing the boundary conditions at the origin, $\psi_1(0) = 0$, and at infinity, $\psi_3(x \longrightarrow \infty) = 0$, the wave functions are given by

$$\psi_1(x) = A \sin(k_1 x) , \qquad \psi_2(x) = B e^{\kappa_2 x} + C e^{-\kappa_2 x} , \qquad \psi_3(x) = D e^{-\kappa_3 x} .$$

The continuity of the wave function and its first derivative at $x = a$ and $x = b$ yield the following relations, respectively,

$$B e^{\kappa_2 a} + C e^{-\kappa_2 a} = A \sin(k_1 a) , \qquad B e^{\kappa_2 a} - C e^{-\kappa_2 a} = A \frac{k_1}{\kappa_2} \cos(k_1 a) ,$$

and

$$B e^{\kappa_2 b} + C e^{-\kappa_2 b} = D e^{-\kappa_3 b} , \qquad B e^{\kappa_2 b} - C e^{-\kappa_2 b} = -D \frac{\kappa_3}{\kappa_2} e^{-\kappa_3 b} .$$

The first two relations give

$$B = A \frac{e^{-\kappa_2 a}}{2} \left[\sin(k_1 a) + \frac{k_1}{\kappa_2} \cos(k_1 a) \right] , \qquad C = A \frac{e^{\kappa_2 a}}{2} \left[\sin(k_1 a) - \frac{k_1}{\kappa_2} \cos(k_1 a) \right] ,$$

while the next two give

$$B = D\,\frac{e^{-(\kappa_2+\kappa_3)b}}{2}\left(1 - \frac{\kappa_3}{\kappa_2}\right), \qquad C = D\,\frac{e^{(\kappa_2-\kappa_3)b}}{2}\left(1 + \frac{\kappa_3}{\kappa_2}\right),$$

so that we have the homogeneous linear system in the unknowns A and D:

$$A\,e^{-\kappa_2 a}\left[\sin(k_1 a) + \frac{k_1}{\kappa_2}\cos(k_1 a)\right] = D\,e^{-(\kappa_2+\kappa_3)b}\left(1 - \frac{\kappa_3}{\kappa_2}\right),$$

$$A\,e^{\kappa_2 a}\left[\sin(k_1 a) - \frac{k_1}{\kappa_2}\cos(k_1 a)\right] = D\,e^{(\kappa_2-\kappa_3)b}\left(1 + \frac{\kappa_3}{\kappa_2}\right).$$

For non-trivial solutions, the determinant of this system must vanish, which yields the condition

$$e^{\kappa_2(b-a)}\left[\sin(k_1 a) + \frac{k_1}{\kappa_2}\cos(k_1 a)\right]\left(1 + \frac{\kappa_3}{\kappa_2}\right) - e^{-\kappa_2(b-a)}\left[\sin(k_1 a) - \frac{k_1}{\kappa_2}\cos(k_1 a)\right]\left(1 - \frac{\kappa_3}{\kappa_2}\right) = 0.$$

It is convenient to expand the products and combine the exponentials into hyperbolic sine and cosine functions, to obtain

$$\sin(k_1 a)\,\sinh[\kappa_2(b-a)] + \frac{k_1}{\kappa_2}\cos(k_1 a)\,\cosh[\kappa_2(b-a)] + \frac{\kappa_3}{\kappa_2}\sin(k_1 a)\,\cosh[\kappa_2(b-a)]$$

$$+ \frac{k_1}{\kappa_2}\frac{\kappa_3}{\kappa_2}\cos(k_1 a)\,\sinh[\kappa_2(b-a)] = 0,$$

which can also be written as (here, $k_1 a$ is assumed not to be an odd multiple of $\pi/2$, otherwise $\cos(k_1 a)$ would vanish)

$$\tan(k_1 a) = -\frac{k_1}{\kappa_2}\frac{1 + (\kappa_3/\kappa_2)\tanh[\kappa_2(b-a)]}{\tanh[\kappa_2(b-a)] + \kappa_3/\kappa_2}.$$

Part 2

To gain insight into the solutions of the above equation, it is convenient to define

$$x = k_1 a \implies |\epsilon| = v_0 - (x/a)^2,$$

and so the variable x must be in the range $0 < x < x_0 \equiv a\sqrt{v_0}$. In terms of x, we also have

$$a\kappa_2 = a\sqrt{w_0 + |\epsilon|} = \sqrt{a^2(v_0 + w_0) - x^2}, \qquad a\kappa_3 = a\sqrt{|\epsilon|} = \sqrt{a^2 v_0 - x^2},$$

and the above equation becomes

$$\tan x = -\underbrace{\frac{x}{\sqrt{x_1^2 - x^2}}\frac{\sqrt{x_1^2 - x^2} + \sqrt{x_0^2 - x^2}\,\tanh(\Delta\sqrt{x_1^2 - x^2})}{\sqrt{x_0^2 - x^2} + \sqrt{x_1^2 - x^2}\,\tanh(\Delta\sqrt{x_1^2 - x^2})}}_{g(x)},$$

where we have introduced the parameters

$$x_0 = a\sqrt{v_0}, \qquad x_1 = a\sqrt{v_0 + w_0}, \qquad \Delta = b/a - 1 > 0.$$

The function $g(x)$ is always negative in the range $0 < x < x_0$, and so a minimum requirement for a solution to exist is that $x_0 > \pi/2$ (if this is not satisfied then $\tan(x)$ is always positive). This will not ensure, however, that a solution does exist: whether this is the case will depend on the potential

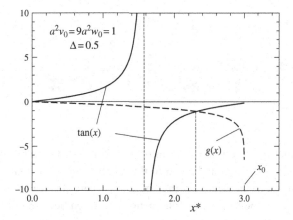

Fig. 4.8 Graphical solution of $\tan(x) = g(x)$ for a specific set of potential parameters (here $x_0 = 3$ and the vertical dashed line is at $\pi/2$). Only one solution, x^*, is found.

parameters contained in the function $g(x)$. The latter is a monotonically decreasing function of x, which vanishes for $x = 0$ and with the value at the end point $x_0 = a\sqrt{v_0}$ of the allowed interval given by

$$g(x_0) = -\sqrt{\frac{v_0}{w_0}} \frac{1}{\tanh(\Delta a \sqrt{w_0})} \ .$$

At least one bound state occurs if the potential parameters are such that (with $x_0 > \pi/2$)

$$\tan(x_0) \geq g(x_0) \implies \tan(a\sqrt{v_0}) \geq -\sqrt{\frac{v_0}{w_0}} \frac{1}{\tanh(\Delta a \sqrt{w_0})} \ ;$$

indeed, when the equality holds, this bound-state energy has vanishing energy. A particular case is illustrated in Fig. 4.8.

Part 3

We examine the two limiting cases (i) $w_0 \longrightarrow 0$ and (ii) $w_0 \gg v_0$. In the former, we have no barrier and

$$\lim_{w_0 \to 0} g(x) = -\frac{x}{\sqrt{x_0^2 - x^2}} \ ,$$

so that n bound states occur if $a\sqrt{v_0} > (2n-1)\pi/2$ with $n = 1, 2, \ldots$, as expected (these are the bound states in a potential $v(x) = \infty$ for $x < 0$ and $v(x) = -v_0\theta(a-x)$ for $x > 0$; see Problem 7 of Chapter 5). Limiting case (ii) corresponds to an infinite barrier at $x = a$, and one would expect to find an infinite set of bound states with energies $\epsilon_n = (n\pi/a)^2 - v_0$. This is indeed the case, since in that limit

$$\lim_{w_0 \to \infty} g(x) = 0 \ ,$$

and the condition for bound states becomes $\tan(x) = 0$ or $x_n = n\pi$, yielding the ϵ_n given above.

5

Scattering in One Dimension

In a scattering experiment a beam of particles is scattered by a target. The interactions between the particles in the beam and those in the target occur in a small region of space. The regions where the particles in the beam are prepared and where the scattered particles are detected are both well outside the region where the interactions between beam particles and target particles are effective. Therefore, the particles in the initial state (before entering the interaction region) and in the final state (after passing through the interaction region) can be considered as free. Of course, we are assuming that the interactions among the beam or target particles can be neglected. Thus, a scattering process essentially involves the transition from an initial free-particle state to a final free-particle state.

Here, we consider the simple but instructive case of a particle under the influence of a potential $v(x)$ in one dimension. The particle can only scatter forward or backward. The question we want to answer is the following: given $v(x)$, what is the probability that the particle is scattered forward? And backward? These probabilities will depend on the initial energy of the particle and, of course, must add up to 1.

As before, we assume that $v(x)$ is piecewise continuous with limits $v(x \longrightarrow \pm\infty) = v_\pm$ and $v_- \geq v_+$. In the asymptotic regions $x \longrightarrow \pm\infty$, the solutions of the Schrödinger equation for $\epsilon > v_-$ are given by Eq. (4.15), with $\alpha_\pm, \beta_\pm$ generally complex constants. The matching of the left, $\psi_-(x)$, and right, $\psi_+(x)$, wave functions determines two of these four constants, say β_- and α_+, which can then be expressed as functions of the remaining two, namely

$$\beta_- = \beta_-(\alpha_-, \beta_+), \qquad \alpha_+ = \alpha_+(\alpha_-, \beta_+). \tag{5.1}$$

Two independent solutions follow by taking, for example, $\beta_+ = 0$ and α_- arbitrary, and $\alpha_- = 0$ and β_+ arbitrary, which result in

$$\psi_-^{(+)}(x) = \alpha_- e^{ik_- x} + \beta_- e^{-ik_- x}, \qquad \psi_-^{(-)}(x) = \beta_- e^{-ik_- x}, \tag{5.2}$$

$$\psi_+^{(+)}(x) = \alpha_+ e^{ik_+ x}, \qquad \psi_+^{(-)}(x) = \alpha_+ e^{ik_+ x} + \beta_+ e^{-ik_+ x}. \tag{5.3}$$

Physically, solution (+) and solution (−) correspond, respectively, to a particle traveling only to the right or only to the left in the regions $x \gg 0$ or $x \ll 0$. The matching for solution (+) and solution (−) leads to a determination of, respectively, the ratios β_-/α_-, α_+/α_- and α_+/β_+, β_-/β_+ as functions of k (that is, of the energy ϵ):

$$\frac{\beta_-}{\alpha_-} = A(k), \qquad \frac{\alpha_+}{\alpha_-} = B(k), \tag{5.4}$$

$$\frac{\alpha_+}{\beta_+} = C(k), \qquad \frac{\beta_-}{\beta_+} = D(k). \tag{5.5}$$

In view of the discussion in terms of wave packets to follow, we will write the solutions as

$$\psi_k^{(+)}(x) = e^{ik_- x} + A(k) e^{-ik_- x} \qquad x \ll 0$$

$$= B(k) e^{ik_+ x} \qquad x \gg 0 \tag{5.6}$$

and

$$\psi_k^{(-)}(x) = D(k)\, e^{-ik_- x} \qquad\qquad x \ll 0$$
$$= e_+^{-ikx} + C(k)\, e_+^{ikx} \qquad x \gg 0\,, \tag{5.7}$$

up to irrelevant overall constants α_- and β_+, as it will become clear below. These wave functions are not normalizable; consequently, as such, they cannot be used to calculate probabilities. In order to remove this difficulty, we introduce wave packets.

5.1 Description in Terms of Wave Packets

We construct the wave packet (we need only to consider one of the two solutions; the treatment is similar for the other)

$$\Psi^{(+)}(x,t) = \int_{k^*}^{\infty} \frac{dk}{\sqrt{2\pi}}\, g(k)\, \underbrace{\psi_k^{(+)}(x)\, e^{-i\omega_k t}}_{\Psi_k^{(+)}(x,t)}\,, \qquad \omega_k = \frac{\hbar k^2}{2m} = \frac{E}{\hbar}\,, \tag{5.8}$$

where the lower integration limit $k^* \equiv \sqrt{v_-}$ is the minimum allowed wave number (we are considering the regime $\epsilon > v_-$) and $g(k)$, which is generally a complex function of k, is taken here to be real and positive for simplicity (this does not affect the following discussion in any significant way). The function $g(k)$ on the left in Eq. (5.8) has the property that it is non-vanishing only in a narrow interval centered at $k_0 = \sqrt{\epsilon_0} > k^*$. Inserting the expressions obtained earlier, we find

$$\Psi^{(+)}(x,t) = \underbrace{\int_{k^*}^{\infty} \frac{dk}{\sqrt{2\pi}}\, g(k)\, e^{i(k_- x - \omega_k t)}}_{\Psi_i^{(+)}(x,t)} + \underbrace{\int_{k^*}^{\infty} \frac{dk}{\sqrt{2\pi}}\, g(k)\, A(k)\, e^{-i(k_- x + \omega_k t)}}_{\Psi_r^{(+)}(x,t)} \qquad x \ll 0 \tag{5.9}$$

$$= \underbrace{\int_{k^*}^{\infty} \frac{dk}{\sqrt{2\pi}}\, g(k)\, B(k)\, e^{i(k_+ x - \omega_k t)}}_{\Psi_t^{(+)}(x,t)} \qquad\qquad x \gg 0\,. \tag{5.10}$$

Note that, since $\psi_k^{(+)}(x)$ is continuous along with its first derivative, the wave packet (linear superposition) $\Psi^{(+)}(x,t)$ will also be continuous; it consists of three terms, labeled as incident (i), reflected (r), and transmitted (t), which live in different regions of space. The incident and reflected wave packets live in $x \ll 0$ and the transmitted wave packet lives in $x \gg 0$. Using the stationary phase method, we find that the incident wave packet is given approximately by

$$\Psi_i^{(+)}(x,t) \approx e^{i(k_-^* x - \omega_{k_0} t)} \int_{k^*}^{\infty} \frac{dk}{\sqrt{2\pi}}\, g(k)\, e^{i(k_0/k_-^*)[x - x_i(t)](k - k_0)} \qquad x \ll 0\,, \tag{5.11}$$

where we have defined $k_-^* = \sqrt{\epsilon_0 - v_-}$. The wave packet is largest when $x \approx x_i(t)$. Its "center" $x_i(t)$ travels according to

$$x_i(t) = \frac{\hbar k_-^*}{m}\, t\,, \tag{5.12}$$

with group velocity $\hbar k_-^*/m$, and therefore is in the region $x \ll 0$ only when $t \ll 0$. We can analyze similarly the reflected and transmitted wave packets. Expressing the generally complex functions $A(k)$ and $B(k)$ as

$$A(k) = |A(k)| \, e^{i\alpha(k)} \,, \qquad B(k) = |B(k)| \, e^{i\beta(k)} \,, \tag{5.13}$$

we find the reflected wave packet to be approximately given by

$$\Psi_r^{(+)}(x,t) \approx e^{-i(k_-^* x + \omega_{k_0} t)} \, e^{i\alpha(k_0)} \int_{k^*}^{\infty} \frac{dk}{\sqrt{2\pi}} \, g(k) \, |A(k)| \, e^{-i(k_0/k_-^*)[x - x_r(t)](k - k_0)}$$

$$\approx e^{-i(k_-^* x + \omega_{k_0} t)} \underbrace{e^{i\alpha(k_0)}|A(k_0)|}_{A(k_0)} \int_{k^*}^{\infty} \frac{dk}{\sqrt{2\pi}} \, g(k) \, e^{-i(k_0/k_-^*)[x - x_r(t)](k - k_0)} \qquad x \ll 0 \,, \tag{5.14}$$

where in the second line we have assumed that $|A(k)|$ does not vary appreciably over the interval centered at k_0 in which $g(k)$ is non-vanishing. The wave packet's center $x_r(t)$ travels with the group velocity $-\hbar k_-^*/m$ according to

$$x_r(t) = -\frac{\hbar k_-^*}{m} \, t + \frac{k_-^*}{k_0} \, \alpha'(k_0) = -\frac{\hbar k_-^*}{m} \, (t - \tau_r) \,, \tag{5.15}$$

showing that the reflected wave packet is in the region $x < 0$ only after a time delay τ_r given by

$$\tau_r = \frac{m}{\hbar k_0} \, \alpha'(k_0) \,. \tag{5.16}$$

Similarly, we obtain for the transmitted wave packet

$$\Psi_t^{(+)}(x,t) \approx e^{i(k_+^* x - \omega_{k_0} t)} \, B(k_0) \int_{k^*}^{\infty} \frac{dk}{\sqrt{2\pi}} \, g(k) \, e^{i(k_0/k_+^*)[x - x_t(t)](k - k_0)} \qquad x \gg 0 \,, \tag{5.17}$$

where $k_+^* = \sqrt{\epsilon_0 - v_+}$ and

$$x_t(t) = \frac{\hbar k_+^*}{m}(t - \tau_t) \,, \qquad\qquad \tau_t = \frac{m}{\hbar k_0} \, \beta'(k_0) \,. \tag{5.18}$$

The transmitted wave packet is in the region $x > 0$ only after a time delay τ_t.

We now observe that the integrals over k can be extended from $-\infty$ to ∞, since the profile function $g(k)$ can be defined to vanish for $k < k^*$, which allows us to write the wave packet as

$$\Psi^{(+)}(x,t) = \underbrace{e^{i(k_-^* x - \omega_{k_0} t)} \, G[(k_0/k_-^*)(x - x_i)]}_{\Psi_i^{(+)}(x,t)} + \underbrace{A(k_0) \, e^{-i(k_-^* x + \omega_{k_0} t)} \, G[(k_0/k_-^*)(x_r - x)]}_{\Psi_r^{(+)}(x,t)} \qquad x \ll 0 \tag{5.19}$$

$$= \underbrace{B(k_0) \, e^{i(k_+^* x - \omega_{k_0} t)} \, G[(k_0/k_+^*)(x - x_t)]}_{\Psi_t^{(+)}(x,t)} \qquad\qquad x \gg 0 \,, \tag{5.20}$$

where

$$\int_{-\infty}^{\infty} \frac{dk}{\sqrt{2\pi}} \, g(k) \, e^{i(k_0/k_-^*)(x - x_i)(k - k_0)} = \int_{-\infty}^{\infty} \frac{dk'}{\sqrt{2\pi}} \, g(k' + k_0) \, e^{i(k_0/k_-^*)(x - x_i)k'} \equiv G[(k_0/k_-^*)(x - x_i)] \,, \tag{5.21}$$

with similar expressions for the reflected and transmitted wave packets. The function $G[\cdots]$ is largest when its argument vanishes. For times $t \ll 0$ (i.e., in the remote past), the center $x_i(t)$ of the incident wave packet is in the region $x \ll 0$ and propagates from left to right with constant velocity $\hbar k_-^*/m$. On the other hand, the position $x_r(t)$ of the center of the reflected wave packet is large and positive, and therefore outside the region where the expression in Eq. (5.19) is valid. Similar considerations hold for the transmitted wave packet. Therefore, at $t \ll 0$, $\Psi^{(+)}(x,t)$ consists only of the incident wave packet:

$$\text{for } t \longrightarrow -\infty, \qquad \Psi^{(+)}(x,t) \approx \Psi_i^{(+)}(x,t) \qquad x \ll 0$$

$$\approx 0 \qquad\qquad x \gg 0 \,. \tag{5.22}$$

By contrast, at $t \gg 0$ (i.e., in the distant future), $\Psi^{(+)}(x, t)$ consists of a reflected wave packet for $x \ll 0$ and of a transmitted wave packet for $x \gg 0$, specifically:

$$\text{for } t \longrightarrow \infty, \qquad \Psi^{(+)}(x, t) \approx \Psi_r^{(+)}(x, t) \qquad x \ll 0$$
$$\approx \Psi_t^{(+)}(x, t) \qquad x \gg 0 . \qquad (5.23)$$

Because of the conservation of probability, we must have

$$\underbrace{\int_{-\infty}^{0} dx \, |\Psi_i^{(+)}(x, t)|^2}_{t \longrightarrow -\infty} = \underbrace{\int_{-\infty}^{0} dx \, |\Psi_r^{(+)}(x, t)|^2 + \int_{0}^{\infty} dx \, |\Psi_t^{(+)}(x, t)|^2}_{t \longrightarrow \infty} , \qquad (5.24)$$

and, since the wave packets are localized at $x_i(t)$, $x_r(t)$, and $x_t(t)$, we can extend the x-integrations from $-\infty$ to ∞, to find

$$\int_{-\infty}^{\infty} dx \, |G[(k_0/k_-^*)(x - x_i)]|^2 = |A(k_0)|^2 \int_{-\infty}^{\infty} dx \, |G[(k_0/k_-^*)(x_r - x)]|^2 + |B(k_0)|^2 \int_{-\infty}^{\infty} dx \, |G[(k_0/k_+^*)(x - x_t)]|^2 , \qquad (5.25)$$

which, after appropriate rescaling of the integration variables, yields

$$\underbrace{|A(k_0)|^2}_{R(k_0)} + \underbrace{\frac{k_+^*}{k_-^*} \, |B(k_0)|^2}_{T(k_0)} = 1 . \qquad (5.26)$$

Here, $R(k_0)$ and $T(k_0)$ represent the probabilities for the particle to be reflected and transmitted, respectively, and

$$R(k_0) + T(k_0) = 1 . \qquad (5.27)$$

Proceeding similarly, we can show that the wave packet based on the solution $(-)$ leads to reflection and transmission coefficients given by

$$\overline{R}(k_0) = |C(k_0)|^2 , \qquad \overline{T}(k_0) = \frac{k_-^*}{k_+^*} \, |D(k_0)|^2 . \qquad (5.28)$$

For a potential that is invariant under parity, it is obvious that $\overline{R}(k_0) = R(k_0)$ and $\overline{T}(k_0) = T(k_0)$, but in fact this remains true for a generic $v(x)$ (it is related to invariance under time reversal; see Problems 14 and 17).

5.2 An Alternative Treatment

Here, we present an alternative treatment of scattering that does not rely directly on the use of wave packets. We begin with the two independent solutions derived earlier,

$$\psi_k(x) = \underbrace{e^{ik_-x}}_{\psi_i(x)} + \underbrace{A(k)e^{-ik_-x}}_{\psi_r(x)} \qquad x \ll 0$$
$$= \underbrace{B(k) \, e^{ik_+x}}_{\psi_t(x)} \qquad x \gg 0 , \qquad (5.29)$$

and the similar one in Eq. (5.7) (we have dropped the superscripts distinguishing these independent wave functions). In view of our previous discussion in terms of wave packets, we can interpret

$\psi_i(x)$, $\psi_r(x)$, and $\psi_t(x)$ as, respectively, the incident, reflected, and transmitted wave functions. The full time-dependent wave function reads $\Psi_k(x,t) = \psi_k(x)e^{-i\omega_k t}$. While $\Psi_k(x,t)$ is not normalizable, we can still evaluate its probability current density. It is time independent (we are dealing with a stationary wave function) and is given by

$$j(x) = \frac{\hbar}{2mi}\left[\psi_k^*(x)\psi_k'(x) - \text{c.c.}\right].\tag{5.30}$$

Inserting the explicit expression for $\psi_k(x)$ yields simply

$$\begin{aligned}
j(x) &= \frac{\hbar k_-}{m}\left(1 - |A(k)|^2\right) = j_i + j_r & x \ll 0 \\
&= \frac{\hbar k_+}{m}|B(k)|^2 = j_t & x \gg 0,
\end{aligned}\tag{5.31}$$

where we have defined the incident, reflected, and transmitted current densities

$$j_i = \frac{\hbar k_-}{m}, \qquad j_r = -\frac{\hbar k_-}{m}|A(k)|^2, \qquad j_t = \frac{\hbar k_+}{m}|B(k)|^2,\tag{5.32}$$

associated, respectively, with the wave functions $\psi_i(x)$, $\psi_r(x)$, and $\psi_t(x)$. The minus sign in j_r is needed because the reflected wave is propagating from right to left. In the absence of time dependence, the continuity equation requires

$$\frac{d}{dx}j(x) = 0,\tag{5.33}$$

that is, $j(x)$ is independent of x and hence $j(\infty) = j(-\infty)$, yielding $j_i + j_r = j_t$. We define

$$R(k) = \frac{|j_r|}{|j_i|} = |A(k)|^2, \qquad T(k) = \frac{|j_t|}{|j_i|} = \frac{k_+}{k_-}|B(k)|^2, \qquad R(k) + T(k) = 1,\tag{5.34}$$

the last relation following from the conservation of the probability current. The expressions for $R(k)$ and $T(k)$ are in agreement with those obtained earlier in the analysis based on wave packets.

5.3 Problems

Problem 1 Scattering in a Repulsive δ-Function Potential

Consider a repulsive δ-function potential located at the origin with

$$v(x) = v_0\,\delta(x), \qquad v_0 > 0.$$

Doubly degenerate continuum solutions exist for any energy $\epsilon > 0$. Introduce wave packets using a profile function $g(k)$ that is real, positive, and non-vanishing in a small interval centered at k_0. Calculate the reflection and transmission coefficients, and obtain the time delays associated with the reflected and transmitted wave packets. Discuss how the quantum mechanical description of the process differs from the classical description (think of $v(x)$ as a very thin and infinitely high wall located at the origin).

Solution

Split the real axis into two regions: $x < 0$ (region $<$) and $x > 0$ (region $>$). In these regions, the Schrödinger equation reads

$$\psi_<''(x) = -\epsilon\psi_<(x) \,, \qquad\qquad \psi_>''(x) = -\epsilon\psi_>(x) \,,$$

with boundary conditions at the origin given by

$$\psi_<(0) = \psi_>(0) \,, \qquad \psi_>'(0) - \psi_<'(0) = v_0\psi_>(0) \,.$$

The general form of these solutions is obtained as

$$\psi_<(x) = \alpha_< \, e^{ikx} + \beta_< \, e^{-ikx} \,, \qquad \psi_>(x) = \alpha_> \, e^{ikx} + \beta_> \, e^{-ikx} \,,$$

where $k = \sqrt{\epsilon} > 0$ and $\alpha_<, \beta_<$ and $\alpha_>, \beta_>$ are constants. The boundary conditions for two of these four constants, say $\beta_<$ and $\alpha_>$, can then be expressed as functions of the remaining two, namely

$$\beta_< = \beta_<(\alpha_<, \beta_>) \,, \qquad \alpha_> = \alpha_>(\alpha_<, \beta_>) \,.$$

Two independent solutions follow by taking $\beta_> = 0$ and $\alpha_<$ arbitrary, and $\alpha_< = 0$ and $\beta_>$ arbitrary, which results in

$$\psi_<^{(+)}(x) = \alpha_< \, e^{ikx} + \beta_< \, e^{-ikx} \,, \qquad \psi_<^{(-)}(x) = \beta_< \, e^{-ikx} \,,$$
$$\psi_>^{(+)}(x) = \alpha_> \, e^{ikx} \,, \qquad\qquad \psi_>^{(-)}(x) = \alpha_> \, e^{ikx} + \beta_> \, e^{-ikx} \,,$$

corresponding to the outgoing waves in $x > 0$ and in $x < 0$, respectively. Next, we impose the boundary conditions at the origin on each of these two solutions; for example, for solution $(+)$ we get

$$\underbrace{\alpha_< + \beta_<}_{\psi_<^{(+)}(0)} = \underbrace{\alpha_>}_{\psi_>^{(+)}(0)} \,, \qquad \underbrace{ik\alpha_>}_{\psi_>^{(+)\prime}(0)} - \underbrace{ik\big(\alpha_< - \beta_<\big)}_{\psi_<^{(+)\prime}(0)} = v_0 \underbrace{\alpha_>}_{\psi_>^{(+)}(0)} \,,$$

which can be solved with respect to $\alpha_<$. We proceed similarly for the solution $(-)$, and finally arrive at

$$\text{solution } (+): \qquad \alpha_> = \frac{2ik}{-v_0 + 2ik} \, \alpha_< \,, \qquad \beta_< = \frac{v_0}{-v_0 + 2ik} \, \alpha_< \,,$$

$$\text{solution } (-): \qquad \beta_< = \frac{2ik}{-v_0 + 2ik} \, \beta_> \,, \qquad \alpha_> = \frac{v_0}{-v_0 + 2ik} \, \beta_> \,.$$

We write the solutions as

$$\psi_k^{(+)}(x) = e^{ikx} + A(k) \, e^{-ikx} \qquad x < 0$$
$$= B(k) \, e^{ikx} \qquad\qquad x > 0$$

and

$$\psi_k^{(-)}(x) = B(k) \, e^{-ikx} \qquad\qquad x < 0$$
$$= e^{-ikx} + A(k) \, e^{ikx} \qquad x > 0$$

up to overall constants $\alpha_<$ and $\beta_>$, where we have defined

$$A(k) = \frac{v_0}{-v_0 + 2ik} \,, \qquad B(k) = \frac{2ik}{-v_0 + 2ik} \,.$$

We form the wave packet and obtain for the solution (+)

$$\Psi^{(+)}(x,t) = \underbrace{\int_0^\infty \frac{dk}{\sqrt{2\pi}} g(k) \, e^{i(kx-\omega_k t)}}_{\Psi_i^{(+)}(x,t)} + \underbrace{\int_0^\infty \frac{dk}{\sqrt{2\pi}} g(k) \, A(k) \, e^{-i(kx+\omega_k t)}}_{\Psi_r^{(+)}(x,t)} \qquad x < 0$$

$$= \underbrace{\int_0^\infty \frac{dk}{\sqrt{2\pi}} g(k) \, B(k) \, e^{i(kx-\omega_k t)}}_{\Psi_t^{(+)}(x,t)} \qquad\qquad\qquad x > 0 \,.$$

Using the stationary phase method for a generic wave packet of the form

$$\Psi(x,t) = \int \frac{dk}{\sqrt{2\pi}} g(k) \, e^{i\varphi(k)} \,,$$

we find that $|\Psi(x,t)|$ is largest when

$$\frac{d}{dk}\varphi(k)\Big|_{k=k_0} \approx 0 \,.$$

By expanding around k_0, we can express the wave packet approximately as

$$\Psi(x,t) \approx e^{i\varphi(k_0)} \int \frac{dk}{\sqrt{2\pi}} g(k) \, e^{i\varphi'(k_0)(k-k_0)} \,.$$

This procedure yields for the incident wave packet

$$\Psi_i^{(+)}(x,t) \approx e^{i(k_0 x - \omega_{k_0} t)} \int_0^\infty \frac{dk}{\sqrt{2\pi}} g(k) \, e^{i[x-x_i(t)](k-k_0)} \qquad x < 0 \,,$$

where $x_i(t) = (\hbar k_0/m)\,t$. The maximum of the wave packet is approximately at $x_i(t)$ and therefore is in the region $x < 0$ only for $t < 0$. Its center reaches the origin, where the δ-function is located, at time $t = 0$.

For the reflected and transmitted wave packets $A(k)$ and $B(k)$ are now complex, and the corresponding phase factors are $\varphi(k) = -kx - \omega_k t + \alpha(k)$ and $\varphi(k) = kx - \omega_k t + \beta(k)$, where

$$\alpha(k) = \tan^{-1}\left(\frac{2k}{v_0}\right) \,, \qquad \beta(k) = -\tan^{-1}\left(\frac{v_0}{2k}\right) \,.$$

For example, the complex function $A(k)$ can be written as

$$A(k) = \frac{v_0}{-v_0 + 2ik} = \frac{v_0}{-v_0 + 2ik}\frac{-v_0 - 2ik}{-v_0 - 2ik} = -\underbrace{\frac{v_0^2}{v_0^2 + 4k^2}}_{A_R(k)} - \underbrace{\frac{2kv_0}{v_0^2 + 4k^2}}_{A_I(k)} i \,,$$

and hence $A(k) = |A(k)|\, e^{i\alpha(k)}$ with $|A(k)| = v_0/\sqrt{v_0^2 + 4k^2}$ and $\alpha(k)$ as given above. We find

$$\Psi_r^{(+)}(x,t) \approx e^{-i(k_0 x + \omega_{k_0} t)} A(k_0) \int_0^\infty \frac{dk}{\sqrt{2\pi}} g(k) \, e^{-i[x-x_r(t)](k-k_0)} \qquad x < 0 \,,$$

where

$$x_r(t) = -\frac{\hbar k_0}{m} t + \frac{d}{dk}\alpha(k)\Big|_{k=k_0} = -\frac{\hbar k_0}{m} t + \frac{2v_0}{v_0^2 + 4k_0^2} = -\frac{\hbar k_0}{m}(t - \tau) \,,$$

and the time delay is

$$\tau = \frac{m}{\hbar k_0} \frac{2 v_0}{v_0^2 + 4 k_0^2} > 0 \,.$$

Similarly, we obtain

$$\Psi_t^{(+)}(x, t) \approx e^{i(k_0 x - \omega_{k_0} t)} B(k_0) \int_0^\infty \frac{dk}{\sqrt{2\pi}} \, g(k) \, e^{i[x - x_t(t)](k - k_0)} \qquad x > 0 \,,$$

where

$$x_t(t) = \frac{\hbar k_0}{m} (t - \tau) \,,$$

and the time delay is the same as above. Collecting these various terms, we express the complete wave packet as

$$\Psi^{(+)}(x, t) = \underbrace{e^{i(k_0 x - \omega_{k_0} t)} \, G[x - x_i(t)]}_{\Psi_i^{(+)}(x, t)} + \underbrace{A(k_0) \, e^{-i(k_0 x + \omega_{k_0} t)} \, G[x_r(t) - x]}_{\Psi_r^{(+)}(x, t)} \qquad x < 0$$

$$= \underbrace{B(k_0) \, e^{i(k_0 x - \omega_{k_0} t)} \, G[x - x_t(t)]}_{\Psi_t^{(+)}(x, t)} \qquad x > 0 \,,$$

where we have introduced the function

$$\int_{-\infty}^\infty \frac{dk}{\sqrt{2\pi}} \, g(k) \, e^{i[x - x_i(t)](k - k_0)} = \int_{-\infty}^\infty \frac{dk'}{\sqrt{2\pi}} \, g(k' + k_0) \, e^{i[x - x_i(t)]k'} \equiv G[x - x_i(t)] \,,$$

by setting $g(k) = 0$ for $k < 0$, which allows one to extend the integration in k over the whole real axis, as illustrated above. We conclude that, in the remote past,

$$\text{for } t \longrightarrow -\infty: \qquad \Psi^{(+)}(x, t) \approx \Psi_i^{(+)}(x, t) \qquad x < 0$$
$$\approx 0 \qquad x > 0 \,,$$

while, in the distant future,

$$\text{for } t \longrightarrow \infty: \qquad \Psi^{(+)}(x, t) \approx \Psi_r^{(+)}(x, t) \qquad x < 0$$
$$\approx \Psi_t^{(+)}(x, t) \qquad x > 0 \,.$$

Conservation of probability implies that

$$\int_{-\infty}^0 dx \, |\Psi_i^{(+)}(x, t)|^2 = \int_{-\infty}^0 dx \, |\Psi_r^{(+)}(x, t)|^2 + \int_0^\infty dx \, |\Psi_t^{(+)}(x, t)|^2 \,,$$

and, since the wave packets are localized at $x_i(t)$, $x_r(t)$, and $x_r(t)$, the x-integrations can be extended to the whole real axis, yielding

$$\int_{-\infty}^\infty dx \, |G[x - x_i(t)]|^2 = |A(k_0)|^2 \int_{-\infty}^\infty dx \, |G[x_r(t) - x]|^2 + |B(k_0)|^2 \int_{-\infty}^\infty dx \, |G[x - x_t(t)]|^2 \,,$$

or, by shifting the integration variables,

$$|A(k_0)|^2 + |B(k_0)|^2 = 1 \,.$$

Inserting the expressions for $A(k_0)$ and $B(k_0)$ obtained earlier, we see that the conservation of probability is satisfied. In particular, the reflection and transmission coefficients are

$$R(k_0) = |A(k_0)|^2 = \frac{v_0^2}{v_0^2 + 4 k_0^2} \,, \qquad T(k_0) = |B(k_0)|^2 = \frac{4 k_0^2}{v_0^2 + 4 k_0^2} \,.$$

It is easily verified that the wave packet formed with the solution $(-)$ leads to the same $R(k_0)$ and $T(k_0)$.

To elucidate further the physical interpretation of what is going on, it is useful to analyze the situation first classically and then quantum mechanically. Classically, a particle approaching from, say, the left, hits the wall at $x=0$, and instantaneously inverts its direction of motion, moving back towards $-\infty$. The magnitude of the momentum before and after hitting the wall is the same, $|p| = \sqrt{2mE}$. The particle cannot penetrate into the $x > 0$ region.

Quantum mechanically, the situation is rather different. The center of the incident wave packet arrives at the wall at time $t=0$. Then, when t is sufficiently large $(t \gg \tau)$, we see that the incident wave packet has disappeared, and we are left with the reflected (for $x < 0$) and transmitted (for $x > 0$) wave packets. The probability that the particle is in the classically forbidden region $x > 0$ is given by $T(k_0)$. For a fixed energy ϵ_0, this probability becomes smaller and smaller as the strength v_0 of the δ-function potential increases. In contrast with the classical case, the particle has a finite probability of penetrating the region $x > 0$ (the tunneling effect).

The reflected and transmitted wave packets propagate towards the left and right with velocities $-\hbar k_0/m$ and $+\hbar k_0/m$, respectively. A time delay is introduced between the reflected and transmitted wave packets. Contrary to what is predicted by classical mechanics, the particle is not instantaneously reflected by the wall. It is as if the particle "hangs around" in a region of approximate size $(\hbar k_0/m)\tau$ before being reflected and transmitted.

Problem 2   Study of the Potential Step

Consider the potential step

$$
\begin{aligned}
V(x) &= 0 & x < 0 \\
&= V_0 & x \geq 0 ,
\end{aligned}
$$

with $V_0 > 0$. Describe the classical motion of a particle of mass m approaching the potential step from the far left $(x < 0)$ with momentum p. Distinguish two cases:

$$
\text{case (a),} \qquad V_0 < E; \qquad \text{case (b),} \qquad 0 < E < V_0 ,
$$

where E is the energy of the particle. What are its velocities in the two cases? Next analyze the problem quantum mechanically.

1. Discuss the expected energy spectrum.
2. Obtain the reflection and transmission probabilities in case (a) from the incident, reflected, and transmitted probability current densities.
3. Do the same for case (b). What are the reflection and transmission probabilities in this case? Discuss the result.
4. Repeat the calculation in part 2 of the problem for case (a), utilizing wave packets defined as

$$
\Psi(x, t) = \int_0^\infty \frac{dk}{\sqrt{2\pi}} g(k)\, \psi_k(x)\, e^{-i\omega_k t} , \qquad \omega_k = \frac{\hbar k^2}{2m} ,
$$

where the profile function $g(k)$ is real and non-vanishing in a small interval centered at $k_* > k_0 = \sqrt{v_0}$, and $\psi_k(x)$ is one of the (two) independent solutions that you found in part 2. It is to be assumed that

$$
\int_{-\infty}^\infty \frac{dk}{\sqrt{2\pi}} |g(k)|^2 = 1 .
$$

5. Work out what happens in case (b), utilizing wave packets for the incident and reflected waves (there is no transmitted wave packet in this case). Assume the profile function to be sharply peaked at k_* with $0 < k_* < k_0$. Note that the reflected wave packet becomes appreciable only after a time delay τ. Calculate τ and provide a physical interpretation.

Solution

In case (a), the (classical) particle will suffer an instantaneous deceleration at $x=0$ upon entering the region $x > 0$ where the potential is effective. In $x > 0$ the energy of the particle will be $mv'^2/2 = E - V_0 = mv^2/2 - V_0$ and its velocity will be $v' = \sqrt{v^2 - 2V_0/m}$. It will continue to travel with constant velocity v' towards $x = \infty$.

In case (b) the particle will be reflected at $x=0$, inverting the direction of motion and traveling towards $x = -\infty$ with the same speed v as it had initially.

Part 1

In both cases only continuum solutions with $E > 0$ are obtained. They are doubly degenerate in case (a) and non-degenerate in case (b).

Part 2

We split the real axis into two regions: $x < 0$ (region $<$) and $x > 0$ (region $>$). In these regions, the Schrödinger equation reads

$$\psi_<''(x) = -k^2 \psi_<(x) \,, \qquad \psi_>''(x) = -K^2 \psi_>(x) \,,$$

where we have defined

$$k^2 = \frac{2mE}{\hbar^2} = \epsilon \,, \qquad K^2 = \frac{2m(E - V_0)}{\hbar^2} = \epsilon - v_0 \,.$$

The wave function and its first derivative must be continuous at the origin, since the potential is non-singular (as a matter of fact, it is piecewise continuous),

$$\psi_<(0) = \psi_>(0) \,, \qquad \psi_<'(0) = \psi_>'(0) \,.$$

We take the two independent solutions as

$$\psi_<^{(+)}(x) = \alpha_< e^{ikx} + \beta_< e^{-ikx} \,, \qquad \psi_<^{(-)}(x) = \beta_< e^{-ikx} \,,$$
$$\psi_>^{(+)}(x) = \alpha_> e^{iKx} \,, \qquad \psi_>^{(-)}(x) = \alpha_> e^{iKx} + \beta_> e^{-iKx} \,,$$

and impose boundary conditions at the origin on each of these two solutions; for example, on solution (+) we get

$$\underbrace{\alpha_< + \beta_<}_{\psi_<^{(+)}(0)} = \underbrace{\alpha_>}_{\psi_>^{(+)}(0)} \,, \qquad \underbrace{ik\left(\alpha_< - \beta_<\right)}_{\psi_<^{(+)\prime}(0)} = \underbrace{iK\alpha_>}_{\psi_>^{(+)\prime}(0)} \,,$$

which can be solved with respect to $\alpha_<$. We proceed similarly for solution $(-)$, and finally arrive at

$$\text{solution } (+): \quad \alpha_> = \frac{2k}{k+K}\alpha_< \,, \qquad \beta_< = \frac{k-K}{k+K}\alpha_< \,,$$
$$\text{solution } (-): \quad \beta_< = \frac{2K}{k+K}\beta_> \,, \qquad \alpha_> = \frac{K-k}{k+K}\beta_> \,.$$

Dropping an irrelevant constant, we express the solutions as

$$\psi_k^{(+)}(x) = \underbrace{e^{ikx}}_{\psi_i^{(+)}(x)} + \underbrace{A(k)\,e^{-ikx}}_{\psi_r^{(+)}(x)} \qquad x < 0$$

$$= \underbrace{B(k)\,e^{iKx}}_{\psi_t^{(+)}(x)} \qquad\qquad x > 0$$

and

$$\psi_k^{(-)}(x) = \underbrace{\overline{B}(k)\,e^{-ikx}}_{\psi_t^{(-)}(x)} \qquad\qquad x < 0$$

$$= \underbrace{e^{-iKx}}_{\psi_i^{(-)}(x)} + \underbrace{\overline{A}(k)\,e^{iKx}}_{\psi_r^{(-)}(x)} \qquad x > 0$$

where

$$A(k) = \frac{k-K}{k+K}, \qquad B(k) = \frac{2k}{k+K},$$
$$\overline{A}(k) = \frac{K-k}{k+K}, \qquad \overline{B}(k) = \frac{2K}{k+K},$$

and

$$K = \sqrt{k^2 - k_0^2} \qquad \text{with} \qquad k_0^2 = v_0 \,.$$

We are now in a position to evaluate the incident, reflected, and transmitted probability currents associated with the corresponding waves. Using

$$j = \frac{\hbar}{2mi}\left[\psi^*(x)\frac{d}{dx}\psi(x) - \psi(x)\frac{d}{dx}\psi^*(x)\right],$$

we find

$$j_i^{(+)} = \frac{\hbar k}{m}\,, \qquad j_r^{(+)} = -\frac{\hbar k}{m}\,|A(k)|^2\,, \qquad j_t^{(+)} = \frac{\hbar K}{m}\,|B(k)|^2\,,$$
$$j_i^{(-)} = -\frac{\hbar K}{m}\,, \qquad j_r^{(-)} = \frac{\hbar K}{m}\,|\overline{A}(k)|^2\,, \qquad j_t^{(-)} = -\frac{\hbar k}{m}\,|\overline{B}(k)|^2\,,$$

where the signs correspond to propagation from left to right (+) or right to left (−). The magnitudes of the incident, reflected, and transmitted probability current densities must satisfy

$$|j_i| = |j_r| + |j_t| \implies \underbrace{\frac{|j_r|}{|j_i|}}_{R(k)} + \underbrace{\frac{|j_t|}{|j_i|}}_{T(k)} = 1\,,$$

which leads to expressions for the reflection and transmission probabilities, $R(k)$ and $T(k)$:

$$R(k) = |A(k)|^2 = \left|\overline{A}(k)\right|^2\,, \qquad T(k) = \frac{K}{k}\,|B(k)|^2 = \frac{k}{K}\,\left|\overline{B}(k)\right|^2\,,$$

and note that they are independent of whether we are considering solution (+) or solution (−) (see Problem 17 for a general proof of this equality). It is easily verified that $R(k) + T(k) = 1$.

Part 3

We have continuum non-degenerate solutions in the energy range: $0 < E < V_0$. The Schrödinger equation reads

$$\psi_<''(x) = -k^2 \psi_<(x), \qquad \psi_>''(x) = \kappa^2 \psi_>(x),$$

where

$$k^2 = \frac{2mE}{\hbar^2}, \qquad \kappa^2 = \frac{2m(V_0 - E)}{\hbar^2} = k_0^2 - k^2.$$

The general form of the solution is obtained as

$$\psi_<(x) = \alpha_< \, e^{ikx} + \beta_< \, e^{-ikx}, \qquad \psi_>(x) = \beta_> \, e^{-\kappa x},$$

since the wave function must vanish at $x \longrightarrow \infty$. Imposing the continuity condition for $\psi(x)$ and $\psi'(x)$ at the origin, we find

$$\beta_< = \frac{k - i\kappa}{k + i\kappa} \alpha_<, \qquad \beta_> = \frac{2k}{k + i\kappa} \alpha_<.$$

Dropping the irrelevant constant $\alpha_<$, we can express the solution as

$$\psi_k(x) = \underbrace{e^{ikx}}_{\psi_i(x)} + \underbrace{C(k) \, e^{-ikx}}_{\psi_r(x)} \qquad x < 0$$

$$= \underbrace{D(k) \, e^{-\kappa x}}_{\psi_t(x)} \qquad x > 0,$$

where

$$C(k) = \frac{k - i\kappa}{k + i\kappa}, \qquad D(k) = \frac{2k}{k + i\kappa}.$$

Note that, apart from the complex constant $D(k)$, the transmitted wave function is real, and therefore its current density vanishes. There is no transmitted wave function. On the other hand, we find

$$j_i = \frac{\hbar k}{m}, \qquad j_r = -\frac{\hbar k}{m} |C(k)|^2 = -\frac{\hbar k}{m},$$

since $C(k)$ has unit magnitude and is just a phase factor. Thus in this case $R(k) = 1$ independently of the energy. As in classical mechanics, the particle is always reflected. However, there is a finite probability for the particle to be found in the region $x > 0$, since

$$|\psi_>(x)|^2 \propto e^{-2\kappa x}.$$

This probability becomes negligible when $x \gg 1/(2\kappa)$.

Part 4

We insert the explicit expression for $\psi_k^{(+)}(x)$ found earlier in part 2 in order to obtain the wave packet; we will consider only the wave packet associated with this solution, as the treatment of the wave packet associated with solution (−) is similar. We have

$$\Psi^{(+)}(x,t) = \underbrace{\int_{k_0}^{\infty} \frac{dk}{\sqrt{2\pi}} g(k) \, e^{i(kx - \omega_k t)}}_{\Psi_i^{(+)}(x,t)} + \underbrace{\int_{k_0}^{\infty} \frac{dk}{\sqrt{2\pi}} g(k) \, A(k) \, e^{-i(kx + \omega_k t)}}_{\Psi_r^{(+)}(x,t)} \qquad x < 0$$

$$= \underbrace{\int_{k_0}^{\infty} \frac{dk}{\sqrt{2\pi}} g(k) \, B(k) \, e^{i(Kx - \omega_k t)}}_{\Psi_t^{(+)}(x,t)} \qquad\qquad\qquad\qquad x > 0 \, .$$

Since $\Psi^{(+)}(x,t)$ is a linear superposition of the $\psi_k^{(+)}(x)$, it will satisfy the boundary conditions at the origin; it consists of three terms, to be labeled incident (i), reflected (r), and transmitted (t). In the present case the constants $A(k)$ and $B(k)$ are real, as is the function $g(k)$. Using the stationary phase method, we find that the centers of the incident, reflected, and transmitted wave packets are, respectively, at

$$x_i(t) = \frac{\hbar k_*}{m} t \, , \qquad x_r(t) = -\frac{\hbar k_*}{m} t \, , \qquad x_t(t) = \frac{\hbar K_*}{m} t \, ,$$

where we have defined $K_* = \sqrt{k_*^2 - k_0^2}$ (note that x_t travels in the $x > 0$ region with the velocity of the classical particle of energy E_*); for example,

$$\frac{d}{dk}(Kx - \omega_k t)\Big|_{k=k_*} = 0 \implies \frac{k_*}{K_*} x - \frac{\hbar k_*}{m} t = 0 \, ,$$

from which we obtain the position $x_t(t)$ of the center of the transmitted wave packet. Furthermore, to linear terms in $k - k_*$ the phase in the transmitted wave packet can be expanded as

$$Kx - \omega_k t = K_* x - \omega_{k_*} t + \overbrace{\left(\frac{k_*}{K_*} x - \frac{\hbar k_*}{m} t \right)}^{\frac{d}{dk}(Kx - \omega_k t)|_{k=k_*}} (k - k_*) + \cdots = K_* x - \omega_{k_*} t + \frac{k_*}{K_*} [x - x_t(t)](k - k_*) + \cdots \, .$$

Expanding the phase factors in the incident, reflected, and transmitted wave packets, we can write

$$\Psi_i^{(+)}(x,t) \approx e^{i(k_* x - \omega_{k_*} t)} \int_{k_0}^{\infty} \frac{dk}{\sqrt{2\pi}} g(k) \, e^{i[x - x_i(t)](k - k_*)} \qquad x < 0 \, ,$$

$$\Psi_r^{(+)}(x,t) \approx e^{-i(k_* x + \omega_{k_*} t)} \int_{k_0}^{\infty} \frac{dk}{\sqrt{2\pi}} g(k) \, A(k) \, e^{-i[x - x_r(t)](k - k_*)} \qquad x < 0 \, ,$$

$$\Psi_t^{(+)}(x,t) \approx e^{i(K_* x - \omega_{k_*} t)} \int_{k_0}^{\infty} \frac{dk}{\sqrt{2\pi}} g(k) \, B(k) \, e^{i(k_*/K_*)[x - x_t(t)](k - k_*)} \qquad x > 0 \, .$$

Since $g(k)$ is sharply peaked at k_*, we can extend the integrations in k from $-\infty$ to ∞, incurring only negligible error. Furthermore, the coefficients $A(k)$ and $B(k)$ are assumed to vary very slowly around k_*, namely

$$\left| \frac{1}{A(k)} \frac{dA(k)}{dk} \right|_{k_*} \ll 1 \, , \qquad \left| \frac{1}{B(k)} \frac{dB(k)}{dk} \right|_{k_*} \ll 1 \, .$$

For example, we find

$$\frac{dB(k)}{dk}\Big|_{k_*} = \frac{2}{k_* + K_*} \left(1 - \frac{k_*}{K_*} \right) \, ,$$

and hence

$$\left| \frac{1}{B(k)} \frac{dB(k)}{dk} \right|_{k_*} = \left| \frac{1}{k_*} - \frac{1}{K_*} \right| \ll 1 \implies \frac{k_0}{k_*} \ll 1 \, ,$$

and the energy E_* is much larger than the potential step. Under these conditions, we have for the wave packets

$$\Psi^{(+)}(x,t) \approx \underbrace{e^{i(k_*x-\omega_{k_*}t)}\,G(x-x_i)}_{\Psi_i^{(+)}(x,t)} + \underbrace{e^{-i(k_*x+\omega_{k_*}t)}\,A(k_*)\,G(x_r-x)}_{\Psi_r^{(+)}(x,t)} \qquad x < 0$$

$$\approx \underbrace{e^{i(K_*x-\omega_{k_*}t)}\,B(k_*)\,G[k_*(x-x_t)/K_*]}_{\Psi_t^{(+)}(x,t)} \qquad\qquad x > 0$$

where the function $G(y)$ is defined as

$$G(y) = \int_{-\infty}^{\infty} \frac{dk}{\sqrt{2\pi}}\,g(k)\,e^{iy(k-k_*)}\,,$$

and we note that the argument of this function in the transmitted wave packet is $k_*(x-x_t)/K_*$.

For $t \ll 0$, only the incident wave packet survives (x_i is negative); by contrast, for $t \gg 0$ only the reflected and transmitted wave packets survive, the former in the $x \ll 0$ region and the latter in the $x \gg 0$ region. Thus, we have

1. $t \ll 0$: $\Psi^{(+)}(x,t) \approx \Psi_i^{(+)}(x,t)$, and

$$\int_{-\infty}^{\infty} dx\,|\Psi^{(+)}(x,t)|^2 \approx \int_{-\infty}^{\infty} dx\,|G(x-x_i)|^2 = \int_{-\infty}^{\infty} dx\,|G(x)|^2 = \int_{-\infty}^{\infty} dk\,|g(k)|^2 = 1\,,$$

where in the last step we have used Parseval's identity for the Fourier transform.

2. $t \gg 0$: $\Psi^{(+)}(x,t) \approx \Psi_r^{(+)}(x,t)$ for $x \ll 0$ and $\Psi^{(+)}(x,t) \approx \Psi_t^{(+)}(x,t)$ for $x \gg 0$, and

$$\int_{-\infty}^{\infty} dx\,|\Psi^{(+)}(x,t)|^2 \approx \int_{-\infty}^{0} dx\,|\Psi_r^{(+)}(x,t)|^2 + \int_{0}^{\infty} dx\,|\Psi_t^{(+)}(x,t)|^2$$

$$= |A(k_*)|^2 \int_{-\infty}^{\infty} dx\,|G(x)|^2 + \frac{K_*}{k_*}\,|B(k_*)|^2 \int_{-\infty}^{\infty} dx\,|G(x)|^2 = |A(k_*)|^2 + \frac{K_*}{k_*}\,|B(k_*)|^2\,,$$

where in the second line we have extended the integration limits to the whole real axis since the reflected (transmitted) wave packet is negligible for $x \gg 0$ ($x \ll 0$), and we have rescaled the integration variable $k_*(x-x_t)/K_*$ to x.

Since the normalization is independent of time (the total probability is conserved), we must have

$$|A(k_*)|^2 + \frac{K_*}{k_*}\,|B(k_*)|^2 = 1\,,$$

and indeed this holds true. We interpret $|A(k_*)|^2 = R(k_*)$ as the probability for the particle to be reflected, and $(K_*/k_*)|B(k_*)|^2 = T(k_*)$ as the probability for it to be transmitted.

Part 5

In this case, we have

$$\Psi(x,t) = \underbrace{\int_{0}^{k_0} \frac{dk}{\sqrt{2\pi}}\,g(k)\,e^{i(kx-\omega_k t)}}_{\Psi_i(x,t)} + \underbrace{\int_{0}^{k_0} \frac{dk}{\sqrt{2\pi}}\,g(k)\,C(k)\,e^{-i(kx+\omega_k t)}}_{\Psi_r(x,t)} \qquad x < 0$$

$$= D(k_*)\,e^{-\kappa_* x} \qquad\qquad\qquad\qquad x > 0\,,$$

where $\kappa_* = \sqrt{k_0^2 - k_*^2}$. We are interested in the reflected wave packet. Recall that $C(k)$ is just a phase factor; noting that

$$k + i\kappa = \sqrt{k^2 + \kappa^2}\, e^{i\delta(k)}\,, \qquad \delta(k) = \tan^{-1}(\kappa/k)\,,$$

$C(k)$ can be expressed as follows:

$$C(k) = \frac{k - i\kappa}{k + i\kappa} = e^{-2i\delta(k)}\,.$$

The reflected wave function is written as

$$\Psi_r(x, t) = \int_0^{k_0} \frac{dk}{\sqrt{2\pi}}\, g(k)\, e^{-i[kx + \omega_k t + 2\delta(k)]}\,.$$

Its center is at

$$x_r(t) = -\frac{\hbar k_*}{m}\, t - 2\frac{d}{dk}\delta(k)\Big|_{k=k_*}\,,$$

and we have

$$\frac{d}{dk}\delta(k) = \frac{d}{dk}\tan^{-1}\frac{\sqrt{k_0^2 - k^2}}{k} = -\frac{1}{\sqrt{k_0^2 - k^2}}\,,$$

which allows us to write

$$x_r(t) = -\frac{\hbar k_*}{m}\, t + \frac{2}{\kappa_*} \implies x_r(t) = -\frac{\hbar k_*}{m}\,(t - \tau)\,, \qquad \tau = \frac{2m}{\hbar \kappa_* k_*}\,.$$

Therefore, the reflected wave packet becomes appreciable only after a time delay τ, that is, its center is at $x < 0$ for $t > \tau$. In the time interval τ, the reflected wave packet tunnels into the (classically forbidden) $x > 0$ region and then tunnels back into the (classically allowed) $x < 0$ region, covering a distance of the order $(\hbar k_*/m)\tau/2 = 1/\kappa_*$. This distance is of the same order as the distance $1/(2\kappa_*)$ over which the probability density for the particle in the classically forbidden region is appreciable.

Problem 3   Linear Potential in Momentum Space

Consider a particle of mass m in one dimension under the influence of a linear potential (classically, a constant force F)

$$v(x) = -\frac{2m}{\hbar^2}\, Fx\,,$$

where F is a constant. The energy spectrum is known to be continuous and non-degenerate.

1. Solve the time-independent Schrödinger equation in momentum space, and denote the solutions as $\widetilde{\psi}_E(p)$. Normalize them by writing

$$\int_{-\infty}^{\infty} dp\, \widetilde{\psi}_E^*(p)\, \widetilde{\psi}_{E'}(p) = \delta(E - E')\,.$$

2. Obtain the coordinate-space representation $\psi_E(x)$. Show that it can be related to the Airy function for which the integral representation is given by

$$\mathrm{Ai}(y) = \frac{1}{2\pi} \int_{-\infty}^{\infty} du\, e^{i(yu + u^3/3)}\,, \qquad y \text{ real}\,.$$

Solution

Part 1

Recalling that, in the p-representation, $x \longrightarrow i\hbar\, d/dp$, the Schrödinger equation reads

$$\frac{p^2}{2m}\,\widetilde{\varphi}_E(p) - i\hbar F\,\frac{d}{dp}\,\widetilde{\varphi}_E(p) = E\,\widetilde{\varphi}_E(p)\,,$$

and this first-order differential equation is easily solved:

$$\frac{d}{dp}\,\widetilde{\varphi}_E(p) = -\frac{i}{\hbar F}\left(\frac{p^2}{2m} - E\right)\widetilde{\varphi}_E(p) \implies \widetilde{\varphi}_E(p) = C\exp\left[-\frac{i}{\hbar F}\int^p dp'\left(\frac{p'^2}{2m} - E\right)\right]\,,$$

where C is a constant. The p-space wave function is given by

$$\widetilde{\varphi}_E(p) = C\,e^{i[Ep - p^3/(6m)]/(\hbar F)}\,.$$

We determine C by imposing the continuum normalization,

$$\int_{-\infty}^{\infty} dp\,\widetilde{\varphi}_E^*(p)\,\widetilde{\varphi}_{E'}(p) = |C|^2\int_{-\infty}^{\infty} dp\, e^{-ip(E-E')/(\hbar F)} = \hbar F|C|^2\int_{-\infty}^{\infty} dy\, e^{-i(E-E')y}$$

$$= 2\pi\hbar F|C|^2\delta(E-E') \implies |C| = \frac{1}{\sqrt{2\pi\hbar F}}\,.$$

Up to a phase factor, we have

$$\widetilde{\varphi}_E(p) = \frac{1}{\sqrt{2\pi\hbar F}}\,e^{i[Ep - p^3/(6m)]/(\hbar F)}\,.$$

Part 2

The coordinate-space wave function is related to the momentum-space wave function by a Fourier transform,

$$\varphi(x) = \int_{-\infty}^{\infty} dp\,\frac{e^{ipx/\hbar}}{\sqrt{2\pi\hbar}}\,\widetilde{\varphi}(p) = \frac{1}{2\pi\hbar\sqrt{F}}\int_{-\infty}^{\infty} dp\, e^{ipx/\hbar}\,e^{i[Ep - p^3/(6m)]/(\hbar F)}\,,$$

which can be related to the integral representation of the Airy function, given by

$$\mathrm{Ai}(y) = \frac{1}{2\pi}\int_{-\infty}^{\infty} du\, e^{i(yu + u^3/3)} = \frac{1}{\pi}\int_0^{\infty} du\,\cos(yu + u^3/3)\,,\qquad y\ \text{real}\,.$$

Indeed, in the Fourier transform introduce the variable

$$u = -\frac{p}{(2m\hbar F)^{1/3}}\,,$$

to obtain

$$\varphi(x) = \frac{1}{2\pi\hbar\sqrt{F}}\,(2m\hbar F)^{1/3}\int_{-\infty}^{\infty} du\, e^{i(yu + u^3/3)} = \frac{1}{\sqrt{\hbar}}\left[\frac{(2m)^2}{\hbar F}\right]^{1/6}\mathrm{Ai}(y),\quad y = -\left[\frac{2m}{(\hbar F)^2}\right]^{1/3}(E+Fx)\,.$$

Problem 4 Finite Barrier: Tunneling, Transmission, and Transit Time

Calculate the transmission probability for a particle in a potential that is given by $v(x) = 0$ for $x < 0$ and $x > a$ and is equal to a constant $v_0 > 0$ for $0 < x < a$. Consider the two cases $0 < \epsilon < v_0$ and $\epsilon > v_0$. Discuss the motion of a classical particle in the two cases.

By introducing wave packets, calculate the transit time – the time needed to traverse the barrier – of the transmitted wave for the two cases $0 < \epsilon < v_0$ and $\epsilon > v_0$ and compare it with that for a classical particle.

Solution

The case $0 < \epsilon < v_0$. We have three regions: region I for $x < 0$, region II for $0 < x < a$, and region III for $x > a$. In these regions the Schrödinger equation reads

$$\psi_I''(x) = -k^2 \psi_I(x) \qquad k^2 = \epsilon > 0 , \qquad\qquad \psi_{II}''(x) = \kappa^2 \psi_{II}(x) \qquad \kappa^2 = v_0 - \epsilon > 0 ,$$

and

$$\psi_{III}''(x) = -k^2 \psi_{III}(x) ,$$

with solutions

$$\psi_I(x) = e^{ikx} + A\, e^{-ikx} , \qquad \psi_{II}(x) = C \cosh(\kappa x) + D \sinh(\kappa x) , \qquad \psi_{III}(x) = B\, e^{ik(x-a)} ,$$

where the incident wave is from the left and the transmitted wave to the right. The transmission coefficient is given by

$$T = |B\, e^{-ika}|^2 = |B|^2 ,$$

which follows from evaluating the incident (j_i) and transmitted (j_t) probability flux $T = |j_t|/|j_i|$. We impose the following boundary conditions at $x = 0$ and $x = a$,

$$1 + A = C$$
$$1 - A = -i\,\frac{\kappa}{k}\, D$$
$$B = C \cosh(\kappa a) + D \sinh(\kappa a)$$
$$i\,\frac{k}{\kappa}\, B = C \sinh(\kappa a) + D \cosh(\kappa a) .$$

We solve the last two equations with respect to C and D,

$$C = \left[\cosh(\kappa a) - i\,\frac{k}{\kappa}\, \sinh(\kappa a) \right] B ,$$

$$D = \left[i\,\frac{k}{\kappa}\, \cosh(\kappa a) - \sinh(\kappa a) \right] B ,$$

and substitute these expressions into the first two of the equations to find

$$1 + A = \left[\cosh(\kappa a) - i\,\frac{k}{\kappa}\, \sinh(\kappa a) \right] B ,$$

$$1 - A = -i\,\frac{\kappa}{k} \left[i\,\frac{k}{\kappa}\, \cosh(\kappa a) - \sinh(\kappa a) \right] B = \left[\cosh(\kappa a) + i\,\frac{\kappa}{k}\, \sinh(\kappa a) \right] B .$$

We solve these with respect to B:

$$B = \frac{2}{2\cosh(\kappa a) + i(\kappa/k - k/\kappa)\sinh(\kappa a)} .$$

Finally, the transmission coefficient is obtained as

$$T = \frac{4}{4\cosh^2(\kappa a) + (\kappa/k - k/\kappa)^2\sinh^2(\kappa a)} = \frac{4}{4 + [(\kappa/k - k/\kappa)^2 + 4]\sinh^2(\kappa a)} ,$$

which can be written as

$$T(\epsilon) = \frac{4\epsilon(v_0 - \epsilon)}{4\epsilon(v_0 - \epsilon) + v_0^2\sinh^2(\sqrt{v_0 - \epsilon}\, a)} ,$$

since $(\kappa/k - k/\kappa)^2 + 4 = v_0/[\epsilon(v_0 - \epsilon)]$. The transmission coefficient vanishes at $\epsilon = 0$, and increases as ϵ increases, reaching a maximum at $\epsilon = v_0$ with

$$T(v_0) = \frac{1}{1 + v_0 a^2/4} ,$$

where we have used $\sinh^2(\sqrt{v_0 - \epsilon}\, a) \approx (v_0 - \epsilon)a^2$ as $\epsilon \longrightarrow v_0$.

The case $\epsilon > v_0$. The procedure above can be repeated in this case. The solution in region II now involves sine and cosine functions. However, we can obtain the transmission coefficient in the present case by observing that

$$\kappa \longrightarrow iK \qquad K = \sqrt{\epsilon - v_0} ,$$

which leads to

$$T = \frac{4}{4 + [(iK/k + ik/K)^2 + 4]\sinh^2(iKa)} .$$

Note that

$$\sinh(ix) = \sum_{k=0}^{\infty} \frac{(ix)^{2k+1}}{(2k+1)!} = i\sum_{k=0}^{\infty} i^{2k}\frac{x^{2k+1}}{(2k+1)!} = i\sin(x)$$

and

$$4 + (iK/k + ik/K)^2 = 4 - (K/k + k/K)^2 = -\frac{v_0^2}{\epsilon(\epsilon - v_0)} .$$

We finally arrive at

$$T(\epsilon) = \frac{4\epsilon(\epsilon - v_0)}{4\epsilon(\epsilon - v_0) + v_0^2\sin^2(\sqrt{\epsilon - v_0}\, a)} ,$$

and we see that $T(\epsilon)$ oscillates between a minimum value equal to $T(v_0)$, obtained earlier, and unity. The transmission is equal to unity at energies such that

$$T(\epsilon_n) = 1 \implies \epsilon_n = v_0 + \frac{n^2\pi^2}{a^2} \qquad n = 1, 2, \ldots$$

It is interesting to compare the motion of a classical particle and that of a quantum particle. If the energy in the former case is less than the height of the barrier, the classical particle rebounds from the barrier without being able to penetrate it. By contrast, the wave packet describing the quantum particle splits into a reflected and a transmitted wave packet, whose intensity never vanishes. As the energy increases from 0 to v_0 the transmission coefficient increases from 0 to the maximum $T(v_0)$. The quantum particle has a finite probability of penetrating the barrier and "materializing" on the other side. This is the *tunneling effect*. Such an effect plays a role, for example, in the theory of α emission by nuclei (α radioactivity).

On the other hand, if the classical particle has energy larger than the height of the barrier, it will traverse it and travel beyond it. Assuming it is approaching the barrier from the left, the particle will suffer a sudden deceleration at the origin and a sudden acceleration at $x = a$, which will restore the original velocity. It will then travel with this velocity to $+\infty$. This is very different from the quantum description, in which the wave packet will almost always be partially reflected, the exception being when the particle energy is ϵ_n, for $n = 1, 2, \ldots$, in which case $T(\epsilon_n) = 1$ and there is no reflected wave packet.

The transmitted wave packet for $k < k_0$ in the region $x > a$ has the form

$$\Psi_t(x, t) = \int_0^{k_0} \frac{dk}{\sqrt{2\pi}} \, g(k) \, B(k) \, e^{i[k(x-a) - \omega_k t]} \qquad\qquad x > a \,,$$

where $g(k)$ (assumed real) is sharply peaked at k_* and

$$B = |B| \, e^{i\delta(k)} \,, \qquad \delta(k) = -\tan^{-1}\left[\frac{1}{2}\left(\frac{\kappa}{k} - \frac{k}{\kappa}\right) \tanh(\kappa a)\right] \,.$$

Using the stationary phase method, where the phase is

$$\varphi(k) = k(x - a) + \delta(k) - \omega_k t \,,$$

we find from $\varphi'(k_*) = 0$ that the center $x_t(t)$ of the transmitted wave packet is given by

$$x_t(t) = a + \frac{\hbar k_*}{m} t - \delta'(k_*) = a + \frac{\hbar k_*}{m}(t - \tau) \,, \qquad \tau = \frac{m}{\hbar k_*} \delta'(k_*) \,.$$

The incident wave packet in the region $x < 0$ has the form

$$\Psi_i(x, t) = \int_0^{k_0} \frac{dk}{\sqrt{2\pi}} \, g(k) \, e^{i(kx - \omega_k t)} \qquad\qquad x < 0 \,,$$

and the center of the incident wave packet reaches the barrier ($x = 0$) at time $t = 0$. Thus we interpret τ as the time for the particle to materialize on the other side of the barrier at $x = a$. We expect τ to become larger and larger as the energy ϵ_* approaches 0. We are left with the evaluation of $\delta'(k_*)$. It is convenient to define

$$\delta(k) = -\tan^{-1}\left[\alpha(k) \, \tanh(\kappa(k)a)\right] \,, \qquad \alpha(k) = \frac{1}{2}\left[\frac{\kappa(k)}{k} - \frac{k}{\kappa(k)}\right] = \frac{k_0^2/2 - k^2}{k\kappa(k)} \,,$$

with

$$\kappa'(k) = -\frac{k}{\kappa(k)} \,, \qquad \alpha'(k) = -\frac{k_0^4/2}{k^2 \kappa^3(k)} \,.$$

We obtain

$$\delta'(k) = -\frac{1}{1 + \alpha^2(k) \tanh^2(\kappa(k)a)}\left[\alpha'(k) \, \tanh(\kappa(k)a) - \frac{ak\alpha(k)}{\kappa(k) \, \cosh^2(\kappa(k)a)}\right]$$

or, expressing $\delta'(k)$ in term of the energy ϵ,

$$\delta'(\epsilon) = \frac{f(\epsilon)}{\epsilon(v_0 - \epsilon)^{3/2} + (v_0/2 - \epsilon)^2(v_0 - \epsilon)^{1/2} \tanh^2(\sqrt{v_0 - \epsilon}\, a)} \,,$$

where

$$f(\epsilon) = \frac{v_0^2}{2}\left[\tanh(\sqrt{v_0 - \epsilon}\,a) + \frac{a}{v_0^2/2}\frac{\epsilon\,(v_0/2 - \epsilon)\sqrt{v_0 - \epsilon}}{\cosh^2(\sqrt{v_0 - \epsilon}\,a)}\right].$$

Using $k_* = \sqrt{\epsilon_*}$, the time delay can be written as

$$\tau = \frac{m}{\hbar\sqrt{\epsilon_*}}\,\delta'(\epsilon_*).$$

Since $\delta'(\epsilon_* \longrightarrow 0)$ is a constant, it takes an infinite time for the transmitted wave to materialize on the other side of the barrier in this limit. By contrast, since $\delta'(\epsilon_* \longrightarrow v_0)$ is a constant, at the other end of the energy range there is a finite time delay given by

$$\tau(v_0) = \frac{m}{\hbar\sqrt{v_0}}\,\delta'(v_0).$$

Indeed, using the following expansions for $x \longrightarrow 0$,

$$\cosh(x) = 1 + \frac{x^2}{2} + \cdots, \qquad\qquad \tanh(x) = x + \frac{x^3}{3} + \cdots,$$

we have

$$f(\epsilon_* \longrightarrow v_0) \approx \frac{2a}{3}\,(av_0)^2\,(v_0 - \epsilon_*)^{3/2} \implies \delta'(v_0) = \frac{2a}{3}\frac{a^2 v_0}{1 + a^2 v_0/4}$$

and

$$\tau(v_0) = \frac{2}{3}\frac{ma}{\hbar\sqrt{v_0}}\frac{a^2 v_0}{1 + a^2 v_0/4}.$$

As noted above, in this range of energies the classical particle does not penetrate the barrier.

For $\epsilon > v_0$ or, in the usual units, $E > V_0$, the classical particle traverses the region of the barrier in a time given by $\tau_{\rm cl} = a/v'$, where $v' = \sqrt{2(E - V_0)/m}$ is the constant velocity in $0 < x < a$. In the quantum case, the time delay of the transmitted wave packet follows from

$$\tau = \frac{m}{\hbar\sqrt{\epsilon_*}}\,\delta'(\epsilon_*),$$

where $\epsilon_* > v_0$,

$$\delta'(\epsilon) = \frac{g(\epsilon)}{\epsilon(\epsilon - v_0)^{3/2} + (v_0/2 - \epsilon)^2(\epsilon - v_0)^{1/2}\tan^2(\sqrt{\epsilon - v_0}\,a)},$$

and

$$g(\epsilon) = \frac{v_0^2}{2}\left[\frac{a}{v_0^2/2}\frac{\epsilon(\epsilon - v_0/2)\sqrt{\epsilon - v_0}}{\cos^2(\sqrt{\epsilon - v_0}\,a)} - \tan(\sqrt{\epsilon - v_0}\,a)\right].$$

At the energies ϵ_n for which $T(\epsilon_n) = 1$, the time delay reduces to

$$\tau(\epsilon_n) = \frac{ma}{\hbar\sqrt{\epsilon_n}}\frac{\epsilon_n - v_0/2}{\epsilon_n - v_0} = \frac{\epsilon_n - v_0/2}{\sqrt{\epsilon_n(\epsilon_n - v_0)}}\,\tau_{\rm cl} = \frac{1}{2}\left(\sqrt{\frac{\epsilon_n}{\epsilon_n - v_0}} + \sqrt{\frac{\epsilon_n - v_0}{\epsilon_n}}\right)\tau_{\rm cl},$$

and it approaches the classical limit as $\epsilon_n \longrightarrow \infty$.

Problem 5 Tunneling in the Limit of a High and/or Wide Barrier

Consider a particle in one dimension under the influence of an even potential given by

$$v(x) = v_0 \qquad |x| \le a, \qquad v(x) = 0 \qquad |x| > a,$$

where $v_0 > 0$. Assume that the particle's energy ϵ is in the range $0 < \epsilon < v_0$. Write the general solution as

$$
\begin{aligned}
\psi(x) &= A\,e^{ikx} + B\,e^{-ikx} & x &< -a, \\
&= C\,e^{-\kappa x} + D\,e^{\kappa x} & -a &< x < a, \\
&= F\,e^{ikx} + G\,e^{-ikx} & x &> a,
\end{aligned}
$$

where

$$k = \sqrt{\epsilon}, \qquad \kappa = \sqrt{v_0 - \epsilon}.$$

1. By imposing the appropriate boundary conditions at $x = \pm a$, show that the coefficients A and B are related to the coefficients F and G by a 2×2 matrix, namely

$$\begin{pmatrix} A \\ B \end{pmatrix} = \underline{M} \begin{pmatrix} F \\ G \end{pmatrix}.$$

Determine $\underline{M}$.
2. From the matrix relation obtained in part 1, show that the transmission coefficient T for a high and wide barrier with $\kappa a \gg 1$ is approximately given by

$$T \approx 16\,e^{-4\kappa a} \left(\frac{k\kappa}{k^2 + \kappa^2} \right)^2.$$

3. Consider the case of a very narrow but very high barrier such that av_0 is finite. Assume that $v_0 \gg \epsilon$, $\kappa \gg k$, and $\kappa a \ll 1$ but $\kappa^2 a$ is finite. Show that the transmission coefficient under these conditions is the same as that obtained in the repulsive δ-function potential $v(x) = 2av_0\,\delta(x)$, namely

$$T = \frac{\epsilon}{\epsilon + (v_0 a)^2}.$$

Solution

Part 1

Impose the continuity of $\psi(x)$ and $\psi'(x)$ at $x = -a$:

$$A\,e^{-ika} + B\,e^{ika} = C\,e^{\kappa a} + D\,e^{-\kappa a}$$

$$ik\left(A\,e^{-ika} - B\,e^{ika} \right) = -\kappa\left(C\,e^{\kappa a} - D\,e^{-\kappa a} \right) \implies A\,e^{-ika} - B\,e^{ika} = i\frac{\kappa}{k}\left(C\,e^{\kappa a} - D\,e^{-\kappa a} \right),$$

and first summing and then subtracting these two relations we arrive at

$$2A\,e^{-ika} = C\left(1 + i\frac{\kappa}{k} \right)e^{\kappa a} + D\left(1 - i\frac{\kappa}{k} \right)e^{-\kappa a}$$

$$2B\,e^{ika} = C\left(1 - i\frac{\kappa}{k} \right)e^{\kappa a} + D\left(1 + i\frac{\kappa}{k} \right)e^{-\kappa a},$$

which can be cast into matrix form as

$$\begin{pmatrix} A \\ B \end{pmatrix} = \frac{1}{2} \begin{bmatrix} (1 + i\kappa/k)\, e^{\kappa a + ika} & (1 - i\kappa/k)\, e^{-\kappa a + ika} \\ (1 - i\kappa/k)\, e^{\kappa a - ika} & (1 + i\kappa/k)\, e^{-\kappa a - ika} \end{bmatrix} \begin{pmatrix} C \\ D \end{pmatrix} .$$

Similarly, we find that

$$F\, e^{ika} + G\, e^{-ika} = C\, e^{-\kappa a} + D\, e^{\kappa a}$$

$$ik\left(F\, e^{ika} - G\, e^{-ika}\right) = -\kappa\left(C\, e^{-\kappa a} - D\, e^{\kappa a}\right) \implies -i\frac{k}{\kappa}\left(F\, e^{ika} - G\, e^{-ika}\right) = C\, e^{-\kappa a} - D\, e^{\kappa a} ,$$

or, in matrix notation,

$$\begin{pmatrix} C \\ D \end{pmatrix} = \frac{1}{2} \begin{bmatrix} (1 - ik/\kappa)\, e^{\kappa a + ika} & (1 + ik/\kappa)\, e^{\kappa a - ika} \\ (1 + ik/\kappa)\, e^{-\kappa a + ika} & (1 - ik/\kappa)\, e^{-\kappa a - ika} \end{bmatrix} \begin{pmatrix} F \\ G \end{pmatrix} .$$

It follows that the matrix connecting A, B to F, G is given by

$$\underline{M} = \frac{1}{4} \begin{bmatrix} (1 + i\kappa/k)\, e^{\kappa a + ika} & (1 - i\kappa/k)\, e^{-\kappa a + ika} \\ (1 - i\kappa/k)\, e^{\kappa a - ika} & (1 + i\kappa/k)\, e^{-\kappa a - ika} \end{bmatrix} \begin{bmatrix} (1 - ik/\kappa)\, e^{\kappa a + ika} & (1 + ik/\kappa)\, e^{\kappa a - ika} \\ (1 + ik/\kappa)\, e^{-\kappa a + ika} & (1 - ik/\kappa)\, e^{-\kappa a - ika} \end{bmatrix} .$$

We find

$$M_{11} = \frac{1}{4}\Big[\, \underbrace{(1 + i\kappa/k)\,(1 - ik/\kappa)}_{2 + i\alpha}\, e^{2\kappa a + 2ika} + \underbrace{(1 - i\kappa/k)\,(1 + ik/\kappa)}_{2 - i\alpha}\, e^{-2\kappa a + 2ika}\,\Big]$$

$$= \left[\cosh 2\kappa a + i(\alpha/2)\, \sinh 2\kappa a\right] e^{2ika} ,$$

$$M_{12} = \frac{1}{4}\Big[\, \underbrace{(1 + i\kappa/k)\,(1 + ik/\kappa)}_{i\beta}\, e^{2\kappa a} + \underbrace{(1 - i\kappa/k)\,(1 - ik/\kappa)}_{-i\beta}\, e^{-2\kappa a}\,\Big]$$

$$= i(\beta/2)\, \sinh 2\kappa a ,$$

and similarly for the remaining two matrix elements, to obtain

$$\underline{M} = \begin{bmatrix} (\cosh 2\kappa a + i(\alpha/2)\, \sinh 2\kappa a)\, e^{2ika} & i(\beta/2)\, \sinh 2\kappa a \\ -i(\beta/2)\, \sinh 2\kappa a & (\cosh 2\kappa a - i(\alpha/2)\, \sinh 2\kappa a)\, e^{-2ika} \end{bmatrix} .$$

Part 2

The transmission coefficient follows from

$$T = \frac{|j_t|}{|j_i|} = \frac{|F|^2}{|A|^2} ,$$

and, setting $G = 0$ in the matrix equation,

$$\begin{pmatrix} A \\ B \end{pmatrix} = \begin{pmatrix} M_{11} & M_{12} \\ M_{21} & M_{22} \end{pmatrix} \begin{pmatrix} F \\ 0 \end{pmatrix} ,$$

we obtain

$$A = M_{11}F \implies T = \frac{1}{|M_{11}|^2} = \frac{1}{\cosh^2(2\kappa a) + (\alpha/2)^2\, \sinh^2(2\kappa a)} .$$

In the limit $\kappa a \gg 1$, we have

$$\cosh(2\kappa a) \approx \frac{1}{2}\, e^{2\kappa a} , \qquad \sinh(2\kappa a) \approx \frac{1}{2}\, e^{2\kappa a}$$

and

$$T \approx \frac{4}{e^{4\kappa a} \left(1 + \alpha^2/4\right)} = \frac{4\,e^{-4\kappa a}}{1 + (\kappa/k - k/\kappa)^2/4} = 16\,e^{-4\kappa a} \left(\frac{k\kappa}{k^2 + \kappa^2}\right)^2 .$$

Part 3

In the limit $\kappa a \ll 1$ we have, neglecting terms quadratic in κa,

$$\cosh(2\kappa a) \approx 1 , \qquad \sinh(2\kappa a) \approx 2\kappa a , \qquad \alpha \approx \kappa/k , \qquad \kappa^2 \approx v_0 ,$$

and

$$T \approx \frac{1}{1 + (\kappa/k)^2 \, (\kappa a)^2} = \frac{k^2}{k^2 + (\kappa^2 a)^2} = \frac{\epsilon}{\epsilon + (v_0 a)^2} ,$$

which is the result obtained for transmission in a δ-function potential with strength $2aV_0$.

Problem 6   Bound- and Scattering-State Problems in a Potential $v(x) = v_0\,\theta(-x) - w_0\delta(x)$

Consider a particle of mass m under the action of a potential given by

$$v(x) = v_0\theta(-x) - w_0\delta(x) ,$$

where $\theta(z) = 1$ if $z > 0$ and 0 otherwise.

1. Show that the presence of the potential step of height v_0 does not alter the boundary conditions that the wave function and its first derivative must satisfy at $x = 0$. Is there a bound state? If there is, what is its energy?
2. Assume $\epsilon > v_0$. Calculate the reflection and transmission coefficients. Find expressions for these coefficients in the limit $\epsilon \gg v_0, w_0$.
3. Now, assume $0 < \epsilon < v_0$. Calculate the reflection and transmission coefficients in this case. Explain why you could have anticipated the result.

Solution

Part 1

In regions 1 and 2, defined, respectively, as $x < 0$ and $x > 0$, the Schrödinger equation reads

$$\psi_1''(x) = (|\epsilon| + v_0)\psi_1(x) , \qquad \psi_2''(x) = |\epsilon|\psi_2(x) ,$$

where $\epsilon < 0$ for bound states. We define $\kappa_1 = \sqrt{|\epsilon| + v_0}$ and $\kappa_2 = \sqrt{|\epsilon|}$, and obtain

$$\psi_1(x) = \alpha\,e^{\kappa_1 x} , \qquad \psi_2(x) = \beta\,e^{-\kappa_2 x} ,$$

with boundary conditions

$$\psi_1(0) = \psi_2(0) , \qquad \psi_2'(0) - \psi_1'(0) = -w_0\psi_2(0) .$$

We note that the presence of the step potential $v_0\theta(-x)$ does not alter the condition on the first derivative; indeed, integrating the Schrödinger equation over a small interval around the origin,

$$\int_{-\eta}^{\eta} dx\, \psi''(x) = \int_{-\eta}^{\eta} [-w_0\delta(x) + v_0\theta(-x) - \epsilon]\psi(x)\,,$$

yields

$$\psi'(\eta) - \psi'(-\eta) = -w_0\psi(0) + v_0\int_{-\eta}^{0} dx\, \psi(x) - \epsilon\int_{-\eta}^{\eta} dx\, \psi(x)$$

and hence, in the limit $\eta \longrightarrow 0$, the jump discontinuity of the first derivative.

The boundary conditions lead to a homogeneous system in α and β given by

$$\alpha - \beta = 0\,, \qquad \alpha + \left(\frac{\kappa_2}{\kappa_1} - \frac{w_0}{\kappa_1}\right)\beta = 0\,,$$

which has non-trivial solutions if

$$\kappa_1 + \kappa_2 = w_0 \implies \sqrt{|\epsilon| + v_0} + \sqrt{|\epsilon|} = w_0\,.$$

Squaring the above expression twice yields the bound-state energy

$$\epsilon = -\frac{(w_0^2 - v_0)^2}{4w_0^2}\,.$$

In the limit $v_0 = 0$, or $w_0 \gg \sqrt{v_0}$, ϵ reduces to the well-known result for the attractive δ-function potential, as expected.

Part 2

It is assumed that $\epsilon > v_0$. We consider the case of an incident wave approaching from the right (from region 2, having $x > 0$). The solution of the Schrödinger equation is given by

$$\psi_1(x) = C\,e^{-ik'x}\,, \qquad \psi_2(x) = A\,e^{-ikx} + B\,e^{ikx}\,,$$

where

$$k = \sqrt{\epsilon}\,, \qquad k' = \sqrt{\epsilon - v_0} = \sqrt{k^2 - v_0}\,.$$

The boundary conditions at the origin yield

$$A + B = C\,, \qquad -ik(A - B) + ik'C = -w_0 C\,,$$

which, when solved with respect to the incident-wave amplitude, give

$$B = \frac{k - k' + iw_0}{k + k' - iw_0}A\,, \qquad C = \frac{2k}{k + k' - iw_0}A\,.$$

The incident, reflected, and transmitted probability current densities are obtained as

$$j_i = -\frac{\hbar k}{m}A\,, \qquad j_r = \frac{\hbar k}{m}B\,, \qquad j_t = -\frac{\hbar k'}{m}C\,,$$

and the reflection and transmission coefficients follow from

$$R = \frac{|j_r|}{|j_i|} = \frac{(k - k')^2 + w_0^2}{(k + k')^2 + w_0^2}\,, \qquad T = \frac{|j_t|}{|j_i|} = \frac{4kk'}{(k + k')^2 + w_0^2}\,, \qquad R + T = 1\,.$$

In the limit $k \gg \sqrt{v_0}$ and $k \gg w_0$, we have

$$k - k' = k\left(1 - \sqrt{1 - v_0/k^2}\right) \approx \frac{v_0}{2k}, \qquad k + k' = k\left(1 + \sqrt{1 - v_0/k^2}\right) \approx 2k - \frac{v_0}{2k},$$

and so

$$R \approx \frac{w_0^2}{4k^2} \longrightarrow 0, \qquad T \approx \frac{4k^2}{4k^2 + w_0^2} \longrightarrow 1.$$

Part 3

Now, it is assumed that $0 < \epsilon < v_0$, which implies the solutions

$$\psi_1(x) = C e^{\kappa x}, \qquad \psi_2(x) = A e^{-ikx} + B e^{ikx},$$

where

$$\kappa = \sqrt{v_0 - \epsilon} = \sqrt{v_0 - k^2}.$$

The probability current density associated with the transmitted wave vanishes, since (up to the possibly complex constant C) $\psi_1(x)$ is real. Hence, we must have $j_i + j_r = 0$ (by the conservation of j). As a consequence, the reflection coefficient is unity for any energy in this range. The values of B and C follow from those of part 2 by simply making the replacement $-ik' \longrightarrow \kappa$ or $k' \longrightarrow i\kappa$, so that

$$B = \frac{k - i(\kappa - w_0)}{k + i(\kappa - w_0)} A, \qquad C = \frac{2k}{k + i(\kappa - w_0)} A,$$

and it should be noted that B differs from A just by a phase factor, since

$$\frac{k - i(\kappa - w_0)}{k + i(\kappa - w_0)} = e^{-2i\varphi}, \qquad \varphi = \tan^{-1}\left(\frac{\kappa - w_0}{k}\right).$$

Problem 7 Wave Functions for a Potential $v(x < 0) = \infty$ and $v(x > 0) = -v_0 \theta(a - x)$

Consider the one-dimensional potential $v(x)$ given by

$$v(x < 0) = \infty, \qquad v(0 < x < a) = -v_0, \qquad v(x > a) = 0,$$

with $v_0 > 0$.

1. Discuss the energy spectrum: for what energies (if any) are there bound states? For what energies are there continuum states? What is the degeneracy of the energy eigenvalues?
2. Assume the energy ϵ of the particle is less than zero. Determine under what conditions there is at least a bound state.
3. Show that for $x > a$ the positive energy solution (up to an overall constant) can be written as

$$\psi_>(x) = e^{i[k(x-a)-2\delta]} + e^{-ik(x-a)}.$$

Determine δ and the solution $\psi_<(x)$ in $0 < x < a$ (up to an overall constant). Calculate the reflection coefficient.

4. Denote by $\psi_k(x)$ the complete positive-energy solution obtained in part 3 and consider the wave packet

$$\Psi(x, t) = \int_0^\infty dk \, g(k) \, \psi_k(x) \, e^{-i\omega_k t} \qquad \text{with } \omega_k = \frac{\hbar k^2}{2m},$$

where the real function $g(k)$ is sharply peaked at k_0. Examine the incident and reflected wave packets in the region $x > a$, and show that the incident wave packet reaches the edge of the potential well at $x = a$ at time $t = 0$, while the reflected wave packet leaves this edge after a time delay τ. Calculate τ and show that it is given by

$$\tau = \frac{2am}{\hbar \sqrt{\epsilon_0}} \frac{1}{\epsilon_0 + v_0 \cos^2(K_0 a)} \left[\epsilon_0 + v_0 \frac{\sin(2K_0 a)}{2K_0 a} \right] ,$$

where $\epsilon_0 = k_0^2$ and $K_0 = \sqrt{k_0^2 + v_0}$. Discuss the limit $\epsilon_0 \gg v_0$. Under what conditions will τ be very long? Explain.

Solution

Part 1

Bound states are possible for energies $-v_0 < \epsilon < 0$; continuum states occur for $\epsilon > 0$. The whole (discrete and continuous) spectrum is non-degenerate.

Part 2

Let $\psi_<(x)$ and $\psi_>(x)$ be the wave functions in the regions $0 < x < a$ and $x > a$, respectively. They satisfy the Schrödinger equations:

$$\psi_<''(x) = -(v_0 - |\epsilon|)\psi_<(x) , \qquad \psi_>''(x) = |\epsilon|\psi_<(x) ,$$

where for bound states $-v_0 < \epsilon < 0$. Define $k = \sqrt{v_0 - |\epsilon|}$ and $\kappa = \sqrt{|\epsilon|}$, so that

$$\psi_<(x) = A \sin(kx) , \qquad \psi_>(x) = B\, e^{-\kappa x} ,$$

where we have imposed the boundary condition $\psi_<(0) = 0$ and $\psi_>(x \longrightarrow \infty) = 0$. The continuity of $\psi(x)$ and its first derivative at $x = a$ require

$$A \sin(ka) = B\, e^{-\kappa a} , \qquad Ak \cos(ka) = -B\kappa\, e^{-\kappa a} ,$$

and the (possible) eigenvalues ϵ are found from

$$\tan(ka) = -\frac{k}{\kappa} .$$

Set $x = ka$ and hence $\kappa a = \sqrt{x_0^2 - x^2}$ with $x_0^2 = a^2 v_0$, and the above equation reduces to

$$\tan x = -\frac{x}{\sqrt{x_0^2 - x^2}} \qquad 0 < x < x_0 .$$

A rough plot of the left- and right-hand sides of this relation shows that for at least a single bound state to occur we must have $x_0 > \pi/2$. As a matter of fact, if x_0 exceeds $(2N - 1)\pi/2$ with N integer there are N bound states. Having determined the solution $x_n^\star$ for the above equation with $1 \le n \le N$, the bound-state energy results from

$$ka = \sqrt{v_0 - |\epsilon|}\, a = x_n^\star \implies \epsilon_n = -v_0 + (x_n^\star/a)^2 .$$

Part 3

For energies $\epsilon > 0$ (the continuum solutions), the Schrödinger equations read

$$\psi''_<(x) = -(\epsilon + v_0)\psi_<(x) , \qquad \psi''_>(x) = -\epsilon\psi_<(x) .$$

They have the solutions

$$\psi_<(x) = A \sin(Kx) , \qquad \psi_>(x) = B\,e^{ik(x-a)} + C\,e^{-ik(x-a)} ,$$

where

$$K = \sqrt{\epsilon + v_0} , \qquad k = \sqrt{\epsilon} ,$$

and the requirement $\psi_<(0) = 0$ has already been imposed. The boundary conditions at a give

$$A \sin(Ka) = B + C , \qquad AK \cos(Ka) = ik(B - C) ,$$

which yield

$$\frac{B}{A} = \frac{1}{2}\left[\sin(Ka) - i\frac{K}{k}\cos(Ka)\right] , \qquad \frac{C}{A} = \frac{1}{2}\left[\sin(Ka) + i\frac{K}{k}\cos(Ka)\right] ,$$

or, in terms of the ratios A/C and B/C,

$$\frac{A}{C} = \frac{2}{\sin(Ka) + i(K/k)\cos(Ka)} , \qquad \frac{B}{C} = \frac{B}{A}\frac{A}{C} = \frac{\sin(Ka) - i(K/k)\cos(Ka)}{\sin(Ka) + i(K/k)\cos(Ka)} .$$

Note that the ratio B/C is just a phase factor, $B/C = e^{-2i\delta}$, and

$$\tan\delta = \frac{K}{k}\cot(Ka) .$$

Up to the overall constant C the solution is obtained as

$$\psi_<(x) = \frac{2\,e^{-i\delta}}{\sqrt{\sin^2(Ka) + (K/k)^2\cos^2(Ka)}}\sin(Kx) , \qquad \psi_>(x) = e^{i[k(x-a)-2\delta]} + e^{-ik(x-a)} .$$

The incident and reflected probability currents in the region $x > 0$ are given by

$$j_i = -\frac{\hbar k}{m}|C|^2 , \qquad j_r = \frac{\hbar k}{m}|B|^2 ,$$

yielding a reflection coefficient

$$R = \frac{|j_r|}{|j_i|} = \frac{|B|^2}{|C|^2} = |e^{-2i\delta}|^2 = 1 .$$

Part 4

The incident and reflected wave packets ($x > a$) are given by

$$\Psi_i(x, t) = \int_0^\infty dk\, g(k)\, e^{-ik(x-a)}\, e^{-i\omega_k t} , \qquad \Psi_r(x, t) = \int_0^\infty dk\, g(k)\, e^{i[k(x-a)-2\delta]}\, e^{-i\omega_k t} .$$

Using the stationary phase method, the center of the incident wave packet moves according to

$$\varphi(k) = -k(x - a) - \omega_k t \quad\text{and}\quad \varphi'(k_0) = 0 \implies x_i(t) = a - \frac{\hbar k_0}{m}t ,$$

and so for $t \ll 0$ it is at $x \gg a$, reaching a at $t=0$. The center of the reflected wave packet moves according to

$$\varphi(k) = k(x - a) - 2\delta(k) - \omega_k t \quad \text{and} \quad \varphi'(k_0) = 0 \implies x_r(t) = a + \frac{\hbar k_0}{m} t + 2\delta'(k_0) \,,$$

which can be expressed as

$$x_r(t) = a + \frac{\hbar k_0}{m}(t - \tau) \,, \qquad \tau = -\frac{2m}{\hbar k_0} \delta'(k_0) \,.$$

Recalling that

$$\delta(k) = \tan^{-1}\left[\frac{K}{k} \cot(Ka)\right] \,,$$

and using

$$\frac{dK}{dk} = \frac{d}{dk}\sqrt{k^2 + v_0} = \frac{k}{K} \,, \qquad \frac{d\cot(Ka)}{dk} = -\frac{1}{\sin^2(Ka)} \frac{ka}{K} \,,$$

we find

$$\delta'(k) = \frac{1}{1 + (K/k)^2 \cot^2(Ka)}\left[-\frac{K}{k^2}\cot(Ka) + \frac{1}{K}\cot(Ka) - \frac{a}{\sin^2(Ka)}\right]$$

$$= -\frac{1}{1 + (K/k)^2 \cot^2(Ka)}\left[\left(\frac{K^2}{k^2} - 1\right)\frac{\cot(Ka)}{K} + \frac{a}{\sin^2(Ka)}\right] \,,$$

which can also be expressed as (substituting $\cot x = \cos x/\sin x$ and $\sin x \cos x = \sin(2x)/2$)

$$\tau = \frac{2am}{\hbar k_0} \frac{1}{k_0^2 \sin^2(K_0 a) + K_0^2 \cos^2(K_0 a)}\left[k_0^2 + \left(K_0^2 - k_0^2\right)\frac{\sin(2K_0 a)}{2K_0 a}\right] \,.$$

Lastly, since $K_0^2 = k_0^2 + v_0$ and $k_0 = \sqrt{\epsilon_0}$, this can also be written as

$$\tau = \frac{2am}{\hbar\sqrt{\epsilon_0}} \frac{1}{\epsilon_0 + v_0 \cos^2(K_0 a)}\left[\epsilon_0 + v_0\frac{\sin(2K_0 a)}{2K_0 a}\right] \,.$$

In the limit $\epsilon_0 \gg v_0$, we have $K_0 \approx \sqrt{\epsilon_0}$ and

$$\frac{1}{1 + (v_0/\epsilon_0) \cos^2(\sqrt{\epsilon_0}\, a)}\left[1 + \frac{v_0}{\epsilon_0}\frac{\sin(2\sqrt{\epsilon_0}\, a)}{2\sqrt{\epsilon_0}\, a}\right] \approx 1 \,,$$

and thus the time delay vanishes as

$$\tau \approx \frac{2am}{\hbar\sqrt{\epsilon_0}} \qquad \text{for} \quad \epsilon_0 \gg v_0 \,.$$

The reflected wave packet starts off without delay to move back towards $x \longrightarrow \infty$.

It is also interesting to examine the case in which

$$K_0 a = \frac{2n - 1}{2}\pi \implies \epsilon_n \equiv \left[\frac{(2n - 1)\pi}{2a}\right]^2 - v_0 \,,$$

and, since ϵ_n must be larger than zero, there is a lower bound $n^\star$, where $n^\star$ is defined by the condition

$$\left[\frac{(2n^\star - 1)\pi}{2a}\right]^2 > v_0 \,,$$

on the allowed n-values. At these resonant energies ϵ_n, $K_0 a$ is an odd multiple of $\pi/2$, and the time delay reduces to

$$\tau = \frac{2am}{\hbar\sqrt{\epsilon_n}} \qquad \text{for} \quad \epsilon_0 = \epsilon_n .$$

If the depth v_0 is very close to (but slightly less than) $(2n^\star - 1)^2 \pi^2/(4a^2)$, then $\epsilon_{n^\star}$ will be very small, and the associated $\tau^\star$ will be very long, that is, the reflected wave packet will take a long time before starting off to move towards $x \longrightarrow \infty$. It is as if the particle hangs around the region of the potential for a long time. As n increases (to $n^\star + 1$, $n^\star + 2$, ...), the time delay will be progressively reduced, but it will still be relatively fairly long if $n^\star \gg 1$, namely if the potential is very deep.

Problem 8   Transmission and Reflection Coefficients in WKB Approximation with Application

This problem deals with the application of the WKB method (developed in Problem 12 of Chapter 4) to the calculation of the transmission probability through a generic potential $v(x)$. The only assumptions made on $v(x)$ is that it vanishes as $|x| \longrightarrow \infty$ and that it has a (single) maximum v_0 at x_0. Continuum solutions of the Schrödinger equation exist for $\epsilon > 0$. Hereafter we restrict the energy ϵ to the range $0 < \epsilon < v_0$. For ϵ in this range, there are two classical inversion points, a and b with $a < b$. Write the WKB solutions as follows:

$$\psi(x \ll a) \approx \underbrace{\frac{B_+}{\sqrt{k(x)}} \exp\left[-i \int_x^a dx'\, k(x')\right]}_{\psi_i(x)} + \underbrace{\frac{B_-}{\sqrt{k(x)}} \exp\left[i \int_x^a dx'\, k(x')\right]}_{\psi_r(x)} ,$$

$$\psi(a \ll x \ll b) \approx \frac{A_-}{\sqrt{\kappa(x)}} \exp\left[-\int_x^b dx'\, \kappa(x')\right] + \frac{A_+}{\sqrt{\kappa(x)}} \exp\left[\int_x^b dx'\, \kappa(x')\right] ,$$

$$\psi(x \gg b) \approx \underbrace{\frac{B'_-}{\sqrt{k(x)}} \exp\left[i \int_b^x dx'\, k(x')\right]}_{\psi_t(x)} ,$$

where $A_\pm$, $B_\pm$, and B'_- are constants, and

$$k(x) = \sqrt{\epsilon - v(x)} \qquad \text{for } \epsilon > v(x) , \qquad\qquad \kappa(x) = \sqrt{v(x) - \epsilon} \qquad \text{for } \epsilon < v(x) .$$

1. Provide a simple argument that justifies the identification of incident, reflected, and transmitted wave functions made above. Show that the probability current densities associated with these wave functions are given by

$$j_i = \frac{\hbar}{m} |B_+|^2 , \qquad j_r = -\frac{\hbar}{m} |B_-|^2 , \qquad j_t = \frac{\hbar}{m} |B'_-|^2 ,$$

and that the transmission probability is obtained as

$$T = \frac{|B'_-|^2}{|B_+|^2} .$$

Hint: Recall that, for the validity of the WKB approximation, $|k'(x)| \ll k^2(x)$.

2. Using the connection formulae derived in Problem 12 of Chapter 4, show that

$$B_+ = \left(\alpha + \frac{1}{4\alpha}\right) B'_- , \qquad B_- = -i\left(\alpha - \frac{1}{4\alpha}\right) B'_- ,$$

where

$$\alpha = \exp\left[\int_a^b dx\,\kappa(x)\right].$$

Obtain the transmission probability T under the assumption that $1/\alpha \ll 1$. What is the reflection probability under the same assumption?

3. Evaluate T for the potential given by

$$v(x) = v_0\left(1 - \frac{|x|}{a}\right) \quad \text{for } |x| \le a; \qquad v(x) = 0 \quad \text{for } |x| > a,$$

where $v_0 \gg 0$ and for energies $\epsilon < v_0$.

Solution

Part 1

In the limit of $k(x)=k$, a constant, the asymptotic wave functions in the regions $x \ll a$ and $x \gg b$ are given by

$$\psi(x \ll a) \approx \frac{B_+}{\sqrt{k}}\,e^{ik(x-a)} + \frac{B_-}{\sqrt{k}}\,e^{-ik(x-a)} = \underbrace{\widetilde{B}_+\,e^{ikx}}_{\psi_i(x)} + \underbrace{\widetilde{B}_-\,e^{-ikx}}_{\psi_r(x)},$$

$$\psi(x \gg b) \approx \frac{B'_-}{\sqrt{k}}\,e^{ik(x-b)} = \underbrace{\widetilde{B}'_-\,e^{ikx}}_{\psi_t(x)},$$

which represent incident, reflected, and transmitted waves, respectively. In order to compute the probability current associated with the WKB solution, we first note that, say for $\psi_i(x)$,

$$\psi'_i(x) = -\frac{B_+}{2}\,k^{-3/2}(x)\,k'(x)\,e^{i\int_a^x dx'k(x')} + iB_+k^{1/2}(x)\,e^{i\int_a^x dx'k(x')}$$

$$= iB_+\,k^{1/2}(x)\,e^{i\int_a^x dx'k(x')}\left[1 + i\frac{k'(x)}{2k^2(x)}\right] \approx iB_+k^{1/2}(x)\,e^{i\int_a^x dx'k(x')},$$

where in the last step we have ignored $k'(x)/k^2(x) \ll 1$. It follows that

$$j_i = \frac{\hbar}{2mi}\left[\psi_i^*(x)\,\psi'_i(x) - \text{c.c.}\right] = \frac{\hbar}{2mi}\left[B_+^*\,k^{-1/2}(x)\,e^{-i\int_a^x dx'k(x')}\,iB_+k^{1/2}(x)\,e^{i\int_a^x dx'k(x')} - \text{c.c.}\right]$$

$$= \frac{\hbar}{2mi}\left(i|B_+|^2 - \text{c.c.}\right) = \frac{\hbar}{m}|B_+|^2.$$

Proceeding in a similar fashion, we find that

$$j_r = -\frac{\hbar}{m}|B_-|^2, \qquad j_t = \frac{\hbar}{m}|B'_-|^2,$$

from which we deduce that the transmission coefficient is given by

$$T = \frac{|j_t|}{|j_i|} = \frac{|B'_-|^2}{|B_+|^2}.$$

Part 2

The first step consists in joining the WKB solution for $x > b$ to that for $x < b$. We use the connection formulae obtained in Problem 12 of Chapter 4,

$$A_- = \frac{1}{2}\left(e^{-i\pi/4}B_+ + e^{i\pi/4}B_-\right), \qquad A_+ = i\left(e^{-i\pi/4}B_+ - e^{i\pi/4}B_-\right),$$

where in the present case $B_+ = 0$ and $B_- = B'_-$, yielding (note that $-i\,e^{i\pi/4} = e^{-i\pi/4}$)

$$A_- = \frac{e^{i\pi/4}}{2}B'_-, \qquad A_+ = e^{-i\pi/4}B'_-.$$

Using

$$\int_x^b dx'\,\kappa(x') = \int_a^b dx\,\kappa(x) - \int_a^x dx'\,\kappa(x'),$$

the WKB solution in the classically forbidden region can be written as

$$\psi(a \ll x \ll b) = A_-\,e^{-\int_a^b dx\,\kappa(x)}\frac{e^{\int_a^x dx'\,\kappa(x')}}{\sqrt{\kappa(x)}} + A_+\,e^{\int_a^b dx\,\kappa(x)}\frac{e^{-\int_a^x dx'\,\kappa(x')}}{\sqrt{\kappa(x)}}$$

$$= \frac{\widetilde{A}_+}{\sqrt{\kappa(x)}}\,e^{\int_a^x dx'\,\kappa(x')} + \frac{\widetilde{A}_-}{\sqrt{\kappa(x)}}\,e^{-\int_a^x dx'\,\kappa(x')},$$

where

$$\widetilde{A}_+ \equiv A_-\,e^{-\int_a^b dx\,\kappa(x)} = \frac{e^{i\pi/4}}{2}B'_-\,e^{-\int_a^b dx\,\kappa(x)},$$

$$\widetilde{A}_- \equiv A_+\,e^{\int_a^b dx\,\kappa(x)} = e^{-i\pi/4}B'_-\,e^{\int_a^b dx\,\kappa(x)}.$$

The WKB solution for $a < x < b$ can now be joined to that for $x < a$ by using the connection formulae

$$B_- = e^{-i\pi/4}\left(\widetilde{A}_- + \frac{i}{2}\widetilde{A}_+\right), \qquad B_+ = e^{i\pi/4}\left(\widetilde{A}_- - \frac{i}{2}\widetilde{A}_+\right).$$

We find that

$$B_- = e^{-i\pi/4}\left[e^{-i\pi/4}B'_-\,e^{\int_a^b dx\,\kappa(x)} + \frac{i}{2}\frac{e^{i\pi/4}}{2}B'_-\,e^{-\int_a^b dx\,\kappa(x)}\right] = -i\left(\alpha - \frac{1}{4\alpha}\right)B'_-$$

$$B_+ = e^{i\pi/4}\left[e^{-i\pi/4}B'_-\,e^{\int_a^b dx\,\kappa(x)} - \frac{i}{2}\frac{e^{i\pi/4}}{2}B'_-\,e^{-\int_a^b dx\,\kappa(x)}\right] = \left(\alpha + \frac{1}{4\alpha}\right)B'_-,$$

where

$$\alpha = e^{\int_a^b dx\,\kappa(x)}.$$

We obtain the transmission coefficient as

$$T = \frac{|B'_-|^2}{|B_+|^2} = \frac{1}{\alpha^2}\frac{1}{[1 + 1/(2\alpha)^2]^2} \approx \frac{1}{\alpha^2},$$

for $1/\alpha \ll 1$, and hence

$$T = \exp\left[-2\int_a^b dx\,\kappa(x)\right].$$

Under the same approximation, the reflection coefficient follows:

$$R = \frac{|j_r|}{|j_i|} = \frac{|B_-|^2}{|B_+|^2} = \left[\frac{\alpha - 1/(4\alpha)}{\alpha + 1/(4\alpha)}\right]^2 = \left[\frac{1 - 1/(2\alpha)^2}{1 + 1/(2\alpha)^2}\right]^2 \approx \left[1 - \frac{1}{(2\alpha)^2}\right]^4 \approx 1 - \frac{1}{\alpha^2},$$

or $R = 1 - T$.

Part 3

For a given energy ϵ in $0 < \epsilon < v_0$, the inversion points are given by

$$\epsilon = v(x) \implies b_\pm = \pm b_0 = \pm a\left(1 - \frac{\epsilon}{v_0}\right),$$

and the transmission coefficient is obtained as

$$T = \exp\left[-2\int_{-b_0}^{b_0} dx\,\sqrt{v_0(1 - |x|/a) - \epsilon}\right] = \exp\left[-4\sqrt{\frac{v_0}{a}}\int_0^{b_0} dx\,\sqrt{b_0 - x}\right]$$

$$= \exp\left[-\frac{8}{3}\sqrt{\frac{v_0}{a}}\,b_0^{3/2}\right] = \exp\left[-\frac{8}{3}\sqrt{v_0 a^2}\left(1 - \frac{\epsilon}{v_0}\right)^{3/2}\right].$$

Problem 9 Scattering and Resonances in a Potential $v(x) = \infty$ for $x < 0$ and $v_0\,\delta(x - a)$ for $x > 0$

Consider a particle of mass m confined in the region $x \geq 0$ and subject to a repulsive δ-function potential located at $x = a$, that is,

$$v(x) = v_0\,\delta(x - a) \qquad \text{for } x \geq 0.$$

1. Without doing any detailed calculations, explain why the reflection coefficient is unity in this case for any $\epsilon > 0$.
2. Obtain the complete solution $\psi_k(x)$ with $k = \sqrt{\epsilon}$ in the whole allowed region $x \geq 0$, thus justifying the above inference.
3. Construct the wave packet

$$\Psi(x, t) = \int_0^\infty \frac{dk}{\sqrt{2\pi}}\,g(k)\,\psi_k(x)\,e^{-i\omega_k t} \qquad \omega_k = \frac{\hbar k^2}{2m},$$

    where the profile function $g(k)$ is assumed to be real and strongly peaked at $k^\star > 0$. Show that the center of the incident wave packet reaches a at time $t = 0$ and that the center of the reflected wave packet moves from a back to infinity with time delay 2τ. Determine τ.
4. Show that the magnitude squared of the wave packet in the region $0 < x < a$ is suppressed by a factor $1/\rho_k^2$. Provide a plot of $1/\rho_k^2$ for $x_0 = v_0 a = 50$ as a function of $x = ka$ for x in the range $0 < x \leq 20$. Comment on the plot.
5. Compute the time delay τ and provide a plot of τ/τ_0, where $\tau_0 = ma^2/\hbar$, for the case where $x_0 = 50$ and x is in the range 0–20, as above. Comment on the plot.

Solution

Part 1

Continuum non-degenerate energies exist for any $\epsilon > 0$. The solutions of the Schrödinger equation in the regions $0 < x < a$ and $x > a$ are

$$\psi_<(x) = A \sin(kx) , \qquad \psi_>(x) = B\, e^{-ik(x-a)} + C\, e^{ik(x-a)} ,$$

with $k = \sqrt{\epsilon}$. The boundary condition at the origin has already been imposed. The probability current density vanishes in the region $0 < x < a$, since the wave functions is real (up the possibly complex constant A). In the region $x > a$ it is given by

$$j = -\frac{i\hbar}{2m} \left[\psi_>^*(x)\, \psi_>'(x) - \text{c.c.} \right] = \underbrace{-\frac{\hbar k}{m} |B|^2}_{j_i} + \underbrace{\frac{\hbar k}{m} |C|^2}_{j_r},$$

where j_i is the probability current density associated with the incident wave (from right to left, hence the minus sign), and j_r is the analogous quantity associated with the reflected wave propagating from left to right. Since j is conserved, we must have $j_i + j_r = 0$. Thus, there is no transmitted wave and the reflection coefficient R must be unity for any energy $\epsilon > 0$,

$$R = \frac{|j_r|}{|j_i|} = \frac{|C|^2}{|B|^2} = 1 ,$$

implying that C must differ from B by at most a phase factor.

Part 2

We now impose the boundary conditions at $x = a$, where the δ-function potential is located, to find

$$\underbrace{A \sin(ka)}_{\psi_<(a)} = \underbrace{B + C}_{\psi_>(a)} , \qquad \underbrace{ik(C - B)}_{\psi_>'(a)} - \underbrace{Ak \cos(ka)}_{\psi_<'(a)} = v_0 \underbrace{(B + C)}_{\psi_>(a)} .$$

We solve the linear system with respect to A and C in terms of B (associated with the incident wave). This linear system has the form

$$A \sin(ka) - C = B , \qquad A \cos(ka) + \left(\frac{v_0}{k} - i \right) C = -\left(\frac{v_0}{k} + i \right) B$$

with solutions given by (using Cramer's rule)

$$A = \frac{\begin{vmatrix} B & -1 \\ -(v_0/k + i)B & v_0/k - i \end{vmatrix}}{\begin{vmatrix} \sin(ka) & -1 \\ \cos(ka) & v_0/k - i \end{vmatrix}} , \qquad C = \frac{\begin{vmatrix} \sin(ka) & B \\ \cos(ka) & -(v_0/k + i)B \end{vmatrix}}{\begin{vmatrix} \sin(ka) & -1 \\ \cos(ka) & v_0/k - i \end{vmatrix}} .$$

We obtain

$$A = -\underbrace{\frac{2i}{(v_0/k - i) \sin(ka) + \cos(ka)}}_{z_A} B , \qquad C = -\underbrace{\frac{(v_0/k + i) \sin(ka) + \cos(ka)}{(v_0/k - i) \sin(ka) + \cos(ka)}}_{z_C} B ,$$

where the complex numbers z_A and z_C are as follows:

$$z_A = -\frac{2i}{\rho} e^{-i\varphi} , \qquad z_C = -e^{-2i\varphi} ,$$

with

$$\rho = \sqrt{1 + (v_0/k) \sin(2ka) + (v_0/k)^2 \sin^2(ka)} , \qquad \varphi = -\tan^{-1}\left[\frac{k}{v_0 + k \cot(ka)}\right] ,$$

and the complex number z_C has magnitude $|z_C| = 1$. It follows that $|C| = |B|$, and hence $R = 1$, as must be the case.

Part 3

The complete continuum wave function for $\epsilon > 0$ follows:

$$\begin{aligned}
\psi_k(x) &= A_k \sin(kx) & 0 < x < a \\
&= e^{-ik(x-a)} + C_k e^{ik(x-a)} & x > a ,
\end{aligned}$$

where the coefficients A_k and C_k are given by the relations in part 2; the coefficient B_k has been set to unity, since it is irrelevant (see below). As a check, we consider the limit $v_0 = 0$, namely the situation when there is no δ-function potential and the particle is free in the region $x > 0$. We have, in this limit,

$$A_k = -2i\, e^{ika} , \qquad C_k = -e^{2ika}$$

and so

$$\begin{aligned}
\psi_k(x) &= -2i\, e^{ika} \sin(kx) = e^{-ik(x-a)} - e^{ik(x+a)} & 0 < x < a \\
&= e^{-ik(x-a)} - e^{ik(x+a)} & x > a ,
\end{aligned}$$

that is, a free-particle solution for $x > 0$ satisfying the boundary condition $\psi_k(0) = 0$.

We now form the wave packet

$$\Psi(x, t) = \int_0^\infty \frac{dk}{\sqrt{2\pi}} g(k)\, \psi_k(x)\, e^{-i\omega_k t} \qquad \omega_k = \frac{\hbar k^2}{2m} ,$$

and the profile function $g(k)$, which is assumed for simplicity to be real, is strongly peaked at $k^\star > 0$. We have

$$\begin{aligned}
\Psi(x, t) &= \int_0^\infty \frac{dk}{\sqrt{2\pi}} g(k)\, \frac{e^{-i(kx + \varphi_k + \omega_k t)}}{\rho_k} - \int_0^\infty \frac{dk}{\sqrt{2\pi}} g(k)\, \frac{e^{i(kx - \varphi_k - \omega_k t)}}{\rho_k} & x < a \\
&= \underbrace{\int_0^\infty \frac{dk}{\sqrt{2\pi}} g(k)\, e^{-i[k(x-a) + \omega_k t]}}_{\Psi_i(x,t)} - \underbrace{\int_0^\infty \frac{dk}{\sqrt{2\pi}} g(k)\, e^{i[k(x-a) - 2\varphi_k - \omega_k t]}}_{\Psi_r(x,t)} & x > a ,
\end{aligned}$$

where we have identified the incident and reflected wave packets and have substituted for A_k and C_k the expressions $-2i\, e^{-i\varphi_k}/\rho_k$ and $e^{-2i\varphi_k}$, respectively. We find that the center of the incident wave packet for $x > a$ moves according to

$$x_i(t) = a - v^\star t \qquad v^\star = \frac{\hbar k^\star}{m} ,$$

and so it reaches a at time $t = 0$; it is approximately given by

$$\Psi_i(x, t) \approx e^{-i[k^\star(x-a)+\omega^\star t]} \int_0^\infty \frac{dk}{\sqrt{2\pi}} \, g(k) \, e^{i(k-k^*)[x_i(t)-x]} \qquad x > a \,,$$

where we have used the stationary phase method and expanded the the phase factor $e^{i\eta}$, with $\eta = -k(x-a) - \omega_k t$, about $k^\star$ in linear terms $k - k^\star$ and have defined $\omega^\star = \omega_{k^\star}$. Similarly, we find for $x > a$

$$\Psi_r(x, t) \approx -e^{i[k^\star(x-a)-2\,\varphi^\star-\omega^\star t]} \int_0^\infty \frac{dk}{\sqrt{2\pi}} \, g(k) \, e^{i(k-k^*)[x-x_r(t)]} \qquad x > a \,,$$

where

$$x_r(t) = a + 2 \left.\frac{d\varphi_k}{dk}\right|_{k^\star} + v^\star t = a + v^\star(t - 2\,\tau) \,,$$

and τ is the time delay, defined as

$$\tau = -\frac{1}{v^\star} \left.\frac{d\varphi_k}{dk}\right|_{k^\star} \,.$$

The reflected wave packet will start on its way back from a to ∞ with a delay 2τ.

Part 4

It is also interesting to examine the region $x < a$, where the incident and reflected wave packets are now, respectively,

$$\Psi_i(x, t) \approx \frac{e^{-i(k^\star x+\varphi^\star+\omega^\star t)}}{\rho^\star} \int_0^\infty \frac{dk}{\sqrt{2\pi}} \, g(k) \, e^{i(k-k^*)[x_i'(t)-x]} \qquad x < a \,,$$

and

$$\Psi_r(x, t) \approx -\frac{e^{i(k^\star x-\varphi^\star-\omega^\star t)}}{\rho^\star} \int_0^\infty \frac{dk}{\sqrt{2\pi}} \, g(k) \, e^{i(k-k^*)[x-x_r'(t)]} \qquad x < a \,,$$

where

$$x_i'(t) = -v^\star(t - \tau) \,, \qquad x_r'(t) = v^\star(t - \tau) \,.$$

Thus, the incident wave packet reaches a (traveling from ∞) at time $t = 0$ and takes a further time τ to reach the origin (where the infinite barrier is located); the reflected wave packet starts off at time $t = \tau$ to travel from the origin to the edge a, which it reaches at time 2τ, and then moves back to ∞. However, the amplitudes of the incident and reflected wave packets in the region $x < a$ are suppressed by the factor $1/\rho^\star$, where

$$\rho^{\star 2} = 1 + \frac{x_0}{x} \sin(2x) + \frac{x_0^2}{x^2} \sin^2 x \qquad \text{with } x_0 = av_0 \,, \qquad x = ak^\star \,.$$

The factor $1/\rho^{\star 2}$ is an oscillating function of x (that is, of the energy $\epsilon = k^2$). However, for $x_0 \gg 1$ this factor is very small everywhere, except for sharp peaks occurring at certain energies. At these energies, the magnitude of the wave packet is relatively large in $x < a$, and the particle has a non-vanishing probability of being in this region; see Fig. 5.1. The states corresponding to these energies are known as resonances.

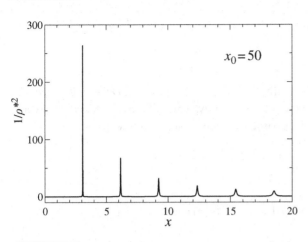 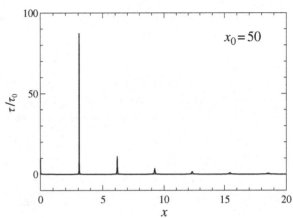

Fig. 5.1 Left-hand panel: The suppression factor $1/\rho^{\star 2}$ as a function of $x = ka$. Right-hand panel: The time delay τ in units $\tau_0 = ma^2/\hbar$.

Part 5

We now investigate the time delay obtained above as a function of energy. We find

$$\frac{d\varphi_k}{dk} = -\frac{1}{1 + k^2/[v_0 + k\cot(ka)]^2}\frac{v_0 + k\cot(ka) - k[\cot(ka) - ka/\sin^2(ka)]}{[v_0 + k\cot(ka)]^2},$$

which can also written as

$$\left.\frac{d\varphi_k}{dk}\right|_{k^\star} = -a\frac{1 + (x_0/x^2)\sin^2 x}{1 + (x_0/x)^2\sin^2 x + (x_0/x)\sin(2x)},$$

where the denominator is just $\rho^{\star 2}$. The time delay is given by

$$\frac{\tau}{\tau_0} = \frac{1}{x}\frac{1 + (x_0/x^2)\sin^2 x}{1 + (x_0/x)^2\sin^2 x + (x_0/x)\sin(2x)} \qquad \tau_0 = \frac{ma^2}{\hbar},$$

and we see in Fig. 5.1 that (again, in the limit $x_0 \gg 1$) it is very large at the resonant energies, in consistency with the notion that the particle "hangs around" the region $x < a$ for a relatively long time before traveling back to ∞.

Problem 10 Scattering in a Potential with an Infinite Barrier at $x = 0$ and a Finite Barrier for $a < x < b$

Consider a particle in a square well potential consisting of an infinite barrier at $x = 0$ and

$$v(x) = -v_0\,\theta(a - x) + w_0\left[\theta(x - a) - \theta(x - b)\right] \qquad x > 0,$$

with $v_0, w_0 > 0$ and $b > a$. In the following, we will be interested only in the continuum solutions corresponding to energy ϵ with $w_0 > \epsilon > 0$.

1. Introduce the wave numbers

$$k_1 = \sqrt{v_0 + \epsilon}, \qquad \kappa_2 = \sqrt{w_0 - \epsilon}, \qquad k_3 = \sqrt{\epsilon},$$

in regions 1, 2, and 3, corresponding, respectively, to $0 < x < a$, $a < x < b$, and $x > b$, and write the wave function in region 3 as

$$\psi_3(x) = e^{-ik_3 x} + D\,e^{ik_3 x} \qquad x > b \,.$$

By using the local conservation of the probability current density, show that D has unit magnitude and conclude that the reflection coefficient R is equal to unity, that is, there is no transmitted wave at $x = 0$, the position of the infinite barrier. In particular, show that D can be expressed as

$$D = \frac{\alpha + i\beta}{-\alpha + i\beta}\, e^{-2ik_3 b} \equiv e^{-2i\varphi}\, e^{-2ik_3 b}\,,$$

where

$$\alpha = \sin(k_1 a)\sinh[\kappa_2(b-a)] + \frac{k_1}{\kappa_2}\cos(k_1 a)\cosh[\kappa_2(b-a)]$$

and

$$\beta = \frac{k_3}{\kappa_2}\left[\sin(k_1 a)\cosh[\kappa_2(b-a)] + \frac{k_1}{\kappa_2}\cos(k_1 a)\sinh[\kappa_2(b-a)]\right]\,.$$

2. Hereafter, set $k = k_3$ and form the wave packet

$$\Psi(x,t) = \int_0^{w_0} \frac{dk}{\sqrt{2\pi}}\, g(k)\,\psi_k(x)\, e^{-i\omega_k t}\,, \qquad\qquad \omega_k = \frac{\hbar k^2}{2m}\,,$$

where

$$\psi_k(x) = e^{ikb}\,\psi(x)\,,$$

and $\psi(x)$ is the complete solution determined in part 1, and the (irrelevant) phase factor in $\psi_k(x)$ is introduced for convenience. The profile function $g(k)$ is taken to be real and strongly peaked at $k^\star$ with $0 < k^\star < \sqrt{w_0}$. Show that the incident wave packet reaches the "edge" b at time $t = 0$, but the reflected wave packet starts traveling from b to ∞ only after a time delay τ. Show that this time delay is given by

$$\tau = -\frac{2m}{\hbar k^\star}\frac{d\varphi(k)}{dk}\bigg|_{k^\star}\,,$$

where

$$\varphi(k) = -\tan^{-1}\left[\frac{k}{\kappa_2}\frac{\sin(k_1 a)\cosh[\kappa_2(b-a)] + (k_1/\kappa_2)\cos(k_1 a)\sinh[\kappa_2(b-a)]}{\sin(k_1 a)\sinh[\kappa_2(b-a)] + (k_1/\kappa_2)\cos(k_1 a)\cosh[\kappa_2(b-a)]}\right]\,.$$

3. Introduce the variable $x = k_1 a$, and show that the time delay can be written as

$$\tau = -\frac{2ma^2}{\hbar x^\star}\frac{d\varphi(x)}{dx}\bigg|_{x^\star}\,,$$

where

$$\varphi(x) = -\tan^{-1}\left[\sqrt{\frac{x^2 - x_0^2}{x_1^2 - x^2}}\,\frac{\tan x + \left(x/\sqrt{x_1^2 - x^2}\right)\tanh(\Delta\sqrt{x_1^2 - x^2})}{\tan x\,\tanh(\Delta\sqrt{x_1^2 - x^2}) + x/\sqrt{x_1^2 - x^2}}\right]\,,$$

with $x_0 = a\sqrt{v_0}$, $x_1 = a\sqrt{v_0 + w_0}$, and $\Delta = b/a - 1$. Here, x is in the interval $x_0 < x < x_1$. Show graphically that $\varphi(x)$ is a decreasing function of x in this interval and hence that $\tau > 0$. Draw the four cases corresponding to the following sets of potential parameters, all having $\Delta = 0.5$: (i) $x_0 = 0.5$ and $x_1 = 1.5$; (ii) $x_0 = 0.5$ and $x_1 = 3.0$; (iii) $x_0 = 2.5$ and $x_1 = 4.0$; and (iv) $x_0 = 2.5$ and $x_1 = 6.0$. Comment on the figures.

Solution

Part 1

We consider only the case $0 < \epsilon < w_0$. Continuum non-degenerate energies exist in this range. The solutions of the Schrödinger equation in regions 1 and 2 are

$$\psi_1(x) = A \sin(k_1 x), \qquad \psi_2(x) = B\, e^{\kappa_2 x} + C\, e^{-\kappa_2 x},$$

with

$$k_1 = \sqrt{v_0 + \epsilon}, \qquad \kappa_2 = \sqrt{w_0 - \epsilon}.$$

In region 3 we have

$$\psi_3(x) = D\, e^{ik_3 x} + E\, e^{-ik_3 x}, \qquad k_3 = \sqrt{\epsilon}.$$

The solutions are not normalizable. The probability current density vanishes in regions 1 and 2, since the wave functions are real (up the possibly complex constants A, B, and C). In region 3, it is given by

$$j = -\frac{i\hbar}{2m} \left[\psi_3^*(x)\, \psi_3'(x) - \text{c.c.} \right] = \underbrace{\frac{\hbar k_3}{m} |D|^2}_{j_r} - \underbrace{\frac{\hbar k_3}{m} |E|^2}_{j_i},$$

where j_i is the probability current density associated with the incident wave (from right to left, hence the minus sign), and j_r is the analogous quantity associated with the reflected wave propagating from left to right. Since j is conserved, we must have $j_i + j_r = 0$. Thus, there is no transmitted wave, and the reflection coefficient R must be unity for any energy in this range (as a matter of fact, for any $\epsilon > 0$),

$$R = \frac{|j_r|}{|j_i|} = \frac{|D|^2}{|E|^2} = 1,$$

which says that D must differ from E only by a phase factor. We now impose the boundary conditions at $x = a$ and $x = b$ and solve for D as function of E. The boundary conditions at $x = a$ yield, as before,

$$B = A\, \frac{e^{-\kappa_2 a}}{2} \left[\sin(k_1 a) + \frac{k_1}{\kappa_2} \cos(k_1 a) \right], \qquad C = A\, \frac{e^{\kappa_2 a}}{2} \left[\sin(k_1 a) - \frac{k_1}{\kappa_2} \cos(k_1 a) \right].$$

The boundary conditions at $x = b$ give

$$B\, e^{\kappa_2 b} + C\, e^{-\kappa_2 b} = D\, e^{ik_3 b} + E\, e^{-ik_3 b}, \qquad B\, e^{\kappa_2 b} - C\, e^{-\kappa_2 b} = i\,\frac{k_3}{\kappa_2} \left(D\, e^{ik_3 b} - E\, e^{-ik_3 b} \right),$$

which lead to

$$B\, e^{\kappa_2 b} = \frac{D}{2}\, e^{ik_3 b} \left(1 + i\frac{k_3}{\kappa_2} \right) + \frac{E}{2}\, e^{-ik_3 b} \left(1 - i\frac{k_3}{\kappa_2} \right),$$

and

$$C\, e^{-\kappa_2 b} = \frac{D}{2}\, e^{ik_3 b} \left(1 - i\frac{k_3}{\kappa_2} \right) + \frac{E}{2}\, e^{-ik_3 b} \left(1 + i\frac{k_3}{\kappa_2} \right).$$

Substituting for B and C in terms of A, we arrive at an inhomogeneous linear system in the unknown A and D:

$$A\, e^{\kappa_2(b-a)} \left[\sin(k_1 a) + \frac{k_1}{\kappa_2} \cos(k_1 a) \right] - D\, e^{ik_3 b} \left(1 + i\frac{k_3}{\kappa_2} \right) = E\, e^{-ik_3 b} \left(1 - i\frac{k_3}{\kappa_2} \right),$$

$$A\, e^{-\kappa_2(b-a)} \left[\sin(k_1 a) - \frac{k_1}{\kappa_2} \cos(k_1 a) \right] - D\, e^{ik_3 b} \left(1 - i\frac{k_3}{\kappa_2} \right) = E\, e^{-ik_3 b} \left(1 + i\frac{k_3}{\kappa_2} \right).$$

This linear system has the form

$$\lambda_{11}A + \lambda_{12}D = \mu_1 , \qquad \lambda_{21}A + \lambda_{22}D = \mu_2 ,$$

with solutions (from Cramer's rule)

$$A = \frac{\mu_1\lambda_{22} - \mu_2\lambda_{12}}{\lambda_{11}\lambda_{22} - \lambda_{12}\lambda_{21}} , \qquad D = \frac{\mu_2\lambda_{11} - \mu_1\lambda_{21}}{\lambda_{11}\lambda_{22} - \lambda_{12}\lambda_{21}} .$$

We obtain

$$A = z_A E , \qquad D = z_D E ,$$

with

$$z_A = \frac{k_3}{\kappa_2} \frac{2}{\beta + i\alpha} , \qquad z_D = e^{-2ik_3 b} \frac{\alpha + i\beta}{-\alpha + i\beta} ,$$

where we define

$$\alpha = \sin(k_1 a) \sinh[\kappa_2(b - a)] + \frac{k_1}{\kappa_2} \cos(k_1 a) \cosh[\kappa_2(b - a)]$$

and

$$\beta = \frac{k_3}{\kappa_2} \left[\sin(k_1 a) \cosh[\kappa_2(b - a)] + \frac{k_1}{\kappa_2} \cos(k_1 a) \sinh[\kappa_2(b - a)] \right] .$$

The complex number z_D has magnitude $|z_D| = 1$,

$$z_D = e^{-2ik_3 b}\, e^{i(\pi - 2\varphi)} , \qquad \varphi = -\tan^{-1}\frac{\beta}{\alpha} ,$$

where φ is given explicitly by

$$\varphi = -\tan^{-1}\left[\frac{k_3}{\kappa_2} \frac{\sin(k_1 a) \cosh[\kappa_2(b - a)] + (k_1/\kappa_2) \cos(k_1 a) \sinh[\kappa_2(b - a)]}{\sin(k_1 a) \sinh[\kappa_2(b - a)] + (k_1/\kappa_2) \cos(k_1 a) \cosh[\kappa_2(b - a)]} \right] .$$

It follows that $|D| = |E|$, and hence $R = 1$, as must be the case.

Part 2

The complete continuum wave function for $0 < \epsilon < w_0$ follows as

$$\begin{aligned}
\psi_k(x) &= e^{ikb} A_k \sin\left(\sqrt{v_0 + k^2}\, x\right) & & 0 < x < a \\
&= e^{ikb} \left(B_k\, e^{\sqrt{w_0 - k^2}\, x} + C_k\, e^{-\sqrt{w_0 - k^2}\, x} \right) & & a < x < b \\
&= e^{ikb} \left(D_k\, e^{ikx} + e^{-ikx} \right) & & x > b ,
\end{aligned}$$

where we have simplified the notation by setting $k = \sqrt{\epsilon} = k_3$, $\kappa_2 = \sqrt{w_0 - k^2}$, $k_1 = \sqrt{v_0 + k^2}$, and have multiplied by the overall phase factor e^{ikb}, as instructed by the text of the problem, for later convenience. The coefficients A_k, B_k, C_k, and D_k are given by the relations in part 3; the coefficient E_k has been set to unity, since it is irrelevant (see below). We now form the wave packet

$$\Psi(x, t) = \int_0^{w_0} \frac{dk}{\sqrt{2\pi}}\, g(k)\, \psi_k(x)\, e^{-i\omega_k t} , \qquad \omega_k = \frac{\hbar k^2}{2m} ,$$

and the profile function $g(k)$, which is assumed for simplicity to be real, is strongly peaked at $k^\star$ with $0 < k^\star < \sqrt{w_0}$. In the region $x > b$ we have

$$\Psi(x,t) = \underbrace{\int_0^{w_0} \frac{dk}{\sqrt{2\pi}} g(k) \, e^{-i[k(x-b)+\omega_k t]}}_{\Psi_i(x,t)} + \underbrace{\int_0^{w_0} \frac{dk}{\sqrt{2\pi}} g(k) \, e^{i[k(x-b)+\pi-2\varphi_k-\omega_k t]}}_{\Psi_r(x,t)} \qquad x > b \; ;$$

and we have identified the incident and reflected wave packets. We find that the center of the incident wave packet moves according to

$$x_i(t) = b - v^\star t \qquad v^\star = \frac{\hbar k^\star}{m} \, ,$$

and so it reaches the edge b at time $t = 0$ (this is the reason for multiplying the wave function by the phase factor e^{ikb}). The incident wave packet is approximately given by

$$\Psi_i(x,t) \approx e^{-i[k^\star(x-b)+\omega^\star t]} \int_0^{w_0} \frac{dk}{\sqrt{2\pi}} g(k) \, e^{i(k-k^\star)[x_i(t)-x]} \, ,$$

where we have used the stationary phase method and expanded the phase factor $e^{i\eta}$, with $\eta = -k(x-b) - \omega_k t$, about $k^\star$ in linear terms $k - k^\star$ and have defined $\omega^\star = \omega_{k^\star}$. Similarly, we find

$$\Psi_r(x,t) \approx e^{i[k^\star(x-b)+\pi-2\varphi^\star-\omega^\star t]} \int_0^{w_0} \frac{dk}{\sqrt{2\pi}} g(k) \, e^{i(k-k^\star)[x-x_r(t)]} \, ,$$

where

$$x_r(t) = b + 2\left.\frac{d\varphi_k}{dk}\right|_{k^\star} + v^\star t = b + v^\star(t - \tau)$$

and τ is the time delay, defined as

$$\tau = -\frac{2}{v^\star} \left.\frac{d\varphi_k}{dk}\right|_{k^\star} .$$

The reflected wave packet will start on its way back from b to ∞ after a delay τ.

Part 3

We now investigate the time delay obtained above as a function of energy for $0 < \epsilon < w_0$. We go back to the expression for φ given in part 3 above, and introduce the following definitions:

$$k_1 a = x \implies \epsilon = (x/a)^2 - v_0 , \qquad k_3 a = a\sqrt{\epsilon} = \sqrt{x^2 - x_0^2} , \qquad x_0 = a\sqrt{v_0} ,$$

and

$$\kappa_2 a = \sqrt{a^2 w_0 - a^2\epsilon} = \sqrt{a^2(v_0 + w_0) - x^2} \equiv \sqrt{x_1^2 - x^2} , \qquad x_1 = a\sqrt{v_0 + w_0} ,$$

so that the allowed range of x is $x_0 < x < x_1$. In terms of these variables, the function $\varphi(x)$ can be written as

$$\varphi(x) = -\tan^{-1}\left[\sqrt{\frac{x^2 - x_0^2}{x_1^2 - x^2}} \, \frac{\tan x + \left(x/\sqrt{x_1^2 - x^2}\right)\tanh(\Delta\sqrt{x_1^2 - x^2})}{\tan x \, \tanh(\Delta\sqrt{x_1^2 - x^2}) + x/\sqrt{x_1^2 - x^2}}\right] ,$$

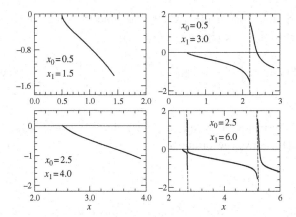

Fig. 5.2 The function $\varphi(x)$ for four different sets of potential parameters x_0 and x_1; in all cases $\Delta = 0.5$. The vertical dashed lines indicate the points at which the argument of the $\tan^{-1}$ function is infinite.

where we have defined $\Delta = b/a - 1$. This function is shown in Fig. 5.2 for four different choices of the potential parameters. In each case $\varphi(x)$ is a decreasing function of x and therefore we have $d\varphi(x)/dx < 0$. The (positive) time delay can be expressed as

$$
\tau = -\frac{2m}{\hbar k^\star}\frac{d\varphi(k)}{dk}\bigg|_{k^\star} = -\frac{2m}{\hbar k^\star}\left[\frac{dx}{dk}\frac{d\varphi(x)}{dx}\bigg|_{x^\star}\right] = -\frac{2ma^2}{\hbar x^\star}\frac{d\varphi(x)}{dx}\bigg|_{x^\star},
$$

where we have used $x = \sqrt{x_0^2 + a^2 k^2}$. The function $\varphi(x)$ vanishes at the lower end point, x_0, and is finite at the higher end point, x_1:

$$
\lim_{x \to x_1} \varphi(x) = -\tan^{-1}\left[\sqrt{x_1^2 - x_0^2}\left(\frac{\tan x_1}{x_1} + \Delta\right)\right].
$$

It is generally smooth, except at the points where its denominator vanishes, namely when

$$
\tan x = -\frac{x}{\sqrt{x_1^2 - x^2}}\frac{1}{\tanh(\Delta\sqrt{x_1^2 - x^2})}.
$$

A necessary requirement for this to happen is that $x_1 > \pi/2$, which ensures that $\tan x$ is negative in a portion of the interval $x_0 < x < x_1$ (otherwise there is no solution). At the points x_n where $\varphi(x)$ changes rapidly (see Fig. 5.2) the slope is large and negative), implying a large time delay. These points x_n (their number will depend on the potential parameters) correspond to energies $\epsilon_n = (x_n/a)^2 - v_0$. So, at these energies the particle "hangs around" the region $0 < x < a$ for a long time, before traveling back to ∞. The states corresponding to the ϵ_n are known as resonances.

Problem 11 Reflection and Transmission for a Particle Confined to the xy-Plane with $v(x, y) = v_0\,\theta(x)$

Suppose that a particle is confined to the xy-plane and is subject to the repulsive potential

$$
v(x) = v_0\,\theta(x) \qquad\qquad v_0 > 0\,.
$$

Assume that the particle's energy $\epsilon > v_0$.

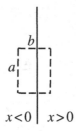

$x<0$ | $x>0$

Fig. 5.3   The rectangle used for the evaluation of the "surface integral."

1. Use separation of variables to solve the Schrödinger equation and show that a solution can be written as

$$\psi_{\mathbf{k}}(x,y) = A\,e^{i(k_x x + k_y y)} + B\,e^{i(-k_x x + k_y y)} \qquad x < 0$$
$$= C\,e^{i(K_x x + k_y y)} \qquad\qquad\quad x > 0 \,,$$

namely, an incident and a reflected wave function for $x < 0$ and a transmitted wave function for $x > 0$. Here, $\epsilon = k_x^2 + k_y^2$ and $K_x^2 = k_x^2 - v_0$. Impose appropriate boundary conditions to relate B and C to A.

2. Introduce the wave numbers $\mathbf{k}_i$, $\mathbf{k}_r$, and $\mathbf{k}_t$, associated with the incident, reflected, and transmitted wave functions, and indicate by θ and φ the angles made by $\mathbf{k}_i$ and $\mathbf{k}_t$, respectively, with the x-axis. Show that

$$\sqrt{\epsilon}\,\sin\theta = \sqrt{\epsilon - v_0}\,\sin\varphi \,,$$

which is Snell's law.

3. Verify that the probability current density $\mathbf{j}$ associated with $\psi_{\mathbf{k}}(\mathbf{r})$ is conserved. To this end, consider Fig. 5.3.

Solution

Part 1

The Schrödinger equation reads

$$-\left(\frac{\partial^2}{\partial x^2} + \frac{\partial^2}{\partial y^2}\right)\psi(x,y) + v_0\theta(x)\psi(x,y) = \epsilon\psi(x,y) \,,$$

with boundary conditions

$$\psi(0^-,y) = \psi(0^+,y)\,, \qquad \frac{\partial}{\partial x}\psi(x,y)\bigg|_{x=0^-} = \frac{\partial}{\partial x}\psi(x,y)\bigg|_{x=0^+} \,.$$

We solve it by the separation of variables $\psi(x,y) = X(x)\,Y(y)$, which, after some rearranging, leads to

$$\frac{X''}{X} - v_0\theta(x) + \epsilon = -\frac{Y''}{Y}$$

and hence

$$\frac{X''}{X} = v_0\theta(x) - k_x^2\,, \qquad \frac{Y''}{Y} = -k_y^2\,, \qquad \epsilon = k_x^2 + k_y^2 \,.$$

These equations have solutions of the type (assuming $\epsilon > v_0$) $X(x) \sim e^{\pm ik_x x}$ for $x < 0$ and $X(x) \sim e^{\pm iK_x x}$ for $x > 0$, where $K_x^2 = k_x^2 - v_0$, and $Y(y) \sim e^{\pm ik_y y}$. We write the complete wave function as

$$\psi_{\mathbf{k}}(x,y) = \underbrace{A\,e^{i(k_x x + k_y y)}}_{\psi_i} + \underbrace{B\,e^{i(-k_x x + k_y y)}}_{\psi_r} \qquad x < 0$$

$$= \underbrace{C\,e^{i(K_x x + k_y y)}}_{\psi_t} \qquad x > 0\,,$$

where ψ_i, ψ_r, and ψ_t indicate the incident, reflected, and transmitted wave functions. Note that the current corresponding to $\psi_{\mathbf{k}}(x,y)$ in the region $x < 0$ is not simply given by the sum of the currents associated with the incident and reflected wave functions; there are interference terms between these waves which do not vanish. This difficulty is avoided by introducing wave packets.

The boundary conditions require (note that the common factors $e^{ik_y y}$ cancel out)

$$A + B = C\,, \qquad A - B = \frac{K_x}{k_x} C\,,$$

with solutions with respect to A given by

$$B = \frac{k_x - K_x}{k_x + K_x} A\,, \qquad C = \frac{2k_x}{k_x + K_x} A\,.$$

Part 2

We introduce the wave-number vectors $\mathbf{k}_i$, $\mathbf{k}_r$, and $\mathbf{k}_t$ associated with the incident, reflected, and transmitted wave functions, respectively:

$$\mathbf{k}_i = (k_x, k_y) = \underbrace{\sqrt{k_x^2 + k_y^2}}_{\sqrt{\epsilon}}\,(\cos\theta, \sin\theta)\,, \qquad \mathbf{k}_r = (-k_x, k_y) = \sqrt{\epsilon}\,(-\cos\theta, \sin\theta)\,,$$

and

$$\mathbf{k}_t = (K_x, k_y) = \sqrt{K_x^2 + k_y^2}\,(\cos\varphi, \sin\varphi)\,,$$

where θ and φ are the angles made by the incident and transmitted wave-number vectors with the x-axis. Recalling that $K_x^2 = k_x^2 - v_0$, we find

$$\sin\varphi = \frac{k_y}{\sqrt{K_x^2 + k_y^2}} = \frac{\sqrt{\epsilon}\,\sin\theta}{\sqrt{\epsilon - v_0}}\,.$$

The above relation is just Snell's law,

$$\sqrt{\epsilon}\,\sin\theta = \sqrt{\epsilon - v_0}\,\sin\varphi\,.$$

Part 3

The incident probability current density is obtained as

$$\mathbf{j}_i = -\frac{i\hbar}{2m}\left(\psi_{\mathbf{k}_i}^* \boldsymbol{\nabla} \psi_{\mathbf{k}_i} - \text{c.c.}\right) = \frac{\hbar\,\mathbf{k}_i}{m}\,|A|^2 \qquad \text{for } x < 0\,,$$

and similarly for the reflected and transmitted probability current densities,

$$\mathbf{j}_r = \frac{\hbar \, \mathbf{k}_r}{m} \, |B|^2 \quad \text{for } x < 0 \,, \qquad \mathbf{j}_t = \frac{\hbar \, \mathbf{k}_t}{m} \, |C|^2 \quad \text{for } x > 0 \,.$$

Since the wave function is a stationary solution, its probability density is time independent, and hence

$$\nabla \cdot \mathbf{j} = 0 \implies \int d\mathbf{r} \, \nabla \cdot \mathbf{j} = \int dS \, \hat{\mathbf{n}} \cdot \mathbf{j} = 0,$$

where $\mathbf{j}$ is the total probability current density, given by

$$\mathbf{j} = \mathbf{j}_i + \mathbf{j}_r \quad \text{for } x < 0 \,, \qquad \mathbf{j} = \mathbf{j}_t \quad \text{for } x > 0 \,.$$

In order to verify the above conservation law, we consider the "surface" integral as in Fig. 5.3. We find

$$-a \, (\mathbf{j}_i + \mathbf{j}_r) \cdot \hat{\mathbf{x}} + a \, \mathbf{j}_t \cdot \hat{\mathbf{x}} = 0$$

and note that the contributions from the upper and lower sides (of length b) of the rectangle cancel each other out in the regions $x < 0$ and $x > 0$. Thus, we find

$$j_{i,x} + j_{r,x} = j_{t,x} \implies k_x \left(|A|^2 - |B|^2 \right) = K_x \, |C|^2 \,,$$

and hence, substituting the expressions previously obtained for B and C in terms of A, we conclude that the above identity is verified, since

$$k_x \left[1 - \frac{(k_x - K_x)^2}{(k_x + K_x)^2} \right] |A|^2 = \frac{4k_x^2 K_x}{(k_x + K_x)^2} \, |A|^2 = K_x \, |C|^2 \,.$$

Problem 12 Reflection and Transmission at Two Repulsive δ-Function Potentials

Consider a particle in one dimension under the influence of a potential consisting of two repulsive δ-functions:

$$v(x) = v_0 \, \delta(x) + v_0 \, \delta(x - a) \qquad v_0 > 0 \,.$$

Calculate the transmission and reflection coefficients and determine for which values of a the transmission coefficient is unity.

Solution

Denoting by I, II, and III the regions $x < 0$, $0 < x < a$, and $x > a$, we have

$$\psi_I(x) = e^{ikx} + A \, e^{-ikx} \,, \qquad \psi_{II}(x) = B \, e^{ikx} + C \, e^{-ikx} \,, \qquad \psi_{III}(x) = D \, e^{ikx} \,.$$

The matching conditions for the wave function and its first derivative lead to the inhomogeneous linear system

$$A - B - C = -1$$
$$\left(1 + i \frac{v_0}{k} \right) A + B - C = 1 - i \frac{v_0}{k}$$
$$B + e^{-2ika} \, C - D = 0$$
$$-B + e^{-2ika} \, C + \left(1 + i \frac{v_0}{k} \right) D = 0 \,.$$

The determinant is

$$\Delta = e^{-2ika}[1 + (1 + iv_0/k)]^2 - [1 + (1 - iv_0/k)]^2 = e^{-2ika}(2 + iv_0/k)^2 + v_0^2/k^2$$
$$= e^{-ika}\left[2i(v_0/k)^2 \sin(ka) + 4 e^{-ika}(1 + iv_0/k)\right] \,.$$

The reflection and transmission coefficients are then given by

$$R = |j_r|/|j_i| = |A|^2 \,, \qquad T = |j_t|/|j_i| = |D|^2 \,.$$

Cramer's rule allows us to obtain

$$A = \frac{1}{\Delta} \det \begin{bmatrix} -1 & -1 & -1 & 0 \\ 1 - iv_0/k & 1 & -1 & 0 \\ 0 & 1 & e^{-2ika} & -1 \\ 0 & -1 & e^{-2ika} & 1 + iv_0/\rho \end{bmatrix} \,,$$

$$D = \frac{1}{\Delta} \det \begin{bmatrix} 1 & -1 & -1 & -1 \\ 1 + iv_0/k & 1 & -1 & 1 - iv_0/k \\ 0 & 1 & e^{-2ika} & 0 \\ 0 & -1 & e^{-2ika} & 0 \end{bmatrix} \,.$$

Evaluating the determinants, we find

$$A = -2i\frac{v_0^2}{k^2}\frac{\exp(-ika)}{\Delta}[\sin(ka) + 2(k/v_0)\cos(ka)] = -\frac{\sin(ka) + 2(k/v_0)\cos(ka)}{\sin(ka) + 2(k/v_0)\exp(-ika)(1 - ik/v_0)} \,,$$

$$D = \frac{4\exp(-2ika)}{\Delta} = -\frac{2i(k/v_0)^2 \exp(-ika)}{\sin(ka) + 2(k/v_0)\exp(-ika)(1 - ik/v_0)} \,,$$

from which we obtain

$$R = \frac{[\sin(ka) + 2(k/v_0)\cos(ka)]^2}{4k^4/v_0^4 + \sin^2(ka) + 4(k/v_0)\sin(ka)\cos(ka) + 4(k^2/v_0^2)\cos^2(ka)} \,,$$

$$T = \frac{4(k^4/v_0^4)}{4k^4/v_0^4 + \sin^2(ka) + 4(k/v_0)\sin(ka)\cos(ka) + 4(k^2/v_0^2)\cos^2(ka)} \,,$$

and $R + T = 1$ as required. It is convenient to write

$$R = \frac{[(v_0/2k)\sin(ka) + \cos(ka)]^2}{1 + [(v_0/2k)\sin(ka) + \cos(ka)]^2} \,,$$

$$T = \frac{1}{1 + [(v_0/2k)\sin(ka) + \cos(ka)]^2} \,.$$

Resonances occur when $T = 1$, which implies that

$$\frac{v_0}{2k}\sin(ka) + \cos(ka) = 0 \implies \tan(ka) = -\frac{2k}{v_0} \quad \text{with } ka \neq (2n + 1)\frac{\pi}{2} \quad n = 0, 1, \ldots$$

This condition is satisfied by

$$a_n = \frac{n\pi}{k} - \frac{1}{k}\tan^{-1}\left(\frac{2k}{v_0}\right) \quad n = 1, 2, \ldots$$

The resonances do not occur at the values $n\pi/k$ or $n\lambda/2$, where λ is the de Broglie wavelength. The reason is that the reflections of the particle at $x = 0$ and $x = a$ occur with a phase shift of the wave function. As a consequence, resonances occur at a_n.

Problem 13   Transmission through Equally Spaced Repulsive δ-Function Potentials

Determine the transmission coefficient for a particle under the influence of a potential given by

$$v(x) = \sum_{n=0}^{N} v_0\, \delta(x - na) \qquad v_0 > 0 \ .$$

Solution

The Schrödinger equation is given by

$$\psi''(x) = \left[v_0 \sum_{n=0}^{N} \delta(x - na) - k^2 \right] \psi(x) \ .$$

For $x < 0$ the solution is given by

$$\psi_0(x) = A_0\, e^{ikx} + B_0\, e^{-ikx} \ ,$$

while in the interval $(n - 1)a < x < na$ with $n = 1, \ldots, N$ it can be written as

$$\psi_n(x) = A_n\, e^{ik(x-na)} + B_n\, e^{-ik(x-na)} \ ,$$

and for $x > Na$ it is

$$\psi_{N+1}(x) = A_{N+1}\, e^{ik[x-(N+1)a]} + B_{N+1}\, e^{-ik[x-(N+1)a]} \ .$$

The boundary conditions at $x = na$ with $n = 0, \ldots, N$ require that

$$\underbrace{A_n + B_n}_{\psi_n(na)} = \underbrace{A_{n+1}\, e^{-ika} + B_{n+1}\, e^{ika}}_{\psi_{n+1}(na)}$$

and

$$\underbrace{ik(A_{n+1}\, e^{-ika} - B_{n+1}\, e^{ika})}_{\psi'_{n+1}(na)} - \underbrace{ik(A_n - B_n)}_{\psi'_n(na)} = v_0 \underbrace{(A_n + B_n)}_{\psi_n(na)} \ .$$

In order to solve this linear system, first define

$$A'_{n+1} = e^{-ika}\, A_{n+1} \ , \qquad B'_{n+1} = e^{ika}\, B_{n+1} \ .$$

In terms of these variables, the system is now given by

$$A'_{n+1} + B'_{n+1} = A_n + B_n \ , \qquad A'_{n+1} - B'_{n+1} = \left(1 - i\frac{v_0}{k}\right) A_n - \left(1 + i\frac{v_0}{k}\right) B_n \ ,$$

yielding the solutions

$$A'_{n+1} = \left(1 - i\frac{v_0}{2k}\right) A_n - i\frac{v_0}{2k}\, B_n \ , \qquad B'_{n+1} = i\frac{v_0}{2k}\, A_n + \left(1 + i\frac{v_0}{2k}\right) B_n \ .$$

In matrix form the relation between A_{n+1} and B_{n+1} and A_n and B_n follows from

$$\begin{pmatrix} A'_{n+1} \\ B'_{n+1} \end{pmatrix} = \underbrace{\begin{pmatrix} e^{-ika} & 0 \\ 0 & e^{ika} \end{pmatrix}}_{\underline{N}} \begin{pmatrix} A_{n+1} \\ B_{n+1} \end{pmatrix} = \underbrace{\begin{pmatrix} 1 - iv_0/2k & -iv_0/2k \\ iv_0/2k & 1 + iv_0/2k \end{pmatrix}}_{\underline{M}} \begin{pmatrix} A_n \\ B_n \end{pmatrix}$$

or

$$\begin{pmatrix} A_{n+1} \\ B_{n+1} \end{pmatrix} = \underline{N}^{-1}\, \underline{M} \begin{pmatrix} A_n \\ B_n \end{pmatrix} = \underline{N}^{\dagger}\, \underline{M} \begin{pmatrix} A_n \\ B_n \end{pmatrix} \ ;$$

$\underline{N}$ is unitary since

$$\underline{N}^{-1} = \begin{pmatrix} e^{ika} & 0 \\ 0 & e^{-ika} \end{pmatrix} = \underline{N}^\dagger \ .$$

We can use the recursion relation established above to relate A_{N+1} and B_{N+1} to A_0 and B_0,

$$\begin{pmatrix} A_{N+1} \\ B_{N+1} \end{pmatrix} = \underline{N}^\dagger \underline{M} \begin{pmatrix} A_N \\ B_N \end{pmatrix} = \left(\underline{N}^\dagger \underline{M} \right) \left(\underline{N}^\dagger \underline{M} \right) \begin{pmatrix} A_{N-1} \\ B_{N-1} \end{pmatrix} = \cdots$$

or, defining $\underline{P} = \underline{N}^\dagger \underline{M}$,

$$\begin{pmatrix} A_{N+1} \\ B_{N+1} \end{pmatrix} = \underbrace{\underline{P} \cdots \underline{P}}_{N+1 \text{ times}} \begin{pmatrix} A_0 \\ B_0 \end{pmatrix} = \underline{P}^{N+1} \begin{pmatrix} A_0 \\ B_0 \end{pmatrix} \ .$$

In order to evaluate this product, assume that $\underline{P}$ is diagonalizable (this hypothesis will be verified *a posteriori* – note that $\underline{P}$ is not hermitian, unitary, or normal):

$$\underline{P} = \underline{S}\,\underline{D}\,\underline{S}^{-1} \ , \qquad \underline{D} = \begin{pmatrix} \lambda_+ & 0 \\ 0 & \lambda_- \end{pmatrix} \ ,$$

where $\lambda_\pm$ are the eigenvalues and $\underline{S}$ is the matrix that transforms the original basis to that of the eigenvectors, and hence

$$\underline{P}^{N+1} = \left(\underline{S}\,\underline{D}\,\underline{S}^{-1} \right) \left(\underline{S}\,\underline{D}\,\underline{S}^{-1} \right) \cdots \left(\underline{S}\,\underline{D}\,\underline{S}^{-1} \right) = \underline{S}\,\underline{D}^{N+1}\,\underline{S}^{-1} \ .$$

The power of the diagonal matrix is easily evaluated:

$$\underline{D}^{N+1} = \begin{pmatrix} \lambda_+ & 0 \\ 0 & \lambda_- \end{pmatrix} \begin{pmatrix} \lambda_+ & 0 \\ 0 & \lambda_- \end{pmatrix} \cdots \begin{pmatrix} \lambda_+ & 0 \\ 0 & \lambda_- \end{pmatrix} = \begin{pmatrix} \lambda_+^{N+1} & 0 \\ 0 & \lambda_-^{N+1} \end{pmatrix} \ .$$

The problem reduces to finding the eigenvalues and eigenvectors of $\underline{P}$. To this end, first note that

$$\underline{P} = \begin{pmatrix} e^{ika} & 0 \\ 0 & e^{-ika} \end{pmatrix} \begin{pmatrix} 1 - iv_0/2k & -iv_0/2k \\ iv_0/2k & 1 + iv_0/2k \end{pmatrix} = \begin{pmatrix} c & 0 \\ 0 & c^* \end{pmatrix} \begin{pmatrix} a & b \\ b^* & a^* \end{pmatrix} = \begin{pmatrix} ac & bc \\ b^*c^* & a^*c^* \end{pmatrix} \ ,$$

and

$$\det \begin{bmatrix} ac - \lambda & bc \\ b^*c^* & a^*c^* - \lambda \end{bmatrix} = 0 \implies \lambda^2 - 2\lambda \, \mathrm{Re}(ac) + |a|^2 - |b|^2 = 0 \ ,$$

since $|c| = 1$. Substituting for a, b, and c yields

$$|a|^2 - |b|^2 = \left| 1 - i\frac{v_0}{2k} \right|^2 - \left| -i\frac{v_0}{2k} \right|^2 = 1 \ ,$$

and so

$$\mathrm{Re}(ac) = \mathrm{Re}\left[\left(1 - i\frac{v_0}{2k} \right) e^{ika} \right] = \cos(ka) + \frac{v_0}{2k} \sin(ka) \ .$$

Define $\tan\alpha = v_0/2k$ with $0 < \alpha < \pi/2$ as the energy varies in $0 < \epsilon < \infty$, so that

$$\mathrm{Re}(ac) = \cos(ka) + \tan\alpha \, \sin(ka) = \frac{\cos\alpha \, \cos(ka) + \sin\alpha \, \sin(ka)}{\cos\alpha} = \frac{\cos(ka - \alpha)}{\cos\alpha} \ .$$

The eigenvalues are then obtained from

$$\lambda^2 - 2\frac{\cos(ka - \alpha)}{\cos\alpha}\lambda + 1 = 0 \implies \lambda_\pm = \frac{\cos(ka - \alpha)}{\cos\alpha} \pm \sqrt{\frac{\cos^2(ka - \alpha)}{\cos^2\alpha} - 1} \ .$$

We write

$$\frac{\cos^2(ka - \alpha)}{\cos^2 \alpha} = \frac{1 + \cos(2ka - 2\alpha)}{1 + \cos(2\alpha)} .$$

The following cases can be distinguished.

1. Real non-degenerate eigenvalues, that is,

$$\cos^2(ka - \alpha)/\cos^2 \alpha > 1 \implies \cos(2ka - 2\alpha) > \cos(2\alpha) ;$$

this is satisfied if

$$p\pi < ka < (p + 1)\pi \quad \text{and} \quad \alpha > (ka - p\pi)/2 \quad p = 0, 1, 2, \ldots$$

2. Real degenerate eigenvalues, that is,

$$\cos^2(ka - \alpha)/\cos^2 \alpha = 1 \implies \cos(2ka - 2\alpha) = \cos(2\alpha) ;$$

this is satisfied if

$$p\pi < ka < (p + 1)\pi \quad \text{and} \quad \alpha = (ka - p\pi)/2 \quad p = 0, 1, 2, \ldots$$

3. Imaginary eigenvalues, that is,

$$\cos^2(ka - \alpha)/\cos^2 \alpha < 1 \implies \cos(2ka - 2\alpha) < \cos(2\alpha) ;$$

this is satisfied if

$$p\pi < ka < (p + 1)\pi \quad \text{and} \quad \alpha < (ka - p\pi)/2 \quad p = 0, 1, 2, \ldots$$

It is convenient to define

$$\xi = \frac{\cos(ka - \alpha)}{\cos \alpha} .$$

Note that in the range where real eigenvalues occur, we have

$$\xi > 1 \quad \text{if} \quad p\pi < ka < (p + 1)\pi \quad \text{and} \quad \alpha > (ka - p\pi)/2 \quad p = 0, 2, 4, \ldots$$
$$\xi < -1 \quad \text{if} \quad p\pi < ka < (p + 1)\pi \quad \text{and} \quad \alpha > (ka - p\pi)/2 \quad p = 1, 3, 5, \ldots .$$

In terms of ξ the eigenvalues are the following:

1. Real for $|\xi| \geq 1$:

$$\lambda_\pm = \xi \pm \sqrt{\xi^2 - 1}, \quad \text{non-degenerate} ; \quad \lambda_\pm = 1, \quad \text{degenerate case}$$

2. Imaginary for $|\xi| < 1$:

$$\lambda_\pm = \xi \pm i \sqrt{1 - \xi^2} ,$$

with $|\lambda_\pm| = 1$, namely, a phase factor.

The incident wave (from the left) is $A_0 e^{ikx}$, the reflected wave is $B_0 e^{-ikx}$, the transmitted wave is $A_{N+1} e^{ik[x-(N+1)a]}$, and the reflection and transmission coefficients are given by

$$R = \frac{|B_0|^2}{|A_0|^2} , \quad T = \frac{|A_{N+1}|^2}{|A_0|^2} .$$

Using

$$\begin{pmatrix} A_{N+1} \\ 0 \end{pmatrix} = \underline{S}\,\underline{D}^{N+1}\,\underline{S}^{-1} \begin{pmatrix} A_0 \\ B_0 \end{pmatrix} = \begin{pmatrix} Q_{11} & Q_{12} \\ Q_{21} & Q_{22} \end{pmatrix} \begin{pmatrix} A_0 \\ B_0 \end{pmatrix} ,$$

where we have set $B_{N+1} = 0$ in order to have only an outgoing wave for $x > (N+1)a$, we obtain

$$Q_{21}A_0 + Q_{22}B_0 = 0 \implies \frac{B_0}{A_0} = -\frac{Q_{21}}{Q_{22}}$$

and

$$A_{N+1} = Q_{11}A_0 + Q_{12}B_0 = \left(Q_{11} - \frac{Q_{12}Q_{21}}{Q_{22}} \right) A_0 \implies \frac{A_{N+1}}{A_0} = \frac{\det \underline{Q}}{Q_{22}} .$$

The determinant of $\underline{Q}$ is equal to 1 since, using $\det(\underline{A}\,\underline{B}) = \det(\underline{A})\det(\underline{B})$ and $\det(\underline{S}^{-1}) = 1/\det(\underline{S})$, it follows that

$$\det \underline{Q} = \det\left(\underline{S}\,\underline{D}^{N+1}\,\underline{S}^{-1} \right) = \det(\underline{D}^{N+1}) = (\lambda_+\,\lambda_-)^{N+1} ,$$

and $\lambda_+\,\lambda_- = 1$ for any ξ. We finally arrive at

$$R = \frac{|Q_{21}|^2}{|Q_{22}|^2} , \qquad T = \frac{1}{|Q_{22}|^2} .$$

In order to calculate $\underline{Q}$, we need to find $\underline{S}$, in other words the eigenvectors of $\underline{P}$. To this end, it is convenient to define $\cos\eta = \xi$ when $|\xi| < 1$, and $\text{sign}(\xi)\cosh\eta = \xi$ when $|\xi| > 1$, where $\text{sign}(\xi) = \pm$ depending on whether $\xi > 1$ or $\xi < -1$, and to rewrite the eigenvalues as

$$\lambda_\pm = e^{\pm i\eta} \quad |\xi| < 1 , \qquad \lambda_\pm = \text{sign}(\xi)\,e^{\pm\eta} \quad |\xi| > 1 .$$

The matrix $\underline{P}$ is given as follows:

$$\underline{P} = \begin{pmatrix} ac & bc \\ b^*c^* & a^*c^* \end{pmatrix} = \begin{bmatrix} (1 - i\tan\alpha)e^{ika} & -i\tan\alpha\,e^{ika} \\ i\tan\alpha\,e^{-ika} & (1 + i\tan\alpha)e^{-ika} \end{bmatrix}$$

$$= \begin{bmatrix} e^{i(ka-\alpha)}/\cos\alpha & -i\tan\alpha\,e^{ika} \\ i\tan\alpha\,e^{-ika} & e^{-i(ka-\alpha)}/\cos\alpha \end{bmatrix} .$$

The Case $|\xi| < 1$. The eigenvectors corresponding to the eigenvalues $\lambda_\pm$ result from

$$(\lambda_\pm \underline{I} - \underline{P}) \begin{pmatrix} a_\pm \\ b_\pm \end{pmatrix} = 0 \implies \left[e^{\pm i\eta} - \frac{e^{i(ka-\alpha)}}{\cos\alpha} \right] a_\pm + i\tan\alpha\,e^{ika}\, b_\pm = 0 ,$$

which can be solved with respect to $b_\pm$ to obtain

$$\frac{b_\pm}{a_\pm} = i\left[e^{\pm i\eta} - \frac{e^{i(ka-\alpha)}}{\cos\alpha} \right] \cot\alpha\,e^{-ika} = i\left[e^{\pm i(\eta\mp ka)}\cos\alpha - e^{-i\alpha} \right] \frac{1}{\sin\alpha} .$$

We set $a_\pm = \sin\alpha$ and obtain

$$a_\pm = \sin\alpha , \qquad b_\pm = i\left[e^{\pm i(\eta\mp ka)}\cos\alpha - e^{-i\alpha} \right] .$$

The matrices $\underline{S}$ and $\underline{S}^{-1}$ read

$$\underline{S} = \begin{pmatrix} a_+ & a_- \\ b_+ & b_- \end{pmatrix} , \qquad \underline{S}^{-1} = \frac{1}{\det(\underline{S})} \begin{pmatrix} b_- & -a_- \\ -b_+ & a_+ \end{pmatrix}$$

or, explicitly,

$$\underline{S} = \begin{bmatrix} \sin\alpha & \sin\alpha \\ i\,e^{i(\eta-ka)}\,\cos\alpha - i\,e^{-i\alpha} & i\,e^{-i(\eta+ka)}\,\cos\alpha - i\,e^{-i\alpha} \end{bmatrix}$$

and

$$\underline{S}^{-1} = \frac{e^{ika}}{2\,\sin\eta\,\sin\alpha\,\cos\alpha} \begin{bmatrix} i\,e^{-i(\eta+ka)}\,\cos\alpha - i\,e^{-i\alpha} & -\sin\alpha \\ -i\,e^{i(\eta-ka)}\,\cos\alpha + i\,e^{-i\alpha} & \sin\alpha \end{bmatrix}.$$

In order to calculate the transmission coefficient, we need the matrix element Q_{22}, that is,

$$Q_{22} = \left(\underline{S}\underline{D}^{N+1}\underline{S}^{-1}\right)_{22} = \frac{a_+ b_- \lambda_-^{N+1} - a_- b_+ \lambda_+^{N+1}}{\det(\underline{S})}$$

or, explicitly,

$$Q_{22} = \frac{\sin[(N+2)\eta] - e^{i(ka-\alpha)}\,\sin[(N+1)\eta]/\cos\alpha}{\sin\eta}.$$

The transmission coefficient is then obtained as

$$T = \frac{\sin^2\eta}{|\sin[(N+2)\eta] - e^{i(ka-\alpha)}\,\sin[(N+1)\eta]/\cos\alpha|^2}.$$

Thus it is an oscillating function of the energy in energy bands defined by the condition $p\pi < ka < (p+1)\pi$ and $\alpha < (ka - p\pi)/2$ with $p = 0, 1, 2, \ldots$.

The Case $|\xi| > 1$. For $\xi > 1$ make the replacements $\eta \longrightarrow -i\eta$ and $\sin\eta \longrightarrow -i\sinh\eta$, corresponding to $e^{\pm i\eta} \longrightarrow e^{\pm\eta}$; for $\xi < -1$ make the replacements $\eta \longrightarrow -i\eta + \pi$ and $\sin\eta \longrightarrow i\sinh\eta$, corresponding to $e^{\pm i\eta} \longrightarrow -e^{\pm\eta}$ (recall that for $|\xi| > 1$ the eigenvalues are $\text{sign}(\xi)\,e^{\pm\eta}$). Thus for $\xi > 1$ the coefficient Q_{22} is given by

$$Q_{22} = \frac{-i\,\sinh[(N+2)\eta] + i\,e^{i(ka-\alpha)}\,\sinh[(N+1)\eta]/\cos\alpha}{-i\,\sinh\eta},$$

and hence

$$T = \frac{1}{|Q_{22}|^2} = \frac{\sinh^2\eta}{|\sinh[(N+2)\eta] - e^{i(ka-\alpha)}\,\sinh[(N+1)\eta]/\cos\alpha|^2}.$$

For $\xi < -1$, we note that

$$\sin[(N+2)\eta] \longrightarrow \sin[-i(N+2)\eta + (N+2)\pi] = -i(-1)^{N+2}\,\sinh[(N+2)\eta],$$

and so

$$Q_{22} = \frac{-i(-1)^{N+2}\,\sinh[(N+2)\eta] + i(-1)^{N+1}\,e^{i(ka-\alpha)}\,\sinh[(N+1)\eta]/\cos\alpha}{i\,\sinh\eta},$$

which yields a similar expression as that above except for a sign switch in the denominator:

$$T = \frac{1}{|Q_{22}|^2} = \frac{\sinh^2\eta}{|\sinh[(N+2)\eta] + e^{i(ka-\alpha)}\,\sinh[(N+1)\eta]/\cos\alpha|^2}.$$

In the limit $N \gg 1$, we can approximate $\sinh[(N+2)\eta]$ by $e^{N\eta}/2$ and obtain that $T \propto e^{-2N\eta}$ for fixed η (energy). Thus, transmission for energies in the bands defined by the condition $p\pi < ka < (p+1)\pi$ and $\alpha > (ka - p\pi)/2$ with $p = 0, 1, 2, \ldots$ is strongly suppressed in the limit $N \gg 1$.

The Case $|\xi| = 1$. The transmission coefficient in this case can be obtained from that corresponding to $|\xi| > 1$ by taking the limits $\xi \longrightarrow \pm 1$ and $\eta \longrightarrow 0$. Note that $|\xi| = 1$ requires energies such that $\alpha = (ka - p\pi)/2$, and hence

$$e^{i(ka-\alpha)}/\cos\alpha \longrightarrow -1 + i\cot(ka/2) \qquad p \text{ odd}$$
$$\longrightarrow 1 + i\tan(ka/2) \qquad p \text{ even}.$$

Using $\sinh x \approx x$ as x goes to zero, we find

$$T = \frac{1}{1 + (N+1)^2 \cot^2(ka/2)} \qquad p \text{ odd},$$
$$= \frac{1}{1 + (N+1)^2 \tan^2(ka/2)} \qquad p \text{ even}.$$

The requirement $\alpha = (ka - p\pi)/2$ puts a condition on the energy since $\alpha = \tan^{-1}(v_0/2k)$, and hence the energy must be such that $\tan(ka/2 - p\pi/2) = v_0/2k$, yielding $\cot(ka/2) = -v_0/2k$ for p odd and $\tan(ka/2) = v_0/2k$ for p even, and hence the transmission coefficients can also be written as

$$T = \frac{1}{1 + (N+1)^2 (v_0/2k)^2} \qquad p = 0, 1, \ldots .$$

It is interesting to note that in this case the transmission coefficient is identical to that corresponding to a single δ-function potential of strength $(N+1)v_0$ (here $N+1$ is the number of individual δ-function potentials present in the problem). We also note that, for fixed energy, T vanishes as $N \gg 1$.

Problem 14   A General Treatment of Scattering: S-Matrix

Consider a potential $v(x)$ that vanishes as $|x| \longrightarrow \infty$. We will be concerned below with the (two-fold degenerate) continuum solutions corresponding to energy $\epsilon > 0$. In the asymptotic regions $x \longrightarrow -\infty$ and $x \longrightarrow \infty$ these solutions satisfy the free-particle Schrödinger equation and can be written as

$$\psi_-(x) = A\,e^{ikx} + B\,e^{-ikx} \qquad x \longrightarrow -\infty,$$
$$\psi_+(x) = C\,e^{ikx} + D\,e^{-ikx} \qquad x \longrightarrow \infty,$$

where $k = \sqrt{\epsilon}$.

1. Evaluate the probability current densities $j_\pm$ corresponding to the wave functions $\psi_\pm(x)$ and use the conservation of probability current to show that

$$|A|^2 - |B|^2 = |C|^2 - |D|^2 .$$

Show that in terms of these coefficients the reflection and transmission probabilities are obtained from either of the following two relations (they are the same, regardless of which set of relations we use; see part 3 below):

$$R = \frac{|B|^2}{|A|^2} \quad \text{and} \quad T = \frac{|C|^2}{|A|^2} \qquad \text{or} \qquad R = \frac{|C|^2}{|D|^2} \quad \text{and} \quad T = \frac{|B|^2}{|D|^2} ,$$

where $R + T = 1$.

2. After joining the $\psi_-(x)$ solution smoothly to the $\psi_+(x)$ solution, relations between the coefficients A and B, on the one hand, and C and D, on the other hand, are obtained. These relations can be written equivalently as relations between the coefficients B and C of the "outgoing waves" e^{-ikx} for $x \longrightarrow -\infty$ and e^{ikx} for $x \longrightarrow \infty$, and the coefficients A and D of the "incoming waves" e^{ikx}, for $x \longrightarrow -\infty$ and e^{-ikx} for $x \longrightarrow \infty$, namely,

$$\begin{pmatrix} B \\ C \end{pmatrix} = \begin{bmatrix} S_{11}(k) & S_{12}(k) \\ S_{21}(k) & S_{22}(k) \end{bmatrix} \begin{pmatrix} A \\ D \end{pmatrix} ,$$

where the matrix elements $S_{ij}(k)$ are complex functions of the wave number (or energy). Use the relation $|A|^2 + |D|^2 = |B|^2 + |C|^2$ implied by the conservation of the probability current to show that the matrix

$$\underline{S}(k) = \begin{bmatrix} S_{11}(k) & S_{12}(k) \\ S_{21}(k) & S_{22}(k) \end{bmatrix}$$

is such that

$$\underline{S}^\dagger(k)\,\underline{S}(k) = \underline{I} \equiv \begin{pmatrix} 1 & 0 \\ 0 & 1 \end{pmatrix} ,$$

that is, $\underline{S}$ is unitary. Obtain the reflection and transmission coefficients in terms of the matrix elements of $\underline{S}(k)$.

3. Since the potential is real, the complex-conjugate solution also satisfies the Schrödinger equation, with asymptotic behavior given by

$$\psi_-^*(x) = A^*\,e^{-ikx} + B^*\,e^{ikx} \qquad\qquad x \longrightarrow -\infty ,$$
$$\psi_+^*(x) = C^*\,e^{-ikx} + D^*\,e^{ikx} \qquad\qquad x \longrightarrow \infty .$$

Show that

$$\begin{pmatrix} A^* \\ D^* \end{pmatrix} = \begin{bmatrix} S_{11}(k) & S_{12}(k) \\ S_{21}(k) & S_{22}(k) \end{bmatrix} \begin{pmatrix} B^* \\ C^* \end{pmatrix} ,$$

and deduce from this that

$$\underline{S}^*(k)\,\underline{S}(k) = \underline{I} ,$$

and hence that $\underline{S}(k)$ must be symmetric (that is, equal to its transpose), in addition to being unitary (this is a consequence of time reversal invariance as we will see in Chapter 16). Conclude from this that the reflection coefficients obtained from either $|B|^2/|A|^2$ or $|C|^2/|D|^2$ are the same, and similarly for the transmission coefficients.

Solution

Part 1

We have

$$\psi_-'(x) = ik \left(A\,e^{ikx} - B\,e^{-ikx} \right) .$$
$$\psi_-^*(x)\,\psi_-'(x) = ik \left(A^*\,e^{-ikx} + B^*\,e^{ikx} \right) \left(A\,e^{ikx} - B\,e^{-ikx} \right)$$
$$= ik \left(|A|^2 - |B|^2 - A^*B\,e^{-2ikx} + AB^*\,e^{2ikx} \right) ,$$

and

$$\psi'^{*}_{-}(x)\,\psi_{-}(x) = \left[\psi^{*}_{-}(x)\,\psi'_{-}(x)\right]^{*} = -ik\left(|A|^2 - |B|^2 - AB^*\,e^{2ikx} + A^*B\,e^{-2ikx}\right) .$$

The j_- probability current is as follows:

$$j_- = \frac{\hbar}{2mi}\left[\psi^{*}_{-}(x)\,\psi'_{-}(x) - \text{c.c.}\right] = \frac{\hbar k}{m}\left(|A|^2 - |B|^2\right) ;$$

similarly,

$$j_+ = \frac{\hbar}{2mi}\left[\psi^{*}_{+}(x)\,\psi'_{+}(x) - \text{c.c.}\right] = \frac{\hbar k}{m}\left(|C|^2 - |D|^2\right) .$$

The probability density for eigenfunctions of the Hamiltonian is time-independent (they are stationary states), and the associated probability current density is uniform (that is, independent of x); hence

$$j_- = j_+ \implies |A|^2 - |B|^2 = |C|^2 - |D|^2 .$$

One solution corresponds to an outgoing wave propagating to $x \longrightarrow \infty$, which is obtained by setting $D = 0$; we then have

$$j_i = \frac{\hbar k}{m}|A|^2 , \qquad j_r = -\frac{\hbar k}{m}|B|^2 , \qquad j_t = \frac{\hbar k}{m}|C|^2,$$

and

$$R = \frac{|j_r|}{|j_i|} = \frac{|B|^2}{|A|^2} , \qquad T = \frac{|j_t|}{|j_i|} = \frac{|C|^2}{|A|^2} .$$

The other, independent, solution has an outgoing wave propagating to $x \longrightarrow -\infty$, which is obtained by setting $A = 0$, yielding

$$j_i = -\frac{\hbar k}{m}|D|^2 , \qquad j_r = \frac{\hbar k}{m}|C|^2 , \qquad j_t = -\frac{\hbar k}{m}|B|^2,$$

and

$$R = \frac{|j_r|}{|j_i|} = \frac{|C|^2}{|D|^2} , \qquad T = \frac{|j_t|}{|j_i|} = \frac{|B|^2}{|D|^2} .$$

The two sets of R and T are identical. This fact is obvious for a parity-invariant potential but also holds for the generic (real) potential that we are considering here (see below as well as Problem 17).

Part 2

From the conservation of probability current, it follows that

$$|B|^2 + |C|^2 = |A|^2 + |D|^2 ,$$

which can be written in matrix form as

$$\begin{pmatrix} B^* & C^* \end{pmatrix}\begin{pmatrix} B \\ C \end{pmatrix} = \begin{pmatrix} A^* & D^* \end{pmatrix}\begin{pmatrix} A \\ D \end{pmatrix} .$$

We have

$$\begin{pmatrix} B \\ C \end{pmatrix} = \begin{pmatrix} S_{11} & S_{12} \\ S_{21} & S_{22} \end{pmatrix}\begin{pmatrix} A \\ D \end{pmatrix} ,$$

and taking the complex conjugate and transpose of both sides gives

$$\begin{pmatrix} B^* & C^* \end{pmatrix} = \begin{pmatrix} B^* \\ C^* \end{pmatrix}^T = \left[\begin{pmatrix} S_{11}^* & S_{12}^* \\ S_{21}^* & S_{22}^* \end{pmatrix} \begin{pmatrix} A^* \\ D^* \end{pmatrix} \right]^T = \begin{pmatrix} A^* \\ D^* \end{pmatrix}^T \begin{pmatrix} S_{11}^* & S_{12}^* \\ S_{21}^* & S_{22}^* \end{pmatrix}^T = \begin{pmatrix} A^* & D^* \end{pmatrix} \begin{pmatrix} S_{11}^* & S_{21}^* \\ S_{12}^* & S_{22}^* \end{pmatrix} .$$

By defining the matrix

$$\underline{S}^\dagger = \left(\underline{S}^* \right)^T = \left(\underline{S}^T \right)^* ,$$

it follows from current conservation that

$$\begin{pmatrix} B^* & C^* \end{pmatrix} \begin{pmatrix} B \\ C \end{pmatrix} = \begin{pmatrix} A^* & D^* \end{pmatrix} \underline{S}^\dagger \, \underline{S} \begin{pmatrix} A \\ D \end{pmatrix} = \begin{pmatrix} A^* & D^* \end{pmatrix} \begin{pmatrix} A \\ D \end{pmatrix} \implies \underline{S}^\dagger \underline{S} = \underline{I}.$$

We first set $D = 0$, obtaining

$$\begin{pmatrix} B \\ C \end{pmatrix} = \begin{pmatrix} S_{11} & S_{12} \\ S_{21} & S_{22} \end{pmatrix} \begin{pmatrix} A \\ 0 \end{pmatrix} \implies B = S_{11} A , \qquad C = S_{21} A \implies R = |S_{11}|^2 , \qquad T = |S_{21}|^2 ,$$

and we then set $A = 0$, obtaining

$$\begin{pmatrix} B \\ C \end{pmatrix} = \begin{pmatrix} S_{11} & S_{12} \\ S_{21} & S_{22} \end{pmatrix} \begin{pmatrix} 0 \\ D \end{pmatrix} \implies B = S_{12} D , \qquad C = S_{22} D \implies R = |S_{22}|^2 , \qquad T = |S_{12}|^2 ,$$

which then lead to

$$|S_{11}|^2 + |S_{21}|^2 = 1 = |S_{22}|^2 + |S_{12}|^2 .$$

Part 3

Using the unitarity of $\underline{S}$, we have

$$\begin{pmatrix} B \\ C \end{pmatrix} = \underline{S} \begin{pmatrix} A \\ D \end{pmatrix} \implies \begin{pmatrix} A \\ D \end{pmatrix} = \underline{S}^\dagger \begin{pmatrix} B \\ C \end{pmatrix} .$$

Now, under complex conjugation the roles of incoming and outgoing waves are inverted, that is, $A \rightleftharpoons B^*$ and $C \rightleftharpoons D^*$, or

$$\begin{pmatrix} B^* \\ C^* \end{pmatrix} = \underline{S}^\dagger \begin{pmatrix} A^* \\ D^* \end{pmatrix} ,$$

and hence taking the complex conjugate of both sides we find

$$\begin{pmatrix} B \\ C \end{pmatrix} = \left(\underline{S}^\dagger \right)^* \begin{pmatrix} A \\ D \end{pmatrix} \implies \left(\underline{S}^\dagger \right)^* = \underline{S} \implies \left[\left(\underline{S}^T \right)^* \right]^* = \underline{S} \implies \underline{S}^T = \underline{S},$$

and so the S-matrix is symmetric (and unitary), namely $S_{12} = S_{21}$. This implies that the reflection and transmission coefficients are the same whether we use the solution with outgoing wave for $x \gg 0$ or that with outgoing wave for $x \ll 0$.

We note in closing that a parity-invariant potential $v(x)$ also yields a symmetric $\underline{S}$-matrix. Indeed, rearranging the matrix gives

$$\begin{pmatrix} B \\ C \end{pmatrix} = \begin{pmatrix} S_{11} & S_{12} \\ S_{21} & S_{22} \end{pmatrix} \begin{pmatrix} A \\ D \end{pmatrix} \implies \begin{pmatrix} C \\ B \end{pmatrix} = \begin{pmatrix} S_{22} & S_{21} \\ S_{12} & S_{11} \end{pmatrix} \begin{pmatrix} D \\ A \end{pmatrix} ,$$

and under space inversion $A \rightleftharpoons D$ and $B \rightleftharpoons C$. This leads to

$$\begin{pmatrix} B \\ C \end{pmatrix} = \begin{pmatrix} S_{22} & S_{21} \\ S_{12} & S_{11} \end{pmatrix} \begin{pmatrix} A \\ D \end{pmatrix} \implies \begin{pmatrix} S_{22} & S_{21} \\ S_{12} & S_{11} \end{pmatrix} = \begin{pmatrix} S_{11} & S_{12} \\ S_{21} & S_{22} \end{pmatrix} \implies S_{11} = S_{22} \qquad S_{12} = S_{21} .$$

Thus, in one dimension parity invariance does not impose any additional constraints on $\underline{S}$ beyond those imposed by time-reversal invariance.

Problem 15   Scattering in a Parity-Invariant Potential: Phase-Shift Method

In this problem we consider the case of a parity-invariant potential $v(x) = v(-x)$, such that $v(|x| \longrightarrow \infty) = 0$, and describe the scattering process in terms of even and odd wave functions. The effect of the potential is to introduce phase shifts in the asymptotic behaviors of these wave functions. The reflection and transmission coefficients are related to these phase shifts, which then contain all relevant information about the scattering process.[1]

1. Write the asymptotic even (e) wave function as

$$\begin{aligned} \psi_k^e(x) &= A\,e^{ikx} + B\,e^{-ikx} & x \ll 0 \\ &= B\,e^{ikx} + A\,e^{-ikx} & x \gg 0 , \end{aligned}$$

and the asymptotic odd (o) wave function as

$$\begin{aligned} \psi_k^o(x) &= C\,e^{ikx} - D\,e^{-ikx} & x \ll 0 \\ &= D\,e^{ikx} - C\,e^{-ikx} & x \gg 0 . \end{aligned}$$

Show that $|A| = |B|$ and $|C| = |D|$. Introduce the phase factors

$$\frac{B}{A} = e^{2i\delta_e} \qquad \text{and} \qquad \frac{D}{C} = e^{2i\delta_o} ,$$

and show that, up to overall factors, these wave functions can be written as

$$\begin{aligned} \psi_k^e(x) &= \cos(kx - \delta_e) & \psi_k^o(x) &= \sin(kx - \delta_o) & x \ll 0 \\ &= \cos(kx + \delta_e) & &= \sin(kx + \delta_o) & x \gg 0 . \end{aligned}$$

2. By taking appropriate linear combinations of $\psi_k^e(x)$ and $\psi_k^o(x)$, construct the wave functions having outgoing waves only in the $x \gg 0$ and $x \ll 0$ regions, respectively. Read off the reflection and transmission coefficients in terms of the phase shifts $\delta_e(k)$ and $\delta_o(k)$, which are of course functions of k. Using the results of Problem 14, obtain the S-matrix in terms of these phase shifts.

Solution

Part 1

Conservation of the probability current yields $j(x \ll 0) = j(x \gg 0)$ and hence, for the even wave function,

$$k(|A|^2 - |B|^2) = k(|B|^2 - |A|^2) \implies |A| = |B| ,$$

---

[1] The phase-shift method is especially useful in the case of a rotationally invariant (central) potential in three dimensions. In one dimension, rotational invariance is equivalent to invariance under parity.

and similarly for the odd wave function. After introducing the phase factors δ_e and δ_o, we have

$$\psi_k^e(x) = A\left(e^{ikx} + e^{-ikx+2i\delta_e}\right) = A\,e^{i\delta_e}\left[e^{i(kx-\delta_e)} + e^{-i(kx-\delta_e)}\right] = 2A\,e^{i\delta_e}\,\cos(kx - \delta_e) \qquad x \ll 0$$
$$= A\left(e^{ikx+2i\delta_e} + e^{-ikx}\right) = A\,e^{i\delta_e}\left[e^{i(kx+\delta_e)} + e^{-i(kx+\delta_e)}\right] = 2A\,e^{i\delta_e}\,\cos(kx + \delta_e) \qquad x \gg 0$$

and

$$\psi_k^o(x) = C\left(e^{ikx} - e^{-ikx+2i\delta_o}\right) = C\,e^{i\delta_o}\left[e^{i(kx-\delta_o)} - e^{-i(kx-\delta_o)}\right] = 2iC\,e^{i\delta_o}\,\sin(kx - \delta_o) \qquad x \ll 0$$
$$= C\left(e^{ikx+2i\delta_o} - e^{-ikx}\right) = C\,e^{i\delta_o}\left[e^{i(kx+\delta_o)} - e^{-i(kx+\delta_e)}\right] = 2iC\,e^{i\delta_o}\,\sin(kx + \delta_o) \qquad x \gg 0 \,,$$

which is the required result.

Part 2

Consider the linear combinations

$$\psi_k^{(\pm)}(x) = e^{i\delta_e(k)}\,\psi_k^e(x) \pm i\,e^{i\delta_o(k)}\,\psi_k^o(x) \,.$$

We find

$$\psi_k^{(+)}(x) = e^{i\delta_e}\frac{e^{i(kx-\delta_e)} + e^{-i(kx-\delta_e)}}{2} + i\,e^{i\delta_o}\frac{e^{i(kx-\delta_o)} - e^{-i(kx-\delta_o)}}{2i} = e^{ikx} + e^{-ikx}\frac{e^{2i\delta_e} - e^{2i\delta_o}}{2} \qquad x \ll 0$$
$$= e^{i\delta_e}\frac{e^{i(kx+\delta_e)} + e^{-i(kx+\delta_e)}}{2} + i\,e^{i\delta_o}\frac{e^{i(kx+\delta_o)} - e^{-i(kx+\delta_o)}}{2i} = e^{ikx}\frac{e^{2i\delta_e} + e^{2i\delta_o}}{2} \qquad x \gg 0 \,,$$

and similarly

$$\psi_k^{(-)}(x) = e^{-ikx}\frac{e^{2i\delta_e} + e^{2i\delta_o}}{2} \qquad x \ll 0$$
$$= e^{-ikx} + e^{ikx}\frac{e^{2i\delta_e} - e^{2i\delta_o}}{2} \qquad x \gg 0 \,.$$

Note that

$$\frac{e^{2i\delta_e} + e^{2i\delta_o}}{2} = e^{i(\delta_e+\delta_o)}\,\cos(\delta_e - \delta_o) \,, \qquad \frac{e^{2i\delta_e} - e^{2i\delta_o}}{2} = i\,e^{i(\delta_e+\delta_o)}\,\sin(\delta_e - \delta_o) \,,$$

and the reflection and transmission coefficients read

$$R(k) = \sin^2[\delta_e(k) - \delta_o(k)] \,, \qquad T(k) = \cos^2[\delta_e(k) - \delta_o(k)] \,.$$

Using the results of the Problem 14, we find that

$$\underline{S} = \begin{bmatrix} i\,e^{i(\delta_e+\delta_o)}\,\sin(\delta_e - \delta_o) & e^{i(\delta_e+\delta_o)}\,\cos(\delta_e - \delta_o) \\ e^{i(\delta_e+\delta_o)}\,\cos(\delta_e - \delta_o) & i\,e^{i(\delta_e+\delta_o)}\,\sin(\delta_e - \delta_o) \end{bmatrix} \,,$$

and it is easily verified that $\underline{S}$ is unitary and symmetric.

Problem 16 Application to Scattering in a Repulsive δ-Function Potential

Consider a repulsive δ-function potential $v(x) = v_0\,\delta(x)$ that is even.

1. Construct the even (e) and odd (o) wave functions for a given energy $\epsilon = k^2$, and show that they can be expressed as

$$\psi_k^e(x) = \cos[k|x| + \delta(k)] \,, \qquad \psi_k^o(x) = \sin(kx) \,,$$

up to irrelevant multiplicative factors. Obtain the phase shift $\delta(k)$.

2. Explain why the odd wave function is unaffected by $v(x)$. What is the effect of $v(x)$ on the nodes of the even wave function?

3. Express the reflection and transmission coefficients in terms of $\delta(k)$.

Solution

Part 1

We write the even (e) wave function as

$$\psi_k^e(x) = \alpha\, e^{ikx} + \beta\, e^{-ikx} \qquad x < 0$$
$$= \beta\, e^{ikx} + \alpha\, e^{-ikx} \qquad x > 0$$

and the odd (o) wave function as

$$\psi_k^o(x) = \alpha'\, e^{ikx} - \beta'\, e^{-ikx} \qquad x < 0$$
$$= \beta'\, e^{ikx} - \alpha'\, e^{-ikx} \qquad x > 0,$$

where α and β, and α' and β', are constants; conservation of the probability current leads to $|\alpha| = |\beta|$ and $|\alpha'| = |\beta'|$. The boundary conditions at the origin imposed by the δ-function potential require

$$\psi_k^e(0^-) = \psi_k^e(0^+) \quad \Longrightarrow \quad \alpha + \beta = \beta + \alpha,$$
$$\psi_k^{e\,\prime}(0^+) - \psi_k^{e\,\prime}(0^-) = v_0\,\psi_k^e(0^+) \quad \Longrightarrow \quad ik\big(\beta - \alpha\big) - ik\big(\alpha - \beta\big) = v_0\big(\alpha + \beta\big)\;;$$

and the first equation is a tautology, while the second gives

$$\beta = \underbrace{\frac{2ik + v_0}{2ik - v_0}}_{z \text{ with } |z|=1}\,\alpha,$$

For the odd wave function we find $\beta' = \alpha'$. We note that the complex number z has unit magnitude, and therefore it can be written as

$$\frac{2ik + v_0}{2ik - v_0} \equiv e^{2i\delta(k)}, \qquad \delta(k) = -\frac{1}{2}\tan^{-1}\left(\frac{4kv_0}{4k^2 - v_0}\right).$$

We can now express the even and odd wave functions as

$$\psi_k^e(x) = \alpha\left[e^{ikx} + e^{-ikx + 2i\delta(k)}\right] = \alpha\, e^{i\delta(k)}\left[e^{i[kx - \delta(k)]} + e^{-i[kx - \delta(k)]}\right] \qquad x < 0$$
$$= \alpha\left[e^{ikx + 2i\delta(k)} + e^{-ikx}\right] = \alpha\, e^{i\delta(k)}\left[e^{i[kx + \delta(k)]} + e^{-i[kx + \delta(k)]}\right] \qquad x > 0$$

and

$$\psi_k^o(x) = \alpha'\left(e^{ikx} - e^{-ikx}\right) \qquad x > 0$$
$$= \alpha'\left(e^{ikx} - e^{-ikx}\right) \qquad x < 0,$$

and thus up to irrelevant (complex) constants, respectively given by $2\alpha\, e^{i\delta(k)}$ and $2i\alpha'$, as

$$\psi_k^e(x) = \cos[k|x| + \delta(k)], \qquad \psi_k^o(x) = \sin(kx),$$

over the whole region $-\infty < x < \infty$.

Part 2

The potential produces a phase shift $\delta(k)$ in the even wave function but has no effect on the odd wave function. Indeed, $\sin(kx)$ is a free solution of the Schrödinger equation. Since the odd wave function $\psi_k^o(x)$ must vanish at the origin, it does not feel the potential, which only acts at $x = 0$. In contrast, the even wave function is shifted as a whole, the nodes in $x > 0$, for example, occurring at

$$x_n = \frac{n\pi}{2k} - \frac{\delta(k)}{k} \qquad n = 1, 3, 5, \ldots,$$

rather than at $n\pi/(2k)$ as in the free-particle case. Whether it is shifted in or out depends on the sign of the phase shift.

Part 3

In order to relate the reflection and transmission coefficients to the phase shift, we need to relate the solutions in terms of standing waves to those in terms of traveling waves. We write the latter as

$$\psi_k^{(+)}(x) = e^{ikx} + A(k)\, e^{-ikx} \qquad x < 0$$
$$= B(k)\, e^{ikx} \qquad x > 0$$

and

$$\psi_k^{(-)}(x) = D(k)\, e^{-ikx} \qquad x < 0$$
$$= e^{-ikx} + C(k)\, e^{ikx} \qquad x > 0\,.$$

The linear combinations are easily found as

$$\psi_k^{(+)}(x) = e^{i\delta(k)}\, \psi_k^e(x) + i\psi_k^o(x)\,,$$
$$\psi_k^{(-)}(x) = e^{i\delta(k)}\, \psi_k^e(x) - i\psi_k^o(x)\,.$$

Indeed, we have

$$\psi_k^{(+)}(x) = e^{i\delta}\, \underbrace{\frac{1}{2}\left(e^{ikx-i\delta} + e^{-ikx+i\delta}\right)}_{\cos(kx-\delta)} + i\underbrace{\frac{1}{2i}\left(e^{ikx} - e^{-ikx}\right)}_{\sin(kx)} \qquad x < 0$$

$$= e^{i\delta}\, \underbrace{\frac{1}{2}\left(e^{ikx+i\delta} + e^{-ikx-i\delta}\right)}_{\cos(kx+\delta)} + i\underbrace{\frac{1}{2i}\left(e^{ikx} - e^{-ikx}\right)}_{\sin(kx)} \qquad x > 0\,,$$

from which we can read off

$$A(k) = \frac{1}{2}\left[e^{2i\delta(k)} - 1\right] = i\,e^{i\delta(k)} \sin\delta(k)\,, \qquad B(k) = \frac{1}{2}\left[e^{2i\delta(k)} + 1\right] = e^{i\delta(k)} \cos\delta(k)\,.$$

The reflection and transmission coefficients are obtained as

$$R(k) = |A(k)|^2 = \sin^2\delta(k) = \frac{1 - \cos[2\,\delta(k)]}{2}\,, \qquad T(k) = |B(k)|^2 = \cos^2\delta(k) = \frac{1 + \cos[2\,\delta(k)]}{2}\,.$$

After making the insertion

$$e^{2i\delta(k)} = \frac{2ik + v_0}{2ik - v_0} \implies \cos[2\,\delta(k)] = \frac{4k^2 - v_0^2}{4k^2 + v_0^2} \quad \text{and} \quad \sin[2\,\delta(k)] = -\frac{4kv_0}{4k^2 + v_0^2}\,,$$

we find as in the treatment based on traveling waves,

$$R(k) = \frac{v_0^2}{4k^2 + v_0^2}, \qquad T(k_0) = \frac{4k^2}{4k^2 + v_0^2}.$$

Problem 17 Reflection and Transmission in a Generic Potential

Consider a generic potential $v(x)$ such that $v(x \longrightarrow \pm\infty) = v_\pm$, where the constants $v_\pm$ are such that $v_- \geq v_+$. We are interested in the scattering-state solutions occurring for energies $\epsilon \geq v_-$.

1. Show that in the asymptotic regions $x \longrightarrow \pm\infty$ the two independent solutions can be written as

$$\psi^{(+)}(x) = A\,e^{ik_-x} + B\,e^{-ik_-x} \qquad x \ll 0$$
$$= C\,e^{ik_+x} \qquad\qquad\quad x \gg 0$$

and

$$\psi^{(-)}(x) = B\,e^{-ik_-x} \qquad\qquad\qquad x \ll 0$$
$$= C\,e^{ik_+x} + D\,e^{-ik_+x} \qquad x \gg 0,$$

where $k_\pm = \sqrt{\epsilon - v_\pm}$. Obtain the reflection and transmission coefficients, $R^{(l)}(k)$ and $T^{(l)}(k)$, respectively, corresponding to the solution $\psi^{(l)}(x)$, with $l = +$ and $-$, in terms of A, B, C, and D. The goal of this problem is to show that

$$R^{(+)}(k) = R^{(-)}(k), \qquad T^{(+)}(k) = T^{(-)}(k).$$

Such a result, while obvious for a parity-invariant $v(x)$, also holds for a generic $v(x)$.

2. Write the general asymptotic solution as

$$\psi(x) = A\,e^{ik_-x} + B\,e^{-ik_-x} \qquad x \ll 0$$
$$= C\,e^{ik_+x} + D\,e^{-ik_+x} \qquad x \gg 0,$$

and, using the conservation of probability, show that

$$|\overline{B}|^2 + |\overline{C}|^2 = |\overline{A}|^2 + |\overline{D}|^2$$

or, in matrix notation,

$$\begin{pmatrix}\overline{B}^* & \overline{C}^*\end{pmatrix}\begin{pmatrix}\overline{B}\\\overline{C}\end{pmatrix} = \begin{pmatrix}\overline{A}^* & \overline{D}^*\end{pmatrix}\begin{pmatrix}\overline{A}\\\overline{D}\end{pmatrix},$$

where the coefficients with overbars are related to the corresponding coefficients introduced above. Determine this relationship.

3. Justify why there must exist a matrix relation of the type

$$\begin{pmatrix}\overline{B}\\\overline{C}\end{pmatrix} = \underbrace{\begin{pmatrix}S_{11} & S_{12}\\S_{21} & S_{22}\end{pmatrix}}_{\underline{S}}\begin{pmatrix}\overline{A}\\\overline{D}\end{pmatrix}$$

between these coefficients. Show that $\underline{S}$ is unitary. Express the reflection and transmission coefficients derived in part 1 in terms of the matrix elements of $\underline{S}$.

4. Noting that $[\psi^{(l)}(x)]^*$ is a solution corresponding to the same energy, deduce that $\underline{S}$, in addition to being unitary, is also symmetric, that is, $\underline{S}^T = \underline{S}$, and that, as a consequence, $R^{(+)}(k) = R^{(-)}(k)$ and $T^{(+)}(k) = T^{(-)}(k)$.

Solution

Part 1

The (asymptotic) probability current densities are obtained, respectively, as

$$j^{(+)}(x \ll 0) = \underbrace{\frac{\hbar k_-}{m} |A|^2}_{j_i^{(+)}} \underbrace{- \frac{\hbar k_-}{m} |B|^2}_{j_r^{(+)}}, \qquad j^{(+)}(x \gg 0) = \underbrace{\frac{\hbar k_+}{m} |C|^2}_{j_t^{(+)}}$$

and

$$j^{(-)}(x \ll 0) = \underbrace{-\frac{\hbar k_-}{m} |B|^2}_{j_t^{(-)}}, \qquad j^{(-)}(x \gg 0) = \underbrace{\frac{\hbar k_+}{m} |C|^2}_{j_r^{(-)}} \underbrace{- \frac{\hbar k_+}{m} |D|^2}_{j_i^{(-)}}.$$

Conservation of the probability current then requires that $j_i^{(\alpha)} + j_r^{(\alpha)} = j_t^{(\alpha)}$ with $\alpha = 1, 2$, from which we deduce the reflection and transmission coefficients,

$$R^{(+)} = \frac{|B|^2}{|A|^2}, \qquad T^{(+)} = \frac{k_+}{k_-} \frac{|C|^2}{|A|^2}$$

and

$$R^{(-)} = \frac{|C|^2}{|D|^2}, \qquad T^{(-)} = \frac{k_-}{k_+} \frac{|B|^2}{|D|^2},$$

where clearly $R^{(\alpha)} + T^{(\alpha)} = 1$.

Part 2

Using the asymptotic form of the solution, conservation of the probability current yields

$$k_-(|A|^2 - |B|^2) = k_+(|C|^2 - |D|^2) \implies k_-|B|^2 + k_+|C|^2 = k_-|A|^2 + k_+|D|^2.$$

By rescaling these constants as

$$\overline{A} = \sqrt{k_-}A, \qquad \overline{B} = \sqrt{k_-}B, \qquad \overline{C} = \sqrt{k_+}A, \qquad \overline{D} = \sqrt{k_+}D,$$

we have

$$|\overline{B}|^2 + |\overline{C}|^2 = |\overline{A}|^2 + |\overline{D}|^2 \implies \left(\overline{B}^* \;\; \overline{C}^*\right)\begin{pmatrix} \overline{B} \\ \overline{C} \end{pmatrix} = \left(\overline{A}^* \;\; \overline{D}^*\right)\begin{pmatrix} \overline{A} \\ \overline{D} \end{pmatrix},$$

where in the last step we have introduced, for later convenience, two-component row and column vectors.

Part 3

The asymptotic solutions in the $x \longrightarrow \pm\infty$ regions must be matched, so that the wave function, along with its first derivative, is continuous everywhere. This implies a relation between the pair $\overline{B}, \overline{C}$ and the pair $\overline{A}, \overline{D}$, that is,

$$\overline{B} = S_{11}\overline{A} + S_{12}\overline{D}, \qquad \overline{C} = S_{21}\overline{A} + S_{22}\overline{D} \implies \begin{pmatrix} \overline{B} \\ \overline{C} \end{pmatrix} = \underbrace{\begin{pmatrix} S_{11} & S_{12} \\ S_{21} & S_{22} \end{pmatrix}}_{\underline{S}}\begin{pmatrix} \overline{A} \\ \overline{D} \end{pmatrix},$$

and the coefficients S_{ij}, which are understood to be functions of the energy ϵ, will depend on the potential. Conservation of probability then yields

$$\left(\overline{A}^* \ \overline{D}^*\right) \underline{S}^\dagger \underline{S} \begin{pmatrix} \overline{A} \\ \overline{D} \end{pmatrix} = \left(\overline{A}^* \ \overline{D}^*\right) \begin{pmatrix} \overline{A} \\ \overline{D} \end{pmatrix} \implies \underline{S}^\dagger \underline{S} = \underline{1} \, ,$$

that is, the matrix $\underline{S}$ must be unitary, or, in terms of matrix elements,

$$|S_{11}|^2 + |S_{21}|^2 = 1 = |S_{12}|^2 + |S_{22}|^2 \, , \qquad S_{11}^* S_{12} + S_{21}^* S_{22} = 0 \, .$$

The previously calculated reflection and transmission coefficients are related to the matrix elements of $\underline{S}$ via

$$R^{(+)} = \frac{|B|^2}{|A|^2} = \frac{|\overline{B}|^2}{|\overline{A}|^2} = |S_{11}|^2 \, , \qquad T^{(+)} = \frac{k_+}{k_-} \frac{|C|^2}{|A|^2} = \frac{|\overline{C}|^2}{|\overline{A}|^2} = |S_{21}|^2 \, ,$$

and similarly

$$R^{(-)} = |S_{22}|^2 \, , \qquad T^{(-)} = |S_{12}|^2 \, .$$

Part 4

In fact, the matrix $\underline{S}$ is not only unitary but also symmetric ($S_{12} = S_{21}$). This can be seen by noting that, since the potential $v(x)$ is real, $\psi^*(x)$ in part 2 is also a solution corresponding to the same energy,

$$\psi^*(x) = A^* \, e^{-ik_- x} + B^* \, e^{ik_- x} \qquad x \ll 0$$
$$= C^* \, e^{-ik_+ x} + D^* \, e^{ik_+ x} \qquad x \gg 0 \, .$$

Complex conjugation converts incoming waves to outgoing waves and vice versa, that is,

$$A^* \rightleftharpoons B \, , \qquad C^* \rightleftharpoons D \, .$$

Exploiting the unitarity of $\underline{S}$, we have

$$\begin{pmatrix} \overline{A} \\ \overline{D} \end{pmatrix} = \underline{S}^\dagger \begin{pmatrix} \overline{B} \\ \overline{C} \end{pmatrix} \, ,$$

and, by virtue of the above correspondence,

$$\begin{pmatrix} \overline{B}^* \\ \overline{C}^* \end{pmatrix} = \underline{S}^\dagger \begin{pmatrix} \overline{A}^* \\ \overline{D}^* \end{pmatrix} \implies \begin{pmatrix} \overline{B} \\ \overline{C} \end{pmatrix} = \left(\underline{S}^\dagger\right)^* \begin{pmatrix} \overline{A} \\ \overline{D} \end{pmatrix} \, .$$

Comparing it with the definition in part 3,

$$\begin{pmatrix} \overline{B} \\ \overline{C} \end{pmatrix} = \underline{S} \begin{pmatrix} \overline{A} \\ \overline{D} \end{pmatrix} \, ,$$

we deduce that

$$\left(\underline{S}^\dagger\right)^* = \underline{S} \implies \underline{S}^T = \underline{S} \, ,$$

where in the last step we have used $\underline{S}^\dagger = \left(\underline{S}^T\right)^*$. From $S_{12} = S_{21}$, we obtain $|S_{11}|^2 = |S_{22}|^2$ from the unitarity conditions on the matrix elements of $\underline{S}$, and hence conclude that

$$R^{(+)} = R^{(-)} \, , \qquad T^{(+)} = T^{(-)} \, ,$$

as anticipated.

6 Mathematical Formulation of Quantum Mechanics

In quantum mechanics, the wave function $\Psi(\mathbf{r}, t)$ plays a central role,[1] as it defines the state of the particle and allows us to calculate the probability of obtaining a given result from the measurement of a given dynamical variable (for example, the particle's position or momentum). Because of the interpretation of $|\Psi(\mathbf{r}, t)|^2$ as a probability density, we must have

$$\int d\mathbf{r} \, |\Psi(\mathbf{r}, t)|^2 < \infty \,. \tag{6.1}$$

Therefore, we are naturally led to consider the set of square integrable functions for which the above integral is convergent. In this section, we begin by reviewing the Hilbert space of square-integrable functions and introducing operators and bases in this space. Next, we generalize these concepts and introduce abstract state spaces and the notion of representations. We conclude with a review of observables and tensor products.

6.1 Hilbert Space of Square-Integrable Functions

The set of square-integrable functions, denoted as $L^2(\mathbb{R})$, forms a Hilbert space. Generally, a Hilbert space is a linear vector space in which a scalar product has been defined. In the specific case of $L^2(\mathbb{R})$, this implies that, if the generally complex-valued functions $\psi_1(\mathbf{r})$ and $\psi_2(\mathbf{r})$ are square-integrable, that is, $\psi_1(\mathbf{r}), \psi_2(\mathbf{r}) \in L^2(\mathbb{R})$, then any linear combination $\lambda_1 \psi_1(\mathbf{r}) + \lambda_2 \psi_2(\mathbf{r})$, where the λ_i are complex numbers, is also square-integrable. Any function belonging to $L^2(\mathbb{R})$ can be thought of as a "vector" in this space.

Given two vectors $\varphi(\mathbf{r}), \psi(\mathbf{r}) \in L^2(\mathbb{R})$, their scalar product (φ, ψ) is defined as

$$(\varphi, \psi) = \int d\mathbf{r} \, \varphi^*(\mathbf{r}) \, \psi(\mathbf{r}) \,. \tag{6.2}$$

This definition associates a complex number, denoted as (φ, ψ), to any two square-integrable functions $\varphi(\mathbf{r})$ and $\psi(\mathbf{r})$. From the above definition, the following properties follow easily:

1. $(\psi, \varphi) = (\varphi, \psi)^*$;
2. $(\varphi, \lambda_1 \psi_1 + \lambda_2 \psi_2) = \lambda_1 (\varphi, \psi_1) + \lambda_2 (\varphi, \psi_2)$;
3. $(\lambda_1 \varphi_1 + \lambda_2 \varphi_2, \psi) = \lambda_1^* (\varphi_1, \psi) + \lambda_2^* (\varphi_2, \psi)$.
4. A norm is associated with any $\psi \in L^2(\mathbb{R})$ and defined as

$$\|\psi\| = \sqrt{(\psi, \psi)} = \left[\int d\mathbf{r} \, |\psi(\mathbf{r})|^2 \right]^{1/2} \geq 0 \,, \tag{6.3}$$

and $\|\psi\| = 0$ iff $\psi(\mathbf{r}) = 0$.

---

[1] Note that the considerations made here for wave functions in the coordinate representation also hold for wave functions in the momentum representation.

An operator is generally any mapping of a Hilbert space onto itself. Specifically, an operator A in $L^2(\mathbb{R})$ takes any vector $\in L^2(\mathbb{R})$ into another vector $\in L^2(\mathbb{R})$:

$$\psi(\mathbf{r}) \longmapsto \psi'(\mathbf{r}) = A\psi(\mathbf{r}) . \tag{6.4}$$

In quantum mechanics we are dealing primarily with linear operators (the exception being the time-reversal operator, which is anti-linear). A linear operator is such that $A\left[\lambda_1\psi_1(\mathbf{r}) + \lambda_2\psi_2(\mathbf{r})\right] = \lambda_1 A\psi_1(\mathbf{r}) + \lambda_2 A\psi_2(\mathbf{r})$. Given two operators A and B, the product operator AB is defined as the mapping $\psi(\mathbf{r}) \longmapsto \psi' = A\left[B\psi(\mathbf{r})\right]$, and generally $AB \neq BA$, so that the ordering in which the operators are applied is important, unless, of course, $AB = BA$, when the operators are said to commute. Of particular importance are adjoint, self-adjoint or hermitian, and unitary operators. Given an operator A, its **adjoint operator** $A^\dagger$, if it exists, is defined as that operator for which

$$(\varphi, A\,\psi) = \int d\mathbf{r}\, \varphi^*(\mathbf{r}) \left[A\psi(\mathbf{r})\right] = \int d\mathbf{r} \left[A^\dagger\varphi(\mathbf{r})\right]^* \psi(\mathbf{r}) = (A^\dagger\varphi, \psi) . \tag{6.5}$$

By using property 1 of the scalar product, we also have $(\varphi, A\psi) = (\psi, A^\dagger\varphi)^*$. The following properties follow from the definition of the adjoint operator:

1. $\left(A^\dagger\right)^\dagger = A$;
2. $(\lambda A)^\dagger = \lambda^* A^\dagger$;
3. $(A + B)^\dagger = A^\dagger + B^\dagger$;
4. $(AB)^\dagger = B^\dagger A^\dagger$.

A **hermitian** or **self-adjoint operator** A is such that $A^\dagger = A$, while a **unitary operator** U is such that, for any two vectors $\varphi(\mathbf{r})$, $\psi(\mathbf{r}) \in L^2(\mathbb{R})$, it preserves the scalar product,

$$(U\varphi, U\psi) = (\varphi, \psi) , \tag{6.6}$$

which also implies that

$$(U\varphi, U\psi) = (U^\dagger U\varphi, \psi) = (\varphi, \psi) \implies U^\dagger U = \mathbb{1} , \tag{6.7}$$

where $\mathbb{1}$ is the identity operator. By multiplying both sides of the last relation from the right by U^{-1}, we see that the adjoint of a unitary operator is equal to its inverse, $U^\dagger = U^{-1}$.

Hermitian and unitary operators play a special role in quantum mechanics. The former are associated with observables – the dynamical properties of the system that we can, at least in principle, measure (position, momentum, and energy, for example). The latter induce symmetry transformations on the system, such as, for example, translations and rotations. The time evolution of a quantum state is also governed by a unitary operator.

Before introducing the concept of a basis, it is useful to recall the concept of linear independence. A set of vectors $\psi_1(\mathbf{r}), \psi_2(\mathbf{r}), \ldots, \psi_p(\mathbf{r})$ is said to be linearly independent when

$$\lambda_1\psi_1(\mathbf{r}) + \lambda_2\psi_2(\mathbf{r}) + \cdots + \lambda_p\psi_p(\mathbf{r}) = 0 , \tag{6.8}$$

where the λ_i are complex numbers, is satisfied iff $\lambda_1 = \lambda_2 = \cdots = \lambda_p = 0$. In particular, if the vectors $\psi_m(\mathbf{r})$ are orthogonal, so that

$$(\psi_m, \psi_n) = 0 \qquad \text{if } m \neq n , \tag{6.9}$$

then they are independent, since taking the scalar product of Eq. (6.8) with any of the ψ_m yields

$$(\psi_m, \lambda_1\psi_1 + \lambda_2\psi_2 + \cdots + \lambda_p\psi_p) = \lambda_m(\psi_m, \psi_m) = 0 \implies \lambda_m = 0 \qquad m = 1, \ldots, p . \tag{6.10}$$

The converse is not true, and the vectors of an independent set are not necessarily orthogonal to each other. However, if p vectors are independent, they can be made orthonormal by the Gram–Schmidt procedure, as illustrated by Problem 4 in Chapter 3.

The infinite set of orthonormal functions $\phi_1(\mathbf{r}), \ldots, \phi_p(\mathbf{r}), \ldots$ with $(\phi_m, \phi_n) = \delta_{mn}$ is a basis in $L^2(\mathbb{R})$ if the series

$$\sum_{m=1}^{\infty} c_m \phi_m(\mathbf{r}), \qquad c_m = (\phi_m, \psi), \tag{6.11}$$

converges to $\psi(\mathbf{r})$, in the sense that $\psi_p(\mathbf{r}) = \psi(\mathbf{r}) - \sum_{m=1}^{p} c_m \phi_m(\mathbf{r})$ and $\lim_{p \to \infty} \|\psi_p\| \longrightarrow 0$. Such a basis is denumerable (the subscript p identifying the individual basis functions takes on integer values) and is made up of functions which are square-integrable and, therefore, themselves in $L^2(\mathbb{R})$. It is known as a discrete basis. However, it is also convenient to introduce bases which are not in $L^2(\mathbb{R})$ but in terms of which any square-integrable function can nonetheless be expanded. These are so-called continuous bases. In a continuous basis, the subscript identifying the basis functions assumes a continuum of values, $\phi_m(\mathbf{r}) \longrightarrow \phi_\alpha(\mathbf{r})$, where α varies over some interval I of the real axis. The "orthonormality condition" now reads

$$(\phi_\alpha, \phi_\beta) = \delta(\alpha - \beta); \tag{6.12}$$

it involves a δ-function rather than a Kronecker δ. The expansion of a vector $\psi(\mathbf{r})$ is given by

$$\psi(\mathbf{r}) = \int_I d\alpha\, c(\alpha)\, \phi_\alpha(\mathbf{r}), \qquad c(\alpha) = (\phi_\alpha, \psi) = \int d\mathbf{r}\, \phi_\alpha^*(\mathbf{r})\, \psi(\mathbf{r}). \tag{6.13}$$

An important example of a continuous basis is that formed by the eigenfunctions of the momentum operator $-i\hbar\boldsymbol{\nabla}$ (a hermitian operator),

$$\phi_\mathbf{p}(\mathbf{r}) = \frac{e^{i\mathbf{p}\cdot\mathbf{r}/\hbar}}{(2\pi\hbar)^{3/2}}, \qquad -i\hbar\,\boldsymbol{\nabla}\phi_\mathbf{p}(\mathbf{r}) = \mathbf{p}\,\phi_\mathbf{p}(\mathbf{r}), \tag{6.14}$$

where the components of $\mathbf{p}$, which vary in $-\infty < p_\alpha < \infty$, are the eigenvalues of the momentum operator and label the basis functions. The orthonormality condition reads, in this case,

$$(\phi_{\mathbf{p}'}, \phi_\mathbf{p}) = \frac{1}{(2\pi\hbar)^3} \int d\mathbf{r}\, e^{i(\mathbf{p}-\mathbf{p}')\cdot\mathbf{r}/\hbar} = \delta(\mathbf{p} - \mathbf{p}'). \tag{6.15}$$

Thus the momentum-space wave function $\widetilde{\psi}(\mathbf{p}) = (\phi_\mathbf{p}, \psi)$ introduced in previous sections can be viewed as the component of the vector $\psi(\mathbf{r})$ in the basis $\phi_\mathbf{p}(\mathbf{r})$.

In general we will be dealing with bases that have discrete and continuum portions. For example, the eigenfunctions of a Hamiltonian for a particle in a one-dimensional attractive δ-function potential form a basis which has a discrete part, consisting of the single bound-state wave function with negative energy, and a continuous part, consisting of the two independent scattering wave functions corresponding to any energy ≥ 0.

6.2 Abstract Hilbert Space

In the previous section we established that the wave function describing the state of a system can be uniquely represented by its components in a basis. The situation is similar to that in ordinary space, where knowledge of a vector is equivalent to knowing its components in a coordinate system. These components depend, of course, on the particular coordinate system we have chosen.

However, we can also think of a vector in a more abstract way, as a quantity with magnitude and direction. Then, we can describe relations between vectors without having to specify a coordinate system. For example, we can say that two vectors are parallel or orthogonal to each other. These statements are independent of the coordinate system. While the representation of a vector in terms of components is often more convenient for doing calculations, thinking of a vector as an entity in itself, while more abstract, has advantages too.

Ket space

Hereafter, the states of a system are represented as vectors ("state vectors") belonging to a Hilbert space, or "state space," which we denote as $\mathbb{E}$. The dimensionality of this space can be finite or infinite, depending on the system under consideration. Following Dirac, a state vector is denoted by a "ket" $|\psi\rangle$, which is assumed to contain all information about the physical state. Kets satisfy the superposition principle and hence, if $|\psi_1\rangle, |\psi_2\rangle \in \mathbb{E}$ then so does any linear combination $\lambda_1|\psi_1\rangle + \lambda_2|\psi_2\rangle$, where the λ_i are complex numbers. It is also postulated that the states $|\psi\rangle$ and $\lambda|\psi\rangle$ with $\lambda \neq 0$ represent the same physical state: only the "direction" of the state in $\mathbb{E}$ matters.

Bra space and inner (or scalar) product

It is convenient to introduce the concept of "bra space", which is dual to "ket space." It is postulated that to every ket $|\psi\rangle$ corresponds a bra denoted as $\langle\psi|$ in this dual space. The ket-bra correspondence is one-to-one, that is, $|\psi\rangle \rightleftharpoons \langle\psi|$. It is also postulated that the correspondence is antilinear, in the sense that

$$\lambda_1|\psi_1\rangle + \lambda_2|\psi_2\rangle \rightleftharpoons \lambda_1^*\langle\psi_1| + \lambda_2^*\langle\psi_2| \,. \tag{6.16}$$

The inner or scalar product associates a complex number with a bra $\langle\varphi|$ and a ket $|\psi\rangle$ (taken in this order); it is indicated as

$$\langle\varphi|\psi\rangle = ((\langle\varphi|) \cdot (|\psi\rangle)) \,. \tag{6.17}$$

The inner product is assumed to satisfy the same properties as the scalar product in $L^2(\mathbb{R})$. The kets $|\varphi\rangle$ and $|\psi\rangle$ are said to be orthogonal if $\langle\varphi|\psi\rangle = 0$ (or, equivalently, if $\langle\psi|\varphi\rangle = 0$).

The operator concept can be carried over from the previous section. Specifically, an operator maps a ket into another ket,

$$|\psi\rangle \longmapsto |\psi'\rangle = \hat{A}|\psi\rangle \,, \tag{6.18}$$

where we denote operators acting in the state space $\mathbb{E}$ with a caret on a capital or lower-case letter. Two operators $\hat{A}$ and $\hat{B}$ are the same if for any ket $|\psi\rangle$ we have $\hat{A}|\psi\rangle = \hat{B}|\psi\rangle$); by contrast, $\hat{A}$ is the null operator if for any $|\psi\rangle$ the state $\hat{A}|\psi\rangle = 0$. Apart from the time-reversal operator, which is antilinear, we are dealing with linear operators, such that $\hat{A}\left(\lambda_1|\psi_1\rangle + \lambda_2|\psi_2\rangle\right) = \lambda_1\hat{A}|\psi_1\rangle + \lambda_2\hat{A}|\psi_2\rangle$). We can also consider the action of an operator $\hat{A}$ on a bra,

$$\langle\psi| \longmapsto \langle\psi'| = \langle\psi|\hat{A} \,, \tag{6.19}$$

where it is understood that $\hat{A}$ acts on $\langle\psi|$ from the right. In general, the ket $\hat{A}|\psi\rangle$ and the bra $\langle\psi|\hat{A}$ are not dual to each other.

The addition of operators is both commutative and associative, namely, $\hat{A} + \hat{B} = \hat{B} + \hat{A}$ and $(\hat{A} + \hat{B}) + \hat{C} = \hat{A} + (\hat{B} + \hat{C})$. However, while the product is associative, $(\hat{A}\hat{B})\hat{C} = \hat{A}(\hat{B}\hat{C}) = \hat{A}\hat{B}\hat{C}$, it is generally non-commutative, $\hat{A}\hat{B} \neq \hat{B}\hat{A}$. In particular, the action of a product on a ket or a bra is defined as

$$\hat{A}\,(\hat{B}\,|\psi\rangle) = (\hat{A}\hat{B})|\psi\rangle = \hat{A}\hat{B}\,|\psi\rangle\,, \qquad (\langle\psi|\,\hat{A})\hat{B} = \langle\psi|(\hat{A}\hat{B}) = \langle\psi|\,\hat{A}\hat{B}\,, \tag{6.20}$$

and so $\hat{B}$ acts on a ket from the left first, and then $\hat{A}$ acts on the result; by contrast, first $\hat{A}$ acts on the bra from the right, and then $\hat{B}$ acts on the result.

An interesting operator can be constructed by "multiplying" (or by taking the so-called outer product of) a ket $|\psi\rangle$ by a bra $\langle\varphi|$ (in this order), namely

$$\hat{A} = |\psi\rangle\langle\varphi|\,. \tag{6.21}$$

When acting on a ket or a bra, this operator yields, respectively,

$$\hat{A}\,|\phi\rangle = (|\psi\rangle\langle\varphi|) \cdot |\phi\rangle = |\psi\rangle\langle\varphi|\phi\rangle\,, \qquad \langle\phi|\,\hat{A} = \langle\phi| \cdot (|\psi\rangle\langle\varphi|) = \langle\phi|\psi\rangle\langle\varphi|\,; \tag{6.22}$$

in the first case the action of $\hat{A}$ consists in multiplying the ket $|\psi\rangle$ by the complex number $\langle\varphi|\phi\rangle$, while in the second case the action of $\hat{A}$ consists in multiplying the bra $\langle\varphi|$ by the complex number $\langle\phi|\psi\rangle$. This example makes it plain that $\langle\phi|\,\hat{A}$ is not dual to $\hat{A}\,|\phi\rangle$.

Products such as $\langle\varphi|\psi\rangle$, $\hat{A}\,|\psi\rangle$, $\langle\varphi|\,\hat{A}$, $|\psi\rangle\langle\varphi|\phi\rangle$, and $\langle\phi|\psi\rangle\langle\varphi|$ are all meaningful and well formed. However, products such as $\hat{A}\langle\psi|$ and $|\psi\rangle\hat{A}$ are simply meaningless.[2] So, in order to avoid any misunderstanding, we should write, for example, $\langle\varphi|\phi\rangle\,|\psi\rangle = (\langle\varphi|\phi\rangle) \cdot |\psi\rangle$ without omitting the parentheses and multiplication dot; by contrast, there is no ambiguity in $|\psi\rangle\langle\varphi|\phi\rangle$ since $(|\psi\rangle\langle\varphi|) \cdot |\phi\rangle$ and $|\psi\rangle \cdot (\langle\varphi|\phi\rangle)$ are well-formed products representing the same state, although the former expression says that this state is obtained by the action of the operator $|\psi\rangle\langle\varphi|$ on $|\phi\rangle$, while the latter says that it is obtained by multiplying $|\psi\rangle$ by the complex number $\langle\varphi|\phi\rangle$. In the notation invented by Dirac, this equality results from the application of a generalized associative axiom. Another illustration of this axiom is that, given an operator $\hat{A}$, we have

$$\langle\varphi| \cdot (\hat{A}\,|\psi\rangle) = (\langle\varphi|\,\hat{A}) \cdot |\psi\rangle = \langle\varphi|\hat{A}|\psi\rangle\,, \tag{6.23}$$

where in the last step we have removed the parentheses, since whether the operator acts to the right on the ket or to the left on the bra is irrelevant; the result is the same.

The **adjoint operator** $\hat{A}^{\dagger}$ of an operator $\hat{A}$ is defined by the following property:

$$\hat{A}\,|\psi\rangle \rightleftharpoons \langle\psi|\,\hat{A}^{\dagger}\,, \tag{6.24}$$

that is, $\langle\psi|\,\hat{A}^{\dagger}$ is the bra dual to the ket $\hat{A}\,|\psi\rangle$. An operator such that $\hat{A}^{\dagger} = \hat{A}$ is known as **self-adjoint** or **hermitian**. The adjoint $(\hat{A}\hat{B})^{\dagger}$, defined as $(\hat{A}\hat{B})\,|\psi\rangle \rightleftharpoons \langle\psi|\,(\hat{A}\hat{B})^{\dagger}$, satisfies

$$(\hat{A}\hat{B})^{\dagger} = \hat{B}^{\dagger}\hat{A}^{\dagger}\,, \tag{6.25}$$

since $\hat{A}\,|\psi'\rangle \rightleftharpoons \langle\psi'|\,\hat{A}^{\dagger}$ and $|\psi'\rangle = B\,|\psi\rangle \rightleftharpoons \langle\psi'| = \langle\psi|\,\hat{B}^{\dagger}$. As in the space $L^2(\mathbb{R})$, the following identities hold:

$$(\hat{A}^{\dagger})^{\dagger} = \hat{A}\,, \qquad (\lambda\hat{A})^{\dagger} = \lambda^*\hat{A}^{\dagger}\,, \qquad (\hat{A} + \hat{B})^{\dagger} = \hat{A}^{\dagger} + \hat{B}^{\dagger}\,. \tag{6.26}$$

---

[2] As a matter of fact, a product such as $|\psi\rangle|\varphi\rangle$ or $\langle\psi|\langle\varphi|$ is also nonsensical, unless the two kets or bras belong to different state or dual spaces; for example, if $|\psi\rangle$ and $|\varphi\rangle$ are in the state spaces of particle 1 and 2 or in the orbital and spin spaces of a particle, respectively. Indeed, in such cases we should more properly write $|\psi\rangle \otimes |\varphi\rangle$ or $\langle\psi| \otimes \langle\varphi|$, that is, a tensor product (this good practice is not always followed in this book!).

The adjoint of $\hat{A} = |\psi\rangle\langle\varphi|$, for example, is obtained as

$$\hat{A}\,|\phi\rangle = |\psi\rangle\langle\varphi|\phi\rangle = |\psi\rangle\underbrace{\langle\phi|\varphi\rangle^*}_{\lambda^*} \rightleftharpoons \underbrace{\langle\phi|\varphi\rangle}_{\lambda}\langle\psi| = \langle\phi| \cdot (|\varphi\rangle\langle\psi|) = \langle\phi|\,\hat{A}^\dagger\,, \qquad (6.27)$$

where we have used the property $\lambda^*|\psi\rangle \rightleftharpoons \lambda\langle\psi|$. We conclude that $\hat{A}^\dagger = |\varphi\rangle\langle\psi|$. Another example is the following:

$$\langle\varphi|\hat{A}\,|\psi\rangle = \langle\varphi| \cdot (\hat{A}\,|\psi\rangle) = [(\langle\psi|\,\hat{A}^\dagger) \cdot |\varphi\rangle]^* = \langle\psi|\,\hat{A}^\dagger|\varphi\rangle^*\,, \qquad (6.28)$$

where we have used the definition of the adjoint operator (that is, $\hat{A}\,|\psi\rangle \rightleftharpoons \langle\psi|\,\hat{A}^\dagger$) and the property of the inner product in order to obtain a relation between the matrix element of $\hat{A}$ and that of its adjoint. In particular, for a hermitian operator we have

$$\langle\varphi|\hat{A}\,|\psi\rangle = \langle\psi|\hat{A}|\varphi\rangle^*\,, \qquad (6.29)$$

which implies, in particular, that $\langle\psi|A|\psi\rangle$ is real.

A **unitary operator** $\hat{U}$ is such that, for any $\hat{U}|\psi\rangle$ and $\langle\varphi|\,\hat{U}^\dagger$, where the latter is the bra dual to the ket $\hat{U}|\varphi\rangle$, we have

$$(\langle\varphi|\,\hat{U}^\dagger) \cdot (\hat{U}|\psi\rangle) = \langle\varphi|\,\hat{U}^\dagger\hat{U}|\psi\rangle = \langle\varphi|\psi\rangle, \qquad (6.30)$$

and so $\hat{U}$ preserves inner products, which in turn implies that $\hat{U}^\dagger = \hat{U}^{-1}$.

As a final remark, hereafter in this text we will primarily use Dirac notation. At times, though, we will revert to the notation, preferred by mathematicians, that we used previously in the context of the $L^2(\mathbb{R})$ Hilbert space. In such a notation, for example, the adjoint of an operator $\hat{A}$ is defined by $(\varphi, \hat{A}\psi) = (\hat{A}^\dagger\varphi, \psi)$, which has no counterpart in Dirac notation; instead, we write $\langle\varphi|\hat{A}|\psi\rangle = \langle\psi|\hat{A}^\dagger|\varphi\rangle^*$.

6.3 Representations

Choosing a representation means choosing a basis in the state space. Once a representation is chosen, kets, bras, and operators are represented by column vectors, row vectors, and square matrices, respectively; the elements are generally complex numbers. To see how this is effected, we restate below the properties of discrete and continuous bases in Dirac notation. A discrete basis $|\phi_m\rangle$ is a denumerable set of state vectors labeled by the subscript m (an integer) with[3]

1. orthonormality, $\langle\phi_m|\phi_n\rangle = \delta_{mn}$;
2. completeness, any state $|\psi\rangle \in \mathbb{E}$ can be expanded as follows:

$$|\psi\rangle = \sum_{m=1}^{\infty} c_m|\phi_m\rangle\,, \qquad c_m = \langle\phi_m|\psi\rangle\,, \qquad (6.31)$$

implying that

$$|\psi\rangle = \sum_{m=1}^{\infty} |\phi_m\rangle\langle\phi_m|\psi\rangle = \Big(\sum_{m=1}^{\infty} \underbrace{|\phi_m\rangle\langle\phi_m|}_{\hat{P}_m}\Big) \cdot |\psi\rangle \qquad \text{or} \qquad \sum_{m=1}^{\infty} \hat{P}_m = \hat{\mathbb{1}}\,, \qquad (6.32)$$

where $\hat{P}_m = |\phi_m\rangle\langle\phi_m|$ is the projection operator onto state $|\phi_m\rangle$.

---

[3] Of course, if the basis if finite, that is, the state space is finite dimensional, then the series are replaced by finite sums.

A continuous basis $|\phi_\alpha\rangle$ (whose vectors are not in $\mathbb{E}$, since they are not normalizable) is a set of vectors labeled by a subscript α which varies continuously in some interval I of the real axis, with the properties of

1. orthonormality, $\langle\phi_\alpha|\phi_\beta\rangle = \delta(\alpha - \beta)$;
2. completeness, any state $|\psi\rangle \in \mathbb{E}$ can be expanded as

$$|\psi\rangle = \int_I d\alpha\, c(\alpha)|\phi_\alpha\rangle, \qquad c(\alpha) = \langle\phi_\alpha|\psi\rangle, \tag{6.33}$$

implying that

$$\int_I d\alpha\, \underbrace{|\phi_\alpha\rangle\langle\phi_\alpha|}_{\hat{P}_\alpha} = \hat{\mathbb{1}} \qquad \text{or} \qquad \int_I d\alpha\, \hat{P}_\alpha = \hat{\mathbb{1}}. \tag{6.34}$$

Having chosen a discrete basis, a state $|\psi\rangle$ and its dual – the bra $\langle\psi|$ – are represented as

$$|\psi\rangle \longrightarrow \begin{pmatrix} c_1 \\ c_2 \\ \vdots \end{pmatrix}, \qquad \langle\psi| \longrightarrow \begin{pmatrix} c_1^* & c_2^* & \cdots \end{pmatrix}, \tag{6.35}$$

since, for example, $\langle\psi| = \langle\psi| \sum_{m=1}^\infty |\phi_m\rangle\langle\phi_m| = \sum_{m=1}^\infty c_m^* \langle\phi_m|$. In the case of a continuous basis, the column and row vectors are replaced by the functions $c(\alpha)$ and $c^*(\alpha)$, respectively. If $|\psi'\rangle = \hat{A}|\psi\rangle$ and $|\psi'\rangle = \sum_{m=1}^\infty c_m' |\phi_m\rangle$, it follows that

$$c_m' = \langle\phi_m|\hat{A}|\psi\rangle = \langle\phi_m|\hat{A} \sum_{n=1}^\infty |\phi_n\rangle\langle\phi_n|\psi\rangle = \sum_{n=1}^\infty \underbrace{\langle\phi_m|\hat{A}|\phi_n\rangle}_{A_{mn}} c_n, \tag{6.36}$$

and the operator $\hat{A}$ is represented by a square matrix, that is, $\hat{A} \longrightarrow \underline{A} = (A_{mn})$. The equivalent relation in the case of a continuous basis is

$$c'(\alpha) = \int_I d\beta\, A(\alpha, \beta)\, c(\beta). \tag{6.37}$$

Given the matrix $\underline{A}$ representing the operator $\hat{A}$, the matrix that represents the adjoint operator $\hat{A}^\dagger$ is obtained by taking the transpose of $\underline{A}$ and the complex conjugate of all matrix elements, since, using the definition of the adjoint, we have

$$\langle\phi_m|\hat{A}^\dagger|\phi_n\rangle = \langle\phi_n|\hat{A}|\phi_m\rangle^* \implies (\underline{A}^\dagger)_{mn} = (\underline{A})_{nm}^*. \tag{6.38}$$

A hermitian operator $\hat{A}^\dagger = \hat{A}$ is represented by a hermitian matrix $\underline{A} = \underline{A}^\dagger$ (note that the diagonal matrix elements are real in this case). Lastly, a unitary operator is represented by a unitary matrix $\underline{U}$ such that

$$\sum_{p=1}^\infty \hat{U}_{mp}^\dagger \hat{U}_{pn} = \sum_{p=1}^\infty \hat{U}_{pm}^* \hat{U}_{pn} = \delta_{mn}. \tag{6.39}$$

Similar relations are obtained in the case of a continuous basis, the only difference being that sums and Kronecker deltas are replaced by integrals and δ-functions, respectively.

Now suppose that $|\phi_m\rangle$ and $|\chi_m\rangle$ are two different bases. The vectors representing states in basis $|\phi_m\rangle$ are related to those representing these same states in basis $|\chi_m\rangle$ in the following way,

$$c'_m = \langle\chi_m|\psi\rangle = \sum_{n=1}^{\infty} \underbrace{\langle\chi_m|\phi_n\rangle}_{S^*_{nm}} \underbrace{\langle\phi_n|\psi\rangle}_{c_n} = \sum_{n=1}^{\infty} S^*_{nm} c_n = \sum_{n=1}^{\infty} S^\dagger_{mn} c_n \,, \qquad (6.40)$$

where we have defined a matrix $\underline{S}$ with matrix elements given by $S_{nm} = \langle\phi_n|\chi_m\rangle$; $\underline{S}$ is unitary. Of course, the above relation can be inverted to obtain the c_m in terms of the c'_m, with $c_m = \sum_{n=1}^{\infty} S_{mn} c'_n$.

To establish the relationship between the matrices $\underline{A}'$ and $\underline{A}$ representing the operator $\hat{A}$ in the two bases, we proceed in a similar fashion:

$$A'_{mn} = \langle\chi_m|\hat{A}|\chi_n\rangle = \sum_{p=1}^{\infty}\sum_{q=1}^{\infty} \langle\chi_m|\phi_p\rangle\langle\phi_p|\hat{A}|\phi_q\rangle\langle\phi_q|\chi_n\rangle = \sum_{p=1}^{\infty}\sum_{q=1}^{\infty} S^*_{pm} A_{pq} S_{qn} = (\underline{S}^\dagger \underline{A}\, \underline{S})_{mn} \,, \quad (6.41)$$

or, in matrix notation,

$$\underline{A}' = \underline{S}^\dagger \underline{A}\, \underline{S} \implies \underline{A} = \underline{S}\, \underline{A}'\, \underline{S}^\dagger \,, \qquad (6.42)$$

where the last relation follows from the unitarity of $\underline{S}$.

6.4 Hermitian Operators and Observables

In quantum mechanics, hermitian operators play a special role: they are associated with quantities that can be measured, that is, observables. The measurement of an observable results in one of the eigenvalues of the associated hermitian operator.[4] It is also postulated that the eigenstates of an observables form a basis, that is,

$$\hat{A}|\psi_i^{(k)}\rangle = \lambda_i|\psi_i^{(k)}\rangle \quad \text{with} \quad k = 1,\ldots,g_i \,, \qquad \sum_i \sum_{k=1}^{g_i} |\psi_i^{(k)}\rangle\langle\psi_i^{(k)}| = \hat{\mathbb{1}} \,, \qquad (6.43)$$

where $g_i \geq 1$ is the degeneracy of eigenvalue λ_i; if $g_i = 1$, the eigenvalue is non-degenerate. There are three important results that hold for observables:

1. If $\hat{A}$ and $\hat{B}$ commute, and $|\psi\rangle$ is an eigenstate of $\hat{A}$ with eigenvalue a, then the state $\hat{B}|\psi\rangle$ is also an eigenstate of $\hat{A}$ corresponding to the same eigenvalue a,

$$\hat{A}(\hat{B}|\psi\rangle) = \hat{A}\hat{B}|\psi\rangle = \hat{B}\hat{A}|\psi\rangle = \hat{B}(a|\psi\rangle) = a\hat{B}|\psi\rangle \,. \qquad (6.44)$$

If the eigenvalue a is non-degenerate then $|\psi\rangle$ is also an eigenstate of $\hat{B}$, since $\hat{B}|\psi\rangle$ must be proportional to $|\psi\rangle$, that is, $\hat{B}|\psi\rangle = b|\psi\rangle$. If the eigenvalue a is g-fold degenerate, then $\hat{B}|\psi^{(k)}\rangle$ is in the subspace spanned by the eigenstates of $\hat{A}$ having the eigenvalue a, that is, it can be expanded as follows:

$$\hat{B}|\psi^{(k)}\rangle = \sum_{m=1}^{g} c_m^{(k)} |\psi^{(m)}\rangle, \qquad c_m^{(k)} = \langle\psi^{(m)}|\hat{B}|\psi^{(k)}\rangle \,, \qquad (6.45)$$

where $\hat{A}|\psi^{(m)}\rangle = a|\psi^{(m)}\rangle$.

---

[4] Recall that hermitian operators have real eigenvalues and that eigenstates belonging to different eigenvalues are orthogonal.

2. If $\hat{A}$ and $\hat{B}$ commute, and $|\psi_1\rangle$ and $|\psi_2\rangle$ are two eigenstates of $\hat{A}$ corresponding to distinct eigenvalues a_1 and a_2, then the matrix element

$$\langle\psi_1|\hat{B}|\psi_2\rangle = 0\,, \tag{6.46}$$

since

$$0 = \langle\psi_1|\underbrace{\hat{A}\hat{B} - \hat{B}\hat{A}}_{[\hat{A},\hat{B}]=0}|\psi_2\rangle = \langle\psi_1|\hat{A}^\dagger\hat{B} - \hat{B}\hat{A}|\psi_2\rangle = \underbrace{(a_1 - a_2)}_{\neq 0}\langle\psi_1|\hat{B}|\psi_2\rangle\,. \tag{6.47}$$

3. If $\hat{A}$ and $\hat{B}$ commute then there exists a basis of common eigenstates of $\hat{A}$ and $\hat{B}$. The proof is straightforward. Let $|\psi_i^{(k)}\rangle$ with $i = 1, 2, \ldots$ and $k = 1, \ldots, g_i$ be the basis consisting of eigenstates of $\hat{A}$. We order it as follows (as an example, we are assuming here that a_1 is g_1-degenerate, a_2 is non-degenerate, etc.):

$$\underbrace{|\psi_1^{(1)}\rangle, \ldots, |\psi_1^{(g_1)}\rangle}_{a_1}, \underbrace{|\psi_2\rangle}_{a_2}, \underbrace{|\psi_1^{(1)}\rangle, \ldots, |\psi_1^{(g_3)}\rangle}_{a_3}, \underbrace{|\psi_4\rangle}_{a_4}, \ldots\,. \tag{6.48}$$

In such a basis, $\hat{B}$ has matrix elements (from properties 1 and 2 above)

$$\begin{aligned}\langle\psi_i^{(k)}|\hat{B}|\psi_j^{(l)}\rangle &= 0 && \text{if } i \neq j\\ &= b_i^{(kl)} && \text{if } i = j\,,\end{aligned} \tag{6.49}$$

that is, the matrix $\underline{B}$ is block-diagonal,

$$\underline{B} = \begin{pmatrix} b_1^{(1,1)} & \cdots & b_1^{(1,g_1)} & 0 & 0 & \cdots & 0 & 0\\ \vdots & & \vdots & \vdots & \vdots & \cdots & \vdots & \vdots\\ b_1^{(g_1,1)} & \cdots & b_1^{(g_1,g_1)} & 0 & 0 & \cdots & 0 & 0\\ 0 & \cdots & 0 & b_2 & 0 & \cdots & 0 & 0\\ 0 & \cdots & 0 & 0 & b_3^{(1,1)} & \cdots & b_3^{(1,g_3)} & 0\\ 0 & \cdots & 0 & \vdots & \vdots & \cdots & \vdots & \vdots\\ 0 & \cdots & 0 & 0 & b_3^{(g_3,1)} & \cdots & b_3^{(g_3,g_3)} & 0\\ 0 & \cdots & 0 & 0 & 0 & \cdots & 0 & b_4 \end{pmatrix}\,, \tag{6.50}$$

where only the first four blocks are illustrated, for simplicity. If a_i is non-degenerate then property 1 above says that $|\psi_i\rangle$ is also an eigenstate of $\hat{B}$ with eigenvalue b_i (this case corresponds to $i = 2$ and $i = 4$ in the matrix above). By contrast, if a_i is degenerate then $\hat{B}$ is represented by a hermitian matrix of dimension $g_i \times g_i$. As such, it can be diagonalized,

$$\underbrace{|\xi_1^{(i)}\rangle, \ldots, |\xi_{g_i}^{(i)}\rangle}_{g_i\text{ eigenstates}}, \qquad \hat{B}|\xi_k^{(i)}\rangle = b_k^{(i)}|\xi_k^{(i)}\rangle\,, \tag{6.51}$$

and some of the above eigenvalues may be degenerate. Each of the eigenstates $|\xi_k^{(i)}\rangle$ consists of a linear combination of the eigenstates $|\psi_i^{(l)}\rangle$ of $\hat{A}$ all having the same eigenvalue a_i,

$$|\xi_k^{(i)}\rangle = \sum_{l=1}^{g_i} c_l^{(k)}|\psi_i^{(l)}\rangle\,, \qquad k = 1, \ldots, g_i\,; \tag{6.52}$$

each $|\xi_k^{(i)}\rangle$ is obviously an eigenstate of $\hat{A}$ with eigenvalue a_i. We can repeat this procedure in each block, and so obtain a basis of common eigenstates of $\hat{A}$ and $\hat{B}$, given by

$$|\xi_1^{(1)}\rangle, \ldots, |\xi_{g_1}^{(1)}\rangle, |\psi_2\rangle, |\xi_1^{(3)}\rangle, \ldots, |\xi_{g_3}^{(3)}\rangle, |\psi_4\rangle, \ldots \tag{6.53}$$

Suppose that $\hat{A}, \hat{B}, \hat{C}, \ldots$ form a set of commuting observables. Then there exists a basis of common eigenstates. Such a set of observables is said to be "complete," that is, a **complete set of commuting observables**, if specifying the eigenvalue of each observable specifies the unique common eigenstate. As an example, consider the case of the Hamiltonian and parity observables, $\hat{H}$ and $\hat{P}$, for a particle in a one-dimensional repulsive δ-function potential located at the origin. The energy eigenvalues are doubly degenerate. However, $\hat{H}$ and $\hat{P}$ commute and specifying the energy *and* parity (either $+$ or $-$) eigenvalues specifies a unique common eigenstate. So, in this case $\hat{H}$ and $\hat{P}$ form a complete set of commuting observables.

6.5 The Coordinate and Momentum Representations

It is interesting to see how wave mechanics emerges in the present abstract formulation. To this end, we introduce the position operator $\hat{x}$, an observable (in one dimension for simplicity but the discussion is easily generalized to the three-dimensional case). It has eigenstates

$$\hat{x}\,|\phi_x\rangle = x\,|\phi_x\rangle \qquad -\infty < x < \infty\,, \tag{6.54}$$

where x is the (non-degenerate) eigenvalue, which takes on a continuum of values, and $|\phi_x\rangle$ represents the state in which the particle is at position x. The eigenstates $|\phi_x\rangle$ form a continuous basis, and hence

$$\langle\phi_x|\phi_{x'}\rangle = \delta(x - x')\,, \qquad \int_{-\infty}^{\infty} dx\,|\phi_x\rangle\langle\phi_x| = \hat{\mathbb{1}}\,. \tag{6.55}$$

A particle in a state $|\psi\rangle$ can be expanded in the basis of eigenstates of $\hat{x}$,

$$|\psi\rangle = \int_{-\infty}^{\infty} dx\,\psi(x)\,|\phi_x\rangle\,, \qquad \psi(x) = \langle\phi_x|\psi\rangle\,, \tag{6.56}$$

and so the wave function $\psi(x)$ is simply the component of $|\psi\rangle$ in the basis of eigenstates of $\hat{x}$. From this perspective, wave mechanics in coordinate space (as opposed to momentum space, see below) consists of using the basis consisting of the eigenstates of $\hat{x}$ to represent states and operators.

An operator that is a function of $\hat{x}$, such as the potential energy operator $\hat{V}(\hat{x})$, has matrix elements

$$\langle\phi_x|\hat{V}(\hat{x})|\phi_{x'}\rangle = \delta(x - x')\,V(x)\,, \tag{6.57}$$

where $V(x)$ is the function obtained by replacing $\hat{x}$ by its eigenvalue x. This can be seen by expanding $\hat{V}(\hat{x})$ as follows:

$$\hat{V}(\hat{x}) = \sum_{m=0}^{\infty} \frac{V^{(m)}(0)}{m!}\,\hat{x}^m\,, \qquad V^{(m)}(0) = \frac{d^m}{dz^m}V(z)\Big|_{z=0}\,, \tag{6.58}$$

and by noting that

$$\langle\phi_x|\hat{V}(\hat{x})|\phi_{x'}\rangle = \sum_{m=0}^{\infty} \frac{V^{(m)}(0)}{m!}\,\langle\phi_x|\hat{x}^m|\phi_{x'}\rangle = \sum_{m=0}^{\infty} \frac{V^{(m)}(0)}{m!}\,x'^m\delta(x - x') = V(x)\delta(x - x')\,, \tag{6.59}$$

since the δ-function enforces $x = x'$ and hence $V(x') = V(x)$, that is, the operator $\hat{V}(\hat{x})$ is diagonal in the $|\phi_x\rangle$ representation. It is shown in Problem 9 that matrix elements of the momentum operator $\hat{p}$ are given by

$$\langle \phi_x | \hat{p} | \phi_{x'} \rangle = -i\hbar \frac{d}{dx} \delta(x - x') , \tag{6.60}$$

or, more generally, that

$$\langle \phi_x | \hat{p}^n | \phi_{x'} \rangle = (-i\hbar)^n \frac{d^n}{dx^n} \delta(x - x') . \tag{6.61}$$

As an example, the eigenvalue problem for a generic observable $\hat{A}$ reads, in this representation,

$$\int_{-\infty}^{\infty} dx' \, A(x, x') \psi(x') = a \psi(x) , \qquad A(x, x') = \langle \phi_x | \hat{A} | \phi_{x'} \rangle , \tag{6.62}$$

which follows by projecting both sides of $\hat{A} | \psi \rangle = a | \psi \rangle$ onto position eigenstates and using completeness.

We can equally well use the momentum representation, namely, we can work with the continuous basis $| \phi_p \rangle$ of momentum eigenstates, for which

$$\langle \phi_p | \phi_{p'} \rangle = \delta(p - p') , \qquad \int_{-\infty}^{\infty} dp \, | \phi_p \rangle \langle \phi_p | = \hat{1} . \tag{6.63}$$

Clearly, an operator function of $\hat{p}$ is diagonal in this basis, since

$$\langle \phi_p | \hat{F}(\hat{p}) | \phi_{p'} \rangle = F(p) \, \delta(p - p') , \tag{6.64}$$

whereas (see Problem 11)

$$\langle \phi_p | \hat{x} | \phi_{p'} \rangle = i\hbar \frac{d}{dp} \delta(p - p') . \tag{6.65}$$

The eigenvalue problem for an operator $\hat{A}$ follows:

$$\int_{-\infty}^{\infty} dp' \, A(p, p') \, \widetilde{\psi}(p') = a \, \widetilde{\psi}(p) , \qquad A(p, p') = \langle \phi_p | \hat{A} | \phi_{p'} \rangle , \tag{6.66}$$

where $\widetilde{\psi}(p)$ is the momentum-space wave function,

$$\widetilde{\psi}(p) = \langle \phi_p | \psi \rangle = \int_{-\infty}^{\infty} dx \, \langle \phi_p | \phi_x \rangle \langle \phi_x | \psi \rangle = \frac{1}{\sqrt{2\pi\hbar}} \int_{-\infty}^{\infty} dx \, e^{-ipx/\hbar} \, \psi(x) , \tag{6.67}$$

where we have used (see Problem 10) $\langle \phi_x | \phi_p \rangle = e^{ipx/\hbar} / \sqrt{2\pi\hbar}$. In particular, $S(x, p) = \langle \phi_x | \phi_p \rangle$ can be viewed as the "transformation matrix" relating the $| \phi_x \rangle$ and $| \phi_p \rangle$ bases,

$$| \phi_p \rangle = \int_{-\infty}^{\infty} dx \, S(x, p) \, | \phi_x \rangle , \qquad | \phi_x \rangle = \int_{-\infty}^{\infty} dp \, S^*(x, p) \, | \phi_p \rangle , \qquad \int dx \, S^*(x, p) \, S(x, p') = \delta(p - p') , \tag{6.68}$$

the last relationship expressing the unitarity of $S(x, p)$.

6.6 Tensor Products

Consider the cases of a single particle in one dimension and in three dimensions. We denote the corresponding state spaces as $\mathbb{E}_x$ and $\mathbb{E}_{\mathbf{r}}$. Possible choices for bases in $\mathbb{E}_x$ and $\mathbb{E}_{\mathbf{r}}$ are those of the eigenstates $| \phi_x \rangle$ and $| \phi_{\mathbf{r}} \rangle$ of the position operator $\hat{x}$ in one dimension and $\hat{\mathbf{r}}$ in three dimensions, respectively. The state space $\mathbb{E}_{\mathbf{r}}$ can also be thought as of consisting of the three state spaces $\mathbb{E}_x$, $\mathbb{E}_y$, and $\mathbb{E}_z$, which we indicate as

$$\mathbb{E}_{\mathbf{r}} = \mathbb{E}_x \otimes \mathbb{E}_y \otimes \mathbb{E}_z \,, \tag{6.69}$$

namely, the tensor product of $\mathbb{E}_x$, $\mathbb{E}_y$, and $\mathbb{E}_z$. Similarly, the basis $|\phi_{\mathbf{r}}\rangle$ can be viewed as the tensor product of the three bases:

$$|\phi_{\mathbf{r}}\rangle = |\phi_x\rangle \otimes |\phi_y\rangle \otimes |\phi_z\rangle = |\phi_y\rangle \otimes |\phi_x\rangle \otimes |\phi_z\rangle = \cdots \,, \tag{6.70}$$

and the relative ordering of the states in the tensor product is irrelevant, since it is understood that $|\phi_x\rangle \in \mathbb{E}_x$, $|\phi_y\rangle \in \mathbb{E}_y$, and $|\phi_z\rangle \in \mathbb{E}_z$. However, it is important to stress that a state $|\psi\rangle \in \mathbb{E}_{\mathbf{r}}$ cannot generally be written as the tensor product of three states $|\varphi_x\rangle$, $|\chi_y\rangle$, and $|\xi_z\rangle$ belonging to, respectively, $\mathbb{E}_x$, $\mathbb{E}_y$, and $\mathbb{E}_z$. While the state $|\psi\rangle$ can be expanded as

$$|\psi\rangle = \int dx \int dy \int dz\, \psi(x,y,z)\, |\phi_x\rangle \otimes |\phi_y\rangle \otimes |\phi_z\rangle = \int d\mathbf{r}\, \psi(\mathbf{r})\, |\phi_{\mathbf{r}}\rangle \,, \tag{6.71}$$

it is only in the special case in which $\psi(x,y,z) = \varphi(x)\,\chi(y)\,\xi(z)$, that is, the expansion coefficients (wave functions) factorize, that we have[5]

$$\begin{aligned}
|\psi\rangle &= \int dx \int dy \int dz\, \varphi(x)\,\chi(y)\,\xi(z)\, |\phi_x\rangle \otimes |\phi_y\rangle \otimes |\phi_z\rangle \\
&= \underbrace{\int dx\, \varphi(x)\, |\phi_x\rangle}_{|\varphi_x\rangle} \otimes \underbrace{\int dy\, \chi(y)\, |\phi_y\rangle}_{|\chi_y\rangle} \otimes \underbrace{\int dz\, \xi(z)\, |\phi_z\rangle}_{|\xi_z\rangle} \,.
\end{aligned} \tag{6.72}$$

To any operator, for example, $\hat{A}_x$, acting in the state space $\mathbb{E}_x$ there corresponds an operator $\hat{A}_x \otimes \mathbb{1}_y \otimes \mathbb{1}_z$ acting in $\mathbb{E}$, such that $\hat{A}_x \otimes \hat{\mathbb{1}}_y \otimes \hat{\mathbb{1}}_z |\phi_{\mathbf{r}}\rangle = (\hat{A}_x |\phi_x\rangle) \otimes |\phi_y\rangle \otimes |\phi_z\rangle$ and hence, on a generic state $|\psi\rangle$,

$$\hat{A}_x \otimes \hat{\mathbb{1}}_y \otimes \hat{\mathbb{1}}_z |\psi\rangle = \int dx \int dy \int dz\, \psi(x,y,z)\, (\hat{A}_x |\phi_x\rangle) \otimes |\phi_y\rangle \otimes |\phi_z\rangle \,. \tag{6.73}$$

It is easily seen that operators acting in different spaces commute, that is, $\hat{A}_x \otimes \hat{B}_y \otimes \hat{\mathbb{1}}_z$ and $\hat{B}_y \otimes \hat{A}_x \otimes \hat{\mathbb{1}}_z$ acting on any state $|\psi\rangle$ yield the same result. In particular, if $\hat{A}_x$, $\hat{B}_y$, and $\hat{C}_z$ are three observables with eigenstates $|a_n\rangle^x$, $|b_p\rangle^y$, and $|c_q\rangle^z$ and corresponding eigenvalues a_n^x, b_p^y, and c_q^z in state spaces $\mathbb{E}_x$, $\mathbb{E}_y$, and $\mathbb{E}_z$, respectively, then the observable

$$\hat{O} = \hat{A}_x \otimes \hat{\mathbb{1}}_y \otimes \hat{\mathbb{1}}_z + \hat{\mathbb{1}}_x \otimes \hat{B}_y \otimes \hat{\mathbb{1}}_z + \hat{\mathbb{1}}_x \otimes \hat{\mathbb{1}}_y \otimes \hat{C}_z \,, \tag{6.74}$$

acting on the tensor product state $|a_n^x; b_p^y; c_q^z\rangle = |a_n\rangle^x \otimes |b_p\rangle^y \otimes |c_q\rangle^z$ has eigenvalues

$$\hat{O} |a_n^x; b_p^y; c_q^z\rangle = (a_n^x + b_p^y + c_q^z) |a_n^x; b_p^y; c_q^z\rangle \,. \tag{6.75}$$

Similarly, the operator $\hat{A}_x \otimes \hat{B}_y \otimes \hat{C}_z$ acting on $|a_n^x; b_p^y; c_q^z\rangle$ in $\mathbb{E}$ has eigenvalues $a_n^x b_p^y c_q^z$ given by the product of the individual eigenvalues. In the following, we will often omit the $\otimes$ symbol in the tensor product, and write simply $|a_n^x; b_p^y; c_q^z\rangle = |a_n\rangle^x |b_p\rangle^y |c_q\rangle^z$ or $\hat{O} = \hat{A}_x + \hat{B}_y + \hat{C}_z$, unless the meaning

---

[5] Generally, let $\mathbb{E}_1$ and $\mathbb{E}_2$ be two state spaces with bases $|\chi_n(1)\rangle$ and $|\xi_p(2)\rangle$, respectively. The basis $|\chi_n(1); \xi_p(2)\rangle = |\chi_n(1)\rangle \otimes |\xi_p(2)\rangle$ spans the space $\mathbb{E} = \mathbb{E}_1 \otimes \mathbb{E}_2$. If both $\mathbb{E}_1$ and $\mathbb{E}_2$ have finite dimensions N_1 and N_2, then $\mathbb{E}$ has dimension $N_1 \times N_2$; if either of $\mathbb{E}_1$ and $\mathbb{E}_2$, or both, are infinite dimensional then $\mathbb{E}$ is also infinite dimensional. A state $|\psi\rangle$ in $\mathbb{E}$ can be generally expanded as $|\psi\rangle = \sum_{np} c_{np} |\chi_n(1); \xi_p(2)\rangle$, and, if the coefficients c_{np} factorize into $c_{np} = a_n b_p$, the state above is not "entangled" and can be written as the tensor product of states $|\phi(1)\rangle = \sum_n a_n |\chi_n\rangle$ and $|\varphi(2)\rangle = \sum_p b_p |\xi_p\rangle$ belonging to, respectively, $\mathbb{E}_1$ and $\mathbb{E}_2$.

is unclear from the context. Tensor products are used, for example, to describe the states of many-particle systems or of a single particle with orbital and spin degrees of freedom.

6.7 Problems

Problem 1 The Set of Square-Integrable Functions Forms a Linear Vector Space

Show that, if $\psi_1(\mathbf{r})$ and $\psi_2(\mathbf{r})$ are square integrable then $\lambda_1 \psi_1(\mathbf{r}) + \lambda_2 \psi_2(\mathbf{r})$ is also square integrable, implying that the set of square-integrable functions forms a linear vector space.

Solution

Consider

$$
\begin{aligned}
|\lambda_1\psi_1(\mathbf{r}) + \lambda_2\psi_2(\mathbf{r})|^2 &= |\lambda_1|^2 |\psi_1(\mathbf{r})|^2 + |\lambda_2|^2 |\psi_2(\mathbf{r})|^2 + 2\,\mathrm{Re}\left[\lambda_1^*\lambda_2\psi_1^*(\mathbf{r})\psi_2(\mathbf{r})\right] \\
&\leq |\lambda_1|^2 |\psi_1(\mathbf{r})|^2 + |\lambda_2|^2 |\psi_2(\mathbf{r})|^2 + 2\,|\lambda_1||\lambda_2|\,|\psi_1(\mathbf{r})||\psi_2(\mathbf{r})| \\
&\leq |\lambda_1|^2 |\psi_1(\mathbf{r})|^2 + |\lambda_2|^2 |\psi_2(\mathbf{r})|^2 + |\lambda_1||\lambda_2|\left[|\psi_1(\mathbf{r})|^2 + |\psi_2(\mathbf{r})|^2\right] ,
\end{aligned}
$$

where in the last step we have used $[\,|\psi_1(\mathbf{r})| - |\psi_2(\mathbf{r})|\,]^2 \geq 0$, which implies that $2\,|\psi_1(\mathbf{r})||\psi_2(\mathbf{r})| \leq |\psi_1(\mathbf{r})|^2 + |\psi_2(\mathbf{r})|^2$. Thus, by integrating over $\mathbf{r}$ on both sides, we obtain the inequality

$$
\int d\mathbf{r}\,|\lambda_1\psi_1(\mathbf{r}) + \lambda_2\psi_2(\mathbf{r})|^2 \leq (|\lambda_1| + |\lambda_2|)\left[|\lambda_1| \int d\mathbf{r}\,|\psi_1(\mathbf{r})|^2 + |\lambda_2| \int d\mathbf{r}\,|\psi_2(\mathbf{r})|^2\right] ,
$$

and the integral on the left-hand side converges.

Problem 2 The Parity Operator as a Hermitian and Unitary Operator

Show that the parity (or space-inversion) operator $Q\psi(\mathbf{r}) = \psi(-\mathbf{r})$ is hermitian and unitary.

Solution

The parity operator Q is hermitian, since

$$
\left(\varphi, Q\psi\right) = \int d\mathbf{r}\,\varphi^*(\mathbf{r})\left[Q\psi(\mathbf{r})\right] = \int d\mathbf{r}\,\varphi^*(\mathbf{r})\,\psi(-\mathbf{r}) = \underbrace{\int d\mathbf{r}\,\varphi^*(-\mathbf{r})\,\psi(\mathbf{r})}_{\text{change variable } \mathbf{r}\longrightarrow-\mathbf{r}}
$$

$$
= \int d\mathbf{r}\,\left[Q\varphi(\mathbf{r})\right]^*\,\psi(\mathbf{r}) = (Q\varphi, \psi) ,
$$

where we have replaced $\mathbf{r}$ by $-\mathbf{r}$ (the Jacobian of the transformation is unity). The parity operator is also unitary, since

$$
(Q\varphi, Q\psi) = \int d\mathbf{r}\,\left[Q\varphi(\mathbf{r})\right]^*\left[Q\psi(\mathbf{r})\right] = \int d\mathbf{r}\,\varphi^*(-\mathbf{r})\,\psi(-\mathbf{r}) = \int d\mathbf{r}\,\varphi^*(\mathbf{r})\,\psi(\mathbf{r}) = (\varphi, \psi) ,
$$

and, therefore, $Q^\dagger = Q^{-1} = Q$.

Problem 3   Properties of the Projection Operator

Consider a normalized state $|\psi\rangle$, and define the operator $\hat{P} = |\psi\rangle\langle\psi|$. Show that $\hat{P}$ is hermitian and that $\hat{P}^2 = \hat{P}$. Show further that, given a state $|\varphi\rangle$, the state $(\hat{\mathbb{1}} - \hat{P})|\varphi\rangle$ is orthogonal to $|\psi\rangle$.

The notion of a projection operator can be generalized to the case of a set of orthonormal states $|\psi_m\rangle$, that is, $\hat{P} = \sum_m |\psi_m\rangle\langle\psi_m|$. Show that $\hat{P}|\varphi\rangle$ is the projection of $|\varphi\rangle$ onto the subspace spanned by the set $|\psi_m\rangle$.

Solution

To show that $\hat{P}$ is hermitian, note that

$$\hat{P}|\varphi\rangle = (|\psi\rangle\langle\psi|) \cdot |\varphi\rangle = |\psi\rangle \underbrace{\langle\varphi|\psi\rangle^*}_{\lambda} \rightleftharpoons \underbrace{\langle\varphi|\psi\rangle}_{\lambda^*}\langle\psi| = \langle\varphi| \cdot (|\psi\rangle\langle\psi|) = \langle\varphi|\hat{P},$$

where we have used the ket–bra correspondence property $\lambda|\varphi\rangle \rightleftharpoons \lambda^*\langle\varphi|$. We also find that

$$\hat{P}^2 = (|\psi\rangle\langle\psi|) \cdot (|\psi\rangle\langle\psi|) = |\psi\rangle\langle\psi| = \hat{P},$$

since $|\psi\rangle$ is normalized, and

$$\langle\psi|(\hat{\mathbb{1}} - \hat{P})|\varphi\rangle = \langle\psi|\varphi\rangle - \langle\psi|\psi\rangle\langle\psi|\varphi\rangle = 0,$$

which says that the state $(\hat{\mathbb{1}} - \hat{P})|\varphi\rangle$ is orthogonal to $|\psi\rangle$. Similarly, for a set of orthonormal states, $\hat{P}$ is hermitian and

$$\hat{P}^2 = \sum_{mn} |\psi_m\rangle \underbrace{\langle\psi_m|\psi_n\rangle}_{\delta_{mn}}\langle\psi_n| = \sum_m |\psi_m\rangle\langle\psi_m| = \hat{P}.$$

Given a state $|\varphi\rangle$, the state $|\varphi\rangle - \hat{P}|\varphi\rangle$ is orthogonal to any of the $|\psi_k\rangle$, and hence $\hat{P}|\varphi\rangle$ lies in the subspace spanned by the $|\psi_m\rangle$; indeed,

$$\langle\psi_k| \cdot (|\varphi\rangle - \hat{P}|\varphi\rangle) = \langle\psi_k|\varphi\rangle - \sum_m \underbrace{\langle\psi_k|\psi_m\rangle}_{\delta_{km}}\langle\psi_m|\varphi\rangle = 0.$$

Problem 4   A Projection Operator onto a State $|\psi\rangle$ in a Three-Dimensional State Space

The state space of a physical system is three dimensional, and the states $|\phi_1\rangle$, $|\phi_2\rangle$, and $|\phi_3\rangle$ form an orthonormal basis in this space. Consider the state $|\psi\rangle$ defined as

$$|\psi\rangle = \frac{1}{\sqrt{2}}|\phi_1\rangle + \frac{i}{2}|\phi_2\rangle + \frac{1}{2}|\phi_3\rangle.$$

Verify explicitly that $|\psi\rangle$ is normalized and calculate the matrix representing the projection operator $\hat{P}$ onto the state $|\psi\rangle$ in the basis $|\phi_1\rangle$, $|\phi_2\rangle$, and $|\phi_3\rangle$. Verify that this matrix satisfies the properties associated with $\hat{P}$, namely that $\hat{P}$ is hermitian and $\hat{P}^2 = \hat{P}$.

Solution

We find

$$\langle\psi|\psi\rangle = \left(\frac{1}{\sqrt{2}}\langle\phi_1| - \frac{i}{2}\langle\phi_2| + \frac{1}{2}\langle\phi_3|\right)\left(\frac{1}{\sqrt{2}}|\phi_1\rangle + \frac{i}{2}|\phi_2\rangle + \frac{1}{2}|\phi_3\rangle\right) = \frac{1}{2} + \frac{1}{4} + \frac{1}{4} = 1,$$

where we have used $\langle\phi_i|\phi_j\rangle = \delta_{ij}$. The projection operator is defined as

$$\hat{P} = |\psi\rangle\langle\psi| = \left(\frac{1}{\sqrt{2}}|\phi_1\rangle + \frac{i}{2}|\phi_2\rangle + \frac{1}{2}|\phi_3\rangle\right)\left(\frac{1}{\sqrt{2}}\langle\phi_1| - \frac{i}{2}\langle\phi_2| + \frac{1}{2}\langle\phi_3|\right)$$

$$= \frac{1}{2}|\phi_1\rangle\langle\phi_1| + \frac{i}{2\sqrt{2}}|\phi_2\rangle\langle\phi_1| + \frac{1}{2\sqrt{2}}|\phi_3\rangle\langle\phi_1| - \frac{i}{2\sqrt{2}}|\phi_1\rangle\langle\phi_2| + \frac{1}{4}|\phi_2\rangle\langle\phi_2| - \frac{i}{4}|\phi_3\rangle\langle\phi_2|$$

$$+ \frac{1}{2\sqrt{2}}|\phi_1\rangle\langle\phi_3| + \frac{i}{4}|\phi_2\rangle\langle\phi_3| + \frac{1}{4}|\phi_3\rangle\langle\phi_3| ,$$

and its matrix elements are given by

$$\underline{P}_{ij} = \langle\phi_i|\hat{P}|\phi_j\rangle \implies \underline{P} = \begin{pmatrix} 1/2 & -i/2\sqrt{2} & 1/2\sqrt{2} \\ i/2\sqrt{2} & 1/4 & i/4 \\ 1/2\sqrt{2} & -i/4 & 1/4 \end{pmatrix} .$$

The matrix $\underline{P}$ is obviously hermitian. Direct multiplication shows that $\underline{P}\,\underline{P} = \underline{P}$.

Problem 5 Properties of the Operator $\hat{O}_{mn} = |\phi_m\rangle\langle\phi_n|$

Let $|\phi_n\rangle$ be a discrete orthonormal basis. Consider the operator $\hat{O}_{mn}$ defined as the outer product of the ket $|\phi_m\rangle$ and bra $\langle\phi_n|$, that is, $\hat{O}_{mn} = |\phi_m\rangle\langle\phi_n|$. Obtain the adjoint $\hat{O}_{mn}^\dagger$ and show that

$$\hat{O}_{mn}\,\hat{O}_{pq}^\dagger = \delta_{nq}\,\hat{O}_{mp} .$$

The trace of an operator $\hat{A}$ is defined as

$$\mathrm{tr}(\hat{A}) = \sum_n \langle\phi_n|\hat{A}|\phi_n\rangle .$$

Calculate the trace of $\hat{O}_{mn}$. If $\hat{A}$ has matrix elements $A_{mn} = \langle\phi_m|\hat{A}|\phi_n\rangle$, show that

$$\hat{A} = \sum_{mn} A_{mn}\,\hat{O}_{mn} , \qquad A_{pq} = \mathrm{tr}\left(\hat{A}\,\hat{O}_{pq}^\dagger\right) .$$

Solution

Using the definition, we have, for any state $|\psi\rangle$,

$$\hat{O}_{mn}|\psi\rangle = |\phi_m\rangle\langle\phi_n|\psi\rangle = \langle\psi|\phi_n\rangle^*|\phi_m\rangle \rightleftharpoons \langle\psi|\phi_n\rangle\langle\phi_m| = \langle\psi|\cdot(|\phi_n\rangle\langle\phi_m|) = \langle\psi|\hat{O}_{nm}^\dagger ,$$

giving

$$\hat{O}_{mn}^\dagger = \hat{O}_{nm} .$$

We also have

$$\hat{O}_{mn}\,\hat{O}_{pq}^\dagger = |\phi_m\rangle\langle\phi_n|\phi_q\rangle\langle\phi_p| = \delta_{nq}\,|\phi_m\rangle\langle\phi_p| = \delta_{nq}\,\hat{O}_{mp} .$$

Using the definition of the trace, we find that

$$\mathrm{tr}(\hat{O}_{mn}) = \sum_p \langle\phi_p|\hat{O}_{mn}|\phi_p\rangle = \sum_p \langle\phi_p|\phi_m\rangle\langle\phi_n|\phi_p\rangle = \sum_p \delta_{pm}\,\delta_{np} = \delta_{mn} .$$

Using the completeness of the basis $|\phi_m\rangle$, we obtain

$$\sum_{mn} A_{mn} \hat{O}_{mn} = \sum_{mn} \underbrace{\langle \phi_m| \hat{A} |\phi_n\rangle}_{\text{complex number}} |\phi_m\rangle\langle\phi_n| = \sum_{mn} |\phi_m\rangle\langle\phi_m| \hat{A} |\phi_n\rangle\langle\phi_n|$$

$$= \left(\sum_m |\phi_m\rangle\langle\phi_m|\right) \hat{A} \left(\sum_n |\phi_n\rangle\langle\phi_n|\right) = \hat{A} \, ,$$

and

$$\text{tr}(\hat{A} \, \hat{O}_{pq}^\dagger) = \sum_n \langle\phi_n|\hat{A} |\phi_q\rangle \underbrace{\langle\phi_p|\phi_n\rangle}_{\delta_{pn}} = \langle\phi_p|\hat{A} |\phi_q\rangle = A_{pq} \, .$$

Problem 6   A Unitary Operator

Suppose that an operator $\hat{A}$ is hermitian. Show that $e^{i\hat{A}}$ is unitary.

Hint: An operator $f(\hat{A})$ is defined through its power series expansion (assumed to exist):

$$f(\hat{A}) = \sum_{m=0}^{\infty} \frac{1}{m!} f^{(m)}(0) \hat{A}^m \, , \qquad f^{(m)}(0) = \frac{d^m}{dz^m} f(z)\Big|_{z=0} \, .$$

Solution

Since $\hat{A}$ is hermitian, we have

$$e^{i\hat{A}} = \sum_{m=0}^{\infty} \frac{i^m}{m!} \hat{A}^m \implies \left(e^{i\hat{A}}\right)^\dagger = \sum_{m=0}^{\infty} \left(\frac{i^m}{m!} \hat{A}^m\right)^\dagger = \sum_{m=0}^{\infty} \frac{(-i)^m}{m!} \hat{A}^m = e^{-i\hat{A}} \, .$$

Now, consider

$$e^{-i\hat{A}} e^{i\hat{A}} = \sum_{m=0}^{\infty} \sum_{n=0}^{\infty} \frac{(-i)^m}{m!} \frac{i^n}{n!} \hat{A}^{m+n} = \sum_{p=0}^{\infty} \sum_{n=0}^{p} \frac{(-i)^{p-n}}{(p-n)!} \frac{i^n}{n!} \hat{A}^p = \sum_{p=0}^{\infty} \frac{(-i)^p}{p!} \hat{A}^p \sum_{n=0}^{p} \frac{p!}{(p-n)! \, n!} (-1)^n \, ,$$

where the sum over unrestricted m and n has been written equivalently as a sum on $p = m + n$ and n, with p unrestricted and n restricted to the range 0 to p. The last summation on n (for $p \geq 1$) is simply $(-1 + 1)^p$, which of course vanishes. This is easily seen by using the binomial expansion

$$(-1 + 1)^p = \sum_{n=0}^{p} \binom{p}{n} (-1)^n \, 1^{p-n} = \sum_{n=0}^{p} \frac{p!}{(p-n)! \, n!} (-1)^n \, .$$

Hence, the only surviving term is that having $p = 0$, yielding $e^{-i\hat{A}} e^{i\hat{A}} = \hat{1}$. Note that, given two operators $\hat{A}$ and $\hat{B}$, generally we have $e^{\hat{A}} e^{\hat{B}} \neq e^{\hat{A}+\hat{B}}$ unless $[\hat{A}, \hat{B}] = 0$, which is the case of interest here since $\hat{A}$ commutes with itself. Therefore, more directly we could have obtained

$$e^{-i\hat{A}} e^{i\hat{A}} = e^{\hat{0}} = \hat{1} \, ,$$

where $\hat{0}$ is the null operator, defined as $\hat{0} |\psi\rangle = 0$ for any $|\psi\rangle$.

Problem 7 Exponentiating the Pauli Matrix σ_y

The matrix $\underline{\sigma}_y$ is defined as

$$\underline{\sigma}_y = \begin{pmatrix} 0 & -i \\ i & 0 \end{pmatrix} .$$

Show that

$$e^{i\alpha\underline{\sigma}_y} = \cos\alpha\,\underline{\mathbb{1}} + i\sin\alpha\,\underline{\sigma}_y ,$$

where $\underline{\mathbb{1}}$ is the 2×2 identity matrix.

Solution

We find that

$$e^{i\alpha\underline{\sigma}_y} = \underline{\mathbb{1}} + i\alpha\,\underline{\sigma}_y + \frac{(i\alpha)^2}{2!}\,\underline{\sigma}_y^2 + \frac{(i\alpha)^3}{3!}\,\underline{\sigma}_y^3 + \cdots ,$$

where the matrix $\underline{\sigma}_y$ is such that

$$\underline{\sigma}_y^2 = \begin{pmatrix} 0 & -i \\ i & 0 \end{pmatrix}\begin{pmatrix} 0 & -i \\ i & 0 \end{pmatrix} = \begin{pmatrix} 1 & 0 \\ 0 & 1 \end{pmatrix} = \underline{\mathbb{1}} .$$

Therefore, even powers of $\underline{\sigma}_y$ give the unit matrix and odd powers of $\underline{\sigma}_y$ are just $\underline{\sigma}_y$, so that

$$\begin{aligned}
e^{i\alpha\underline{\sigma}_y} &= \underline{\mathbb{1}} + i\alpha\,\underline{\sigma}_y + \frac{(i\alpha)^2}{2!}\,\underline{\sigma}_y^2 + \frac{(i\alpha)^3}{3!}\,\underline{\sigma}_y^3 + \frac{(i\alpha)^4}{4!}\,\underline{\sigma}_y^4 + \frac{(i\alpha)^5}{5!}\,\underline{\sigma}_y^5 + \frac{(i\alpha)^6}{6!}\,\underline{\sigma}_y^6 + \cdots \\
&= \underbrace{\left(1 - \frac{\alpha^2}{2!} + \frac{\alpha^4}{4!} - \frac{\alpha^6}{6!} + \cdots\right)}_{\cos\alpha}\underline{\mathbb{1}} + i\underbrace{\left(\alpha - \frac{\alpha^3}{3!} + \frac{\alpha^5}{5!} + \cdots\right)}_{\sin\alpha}\underline{\sigma}_y = \cos\alpha\,\underline{\mathbb{1}} + i\sin\alpha\,\underline{\sigma}_y .
\end{aligned}$$

Problem 8 The Transformation Matrix Relating Two Bases is Unitary

Show that the transformation matrix relating two bases is unitary.

Solution

The components $c_m = \langle\phi_m|\psi\rangle$ representing a state $|\psi\rangle$ in basis $|\phi_m\rangle$ are related to those representing the same state in basis $|\chi_m\rangle$ via

$$c_m' = \sum_{n=1}^{\infty} S_{mn}^{\dagger}\, c_n .$$

It follows that

$$c_m = \sum_{n=1}^{\infty} S_{mn}\, c_n' = \sum_{n=1}^{\infty} S_{mn} \underbrace{\sum_{p=1}^{\infty} S_{np}^{\dagger}\, c_p}_{c_n'} = \sum_{p=1}^{\infty}\left(\sum_{n=1}^{\infty} S_{mn}\, S_{np}^{\dagger}\right) c_p = \sum_{p=1}^{\infty}\underbrace{\left(\underline{S}\,\underline{S}^{\dagger}\right)_{mp}}_{\delta_{mp}} c_p ,$$

which implies $\underline{S}\,\underline{S}^{\dagger} = \underline{\mathbb{1}}$, where $\underline{\mathbb{1}}$ is the identity matrix.

Problem 9 The Momentum Operator in the Coordinate Representation

Derive the coordinate representation of the momentum operator $\hat{p}$ using only the commutation relation $[\hat{x}, \hat{p}] = i\hbar$ (in one dimension for simplicity). Proceed as follows:

1. Consider the operator $\hat{T} = \hat{\mathbb{1}} - (i/\hbar)\eta\hat{p}$ with the parameter $\eta \longrightarrow 0$. Show that, to linear terms in η included,

$$\hat{x}\,\hat{T}|\phi_x\rangle = (x + \eta)\,\hat{T}|\phi_x\rangle\,,$$

   where $|\phi_x\rangle$ is the position eigenstate with eigenvalue x.
2. Deduce that $\hat{T}|\phi_x\rangle$ is proportional to $|\phi_{x+\eta}\rangle$, and show that the proportionality constant can be, at most, a phase factor (note that $\hat{T}$ is unitary to linear terms in η).
3. Consider the matrix element $\langle\phi_x|\hat{T}|\phi_{x'}\rangle$ and deduce that, in the limit $\eta \longrightarrow 0$,

$$\langle\phi_x|\hat{p}\,|\phi_{x'}\rangle = -i\hbar\frac{d}{dx}\delta(x - x')\,.$$

Show that

$$\langle\phi_x|\hat{p}^n\,|\phi_{x'}\rangle = (-i\hbar)^n\frac{d^n}{dx^n}\delta(x - x')\,.$$

Solution

Part 1

We have

$$\hat{x}\,\hat{T}|\phi_x\rangle = \left(\hat{T}\hat{x} + [\hat{x},\,\hat{T}]\right)|\phi_x\rangle = x\,\hat{T}|\phi_x\rangle + [\hat{x},\,\hat{T}]|\phi_x\rangle\,,$$

where in the next to last term we have used the fact that $|\phi_x\rangle$ is an eigenstate of $\hat{x}$ (and that $\hat{T}$ is linear, of course). Evaluation of the commutator gives

$$[\hat{x},\,\hat{T}] = \left[\hat{x},\,\hat{\mathbb{1}} - \frac{i}{\hbar}\eta\hat{p}\right] = -\frac{i}{\hbar}\eta\,[\hat{x},\,\hat{p}] = \eta\,,$$

where the identity operator is understood. Therefore, we find

$$\hat{x}\,\hat{T}|\phi_x\rangle = x\,\hat{T}|\phi_x\rangle + \eta\,|\phi_x\rangle = x\,\hat{T}|\phi_x\rangle + \eta\,\hat{T}|\phi_x\rangle = (x + \eta)\,\hat{T}|\phi_x\rangle\,,$$

where in the next to last step we have ignored terms quadratic in η and have used the relation $\eta\,|\phi_x\rangle = \eta\,\hat{T}|\phi_x\rangle$. We conclude that (in the limit $\eta \longrightarrow 0$) the state $\hat{T}|\phi_x\rangle$ is an eigenstate of $\hat{x}$ with eigenvalue $x + \eta$.

Part 2

Since the eigenvalues of $\hat{x}$ are non-degenerate, it is clear that $\hat{T}|\phi_x\rangle = c\,|\phi_{x+\eta}\rangle$, where c is constant. However, this constant has unit magnitude, since $\hat{T}$ is unitary:

$$\hat{T}^\dagger\,\hat{T} = \left(\hat{\mathbb{1}} + i\eta\hat{p}/\hbar\right)\left(\hat{\mathbb{1}} - i\eta\hat{p}/\hbar\right) = \hat{\mathbb{1}} + \text{terms quadratic in } \eta\,,$$

thus yielding

$$\delta(x - x') = \langle\phi_{x+\eta}|\phi_{x'+\eta}\rangle = \frac{1}{|c|^2}\langle\phi_x|\,\hat{T}^\dagger\,\hat{T}|\phi_{x'}\rangle = \frac{1}{|c|^2}\langle\phi_x|\phi_{x'}\rangle = \frac{1}{|c|^2}\delta(x - x') \implies |c|^2 = 1\,.$$

Part 3

The matrix elements of $\hat{T}$ read

$$\langle\phi_x|\hat{T}|\phi_{x'}\rangle = \langle\phi_x|\phi_{x'+\eta}\rangle = \delta(x-x'-\eta) = \langle\phi_x|\hat{\mathbb{1}} - \frac{i}{\hbar}\,\eta\hat{p}\,|\phi_{x'}\rangle = \delta(x-x') - \frac{i}{\hbar}\,\eta\,\langle\phi_x|\hat{p}\,|\phi_{x'}\rangle\,,$$

which leads to the required result,

$$\langle\phi_x|\hat{p}\,|\phi_{x'}\rangle = i\hbar\,\lim_{\eta\to 0}\,\frac{\delta(x-x'-\eta) - \delta(x-x')}{\eta} = -i\hbar\,\frac{d}{dx}\delta(x-x')\,,$$

where we have used the definition of the derivative,

$$\frac{d}{dx}f(x) = \lim_{\eta\to 0}\frac{f(x+\eta) - f(x)}{\eta} = -\lim_{\eta\to 0}\frac{f(x-\eta) - f(x)}{\eta}\,.$$

To show the last part, proceed by induction and assume that

$$\langle\phi_x|\hat{p}^{n-1}\,|\phi_{x'}\rangle = (-i\hbar)^{n-1}\frac{d^{n-1}}{dx^{n-1}}\delta(x-x')\,.$$

It then follows that

$$\langle\phi_x|\hat{p}^n\,|\phi_{x'}\rangle = \underbrace{\int_{-\infty}^{\infty}dy\,\langle\phi_x|\hat{p}^{n-1}\,|\phi_y\rangle\langle\phi_y|\hat{p}\,|\phi_{x'}\rangle}_{\text{insert completeness between } \hat{p}^{n-1} \text{ and } \hat{p}} = \int_{-\infty}^{\infty}dy\,(-i\hbar)^{n-1}\frac{d^{n-1}}{dx^{n-1}}\delta(x-y)\,(-i\hbar)\frac{d}{dy}\delta(y-x')$$

$$= (-i\hbar)^n\frac{d^{n-1}}{dx^{n-1}}\underbrace{\int_{-\infty}^{\infty}dy\,\delta(x-y)\frac{d}{dy}\delta(y-x')}_{\text{integrate out }\delta(x-y)} = (-i\hbar)^n\frac{d^n}{dx^n}\delta(x-x')\,.$$

Problem 10 Momentum and Hamiltonian Eigenvalue Problems in the Coordinate Representation

Solve the eigenvalue problem $\hat{p}\,|\phi_p\rangle = p\,|\phi_p\rangle$ in the coordinate representation, where $\hat{p}$ is the momentum operator and the eigenstates $|\phi_p\rangle$ are normalized as $\langle\phi_p|\phi_{p'}\rangle = \delta(p-p')$. Also show that the eigenvalue problem for the Hamiltonian operator,

$$\hat{H} = \frac{\hat{p}^2}{2m} + \hat{V}(\hat{x})\,,$$

just yields the (time-independent) Schrödinger equation.

Solution

We have

$$\int_{-\infty}^{\infty}dx'\,\underbrace{p(x,x')\,\phi_p(x')}_{\langle\phi_x|\hat{p}|\phi_{x'}\rangle\langle\phi_{x'}|\phi_p\rangle} = -i\hbar\int_{-\infty}^{\infty}dx'\,\frac{d}{dx}\delta(x-x')\,\phi_p(x') = -i\hbar\frac{d}{dx}\phi_p(x) = p\langle\phi_x|\phi_p\rangle = p\phi_p(x)\,,$$

and the eigenvalue problem is reduced to solving the first-order differential equation $\phi_p'(x) = i(p/\hbar)\phi_p(x)$, which has the solution

$$\phi_p(x) = \langle\phi_x|\phi_p\rangle = c\,e^{ipx/\hbar}\,,$$

where c is constant. Its magnitude is fixed by requiring that the eigenstates $|\phi_p\rangle$ satisfy the continuum normalization:

$$\langle \phi_p | \phi_{p'} \rangle = \int_{-\infty}^{\infty} dx \, \langle \phi_p | \phi_x \rangle \langle \phi_x | \phi_{p'} \rangle = |c|^2 \int_{-\infty}^{\infty} dx \, e^{-i(p-p')x/\hbar} = |c|^2 \, 2\pi\hbar \, \delta(p-p') \implies |c| = \frac{1}{\sqrt{2\pi\hbar}} \, .$$

In the case of the Hamiltonian, the eigenvalue problem $\hat{H}|\psi\rangle = E|\psi\rangle$ reads

$$\int_{-\infty}^{\infty} dx' \, \underbrace{H(x,x') \, \psi(x')}_{\langle \phi_x | \hat{H} | \phi_{x'} \rangle \langle \phi_{x'} | \psi \rangle} = \int_{-\infty}^{\infty} dx' \left[-\frac{\hbar^2}{2m} \frac{d^2}{dx^2} + V(x) \right] \delta(x - x') \, \psi(x') = E\psi(x) \, ,$$

or, integrating out the δ-function,

$$\left[-\frac{\hbar^2}{2m} \frac{d^2}{dx^2} + V(x) \right] \psi(x) = E\psi(x) \, .$$

Problem 11   The Position Operator in the Momentum Representation

Obtain the momentum representation of the position operator $\hat{x}$.

Solution

The position-operator matrix elements are given by

$$\langle \phi_p | \hat{x} | \phi_{p'} \rangle = \underbrace{\int_{-\infty}^{\infty} dx \, \langle \phi_p | \hat{x} | \phi_x \rangle \langle \phi_x | \phi_{p'} \rangle}_{\text{completeness of } \hat{x} \text{ eigenstates}} = \int_{-\infty}^{\infty} dx \, x \, \langle \phi_p | \phi_x \rangle \langle \phi_x | \phi_{p'} \rangle = \frac{1}{2\pi\hbar} \int_{-\infty}^{\infty} dx \, x \, e^{-i(p-p')x/\hbar}$$

$$= \, = i\hbar \frac{d}{dp} \underbrace{\frac{1}{2\pi\hbar} \int_{-\infty}^{\infty} dx \, e^{-i(p-p')x/\hbar}}_{\delta(p-p')} = i\hbar \frac{d}{dp} \delta(p - p') \, ,$$

as expected.

Problem 12   The Hamiltonian Eigenvalue Problem in the Momentum Representation

Show that the Hamiltonian eigenvalue problem $\hat{H}|\psi\rangle = E|\psi\rangle$ in the momentum representation yields the momentum-space Schrödinger equation,

$$\left(E - \frac{p^2}{2m} \right) \widetilde{\psi}(p) = \frac{1}{\sqrt{2\pi\hbar}} \int_{-\infty}^{\infty} dp' \, \widetilde{V}(p - p') \, \widetilde{\psi}(p') \, , \qquad \widetilde{V}(p - p') = \frac{1}{\sqrt{2\pi\hbar}} \int_{-\infty}^{\infty} dx \, V(x) \, e^{-i(p-p')x/\hbar} \, .$$

Solution

We find that

$$\int_{-\infty}^{\infty} dp' \, \underbrace{H(p,p') \, \widetilde{\psi}(p')}_{\langle \phi_p | \hat{H} | \phi_{p'} \rangle \langle \phi_{p'} | \psi \rangle} = \int_{-\infty}^{\infty} dp' \left[\frac{p^2}{2m} \delta(p - p') + \langle \phi_p | \hat{V}(\hat{x}) | \phi_{p'} \rangle \right] \widetilde{\psi}(p') = E\widetilde{\psi}(p) \, ,$$

which, after integrating out the δ-function, can be expressed as

$$\left(E - \frac{p^2}{2m}\right) \widetilde{\psi}(p) = \int_{-\infty}^{\infty} dp' \, \langle \phi_p | \hat{V}(\hat{x}) | \phi_{p'} \rangle \, \widetilde{\psi}(p') \, .$$

Evaluation of the potential energy operator matrix element is facilitated by inserting the completeness relation for the position eigenstates as follows:

$$\langle \phi_p | \hat{V}(\hat{x}) | \phi_{p'} \rangle = \int_{-\infty}^{\infty} dx \, \langle \phi_p | \hat{V}(\hat{x}) | \phi_x \rangle \langle \phi_x | \phi_{p'} \rangle = \int_{-\infty}^{\infty} dx \, V(x) \, \langle \phi_p | \phi_x \rangle \langle \phi_x | \phi_{p'} \rangle$$

$$= \underbrace{\frac{1}{2\pi\hbar} \int_{-\infty}^{\infty} dx \, V(x) \, e^{-i(p-p')x/\hbar}}_{\widetilde{V}(p-p')/\sqrt{2\pi\hbar}} \, ,$$

which then yields the expression provided in the text of the problem.

Problem 13 Some Consequences of the Commutation Relation between $\hat{x}$ and $\hat{p}$

The Hamiltonian $\hat{H}$ of a particle in a one dimension is given by

$$\hat{H} = \frac{\hat{p}^2}{2m} + \hat{V}(\hat{x}) \, .$$

Let $|\phi_n\rangle$ be the eigenstates of $\hat{H}$ with $\hat{H}|\phi_n\rangle = E_n|\phi_n\rangle$, where n is a discrete index. Evaluate the commutator $[\hat{x}, \hat{H}]$ and deduce that

$$\langle \phi_m | \hat{p} | \phi_n \rangle = \alpha_{mn} \langle \phi_m | \hat{x} | \phi_n \rangle \, ,$$

where the coefficient α_{mn} depends on the energy difference $E_m - E_n$. Using the closure relation satisfied by the eigenstates of $\hat{H}$ and the previous result, obtain the following sum rule:

$$\sum_n (E_m - E_n)^2 \, |\langle \phi_m | \hat{x} | \phi_n \rangle|^2 = \frac{\hbar^2}{m^2} \, \langle \phi_m | \hat{p}^2 | \phi_m \rangle \, .$$

Solution

We evaluate the commutator

$$[\hat{x}, \hat{H}] = \frac{1}{2m} [\hat{x}, \hat{p}^2] = \frac{i\hbar}{m} \hat{p} \, ,$$

where we have used the fact that $\hat{x}$ commutes with any function of $\hat{x}$, such as $\hat{V}(\hat{x})$, and the commutator property valid for any three operators $\hat{A}$, $\hat{B}$, and $\hat{C}$,

$$[\hat{A}, \hat{B}\hat{C}] = \hat{B}\hat{A}\hat{C} \underbrace{-\hat{B}\hat{C}\hat{A} + \hat{A}\hat{B}\hat{C}}_{[\hat{A}, \hat{B}\hat{C}]} - \hat{B}\hat{A}\hat{C} = \hat{B}[\hat{A}, \hat{C}] + [\hat{A}, \hat{B}]\hat{C} \, ,$$

where we have added and subtracted $\hat{B}\hat{A}\hat{C}$. Therefore, it follows that

$$\langle \phi_m | \hat{p} | \phi_n \rangle = -i \frac{m}{\hbar} \langle \phi_m | [\hat{x}, \hat{H}] | \phi_n \rangle = -i \frac{m}{\hbar} \langle \phi_m | \hat{x}\hat{H} - \hat{H}\hat{x} | \phi_n \rangle = i \underbrace{\frac{m}{\hbar} (E_m - E_n)}_{\alpha_{mn}} \langle \phi_m | \hat{x} | \phi_n \rangle \, .$$

Next, we consider

$$
\frac{\hbar^2}{m^2} \langle \phi_n | \hat{p}^2 | \phi_n \rangle = \frac{\hbar^2}{m^2} \sum_m \underbrace{\langle \phi_n | \hat{p} | \phi_m \rangle}_{\langle \phi_m | \hat{p} | \phi_n \rangle^*} \langle \phi_m | \hat{p} | \phi_n \rangle = \frac{\hbar^2}{m^2} \sum_m \alpha^*_{mn} \underbrace{\langle \phi_n | \hat{x} | \phi_m \rangle}_{\langle \phi_m | \hat{x} | \phi_n \rangle^*} \alpha_{mn} \langle \phi_m | \hat{x} | \phi_n \rangle
$$

$$
= \sum_m (E_m - E_n)^2 \, |\langle \phi_m | \hat{x} | \phi_n \rangle|^2 \, ,
$$

where we have used the completeness of the states $|\phi_m\rangle$ and the result derived previously.

Problem 14 Trace of an Operator

Prove or evaluate the following relations:

1. Given an operator $\hat{A}$ and a basis $|\phi_n\rangle$, show that $\mathrm{tr}(\hat{A}\hat{B}) = \mathrm{tr}(\hat{B}\hat{A})$, where $\hat{A}$ and $\hat{B}$ are non-commuting operators.
2. Show that the trace of an operator is independent of the basis $|\phi_n\rangle$ used to evaluate it.
3. Let $\hat{A}$ be a hermitian operator and let $|\phi_n\rangle$ be the basis of eigenstates of $\hat{A}$, that is, $\hat{A}|\phi_n\rangle = a_n|\phi_n\rangle$. Evaluate $\exp[if(\hat{A})]$.

Solution

Part 1

Using the definition of the trace and the completeness of the basis, we find

$$
\mathrm{tr}(\hat{A}\,\hat{B}) = \sum_m \langle \phi_m | \hat{A}\hat{B} | \phi_m \rangle = \sum_{n,m} \underbrace{\langle \phi_m | \hat{A} | \phi_n \rangle}_{c-number} \underbrace{\langle \phi_n | \hat{B} | \phi_m \rangle}_{c-number} = \sum_{n,m} \langle \phi_n | \hat{B} | \phi_m \rangle \langle \phi_m | \hat{A} | \phi_n \rangle
$$

$$
= \sum_n \langle \phi_n | \hat{B}\hat{A} | \phi_n \rangle = \mathrm{tr}(\hat{B}\hat{A}) \, .
$$

Part 2

Consider two bases $\{|\phi_n\rangle\}$ and $\{|\chi_n\rangle\}$, and define $S_{nm} = \langle \phi_n | \chi_m \rangle$. We have

$$
\mathrm{tr}(\hat{A}) = \sum_n \langle \phi_n | \hat{A} | \phi_n \rangle = \sum_{n,m,p} \langle \phi_n | \chi_m \rangle \langle \chi_m | \hat{A} | \chi_p \rangle \langle \chi_p | \phi_n \rangle = \sum_{n,m,p} S_{nm} S^*_{np} \langle \chi_m | \hat{A} | \chi_p \rangle
$$

$$
= \sum_{m,p} \delta_{m,p} \langle \chi_m | \hat{A} | \chi_p \rangle = \sum_m \langle \chi_m | \hat{A} | \chi_m \rangle = \mathrm{tr}(\hat{A}) \, ,
$$

where we have used the unitarity of $\underline{S}$. In matrix notation we have

$$
\underline{A} = \underline{S}\,\underline{A}'\,\underline{S}^\dagger \, ,
$$

and, using the property derived in part 1 above, we obtain immediately

$$
\mathrm{tr}(\underline{A}) = \mathrm{tr}(\underline{S}\,\underline{A}'\,\underline{S}^\dagger) = \mathrm{tr}(\underline{A}'\,\underbrace{\underline{S}^\dagger\,\underline{S}}_{\underline{I}}) = \mathrm{tr}(\underline{A}') \, .
$$

Part 3

We have $f(\hat{A})\,|\phi_n\rangle = f(a_n)\,|\phi_n\rangle$ and

$$e^{if(\hat{A})}\,|\phi_n\rangle = e^{if(a_n)}\,|\phi_n\rangle\,.$$

Problem 15 Properties of Eigenvalues and Eigenstates of a Hermitian Operator

Show that the eigenvalues of a hermitian operator $\hat{A}$ are real and that eigenstates belonging to different eigenvalues are orthogonal.

Solution

Since $\hat{A}$ is hermitian, we have

$$\hat{A}\,|\psi\rangle = \lambda\,|\psi\rangle \rightleftharpoons \langle\psi|\,\hat{A} = \lambda^*\langle\psi|\,,$$

and therefore by taking the scalar product with $\langle\psi|$ in the first relation and $|\psi\rangle$ in the second it follows that

$$\lambda\langle\psi|\psi\rangle = \langle\psi|\hat{A}|\psi\rangle = \lambda^*\langle\psi|\psi\rangle\,,$$

and so either $\langle\psi|\psi\rangle \neq 0$ and $\lambda = \lambda^*$, or $\langle\psi|\psi\rangle = 0$ and $|\psi\rangle$ is the null state. Eigenstates belonging to different eigenvalues are orthogonal, since if $\hat{A}\,|\psi_1\rangle = \lambda_1|\psi_1\rangle$ and $\hat{A}\,|\psi_2\rangle = \lambda_2|\psi_2\rangle$ with $\lambda_1 \neq \lambda_2$ then

$$\langle\psi_1|\hat{A}|\psi_2\rangle = \langle\psi_1|\cdot(\hat{A}|\psi_2\rangle) = \lambda_2\langle\psi_1|\psi_2\rangle \qquad \text{and also} \qquad \langle\psi_1|\hat{A}|\psi_2\rangle = (\langle\psi_1|\hat{A}^\dagger)\cdot|\psi_2\rangle = \lambda_1\langle\psi_1|\psi_2\rangle\,;$$

then subtracting one from the other we arrive at

$$\underbrace{(\lambda_1 - \lambda_2)}_{\neq 0}\langle\psi_1|\psi_2\rangle = 0 \implies \langle\psi_1|\psi_2\rangle = 0\,.$$

Problem 16 Decompositions of Hermitian Operator in Terms of Its Eigenvalues or Eigenstates

Let $\hat{A}$ be a hermitian operator and let $|\phi_n\rangle$ be the basis consisting of the eigenstates of $\hat{A}$, that is, $\hat{A}\,|\phi_n\rangle = a_n|\phi_n\rangle$. Assume that the eigenvalues are non-degenerate and that the basis can be either finite or infinite dimensional.

1. Prove that

$$\prod_n (\hat{A} - a_n)$$

   is the null operator.
2. Explain the significance of the following operator:

$$\prod_{n(n\neq m)}^{\infty} \frac{\hat{A} - a_n}{a_m - a_n}\,.$$

(Adapted from J. J. Sakurai and J. Napolitano 2020, *Modern Quantum Mechanics*, Cambridge University Press.)

Solution

Part 1

For any state $|\psi\rangle$ we have

$$|\psi\rangle = \sum_m c_m |\phi_m\rangle \,, \qquad c_m = \langle \phi_m | \psi \rangle \,,$$

and therefore

$$\left[\prod_n (\hat{A} - a_n)\right] |\psi\rangle = \sum_m c_m \left[\prod_n (\hat{A} - a_n)\right] |\phi_m\rangle = \sum_m c_m \left[\prod_n (a_m - a_n)\right] |\phi_m\rangle = 0 \,,$$

since

$$\prod_n (a_m - a_n) = (a_m - a_1) \cdots (a_m - a_{m-1})(a_m - a_m)(a_m - a_{m+1}) \cdots = 0 \,.$$

Part 2

For any state $|\psi\rangle$ we have

$$\left(\prod_{n(n \neq m)}^{\infty} \frac{\hat{A} - a_n}{a_m - a_n}\right) |\psi\rangle = \sum_p c_p \left(\prod_{n(n \neq m)} \frac{\hat{A} - a_n}{a_m - a_n}\right) |\phi_p\rangle = \sum_p c_p \underbrace{\left(\prod_{n(n \neq m)} \frac{a_p - a_n}{a_m - a_n}\right)}_{\delta_{p,m}} |\phi_p\rangle = c_m |\phi_m\rangle \,,$$

since, for any $p \neq m$,

$$\prod_{n(n \neq m)} \frac{a_p - a_n}{a_m - a_n} = \frac{a_p - a_1}{a_m - a_1} \cdots \frac{a_p - a_p}{a_m - a_p} \cdots = 0 \,,$$

and, for $p = m$,

$$\prod_{n(n \neq m)} \frac{a_m - a_n}{a_m - a_n} = 1 \,.$$

Thus the operator projects $|\psi\rangle$ along the eigenstate $|\phi_m\rangle$.

Problem 17 Basis of Simultaneous Eigenstates for Two Observables, Commutativity

Consider two observables $\hat{A}$ and $\hat{B}$. Assume that there are simultaneous eigenstates $|\psi_{a_m,b_n}\rangle$ of $\hat{A}$ and $\hat{B}$ with

$$\hat{A} |\psi_{a_m,b_n}\rangle = a_m |\psi_{a_m,b_n}\rangle \,, \qquad \hat{B} |\psi_{a_m,b_n}\rangle = b_n |\psi_{a_m,b_n}\rangle \,,$$

which form an orthonormal basis. Can you conclude that the two observables commute? Justify your answer. (Adapted from *Modern Quantum Mechanics*, J. J. Sakurai and J. Napolitano, Cambridge University Press.)

Solution

Consider a generic state $|\psi\rangle$. Since the eigenstates $|\psi_{a_m,b_n}\rangle$ form a basis, it can be expanded as follows:

$$|\psi\rangle = \sum_{m,n} c_{m,n} |\psi_{a_m,b_n}\rangle \,, \qquad c_{m,n} = \langle \psi_{a_m,b_n} | \psi \rangle \,.$$

Now, acting on $|\psi\rangle$ with $\hat{A}\hat{B}$ and $\hat{B}\hat{A}$ yields, respectively,

$$\hat{A}\hat{B} |\psi\rangle = \sum_{m,n} c_{m,n} \hat{A}\hat{B} |\psi_{a_m,b_n}\rangle = \sum_{m,n} c_{m,n} b_n a_m |\psi_{a_m,b_n}\rangle$$

and

$$\hat{B}\hat{A} |\psi\rangle = \sum_{m,n} c_{m,n} \hat{B}\hat{A} |\psi_{a_m,b_n}\rangle = \sum_{m,n} c_{m,n} a_m b_n |\psi_{a_m,b_n}\rangle \,,$$

that is, the same result. We conclude that $(\hat{A}\hat{B} - \hat{B}\hat{A})|\psi\rangle = 0$ for any $|\psi\rangle$, implying that $\hat{A}$ and $\hat{B}$ commute.

Problem 18 Simultaneous Eigenstate of Two Anticommuting Observables

Suppose that $\hat{A}$ and $\hat{B}$ are two observables that anticommute, that is, $\{\hat{A},\hat{B}\} \equiv \hat{A}\hat{B} + \hat{B}\hat{A} = 0$. Can there be a common eigenstate of $\hat{A}$ and $\hat{B}$? Justify your answer. (Adapted from J. J. Sakurai and J. Napolitano, *Modern Quantum Mechanics*, Cambridge University Press.)

Solution

Since $\hat{A}$ is an observable it can be diagonalized, so that in the basis consisting of eigenstates of $\hat{A}$, the associated matrix is given by $(\underline{A})_{ij} = a_i \delta_{ij}$. In this basis, observable $\hat{B}$ is represented by a hermitian matrix $(\underline{B})_{ij} = b_{ij}$. The requirement that $\hat{A}$ and $\hat{B}$ anticommute leads to the condition

$$(\underline{A}\,\underline{B} + \underline{B}\,\underline{A})_{ij} = \sum_k \left(a_i \delta_{ik} b_{kj} + b_{ik} \delta_{kj} a_j \right) = b_{ij} \left(a_i + a_j \right) = 0 \,.$$

Suppose the state space is two dimensional, in which case the above condition requires that

$$a_1 b_{11} = 0 \,, \qquad (a_1 + a_2) b_{12} = 0 \,, \qquad a_2 b_{22} = 0 \,;$$

and note that $(a_1 + a_2) b_{21} = 0$ is redundant since $(a_1 + a_2) b_{21} = [(a_1 + a_2) b_{12}]^*$ (the eigenvalues a_i are real and $\underline{B}$ is hermitian). The coefficients a_i cannot both be zero, otherwise we would have the null matrix. We take $a_2 = 0$; then $b_{11} = 0$ and $b_{12} = 0$, and hence the matrices representing $\hat{A}$ and $\hat{B}$ must have the forms

$$\underline{A} = \begin{pmatrix} a & 0 \\ 0 & 0 \end{pmatrix} \,, \qquad \underline{B} = \begin{pmatrix} 0 & 0 \\ 0 & b \end{pmatrix} \,,$$

with eigenvalues a and 0, and 0 and b, respectively. The common eigenvectors are

$$\underline{v}_1 = \begin{pmatrix} 1 \\ 0 \end{pmatrix} \,, \qquad \underline{v}_2 = \begin{pmatrix} 0 \\ 1 \end{pmatrix} \,,$$

and

$$\underline{A}\,\underline{B} = \begin{pmatrix} a & 0 \\ 0 & 0 \end{pmatrix} \begin{pmatrix} 0 & 0 \\ 0 & b \end{pmatrix} = \begin{pmatrix} 0 & 0 \\ 0 & 0 \end{pmatrix} = \begin{pmatrix} 0 & 0 \\ 0 & b \end{pmatrix} \begin{pmatrix} a & 0 \\ 0 & 0 \end{pmatrix} = \underline{B}\,\underline{A} \,,$$

and $\underline{A}$ and $\underline{B}$ certainly anticommute (note that they also commute!). For a less trivial case, consider a three-dimensional state space with

$$\underline{A} = \begin{pmatrix} a & 0 & 0 \\ 0 & 0 & 0 \\ 0 & 0 & -a \end{pmatrix}, \qquad \underline{B} = \begin{pmatrix} 0 & 0 & b \\ 0 & 0 & 0 \\ b^* & 0 & 0 \end{pmatrix}.$$

It is easily verified that the above matrices anticommute and have the common eigenvector

$$\underline{v} = \begin{pmatrix} 0 \\ 1 \\ 0 \end{pmatrix},$$

both with eigenvalue 0.

Problem 19　Normal Operators and Associated Eigenvalues

A *normal* operator is one that commutes with its adjoint. A linear algebra theorem assures us that (at least for a finite-dimensional state space) a normal operator can be diagonalized and that the eigenvectors form a basis in the state space.

1. Prove the converse statement, that an operator whose eigenstates form a basis in the state space must be normal.
2. A hermitian operator is an obvious example of a normal operator. Is a unitary operator also a normal operator? Show that the eigenvalues of a unitary operator all have unit magnitude.
3. Show that the eigenvalues of any operator are independent of the basis used to represent this operator.
4. Prove that if $\underline{A}$ is a normal matrix representing the (normal) operator $\hat{A}$ then

$$\det\left(e^{\underline{A}}\right) = e^{\mathrm{tr}\underline{A}}, \qquad \mathrm{tr}\left(\underline{A}\right) = \sum_i A_{ii}.$$

Solution

Part 1

Since the eigenstates of an operator $\hat{A}$ form a basis in the state space, we have the completeness relation

$$\sum_n \sum_{k=1}^{g_n} |\phi_n^{(k)}\rangle\langle\phi_n^{(k)}| = \hat{\mathbb{1}},$$

where a_n (with degeneracy g_n) is the eigenvalue (not necessarily real) corresponding to the eigenstates $|\phi_n^{(k)}\rangle$. We obtain

$$\hat{A} = \sum_n \sum_{k=1}^{g_n} a_n |\phi_n^{(k)}\rangle\langle\phi_n^{(k)}|,$$

where we have multiplied $\hat{A}$ on the right by the identity operator $\hat{\mathbb{1}}$ and exploited the completeness relation. Now, taking the adjoint yields

$$\hat{A}^\dagger = \sum_n \sum_{k=1}^{g_n} a_n^* |\phi_n^{(k)}\rangle\langle\phi_n^{(k)}|,$$

and hence, using the orthonormality property of the eigenstates, $\langle\phi_n^{(k)}|\phi_p^{(l)}\rangle = \delta_{np}\,\delta_{kl}$, we have

$$\hat{A}\,\hat{A}^\dagger = \sum_n \sum_{k=1}^{g_n} |a_n|^2\, |\phi_n^{(k)}\rangle\langle\phi_n^{(k)}| = \hat{A}^\dagger\,\hat{A}$$

and $\hat{A}$ commutes with $\hat{A}^\dagger$.

Part 2

Let $\hat{U}$ be a unitary operator. Since $\hat{U}^\dagger = \hat{U}^{-1}$, we see that $\hat{U}^\dagger\hat{U} = \hat{\mathbb{1}} = \hat{U}\hat{U}^\dagger$ and hence $\hat{U}$ is normal. Now, if $|\psi\rangle$ is an eigenstate of $\hat{U}$ with eigenvalue u, then

$$\underbrace{\hat{U}|\psi\rangle}_{|\psi'\rangle} = u|\psi\rangle\,,$$

with $\langle\psi'|\psi'\rangle = |u|^2\langle\psi|\psi\rangle$. Exploiting unitarity leads to

$$\langle\psi'|\psi'\rangle = \langle\psi|\psi\rangle \implies |u|^2 = 1\,,$$

that is, the eigenvalue u is a phase factor.

Part 3

The characteristic equation determining the eigenvalues is given by

$$\det\left(\underline{A} - \lambda\,\underline{\mathbb{1}}\right)\,.$$

Under a basis transformation, we have

$$\underline{A} \longrightarrow \underline{A}' = \underline{S}^\dagger\,\underline{A}\,\underline{S}\,,$$

and $\underline{S}$ is unitary. Using Binet's theorem, that $\det(\underline{A}\,\underline{B}) = \det(\underline{A})\,\det(\underline{B})$, it follows that

$$\det\left(\underline{A}' - \lambda\,\underline{\mathbb{1}}\right) = \det\left[\underline{S}^\dagger\,(\underline{A} - \lambda\,\underline{\mathbb{1}})\,\underline{S}\right] = \det\left(\underline{S}^\dagger\right)\det\left(\underline{A} - \lambda\,\underline{\mathbb{1}}\right)\det\left(\underline{S}\right) = \det\left(\underline{A} - \lambda\,\underline{\mathbb{1}}\right)\,,$$

since $\det\left(\underline{S}^\dagger\right) = \det\left(\underline{S}^{-1}\right) = \left[\det\left(\underline{S}\right)\right]^{-1}$. We conclude that the characteristic equation and therefore its roots are invariant under a basis transformation.

Part 4

Given that $\hat{A}$ is normal, it can be diagonalized. In the basis of eigenstates, we have

$$\underline{A} = \underline{S}\,\underline{A}_D\,\underline{S}^\dagger\,,$$

where $\underline{A}_D$ is the matrix with eigenvalues on its diagonal. We also have

$$e^{\underline{A}} = \sum_{n=0}^\infty \frac{1}{n!}\,\underline{A}^n = \sum_{n=0}^\infty \frac{1}{n!}\,\left(\underline{S}\,\underline{A}_D\,\underline{S}^\dagger\right)^n = \underline{S}\left[\sum_{n=0}^\infty \frac{1}{n!}\,\underline{A}_D^n\right]\underline{S}^\dagger\,.$$

Now, being diagonal, the matrix $\underline{A}_D^n$ has the form (we are assuming it has dimension p)

$$\underline{A}_D^n = \begin{pmatrix} a_1^n & 0 & \cdots & 0 \\ 0 & a_2^n & \cdots & 0 \\ \vdots & \vdots & \cdots & \vdots \\ 0 & 0 & \cdots & a_p^n \end{pmatrix},$$

so that

$$e^{\underline{A}_D} = \sum_{n=0}^{\infty} \frac{1}{n!} \underline{A}_D^n = \underline{A}_D^n = \begin{pmatrix} \sum_n a_1^n/n! & 0 & \cdots & 0 \\ 0 & \sum_n a_2^n/n! & \cdots & 0 \\ \vdots & \vdots & \cdots & \vdots \\ 0 & 0 & \cdots & \sum_n a_p^n/n! \end{pmatrix} = \begin{pmatrix} e^{a_1} & 0 & \cdots & 0 \\ 0 & e^{a_2} & \cdots & 0 \\ \vdots & \vdots & \cdots & \vdots \\ 0 & 0 & \cdots & e^{a_p} \end{pmatrix}.$$

Using Binet's theorem, we have

$$\det\left(e^{\underline{A}}\right) = \det\left(e^{\underline{A}_D}\right) = \prod_n e^{a_n} = e^{\sum_n a_n} = e^{\operatorname{tr}(\underline{A}_D)} = e^{\operatorname{tr}(\underline{A})},$$

where in the last step we have exploited the invariance of the trace under a change of basis,

$$\operatorname{tr}(\underline{A}) = \operatorname{tr}(\underline{S}\,\underline{A}_D\,\underline{S}^\dagger) = \operatorname{tr}(\underline{A}_D\,\underline{S}^\dagger\,\underline{S}) = \operatorname{tr}(\underline{A}_D).$$

Problem 20 A Model Hamiltonian

Suppose that a linear operator $\hat{S}$ satisfies the two equations

$$\hat{S}^\dagger \hat{S}^\dagger = 0, \qquad \hat{S}\hat{S}^\dagger + \hat{S}^\dagger \hat{S} = \hat{\mathbb{1}}.$$

The Hamiltonian $\hat{H}$ of the system is given by

$$\hat{H} = \lambda \hat{S}\hat{S}^\dagger, \qquad \lambda \text{ real}.$$

1. Show that $\hat{H}$ is hermitian.
2. Express the operator $\hat{H}^2$ in terms of $\hat{H}$.
3. Determine the eigenvalues of $\hat{H}$.

Solution

Part 1

Since λ is real, we have

$$\hat{H}^\dagger = \lambda\left(\hat{S}\hat{S}^\dagger\right)^\dagger = \lambda\left(\hat{S}^\dagger\right)^\dagger \hat{S}^\dagger = \lambda\hat{S}\hat{S}^\dagger = \hat{H}.$$

Part 2

Using the given properties of $\hat{S}$, we find

$$\hat{H}^2 = \lambda^2 \hat{S}\hat{S}^\dagger\hat{S}\hat{S}^\dagger = \lambda^2 \hat{S}(\hat{\mathbb{1}} - \hat{S}\hat{S}^\dagger)\hat{S}^\dagger = \lambda\,\hat{H},$$

since $\hat{S}^\dagger \hat{S}^\dagger = 0$.

Part 3

Let $|\psi_E\rangle$ be an eigenstate of $\hat{H}$ with eigenvalue E, namely $\hat{H}|\psi_E\rangle = E|\psi_E\rangle$. We have

$$E^2|\psi_E\rangle = \hat{H}^2|\psi_E\rangle = \lambda\,\hat{H}|\psi_E\rangle = \lambda E\,|\psi_E\rangle\,,$$

yielding $E^2 = \lambda E$. The eigenvalues E are 0 or λ.

Problem 21 A Simple Two-State Hamiltonian

The Hamiltonian operator for a two-state system is given in some basis $|1\rangle, |2\rangle$ by

$$\hat{H} = E(|1\rangle\langle 1| - |2\rangle\langle 2| + |1\rangle\langle 2| + |2\rangle\langle 1|)\,,$$

where E is a constant with the dimensions of energy. Find the energy eigenvalues and corresponding eigenstates (as linear combinations of $|1\rangle$ and $2\rangle$). (Adapted from J. J. Sakurai and J. Napolitano (2020), *Modern Quantum Mechanics*, Cambridge University Press.)

Solution

In the basis $|1\rangle, |2\rangle$, the matrix representing $\hat{H}$ has matrix elements given by

$$\langle 1|\hat{H}|1\rangle = E\,, \qquad \langle 2|\hat{H}|2\rangle = -E\,, \qquad \langle 1|\hat{H}|2\rangle = \langle 2|\hat{H}|1\rangle = E\,,$$

so that

$$\underline{H} = E\begin{pmatrix} 1 & 1 \\ 1 & -1 \end{pmatrix}\,.$$

We find that

$$\det\begin{bmatrix} E - \lambda & E \\ E & -E - \lambda \end{bmatrix} = \lambda^2 - 2E^2 = 0 \implies \lambda_\pm = \pm\sqrt{2}\,E\,.$$

The eigenstates corresponding to $\lambda_\pm$ have the following expansion in the basis $\{|1\rangle, |2\rangle\}$:

$$|\pm\rangle = \sum_{i=1,2} c_i^{(\pm)}|i\rangle\,,$$

where the expansion parameters $c_i^{(\pm)}$ are found from the (linearly dependent) linear system

$$\begin{bmatrix} E - \lambda_\pm & E \\ E & -E - \lambda_\pm \end{bmatrix}\begin{bmatrix} c_1^{(\pm)} \\ c_2^{(\pm)} \end{bmatrix} = 0\,;$$

utilizing the first equation we have

$$(E - \lambda_\pm)c_1^{(\pm)} + Ec_2^{(\pm)} = 0 \implies c_2^{(\pm)} = (\pm\sqrt{2} - 1)c_1^{(\pm)}$$

or

$$|+\rangle \longrightarrow c_+\begin{pmatrix} 1 \\ \sqrt{2} - 1 \end{pmatrix}\,, \qquad |-\rangle \longrightarrow c_-\begin{pmatrix} 1 \\ -\sqrt{2} - 1 \end{pmatrix}\,,$$

with normalization constants $c_\pm$.

Problem 22 Eigenvalues and Eigenvectors of a Two-State Hamiltonian

Consider a system with a two-dimensional state space. In this space, the states $|1\rangle$ and $|2\rangle$ form an orthonormal basis. The Hamiltonian describing the system in this basis has the form

$$\hat{H} = H_{11}|1\rangle\langle 1| + H_{22}|2\rangle\langle 2| + H_{12}(|1\rangle\langle 2| + |2\rangle\langle 1|),$$

where H_{11}, H_{22}, and H_{12} are real parameters with dimensions of energy. Obtain the eigenvalues and corresponding eigenvectors of $\hat{H}$. Make sure that they reduce to the expected (and obvious!) eigenvalues and eigenvectors in the limit $H_{12} \longrightarrow 0$.

It is convenient to introduce the parameter

$$\eta = \frac{2H_{12}}{H_{11} - H_{22}} \qquad \text{with } H_{11} \neq H_{22},$$

and to express the results in terms of η.

Solution

The eigenvalues result from

$$\det \begin{bmatrix} H_{11} - E & H_{12} \\ H_{12} & H_{22} - E \end{bmatrix} = 0,$$

where H_{11}, H_{22}, and H_{12} are real. We find that

$$(H_{11} - E)(H_{22} - E) - H_{12}^2 = 0 \implies E_\pm = \frac{H_{11} + H_{22} \pm \sqrt{(H_{11} - H_{22})^2 + 4H_{12}^2}}{2}.$$

We now introduce the parameter

$$\eta = 2\frac{H_{12}}{H_{11} - H_{22}},$$

so that

$$E_\pm = \frac{H_{11} + H_{22}}{2} \pm \frac{H_{11} - H_{22}}{2}\sqrt{1 + \eta^2}.$$

In the basis consisting of $|1\rangle$ and $|2\rangle$, we write the corresponding eigenvectors as column vectors $(a_\pm, b_\pm)^T$ and obtain

$$H_{12}a_+ + (H_{22} - E_+)b_+ = 0, \qquad (H_{11} - E_-)a_- + H_{12}b_- = 0,$$

where we consider only the second and first equations for the eigenvector, corresponding to E_+ and E_-, respectively. Inserting the expressions for the eigenvalues yields the eigenvectors

$$H_{12}a_+ - \frac{H_{11} - H_{22}}{2}\left(1 + \sqrt{1 + \eta^2}\right)b_+ = 0 \implies a_+ \begin{bmatrix} 1 \\ \eta/\left(1 + \sqrt{1 + \eta^2}\right) \end{bmatrix}$$

and

$$\frac{H_{11} - H_{22}}{2}\left(1 + \sqrt{1 + \eta^2}\right)a_- + H_{12}b_- = 0 \implies b_- \begin{bmatrix} -\eta/\left(1 + \sqrt{1 + \eta^2}\right) \\ 1 \end{bmatrix}.$$

The normalization condition gives

$$|a_+|^2 \left[1 + \frac{\eta^2}{(1 + \sqrt{1 + \eta^2})^2} \right] = 1 \implies |a_+| = \frac{1}{\sqrt{2}} \left(1 + \frac{1}{\sqrt{1 + \eta^2}} \right)^{1/2},$$

with an identical expression for $|b_-|$. Note that in the limit $H_{12} \longrightarrow 0$, we have $\eta \longrightarrow 0$ and

$$E_+ \longrightarrow H_{11}, \qquad E_- \longrightarrow H_{22},$$

with corresponding eigenvectors $(1, 0)^T$ and $(0, 1)^T$.

Problem 23 Two Observables in a Three-Dimensional State Space: An Example

Consider a three-dimensional state space. If a certain set of orthonormal kets $|\phi_1\rangle$, $|\phi_2\rangle$, and $|\phi_3\rangle$ are used as the basis kets, the operators $\hat{A}$ and $\hat{B}$ are represented by

$$\underline{A} = \begin{pmatrix} a & 0 & 0 \\ 0 & -a & 0 \\ 0 & 0 & -a \end{pmatrix}, \qquad \underline{B} = \begin{pmatrix} b & 0 & 0 \\ 0 & 0 & -ib \\ 0 & ib & 0 \end{pmatrix},$$

where a and b are real.

1. It is obvious that $\hat{A}$ has a degenerate spectrum. Is the spectrum of $\hat{B}$ also degenerate?
2. Show that $\hat{A}$ and $\hat{B}$ commute.
3. Find a new set of orthonormal kets which are simultaneous eigenstates of both $\hat{A}$ and $\hat{B}$. Specify the eigenvalues of $\hat{A}$ and $\hat{B}$ for each of these three eigenstates. Does specifying these eigenvalues uniquely identify the relative common eigenstate? That is, do $\hat{A}$ and $\hat{B}$ form a complete set of commuting observables? (Adapted from J. J. Sakurai and J. Napolitano 2020, *Modern Quantum Mechanics*, Cambridge University Press.)

Solution

Part 1

The eigenvalues of $\underline{B}$ follow from

$$\det|\underline{B} - \lambda \underline{I}| = \det \begin{bmatrix} b - \lambda & 0 & 0 \\ 0 & -\lambda & -ib \\ 0 & ib & -\lambda \end{bmatrix} = (b - \lambda)(\lambda^2 - b^2) = 0,$$

namely $\lambda_1 = b$ with degeneracy $g_1 = 2$ and $\lambda_2 = -b$ non-degenerate, $g_2 = 1$.

Part 2

We have by direct evaluation

$$\underbrace{\begin{pmatrix} a & 0 & 0 \\ 0 & -a & 0 \\ 0 & 0 & -a \end{pmatrix}}_{\underline{A}} \underbrace{\begin{pmatrix} b & 0 & 0 \\ 0 & 0 & -ib \\ 0 & ib & 0 \end{pmatrix}}_{\underline{B}} = \begin{pmatrix} ab & 0 & 0 \\ 0 & 0 & iab \\ 0 & -iab & 0 \end{pmatrix},$$

and

$$\underbrace{\begin{pmatrix} b & 0 & 0 \\ 0 & 0 & -ib \\ 0 & ib & 0 \end{pmatrix}}_{B} \underbrace{\begin{pmatrix} a & 0 & 0 \\ 0 & -a & 0 \\ 0 & 0 & -a \end{pmatrix}}_{A} = \begin{pmatrix} ab & 0 & 0 \\ 0 & 0 & iab \\ 0 & -iab & 0 \end{pmatrix},$$

and so $\underline{A}$ and $\underline{B}$ commute.

Part 3

The matrix $\underline{A}$ is diagonal and so

$$\hat{A}|\phi_1\rangle = a|\phi_1\rangle, \qquad \hat{A}|\phi_2\rangle = -a|\phi_2\rangle, \qquad \hat{A}|\phi_3\rangle = -a|\phi_3\rangle.$$

Since $\hat{A}$ and $\hat{B}$ commute and the eigenvalue a is non-degenerate, $|\phi_1\rangle$ is also an eigenstate of $\hat{B}$ with eigenvalue b; indeed,

$$\begin{pmatrix} b & 0 & 0 \\ 0 & 0 & -ib \\ 0 & ib & 0 \end{pmatrix} \begin{pmatrix} 1 \\ 0 \\ 0 \end{pmatrix} = b \begin{pmatrix} 1 \\ 0 \\ 0 \end{pmatrix}.$$

In the degenerate subspace spanned by $|\phi_2\rangle$ and $|\phi_3\rangle$, $\hat{B}$ has the representation

$$\begin{pmatrix} 0 & -ib \\ ib & 0 \end{pmatrix}$$

with eigenvalues $\lambda_1 = b$ and $\lambda_2 = -b$. The relative eigenstates are obtained from

$$\begin{pmatrix} -b & -ib \\ ib & -b \end{pmatrix} \begin{pmatrix} c_1^{(1)} \\ c_2^{(1)} \end{pmatrix} = 0 \qquad c_1^{(1)} = -ic_2^{(1)}$$

and

$$\begin{pmatrix} b & -ib \\ ib & b \end{pmatrix} \begin{pmatrix} c_1^{(2)} \\ c_2^{(2)} \end{pmatrix} = 0 \qquad c_1^{(2)} = ic_2^{(2)}.$$

After normalization, the eigenstates read

$$\underline{c}^{(1)} = \frac{1}{\sqrt{2}} \begin{pmatrix} 1 \\ i \end{pmatrix}, \qquad \underline{c}^{(2)} = \frac{1}{\sqrt{2}} \begin{pmatrix} 1 \\ -i \end{pmatrix}.$$

In terms of $|\phi_2\rangle$ and $|\phi_3\rangle$, we have

$$|\psi_2\rangle = \frac{1}{\sqrt{2}} (|\phi_2\rangle + i|\phi_3\rangle), \qquad |\psi_3\rangle = \frac{1}{\sqrt{2}} (|\phi_2\rangle - i|\phi_3\rangle).$$

Note that

$$\hat{A}|\psi_2\rangle = -a|\psi_2\rangle, \qquad \hat{A}|\psi_3\rangle = -a|\psi_3\rangle.$$

The common basis of simultaneous eigenstates consists of $|\phi_1\rangle$, $|\psi_2\rangle$, and $|\psi_3\rangle$ with corresponding eigenvalues of $\hat{A}$, $\hat{B}$ given by a, b, $-a, b$, and $-a, -b$. Thus, specifying the pair of simultaneous eigenvalues uniquely identifies the eigenstate, and $\hat{A}$ and $\hat{B}$ form a complete set of commuting observables.

Problem 24   Model for a Planar Molecule

A certain molecule is composed of six identical atoms $A_1, A_2, \ldots, A_6$ which form a regular hexagon. Consider an electron which can be localized on each of the atoms. Denote by $|\psi_n\rangle$ the state in which the electron is localized on the nth atom ($n = 1, \ldots, 6$). The electron states are taken to be limited to the space spanned by the $|\psi_n\rangle$, assumed to be orthonormal $\langle\psi_m|\psi_n\rangle = \delta_{mn}$; in other words, these six states form a basis.

1. Define the operator $\hat{R}$ by the following relations:

$$\hat{R}|\psi_1\rangle = |\psi_2\rangle, \qquad \hat{R}|\psi_2\rangle = |\psi_3\rangle, \qquad \ldots, \qquad \hat{R}|\psi_6\rangle = |\psi_1\rangle.$$

   Find the eigenvalues and eigenstates of $\hat{R}$. Show that the eigenvectors form an orthonormal set (i.e., they form a basis).

2. Show that the adjoint operator $\hat{R}^\dagger$ acts as follows:

$$\hat{R}^\dagger|\psi_1\rangle = |\psi_6\rangle, \qquad \hat{R}^\dagger|\psi_2\rangle = |\psi_1\rangle, \qquad \ldots, \qquad \hat{R}^\dagger|\psi_6\rangle = |\psi_5\rangle.$$

   Also, show that $\hat{R}$ is unitary.

3. When the probability of the electron jumping from one site to a contiguous one to the left or right is neglected, its energy is described by the Hamiltonian $\hat{H}_0$, whose eigenstates are the six states $|\psi_n\rangle$ all having the same eigenvalue E_0, namely,

$$\hat{H}_0|\psi_n\rangle = E_0|\psi_n\rangle.$$

   The possibility that the electron will jump from one site to another is modeled by adding to the Hamiltonian $\hat{H}_0$ a perturbation $\hat{V}$ such that

$$\hat{V}|\psi_1\rangle = -a|\psi_6\rangle - a|\psi_2\rangle, \qquad \hat{V}|\psi_2\rangle = -a|\psi_1\rangle - a|\psi_3\rangle, \qquad \ldots,$$

$$\hat{V}|\psi_6\rangle = -a|\psi_5\rangle - a|\psi_1\rangle.$$

   Show that $\hat{R}$ commutes with the total Hamiltonian $\hat{H} = \hat{H}_0 + \hat{V}$. From this deduce the eigenstates and eigenvalues of $\hat{H}$. In these eigenstates is the electron localized?

   Hint: The N distinct complex roots of $z^N = 1$ are given by

$$z_n = e^{i2\pi n/N} \qquad n = 1, 2, \ldots, N,$$

and the following identity holds:

$$\sum_{n=0}^{N} z^n = \frac{1 - z^{N+1}}{1 - z}, \qquad z \text{ complex}.$$

(Adapted from C. Cohen-Tannoudji, B. Diu, and F. Laloë 1997, *Quantum Mechanics*, vol. 1, Wiley.)

Solution

Part 1

The "hopping" operator $\hat{R}$ satisfies

$$\hat{R}|\psi_i\rangle = |\psi_{i+1}\rangle \qquad i = 1, \ldots, 6, \qquad \text{and} \qquad |\psi_7\rangle = |\psi_1\rangle,$$

yielding $\hat{R}^6 = \hat{\mathbb{1}}$. Let $|r\rangle$ be an eigenstate of $\hat{R}$ with eigenvalue λ, namely $\hat{R}|r\rangle = \lambda|r\rangle$. We find

$$|r\rangle = \hat{R}^6|r\rangle = \lambda\hat{R}^5|r\rangle = \lambda^2\hat{R}^4|r\rangle = \cdots = \lambda^6|r\rangle \implies \lambda^6 = 1 \,,$$

which has the solutions

$$\lambda_n = e^{2\pi i n/6} = e^{i\pi n/3} \qquad n = 1,\ldots,6 \,.$$

To determine the eigenstate corresponding to eigenvalue λ_n, we first obtain the representation of $\hat{R}$ in the basis $|\psi_i\rangle$,

$$R_{ij} = \langle\psi_i|\hat{R}|\psi_j\rangle = \langle\psi_i|\psi_{j+1}\rangle = \delta_{i,j+1} \,,$$

and then solve the linear system

$$(\underline{R} - \lambda_n\,\underline{\mathbb{1}})\underline{C} = \begin{pmatrix} -\lambda_n & 0 & 0 & 0 & 0 & 1 \\ 1 & -\lambda_n & 0 & 0 & 0 & 0 \\ 0 & 1 & -\lambda_n & 0 & 0 & 0 \\ 0 & 0 & 1 & -\lambda_n & 0 & 0 \\ 0 & 0 & 0 & 1 & -\lambda_n & 0 \\ 0 & 0 & 0 & 0 & 1 & -\lambda_n \end{pmatrix}\begin{pmatrix} c_1 \\ c_2 \\ c_3 \\ c_4 \\ c_5 \\ c_6 \end{pmatrix} = 0 \,,$$

where, in the basis $|\psi_i\rangle$,

$$|r\rangle = \sum_{i=1}^{6} c_i|\psi_i\rangle \,.$$

Explicitly, the linear system reads

$$c_6 = \lambda_n c_1\,, \quad c_1 = \lambda_n c_2\,, \quad c_2 = \lambda_n c_3\,, \quad c_3 = \lambda_n c_4\,, \quad c_4 = \lambda_n c_5\,, \quad c_5 = \lambda_n c_6\,,$$

and, since $|\lambda_n| = 1$, it can also be written as

$$c_1 = \lambda_n^* c_6\,, \quad c_2 = \lambda_n^* c_1\,, \quad c_3 = \lambda_n^* c_2\,, \quad c_4 = \lambda_n^* c_3\,, \quad c_5 = \lambda_n^* c_4\,, \quad c_6 = \lambda_n^* c_5\,,$$

which can be easily solved to yield

$$c_1 = \lambda_n^* c_6\,, \quad c_2 = (\lambda_n^*)^2 c_6\,, \quad c_3 = (\lambda_n^*)^3 c_6\,, \quad c_4 = (\lambda_n^*)^4 c_6\,, \quad c_5 = (\lambda_n^*)^5 c_6\,, \quad c_6 = (\lambda_n^*)^6 c_6\,.$$

The last equation is a tautology since $(\lambda_n^*)^6 = 1$. Thus, the eigenstate $|r_n\rangle$ corresponding to the eigenvalue λ_n is given by

$$|r_n\rangle = c_6 \sum_{i=1}^{6} (\lambda_n^*)^i |\psi_i\rangle \,.$$

The constant $|c_6|$ is fixed by requiring that $|r_n\rangle$ be normalized,

$$1 = \langle r_n|r_n\rangle = |c_6|^2 \sum_{i,j=1}^{6} \lambda_n^i\,\lambda_n^{*j}\,\underbrace{\langle\psi_i|\psi_j\rangle}_{\delta_{ij}} = |c_6|^2 \sum_{i=1}^{6} (\lambda_n\lambda_n^*)^i = 6|c_6|^2 \implies |c_6| = \frac{1}{\sqrt{6}} \,.$$

Note that $\hat{R}$ cannot be hermitian since some of its eigenvalues are complex. However, it is still true that the eigenstates are orthonormal; indeed, for $n \neq m$ we have

$$\langle r_n|r_m\rangle = \frac{1}{6} \sum_{i,j=1}^{6} \lambda_n^i\,\lambda_m^{*j}\langle\psi_i|\psi_j\rangle = \frac{1}{6} \sum_{i=1}^{6} (\lambda_n\lambda_m^*)^i \,.$$

Now set $z \equiv \lambda_n \lambda_m^* = e^{i\pi(n-m)/3}$ with $z^6 = 1$, and hence

$$\langle r_n | r_m \rangle \propto \sum_{i=1}^{6} z^i = \sum_{i=0}^{6} z^i - 1 = \frac{1-z^7}{1-z} - 1 = \frac{1-zz^6}{1-z} - 1 = \frac{1-z}{1-z} - 1 = 0 .$$

We conclude that the six states $|r_n\rangle$ form an orthonormal set in the six-dimensional Hilbert space, and therefore constitute a basis.

Part 2

To determine the action of $\hat{R}^\dagger$, we use the fact that the matrix elements of $\hat{R}$ are real and hence

$$\langle \psi_i | \hat{R} | \psi_j \rangle = \delta_{i,j+1} = \langle \psi_i | \hat{R} | \psi_j \rangle^* = \langle \psi_j | \hat{R}^\dagger | \psi_i \rangle \implies \langle \psi_i | \hat{R}^\dagger | \psi_j \rangle = \delta_{i+1,j} ,$$

and the matrix representing $\hat{R}^\dagger$ is given by

$$\underline{R}^\dagger = \begin{pmatrix} 0 & 1 & 0 & 0 & 0 & 0 \\ 0 & 0 & 1 & 0 & 0 & 0 \\ 0 & 0 & 0 & 1 & 0 & 0 \\ 0 & 0 & 0 & 0 & 1 & 0 \\ 0 & 0 & 0 & 0 & 0 & 1 \\ 1 & 0 & 0 & 0 & 0 & 0 \end{pmatrix} \implies \hat{R}^\dagger | \psi_i \rangle = | \psi_{i-1} \rangle \qquad \text{with } i = 1, \ldots, 6 \text{ and } |\psi_0\rangle = |\psi_6\rangle .$$

In fact, $\hat{R}$ is unitary, since

$$\langle \psi_j | \hat{R} \hat{R}^\dagger | \psi_i \rangle = \langle \psi_j | \hat{R} | \psi_{i-1} \rangle = \langle \psi_j | \psi_i \rangle = \delta_{ij} \implies \hat{R} \hat{R}^\dagger = \hat{1} ,$$

and similarly $\hat{R}^\dagger \hat{R} = \hat{1}$, which also implies that $[\hat{R}, \hat{R}^\dagger] = 0$, that is, $\hat{R}$ is a normal operator. The unitarity of $\hat{R}$ can also be shown by multiplying out the matrices representing $\hat{R}$ and $\hat{R}^\dagger$.

Part 3

The perturbation can be expressed as

$$\hat{V} = -a(\hat{R} + \hat{R}^\dagger) ,$$

since $\hat{V}|\psi_i\rangle = -a|\psi_{i+1}\rangle - a|\psi_{i-1}\rangle$. By contrast, $\hat{H}_0$ is given by $\hat{H}_0 = E_0 \hat{1}$, and $\hat{H} = \hat{H}_0 + \hat{V}$ obviously commutes with $\hat{R}$ (or $\hat{R}^\dagger$). Note that, given $\hat{R}|r_n\rangle = \lambda_n|r_n\rangle$, it follows that

$$|r_n\rangle = \underbrace{\hat{R}^\dagger \hat{R}}_{\hat{1}} |r_n\rangle = \lambda_n \hat{R}^\dagger |r_n\rangle \implies \hat{R}^\dagger |r_n\rangle = \lambda_n^* |r_n\rangle ,$$

after multiplying both sides of the first relation by λ_n^* and using $\lambda_n \lambda_n^* = 1$. The $|r_n\rangle$ are also eigenstates of $\hat{H}$, since

$$\hat{H}|r_n\rangle = E_0|r_n\rangle - a(\lambda_n + \lambda_n^*)|r_n\rangle \implies E_n = E_0 - 2a\,\mathrm{Re}(\lambda_n) = E_0 - 2a\cos(n\pi/3) .$$

The eigenvalues E_n are partly degenerate, since

$$E_1 = E_5 = E_0 - a , \qquad E_2 = E_4 = E_0 + a , \qquad E_3 = E_0 + 2a , \qquad E_6 = E_0 - 2a .$$

In each of these energy eigenstates the electron is delocalized. As a matter of fact, the probability that the electron is at a given site i is

$$P_{i,n} = |\langle \psi_i | r_n \rangle|^2 = \frac{1}{6} |(\lambda_n^*)^i|^2 = \frac{1}{6} ,$$

and so it is the same for any site, regardless of the state $|r_n\rangle$.

Problem 25   Derivation of Formulae Relating to Exponentials of Operators

Let $\hat{A}$ and $\hat{B}$ be two generic operators.

1. Show that

$$e^{\lambda \hat{B}} \hat{A} e^{-\lambda \hat{B}} = \hat{A} + \lambda [\hat{B}, \hat{A}] + \frac{\lambda^2}{2!} [\hat{B}, [\hat{B}, \hat{A}]] + \cdots + \frac{\lambda^n}{n!} [\hat{B}, [\hat{B}, \cdots [\hat{B}, \hat{A}] \cdots]] + \cdots .$$

2. Show that, if $\hat{A}$ and $\hat{B}$ each commute with the operator resulting from $[\hat{A}, \hat{B}]$, then the following identity (known as the Baker–Campbell–Hausdorff formula) holds:

$$e^{\hat{A}} e^{\hat{B}} = e^{\hat{A} + \hat{B} + [\hat{A}, \hat{B}]/2} .$$

Solution

Part 1

Define

$$\hat{F}(\lambda) = e^{\lambda \hat{B}} \hat{A} e^{-\lambda \hat{B}} ,$$

and differentiate with respect to λ to obtain

$$\frac{d}{d\lambda} \hat{F}(\lambda) = \hat{B} e^{\lambda \hat{B}} \hat{A} e^{-\lambda \hat{B}} - e^{\lambda \hat{B}} \hat{A} e^{-\lambda \hat{B}} \hat{B} = [\hat{B}, \hat{F}(\lambda)] ,$$

since $\hat{B}$ and $e^{\pm \lambda \hat{B}}$ commute with each other. Note that the initial condition at $\lambda = 0$ is $\hat{F}(0) = \hat{A}$. A formal solution to the differential equation including the initial condition is given by

$$\hat{F}(\lambda) = \hat{A} + \int_0^\lambda d\lambda' \, [\hat{B}, \hat{F}(\lambda')] ,$$

which we solve by iteration:

$$\hat{F}(\lambda) = \hat{A} + \int_0^\lambda d\lambda_1 [\hat{B}, \hat{A}] + \int_0^\lambda d\lambda_1 \int_0^{\lambda_1} d\lambda_2 [\hat{B}, [\hat{B}, \hat{A}]]$$

$$+ \int_0^\lambda d\lambda_1 \int_0^{\lambda_1} d\lambda_2 \int_0^{\lambda_2} d\lambda_3 [\hat{B}, [\hat{B}, [\hat{B}, \hat{A}]]] + \cdots$$

$$= \hat{A} + [\hat{B}, \hat{A}] I_1(\lambda) + [\hat{B}, [\hat{B}, \hat{A}]] I_2(\lambda) + [\hat{B}, [\hat{B}, [\hat{B}, \hat{A}]]] I_3(\lambda) + \cdots ,$$

since the commutators are independent of λ and where we define

$$I_n(\lambda) = \int_0^\lambda d\lambda_1 \int_0^{\lambda_1} d\lambda_2 \cdots \int_0^{\lambda_{n-1}} d\lambda_n .$$

These integrals are easily performed,

$$I_1(\lambda) = \int_0^\lambda d\lambda_1 = \lambda \,, \qquad I_2(\lambda) = \int_0^\lambda d\lambda_1 \int_0^{\lambda_1} d\lambda_2 = \frac{\lambda^2}{2!} \,,$$

and, generally, $I_n(\lambda) = \lambda^n/n!$, which leads to the formula given in the text.

Part 2

Define $\hat{F}(\lambda) = e^{\lambda \hat{A}} \, e^{\lambda \hat{B}}$, so that

$$\hat{F}'(\lambda) = e^{\lambda \hat{A}} \, \hat{A} \, e^{\lambda \hat{B}} + e^{\lambda \hat{A}} \, \hat{B} \, e^{\lambda \hat{B}} = e^{\lambda \hat{A}} \, e^{\lambda \hat{B}} \, e^{-\lambda \hat{B}} \, \hat{A} \, e^{\lambda \hat{B}} + e^{\lambda \hat{A}} \, e^{\lambda \hat{B}} \, \hat{B} = \hat{F}(\lambda) \left[e^{-\lambda \hat{B}} \, \hat{A} \, e^{\lambda \hat{B}} + \hat{B} \right] \,.$$

We use

$$e^{-\lambda \hat{B}} \, \hat{A} \, e^{\lambda \hat{B}} = \hat{A} - \lambda \, [\hat{B}, \hat{A}] + \frac{\lambda^2}{2!} \, [\hat{B}, [\hat{B}, \hat{A}]] + \cdots = \hat{A} + \lambda \, [\hat{A}, \hat{B}] \,,$$

since by assumption $\hat{B}$ commutes with the commutator $[\hat{A}, \hat{B}]$. We conclude that

$$\hat{F}'(\lambda) = \hat{F}(\lambda) \left(\hat{A} + \hat{B} + \lambda \, [\hat{A}, \hat{B}] \right) = \hat{F}(\lambda) \, \hat{G}(\lambda) \,, \qquad \hat{G}(\lambda) \equiv \hat{A} + \hat{B} + \lambda \, [\hat{A}, \hat{B}] \,.$$

We can also write

$$\hat{F}'(\lambda) = \hat{A} \, e^{\lambda \hat{A}} \, e^{\lambda \hat{B}} + e^{\lambda \hat{A}} \, \hat{B} \, e^{\lambda \hat{B}} = \hat{A} \, e^{\lambda \hat{A}} \, e^{\lambda \hat{B}} + e^{\lambda \hat{A}} \, \hat{B} \, e^{-\lambda \hat{A}} \, e^{\lambda \hat{A}} \, e^{\lambda \hat{B}} = \underbrace{\left[\hat{A} + e^{\lambda \hat{A}} \, \hat{B} \, e^{-\lambda \hat{A}} \right] \hat{F}(\lambda)}_{\hat{G}(\lambda) \hat{F}(\lambda)} \,,$$

where the last step follows because $\hat{A}$ also commutes with the commutator $[\hat{A}, \hat{B}]$. Thus, we have

$$\hat{F}'(\lambda) = \hat{F}(\lambda) \, \hat{G}(\lambda) = \hat{G}(\lambda) \, \hat{F}(\lambda) \implies [\hat{F}(\lambda), \hat{G}(\lambda)] = 0 \,,$$

and the solution of the differential equation is simply obtained as

$$\hat{F}(\lambda) = \exp \left[\int_0^\lambda d\lambda' \, \hat{G}(\lambda') \right] \implies \hat{F}(\lambda) = e^{(\hat{A} + \hat{B})\lambda + [\hat{A}, \hat{B}]\lambda^2/2} \,,$$

which for $\lambda = 1$ yields the required identity. We emphasize that the above solution holds true only if $\hat{F}(\lambda)$ and $\hat{G}(\lambda)$ commute with each other, as is the case here.

7 Physical Interpretation: Postulates of Quantum Mechanics

The physical interpretation of the mathematical framework outlined in the previous chapter rests on a number of "postulates", which are introduced here:[1]

- The physical state of a system is represented by a vector – more precisely, a "ray", a normalized vector – in the state space $\mathbb{E}$, a Hilbert space; as a consequence, a linear combination of state vectors is also a state vector (linear superposition).
- Measurable physical quantities – observables – are represented by hermitian operators acting in $\mathbb{E}$; if $\hat{A}$ is an observable then there exists a basis consisting of eigenstates of $\hat{A}$.
- The only possible result of a measurement of an observable $\hat{A}$ is one of its eigenvalues. If the spectrum of $\hat{A}$ is discrete, the results obtained by measuring $\hat{A}$ are quantized.
- When the observable $\hat{A}$ is measured on a system in state $|\psi\rangle$, the probability of obtaining the discrete eigenvalue a_i is

$$p(a_i) = \sum_{k=1}^{g_i} |\langle \phi_i^{(k)}|\psi\rangle|^2 = \langle \psi|\hat{P}_i|\psi\rangle \quad \text{with } \hat{P}_i = \sum_{k=1}^{g_i} |\phi_i^{(k)}\rangle\langle\phi_i^{(k)}| \text{ and } \langle\psi|\psi\rangle = 1 , \quad (7.1)$$

where $|\phi_i^{(k)}\rangle$ are the (normalized) eigenstates of $\hat{A}$ corresponding to eigenvalue a_i, having degeneracy g_i,

$$\hat{A}|\phi_i^{(k)}\rangle = a_i|\phi_i^{(k)}\rangle , \qquad k = 1,\ldots,g_i , \quad (7.2)$$

and $\hat{P}_i$ is the projection operator onto the subspace spanned by the eigenstates $|\phi_i^{(k)}\rangle$. Of course $\sum_i p(a_i) = 1$, which follows from the completeness of the basis of $\hat{A}$ eigenstates. Note that when a_i is degenerate, there is an infinity of possible bases in the subspace of $\mathbb{E}$ belonging to the eigenvalue a_i; however, the probability $p(a_i)$ is independent of the basis chosen.

If the observable $\hat{A}$ has a continuous spectrum then the probability that we measure an eigenvalue centered at α in an interval $d\alpha$ is obtained as

$$p(\alpha)d\alpha = \langle \psi|\hat{P}_\alpha|\psi\rangle \, d\alpha = \left[\sum_{k=1}^{g_\alpha} |\langle\phi_\alpha^{(k)}|\psi\rangle|^2 \right] d\alpha , \qquad \hat{P}_\alpha = \sum_{k=1}^{g_\alpha} |\phi_\alpha^{(k)}\rangle\langle\phi_\alpha^{(k)}| , \quad (7.3)$$

and $p(\alpha)$ can be thought of as a probability density. This interpretation is in line with that of $|\psi(\mathbf{r})|^2$ as the probability density for the particle to be found at position $\mathbf{r}$. Indeed, the position operator has a (non-degenerate) continuous spectrum, $\hat{\mathbf{r}}|\phi_{\mathbf{r}}\rangle = \mathbf{r}|\phi_{\mathbf{r}}\rangle$, and the probability density that the system will be found at $\mathbf{r}$ is $|\langle\phi_{\mathbf{r}}|\psi\rangle|^2 = |\psi(\mathbf{r})|^2$.

If many measurements of the observable $\hat{A}$ are performed with the system always in the same state $|\psi\rangle$, then the average value, denoted as $\langle\hat{A}\rangle$, of the results obtained in these measurements is given by

---

[1] There is a large amount of literature on the interpretation of quantum mechanics and its conceptual difficulties. A good summary is given in S. Weinberg 2012, *Lectures on Quantum Mechanics*, Cambridge University Press, Section 3.7.

$$\langle \hat{A} \rangle = \sum_i a_i p(a_i) \,, \tag{7.4}$$

where the sum is over all possible results. This average can also be expressed as the expectation value of $\hat{A}$ on the state $|\psi\rangle$,

$$\langle \hat{A} \rangle = \langle \psi | \hat{A} | \psi \rangle \,, \tag{7.5}$$

as implied by the completeness of the basis of eigenstates of $\hat{A}$,

$$\hat{A} = \hat{A} \underbrace{\sum_i \sum_{k=1}^{g_i} |\phi_i^{(k)}\rangle\langle\phi_i^{(k)}|}_{\hat{1}} = \sum_i \sum_{k=1}^{g_i} a_i |\phi_i^{(k)}\rangle\langle\phi_i^{(k)}| \,. \tag{7.6}$$

- If the measurement of the observable $\hat{A}$ on the system in state $|\psi\rangle$ gives the result a_i then the state of the system immediately after the measurement is either $|\phi_i\rangle$, if a_i is non-degenerate, or a normalized linear combination obtained by projecting $|\psi\rangle$ onto the subspace spanned by the eigenstates $|\phi_i^{(k)}\rangle$, if a_i is degenerate. We can encompass both cases by writing

$$\text{normalized state directly after measurement} = \frac{\hat{P}_i |\psi\rangle}{\sqrt{\langle\psi|\hat{P}_i|\psi\rangle}} \,, \tag{7.7}$$

where $\hat{P}_i$ is the projection operator onto the subspace corresponding to eigenvalue a_i.

If the observable $\hat{A}$ has a continuous spectrum then its measurement on a system described by the state $|\psi\rangle$ will yield a result α_0 to within $\Delta\alpha$, with probability

$$\int_{\alpha_0 - \Delta\alpha/2}^{\alpha_0 + \Delta\alpha/2} d\alpha \sum_{k=1}^{g_\alpha} |\langle\phi_\alpha^{(k)}|\psi\rangle|^2 \,. \tag{7.8}$$

The state immediately after the measurement is given as follows:

$$\text{normalized state directly after measurement} = \frac{\hat{P}_{\Delta\alpha}(\alpha_0) |\psi\rangle}{\sqrt{\langle\psi|\hat{P}_{\Delta\alpha}(\alpha_0)|\psi\rangle}} \,, \tag{7.9}$$

where we define the projection operator

$$\hat{P}_{\Delta\alpha}(\alpha_0) = \int_{\alpha_0 - \Delta\alpha/2}^{\alpha_0 + \Delta\alpha/2} d\alpha \sum_{k=1}^{g_\alpha} |\phi_\alpha^{(k)}\rangle\langle\phi_\alpha^{(k)}| \,. \tag{7.10}$$

Given that in practice any measuring apparatus has a finite resolution, the definition above is physically sensible.

The present postulate implies an interaction between the system and the measuring apparatus, in consequence of which the state of the system is changed; indeed, the measurement results in the *collapse* of the state $|\psi\rangle$ into the state $\hat{P}_i|\psi\rangle$ or $\hat{P}_{\Delta\alpha}(\alpha_0)|\psi\rangle$, up to a normalization constant. It should be also clear that, if immediately after the first measurement, we carry out a second measurement of the same observable $\hat{A}$, we are bound to obtain the result a_i or α_0 within $\Delta\alpha$ with certainty, that is, with probability 1. Here, *immediately after the first measurement* means that the system has not had time to evolve (time evolution in quantum mechanics is the subject of the next postulate).

- The time evolution of the state vector $|\psi(t)\rangle$ is governed by the Schrödinger equation

$$i\hbar \frac{d}{dt}|\psi(t)\rangle = \hat{H}(t)|\psi(t)\rangle \,, \tag{7.11}$$

where $\hat{H}(t)$ is the (possibly time-dependent) Hamiltonian. Time evolution in quantum mechanics is deterministic: given the state $|\psi(t_0)\rangle$ at time t_0, the (unique) state at time $t > t_0$ is obtained by solving a first-order differential equation. This determinism is at odds with the process of measurement. If the time-dependent Schrödinger equation were to describe this process then we could predict that the system would be in some definite state after the measurement, rather than in any of a number of possible states, each characterized by a specific probability.

7.1 Time Evolution Operator and Time Dependence of Expectation Values

The time evolution of the state can also be described by an operator (a unitary operator, as it turns out),

$$|\psi(t)\rangle = \hat{U}(t, t_0)|\psi(t_0)\rangle \qquad t \geq t_0 \,, \tag{7.12}$$

where $\hat{U}(t, t_0)$ satisfies

$$i\hbar \frac{d}{dt}\hat{U}(t, t_0) = \hat{H}(t)\,\hat{U}(t, t_0) \,, \qquad \hat{U}(t_0, t_0) = \hat{1} \,. \tag{7.13}$$

When $\hat{H}$ is time independent, this equation is easily solved to yield

$$\hat{U}(t, t_0) = e^{-i\hat{H}(t-t_0)/\hbar} \,. \tag{7.14}$$

However, when $\hat{H}$ is time dependent, a formal solution can be obtained by first converting the differential equation into an integral equation,

$$\hat{U}(t, t_0) = \hat{1} - \frac{i}{\hbar}\int_{t_0}^{t} dt' \, \hat{H}(t') \, \hat{U}(t', t_0) \,, \tag{7.15}$$

and then solving the latter by iteration to find

$$\hat{U}(t, t_0) = \hat{1} - \frac{i}{\hbar}\int_{t_0}^{t} dt_1 \, \hat{H}(t_1) + \left(-\frac{i}{\hbar}\right)^2 \int_{t_0}^{t} dt_1 \int_{t_0}^{t_1} dt_2 \, \hat{H}(t_1)\,\hat{H}(t_2) + \cdots$$
$$+ \left(-\frac{i}{\hbar}\right)^n \int_{t_0}^{t} dt_1 \cdots \int_{t_0}^{t_{n-1}} dt_n \, \hat{H}(t_1)\cdots\hat{H}(t_n) + \cdots \,; \tag{7.16}$$

the nested time-integration limits should be noted. In general, the Hamiltonians at different times will not commute, so the ordering in the expansion above is important.

The unitarity of $\hat{U}(t, t_0)$ is obvious when the Hamiltonian is time independent. When it is time dependent, we observe that the evolution from t to $t + \eta$, where η is a small increment, is governed by Eq. (7.15):

$$\hat{U}(t + \eta, t) = \hat{1} - \frac{i}{\hbar}\int_{t}^{t+\eta} dt' \, \hat{H}(t') \, \hat{U}(t', t) \implies \hat{U}(t + \eta, t) = \hat{1} - i\frac{\eta}{\hbar}\hat{H}(t) + O(\eta^2) \,, \tag{7.17}$$

where $\hat{U}(t + \eta, t)$ is unitary up to linear terms in η. For a finite time interval, $\hat{U}(t, t_0)$ can be written as the product

$$\hat{U}(t, t_0) = \hat{U}(t, t - \eta) \, \hat{U}(t - \eta, t - 2\eta) \cdots \hat{U}(t_0 + 2\eta, t_0 + \eta) \, \hat{U}(t_0 + \eta, t_0) \; ; \qquad (7.18)$$

we note that the product of unitary operators is unitary. Thus time evolution preserves the normalization of the state, that is, $\langle \psi(t) | \psi(t) \rangle = \langle \psi(t_0) | \psi(t_0) \rangle$.

Having established how states evolve in time, we can deduce how expectation values change in time. Let $\hat{A}$ be an observable and let $\hat{H}$ be the system Hamiltonian, both assumed to be time independent. The expectation value of $\hat{A}$ will generally depend on time,

$$\langle \hat{A}(t) \rangle = \langle \psi(t) | \, \hat{A} \, | \psi(t) \rangle \, , \qquad (7.19)$$

and this time dependence can be made explicit by expanding $|\psi(t)\rangle$ on a basis consisting of eigenstates of $\hat{H}$:

$$|\psi(t)\rangle = \sum_i \sum_{k=1}^{g_i} e^{-iE_i(t-t_0)/\hbar} \, c_i^{(k)}(t_0) |\phi_i^{(k)}\rangle \, , \qquad c_i^{(k)}(t_0) = \langle \phi_i^{(k)} | \psi(t_0) \rangle \, . \qquad (7.20)$$

Using the Schrödinger equation, the rate of change of the expectation value can be expressed as follows:

$$\frac{d}{dt} \langle \hat{A}(t) \rangle = \underbrace{\left[\frac{d}{dt} \langle \psi(t) | \right]}_{(i/\hbar)\langle \psi(t) | \hat{H}} \hat{A} | \psi(t) \rangle + \langle \psi(t) | \hat{A} \underbrace{\left[\frac{d}{dt} | \psi(t) \rangle \right]}_{-(i/\hbar)\hat{H} | \psi(t) \rangle} = \frac{i}{\hbar} \langle \psi(t) | [\hat{H}, \hat{A}] | \psi(t) \rangle \, . \qquad (7.21)$$

We conclude that the expectation value is time independent, namely, the observable is a constant of motion if it commutes with the Hamiltonian,[2]

$$[\hat{H}, \hat{A}] = 0 \implies \hat{A} \text{ is a constant of motion} \, . \qquad (7.22)$$

Obviously, since $\hat{H}$ commutes with itself, the average value of the energy for a system in a state $|\psi(t)\rangle$ is constant in time. The above results can also be used to derive Ehrenfest's relations (see Problems 10 and 12 in Chapter 3; the latter relates to Ehrenfest's relations for a charged particle in an electromagnetic field),

$$\frac{d}{dt} \langle \hat{\mathbf{r}}(t) \rangle = \frac{1}{m} \langle \psi(t) | \, \hat{\mathbf{p}} \, | \psi(t) \rangle \, , \qquad \frac{d}{dt} \langle \hat{\mathbf{p}}(t) \rangle = -\langle \psi(t) | \, \boldsymbol{\nabla} \hat{V}(\hat{\mathbf{r}}) \, | \psi(t) \rangle \, , \qquad (7.23)$$

for a system described by the Hamiltonian

$$\hat{H} = \frac{\hat{\mathbf{p}}^2}{2m} + \hat{V}(\hat{\mathbf{r}}) \, . \qquad (7.24)$$

The above relations resemble Hamilton's equations of classical mechanics. However, they do not imply that the averages $\langle \hat{\mathbf{r}}(t) \rangle$ and $\langle \hat{\mathbf{p}}(t) \rangle$ generally follow the laws of classical mechanics, except for the special cases in which the particle is free or the potential energy operator $\hat{V}(\hat{\mathbf{r}})$ is linear or quadratic in $\hat{\mathbf{r}}$.

---

[2] The above relation has a counterpart in classical mechanics, where a quantity $A(q_i, p_i)$ is a function of the generalized coordinates and momenta q_i and p_i with $i = 1, \ldots, n$ (n is the number of degrees of freedom) is conserved, that is, a constant of motion, if it has vanishing Poisson brackets with the Hamiltonian $H(q_i, p_i)$: $\{A, H\} = 0$. Poisson brackets are generally defined for any two functions B and C of the q_i and p_i as

$$\{B, C\} = \sum_{k=1}^{n} \left[\left(\frac{\partial B}{\partial q_k} \right) \frac{\partial C}{\partial p_k} - \left(\frac{\partial B}{\partial p_k} \right) \frac{\partial C}{\partial q_k} \right] \, .$$

In particular, note the fundamental Poisson brackets $\{q_i, p_j\} = \delta_{ij}$. The transition from classical to quantum mechanics is implemented via the replacement $q_i, p_i \longrightarrow \hat{q}_i, \hat{p}_i$ and by requiring that $[\hat{q}_i, \hat{p}_j] = i\hbar \{q_i, p_j\}$. Indeed, we have the general correspondence $[\hat{B}, \hat{C}] \longrightarrow i\hbar \{B, C\}$.

7.2 Schrödinger and Heisenberg Pictures

In the Schrödinger picture (the S-picture, denoted by the subscript S below), states carry the time dependence,

$$|\psi_S(t)\rangle = \hat{U}(t, t_0) \, |\psi_S(t_0)\rangle \,. \tag{7.25}$$

The time dependence can be transferred from the states to the operators by defining

$$|\psi_H\rangle = \hat{U}^\dagger(t, t_0) \, |\psi_S(t)\rangle = |\psi_S(t_0)\rangle \,, \tag{7.26}$$

and it follows that

$$\hat{A}_H(t) = \hat{U}^\dagger(t, t_0) \, \hat{A}_S(t) \, \hat{U}(t, t_0) \implies \hat{A}_S(t) = \hat{U}(t, t_0) \, \hat{A}_H(t) \, \hat{U}^\dagger(t, t_0) \,, \tag{7.27}$$

where the subscript H stands for the Heisenberg picture (H-picture).[3] In the H-picture, the states are time independent and the operators carry the time dependence. The definitions are such that matrix elements of H-picture operators between H-picture states at time t are the same as those of S-picture operators between S-picture states at the same time, namely

$$\langle \psi_H | \hat{A}_H(t) | \phi_H \rangle = \langle \psi_S(t) | \hat{A}_S(t) | \phi_S(t) \rangle \,. \tag{7.28}$$

We can determine how $\hat{A}_H(t)$ evolves in time. Recalling the general relation

$$i\hbar \frac{d}{dt} \hat{U}(t, t_0) = \hat{H}_S(t) \, \hat{U}(t, t_0) \,, \tag{7.29}$$

it is easily seen that

$$i\hbar \frac{d}{dt} \hat{A}_H(t) = \left[\hat{A}_H(t) , \, \hat{H}_H(t) \right] + i\hbar \, \hat{U}^\dagger(t, t_0) \left[\frac{d}{dt} \hat{A}_S(t) \right] \hat{U}(t, t_0) \,. \tag{7.30}$$

Note that if the Hamiltonian in the S-picture is time independent then

$$\hat{U}(t, t_0) = e^{-i\hat{H}_S(t-t_0)/\hbar} \implies \hat{H}_H(t) = e^{i\hat{H}_S(t-t_0)/\hbar} \, H_S \, e^{-i\hat{H}_S(t-t_0)/\hbar} = H_S \,, \tag{7.31}$$

and the matrix elements of the Hamiltonian in the H-picture are time independent, as they are in the S-picture, since

$$\langle \psi_H | \hat{H}_H(t) | \phi_H \rangle = \langle \psi_S(t) | \hat{H}_S | \phi_S(t) \rangle = \langle \psi_S(t_0) | \hat{H}_S | \phi_S(t_0) \rangle \,. \tag{7.32}$$

7.3 Problems

Problem 1 The Probability $p(a_i)$ is Independent of the Basis Adopted in a Degenerate Subspace

Let a_i be a degenerate eigenvalue of observable $\hat{A}$ and let $|\psi\rangle$ be the (normalized) state describing the system. Show that the probability $p(a_i)$ that the result of a measurement of $\hat{A}$ will be a_i is independent of the basis of $\hat{A}$ eigenstates chosen in the degenerate subspace.

---

[3] In contrast with the previous section, it is assumed that the generic S-picture operator $A_S(t)$ may itself be time dependent.

Solution

Let $|\phi_i^{(k)}\rangle$ and $|\chi_i^{(k)}\rangle$ be two bases in the subspace of $\mathbb{E}$ corresponding to the eigenvalue a_i, where $k = 1, \ldots, g_i$. There is a unitary matrix $\underline{S}$ of dimensions $g_i \times g_i$ that transforms one basis into the other. Its matrix elements are given by

$$S_{lm} = \langle \phi_i^{(l)} | \chi_i^{(m)} \rangle \,,$$

such that

$$|\phi_i^{(l)}\rangle = \sum_{m=1}^{g_i} |\chi_i^{(m)}\rangle\langle\chi_i^{(m)}|\phi_i^{(l)}\rangle = \sum_{m=1}^{g_i} S_{lm}^* |\chi_i^{(m)}\rangle \,, \qquad \langle\phi_i^{(l)}| = \sum_{m=1}^{g_i} S_{lm} \langle\chi_i^{(m)}| \,.$$

It follows that

$$\underbrace{\sum_{k=1}^{g_i} |\phi_i^{(k)}\rangle\langle\phi_i^{(k)}|}_{\hat{P}_i} = \sum_{k=1}^{g_i}\sum_{l=1}^{g_i}\sum_{m=1}^{g_i} S_{km} \underbrace{S_{kl}^*}_{\underline{S}_{lk}^\dagger} |\chi_i^{(l)}\rangle\langle\chi_i^{(m)}| = \sum_{l=1}^{g_i}\sum_{m=1}^{g_i} \underbrace{(\underline{S}^\dagger\,\underline{S})_{lm}}_{\delta_{lm}} |\chi_i^{(l)}\rangle\langle\chi_i^{(m)}| = \underbrace{\sum_{l=1}^{g_i} |\chi_i^{(l)}\rangle\langle\chi_i^{(l)}|}_{\hat{P}_i} \,,$$

and hence $p(a_i) = \langle\psi|\hat{P}_i|\psi\rangle$ is invariant with respect to a change of basis.

Problem 2 Explicit Time Dependence of $\langle\hat{A}(t)\rangle$

Using the basis of Hamiltonian eigenstates (stationary states), make explicit the time dependence of the expectation value $\langle\hat{A}(t)\rangle$ for a generic observable $\hat{A}$. Under what condition is $\langle\hat{A}(t)\rangle$ time independent? Justify your answer.

Solution

Expanding $|\psi(t)\rangle$ on a basis of eigenstates $|\phi_i^{(k)}\rangle$ of $\hat{H}$ yields

$$|\psi(t)\rangle = \sum_i\sum_{k=1}^{g_i} |\phi_i^{(k)}\rangle\langle\phi_i^{(k)}| \underbrace{e^{-i\hat{H}(t-t_0)/\hbar} |\psi(t_0)\rangle}_{|\psi(t)\rangle} = \sum_i\sum_{k=1}^{g_i} e^{-iE_i(t-t_0)/\hbar} c_i^{(k)}(t_0) |\phi_i^{(k)}\rangle \,, \qquad c_i^{(k)}(t_0) = \langle\phi_i^{(k)}|\psi(t_0)\rangle \,,$$

and inserting the above expression into that for $\langle\hat{A}(t)\rangle$ leads to

$$\langle\hat{A}(t)\rangle = \langle\psi(t)|\hat{A}|\psi(t)\rangle = \sum_{i,j}\sum_{k=1}^{g_i}\sum_{l=1}^{g_j} \underbrace{e^{-i(E_j-E_i)(t-t_0)/\hbar}}_{\text{explicit time dependence}} c_i^{(k)*}(t_0)\, c_j^{(l)}(t_0)\langle\phi_i^{(k)}|\hat{A}|\phi_j^{(l)}\rangle \,.$$

If $\hat{A}$ commutes with $\hat{H}$, then matrix elements of $\hat{A}$ between eigenstates of $\hat{H}$ corresponding to different eigenvalues vanish, that is, $\langle\phi_i^{(k)}|\hat{A}|\phi_j^{(l)}\rangle = 0$ if $i \neq j$, and hence

$$\langle\hat{A}(t)\rangle = \underbrace{\sum_i\sum_{k=1}^{g_i}\sum_{l=1}^{g_i} c_i^{(k)*}(t_0)\, c_i^{(l)}(t_0)\langle\phi_i^{(k)}|\hat{A}|\phi_i^{(l)}\rangle}_{\text{independent of time}} \qquad \text{if } [\hat{A},\hat{H}] = 0 \,,$$

of course, in agreement with the direct result that the rate of change of $\langle\hat{A}(t)\rangle$, being proportional to the commutator of $\hat{H}$ with $\hat{A}$, vanishes in this case.

Problem 3 Time–Energy Relation

Suppose that $\hat{A}$ is an observable for a system governed by the Hamiltonian $\hat{H}$. Define the time

$$\tau_A = \frac{\Delta A}{|d\langle\hat{A}(t)\rangle/dt|} \, ,$$

where ΔA is the standard deviation

$$\Delta A = \left[\langle\psi(t)|\hat{A}^2|\psi(t)\rangle - \langle\psi(t)|\hat{A}|\psi(t)\rangle^2 \right]^{1/2} \, .$$

Show that $\tau_A \, \Delta H \geq \hbar/2$. Note that, while ΔA is generally time dependent, ΔH, similarly defined, is not. Provide a physical interpretation for the time τ_A.

Solution

Heisenberg's uncertainty relation gives

$$\Delta A \, \Delta H \geq \frac{1}{2}\left| \langle\psi(t)|[\hat{A}, \hat{H}]|\psi(t)\rangle \right| \, .$$

However, we also have that the rate of change of $\langle\hat{A}(t)\rangle$ can be expressed as

$$i\hbar\frac{d}{dt}\langle\psi(t)|\hat{A}|\psi(t)\rangle = \langle\psi(t)|[\hat{A}, \hat{H}]|\psi(t)\rangle \, ,$$

and therefore

$$\Delta A \, \Delta H \geq \frac{\hbar}{2}\left| \frac{d}{dt}\langle\psi(t)|\hat{A}|\psi(t)\rangle \right| \, ,$$

which yields the required relation. In particular, τ_A can be interpreted as the time required for the average of $\hat{A}$ to change by an amount equal to its uncertainty ΔA; equivalently, $\langle\hat{A}(t_1)\rangle \simeq \langle\hat{A}(t_0)\rangle$ for $t_1 > t_0$, if the time interval $t_1 - t_0$ is small compared with τ_A, which characterizes the rate of change of the physical property represented by $\hat{A}$.

Note that, if the system is in one of the eigenstates of the Hamiltonian, i.e., in a "stationary" state, so that

$$|\psi(t_0)\rangle = |\phi_i^{(k)}\rangle \implies |\psi(t)\rangle = e^{-iE_i(t-t_0)/\hbar}|\phi_i^{(k)}\rangle \, ,$$

then $d\langle\hat{A}(t)\rangle/dt = 0$ regardless of whether $\hat{A}$ is a constant of motion or not, that is, whether it commutes with $\hat{H}$ or not. The characteristic time τ_A becomes infinite in this case.

Problem 4 Measurements on a Generic Wave Function $\psi(\mathbf{r})$

A particle is in a state represented by the normalized wave function $\psi(\mathbf{r})$. Calculate:

1. The probability that a measurement of the momentum component $\hat{p}_x$ will yield a result between p_1 and p_2;
2. The probability that a simultaneous measurement of the position component $\hat{x}$ and momentum component $\hat{p}_z$ will yield results $x_1 \leq x \leq x_2$ and $p_z \geq 0$, respectively;
3. The probability that a measurement of the observable $\hat{u} = \hat{x} + \hat{y} + \hat{z}$ will yield a result between u_1 and u_2.

Solution

Part 1

One way to obtain this probability is to compute the momentum-space wave function,

$$\widetilde{\psi}(\mathbf{p}) = \langle\phi_{\mathbf{p}}|\psi\rangle = \int d\mathbf{r}\,\langle\phi_{\mathbf{p}}|\phi_{\mathbf{r}}\rangle\langle\phi_{\mathbf{r}}|\psi\rangle = \int \frac{d\mathbf{r}}{(2\pi\hbar)^{3/2}}\, e^{-i\mathbf{p}\cdot\mathbf{r}/\hbar}\,\psi(\mathbf{r})\,,$$

where $|\phi_{\mathbf{p}}\rangle$ and $|\phi_{\mathbf{r}}\rangle$ are the momentum and position eigenstates, respectively, with

$$\hat{\mathbf{p}}\,|\phi_{\mathbf{p}}\rangle = \mathbf{p}\,|\phi_{\mathbf{p}}\rangle\,,\qquad \hat{\mathbf{r}}\,|\phi_{\mathbf{r}}\rangle = \mathbf{r}\,|\phi_{\mathbf{r}}\rangle\,,\qquad \langle\phi_{\mathbf{r}}|\psi_{\mathbf{p}}\rangle = \frac{e^{i\mathbf{p}\cdot\mathbf{r}/\hbar}}{(2\pi\hbar)^{3/2}}\,.$$

The probability of measuring a p_x value between p_1 and p_2 is then obtained as

$$\mathcal{P}(p_1 \leq p_x \leq p_2) = \int_{p_1}^{p_2} dp_x \int_{-\infty}^{\infty} dp_y \int_{-\infty}^{\infty} dp_z\,|\widetilde{\psi}(\mathbf{p})|^2\,.$$

Another way to obtain the probability is to introduce the projection operator

$$\hat{P}_{p_x\in[p_1,p_2]} = \left(\int_{p_1}^{p_2} dp_x\,|\phi_{p_x}\rangle\langle\phi_{p_x}|\right)\otimes\hat{1}_y\otimes\hat{1}_z = \left(\int_{p_1}^{p_2} dp_x\,|\phi_{p_x}\rangle\langle\phi_{p_x}|\right)\otimes\left(\int_{-\infty}^{\infty} dy\,|\phi_y\rangle\langle\phi_y|\right)\otimes\left(\int_{-\infty}^{\infty} dz\,|\phi_z\rangle\langle\phi_z|\right)$$

$$= \int_{p_1}^{p_2} dp_x \int_{-\infty}^{\infty} dy \int_{-\infty}^{\infty} dz\,\underbrace{\left(|\phi_{p_x}\rangle\otimes|\phi_y\rangle\otimes|\phi_z\rangle\right)}_{|\phi_{p_x,y,z}\rangle}\underbrace{\left(\langle\phi_{p_x}|\otimes\langle\phi_y|\otimes\langle\phi_z|\right)}_{\langle\phi_{p_x,y,z}|}\,,$$

where we have expressed the completeness in the state spaces $\mathbb{E}_y$ and $\mathbb{E}_z$ in terms of the $\hat{y}$ and $\hat{z}$ eigenstates, respectively. Of course, we are free to choose any bases we like in $\mathbb{E}_y$ and $\mathbb{E}_z$, such as, for example, those of the $\hat{p}_y$ and $\hat{p}_z$ eigenstates, as we did previously. The probability then follows by evaluating the expectation value of $\hat{P}_{p_x\in[p_1,p_2]}$ for the state $|\psi\rangle$,

$$\mathcal{P}(p_1 \leq p_x \leq p_2) = \langle\psi|\hat{P}_{p_x\in[p_1,p_2]}|\psi\rangle = \int_{p_1}^{p_2} dp_x \int_{-\infty}^{\infty} dy \int_{-\infty}^{\infty} dz\,\langle\psi|\phi_{p_x,y,z}\rangle\langle\phi_{p_x,y,z}|\psi\rangle$$

$$= \int_{p_1}^{p_2} dp_x \int_{-\infty}^{\infty} dy \int_{-\infty}^{\infty} dz\,|\widetilde{\psi}(p_x,y,z)|^2\,,$$

where

$$\widetilde{\psi}(p_x,y,z) = \frac{1}{(2\pi\hbar)^{1/2}}\int_{-\infty}^{\infty} dx\, e^{-ip_x x/\hbar}\,\psi(x,y,z)\,.$$

This probability is easily verified to be same as that obtained previously.

Part 2

Following the discussion above, we choose the basis $|\phi_{x,y,p_z}\rangle = |\phi_x\rangle\otimes|\phi_y\rangle\otimes|\phi_{p_z}\rangle$ in the state space $\mathbb{E} = \mathbb{E}_x\otimes\mathbb{E}_y\otimes\mathbb{E}_z$, and the relevant projection operator is written in the present case as

$$\hat{P}_{x\in[x_1,x_2],p_z\in[0,\infty[} = \int_{x_1}^{x_2} dx \int_{-\infty}^{\infty} dy \int_{0}^{\infty} dp_z\,|\phi_{x,y,p_z}\rangle\langle\phi_{x,y,p_z}|\,.$$

The probability $\mathcal{P}(x_1 \leq x \leq x_2; p_z \geq 0)$ follows:

$$\mathcal{P}(x_1 \leq x \leq x_2; p_z \geq 0) = \int_{x_1}^{x_2} dx \int_{-\infty}^{\infty} dy \int_{0}^{\infty} dp_z\,|\widetilde{\psi}(x,y,p_z)|^2\,,$$

where

$$\widetilde{\psi}(x,y,p_z) = \frac{1}{(2\pi\hbar)^{1/2}} \int_{-\infty}^{\infty} dz\, e^{-ip_z z/\hbar}\, \psi(x,y,z) \;.$$

Part 3

The observable $\hat{u}$ commutes with any of the components of the position observable $\hat{\mathbf{r}}$, and hence

$$\hat{u}\,|\phi_x\rangle \otimes |\phi_y\rangle \otimes |\phi_z\rangle = (x+y+z)\,|\phi_x\rangle \otimes |\phi_y\rangle \otimes |\phi_z\rangle \;.$$

In particular, the states $|\phi_{u,y,z}\rangle = |\phi_{u-y-z}\rangle \otimes |\phi_y\rangle \otimes |\phi_z\rangle$ are infinitely degenerate eigenstates of $\hat{u}$ corresponding to the eigenvalue u (u assumes continuous values), since for any y and z we have $\hat{u}\,|\phi_{u,y,z}\rangle = u|\phi_{u,y,z}\rangle$. However, $\hat{u}$, $\hat{y}$, and $\hat{z}$ form a complete set of commuting observables. We introduce the projection operator

$$\hat{P}_{u\in[u_1,u_2]} = \int_{u_1}^{u_2} du \int_{-\infty}^{\infty} dy \int_{-\infty}^{\infty} dz\, |\phi_{u,y,z}\rangle\langle\phi_{u,y,z}| \;,$$

from which we obtain

$$\mathcal{P}(u_1 \le u \le u_2) = \int_{u_1}^{u_2} du \int_{-\infty}^{\infty} dy \int_{-\infty}^{\infty} dz\, |\langle\phi_{u,y,z}|\psi\rangle|^2 = \int_{u_1}^{u_2} du \underbrace{\int_{-\infty}^{\infty} dy \int_{-\infty}^{\infty} dz\, |\psi(u-y-z,y,z)|^2}_{\rho(u)} \;.$$

Here $\rho(u)$ is the probability density of measuring u in du; by exploiting the δ-function property, it can also be written as follows:

$$\rho(u) = \int_{-\infty}^{\infty} dy \int_{-\infty}^{\infty} dz\, |\psi(u-y-z,y,z)|^2 = \int d\mathbf{r}\, \delta(x+y+z-u)\,|\psi(\mathbf{r})|^2 \;,$$

which makes it obvious that

$$\int_{-\infty}^{\infty} du\, \rho(u) = \int d\mathbf{r}\, |\psi(\mathbf{r})|^2 = 1 \;.$$

Of course, we could have taken either $\hat{u}$, $\hat{x}$, $\hat{z}$ or $\hat{u}$, $\hat{x}$, $\hat{y}$ as complete sets of commuting observables. The probability $\mathcal{P}(u_1 \le u \le u_2)$ calculated above would be the same.

Problem 5   Measurements of Non-Commuting Observables

Consider three non-commuting observables $\hat{A}$, $\hat{B}$, and $\hat{C}$. Observable $\hat{A}$ is measured, yielding the non-degenerate eigenvalue a with corresponding eigenstate $|\psi_a\rangle$. Then, the following two different experiments are carried out.

1. In the first experiment, before the system has had any time to evolve after the measurement of $\hat{A}$, observable $\hat{C}$ is measured, yielding the non-degenerate eigenvalue c with corresponding eigenstate $|\chi_c\rangle$. Calculate the probability $P_a(c)$ of measuring c.
2. In the second experiment, observables $\hat{B}$ and $\hat{C}$ are measured in rapid succession so that the system has not had any time to evolve between the measurements of $\hat{A}$ and $\hat{B}$ and those of $\hat{B}$ and $\hat{C}$.

Suppose these measurements yield, respectively, the non-degenerate eigenvalues b and c, where $|\varphi_b\rangle$ is the eigenstate corresponding to the eigenvalue b. Calculate the probability $P_a(b,c)$ of measuring b and c.

3. Assuming that the eigenvalues of $\hat{B}$ are all non-degenerate, show that

$$P_a(c) = \sum_b P_a(b,c) + \text{something else},$$

and provide an expression for the "something else." What is then the essential difference between the two experiments? What conclusions can you draw?

Solution

Part 1

We find

$$P_a(c) = |\langle \chi_c | \psi_a \rangle|^2.$$

Part 2

Given that the system is in state $|\psi_a\rangle$ (after the measurement of $\hat{A}$ yielding a), we find, for the probability of measuring b,

$$P_a(b) = |\langle \varphi_b | \psi_a \rangle|^2.$$

Then, the probability of measuring c with the system in state $|\varphi_b\rangle$ is given by

$$P_b(c) = |\langle \chi_c | \varphi_b \rangle|^2,$$

and hence the probability of measuring b and c with the system in state $|\psi_a\rangle$ reads

$$P_a(b,c) = P_b(c)P_a(b) = |\langle \chi_c | \varphi_b \rangle|^2 \, |\langle \varphi_b | \psi_a \rangle|^2.$$

Part 3

Using the completeness of the eigenstates of $\hat{B}$ (the corresponding eigenvalues are assumed to be all non-degenerate), we can express the *amplitude* $\langle \chi_c | \psi_a \rangle$ as follows:

$$\langle \chi_c | \psi_a \rangle = \sum_b \langle \chi_c | \varphi_b \rangle \langle \varphi_b | \psi_a \rangle,$$

from which we deduce

$$P_a(c) = \left| \sum_b \langle \chi_c | \varphi_b \rangle \langle \varphi_b | \psi_a \rangle \right|^2 = \left(\sum_b \langle \chi_c | \varphi_b \rangle \langle \varphi_b | \psi_a \rangle \right) \left(\sum_{b'} \langle \chi_c | \varphi_{b'} \rangle \langle \varphi_{b'} | \psi_a \rangle \right)^*$$

$$= \underbrace{\sum_b |\langle \chi_c | \varphi_b \rangle|^2 \, |\langle \varphi_b | \psi_a \rangle|^2}_{\sum_b P_a(b,c)} + \underbrace{\sum_{b \neq b'} \langle \chi_c | \varphi_b \rangle \langle \chi_c | \varphi_{b'} \rangle^* \langle \varphi_b | \psi_a \rangle \langle \varphi_{b'} | \psi_a \rangle^*}_{\text{something else}}.$$

In the second experiment, the measurement of $\hat{B}$ yielding the result b forces the system into the state $|\varphi_b\rangle$, while in the first experiment, when $\hat{B}$ is not measured, the system can transition into the state $|\chi_c\rangle$ from any of the states $|\varphi_b\rangle$. However, we must account for this possibility at the level of amplitudes rather than probabilities, otherwise crucial interference effects are missed.

Problem 6   Energy Measurements for a Particle in a One-Dimensional Infinitely Deep Well

A particle in an infinitely deep one-dimensional potential well $V(x)$, for which $V(x) = 0$ for $|x| \leq a/2$ and $V(x) = \infty$ for $|x| > a/2$, is at time $t = 0$ in a state

$$|\psi(0)\rangle = c_1|\psi_1\rangle + c_2|\psi_2\rangle + c_3|\psi_3\rangle + c_4|\psi_4\rangle ,$$

where the $|\psi_n\rangle$ are the eigenstates of the Hamiltonian $H = p^2/(2m) + V(x)$. Suppose an energy measurement is performed: what is the probability of finding a value smaller than $3\pi^2\hbar^2/(ma^2)$? What are the mean value and root-mean-square deviation of the energy in state $|\psi(0)\rangle$? Obtain the state $|\psi(t)\rangle$ at time t and show that the results above remain valid for any t: why is this?

The energy is measured and the result $8\pi^2\hbar^2/(ma^2)$ is obtained. After the measurement, what is the state of the system? What is the result if the energy is measured again?

Solution

The eigenvalues and eigenstates of the Hamiltonian of a particle in an infinitely deep potential well are

$$\hat{H}|\psi_n\rangle = E_n|\psi_n\rangle , \qquad E_n = \frac{\pi^2\hbar^2}{2ma^2}\, n^2 , \qquad n = 1, 2, \ldots$$

The state $|\psi(0)\rangle$ is assumed to be normalized. The probability of measuring energy E_m is given by

$$p(E_m) = |\langle\psi_m|\psi(0)\rangle|^2 ,$$

where

$$\langle\psi_m|\psi(0)\rangle = c_1\,\delta_{m,1} + c_2\,\delta_{m,2} + c_3\,\delta_{m,3} + c_4\,\delta_{m,4} .$$

We define

$$\epsilon_0 = \frac{\pi^2\hbar^2}{2ma^2} ,$$

so that the eigenvalues can be written as $E_n = \epsilon_0 n^2$, and the probability of measuring energy,

$$E_m < 3\,\frac{\pi^2\hbar^2}{ma^2} = 6\epsilon_0 ,$$

is given by

$$p(E < 6\epsilon_0) = p(E_1) + p(E_2) = |c_1|^2 + |c_2|^2 .$$

The expectation value of H is given by

$$\langle H \rangle = \langle\psi(0)|\hat{H}|\psi(0)\rangle = \sum_{i=1}^{4} E_i|c_i|^2 = \sum_{i=1}^{4} E_i p(E_i) .$$

The variance of the energy is given by

$$(\Delta H)^2 = \langle\psi(0)|\hat{H}^2|\psi(0)\rangle - (\langle H\rangle)^2 = \sum_{i=1}^{4} E_i^2 |c_i|^2 - \left(\sum_{i=1}^{4} E_i |c_i|^2\right)^2$$

$$= \sum_{i=1}^{4} E_i^2\, p(E_i)[1 - p(E_i)] - \sum_{i\neq j=1}^{4} E_i E_j\, p(E_i)\, p(E_j)\,,$$

and the root-mean-square deviation is $\sqrt{\langle(\Delta H)^2\rangle}$.

The state at time t follows from

$$|\psi(t)\rangle = \mathrm{e}^{-i\hat{H}t/\hbar}\,|\psi(0)\rangle = \sum_{i=1}^{4} c_i\, \mathrm{e}^{-i\hat{H}t/\hbar}|\psi_i\rangle = \sum_{i=1}^{4} c_i\, \mathrm{e}^{-iE_i t/\hbar}|\psi_i\rangle\,.$$

Note that $p(E_m)$ is constant in time, since

$$p(E_m, t) = \langle\psi(t)|\hat{P}_m|\psi(t)\rangle = \langle\psi(0)|\mathrm{e}^{i\hat{H}t/\hbar}\,\hat{P}_m\,\mathrm{e}^{-i\hat{H}t/\hbar}|\psi(0)\rangle = \langle\psi(0)|\hat{P}_m|\psi(0)\rangle = p(E_m, 0)\,,$$

where we have defined the projector onto eigenstate $|\psi_m\rangle$ as

$$\hat{P}_m = |\psi_m\rangle\langle\psi_m|\,,$$

and have used the fact that it commutes with $\hat{H}$ and hence with the time evolution operator. Therefore the probability of measuring an energy less than $6\epsilon_0$ is the same as before. A similar argument shows that the variance is time independent; indeed,

$$\langle\psi(t)|(\hat{H} - \langle H\rangle)^2|\psi(t)\rangle = \langle\psi(0)|\mathrm{e}^{i\hat{H}t/\hbar}(\hat{H} - \langle H\rangle)^2\mathrm{e}^{-i\hat{H}t/\hbar}|\psi(0)\rangle = \langle\psi(0)|(\hat{H} - \langle H\rangle)^2|\psi(0)\rangle\,.$$

The energy $16\epsilon_0$ is that of the third excited state, having $n=4$. Right after the measurement, say at time t_1, the state $|\psi(t_1)\rangle$ collapses into the state $|\psi_4\rangle$. The latter then evolves as

$$|\psi(t > t_1)\rangle = \mathrm{e}^{-i\hat{H}(t-t_1)/\hbar}\,|\psi_4\rangle = \mathrm{e}^{-iE_4(t-t_1)/\hbar}\,|\psi_4\rangle\,.$$

If the energy is measured again at some later time $t > t_1$, the result E_4 will be obtained with certainty, namely $p(E_4) = 1$.

Problem 7   Energy Measurements for a Particle in a Two-Dimensional Infinitely Deep Well

Consider a particle of mass m in two dimensions with Hamiltonian given by

$$\hat{H} = \hat{H}_x + \hat{H}_y\,, \qquad \hat{H}_x = \frac{\hat{p}_x^2}{2m} + \hat{V}(\hat{x})\,, \qquad \hat{H}_y = \frac{\hat{p}_y^2}{2m} + \hat{V}(\hat{y})\,,$$

where the potentials $V(x)$ and $V(y)$ vanish when $0 \le x \le a$ and $0 \le y \le a$ and are infinite otherwise.

1. Consider the four sets of observables $\hat{H}$, $\hat{H}_x$, $(\hat{H}_x, \hat{H}_y)$, and $(\hat{H}, \hat{H}_x)$. Which of them constitutes a complete set of commuting observables?
2. Consider a particle whose wave function is given by

$$\psi(x, y) = N \cos(\pi x/a)\, \cos(\pi y/a)\, \sin(2\pi x/a)\, \sin(2\pi y/a)\,,$$

where N is a normalization constant and $0 \le x \le a$ and $0 \le y \le a$.

a. What is the mean value of $\hat{H}$? If $\hat{H}$ is measured, what results can be found and with what probabilities?

b. The observable $\hat{H}_x$ is measured: what results can be found and with what probabilities? If this measurement yields the result $\pi^2\hbar^2/(2ma^2)$, what will the results of a subsequent measurement of $\hat{H}_y$ be? With what probabilities?

c. Instead of performing the preceding measurement, a simultaneous measurement of $\hat{H}_x$ and $\hat{p}_y$ is carried out. What are the probabilities of obtaining $E_x = 9\pi^2\hbar^2/(2ma^2)$ and p_y at p_0 in dp?

(Adapted from C. Cohen-Tannoudji, B. Diu, and F. Laloë 1997, *Quantum Mechanics*, vol. 1, Wiley.)

Solution

We consider the state space $E = E_x \otimes E_y$. The eigenstates of $\hat{H}$ in this space are the tensor product of those of $\hat{H}_x$ and $\hat{H}_y$ in state spaces E_x and E_y, respectively; that is, if $\hat{H}_x|\psi_{n_x}\rangle = E_{n_x}|\psi_{n_x}\rangle$ and $\hat{H}_y|\psi_{n_y}\rangle = E_{n_y}|\psi_{n_y}\rangle$ then

$$\hat{H}\left[|\psi_{n_x}\rangle \otimes |\psi_{n_y}\rangle\right] = \left[\hat{H}_x|\psi_{n_x}\rangle\right] \otimes |\psi_{n_y}\rangle + |\psi_{n_x}\rangle \otimes \left[\hat{H}_y|\psi_{n_y}\rangle\right] = (E_{n_x} + E_{n_y})|\psi_{n_x}\rangle \otimes |\psi_{n_y}\rangle .$$

The eigenfunctions of $\hat{H}_x$ (or $\hat{H}_y$) are easily obtained from solutions of the Schrödinger equation

$$-\frac{\hbar^2}{2m}\frac{d^2\psi(x)}{dx^2} = E\psi(x) \qquad \text{with } \psi(0) = \psi(a) = 0 ,$$

and the (normalized) eigenfunctions and corresponding eigenvalues are given by

$$\psi_n(x) = \sqrt{\frac{2}{a}}\sin(n\pi x/a) \qquad E_n = \frac{(n\pi\hbar)^2}{2ma^2} ,$$

where $n = 1, 2, \ldots$. The (normalized) eigenfunctions of the total Hamiltonian $\hat{H}$ are

$$\psi_{n_x,n_y}(x,y) = \frac{2}{a}\sin(n_x\pi x/a)\sin(n_y\pi y/a) , \qquad E_{n_x,n_y} = \frac{(\pi\hbar)^2}{2ma^2}\left(n_x^2 + n_y^2\right) .$$

For given n_x and n_y we list in Table 7.1 a few of the resulting $n_x^2 + n_y^2$ (in increasing order). We see that generally when $n_x = n_y$ there is no degeneracy, while when $n_x \neq n_y$ the degeneracy is two-fold. However, additional degeneracy can occur in special cases; for example, the states $\psi_{5,5}$, $\psi_{7,1}$, and $\psi_{1,7}$ are degenerate, with energy $25(\pi\hbar)^2/(ma^2)$.

Part 1

We see that $\hat{H}$ and $\hat{H}_x$ do not constitute a complete set of commuting observables. For example, specifying the eigenvalue $10(\pi\hbar)^2/(2ma^2)$ of $\hat{H}$ does not uniquely specify the state of the particle, since this eigenvalue is two-fold degenerate with eigenfunctions $\psi_{3,1}(x,y)$ and $\psi_{1,3}(x,y)$. Similarly, each eigenvalue $n_x^2(\pi\hbar)^2/(2ma^2)$ of $\hat{H}_x$ is infinitely degenerate in the state space E, since $\psi_{n_x,1}(x,y)$, $\psi_{n_x,2}(x,y)$, $\ldots$ are all eigenfunctions of $\hat{H}_x$ corresponding to this same eigenvalue. By contrast, $(\hat{H}_x, \hat{H}_y)$ and $(\hat{H}, \hat{H}_x)$ constitute complete sets of commuting observables.

Table 7.1 States corresponding to energies levels E_{n_x,n_y}.			
State	n_x	n_y	$n_x^2 + n_y^2$
0	1	1	2
1a	1	2	5
1b	2	1	5
2	2	2	8
3a	1	3	10
3b	3	1	10
4a	2	3	13
4b	3	2	13
5a	4	1	17
5b	1	4	17
6	3	3	18

Part 2a

The following formula is useful:

$$\sin\alpha\,\cos\beta = \frac{1}{2}\left[\sin(\alpha+\beta) + \sin(\alpha-\beta)\right] \; ;$$

so that thus we can write

$$\sin(2\pi x/a)\,\cos(\pi x/a) = \frac{1}{2}\left[\sin(3\pi x/a) + \sin(\pi x/a)\right] \; .$$

The wave function $\psi(x,y)$ can be expanded in the basis of eigenfunctions $\psi_{n_x,n_y}(x,y)$ as

$$\psi(x,y) = N\frac{a}{8}\left[\psi_{1,1}(x,y) + \psi_{3,1}(x,y) + \psi_{1,3}(x,y) + \psi_{3,3}(x,y)\right] \; .$$

Using the orthonormality of the basis, the normalization factor is easily calculated;

$$\int_0^a dx \int_0^a dy\,|\psi(x,y)|^2 = |N|^2\frac{a^2}{16} = 1 \implies |N| = \frac{4}{a} \; ,$$

and up to a phase factor the normalized state is given by

$$|\psi\rangle = \frac{1}{2}\left(|\psi_{1,1}\rangle + |\psi_{3,1}\rangle + |\psi_{1,3}\rangle + |\psi_{3,3}\rangle\right) \; .$$

In order to calculate the mean value of $\hat{H}$, we first note that

$$\hat{H}|\psi\rangle = \frac{1}{2}\left[E_{1,1}|\psi_{1,1}\rangle + E_{3,1}(|\psi_{3,1}\rangle + |\psi_{1,3}\rangle) + E_{3,3}|\psi_{3,3}\rangle\right] \; ,$$

and hence

$$\langle\psi|\hat{H}|\psi\rangle = \frac{1}{4}\left(E_{1,1} + 2E_{3,1} + E_{3,3}\right) = 5\frac{(\pi\hbar)^2}{ma^2} \; .$$

The probability of measuring the non-degenerate eigenvalue $E_{m,m}$ with $m = 1, 2, \ldots$ is given by

$$p(E_{m,m}) = |\langle\psi_{m,m}|\psi\rangle|^2 = \frac{1}{4}\left(\delta_{m,1} + \delta_{m,3}\right) \; ,$$

while the probability of measuring the (ordinarily two-fold) degenerate eigenvalue $E_{m,n} = E_{n,m}$ with $m \neq n$ is given by

$$p(E_{m,n}) = |\langle \psi_{m,n} | \psi \rangle|^2 + |\langle \psi_{n,m} | \psi \rangle|^2 = \frac{1}{2} \delta_{m,3} \delta_{n,1} .$$

Referring to the table above, we conclude that the ground-state and sixth-excited-state energies will be measured each with probability 1/4, while the third-excited-state energy will be measured with probability 1/2.

Part 2b

The probability of measuring the energy E_m is obtained from

$$p(E_m) = \sum_{n=1}^{\infty} |\langle \psi_{m,n} | \psi \rangle|^2 ,$$

since, in the full state space $\mathbb{E}$, each eigenvalue of $\hat{H}_x$ is infinitely degenerate; hence, we find

$$p(E_m) = \frac{1}{4} \sum_{n=1}^{\infty} (\delta_{m,1} \delta_{n,1} + \delta_{m,1} \delta_{n,3} + \delta_{m,3} \delta_{n,1} + \delta_{m,3} \delta_{n,3})^2 = \frac{1}{2} (\delta_{m,1} + \delta_{m,3}) ,$$

and so only the energies E_1 and E_3 will be measured, each with probability 1/2. If the measurement is carried out and E_1 is measured, then directly after the measurement the normalized state of the system reads

$$|\varphi\rangle = N \sum_{n=1}^{\infty} |\psi_{1,n}\rangle \langle \psi_{1,n} | \psi \rangle \qquad \text{with} \quad |N| = \left[\sum_{n=1}^{\infty} |\langle \psi_{1,n} | \psi \rangle|^2 \right]^{-1/2} ,$$

yielding

$$|\varphi\rangle = \frac{1}{\sqrt{2}} \left(|\psi_{1,1}\rangle + |\psi_{1,3}\rangle \right) .$$

The probability of measuring the energy E_m for $\hat{H}_y$ follows from

$$p(E_m) = \sum_{n=1}^{\infty} |\langle \psi_{n,m} | \varphi \rangle|^2 = \frac{1}{2} (\delta_{m,1} + \delta_{m,3}) ,$$

and only the energies E_1 and E_3 will be measured, each with probability 1/2.

Part 2c

Let $|\psi_{n,p}\rangle$ be the basis of simultaneous eigenstates of $\hat{H}_x$ and $\hat{p}_y$, where n assumes discrete values and p continuous values. We have

$$\hat{H}_x |\psi_{n,p}\rangle = \left[\hat{H}_x |\psi_n\rangle \right] \otimes |\phi_p\rangle = E_n |\psi_n\rangle \otimes |\phi_p\rangle = E_n |\psi_{n,p}\rangle ,$$

and similarly for $\hat{p}_y$. In the coordinate representation, the associated wave functions read

$$\psi_{n,p}(x, y) = \langle \phi_{x,y} | \psi_{n,p} \rangle = \sqrt{\frac{2}{a}} \sin(n\pi x/a) \frac{e^{ipy/\hbar}}{(2\pi\hbar)^{1/2}} .$$

Generally, the probability density of measuring a given energy E_n and a given momentum p follows from

$$\rho_n(p) = |\langle \psi_{n,p} | \psi \rangle|^2 ,$$

and in our case this probability is given by $p(E_3, p_0) = \rho_3(p_0)\, dp$. The probability density can be calculated explicitly. We have (for $n = 3$)

$$\langle \psi_{3,p} | \psi \rangle = \int_0^a dx \int_0^a dy\, \psi_3(x)\, \frac{e^{-ipy/\hbar}}{(2\pi\hbar)^{1/2}}\, \psi(x,y)$$

or

$$\langle \psi_{3,p} | \psi \rangle = \frac{4}{a} \left[\int_0^a dx\, \psi_3(x)\, \cos(\pi x/a)\, \sin(2\pi x/a) \right] \left[\frac{1}{(2\pi\hbar)^{1/2}} \int_0^a dy\, e^{-ipy/\hbar} \cos(\pi y/a) \sin(2\pi y/a) \right] .$$

The x- and y-integrals yield, respectively,

$$[\ldots]_x = \frac{1}{2}\sqrt{\frac{a}{2}} \int_0^a dx\, \psi_3(x) \left[\psi_1(x) + \psi_3(x) \right] = \frac{1}{2}\sqrt{\frac{a}{2}} ,$$

and

$$[\ldots]_y = \frac{1}{2\,(2\pi\hbar)^{1/2}} \int_0^a dy\, e^{-ipy/\hbar} \left[\sin(\pi y/a) + \sin(3\pi y/a) \right] .$$

Consider first the integral for a generic n:

$$\int_0^a dy\, e^{-ipy/\hbar} \sin(n\pi y/a) = \frac{1}{2i} \int_0^a dy \left[e^{i(n\pi/a - p/\hbar)y} - e^{-i(n\pi/a + p/\hbar)y} \right]$$

$$= \frac{1}{2i} \left[\frac{e^{i(n\pi - pa/\hbar)} - 1}{i(n\pi/a - p/\hbar)} + \frac{e^{-i(n\pi + pa/\hbar)} - 1}{i(n\pi/a + p/\hbar)} \right]$$

$$= n\pi a \frac{e^{-ipa/2\hbar}}{(n\pi)^2 - (pa/\hbar)^2} \left[e^{ipa/2\hbar} + (-)^{n+1}\, e^{-ipa/2\hbar} \right] ,$$

yielding finally

$$\int_0^a dy\, e^{-ipy/\hbar} \sin(n\pi y/a) = 2n\pi a \frac{e^{-ipa/2\hbar}}{(n\pi)^2 - (pa/\hbar)^2} \cos(pa/2\hbar) \qquad n = \text{odd}$$

$$= 2n\pi a \frac{e^{-i(pa/2\hbar - \pi/2)}}{(n\pi)^2 - (pa/\hbar)^2} \sin(pa/2\hbar) \qquad n = \text{even} .$$

We obtain

$$\langle \psi_{3,p} | \psi \rangle = \frac{4}{a} \left(\frac{1}{2}\sqrt{\frac{a}{2}} \right) \frac{\pi a}{(2\pi\hbar)^{1/2}}\, e^{-ipa/2\hbar} \cos(pa/2\hbar) \left[\frac{1}{\pi^2 - (pa/\hbar)^2} + \frac{3}{9\pi^2 - (pa/\hbar)^2} \right] ,$$

and hence

$$p(E_3, p_0) = \frac{\pi a}{\hbar} \cos^2(ap_0/2\hbar) \left[\frac{1}{\pi^2 - (ap_0/\hbar)^2} + \frac{3}{9\pi^2 - (ap_0/\hbar)^2} \right]^2 .$$

Problem 8　Measurements of Three Observables in a Three-Dimensional State Space: Example

Consider a system with a three-dimensional state space. The Hamiltonian $\hat{H}$ has a non-degenerate eigenvalue $E_1 = E_0$ with (normalized) eigenstate $|\phi_1\rangle$ and a degenerate eigenvalue $E_2 = 2E_0$ with (orthonormal) eigenstates $|\phi_2\rangle$ and $|\phi_3\rangle$. Suppose at time $t=0$ the system is in the normalized state $|\psi(0)\rangle$ given by

$$|\psi(0)\rangle = \frac{1}{\sqrt{2}}|\phi_1\rangle + \frac{1}{2}\left(|\phi_2\rangle + |\phi_3\rangle\right) .$$

1. At $t=0$ the energy of the system is measured. What values can be found and with what probabilities? Calculate $\langle H \rangle$ and the root-mean-square deviation ΔH for the system in the state $|\psi(0)\rangle$.
2. Suppose that at $t=0$, instead of $\hat{H}$, we measure the observable $\hat{A}$, which in the basis $|\phi_1\rangle$, $|\phi_2\rangle$, and $|\phi_3\rangle$ is represented by the following matrix:

$$\underline{A} = a \begin{pmatrix} 1 & 0 & 0 \\ 0 & 0 & 1 \\ 0 & 1 & 0 \end{pmatrix} .$$

Here, a is real and positive. What results can be found and with what probabilities?
3. Obtain the state vector $|\psi(t)\rangle$ at time t.
4. In addition to $\hat{A}$ as above, consider another observable $\hat{B}$ with representation (in the same basis of Hamiltonian eigenstates) given by

$$\underline{B} = b \begin{pmatrix} 0 & 1 & 0 \\ 1 & 0 & 0 \\ 0 & 0 & 1 \end{pmatrix} ,$$

where b is real and positive. What are the mean values $\langle A(t) \rangle$ and $\langle B(t) \rangle$? Any comments?
5. What results are obtained if $\hat{A}$ is measured at time t? Answer the same question for $\hat{B}$ and interpret these results.

Solution

Part 1

We have

$$p(E_1) = |\langle\phi_1|\psi(0)\rangle|^2 = \frac{1}{2}, \qquad p(E_2) = |\langle\phi_2|\psi(0)\rangle|^2 + |\langle\phi_3|\psi(0)\rangle|^2 = \frac{1}{2} .$$

The average energy is

$$\langle\hat{H}\rangle = \left(1/\sqrt{2}\ \ 1/2\ \ 1/2\right) \underbrace{\begin{pmatrix} E_0 & 0 & 0 \\ 0 & 2E_0 & 0 \\ 0 & 0 & 2E_0 \end{pmatrix} \begin{pmatrix} 1/\sqrt{2} \\ 1/2 \\ 1/2 \end{pmatrix}}_{\langle\psi(0)|\hat{H}|\psi(0)\rangle} = E_0 \left(1/\sqrt{2}\ \ 1/2\ \ 1/2\right) \begin{pmatrix} 1/\sqrt{2} \\ 1 \\ 1 \end{pmatrix} = \frac{3E_0}{2} ,$$

and similarly

$$\langle\hat{H}^2\rangle = \left(1/\sqrt{2}\ \ 1/2\ \ 1/2\right) \underbrace{\begin{pmatrix} E_0 & 0 & 0 \\ 0 & 2E_0 & 0 \\ 0 & 0 & 2E_0 \end{pmatrix} \begin{pmatrix} E_0 & 0 & 0 \\ 0 & 2E_0 & 0 \\ 0 & 0 & 2E_0 \end{pmatrix} \begin{pmatrix} 1/\sqrt{2} \\ 1/2 \\ 1/2 \end{pmatrix}}_{\langle\psi(0)|\hat{H}^2|\psi(0)\rangle} = \frac{5E_0^2}{2} ,$$

so that the root-mean-square deviation is given by

$$\Delta H = \sqrt{\langle \hat{H}^2 \rangle - \langle \hat{H} \rangle^2} = \frac{E_0}{2} \; .$$

Part 2

We need to find the eigenvalues and eigenstates of $\hat{A}$. It is easy to verify that $\hat{H}$ and $\hat{A}$ commute; hence, $\hat{A} |\phi_1\rangle$ is an eigenstate of $\hat{H}$ with eigenvalue E_1. Since E_1 is non-degenerate, it follows that $|\phi_1\rangle$ must be an eigenstate of $\hat{A}$; it is indeed such an eigenstate, with eigenvalue a, as is obvious from the matrix representation of $\hat{A}$. It is only necessary to diagonalize $\hat{A}$ in the degenerate subspace belonging to the eigenvalue E_2 of $\hat{H}$. We find

$$\det \begin{pmatrix} -\lambda & a \\ a & -\lambda \end{pmatrix} = 0 \implies \lambda_\pm = \pm a \; ,$$

with corresponding eigenstates

$$|\varphi_\pm\rangle = \frac{1}{\sqrt{2}} \left(|\phi_2\rangle \pm |\phi_3\rangle \right) \; .$$

We see that the eigenvalue a is doubly degenerate, with eigenstates $|\phi_1\rangle$ and $|\varphi_+\rangle$, while the eigenvalue $-a$ is non-degenerate with eigenstate $|\varphi_-\rangle$. We obtain for the probabilities of measuring a and $-a$,

$$p_a = |\langle \phi_1 | \psi(0) \rangle|^2 + |\langle \varphi_+ | \psi(0) \rangle|^2 = 1 \; , \qquad p_{-a} = |\langle \varphi_- | \psi(0) \rangle|^2 = 0 \; .$$

Note that the states $|\phi_1\rangle$, $|\varphi_+\rangle$, and $|\varphi_-\rangle$ are simultaneous eigenstates of $\hat{H}$ and $\hat{A}$ with eigenvalues, respectively, E_0, $2E_0$, and $2E_0$, and a, a, and $-a$. As a matter of fact, $\hat{H}$ and $\hat{A}$ form a complete set of commuting observables.

Part 3

We have

$$|\psi(t)\rangle = \mathrm{e}^{-i\hat{H}t/\hbar} |\psi(0)\rangle = \mathrm{e}^{-iE_0 t/\hbar} \left[\frac{1}{\sqrt{2}} |\phi_1\rangle + \frac{1}{2} \mathrm{e}^{-iE_0 t/\hbar} \left(|\phi_2\rangle + |\phi_3\rangle \right) \right] \; .$$

Part 4

Since $\hat{A}$ commutes with $\hat{H}$, the expectation value $\langle \psi(t) | \hat{A} | \psi(t) \rangle$ is time independent and hence equal to a, given that a measurement of $\hat{A}$ with the system in state $|\psi(0)\rangle$ yields with certainty the result a. By contrast, it is easily verified that $\hat{B}$ does not commute with $\hat{H}$, and hence

$$\langle \hat{B}(t)\rangle = \left(1/\sqrt{2} \ e^{iE_0 t/\hbar}/2 \ \ e^{iE_0 t/\hbar}/2\right)\underbrace{\begin{pmatrix} 0 & b & 0 \\ b & 0 & 0 \\ 0 & 0 & b \end{pmatrix}\begin{pmatrix} 1/\sqrt{2} \\ e^{-iE_0 t/\hbar}/2 \\ e^{-iE_0 t/\hbar}/2 \end{pmatrix}}_{\langle\psi(t)|\hat{B}|\psi(t)\rangle} = \frac{b}{4}\left[1 + 2\sqrt{2} \ \cos(E_0 t/\hbar)\right] \ ,$$

and so the expectation value, as expected, is time dependent.

Part 5

The observables $\hat{H}$ and $\hat{A}$ commute and the probabilities of measuring a and $-a$ at time t are the same as those at time $t=0$, since, for example,

$$p_a(t) = \langle\psi(t)|\hat{P}_a|\psi(t)\rangle = \langle\psi(0)| e^{i\hat{H}t/\hbar} \ \hat{P}_a \ e^{-i\hat{H}t/\hbar} \ |\psi(0)\rangle = \langle\psi(0)|\hat{P}_a|\psi(0)\rangle = P_a(0) \ ,$$

where we have defined the projection operator

$$\hat{P}_a = |\phi_1\rangle\langle\phi_1| + |\varphi_+\rangle\langle\varphi_+| \ ,$$

and have used the fact that

$$[\hat{P}_a, \hat{H}] = 0 \implies e^{i\hat{H}t/\hbar} \ \hat{P}_a \ e^{-i\hat{H}t/\hbar} = \hat{P}_a \ .$$

By contrast, the observables $\hat{B}$ and $\hat{H}$ do not commute. The eigenvalues of $\hat{B}$ are $-b$ (non-degenerate) with eigenstate

$$|\chi_-\rangle = \frac{1}{\sqrt{2}} \left(|\phi_1\rangle - |\phi_2\rangle\right) \ ,$$

and b (doubly degenerate) with eigenstates $|\phi_3\rangle$ and

$$|\chi_+\rangle = \frac{1}{\sqrt{2}} \left(|\phi_1\rangle + |\phi_2\rangle\right) \ .$$

The probability of measuring $-b$ at time t is then given by

$$p_{-b}(t) = |\langle\chi_-|\psi(t)\rangle|^2 = \frac{1}{4}\left|1 - \frac{1}{\sqrt{2}} \ e^{-iE_0 t/\hbar}\right|^2 = \frac{3}{8}\left[1 - \frac{2\sqrt{2}}{3} \ \cos(E_0 t/\hbar)\right] \ ,$$

and the probability of measuring b at time t is

$$p_b(t) = |\langle\chi_+|\psi(t)\rangle|^2 + |\langle\phi_3|\psi(t)\rangle|^2 = \frac{1}{4}\left|1 + \frac{1}{\sqrt{2}} \ e^{-iE_0 t/\hbar}\right|^2 + \frac{1}{4} = \frac{5}{8}\left[1 + \frac{2\sqrt{2}}{5} \ \cos(E_0 t/\hbar)\right] \ .$$

As a check, note that

$$\langle\hat{B}(t)\rangle = bp_b(t) - bp_{-b}(t) = \frac{b}{4}\left[1 + 2\sqrt{2} \ \cos(E_0 t/\hbar)\right] \ ,$$

in agreement with the result found earlier.

Problem 9 Consequences of a Sudden Change in the Potential

In a nuclear β decay, a proton is instantaneously changed by the weak interaction into a neutron, positron (e^+), and electron neutrino (ν_e). The positron and electron neutrino leave the nucleus without suffering any interactions on their way out (or, rather, minor ones in the case of the positron). However, the average attractive potential felt by the proton is different from that felt by the neutron (for one thing, the neutron does not carry any charge and is therefore unaffected by the repulsive Coulomb potential due to the remaining protons in the nucleus). This is the motivation for the following problem.

Consider a particle of mass m subject to an attractive δ-function potential given by $-v_0\,\delta(x)$ (in the notation of Chapter 4).

1. Show that the complete set of energy eigenfunctions can be written as

$$\psi_{\epsilon_0}(x) = \sqrt{\frac{v_0}{2}}\, e^{-v_0|x|/2}\,, \qquad \epsilon_0 = -\frac{v_0^2}{4}\,,$$

for the bound state, and

$$\psi_{k,\pm}(x) = \frac{1}{\sqrt{2\pi}}\left(e^{\pm ikx} - \frac{v_0}{v_0 + 2ik}\, e^{ik|x|}\right),$$

for the doubly degenerate $\pm$ scattering states corresponding to incident wave from, respectively, the left and the right.

2. Show that the bound-state wave function is orthogonal to the scattering-state wave functions. Also, show that the wave functions $\psi_{k,\pm}(x)$ satisfy the continuum normalization

$$\int_{-\infty}^{\infty} dx\, \psi_{k,\alpha}^*(x)\, \psi_{k',\alpha'}(x) = \delta(k-k')\,\delta_{\alpha,\alpha'}\,, \qquad \alpha,\alpha' = \pm\,.$$

The following formula is helpful:

$$\int_0^{\infty} dx\, e^{-iqx} = \int_{-\infty}^0 dx\, e^{iqx} = \lim_{\eta\to 0}\int_{-\infty}^0 dx\, e^{iqx+\eta x} = \lim_{\eta\to 0}\frac{1}{\eta + iq} = \pi\,\delta(q) - i\,\mathrm{PV}\left(\frac{1}{q}\right),$$

where PV indicates the Cauchy principal-value integration, which, given a function $f(q)$, is defined as

$$\mathrm{PV}\left(\frac{1}{q}\right) f(q) \equiv \lim_{\eta\to 0}\left[\int_{-\infty}^{-\eta} dq\,\frac{f(q)}{q} + \int_{\eta}^{\infty} dq\,\frac{f(q)}{q}\right].$$

3. Suppose that the particle is initially in the bound state. Suddenly, the strength of the potential v_0 is changed to $\bar v_0 > v_0$. Assume that this sudden change does not affect the state of the particle. What is the wave function $\psi(x,t)$ at time t, given that it is

$$\psi(x,0) = \sqrt{\frac{v_0}{2}}\, e^{-v_0|x|/2}\,,$$

at the initial time $t=0$? Compute the probabilities that the particle is found in the ground state and $\pm$ scattering states corresponding to the $-\bar v_0\,\delta(x)$ potential.

4. Evaluate the expectation value of the Hamiltonian with the $-\bar v_0\,\delta(x)$ potential and hence obtain the energy required to change v_0 to $\bar v_0$.

Solution

Part 1

The bound-state wave function is as obtained in Problem 4 of Chapter 4. The continuum wave functions read

$$\psi_{k,+}(x) = \frac{1}{\sqrt{2\pi}}\left(e^{ikx} - \frac{v_0}{v_0 + 2ik}\,e^{-ikx}\right) = \frac{1}{\sqrt{2\pi}}\left[e^{ikx} + A(k)\,e^{-ikx}\right] \qquad x < 0$$

$$= \frac{1}{\sqrt{2\pi}}\left(1 - \frac{v_0}{v_0 + 2ik}\right)e^{ikx} = \frac{1}{\sqrt{2\pi}}\,B(k)\,e^{ikx} \qquad\qquad x > 0$$

and

$$\psi_{k,-}(x) = \frac{1}{\sqrt{2\pi}}\left(1 - \frac{v_0}{v_0 + 2ik}\right)e^{-ikx} = \frac{1}{\sqrt{2\pi}}\,B(k)\,e^{-ikx} \qquad\qquad x < 0$$

$$= \frac{1}{\sqrt{2\pi}}\left(e^{-ikx} - \frac{v_0}{v_0 + 2ik}\,e^{ikx}\right) = \frac{1}{\sqrt{2\pi}}\left[e^{-ikx} + A(k)\,e^{ikx}\right] \qquad x > 0$$

and are in agreement with those derived in Problem 1 of Chapter 5 for a repulsive δ-function potential, provided that $v_0 \longrightarrow -v_0$. Here, we have defined

$$A(k) = -\frac{v_0}{v_0 + 2ik}\,, \qquad B(k) = \frac{2ik}{v_0 + 2ik}\,.$$

Part 2

It is convenient to introduce the parameter $a = 2/v_0$. Orthogonality of the bound and continuum states follows from

$$\langle \psi_{k,+} | \psi_{\epsilon_0} \rangle = \frac{1}{\sqrt{2\pi a}}\left[\int_{-\infty}^{0} dx\left[e^{-ikx} - \frac{e^{ikx}}{1 - iak}\right]e^{x/a} - \frac{iak}{1 - iak}\int_{0}^{\infty} dx\,e^{-ikx}\,e^{-x/a}\right]$$

$$= \frac{1}{\sqrt{2\pi a}}\left[\frac{1}{1/a - ik} - \frac{1}{(1/a + ik)(1 - iak)} - \frac{iak}{(1/a + ik)(1 - iak)}\right]$$

$$= \frac{1}{\sqrt{2\pi a}}\left[\frac{a}{1 - iak} - \frac{a}{1 - iak}\right] = 0\,,$$

and similarly for $\langle \psi_{k,-} | \psi_{\epsilon_0} \rangle$. The orthogonality of the continuum states follows from (we only consider the $+$ solution, for illustration; the proof of orthogonality between $+$ and $-$ or between $-$ and $-$ is identical)

$$\langle \psi_{k,+} | \psi_{k',+} \rangle = \int_{-\infty}^{\infty} dx\,\psi_{k,+}^{*}(x)\,\psi_{k',+}(x) = \frac{1}{2\pi}\int_{-\infty}^{0} dx\left[e^{-ikx} - \frac{1}{1 - iak}\,e^{ikx}\right]\left[e^{ik'x} - \frac{1}{1 + iak'}\,e^{-ik'x}\right]$$

$$- \frac{1}{2\pi}\int_{0}^{\infty} dx\,\frac{iak}{1 - iak}\,e^{-ikx}\,\frac{iak'}{1 + iak'}\,e^{ik'x}\,.$$

Using the result given in the text, we have

$$
\langle \psi_{k,+} | \psi_{k',+} \rangle = \frac{1}{2\pi} \left(\pi \underbrace{\delta(k'-k)}_{1} - i \underbrace{\frac{PV}{k'-k}}_{2} - \frac{1}{1-iak} \left[\pi \underbrace{\delta(k'+k)}_{3} - i \underbrace{\frac{PV}{k'+k}}_{4} \right] \right.
$$

$$
- \frac{1}{1+iak'} \left[\pi \underbrace{\delta(k'+k)}_{5} + i \underbrace{\frac{PV}{k'+k}}_{6} \right] + \frac{1}{1-iak}\frac{1}{1+iak'} \left[\pi \underbrace{\delta(k'-k)}_{7} + i \underbrace{\frac{PV}{k'-k}}_{8} \right]
$$

$$
\left. + \frac{ak}{1-iak}\frac{ak'}{1+iak'} \left[\pi \underbrace{\delta(k'-k)}_{9} + i \underbrace{\frac{PV}{k'-k}}_{10} \right] \right) .
$$

Consider the terms proportional to $\delta(k'-k)$,

$$
1 + 7 + 9 = \frac{1}{2\pi} \pi \left[1 + \frac{1}{1+(ak)^2} + \frac{(ak)^2}{1+(ak)^2} \right] \delta(k'-k) = \delta(k'-k) ,
$$

where we have enforced $k' = k$ as dictated by the δ-function. The other terms give

$$
4 + 6 = \frac{1}{2\pi} \left[\frac{i}{1-iak} - \frac{i}{1+iak'} \right] \frac{PV}{k'+k} = -\frac{a}{2\pi} \frac{k'+k}{(1-iak)(1+iak')} \frac{PV}{k'+k} = -\frac{a/(2\pi)}{(1-iak)(1+iak')}
$$

and

$$
2 + 8 + 10 = \frac{1}{2\pi} \left[-i + \frac{i + ia^2 kk'}{1+iak'} \right] \frac{PV}{k'-k} = \frac{i}{2\pi} \frac{-1 - a^2 kk' + iak - iak' + 1 + a^2 kk'}{(1-iak)(1+iak')} \frac{PV}{k'-k}
$$

$$
= \frac{a}{2\pi} \frac{k'-k}{(1-iak)(1+iak')} \frac{PV}{k'-k} = \frac{a/(2\pi)}{(1-iak)(1+iak')} ,
$$

which exactly cancels the contribution from terms $4 + 6$. Lastly, the contribution $3 + 5$ is proportional to $\delta(k+k')$, which can never be satisfied, since $k, k' > 0$. Thus, we obtain

$$
\langle \psi_{k,+} | \psi_{k',+} \rangle = \delta(k-k') .
$$

Part 3

We use the completeness of the $\hat{\overline{H}}$ eigenstates, where quantities with an overbar correspond to the new potential,

$$
|\overline{\psi}_{\overline{\epsilon}_0}\rangle\langle\overline{\psi}_{\overline{\epsilon}_0}| + \sum_{\alpha=\pm} \int_0^\infty dk \, |\overline{\psi}_{k,\alpha}\rangle\langle\overline{\psi}_{k,\alpha}| = \hat{\mathbb{1}} ,
$$

to expand the initial state $|\psi(0)\rangle$ as

$$
|\psi(0)\rangle = C_{\overline{\epsilon}_0} |\overline{\psi}_{\overline{\epsilon}_0}\rangle + \sum_{\alpha=\pm} \int_0^\infty dk \, C_\alpha(k) \, |\overline{\psi}_{k,\alpha}\rangle ,
$$

where

$$
C_{\overline{\epsilon}_0} = \langle\overline{\psi}_{\overline{\epsilon}_0}|\psi(0)\rangle , \qquad C_\alpha(k) = \langle\overline{\psi}_{k,\alpha}|\psi(0)\rangle .
$$

At time t the state is given by

$$|\psi(t)\rangle = e^{-i\hat{H}t/\hbar}|\psi(0)\rangle = C_{\overline{\epsilon}_0}\, e^{-iE_0 t/\hbar}|\overline{\psi}_{\overline{\epsilon}_0}\rangle + \sum_{\alpha=\pm}\int_0^\infty dk\, C_\alpha(k)\, e^{-iE_k t/\hbar}|\overline{\psi}_{k,\alpha}\rangle ,$$

where

$$E_0 = \frac{\hbar^2\,\overline{\epsilon}_0}{2m} , \qquad E_k = \frac{\hbar^2\,k^2}{2m} .$$

In the coordinate representation, the wave function $\psi(x,t) = \langle\phi_x|\psi(t)\rangle$ reads

$$\psi(x,t) = C_{\overline{\epsilon}_0}\, e^{-iE_0 t/\hbar}\,\overline{\psi}_{\overline{\epsilon}_0}(x) + \sum_{\alpha=\pm}\int_0^\infty dk\, C_\alpha(k)\, e^{-iE_k t/\hbar}\,\overline{\psi}_{k,\alpha}(x) .$$

The probabilities that the particle will be found in the ground state or in the $\pm$ scattering states (regardless of the momentum value $\hbar k$) are given by

$$p_0 = |C_{\overline{\epsilon}_0}|^2 , \qquad p_\alpha = \int_0^\infty dk\, |C_\alpha(k)|^2 \qquad \alpha = \pm .$$

We find that

$$C_{\overline{\epsilon}_0} = \sqrt{\frac{\overline{v}_0}{2}}\sqrt{\frac{v_0}{2}}\int_{-\infty}^\infty dx\, e^{-\overline{v}_0|x|/2}\, e^{-v_0|x|/2} = \frac{2\sqrt{\overline{v}_0 v_0}}{\overline{v}_0 + v_0}$$

and

$$
\begin{aligned}
C_+(k) &= \frac{1}{\sqrt{2\pi}}\sqrt{\frac{v_0}{2}}\left[\int_{-\infty}^0 dx\,\left[e^{-ikx} + \overline{A}^*(k)\, e^{ikx}\right] e^{v_0 x/2} + \overline{B}^*(k)\int_0^\infty dx\, e^{-ikx}\, e^{-v_0 x/2}\right]\\
&= \frac{1}{\sqrt{2\pi}}\sqrt{\frac{v_0}{2}}\left[\frac{1}{v_0/2 - ik} + \frac{\overline{A}^*(k)}{v_0/2 + ik} + \frac{\overline{B}^*(k)}{v_0/2 + ik}\right]\\
&= \frac{2}{\sqrt{2\pi}}\sqrt{\frac{v_0}{2}}\left[\frac{1}{v_0 - 2ik} - \frac{\overline{v}_0 + 2ik}{(v_0 + 2ik)(\overline{v}_0 - 2ik)}\right]\\
&= \sqrt{\frac{v_0}{\pi}}\,\frac{4ik\,(\overline{v}_0 - v_0)}{(v_0^2 + 4k^2)(\overline{v}_0 - 2ik)} ,
\end{aligned}
$$

and an identical result for $C_-(k)$, of course (simply change x to $-x$ in the first line). Thus we obtain, for the probabilities,

$$P_0 = \frac{4\,\overline{v}_0 v_0}{(\overline{v}_0 + v_0)^2} , \qquad P_\alpha = \frac{v_0(\overline{v}_0 - v_0)^2}{4\pi}\underbrace{\int_0^\infty dk\,\frac{k^2}{(v_0^2/4 + k^2)^2\,(\overline{v}_0^2/4 + k^2)}}_{I} .$$

The integral can be worked out in the following way:

$$I = \int_0^\infty dk \, \frac{v_0^2/4 + k^2 - v_0^2/4}{(v_0^2/4 + k^2)^2 \, (\bar{v}_0^2/4 + k^2)} = \underbrace{\int_0^\infty dk \, \frac{1}{(v_0^2/4 + k^2) \, (\bar{v}_0^2/4 + k^2)}}_{I_1}$$

$$- \frac{v_0^2}{\bar{v}_0^2 - v_0^2} \int_0^\infty dk \, \frac{1}{v_0^2/4 + k^2} \left[\frac{1}{v_0^2/4 + k^2} - \frac{1}{\bar{v}_0^2/4 + k^2} \right]$$

$$= \left(1 + \frac{v_0^2}{\bar{v}_0^2 - v_0^2} \right) I_1 - \frac{v_0^2}{\bar{v}_0^2 - v_0^2} \underbrace{\int_0^\infty dk \, \frac{1}{(v_0^2/4 + k^2)^2}}_{I_2} = \frac{\bar{v}_0^2}{\bar{v}_0^2 - v_0^2} I_1 - \frac{v_0^2}{\bar{v}_0^2 - v_0^2} I_2 \, .$$

The integrals I_1 and I_2 can be carried out by contour integration to find

$$I_1 = i\pi \left(\frac{1}{iv_0} \frac{4}{\bar{v}_0^2 - v_0^2} + \frac{1}{i\bar{v}_0} \frac{4}{v_0^2 - \bar{v}_0^2} \right) = \frac{4\pi}{\bar{v}_0 v_0} \frac{1}{\bar{v}_0 + v_0}$$

and

$$I_2 = i\pi \frac{d}{dk} \frac{1}{(k + iv_0/2)^2} \bigg|_{k=iv_0/2} = \frac{2\pi}{v_0^3} \, ,$$

and hence

$$I = \frac{4\pi}{\bar{v}_0 v_0} \frac{1}{\bar{v}_0 + v_0} \frac{\bar{v}_0^2}{\bar{v}_0^2 - v_0^2} - \frac{2\pi}{v_0^3} \frac{v_0^2}{\bar{v}_0^2 - v_0^2} = \frac{2\pi}{v_0} \frac{1}{(\bar{v}_0 + v_0)^2} \, .$$

The probability is then obtained as

$$P_\alpha = \frac{1}{2} \frac{(\bar{v}_0 - v_0)^2}{(\bar{v}_0 + v_0)^2} \, ,$$

and, as expected,

$$P_0 + P_+ + P_- = \frac{4 \bar{v}_0 v_0}{(\bar{v}_0 + v_0)^2} + \frac{(\bar{v}_0 - v_0)^2}{(\bar{v}_0 + v_0)^2} = 1 \, .$$

Part 4

In order to evaluate the expectation value of $\overline{H}$, it is convenient to write

$$\langle \psi(0) | \hat{\overline{H}} | \psi(0) \rangle = \underbrace{\langle \psi(0) | \hat{H} | \psi(0) \rangle}_{E_0} + \langle \psi(0) | \hat{\overline{H}} - \hat{H} | \psi(0) \rangle = \underbrace{-\frac{\hbar^2 v_0^2}{8m}}_{E_0} + \underbrace{\frac{\hbar^2}{4m}}_{|\psi_{\epsilon_0}(0)|^2} v_0 \, (v_0 - \bar{v}_0) = E_0 \left(\frac{2 \bar{v}_0}{v_0} - 1 \right) ,$$

where we have used the fact that $|\psi(0)\rangle$ is an eigenstate of $\hat{H}$. The energy required to change v_0 to $\bar{v}_0$ follows from

$$\Delta E = E_0 - \langle \psi(0) | \hat{\overline{H}} | \psi(0) \rangle = -2 E_0 \left(\frac{\bar{v}_0}{v_0} - 1 \right) > 0 \, .$$

Problem 10   Time Evolution of $\langle x(t) \rangle$ and $\langle p(t) \rangle$ in a One-Dimensional Linear Potential

Consider a particle in a one-dimensional potential given by

$$\hat{V}(\hat{x}) = -F_0 \hat{x} \,, \qquad F_0 > 0 \,.$$

1. Write down Ehrenfest's relations for the mean values of $\hat{x}$ and $\hat{p}$. Integrate the equations and compare the results with those in classical mechanics.
2. Show that the root-mean-square deviation of $\hat{p}$ does not change in time.

Solution

Part 1

Ehrenfest's relations read

$$\frac{d}{dt} \langle \hat{x}(t) \rangle = \frac{1}{m} \langle \psi(t) | \hat{p} | \psi(t) \rangle \,, \qquad \frac{d}{dt} \langle \hat{p}(t) \rangle = -\langle \psi(t) | \hat{V}'(\hat{x}) | \psi(t) \rangle \,,$$

where $\hat{V}'(\hat{x}) = -F_0$ is a constant. We therefore find that

$$\frac{d}{dt} \langle \hat{p}(t) \rangle = F_0 \implies \langle \hat{p}(t) \rangle = p_0 + F_0(t - t_0) \,.$$

Inserting this result into the right-hand side of the equation for $\hat{x}$, it follows that

$$\langle \hat{x}(t) \rangle = x_0 + \frac{p_0}{m}(t - t_0) + \frac{F_0}{2m}(t - t_0)^2 \,,$$

where the constants x_0 and p_0 are given by

$$x_0 = \langle \hat{x}(t_0) \rangle \,, \qquad p_0 = \langle \hat{p}(t_0) \rangle \,.$$

The expressions above for $\langle \hat{x}(t) \rangle$ and $\langle \hat{p}(t) \rangle$ are identical to those obtained in classical mechanics for a particle under the action of a constant force F_0.

Part 2

The variance,

$$(\Delta p)^2 = \langle \psi(t) | (\hat{p} - \langle p(t) \rangle)^2 | \psi(t) \rangle \,,$$

does not change in time because its time derivative vanishes. We consider the time dependence of the expectation value of the operator $\hat{A} = (\hat{p} - \langle p(t) \rangle)^2$ to find

$$\frac{d}{dt} \langle \hat{A} \rangle = \frac{i}{\hbar} \langle \psi(t) | [\hat{H}, (\hat{p} - \langle p(t) \rangle)^2] | \psi(t) \rangle \,,$$

and the commutator is given by

$$\begin{aligned}
[\hat{H}, (\hat{p} - \langle p(t) \rangle)^2] &= -F_0 [\hat{x}, (\hat{p} - \langle p(t) \rangle)^2] \\
&= -F_0 (\hat{p} - \langle p(t) \rangle) [\hat{x}, \hat{p} - \langle p(t) \rangle] - F_0 [\hat{x}, \hat{p} - \langle p(t) \rangle] (\hat{p} - \langle p(t) \rangle) \\
&= -2i\hbar F_0 (\hat{p} - \langle p(t) \rangle) \,.
\end{aligned}$$

It follows that

$$\frac{d}{dt}\langle\hat{A}\rangle = \frac{i}{\hbar}\,(-2i\hbar\,F_0)\,\langle\psi(t)|\,(\hat{p}-\langle p(t)\rangle)\,|\psi(t)\rangle = 2F_0\,[\underbrace{\langle\psi(t)|\,\hat{p}\,|\psi(t)\rangle}_{\langle p(t)\rangle}-\langle p(t)\rangle] = 0\ .$$

Since the variance is constant in time, so is the root-mean-square deviation $\sqrt{(\Delta p)^2}$.

Problem 11 The Virial Theorem in One Dimension

Let $|\phi_n\rangle$ be the eigenstates of a Hamiltonian $\hat{H}$ with $\hat{H}|\phi_n\rangle = E_n|\phi_n\rangle$, where n is a discrete index. Given an arbitrary operator $\hat{A}$, show the expectation value of its commutator with $\hat{H}$ on any state $|\phi_n\rangle$ vanishes. Next, suppose that $\hat{H}$ describes a particle in one dimension and subject to a potential $\hat{V}(\hat{x})$,

$$\hat{H} = \frac{\hat{p}^2}{2m} + \hat{V}(\hat{x})\ .$$

Evaluate the commutators $[\hat{H},\hat{x}]$, $[\hat{H},\hat{p}]$, and $[\hat{H},\hat{x}\hat{p}]$, and deduce that the expectation values of $\hat{p}$ on the eigenstates $|\phi_n\rangle$ vanish. Establish a relation between the expectation values on $|\phi_n\rangle$ of the kinetic energy operator and the expectation values of the operator $\hat{x}\,\hat{V}'(\hat{x})$, where $\hat{V}'(\hat{x})$ is the derivative of the operator $\hat{V}(\hat{x})$. Assuming that $\hat{V}(\hat{x}) = V_0\,\hat{x}^n$ with n integer ($n = 2, 4, \ldots$) and $V_0 > 0$, establish a relation between the expectation values of the kinetic energy and potential energy.

Solution

We have

$$\langle\phi_n|\underbrace{\hat{A}\hat{H}-\hat{H}\hat{A}}_{[\hat{A},\hat{H}]}|\phi_n\rangle = E_n\langle\phi_n|\hat{A}|\phi_n\rangle - E_n\langle\phi_n|\hat{A}|\phi_n\rangle = 0\ ,$$

where in the second step we have used the hermiticity of $\hat{H}$. We also have

$$[\hat{H},\hat{x}] = \frac{1}{2m}\,[\hat{p}^2,\hat{x}] = \frac{\hat{p}}{2m}\,[\hat{p},\hat{x}] + [\hat{p},\hat{x}]\,\frac{\hat{p}}{2m} = -i\frac{\hbar}{m}\,\hat{p}\ ,$$

since $\hat{V}(\hat{x})$ commutes with $\hat{x}$. Similarly, it follows that

$$[\hat{H},\hat{p}] = [\hat{V}(\hat{x}),\hat{p}] = i\hbar\frac{dV(x)}{dx}\bigg|_{x\to\hat{x}} = i\hbar\hat{V}'(\hat{x})\ ,$$

and

$$[\hat{H},\hat{x}\hat{p}] = \frac{1}{2m}\,[\hat{p}^2,\hat{x}\hat{p}] + [\hat{V}(\hat{x}),\hat{x}\hat{p}] = \frac{1}{2m}\,[\hat{p}^2,\hat{x}]\hat{p} + \hat{x}\,[\hat{V}(\hat{x}),\hat{p}] = i\hbar\,\hat{x}\,\hat{V}'(\hat{x}) - i\frac{\hbar}{m}\,\hat{p}^2\ .$$

Using the above result,

$$\hat{p} = i\frac{m}{\hbar}\,[\hat{H},\hat{x}]\ ,$$

we deduce

$$\langle\phi_n|\hat{p}|\phi_n\rangle = i\frac{m}{\hbar}\,\langle\phi_n|[\hat{H},\hat{x}]|\phi_n\rangle = 0\ .$$

From the relation

$$\langle\phi_n|[\hat{H},\hat{x}\hat{p}]|\phi_n\rangle = 0\ ,$$

and evaluating the commutator, we obtain

$$i\hbar \langle \phi_n | \hat{x} \, \hat{V}'(\hat{x}) | \phi_n \rangle - i \frac{\hbar}{m} \langle \phi_n | \hat{p}^2 | \phi_n \rangle = 0 \implies \langle \phi_n | \hat{T} | \phi_n \rangle = \frac{1}{2} \langle \phi_n | \hat{x} \, \hat{V}'(\hat{x}) | \phi_n \rangle \,,$$

where $\hat{T} = \hat{p}^2/(2m)$ is the kinetic energy operator. Assuming $\hat{V}(\hat{x}) = V_0 \, \hat{x}^n$, it follows that

$$\hat{x} \, \hat{V}'(\hat{x}) = n \, V_0 \, \hat{x}^n = n \, \hat{V}(\hat{x}) \implies \langle \phi_n | \hat{T} | \phi_n \rangle = \frac{n}{2} \langle \phi_n | \hat{V} | \phi_n \rangle \,.$$

Problem 12   Interaction Representation

Consider a system governed by the Hamiltonian $\hat{H}_0(t)$ and the corresponding time evolution operator $\hat{U}_0(t, t_0)$, satisfying

$$i\hbar \frac{d}{dt} \hat{U}_0(t, t_0) = \hat{H}_0(t) \, \hat{U}_0(t, t_0) \,, \qquad \hat{U}_0(t_0, t_0) = \hat{\mathbb{1}} \,.$$

Now, assume that the system is perturbed in such a way that its Hamiltonian becomes $\hat{H}(t)$ with

$$\hat{H}(t) = \hat{H}_0(t) + \hat{V}(t) \,.$$

The state vector $|\psi_I(t)\rangle$ of the system in the "interaction" picture is obtained from the Schrödiner-picture state vector $|\psi_S(t)\rangle$ via the unitary transformation

$$|\psi_I(t)\rangle = \hat{U}_0^\dagger(t, t_0) \, |\psi_S(t)\rangle \,.$$

1. Show that the time evolution of $|\psi_I(t)\rangle$ is given by

$$i\hbar \frac{d}{dt} |\psi_I(t)\rangle = \hat{V}_I(t) \, |\psi_I(t)\rangle \,,$$

   where $\hat{V}_I(t)$ is the representation of the perturbation operator $\hat{V}(t)$ in the interaction picture,

$$\hat{V}_I(t) = \hat{U}_0^\dagger(t, t_0) \, \hat{V}(t) \, \hat{U}_0(t, t_0) \,.$$

2. By converting the above differential equation into an integral equation,

$$|\psi_I(t)\rangle = |\psi_I(t_0)\rangle - \frac{i}{\hbar} \int_{t_0}^{t} dt' \, \hat{V}_I(t') | \psi_I(t')\rangle \,,$$

   where $|\psi_I(t_0)\rangle = |\psi_S(t_0)\rangle$, show that the state vector $|\psi_I(t)\rangle$ can be expanded in a power series of $\hat{V}_I(t)$.

Solution

Part 1

Using the unitarity of $\hat{U}_0(t, t_0)$, we express the Schrödinger-picture state as

$$|\psi_S(t)\rangle = \hat{U}_0(t, t_0) \, |\psi_I(t)\rangle \,.$$

It satisfies the time-dependent Schrödinger equation

$$i\hbar \frac{d}{dt} |\psi_S(t)\rangle = [\hat{H}_0(t) + \hat{V}(t)] \, |\psi_S(t)\rangle \,,$$

which gives

$$i\hbar \left[\frac{d}{dt} \hat{U}_0(t,t_0) \right] |\psi_I(t)\rangle + i\hbar \, \hat{U}_0(t,t_0) \frac{d}{dt} |\psi_I(t)\rangle = [\hat{H}_0(t) + \hat{V}(t)] \, \hat{U}_0(t,t_0) \, |\psi_I(t)\rangle \, .$$

We now note that, since $\hat{U}_0(t,t_0)$ is the time evolution operator corresponding to $\hat{H}_0(t)$, it satisfies

$$i\hbar \frac{d}{dt} \hat{U}_0(t,t_0) = \hat{H}_0(t) \, \hat{U}_0(t,t_0) \, ,$$

which, when inserted into the above equation, yields

$$i\hbar \, \hat{U}_0(t,t_0) \frac{d}{dt} |\psi_I(t)\rangle = \hat{V}(t) \, \hat{U}_0(t,t_0) \, |\psi_I(t)\rangle \implies i\hbar \frac{d}{dt} |\psi_I(t)\rangle = \underbrace{\hat{U}_0^\dagger(t,t_0) \, \hat{V}(t) \, \hat{U}_0(t,t_0)}_{V_I(t)} |\psi_I(t)\rangle \, .$$

Part 2

At time t_0 we have $U_0(t_0,t_0) = \hat{\mathbb{1}}$ and hence $|\psi_I(t)\rangle$ satisfies the initial condition $|\psi_I(t_0)\rangle = |\psi_S(t_0)\rangle$. By formally integrating over time both sides of

$$\frac{d}{dt} |\psi_I(t)\rangle = -\frac{i}{\hbar} V_I(t) \, |\psi_I(t)\rangle \, ,$$

we find

$$\int_{t_0}^t dt' \frac{d}{dt'} |\psi_I(t')\rangle = -\frac{i}{\hbar} \int_{t_0}^t dt' \, V_I(t') \, |\psi_I(t')\rangle \, ,$$

which leads to

$$|\psi_I(t)\rangle = |\psi_I(t_0)\rangle - \frac{i}{\hbar} \int_{t_0}^t dt' \, V_I(t') \, |\psi_I(t')\rangle \, .$$

This integral equation can be solved by iteration. We set the zeroth-order correction as $|\psi_I^{(0)}(t)\rangle = |\psi_I(t_0)\rangle = |\psi_S(t_0)\rangle$ and substitute it into the right-hand side of the integral equation, to obtain the first-order correction as follows:

$$|\psi_I^{(1)}(t)\rangle = |\psi_I(t_0)\rangle - \frac{i}{\hbar} \int_{t_0}^t dt_1 \, \hat{V}_I(t_1) \, |\psi_I^{(0)}(t_1)\rangle$$

$$= |\psi_S(t_0)\rangle - \frac{i}{\hbar} \int_{t_0}^t dt_1 \, \hat{V}_I(t_1) \, |\psi_S(t_0)\rangle \, .$$

We repeat the procedure to obtain the second-order correction as follows:

$$|\psi_I^{(2)}(t)\rangle = |\psi_I(t_0)\rangle - \frac{i}{\hbar} \int_{t_0}^t dt_1 \, \hat{V}_I(t_1) \, |\psi_I^{(1)}(t_1)\rangle$$

$$= |\psi_S(t_0)\rangle - \frac{i}{\hbar} \int_{t_0}^t dt_1 \, \hat{V}_I(t_1) \left[|\psi_S(t_0)\rangle - \frac{i}{\hbar} \int_{t_0}^{t_1} dt_2 \, \hat{V}_I(t_2) \, |\psi_S(t_0)\rangle \right]$$

$$= |\psi_S(t_0)\rangle - \frac{i}{\hbar} \int_{t_0}^t dt_1 \, \hat{V}_I(t_1) \, |\psi_S(t_0)\rangle + \left(-\frac{i}{\hbar} \right)^2 \int_{t_0}^t dt_1 \int_{t_0}^{t_1} dt_2 \, \hat{V}_I(t_1) \, \hat{V}_I(t_2) \, |\psi_S(t_0)\rangle \, ,$$

and so on (note the integration limits). Thus we have the expansion

$$|\psi_I(t)\rangle = \left[\hat{\mathbb{1}} - \frac{i}{\hbar}\int_{t_0}^t dt_1\,\hat{V}_I(t_1) + \left(-\frac{i}{\hbar}\right)^2 \int_{t_0}^t dt_1 \int_{t_0}^{t_1} dt_2\,\hat{V}_I(t_1)\,\hat{V}_I(t_2) + \cdots \right.$$
$$\left. + \left(-\frac{i}{\hbar}\right)^n \int_{t_0}^t dt_1 \cdots \int_{t_0}^{t_{n-1}} dt_n\,\hat{V}_I(t_1)\cdots\hat{V}_I(t_n) + \cdots \right]|\psi_S(t_0)\rangle\,,$$

which can be interpreted as an expansion for $|\psi_S(t)\rangle$ by recalling that

$$|\psi_S(t)\rangle = U_0(t,t_0)|\psi_I(t)\rangle = U_0(t,t_0)\,U_I(t,t_0)\,|\psi_S(t_0)\rangle = U(t,t_0)\,|\psi_S(t_0)\rangle\,,$$

where we defined $U_I(t,t_0)$ as the operator within the brackets $[\cdots]$, and the time evolution operator $U(t,t_0)$ as follows:

$$U(t,t_0) = U_0(t,t_0)\,U_I(t,t_0)\,,$$

corresponding to the full Hamiltonian $\hat{H}_0(t) + \hat{V}(t)$.

Problem 13   Two-Flavor Neutrino Oscillations

This problem deals with the phenomenon of neutrino oscillations in a simplified scenario, in which the neutrino flavors are two (the electron and muon neutrinos), rather than three (the electron, muon, and tau neutrinos) observed in nature. We denote these two flavors by ν_e and ν_μ, respectively. It turns out that an electron neutrino of momentum $\mathbf{p}$ is given by

$$|\nu_e;\mathbf{p}\rangle = \cos\theta\,|\nu_1;\mathbf{p}\rangle + \sin\theta\,|\nu_2;\mathbf{p}\rangle\,,$$

where $|\nu_k;\mathbf{p}\rangle$ with $k=1,2$ is an eigenstate of the relativistic free-particle Hamiltonian

$$\hat{H}_k = c\sqrt{\hat{\mathbf{p}}^2 + (m_k c)^2}\,,$$

that is,

$$\hat{H}_k|\nu_k;\mathbf{p}\rangle = E_k|\nu_k;\mathbf{p}\rangle\,, \qquad E_k = c\sqrt{p^2 + (m_k c)^2}\,,$$

where m_k and c are, respectively, the mass and speed of light, and

$$\langle\nu_k;\mathbf{p}|\nu_l;\mathbf{q}\rangle = \delta_{kl}\,\delta(\mathbf{p}-\mathbf{q})\,.$$

A muon neutrino of momentum $\mathbf{p}$ is given, in terms of the Hamiltonian eigenstates defined above, by the orthogonal linear combination

$$|\nu_\mu;\mathbf{p}\rangle = -\sin\theta\,|\nu_1,;\mathbf{p}\rangle + \cos\theta\,|\nu_2;\mathbf{p}\rangle\,.$$

1. Assume that at time $t=0$ the neutrino state is $|\psi(0)\rangle = |\nu_e;\mathbf{p}\rangle$. Calculate the probability $p_{\nu_e\to\nu_\mu}(t)$ that at time t the neutrino is in state $|\psi(t)\rangle = |\nu_\mu;\mathbf{p}\rangle$, that is, the neutrino has undergone the flavor oscillation $\nu_e \longrightarrow \nu_\mu$. Calculate also the survival probability $p_{\nu_e\to\nu_e}(t)$. Alternatively, assume that the neutrino state at time $t=0$ is $|\psi(0)\rangle = |\nu_\mu;\mathbf{p}\rangle$, and calculate $p_{\nu_\mu\to\nu_e}(t)$ and $p_{\nu_\mu\to\nu_\mu}(t)$.

2. Assume that the momentum $\mathbf{p}$ (that is, the eigenvalue of the momentum operator) is such that $|\mathbf{p}| \gg m_k c$. Show that, to leading order in $\Delta m^2 = m_1^2 - m_2^2$, the probability for the conversion $\nu_e \longrightarrow \nu_\mu$ can be written as

$$p_{\nu_e\to\nu_\mu}(L) = \sin^2(2\theta)\,\sin^2\left(\frac{\Delta m^2 L c^2}{4\hbar p}\right)\,,$$

where L is the distance traveled by the neutrino in time t.

Solution

Part 1

The Hamiltonian eigenstates evolve in time according to

$$e^{-i\hat{H}_k t/\hbar} |\nu_k; \mathbf{p}\rangle = e^{-iE_k t/\hbar} |\nu_k; \mathbf{p}\rangle \,,$$

and, since $|\psi(0)\rangle = |\nu_e; \mathbf{p}\rangle$, the evolved state at time t is obtained as

$$|\psi(t)\rangle = \cos\theta \, e^{-iE_1 t/\hbar} |\nu_1; \mathbf{p}\rangle + \sin\theta \, e^{-iE_2 t/\hbar} |\nu_2; \mathbf{p}\rangle \,.$$

It is convenient to define

$$E_\pm = \frac{E_1 \pm E_2}{2} \,.$$

The amplitude for the transition $\nu_e \longrightarrow \nu_\mu$ reads

$$\langle \nu_\mu; \mathbf{p} | \psi(t)\rangle = \left(-\sin\theta \, \langle \nu_1; \mathbf{p}| + \cos\theta \, \langle \nu_2; \mathbf{p}| \right) \left(\cos\theta \, e^{-iE_1 t/\hbar} |\nu_1; \mathbf{p}\rangle + \sin\theta \, e^{-iE_2 t/\hbar} |\nu_2; \mathbf{p}\rangle \right)$$

$$= -\sin\theta \cos\theta \left(e^{-iE_1 t/\hbar} - e^{-iE_2 t/\hbar} \right) = -\sin\theta \cos\theta \, e^{-iE_+ t/\hbar} \underbrace{\left[e^{-iE_- t/\hbar} - e^{iE_- t/\hbar} \right]}_{-2i\sin(E_- t/\hbar)} \,,$$

yielding

$$\langle \nu_\mu; \mathbf{p} | \psi(t)\rangle = i \sin(2\theta) \, e^{-iE_+ t/\hbar} \sin\left(\frac{E_- t}{\hbar} \right) \,,$$

and hence the transition probability

$$p_{\nu_e \to \nu_\mu}(t) = |\langle \nu_\mu; \mathbf{p} | \psi(t)\rangle|^2 = \sin^2(2\theta) \, \sin^2\left[\frac{(E_1 - E_2)t}{2\hbar} \right] \,.$$

The survival probability is as follows:

$$p_{\nu_e \to \nu_e}(t) = 1 - p_{\nu_e \to \nu_\mu}(t) = 1 - \sin^2(2\theta) \, \sin^2\left[\frac{(E_1 - E_2)t}{2\hbar} \right] \,,$$

which can also be obtained directly. Indeed, we find

$$\langle \nu_e; \mathbf{p} | \psi(t)\rangle = \left(\cos\theta \, \langle \nu_1; \mathbf{p}| + \sin\theta \, \langle \nu_2; \mathbf{p}| \right) \left(\cos\theta \, e^{-iE_1 t/\hbar} |\nu_1; \mathbf{p}\rangle + \sin\theta \, e^{-iE_2 t/\hbar} |\nu_2; \mathbf{p}\rangle \right)$$

$$= \cos^2\theta \, e^{-iE_1 t/\hbar} + \sin^2\theta \, e^{-iE_2 t/\hbar}$$

$$= e^{-iE_+ t/\hbar} \left[\frac{1 + \cos(2\theta)}{2} e^{-iE_- t/\hbar} + \frac{1 - \cos(2\theta)}{2} e^{iE_- t/\hbar} \right]$$

$$= e^{-iE_+ t/\hbar} \left[\cos\left(\frac{E_- t}{\hbar} \right) - i \cos(2\theta) \sin\left(\frac{E_- t}{\hbar} \right) \right] \,,$$

and hence

$$p_{\nu_e \to \nu_e}(t) = \cos^2\left[\frac{(E_1 - E_2)t}{2\hbar} \right] + \cos^2(2\theta) \, \sin^2\left[\frac{(E_1 - E_2)t}{2\hbar} \right]$$

$$= 1 - \sin^2\left[\frac{(E_1 - E_2)t}{2\hbar} \right] \underbrace{\left[1 - \cos^2(2\theta) \right]}_{\sin^2(2\theta)} \,,$$

as expected. By a calculation as similar to that above, it is straightforward to verify that

$$p_{\nu_\mu \to \nu_e}(t) = p_{\nu_e \to \nu_\mu}(t) \, ,$$

and hence that the muon and electron neutrino survival probabilities are the same.

Before moving on, we note that we can also obtain these probabilities by evaluating the time evolution operator directly in the flavor basis. To this end, we introduce the orthogonal matrix

$$\underline{S} = \begin{pmatrix} \cos\theta & \sin\theta \\ -\sin\theta & \cos\theta \end{pmatrix} \, , \qquad \underline{S}^T = \underline{S}^{-1} \, ,$$

so that

$$\begin{pmatrix} |\nu_e; \mathbf{p}\rangle \\ |\nu_\mu; \mathbf{p}\rangle \end{pmatrix} = \underline{S} \begin{pmatrix} |\nu_1; \mathbf{p}\rangle \\ |\nu_2; \mathbf{p}\rangle \end{pmatrix} \, , \qquad \begin{pmatrix} |\nu_1; \mathbf{p}\rangle \\ |\nu_2; \mathbf{p}\rangle \end{pmatrix} = \underline{S}^T \begin{pmatrix} |\nu_e; \mathbf{p}\rangle \\ |\nu_\mu; \mathbf{p}\rangle \end{pmatrix} \, .$$

The time evolution operator in the flavor basis follows from that in the mass basis as

$$\underbrace{\langle \nu_f; \mathbf{p} | \hat{U}(t) | \nu_{f'}; \mathbf{p} \rangle}_{U^{\text{flavor}}_{f,f'}(t)} = \sum_{kk'} \underbrace{\langle \nu_f; \mathbf{p} | \nu_k; \mathbf{p} \rangle}_{(\underline{S}^T)_{k,f}} \underbrace{\langle \nu_k; \mathbf{p} | \hat{U}(t) | \nu_{k'}; \mathbf{p} \rangle}_{U^{\text{mass}}_{k,k'}(t)} \underbrace{\langle \nu_{k'}; \mathbf{p} | \nu_{f'}; \mathbf{p} \rangle}_{S_{f',k'}}$$

$$= \sum_{k,k'} S_{f,k} \, U^{\text{mass}}_{k,k'}(t) \, (\underline{S}^T)_{k',f'} \, ,$$

or, in matrix notation,

$$\underline{U}^{\text{flavor}} = \underline{S}\, \underline{U}^{\text{mass}}\, \underline{S}^T = \begin{pmatrix} \cos\theta & \sin\theta \\ -\sin\theta & \cos\theta \end{pmatrix} \begin{pmatrix} e^{-iE_1 t/\hbar} & 0 \\ 0 & e^{-iE_2 t/\hbar} \end{pmatrix} \begin{pmatrix} \cos\theta & -\sin\theta \\ \sin\theta & \cos\theta \end{pmatrix} \, ,$$

yielding the unitary (and symmetric) flavor matrix

$$\underline{U}^{\text{flavor}} = \begin{pmatrix} \cos^2\theta\, e^{-iE_1 t/\hbar} + \sin^2\theta\, e^{-iE_2 t/\hbar} & -\sin\theta\cos\theta \left(e^{-iE_1 t/\hbar} - e^{-iE_2 t/\hbar} \right) \\ -\sin\theta\cos\theta \left(e^{-iE_1 t/\hbar} - e^{-iE_2 t/\hbar} \right) & \cos^2\theta\, e^{-iE_1 t/\hbar} + \sin^2\theta\, e^{-iE_2 t/\hbar} \end{pmatrix} \, .$$

The transition amplitude $f \longrightarrow f'$ reads simply

$$\langle \nu_{f'}; \mathbf{p} | \hat{U}(t) | \nu_f; \mathbf{p} \rangle = U^{\text{flavor}}_{f',f}(t) \, ,$$

with the corresponding probability given by $p_{f \to f'}(t) = |U^{\text{flavor}}_{f',f}(t)|^2$ as obtained earlier.

Part 2

In the limit $p \gg m_k c$, we have

$$E_k = cp\sqrt{1 + (m_k c/p)^2} \approx cp \left[1 + \frac{(m_k c/p)^2}{2} + \cdots \right] = cp + \frac{m_k^2 c^3}{2p} + \cdots \, ,$$

and the energy difference to leading order in Δm^2 is given by

$$E_1 - E_2 = \frac{(m_1^2 - m_2^2)c^3}{2p} = \frac{\Delta m^2 c^3}{2p} \, .$$

In this limit we also have $L = c\, t$, and hence

$$p_{\nu_e \to \nu_\mu}(L) = \sin^2(2\theta)\, \sin^2\left(\frac{\Delta m^2 L c^2}{4\hbar p} \right) \, .$$

Approximating the neutrino energy as $E \approx cp$, the above probability can be written as

$$p_{\nu_e \to \nu_\mu}(L) = \sin^2(2\theta) \, \sin^2\left(\frac{\Delta m^2 L c^4}{4\hbar c E}\right) .$$

Expressing E in eV, $\Delta m^2 c^4$ in eV^2, and L in km gives (here, $\hbar c \approx 197.33$ MeV fm or $\hbar c \approx 197.33 \times 10^{-12}$ eV km)

$$p_{\nu_e \to \nu_\mu}(L) = \sin^2(2\theta) \, \sin^2\left(1.27 \times 10^9 \, \frac{\Delta m^2 \, [\text{eV}^2/c^4] \, L \, [\text{km}]}{E[\text{eV}]}\right) ,$$

which provides an estimate of the scales involved in the sinusoidal dependence on the parameters Δm^2, L, and E. For example, long-baseline accelerator neutrino experiments involve E values of the order of GeV (10^9 eV) and L values (the distance between the neutrino source and neutrino detector) of the order of 10^3 km; they are sensitive to Δm^2 values of the order of 10^{-2}–10^{-3} eV^2/c^4.

The Harmonic Oscillator

The harmonic oscillator is of a great relevance in quantum mechanics: quadratic Hamiltonians enter all problems involving quantized oscillations, for example, in the theory of molecular and crystalline vibrations. The harmonic oscillator also plays a central role in quantum field theory.

8.1 Lowering and Raising Operators and the Number Operator Representation

The harmonic oscillator Hamiltonian (in one dimension) is given by

$$\hat{H} = \frac{\hat{p}^2}{2m} + \frac{m\omega^2}{2}\hat{x}^2 , \tag{8.1}$$

where $\hat{x}$ and $\hat{p}$ are the position and momentum operators and ω is the natural angular frequency of the oscillator. It is convenient to introduce the following combinations:

$$\hat{a} = \frac{1}{\sqrt{2m\hbar\omega}}\left(i\hat{p} + m\omega\hat{x}\right) \qquad \hat{a}^\dagger = \frac{1}{\sqrt{2m\hbar\omega}}\left(-i\hat{p} + m\omega\hat{x}\right) , \tag{8.2}$$

known, respectively, as the lowering and raising operator. They satisfy the commutation relations

$$[\hat{a}, \hat{a}] = 0 = \left[\hat{a}^\dagger, \hat{a}^\dagger\right] , \qquad \left[\hat{a}, \hat{a}^\dagger\right] = \hat{\mathbb{1}} , \tag{8.3}$$

where $\hat{\mathbb{1}}$ denotes the identity operator (hereafter, the latter will be simply denoted by 1). In terms of $\hat{a}$ and $\hat{a}^\dagger$, the position, momentum, and Hamiltonian operators read

$$\hat{x} = \sqrt{\frac{\hbar}{2m\omega}}\,(\hat{a}^\dagger + \hat{a}) , \qquad \hat{p} = i\sqrt{\frac{2}{m\hbar\omega}}\,(\hat{a}^\dagger - \hat{a}) , \qquad \hat{H} = \hbar\omega\left(\hat{a}^\dagger\hat{a} + \frac{1}{2}\right) . \tag{8.4}$$

The eigenstates of $\hat{H}$ are the same as those of the *number operator* $\hat{N} = \hat{a}^\dagger\hat{a}$. This operator satisfies a number of properties:

1. $\hat{N}\hat{a} = \hat{a}\,(\hat{N} - 1)$ and $\hat{N}\hat{a}^\dagger = \hat{a}^\dagger\,(\hat{N} + 1)$; for example,

$$\hat{N}\hat{a} = \hat{a}^\dagger\hat{a}\hat{a} = \underbrace{\left(\hat{a}\hat{a}^\dagger + \left[\hat{a}^\dagger, \hat{a}\right]\right)}_{\text{identity for } \hat{a}^\dagger\hat{a}}\hat{a} = \hat{a}\hat{a}^\dagger\hat{a} - \hat{a} = \hat{a}\,(\hat{N} - 1) . \tag{8.5}$$

2. If $|\psi\rangle$ is an eigenstate of $\hat{N}$ with eigenvalue λ then $\lambda \geq 0$; indeed,

$$\langle\psi|\hat{N}|\psi\rangle = \lambda\,\langle\psi|\psi\rangle \qquad \text{and} \qquad \langle\psi|\hat{N}|\psi\rangle = \underbrace{\langle\psi|\hat{a}^\dagger\hat{a}|\psi\rangle}_{\langle\phi|\phi\rangle} \geq 0 , \tag{8.6}$$

where the inequality is just the statement that the norm squared of the state $|\phi\rangle = \hat{a}|\psi\rangle$ is larger than or equal to 0. We conclude that $\lambda = \langle\phi|\phi\rangle/\langle\psi|\psi\rangle \geq 0$; further, if $\lambda = 0$ then necessarily $\hat{a}|\psi\rangle = 0$.

3. If $|\psi\rangle$ is an eigenstate of $\hat{N}$ with eigenvalue λ then $\hat{a}|\psi\rangle$ and $\hat{a}^\dagger|\psi\rangle$ are also eigenstates of $\hat{N}$, with eigenvalues $\lambda - 1$ and $\lambda + 1$, respectively; for example,

$$\hat{N}\hat{a}|\psi\rangle = \hat{a}(\hat{N} - 1)|\psi\rangle = \hat{a}(\lambda - 1)|\psi\rangle = (\lambda - 1)\hat{a}|\psi\rangle. \tag{8.7}$$

From property 3 we see that if $|\psi\rangle$ is an eigenstate of $\hat{N}$ with eigenvalue λ then the sequence of states

$$\hat{a}|\psi\rangle, \hat{a}^2|\psi\rangle, \ldots, \hat{a}^p|\psi\rangle, \tag{8.8}$$

consists of eigenstates of $\hat{N}$ with corresponding eigenvalues

$$\lambda - 1, \lambda - 2, \ldots, \lambda - p. \tag{8.9}$$

This set is finite, since the eigenvalues of $\hat{N}$ are necessarily positive or zero. Further, according to property 2, if $\lambda = 0$ then $\hat{a}|\psi\rangle = 0$. Multiplying both sides of this relation by $\hat{a}^\dagger$ leads to $\hat{a}^\dagger\hat{a}|\psi\rangle = \hat{N}|\psi\rangle = 0$, and $|\psi\rangle$ is an eigenstate of $\hat{N}$ with eigenvalue 0. Thus there must exist an integer n such that the action of $\hat{a}$ on a non-zero eigenstate $\hat{a}^n|\psi\rangle$ with eigenvalue $\lambda - n$ yields zero, and hence $\lambda = n$.

By contrast, the sequence of states

$$\hat{a}^\dagger|\psi\rangle, \hat{a}^{\dagger 2}|\psi\rangle, \ldots, \hat{a}^{\dagger p}|\psi\rangle, \ldots \tag{8.10}$$

consists of eigenstates of $\hat{N}$ with eigenvalues, respectively,

$$\lambda + 1, \lambda + 2, \ldots, \lambda + p, \ldots \tag{8.11}$$

This sequence never terminates, and so the eigenvalues of $\hat{N}$ are the non-negative integers $n = 0, 1, 2 \ldots$. We denote the corresponding normalized eigenstates by $|n\rangle$, and obtain for the Hamiltonian

$$\hat{H}|n\rangle = \underbrace{\hbar\omega(n + 1/2)}_{E_n}|n\rangle, \qquad n = 0, 1, 2, \ldots, \tag{8.12}$$

where the eigenvalues E_n are non-degenerate. The action of $\hat{a}$ and $\hat{a}^\dagger$ on $|n\rangle$ results in[1]

$$\hat{a}|n\rangle = \sqrt{n}|n - 1\rangle, \qquad \hat{a}|0\rangle = 0, \qquad \hat{a}^\dagger|n\rangle = \sqrt{n + 1}|n + 1\rangle, \tag{8.13}$$

the last of which yields the recursive relation

$$|n\rangle = \frac{1}{\sqrt{n}}\hat{a}^\dagger|n - 1\rangle \implies |n\rangle = \frac{1}{\sqrt{n!}}\hat{a}^{\dagger n}|0\rangle. \tag{8.14}$$

---

[1] We will show this for $\hat{a}^\dagger$. Since $\hat{a}^\dagger|n\rangle$ is an eigenstate of $\hat{N}$ with eigenvalue $n + 1$, it must be proportional to $|n + 1\rangle$, that is, $\hat{a}^\dagger|n\rangle = \lambda|n + 1\rangle$ (λ constant), given that the eigenvalue is non-degenerate. Now, consider the square of the normalized state $|\phi\rangle = \hat{a}^\dagger|n\rangle$ with dual bra $\langle n|\hat{a} = \langle\phi|$,

$$\langle\phi|\phi\rangle = \langle n|\hat{a}\hat{a}^\dagger|n\rangle = |\lambda|^2\langle n + 1|n + 1\rangle = |\lambda|^2.$$

On the left-hand side we can express $\hat{a}\hat{a}^\dagger$ in terms of the number operator by noting that

$$\hat{a}\hat{a}^\dagger = \left[\hat{a}, \hat{a}^\dagger\right] + \hat{a}^\dagger\hat{a} = 1 + \hat{N},$$

which then leads to

$$|\lambda|^2 = \langle n|\hat{a}\hat{a}^\dagger|n\rangle = \langle n|\hat{N} + 1|n\rangle = (n + 1)\langle n|n\rangle = n + 1,$$

implying that $|\lambda| = \sqrt{n + 1}$ (up to an irrelevant phase factor).

These eigenstates form a basis in the Hilbert space of single-particle states in one dimension,

$$\sum_{n=0}^{\infty} |n\rangle\langle n| = 1 , \qquad \langle m|n\rangle = \delta_{mn} . \tag{8.15}$$

The matrix elements of $\hat{a}$ and $\hat{a}^\dagger$ in this basis follow easily as

$$(a)_{mn} = \langle m|\hat{a}|n\rangle = \sqrt{n}\,\delta_{m,n-1} , \qquad (a^\dagger)_{mn} = \langle m|\hat{a}^\dagger|n\rangle = \sqrt{n+1}\,\delta_{m,n+1} . \tag{8.16}$$

Lastly, it is interesting to point out an alternative interpretation of the operators $\hat{a}$, $\hat{a}^\dagger$, and $\hbar\omega\,\hat{N}$. Since the eigenvalues of $\hat{N}$ are the non-negative integers, we may consider $\hbar\omega\,\hat{N}$ as the Hamiltonian of a system of particles all in one and the same single-particle state of energy $\hbar\omega$. Then, $|n\rangle$ represents the state of the system consisting of n particles while $|0\rangle$ is the vacuum state, in which no particles are present. The operators $\hat{a}$ and $\hat{a}^\dagger$ destroy and create particles, respectively, since $\hat{a}|n\rangle$ and $a^\dagger|n\rangle$ are proportional to the states of the system with $n-1$ and $n+1$ particles. Such an interpretation is used in quantum field theory. For example, the Hamiltonian of the electromagnetic field in a box is written as

$$\hat{H}_\gamma = \sum_{\mathbf{k}} \sum_{\lambda=1,2} \hbar\omega_k\,\hat{a}_{\mathbf{k}\lambda}^\dagger\,\hat{a}_{\mathbf{k}\lambda} , \tag{8.17}$$

where $\hat{a}_{\mathbf{k}\lambda}$ and $\hat{a}_{\mathbf{k}\lambda}^\dagger$ destroy and create photons of wave number $\mathbf{k}$, polarization state λ, and energy $\omega_k = c\,|\mathbf{k}|$ (c is the speed of light).

8.2 Wave Functions of the Harmonic Oscillator

In this section, we construct the harmonic oscillator eigenfunctions $\psi_n(x) = \langle\phi_x|n\rangle$. We begin by considering the ground state $|0\rangle$ with the property $\hat{a}|0\rangle = 0$, which in the coordinate representation is expressed as

$$0 = \langle\phi_x|\,\hat{a}\,|0\rangle = \int_{-\infty}^{\infty} dx'\,\langle\phi_x|\,\hat{a}\,|\phi_{x'}\rangle\langle\phi_{x'}|0\rangle = \int_{-\infty}^{\infty} dx'\,\langle\phi_x|\,\hat{a}\,|\phi_{x'}\rangle\,\psi_0(x') , \tag{8.18}$$

where the matrix element of a reads

$$\langle\phi_x|\hat{a}|\phi_{x'}\rangle = \frac{1}{\sqrt{2m\hbar\omega}}\langle\phi_x|\,i\hat{p} + m\omega\hat{x}\,|\phi_{x'}\rangle = \frac{1}{\sqrt{2m\hbar\omega}}\left(\hbar\frac{d}{dx} + m\omega x\right)\delta(x-x') . \tag{8.19}$$

Inserting the above result into Eq. (8.18) and integrating out the δ-function yields the first-order differential equation

$$\psi_0'(x) + \frac{m\omega}{\hbar}\,x\psi_0(x) = 0 , \tag{8.20}$$

which has the normalized solution

$$\psi_0(x) = c_0\,e^{-m\omega x^2/(2\hbar)} , \qquad c_0 = \left(\frac{m\omega}{\pi\hbar}\right)^{1/4} , \tag{8.21}$$

a Gaussian centered at the origin. The excited-state wave functions follow from

$$\psi_n(x) = \langle\phi_x|n\rangle = \frac{1}{\sqrt{n!}}\int_{-\infty}^{\infty} dx'\,\langle\phi_x|\,\hat{a}^{\dagger\,n}\,|\phi_{x'}\rangle\,\psi_0(x') . \tag{8.22}$$

The operator $\hat{a}^\dagger$ has the coordinate representation

$$\langle \phi_x | \hat{a}^\dagger | \phi_{x'} \rangle = \frac{1}{\sqrt{2m\hbar\omega}} \langle \phi_x | - i\hat{p} + m\omega\,\hat{x} | \phi_{x'} \rangle = \sqrt{\frac{m\omega}{2\hbar}} \left(x - \frac{\hbar}{m\omega} \frac{d}{dx} \right) \delta(x - x') \equiv O(x)\,\delta(x - x') . \tag{8.23}$$

By induction, the matrix elements of $\hat{a}^{\dagger\,n}$ are as follows:

$$\langle \phi_x | \hat{a}^{\dagger\,n} | \phi_{x'} \rangle = \int_{-\infty}^{\infty} dy \, \langle \phi_x | \hat{a}^\dagger | \phi_y \rangle \langle \phi_y | \hat{a}^{\dagger\,n-1} | \phi_{x'} \rangle = O(x) \int_{-\infty}^{\infty} dy \, \delta(x - y)\,[O(y)]^{n-1}\,\delta(y - x') = [O(x)]^n\,\delta(x - x') , \tag{8.24}$$

and inserting the above relation into Eq. (8.22) yields

$$\psi_n(x) = \frac{1}{\sqrt{n!}} \left(\frac{m\omega}{2\hbar} \right)^{n/2} \left(x - \frac{\hbar}{m\omega} \frac{d}{dx} \right)^n \psi_0(x) . \tag{8.25}$$

These wave functions can be written in standard form as

$$\psi_n(x) = c_n h_n(\xi)\,e^{-\xi^2/2} , \qquad \xi = \left(\frac{m\omega}{\hbar} \right)^{1/2} x , \tag{8.26}$$

where the $h_n(\xi)$ are (real) polynomials of order n in ξ, known as Hermite polynomials, and the c_n are normalization factors, given by

$$c_n = \left(\frac{1}{\sqrt{\pi}\,2^n n!} \right)^{1/2} \left(\frac{m\omega}{\hbar} \right)^{1/4} . \tag{8.27}$$

The Hamiltonian of the harmonic oscillator is invariant under space inversion. As a consequence, the (non-degenerate) eigenstates have definite parity. In particular, as expected, the ground-state wave function $\psi_0(x)$ has parity $+$, while the excited-state wave functions $\psi_n(x)$ have parity $(-)^n$, as is obvious from Eq. (8.25), which under the mapping $x \longmapsto -x$ becomes

$$\psi_n(x) \longmapsto \psi_n(-x) = \frac{1}{\sqrt{n!}} \left(\frac{m\omega}{2\hbar} \right)^{n/2} \left[-x - \frac{\hbar}{m\omega} \frac{d}{d(-x)} \right]^n \psi_0(-x) = (-1)^n\,\psi_n(x) . \tag{8.28}$$

Of course, this property also follows from Eq. (8.26) by observing that $h_n(-\xi) = (-1)^n\,h_n(\xi)$.

8.3 Problems

Problem 1 Some Properties of the Raising and Lowering Operators

Let $\hat{a}$ and $\hat{a}^\dagger$ be the lowering and raising operators for a harmonic oscillator with $[\hat{a}, \hat{a}^\dagger] = 1$ and $\hat{N} = \hat{a}^\dagger \hat{a}$. Show that

1. the operators $\hat{a}$ and $\hat{a}^\dagger$ have no inverse;
2. the relations $[\hat{N}, \hat{a}^p] = -p\,\hat{a}^p$ and $[\hat{N}, \hat{a}^{\dagger p}] = p\,\hat{a}^{\dagger p}$, where p is a positive integer, hold.

Solution

Part 1

There is a non-null state $|0\rangle$ such that $\hat{a}\,|0\rangle = 0$. Suppose that $\hat{a}^{-1}$ exists; then, we would have

$$\hat{a}^{-1}(\hat{a}\,|0\rangle) = 0 = (\hat{a}^{-1}\,\hat{a})|0\rangle = |0\rangle \implies |0\rangle = 0 ,$$

a contradiction. Therefore $\hat{a}^{-1}$ does not exist. A similar argument for $\hat{a}^\dagger$, which is defined in such a way that $\langle 0|\hat{a}^\dagger = 0$, shows that its inverse does not exist.

Part 2

Proceed by induction. The relations hold for $p = 1$, since

$$[\hat{N}, \hat{a}] = \underbrace{\hat{N}\hat{a}}_{\hat{a}\hat{N}-\hat{a}} - \hat{a}\hat{N} = -\hat{a}, \qquad [\hat{N}, \hat{a}^\dagger] = \underbrace{\hat{N}\hat{a}^\dagger}_{\hat{a}^\dagger\hat{N}+\hat{a}^\dagger} - \hat{a}^\dagger\hat{N} = \hat{a}^\dagger.$$

Assume the relations also hold for p; then for $p + 1$ we have

$$[\hat{N}, \hat{a}^{p+1}] = \hat{a}^p [\hat{N}, \hat{a}] + [\hat{N}, \hat{a}^p]\hat{a} = -\hat{a}^{p+1} - p\,\hat{a}^{p+1} = -(p+1)\,\hat{a}^{p+1},$$

and similarly for $\hat{a}^\dagger$.

Problem 2   Eigenvalue Problem for $\hat{x}$ in the N-Representation

Form the matrices representing the position and momentum observables, $\hat{x}$ and $\hat{p}$, in the representation $|n\rangle$ of eigenstates of $\hat{N}$. Verify that they are hermitian and satisfy the commutation relation $[\hat{x}, \hat{p}] = i\hbar$. Set up the eigenvalue problem of $\hat{x}$ in this representation; show that the spectrum of $\hat{x}$ is continuous and non-degenerate and extends from $-\infty$ to ∞. Form explicitly the eigenstate of $\hat{x}$ corresponding to the eigenvalue 0.

Solution

The position and momentum operators in terms of $\hat{a}$ and $\hat{a}^\dagger$ are

$$\hat{x} = \sqrt{\frac{\hbar}{2m\omega}}\left(\hat{a}^\dagger + \hat{a}\right), \qquad \hat{p} = i\sqrt{\frac{m\hbar\omega}{2}}\left(\hat{a}^\dagger - \hat{a}\right).$$

In the basis consisting of eigenstates of $\hat{N}$, the matrices representing $\hat{a}$ and $\hat{a}^\dagger$ have the form

$$\underline{a} = \begin{pmatrix} 0 & \sqrt{1} & 0 & 0 & 0 & \cdots \\ 0 & 0 & \sqrt{2} & 0 & 0 & \cdots \\ 0 & 0 & 0 & \sqrt{3} & 0 & \cdots \\ 0 & 0 & 0 & 0 & \sqrt{4} & \cdots \\ \vdots & \vdots & \vdots & \vdots & \vdots & \vdots \end{pmatrix}, \qquad \underline{a}^\dagger = \begin{pmatrix} 0 & 0 & 0 & 0 & \cdots \\ \sqrt{1} & 0 & 0 & 0 & \cdots \\ 0 & \sqrt{2} & 0 & 0 & \cdots \\ 0 & 0 & \sqrt{3} & 0 & \cdots \\ 0 & 0 & 0 & \sqrt{4} & \cdots \\ \vdots & \vdots & \vdots & \vdots & \vdots \end{pmatrix}.$$

and hence

$$\underline{x} = \sqrt{\frac{\hbar}{2m\omega}} \begin{pmatrix} 0 & \sqrt{1} & 0 & 0 & 0 & \cdots \\ \sqrt{1} & 0 & \sqrt{2} & 0 & 0 & \cdots \\ 0 & \sqrt{2} & 0 & \sqrt{3} & 0 & \cdots \\ 0 & 0 & \sqrt{3} & 0 & \sqrt{4} & \cdots \\ \vdots & \vdots & \vdots & \vdots & \vdots & \vdots \end{pmatrix}, \qquad \underline{p} = \sqrt{\frac{m\hbar\omega}{2}} \begin{pmatrix} 0 & -i\sqrt{1} & 0 & 0 & \cdots \\ i\sqrt{1} & 0 & -i\sqrt{2} & 0 & \cdots \\ 0 & i\sqrt{2} & 0 & -i\sqrt{3} & \cdots \\ 0 & 0 & i\sqrt{3} & 0 & \cdots \\ 0 & 0 & 0 & i\sqrt{4} & \cdots \\ \vdots & \vdots & \vdots & \vdots & \vdots \end{pmatrix},$$

which are manifestly hermitian. By direct multiplication we find

$$
\underline{x}\,\underline{p} = \frac{\hbar}{2}
\begin{pmatrix}
i & 0 & -i\sqrt{2} & 0 & \cdots \\
0 & i & 0 & -i\sqrt{6} & \cdots \\
i\sqrt{2} & 0 & i & 0 & \cdots \\
0 & i\sqrt{6} & 0 & i & \cdots \\
\vdots & \vdots & \vdots & \vdots & \vdots
\end{pmatrix},
\qquad
\underline{p}\,\underline{x} = \frac{\hbar}{2}
\begin{pmatrix}
-i & 0 & -i\sqrt{2} & 0 & \cdots \\
0 & -i & 0 & -i\sqrt{6} & \cdots \\
i\sqrt{2} & 0 & -i & 0 & \cdots \\
0 & i\sqrt{6} & 0 & -i & \cdots \\
\vdots & \vdots & \vdots & \vdots & \vdots
\end{pmatrix},
$$

and hence

$$
\underline{x}\,\underline{p} - \underline{p}\,\underline{x} = \frac{\hbar}{2}
\begin{pmatrix}
2i & 0 & 0 & 0 & \cdots \\
0 & 2i & 0 & 0 & \cdots \\
0 & 0 & 2i & 0 & \cdots \\
0 & 0 & 0 & 2i & \cdots \\
\vdots & \vdots & \vdots & \vdots & \vdots
\end{pmatrix}.
$$

We could also have evaluated the matrix element

$$
\begin{aligned}
\langle m|\hat{x}\hat{p}|n\rangle &= \sum_{k=0}^{\infty} \langle m|\hat{x}|k\rangle\langle k|\hat{p}|n\rangle = i\frac{\hbar}{2}\sum_{k=0}^{n}\langle m|\hat{a}^{\dagger}+\hat{a}|k\rangle\langle k|\hat{a}^{\dagger}-\hat{a}|n\rangle \\
&= i\frac{\hbar}{2}\sum_{k=0}^{\infty}\left(\sqrt{k+1}\,\delta_{m,k+1}+\sqrt{k}\,\delta_{m,k-1}\right)\left(\sqrt{n+1}\,\delta_{k,n+1}-\sqrt{n}\,\delta_{k,n-1}\right) \\
&= i\frac{\hbar}{2}\left[\sqrt{(n+1)(n+2)}\,\delta_{m,n+2}+\delta_{m,n}-\sqrt{n(n-1)}\,\delta_{m,n-2}\right]
\end{aligned}
$$

and similarly

$$
\langle m|\hat{p}\hat{x}|n\rangle = i\frac{\hbar}{2}\left[\sqrt{(n+1)(n+2)}\,\delta_{m,n+2}-\delta_{m,n}-\sqrt{n(n-1)}\,\delta_{m,n-2}\right],
$$

yielding

$$
\langle m|\hat{x}\hat{p}|n\rangle - \langle m|\hat{p}\hat{x}|n\rangle = i\hbar\delta_{m,n}.
$$

Let $|\phi_x\rangle$ be an eigenstate of $\hat{x}$ with eigenvalue x, that is, $\hat{x}|\phi_x\rangle = x|\phi_x\rangle$. In the basis $|n\rangle$ the eigenvalue equation reads

$$
\langle m|\hat{x}|\phi_x\rangle = x\langle m|\phi_x\rangle \qquad \text{or} \qquad \sum_{k=0}^{\infty}\langle m|\hat{x}|k\rangle\langle k|\phi_x\rangle = x\langle m|\phi_x\rangle.
$$

We set $c_m^{(x)} = \langle m|\phi_x\rangle$ and

$$
\sqrt{\frac{\hbar}{2m\omega}}\sum_{k=0}^{\infty}\left(\sqrt{k+1}\,\delta_{m,k+1}+\sqrt{k}\,\delta_{m,k-1}\right)c_k^{(x)} = x\,c_m^{(x)},
$$

implying

$$
\sqrt{m}\,c_{m-1}^{(x)} + \sqrt{m+1}\,c_{m+1}^{(x)} = \underbrace{\sqrt{\frac{2m\omega}{\hbar}}\,x}_{\lambda}\,c_m^{(x)}
$$

or explicitly (here λ is real, since the eigenvalues of a hermitian operator, such as $\hat{x}$, are real)

$$\lambda c_0^{(x)} = \sqrt{1}\, c_1^{(x)}$$
$$\lambda c_1^{(x)} = \sqrt{1}\, c_0^{(x)} + \sqrt{2}\, c_2^{(x)}$$
$$\lambda c_2^{(x)} = \sqrt{2}\, c_1^{(x)} + \sqrt{3}\, c_3^{(x)}$$
$$\lambda c_3^{(x)} = \sqrt{3}\, c_2^{(x)} + \sqrt{4}\, c_4^{(x)}$$
$$\vdots$$

This system has a non-trivial unique solution for any λ in $-\infty < \lambda < \infty$ as long as $c_0^{(x)} \neq 0$; indeed, we have

$$c_1^{(x)} = \frac{\lambda}{\sqrt{1!}}\, c_0^{(x)}$$

$$c_2^{(x)} = \frac{\lambda^2 - 1}{\sqrt{2!}}\, c_0^{(x)}$$

$$c_3^{(x)} = \frac{\lambda^3 - 3\lambda}{\sqrt{3!}}\, c_0^{(x)}$$

$$c_4^{(x)} = \frac{\lambda^4 - 6\lambda^2 + 3}{\sqrt{4!}}\, c_0^{(x)}$$

$$\vdots$$

We conclude that, as expected, $\hat{x}$ has a continuous and non-degenerate spectrum (note that the spectrum is continuous owing to the fact that we are dealing here with a linear system consisting of an infinite number of equations). The components of the $x = 0$ eigenstate are easily obtained by setting $\lambda = 0$ in the linear system. This yields $c_m^{(0)} = 0$ for m odd, and

$$c_2^{(0)} = -\sqrt{\frac{1}{2}}\, c_0^{(0)}\,, \qquad c_4^{(0)} = \sqrt{\frac{1 \times 3}{2 \times 4}}\, c_0^{(0)}\,, \qquad c_4^{(0)} = -\sqrt{\frac{1 \times 3 \times 5}{2 \times 4 \times 6}}\, c_0^{(0)}\,, \qquad \cdots\,.$$

The coefficient $c_0^{(0)}$ follows from

$$c_0^{(0)} = \langle 0 | \phi_0 \rangle = \langle \phi_0 | 0 \rangle^* = \psi_0^*(0) = \left(\frac{m\omega}{\pi\hbar}\right)^{1/4}\,,$$

and $\psi_0(x) = \langle \phi_x | 0 \rangle$ is the ground-state wave function.

By inspection, we see that the solution $c_m^{(x)}$ is proportional to the Hermite polynomial of order m, that is,

$$H_m(\lambda/\sqrt{2}) = H_m(\sqrt{m\omega/\hbar}\, x) = 2^{m/2}\,\sqrt{m!}\, c_m^{(x)}/c_0^{(x)}\,.$$

Insertion of this into the recurrence relation for the $c_m^{(x)}$ (rescaled by c_0),

$$\sqrt{2}\sqrt{\frac{m\omega}{\hbar}}\, x c_m^{(x)}/c_0^{(x)} = \sqrt{m}\, c_{m-1}^{(x)}/c_0^{(x)} + \sqrt{m+1}\, c_{m+1}^{(x)}/c_0^{(x)}\,,$$

yields

$$\sqrt{2}\sqrt{\frac{m\omega}{\hbar}}\, x \frac{H_m(\sqrt{m\omega/\hbar}\, x)}{2^{m/2}\,\sqrt{m!}} = \sqrt{m}\, \frac{H_{m-1}(\sqrt{m\omega/\hbar}\, x)}{2^{(m-1)/2}\,\sqrt{(m-1)!}} + \sqrt{m+1}\, \frac{H_{m+1}(\sqrt{m\omega/\hbar}\, x)}{2^{(m+1)/2}\,\sqrt{(m+1)!}}\,,$$

or, setting $\xi = \sqrt{m\omega/\hbar}\, x$,

$$2\xi H_m(\xi) = 2m H_{m-1}(\xi) + H_{m+1}(\xi) ,$$

a recurrence relation satisfied by the Hermite polynomials. Thus, we find

$$\frac{c_m^{(x)}}{c_0^{(x)}} = \frac{H_m(\xi)}{2^{m/2}\sqrt{m!}} \implies \langle m|\phi_x\rangle = \frac{H_m(\xi)}{2^{m/2}\sqrt{m!}}\, c_0^{(x)} .$$

We could have obtained the result for $c_m^{(0)}$ directly by noting that

$$\langle \phi_x|m\rangle = \langle m|\phi_x\rangle^* = N_m H_m(\xi)\, e^{-\xi^2/2} = \frac{H_m(\xi)}{2^{m/2}\sqrt{m!}} \underbrace{\left(\frac{m\omega}{\pi\hbar}\right)^{1/4} e^{-\xi^2/2}}_{c_0^{(x)}=\langle 0|\phi_x\rangle=\langle\phi_x|0\rangle^*} ,$$

where we have inserted the normalization factor

$$N_m = \left(\frac{1}{\sqrt{\pi}\, 2^m m!}\right)^{1/2} \left(\frac{m\omega}{\hbar}\right)^{1/4} .$$

The eigenstate corresponding to the eigenvalue $x = 0$ is easily obtained as

$$c_m^{(0)} = \frac{H_m(0)}{2^{m/2}\sqrt{m!}}\, c_0^{(0)} ,$$

and since the parity of $H_m(\xi)$ is $(-1)^m$ only those for even m are non-zero at the origin,

$$c_{2m}^{(0)} = \frac{H_{2m}(0)}{2^m\sqrt{(2m)!}}\, c_0^{(0)} = \frac{(-1)^m\sqrt{(2m)!}}{2^m\, m!}\, c_0^{(0)} \qquad m = 0, 1, 2, \dots ,$$

where we have used $H_{2m}(0) = (-1)^m\, (2m)!\, /m!$; this agrees with the result obtained earlier.

Problem 3 Position Measurement and Subsequent Time Evolution of a Harmonic Oscillator

A harmonic oscillator of mass m and angular frequency ω is in the ground state with normalized wave function given by

$$\psi_0(x) = \left(\frac{m\omega}{\pi\hbar}\right)^{1/4} e^{-m\omega x^2/(2\hbar)} .$$

1. At $t = 0$ a measurement of the particle's position is made. What is the probability that the particle will be found in an interval dx centered at $x = 0$?
2. Assume that the measurement of the particle's position at $t = 0$ does indeed yield the result $x = 0$. What is the particle's wave function $\psi(x, 0)$ immediately after this measurement?
3. What is the particle's wave function $\psi(x, t)$ at a later time t?
4. What is the probability that a measurement of the energy at time t yields the result $\hbar\omega/2$?
5. Without doing any detailed calculation, explain why the probability of measuring an energy $(n + 1/2)\hbar\omega$ with n odd vanishes.

Solution

Part 1

The probability that the particle will be in an interval dx centered at the origin is

$$p_0 = |\psi_0(0)|^2\, dx = \left(\frac{m\omega}{\pi\hbar}\right)^{1/2} dx\,.$$

Part 2

Directly after the measurement the particle is in the eigenstate $|\phi_x\rangle$ of the position operator with eigenvalue $x=0$, namely $|\phi_0\rangle$. Recalling that

$$\langle \phi_x|\phi_{x'}\rangle = \delta(x - x')\,,$$

we obtain

$$\psi(x,0) = \langle \phi_x|\phi_0\rangle = \delta(x)\,.$$

Part 3

The time evolution of a state is governed by the Hamiltonian,

$$|\psi(t)\rangle = \mathrm{e}^{-iHt/\hbar}\,|\psi(0)\rangle\,.$$

We denote by $|n\rangle$ the eigenstates of the harmonic oscillator Hamiltonian with eigenvalues $\hbar\omega(n+1/2)$. Using the completeness of the basis $|n\rangle$, we can express $|\psi(t)\rangle$ as

$$|\psi(t)\rangle = \sum_{n=0}^{\infty} \mathrm{e}^{-i\omega(n+1/2)t}\, c_n|n\rangle\,, \qquad c_n = \langle n|\psi(0)\rangle\,,$$

or, in the coordinate representation,

$$\psi(x,t) = \langle \phi_x|\psi(t)\rangle = \sum_{n=0}^{\infty} \mathrm{e}^{-i\omega(n+1/2)t}\, c_n\langle\phi_x|n\rangle = \sum_{n=0}^{\infty} \mathrm{e}^{-i\omega(n+1/2)t}\, c_n\psi_n(x)\,,$$

where the $\psi_n(x)$ are the harmonic oscillator wave functions. In our case we have $|\psi(0)\rangle = |\phi_0\rangle$ and

$$c_n = \langle n|\phi_0\rangle = \int_{-\infty}^{\infty} dx\, \langle n|\phi_x\rangle\langle\phi_x|\phi_0\rangle = \int_{-\infty}^{\infty} dx\, \psi_n^*(x)\,\delta(x) = \psi_n^*(0) = \psi_n(0)\,,$$

since the harmonic oscillator wave functions are taken to be real. The $\psi_n(x)$ have definite parity,

$$\psi_n(-x) = (-1)^n\psi_n(x)\,,$$

implying that

$$\psi(x,t) = \sum_{\text{even } n=0}^{\infty} \mathrm{e}^{-i\omega(n+1/2)t}\, c_n\psi_n(x)\,.$$

Part 4

The probability of measuring the ground-state energy is given by

$$p_{\text{gs}} = |\langle 0|\psi(t)\rangle|^2 = |c_0|^2 = |\psi_0(0)|^2 = \left(\frac{m\omega}{\pi\hbar}\right)^{1/2}\,.$$

Part 5

The state $|\psi(t)\rangle$ is expanded in terms of eigenstates $|n\rangle$ with n even only; hence, $\langle m|\psi(t)\rangle$ for any odd m vanishes.

Problem 4 *N* Uncoupled Harmonic Oscillators

A system consists of N non-interacting particles, each of mass m and in a harmonic potential of angular frequency ω. The Hamiltonian is given by

$$\hat{H} = \sum_{i=1}^{N} \hat{H}_i , \qquad \hat{H}_i = \frac{\hat{p}_i^2}{2m} + \frac{m\omega^2}{2} \hat{x}_i^2 ,$$

where $\hat{x}_i$ and $\hat{p}_i$ are the position and momentum operators for particle i.

1. Calculate the energies of this system, their degeneracy, and the corresponding eigenstates. Which sets of observables constitute a complete set of commuting observables? Write down the orthonormalization and closure relations for the eigenstates $|n_1, n_2, \ldots, n_N\rangle$ of $\hat{H}$.
2. The system is in the state

$$|\psi\rangle = \frac{1}{2} (|0, 0, 0, \underbrace{0, \ldots}_{\text{all } 0}\rangle + |1, 0, 0, \underbrace{0, \ldots}_{\text{all } 0}\rangle + |0, 1, 0, \underbrace{0, \ldots}_{\text{all } 0}\rangle + |0, 0, 1, \underbrace{0, \ldots}_{\text{all } 0}\rangle) .$$

What results can be found, and with what probabilities, if a measurement is made of (i) the total energy of the system, (ii) the energy of particle 1, (iii) the position or momentum of particle 1?

Solution

Part 1

The eigenvalues and eigenstates of $\hat{H}_i$ are, respectively,

$$\hat{H}_i |n_i\rangle = \underbrace{\hbar\omega(n_i + 1/2)}_{E_{n_i}} |n_i\rangle , \qquad n_i = 0, 1, 2, \ldots ,$$

and

$$|n_i\rangle = \frac{1}{\sqrt{n_i!}} \hat{a}_i^{\dagger\, n_i} |0_i\rangle .$$

The eigenstates of the total Hamiltonian are

$$|n_1, n_2, \ldots, n_N\rangle = |n_1\rangle \otimes |n_2\rangle \otimes \cdots \otimes |n_N\rangle ,$$

where $\otimes$ denotes the tensor product; we have

$$\hat{H}|n_1, \ldots, n_N\rangle = \underbrace{(\hat{H}_1 |n_1\rangle)}_{E_{n_1}|n_1\rangle} \otimes \cdots \otimes |n_N\rangle + \cdots + |n_1\rangle \otimes \cdots \otimes \underbrace{(\hat{H}_N |n_N\rangle)}_{E_{n_N}|n_N\rangle} = (E_{n_1} + \cdots + E_{n_N}) |n_1, \ldots, n_N\rangle .$$

We denote the eigenvalues of $\hat{H}$ as

$$E_n(N) = \hbar\omega(n + N/2) , \qquad n = n_1 + n_2 + \cdots + n_N = 0, 1, 2, \ldots$$

The degeneracy of a generic level n (for N harmonic oscillators) is given by

$$g_n(N) = \sum_{n_1=0}^{n} \sum_{n_2=0}^{n} \cdots \sum_{n_N=0}^{n} \delta_{n_1+n_2+\cdots+n_N,n} \; .$$

Using the identity

$$\frac{1}{2\pi} \int_0^{2\pi} d\theta \, e^{i\theta(p-q)} = \delta_{p,q}, \qquad p,q \text{ integers} \; ,$$

the above degeneracy can also be expressed as follows:

$$g_n(N) = \frac{1}{2\pi} \int_0^{2\pi} d\theta \sum_{n_1=0}^{n} \sum_{n_2=0}^{n} \cdots \sum_{n_N=0}^{n} e^{i\theta(n_1+n_2+\cdots+n_N-n)} = \frac{1}{2\pi} \int_0^{2\pi} d\theta \, e^{-in\theta} \left(\sum_{p=0}^{n} e^{ip\theta} \right)^N .$$

We introduce the variable $z = e^{i\theta}$, so that the above integral is converted into an integral in the complex plane along a closed path consisting of a circle of unit radius; given $d\theta = dz/(iz)$, we find

$$g_n(N) = \frac{1}{2\pi i} \oint dz \, \frac{1}{z^{n+1}} \left(\sum_{p=0}^{n} z^p \right)^N ,$$

and the integrand is an analytic function in the complex plane, except for a pole of order $n+1$ at the origin. Using the residue theorem, we find that the contour integral is given by

$$g_n(N) = \frac{1}{n!} \frac{d^n}{dz^n} \left(\sum_{p=0}^{n} z^p \right)^N \bigg|_{z=0} = \frac{1}{n!} \frac{d^n}{dz^n} \left(\frac{1-z^{n+1}}{1-z} \right)^N \bigg|_{z=0} ,$$

which can be written as

$$g_n(N) = \frac{1}{n!} \sum_{k=0}^{n} \frac{n!}{k! \, (n-k)!} \left[\frac{d^k}{dz^k} (1-z^{n+1})^N \right]_{z=0} \left[\frac{d^{n-k}}{dz^{n-k}} \frac{1}{(1-z)^N} \right]_{z=0} .$$

Using the binomial expansion, we have, for the first factor in square brackets,

$$\frac{d^k}{dz^k} (1-z^{n+1})^N \bigg|_{z=0} = \frac{d^k}{dz^k} \sum_{p=0}^{N} \frac{N!}{p! \, (N-p)!} \, (-)^p \left(z^{n+1} \right)^p \bigg|_{z=0} .$$

However, since k is an integer between 0 and n, this factor vanishes unless p and k are both 0, which yields

$$g_n(N) = \frac{1}{n!} \frac{d^n}{dz^n} \frac{1}{(1-z)^N} \bigg|_{z=0} = \frac{N(N+1)\cdots(N+n-1)}{n!} \frac{1}{(1-z)^{N+n}} \bigg|_{z=0} = \frac{(n+N-1)!}{n! \, (N-1)!} \; .$$

We label the corresponding eigenstates as $|n^{(k)}\rangle$, where the superscript k runs over the degenerate eigenstates; for example, for $n=1$ the degeneracy is N and the eigenstates are

$$|1^{(1)}\rangle = |1,0,0,\ldots,0,0\rangle , \qquad |1^{(2)}\rangle = |0,1,0,\ldots 0,0\rangle , \qquad \ldots , \qquad |1^{(N)}\rangle = |0,0,0,\ldots,0,1\rangle .$$

Obviously, $\hat{H}$ on its own does not form a complete set of commuting observables, since its eigenvalues are degenerate. However, the N single-particle Hamiltonians $\hat{H}_1, \ldots \hat{H}_N$, as well as, for example, $\hat{H}, \hat{H}_1, \ldots \hat{H}_{N-1}$ (or any other set containing $\hat{H}$ and $N-1$ single-particle Hamiltonians), do form a complete set of commuting observables, since the corresponding eigenvalues uniquely identify the state of the system. In the basis $|n_1, \ldots, n_N\rangle$, the orthonormalization and closure relations read, respectively,

$$\langle n_1, \ldots, n_N | n_1', \ldots, n_N'\rangle = \langle n_1 | n_1'\rangle \cdots \langle n_N | n_N'\rangle = \delta_{n_1, n_1'} \cdots \delta_{n_N, n_N'}$$

and

$$\sum_{n_1=0}^{\infty} \cdots \sum_{n_N=0}^{\infty} |n_1, \ldots, n_N\rangle \langle n_1, \ldots, n_N| = \left(\sum_{n_1=0}^{\infty} |n_1\rangle\langle n_1| \right) \otimes \cdots \otimes \left(\sum_{n_N=0}^{\infty} |n_N\rangle\langle n_N| \right) = \hat{1}_1 \otimes \cdots \otimes \hat{1}_N = \hat{1} \,.$$

Note that in the basis $|n^{(k)}\rangle$ we have the corresponding relations

$$\langle m^{(k)} | n^{(l)}\rangle = \delta_{mn}\, \delta_{kl} \,, \qquad \sum_{n=0}^{\infty} \sum_{k=1}^{g_n} |n^{(k)}\rangle \langle n^{(k)}| = \hat{1} \,.$$

Part 2

In the basis $|n^{(k)}\rangle$ the state $|\psi\rangle$ can be written as a linear combination:

$$|\psi\rangle = \frac{1}{2} \left(|0\rangle + |1^{(1)}\rangle + |1^{(2)}\rangle + |1^{(3)}\rangle \right) \,.$$

The probability of measuring the total energy $E_n(N)$ is given by

$$P(E_n) = \sum_{k=1}^{g_n} |\langle n^{(k)} | \psi\rangle|^2 \,,$$

where

$$\langle n^{(k)} | \psi\rangle = \frac{1}{2} \left(\delta_{n,0} + \delta_{n,1}\, \delta_{k,1} + \delta_{n,1}\, \delta_{k,2} + \delta_{n,1}\, \delta_{k,3} \right) \,,$$

from which we conclude that

$$P(E_0) = \frac{1}{4} \,, \qquad P(E_1) = \frac{3}{4} \,.$$

The probability of measuring an energy E_{n_1} for oscillator 1 (regardless of what the energies are for the remaining $N-1$ oscillators) is given by

$$P(E_{n_1}) = \sum_{n_2=0}^{\infty} \cdots \sum_{n_N=0}^{\infty} |\langle n_1, n_2, \ldots, n_N | \psi\rangle|^2 \,,$$

where it is convenient in this case to use the basis $|n_1, n_2, \ldots, n_N\rangle$. We find, for the amplitude in the above expression,

$$\langle n_1, n_2, \ldots, n_N | \psi\rangle = \frac{1}{2} \left(\delta_{n_1,0}\, \delta_{n_2,0}\, \delta_{n_3,0} \cdots \delta_{n_N,0} + \delta_{n_1,1}\, \delta_{n_2,0}\, \delta_{n_3,0} \cdots \delta_{n_N,0} \right.$$
$$\left. + \delta_{n_1,0}\, \delta_{n_2,1}\, \delta_{n_3,0} \cdots \delta_{n_N,0} + \delta_{n_1,0}\, \delta_{n_2,0}\, \delta_{n_3,1} \cdots \delta_{n_N,0} \right) \,,$$

yielding

$$P(\hbar\omega/2) = \frac{3}{4} \,, \qquad P(3\hbar\omega/2) = \frac{1}{4} \,.$$

The probability that measuring the position of particle 1 will give a value that lies in an interval dx_1 centered at x_1 is given by $p(x_1)\,dx_1$, where

$$P(x_1) = \int_{-\infty}^{\infty} dx_2 \cdots dx_N \, |\langle \phi_{x_1,x_2,\ldots,x_N}|\psi\rangle|^2 \,,$$

and the state $|\phi_{x_1,\ldots,x_N}\rangle = |\phi_{x_1}\rangle \otimes \cdots \otimes |\phi_{x_N}\rangle$ is the tensor product of the eigenstates of the position operators $\hat{x}_1, \ldots, \hat{x}_N$. We find that

$$\langle \phi_{x_1,\ldots,x_N}|\psi\rangle = \frac{1}{2}[\psi_0(x_1)\,\psi_0(x_2)\,\psi_0(x_3) + \psi_1(x_1)\,\psi_0(x_2)\,\psi_0(x_3) + \psi_0(x_1)\,\psi_1(x_2)\,\psi_0(x_3)$$

$$+ \psi_0(x_1)\,\psi_0(x_2)\,\psi_1(x_3)]\prod_{i=4}^{N}\psi_0(x_i)\,,$$

where $\psi_n(x)$ are the normalized wave functions of the harmonic oscillator energy eigenstates, that is, $\psi_n(x) = \langle \phi_x|n\rangle$. Squaring the above matrix element, integrating over $x_2, \ldots, x_N$, and using the orthonormality of the wave functions,

$$\int_{-\infty}^{\infty} dx_i \, \psi_m^*(x_i)\,\psi_n(x_i) = \delta_{mn}\,,$$

yields

$$P(x_1) = \frac{3}{4}\,\psi_0^2(x_1) + \frac{1}{4}\,\psi_1^2(x_1) + \frac{1}{2}\,\mathrm{Re}\big[\psi_0^*(x_1)\,\psi_1(x_1)\big]\,.$$

The probability that measuring the momentum of particle 1 will give a value in an interval dp_1 centered at p_1 is $p(p_1)\,dp_1$, where

$$P(p_1) = \int_{-\infty}^{\infty} dp_2 \cdots dp_N \, |\langle \phi_{p_1,\ldots,p_N}|\psi\rangle|^2 \,,$$

and where the state $|\phi_{p_1,\ldots,p_2}\rangle = |\phi_{p_1}\rangle \otimes \cdots \otimes |\phi_{p_N}\rangle$ is the tensor product of the eigenstates of the momentum operators $\hat{p}_1, \ldots \hat{p}_N$. The amplitude reads

$$\langle \phi_{p_1,\ldots,p_N}|\psi\rangle = \frac{1}{2}[\widetilde{\psi}_0(p_1)\,\widetilde{\psi}_0(p_2)\,\widetilde{\psi}_0(p_3) + \widetilde{\psi}_1(p_1)\,\widetilde{\psi}_0(p_2)\,\widetilde{\psi}_0(p_3) + \widetilde{\psi}_0(p_1)\,\widetilde{\psi}_1(p_2)\,\widetilde{\psi}_0(p_3)$$

$$+ \widetilde{\psi}_0(p_1)\,\widetilde{\psi}_0(p_2)\,\widetilde{\psi}_1(p_3)]\prod_{i=4}^{N}\widetilde{\psi}_0(p_i)\,,$$

where $\widetilde{\psi}_n(p)$ are the (normalized) momentum-space wave functions of the eigenstates $|n\rangle$, namely $\widetilde{\psi}_n(p) = \langle \phi_p|n\rangle$. These are related to the wave functions $\psi_n(x)$ via

$$\widetilde{\psi}_n(p) = \langle \phi_p|n\rangle = \int_{-\infty}^{\infty} dx \, \langle \phi_p|\phi_x\rangle\langle \phi_x|n\rangle = \frac{1}{\sqrt{2\pi\hbar}}\int_{-\infty}^{\infty} dx \, e^{-ipx/\hbar}\,\psi_n(x)\,.$$

A calculation identical to that above leads to

$$P(p_1) = \frac{3}{4}\,|\widetilde{\psi}_0(p_1)|^2 + \frac{1}{4}\,|\widetilde{\psi}_1(p_1)|^2 + \frac{1}{2}\,\mathrm{Re}\big[\widetilde{\psi}_0^*(p_1)\widetilde{\psi}_1(p_1)\big]\,.$$

Problem 5 Energy Measurements on Two Uncoupled Harmonic Oscillators

The Hamiltonian of a two-particle system is given by

$$\hat{H} = \hat{H}_1 + \hat{H}_2\,, \qquad \hat{H}_i = \frac{\hat{p}_i^2}{2m} + \frac{m\omega^2}{2}\,\hat{x}_i^2\,,$$

where $\hat{x}_i$ and $\hat{p}_i$ are the position and momentum operator of particle i, respectively. Write down the eigenvalues and eigenstates of $\hat{H}$ (see Problem 4). What is the degeneracy of the eigenvalues? Suppose at time $t = 0$ the system is in the state

$$|\psi(0)\rangle = \frac{1}{2}(|0,0\rangle + |1,0\rangle + |0,1\rangle + |1,1\rangle) \ .$$

1. Calculate the mean values of the energy and position of particle 1, after a measurement of the total energy at $t = 0$ has yielded the result $2\hbar\omega$. After this measurement, the system evolves undisturbed for a time t. If a measurement of the energy of particle 1 is then carried out at time t, what are the possible results and with what probabilities? If, instead of the energy of particle 1, its position is measured, what results could be found and with what probabilities?
2. Suppose that at $t = 0$, instead of the total energy, the energy of particle 2 is measured, yielding the result $\hbar\omega/2$. How do the answers to the questions of part 1 change?

Solution

We label the eigenstates of the total Hamiltonian as $|E_n^{(k)}\rangle$ with $n = 0, 1, \ldots$ and $k = 1, \ldots, g_n$ (here, $g_n = n + 1$ is the degeneracy), namely

$$\hat{H}|E_n^{(k)}\rangle = E_n|E_n^{(k)}\rangle \ , \qquad E_n = \hbar\omega(n + 1) \ , \qquad \langle E_m^{(k)}|E_n^{(l)}\rangle = \delta_{mn}\,\delta_{kl} \ ,$$

and

$$|E_n^{(1)}\rangle = |n,0\rangle \ , \ |E_n^{(2)}\rangle = |n-1,1\rangle \ , \ \ldots \ , \ |E_n^{(g_n)}\rangle = |0,n\rangle \ ;$$

see Problem 4. The initial state is written as

$$|\psi(0)\rangle = \frac{1}{2}(|E_0\rangle + |E_1^{(1)}\rangle + |E_1^{(2)}\rangle + |E_2^{(2)}\rangle) \ ,$$

and note that the ground state $|E_0\rangle$ is non-degenerate.

Part 1

We define the projection operator over the subspace of states with total energy E_n as

$$\hat{P}_n = \sum_{k=1}^{g_n} |E_n^{(k)}\rangle\langle E_n^{(k)}| \ .$$

Since the total energy $E_1 = 2\hbar\omega$ is measured, after the measurement the normalized state of the system is given by

$$|\psi'(0)\rangle = \frac{1}{[\langle\psi(0)|\hat{P}_1|\psi(0)\rangle]^{1/2}}\,\hat{P}_1|\psi(0)\rangle \ .$$

Using the orthonormality of the basis $|E_n^{(k)}\rangle$, it follows that

$$\hat{P}_1|\psi(0)\rangle = \sum_{k=1}^{2} |E_1^{(k)}\rangle\langle E_1^{(k)}|\psi(0)\rangle = \frac{1}{2}(|E_1^{(1)}\rangle + |E_1^{(2)}\rangle)$$

and

$$\langle\psi(0)|\hat{P}_1|\psi(0)\rangle = \langle\psi(0)|\hat{P}_1\hat{P}_1|\psi(0)\rangle = \frac{1}{4}((\langle E_1^{(1)}| + \langle E_1^{(2)}|)(|E_1^{(1)}\rangle + |E_1^{(2)}\rangle)) = \frac{1}{2} \ ,$$

where we have used the projector property $\hat{P}_n^2 = \hat{P}_n$. The normalized state is obtained as

$$|\psi'(0)\rangle = \frac{1}{\sqrt{2}}(|E_1^{(1)}\rangle + |E_1^{(2)}\rangle) = \frac{1}{\sqrt{2}}(|1,0\rangle + |0,1\rangle) \,,$$

and the mean value of $\hat{x}_1$ is

$$\langle\psi'(0)|\hat{x}_1|\psi'(0)\rangle = \frac{1}{2}(\langle 1,0| + \langle 0,1|)\hat{x}_1(|1,0\rangle + |0,1\rangle) = \frac{1}{2}(\langle 1|\hat{x}_1|1\rangle + \langle 0|\hat{x}_1|0\rangle) = 0 \,,$$

where we have used

$$\langle m_1, m_2|\hat{x}_1|n_1, n_2\rangle = (\langle m_1| \otimes \langle m_2|)\hat{x}_1(|n_1\rangle \otimes |n_2\rangle) = \langle m_1|\hat{x}_1|n_1\rangle\langle m_2|n_2\rangle = \langle m_1|\hat{x}_1|n_1\rangle\,\delta_{m_2,n_2} \,,$$

and have noted that expectation values of $\hat{x}_1$ in eigenstates of $\hat{H}_1$ vanish, since $\hat{x}_1 \propto \hat{a}_1 + \hat{a}_1^\dagger$, $\langle m_1|\hat{a}_1|m_1\rangle = 0$, and $\langle m_1|\hat{a}_1^\dagger|m_1\rangle = 0$. For the expectation value of $\hat{H}_1$ we find

$$\langle\psi'(0)|\hat{H}_1|\psi'(0)\rangle = \frac{1}{2}(\underbrace{\langle 1|\hat{H}_1|1\rangle}_{3\hbar\omega/2} + \underbrace{\langle 0|\hat{H}_1|0\rangle}_{\hbar\omega/2}) = \hbar\omega \,.$$

After the total energy measurement, the state evolves according to

$$|\psi'(t)\rangle = e^{-i\hat{H}t/\hbar}|\psi'(0)\rangle = e^{-i\hat{H}t/\hbar}\frac{1}{\sqrt{2}}(|E_1^{(1)}\rangle + |E_1^{(2)}\rangle) = e^{-iE_1 t/\hbar}|\psi'(0)\rangle \,,$$

since $\hat{H}|E_1^{(k)}\rangle = E_1|E_1^{(k)}\rangle$. The probability of measuring energy E_{n_1} for oscillator 1 and energy E_{n_2} for oscillator 2 is given by

$$P_{n_1,n_2}(t) = |\langle n_1, n_2|\psi'(t)\rangle|^2 \,,$$

where $\hat{H}_1|n_1, n_2\rangle = E_{n_1}|n_1, n_2\rangle$ and similarly for $\hat{H}_2$. The probability of measuring energy E_{n_1} for oscillator 1 is then obtained as

$$P_{n_1}(t) = \sum_{n_2=0}^{\infty} P_{n_1,n_2}(t) = \sum_{n_2=0}^{\infty} |\langle n_1, n_2|\psi'(t)\rangle|^2 = \sum_{n_2=0}^{\infty} |\langle n_1, n_2|\psi'(0)\rangle|^2 \,,$$

since $|\psi'(t)\rangle$ differs from $|\psi'(0)\rangle$ only by an overall phase factor. Hence, $P_{n_1}(t)$ is time independent. We find

$$\langle n_1, n_2|\psi'(0)\rangle = \frac{1}{\sqrt{2}}(\delta_{n_1,1}\,\delta_{n_2,0} + \delta_{n_1,0}\,\delta_{n_2,1}) \implies P_{n_1}(t) = P_{n_1}(0) = \frac{1}{2}(\delta_{n_1,1} + \delta_{n_1,0}) \,,$$

and so the energies which can be measured are either $E_{n_1=0} = \hbar\omega/2$ or $E_{n_1=1} = 3\hbar\omega/2$, each with probability 1/2. Considering next the measurement of the position of particle 1, we note that the probability density that particles 1 and 2 are found at positions x_1 and x_2, respectively, is given by

$$P(x_1, x_2; t) = |\langle\phi_{x_1,x_2}|\psi'(t)\rangle|^2 \,,$$

where $|\phi_{x_1,x_2}\rangle$ are the (continuum) eigenstates of the position operators $\hat{x}_1$ and $\hat{x}_2$,

$$|\phi_{x_1,x_2}\rangle = |\phi_{x_1}\rangle \otimes |\phi_{x_2}\rangle \,, \qquad \hat{x}_i|\phi_{x_1,x_2}\rangle = x_i|\phi_{x_1,x_2}\rangle \,.$$

It follows that

$$P(x_1; t) = \int_{-\infty}^{\infty} dx_2\, P(x_1, x_2; t) = \int_{-\infty}^{\infty} dx_2\, |\langle\phi_{x_1,x_2}|\psi'(t)\rangle|^2 = P(x_1; 0)$$

and

$$\langle \phi_{x_1, x_2} | \psi'(0) \rangle = \frac{1}{\sqrt{2}} [\psi_1(x_1)\, \psi_0(x_2) + \psi_0(x_1)\, \psi_1(x_2)] \,,$$

where $\psi_n(x)$ are the orthonormal oscillator wave functions. The probability density $p(x_1; t)$ is simply given by

$$P(x_1; t) = \frac{1}{2} [|\psi_1(x_1)|^2 + |\psi_0(x_1)|^2] \,,$$

where we have used

$$\int_{-\infty}^{\infty} dx_2\, \psi_m^*(x_2)\, \psi_n(x_2) = \delta_{m,n} \,.$$

Part 2

The projection operator onto the eigenstate $|n_1, n_2\rangle$ of $\hat{H}_1$ and $\hat{H}_2$ with eigenvalues E_{n_1} and E_{n_2}, respectively, is given by

$$\hat{P}_{n_1, n_2} = |n_1, n_2\rangle\langle n_1, n_2| \,.$$

Since the energy $E_{n_2=0} = \hbar\omega/2$ is measured at time $t=0$, after the measurement the system is in the state

$$|\phi(0)\rangle = \frac{1}{[\langle \psi(0) | \sum_{n_1} \hat{P}_{n_1, 0} | \psi(0)\rangle]^{1/2}} \sum_{n_1} \hat{P}_{n_1, 0} |\psi(0)\rangle \,,$$

where

$$\sum_{n_1} \hat{P}_{n_1, 0} |\psi(0)\rangle = \sum_{n_1} |n_1, 0\rangle\langle n_1, 0|\psi(0)\rangle = \frac{1}{2}(|0, 0\rangle + |1, 0\rangle) \,;$$

this also follows by noting that generally

$$\sum_{n_1} \hat{P}_{n_1, n2} = \Big(\sum_{n_1} |n_1\rangle\langle n_1| \Big) \otimes |n_2\rangle\langle n_2| = \hat{\mathbb{1}}_1 \otimes |n_2\rangle\langle n_2| \,.$$

The normalized state reads

$$|\phi(0)\rangle = \frac{1}{\sqrt{2}}(|0, 0\rangle + |1, 0\rangle) \,.$$

The mean value of $\hat{x}_1$ follows:

$$\langle \phi(0)|\hat{x}_1|\phi(0)\rangle = \frac{1}{2}(\langle 0, 0| + \langle 1, 0|)\hat{x}_1(|0, 0\rangle + |1, 0\rangle) = \frac{1}{2}(\langle 1|\hat{x}_1|0\rangle + \langle 0|\hat{x}_1|1\rangle) \,,$$

since the expectation value $\langle n_1|\hat{x}_1|n_1\rangle = 0$. To evaluate the matrix element, it is convenient to express $\hat{x}_1$ as

$$\hat{x}_1 = \sqrt{\frac{\hbar}{2m\omega}}(\hat{a}_1 + \hat{a}_1^\dagger) \,,$$

so that

$$\langle \phi(0)|\hat{x}_1|\phi(0)\rangle = \frac{1}{2} \sqrt{\frac{\hbar}{2m\omega}}(\langle 1|\hat{a}_1^\dagger|0\rangle + \langle 0|\hat{a}_1|1\rangle) = \sqrt{\frac{\hbar}{2m\omega}} \,.$$

The mean value of H_1 is obtained as

$$\langle \phi(0)|\hat{H}_1|\phi(0)\rangle = \frac{1}{2}(\langle 0, 0| + \langle 1, 0|)\hat{H}_1(|0, 0\rangle + |1, 0\rangle) = \frac{1}{2}(\langle 0|\hat{H}_1|0\rangle + \langle 1|\hat{H}_1|1\rangle) = \hbar\omega \,.$$

After the measurement of the energy of particle 2, the state evolves according to

$$|\phi(t)\rangle = e^{-i\hat{H}t/\hbar}|\phi(0)\rangle = e^{-i\hat{H}t/\hbar}\frac{1}{\sqrt{2}}(|E_0\rangle + |E_1^{(1)}\rangle)$$

$$= \frac{e^{-iE_0t/\hbar}}{\sqrt{2}}(|E_0\rangle + e^{-i\Delta Et/\hbar}|E_1^{(1)}\rangle) = \frac{e^{-i\omega t}}{\sqrt{2}}(|0,0\rangle + e^{-i\omega t}|1,0\rangle),$$

where $E_0 = \hbar\omega$ and $\Delta E = E_1 - E_0 = \hbar\omega$. The probability of measuring energy E_{n_1} for oscillator 1 is given by

$$P_{n_1}(t) = \sum_{n_2=0}^{\infty} P_{n_1,n_2}(t) = \sum_{n_2=0}^{\infty} |\langle n_1, n_2|\phi(t)\rangle|^2,$$

and

$$\langle n_1, n_2|\phi(t)\rangle = \frac{e^{-i\omega t}}{\sqrt{2}}(\delta_{n_1,0}\delta_{n_2,0} + e^{-i\omega t}\delta_{n_1 1}\delta_{n_2,0}) \implies P_{n_1}(t) = P_{n_1}(0) = \frac{1}{2}(\delta_{n_1,0} + \delta_{n_1,1}),$$

and thus the energies which can be measured are either $E_{n_1=0} = \hbar\omega/2$ or $E_{n_1=1} = 3\hbar\omega/2$, each with probability 1/2. The probability density for oscillator 1 to be at position x_1 and oscillator 2 to be at position x_2 is given by

$$P(x_1, x_2; t) = |\langle\phi_{x_1,x_2}|\phi(t)\rangle|^2,$$

from which it follows that

$$P(x_1; t) = \int_{-\infty}^{\infty} dx_2\, P(x_1, x_2; t) = \int_{-\infty}^{\infty} dx_2\, |\langle\phi_{x_1,x_2}|\phi(t)\rangle|^2$$

and

$$\langle\phi_{x_1,x_2}|\phi(t)\rangle = \frac{e^{-i\omega t}}{\sqrt{2}}[\psi_0(x_1) + e^{-i\omega t}\psi_1(x_1)]\psi_0(x_2).$$

We finally obtain the probability density $P(x_1; t)$ as follows:

$$P(x_1; t) = \frac{1}{2}[|\psi_0(x_1)|^2 + |\psi_1(x_1)|^2 + 2\cos(\omega t)\,\psi_0(x_1)\,\psi_1(x_1)],$$

where we have made use of the fact that the eigenfunctions of the harmonic oscillator can be chosen to be real.

Problem 6 Time Evolution of Position and Momentum Operators in Heisenberg Picture

Consider a one-dimensional harmonic oscillator with Hamiltonian

$$\hat{H} = \frac{\hat{p}^2}{2m} + \frac{m\omega^2\hat{x}^2}{2}.$$

1. Solve Heisenberg's equations of motion for $d\hat{x}_H(t)/dt$ and $d\hat{p}_H(t)/dt$, where $\hat{x}_H(t)$ and $\hat{p}_H(t)$ are the position and momentum operators in the Heisenberg picture,

$$\hat{x}_H(t) = e^{i\hat{H}t/\hbar}\,\hat{x}\,e^{-i\hat{H}t/\hbar},$$

and similarly for $\hat{p}_H(t)$. Denote by $\hat{x}_H(0) = \hat{x}$ and $\hat{p}_H(0) = \hat{p}$ the operators at the initial time $t = 0$; they coincide with the Schrödinger picture operators $\hat{x}$ and $\hat{p}$. It may be convenient first to solve Heisenberg's equations of motion for $\hat{a}_H(t)$ and $\hat{a}_H^\dagger(t)$.

2. Using the above results, compute the expectation values $\langle \hat{x}_H^2(t) \rangle$ and $\langle \hat{p}_H^2(t) \rangle$ on a generic state $|\psi_H\rangle = |\psi(0)\rangle$ in terms of those at the initial time; here, $|\psi(0)\rangle$ is the Schrödinger-picture state at the initial time.

3. Suppose the Schrödinger-picture state $|\psi(0)\rangle$ is represented by a real and even wave function in the coordinate representation. Show that the variances of $\hat{x}_H(t)$ and $\hat{p}_H(t)$ at time t are given by

$$[\Delta x(t)]^2 = \langle \hat{x}^2 \rangle \cos^2(\omega t) + \frac{\langle \hat{p}^2 \rangle}{(m\omega)^2} \sin^2(\omega t)$$

and

$$[\Delta p(t)]^2 = \langle \hat{p}^2 \rangle \cos^2(\omega t) + (m\omega)^2 \langle \hat{x}^2 \rangle \sin^2(\omega t) ,$$

where $\langle \hat{x}^2 \rangle = \langle \psi(0)|\hat{x}^2|\psi(0)\rangle$ and $\langle \hat{p}^2 \rangle = \langle \psi(0)|\hat{p}^2|\psi(0)\rangle$.

Solution

Part 1

Since $\hat{x}$ and $\hat{p}$ are linear combinations of $\hat{a}$ and $\hat{a}^\dagger$, we first work out Heisenberg's equations of motion for $\hat{a}_H(t)$ and its adjoint $\hat{a}_H^\dagger(t)$. We have

$$\frac{d}{dt}\hat{a}_H(t) = \frac{d}{dt}\left[e^{i\hat{H}t/\hbar}\,\hat{a}\,e^{-i\hat{H}t/\hbar} \right] = \frac{i}{\hbar}e^{i\hat{H}t/\hbar}\left[\hat{H},\hat{a}\right]e^{-i\hat{H}t/\hbar} ,$$

with the commutator yielding

$$\left[\hat{H},\hat{a}\right] = \hbar\omega\left[\hat{a}^\dagger\hat{a} + 1/2,\hat{a}\right] = \hbar\omega\,\hat{a}^\dagger\left[\hat{a},\hat{a}\right] + \hbar\omega\left[\hat{a}^\dagger,\hat{a}\right]\hat{a} = -\hbar\omega\,\hat{a} .$$

Therefore, we have

$$\frac{d}{dt}\hat{a}_H(t) = -i\omega\,e^{i\hat{H}t/\hbar}\,\hat{a}\,e^{-i\hat{H}t/\hbar} = -i\omega\,\hat{a}_H(t) \implies \hat{a}_H(t) = e^{-i\omega t}\,\hat{a}_H(0) = e^{-i\omega t}\,\hat{a}$$

and

$$\hat{a}_H^\dagger(t) = e^{i\omega t}\,\hat{a}^\dagger .$$

Recalling that

$$\hat{x} = \sqrt{\frac{\hbar}{2m\omega}}\,(\hat{a} + \hat{a}^\dagger) , \qquad \hat{p} = i\sqrt{\frac{m\hbar\omega}{2}}\,(\hat{a}^\dagger - \hat{a}) ,$$

we find that

$$\hat{x}_H(t) = \sqrt{\frac{\hbar}{2m\omega}}\left[e^{-i\omega t}\,\hat{a} + e^{i\omega t}\,\hat{a}^\dagger \right] = \sqrt{\frac{\hbar}{2m\omega}}\left[\cos(\omega t)\,(\hat{a} + \hat{a}^\dagger) + i\sin(\omega t)\,(\hat{a}^\dagger - \hat{a}) \right]$$

$$= \hat{x}\,\cos(\omega t) + \frac{\hat{p}}{m\omega}\,\sin(\omega t) ,$$

and

$$\hat{p}_H(t) = i\sqrt{\frac{\hbar}{2m\omega}}\left[-e^{-i\omega t}\,\hat{a} + e^{i\omega t}\,\hat{a}^\dagger \right] = i\sqrt{\frac{\hbar}{2m\omega}}\left[\cos(\omega t)\,(\hat{a}^\dagger - \hat{a}) + i\sin(\omega t)\,(\hat{a}^\dagger + \hat{a}) \right]$$

$$= \hat{p}\,\cos(\omega t) - m\omega\hat{x}\,\sin(\omega t) .$$

Part 2

Squaring the Heisenberg-picture operators, we obtain

$$\hat{x}_H^2(t) = \left[\hat{x}\,\cos(\omega t) + \frac{\hat{p}}{m\omega}\,\sin(\omega t)\right]^2 = \hat{x}^2\,\cos^2(\omega t) + \frac{\hat{p}^2}{(m\omega)^2}\,\sin^2(\omega t) + \frac{\hat{x}\hat{p}+\hat{p}\hat{x}}{2m\omega}\,\sin(2\omega t)\,,$$

and similarly

$$\hat{p}_H^2(t) = \hat{p}^2\,\cos^2(\omega t) + (m\omega)^2\,\hat{x}^2\,\sin^2(\omega t) - \frac{m\omega}{2}\,(\hat{x}\hat{p}+\hat{p}\hat{x})\,\sin(2\omega t)\,.$$

Expectation values are as follows:

$$\langle\psi_H|\hat{x}_H^2(t)|\psi_H\rangle = \langle\hat{x}^2\rangle\,\cos^2(\omega t) + \frac{\langle\hat{p}^2\rangle}{(m\omega)^2}\,\sin^2(\omega t) + \frac{\langle\hat{x}\hat{p}\rangle+\langle\hat{p}\hat{x}\rangle}{2m\omega}\,\sin(2\omega t)$$

and

$$\langle\psi_H|\hat{p}_H^2(t)|\psi_H\rangle = \langle\hat{p}^2\rangle\,\cos^2(\omega t) + (m\omega)^2\,\langle\hat{x}^2\rangle\,\sin^2(\omega t) - \frac{m\omega}{2}\,(\langle\hat{x}\hat{p}\rangle+\langle\hat{p}\hat{x}\rangle)\,\sin(2\omega t)\,,$$

where here $\langle\hat{x}^2\rangle$ is short-hand for $\langle\psi(0)|\hat{x}^2|\psi(0)\rangle$, and so on.

Part 3

Since $|\psi(0)\rangle$ is even under parity, the expectation values $\langle\hat{x}\rangle$ and $\langle\hat{p}\rangle$ vanish since the observables $\hat{x}$ and $\hat{p}$ are odd under parity; specifically,

$$\underbrace{\int_{-\infty}^{\infty} dx\,\psi^*(x;0)\,x\,\psi(x;0)}_{\langle x\rangle} = \int_{-\infty}^{\infty} dx\,\psi^*(-x;0)\,x\,\psi(-x;0) = -\underbrace{\int_{-\infty}^{\infty} dx\,\psi^*(x;0)\,x\,\psi(x;0)}_{\langle x\rangle}\,,$$

where in the last step we have replaced x by $-x$; so $\langle\hat{x}\rangle = -\langle\hat{x}\rangle$ and so $\langle\hat{x}\rangle = 0$. A similar argument holds for $\langle\hat{p}\rangle$, showing that $\langle\hat{p}\rangle = 0$. It should be noted that this argument proving that the expectation values of $\hat{x}$ and $\hat{p}$ vanish requires only that $|\psi(0)\rangle$ has definite parity (either even or odd). Of course, we obviously have $\langle\psi_H|\hat{x}_H(t)|\psi_H\rangle = \langle\psi_H|\hat{p}_H(t)|\psi_H\rangle = 0$.

Now, we further assume that the wave function $\psi(x;0)$ is real and show that the expectation value of $\hat{x}\hat{p}+\hat{p}\hat{x}$ vanishes in this case; indeed, consider

$$\langle\hat{p}\hat{x}\rangle = \langle\hat{x}\hat{p}\rangle^* = \left[-i\hbar\int_{-\infty}^{\infty} dx\,\psi(x;0)\,x\,\psi'(x;0)\right]^* = i\hbar\int_{-\infty}^{\infty} dx\,\psi(x;0)\,x\,\psi'(x;0) = -\langle\hat{x}\hat{p}\rangle\,.$$

Therefore, the variance $[\Delta x(t)]^2$ is obtained as

$$[\Delta x(t)]^2 = \langle\hat{x}_H^2(t)\rangle - \langle\hat{x}_H(t)\rangle^2 = \langle\hat{x}^2\rangle\,\cos^2(\omega t) + \frac{\langle\hat{p}^2\rangle}{(m\omega)^2}\,\sin^2(\omega t)\,,$$

and similarly

$$[\Delta p(t)]^2 = \langle\hat{p}^2\rangle\,\cos^2(\omega t) + (m\omega)^2\,\langle\hat{x}^2\rangle\,\sin^2(\omega t)\,.$$

Problem 7   Two Uncoupled Harmonic Oscillators and the Exchange Operator

The Hamiltonian of a two-particle system of uncoupled harmonic oscillators is given by

$$\hat{H} = \hat{H}_1 + \hat{H}_2 , \qquad \hat{H}_i = \frac{\hat{p}_i^2}{2m} + \frac{m\omega^2}{2}\,\hat{x}_i^2 ,$$

where $\hat{x}_i$ and $\hat{p}_i$ are the position and momentum operators for particle i. Let $|n_1, n_2\rangle = |n_1\rangle \otimes |n_2\rangle$ denote the common eigenstates of $\hat{H}_1$ and $\hat{H}_2$, with eigenvalues $\hbar\omega(n_1 + 1/2)$ and $\hbar\omega(n_2 + 1/2)$, respectively.

1. Define the two-particle exchange operator as

$$\hat{P}_e\,|n_1, n_2\rangle = |n_2, n_1\rangle .$$

   Show that $\hat{P}_e$ is both hermitian and unitary, and that its eigenvalues are ± 1. Further, show that under the action of $\hat{P}_e$ the observables $\hat{H}_1$, $\hat{x}_1$, and $\hat{p}_1$ transform as follows:

$$\hat{P}_e\,\hat{H}_1\,\hat{P}_e^\dagger = \hat{H}_2 , \qquad \hat{P}_e\,\hat{x}_1\,\hat{P}_e^\dagger = \hat{x}_2 , \qquad \hat{P}_e\,\hat{p}_1\,\hat{P}_e^\dagger = \hat{p}_2 ,$$

   and vice versa.

2. Since $\hat{H}$ is invariant under the action of $\hat{P}_e$, show that there exists a basis of common eigenstates of $\hat{H}$ and $\hat{P}_e$. Construct such a basis. Do $\hat{H}$ and $\hat{P}_e$ form a complete set of commuting observables?

Solution

Part 1

Using the definition, it follows that, on the basis $|n_1, n_2\rangle$,

$$\hat{P}_e^2\,|n_1, n_2\rangle = \hat{P}_e\,|n_2, n_1\rangle = |n_1, n_2\rangle \implies \hat{P}_e^2 = \hat{\mathbb{1}} ,$$

and therefore, multiplying both sides of the last relation by $\hat{P}_e^{-1}$, we arrive at

$$\hat{P}_e = \hat{P}_e^{-1} .$$

In fact the exchange operator is also hermitian, since

$$\langle n_1', n_2'|\hat{P}_e|n_1, n_2\rangle = \langle n_1', n_2'|n_2, n_1\rangle = \delta_{n_1', n_2}\,\delta_{n_2', n_1}$$

and

$$[\langle n_1, n_2|\hat{P}_e|n_1', n_2'\rangle]^* = [\langle n_1, n_2|n_2', n_1'\rangle]^* = \delta_{n_1, n_2'}\,\delta_{n_2, n_1'} = \langle n_1', n_2'|\hat{P}_e|n_1, n_2\rangle \implies \hat{P}_e^\dagger = \hat{P}_e .$$

Since $\hat{P}_e = \hat{P}_e^{-1}$, we conclude that $\hat{P}_e$ is also unitary: $\hat{P}_e^\dagger = \hat{P}_e^{-1}$. Let $|\psi\rangle$ be an eigenstate of $\hat{P}_e$ with eigenvalue λ, that is, $\hat{P}_e\,|\psi\rangle = \lambda\,|\psi\rangle$; then

$$|\psi\rangle = \hat{P}_e^2|\psi\rangle = \lambda\,\hat{P}_e\,|\psi\rangle = \lambda^2|\psi\rangle \implies \lambda = \pm 1 .$$

On the basis $|n_1, n_2\rangle$, where it is understood that the first and second labels specify the states in the state spaces of, respectively, oscillators 1 and 2, that is,

$$|n_1, n_2\rangle = |n_1\rangle \otimes |n_2\rangle ,$$

we have

$$\langle n_1', n_2'|\hat{P}_e\,\hat{H}_1\,\hat{P}_e^\dagger|n_1, n_2\rangle = \langle n_2', n_1'|\hat{H}_1|n_2, n_1\rangle = E_{n_2}\langle n_2', n_1'|n_2, n_1\rangle = E_{n_2}\,\delta_{n_2', n_2}\,\delta_{n_1', n_1} .$$

On the other hand, we also find

$$\langle n_1', n_2' | \hat{H}_2 | n_1, n_2 \rangle = E_{n_2} \delta_{n_1', n_1} \delta_{n_2', n_2} = \langle n_1', n_2' | \hat{P}_e \hat{H}_1 \hat{P}_e^\dagger | n_1, n_2 \rangle \, ,$$

which implies

$$\hat{P}_e \hat{H}_1 \hat{P}_e^\dagger = \hat{H}_2 \, .$$

To show that $\hat{P}_e \hat{x}_1 \hat{P}_e^\dagger = \hat{x}_2$ and vice versa, we can proceed in a similar way to that above. However, rather than using a basis of eigenstates of $\hat{H}_1$ and $\hat{H}_2$, it is convenient to use the basis $|\phi_{x_1, x_2}\rangle$ of continuum eigenstates of $\hat{x}_1$ and $\hat{x}_2$, namely

$$|\phi_{x_1, x_2}\rangle = |\phi_{x_1}\rangle \otimes |\phi_{x_2}\rangle \, , \qquad \hat{x}_i |\phi_{x_1, x_2}\rangle = x_i |\phi_{x_1, x_2}\rangle \, .$$

The proof is then identical to that above, the only difference being that the Kronecker δs are replaced by δ-functions. In the case of the momentum observable, we use the basis $|\phi_{p_1, p_2}\rangle$ of continuum eigenstates of $\hat{p}_1$ and $\hat{p}_2$, obtaining $\hat{P}_e \hat{p}_1 \hat{P}_e^\dagger = \hat{p}_2$ and vice versa.

Of course, we could also have shown this directly by noting that

$$\langle \phi_{x_1, x_2} | n_1, n_2 \rangle = \psi_{n_1}(x_1) \, \psi_{n_2}(x_2) \, ,$$

and therefore

$$
\begin{aligned}
\langle n_1', n_2' | \hat{P}_e \hat{x}_1 \hat{P}_e^\dagger | n_1, n_2 \rangle = \langle n_2', n_1' | \hat{x}_1 | n_2, n_1 \rangle &= \int_{-\infty}^{\infty} \int_{-\infty}^{\infty} dx_1 dx_2 \, \psi_{n_2'}(x_1) \, \psi_{n_1'}(x_2) \, x_1 \, \psi_{n_2}(x_1) \, \psi_{n_1}(x_2) \\
&= \int_{-\infty}^{\infty} \int_{-\infty}^{\infty} dx_1 dx_2 \, \psi_{n_2'}(x_2) \, \psi_{n_1'}(x_1) \, x_2 \, \psi_{n_2}(x_2) \, \psi_{n_1}(x_1) \\
&= \int_{-\infty}^{\infty} \int_{-\infty}^{\infty} dx_1 dx_2 \, x_2 \, \langle n_1', n_2' | \phi_{x_1, x_2} \rangle \langle \phi_{x_1, x_2} | n_1, n_2 \rangle \\
&= \int_{-\infty}^{\infty} \int_{-\infty}^{\infty} dx_1 dx_2 \, \langle n_1', n_2' | \hat{x}_2 | \phi_{x_1, x_2} \rangle \langle \phi_{x_1, x_2} | n_1, n_2 \rangle = \langle n_1', n_2' | \hat{x}_2 | n_1, n_2 \rangle \, ,
\end{aligned}
$$

where in the second line we made the change of variables $x_1 \rightleftharpoons x_2$ in the integral (the Jacobian is 1), and in the last line we made use of the completeness relation for the position eigenstates.

Part 2

The observable $\hat{H} = \hat{H}_1 + \hat{H}_2$ is obviously invariant under the action of $\hat{P}_e$,

$$\hat{P}_e \hat{H} \hat{P}_e^\dagger = \hat{H} \implies \hat{P}_e \hat{H} = \hat{H} \hat{P}_e \, ,$$

and $\hat{H}$ and $\hat{P}_e$ commute. As a consequence there exists a basis of common eigenstates. To construct such a basis is straightforward:

$$\text{ground state,} \quad |0, 0\rangle$$

$$\text{1st excited states,} \quad \frac{1}{\sqrt{2}}(|1, 0\rangle \pm |0, 1\rangle)$$

$$\text{2nd excited states,} \quad \frac{1}{\sqrt{2}}(|2, 0\rangle \pm |0, 2\rangle) \, , \qquad |1, 1\rangle$$

$$\text{3rd excited states,} \quad \frac{1}{\sqrt{2}}(|3, 0\rangle \pm |0, 3\rangle) \, , \qquad \frac{1}{\sqrt{2}}(|2, 1\rangle \pm |1, 2\rangle) \, ,$$

and so on. Clearly, the states of the type $|n/2, n/2\rangle$ with n even are eigenstates of $\hat{P}_e$ with eigenvalue $+1$; the states $(|n - k, k\rangle \pm |k, n - k\rangle)/\sqrt{2}$ with $k = 0, \ldots, n$ are eigenstates of $\hat{P}_e$ with eigenvalues ± 1. The observables $\hat{H}$ and $\hat{P}_e$ do not form a complete set of commuting observables, since specifying their eigenvalues does not uniquely identify the state; for example, the respective eigenvalues E_2 and $+1$ correspond to the two states

$$\frac{1}{\sqrt{2}}(|2, 0\rangle + |0, 2\rangle), \qquad |1, 1\rangle.$$

Problem 8 Momentum-Space Eigenfunctions of the Harmonic Oscillator

This problem deals with the harmonic oscillator in the momentum representation and, in particular, with the relation between the coordinate- and momentum-representation eigenfunctions of its Hamiltonian.

1. As a preliminary, consider the operator $\hat{y} = \alpha \hat{x}$ with α a real constant, that is, it provides a rescaling of the position operator $\hat{x}$. Denote by $|\phi_x\rangle$ and $|\varphi_y\rangle$ the bases consisting of eigenstates of $\hat{x}$ and $\hat{y}$, respectively, which satisfy the completeness relations

$$\int_{-\infty}^{\infty} dx \, |\phi_x\rangle\langle\phi_x| = \hat{\mathbb{1}}, \qquad \int_{-\infty}^{\infty} dy \, |\varphi_y\rangle\langle\varphi_y| = \hat{\mathbb{1}}.$$

Clearly, we have

$$|\varphi_{\alpha x}\rangle = c |\phi_x\rangle \qquad \text{or} \qquad |\phi_{y/\alpha}\rangle = \frac{1}{c} |\varphi_y\rangle.$$

Exploiting the continuum normalization $\langle\phi_x|\phi_{x'}\rangle = \delta(x - x')$, determine the constant $|c|$. Now, consider the momentum operator $\hat{p}$ and its rescaled version $\hat{q} = \beta \hat{p}$, with β a real constant, and denote by $|\psi_p\rangle$ and $|\chi_q\rangle$ the corresponding bases of eigenstates such that

$$|\chi_{\beta p}\rangle = d|\psi_p\rangle \qquad \text{or} \qquad |\psi_{q/\beta}\rangle = \frac{1}{d} |\chi_q\rangle.$$

Determine $|d|$.

2. By appropriate rescaling of the Hamiltonians in the coordinate and momentum representations, show that the Schrödinger equations satisfied by the coordinate- and momentum-space eigenfunctions can be written as

$$\frac{1}{2}\left(-\frac{d^2}{dy^2} + y^2\right)\xi_n(y) = (n + 1/2)\xi_n(y), \qquad \frac{1}{2}\left(q^2 - \frac{d^2}{dq^2}\right)\widetilde{\xi}_n(q) = (n + 1/2)\widetilde{\xi}_n(q),$$

where

$$\xi_n(y) = \langle\varphi_y|\xi_n\rangle, \qquad \widetilde{\xi}_n(q) = \langle\chi_q|\xi_n\rangle,$$

and $|\xi_n\rangle$ are the harmonic oscillator eigenstates:

$$\left(\frac{\hat{p}^2}{2m} + \frac{m\omega^2}{2}\hat{x}^2\right)|\xi_n\rangle = (n + 1/2)\hbar\omega |\xi_n\rangle.$$

3. Explain why the wave functions $\xi_n(y)$ and $\widetilde{\xi}_n(q)$ can differ by at the most a phase factor, that is,

$$\widetilde{\xi}_n(q) = e^{i\eta_n} \xi_n(y = q).$$

By expressing the relation $a^\dagger |\xi_n\rangle = \sqrt{n+1} |\xi_{n+1}\rangle$ in the (rescaled) position and momentum representations, show that the phase factors satisfy the requirement

$$e^{i\eta_{n+1}} = -i\, e^{i\eta_n} \implies e^{i\eta_n} = (-i)^n\, e^{i\eta_0}\,,$$

where $e^{i\eta_0}$ is the phase factor for the ground state $|\xi_0\rangle$. Using the relation $a|\xi_0\rangle = 0$, show that this phase factor is in fact unity.

4. Show that the wave functions in the coordinate and momentum representations are related via

$$\widetilde{\xi}_n(p) = \frac{(-i)^n}{\sqrt{m\omega}} \, \xi_n[x = p/(m\omega)]\,.$$

Of course, we could have obtained the momentum-space wave function by evaluating the Fourier transform

$$\widetilde{\xi}_n(p) = \int_{-\infty}^{\infty} dx \, \frac{e^{-ipx/\hbar}}{\sqrt{2\pi\hbar}} \, \xi_n(x)\,,$$

but the relation determined above avoids the carrying out of these Fourier transforms and shows that, for example, the probability densities in coordinate and momentum space have the same shape.

Solution

Part 1

We consider

$$\langle \phi_x | \phi_{x'} \rangle = \delta(x - x') = \delta(y/\alpha - y'/\alpha) = |\alpha|\,\delta(y - y') = |\alpha|\,\langle \varphi_y | \varphi_{y'} \rangle\,,$$

from which we deduce $|c| = 1/\sqrt{|\alpha|}$, that is,

$$|\phi_{y/\alpha}\rangle = \sqrt{|\alpha|}\,|\varphi_y\rangle \qquad \text{and} \qquad |\varphi_{\alpha x}\rangle = \frac{1}{\sqrt{|\alpha|}}\,|\phi_x\rangle\,.$$

Proceeding similarly in momentum space, we find

$$\langle \psi_p | \psi_{p'} \rangle = \delta(p - p') = \delta(q/\beta - q'/\beta) = |\beta|\,\delta(q - q') = |\beta|\,\langle \chi_q | \chi_{q'} \rangle\,,$$

and so

$$|d| = 1/\sqrt{|\beta|}\,, \qquad |\psi_{q/\beta}\rangle = \sqrt{|\beta|}\,|\chi_q\rangle\,, \qquad \text{and} \qquad |\chi_{\beta p}\rangle = \frac{1}{\sqrt{|\beta|}}\,|\psi_p\rangle\,.$$

Part 2

In the coordinate representation the Schrödinger equation satisfied by the eigenfunction $\xi_n(x)$ reads (after dividing both sides by $\hbar\omega$)

$$\frac{1}{2}\left(-\frac{\hbar}{m\omega}\frac{d^2}{dx^2} + \frac{m\omega}{\hbar}x^2\right)\xi_n(x) = (n + 1/2)\xi_n(x)\,.$$

We rescale x as follows:

$$y = \underbrace{\sqrt{\frac{m\omega}{\hbar}}}_{\alpha}\, x\,, \qquad \frac{d}{dx} = \frac{dy}{dx}\frac{d}{dy} = \sqrt{\frac{m\omega}{\hbar}}\frac{d}{dy}$$

and hence

$$\frac{1}{2}\left(-\frac{d^2}{dy^2} + y^2\right)\xi_n(y) = (n+1/2)\xi_n(y)\,.$$

Note that the wave functions in the two bases $|\phi_x\rangle$ and $|\varphi_y\rangle$ are related via

$$\xi_n(x) = \langle\phi_x|\xi_n\rangle = \underbrace{\left(\frac{m\omega}{\hbar}\right)^{1/4}}_{\sqrt{|\alpha|}}\langle\varphi_y|\xi_n\rangle = \left(\frac{m\omega}{\hbar}\right)^{1/4}\xi_n(y)\,.$$

Similarly, in the momentum representation the Schrödinger equation is given by (recall that $\hat{x} \longrightarrow i\hbar\,d/dp$)

$$\frac{1}{2}\left(\frac{1}{m\hbar\omega}p^2 - m\hbar\omega\frac{d^2}{dp^2}\right)\widetilde{\xi}_n(p) = (n+1/2)\widetilde{\xi}_n(p)\,,$$

which reduces to

$$\frac{1}{2}\left(q^2 - \frac{d^2}{dq^2}\right)\widetilde{\xi}_n(q) = (n+1/2)\widetilde{\xi}_n(q)\,,$$

after rescaling by

$$q = \underbrace{\frac{1}{\sqrt{m\hbar\omega}}}_{\beta}p\,, \qquad \frac{d}{dp} = \frac{1}{\sqrt{m\hbar\omega}}\frac{d}{dq}\,.$$

The momentum-space wave functions are related via

$$\widetilde{\xi}_n(p) = \langle\psi_p|\xi_n\rangle = \underbrace{(m\hbar\omega)^{-1/4}}_{\sqrt{|\beta|}}\langle\chi_q|\xi_n\rangle = (m\hbar\omega)^{-1/4}\widetilde{\xi}_n(q)\,.$$

Part 3

The wave functions $\xi_n(y)$ and $\widetilde{\xi}_n(q)$ satisfy identical differential equations and are both normalized. Given that the eigenvalues $n+1/2$ are all non-degenerate, we conclude that these wave functions can differ by at most a phase factor. We can determine this phase factor by noting that the raising operator,

$$\hat{a}^\dagger = \frac{1}{\sqrt{2m\hbar\omega}}\left(-i\hat{p} + m\omega\hat{x}\right) = \frac{1}{\sqrt{2}}\left(-\frac{i}{\alpha\hbar}\hat{p} + \alpha\hat{x}\right) \qquad \text{with } \alpha = \sqrt{\frac{m\omega}{\hbar}}$$

$$= \frac{1}{\sqrt{2}}\left(-i\beta\hat{p} + \frac{1}{\beta\hbar}\hat{x}\right) \qquad \text{with } \beta = \frac{1}{\sqrt{m\hbar\omega}}\,,$$

reads in the rescaled position and momentum bases (we insert below the completeness relative to the basis $|\phi_x\rangle$)

$$\langle\varphi_y|\hat{a}^\dagger|\varphi_{y'}\rangle = \int_{-\infty}^{\infty}dx\int_{-\infty}^{\infty}dx'\,\underbrace{\frac{1}{\sqrt{|\alpha|}}\delta(x-y/\alpha)}_{\langle\varphi_y|\phi_x\rangle}\underbrace{\frac{1}{\sqrt{2}}\left(-\frac{1}{\alpha}\frac{d}{dx} + \alpha x\right)\delta(x-x')}_{\langle\phi_x|\hat{a}^\dagger|\phi_{x'}\rangle}\underbrace{\frac{1}{\sqrt{|\alpha|}}\delta(x'-y'/\alpha)}_{\langle\phi_{x'}|\varphi_{y'}\rangle}$$

$$= \frac{1}{|\alpha|}\frac{1}{\sqrt{2}}\left[-\frac{1}{\alpha}\frac{d}{d(y/\alpha)} + \alpha\frac{y}{\alpha}\right]\delta(y/\alpha - y'/\alpha) = \frac{1}{\sqrt{2}}\left(-\frac{d}{dy} + y\right)\delta(y-y')\,,$$

and similarly

$$\langle \chi_q | a^\dagger | \chi_{q'} \rangle = -\frac{i}{\sqrt{2}} \left(-\frac{d}{dq} + q \right) \delta(q - q') \ .$$

Using $|\xi_{n+1}\rangle = (1/\sqrt{n+1}) \, \hat{a}^\dagger \, |\xi_n\rangle$, we have

$$\xi_{n+1}(y) = \frac{1}{\sqrt{2(n+1)}} \left(-\frac{d}{dy} + y \right) \xi_n(y) \ , \qquad \widetilde{\xi}_{n+1}(q) = -\frac{i}{\sqrt{2(n+1)}} \left(-\frac{d}{dq} + q \right) \widetilde{\xi}_n(q) \ ,$$

where the last relation implies that

$$e^{i\eta_{n+1}} \, \xi_{n+1}(y) = -i \, \frac{1}{\sqrt{2(n+1)}} \left(-\frac{d}{dy} + y \right) e^{i\eta_n} \, \xi_n(y) = -i \, e^{i\eta_n} \, \xi_{n+1}(y)$$

and hence

$$e^{i\eta_{n+1}} = -i \, e^{i\eta_n} \implies e^{i\eta_{n+1}} = (-i)(-i) \, e^{i\eta_{n-1}} = \cdots = (-i)^{n+1} \, e^{i\eta_0} \ .$$

Lastly, to obtain the phase factor for the ground state, we use the requirement that $\hat{a} \, |\xi_0\rangle = 0$, with

$$\langle \varphi_y | \hat{a} | \varphi_{y'} \rangle = \frac{1}{\sqrt{2}} \left(\frac{d}{dy} + y \right) \delta(y - y') \ , \qquad \langle \chi_q | \hat{a} | \chi_{q'} \rangle = \frac{i}{\sqrt{2}} \left(\frac{d}{dq} + q \right) \delta(q - q') \ ,$$

which imply that

$$\left(\frac{d}{dy} + y \right) \xi_0(y) = 0 \ , \qquad \left(\frac{d}{dq} + q \right) \widetilde{\xi}_0(q) = 0 \ ,$$

and hence $e^{i\eta_0}$ can be set to unity. We conclude that

$$\widetilde{\xi}_n(q) = (-i)^n \, \xi_n(y = q) \ .$$

Part 4

Using the following relations, obtained earlier,

$$\xi_n(y) = \left(\frac{m\omega}{\hbar} \right)^{-1/4} \xi_n(x) \ , \qquad \widetilde{\xi}_n(p) = (m\hbar\omega)^{-1/4} \, \widetilde{\xi}_n(q) \ ,$$

it follows that

$$\widetilde{\xi}_n(p/\sqrt{m\hbar\omega}) = (m\hbar\omega)^{-1/4} \, \widetilde{\xi}_n(q) = (-i)^n \, (m\hbar\omega)^{-1/4} \, \xi_n(y) = \frac{(-i)^n}{\sqrt{m\omega}} \, \xi_n(\sqrt{m\omega/\hbar} \, x) \ ,$$

which can also be written as

$$\widetilde{\xi}_n(p) = \frac{(-i)^n}{\sqrt{m\omega}} \, \xi_n[x = p/(m\omega)] \ .$$

Problem 9   A Hamiltonian Consisting of a Harmonic Oscillator Plus a δ-Function Potential

Consider the following one-dimensional Hamiltonian:

$$\hat{H} = \frac{\hat{p}^2}{2m} + \frac{m\omega^2 \hat{x}^2}{2} + V_0 \, \delta(\hat{x}) \ ,$$

where V_0 is a real parameter. Note that the potential operator is invariant under parity.

1. Obtain the Schrödinger equation in the momentum representation.
2. Explain why the energies of the odd eigenfunctions, as well as the eigenfunctions themselves, coincide with those of the simple harmonic oscillator Hamiltonian.
3. By expanding the even eigenfunctions $\widetilde{\psi}_E(p)$ of the above Hamiltonian on a basis consisting of eigenfunctions $\widetilde{\xi}_n(p)$ of the harmonic oscillator, show that the energies of these eigenfunctions are given by

$$\underbrace{\sum_{p=0}^{\infty} \frac{[(2p-1)!!]^2}{(2p)!} \frac{1}{\epsilon - 2p - 1/2}}_{f(\epsilon)} = \frac{1}{v_0} \sqrt{\frac{4\pi m\omega}{\hbar}},$$

where

$$\epsilon = \frac{E}{\hbar\omega}, \qquad v_0 = \frac{2m}{\hbar^2} V_0.$$

4. Noting that the function $f(\epsilon)$ defined above can be expressed as

$$f(\epsilon) = \frac{2\sqrt{\pi}}{2\epsilon - 1} \frac{\Gamma(5/4 - \epsilon/2)}{\Gamma(3/4 - \epsilon/2)},$$

involving a ratio of Γ-functions, discuss the type of solutions for ϵ resulting when the parameter v_0 is positive or negative, corresponding to a repulsive or attractive δ-function potential, respectively.

Solution

Part 1

The eigenvalue equation for the Hamiltonian in the basis $|\phi_p\rangle$ of momentum eigenstates can be written as

$$\hat{H}|\psi_E\rangle = E|\psi_E\rangle \implies \int_{-\infty}^{\infty} dp' \, \langle\phi_p|\hat{H}|\phi_{p'}\rangle \, \widetilde{\psi}_E(p') = E\widetilde{\psi}_E(p),$$

where the Hamiltonian matrix elements are given by

$$\langle\phi_p|\hat{H}|\phi_{p'}\rangle = \left[\frac{p^2}{2m} - \frac{m(\hbar\omega)^2}{2} \frac{d^2}{dp^2}\right] \delta(p - p') + \frac{V_0}{2\pi\hbar},$$

and we have used

$$\langle\phi_p|\delta(\hat{x})|\phi_{p'}\rangle = \int_{-\infty}^{\infty} dx \, \underbrace{\langle\phi_p|\delta(\hat{x})|\phi_x\rangle\langle\phi_x|\phi_{p'}\rangle}_{\delta(x)\,\langle\phi_p|\phi_x\rangle\langle\phi_x|\phi_{p'}\rangle} = \int_{-\infty}^{\infty} dx \, \delta(x) \frac{e^{-i(p-p')x/\hbar}}{2\pi\hbar} = \frac{1}{2\pi\hbar}.$$

Therefore, we find for the eigenvalue equation

$$\left[\frac{p^2}{2m} - \frac{m(\hbar\omega)^2}{2} \frac{d^2}{dp^2}\right] \widetilde{\psi}_E(p) + \frac{V_0}{2\pi\hbar} \int_{-\infty}^{\infty} dp \, \widetilde{\psi}_E(p) = E\widetilde{\psi}_E(p).$$

Part 2

Since the Hamiltonian is invariant under parity, the eigenstates are either even or odd under parity (only bound states are possible for the present Hamiltonian). Note that this property of the eigenstates is independent of the representation adopted; for example, using the coordinate representation, we have for an even or odd wave function the general relation $\psi_E(x) = \pm\psi_E(-x)$, and

$$\widetilde{\psi}_E(p) = \int_{-\infty}^{\infty} dx \frac{e^{-ipx/\hbar}}{\sqrt{2\pi\hbar}} \psi_E(x) = \pm \int_{-\infty}^{\infty} dx \frac{e^{-ipx/\hbar}}{\sqrt{2\pi\hbar}} \psi_E(-x) = \pm \int_{-\infty}^{\infty} dx \frac{e^{ipx/\hbar}}{\sqrt{2\pi\hbar}} \psi_E(x) = \pm \widetilde{\psi}(-p) \,,$$

where in the next-to-last step we have replaced x by $-x$. The integral

$$\int_{-\infty}^{\infty} dp \, \widetilde{\psi}_E(p) = 0$$

vanishes for odd $\widetilde{\psi}_E(p)$. We conclude that the odd eigenstates of the harmonic oscillator Hamiltonian are also eigenstates of the present Hamiltonian. This, of course, is immediately obvious in the coordinate representation, since odd wave functions vanish at the origin and are unaffected by the δ-function potential. Thus, in this case we have

$$E \longrightarrow \hbar\omega(2n - 1/2) \,, \qquad \widetilde{\psi}_E(p) \longrightarrow \widetilde{\xi}_{2n-1}(p) \,,$$

where $n = 1, 2, \ldots$ and $\widetilde{\xi}_m(p)$ are the momentum-space harmonic oscillator wave functions (discussed in Problem 8).

Part 3

In the remainder, we are concerned with the even solutions of the eigenvalue problem. First, we note that

$$\int_{-\infty}^{\infty} dp \, \widetilde{\psi}_E(p) = \sqrt{2\pi\hbar} \int_{-\infty}^{\infty} dp \frac{e^{ipx/\hbar}}{\sqrt{2\pi\hbar}} \widetilde{\psi}_E(p) \bigg|_{x=0} = \sqrt{2\pi\hbar} \, \psi_E(0) \,,$$

where $\psi_E(0)$ is the coordinate-space wave function evaluated at the origin. Second, it is convenient to expand $\widetilde{\psi}_E(p)$ on the basis of harmonic oscillator momentum-space wave functions $\widetilde{\xi}_m(p)$:

$$\widetilde{\psi}_E(p) = \sum_{n=0}^{\infty} c_{2n} \widetilde{\xi}_{2n}(p) \,, \qquad c_{2n} = \int_{-\infty}^{\infty} dp \, \widetilde{\xi}_{2n}^*(p) \, \widetilde{\psi}_E(p) \,,$$

where only the even basis functions occur in the expansion. Inserting this expansion into the eigenvalue equation, we arrive at

$$\sum_{n=0}^{\infty} c_{2n} \left[E - \hbar\omega(2n + 1/2) \right] \widetilde{\xi}_{2n}(p) = \underbrace{\frac{V_0}{\sqrt{2\pi\hbar}} \psi_E(0)}_{C} \,.$$

Multiplying both sides by $\widetilde{\xi}_{2m}^*(p)$ and integrating over p yields

$$c_{2m} \left[E - \hbar\omega(2m + 1/2) \right] = C \int_{-\infty}^{\infty} dp \, \widetilde{\xi}_{2m}^*(p) = \sqrt{2\pi\hbar} C \xi_{2m}(0) \,,$$

and have used the fact that the coordinate-space $\xi_m(x)$ are real. We find the solution

$$c_{2m} = \frac{V_0 \psi_E(0)}{E - \hbar\omega(2m + 1/2)} \xi_{2m}(0) \,, \qquad m = 0, 1, 2, \ldots$$

Now, consider

$$\psi_E(0) = \sum_{p=0}^{\infty} c_{2p} \xi_{2p}(0) = V_0 \psi_E(0) \sum_{p=0}^{\infty} \frac{\xi_{2p}^2(0)}{E - \hbar\omega(2p + 1/2)} \,;$$

the eigenenergies result from

$$\sum_{p=0}^{\infty} \frac{\xi_{2p}^2(0)}{E - \hbar\omega(2p + 1/2)} = \frac{1}{V_0} .$$

The squared value of the harmonic oscillator coordinate-space wave function at the origin is given by[2]

$$\xi_{2p}^2(0) = \sqrt{\frac{m\omega}{\pi\hbar}} \frac{[(2p-1)!!]^2}{(2p)!} ,$$

where we define $(-1)!! \equiv 1$ corresponding to $p = 0$. Expressing

$$E = \hbar\omega\epsilon , \qquad V_0 = \frac{\hbar^2}{2m} v_0 ,$$

the eigenvalue equation can be written as

$$\underbrace{\sum_{p=0}^{\infty} \frac{[(2p-1)!!]^2}{(2p)!} \frac{1}{\epsilon - 2p - 1/2}}_{f(\epsilon)} = \frac{1}{v_0} \sqrt{\frac{4\pi m\omega}{\hbar}} ,$$

and the left-hand side defines a function $f(\epsilon)$. This function has singularities (simple poles) at $\epsilon_p = 2p + 1/2$ with $p = 0, 1, 2, \ldots$

Part 4

A Mathematica program gives for the function $f(\epsilon)$ defined above the result

$$f(\epsilon) = \frac{2\sqrt{\pi}}{2\epsilon - 1} \frac{\Gamma(5/4 - \epsilon/2)}{\Gamma(3/4 - \epsilon/2)} = -\frac{\sqrt{\pi}}{2} \frac{\Gamma(1/4 - \epsilon/2)}{\Gamma(3/4 - \epsilon/2)} ,$$

where in the last step we have used the property $\Gamma(z + 1) = z\Gamma(z)$, so that $\Gamma(5/4 - \epsilon/2) = (1/4 - \epsilon/2)\Gamma(1/4 - \epsilon/2)$. The eigenvalue equation reduces to

$$g(\epsilon) \equiv \frac{\Gamma(1/4 - \epsilon/2)}{\Gamma(3/4 - \epsilon/2)} = \underbrace{-\frac{4}{v_0} \sqrt{\frac{m\omega}{\hbar}}}_{\epsilon^\star}$$

The Γ-function $\Gamma(z)$ has simple poles at $z = 0, -1, -2, \ldots$; as a consequence, for $\epsilon > 0$ the function $g(\epsilon)$ has zeros at

$$\frac{3}{4} - \frac{\epsilon}{2} = -n \implies \epsilon_n = 2n + \frac{3}{2} ,$$

and simple poles at

$$\frac{1}{4} - \frac{\epsilon}{2} = -n \implies \epsilon_n = 2n + \frac{1}{2} ,$$

---

[2] The normalized harmonic oscillator wave function evaluated at the origin is given by $\xi_{2p}(0) = c_{2p} h_{2p}(0)$ with

$$c_{2p} = \left[\frac{1}{2^{2p}(2p)!} \right]^{1/2} \left(\frac{m\omega}{\pi\hbar} \right)^{1/4} ,$$

and the Hermite polynomial satisfies $h_{2p+2}(0) = -2(2p + 1)h_{2p}(0)$, yielding $h_{2p}(0) = (-2)^p(2p - 1)!!$.

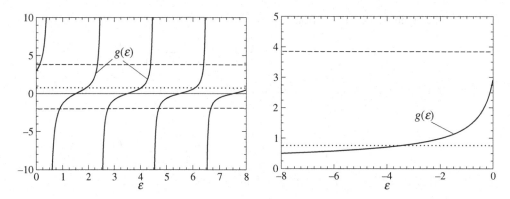

Fig. 8.1 Graphical solution of the eigenvalue equation: the regions $\epsilon > 0$ (left-hand panel) and $\epsilon < 0$ (right-hand panel).

where n is a non-negative integer (the poles correspond to the singularities seen in the left-hand panel of Fig. 8.1). For negative ϵ, the function $g(\epsilon)$, given by

$$g(\epsilon \leq 0) = \frac{\Gamma(1/4 + |\epsilon|/2)}{\Gamma(3/4 + |\epsilon|/2)},$$

is a monotonically decreasing function of $|\epsilon|$ (see the right-hand panel of Fig. 8.1), which vanishes as $\sqrt{2/|\epsilon|}$ in the limit $|\epsilon| \longrightarrow \infty$.

For a repulsive δ-function potential ($v_0 > 0$), the horizontal line $-\epsilon^\star < 0$ (shown as the lower dashed line in the left-hand panel of Fig. 8.1) intersects $g(\epsilon)$ at ϵ values larger than those at which the poles $\epsilon_n = 1/2, 5/2, 9/2, \ldots$ occur, for any $\epsilon^\star$ or, equivalently, any $v_0 > 0$. There is a one-to-one correspondence between these energy eigenvalues and those of the simple harmonic oscillator for even n.

For an attractive δ-function potential ($v_0 < 0$), the cases $\epsilon > 0$ and $\epsilon < 0$ need to be considered separately (bound states may also exist for $\epsilon < 0$). We define $g(0) = -\epsilon_0^\star$, yielding

$$\epsilon_0^\star = -\frac{\Gamma(1/4)}{\Gamma(3/4)} \implies v_0^\star = -4\frac{\Gamma(3/4)}{\Gamma(1/4)}\sqrt{\frac{m\omega}{\hbar}}.$$

If $0 > v_0 > v_0^\star$ then solutions exist only for $\epsilon > 0$, and the horizontal line $-\epsilon^\star > 0$ (shown in the left-hand panel of Fig. 8.1 by the upper dashed line) intersects $g(\epsilon)$ at ϵ values smaller than $\epsilon_n = 1/2, 5/2, 9/2, \ldots$. Again, there is a one-to-one correspondence between these energy eigenvalues and those of the simple harmonic oscillator for even n. By contrast, if $v_0 < v_0^\star$ then there is a single energy eigenvalue less than zero (the intersection of the dotted and solid lines in the right-hand panel of Fig. 8.1), in addition to energy eigenvalues occurring at ϵ values smaller than $\epsilon_n = 5/2, 9/2, 13/2, \ldots$. Note that in this case $\epsilon_0 = 1/2$, corresponding to the ground-state energy $\hbar\omega/2$ of the simple harmonic oscillator, has migrated to negative energy.

Problem 10 Coherent States

This problem deals with a special class of harmonic oscillator states: that of coherent states. These states are defined as the eigenstates of the lowering operator $\hat{a}$,

$$\hat{a}|\chi_\alpha\rangle = \alpha|\chi_\alpha\rangle,$$

where α is a generally complex number; note that, since $\hat{a}$ is not hermitian, its eigenvalues need not be real.[3]

1. Consider the operator $\hat{U}_\alpha$ defined as

$$\hat{U}_\alpha = e^{\alpha\,\hat{a}^\dagger - \alpha^*\hat{a}}\,,$$

where α is a complex constant, and $\hat{a}$ and $\hat{a}^\dagger$ are the lowering and raising operators. Show that $\hat{U}_\alpha$ is unitary.

2. Using the formula (see Problem 25 in Chapter 6 for a derivation)

$$e^{\lambda\hat{B}}\,\hat{A}\,e^{-\lambda\hat{B}} = \hat{A} + \lambda\,[\hat{B},\hat{A}] + \frac{\lambda^2}{2!}\,[\hat{B},[\hat{B},\hat{A}]] + \cdots + \frac{\lambda^n}{n!}\,[\hat{B},[\hat{B},\cdots[\hat{B},\hat{A}]\cdots]] + \cdots,$$

where $\hat{A}$ and $\hat{B}$ are two generic operators, show that

$$\hat{U}_\alpha^\dagger\,\hat{a}\,\hat{U}_\alpha = \hat{a} + \alpha\,\hat{\mathbb{1}}\,.$$

Deduce from this that

$$\hat{a}\,\hat{U}_\alpha|\chi_0\rangle = \alpha\,\hat{U}_\alpha\,|0\rangle\,,$$

which implies that the eigenstates of $\hat{a}$ are of the form $|\chi_a\rangle = \hat{U}_\alpha\,|0\rangle$ with α a complex number. Here $|0\rangle$ is the ground state of the harmonic oscillator Hamiltonian. Thus, $\hat{a}$ has a continuous spectrum. Are these eigenstates normalized to one? If they are, why is this surprising?

3. For two generic operators $\hat{A}$ and $\hat{B}$, each of which commutes with the operator resulting from $[\hat{A},\hat{B}]$, the following relation holds (see Problem 25 in Chapter 6):

$$e^{\hat{A}}\,e^{\hat{B}} = e^{\hat{A}+\hat{B}+[\hat{A},\hat{B}]/2}\,.$$

Express $\hat{U}_\alpha$ as the product of two exponential operators. Using this result, show that

$$|\chi_a\rangle = e^{-|\alpha|^2/2}\,e^{\alpha\hat{a}^\dagger}\,|0\rangle\,.$$

4. Show that if $|\chi_\alpha\rangle$ and $|\chi_\beta\rangle$ are two eigenstates of $\hat{a}$; then their scalar product is given by

$$\langle\chi_\beta|\chi_\alpha\rangle = e^{-|\alpha|^2/2 - |\beta|^2/2 + \alpha\beta^*}\,,$$

that is, the eigenstates are not orthogonal to each other (which is not surprising, since $\hat{a}$ is not hermitian).

   Hint: Use twice the operator identity in part 3 to move $e^{\beta^*\hat{a}}$ to the right of $e^{\alpha\hat{a}^\dagger}$.

5. Show that $|\chi_\alpha\rangle$ can be expanded as

$$|\chi_\alpha\rangle = e^{-|\alpha|^2/2}\sum_{n=0}^{\infty}\frac{\alpha^n}{\sqrt{n!}}\,|n\rangle$$

---

[3] Generally, only a normal operator $\hat{A}$, which by definition commutes with its adjoint, that is, $[\hat{A},\hat{A}^\dagger]=0$, is guaranteed to be diagonalizable. In particular, hermitian and unitary operators are examples of normal (and hence diagonalizable) operators. An operator that is not normal may not be diagonalizable. As an example, consider an operator $\hat{A}$ that in a two-dimensional state space is represented by a matrix,

$$\underline{A} = \begin{pmatrix} 0 & \alpha \\ 0 & 0 \end{pmatrix}\,,$$

where α is a complex number. This matrix has a single (two-fold degenerate) eigenvalue equal to zero; however, two independent eigenstates corresponding to this eigenvalue do not exist.

in the orthonormal basis of the harmonic oscillator eigenstates. Calculate the probability that the state $|\chi_\alpha\rangle$ has energy E_n. What is the average energy?

6. By integrating over the real and imaginary parts of α, show that the states $|\chi_\alpha\rangle$ satisfy the following completeness relation:

$$\frac{1}{\pi} \int_{-\infty}^{\infty} d\alpha_R \int_{-\infty}^{\infty} d\alpha_I \, |\chi_\alpha\rangle\langle\chi_\alpha| = \hat{\mathbb{1}} \, ,$$

where $\alpha = \alpha_R + i\alpha_I$.

   Hint: Insert expansions in terms of Hamiltonian eigenstates and use polar coordinates, setting $\alpha_R = \rho \cos\phi$ and $\alpha_I = \rho \sin\phi$.

7. Show that the raising operator has no normalizable eigenstates.

   Hint: Assume there is an eigenstate of $\hat{a}^\dagger$ such that $\hat{a}^\dagger |\varphi\rangle = \gamma |\varphi\rangle$; expand $|\varphi\rangle$ in the basis of Hamiltonian eigenstates, and obtain a recursion relation for the expansion coefficients from the eigenvalue equation . . . .

8. Recalling that

$$\hat{x} = \sqrt{\frac{\hbar}{2m\omega}} \, (\hat{a}^\dagger + \hat{a}) \, , \qquad \hat{p} = i\sqrt{\frac{m\hbar\omega}{2}} \, (\hat{a}^\dagger - \hat{a}) \, ,$$

calculate the expectation values of $\hat{x}$ and $\hat{x}^2$, and those of $\hat{p}$ and $\hat{p}^2$, for the state $|\chi_\alpha\rangle$; what is $\Delta x \, \Delta p$ for these states?

9. Obtain the wave function of the state $|\chi_\alpha\rangle$ and show that it can be written in the following way:

$$\chi_\alpha(x) = \langle\phi_x|\chi_\alpha\rangle = N \exp\left[-\frac{(x - \langle\hat{x}\rangle)^2}{4(\Delta x)^2} + \frac{i}{\hbar} \langle\hat{p}\rangle x \right] \, ,$$

where N is a normalization constant and $\langle\hat{x}\rangle$ and $\langle\hat{p}\rangle$ are the expectation values calculated above.

Solution

Part 1

Define $\hat{B} = \alpha\hat{a}^\dagger - \alpha^*\hat{a}$, so that $\hat{U}_\alpha = e^{\hat{B}}$. We then have

$$\hat{B}^\dagger = (\alpha \, \hat{a}^\dagger - \alpha^* \hat{a})^\dagger = \alpha^* \hat{a} - \alpha \, \hat{a}^\dagger = -\hat{B} \implies \hat{U}_\alpha^\dagger = e^{-\hat{B}} \, ,$$

and $\hat{U}_\alpha^\dagger \hat{U}_\alpha = e^{-\hat{B}} e^{\hat{B}} = \hat{\mathbb{1}}$.

Part 2

We first evaluate the commutator

$$[\hat{B}, \, \hat{a}] = \alpha[\hat{a}^\dagger, \, \hat{a}] - \alpha^*[\hat{a}, \, \hat{a}] = -\alpha \, \hat{\mathbb{1}} \, ,$$

and, since this commutator is proportional to the identity operator, it follows that higher-order nested commutators all vanish,

$$[\hat{B}, \, [\hat{B}, \, \hat{a}]] = \cdots = [\hat{B}, \, [\hat{B}, \, \cdots \, [\hat{B}, \, \hat{a}] \cdots]] = \cdots = 0 \, .$$

We conclude that (with $\lambda = -1$)

$$\hat{U}_\alpha^\dagger \, \hat{a} \, \hat{U}_\alpha = e^{-\hat{B}} \, \hat{a} \, e^{\hat{B}} = \hat{a} - [\hat{B}, \, \hat{a}] = \hat{a} + \alpha \, \hat{\mathbb{1}} \, .$$

On multiplying both sides on the left by $\hat{U}_\alpha$ and using the unitarity of $\hat{U}_\alpha$ we obtain that

$$\hat{U}_\alpha \, \hat{U}_\alpha^\dagger \, \hat{a} \, \hat{U}_\alpha = \hat{a} \, \hat{U}_\alpha = \hat{U}_\alpha(\hat{a} + \alpha \, \hat{\mathbb{1}}) = \hat{U}_\alpha \, \hat{a} + \alpha \, \hat{U}_\alpha \,.$$

Let $|\chi_\alpha\rangle$ be an eigenstate of $\hat{a}$ with eigenvalue α (generally a complex number); then, $\hat{U}_\beta \, |\chi_\alpha\rangle$ is an eigenstate of $\hat{a}$ with eigenvalue $\alpha + \beta$ since

$$\hat{a} \, \hat{U}_\beta \, |\chi_\alpha\rangle = (\hat{U}_\beta \, \hat{a} + \beta \, \hat{U}_\beta) \, |\chi_\alpha\rangle = (\alpha + \beta) \, \hat{U}_\beta \, |\chi_\alpha\rangle \,.$$

Thus we can generate all the eigenstates by starting with the state $|\chi_0\rangle$ corresponding to $\alpha = 0$ and applying $\hat{U}_\alpha$ to it, that is, $|\chi_\alpha\rangle = \hat{U}_\alpha \, |\chi_0\rangle$, where α is any complex number. In particular, since $|\chi_0\rangle$ is an eigenstate of $\hat{a}$ with zero eigenvalue, we have

$$\hat{a} \, |\chi_0\rangle = 0 \implies |\chi_0\rangle = e^{i\delta} \, |0\rangle \,,$$

and so, up to an irrelevant phase factor (which we set to 1), the state $|\chi_0\rangle$ coincides with the ground state of the harmonic oscillator Hamiltonian, assumed to be normalized to 1. We see that the eigenvalues of $\hat{a}$ form a continuum (the complex numbers). The unitarity of $\hat{U}_\alpha$ ensures that $|\chi_\alpha\rangle = \hat{U}_\alpha \, |0\rangle$ is also normalized to 1. Here, then, is an example of an operator with a continuum spectrum, which nevertheless has normalizable eigenstates (rather than eigenstates obeying a continuum normalization condition).

Part 3

Define

$$\hat{A} = \alpha \, \hat{a}^\dagger \,, \qquad \hat{B} = -\alpha^* \, \hat{a} \,, \qquad [\hat{A}, \, \hat{B}] = -\alpha \alpha^* \, [\hat{a}^\dagger, \, \hat{a}] = |\alpha|^2 \, \hat{\mathbb{1}} \,,$$

yielding

$$\hat{U}_\alpha = e^{\hat{A} + \hat{B} + [\hat{A}, \hat{B}]/2 - [\hat{A}, \hat{B}]/2} = e^{\hat{A} + \hat{B} + [\hat{A}, \hat{B}]/2} \, e^{-[\hat{A}, \hat{B}]/2} = e^{-|\alpha|^2/2} \, e^{\hat{A}} \, e^{\hat{B}} = e^{-|\alpha|^2/2} \, e^{\alpha \, \hat{a}^\dagger} \, e^{-\alpha^* \, \hat{a}} \,,$$

where in the second step we have used the fact $[\hat{A}, \, \hat{B}]$ is proportional to the identity and hence for any operator $\hat{C}$ we have $e^{\hat{C} + \gamma \hat{\mathbb{1}}} = e^{\gamma \hat{\mathbb{1}}} \, e^{\hat{C}} = e^{\hat{C}} \, e^{\gamma \hat{\mathbb{1}}} = e^\gamma \, e^{\hat{C}}$ (and the identity operator is understood). Using $\hat{a} \, |0\rangle = 0$, we see that

$$e^{-\alpha^* \hat{a}} \, |0\rangle = \sum_{n=0}^\infty \frac{(-\alpha^*)^n}{n!} \, \hat{a}^n \, |0\rangle = |0\rangle \,,$$

since in the series only the term with $n = 0$ is left. We conclude that

$$|\chi_\alpha\rangle = \hat{U}_\alpha \, |0\rangle = e^{-|\alpha|^2/2} \, e^{\alpha \, \hat{a}^\dagger} \, |0\rangle \,.$$

Part 4

We find that

$$\langle \chi_\beta | \chi_\alpha \rangle = e^{-|\alpha|^2/2 - |\beta|^2/2} \, \langle 0| \, e^{\beta^* \hat{a}} e^{\alpha \hat{a}^\dagger} \, |0\rangle \,,$$

and the evaluation of the matrix element is straightforward if $e^{\beta^* \, \hat{a}}$ can be passed to the right of $e^{\alpha \, \hat{a}^\dagger}$, since then

$$e^{\beta^* \hat{a}} \, |0\rangle = |0\rangle \,, \qquad \langle 0| e^{\alpha \, \hat{a}^\dagger} = \left(e^{\alpha^* \hat{a}} \, |0\rangle \right)^\dagger = \langle 0| \implies \langle 0| e^{\alpha \, \hat{a}^\dagger} e^{\beta^* \hat{a}} \, |0\rangle = \langle 0|0\rangle = 1 \,.$$

This can be achieved by using the operator identity in part 3. Setting

$$\hat{A} = \beta^* \, \hat{a}, \qquad \hat{B} = \alpha \, \hat{a}^\dagger,$$

we have

$$e^{\hat{A}} \, e^{\hat{B}} = e^{\hat{A}+\hat{B}} \, e^{[\hat{A},\hat{B}]/2},$$

where we used the fact that the commutator is proportional to the identity; indeed,

$$[\hat{A}, \hat{B}] = \alpha \beta^*.$$

Under the exchange $\hat{A} \rightleftharpoons \hat{B}$ the operator identity reads

$$e^{\hat{B}} \, e^{\hat{A}} = e^{\hat{B}+\hat{A}+[\hat{B},\hat{A}]/2} = e^{\hat{A}+\hat{B}-[\hat{A},\hat{B}]/2} \implies e^{\hat{B}} \, e^{\hat{A}} = e^{\hat{A}} \, e^{\hat{B}} \, e^{-[\hat{A},\hat{B}]}$$

or

$$e^{\beta^* \hat{a}} \, e^{\alpha \, \hat{a}^\dagger} = e^{\alpha \, \hat{a}^\dagger} \, e^{\beta^* \hat{a}} \, e^{\alpha \beta^*},$$

and hence

$$\langle \chi_\beta | \chi_\alpha \rangle = e^{-|\alpha|^2/2 - |\beta|^2/2} \, \langle 0 | e^{\beta^* \hat{a}} \, e^{\alpha \hat{a}^\dagger} | 0 \rangle = e^{-|\alpha|^2/2 - |\beta|^2/2 + \alpha\beta^*} \, \langle 0 | e^{\alpha \hat{a}^\dagger} \, e^{\beta^* \hat{a}} | 0 \rangle = e^{-|\alpha|^2/2 - |\beta|^2/2 + \alpha\beta^*};$$

in particular

$$|\langle \chi_\beta | \chi_\alpha \rangle|^2 = e^{-|\alpha|^2 - |\beta|^2 + \alpha\beta^* + \alpha^*\beta} = e^{-|\alpha-\beta|^2}.$$

Part 5

We find that

$$|\chi_\alpha\rangle = e^{-|\alpha|^2/2} \, e^{\alpha \hat{a}^\dagger} | 0 \rangle = e^{-|\alpha|^2/2} \sum_{n=0}^{\infty} \frac{\alpha^n}{n!} \, \hat{a}^{\dagger n} | 0 \rangle = e^{-|\alpha|^2/2} \sum_{n=0}^{\infty} \frac{\alpha^n}{\sqrt{n!}} | n \rangle,$$

where the last step follows since

$$|n\rangle = \frac{\hat{a}^{\dagger n}}{\sqrt{n!}} | 0 \rangle.$$

The probability that $|\chi_\alpha\rangle$ has energy $E_m = (m + 1/2)\hbar\omega$ is given by

$$p(E_m) = |\langle m | \chi_\alpha \rangle|^2 = e^{-|\alpha|^2} \frac{|\alpha|^{2m}}{m!},$$

and the average energy is obtained as

$$\langle \chi_\alpha | \hat{H} | \chi_\alpha \rangle = \sum_{m=0}^{\infty} E_m p(E_m) = e^{-|\alpha|^2} \hbar\omega \sum_{m=0}^{\infty} \left(m + \frac{1}{2} \right) \frac{|\alpha|^{2m}}{m!}.$$

The series is easily summed as follows:

$$\sum_{m=0}^{\infty} \left(m + \frac{1}{2} \right) \frac{|\alpha|^{2m}}{m!} = \sum_{m=1}^{\infty} \frac{(|\alpha|^2)^m}{(m-1)!} + \frac{1}{2} \sum_{m=0}^{\infty} \frac{(|\alpha|^2)^m}{m!} = \left(|\alpha|^2 + \frac{1}{2} \right) e^{|\alpha|^2},$$

where we have used the relations

$$\sum_{m=1}^{\infty} \frac{(|\alpha|^2)^m}{(m-1)!} = |\alpha|^2 \sum_{m=1}^{\infty} \frac{(|\alpha|^2)^{m-1}}{(m-1)!} = |\alpha|^2 \, e^{|\alpha|^2}.$$

We finally obtain

$$\langle \chi_\alpha | \hat{H} | \chi_\alpha \rangle = \hbar \omega \left(|\alpha|^2 + \frac{1}{2} \right) .$$

Part 6

We introduce plane polar coordinates by setting $\alpha_R = \rho \cos \theta$ and $\alpha_I = \rho \sin \theta$, and hence $\alpha = \rho\, e^{i\theta}$. We then have, inserting the expansions in terms of oscillator states,

$$\frac{1}{\pi} \int_{-\infty}^{\infty} d\alpha_R \int_{-\infty}^{\infty} d\alpha_I |\chi_\alpha\rangle\langle\chi_\alpha| = \frac{1}{\pi} \int_0^{\infty} d\rho\, \rho \int_0^{2\pi} d\theta\, e^{-|\alpha|^2} \sum_{m,n=0}^{\infty} \frac{\alpha^m}{\sqrt{m!}} |m\rangle\langle n| \frac{(\alpha^*)^n}{\sqrt{n!}}$$

$$= \frac{1}{\pi} \int_0^{\infty} d\rho\, \rho \int_0^{2\pi} d\theta\, e^{-\rho^2} \sum_{m,n=0}^{\infty} \rho^{m+n} \frac{e^{i(m-n)\theta}}{\sqrt{m!\, n!}} |m\rangle\langle n| .$$

The integration over angles is as follows:

$$\frac{1}{\pi} \int_0^{2\pi} d\theta\, e^{i(m-n)\theta} = 2\,\delta_{m,n} ,$$

since, for $m \neq n$,

$$\int_0^{2\pi} d\theta\, e^{i(m-n)\theta} = \left. \frac{e^{i(m-n)\theta}}{i(m-n)} \right|_0^{2\pi} = 0 .$$

We are left with

$$\frac{1}{\pi} \int_{-\infty}^{\infty} d\alpha_R \int_{-\infty}^{\infty} d\alpha_I |\chi_\alpha\rangle\langle\chi_\alpha| = 2 \sum_{n=0}^{\infty} \frac{|n\rangle\langle n|}{n!} \int_0^{\infty} d\rho\, e^{-\rho^2} \rho^{2n+1} .$$

The last task is evaluation of the ρ-integral. We note that

$$2 \int_0^{\infty} d\rho\, e^{-\lambda\rho^2} \rho^{2n+1} = (-1)^n\, 2 \frac{d^n}{d\lambda^n} \int_0^{\infty} d\rho\, \rho\, e^{-\lambda\rho^2} = (-1)^n\, 2 \frac{d^n}{d\lambda^n} \left(-\frac{e^{-\lambda\rho^2}}{2\lambda} \right) \Big|_0^{\infty}$$

$$= (-1)^n \frac{d^n}{d\lambda^n} \frac{1}{\lambda} = \frac{n!}{\lambda^{n+1}} .$$

Inserting this result (evaluated at $\lambda = 1$) into the previous expansion we finally arrive at

$$\frac{1}{\pi} \int_{-\infty}^{\infty} d\alpha_R \int_{-\infty}^{\infty} d\alpha_I |\chi_\alpha\rangle\langle\chi_\alpha| = \sum_{n=0}^{\infty} |n\rangle\langle n| = \hat{\mathbb{1}} ,$$

and the states $|\chi_\alpha\rangle$ form a complete but non-orthogonal set of states.

Part 7

Suppose that $|\varphi\rangle$ is an eigenstate of $\hat{a}^\dagger$ with eigenvalue $\gamma \neq 0$,

$$\hat{a}^\dagger |\varphi\rangle = \gamma |\varphi\rangle .$$

Expand $|\varphi\rangle$ in the basis of oscillator eigenstates:

$$|\varphi\rangle = \sum_{n=0}^{\infty} c_n |n\rangle , \qquad c_n = \langle n|\varphi\rangle ,$$

and insert it into the eigenvalue equation above, to find

$$\sum_{n=0}^{\infty} c_n \, \hat{a}^\dagger \, |n\rangle = \sum_{n=0}^{\infty} c_n \, \sqrt{n+1} \, |n+1\rangle = \gamma \sum_{n=0}^{\infty} c_n \, |n\rangle \, .$$

Note that for $\gamma = 0$ the only possible solution is $c_m = 0$ for any m, that is, the null state. Projecting the previous equation onto state $|m\rangle$ yields the recursion relation

$$\sum_{n=0}^{\infty} c_n \, \sqrt{n+1} \, \underbrace{\langle m|n+1\rangle}_{\delta_{m,n+1}} = \gamma \sum_{n=0}^{\infty} c_n \, \underbrace{\langle m|n\rangle}_{\delta_{m,n}} \implies c_{m-1} \, \sqrt{m} = \gamma c_m \, ,$$

which can be easily solved as follows:

$$c_1 = \frac{\sqrt{1}}{\gamma} \, c_0 \, , \qquad c_2 = \frac{\sqrt{2}}{\gamma} \, c_1 = \frac{\sqrt{2}}{\gamma} \frac{\sqrt{1}}{\gamma} \, c_0 \, , \, \dots \, ;$$

the general coefficient is given by

$$c_m = \frac{\sqrt{m!}}{\gamma^m} \, c_0 \, .$$

It then follows that

$$\langle \varphi | \varphi \rangle = |c_0|^2 \sum_{m=0}^{\infty} \frac{m!}{|\gamma|^{2m}} \implies \text{divergent series} \, ,$$

since, denoting the generic term as $a_m = m! \, / |\gamma|^{2m}$, we have

$$\lim_{m \to \infty} \frac{a_{m+1}}{a_m} = \lim_{m \to \infty} \frac{m+1}{|\gamma|^2} = \infty \, .$$

Therefore the eigenstates of $\hat{a}^\dagger$ are not normalizable.

Part 8

Given that $|\chi_\alpha\rangle$ are eigenstates of $\hat{a}$, we have

$$\hat{a} \, |\chi_\alpha\rangle = \alpha |\chi_\alpha\rangle \, , \qquad \langle \chi_\alpha| \, \hat{a}^\dagger = \alpha^* \langle \chi_\alpha| \, ;$$

we thus obtain

$$\langle \hat{x} \rangle = \sqrt{\frac{\hbar}{2m\omega}} \, \langle \chi_\alpha| \hat{a}^\dagger + \hat{a} |\chi_\alpha\rangle = \sqrt{\frac{\hbar}{2m\omega}} \, (\alpha^* + \alpha)$$

and

$$\langle \hat{p} \rangle = i \sqrt{\frac{m\hbar\omega}{2}} \, \langle \chi_\alpha| \hat{a}^\dagger - \hat{a} |\chi_\alpha\rangle = i \sqrt{\frac{m\hbar\omega}{2}} \, (\alpha^* - \alpha) \, .$$

The operator $\hat{x}^2$ is given by

$$\hat{x}^2 = \frac{\hbar}{2m\omega} (\hat{a}^\dagger + \hat{a})^2 = \frac{\hbar}{2m\omega} (\hat{a}^{\dagger 2} + \hat{a}^2 + \hat{a}^\dagger \, \hat{a} + \hat{a} \, \hat{a}^\dagger) = \frac{\hbar}{2m\omega} (\hat{a}^{\dagger 2} + \hat{a}^2 + 2\hat{a}^\dagger \, \hat{a} + \hat{\mathbb{1}}) \, ,$$

where in the last step we have moved the raising operator to the left of the lowering operator by making use of the commutation relation, that is, $\hat{a} \, \hat{a}^\dagger = [\hat{a}, \, \hat{a}^\dagger] + \hat{a}^\dagger \, \hat{a} = \hat{a}^\dagger \, \hat{a} + \hat{\mathbb{1}}$ (this operation is known as "normal ordering"; a normal-ordered product of $\hat{a}$ and $\hat{a}^\dagger$ is a product in which all the $\hat{a}^\dagger$ are to the left of the $\hat{a}$). Similarly, we write

$$\hat{p}^2 = -\frac{m\hbar\omega}{2}(\hat{a}^{\dagger 2} + \hat{a}^2 - 2\,\hat{a}^\dagger\,\hat{a} - \hat{1}).$$

Using

$$\hat{a}^2\,|\chi_\alpha\rangle = \alpha^2|\chi_\alpha\rangle\,, \qquad \langle\chi_\alpha|\,\hat{a}^{\dagger 2} = \alpha^{*2}\langle\chi_\alpha|\,,$$

we find

$$\langle\hat{x}^2\rangle = \frac{\hbar}{2m\omega}(\alpha^{*2} + \alpha^2 + 2\alpha^*\alpha + 1)\,, \qquad \langle\hat{p}^2\rangle = -\frac{m\hbar\omega}{2}(\alpha^{*2} + \alpha^2 - 2\alpha^*\alpha - 1)\,.$$

The variances are thereby obtained:

$$(\Delta x)^2 = \langle\hat{x}^2\rangle - (\langle\hat{x}\rangle)^2 = \frac{\hbar}{2m\omega}\left[\alpha^{*2} + \alpha^2 + 2\alpha^*\alpha + 1 - (\alpha^* + \alpha)^2\right] = \frac{\hbar}{2m\omega}\,,$$

$$(\Delta p)^2 = \langle\hat{p}^2\rangle - (\langle\hat{p}\rangle)^2 = \frac{m\hbar\omega}{2}\left[-\alpha^{*2} - \alpha^2 + 2\,\alpha^*\,\alpha + 1 + (\alpha^* - \alpha)^2\right] = \frac{m\hbar\omega}{2}\,.$$

Hence

$$\Delta x\,\Delta p = \frac{\hbar}{2}\,.$$

The states $|\chi_\alpha\rangle$ have minimum uncertainty in the coordinate and momentum observables.

Part 9

Project the eigenvalue equation for $\hat{a}$ onto the position eigenstates:

$$\langle\phi_x|\hat{a}|\chi_\alpha\rangle = \alpha\,\langle\phi_x|\chi_\alpha\rangle \implies \int_{-\infty}^{\infty} dx'\,\langle\phi_x|\hat{a}|\phi_{x'}\rangle\langle\phi_{x'}|\chi_\alpha\rangle = \alpha\,\langle\phi_x|\chi_\alpha\rangle\,.$$

Recalling that

$$\hat{a} = \frac{1}{\sqrt{2m\hbar\omega}}\,(i\hat{p} + m\omega\,\hat{x})\,,$$

and inserting the x-representation of $\hat{x}$ and $\hat{p}$, we find

$$\frac{1}{\sqrt{2m\hbar\omega}}\int_{-\infty}^{\infty} dx'\,\left(\hbar\frac{d}{dx} + m\omega x\right)\delta(x - x')\chi_\alpha(x') = \alpha\chi_\alpha(x)\,.$$

Integrating out the δ-function, we find the first-order differential equation

$$\chi'_\alpha(x) + \frac{m\omega}{\hbar}x\chi_\alpha(x) = \sqrt{\frac{2m\omega}{\hbar}}\,\alpha\chi_\alpha(x)\,,$$

which has the solution

$$\chi_\alpha(x) = N\exp\left(\sqrt{\frac{2m\omega}{\hbar}}\,\alpha\,x - \frac{m\omega}{2\hbar}\,x^2\right)\,.$$

Note that

$$\alpha = \frac{\alpha + \alpha^*}{2} + \frac{\alpha - \alpha^*}{2} = \sqrt{\frac{m\omega}{2\hbar}}\,\langle\hat{x}\rangle + \frac{i}{\sqrt{2m\hbar\omega}}\,\langle\hat{p}\rangle$$

and hence

$$\sqrt{\frac{2m\omega}{\hbar}}\,\alpha x = \frac{m\omega}{\hbar}\,\langle x\rangle x + i\,\frac{\langle\hat{p}\rangle}{\hbar}\,x \qquad \text{and} \qquad \frac{m\omega}{2\hbar} = \frac{1}{4(\Delta x)^2}\,,$$

so that

$$\chi_\alpha(x) = N\exp\left[\frac{m\omega}{\hbar}\,\langle x\rangle x + i\,\frac{\langle\hat{p}\rangle}{\hbar}\,x - \frac{x^2}{4(\Delta x)^2}\right]\,.$$

As a last step, we introduce two constants a and b:

$$\frac{m\omega}{\hbar}\,\langle\hat{x}\rangle x - \frac{x^2}{4(\Delta x)^2} = -\frac{(x - a\langle x\rangle)^2}{4(\Delta x)^2} + b \implies b = \frac{a^2}{4}\,\frac{\langle x\rangle^2}{(\Delta x)^2} \qquad \text{and} \qquad a = \frac{2m\omega}{\hbar}\,(\Delta x)^2 = 1\,,$$

which then yields

$$\chi_\alpha(x) = \underbrace{N\exp\left[\frac{\langle x\rangle^2}{4(\Delta x)^2}\right]}_{\text{normalization constant}}\exp\left[-\frac{(x - \langle\hat{x}\rangle)^2}{4(\Delta x)^2} + \frac{i}{\hbar}\,\langle\hat{p}\rangle x\right]\,,$$

where N is a normalization constant, and $\langle\hat{x}\rangle$ and $\langle\hat{p}\rangle$ are the expectation values calculated above. The wave function has the form of a "minimum" wave packet.

Problem 11 Model of a One-Dimensional Crystal

This problem deals with the small oscillations of a one-dimensional system consisting of N particles, each of mass m and each connected to its nearest neighbors by springs with spring constant k. The unstretched springs have length a and the equilibrium positions x_l^0 of the particles are $x_l^0 = la$ with $l = 1, 2, \ldots, N$. Springs connect the first and last particle to the points $x_0^0 = 0$ and $x_{N+1}^0 = (N + 1)a$. This is a simple model for a one-dimensional crystal of length $L = (N + 1)a$.

1. Introduce displacements from equilibrium as generalized coordinates $\eta_l(t) = x_l(t) - x_l^0$, and show that the classical Lagrangian for this system can be written as (the time dependence is understood)

$$L = T - V = \frac{m}{2}\sum_{l=1}^{N}\dot{\eta}_l^2 - \frac{k}{2}\sum_{l=1}^{N+1}(\eta_l - \eta_{l-1})^2\,,$$

where T and V are, respectively, the kinetic and potential energies, and we define

$$\eta_0 = 0 = \eta_{N+1}\,.$$

Obtain the classical equations of motion for this system via the Euler–Lagrange equations

$$\frac{d}{dt}\frac{\partial L}{\partial\dot{\eta}_k} - \frac{\partial L}{\partial\eta_k} = 0\,, \qquad k = 1, 2, \ldots, N\,.$$

2. By positing that the solutions are of the form (here, $x_k^0 = ka$)

$$\eta_k(t) = \text{Re}\left[A_+\,e^{i(qx_k^0 - \omega t)} + A_-\,e^{i(-qx_k^0 - \omega t)}\right]\,,$$

namely a superposition of traveling waves with wave numbers q and $-q$ and angular frequency ω, substitute them into the equations of motion derived above and determine the relation between ω and q, the dispersion relation. By imposing the conditions $\eta_0(t) = 0 = \eta_{N+1}(t)$ determine the

N allowed values of q_n and the corresponding ω_n. Show that the general solutions $\eta_k(t)$ can be written as the superposition

$$\eta_k(t) = \sum_{n=1}^{N} v_k^{(n)} A_n \cos(\omega_n t + \varphi_n) ,$$

where the A_n and φ_n with $n = 1, 2, \ldots, N$ are $2N$ real constants fixed by the initial conditions $\eta_k(0)$ and $\dot{\eta}_k(0)$, and

$$v_k^{(n)} = \alpha_n \sin(q_n a k) ,$$

with α_n a real constant to be determined below.

3. Introduce the vectors

$$\underline{\eta} = \begin{pmatrix} \eta_1 \\ \eta_2 \\ \vdots \\ \eta_N \end{pmatrix}, \qquad \underline{\dot{\eta}} = \begin{pmatrix} \dot{\eta}_1 \\ \dot{\eta}_2 \\ \vdots \\ \dot{\eta}_N \end{pmatrix},$$

and express the Lagrangian as

$$L = \frac{1}{2} \underline{\dot{\eta}}^T \underline{M} \underline{\dot{\eta}} - \frac{1}{2} \underline{\eta}^T \underline{V} \underline{\eta} .$$

Construct the matrices $\underline{M}$ and $\underline{V}$. Show that the vectors defined in part 2,

$$\underline{v}^{(n)} = \alpha_n \begin{pmatrix} \sin(q_n a) \\ \sin(2 q_n a) \\ \sin(3 q_n a) \\ \vdots \\ \sin(N q_n a) \end{pmatrix}, \qquad n = 1, 2, \ldots, N,$$

are in fact eigenvectors of both $\underline{M}$ and $\underline{V}$, and determine the eigenvalues.

4. Show by explicit calculation that the scalar products $\underline{v}^{(m)\,T} \underline{v}^{(n)}$ vanish for $m \neq n$, and that the normalization constant $\alpha_n = \sqrt{2/(N+1)}$. Write down the matrix $\underline{S}$ that diagonalizes both $\underline{M}$ and $\underline{V}$. The following results may be useful:

$$\cos \alpha = \mathrm{Re}\left(e^{\pm i\alpha}\right) , \qquad \sum_{n=1}^{N} z^n = \frac{1 - z^{N+1}}{1 - z} - 1, \quad \text{with } z \text{ complex} .$$

5. Introduce a new set of generalized coordinates via

$$\underline{\eta}(t) = \underline{S}\, \underline{\xi}(t), \qquad \text{where} \qquad \eta_k(t) = \sum_{l=1}^{N} S_{kl}\, \xi_l(t) ,$$

and solve the equations of motion for the $\xi_k(t)$ with $k = 1, 2, \ldots, N$.

6. Quantize the system by requiring that $[\hat{\eta}_l, \hat{p}_m] = i\hbar\, \delta_{lm}$, where $\hat{p}_l$ is the momentum conjugate to η_l. Introduce a new set of position and momentum operators via the transformation

$$\hat{\xi}_l = \sum_{lm} \left(\underline{S}^T\right)_{lm} \hat{\eta}_m , \qquad \hat{q}_l = \sum_{lm} \left(\underline{S}^T\right)_{lm} \hat{p}_m ,$$

and show that

$$\left[\hat{\xi}_k, \hat{q}_l\right] = i\hbar\, \delta_{kl} .$$

Express the Hamiltonian in terms of $\hat{\xi}_k$ and $\hat{q}_k$, and write down its eigenstates and eigenvalues.

7. Knowing that, for a system held in equilibrium at temperature T, the probability that it will be in a state of energy E is proportional to $e^{-\beta E}$ with $\beta^{-1} = k_B T$ (k_B is Boltzmann's constant), show that the average energy of the system at temperature T is given by

$$\langle E \rangle_T = \sum_{k=1}^{N} \left(\frac{\hbar \omega_k}{2} + \frac{\hbar \omega_k}{e^{\beta \hbar \omega_k} - 1} \right) .$$

Consider the thermodynamic limit (that is, the limit in which $N, L \longrightarrow \infty$ with the spacing a held fixed), and show that in the low-temperature limit the heat capacity c_V (that is, the derivative of $\langle E \rangle_T$ with respect to T) is linear in T. Does the high-temperature limit for c_V conform to expectations?

Solution

Part 1

The equilibrium positions are $x_l^0 = l a$ and the displacements from these equilibrium positions are $\eta_l = x_l - x_l^0$. The kinetic energy of the system is given by

$$T = \frac{m}{2} \sum_{l=1}^{N} \dot{\eta}_l^2 ,$$

since $\dot{x}_l = \dot{\eta}_l$, while the potential energy follows as

$$V = \frac{k}{2} (x_1 - x_0^0 - a)^2 + \frac{k}{2} (x_2 - x_1 - a)^2 + \cdots + \frac{k}{2} (x_{N-1} - x_N - a)^2 + \frac{k}{2} (x_{N+1}^0 - x_N - a)^2 ,$$

and a is the length of the unstretched spring. In terms of the displacements, V reads

$$V = \frac{k}{2} \left[\eta_1^2 + \sum_{l=2}^{N} (\eta_l - \eta_{l-1})^2 + \eta_N^2 \right] ,$$

which can also be written as

$$V = \frac{k}{2} \sum_{l=1}^{N+1} (\eta_l - \eta_{l-1})^2 \qquad \text{with} \ \ \eta_0 = 0 = \eta_{N+1} .$$

The Lagrangian of the system is then obtained as

$$L = \frac{m}{2} \sum_{l=1}^{N} \dot{\eta}_l^2 - \frac{k}{2} \sum_{l=1}^{N+1} (\eta_l - \eta_{l-1})^2 ,$$

leading to the equations of motion

$$\frac{d}{dt} \frac{\partial L}{\partial \dot{\eta}_k} = m \ddot{\eta}_k , \qquad \frac{\partial L}{\partial \eta_k} = -k \sum_{l} (\eta_l - \eta_{l-1}) (\delta_{l,k} - \delta_{l-1,k}) = -k (2\eta_k - \eta_{k-1} - \eta_{k+1}) .$$

Combining these results, we arrive at

$$\ddot{\eta}_k(t) = -\omega_0^2 [2\eta_k(t) - \eta_{k-1}(t) - \eta_{k+1}(t)] \qquad \text{with} \qquad \eta_0(t) = 0 = \eta_{N+1}(t) \ \text{and} \ \omega_0 = \sqrt{\frac{k}{m}} ,$$

a set of coupled differential equations.

As an aside, we note that the above set of differential equations also describes the transverse (as opposed to longitudinal) oscillations of N identical masses, equally spaced, on a stretched massless string. Let T be the uniform tension in the string; then, we have

$$m\ddot{\eta}_k = T\sin\theta - T\sin\phi \approx T(\theta - \phi) \,,$$

where η_k now denotes the *transverse* displacement of the mass located at x_k^0, $\sin\theta \approx \theta \approx (\eta_{k-1} - \eta_k)/a$, and $\sin\phi \approx \phi \approx (\eta_k - \eta_{k+1})/a$. We find that

$$\ddot{\eta}_k(t) = -\frac{T}{ma}\left[2\eta_k(t) - \eta_{k-1}(t) - \eta_{k+1}(t)\right]\,.$$

It is easier to visualize the transverse oscillations of the string than the longitudinal oscillations of the spring-connected masses.

Part 2

We substitute the ansatz

$$\eta_k(t) = \mathrm{Re}\left[A\,e^{i(qx_k^0 - \omega t)}\right]\,,$$

where the wave number q and angular frequency ω are as yet unknown and $x_k^0 = ka$, into the equations of motion, to obtain

$$-\omega^2 A\,e^{i(qx_k^0 - \omega t)} = -\omega_0^2 A\,e^{i(qx_k^0 - \omega t)}\left[2 - e^{iq(x_{k-1}^0 - x_k^0)} - e^{iq(x_{k+1}^0 - x_k^0)}\right]\,,$$

and the real part can be taken at the end (the differential equations are linear). Therefore, we have the dispersion relation

$$\omega^2 = \omega_0^2\left(2 - e^{-iqa} - e^{iqa}\right) = 4\omega_0^2\sin^2(qa/2)\,.$$

In order to determine q, we impose the boundary conditions $\eta_0(t) = 0 = \eta_{N+1}(t)$. To this end first note that, since the dispersion relation is even under q, the solution consists of a superposition of traveling waves:

$$\eta_k(t) = \left[A_+\,e^{iqx_k^0} + A_-\,e^{-iqx_k^0}\right]e^{-i\omega t}\,,$$

and hence

$$\eta_0(t) = 0 \implies (A_+ + A_-)\,e^{-i\omega t} = 0 \implies A_- = -A_+$$

and

$$\eta_{N+1}(t) = 0 \implies \left[A_+\,e^{iq(N+1)a} + A_-\,e^{-iq(N+1)a}\right]e^{-i\omega t} = 0 \implies \sin[qa(N+1)] = 0\,,$$

this last relation yielding the allowed values q_n,

$$q_n = \frac{n}{N+1}\frac{\pi}{a} \qquad n = 1, 2, \ldots, N\,,$$

with corresponding eigenfrequencies

$$\omega_n = 2\omega_0\sin\left(\frac{n}{N+1}\frac{\pi}{2}\right)\,.$$

We have just determined N independent solutions, and the general solution for $\eta_k(t)$ is obtained by linear superposition as follows:

$$\eta_k(t) = \mathrm{Re}\left[\sum_{n=1}^{N} C_n\left(e^{iq_n x_k^0} - e^{-iq_n x_k^0}\right)e^{-i\omega_n t}\right]\,,$$

where the generally complex constants $C_n = |C_n|\, e^{i\varphi_n}$ are determined by the $2N$ initial conditions on $\eta_k(0)$ and $\dot{\eta}_k(0)$. After the redefinition of the constants $2|C_n| \longrightarrow A_n$ and $\varphi_n + \pi/2 \longrightarrow -\varphi_n$, the solution above can be simply written as

$$\eta_k(t) = \sum_{n=1}^{N} A_n \, \sin\left(aq_n k\right) \, \cos(\omega_n t + \varphi_n) = \sum_{n=1}^{N} v_k^{(n)} \, A_n \, \cos(\omega_n t + \varphi_n) \,,$$

where in the last step we introduced

$$v_k^{(n)} = \alpha_n \, \sin(q_n a\, k)$$

and have relabeled A_n/α_n as A_n. It can be easily verified that the above $\eta_k(t)$ satisfy the equations of motion, by direct substitution.

Part 3

We introduce the vectors

$$\underline{\eta} = \begin{pmatrix} \eta_1 \\ \eta_2 \\ \vdots \\ \eta_N \end{pmatrix}, \qquad \underline{\dot{\eta}} = \begin{pmatrix} \dot{\eta}_1 \\ \dot{\eta}_2 \\ \vdots \\ \dot{\eta}_N \end{pmatrix},$$

and express the Lagrangian as

$$L = \frac{1}{2}\, \underline{\dot{\eta}}^T \underline{M}\, \underline{\dot{\eta}} - \frac{1}{2}\, \underline{\eta}^T \underline{V} \underline{\eta} \,,$$

where the matrices $\underline{M}$ and $\underline{V}$ are given by

$$\underline{M} = \begin{pmatrix} m & 0 & 0 & \cdots & 0 \\ 0 & m & 0 & \cdots & 0 \\ \vdots & \vdots & \vdots & & \vdots \\ 0 & 0 & 0 & \cdots & m \end{pmatrix} = m\,\underline{I}\,, \qquad \underline{V} = \begin{pmatrix} 2k & -k & 0 & 0 & 0 & \cdots & 0 & 0 \\ -k & 2k & -k & 0 & 0 & \cdots & 0 & 0 \\ 0 & -k & 2k & -k & 0 & \cdots & 0 & 0 \\ \vdots & \vdots & \vdots & \vdots & \vdots & & \vdots & \vdots \\ 0 & 0 & 0 & 0 & 0 & \cdots & -k & 2k \end{pmatrix}.$$

We see that $\underline{M}$ is proportional to the identity matrix, while $\underline{V}$ has vanishing matrix elements except for those on the diagonal (all equal to twice the spring constant) and to the immediate left and right of the diagonal (all equal to minus the spring constant). Note that $\underline{V}$ is a real symmetric matrix, and as such it can be diagonalized. Indeed, the eigenvalues are $m\omega_n^2$ with corresponding eigenvectors

$$\underline{v}^{(n)} = \alpha_n \begin{pmatrix} \sin(q_n a) \\ \sin(2q_n a) \\ \sin(3q_n a) \\ \vdots \\ \sin(Nq_n a) \end{pmatrix}, \qquad n = 1, 2, \ldots, N,$$

where the $v_k^{(n)}$ is the coefficient in the solution $\eta_k(t)$ obtained in part 2. This is easily verified,

$$\underline{V}\underline{v}^{(n)} = k\alpha_n \begin{pmatrix} 2\sin(q_n a) - \sin(2q_n a) \\ -\sin(q_n a) + 2\sin(2q_n a) - \sin(3q_n a) \\ \vdots \\ -\sin[(l-1)q_n a] + 2\sin(lq_n a) - \sin[(l+1)q_n a] \\ \vdots \\ 2\sin(Nq_n a) - \sin[(N-1)q_n a] \end{pmatrix} = \underbrace{k[2 - \cos(q_n a)]}_{m\omega_n^2}\,\underline{v}^{(n)}\,,$$

since, using $\sin(x \pm q_n a) = \sin x \cos(q_n a) \pm \cos x \sin(q_n a)$ with $x = lq_n a$,

$$2\sin(q_n a) - \sin(2q_n a) = [2 - 2\cos(q_n a)]\sin(q_n a)$$

$$\vdots$$

$$-\sin[(l-1)q_n a] + 2\sin(lq_n a) - \sin[(l+1)q_n a] = [2 - 2\cos(q_n a)]\sin(lq_n a)$$

$$\vdots$$

where the last component also gives

$$2\sin(Nq_n a) - \sin[(N-1)q_n a] = 2\sin(Nq_n a) - \sin[(N-1)q_n a] - \underbrace{\sin[(N+1)q_n a]}_{\sin(n\pi)=0}$$

$$= [2 - 2\cos(q_n a)]\sin(Nq_n a)\,;$$

in the first line we have added 0 as $\sin[(N+1)q_n a]$.

Part 4

Consider $m \neq n$ and then the scalar product of $\underline{v}^{(m)}$ and $\underline{v}^{(n)}$ vanishes; indeed, we have

$$\underline{v}^{(m)\,T}\underline{v}^{(n)} \propto \sum_{k=1}^{N} \sin(q_m a k)\sin(q_n a\, k) \propto \mathrm{Re}\sum_{k=1}^{N}\left[e^{i(q_m - q_n)ak} - e^{i(q_m + q_n)ak}\right]\,,$$

where we have set $2\sin(q_m ak)\sin(q_n ak) = \cos[(q_m - q_n)ak] - \cos[(q_m + q_n)ak]$. The sums can be carried out explicitly (they are geometric sums) to obtain

$$\underline{v}^{(m)\,T}\underline{v}^{(n)} \propto \mathrm{Re}\left[\frac{1 - e^{i(q_m - q_n)a(N+1)}}{1 - e^{i(q_m - q_n)a}} - \frac{1 - e^{i(q_m + q_n)a(N+1)}}{1 - e^{i(q_m + q_n)a}}\right]$$

$$\propto \mathrm{Re}\left[\frac{1 - (-1)^{m-n}}{1 - e^{i(q_m - q_n)a}} - \frac{1 - (-1)^{m+n}}{1 - e^{i(q_m + q_n)a}}\right]\,,$$

and it is now clear that if m and n are both either even or odd the right-hand side is zero, since $(-1)^{m-n} = 1 = (-1)^{m+n}$. On the other hand, if m is even and n is odd, or vice versa, we note that

$$\frac{1 - (-1)^{m+n}}{1 - e^{i(q_m \mp q_n)a}} = \frac{2\,e^{-i(q_m \mp q_n)a/2}}{e^{-i(q_m \mp q_n)a/2} - e^{i(q_m \mp q_n)a/2}} = \frac{i\cos[(q_m \mp q_n)a/2] + \sin[(q_m \mp q_n)a/2]}{\sin[(q_m \mp q_n)a/2]}\,,$$

and so

$$\underline{v}^{(m)\,T}\underline{v}^{(n)} \propto \mathrm{Re}\left[i\cot[(q_m - q_n)a/2] + 1 - i\cot[(q_m + q_n)a/2] - 1\right] = 0\,.$$

The normalization constant α_n follows:

$$\underline{v}^{(n)T}\underline{v}^{(n)} = \alpha_n^2 \sum_{k=1}^{N} \sin^2(q_n a k) = \frac{\alpha_n^2}{2} \sum_{k=1}^{N} [1 - \cos(2q_n a k)]$$

$$= \frac{\alpha_n^2}{2} \left(N - \text{Re} \sum_{k=1}^{N} e^{2iq_n a k} \right) = \frac{\alpha_n^2}{2} (N+1) \,,$$

since

$$\sum_{k=1}^{N} e^{2iq_n a k} = \sum_{k=0}^{N} e^{2iq_n a k} - 1 = \frac{1 - e^{2iq_n a(N+1)}}{1 - e^{2iq_n a}} - 1 = \frac{1 - e^{2in\pi}}{1 - e^{2in\pi/(N+1)}} - 1 = -1 \,.$$

The normalized eigenvectors are then given by

$$\underline{v}^{(n)} = \sqrt{\frac{2}{N+1}} \begin{pmatrix} \sin(q_n a) \\ \sin(2q_n a) \\ \sin(3q_n a) \\ \vdots \\ \sin(Nq_n a) \end{pmatrix} \,, \qquad n = 1, 2, \ldots, N \,,$$

and the orthogonal matrix – modal matrix – that diagonalizes $\underline{V}$ is obtained as

$$\underline{S} = \left[\underline{v}^{(1)} \ \underline{v}^{(2)} \ \cdots \ \underline{v}^{(N)} \right] = \sqrt{\frac{2}{N+1}} \begin{pmatrix} \sin(q_1 a) & \sin(q_2 a) & \ldots & \sin(q_N a) \\ \sin(2q_1 a) & \sin(2q_2 a) & \ldots & \sin(2q_N a) \\ \sin(3q_1 a) & \sin(3q_2 a) & \ldots & \sin(3q_N a) \\ \vdots & \vdots & \ldots & \vdots \\ \sin(Nq_1 a) & \sin(Nq_2 a) & \ldots & \sin(Nq_N a) \end{pmatrix} \,,$$

with matrix element $S_{mn} = v_m^{(n)}$; hence, we have

$$\left(\underline{S}^T \underline{V} \underline{S} \right)_{mn} = \sum_{kl} \left(\underline{S}^T \right)_{mk} V_{kl} S_{ln} = \sum_{k} \left(\underline{S}^T \right)_{mk} \underbrace{\sum_{l} V_{kl} v_l^{(n)}}_{\underline{V}\underline{v}^{(n)} = m\omega_n^2 \underline{v}^{(n)}} = \sum_{k} \left(\underline{S}^T \right)_{mk} m\omega_n^2 v_k^{(n)}$$

$$= m\omega_n^2 \sum_{k} \left(\underline{S}^T \right)_{mk} S_{kn} = m\omega_n^2 \left(\underline{S}^T \underline{S} \right)_{mn} = m\omega_n^2 \delta_{mn} \,.$$

Part 5

We introduce a new set of coordinates (the normal coordinates) $\xi_l(t)$ via

$$\underline{\eta}(t) = \underline{S}\,\underline{\xi}(t) \qquad \text{or} \qquad \eta_k(t) = \sum_{l=1}^{N} S_{kl}\,\xi_l(t) \,,$$

and the transformation is simply inverted, $\underline{\xi}(t) = \underline{S}^T \underline{\eta}$. The Lagrangian in these new coordinates reads

$$L = \frac{1}{2} \underline{\dot{\xi}}^T \underline{S}^T \underline{M} \underline{S} \underline{\dot{\xi}} - \frac{1}{2} \underline{\xi}^T \underline{S}^T \underline{V} \underline{S} \underline{\xi} = \frac{m}{2} \sum_{k=1}^{N} \left(\dot{\xi}_k^2 - \omega_k^2 \xi_k^2 \right) \,,$$

since $\underline{M} = m\,\underline{I}$, and the oscillators decouple in the normal coordinates,

$$\frac{d}{dt}\frac{\partial L}{\partial \dot{\xi}_l} - \frac{\partial L}{\partial \xi_l} = 0 \implies \ddot{\xi}_l(t) + \omega_l^2\,\xi_l(t)\,,$$

with solutions $\xi_l(t) = A_l\,\cos(\omega_l t + \varphi_l)$. The $2N$ constants A_l and φ_l follow from the initial conditions $\eta_l(0)$ and $\dot{\eta}_l(0)$ as

$$\underline{\xi}(0) = \begin{pmatrix} A_1\,\cos\varphi_1 \\ A_2\,\cos\varphi_2 \\ \vdots \\ A_N\,\cos\varphi_N \end{pmatrix} = \underline{S}^T \begin{pmatrix} \eta_1(0) \\ \eta_2(0) \\ \vdots \\ \eta_N(0) \end{pmatrix}\,, \qquad \underline{\dot{\xi}}(0) = \begin{pmatrix} -\omega_1 A_1\,\sin\varphi_1 \\ -\omega_2 A_2\,\sin\varphi_2 \\ \vdots \\ -\omega_N A_N\,\sin\varphi_N \end{pmatrix} = \underline{S}^T \begin{pmatrix} \dot{\eta}_1(0) \\ \dot{\eta}_2(0) \\ \vdots \\ \dot{\eta}_N(0) \end{pmatrix}\,,$$

and the classical problem is completely solved.

Part 6

We now quantize the system by requiring that $\eta_l \longrightarrow \hat{\eta}_l$ and $m\dot{\eta}_l = p_l \longrightarrow \hat{p}_l$ with $[\hat{\eta}_l, \hat{p}_m] = i\hbar\,\delta_{lm}$ with all other commutators vanishing. The Hamiltonian of this system is then given by

$$\hat{H} = \sum_{k=1}^{N} \frac{\hat{p}_k^2}{2m} + \frac{k}{2}\left[\hat{\eta}_1^2 + \sum_{k=2}^{N}(\hat{\eta}_k - \hat{\eta}_{k-1})^2 + \hat{\eta}_N^2\right]\,,$$

a set of N coupled harmonic oscillators. Following our classical treatment above, we introduce a set of position and momentum operators $\hat{\xi}_l$ and $\hat{q}_l$ via the transformation

$$\hat{\xi}_l = \sum_m \left(\underline{S}^T\right)_{lm} \hat{\eta}_m\,, \qquad \hat{q}_l = \sum_m \left(\underline{S}^T\right)_{lm} \hat{p}_m\,,$$

where $\underline{S}$ is the modal matrix defined earlier. Such transformation is canonical since it preserves the commutation relations; indeed, we have

$$\left[\hat{\xi}_k, \hat{q}_l\right] = \sum_{m,n=1}^{N} \left(\underline{S}^T\right)_{km} \left(\underline{S}^T\right)_{ln} [\hat{\eta}_m, \hat{p}_n] = i\hbar \sum_{m,n=1}^{N} \left(\underline{S}^T\right)_{km} \left(\underline{S}^T\right)_{ln} \delta_{mn}$$

$$= i\hbar \sum_{m=1}^{N} \left(\underline{S}^T\right)_{km} \underline{S}_{ml} = i\hbar\,\delta_{kl}\,.$$

In terms of the new observables (note that, since the modal matrix is real, the new position and momentum operators are hermitian), the Hamiltonian is given by

$$\hat{H} = \sum_{k=1}^{N} \underbrace{\left(\frac{\hat{q}_k^2}{2m} + \frac{m\omega_k^2}{2}\hat{\xi}_k^2\right)}_{\hat{H}_k}\,, \qquad \omega_k = 2\omega_0\,\sin\left(\frac{k}{N+1}\frac{\pi}{2}\right)\,,$$

the sum of N independent (and commuting) harmonic oscillator Hamiltonians. We denote by $E_k = \hbar\omega_k(n_k + 1/2)$ and $|n_k\rangle$ the eigenvalues and corresponding eigenvectors of $\hat{H}_k$. The eigenstates of the full Hamiltonian consist of a tensor product:

$$|n_1, n_2, \ldots, n_N\rangle = |n_1\rangle \otimes |n_2\rangle \otimes \cdots \otimes |n_N\rangle$$

with eigenvalues

$$\hat{H}|n_1, n_2, \ldots, n_N\rangle = \underbrace{\sum_{k=1}^{N} \hbar\omega_k(n_k + 1/2)}_{E_{n_1 n_2 \ldots n_N}} |n_1, n_2, \ldots, n_N\rangle .$$

The ground state $|0\rangle \equiv |00\cdots0\rangle$ has energy $E_0 = \sum_k \hbar\omega_k/2$.

Part 7

Assuming the system is kept in equilibrium at temperature T, the probability that it is in the state $|n_1, n_2, \ldots, n_N\rangle$ with energy $E_{n_1 n_2 \ldots n_N}$ is proportional to

$$p(E_{n_1 n_2 \ldots n_N}) \propto e^{-\beta E_{n_1 n_2 \ldots n_N}} \qquad \text{with} \quad \beta = \frac{1}{k_B T} ,$$

resulting in an average energy

$$\langle E \rangle_T = \sum_{n_1, n_2, \ldots, n_N=0}^{\infty} E_{n_1 n_2 \ldots n_N} p(E_{n_1 n_2 \ldots n_N}) = -\frac{\partial}{\partial\beta} \ln \underbrace{\left[\sum_{n_1, n_2, \ldots, n_N=0}^{\infty} e^{-\beta E_{n_1 n_2 \ldots n_N}} \right]}_{Z(\beta)} ,$$

where $Z(\beta)$ is known as the partition function. It can be easily evaluated in this case:

$$Z(\beta) = \sum_{n_1, n_2, \ldots, n_N=0}^{\infty} e^{-\beta E_{n_1 n_2 \ldots n_N}} = \sum_{n_1, n_2, \ldots, n_N=0}^{\infty} e^{-\beta \sum_{k=1}^{N} \hbar\omega_k(n_k+1/2)} = \sum_{n_1, n_2, \ldots, n_N=0}^{\infty} \left[\prod_{k=1}^{N} e^{-\beta\hbar\omega_k(n_k+1/2)} \right]$$

$$= \left[\sum_{n_1=0}^{\infty} e^{-\beta\hbar\omega_1(n_1+1/2)} \right] \left[\sum_{n_2=0}^{\infty} e^{-\beta\hbar\omega_2(n_2+1/2)} \right] \cdots \left[\sum_{n_N=0}^{\infty} e^{-\beta\hbar\omega_N(n_N+1/2)} \right]$$

$$= \prod_{k=1}^{N} \left[\sum_{n_k=0}^{\infty} e^{-\beta\hbar\omega_k(n_k+1/2)} \right] = \prod_{k=1}^{N} \frac{e^{-\beta\hbar\omega_k/2}}{1 - e^{-\beta\hbar\omega_k}} ,$$

from which we obtain the average energy as

$$\langle E \rangle_T = -\frac{\partial}{\partial\beta} \ln Z(\beta) = -\frac{\partial}{\partial\beta} \sum_{k=1}^{N} \ln \left(\frac{e^{-\beta\hbar\omega_k/2}}{1 - e^{-\beta\hbar\omega_k}} \right) = \sum_{k=1}^{N} \left(\frac{\hbar\omega_k}{2} + \frac{\hbar\omega_k}{e^{\beta\hbar\omega_k} - 1} \right) .$$

The allowed wave numbers are given by $q_k = [k/(N+1)](\pi/a) = (\pi/L)k$ in terms of the total length $L = (N+1)a$ of the crystal. In the thermodynamic limit $L \longrightarrow \infty$, these wave numbers are densely distributed. There is a single mode in each segment of length π/L and therefore $\rho\,(\pi/L) = 1$, where ρ is the density of the normal modes. We convert the sum over k to an integral over the wave number q,

$$\sum_{k=1}^{N} \longrightarrow \frac{L}{\pi} \int_{\pi/L}^{N\pi/L} dq \longrightarrow \frac{L}{\pi} \int_0^{\pi/a} dq ,$$

where in the last step we have replaced the upper integration limit by $N\pi/L = (\pi/a)N/(N+1) \longrightarrow \pi/a$. We obtain for the average energy

$$\langle E \rangle_T = \frac{L}{\pi} \int_0^{\pi/a} dq \left(\frac{\hbar\omega_q}{2} + \frac{\hbar\omega_q}{e^{\beta\hbar\omega_q} - 1} \right) \qquad \text{with} \quad \omega_q = 2\omega_0 \sin(qa/2) .$$

In the low-temperature limit, the dominant contribution to the energy is from the low-frequency modes having $\omega_q \approx qa\omega_0$. The parameter $a\omega_0$ has dimensions of velocity (it is the sound velocity). Ignoring the zero-point contribution, the average energy follows in this limit as

$$\langle E \rangle_T = \frac{L}{\pi} \int_0^{\pi/a} dq \, \frac{qa\hbar\omega_0}{e^{\beta qa\hbar\omega_0} - 1} \, ,$$

or, changing variables,

$$\langle E \rangle_T = \frac{L}{a\pi} \frac{(k_B T)^2}{\hbar\omega_0} \int_0^{T_D/T} dx \, \frac{x}{e^x - 1} \qquad \text{with } T_D = \frac{\pi\hbar\omega_0}{k_B} \, .$$

In the low-temperature limit, the average energy and heat capacity are given by

$$\langle E \rangle_{T \ll T_D} = \frac{L}{a\pi} \frac{(k_B T)^2}{\hbar\omega_0} \int_0^\infty dx \, \frac{x}{e^x - 1} \implies c_V = \frac{\partial}{\partial T} \langle E \rangle_{T \ll T_D} \propto T \, .$$

In the high-temperature limit, $T_D/T \ll 1$, the integrand $x/(e^x - 1) \approx 1$ in the expression for $\langle E \rangle_T$, and hence, to leading order,

$$\langle E \rangle_{T \gg T_D} = \frac{L}{a\pi} \frac{(k_B T)^2}{\hbar\omega_0} \frac{T_D}{T} = \frac{L}{a} k_B T \implies c_V = \frac{\partial}{\partial T} \langle E \rangle_{T \gg T_D} \propto N k_B \, ,$$

where $L/a = N$ (recall that $N \gg 1$ in the thermodynamic limit). The result above is in agreement with the equipartition theorem of classical statistical mechanics.

9 Particle in a Central Potential; Orbital Angular Momentum

In the present and next chapters, we consider the important problem of a particle under the action of a potential which depends only on the distance from a center of force (a so-called central potential). Physically relevant examples are an electron in the Coulomb field of the proton (the hydrogen atom) and the classic problem of a particle in a three-dimensional harmonic potential. Unless stated otherwise, hereafter we adopt the coordinate representation. The state is represented by the wave function $\Psi(\mathbf{r}, t) = \langle \phi_{\mathbf{r}} | \Psi(t) \rangle$, and

$$\Psi(\mathbf{r}, t) = e^{-iE(t-t_0)/\hbar} \, \psi(\mathbf{r}) \,, \qquad \underbrace{\left[-\frac{\hbar^2}{2\mu} \nabla^2 + V(r) \right] \psi(\mathbf{r}) = E\psi(\mathbf{r}) \,,}_{H(\mathbf{r}, -i\hbar\nabla)} \tag{9.1}$$

where μ denotes the particle's mass and the potential $V(r)$ depends only on the magnitude $|\mathbf{r}| \equiv r$. The problem reduces to finding the eigenvalues and eigenfunctions of the Hamiltonian $H(\mathbf{r}, \mathbf{p})$ with $\mathbf{p} = -i\hbar\nabla$.

9.1 Orbital Angular Momentum

In classical mechanics, the solution of the equations of motion for a particle in a central potential is greatly facilitated by introducing the angular momentum $\mathbf{L} = \mathbf{r} \times \mathbf{p}$ (relative to the center of force), a constant of motion. As a consequence, it can be seen that the motion of the particle occurs in a plane perpendicular to $\mathbf{L}$ and that its position vector sweeps equal areas in equal times. It is therefore natural to introduce the angular momentum in quantum mechanics as the operator

$$\hat{\mathbf{L}} = \hat{\mathbf{r}} \times \hat{\mathbf{p}} \,; \tag{9.2}$$

in the coordinate representation $\mathbf{L} = -i\hbar \, \mathbf{r} \times \nabla$. Its components can be conveniently expressed as

$$L_i = -i\hbar \sum_{jk} \epsilon_{ijk} r_j \partial_k \,, \qquad \partial_k \equiv \frac{\partial}{\partial r_k} \,, \tag{9.3}$$

where ϵ_{ijk} is the Levi–Civita tensor, with the property that

$$\epsilon_{ijk} = \begin{array}{ll} +1 & \text{if } ijk = 123 \text{ or } 231 \text{ or } 312 \\ -1 & \text{if } ijk = 132 \text{ or } 213 \text{ or } 321 \\ 0 & \text{otherwise} \,. \end{array} \tag{9.4}$$

Thus ϵ_{ijk} is a totally antisymmetric tensor that is equal to $+1$ for a cyclic permutation of 123, to -1 for an anticyclic permutation of 123, and to zero if any two or more indices are the same; it changes sign under the the exchange of any two indices. The following identity for the product of two Levi–Civita tensors is useful in applications,[1]

$$\sum_l \epsilon_{ijl} \, \epsilon_{mnl} = \delta_{im} \, \delta_{jn} - \delta_{in} \, \delta_{jm} \, . \tag{9.5}$$

The canonical commutation relations satisfied by the components of $\mathbf{r}$ and $\mathbf{p}$ imply that

$$\left[L_i , r_j \right] = \sum_{lm} \epsilon_{ilm} \left[r_l p_m , r_j \right] = \sum_{lm} \epsilon_{ilm} \, r_l \left[p_m , r_j \right] = -i\hbar \sum_l \epsilon_{ilj} \, r_l = i\hbar \sum_l \epsilon_{ijl} \, r_l \, , \tag{9.6}$$

where we have used the identity, valid for any three operators, $[BC, A] = B[C, A] + [B, A]C$ (or, equivalently, $[A, BC] = B[A, C] + [A, B]C$) and the property that $\epsilon_{ilj} = -\epsilon_{ijl}$. Similarly, we find that

$$\left[L_i , p_j \right] = i\hbar \sum_l \epsilon_{ijl} \, p_l \, . \tag{9.7}$$

As a matter of fact, any vector operator $\mathbf{V}$ satisfies the following commutation relations with $\mathbf{L}$,

$$\left[L_i , V_j \right] = i\hbar \sum_l \epsilon_{ijl} \, V_l \, , \tag{9.8}$$

as a consequence of the transformation properties of vector operators under rotations. We will return to this topic when we discuss symmetry transformations (translations, rotations, etc.) in quantum mechanics. Here, we just note that the angular momentum, being itself a vector operator, satisfies the commutation relations (see Problem 1 for an explicit derivation)

$$\left[L_i , L_j \right] = i\hbar \sum_l \epsilon_{ijl} \, L_l \, ; \tag{9.9}$$

different components of the angular momentum do not commute with each other and, therefore, cannot be measured simultaneously.

The commutation relations of $\mathbf{L}$ and $\mathbf{V}$ also imply that

$$\left[L_i , \mathbf{V}^2 \right] = 0 \, , \tag{9.10}$$

that is, $\mathbf{L}$ commutes with the scalar operator $\mathbf{V}^2$; in particular, it commute with $\mathbf{p}^2$ and $r = \sqrt{\mathbf{r}^2}$, and hence with the Hamiltonian of a particle in a central potential. The Hamiltonian, $\mathbf{L}^2$, and one of the components of $\mathbf{L}$, which conventionally is taken as the z-component, form a set of commuting observables. As a consequence, there exists a basis of simultaneous eigenfunctions of H, $\mathbf{L}^2$, and L_z.

In order to fully exploit the rotational symmetry of the Hamiltonian of a particle in a central potential (that is, the fact that H commutes with $\mathbf{L}$), first we note that

$$\mathbf{L}^2 = \sum_i L_i L_i = -\hbar^2 \sum_{jklm} \sum_i \epsilon_{ijk} \, \epsilon_{ilm} r_j \partial_k r_l \partial_m = -\hbar^2 \sum_{jklm} \left(\delta_{jl} \, \delta_{km} - \delta_{jm} \, \delta_{kl} \right) r_j \partial_k r_l \partial_m$$

$$= -\hbar^2 \left(\underbrace{\sum_{jk} r_j \partial_k r_j \partial_k}_{\text{term A}} - \underbrace{\sum_{jk} r_j \partial_k r_k \partial_j}_{\text{term B}} \right) , \tag{9.11}$$

----

[1] We write (note the positions of the subscripts): $\sum_l \epsilon_{ijl} \, \epsilon_{mnl} = \epsilon_{ij1} \, \epsilon_{mn1} + \epsilon_{ij2} \, \epsilon_{mn2} + \epsilon_{ij3} \, \epsilon_{mn3}$; the first term requires ij and mn to be equal to 23 or 32 (namely, the four possibilities $ij = 23$ and $mn = 23$ or 32, or $ij = 32$ and $mn = 23$ or 32), the second term requires ij and mn to be equal to 13 or 31, and the last term requires ij and mn to be equal to 12 or 21. Thus, for any of the allowed possibilities, only one of these three terms survives. Consider the first term, $\epsilon_{ij1} \, \epsilon_{mn1}$: for $ij = mn = 23$ or 32 the product $\epsilon_{231} \, \epsilon_{231} = \epsilon_{321} \, \epsilon_{321} = 1$, since $\epsilon_{231} = 1$ and $\epsilon_{321} = -1$; on the other hand, $ij = 23$ and $mn = 32$, or $ij = 32$ and $mn = 23$, the product $\epsilon_{231} \, \epsilon_{321} = -1$. We can summarize these results in the following way:

$$\sum_l \epsilon_{ijl} \, \epsilon_{mnl} = \delta_{im} \, \delta_{jn} - \delta_{in} \, \delta_{jm} \, .$$

where it is important to stress that the partial derivatives act on everything to their right. Next, we consider the two terms A and B separately:[2]

$$\text{term A} = \sum_{jk} r_j (\delta_{jk} + r_j \partial_k) \partial_k = \sum_j r_j \partial_j + \sum_j r_j r_j \sum_k \partial_k \partial_k = \mathbf{r} \cdot \nabla + r^2 \nabla^2 \; ; \tag{9.12}$$

$$\text{term B} = \sum_{jk} r_j (\delta_{kk} + r_k \partial_k) \partial_j = 3 \sum_j r_j \partial_j + \sum_{jk} r_j r_k \partial_j \partial_k$$

$$= 3 \mathbf{r} \cdot \nabla + \sum_{jk} r_j (\partial_j r_k - \delta_{jk}) \partial_k = 2 \mathbf{r} \cdot \nabla + (\mathbf{r} \cdot \nabla)^2 \; , \tag{9.13}$$

to arrive at

$$\mathbf{L}^2 = -\hbar^2 \left[r^2 \nabla^2 - (\mathbf{r} \cdot \nabla)^2 - \mathbf{r} \cdot \nabla \right] \; , \tag{9.14}$$

which can be further simplified by expressing $\mathbf{r} \cdot \nabla$ in spherical coordinates as $r \partial / \partial r$. Therefore, we obtain

$$\mathbf{L}^2 = r^2 \mathbf{p}^2 + \hbar^2 r \frac{\partial}{\partial r} r \frac{\partial}{\partial r} + \hbar^2 r \frac{\partial}{\partial r} \; , \tag{9.15}$$

which we can "solve" for $\mathbf{p}^2$ to find

$$\mathbf{p}^2 = -\frac{\hbar^2}{r^2} \left(r \frac{\partial}{\partial r} r \frac{\partial}{\partial r} + r \frac{\partial}{\partial r} \right) + \frac{\mathbf{L}^2}{r^2} = -\frac{\hbar^2}{r^2} \frac{\partial}{\partial r} r^2 \frac{\partial}{\partial r} + \frac{\mathbf{L}^2}{r^2} \; . \tag{9.16}$$

The terms involving the partial derivatives with respect to r can be rewritten as follows:

$$r \frac{\partial}{\partial r} r \frac{\partial}{\partial r} + r \frac{\partial}{\partial r} = 2r \frac{\partial}{\partial r} + r^2 \frac{\partial^2}{\partial r^2} = \frac{\partial}{\partial r} r^2 \frac{\partial}{\partial r} \; . \tag{9.17}$$

In solving for $\mathbf{p}^2$, we have multiplied both sides on the left by $1/r^2$, the inverse operator of r^2. This operator does not commute with $\partial_r r^2 \partial_r$, and so the relative ordering of $1/r^2$ and $\partial_r r^2 \partial_r$ is important ($1/r^2$ is on the left, as it must); on the other hand, $\mathbf{L}^2$ and $1/r^2$ do commute, and their relative ordering is irrelevant.

9.2 The Spectrum of $\mathbf{L}^2$ and L_z

The Schrödinger equation for a particle in a central potential is conveniently written in spherical coordinates as

$$\underbrace{\left[-\frac{\hbar^2}{2\mu r^2} \frac{\partial}{\partial r} r^2 \frac{\partial}{\partial r} + \frac{\mathbf{L}^2}{2\mu r^2} + V(r) \right]}_{H} \psi(r, \theta, \phi) = E \psi(r, \theta, \phi) \; , \tag{9.18}$$

which makes it obvious that the Hamiltonian and $\mathbf{L}^2$ commute (the square of the angular momentum operator acts only on the polar angle θ and azimuthal angle ϕ, see Problem 2). The next task is to determine the eigenvalues and eigenfunctions of $\mathbf{L}^2$ and L_z. We will see in a later section how this can be done on general grounds by using only the commutation relations satisfied by the L_i.

---

[2] The term $\partial_k r_j$ can be written as $\delta_{jk} + r_j \partial_k$, since for any $f(\mathbf{r})$ we have $\partial_k [r_j f(\mathbf{r})] = \delta_{jk} f(\mathbf{r}) + r_j \partial_k f(\mathbf{r})$; this also yields $r_j \partial_k = \partial_k r_j - \delta_{jk}$.

We assume that the wave function $\psi(\mathbf{r})$ is a smooth (continuous) function of $\mathbf{r}$, in particular near the origin; such an assumption is justified if the potential $V(r)$ is not too singular at $r = 0$ (see below). Under this assumption, $\psi(\mathbf{r})$ can be expanded in a Taylor series around $\mathbf{r} = 0$,

$$\psi(\mathbf{r}) = \psi(0) + \sum_i r_i \partial_i \psi(\mathbf{r})|_{\mathbf{r}=0} + \frac{1}{2!} \sum_{ij} r_i r_j \partial_i \partial_j \psi(\mathbf{r})|_{\mathbf{r}=0} + \frac{1}{3!} \sum_{ijk} r_i r_j r_k \partial_i \partial_j \partial_k \psi(\mathbf{r})|_{\mathbf{r}=0} + \cdots . \tag{9.19}$$

It may happen, for a given wave function, that some terms in this expansion vanish. Suppose that the non-vanishing term with the smallest number of r_i factors has $l \geq 0$ such factors; for example, if $\psi(0) \neq 0$ then $l = 0$; if $\psi(0) = 0$ and $\partial_i \psi(\mathbf{r})|_{\mathbf{r}=0} = 0$ and some of the second partial derivatives are non-vanishing then $l = 2$. Thus, generally we have that

$$\psi(\mathbf{r}) = P_l(r_x, r_y, r_z) + P_{l+1}(r_x, r_y, r_z) + \cdots , \tag{9.20}$$

where $P_n(r_x, r_y, r_z)$ with $n = l, l + 1, \ldots, \infty$ is a homogenous polynomial of order n: it consists of a sum of terms, each of which contains n factors of the r_i; for example, $P_2(r_x, r_y, r_z) = ar_x^2 + br_y r_z$ with a, b constants. When expressed in spherical coordinates, we see that

$$\psi(r, \theta, \phi) = r^l P_l(\theta, \phi) + r^{l+1} P_{l+1}(\theta, \phi) + \cdots , \tag{9.21}$$

since each of the r_i is proportional to r. The Schrödinger equation implies that

$$\mathbf{L}^2 \psi(r, \theta, \phi) = \hbar^2 \frac{\partial}{\partial r} r^2 \frac{\partial}{\partial r} \psi(r, \theta, \phi) + 2\mu r^2 [E - V(r)] \psi(r, \theta, \phi) . \tag{9.22}$$

In the limit $r \longrightarrow 0$, the dominant term in $\psi(r, \theta, \phi)$ is $r^l P_l(\theta, \phi)$. Assuming that $r^2 V(r) \longrightarrow 0$ in this limit (this requirement is satisfied by the Coulomb potential in the physically relevant case of the hydrogen atom), we conclude that

$$\mathbf{L}^2 \psi(r, \theta, \phi) = \hbar^2 \underbrace{\frac{\partial}{\partial r} r^2 \frac{\partial}{\partial r} r^l}_{l(l+1)r^l} P_l(\theta, \phi) = \hbar^2 l(l + 1) \psi(r, \theta, \phi) . \tag{9.23}$$

Even though the above result has been derived in the limit $r \longrightarrow 0$, it must hold for any r; since $\mathbf{L}^2$ acts only on θ and ϕ, the eigenfunctions of $\mathbf{L}^2$ must be proportional to functions of these angles – we denote them generically as $Y(\theta, \phi)$. The coefficient of proportionality can be a function of r, and hence

$$\psi(r, \theta, \phi) = R(r) Y(\theta, \phi) , \qquad \mathbf{L}^2 Y(\theta, \phi) = \hbar^2 l(l + 1) Y(\theta, \phi) . \tag{9.24}$$

We also require that $\psi(r, \theta, \phi)$ be an eigenfunction of L_z,

$$L_z \psi(r, \theta, \phi) = \hbar m \psi(r, \theta, \phi) , \tag{9.25}$$

where we take m to be a real number (L_z is a hermitian operator, and its eigenvalues are real), and the factor $\hbar$ is included for convenience. Since L_z, given by

$$L_z = -i\hbar \frac{\partial}{\partial \phi} , \tag{9.26}$$

acts only on the ϕ dependence, we must have

$$L_z Y(\theta, \phi) = \hbar m Y(\theta, \phi) ; \tag{9.27}$$

expressing $Y(\theta, \phi) = P(\theta)\,\Phi(\phi)$ as a product (using the separability of the variables) of a function of θ times a function of ϕ, we arrive at

$$-i\frac{\partial}{\partial\phi}\Phi(\phi) = m\Phi(\phi)\,, \qquad \Phi(\phi + 2\pi) = \Phi(\phi)\,, \tag{9.28}$$

where the boundary condition ensures that the $\Phi(\phi)$, and hence $Y(\theta, \phi)$, assume the same value as $\phi \longrightarrow \phi + 2\pi$. The solution is given by $\Phi(\phi) \propto e^{im\phi}$ with $m = 0, \pm 1, \pm 2, \ldots$. At this stage, there is no upper limit on $|m|$. In fact, it will turn out that $|m| \leq l$. How this comes about will become clear when we construct the spherical harmonics in the next subsection.

Inserting Eq. (9.24) into the Schrödinger equation, we obtain

$$\left[-\frac{\hbar^2}{2\mu r^2}\frac{d}{dr}r^2\frac{d}{dr} + \frac{l(l+1)\hbar^2}{2\mu r^2} + V(r)\right]R(r) = ER(r)\,, \tag{9.29}$$

where in the limit $r \longrightarrow 0$ the radial function $R(r) \propto r^l$ (of course, we are assuming here that $r^2 V(r) \longrightarrow 0$ in this limit). In addition, for bound-state solutions the wave functions must be normalizable:

$$\int_0^{2\pi} d\phi \int_0^{\pi} d\theta\,\sin\theta\,|Y(\theta, \phi)|^2 \int_0^{\infty} dr\,r^2\,|R(r)|^2 < \infty\,, \tag{9.30}$$

and so the integral over θ and ϕ and the integral over r must each be convergent. The differential equation (9.29) can be cast into a one-dimensional Schrödinger equation for a particle of mass μ under the influence of an *effective* potential $V(r) + l(l+1)\hbar^2/(2\mu r^2)$ by introducing an auxiliary function $u(r)$:

$$R(r) = \frac{u(r)}{r} \qquad \text{with } u(r) \propto r^{l+1} \text{ as } r \longrightarrow 0\,. \tag{9.31}$$

In terms of $u(r)$, the radial Schrödinger equation in the region $0 < r < \infty$ reduces to

$$-\frac{\hbar^2}{2\mu}u''(r) + \left[\frac{l(l+1)\hbar^2}{2\mu r^2} + V(r)\right]u(r) = Eu(r)\,, \tag{9.32}$$

with the requirement (for bound states)

$$\int_0^{\infty} dr\,|u(r)|^2 < \infty\,. \tag{9.33}$$

9.3 Spherical Harmonics

Here, we construct the eigenfunctions of $\mathbf{L}^2$ and L_z, known as spherical harmonics. We have already determined their eigenvalues, respectively $l(l+1)\hbar^2$ and $m\hbar$ with $l = 0, 1, 2, \ldots$ and $m = 0, \pm 1, \pm 2, \ldots$, albeit with no restriction on $|m|$ as of yet. We write the eigenvalue equation for the square of the orbital angular momentum as (see Problem 2 for a derivation of $\mathbf{L}^2$ in spherical coordinates)

$$\underbrace{-\hbar^2\left[\frac{1}{\sin\theta}\frac{\partial}{\partial\theta}\left(\sin\theta\frac{\partial}{\partial\theta}\right) + \frac{1}{\sin^2\theta}\frac{\partial^2}{\partial\phi^2}\right]}_{\mathbf{L}^2}\underbrace{P(\theta)\,e^{im\phi}}_{Y(\theta,\phi)} = l(l+1)\hbar^2\,\underbrace{P(\theta)\,e^{im\phi}}_{Y(\theta,\phi)}\,, \tag{9.34}$$

where $P(\theta)$ is a function of θ to be determined. By acting with the partial derivatives on ϕ, the above equation can be simplified to

$$-\left[\frac{1}{\sin\theta}\frac{d}{d\theta}\left(\sin\theta\frac{d}{d\theta}\right)-\frac{m^2}{\sin^2\theta}\right]P(\theta)=l(l+1)P(\theta)\,. \tag{9.35}$$

Introducing the variable x, where

$$x=\cos\theta \qquad \text{with} \qquad -1\le x\le 1 \qquad \text{and} \qquad \frac{d}{d\theta}=\frac{dx}{d\theta}\frac{d}{dx}=-\sin\theta\frac{d}{dx}\,, \tag{9.36}$$

so that

$$\frac{1}{\sin\theta}\frac{d}{d\theta}\left(\sin\theta\frac{d}{d\theta}\right)=\frac{1}{\sin\theta}\underbrace{\left(-\sin\theta\frac{d}{dx}\right)}_{d/d\theta}\left[\sin\theta\underbrace{\left(-\sin\theta\frac{d}{dx}\right)}_{d/d\theta}\right]=\frac{d}{dx}\left[\underbrace{(1-x^2)}_{\sin^2\theta}\frac{d}{dx}\right]\,, \tag{9.37}$$

leads to the equation

$$\frac{d}{dx}(1-x^2)\frac{dP_l^m(x)}{dx}+\left[l(l+1)-\frac{m^2}{1-x^2}\right]P_l^m(x)=0\,, \tag{9.38}$$

where we have made explicit the dependence of $P(x)$ on the integers l and m – these are known as the associated Legendre functions. If $P_l^{m=0}(x)\equiv P_l(x)$ satisfies the Legendre equation,

$$\frac{d}{dx}(1-x^2)\frac{dP_l(x)}{dx}+l(l+1)\,P_l(x)=0\,, \tag{9.39}$$

then

$$P_l^m(x)=(1-x^2)^{|m|/2}\frac{d^{|m|}P_l(x)}{dx^{|m|}}\,, \tag{9.40}$$

is a solution of Eq. (9.38) ($d^{|m|}/dx^{|m|}$ represents the $|m|$th derivative). The Legendre equation is a second-order differential equation, and as such it has two independent solutions, $P_l(x)$ and $Q_l(x)$. For integer $l\ge 0$, the solution $P_l(x)$ can be expressed as Rodrigues' formula,

$$P_l(x)=\frac{1}{2^l\,l!}\frac{d^l}{dx^l}(x^2-1)^l\,, \tag{9.41}$$

that is, a polynomial of order l in x (known as a Legendre polynomial), whereas the solution $Q_l(x)$ has a singular logarithmic behavior at the end points $x=\pm1$ (it contains a term proportional to $\ln[(1+x)/(1-x)]$) and is therefore not an acceptable solution. Since $P_l(x)=a_lx^l+a_{l-1}x^{l-1}+\cdots+a_0$, it follows that

$$\frac{d^{|m|}P_l(x)}{dx^{|m|}}=0 \qquad \text{if} \qquad |m|>l\,, \tag{9.42}$$

and the allowed integer-m values are such that $|m|\le l$.

The θ and ϕ solutions above are combined into the spherical harmonics $Y_{lm}(\theta,\phi)$, defined as (the pre-factor below is for normalization)

$$Y_{lm}(\theta,\phi)=(-1)^m\sqrt{\frac{2l+1}{4\pi}\frac{(l-m)!}{(l+m)!}}\,P_l^m(\cos\theta)\,e^{im\phi} \qquad 0\le m\le l\,, \tag{9.43}$$

and we have

$$Y_{lm}(\theta,\phi)=(-1)^m\,Y_{l,-m}^*(\theta,\phi) \qquad -l\le m\le -1\,. \tag{9.44}$$

The spherical harmonics are eigenfunctions of $\mathbf{L}^2$ and L_z with eigenvalues, respectively, $l(l+1)\hbar^2$ and $m\hbar$, where $l = 0, 1, 2, \ldots$ and $m = -l, \ldots, l$ (and hence, for fixed l, there are $2l+1$ allowed values of m). They can be shown to be orthonormal (see Problem 3),

$$\int_0^\pi d\theta \, \sin\theta \int_0^{2\pi} d\phi \, Y_{l'm'}^*(\theta, \phi) \, Y_{lm}(\theta, \phi) = \delta_{ll'}\delta_{mm'} , \tag{9.45}$$

and to form a complete set – a basis – on the sphere of unit radius. A function $f(\theta, \phi)$ can be expanded as

$$f(\theta, \phi) = \sum_{l=0}^\infty \sum_{m=-l}^l c_{lm} Y_{lm}(\theta, \phi) , \qquad c_{lm} = \int_0^\pi d\theta \, \sin\theta \int_0^{2\pi} d\phi \, Y_{lm}^*(\theta, \phi) \, f(\theta, \phi) . \tag{9.46}$$

9.4 Problems

Problem 1   Commutation Relations Satisfied by the Components of $\mathbf{L}$

Verify explicitly that $\left[L_i, L_j\right] = i\hbar \sum_k \epsilon_{ijk} L_k$, and that $\left[L_i, \mathbf{V}^2\right] = 0$ for a generic vector operator $\mathbf{V}$.

Solution

Consider

$$\left[L_i, L_j\right] = \sum_{lm} \epsilon_{jlm}\left[L_i, r_l p_m\right] = \sum_{lm} \epsilon_{jlm}\left(r_l\left[L_i, p_m\right] + \left[L_i, r_l\right]p_m\right) = i\hbar \sum_{lm}\sum_n \epsilon_{jlm}\left(\epsilon_{imn} r_l p_n + \epsilon_{iln} r_n p_m\right) .$$

The sums over products of Levi–Civita tensors satisfy

$$\sum_m \epsilon_{jlm}\epsilon_{imn} = -\sum_m \epsilon_{jlm}\epsilon_{inm} \stackrel{?}{=} -\delta_{ji}\delta_{ln} + \delta_{jn}\delta_{li} , \qquad \sum_l \epsilon_{jlm}\epsilon_{iln} = \sum_m \epsilon_{jml}\epsilon_{inl} = \delta_{ji}\delta_{mn} - \delta_{jn}\delta_{mi} .$$

We obtain

$$\left[L_i, L_j\right] = i\hbar \sum_{ln}\left(-\delta_{ji}\delta_{ln} + \delta_{jn}\delta_{li}\right) r_l p_n + i\hbar \sum_{mn}\left(\delta_{ji}\delta_{mn} - \delta_{jn}\delta_{mi}\right) r_n p_m$$

$$= i\hbar\left(-\delta_{ij}\,\mathbf{r}\cdot\mathbf{p} + r_i p_j\right) + i\hbar\left(\delta_{ij}\,\mathbf{r}\cdot\mathbf{p} - r_j p_i\right) = i\hbar\left(r_i p_j - r_j p_i\right) = i\hbar \sum_k \epsilon_{ijk} L_k .$$

Also, $V^2 = |\mathbf{V}|^2 = \mathbf{V}\cdot\mathbf{V}$ has a vanishing commutator with $\mathbf{L}$, since

$$\left[L_i, V^2\right] = \sum_j\left[L_i, V_j^2\right] = \sum_j\left(V_j\left[L_i, V_j\right] + \left[L_i, V_j\right]V_j\right) = i\hbar \sum_{jk}\left(\epsilon_{ijk} V_j V_k + \underbrace{\epsilon_{ijk} V_k V_j}_{\text{exchange } j \rightleftarrows k}\right)$$

$$= i\hbar \sum_{jk}\left(\epsilon_{ijk} V_j V_k + \epsilon_{ikj} V_j V_k\right) = i\hbar \sum_{jk}\left(\epsilon_{ijk} - \epsilon_{ijk}\right) V_j V_k = 0 ,$$

where we emphasize that $\left[L_i, V^2\right] = 0$ remains valid even if the components V_i do not commute among themselves (as is the case for $\mathbf{L}$).

Problem 2 Explicit Expressions for L_i and $\mathbf{L}^2$ in Spherical Coordinates

Obtain explicit expressions for the components of $\mathbf{L}$ and of $\mathbf{L}^2$ in spherical coordinates.

Solution

Spherical coordinates are related to cartesian coordinates via

$$x = r \sin \theta \cos \phi, \qquad y = r \sin \theta \sin \phi, \qquad z = r \cos \theta,$$

where the angles θ and ϕ vary in the ranges $0 \leq \theta \leq \pi$ and $0 \leq \phi \leq 2\pi$. These relations can be inverted to give

$$r = \sqrt{x^2 + y^2 + z^2}, \qquad \theta = \tan^{-1}\left(\sqrt{x^2 + y^2}/z\right), \qquad \phi = \tan^{-1}(y/x) .$$

Using the chain rule for derivatives,

$$\frac{\partial}{\partial x} = \frac{\partial r}{\partial x}\frac{\partial}{\partial r} + \frac{\partial \theta}{\partial x}\frac{\partial}{\partial \theta} + \frac{\partial \phi}{\partial x}\frac{\partial}{\partial \phi},$$

and similarly for the partial derivatives with respect to y and z, we find

$$\frac{\partial r}{\partial x} = \frac{x}{\sqrt{x^2 + y^2 + z^2}} = \sin \theta \cos \phi,$$

$$\frac{\partial \theta}{\partial x} = \frac{1}{1 + (x^2 + y^2)/z^2}\frac{x}{z\sqrt{x^2 + y^2}} = \frac{xz}{r^2\sqrt{x^2 + y^2}} = \frac{1}{r}\cos \theta \cos \phi,$$

$$\frac{\partial \phi}{\partial x} = \frac{1}{1 + y^2/x^2}\left(-\frac{y}{x^2}\right) = -\frac{y}{x^2 + y^2} = -\frac{1}{r}\frac{\sin \phi}{\sin \theta},$$

and similarly

$$\frac{\partial r}{\partial y} = \sin \theta \sin \phi, \qquad\qquad \frac{\partial r}{\partial z} = \cos \theta,$$

$$\frac{\partial \theta}{\partial y} = \frac{1}{r}\cos \theta \sin \phi, \qquad\qquad \frac{\partial \theta}{\partial z} = -\frac{1}{r}\sin \theta,$$

$$\frac{\partial \phi}{\partial y} = \frac{1}{r}\frac{\cos \phi}{\sin \theta}, \qquad\qquad \frac{\partial \phi}{\partial z} = 0.$$

The relationship between the partial derivatives $(\partial_x, \partial_y, \partial_z)$ and $(\partial_r, \partial_\theta, \partial_\phi)$ can be put into matrix form:

$$\begin{pmatrix} \partial_x \\ \partial_y \\ \partial_z \end{pmatrix} = \begin{pmatrix} \sin \theta \cos \phi & \cos \theta \cos \phi/r & -\sin \phi/\sin \theta/r \\ \sin \theta \sin \phi & \cos \theta \sin \phi/r & \cos \phi/\sin \theta/r \\ \cos \theta & -\sin \theta/r & 0 \end{pmatrix} \begin{pmatrix} \partial_r \\ \partial_\theta \\ \partial_\phi \end{pmatrix} ;$$

the matrix, having determinant $(r^2 \sin \theta)^{-1}$, is well defined, if $r > 0$ and $\theta \neq 0$ or π (it can be inverted so as to express the partial derivatives with respect to spherical coordinates in terms of those with respect to cartesian coordinates). For $\theta = 0$ and π, regardless of ϕ, the vector $\mathbf{r}$ lies along the z-axis. Thus, we obtain

$$
\begin{aligned}
L_x &= -i\hbar \left(y\,\frac{\partial}{\partial z} - z\,\frac{\partial}{\partial y} \right) \\
&= -i\hbar \Big[r \sin\theta\,\sin\phi \left(\cos\theta\,\frac{\partial}{\partial r} - \frac{1}{r}\,\sin\theta\,\frac{\partial}{\partial\theta} \right) \\
&\qquad - r\cos\theta \left(\sin\theta\,\sin\phi\,\frac{\partial}{\partial r} + \frac{1}{r}\,\cos\theta\,\sin\phi\,\frac{\partial}{\partial\theta} + \frac{1}{r}\,\frac{\cos\phi}{\sin\theta}\,\frac{\partial}{\partial\phi} \right) \Big] \\
&= -i\hbar \left(-\sin^2\theta\,\sin\phi\,\frac{\partial}{\partial\theta} - \cos^2\theta\,\sin\phi\,\frac{\partial}{\partial\theta} - \cot\theta\,\cos\phi\,\frac{\partial}{\partial\phi} \right) ,
\end{aligned}
$$

or

$$
L_x = i\hbar \left(\sin\phi\,\frac{\partial}{\partial\theta} + \cot\theta\,\cos\phi\,\frac{\partial}{\partial\phi} \right) ,
$$

and similarly

$$
L_y = i\hbar \left(-\cos\phi\,\frac{\partial}{\partial\theta} + \cot\theta\,\sin\phi\,\frac{\partial}{\partial\phi} \right) , \qquad L_z = -i\hbar\,\frac{\partial}{\partial\phi} .
$$

As expected, since $\mathbf{L}$ commutes with r the components L_i act only on the angular dependence; in particular, when L_i is applied to a spherically symmetric wave function $\psi(r)$, it gives $L_i\psi(r)=0$.

We insert the expressions for the angular momentum components to obtain

$$
\begin{aligned}
\mathbf{L}^2 = -\hbar^2 \Big[&\underbrace{\left(\sin\phi\,\frac{\partial}{\partial\theta} + \cot\theta\,\cos\phi\,\frac{\partial}{\partial\phi} \right)\left(\sin\phi\,\frac{\partial}{\partial\theta} + \cot\theta\,\cos\phi\,\frac{\partial}{\partial\phi} \right)}_{L_x L_x} \\
&+ \underbrace{\left(-\cos\phi\,\frac{\partial}{\partial\theta} + \cot\theta\,\sin\phi\,\frac{\partial}{\partial\phi} \right)\left(-\cos\phi\,\frac{\partial}{\partial\theta} + \cot\theta\,\sin\phi\,\frac{\partial}{\partial\phi} \right)}_{L_y L_y} + \underbrace{\frac{\partial^2}{\partial\phi^2}}_{L_z L_z} \Big] \\
= -\hbar^2 \Big(&\sin^2\phi\,\frac{\partial^2}{\partial\theta^2} + \cot^2\theta\,\cos\phi\,\frac{\partial}{\partial\phi}\,\cos\phi\,\frac{\partial}{\partial\phi} + \underbrace{\sin\phi\,\cos\phi\,\frac{\partial}{\partial\theta}\,\cot\theta\,\frac{\partial}{\partial\phi}}_{\text{canceled out}} \\
&+ \cot\theta\,\cos\phi\,\frac{\partial}{\partial\phi}\,\sin\phi\,\frac{\partial}{\partial\theta} + \cos^2\phi\,\frac{\partial^2}{\partial\theta^2} + \cot^2\theta\,\sin\phi\,\frac{\partial}{\partial\phi}\,\sin\phi\,\frac{\partial}{\partial\phi} \\
&- \underbrace{\sin\phi\,\cos\phi\,\frac{\partial}{\partial\theta}\,\cot\theta\,\frac{\partial}{\partial\phi}}_{\text{canceled out}} - \cot\theta\,\sin\phi\,\frac{\partial}{\partial\phi}\,\cos\phi\,\frac{\partial}{\partial\theta} + \frac{\partial^2}{\partial\phi^2} \Big) \\
= -\hbar^2 \Big[&\frac{\partial^2}{\partial\theta^2} + \cot^2\theta\,\underbrace{\left(\cos\phi\,\frac{\partial}{\partial\phi}\,\cos\phi\,\frac{\partial}{\partial\phi} + \sin\phi\,\frac{\partial}{\partial\phi}\,\sin\phi\,\frac{\partial}{\partial\phi} \right)}_{\frac{\partial^2}{\partial\phi^2}} \\
&+ \cot\theta\,\underbrace{\left(\cos\phi\,\frac{\partial}{\partial\phi}\,\sin\phi\,\frac{\partial}{\partial\theta} - \sin\phi\,\frac{\partial}{\partial\phi}\,\cos\phi\,\frac{\partial}{\partial\theta} \right)}_{\frac{\partial}{\partial\theta}} + \frac{\partial^2}{\partial\phi^2} \Big] \\
= -\hbar^2 \Big(&\frac{\partial^2}{\partial\theta^2} + \cot^2\theta\,\frac{\partial^2}{\partial\phi^2} + \cot\theta\,\frac{\partial}{\partial\theta} + \frac{\partial^2}{\partial\phi^2} \Big) ,
\end{aligned}
$$

which can be cast into the standard form

$$\mathbf{L}^2 = -\hbar^2 \left[\frac{1}{\sin\theta} \frac{\partial}{\partial\theta} \left(\sin\theta \frac{\partial}{\partial\theta} \right) + \frac{1}{\sin^2\theta} \frac{\partial^2}{\partial\phi^2} \right] .$$

Problem 3 Orthogonality of Spherical Harmonics

Show that $\int d\Omega\, Y^*_{l'm'}(\Omega)\, Y_{lm}(\Omega) = 0$ for $l, m \neq l', m'$.

Solution

We write the spherical harmonics as

$$Y_{lm}(\theta, \phi) = e^{i\delta_m}\, c_{l|m|} P_l^{|m|}(\cos\theta)\, e^{im\phi}, \qquad c_{l|m|} = \sqrt{\frac{2l+1}{4\pi} \frac{(l-|m|)!}{(l+|m|)!}},$$

where $|m| \leq l$ and the phase factor $e^{i\delta_m}$ is $(-1)^m$ or 1 depending on whether $m \geq 0$ or $m < 0$. We find that

$$\int d\Omega\, Y^*_{l'm'}(\Omega)\, Y_{lm}(\Omega) = e^{-i\delta_{m'} + i\delta_m} c_{l'|m'|} c_{l|m|} \int_{-1}^{1} d\cos\theta\, P_{l'}^{|m'|}(\cos\theta) P_l^{|m|}(\cos\theta) \underbrace{\int_0^{2\pi} d\phi\, e^{-i(m'-m)\phi}}_{2\pi\,\delta_{mm'}}$$

$$= 2\pi c_{l'|m|} c_{l|m|}\, \delta_{mm'} \int_{-1}^{1} dx\, P_{l'}^{|m|}(x)\, P_l^{|m|}(x) .$$

The integral in the second line vanishes unless $l = l'$. To verify this, start from Eq. (9.38) and multiply both sides on the left by $P_{l'm}(x)$, obtaining

$$P_{l'm}(x) \frac{d}{dx}(1-x^2) \frac{dP_l^m(x)}{dx} = P_{l'm}(x) \left[\frac{m^2}{1-x^2} - l(l+1) \right] P_l^m(x) ,$$

and subtract from it the same expression with $l' \rightleftharpoons l$ to obtain

$$\text{l.h.s.} = P_{l'm}(x) \frac{d}{dx}(1-x^2) \frac{dP_l^m(x)}{dx} - P_{lm}(x) \frac{d}{dx}(1-x^2) \frac{dP_{l'}^m(x)}{dx} ,$$
$$\text{r.h.s.} = \left[l'(l'+1) - l(l+1) \right] P_{l'}^m(x) P_l^m(x) .$$

Now, integrate both sides; for the left-hand side, integrating by parts we find

$$\int_{-1}^{1} dx\,(\text{l.h.s.}) = \underbrace{P_{l'm}(x)\,(1-x^2) \frac{dP_l^m(x)}{dx} \Big|_{-1}^{1}}_{A} - \underbrace{\int_{-1}^{1} dx\,(1-x^2) \left[\frac{dP_{l'}^m(x)}{dx} \right] \frac{dP_l^m(x)}{dx}}_{B}$$

$$\underbrace{- P_{lm}(x)\,(1-x^2) \frac{dP_{l'}^m(x)}{dx} \Big|_{-1}^{1}}_{C} + \underbrace{\int_{-1}^{1} dx\,(1-x^2) \left[\frac{dP_l^m(x)}{dx} \right] \frac{dP_{l'}^m(x)}{dx}}_{D} ,$$

and terms A and C vanish when evaluated at the end points $x = \pm 1$, while terms B and D cancel, since they amount to the same integral. We therefore arrive at

$$\left[l'(l'+1) - l(l+1) \right] \int_{-1}^{1} dx\, P_{l'}^m(x) P_l^m(x) = 0 ,$$

which requires $l' = l$ or $l' = -(l + 1)$; however, since l and l' are zero or positive integers, we must have $l' = l$. We could have guessed this result, since $Y_{l'm}$ and Y_{lm} are eigenfunctions of a hermitian operator belonging to different eigenvalues. We therefore obtain

$$\int d\Omega\, Y^*_{l'm'}(\Omega)\, Y_{lm}(\Omega) = \delta_{l'l}\delta_{m'm} \frac{2l+1}{2} \frac{(l-|m|)!}{(l+|m|)!} \int_{-1}^{1} dx\, \left[P_l^{|m|}(x)\right]^2 = \delta_{l'l}\,\delta_{m'm}\;;$$

it can be shown that

$$\int_{-1}^{1} dx\, \left[P_l^{|m|}(x)\right]^2 = \frac{2}{2l+1} \frac{(l+|m|)!}{(l-|m|)!}\,.$$

Problem 4 Commutation Relations of a Vector Operator and L

Prove the relation

$$[\hat{\mathbf{n}} \cdot \mathbf{L},\, \mathbf{V}] = i\hbar\mathbf{V} \times \hat{\mathbf{n}}\,,$$

where $\hat{\mathbf{n}}$ is a unit vector and $\mathbf{V}$ is a generic vector operator.

Solution

We have

$$[\hat{\mathbf{n}} \cdot \mathbf{L},\, V_j] = \sum_i \hat{n}_i\, [L_i,\, V_j] = i\hbar \sum_{ik} \hat{n}_i \underbrace{\epsilon_{ijk}}_{\epsilon_{kij}}\, V_k = i\hbar\, (\mathbf{V} \times \hat{\mathbf{n}})_j\,,$$

where we have used the commutation relations of a vector operator with $\mathbf{L}$.

Problem 5 The Functions $\mathcal{Y}_{lm}(r, \Omega) = r^l\, Y_{lm}(\Omega)$ are Solutions of the Laplace Equation

Define functions $\mathcal{Y}_{lm}(r, \Omega) = r^l\, Y_{lm}(\Omega)$, where the Y_{lm} are the spherical harmonics. Show that

$$\nabla^2\, \mathcal{Y}_{lm}(r, \Omega) = 0\,.$$

The "harmonic" functions $\mathcal{Y}_{lm}(r, \Omega)$ are homogeneous polynomials of order l in $r_\pm$ and r_0, where

$$r_\pm = r_x \pm ir_y = r\,\sin\theta\,e^{\pm i\phi}\,,\qquad r_0 = r_z = r\cos\theta\,.$$

Knowing these facts, obtain the spherical harmonics Y_{33}, Y_{32}, Y_{31}, and Y_{30} up to normalization factors. Check your results for the angular dependence with available tabulations.

Solution

We have

$$\nabla^2\, \mathcal{Y}(r, \Omega) = \left(\frac{1}{r^2} \frac{\partial}{\partial r} r^2 \frac{\partial}{\partial r} - \frac{\mathbf{L}^2}{\hbar^2\, r^2}\right) r^l\, Y_{lm}(\Omega) = l(l+1)r^{l-2}\, Y_{lm}(\Omega) - l(l+1)r^{l-2}\, Y_{lm}(\Omega) = 0\,.$$

The $\mathcal{Y}_{lm}(r, \Omega)$ has a ϕ dependence given by $e^{im\phi}$, and is a homogenous polynomial of order l in $r_\pm$ and r_0,

$$r_\pm = r_x \pm ir_y = r\,\sin\theta\,e^{\pm i\phi}\,,\qquad r_0 = r_z = r\cos\theta\,.$$

The only possibilities are

$$\mathcal{Y}_{33}(r,\Omega) \propto r_+^3 \propto r^3 \sin^3\theta\,e^{3i\phi}\,, \qquad \mathcal{Y}_{32}(r,\Omega) \propto r_+^2\,r_0 \propto r^3\,\sin^2\theta\,\cos\theta\,e^{2i\phi}\,,$$

$$\mathcal{Y}_{31}(r,\Omega) \propto r_+^2 r_- + a r_+ r_0^2\,, \qquad \mathcal{Y}_{30}(r,\Omega) \propto r_+ r_- r_0 + b r_0^3\,,$$

and the first two agree with available tabulations (up to normalization factors). The coefficients a and b can be determined by using the Laplace equation condition. We find

$$\mathcal{Y}_{31}(r,\Omega) \propto r_+(r_+ r_- + a r_0^2) = r_x^3 + r_x r_y^2 + a r_x r_z^2 + i(r_x^2 r_y + r_y^3 + a r_y r_z^2)\,,$$

and

$$(\partial_x^2 + \partial_y^2 + \partial_z^2)[r_x^3 + r_x r_y^2 + a r_x r_z^2 + i(r_x^2 r_y + r_y^3 + a r_y r_z^2)] = 2(4+a)(r_x + i r_y) = 0 \implies a = -4\,;$$

similarly,

$$\mathcal{Y}_{30}(r,\Omega) \propto r_0\,(r_+ r_- + b r_0^2) = r_x^2 r_z + r_y^2 r_z + b r_z^3\,,$$

and

$$(\partial_x^2 + \partial_y^2 + \partial_z^2)[r_x^2 r_z + r_y^2 r_z + b r_z^3] = 2(2+3b)r_z = 0 \implies b = -2/3\,.$$

Thus it follows that

$$\mathcal{Y}_{31}(r,\Omega) \propto r_+^2\,r_- - 4 r_+\,r_0^2 = r^3\,(\sin^3\theta - 4\sin\theta\,\cos^2\theta)\,e^{i\phi} = r^3\,\sin\theta\,(1 - 5\cos^2\theta)\,e^{i\phi}$$

and

$$\mathcal{Y}_{30}(r,\Omega) \propto r_+ r_- r_0 - \frac{2}{3}r_0^3 = r^3\,\cos\theta\left(\sin^2\theta - \frac{2}{3}\cos^2\theta\right) \propto r^3\,(3\cos\theta - 5\cos^3\theta)\,,$$

both of which are in agreement with available tabulations.

Problem 6 Measurements of $\mathbf{L}^2$ and L_z: An Example

The state of a certain particle is represented by the wave function

$$\psi(x,y,z) = N(x + y + z)\,e^{-r^2/\alpha^2}\,,$$

where α is real and N is a normalization constant. The observables $\mathbf{L}^2$ and L_z are measured; what are the probabilities of obtaining, respectively, $2\hbar^2$ and 0? (Adapted from J. J. Sakurai and J. Napolitano 2020, *Modern Quantum Mechanics*, Cambridge University Press.)

Solution

In spherical coordinates the wave function reads

$$\psi(r,\Omega) = N e^{-r^2/\alpha^2} r(\sin\theta\,\cos\phi + \sin\theta\,\sin\phi + \cos\theta)\,.$$

The dependence on θ and ϕ can be expressed as a linear combination of spherical harmonics $Y_{1,m}(\Omega)$ by noting that

$$Y_{1,\pm1}(\Omega) = \mp\sqrt{\frac{3}{8\pi}}\,\sin\theta\,e^{\pm i\phi}\,, \qquad Y_{1,0}(\Omega) = \sqrt{\frac{3}{4\pi}}\,\cos\theta\,,$$

yielding

$$\sin\theta\cos\phi = \sqrt{\frac{2\pi}{3}}\,[Y_{1,-1}(\Omega) - Y_{1,1}(\Omega)]\,,\qquad \sin\theta\sin\phi = i\sqrt{\frac{2\pi}{3}}\,[Y_{1,-1}(\Omega) + Y_{1,1}(\Omega)]$$

and

$$\cos\theta = \sqrt{\frac{4\pi}{3}}\,Y_{1,0}(\Omega)\,.$$

We therefore find that

$$\psi(r,\Omega) = N\sqrt{\frac{4\pi}{3}}\left[\frac{i+1}{\sqrt{2}}\,Y_{1,-1}(\Omega) + \frac{i-1}{\sqrt{2}}\,Y_{1,1}(\Omega) + Y_{1,0}(\Omega)\right] r\,e^{-r^2/\alpha^2}\,,$$

which we write simply as

$$\psi(r,\Omega) = f(r)\sum_{m=\pm1,0} c_{1m}\,Y_{1m}(\Omega)\,,$$

where we define

$$f(r) = N\sqrt{\frac{4\pi}{3}}\,r\,e^{-r^2/\alpha^2}\,,\qquad c_{1\pm1} = \frac{i\mp1}{\sqrt{2}}\,,\qquad c_{10} = 1\,.$$

Note that $|c_{1m}|^2 = 1$. We first obtain the normalization constant:

$$\int d\mathbf{r}\,|\psi(\mathbf{r})|^2 = \int_0^\infty dr\,r^2\,|f(r)|^2 \sum_{m,m'} c^*_{1m'}c_{1m} \underbrace{\int d\Omega\,Y^*_{1m'}(\Omega)\,Y_{1m}(\Omega)}_{\delta_{mm'}} = 3\int_0^\infty dr\,r^2\,|f(r)|^2\,.$$

The normalized wave function can then be written as

$$\overline{\psi}(r,\Omega) = \frac{1}{\sqrt{3}}\,\overline{f}(r)\sum_{m=\pm1,0} c_{1m}\,Y_{1m}(\Omega)\,,\qquad \overline{f}(r) = \frac{1}{\sqrt{\int_0^\infty dr\,r^2\,|f(r)|^2}}\,f(r)\,.$$

Given this wave function, we are asked to calculate the probability of measuring specific values of l and m. In order to answer this question, we introduce the basis consisting of eigenstates of the complete set of commuting observables $\hat{r}$ (the magnitude of the position operator), $\hat{\mathbf{L}}^2$, and $\hat{L}_z$. We denote these eigenstates as $|\varphi_{rlm}\rangle$ and we have

$$\hat{r}\,|\varphi_{rlm}\rangle = r\,|\varphi_{rlm}\rangle\,,\qquad \hat{\mathbf{L}}^2\,|\varphi_{rlm}\rangle = l(l+1)\hbar^2\,|\varphi_{rlm}\rangle\,,\qquad \hat{L}_z\,|\varphi_{rlm}\rangle = m\hbar\,|\varphi_{rlm}\rangle\,.$$

We have reinstated the hat on top of the symbols representing operators for clarity. The basis of eigenstates $|\varphi_{rlm}\rangle$ is continuous. We choose to normalize these states as follows:

$$\langle\varphi_{rlm}|\varphi_{r'l'm'}\rangle = \frac{\delta(r-r')}{r^2}\,\delta_{l,l'}\,\delta_{m,m'}\,,$$

so that the completeness relation is given by

$$\sum_{l=0}^\infty \sum_{m=-l}^l \int_0^\infty dr\,r^2\,|\varphi_{rlm}\rangle\langle\varphi_{rlm}| = \hat{\mathbb{1}}\,.$$

The wave function (in spherical coordinates) corresponding to $|\varphi_{rlm}\rangle$ reads

$$\langle\phi_{\mathbf{r}'}|\varphi_{rlm}\rangle = \frac{\delta(r-r')}{r^2}\,Y_{lm}(\Omega')\,.$$

Given the (normalized) state $|\overline{\psi}\rangle$ above, the probability of measuring l and m values regardless of the r position of the particle is obtained from

$$P_{lm} = \int_0^\infty dr\, r^2\, |\langle \varphi_{rlm}|\overline{\psi}\rangle|^2 \ .$$

It is convenient to evaluate the matrix element in the (spherical) coordinate representation:

$$\langle \varphi_{rlm}|\overline{\psi}\rangle = \int d\mathbf{r}'\langle \varphi_{rlm}|\phi_{\mathbf{r}'}\rangle\langle\phi_{\mathbf{r}'}|\overline{\psi}\rangle = \int_0^\infty dr'\, r'^2 \int d\Omega' \frac{\delta(r-r')}{r^2} Y_{lm}^*(\Omega')\,\overline{\psi}(r',\Omega')$$

$$= \int d\Omega'\, Y_{lm}^*(\Omega')\,\overline{\psi}(r,\Omega') = \frac{1}{\sqrt{3}}\overline{f}(r)\,\delta_{l1}c_{1m}\ ,$$

yielding

$$P_{lm} = \frac{1}{3}\delta_{l1}\,|c_{1m}|^2 \int_0^\infty dr\, r^2\, |\overline{f}(r)|^2 = \frac{1}{3}\delta_{l1}\ .$$

Therefore, we measure the eigenvalue $2\hbar^2$ of $\mathbf{L}^2$ with certainty, and the eigenvalue $m\hbar$ of L_z with probability $1/3$ for $m = \pm 1$ or 0.

Problem 7 Measurement of $\mathbf{L}^2$: An Example

A particle of mass μ is under the influence of a central potential $V(r)$. Its wave function is given by

$$\psi(\mathbf{r}) = (x + y + 3z)f(r)\ .$$

1. Is $\psi(\mathbf{r})$ an eigenfunction of $\mathbf{L}^2$? If so, what is the l-value? If not, what are the possible values of l we may obtain if $\mathbf{L}^2$ is measured?
2. What are the probabilities for the particle to be found in the various m states?
3. Suppose it is known that $\psi(\mathbf{r})$ is an energy eigenfunction with eigenvalue E. Indicate how we may determine the potential $V(r)$.

(Adapted from J. J. Sakurai and J. Napolitano 2020, *Modern Quantum Mechanics*, Cambridge University Press.)

Solution
Part 1

In spherical coordinates the wave function reads

$$\psi(r,\Omega) = rf(r)(\sin\theta\cos\phi + \sin\theta\sin\phi + 3\cos\theta)\ .$$

The dependence on θ and ϕ can be expressed as a linear combination of spherical harmonics $Y_{1,m}(\Omega)$ by noting that

$$Y_{1,\pm1}(\Omega) = \mp\sqrt{\frac{3}{8\pi}}\sin\theta\, e^{\pm i\phi}\ , \qquad Y_{1,0}(\Omega) = \sqrt{\frac{3}{4\pi}}\cos\theta\ ,$$

yielding

$$\sin\theta\cos\phi = \sqrt{\frac{2\pi}{3}}\,[Y_{1,-1}(\Omega) - Y_{1,1}(\Omega)]\ , \qquad \sin\theta\sin\phi = i\sqrt{\frac{2\pi}{3}}\,[Y_{1,-1}(\Omega) + Y_{1,1}(\Omega)]\ ,$$

and

$$\cos \theta = \sqrt{\frac{4\pi}{3}} \, Y_{1,0}(\Omega) \, .$$

We therefore find

$$\psi(r,\Omega) = \sqrt{\frac{2\pi}{3}} \, rf(r) \left[(1+i)Y_{1,-1}(\Omega) - (1-i)Y_{1,1}(\Omega) + 3\sqrt{2} \, Y_{1,0}(\Omega) \right] \, .$$

Since $\psi(r,\Omega)$ consists of a linear combination of spherical harmonics of order $l = 1$, it must be an eigenfunction of $\mathbf{L}^2$ with eigenvalue $2\hbar^2$. A measurement of $\mathbf{L}^2$ would yield this value with certainty. The norm squared of the wave function is given by

$$\int dr \, r^2 \int d\Omega \, |\psi(r,\Omega)|^2 = \frac{2\pi}{3} \underbrace{\int_0^\infty dr \, r^4 \, |f(r)|^2}_{I} \int d\Omega [\cdots]^*[\cdots] = \frac{44\pi}{3} I \, ,$$

making use of the orthonormality of the spherical harmonics, and the normalized wave function is then

$$\psi(r,\Omega) = \frac{1}{\sqrt{22I}} \, rf(r) \left[(1+i)Y_{1,-1}(\Omega) - (1-i)Y_{1,1}(\Omega) + 3\sqrt{2} \, Y_{1,0}(\Omega) \right] \, .$$

Part 2

As in Problem 6, we denote by $|\varphi_{rlm}\rangle$ the eigenstates of the complete set of commuting observables $\hat{r}$ (the magnitude of the position operator), $\hat{\mathbf{L}}^2$, and $\hat{L}_z$,

$$\hat{r} \, |\varphi_{rlm}\rangle = r|\varphi_{rlm}\rangle \, , \qquad \hat{\mathbf{L}}^2 \, |\varphi_{rlm}\rangle = l(l+1)\hbar^2 |\varphi_{rlm}\rangle \, , \qquad \hat{L}_z \, |\varphi_{rlm}\rangle = m\hbar \, |\varphi_{rlm}\rangle \, .$$

Given a normalized state $|\psi\rangle$, the probability of measuring particular l and m values regardless of the r position of the particle is obtained from

$$p_{lm} = \int_0^\infty dr \, r^2 |\langle \varphi_{rlm}|\psi\rangle|^2 \, .$$

Evaluating the matrix element in the (spherical) coordinate representation gives

$$\langle \varphi_{rlm}|\psi\rangle = \int d\Omega Y_{lm}^*(\Omega) \, \psi(r,\Omega) = \frac{1}{\sqrt{22I}} rf(r) \left[(1+i) \, \delta_{l,1} \, \delta_{m,-1} - (1-i) \, \delta_{l,1} \delta_{m,1} + 3\sqrt{2} \, \delta_{l,1} \, \delta_{m,0} \right] \, ,$$

yielding

$$p_{lm} = \frac{1}{22I} \int_0^\infty dr \, r^4 \, |f(r)|^2 \, (2 \, \delta_{l,1} \, \delta_{m,-1} + 2 \, \delta_{l,1} \, \delta_{m,1} + 18 \, \delta_{l,1} \, \delta_{m,0})$$

$$= \frac{1}{11} \, \delta_{l,1} \, \delta_{m,-1} + \frac{1}{11} \, \delta_{l,1} \, \delta_{m,1} + \frac{9}{11} \, \delta_{l,1} \, \delta_{m,0} \, .$$

Therefore, the probability of measuring $m = \pm 1$ is $1/11$, and the probability of measuring $m = 0$ is $9/11$.

Part 3

It is assumed that $\psi(r,\Omega)$ is an eigenfunction of the Hamiltonian corresponding to eigenvalue E; hence

$$\left[-\frac{\hbar^2}{2\mu r} \frac{d^2}{dr^2} r + \frac{\mathbf{L}^2}{2\mu r^2} + V(r) \right] \psi(r,\Omega) = E\psi(r,\Omega) \, ,$$

or, using the fact that $\psi(r, \Omega)$ is also an eigenfunction of $\mathbf{L}^2$ with $l = 1$,

$$\left[-\frac{\hbar^2}{2\mu r} \frac{d^2}{dr^2} r + \frac{\hbar^2}{\mu r^2} + V(r) \right] \psi(r, \Omega) = E \psi(r, \Omega) .$$

Projecting both sides onto $Y_{1,m}(\Omega)$ yields

$$\int d\Omega\, Y_{1,m}^*(\Omega) \left[-\frac{\hbar^2}{2\mu r} \frac{d^2}{dr^2} r + \frac{\hbar^2}{\mu r^2} + V(r) \right] \psi(r, \Omega) = E \int d\Omega\, Y_{1,m}^*(\Omega)\, \psi(r, \Omega) ,$$

and, carrying out the solid angle integrations (common factors resulting from the spherical harmonics drop out),

$$\left[-\frac{\hbar^2}{2\mu r} \frac{d^2}{dr^2} r + \frac{\hbar^2}{\mu r^2} + V(r) \right] r f(r) = E r f(r) ,$$

yielding

$$V(r) = E - \frac{\hbar^2}{\mu r^2} + \frac{\hbar^2}{2\mu r^2} \frac{1}{f(r)} \frac{d^2}{dr^2} r^2 f(r) = E - \frac{\hbar^2}{\mu r^2} + \frac{\hbar^2}{2\mu r^2} \frac{r^2 f''(r) + 4r f'(r) + 2 f(r)}{f(r)}$$

$$= E + \frac{\hbar^2}{2\mu} \frac{r f''(r) + 4 f'(r)}{r f(r)} ,$$

and the function $f(r)$ must be real.

Problem 8   Ehrenfest's Relations for $\langle \mathbf{L}(t) \rangle$

Consider a particle in three dimensions with Hamiltonian given by

$$\hat{H} = \frac{\hat{\mathbf{p}}^2}{2m} + \hat{V}(\hat{\mathbf{r}}) .$$

Show that the time derivative of the expectation value of the orbital angular momentum operator $\hat{\mathbf{L}} = \hat{\mathbf{r}} \times \hat{\mathbf{p}}$ is given by

$$\frac{d}{dt} \langle \psi(t) | \hat{\mathbf{L}} | \psi(t) \rangle = -\langle \psi(t) | \hat{\mathbf{r}} \times \boldsymbol{\nabla} \hat{V}(\hat{\mathbf{r}}) | \psi(t) \rangle .$$

Does this equation have a classical counterpart?

Solution

The rate of change in time of the expectation value of an operator – in this case the orbital angular momentum $\hat{\mathbf{L}}$ – is given by

$$\frac{d}{dt} \langle \psi(t) | \hat{\mathbf{L}} | \psi(t) \rangle = \frac{i}{\hbar} \langle \psi(t) | [\hat{H}, \hat{\mathbf{L}}] | \psi(t) \rangle .$$

Consider the commutator

$$[\hat{H}, \hat{\mathbf{L}}] = \frac{1}{2m} \underbrace{[\hat{\mathbf{p}}^2, \hat{\mathbf{L}}]}_{\text{term 1}} + \underbrace{[\hat{V}(\hat{\mathbf{r}}), \hat{\mathbf{L}}]}_{\text{term 2}} .$$

Term 1 is immediately seen to vanish ($\hat{\mathbf{p}}^2$ is a scalar operator, and as such it commutes with the angular momentum; see Problem 1). For term 2, consider the x-component,

$$[\hat{V}, \hat{L}_x] = [\hat{V}, \hat{y}\hat{p}_z - \hat{z}\hat{p}_y] = \hat{y}[\hat{V}, \hat{p}_z] - \hat{z}[\hat{V}, \hat{p}_y] = i\hbar\hat{y}\frac{\partial\hat{V}}{\partial z} - i\hbar\hat{z}\frac{\partial\hat{V}}{\partial y},$$

and similarly for the other components (here, $\partial\hat{V}/\partial z$ is to be understood as the operator obtained by replacing the vector $\mathbf{r}$ by the vector operator $\hat{\mathbf{r}}$ in the function $\partial V(\mathbf{r})/\partial z$; hence, we have

$$[\hat{V}(\hat{\mathbf{r}}), \hat{\mathbf{L}}] = i\hbar\,\hat{\mathbf{r}} \times \mathbf{\nabla}\hat{V}(\hat{\mathbf{r}}),$$

yielding the required result,

$$\frac{d}{dt}\langle\psi(t)|\hat{\mathbf{L}}|\psi(t)\rangle = -\langle\psi(t)|\hat{\mathbf{r}} \times \mathbf{\nabla}\hat{V}(\hat{\mathbf{r}})|\psi(t)\rangle.$$

This is the quantum mechanical counterpart to the equation of classical mechanics $\dot{\mathbf{L}} = \boldsymbol{\tau}$, where $\boldsymbol{\tau}$ is the torque, defined as $\boldsymbol{\tau} = \mathbf{r} \times \mathbf{F} = -\mathbf{r} \times \mathbf{\nabla}V$.

Problem 9   Some Identities Involving r, p, and L

Show that the following properties relating to the orbital angular momentum operator $\mathbf{L} = \mathbf{r} \times \mathbf{p}$ are satisfied:

1. $\mathbf{r} \cdot \mathbf{L}$ and $\mathbf{L} \cdot \mathbf{r}$, and similarly $\mathbf{p} \cdot \mathbf{L}$ and $\mathbf{L} \cdot \mathbf{p}$, are null operators;
2. $\mathbf{L}^2 = -\mathbf{r} \cdot [\mathbf{p}(\mathbf{p} \cdot \mathbf{r}) - \mathbf{p}^2\,\mathbf{r}]$ (consider the order of the operators); next show that

$$[\mathbf{r}, \mathbf{p}^2] = 2i\hbar\,\mathbf{p}, \qquad \mathbf{r} \cdot \mathbf{p} - \mathbf{p} \cdot \mathbf{r} = 3i\hbar,$$

and hence obtain

$$\mathbf{L}^2 = r^2\,\mathbf{p}^2 + i\hbar\,\mathbf{r} \cdot \mathbf{p} - (\mathbf{r} \cdot \mathbf{p})^2;$$

3. By direct calculation show that in spherical coordinates

$$\mathbf{r} \cdot \mathbf{p} = -i\hbar r\frac{\partial}{\partial r},$$

and using the result in part 2 obtain

$$\mathbf{L}^2 = r^2\,\mathbf{p}^2 + \hbar^2\frac{\partial}{\partial r}r^2\frac{\partial}{\partial r}.$$

Solution

Part 1

We find

$$\mathbf{r} \cdot \mathbf{L} = \sum_i r_i L_i = \sum_{ijk}\epsilon_{ijk}r_ir_jp_k = \sum_k\underbrace{\left(\sum_{ij}\epsilon_{ijk}r_ir_j\right)}_{(\mathbf{r}\times\mathbf{r})_k}p_k = 0,$$

and similarly

$$\mathbf{L} \cdot \mathbf{r} = \sum_i L_ir_i = \sum_{ijk}\epsilon_{ijk}r_jp_kr_i = \sum_{ijk}\epsilon_{ijk}p_kr_ir_j = \sum_k p_k\underbrace{\left(\sum_{ij}\epsilon_{kij}r_ir_j\right)}_{(\mathbf{r}\times\mathbf{r})_k} = 0,$$

where in the third step we have used the fact that, owing to ϵ_{ijk}, the subscripts must satisfy $j \neq k$ and hence the operators r_j and p_k commute. A similar argument shows that $\mathbf{p} \cdot \mathbf{L}$ and $\mathbf{L} \cdot \mathbf{p}$ vanish, since the operator $\mathbf{p} \times \mathbf{p}$ is the null operator.

Part 2

First note that

$$\mathbf{r} \times \mathbf{p} = -\mathbf{p} \times \mathbf{r}$$

and hence

$$\mathbf{L}^2 = -(\mathbf{r} \times \mathbf{p}) \cdot (\mathbf{p} \times \mathbf{r}) = -\sum_{ijklm} \epsilon_{ijk}\, \epsilon_{ilm}\, r_j p_k p_l r_m = -\sum_{jklm} \left(\delta_{jl}\, \delta_{km} - \delta_{jm}\, \delta_{kl} \right) r_j p_k p_l r_m$$

$$= -(\mathbf{r} \cdot \mathbf{p})\,(\mathbf{p} \cdot \mathbf{r}) + \sum_j r_j\, \mathbf{p}^2\, r_j = -\mathbf{r} \cdot [\, \mathbf{p}\,(\mathbf{p} \cdot \mathbf{r}) - \mathbf{p}^2\, \mathbf{r}\,]\,.$$

We find

$$[r_i, \mathbf{p}^2] = \sum_j [r_i, p_j^2] = \sum_j \left(p_j\, [r_i, p_j] + [r_i, p_j] p_j \right) = 2i\hbar \sum_j \delta_{ij}\, p_j = 2i\hbar\, p_i\,,$$

and

$$\mathbf{r} \cdot \mathbf{p} - \mathbf{p} \cdot \mathbf{r} = \sum_i (r_i p_i - p_i r_i) = \sum_i [r_i, p_i] = i\hbar \sum_i \delta_{ii} = 3i\hbar\,,$$

from which we deduce that

$$\mathbf{L}^2 = -\mathbf{r} \cdot \mathbf{p}\, \underbrace{(\mathbf{r} \cdot \mathbf{p} - 3i\hbar)}_{\mathbf{p} \cdot \mathbf{r}} + \mathbf{r} \cdot \underbrace{(\mathbf{r}\,\mathbf{p}^2 - 2i\hbar\,\mathbf{p})}_{\mathbf{p}^2\,\mathbf{r}} = r^2\,\mathbf{p}^2 + i\hbar\,\mathbf{r} \cdot \mathbf{p} - (\mathbf{r} \cdot \mathbf{p})^2\,.$$

Part 3

Using the results of Problem 2, it follows that

$$x\,\partial_x + y\,\partial_y + z\,\partial_z = r \sin\theta \cos\phi\, [\sin\theta \cos\phi\, \partial_r + (\cos\theta \cos\phi/r)\, \partial_\theta - (\sin\phi/\sin\theta/r)\, \partial_\phi]$$
$$+ r \sin\theta \sin\phi\, [\sin\theta \sin\phi\, \partial_r + (\cos\theta \sin\phi/r)\, \partial_\theta + (\cos\phi/\sin\theta/r)\, \partial_\phi]$$
$$+ r \cos\theta\, [\cos\theta\, \partial_r - (\sin\theta/r)\, \partial_\theta] = r\,\partial_r\,,$$

which could have been guessed given that $\mathbf{r} \cdot \mathbf{p}$, as a scalar operator, commutes with $\mathbf{L}$; since the components L_i depend on θ and ϕ, and derivatives with respect to these variables, it follows that $\mathbf{r} \cdot \mathbf{p}$ cannot have any dependence on θ, ∂_θ or ϕ, ∂_ϕ if it has to commute with each L_i. We arrive at

$$\mathbf{L}^2 = r^2\,\mathbf{p}^2 + \hbar^2 r\partial_r + \hbar^2 (r\partial_r)^2 = r^2\,\mathbf{p}^2 + \hbar^2 [2r\partial_r + r^2 \partial_r^2] = r^2\,\mathbf{p}^2 + \hbar^2 \partial_r r^2 \partial_r\,.$$

Problem 10   More Identities Involving $\hat{\mathbf{L}}$ for a System of N Particles

A system consists of N particles with position and momentum operators given by $\hat{\mathbf{r}}_i$ and $\hat{\mathbf{p}}_i$ with $i = 1, 2, \ldots, N$. The $\hat{\mathbf{r}}_i$ and $\hat{\mathbf{p}}_i$ operators for different particles commute. Define the operator

$$\hat{\mathbf{L}} = \sum_{i=1}^{N} \hat{\mathbf{L}}_i\,, \qquad \hat{\mathbf{L}}_i = \hat{\mathbf{r}}_i \times \hat{\mathbf{p}}_i\,.$$

1. Show that the components of $\hat{\mathbf{L}}$ satisfy the commutation relations that define the angular momentum operator. Deduce from this that if $\mathbf{v}$ and $\mathbf{v}'$ are two generic vectors (not vector operators) then

$$\left[\hat{\mathbf{L}} \cdot \mathbf{v}, \hat{\mathbf{L}} \cdot \mathbf{v}'\right] = i\hbar\,(\mathbf{v} \times \mathbf{v}') \cdot \hat{\mathbf{L}}\,.$$

2. Evaluate the commutators of the components of $\hat{\mathbf{L}}$ with those of $\hat{\mathbf{r}}_i$ and $\hat{\mathbf{p}}_i$ and obtain the following three identities:

$$\left[\hat{\mathbf{L}}, \hat{\mathbf{r}}_i \cdot \hat{\mathbf{r}}_j\right] = 0\,, \qquad \left[\hat{\mathbf{L}}, \hat{\mathbf{L}} \cdot \hat{\mathbf{r}}_i\right] = 0\,, \qquad \left[\hat{\mathbf{L}} \cdot \hat{\mathbf{r}}_i, \hat{\mathbf{L}} \cdot \hat{\mathbf{r}}_j\right] = -i\hbar\,(\hat{\mathbf{r}}_i \times \hat{\mathbf{r}}_j) \cdot \hat{\mathbf{L}} = -i\hbar\,\hat{\mathbf{L}} \cdot (\hat{\mathbf{r}}_i \times \hat{\mathbf{r}}_j)\,.$$

Now, introduce the operators

$$\hat{V} = \sum_{i=1}^n a_i\,\hat{\mathbf{r}}_i\,, \qquad \hat{V}' = \sum_{i=1}^n a_i'\,\hat{\mathbf{r}}_i\,,$$

where the a_i and a_i' are given real coefficients. Show that

$$\left[\hat{\mathbf{L}} \cdot \hat{V}, \hat{\mathbf{L}} \cdot \hat{V}'\right] = -i\hbar\,(\hat{V} \times \hat{V}') \cdot \hat{\mathbf{L}}\,.$$

Solution

Part 1

Since angular momentum components of different particles commute, we find

$$\left[\hat{L}_\alpha, \hat{L}_\beta\right] = \sum_{i,j=1}^N \left[\hat{L}_{i,\alpha}, \hat{L}_{j,\beta}\right] = \sum_{i=1}^N \left[\hat{L}_{i,\alpha}, \hat{L}_{i,\beta}\right] = i\hbar \sum_{i=1}^N \sum_\gamma \epsilon_{\alpha\beta\gamma}\,\hat{L}_{i,\gamma} = i\hbar \sum_\gamma \epsilon_{\alpha\beta\gamma}\,\hat{L}_\gamma\,,$$

and so $\hat{\mathbf{L}}$ (the total orbital angular momentum) satisfies the expected commutation relations. Using this result, it immediately follows that

$$\left[\hat{\mathbf{L}} \cdot \mathbf{v}, \hat{\mathbf{L}} \cdot \mathbf{v}'\right] = \sum_{\alpha\beta} v_\alpha v_\beta' \left[\hat{L}_\alpha, \hat{L}_\beta\right] = i\hbar \sum_{\alpha\beta\gamma} \epsilon_{\alpha\beta\gamma}\,v_\alpha v_\beta'\,\hat{L}_\gamma = i\hbar\,(\mathbf{v} \times \mathbf{v}') \cdot \hat{\mathbf{L}}\,.$$

Part 2

The operators $\hat{\mathbf{r}}_i \cdot \hat{\mathbf{r}}_j$ and $\hat{\mathbf{L}} \cdot \hat{\mathbf{r}}_i$ are scalar operators with respect to rotations of the whole system, and hence commute with the generator $\mathbf{L}$ of these rotations. This is easily verified. Since $\hat{\mathbf{r}}_i$ and $\hat{\mathbf{p}}_i$ are vector operators, their commutation relations with $\hat{L}_i$ are given by

$$\left[\hat{L}_{i,\alpha}, \hat{r}_{i,\beta}\right] = i\hbar \sum_\gamma \epsilon_{\alpha\beta\gamma}\,\hat{r}_{i,\gamma}\,, \qquad \left[\hat{L}_{i,\alpha}, \hat{p}_{i,\beta}\right] = i\hbar \sum_\gamma \epsilon_{\alpha\beta\gamma}\,\hat{p}_{i,\gamma}\,,$$

and hence

$$\left[\hat{L}_\alpha, \hat{r}_{i,\beta}\right] = \sum_{k=1}^N \left[\hat{L}_{k,\alpha}, \hat{r}_{i,\beta}\right] = \left[\hat{L}_{i,\alpha}, \hat{r}_{i,\beta}\right] = i\hbar \sum_\gamma \epsilon_{\alpha\beta\gamma}\,\hat{r}_{i,\gamma}\,,$$

and similarly for $\left[\hat{L}_\alpha, \hat{p}_{i,\beta}\right] = i\hbar \sum_\gamma \epsilon_{\alpha\beta\gamma} \hat{p}_{i,\gamma}$. We also obtain

$$
\begin{aligned}
\left[\hat{L}_\alpha, \hat{\mathbf{r}}_i \cdot \hat{\mathbf{r}}_j\right] &= \sum_\beta \left[\hat{L}_\alpha, \hat{r}_{i,\beta}\,\hat{r}_{j,\beta}\right] = \sum_\beta \left(\hat{r}_{i,\beta}\left[\hat{L}_\alpha, \hat{r}_{j,\beta}\right] + \left[\hat{L}_\alpha, \hat{r}_{i,\beta}\right]\hat{r}_{j,\beta}\right) \\
&= i\hbar \sum_{\beta\gamma} \epsilon_{\alpha\beta\gamma}\left(\hat{r}_{i,\beta}\,\hat{r}_{j,\gamma} + \hat{r}_{i,\gamma}\,\hat{r}_{j,\beta}\right) = i\hbar \left(\hat{\mathbf{r}}_i \times \hat{\mathbf{r}}_j + \hat{\mathbf{r}}_j \times \hat{\mathbf{r}}_i\right)_\alpha = 0 \,,
\end{aligned}
$$

since the components of $\hat{r}_{i,\beta}$ and $\hat{r}_{j,\gamma}$ commute for any i and j. Similarly, we have

$$
\sum_\beta \left[\hat{L}_\alpha, \hat{L}_\beta\,\hat{r}_{i,\beta}\right] = i\hbar \sum_{\beta\gamma}\left(\epsilon_{\alpha\beta\gamma}\,\hat{L}_\beta\,\hat{r}_{i,\gamma} + \underbrace{\epsilon_{\alpha\beta\gamma}}_{-\epsilon_{\alpha\gamma\beta}}\,\hat{L}_\gamma\,\hat{r}_{i,\beta}\right) = i\hbar\left(\hat{\mathbf{L}} \times \hat{\mathbf{r}}_i - \hat{\mathbf{L}} \times \hat{\mathbf{r}}_i\right)_\alpha = 0 \,.
$$

Now, consider

$$
\begin{aligned}
\left[\hat{\mathbf{L}} \cdot \hat{\mathbf{r}}_i, \hat{\mathbf{L}} \cdot \hat{\mathbf{r}}_j\right] &= \sum_\alpha \left(L_\alpha\left[\hat{r}_{i,\alpha}, \hat{\mathbf{L}} \cdot \hat{\mathbf{r}}_j\right] + \left[\hat{L}_\alpha, \hat{\mathbf{L}} \cdot \hat{\mathbf{r}}_j\right]\hat{r}_{i,\alpha}\right) = \sum_{\alpha\beta}\hat{L}_\alpha\left[\hat{r}_{i,\alpha}, \hat{L}_\beta\,\hat{r}_{j,\beta}\right] \\
&= \sum_{\alpha\beta}\left(\hat{L}_\alpha\hat{L}_\beta\left[\hat{r}_{i,\alpha}, \hat{r}_{j,\beta}\right] + \hat{L}_\alpha\left[\hat{r}_{i,\alpha}, \hat{L}_\beta\right]\hat{r}_{j,\beta}\right) = -i\hbar \sum_{\alpha\beta\gamma}\underbrace{\epsilon_{\beta\alpha\gamma}}_{\epsilon_{\alpha\gamma\beta}}\hat{L}_\alpha\,\hat{r}_{i,\gamma}\,\hat{r}_{j,\beta} \\
&= -i\hbar\,\hat{\mathbf{L}} \cdot (\hat{\mathbf{r}}_i \times \hat{\mathbf{r}}_j) = -i\hbar(\hat{\mathbf{r}}_i \times \hat{\mathbf{r}}_j) \cdot \hat{\mathbf{L}} \,,
\end{aligned}
$$

where the last step follows on noting that, since $\hat{\mathbf{r}}_i \times \hat{\mathbf{r}}_j$ is a vector operator, it has commutation relations $[\hat{L}_\alpha, (\hat{\mathbf{r}}_i \times \hat{\mathbf{r}}_j)_\beta] = i\hbar \sum_\gamma \epsilon_{\alpha\beta\gamma}(\hat{\mathbf{r}}_i \times \hat{\mathbf{r}}_j)_\gamma$, which vanish when $\alpha = \beta$, yielding $\hat{\mathbf{L}} \cdot (\hat{\mathbf{r}}_i \times \hat{\mathbf{r}}_j) = \sum_\alpha \hat{L}_\alpha(\hat{\mathbf{r}}_i \times \hat{\mathbf{r}}_j)_\alpha = \sum_\alpha(\hat{\mathbf{r}}_i \times \hat{\mathbf{r}}_j)_\alpha \hat{L}_\alpha = (\hat{\mathbf{r}}_i \times \hat{\mathbf{r}}_j) \cdot \hat{\mathbf{L}}$. As a consequence of the previous commutator, we also obtain

$$
\left[\hat{\mathbf{L}} \cdot \hat{\mathbf{V}}, \hat{\mathbf{L}} \cdot \hat{\mathbf{V}}'\right] = \sum_{ij} a_i a_j'\left[\hat{\mathbf{L}} \cdot \hat{\mathbf{r}}_i, \hat{\mathbf{L}} \cdot \hat{\mathbf{r}}_j\right] = -i\hbar \sum_{ij} a_i a_j'(\hat{\mathbf{r}}_i \times \hat{\mathbf{r}}_j) \cdot \hat{\mathbf{L}} = -i\hbar\,(\hat{\mathbf{V}} \times \hat{\mathbf{V}}') \cdot \hat{\mathbf{L}} \,,
$$

which should be contrasted with the relation found in part 1; that relation has the opposite sign on the right-hand side when $\mathbf{v}$ and $\mathbf{v}'$ are ordinary vectors, for example, when they are the unit vectors specifying the directions of a right-handed cartesian coordinate system.

Problem 11 Expectation Values of $\langle\hat{L}_\alpha\rangle_{lm}$ and $\langle\hat{L}_\alpha^2\rangle_{lm}$

Suppose that a particle is in an eigenstate of $\hat{\mathbf{L}}^2$ and $\hat{L}_z$ with eigenvalues $l(l+1)\hbar^2$ and $m\hbar$, respectively. Show that the expectation values $\langle\hat{L}_x\rangle$ and $\langle\hat{L}_y\rangle$ vanish. Using the explicit expressions of $\hat{L}_x$ and $\hat{L}_y$ in spherical coordinates (see Problem 2), show also that

$$
\langle\hat{L}_x^2\rangle = \langle\hat{L}_y^2\rangle \,,
$$

and hence that

$$
\langle\hat{L}_x^2\rangle = \langle\hat{L}_y^2\rangle = \frac{\hbar^2}{2}\left[l(l+1) - m^2\right] \,.
$$

Solution

Using the commutation relations satisfied by the components of $\hat{\mathbf{L}}$, we have

$$
i\hbar\hat{L}_x = [\hat{L}_y, \hat{L}_z] \,, \qquad i\hbar\hat{L}_y = [\hat{L}_z, \hat{L}_x] \,,
$$

and

$$\langle Y_{lm}|\hat{L}_x|Y_{lm}\rangle = -\frac{i}{\hbar}\langle Y_{lm}|\hat{L}_y\hat{L}_z - \hat{L}_z\hat{L}_y|Y_{lm}\rangle = -i(m-m)\langle Y_{lm}|\hat{L}_y|Y_{lm}\rangle = 0 \,,$$

where we have used $\hat{L}_z|Y_{lm}\rangle = m\hbar|Y_{lm}\rangle$); similarly, we find $\langle Y_{lm}|\hat{L}_y|Y_{lm}\rangle = 0$.

We recall that the ϕ dependence of the spherical harmonics is given by

$$Y_{lm}(\theta, \phi) = e^{im\phi}\, Y_{lm}(\theta, 0) \,,$$

so that

$$\hat{L}_x|Y_{lm}\rangle \longrightarrow i\hbar\left(\sin\phi\,\frac{\partial}{\partial\theta} + \cot\theta\,\cos\phi\,\frac{\partial}{\partial\phi}\right)e^{im\phi}\,Y_{lm}(\theta,0) = i\hbar\,e^{im\phi}\left(\sin\phi\,\frac{\partial}{\partial\theta} + im\cot\theta\,\cos\phi\right)Y_{lm}(\theta,0) \,,$$

and similarly

$$\hat{L}_y|Y_{lm}\rangle \longrightarrow i\hbar\left(-\cos\phi\,\frac{\partial}{\partial\theta} + \cot\theta\,\sin\phi\,\frac{\partial}{\partial\phi}\right)e^{im\phi}\,Y_{lm}(\theta,0) = i\hbar\,e^{im\phi}\left(-\cos\phi\,\frac{\partial}{\partial\theta} + im\cot\theta\,\sin\phi\right)Y_{lm}(\theta,0) \,,$$

Now, consider the expectation value

$$\langle Y_{lm}|\hat{L}_x^2|Y_{lm}\rangle = \hbar^2\int_0^\pi d\theta\,\sin\theta\int_0^{2\pi}d\phi\,\left|\left(\sin\phi\,\frac{\partial}{\partial\theta} + im\cot\theta\,\cos\phi\right)Y_{lm}(\theta,0)\right|^2 \,,$$

and the ϕ-integrations are easily performed,

$$\int_0^{2\pi}d\phi\,\sin^2\phi = \int_0^{2\pi}d\phi\,\cos^2\phi = \pi \,, \qquad \int_0^{2\pi}d\phi\,\sin\phi\,\cos\phi = 0 \,,$$

to yield

$$\langle Y_{lm}|\hat{L}_x^2|Y_{lm}\rangle = \pi\hbar^2\int_0^\pi d\theta\,\sin\theta\left[\left|\frac{\partial}{\partial\theta}Y_{lm}(\theta,0)\right|^2 + m^2\,\cot^2\theta\,|Y_{lm}(\theta,0)|^2\right] = \langle Y_{lm}|\hat{L}_y^2|Y_{lm}\rangle \,.$$

We also have $\hat{L}_x^2 + \hat{L}_y^2 = \hat{\mathbf{L}}^2 - \hat{L}_z^2$, and hence

$$\langle Y_{lm}|\hat{L}_x^2|Y_{lm}\rangle + \langle Y_{lm}|\hat{L}_y^2|Y_{lm}\rangle = \langle Y_{lm}|\hat{\mathbf{L}}^2|Y_{lm}\rangle - \langle Y_{lm}|\hat{L}_z^2|Y_{lm}\rangle = [l(l+1) - m(m+1)]\hbar^2 \,,$$

so that

$$\langle Y_{lm}|\hat{L}_x^2|Y_{lm}\rangle = \langle Y_{lm}|\hat{L}_y^2|Y_{lm}\rangle = [l(l+1) - m(m+1)]\frac{\hbar^2}{2} \,.$$

10 Bound States in a Central Potential: Applications

As we have seen, the wave functions in a central potential can be written as

$$\psi_{Elm}(r,\Omega) = \frac{u_{El}(r)}{r} Y_{lm}(\Omega) , \qquad (10.1)$$

and are eigenfunctions of H, $\mathbf{L}^2$, and L_z with eigenvalues E, $l(l+1)\hbar^2$, and $m\hbar$, respectively. The reduced radial wave functions $u_{El}(r)$ in the region $0 < r < \infty$ satisfy (for each l) the one-dimensional Schrödinger equation

$$-\frac{\hbar^2}{2\mu} u_{El}''(r) + \left[\frac{l(l+1)\hbar^2}{2\mu r^2} + V(r)\right] u_{El}(r) = E u_{El}(r) . \qquad (10.2)$$

The boundary conditions are $u_{El}(r) \propto r^{l+1}$ as $r \longrightarrow 0$ and $u_{El}(r) \longrightarrow 0$ as $r \longrightarrow \infty$ for bound states (scattering, or continuum, states require only that $u_{El}(r)$ is bounded in $0 < r < \infty$; we will return to these later). Note that the subscript E specifying the energy eigenvalue is not explicitly included in most of the formulae below.

10.1 Hydrogen-Like Atoms

We consider an electron of charge $-e$ ($e > 0$) in the attractive Coulomb field of a nucleus of charge Ze (for the hydrogen atom $Z=1$), fixed at the origin. The Schrödinger equation for the reduced radial wave function $u_l(r)$ reads

$$\frac{d^2 u_l(r)}{dr^2} + \left[\underbrace{\bar{\epsilon} + \frac{2\mu}{\hbar^2}\frac{Ze^2}{r} - \frac{l(l+1)}{r^2}}_{-v_{\text{eff}}(r)}\right] u_l(r) = 0 , \qquad \bar{\epsilon} = \frac{2\mu E}{\hbar^2} , \qquad (10.3)$$

which corresponds to a one-dimensional Schrödinger equation for an electron constrained in the region in $0 \leq r < \infty$ and under the action of an effective potential given by

$$v_{\text{eff}}(r) = -\frac{2\mu}{\hbar^2}\frac{Ze^2}{r} + \frac{l(l+1)}{r^2} . \qquad (10.4)$$

This effective potential is repulsive at short distances because of the centrifugal barrier (for $l \geq 1$), and is attractive at large distances because of the Coulomb potential. For $l \geq 1$, $v_{\text{eff}}(r)$ has a minimum at

$$r_{\min} = l(l+1)\frac{\hbar^2}{Ze^2\mu} , \qquad v_{\text{eff}}(r_{\min}) = -\frac{(Ze^2\mu)^2}{l(l+1)\hbar^4} . \qquad (10.5)$$

Therefore, if bound states occur (as they do) for $l \geq 1$ then the corresponding energies must be in the range $v_{\text{eff}}(r_{\min}) \leq \epsilon < 0$. For the S-wave ($l=0$), the only requirement is $\epsilon < 0$.

We introduce the following convenient variables:

$$\alpha = \frac{e^2}{\hbar c} \approx \frac{1}{137} \text{ (fine structure constant)}, \quad a_0 = \frac{\hbar^2}{\mu e^2} \approx 0.529 \times 10^{-8} \text{ cm (Bohr radius)}, \quad r_0 = \frac{a_0}{Z}. \tag{10.6}$$

Rescaling r as $x = r/r_0$ yields

$$\frac{d^2 u_l(x)}{dx^2} + \left[-\epsilon + \frac{2}{x} - \frac{l(l+1)}{x^2} \right] u_l(x) = 0, \tag{10.7}$$

where, taking into account that $\bar{\epsilon}$ (or E) is less than zero for bound states, we define

$$\epsilon = r_0^2 |\bar{\epsilon}| = \frac{a_0^2}{Z^2} \frac{2\mu}{\hbar^2} |E| = \frac{a_0^2}{Z^2} \underbrace{\frac{e^2 \mu}{\hbar^2}}_{1/a_0} \frac{2}{e^2} |E| = \frac{|E|}{E_0}. \tag{10.8}$$

The energy E_0 is given by

$$E_0 = \frac{Z^2 e^2}{2 a_0} = Z^2 \underbrace{\frac{\alpha^2 \mu c^2}{2}}_{\text{Rydberg}} \approx Z^2 \times 13.6 \text{ eV}. \tag{10.9}$$

Next, we remove the known short- and long-range behavior of $u_l(x)$ by introducing a function $F_l(x)$ such that

$$u_l(x) = x^{l+1} e^{-\sqrt{\epsilon} x} F_l(x), \tag{10.10}$$

which leads to the following equation for $F_l(x)$:

$$x F_l''(x) + \left[2(l+1) - 2\sqrt{\epsilon} x \right] F_l'(x) + \left[2 - 2\sqrt{\epsilon}(l+1) \right] F_l(x) = 0. \tag{10.11}$$

As a last step, we rescale x by defining $z = 2\sqrt{\epsilon} x$ to arrive at the hypergeometric confluent equation

$$z F_l''(z) + (2l + 2 - z) F_l'(z) + (\xi - l - 1) F_l(z) = 0, \qquad \xi = \frac{1}{\sqrt{\epsilon}}. \tag{10.12}$$

and, in order for $F_l(z)$ to be an acceptable solution, we must have (see Problem 2)

$$l + 1 - \xi = -\nu \implies \epsilon = \frac{1}{(\nu + l + 1)^2}, \qquad \nu = 0, 1, 2, \ldots, \tag{10.13}$$

which then implies that $F_l(z)$ is a polynomial in z. We define $n = \nu + l + 1$, and express the bound-state energies (recall that $\epsilon = -E/E_0$ with $E < 0$) as

$$E_n = -\frac{E_0}{n^2} = -\frac{Z^2}{2n^2} \frac{e^2}{a_0} = -\frac{(Z\alpha)^2}{2n^2} \mu c^2 = -\frac{Z^2}{n^2} \times 13.6 \text{ eV}, \qquad n \geq 1, \tag{10.14}$$

where, for a given n,

$$l = 0, 1, \ldots, n - 1. \tag{10.15}$$

The energies are degenerate, and the degree g_n of degeneracy is given by

$$g_n = \sum_{l=0}^{n-1} \underbrace{(2l+1)}_{(2l+1) \ m \text{ values}} = 2 \frac{n(n-1)}{2} + n = n^2. \tag{10.16}$$

where we have used $\sum_{k=0}^{p} k = p(p+1)/2$ and $\sum_{k=0}^{p} 1 = p + 1$. Hence, the ground state has $n = 1$ and is non-degenerate (of course, we are ignoring here the fact that the electron and nucleus have intrinsic angular momenta, i.e., spin angular momenta; we will come back to this later). Collecting results, and keeping track of the various definitions, the corresponding wave functions can be written as

$$\psi_{nlm}(r, \Omega) = \underbrace{N_{nl}\rho_n^l L_{n-l-1}^{2l+1}(\rho_n)\, e^{-\rho_n/2}}_{R_{nl}(r)} \; Y_{lm}(\Omega)\,, \tag{10.17}$$

where N_{nl} is a normalization constant, the variable ρ_n is defined as

$$\rho_n = 2\sqrt{\epsilon}\, x = \frac{2}{n}\frac{r}{r_0} = \frac{2}{n}\frac{r}{a_0/Z}\,, \tag{10.18}$$

and the $L_p^k(x)$ are polynomials of order p in x, known as Laguerre polynomials. They are related to the hypergeometric-confluent solutions of Eq. (10.12), and can be obtained from

$$L_p^0(x) = e^x \frac{d^p}{dx^p}\left(e^{-x} x^p\right)\,, \qquad L_p^k(x) = (-1)^k \frac{d^k}{dx^k} L_{p+k}^0(x)\,, \tag{10.19}$$

with explicit expressions given by

$$L_p^k(x) = \sum_{q=0}^{p}(-)^q \frac{[(p+k)!\,]^2}{(p-q)!\,(k+q)!\,q!}\, x^q\,. \tag{10.20}$$

The $L_p^k(x)$ can be shown to have p zeros in $0 < x < \infty$ and to satisfy the "orthogonality" condition

$$\int_0^\infty dx\, x^k\, e^{-x}\, L_p^k(x)\, L_q^k(x) = \frac{[(p+k)!\,]^3}{p!}\,\delta_{pq}\,. \tag{10.21}$$

It can be seen that, the wave function decreases at large ρ_n as $\rho_n^l\, \rho_n^{n-l-1}\, e^{-\rho_n/2}$, or, in terms of r,

$$\psi(r, \Omega) \sim r^{n-1}\, e^{-(Z/n)r/a_0}\,, \tag{10.22}$$

and the electron is localized within a region of radius na_0/Z. The results we have derived here are for a single-electron atom in the Coulomb field of a nucleus of charge Ze, and so they are valid for the hydrogen atom, the singly ionized helium atom (which has $Z=2$), or the doubly ionized lithium atom ($Z=3$), and so forth. We have already noted that the degeneracy of the energy levels is $g_n = n^2$. This is due to the $1/r$ behavior of the Coulomb potential,[1] and the accidental extra degeneracy on l (beyond the $2l+1$ degeneracy from rotational symmetry) originates from an additional symmetry of the hydrogen-like Hamiltonian (see Problem 14 in Chapter 11). In multielectron atoms the energies depend on both n and l, since in that case the electrostatic potential ceases to be strictly proportional to $1/r$.

According to the standard notation, the states with $l = 0, 1, 2, 3, 4, \ldots$ are labeled as s, p, d, f, g, $\ldots$, (the letters stand for "sharp", "principal", "diffuse", and so on, for reasons which have to do with the appearance of the spectral lines). Hydrogen-like atom states are often labeled as 1s, 2s, 2p, 3s, 3p, 3d, and so forth, where 1, 2, 3, $\ldots$ specify the "principal" quantum number n.

10.2 Harmonic Oscillator in Three Dimensions

We consider a particle in a harmonic oscillator potential of natural angular frequency ω,

$$V(r) = \frac{\mu\,\omega^2}{2}\, r^2\,. \tag{10.23}$$

---

[1] It turns out that the harmonic oscillator potential too shares this feature, in that an accidental extra degeneracy on l appears, as we will see.

The Schrödinger equation for the reduced radial wave function $u_l(r)$ is given by

$$\frac{d^2 u_l(r)}{dr^2} + \left[\bar{\epsilon} - \frac{\mu^2 \omega^2}{\hbar^2} r^2 - \frac{l(l+1)}{r^2}\right] u_l(r) = 0 , \qquad \bar{\epsilon} = \frac{2\mu E}{\hbar^2} . \tag{10.24}$$

Only bound states are possible, as it will be apparent below from the asymptotic behavior of $u_l(r)$. It is convenient to define the variable $r_0 = \sqrt{\hbar/(\mu\omega)}$ with dimensions of length, and to introduce the adimensional variable $x = r/r_0$, in terms of which the above equation is written as

$$\frac{d^2 u_l(x)}{dx^2} + \left[\epsilon - x^2 - \frac{l(l+1)}{x^2}\right] u_l(x) = 0 , \tag{10.25}$$

where we have defined

$$\epsilon = r_0^2 \bar{\epsilon} = \underbrace{\frac{\hbar}{\mu\omega} \frac{2\mu}{\hbar^2}}_{E_0} E = \frac{E}{E_0} , \qquad E_0 = \frac{\hbar\omega}{2} . \tag{10.26}$$

In the limit $x \longrightarrow \infty$, Eq. (10.25) reduces to

$$u_l''(x) - x^2 u_l(x) = 0 \implies u_l(x) \propto e^{-x^2/2} , \tag{10.27}$$

since $(e^{-x^2/2})'' = (x^2 - 1) e^{-x^2/2} \approx x^2 e^{-x^2/2}$. We see that all solutions vanish in the asymptotic region and, as a consequence, no continuum solutions can exist in a harmonic oscillator potential. We remove the short- and long-range behavior of $u_l(x)$ by introducing the auxiliary function $g_l(x)$,

$$u_l(x) = x^{l+1} e^{-x^2/2} g_l(x) , \tag{10.28}$$

so that

$$x g_l''(x) + 2\left(l + 1 - x^2\right) g_l'(x) + \xi x g_l(x) = 0 , \qquad \xi = \epsilon - 2l - 3 . \tag{10.29}$$

It can be shown (see Problem 3) that, in order for $u_l(x)$ not to be exponentially increasing in the limit $x \longrightarrow \infty$, we must have ξ equal to $2p$ with p an even integer,

$$\xi = 2p \implies \epsilon = 2p + 2l + 3 \qquad \text{and} \qquad p = 0, 2, 4, \ldots , \tag{10.30}$$

which in turn implies that $g_l(x)$ is a polynomial in x. We then obtain the well-known eigenvalues,

$$E_n = \hbar\omega(n + 3/2) , \qquad n = 0, 1, 2, \ldots , \tag{10.31}$$

where $n = p + l$ with $l = 0, 1, 2, \ldots$. The energy level with $n = 0$ is non-degenerate, since p and l must be both zero; however, the first excited level has $n = 1$, which necessarily implies $p = 0$ and $l = 1$, and is three-fold degenerate; the second excited level has $n = 2$, which is satisfied by either $p = 0$ and $l = 2$ or $p = 2$ and $l = 0$ (p must be an even integer) and is therefore six-fold degenerate, and so on. The degeneracy for a given n is

$$g_n = \frac{(n+1)(n+2)}{2} . \tag{10.32}$$

The wave functions are written as

$$\psi_{nlm}(r, \Omega) = \underbrace{N_{nl} x^l g_{nl}(x) e^{-x^2/2}}_{R_{nl}(r)} Y_{lm}(\Omega) , \qquad x = \sqrt{\frac{\mu\omega}{\hbar}} r , \tag{10.33}$$

where N_{nl} is a normalization constant and $g_{nl}(x)$ are polynomials in x which can be related to the Hermite polynomials.

10.3 The Two-Body Problem

In this section we consider the quantum mechanical treatment of two particles. In classical mechanics the two-body problem with a potential that depends only on the relative positions of the two particles is equivalent to a one-body problem. This equivalence also holds in quantum mechanics. Let $\mathbf{r}_1$ and $\mathbf{p}_1 = -i\hbar\boldsymbol{\nabla}_1$, and $\mathbf{r}_2$ and $\mathbf{p}_2 = -i\hbar\boldsymbol{\nabla}_2$, be the position and momentum operators for particles 1 and 2 of masses m_1 and m_2. They satisfy the commutation relations

$$[\, r_{i\alpha}\,,\, r_{j\beta}\,] = 0 = [\, p_{i\alpha}\,,\, p_{j\beta}\,]\,, \qquad [\, r_{i\alpha}\,,\, p_{j\beta}\,] = i\hbar\,\delta_{ij}\delta_{\alpha\beta}\,, \tag{10.34}$$

and note that the position and momentum operators of different particles ($i \neq j$) commute. We now proceed as in classical mechanics and introduce the center-of-mass and relative position operators, respectively, $\mathbf{R}$ and $\mathbf{r}$,

$$\mathbf{R} = \frac{m_1\,\mathbf{r}_1 + m_2\,\mathbf{r}_2}{m_1 + m_2}\,, \qquad \mathbf{r} = \mathbf{r}_1 - \mathbf{r}_2\,, \tag{10.35}$$

and the conjugate momentum operators,

$$\mathbf{P} = \mathbf{p}_1 + \mathbf{p}_2\,, \qquad \mathbf{p} = \frac{m_2\,\mathbf{p}_1 - m_1\,\mathbf{p}_2}{m_1 + m_2}\,. \tag{10.36}$$

We define the total mass M and reduced mass μ as

$$M = m_1 + m_2\,, \qquad \mu = \frac{m_1 m_2}{m_1 + m_2}\,. \tag{10.37}$$

The above relations are easily inverted to yield

$$\mathbf{r}_1 = \mathbf{R} + \frac{\mu}{m_1}\,\mathbf{r}\,, \qquad \mathbf{p}_1 = \frac{m_1}{M}\,\mathbf{P} + \mathbf{p}\,, \tag{10.38}$$

$$\mathbf{r}_2 = \mathbf{R} - \frac{\mu}{m_2}\,\mathbf{r}\,, \qquad \mathbf{p}_2 = \frac{m_2}{M}\,\mathbf{P} - \mathbf{p}\,. \tag{10.39}$$

The center-of-mass and relative variables satisfy canonical commutation relations, that is,

$$[\, R_\alpha\,,\, P_\beta\,] = i\hbar\delta_{\alpha\beta}\,, \qquad [\, r_\alpha\,,\, p_\beta\,] = i\hbar\delta_{\alpha\beta}\,, \tag{10.40}$$

with all remaining commutators vanishing, for example $[\, R_\alpha\,,\, p_\beta\,] = 0 = [\, r_\alpha\,,\, P_\beta\,]$. The Hamiltonian operator, which in the original variables reads

$$H = \frac{\mathbf{p}_1^2}{2m_1} + \frac{\mathbf{p}_2^2}{2m_2} + V(\mathbf{r}_1 - \mathbf{r}_2)\,, \tag{10.41}$$

in terms of the center-of-mass and relative variables simplifies to

$$H = \underbrace{\frac{\mathbf{P}^2}{2M}}_{H_{\text{cm}}} + \underbrace{\frac{\mathbf{p}^2}{2\mu} + V(\mathbf{r})}_{H_{\text{rel}}}\,; \tag{10.42}$$

thus it separates into a center-of-mass and relative Hamiltonian. The center-of-mass Hamiltonian is that of a free particle of mass M, and its eigenstates are those of the center-of-mass momentum operator, $\hat{\mathbf{P}}\,|\phi_{\mathbf{P}}\rangle = \mathbf{P}\,|\phi_{\mathbf{P}}\rangle$, that is,

$$H_{\text{cm}}|\phi_{\mathbf{P}}\rangle = \frac{\mathbf{P}^2}{2M}\,|\phi_{\mathbf{P}}\rangle\,. \tag{10.43}$$

The relative Hamiltonian is that of a particle of mass μ under the influence of a potential $V(\mathbf{r})$ (we are assuming that $V(\mathbf{r})$ vanishes as $|\mathbf{r}| \longrightarrow \infty$ but not that it is a central potential). This Hamiltonian might have discrete eigenvalues $E_i < 0$ with eigenstates $|\psi_i^{(k)}\rangle$,

$$H_{\text{rel}} |\psi_i^{(k)}\rangle = E_i |\psi_i^{(k)}\rangle \, , \qquad k = 1, \ldots, g_i \, , \qquad \langle \psi_i^{(k)} | \psi_j^{(l)} \rangle = \delta_{ij} \, \delta_{kl} \, ; \tag{10.44}$$

it will generally have continuous eigenvalues $E \geq 0$ with corresponding eigenstates $|\psi_E^{(k)}\rangle$,

$$H_{\text{rel}} |\psi_E^{(k)}\rangle = E |\psi_E^{(k)}\rangle \, , \qquad k = 1, \ldots, g_E \, , \qquad \langle \psi_E^{(k)} | \psi_{E'}^{(l)} \rangle = \delta(E - E') \, \delta_{kl} \, . \tag{10.45}$$

The introduction of center-of-mass and relative variables has allowed us to reduce the two-body problem to two independent one-body problems, a significant simplification. The eigenstates of the complete Hamiltonian $H = H_{\text{cm}} + H_{\text{rel}}$ are simply given by the tensor product of eigenstates of H_{cm} and H_{rel},

$$|\phi_{\mathbf{P}}; \psi_i^{(k)}\rangle \equiv |\phi_{\mathbf{P}}\rangle \otimes |\psi_i^{(k)}\rangle \, , \tag{10.46}$$

with corresponding wave functions

$$\langle \phi_{\mathbf{R},\mathbf{r}} | \phi_{\mathbf{P}}; \psi_i^{(k)} \rangle = \frac{e^{i\mathbf{P}\cdot\mathbf{R}/\hbar}}{(2\pi\hbar)^{3/2}} \, \psi_i^{(k)}(\mathbf{r}) \, , \tag{10.47}$$

and similarly for the continuous eigenstates of H_{rel}. We then have

$$H |\phi_{\mathbf{P}}; \psi_i^{(k)}\rangle = (H_{\text{cm}} |\phi_{\mathbf{P}}\rangle) \otimes |\psi_i^{(k)}\rangle + |\phi_{\mathbf{P}}\rangle \otimes (H_{\text{rel}} |\psi_i^{(k)}\rangle) = \left(\frac{\mathbf{P}^2}{2M} + E_i \right) |\phi_{\mathbf{P}}; \psi_i^{(k)}\rangle \, , \tag{10.48}$$

and the eigenvalues of H are the sums of the eigenvalues of H_{cm} and H_{rel}.

10.4 Problems

Problem 1 Solution of the Schrödinger Equation by Separation of Variables

Consider the Schrödinger equation in a central potential $V(r)$,

$$\left[\boldsymbol{\nabla}^2 + \epsilon - v(r) \right] \psi(\mathbf{r}) = 0 \, ,$$

where we have define

$$v(r) = \frac{2\mu}{\hbar^2} \, V(r) \, , \qquad \epsilon = \frac{2\mu E}{\hbar^2} \, .$$

1. Assume the separation of variables (in spherical coordinates), that is,

$$\psi(\mathbf{r}) = R(r) \, P(\theta) \, \Phi(\phi) \, ,$$

and show that

$$r^2 \underbrace{\left[\frac{1}{r^2 R} \frac{d}{dr} \left(r^2 \frac{dR}{dr} \right) + \epsilon - v \right]}_{\text{function of } r} + \underbrace{\left[\frac{1}{\sin\theta \, P} \frac{d}{d\theta} \left(\sin\theta \frac{dP}{d\theta} \right) + \frac{1}{\sin^2\theta \, \Phi} \frac{d^2\Phi}{d\phi^2} \right]}_{\text{function of } \theta \text{ and } \phi} = 0 \, .$$

2. For the above to be satisfied, the function of r must be equal to a constant, which for convenience we call $\alpha(\alpha + 1)$, namely

$$r^2 \left[\frac{1}{r^2 R} \frac{d}{dr} \left(r^2 \frac{dR}{dr} \right) + \epsilon - v \right] = \alpha(\alpha + 1) \,,$$

so that

$$\left[\frac{1}{\sin\theta \, P} \frac{d}{d\theta} \left(\sin\theta \frac{dP}{d\theta} \right) + \frac{1}{\sin^2\theta \, \Phi} \frac{d^2\Phi}{d\phi^2} \right] = -\alpha(\alpha + 1) \,.$$

Impose periodic boundary conditions on $\Phi(\phi)$ and its derivative, that is,

$$\Phi(2\pi) = \Phi(0) \,, \qquad \Phi'(2\pi) = \Phi'(0) \,,$$

to show that

$$\Phi_m(\phi) \propto e^{im\phi} \qquad \text{with} \qquad m = 0, \pm 1, \pm 2, \dots \,,$$

and that $P(\theta)$ must satisfy the equation

$$\frac{1}{\sin\theta} \frac{d}{d\theta} \left(\sin\theta \frac{dP}{d\theta} \right) + \left[\alpha(\alpha + 1) - \frac{m^2}{\sin^2\theta} \right] P = 0 \,.$$

3. Make the change of variable $x = \cos\theta$ and show that the equation for $P(\theta)$ reduces to

$$\frac{d}{dx}(1 - x^2) \frac{dP(x)}{dx} + \left[\alpha(\alpha + 1) - \frac{m^2}{1 - x^2} \right] P(x) = 0 \,,$$

where $-1 \leq x \leq 1$.

4. Knowing that (i) if

$$\frac{d}{dx}(1 - x^2) \frac{dP_\alpha(x)}{dx} + \alpha(\alpha + 1) P_\alpha(x) = 0 \,,$$

then the general solution for $m \neq 0$ follows,

$$P_\alpha^m(x) = (1 - x^2)^{|m|/2} \frac{d^{|m|} P_\alpha(x)}{dx^{|m|}} \,,$$

and that (ii) the equation for $P_\alpha(x)$ has a regular solution in $-1 \leq x \leq 1$ (including the end points) only if α is a non-negative integer $\alpha = l$ with $l = 0, 1, 2, \dots$, in which case $P_l(x)$ reduces to the Legendre polynomial

$$P_l(x) = \frac{1}{2^l} \frac{1}{l!} \frac{d^l}{dx^l} (x^2 - 1)^l \,,$$

deduce that it must be the case that $|m| \leq l$.

Solution

Part 1

We use

$$\underbrace{-\hbar^2 \nabla^2}_{p^2} = -\frac{\hbar^2}{r^2} \frac{\partial}{\partial r} r^2 \frac{\partial}{\partial r} + \frac{\mathbf{L}^2}{r^2} = -\frac{\hbar^2}{r^2} \frac{\partial}{\partial r} r^2 \frac{\partial}{\partial r} - \frac{\hbar^2}{r^2} \left[\frac{1}{\sin\theta} \frac{\partial}{\partial\theta} \left(\sin\theta \frac{\partial}{\partial\theta} \right) + \frac{1}{\sin^2\theta} \frac{\partial^2}{\partial\phi^2} \right] \,,$$

so that the Schrödinger equation is given as follows (for brevity, we have suppressed the dependence on r, θ, and ϕ of, respectively, R, P, and Φ):

$$P\Phi \left[\frac{1}{r^2} \frac{\partial}{\partial r} r^2 \frac{\partial}{\partial r} + \epsilon - v \right] R + \frac{R}{r^2} \left[\frac{1}{\sin\theta} \frac{\partial}{\partial\theta} \left(\sin\theta \frac{\partial}{\partial\theta} \right) + \frac{1}{\sin^2\theta} \frac{\partial^2}{\partial\phi^2} \right] P\Phi = 0 \, .$$

Multiplying by r^2 and dividing both sides of the equation above by the product $RP\Phi$, we arrive at

$$r^2 \left[\frac{1}{r^2 R} \frac{d}{dr} \left(r^2 \frac{dR}{dr} \right) + \epsilon - v \right] + \left[\frac{1}{\sin\theta \, P} \frac{d}{d\theta} \left(\sin\theta \frac{dP}{d\theta} \right) + \frac{1}{\sin^2\theta \, \Phi} \frac{d^2\Phi}{d\phi^2} \right] = 0 \, .$$

Part 2

We require

$$r^2 \left[\frac{1}{r^2 R} \frac{d}{dr} \left(r^2 \frac{dR}{dr} \right) + \epsilon - v \right] = \alpha(\alpha + 1) \, ,$$

where we take α to be a real constant for the time being. Hence, we have

$$\left[\frac{1}{\sin\theta \, P} \frac{d}{d\theta} \left(\sin\theta \frac{dP}{d\theta} \right) + \frac{1}{\sin^2\theta \, \Phi} \frac{d^2\Phi}{d\phi^2} \right] = -\alpha(\alpha + 1) \, .$$

Multiplying both sides of the equation above by $\sin^2\theta$, we obtain

$$\underbrace{\frac{\sin\theta}{P} \left(\frac{d}{d\theta} \sin\theta \frac{dP}{d\theta} \right) + \alpha(\alpha + 1) \sin^2\theta}_{\text{function of } \theta} + \underbrace{\frac{1}{\Phi} \frac{d^2\Phi}{d\phi^2}}_{\text{function of } \phi} = 0 \, ,$$

yielding (here m^2 is a real constant)

$$\frac{\sin\theta}{P} \left(\frac{d}{d\theta} \sin\theta \frac{dP}{d\theta} \right) + \alpha(\alpha + 1) \sin^2\theta = m^2 \, , \qquad \frac{1}{\Phi} \frac{d^2\Phi}{d\phi^2} = -m^2 \, ,$$

which can also be written as

$$\frac{1}{\sin\theta} \frac{d}{d\theta} \left(\sin\theta \frac{dP}{d\theta} \right) + \left[\alpha(\alpha + 1) - \frac{m^2}{\sin^2\theta} \right] P = 0 \, , \qquad \frac{d^2\Phi}{d\phi^2} = -m^2\Phi \, .$$

The second equation is easily solved to give

$$\Phi(\phi) = A\,e^{i|m|\phi} + B\,e^{-i|m|\phi} \, ,$$

and imposing periodic boundary conditions on $\Phi(\phi)$ and on its derivative leads to

$$A\,e^{2\pi|m|i} + B\,e^{-2\pi|m|i} = A + B \, , \qquad i|m|A\,e^{2\pi|m|i} - i|m|B\,e^{-2\pi|m|i} = i|m|A - i|m|B \, ,$$

which can be conveniently written as

$$A\left(e^{2\pi|m|i} - 1 \right) + B\left(e^{-2\pi|m|i} - 1 \right) = 0 \, , \qquad A\left(e^{2\pi|m|i} - 1 \right) - B\left(e^{-2\pi|m|i} - 1 \right) = 0 \, .$$

These relations yield, for example,

$$2A\left(e^{2\pi|m|i} - 1 \right) = 0 \, ,$$

which is satisfied by setting $|m| = 0, 1, 2, \ldots$. (Setting $A = 0$ with $|m|$ non-integer gives $B = 0$, that is, the null or trivial solution.) We simply write $\Phi(\phi)$ as

$$\Phi(\phi) \propto e^{im\phi} \, , \qquad m = 0, \pm 1, \pm 2, \ldots \, ,$$

and there is no upper bound on $|m|$ as of yet.

Part 3

Introduce the variable $x = \cos\theta$ (with $-1 \le x \le 1$), so that

$$\frac{d}{d\theta} = \frac{dx}{d\theta}\frac{d}{dx} = -\sin\theta\,\frac{d}{dx}\,,$$

and hence

$$\frac{1}{\sin\theta}\frac{d}{d\theta}\left(\sin\theta\frac{d}{d\theta}\right) = \frac{1}{\sin\theta}\underbrace{\left(-\sin\theta\frac{d}{dx}\right)}_{d/d\theta}\left[\sin\theta\underbrace{\left(-\sin\theta\frac{d}{dx}\right)}_{d/d\theta}\right] = \frac{d}{dx}\left[\underbrace{(1-x^2)}_{\sin^2\theta}\frac{d}{dx}\right]\,,$$

to obtain the associated Legendre equation

$$\frac{d}{dx}(1-x^2)\frac{dP_l^m(x)}{dx} + \left[l(l+1) - \frac{m^2}{1-x^2}\right]P_l^m(x) = 0\,,$$

where we have made explicit the dependence of $P(x)$ on the integers l and m.

Part 4

Using the binomial expansion, we find

$$P_l(x) = \frac{1}{2^l\,l!}\frac{d^l}{dx^l}(x^2-1)^l = \frac{1}{2^l\,l!}\frac{d^l}{dx^l}\sum_{k=0}^{l}\binom{l}{k}(-)^{l-k}x^{2k}\,.$$

The highest power x in the binomial expansion is x^{2l}:

$$\sum_{k=0}^{l}\binom{l}{k}(-1)^{l-k}x^{2k} = x^{2l} - lx^{2l-2} + \cdots + (-1)^l\,,$$

and taking l derivatives yields

$$\frac{d^l}{dx^l}(x^2-1)^l = 2l(2l-1)\cdots(l+1)x^l + \text{lower powers of } x\,.$$

Thus, we see that $P_l(x)$ is a polynomial of order l. Since

$$P_l^m(x) = (1-x^2)^{|m|/2}\frac{d^{|m|}P_l(x)}{dx^{|m|}}\,,$$

we conclude that

$$\frac{d^{|m|}P_l(x)}{dx^{|m|}} = 0 \qquad \text{if } |m| > l\,,$$

and therefore we arrive at the constraint $|m| \le l$.

Problem 2   Solution of the Radial Equation for the Hydrogen Atom

Start from

$$\frac{d^2 u_l(x)}{dx^2} + \left[-\epsilon + \frac{2}{x} - \frac{l(l+1)}{x^2}\right]u_l(x) = 0\,,$$

and show that bound-state solutions result from the condition $\epsilon = 1/(\nu + l + 1)^2$ with $\nu = 0, 1, 2, \ldots$ and $l = 0, 1, 2, \ldots$.

Solution

In the limit $x \longrightarrow \infty$, Eq. (10.7) reduces to

$$\frac{d^2 u_l(x)}{dx^2} - \epsilon u_l(x) = 0 \implies u_l(x) \propto e^{-\sqrt{\epsilon} x} .$$

This asymptotic behavior can be removed by introducing the auxiliary function $f_l(x)$ via $u_l(x) = e^{-\sqrt{\epsilon} x} f_l(x)$, so that

$$u_l''(x) = \left[\epsilon f_l(x) - 2 \sqrt{\epsilon} f_l'(x) + f_l''(x) \right] e^{-\sqrt{\epsilon} x} ,$$

which yields the following equation for $f_l(x)$:

$$f_l''(x) - 2 \sqrt{\epsilon} f'(x) + \left[\frac{2}{x} - \frac{l(l+1)}{x^2} \right] f_l(x) = 0 .$$

We remove the effect of the centrifugal barrier on the short-range behavior by defining $f_l(x) = x^{l+1} F_l(x)$; we find

$$f_l'(x) = (l+1)x^l F_l(x) + x^{l+1} F_l'(x) , \qquad f_l''(x) = l(l+1)x^{l-1} F_l(x) + 2(l+1)x^l F_l'(x) + x^{l+1} F_l''(x) .$$

After dividing by x^l, we are left with the following equation for $F_l(x)$:

$$xF_l''(x) + \left[2(l+1) - 2\sqrt{\epsilon} x \right] F_l'(x) + \left[2 - 2\sqrt{\epsilon}(l+1) \right] F_l(x) = 0 .$$

As a last step, we rescale x by defining

$$z = 2\sqrt{\epsilon} x , \qquad \frac{d}{dx} = 2\sqrt{\epsilon} \frac{d}{dz} ,$$

to arrive at (on dividing by $2\sqrt{\epsilon}$)

$$zF_l''(z) + (2l + 2 - z) F_l'(z) + (\xi - l - 1)F_l(z) = 0 , \qquad \xi = \frac{1}{\sqrt{\epsilon}} ,$$

which can be rewritten as

$$zF_l''(z) + (b - z) F_l'(z) - aF_l(z) = 0 , \qquad a = l + 1 - \xi , \qquad b = 2l + 2 .$$

We look for a power series solution,

$$F_l(z) = z^s \sum_{q=0}^{\infty} c_q^{(l)} z^q , \qquad c_0^{(l)} \neq 0 .$$

Substituting this expansion into the equation above and lumping together terms with the same power of z leads to

$$\sum_{q=0}^{\infty} [(s+q)(s+q-1) + b(s+q)] c_q^{(l)} z^{q-1} = \sum_{q=0}^{\infty} (s+q+a) c_q^{(l)} z^q .$$

For the power series to be a solution, each coefficient of z^p must vanish:

$$z^{-1}: \qquad [s(s-1) + bs] c_0^{(l)} = 0 \implies s(s+b-1) = 0 ,$$

$$z^0: [s(s+1) + b(s+1)]c_1^{(l)} = (s+a)c_0^{(l)} \implies c_1^{(l)} = \frac{s+a}{(s+1)(s+b)} c_0^{(l)} ,$$

and so on. For the generic term proportional to z^q, we find

$$z^q: \ [(s+q)(s+q+1) + b(s+q+1)] c_{q+1}^{(l)} = (s+q+a)c_q^{(l)} ,$$

which yields the recurrence relation

$$c_{q+1}^{(l)} = \frac{s+q+a}{(s+q+1)(s+b+q)}\, c_q^{(l)}\,, \qquad q = 0,1,2,\dots,$$

with s determined from

$$s(s+b-1) = 0 \implies s = 0 \qquad \text{or} \qquad s = 1-b.$$

The solution corresponding to $s = 1-b = -2l-1$ is the irregular function $G_l(z)$, which is unacceptable, since it behaves in the limit $z \longrightarrow 0$ as $G_l(z) \propto z^{-2l-1}$, and would lead to $u_l(r) \propto r^{-l}$ with incorrect behavior for small r. Thus the only acceptable solution has $s = 0$, which gives

$$c_{q+1}^{(l)} = \underbrace{\frac{q+a}{(q+1)(q+b)}}_{\gamma_q}\, c_q^{(l)}\,, \qquad q = 0,1,2,\dots,$$

where $c_0^{(l)}$ is an arbitrary constant ($\neq 0$) to be ultimately fixed by the normalization condition. The recurrence relation yields $c_{q+1}^{(l)} = \gamma_q\,\gamma_{q-1}\cdots\gamma_0\, c_0^{(l)}$ and the solution has the form

$$F_l(z) = c_0^{(l)} + c_1^{(l)} z + c_2^{(l)} z^2 + c_3^{(l)} z^3 + \cdots = c_0^{(l)} \left[1 + \underbrace{\frac{a}{b}}_{A_1} z + \underbrace{\frac{1}{2}\frac{a(a+1)}{b(b+1)}}_{A_2} z^2 + \underbrace{\frac{1}{6}\frac{a(a+1)(a+2)}{b(b+1)(b+2)}}_{A_3} z^3 + \cdots \right].$$

This is known as the confluent hypergeometric series. The series must terminate, that is, $F_l(z)$ must be a polynomial, because otherwise it would lead to a solution $u_l(r)$ which would diverge exponentially at large r and therefore could not describe a bound state. This can be seen by noting that for $q \gg 1$ the ratio $c_{q+1}^{(l)}/c_q^{(l)}$ behaves as $1/q$. The series can be written, up to the overall normalization factor $c_0^{(l)}$, as

$$F_l(z) = 1 + A_1 z + \cdots + A_{m-1}z^{m-1} + A_m z^m \underbrace{\left[1 + \frac{A_{m+1}}{A_m}z + \frac{A_{m+2}}{A_m}z^2 + \cdots \right]}_{H(z)},$$

where the integer $m \gg 1$, so that

$$\frac{A_{m+1}}{A_m} = \frac{1}{m}\,, \qquad \frac{A_{m+2}}{A_m} = \frac{A_{m+1}}{A_m}\frac{A_{m+2}}{A_{m+1}} = \frac{1}{m(m+1)}\,,$$

and so on. The series for $H(z)$ is given by

$$\begin{aligned}
H(z) &= 1 + \frac{1}{m} z + \frac{1}{m(m+1)} z^2 + \frac{1}{m(m+1)(m+2)} z^3 + \cdots \\
&= 1 + \frac{(m-1)!}{m!} z + \frac{(m-1)!}{(m+1)!} z^2 + \frac{(m-1)!}{(m+2)!} z^3 + \cdots \\
&= (m-1)! \sum_{p=0}^{\infty} \frac{z^p}{(p+m-1)!} = (m-1)! \sum_{n=m-1}^{\infty} \frac{z^{n-m+1}}{n!} \\
&= \frac{(m-1)!}{z^{m-1}} \left(\sum_{n=0}^{\infty} \frac{z^n}{n!} - \sum_{n=0}^{m-2} \frac{z^n}{n!} \right) = \frac{(m-1)!}{z^{m-1}} \left(e^z - \sum_{n=0}^{m-2} \frac{z^n}{n!} \right).
\end{aligned}$$

We therefore conclude that $F_l(z) = (\text{polynomial in } z) + (m-1)!\, A_m z\, e^z$. Since

$$u_l(x) = x^{l+1}\, e^{-\sqrt{\epsilon}\,x}\, F_l(\underbrace{2\sqrt{\epsilon}x}_{z})\,,$$

we find that $u_l(x)$ for large x (or, equivalently, for large r as $x = r/r_0$) behaves as $e^{\sqrt{\epsilon}x}$ and is not normalizable. Thus the series must terminate, and $F_l(z)$ must be a polynomial. This can only happen if in the recurrence relation for $c_q^{(l)}$ we have that the parameter a is 0 or a negative integer, yielding

$$l + 1 - \frac{1}{\sqrt{\epsilon}} = -v \implies \epsilon = \frac{1}{(v + l + 1)^2}, \qquad v = 0, 1, 2, \ldots .$$

Problem 3   The Radial Equation for the Isotropic Harmonic Oscillator in Three Dimensions

Start from

$$x g_l''(x) + 2\left(l + 1 - x^2\right) g_l'(x) + \xi x g_l(x) = 0, \qquad \xi = \epsilon - 2l - 3,$$

and show that bound-state solutions result from the conditions $\xi = 2p$ or $\epsilon = 2p + 2l + 3$, where p is a non-negative even integer. Define $n = p + l$ and show that the degeneracy of energy level n is $g_n = (n + 1)(n + 2)/2$.

Solution

We posit

$$g_l(x) = x^s \sum_{q=0}^{\infty} c_q^{(l)} x^q, \qquad c_0^{(l)} \neq 0.$$

Inserting the above expansion into the radial equation and lumping together terms with the same power of x, we obtain

$$\sum_{q=0}^{\infty} [(s + q)(s + q - 1) + 2(l + 1)(s + q)] c_q^{(l)} x^{q-1} = \sum_{q=0}^{\infty} [2(s + q) - \xi] c_q^{(l)} x^{q+1}.$$

For the power series to be a solution, each coefficient of x^p must vanish. Consider first the terms proportional to x^{-1} and x^0,

$$x^{-1}: \quad [s(s - 1) + 2(l + 1)s] c_0^{(l)} = 0 \implies s = 0, -(2l + 1) \quad (\text{for } c_0^{(l)} \neq 0),$$

$$x^0: \quad \underbrace{[s(s + 1) + 2(l + 1)(s + 1)]}_{\neq 0 \text{ for } s=0 \text{ or } s=-(2l+1)} c_1^{(l)} = 0 \implies c_1^{(l)} = 0,$$

where in the last relation $c_1^{(l)}$ must vanish, since the prefactor $[\cdots]$ is $\neq 0$ when $s = 0$ or $s = -(2l + 1)$. For a general term proportional to x^{p+1} with $p \geq 0$, we find

$$x^{p+1}: [(s + p + 2)(s + p + 1) + 2(l + 1)(s + p + 2)] c_{p+2}^{(l)} = [2(s + p) - \xi] c_p^{(l)},$$

which implies the following recurrence relation:

$$c_{p+2}^{(l)} = \frac{2(s + p) - \xi}{(s + p + 2)(s + p + 1) + 2(l + 1)(s + p + 2)} c_p^{(l)} \qquad \text{with } p = 0, 1, \ldots .$$

Because of the condition $c_1^{(l)} = 0$, it immediately follows that $c_p^{(l)} = 0$ for p odd. Furthermore, only the solution corresponding to $s = 0$ is acceptable, since that corresponding to $s = -(2l + 1)$ is too singular at the origin as it would lead to $u_l(r) \propto r^{-l}$. Thus, the series solution is of the form

$$g_l(x) = \sum_{p=0}^{\infty} c_p^{(l)} x^p \,, \qquad c_{p+2}^{(l)} = \frac{2p - \xi}{(p+2)(p+2l+3)} c_p^{(l)} \,, \qquad c_1^{(l)} = 0 \,.$$

An analysis similar to that carried out for the equation relative to the hydrogen atom in Problem 2 shows that, for the solution to be normalizable (that is, vanishing at $x \pm \infty$), the series must terminate, namely $g_l(x)$ must be a polynomial. This can only happen if $\xi = 2p$ with p even, yielding $\epsilon = 2p+2l+3$ or, in physical units,

$$E_n = \hbar\omega(n + 3/2) \,, \qquad n = p + l \ \text{ with } \ p = 0, 2, 4, \ldots \ \text{ and } \ l = 0, 1, 2, \ldots$$

In order to calculate the degeneracy we observe that, for n even, l must be even and hence (setting $l = 2k$)

$$\sum_{l \text{ even}=0}^{n} (2l + 1) = \sum_{k=0}^{n/2} (4k + 1) = 4\frac{(n/2 + 1)n/2}{2} + (n/2 + 1) = \frac{(n+1)(n+2)}{2} \,,$$

and, for n odd, l must be odd ($l = 2k + 1$), so that

$$\sum_{l \text{ odd}=1}^{n} (2l + 1) = \sum_{k=0}^{(n-1)/2} (4k + 3) = 4\frac{[(n-1)/2 + 1](n-1)/2}{2} + 3[(n-1)/2 + 1] = \frac{(n+1)(n+2)}{2} \,,$$

as expected.

Problem 4 Alternative Solution of the Isotropic Harmonic Oscillator in Three Dimensions

Consider a particle of mass μ under the action of an isotropic harmonic potential of angular frequency ω in three dimensions, with Hamiltonian given by

$$\hat{H} = \frac{\hat{\mathbf{p}}^2}{2\mu} + \frac{\mu\omega^2}{2} \hat{\mathbf{r}}^2 \,,$$

Obtain the eigenvalues and corresponding eigenfunctions, and determine the degeneracy of the eigenvalues.

Solution

The Hamiltonian describing this system consists of three independent harmonic oscillator Hamiltonians

$$\hat{H} = \sum_{i=1}^{3} \hat{H}_i \,, \qquad \hat{H}_i = \frac{\hat{p}_i^2}{2\mu} + \frac{\mu\omega^2}{2} \hat{r}_i^2 \,,$$

where the components of the position and momentum operators satisfy the commutation relations $[\hat{r}_i, \hat{p}_j] = i\hbar\delta_{ij}$ (all other commutators vanish). In the number-operator representation, the eigenstates of the full Hamiltonian $\hat{H}$ consist of tensor products $|n_1\rangle \otimes |n_2\rangle \otimes |n_3\rangle$, such that

$$\hat{H}(|n_1\rangle \otimes |n_2\rangle \otimes |n_3\rangle) = \underbrace{(\hat{H}_1|n_1\rangle)}_{E_1|n_1\rangle} \otimes |n_2\rangle \otimes |n_3\rangle + |n_1\rangle \otimes \underbrace{(\hat{H}_2|n_2\rangle)}_{E_2|n_2\rangle} \otimes |n_3\rangle + \cdots + |n_1\rangle \otimes |n_2\rangle \otimes \underbrace{(\hat{H}_3|n_3\rangle)}_{E_3|n_3\rangle}$$

$$= \underbrace{(E_1 + E_2 + E_3)}_{E} |n_1\rangle \otimes |n_2\rangle \otimes |n_3\rangle \,,$$

where $E_i = \hbar\omega(n_i + 1/2)$. The eigenfunctions are obtained as

$$\psi_{n_1,n_2,n_3}(\mathbf{r}) = \langle\phi_{r_1}|n_1\rangle\langle\phi_{r_2}|n_2\rangle\langle\phi_{r_3}|n_3\rangle = \prod_{i=1}^{3}\psi_{n_i}(r_i),$$

where $\psi_n(x)$ is the single-harmonic-oscillator wave function defined in Eq. (8.26) (with m replaced by μ). The eigenvalues are given by

$$E_n = \sum_{i=1}^{d}E_i = \hbar\omega(\underbrace{n_1 + n_2 + n_3}_{n} + 3/2) = \hbar\omega(n + 3/2).$$

In order to determine the degeneracy g_n, we need to calculate the number of ways in which three non-negative integers n_1, n_2, and n_3 can add up to a given integer n. This number is

$$g_n = \sum_{n_1=0}^{n}\sum_{n_2=0}^{n}\sum_{n_3=0}^{n}\delta_{n,n_1+n_2+n_3} = \sum_{n_1=0}^{n}\sum_{n_2=0}^{n-n_1}1 = \sum_{n_1=0}^{n}(n - n_1 + 1) = (n+1)\sum_{n_1=0}^{n}1 - \sum_{n_1=0}^{n}n_1$$

$$= (n+1)^2 - n(n+1)/2 = (n+1)(n+2)/2.$$

The above procedure is obviously generalizable to the case of the isotropic harmonic oscillator in d dimensions. Eigenstates consists of tensor products $|n_1\rangle \otimes |n_2\rangle \otimes \cdots \otimes |n_d\rangle$ with eigenvalues $E_n = \hbar\omega(n + d/2)$, where $n = n_1 + \cdots + n_d$. The degeneracy of these eigenvalues is given by (see Problem 4 in Chapter 8)

$$g_n = \frac{(n + d - 1)!}{n! \, (d - 1)!}.$$

Problem 5   Differences in the Energy Spectra of Hydrogen and Deuterium

In nature there are two stable isotopes of the hydrogen nucleus: the proton with mass $m_p \approx 1836\, m_e$ and the deuteron (a bound state of a proton and a neutron) with mass $m_d \approx 3670\, m_e$. How does this difference in mass impact the energy spectra of the hydrogen and deuterium atoms?

Solution

The eigenvalues of the relative Hamiltonian of the hydrogen atom are given by

$$E_n(\text{hydrogen}) = -\frac{\alpha^2}{2n^2}\,\mu_{ep}c^2,$$

where α is the fine structure constant and μ_{ep} is the reduced electron–proton mass,

$$\mu_{ep} = \frac{m_e m_p}{m_e + m_p} \approx m_e\left(1 - \frac{m_e}{m_p}\right) \approx 0.99945\, m_e.$$

In the case of deuterium, the only change in the expression for the eigenvalues is the replacement of μ_{ep} by μ_{ed}, the reduced mass of the electron–deuteron system; $\mu_{ed} \approx 0.99973\, m_e$. This difference leads to an overall tiny shift proportional to $\mu_{ed} - \mu_{ep}$ in the frequencies of the light emitted in transitions between energy levels in hydrogen and in deuterium, and consequently a tiny shift in the positions of the corresponding spectral lines. This shift can be detected. From the intensity of the spectral lines astronomers are able to infer the relative abundance of hydrogen and deuterium in the interstellar medium, which in turn reveals conditions in the early universe, when a small fraction of matter consisted of deuterons.

Problem 6. The Schrödinger Equation in Two Dimensions

Consider the two-dimensional Schrödinger equation for the case where the potential energy depends only on the radial variable.

1. Introduce polar coordinates

$$x = \rho \cos \theta , \qquad y = \rho \sin \theta ,$$

   and derive the identity

$$\frac{\partial^2}{\partial x^2} + \frac{\partial^2}{\partial y^2} = \frac{\partial^2}{\partial \rho^2} + \frac{1}{\rho} \frac{\partial}{\partial \rho} + \frac{1}{\rho^2} \frac{\partial^2}{\partial \theta^2} .$$

2. Deduce from this, using the separation of variables, that there is a complete set of eigenfunctions of the form

$$\psi_{E,n}(\rho, \theta) = f_{E,n}(\rho) \, e^{in\theta} \qquad n = 0, \pm 1, \pm 2, \dots ,$$

   where $f_{E,n}(\rho)$ is the solution of the radial equation

$$\left[\frac{\partial^2}{\partial \rho^2} + \frac{1}{\rho} \frac{\partial}{\partial \rho} - \frac{n^2}{\rho^2} - v(\rho) + \epsilon \right] f_{E,n}(\rho) = 0 , \qquad \epsilon = \frac{2mE}{\hbar^2} , \qquad v(\rho) = \frac{2m}{\hbar^2} V(\rho) .$$

3. Consider the free-particle case $v(\rho) = 0$. Define

$$x = k\rho \qquad \text{with } k^2 = \frac{2mE}{\hbar^2} > 0 ,$$

   and show that the radial equation reduces to

$$\left[\frac{d^2}{dx^2} + \frac{1}{x} \frac{d}{dx} - \frac{n^2}{x^2} + 1 \right] f_{k,n}(x) = 0 .$$

4. Solve the free-particle radial equation by assuming

$$f_{k,n}(x) = x^s \sum_{p=0}^{\infty} a_p x^p \qquad \text{with } a_0 \neq 0 ,$$

   where the dependence of the coefficients a_p on the quantum number n is understood. Verify that the regular solution (well behaved in the limit $x \longrightarrow 0$) coincides with the power series expansion of the regular Bessel function.

Solution

Part 1

We introduce polar coordinates

$$x = \rho \cos \theta \qquad y = \rho \sin \theta , \qquad \rho = \sqrt{x^2 + y^2} \qquad \theta = \tan^{-1}(y/x) ,$$

for which

$$\frac{\partial}{\partial x} = \frac{\partial \rho}{\partial x} \frac{\partial}{\partial \rho} + \frac{\partial \theta}{\partial x} \frac{\partial}{\partial \theta} = \frac{x}{\sqrt{x^2 + y^2}} \frac{\partial}{\partial \rho} - \frac{y/x^2}{1 + (y/x)^2} \frac{\partial}{\partial \theta} = \cos \theta \frac{\partial}{\partial \rho} - \frac{\sin \theta}{\rho} \frac{\partial}{\partial \theta} ,$$

and similarly

$$\frac{\partial}{\partial y} = \sin \theta \frac{\partial}{\partial \rho} + \frac{\cos \theta}{\rho} \frac{\partial}{\partial \theta} .$$

Next, consider

$$\frac{\partial^2}{\partial x^2} = \left(\cos \theta \frac{\partial}{\partial \rho} - \frac{\sin \theta}{\rho} \frac{\partial}{\partial \theta} \right) \left(\cos \theta \frac{\partial}{\partial \rho} - \frac{\sin \theta}{\rho} \frac{\partial}{\partial \theta} \right)$$

$$= \cos^2 \theta \frac{\partial^2}{\partial \rho^2} - \frac{\sin \theta}{\rho} \frac{\partial}{\partial \theta} \cos \theta \frac{\partial}{\partial \rho} - \cos \theta \frac{\partial}{\partial \rho} \frac{\sin \theta}{\rho} \frac{\partial}{\partial \theta} + \frac{\sin \theta}{\rho} \frac{\partial}{\partial \theta} \frac{\sin \theta}{\rho} \frac{\partial}{\partial \theta}$$

$$= \cos^2 \theta \frac{\partial^2}{\partial \rho^2} + \frac{\sin^2 \theta}{\rho} \frac{\partial}{\partial \rho} - \frac{\sin \theta \cos \theta}{\rho} \frac{\partial^2}{\partial \theta \partial \rho} + \frac{\sin \theta \cos \theta}{\rho^2} \frac{\partial}{\partial \theta}$$

$$- \frac{\sin \theta \cos \theta}{\rho} \frac{\partial^2}{\partial \theta \partial \rho} + \frac{\sin \theta \cos \theta}{\rho^2} \frac{\partial}{\partial \theta} + \frac{\sin^2 \theta}{\rho^2} \frac{\partial^2}{\partial \theta^2}$$

and

$$\frac{\partial^2}{\partial y^2} = \left(\sin \theta \frac{\partial}{\partial \rho} + \frac{\cos \theta}{\rho} \frac{\partial}{\partial \theta} \right) \left(\sin \theta \frac{\partial}{\partial \rho} + \frac{\cos \theta}{\rho} \frac{\partial}{\partial \theta} \right)$$

$$= \sin^2 \theta \frac{\partial^2}{\partial \rho^2} + \frac{\cos \theta}{\rho} \frac{\partial}{\partial \theta} \sin \theta \frac{\partial}{\partial \rho} + \sin \theta \frac{\partial}{\partial \rho} \frac{\cos \theta}{\rho} \frac{\partial}{\partial \theta} + \frac{\cos \theta}{\rho} \frac{\partial}{\partial \theta} \frac{\cos \theta}{\rho} \frac{\partial}{\partial \theta}$$

$$= \sin^2 \theta \frac{\partial^2}{\partial \rho^2} + \frac{\cos^2 \theta}{\rho} \frac{\partial}{\partial \rho} + \frac{\sin \theta \cos \theta}{\rho} \frac{\partial^2}{\partial \theta \partial \rho} - \frac{\sin \theta \cos \theta}{\rho^2} \frac{\partial}{\partial \theta}$$

$$+ \frac{\sin \theta \cos \theta}{\rho} \frac{\partial^2}{\partial \theta \partial \rho} - \frac{\sin \theta \cos \theta}{\rho^2} \frac{\partial}{\partial \theta} + \frac{\cos^2 \theta}{\rho^2} \frac{\partial^2}{\partial \theta^2} .$$

Combining the above relations yields

$$\frac{\partial^2}{\partial x^2} + \frac{\partial^2}{\partial y^2} = \frac{\partial^2}{\partial \rho^2} + \frac{1}{\rho} \frac{\partial}{\partial \rho} + \frac{1}{\rho^2} \frac{\partial^2}{\partial \theta^2} .$$

Part 2

The Schrödinger equation in polar coordinates reads

$$\left[-\frac{\hbar^2}{2m} \left(\frac{\partial^2}{\partial \rho^2} + \frac{1}{\rho} \frac{\partial}{\partial \rho} + \frac{1}{\rho^2} \frac{\partial^2}{\partial \theta^2} \right) + V(\rho) \right] \psi(\rho, \theta) = E \psi(\rho, \theta) ,$$

which we write as

$$\left[\frac{\partial^2}{\partial \rho^2} + \frac{1}{\rho} \frac{\partial}{\partial \rho} + \frac{1}{\rho^2} \frac{\partial^2}{\partial \theta^2} - v(\rho) + \epsilon \right] \psi(\rho, \theta) = 0 .$$

We posit

$$\psi(\rho, \theta) = f(\rho) \, t(\theta) ,$$

which leads to

$$f'' t + \frac{f' t}{\rho} + \frac{f t''}{\rho^2} + [\epsilon - v(\rho)] f t = 0 \implies \rho^2 \left[\frac{f''}{f} + \frac{1}{\rho} \frac{f'}{f} + \epsilon - v(\rho) \right] = -\frac{t''}{t} .$$

The left-hand side is a function only of ρ and the right-hand side is a function only of θ. In order to satisfy the above equation for any ρ and θ, we must have

$$\frac{t''}{t} = -\mu^2 , \qquad \frac{f''}{f} + \frac{1}{\rho}\frac{f'}{f} + \epsilon - v(\rho) = \frac{\mu^2}{\rho^2} ,$$

where μ is a constant (at this stage, μ can be complex). The function $t(\theta)$ must satisfy the boundary condition $t(\theta + 2\pi) = t(\theta)$. The solution of the differential equation for $t(\theta)$ is $t(\theta) \sim e^{i\mu\theta}$ and the boundary condition requires

$$e^{2\pi\mu i} = 1 \implies \mu = n = 0, \pm 1, \pm 2, \ldots$$

The differential equation for $f_{\epsilon,n}(\rho)$ (we have made explicit its dependence on ϵ and n) is given by

$$\left[\frac{\partial^2}{\partial\rho^2} + \frac{1}{\rho}\frac{\partial}{\partial\rho} - \frac{n^2}{\rho^2} - v(\rho) + \epsilon \right] f_{\epsilon,n}(\rho) = 0 .$$

The wave function is written as

$$\psi_{\epsilon,n}(\rho, \theta) = f_{\epsilon,n}(\rho)\, e^{in\theta} ,$$

and the boundary conditions for $\rho \longrightarrow \infty$ will depend on the problem of interest (for example, in the case of a bound-state solution we must have $f_{\epsilon,n}(\rho \longrightarrow \infty) = 0$). The boundary condition at the origin can be obtained by considering the differential equation in the limit $\rho \longrightarrow 0$. Assuming that $\rho^2 v(\rho) \longrightarrow 0$ in this limit, we find

$$\left[\frac{\partial^2}{\partial\rho^2} + \frac{1}{\rho}\frac{\partial}{\partial\rho} - \frac{n^2}{\rho^2} \right] f_{\epsilon,n}(\rho) = 0 \qquad \text{for } \rho \longrightarrow 0 .$$

Taking $f_{\epsilon,n}(\rho) \propto \rho^\alpha$, it follows that

$$\alpha(\alpha - 1)\rho^{\alpha-2} + \alpha\rho^{\alpha-2} - n^2\rho^{\alpha-2} = 0 \implies \alpha^2 = n^2 ,$$

which has the solution $\alpha = |n|$, yielding

$$f_{\epsilon,n}(\rho \longrightarrow 0) \propto \rho^{|n|} .$$

Part 3

In the free-particle case the radial equation becomes

$$\left[\frac{\partial^2}{\partial\rho^2} + \frac{1}{\rho}\frac{\partial}{\partial\rho} - \frac{n^2}{\rho^2} + k^2 \right] f_{k,n}(\rho) = 0 ,$$

where we have introduced $\epsilon = 2mE/\hbar^2 = k^2$. We define $x = k\rho$, so that

$$\frac{\partial}{\partial\rho} = k\frac{d}{dx} , \qquad \frac{\partial^2}{\partial\rho^2} = k^2\frac{d^2}{dx^2} ,$$

and the radial equation reduces to

$$\left[k^2\frac{d^2}{dx^2} + k^2\frac{1}{x}\frac{d}{dx} - k^2\frac{n^2}{x^2} + k^2 \right] f_{k,n}(x) = 0 \implies \left[\frac{d^2}{dx^2} + \frac{1}{x}\frac{d}{dx} - \frac{n^2}{x^2} + 1 \right] f_{k,n}(x) = 0 .$$

Part 4

We assume the power series solution

$$f_{k,n}(x) = \sum_{p=0}^{\infty} a_p x^{s+p} ,$$

which, when inserted into the Bessel equation, leads to

$$\sum_{p=0}^{\infty} a_p[(s+p)(s+p-1) + (s+p) - n^2]x^{s+p-2} = -\sum_{p=0}^{\infty} a_p x^{s+p} ,$$

or, more conveniently, removing the common x^s factor on the left- and right-hand sides,

$$\sum_{p=0}^{\infty} a_p[(s+p)^2 - n^2]x^{p-2} = -\sum_{p=0}^{\infty} a_p x^p .$$

We compare powers of x on the left- and right-hand sides:

$$x^{-2}: a_0[s^2 - n^2] = 0 , \qquad x^{-1}: a_1[(s+1)^2 - n^2] = 0 ,$$

and

$$x^0: a_2[(s+2)^2 - n^2] = -a_0 , \qquad x^1: a_3[(s+3)^2 - n^2] = -a_1 , \ \ldots$$

The general form of the recurrence relation is

$$a_p = -\frac{1}{[(s+p)^2 - n^2]} a_{p-2} .$$

Since $a_0 \neq 0$, the x^{-2} equation gives the condition $s = \pm|n|$. Then the x^{-1} equation requires $a_1 = 0$, in order to satisfy the condition

$$a_1[(1 \pm |n|)^2 - n^2] = a_1[1 \pm 2|n|] = 0 ,$$

since n is an integer. As a consequence, all odd powers of x vanish. The two independent solutions are that corresponding to $s = |n|$ – the regular solution, well-behaved at the origin – and that corresponding to $s = -|n|$ – the irregular solution, singular at the origin. For the regular solution we find

$$p = 2k, \qquad k = 0, 1, 2, \ldots \implies f_{k,n}(x) = x^{|n|} \sum_{k=0}^{\infty} a_{2k}x^{2k} , \qquad a_{2k} = -\frac{1}{4k(k+|n|)} a_{2k-2} ,$$

or, more explicitly (recall that $a_0 \neq 0$),

$$a_2 = -\frac{1}{4(1+|n|)} a_0 ,$$

$$a_4 = \frac{1}{4(1+|n|)} \frac{1}{8(2+|n|)} a_0 ,$$

$$a_6 = -\frac{1}{4(1+|n|)} \frac{1}{8(2+|n|)} \frac{1}{12(3+|n|)} a_0 ,$$

from which we infer the general form for the coefficient:

$$a_{2k} = \frac{(-1)^k}{4^k k!} \frac{1}{(1+|n|)(2+|n|)\cdots(k+|n|)} a_0 ,$$

and the general solution

$$f_n(x) = |n|! \, x^{|n|} a_0 \sum_{k=0}^{\infty} \frac{(-1)^k}{k!} \frac{(x^2/4)^k}{(|n| + k)!} \; .$$

By choosing[2]

$$a_0 = \frac{1}{2^{|n|} \, |n|!} \; ,$$

the expansion above coincides with that given for the Bessel function $J_n(x)$ in Section 9.1.10 of M. Abramowitz and I. A. Stegun eds., 1965, *Handbook of Mathematical Functions*, Dover, US National Bureau of Standards.

Problem 7 S-Wave Bound States in a Spherical Potential Well

Consider a particle in a spherical potential well defined as

$$V(r) = -V_0 \quad r < a \,, \qquad V(r) = 0 \quad r > a \,,$$

where $V_0 > 0$. Assume the particle has orbital angular momentum $l = 0$ (that is, it is in the S-wave state).

1. Show that there are no bound states unless

$$V_0 a^2 > \frac{\pi^2 \hbar^2}{8\mu} \; .$$

2. If the ground state of the particle is only just bound, show that the well depth V_0 and radius a are related to the binding energy by the expansion

$$V_0 a^2 = \frac{\hbar^2}{2\mu} \left[\frac{\pi^2}{4} + 2\kappa a - \frac{4}{\pi^2} (\kappa a)^2 + \cdots \right] \,,$$

where $\kappa = \sqrt{2\mu|E|/\hbar^2}$. The simplest compound nucleus is the deuteron, which consists of a proton and neutron. They are bound with an energy of 2.226 MeV and there are no discrete excited states. If the deuteron is represented by a particle with reduced mass μ given by

$$\mu = \frac{m_p m_n}{m_p + m_n} \,,$$

where m_p and m_n are, respectively, the proton and neutron mass, moving in a square well potential of radius $a = 1.5$ fm, estimate the depth V_0 of the potential.

Solution

Part 1

Bound states (if they occur) must have energies in the range $-V_0 < E < 0$. The radial Schrödinger equation in the S-wave case for the reduced wave function $u(r)$ reads

$$u_<''(r) + v_0 u_<(r) = |\epsilon| \, u_<(r) \quad 0 \le r < a \,, \qquad u_>''(r) = |\epsilon| u_>(r) \quad r > a \,,$$

---

[2] Since the differential equation defining $f_{k,n}(x)$ is homogeneous, the overall normalization factor is arbitrary.

where we define $v_0 = 2\mu V_0/\hbar^2$ and $\epsilon = 2\mu E/\hbar^2$ (and take $E < 0$). The boundary conditions are

$$u_<(r) \propto r \qquad \text{as } r \longrightarrow 0, \qquad \lim_{r \to \infty} u_>(r) = 0,$$

and $u(r)$ is continuous along with its first derivative at $r = a$. We find

$$u_<(r) = A \sin(Kr), \qquad u_>(r) = B\,e^{-\kappa r},$$

where

$$K = \sqrt{v_0 - |\epsilon|}, \qquad \kappa = \sqrt{|\epsilon|},$$

and note that $0 < |\epsilon| < v_0$. Imposing the continuity of $u(r)$ and $u'(r)$ yields

$$A \sin(Ka) = B\,e^{-\kappa a}, \qquad AK \cos(Ka) = -\kappa\,e^{-\kappa a},$$

from which we obtain the eigenvalue equation

$$\tan(Ka) = -\frac{K}{\kappa}.$$

Set $x = Ka$ and hence $\kappa a = \sqrt{x_0^2 - x^2}$ with $x_0^2 = a^2 v_0$, and the equation above reduces to

$$\tan x = -\frac{x}{\sqrt{x_0^2 - x^2}} \qquad 0 < x < x_0.$$

A rough plot of the left- and right-hand sides of this relation shows that, for at least a single bound state to occur, we must have $x_0 > \pi/2$, and so

$$x_0^2 > \frac{\pi^2}{4} \implies V_0 a^2 > \frac{\pi^2 \hbar^2}{8\mu}.$$

Part 2

It is convenient to start from

$$\tan\sqrt{x_0^2 - z^2} = -\frac{\sqrt{x_0^2 - z^2}}{z} \implies \underbrace{z \tan\sqrt{x_0^2 - z^2}}_{f(z)} = -\sqrt{x_0^2 - z^2}, \qquad z = \kappa a.$$

Since the particle is only just bound, its energy is close to zero (in magnitude). It is also clear that x_0 must be close to $\pi/2$ (that is, $x_0 \longrightarrow \pi^+/2$). We expand $f(z)$ for small z as

$$f(z) = z \tan x_0 + O(z^3),$$

and hence we have, by expanding $\sqrt{x_0^2 - z^2}$ to order z^2,

$$z \tan x_0 \approx -x_0\left(1 - \frac{z^2}{2x_0^2}\right) \implies z \approx -\frac{x_0}{\tan x_0} + \frac{z^2}{2x_0 \tan x_0}.$$

We can solve this algebraic equation for z approximately as

$$z \approx -\frac{x_0}{\tan x_0} + \frac{x_0}{2 \tan^3 x_0} = \underbrace{-\frac{x_0}{\tan x_0}\left(1 - \frac{1}{2 \tan^2 x_0}\right)}_{g(x_0)},$$

and recall that x_0 is close to $\pi^+/2$ where $\tan x_0 < 0$ and very large in magnitude. We expand $g(x_0)$ around $\pi/2$. We define $u = x_0 - \pi/2$ with $|u| \ll 1$, and note that

$$x_0 = \pi/2 + u \, , \qquad \tan x_0 = -\frac{1}{x_0 - \pi/2} + \frac{x_0 - \pi/2}{3} + \cdots = -\frac{1}{u} + \frac{u}{3} + \cdots \, .$$

We insert these expansions into $g(u)$ to find

$$g(u) \approx -\frac{\pi/2 + u}{-1/u + u/3}\left[1 - \frac{1}{2\,(-1/u + u/3)^2}\right] = \frac{u(\pi/2 + u)}{1 - u^2/3}\left[1 - \frac{u^2}{2\,(1 - u^2/3)^2}\right] \, ,$$

and so up, to cubic terms in u,

$$g(u) \approx u(\pi/2 + u)(1 + u^2/3)(1 - u^2/2) \approx u(\pi/2 + u)(1 - u^2/6) \approx u(\pi/2 + u - \pi\,u^2/12) \, .$$

Thus, the equation relating z to $g(x_0)$ reads, in this limit,

$$z \approx u\left(\frac{\pi}{2} + u - \frac{\pi}{12}u^2\right) \, .$$

Since we need the solution for u as function of z, we write

$$u \approx \frac{z}{\pi/2 + u - \pi u^2/12} = \frac{2}{\pi}\frac{z}{1 + 2u/\pi - u^2/6} \approx \frac{2z}{\pi}\left[1 - \frac{2u}{\pi} + \left(\frac{1}{6} + \frac{4}{\pi^2}\right)u^2\right] \, ,$$

or

$$u \approx \frac{2z}{\pi} - \frac{4zu}{\pi^2} + \frac{2zu^2}{\pi}\left(\frac{1}{6} + \frac{4}{\pi^2}\right) \, ,$$

which can be solved by iteration to yield

$$u \approx \frac{2z}{\pi} - \frac{8z^2}{\pi^3} \, ,$$

where only terms up to quadratic in z have been included. We finally have, in terms of x_0,

$$x_0 \approx \frac{\pi}{2} + \frac{2z}{\pi} - \frac{8z^2}{\pi^3} \implies x_0^2 \approx \frac{\pi^2}{4} + 2z - \frac{4z^2}{\pi^2} \, .$$

Recalling the definitions of x_0 and z, the above relation reads

$$V_0 a^2 = \frac{\hbar^2}{2\mu}\left[\frac{\pi^2}{4} + 2\kappa a - \frac{4}{\pi^2}\,(\kappa a)^2 + \cdots\right] \, .$$

We have

$$\frac{\hbar^2}{2\mu} = \frac{(\hbar c)^2}{mc^2} \approx 41.47 \text{ MeV fm}^2$$

where we have taken the nucleon mass to be the average of the proton and neutron masses, $m = (m_p + m_n)/2$ (with $mc^2 \approx 938.9$ MeV), and hence $\mu = m/2$ (also, $\hbar c \approx 197.33$ MeV fm). Using the deuteron binding energy of 2.226 MeV and $a = 1.5$ fm, the expansion parameter is

$$\kappa a \approx 0.348 \implies V_0 \approx 58.3 \text{ MeV} \, ,$$

where we have kept only the linear term κa in the expansion of V_0.

Problem 8 Particle Constrained in a Spherical Potential Well

Consider a particle in a spherical potential well $V(r)$ given by $V(r) = -V_0$ $(V_0 > 0)$ for $r \leq r_0$, and $V(r) = \infty$ for $r > r_0$.

1. Obtain the S-wave energy eigenvalues.
2. Obtain an expression for the energy eigenvalues for a generic orbital angular momentum l and the corresponding eigenfunctions.

Solution

Part 1

The radial equation in $0 \leq r \leq r_0$ reads

$$u_l''(r) + \left[\epsilon + v_0 - \frac{l(l+1)}{r^2} \right] u_l(r) = 0, \qquad v_0 = \frac{2\mu V_0}{\hbar^2} > 0 \text{ and } \epsilon = \frac{2\mu E}{\hbar^2},$$

with the boundary conditions $u_l(r \longrightarrow 0) \propto r^{l+1}$ and $u_l(r_0) = 0$. We must have $\epsilon \geq -v_0$ for acceptable solutions. The case of S-waves (corresponding to $l = 0$) is especially simple, since then

$$u_0''(r) + (\epsilon + v_0)u_0(r) = 0,$$

which (for $\epsilon > -v_0$) has as solutions linear combinations of $\sin(kr)$ and $\cos(kr)$ with

$$k = \sqrt{\epsilon + v_0}, \qquad \epsilon > -v_0.$$

However, $u_0(r)$ must vanish at the origin and also at r_0, and therefore the only allowed solutions are proportional to $\sin(kr)$ with

$$kr_0 = n\pi \implies E_n = -V_0 + \frac{(n\pi)^2 \hbar^2}{2\mu r_0^2}, \qquad n = 1, 2, \dots$$

It can be seen that there is an infinite number of S-wave bound states.

Part 2

For a generic l, the radial equation can be cast into the form of a Bessel equation. To this end, it is convenient to work with $R_l(r)$ rather than $u_l(r)$, so that

$$\frac{1}{r^2} \frac{d}{dr} r^2 \frac{dR_l(r)}{dr} + \left[\epsilon + v_0 - \frac{l(l+1)}{r^2} \right] R_l(r) = 0,$$

and to introduce the non-dimensional variable $\rho = kr$. In terms of ρ the equation above becomes

$$\frac{1}{\rho^2} \frac{d}{d\rho} \rho^2 \frac{dR_l(\rho)}{d\rho} + \left[1 - \frac{l(l+1)}{\rho^2} \right] R_l(\rho) = 0 \qquad \text{or} \qquad R_l''(\rho) + \frac{2}{\rho} R_l'(\rho) + \left[1 - \frac{l(l+1)}{\rho^2} \right] R_l(\rho) = 0.$$

We introduce the auxiliary function $R_l(\rho) \equiv J_l(\rho)/\sqrt{\rho}$, which then leads to

$$R_l'(\rho) = -\frac{J_l(\rho)}{2\rho^{3/2}} + \frac{J_l'(\rho)}{\rho^{1/2}}, \qquad R_l''(\rho) = \frac{3J_l(\rho)}{4\rho^{5/2}} - \frac{J_l'(\rho)}{\rho^{3/2}} + \frac{J_l''(\rho)}{\rho^{1/2}}$$

and therefore

$$\frac{3J_l(\rho)}{4\rho^{5/2}} - \frac{J_l'(\rho)}{\rho^{3/2}} + \frac{J_l''(\rho)}{\rho^{1/2}} + \frac{2}{\rho}\left[-\frac{J_l(\rho)}{2\rho^{3/2}} + \frac{J_l'(\rho)}{\rho^{1/2}}\right] + \left[1 - \frac{(l+1/2)^2 - 1/4}{\rho^2}\right]\frac{J_l(\rho)}{\rho^{1/2}} = 0\,.$$

Canceling terms and multiplying both sides by $\rho^{5/2}$, we are finally left with

$$\rho^2 J_l''(\rho) + \rho J_l'(\rho) + [\rho^2 - \underbrace{(l+1/2)^2}_{\nu^2}]J_l(\rho) = 0\,,$$

which is in the form of a Bessel equation with $\nu = 1/2, 3/2, 5/2, \ldots$, with solutions

$$R_l(\rho) = \frac{J_l(\rho)}{\rho^{1/2}} = \alpha j_l(\rho) + \beta n_l(\rho)\,,$$

where $j_l(\rho)$ and $n_l(\rho)$ are known as the regular and irregular spherical Bessel functions, respectively;[3] for illustration, we list a few of them:

$$j_0(\rho) = \frac{\sin\rho}{\rho}\,, \qquad j_1(\rho) = \frac{\sin\rho}{\rho^2} - \frac{\cos\rho}{\rho}\,, \qquad j_2(\rho) = 3\frac{\sin\rho}{\rho^3} - 3\frac{\cos\rho}{\rho^2} - \frac{\sin\rho}{\rho}\,,$$

$$n_0(\rho) = -\frac{\cos\rho}{\rho}\,, \qquad n_1(\rho) = -\frac{\cos\rho}{\rho^2} - \frac{\sin\rho}{\rho}\,, \qquad n_2(\rho) = -3\frac{\cos\rho}{\rho^3} - 3\frac{\sin\rho}{\rho^2} + \frac{\cos\rho}{\rho}\,.$$

In the limit $\rho \longrightarrow 0$, the Bessel functions behave as

$$j_l(\rho) \propto \rho^l\,, \qquad n_l(\rho) \propto \rho^{-(l+1)}\,.$$

Again, because of the boundary conditions, we must have $\beta = 0$ and $j_l(kr_0) = 0$, that is, kr_0 must be a node of $j_l(\rho)$. There is an infinite number of nodes for any fixed l, and we label them ρ_{nl} with $j_l(\rho_{nl}) = 0$,

$$kr_0 = \rho_{nl} \implies E_{nl} = -V_0 + \frac{\rho_{nl}^2 \hbar^2}{2\mu r_0^2}\,, \qquad n = 1, 2, 3, \ldots$$

The bound-state energies E_{nl} depend on n and l but are independent of the azimuthal quantum number m. Therefore, each eigenvalue E_{nl} has degeneracy $g_{nl} = 2l + 1$; in particular, the ground state, for which $nl = 1s$, is non-degenerate. No analytical expression is available for the nodes ρ_{nl}, except for $l = 0$. The resulting level ordering for the lowest states can be easily obtained from Table 10.1. The eigenfunctions are obtained as

Table 10.1 The first few zeros ρ_{nl} of the spherical Bessel functions $j_l(\rho)$

l/n	1	2	3	4
S	3.142	6.283	9.425	12.566
P	4.493	7.725	10.904	14.066
D	5.763	9.095	12.323	15.515
F	6.988	10.417	13.698	16.924
G	8.183	11.705	15.040	18.301

---

[3] Incidentally, the above procedure also provides the solutions of the radial equation for a free particle. In that case $v_0 = 0$, the range of ρ (or r) is $0 \leq \rho < \infty$, and the condition of regularity at the origin imposes $\beta = 0$ in the linear combination above. The spherical Bessel function $j_l(\rho)$ is bounded (but is not normalizable).

$$\psi_{nlm}(r,\Omega) = \alpha_{nl} j_l(\rho_{nl} r/r_0) Y_{lm}(\Omega) \,,$$

where k has been expressed as ρ_{nl}/r_0 and α_{nl} is a normalization constant.

Problem 9   Given $\psi_{lm}(\mathbf{r}) = r^\beta\, e^{-\gamma r}\, Y_{lm}(\Omega)$, Can All H-Atom Eigenfunctions be Determined?

Assuming that the eigenfunctions for the hydrogen atom are of the form $r^\beta\, e^{-\gamma r}\, Y_{lm}(\Omega)$ with undetermined parameters β and γ, solve the Schrödinger equation. Are all eigenfunctions and eigenvalues obtained in this way?

Solution

By assumption we have

$$\psi(r,\Omega) = R(r)\, Y_{lm}(\Omega) \qquad \text{with } R(r) = r^\beta\, e^{-\gamma r} \,,$$

so that $u(r) = r R(r) = r^{\beta+1} e^{-\gamma r}$. Inserting this expression into the radial equation

$$u''(r) + \left[\epsilon + \frac{2\mu}{\hbar^2} \frac{Ze^2}{r} - \frac{l(l+1)}{r^2} \right] u(r) = 0 \,,$$

with

$$u''(r) = \left[\gamma^2 - \frac{2\gamma(\beta+1)}{r} + \frac{\beta(\beta+1)}{r^2} \right] \underbrace{r^{\beta+1}\, e^{-\gamma r}}_{u(r)} \,,$$

we obtain

$$\left[\gamma^2 - \frac{2\gamma(\beta+1)}{r} + \frac{\beta(\beta+1)}{r^2} \right] u(r) + \left[\epsilon + \frac{2\mu}{\hbar^2} \frac{Ze^2}{r} - \frac{l(l+1)}{r^2} \right] u(r) = 0 \,.$$

Matching the constants yields

$$\beta = l\,, \qquad \gamma(\beta+1) = \frac{\mu Ze^2}{\hbar^2} \,, \qquad \gamma^2 = -\epsilon \,,$$

since the solution $\beta = -l - 1$ is not physically acceptable. Thus we obtain

$$\gamma = \frac{\mu Ze^2}{(l+1)\hbar^2} \implies E = -\frac{\hbar^2}{2\mu} \left[\frac{\mu Ze^2}{(l+1)\hbar^2} \right]^2 = -\frac{(Z\alpha)^2}{2(l+1)^2} \mu c^2 \,,$$

where $\alpha = e^2/(\hbar c)$ is the fine-structure constant. Setting $n = l + 1$, we find

$$E_n = -\frac{(Z\alpha)^2}{2n^2} \mu c^2 \qquad n = 1, 2, \dots \,,$$

which coincides with the exact result. However, for fixed n, we obtain only the wave function with $l = n - 1$, that is the wave function with the highest possible l compatible with the given n: ψ_{1s}, ψ_{2p}, ψ_{3d}, and so on.

Problem 10   Probability for Electron in H Atom to be in Classically Forbidden Region

Compute the probability that the electron in the ground state of a hydrogen atom will be found at a distance from the nucleus greater than its energy would permit in the classical theory. (Adapted from J. J. Sakurai and J. Napolitano (2020), *Modern Quantum Mechanics*, Cambridge University Press.)

Solution

For a given $\epsilon = 2\mu E/\hbar^2 < 0$, the classical inversion points for vanishing angular momentum are given by

$$-\frac{2\gamma}{r} = \epsilon \implies r_< = 0 \text{ and } r_> = \frac{2\gamma}{|\epsilon|} ,$$

where we define

$$\gamma = \frac{e^2 \mu}{\hbar^2} = \frac{1}{a_0} ,$$

and a_0 is the Bohr radius $a_0 = \hbar^2/(\mu e^2)$. In the hydrogen atom, the normalized 1s wave function is given by

$$\psi_{1s}(r) = \underbrace{\frac{2}{a_0^{3/2}} e^{-r/a_0}}_{R_{1s}(r)} Y_{00}(\Omega) ,$$

with corresponding energy

$$E_{1s} = -\frac{e^2}{2a_0} \implies |\epsilon_{1s}| = \frac{1}{a_0^2} \quad \text{and} \quad r_> = 2a_0 .$$

The probability that the electron is found at a distance from the nucleus greater that its energy would permit in the classical theory is obtained as

$$p(1s) = \int_{2a_0}^{\infty} dr\, r^2\, |R_{1s}(r)|^2 \underbrace{\int d\Omega\, |Y_{00}(\Omega)|^2}_{\text{unity}} .$$

Carrying out the integration, we find

$$p(1s) = \int_{2a_0}^{\infty} dr\, r^2\, \frac{4}{a_0^3} e^{-2r/a_0} = \frac{1}{2} \int_4^{\infty} dx\, x^2\, e^{-x} = 13e^{-4},$$

where we have used the formula

$$I_n = \int_b^{\infty} dx\, x^n\, e^{-\gamma x} = (-1)^n \frac{d^n}{d\gamma^n} \int_b^{\infty} dx\, e^{-\gamma x} = (-1)^n \frac{d^n}{d\gamma^n} \frac{e^{-\gamma b}}{\gamma} ,$$

which for $n = 2$ gives

$$I_2 = \left(\frac{b^2}{\gamma} + \frac{2}{\gamma^3} + \frac{2b}{\gamma^2} \right) e^{-\gamma b} \implies I_2 = 26e^{-4} \approx 0.48 \quad \text{for } b = 4 \text{ and } \gamma = 1 .$$

Problem 11 Relation Between the Radial Equations of the Hydrogen Atom and the Harmonic Oscillator

There is a relationship between the radial equations of a particle in a $1/r$ potential and in a harmonic oscillator potential. Consider the radial equation for the hydrogen-like atom,

$$\frac{d^2u(r)}{dr^2} + \left[\epsilon + \frac{2\mu}{\hbar^2} \frac{Ze^2}{r} - \frac{l(l+1)}{r^2} \right] u(r) = 0 \quad \text{with } \epsilon = \frac{2\mu E}{\hbar^2} ,$$

and carry out the following transformation:

$$r = \gamma \bar{r}^2 , \qquad u = \bar{r}^{1/2} \bar{u} ,$$

where γ is a constant. Show that the resulting radial equation can be transformed into that for the harmonic oscillator by an appropriate choice of the various constants. What is the relationship between the energy eigenvalues and quantum numbers of the two systems?

Solution

We have

$$\frac{d}{dr} = \frac{d\bar{r}}{dr}\frac{d}{d\bar{r}} = \frac{1}{2\gamma\bar{r}}\frac{d}{d\bar{r}} \implies \frac{d^2}{dr^2} = \frac{1}{4\gamma^2}\frac{1}{\bar{r}}\frac{d}{d\bar{r}}\frac{1}{\bar{r}}\frac{d}{d\bar{r}} ,$$

and hence

$$\frac{d^2 u}{dr^2} = \frac{1}{4\gamma^2}\frac{1}{\bar{r}}\frac{d}{d\bar{r}}\frac{1}{\bar{r}}\frac{d}{d\bar{r}}\bar{r}^{1/2}\bar{u} = \frac{1}{4\gamma^2}\frac{1}{\bar{r}}\frac{d}{d\bar{r}}\left(\frac{1}{2}\bar{r}^{-3/2}\bar{u} + \bar{r}^{-1/2}\bar{u}'\right)$$

$$= \frac{1}{4\gamma^2}\left(-\frac{3}{4}\bar{r}^{-7/2}\bar{u} + \underbrace{\frac{1}{2}\bar{r}^{-5/2}\bar{u}' - \frac{1}{2}\bar{r}^{-5/2}\bar{u}'}_{\text{cancel}} + \bar{r}^{-3/2}\bar{u}''\right) ,$$

where the primes indicate derivatives of $\bar{u}$ with respect to $\bar{r}$. In terms of $\bar{r}$ and $\bar{u}$, the radial equation reads

$$\frac{\bar{r}^{-3/2}}{4\gamma^2}\left(\bar{u}'' - \frac{3}{4\bar{r}^2}\bar{u}\right) + \left[\epsilon + \frac{2}{\gamma\bar{r}_0}\frac{1}{\bar{r}^2} - \frac{l(l+1)}{\gamma^2\bar{r}^4}\right]\bar{r}^{1/2}\bar{u} = 0 ,$$

where we define $\bar{r}_0 = a_0/Z$ and a_0 is the Bohr radius $a_0 = \hbar^2/(\mu e^2)$. Multiplying both sides by $4\gamma^2\bar{r}^{3/2}$ yields

$$\bar{u}'' + \left[\frac{8\gamma}{\bar{r}_0} - 4\gamma^2|\epsilon|\bar{r}^2 - \frac{4l^2 + 4l + 3/4}{\bar{r}^2}\right]\bar{u} = 0 ,$$

where we have taken ϵ to be negative, since bound states in hydrogen-like atoms occur for $\epsilon < 0$. Note that the centrifugal term can be written as

$$\frac{4l^2 + 4l + 3/4}{\bar{r}^2} = \frac{\bar{l}(\bar{l}+1)}{\bar{r}^2} \qquad\qquad \text{with } \bar{l} = 2l + 1/2 ,$$

and the quadratic potential can be written as the harmonic potential $(\mu^2\omega^2/\hbar^2)\bar{r}^2$ by choosing γ such that

$$4\gamma^2|\epsilon| = \frac{\mu^2\omega^2}{\hbar^2} \implies \gamma = \frac{\mu\omega}{2\hbar\sqrt{|\epsilon|}} .$$

With these substitutions, the equation for $\bar{u}$ takes the form

$$\bar{u}'' + \left[\bar{\epsilon} - \frac{\mu^2\omega^2}{\hbar^2}\bar{r}^2 - \frac{\bar{l}(\bar{l}+1)}{\bar{r}^2}\right]\bar{u} = 0 ,$$

where

$$\bar{\epsilon} = \frac{4\mu\omega}{\hbar\bar{r}_0\sqrt{|\epsilon|}} .$$

The above equation has precisely the form of the Schrödinger equation for the isotropic harmonic oscillator. It has eigenvalues given by Eq. (10.24) and the relation between $\bar{\epsilon}$ and ϵ in Eq. (10.26),

$$\bar{\epsilon} = \overbrace{\frac{\mu\,\omega}{\hbar}}^{1/r_0^2} \underbrace{\left(2p + 2\bar{l} + 3\right)}_{\epsilon} = \frac{\mu\omega}{\hbar}\left(2p + 4l + 4\right) ,$$

where p is an *even* non-negative integer, so that $p = 2\nu$ with $\nu = 0, 1, 2, \ldots$ (and of course $l = 0, 1, 2, \ldots$). Thus we have

$$\bar{\epsilon} = 4\,\frac{\mu\omega}{\hbar}(\nu + l + 1) ,$$

from which we deduce

$$4\frac{\mu\omega}{\hbar}(\nu + l + 1) = \frac{4\mu\omega}{\hbar\bar{r}_0\sqrt{|\epsilon|}} \implies \epsilon_n = -\frac{1}{n^2\bar{r}_0^2} \qquad n = \nu + l + 1 ,$$

or

$$E_n = \frac{\hbar^2}{2\mu}\,\epsilon_n = -\frac{\hbar^2}{2\mu}\frac{Z^2}{n^2 a_0^2} = -\frac{Z^2 e^4 \mu}{2n^2\hbar^2} = -\frac{(Z\alpha)^2}{2n^2}\mu c^2 ,$$

as expected.

Problem 12   WKB Approximation of the Hydrogen-like Atom Spectrum

Apply the WKB formula derived in Problem 13 in Chapter 4 to obtain the (bound-state) energy levels of a S-wave electron (charge $-e$) in the Coulomb field of a nucleus with charge Ze (the hydrogen-like system).

Derive the energy levels for a generic angular momentum l.

The application of the WKB method for the radial equation needs to be discussed. The radial coordinate r is in the range $0 < r < \infty$, and there is no boundary condition as r goes to $-\infty$ (in contrast with the one-dimensional case). In the specific case of the Coulomb attractive potential, the formula as derived in Problem 13 in Chapter 4 can be applied when the orbital angular momentum $l \gg 1$, since only then is the smallest inversion point sufficiently far from the origin. We will ignore this issue here, and use the WKB formula without concerning ourselves with such a difficulty.

Solution

The Schrödinger equation for the reduced radial wave function $u_l(r)$ reads

$$u_l''(r) = \left[\underbrace{\frac{l(l+1)}{r^2} - \frac{2\mu}{\hbar^2}\frac{Ze^2}{r} - \epsilon}_{v(r)}\right]u_l(r) , \qquad \epsilon = \frac{2\mu E}{\hbar^2} ,$$

and bound-state solutions are possible only for $E < 0$. The WKB quantization condition for the energy (see Problem 13 in Chapter 4) requires

$$\int_{r_-}^{r_+} dr \sqrt{\epsilon - v(r)} = (n - 1/2)\pi , \qquad n = 1, 2, 3, \ldots ,$$

where in the classically allowed region $\epsilon > v(r)$. Here, $r_\pm$ are the classical inversion points (with $\epsilon < 0$). Setting

$$\gamma = \frac{Ze^2\mu}{\hbar^2} \,,$$

we find the inversion points from $v(r) = \epsilon$, which yields

$$-\frac{2\gamma}{r} = \epsilon \implies r_- = 0 \text{ and } r_+ = -\frac{2\gamma}{\epsilon} \qquad \text{for } l = 0$$

and

$$\frac{l(l+1)}{r^2} - \frac{2\gamma}{r} = \epsilon \implies r_\pm = -\frac{\gamma \pm \sqrt{\gamma^2 + l(l+1)\epsilon}}{\epsilon} \qquad \text{for } l \geq 1 \,.$$

When $l = 0$, the potential has no minimum. However, when $l \geq 1$, the potential $v(r)$ does have a minimum, at $r_m = l(l+1)/\gamma$, where it assumes the value

$$v_m \equiv v(r_m) = -\frac{\gamma^2}{l(l+1)} < 0 \,,$$

and so

$$r_\pm = -\frac{\gamma \pm \sqrt{l(l+1)(\epsilon - v_m)}}{\epsilon} = \frac{\gamma \pm \sqrt{l(l+1)(|v_m| - |\epsilon|)}}{|\epsilon|} \,,$$

where $0 > \epsilon > v_m$ or $0 < |\epsilon| < |v_m|$.

The case $l = 0$

The integral in the WKB formula is given by (setting $x = r_+/r$ below)

$$I_0 = \int_{r_-}^{r_+} dr \sqrt{-|\epsilon| + \frac{2\gamma}{r}} = \sqrt{|\epsilon|} \int_0^{r_+} dr \sqrt{\frac{r_+}{r} - 1} = r_+ \sqrt{|\epsilon|} \int_1^\infty dx \, \frac{\sqrt{x-1}}{x^2} \,.$$

We change variable to

$$x = \cosh^2 u \,, \qquad dx = 2 \sinh u \cosh u \, du \,,$$

and so

$$I_0 = 2\, r_+ \sqrt{|\epsilon|} \int_0^\infty du \, \frac{\sinh^2 u}{\cosh^3 u} = 2\, r_+ \sqrt{|\epsilon|} \int_0^\infty du \, \sinh u \left(-\frac{1}{2}\frac{d}{du}\cosh^{-2}u\right) = r_+ \sqrt{|\epsilon|} \int_0^\infty du \, \frac{1}{\cosh u} \,,$$

after an integration by parts. We further set

$$t = \tanh(u/2) \implies \cosh u = \frac{1+t^2}{1-t^2} \text{ and } du = \frac{2}{1-t^2} dt \,,$$

which finally yields

$$I_0 = r_+ \sqrt{|\epsilon|} \int_0^\infty du \, \frac{1}{\cosh u} = 2\, r_+ \sqrt{|\epsilon|} \int_0^1 dt \, \frac{1}{1+t^2} = \frac{\pi}{2} r_+ \sqrt{|\epsilon|} \,.$$

The WKB condition then gives

$$\frac{\pi}{2} r_+ \sqrt{|\epsilon|} = (n - 1/2)\pi \implies \frac{\gamma}{\sqrt{|\epsilon|}} = n - 1/2 \implies |\epsilon_{n0}| = \frac{\gamma^2}{(n - 1/2)^2} \,,$$

or

$$E_{n0} = -\frac{1}{2(n-1/2)^2} \frac{(Ze^2)^2}{\hbar^2} \mu = -\frac{(Z\alpha)^2}{2(n-1/2)^2} \mu c^2 \,,$$

where we have introduced the fine structure constant $\alpha = e^2/(\hbar c)$ (c is the speed of light). Note that, in the exact result, $n - 1/2$ is replaced by n with $n = 1, 2, \ldots$

The case $l \geq 1$

The integral in the WKB formula is now given by

$$I_l = \int_{r_-}^{r_+} dr \sqrt{-|\epsilon| + \frac{2\gamma}{r} - \frac{l(l+1)}{r^2}} = \int_{r_-}^{r_+} dr \frac{1}{r} \sqrt{-|\epsilon|r^2 + 2\gamma r - l(l+1)} \,,$$

since $\epsilon < 0$. We write

$$-|\epsilon|r^2 + 2\gamma r - l(l+1) = -\frac{1}{|\epsilon|}(|\epsilon|r - \gamma)^2 + \frac{\gamma^2 - l(l+1)|\epsilon|}{|\epsilon|} = \frac{\gamma^2 - l(l+1)|\epsilon|}{|\epsilon|} \left[1 - \frac{(|\epsilon|r - \gamma)^2}{\gamma^2 - l(l+1)|\epsilon|}\right] \,,$$

and introduce the new variable

$$x = \frac{|\epsilon|r - \gamma}{\sqrt{\gamma^2 - l(l+1)|\epsilon|}} \,, \qquad r = \frac{x\sqrt{\gamma^2 - l(l+1)|\epsilon|} + \gamma}{|\epsilon|} \,,$$

so that

$$\sqrt{-|\epsilon|r^2 + 2\gamma r - l(l+1)} = \sqrt{\frac{\gamma^2 - l(l+1)|\epsilon|}{|\epsilon|}} \sqrt{1 - x^2} \,.$$

The integral becomes

$$I_l = \sqrt{\frac{\gamma^2 - l(l+1)|\epsilon|}{|\epsilon|}} \int_{x_-}^{x_+} dx \frac{\sqrt{1 - x^2}}{x + x_0} \,,$$

where

$$x_0 = \frac{\gamma}{\sqrt{\gamma^2 - l(l+1)|\epsilon|}} = \frac{1}{\sqrt{1 - |\epsilon|/|v_m|}} > 1 \,, \qquad x_\pm = \frac{|\epsilon|r_\pm - \gamma}{\sqrt{\gamma^2 - l(l+1)|\epsilon|}} = \pm 1 \,.$$

We finally arrive at

$$I_l = \sqrt{\frac{\gamma^2 - l(l+1)|\epsilon|}{|\epsilon|}} \int_{-1}^{1} dx \frac{\sqrt{1 - x^2}}{x + x_0} \equiv \sqrt{\frac{\gamma^2 - l(l+1)|\epsilon|}{|\epsilon|}} J_l \,.$$

The integral J_l can be further manipulated to obtain a rational integrand. Begin by making the substitution

$$x = \cos t \quad \text{and} \quad dx = -\sin t \, dt \implies J_l = \int_0^{\pi} dt \frac{\sin^2 t}{\cos t + x_0} \,;$$

then write

$$\sin^2 t = 4 \sin^2(t/2) \cos^2(t/2) \,, \qquad \cos t = \cos^2(t/2) - \sin^2(t/2) \,,$$

and recall that

$$\cos^2(t/2) = \frac{1}{1 + \tan^2(t/2)} \ , \qquad \sin^2(t/2) = \frac{\tan^2(t/2)}{1 + \tan^2(t/2)} \ .$$

Now, make the following substitution:

$$u = \tan(t/2) \qquad du = \frac{1}{2\cos^2(t/2)} \, dt \implies dt = \frac{2}{1 + u^2} \, du$$

to arrive at

$$J_l = 8 \int_0^\infty du \, \frac{u^2}{(1 + u^2)^2 \, [x_0 + 1 + (x_0 - 1)u^2]} = 4 \int_{-\infty}^\infty du \, \frac{u^2}{(1 + u^2)^2 \, [x_0 + 1 + (x_0 - 1)u^2]} \ ,$$

which can be evaluated by contour integration. The integrand has poles at

$$\pm i \ \text{(double poles)} \qquad \text{and} \qquad \pm i \sqrt{\frac{x_0 + 1}{x_0 - 1}} \equiv \pm i u_0 \ \text{(simple poles)} \ ,$$

and

$$J_l = \frac{4}{x_0 - 1} \oint dz \, f(z) \ , \qquad f(z) = \frac{z^2}{(z + i)^2 \, (z - i)^2 \, (z + iu_0) \, (z - iu_0)} \ .$$

We close the contour in the upper half-plane to find

$$J_l = \frac{4}{x_0 - 1} \, 2\pi i \, [\mathrm{Res}(i) + \mathrm{Res}(iu_0)] \ .$$

The residue (times $2\pi i$) at the simple pole iu_0 is given by

$$2\pi i \, \mathrm{Res}(iu_0) = 2\pi i \frac{(iu_0)^2}{2iu_0 \, (iu_0 + i)^2 \, (iu_0 - i)^2} = -\frac{\pi u_0}{(u_0^2 - 1)^2} \ ,$$

while the residue at the double pole i is given by

$$2\pi i \, \mathrm{Res}(i) = 2\pi i \frac{d}{dz} \left[(z - i)^2 f(z) \right] \Big|_{z=i} = \frac{\pi}{u_0^2 - 1} \left(\frac{1}{2} + \frac{1}{u_0^2 - 1} \right) \ .$$

Collecting results, we obtain

$$J_l = \frac{2\pi}{x_0 - 1} \frac{1}{u_0^2 - 1} \left(1 - 2 \frac{u_0 - 1}{u_0^2 - 1} \right) = \frac{2\pi}{(x_0 - 1)(u_0 + 1)^2} \ ,$$

and the original integral I_l finally reads

$$I_l = 2\pi \sqrt{\frac{\gamma^2 - l(l+1)|\epsilon|}{|\epsilon|}} \, \frac{1}{(x_0 - 1)(u_0 + 1)^2} = \pi \sqrt{\frac{\gamma^2 - l(l+1)|\epsilon|}{|\epsilon|}} \, \frac{1}{x_0 + \sqrt{x_0^2 - 1}} \ ,$$

yielding the quantization condition

$$\sqrt{\frac{\gamma^2 - l(l+1)|\epsilon|}{|\epsilon|}} \, \frac{1}{x_0 + \sqrt{x_0^2 - 1}} = n - 1/2 \ , \qquad n = 1, 2, \dots$$

Now, we have

$$\sqrt{x_0^2 - 1} = \sqrt{\frac{\gamma^2}{\gamma^2 - l(l+1)|\epsilon|} - 1} = \sqrt{\frac{l(l+1)|\epsilon|}{\gamma^2 - l(l+1)|\epsilon|}} \ ,$$

so that

$$\sqrt{\frac{\gamma^2 - l(l+1)|\epsilon|}{|\epsilon|}} \frac{1}{x_0 + \sqrt{x_0^2 - 1}} = \sqrt{\frac{\gamma^2 - l(l+1)|\epsilon|}{|\epsilon|}} \frac{\sqrt{\gamma^2 - l(l+1)|\epsilon|}}{\gamma + \sqrt{l(l+1)|\epsilon|}}$$

$$= \frac{1}{\sqrt{|\epsilon|}} \frac{\gamma^2 - l(l+1)|\epsilon|}{\gamma + \sqrt{l(l+1)|\epsilon|}} = \frac{\gamma - \sqrt{l(l+1)|\epsilon|}}{\sqrt{|\epsilon|}}$$

yielding finally

$$\frac{\gamma}{\sqrt{|\epsilon_{nl}|}} = n - 1/2 + \sqrt{l(l+1)} \implies E_{nl} = -\frac{(Z\alpha)^2}{2[n - 1/2 + \sqrt{l(l+1)}]^2} \mu c^2 .$$

Note that if the factor $l(l+1)$ is changed to $(l+1/2)^2$ then the energies above coincide with the exact values; we set $k = n + l$, and obtain

$$E_k = -\frac{(Z\alpha)^2}{2k^2} \mu c^2 , \qquad k = 1, 2, \ldots ,$$

and for a given k there is a degeneracy on $l = 0, 1, \ldots, k - 1$ (in addition to the degeneracy $2l + 1$ on the azimuthal quantum number).

Comment

In reference to the cautionary note in the text of the problem, we observe the following. For S-waves (with an attractive Coulomb potential), the smallest inversion point is the origin, where $u(r)$ must vanish. We can use the WKB formulae as follows. For $r \gg r_+$ (here r_+ is the largest inversion point), we must have a decreasing exponential, namely

$$u_0(r) \approx \frac{A_-}{\sqrt{\kappa(r)}} \exp\left[-\int_{r_+}^r dr' \, \kappa(r')\right] ,$$

while for $r \ll r_+$, in the classically allowed region,

$$u_0(r) \approx \frac{B_-}{\sqrt{k(r)}} \exp\left[i \int_r^{r_+} dr' \, k(r')\right] + \frac{B_+}{\sqrt{k(r)}} \exp\left[-i \int_r^{r_+} dr' \, k(r')\right] .$$

The connection formulae in this case give (here, A_+ is zero)

$$B_- = e^{-i\pi/4} A_- , \qquad B_+ = e^{i\pi/4} A_- ,$$

and the wave function in the classically allowed region is

$$u_0(r) \approx \frac{2A_-}{\sqrt{k(r)}} \cos\left[\int_r^{r_+} dr' \, k(r') - \frac{\pi}{4}\right] .$$

Requiring that the wave function vanishes at the origin yields

$$u_0(0) = 0 \implies \frac{1}{\sqrt{k(r \longrightarrow 0)}} \cos\left[\int_0^{r_+} dr' \, k(r') - \frac{\pi}{4}\right] = 0 .$$

We conclude that

$$\int_0^{r_+} dr' \, k(r') = \left(n - \frac{1}{4}\right) \pi \qquad \text{with } n = 1, 2, \ldots ,$$

which is the correct WKB quantization formula in the present case. It would lead to the following S-wave bound-state energies:

$$E_n = -\frac{(Z\alpha)^2}{2(n - 1/4)^2} \mu c^2 \ .$$

Problem 13 S-Wave Particle in Central Potential $V(r) = -V_0\, e^{-r/a}$

Consider a particle of mass μ under the influence of a central potential given by

$$V(r) = -V_0\, e^{-r/a} \ ,$$

where $V_0, a > 0$.

1. Change variable from r to $x = e^{-r/(2a)}$ and show that the S-wave Schrödinger equation for the reduced radial wave function $u_0(r)$ reduces to the Bessel equation,

$$\chi''(\xi) + \frac{1}{\xi}\, \chi'(\xi) + \left(1 - \frac{\alpha^2}{\xi^2}\right) \chi(\xi) = 0 \ ,$$

   after appropriate rescaling of the variable $x \longrightarrow \xi$.

   Hint: Consult a book on special functions to learn about the properties of the solutions of the above equation, in particular their small-ξ behavior.
2. When the radial wave function u_0 is viewed as a function of ξ (or x), what boundary conditions must it satisfy? Show how these boundary conditions can be used to determine the energy levels.
3. What is the lower limit of V_0 for which at least one bound state exists?

Solution

Part 1

The radial equation for an S-wave reads

$$\left[-\frac{\hbar^2}{2\mu}\frac{d^2}{dr^2} + V(r)\right] u_0(r) = E u_0(r) \ ,$$

and bound states are possible for $E < 0$. The above equation can be written as

$$u_0''(r) + \left[v_0\, e^{-r/a} - |\epsilon|\right] u_0(r) = 0 \ .$$

Under the replacement $x = e^{-r/(2a)}$, we have

$$\frac{d}{dr} = \frac{dx}{dr}\frac{d}{dx} = -\frac{1}{2a}\, e^{-r/(2a)}\frac{d}{dx} = -\frac{x}{2a}\frac{d}{dx}$$

and

$$\frac{d^2}{dr^2} = -\frac{x}{2a}\frac{d}{dx}\left(-\frac{x}{2a}\frac{d}{dx}\right) = \frac{1}{4a^2}\, x\frac{d}{dx}\left(x\frac{d}{dx}\right) = \frac{1}{4a^2}\left(x^2\frac{d^2}{dx^2} + x\frac{d}{dx}\right) \ .$$

In terms of x the radial equation now reads

$$\left[\frac{1}{4a^2}\left(x^2\frac{d^2}{dx^2} + x\frac{d}{dx}\right) u_0(x) + v_0 x^2 - |\epsilon|\right] u_0(x) = 0 \ ,$$

or, multiplying both sides by $4a^2/x^2$,

$$u_0''(x) + \frac{1}{x} u_0'(x) + 4a^2 \left(v_0 - \frac{|\epsilon|}{x^2} \right) = 0 \ .$$

In order to cast this as the Bessel equation, consider the further change of variable $\xi = cx$, with the constant c to be determined. We have

$$\frac{d}{dx} = c \frac{d}{d\xi} \ , \qquad \frac{d^2}{dx^2} = c^2 \frac{d^2}{d\xi^2} \ ,$$

and

$$c^2 u''(\xi) + c^2 \frac{1}{\xi} u_0'(\xi) + 4a^2 v_0 \left(1 - \frac{c^2 |\epsilon|/v_0}{\xi^2} \right) = 0 \ .$$

Choosing $c^2 = 4a^2 v_0$, we obtain the Bessel equation

$$u_0''(\xi) + \frac{1}{\xi} u_0'(\xi) + \left(1 - \frac{\alpha^2}{\xi^2} \right) u_0(\xi) = 0 \ , \qquad \alpha = 2a \sqrt{|\epsilon|} \ ,$$

where we note that here α is a real number.

Part 2

The boundary conditions on $u_0(r)$ are:

$$\lim_{r \to 0} u_0(r) = 0 \ , \qquad \lim_{r \to \infty} u_0(r) = 0 \ .$$

The general solution of the Bessel equation is the linear combination

$$u_0(\xi) = A J_{|\alpha|}(\xi) + B J_{-|\alpha|}(\xi) \ ,$$

where $J_{|\alpha|}(\xi)$ and $J_{-|\alpha|}(\xi)$ are Bessel functions. In terms of ξ the boundary condition at $r \longrightarrow \infty$ implies $\xi \longrightarrow 0$; since $J_{-|\alpha|}(\xi)$ is divergent in this limit, we must have $B = 0$. Hence, we are left with

$$u_0(r) = A J_{|\alpha|} \Big[\underbrace{2a\sqrt{v_0}\, e^{-r/(2a)}}_{\xi} \Big] \ ,$$

and the boundary condition at the origin yields

$$J_{2a\sqrt{|\epsilon|}}(2a\sqrt{v_0}) = 0 \ .$$

The energies result from solving the equation above for given a and v_0.

Part 3

If the particle is weakly bound then its binding energy $|\epsilon|$ is close to zero, which implies $|\alpha| = 0$. The first node of $J_0(\xi)$ occurs at $\xi_0 \approx 2.4$, yielding the minimum value of V_0 for a bound state to exist as

$$V_0 = \frac{\hbar^2}{8\mu} \frac{\xi_0^2}{a^2} \ .$$

Problem 14   Relation Between S-Wave Functions Evaluated at the Origin and $V'(r)$

Consider a central potential $V(r)$ such that $r^2 V(r)$ vanishes as $r \longrightarrow 0$. Show that the (real) bound-state wave functions in an S-wave $\psi_{ns}(r)$,

$$\psi_{ns}(r) = \frac{1}{\sqrt{4\pi}} \frac{u_{ns}(r)}{r} ,$$

evaluated at the origin satisfy the following relation:

$$[\psi_{ns}(0)]^2 = \frac{m}{2\pi\hbar^2} \langle \psi_{ns}| \frac{dV(r)}{dr} |\psi_{ns}\rangle .$$

Verify this relation for the ground states of the hydrogen atom and the isotropic three-dimensional harmonic oscillator.

Solution

The wave function $\psi_{ns}(r)$ is real, and hence

$$\left\langle \frac{dV(r)}{dr} \right\rangle_{ns} = \frac{\hbar^2}{2m} \int d\mathbf{r}\, v'(r)\, [\psi_{ns}(r)]^2 = \frac{\hbar^2}{2m} \int_0^\infty dr\, v'(r)\, u_{ns}^2(r) ,$$

where we define $v(r) = (2m/\hbar^2) V(r)$. An integration by parts and use of the radial equation to re-express $v(r) u_{ns}(r)$ as $u_{ns}''(r) + \epsilon_{ns} u_{ns}(r)$ allows one to obtain

$$\left\langle \frac{dV(r)}{dr} \right\rangle_{ns} = \frac{\hbar^2}{2m} \Big[\underbrace{v(r) u_{ns}^2(r) \big|_0^\infty}_{\text{vanishes}} - 2 \int_0^\infty dr\, \underbrace{v(r) u_{ns}(r)}_{\text{use } u_{ns}''(r) + [\epsilon - v(r)]u_{ns}(r)=0} u_{ns}'(r) \Big]$$

$$= -\frac{\hbar^2}{m} \int_0^\infty dr\, [u_{ns}''(r) + \epsilon_{ns} u_{ns}(r)]\, u_{ns}'(r)$$

$$= -\frac{\hbar^2}{2m} \Big[u_{ns}'^2(r) + \epsilon_{ns} u_{ns}^2(r) \Big] \Big|_0^\infty = \frac{\hbar^2}{2m} u_{ns}'^2(0) ,$$

since $u_{ns}(r \longrightarrow 0) = c\, r$ with c constant and $u_{ns}(r \longrightarrow \infty) = 0$; note that $u_{ns}'(0) = c$ and $u_{ns}'(r \longrightarrow \infty) = 0$. We conclude that

$$\left\langle \frac{dV(r)}{dr} \right\rangle_{ns} = \frac{\hbar^2}{2m} \left[\frac{u_{ns}(r)}{r} \right]^2 \Big|_{r=0} = \frac{2\pi\hbar^2}{m} [\psi_{ns}(0)]^2 ,$$

as expected. In particular, for the ground state of the hydrogen atom we have

$$\psi_{1s}(r) = \frac{1}{\sqrt{\pi a_0^3}}\, e^{-r/a_0} , \qquad a_0 = \frac{\hbar^2}{\mu e^2} ,$$

and the expectation value of $V'(r)$ follows:

$$\langle V'(r)\rangle_{1s} = \int d\Omega \int_0^\infty dr\, r^2\, \underbrace{\frac{e^2}{r^2}}_{V'(r)}\, \underbrace{\frac{1}{\pi a_0^3}\, e^{-2r/a_0}}_{\psi_{1s}^2(r)} = 2\, \frac{e^2}{a_0^2} ,$$

which leads to

$$\frac{\mu}{2\pi\hbar^2}\, \langle V'(r)\rangle_{1s} = \frac{\mu}{\pi\hbar^2}\, \frac{e^2}{a_0^2} = \frac{1}{\pi a_0^3} = \psi_{1s}^2(0) .$$

In the case of the isotropic three-dimensional harmonic oscillator, the ground-state wave function reads

$$\psi_{0s}(r) = \left(\frac{\mu\omega}{\pi\hbar}\right)^{3/4} e^{-\mu\omega r^2/(2\hbar)} ,$$

and

$$\langle V'(r)\rangle_{0s} = \int d\Omega \int_0^\infty dr\, r^2 \underbrace{\mu\omega^2 r}_{V'(r)} \underbrace{\left(\frac{\mu\omega}{\pi\hbar}\right)^{3/2} e^{-\mu\omega r^2/\hbar}}_{\psi_{0s}^2(r)} = \frac{4}{\sqrt{\pi}}\mu\omega^2 \left(\frac{\hbar}{\mu\omega}\right)^{1/2} \underbrace{\int_0^\infty dx\, x^3\, e^{-x^2}}_{1/2} ,$$

where we define $x = \sqrt{\mu\omega/\hbar}\, r$. The last integral is easily done by parts – note that $x\, e^{-x^2} = -(e^{-x^2})'/2$ – and

$$\frac{\mu}{2\pi\hbar^2}\langle V'(r)\rangle_{0s} = \frac{\mu}{2\pi\hbar^2}\frac{2}{\sqrt{\pi}}\mu\omega^2 \left(\frac{\hbar}{\mu\omega}\right)^{1/2} = \left(\frac{\mu\omega}{\pi\hbar}\right)^{3/2} = \psi_{0s}^2(0) .$$

Problem 15 Virial Theorem in Three Dimensions

Consider a particle of mass μ in a central potential which admits bound states.

1. Show that the operator $\hat{O}$, defined as

$$\hat{O} = \hat{\mathbf{r}} \cdot \hat{\mathbf{p}} + \hat{\mathbf{p}} \cdot \hat{\mathbf{r}} ,$$

   has vanishing expectation values on the eigenstates $|\psi_{Elm}\rangle$ of the Hamiltonian.
2. Show that if $|\psi(t)\rangle$ is a generic state at time t then

$$\frac{d}{dt}\langle\psi(t)|\hat{O}|\psi(t)\rangle = 2\langle\psi(t)|\hat{\mathbf{p}}^2/\mu - \hat{\mathbf{r}}\cdot\boldsymbol{\nabla}\,\hat{V}(\hat{r})|\psi(t)\rangle .$$

3. Specialize to the case in which $|\psi(t)\rangle \longrightarrow e^{-iEt/\hbar}|\psi_{Elm}\rangle$. What does the relation derived in part 2 imply in this case? Show that, if the potential is proportional to a power $\hat{r}^\alpha$, this relation allows us to determine the expectation values of the kinetic and potential energy operators independently. Determine these expectation values in the case of the hydrogen atom and isotropic three-dimensional oscillator.

Solution

Part 1

We note the following identity,

$$[\hat{H}, \hat{r}_\alpha] = \frac{1}{2\mu}[\hat{\mathbf{p}}^2, \hat{r}_\alpha] + [\hat{V}(\hat{r}), \hat{r}_\alpha] = \frac{1}{2\mu}\sum_\beta (\hat{p}_\beta \underbrace{[\hat{p}_\beta, \hat{r}_\alpha]}_{-i\hbar\delta_{\alpha\beta}} + [\hat{p}_\beta, \hat{r}_\alpha]\hat{p}_\beta) = -\frac{i\hbar}{\mu}\hat{p}_\alpha ,$$

and hence

$$\hat{O} = \sum_\alpha (\hat{r}_\alpha\hat{p}_\alpha + \hat{p}_\alpha\hat{r}_\alpha) = \frac{i\mu}{\hbar}\sum_\alpha \left(\hat{r}_\alpha[\hat{H}, \hat{r}_\alpha] + [\hat{H}, \hat{r}_\alpha]\hat{r}_\alpha\right) = \frac{i\mu}{\hbar}\left(\hat{H}\hat{\mathbf{r}}^2 - \hat{\mathbf{r}}^2\hat{H}\right) ,$$

where, expanding the two commutators, we see that the term $\hat{r}_\alpha\hat{H}\hat{r}_\alpha$ enters with a plus sign in the first commutator and a minus in the second, canceling out. The expectation value of $\hat{O}$ on a bound eigenstate of $\hat{H}$ vanishes, since

$$\langle \psi_{Elm} | \hat{O} | \psi_{Elm} \rangle = \frac{i\mu}{\hbar} \langle \psi_{Elm} | \hat{H}\hat{\mathbf{r}}^2 - \hat{\mathbf{r}}^2 \hat{H} | \psi_{Elm} \rangle = \frac{i\mu}{\hbar} (E - E) \langle \psi_{Elm} | \hat{\mathbf{r}}^2 | \psi_{Elm} \rangle = 0 \ .$$

It is instructive to derive this result directly. To this end, we note that

$$\hat{\mathbf{r}} \cdot \hat{\mathbf{p}} \longrightarrow -i\hbar \mathbf{r} \cdot \boldsymbol{\nabla} = -i\hbar r \frac{\partial}{\partial r} \ ,$$

so that

$$\langle \psi_{Elm} | \hat{\mathbf{r}} \cdot \hat{\mathbf{p}} | \psi_{Elm} \rangle = -i\hbar \int d\Omega \int dr \, r^2 \, \psi_{Elm}^*(\mathbf{r}) \, r \frac{\partial}{\partial r} \psi_{Elm}(\mathbf{r}) = -i\hbar \int dr \, r^3 R_{El}(r) R'_{El}(r) \ ,$$

where in the last step we have expressed the bound-state wave function as

$$\psi_{Elm}(\mathbf{r}) = R_{El}(r) \, Y_{lm}(\Omega) \ ,$$

and have carried out the angular integration. Note that the (bound-state) radial wave function is real. We also have

$$\langle \psi_{Elm} | \hat{\mathbf{r}} \cdot \hat{\mathbf{p}} | \psi_{Elm} \rangle^* = \langle \psi_{Elm} | \hat{\mathbf{p}} \cdot \hat{\mathbf{r}} | \psi_{Elm} \rangle = i\hbar \int dr \, r^3 R_{El}(r) R'_{El}(r) \ ,$$

which implies that the expectation value of $\hat{O}$ vanishes.

Part 2

For the expectation value on a generic state $|\psi(t)\rangle$, we have

$$\frac{d}{dt} \langle \psi(t) | \hat{O} | \psi(t) \rangle = \frac{i}{\hbar} \, \psi(t) | [\hat{H}, \, \hat{O}] | \psi(t) \rangle \ ,$$

where

$$[\hat{H}, \, \hat{O}] = \sum_\alpha \left(\hat{r}_\alpha \left[\hat{H}, \, \hat{p}_\alpha \right] + \left[\hat{H}, \, \hat{r}_\alpha \right] \hat{p}_\alpha + \hat{p}_\alpha \left[\hat{H}, \, \hat{r}_\alpha \right] + \left[\hat{H}, \, \hat{p}_\alpha \right] \hat{r}_\alpha \right) \ ,$$

and we have already obtained $[\hat{H}, \, \hat{r}_\alpha] = -(i\hbar/\mu)\hat{p}_\alpha$. We find for the remaining commutator

$$[\hat{H}, \, \hat{p}_\alpha] = [\hat{V}(\hat{r}), \, \hat{p}_\alpha] = i\hbar \, \boldsymbol{\nabla}_\alpha \hat{V}(\hat{r}) = i\hbar \, \hat{r}_\alpha \frac{\hat{V}'(\hat{r})}{\hat{r}} \ ,$$

and hence

$$[\hat{H}, \, \hat{O}] = 2i\hbar \sum_\alpha \left[\hat{r}_\alpha^2 \frac{\hat{V}'(\hat{r})}{\hat{r}} - \frac{1}{\mu} \hat{p}_\alpha^2 \right] = 2i\hbar \left[\hat{r} \hat{V}'(\hat{r}) - \frac{\hat{\mathbf{p}}^2}{\mu} \right] = 2i\hbar \left[\hat{\mathbf{r}} \cdot \boldsymbol{\nabla} \hat{V}(\hat{r}) - \frac{\hat{\mathbf{p}}^2}{\mu} \right] \ .$$

We conclude that

$$\frac{d}{dt} \langle \psi(t) | \hat{O} | \psi(t) \rangle = 2 \langle \psi(t) | \hat{\mathbf{p}}^2/\mu - \hat{\mathbf{r}} \cdot \boldsymbol{\nabla} \hat{V}(\hat{r}) | \psi(t) \rangle \ .$$

Part 3

The expectation value of a generic operator on a stationary state (an eigenstate of the Hamiltonian),

$$|\psi_{Elm}(t)\rangle = \mathrm{e}^{-iEt/\hbar} |\psi_{Elm}\rangle \ ,$$

is time independent; in our particular case, we have

$$\langle \psi_{Elm}(t) | \hat{O} | \psi_{Elm}(t) \rangle = \langle \psi_{Elm}(0) | \hat{O} | \psi_{Elm}(0) \rangle = \langle \psi_{Elm} | \hat{O} | \psi_{Elm} \rangle = 0 \ ,$$

which in turn yields

$$\langle \psi_{Elm} | \hat{\mathbf{p}}^2 / \mu - \hat{\mathbf{r}} \cdot \nabla \hat{V}(\hat{r}) | \psi_{Elm} \rangle = 0 \implies 2\langle \psi_{Elm} | \hat{T} | \psi_{Elm} \rangle = \langle \psi_{Elm} | \hat{\mathbf{r}} \cdot \nabla \hat{V}(\hat{r}) | \psi_{Elm} \rangle \,,$$

where $\hat{T}$ is the kinetic energy operator. If the potential energy operator is proportional to $\hat{r}^\alpha$, where α is a real number, then $\hat{\mathbf{r}} \cdot \nabla \hat{V}(\hat{r}) = \alpha \hat{V}(\hat{r})$ and the relation above leads to

$$\langle \hat{T} \rangle_{Elm} = \frac{\alpha}{2} \langle \hat{V} \rangle_{Elm} \,.$$

For the hydrogen atom or isotropic three-dimensional harmonic oscillator this gives, respectively,

$$\langle \hat{T} \rangle_{nlm} = -\frac{1}{2} \left\langle \frac{Ze^2}{\hat{r}} \right\rangle_{nlm} \,, \qquad \langle \hat{T} \rangle_{nlm} = \left\langle \frac{\mu\omega^2}{2} \hat{r}^2 \right\rangle_{nlm} \,.$$

Therefore, since $\langle \hat{T} + \hat{V} \rangle_{nlm} = E_n$, we can determine the expectation values of the kinetic and potential energy operators independently, as

$$\langle \hat{V} \rangle_{nlm} = 2E_n = -\frac{(Z\alpha)^2}{n^2} \mu c^2 \,, \qquad \langle \hat{T} \rangle_{nlm} = \frac{(Z\alpha)^2}{2n^2} \mu c^2$$

for the hydrogen atom, and as

$$\langle \hat{V} \rangle_{nlm} = \frac{E_n}{2} = \frac{\hbar\omega}{2}(n + 3/2) \,, \qquad \langle \hat{T} \rangle_{nlm} = \frac{\hbar\omega}{2}(n + 3/2)$$

for the harmonic oscillator.

Problem 16 Harmonic Oscillator in Two Dimensions in Plane Polar Coordinates

First review Problem 6 and its solution. Then consider the isotropic two-dimensional harmonic oscillator potential.

1. After introducing the non-dimensional variable

$$x = \sqrt{\frac{m\omega}{\hbar}} \rho \,,$$

show that the radial equation reads

$$\left[\frac{d^2}{dx^2} + \frac{1}{x}\frac{d}{dx} - \frac{n^2}{x^2} - x^2 + \epsilon \right] f_{\epsilon n}(x) = 0 \,, \qquad \epsilon = \frac{E}{\hbar\omega/2} \,.$$

Introduce the auxiliary function $g_{\epsilon n}(x)$, defined as

$$f_{\epsilon n}(x) = x^{|n|} e^{-x^2/2} g_{\epsilon n}(x) \,,$$

and show that it satisfies the following differential equation:

$$xg''_{\epsilon n} + \left(b - 2x^2 \right) g'_{\epsilon n} + axg_{\epsilon n} = 0 \,,$$

where

$$a = \epsilon - 2(|n| + 1) \,, \qquad b = 2|n| + 1 \,.$$

2. Posit the following power series solution for $g_{\epsilon n}(x)$:

$$g(x) = x^s \sum_{q=0}^{\infty} c_q x^q \,, \qquad c_0 \neq 0 \,,$$

where the subscripts ϵn on both g and the coefficients c_q have been dropped for brevity. Assume that the series must terminate (that is, $g(x)$ is a polynomial) for the solution to be acceptable. Show that this happens if

$$\epsilon - 2(|n| + 1) = 2p \qquad \text{or} \qquad E_m = \hbar\omega(m+1) \text{ with } m = p + |n| ,$$

where p is even and $|n| = 0, 1, 2, \ldots$ Determine the degeneracy of the energy eigenvalues E_m.

3. Justify the assumption above, that is, show that if the series does not terminate then $g(x)$ behaves asymptotically as $x^2 e^{x^2}$, and hence $f(x)$ behaves as $x^{|n|+2} e^{x^2/2}$.

Solution

Part 1

Using the results of Problem 6, the wave function is given by

$$\psi_{\epsilon n}(\rho, \theta) = f_{\epsilon n}(\rho) e^{in\theta} ,$$

where $n = 0, \pm 1, \pm 2, \ldots$ and the function $f_{\epsilon n}(\rho)$ satisfies the radial equation for $\rho \geq 0$,

$$\left[\frac{d^2}{d\rho^2} + \frac{1}{\rho} \frac{d}{d\rho} - \frac{n^2}{\rho^2} - v(\rho) + \epsilon \right] f_{\epsilon n}(\rho) = 0 .$$

The boundary conditions on $f_{\epsilon n}(\rho)$ are that it vanishes for $\rho \longrightarrow \infty$ for bound states and that it is proportional to $\rho^{|n|}$ as $\rho \longrightarrow 0$ (assuming that $\rho^2 v(\rho) \longrightarrow 0$ in this limit). In the specific case of the the (isotropic) harmonic oscillator in two dimensions, we have

$$\left[\frac{d^2}{d\rho^2} + \frac{1}{\rho} \frac{d}{d\rho} - \frac{n^2}{\rho^2} - \left(\frac{m\omega}{\hbar}\right)^2 \rho^2 + \epsilon \right] f_{\epsilon n}(\rho) = 0 .$$

It is convenient to define the non-dimensional variable

$$x = \sqrt{\frac{m\omega}{\hbar}} \, \rho , \qquad \frac{d}{d\rho} = \frac{dx}{d\rho} \frac{d}{dx} = \sqrt{\frac{m\omega}{\hbar}} \frac{d}{dx} ,$$

in terms of which the radial equation now reads

$$\left[\frac{d^2}{dx^2} + \frac{1}{x} \frac{d}{dx} - \frac{n^2}{x^2} - x^2 + \bar{\epsilon} \right] f_{\epsilon n}(x) = 0 , \qquad \bar{\epsilon} = \frac{\hbar}{m\omega} \epsilon = \frac{E}{\hbar\omega/2} .$$

In the asymptotic region $x \longrightarrow \infty$, $f_{\epsilon n}$ falls off as $e^{-x^2/2}$ (see Section 10.2), and it is convenient to define an auxiliary function $g_{\epsilon n}(x)$:

$$f_{\epsilon n}(x) = x^{|n|} e^{-x^2/2} g_{\epsilon n}(x) .$$

in order to remove the short- and long-range behavior. We find

$$f'_{\epsilon n} = x^{|n|} e^{-x^2/2} \left[\frac{|n|}{x} g_{\epsilon n} - x g_{\epsilon n} + g'_{\epsilon n} \right]$$

and

$$f''_{\epsilon n} = x^{|n|} e^{-x^2/2} \left[\left(\frac{|n|}{x} - x \right) \left(\frac{|n|}{x} g_{\epsilon n} - x g_{\epsilon n} + g'_{\epsilon n} \right) - \frac{|n|}{x^2} g_{\epsilon n} + \frac{|n|}{x} g'_{\epsilon n} - g_{\epsilon n} - x g'_{\epsilon n} + g''_{\epsilon n} \right]$$

$$= x^{|n|} e^{-x^2/2} \left[\left(\frac{n^2}{x^2} - \frac{|n|}{x^2} + x^2 - 2|n| - 1 \right) g_{\epsilon n} + 2 \left(\frac{|n|}{x} - x \right) g'_{\epsilon n} + g''_{\epsilon n} \right] .$$

Inserting this into the differential equation for $f_{\epsilon n}$, we arrive at

$$g_{\epsilon n}'' + \left(\frac{2|n| + 1}{x} - 2x\right) g_{\epsilon n}' + (\bar{\epsilon} - 2|n| - 2) g_{\epsilon n} = 0 \,,$$

or, multiplying both sides by x,

$$x g_{\epsilon n}'' + \left(b - 2x^2\right) g_{\epsilon n}' + ax g_{\epsilon n} = 0 \,,$$

where we define

$$a = \bar{\epsilon} - 2(|n| + 1) \,, \qquad b = 2|n| + 1 \,.$$

Part 2

Inserting power series into the differential equation for $g(x)$, and lumping together terms with the same power of x, we obtain

$$\sum_{q=0}^{\infty} [(s + q)(s + q - 1) + b(s + q)] c_q x^{q-1} = \sum_{q=0}^{\infty} [2(s + q) - a] c_q x^{q+1} \,.$$

For the power series to be a solution, each coefficient of x^p must vanish. Consider first the terms proportional to x^{-1} and x^0,

$$x^{-1}: \ [s(s - 1) + bs] c_0 = 0 \implies s = 0, 1 - b \qquad \text{(for } c_0 \neq 0) \,,$$
$$x^0: \quad \underbrace{[s(s + 1) + b(s + 1)]}_{\neq 0 \text{ for } s=0 \text{ or } s=1-b} c_1 = 0 \implies c_1 = 0 \,,$$

where in the last relation c_1 must vanish, since the pre-factor $[\cdots]$ is $\neq 0$ when s is either 0 or $1 - b$. We also observe that the solution corresponding to $s = 1 - b = -2|n|$ would lead to singular behavior for $x \longrightarrow 0$ – that is, $g_{\epsilon n} \propto x^{-2|n|}$ in turn implying $f_{\epsilon n}(x) \propto x^{-|n|}$ – which is unacceptable. Therefore, the only allowed solution is $s = 0$. Assuming the latter, we find for a generic term proportional to x^{p+1}, with $p \geq 0$,

$$x^{p+1}: \ [(p + 2)(p + 1) + b(p + 2)] c_{p+2} = [2p - a] c_p \,,$$

which implies the following recurrence relation:

$$c_{p+2} = \frac{2p - a}{(p + 2)(p + 1) + b(p + 2)} c_p \qquad \text{with } p = 0, 2, 4, \ldots$$

Because of the condition $c_1 = 0$, it follows that $c_p = 0$ for any odd p. Thus, the series solution is of the form

$$g(x) = \sum_{p \text{ even}=0}^{\infty} c_p x^p \,, \qquad c_{p+2} = \frac{2p - a}{(p + 2)(2|n| + p + 2)} c_p \,,$$

where we have inserted the value for b. Clearly, the series terminates if $a = 2p$ with p even, which yields the eigenvalue

$$\bar{\epsilon} - 2(|n| + 1) = 2p \qquad \text{or} \qquad E_m = \hbar\omega(m + 1), \ \text{with} \ m = p + |n| \,,$$

where p is even and $|n| = 0, 1, 2, \ldots$ The degeneracies of the lowest energy levels are indicated in Table 10.2. It is easily seen that in general E_m has degeneracy $g_m = m + 1$, as expected (see Problem 4 in Chapter 8).

Table 10.2 Degeneracies for the lowest levels of the two-dimensional harmonic oscillator.

m	g_m	p	n
0	1	0	0
1	2	0	1
		0	-1
2	3	2	0
		0	2
		0	-2
3	4	2	1
		2	-1
		0	3
		0	-3

Part 3

The series must terminate, that is, $g(x)$ is a polynomial, because otherwise the solution $f(x)$ would diverge at large x and therefore could not possibly describe a bound state. To see this fact, note that for large $q \gg 1$ we have

$$\frac{c_{2q+2}}{c_{2q}} = \frac{4q - a}{(2q + 2)(2|n| + 2q + 2)} \longrightarrow \frac{1}{q}$$

The series can be written, up to the overall normalization factor c_0, as

$$g(x) = 1 + c_2 x^2 + \cdots + c_{2q-2} x^{2q-2} + c_{2q} x^{2q} \underbrace{\left[1 + \frac{c_{2q+2}}{c_{2q}} x^2 + \frac{c_{2q+4}}{c_{2q}} x^4 + \cdots \right]}_{h(x)},$$

where the integer $q \gg 1$. In this limit, we have

$$\frac{c_{2q+2}}{c_{2q}} = \frac{1}{q}, \qquad \frac{c_{2q+4}}{c_{2q}} = \frac{c_{2q+2}}{c_{2q}} \frac{c_{2q+4}}{c_{2q+2}} = \frac{1}{q(q+1)}, \qquad \cdots,$$

and the series for $H(z)$ is given by

$$\begin{aligned}
h(x) &= 1 + \frac{1}{q} x^2 + \frac{1}{q(q+1)} x^4 + \frac{1}{q(q+1)(q+2)} x^6 + \cdots \\
&= 1 + \frac{(q-1)!}{q!} x^2 + \frac{(q-1)!}{(q+1)!} x^4 + \frac{(q-1)!}{(q+2)!} x^6 + \cdots \\
&= (q-1)! \sum_{p=0}^{\infty} \frac{1}{(p+q-1)!} x^{2p} = (q-1)! \sum_{n=q-1}^{\infty} \frac{1}{n!} x^{2n-2q+2} = \frac{(q-1)!}{x^{2q-2}} \left(\sum_{n=0}^{\infty} \frac{x^{2n}}{n!} - \sum_{n=0}^{q-2} \frac{x^{2n}}{n!} \right) \\
&= \frac{(q-1)!}{x^{2q-2}} \left(e^{x^2} - \sum_{n=0}^{q-2} \frac{x^{2n}}{n!} \right).
\end{aligned}$$

We therefore conclude that

$$g(x) = (\text{polynomial in } x^2) + (q-1)!\, c_{2q} x^2\, e^{x^2} ,$$

which in turn would imply that, in the asymptotic region,

$$f(x) \propto x^{|n|+2}\, e^{x^2/2} ,$$

an exponentially diverging $f(x)$. Hence, the series must terminate.

Angular Momentum: General Properties

We have already introduced the orbital angular momentum operator $\hat{\mathbf{L}}$ in our discussion of a particle in a central potential. In analogy to classical mechanics, it is defined as $\hat{\mathbf{L}} = \hat{\mathbf{r}} \times \hat{\mathbf{p}}$, and satisfies the commutation relations $[\hat{L}_i, \hat{L}_j] = i\hbar \sum_k \epsilon_{ijk} \hat{L}_k$, which follow from the commutation relations of the position and momentum operators. In fact, as we will see later, the angular momentum commutation relations can be derived from the transformation properties of states and vector operators under rotations; in particular, invariance under rotations leads to a conserved angular momentum. This point of view, which is based on symmetry considerations, avoids having to rely on the classical analogy.

In the treatment that follows, we assume only that there exist three hermitian operators $\hat{J}_i$ which satisfy the commutation relations

$$[\hat{J}_i, \hat{J}_j] = i\hbar \sum_k \epsilon_{ijk} \hat{J}_k . \tag{11.1}$$

Using these commutation relations, we will obtain the eigenvalues and eigenstates of $\hat{\mathbf{J}}^2 = \hat{J}_1^2 + \hat{J}_2^2 + \hat{J}_3^2$ and $\hat{J}_3$. We will find that the eigenvalues of $\hat{\mathbf{J}}^2$ and $\hat{J}_3$ are, respectively, $j(j+1)\hbar^2$ and $\mu\hbar$, with $j = 0, 1/2, 1, 3/2, \ldots$ and $\mu = -j, \ldots, j$, and therefore, in addition to integer values, half-integer values are allowed.

11.1 Raising and Lowering Operators: Definitions and Properties

To begin, we note that

$$[\hat{J}_i, \hat{\mathbf{J}}^2] = 0 , \tag{11.2}$$

see Problem 1 in Chapter 9 relating to the orbital angular momentum operator (any property derived using only the commutation relations of the $\hat{L}_i$ clearly also holds in the present case). Next, we define the raising (+) and lowering (−) angular momentum operators as

$$\hat{J}_+ = \hat{J}_1 + i\hat{J}_2 , \qquad \hat{J}_- = \hat{J}_1 - i\hat{J}_2 , \qquad \hat{J}_+^\dagger = \hat{J}_- , \tag{11.3}$$

which obviously commute with $\hat{\mathbf{J}}^2$. We find that

$$[\hat{J}_+, \hat{J}_3] = \underbrace{[\hat{J}_1, \hat{J}_3]}_{-i\hbar\hat{J}_2} + i \underbrace{[\hat{J}_2, \hat{J}_3]}_{i\hbar\hat{J}_1} = -i\hbar\hat{J}_2 - \hbar\hat{J}_1 = -\hbar\hat{J}_+ , \tag{11.4}$$

and[1]

$$[\hat{J}_-, \hat{J}_3] = [\hat{J}_+^\dagger, \hat{J}_3] = \left(\underbrace{[\hat{J}_3, \hat{J}_+]}_{\hbar\hat{J}_+} \right)^\dagger = \hbar\hat{J}_- . \tag{11.5}$$

---

[1] Recall that $[\hat{A}^\dagger, \hat{B}^\dagger] = \hat{A}^\dagger \hat{B}^\dagger - \hat{B}^\dagger \hat{A}^\dagger = (\hat{B}\hat{A} - \hat{A}\hat{B})^\dagger = [\hat{B}, \hat{A}]^\dagger$.

We also find that

$$[\hat{J}_+, \hat{J}_-] = i[\hat{J}_2, \hat{J}_1] - i[\hat{J}_1, \hat{J}_2] = -2i[\hat{J}_1, \hat{J}_2] = 2\hbar \hat{J}_3 . \tag{11.6}$$

To summarize, we have

$$[\hat{J}_\pm, \hat{J}_3] = \mp \hbar \hat{J}_\pm , \qquad [\hat{J}_+, \hat{J}_-] = 2\hbar \hat{J}_3 . \tag{11.7}$$

Since $\hat{\mathbf{J}}^2$ and $\hat{J}_3$ commute, they can be diagonalized simultaneously.

11.2 Determining the Eigenvalues of $\hat{\mathbf{J}}^2$ and $\hat{J}_3$

Let $|\psi_{\lambda\mu}\rangle$ be an eigenstate of $\hat{\mathbf{J}}^2$ and $\hat{J}_3$ with (real) eigenvalues λ and μ (the factors $\hbar$ are included for convenience, to make λ and μ non-dimensional):

$$\hat{\mathbf{J}}^2 |\psi_{\lambda\mu}\rangle = \lambda\hbar^2 |\psi_{\lambda\mu}\rangle , \qquad \hat{J}_3 |\psi_{\lambda\mu}\rangle = \mu\hbar |\psi_{\lambda\mu}\rangle . \tag{11.8}$$

The set $|\psi_{\lambda\mu}\rangle$ forms a basis,

$$\langle \psi_{\lambda\mu} | \psi_{\lambda'\mu'} \rangle = \delta_{\lambda\lambda'} \delta_{\mu\mu'} , \qquad \sum_{\lambda\mu} |\psi_{\lambda\mu}\rangle\langle\psi_{\lambda\mu}| = \hat{\mathbb{1}} . \tag{11.9}$$

The following results follow from the commutation relations of $\hat{J}_\pm$ and $\hat{J}_3$:

1. The matrix elements of $\hat{J}_i$ in the basis $|\psi_{\lambda\mu}\rangle$ vanish unless $\lambda = \lambda'$, since

$$0 = \langle \psi_{\lambda\mu} | \underbrace{\hat{\mathbf{J}}^2 \hat{J}_i - \hat{J}_i \hat{\mathbf{J}}^2}_{[\hat{\mathbf{J}}^2, \hat{J}_i]=0} | \psi_{\lambda'\mu'} \rangle = (\lambda - \lambda')\hbar^2 \langle \psi_{\lambda\mu} | \hat{J}_i | \psi_{\lambda'\mu'} \rangle , \tag{11.10}$$

and, therefore, if $\lambda \neq \lambda'$ then $\langle \psi_{\lambda\mu} | \hat{J}_i | \psi_{\lambda'\mu'} \rangle = 0$. As a consequence the state $\hat{J}_i |\psi_{\lambda\mu}\rangle$ can be expanded as follows:

$$\hat{J}_i |\psi_{\lambda\mu}\rangle = \sum_{\mu'} |\psi_{\lambda\mu'}\rangle\langle\psi_{\lambda\mu'} | \hat{J}_i | \psi_{\lambda\mu}\rangle , \tag{11.11}$$

namely, the expansion is in the subspace spanned by the eigenstates having the same eigenvalue $\lambda\hbar^2$ of $\hat{\mathbf{J}}^2$.

2. The states $\hat{J}_\pm |\psi_{\lambda\mu}\rangle$ are proportional to the eigenstates of $\hat{J}_3$ having eigenvalues $(\mu \pm 1)\hbar$ (that is, $\hat{J}_+$ raises, and $\hat{J}_-$ lowers, μ by one unit). Using the first of equations (11.7) from right to left, we have

$$-\hbar\langle \psi_{\lambda\mu} | \hat{J}_+ | \psi_{\lambda\mu'} \rangle = \langle \psi_{\lambda\mu} | \underbrace{\hat{J}_+ \hat{J}_3 - \hat{J}_3 \hat{J}_+}_{[\hat{J}_+, \hat{J}_3]} | \psi_{\lambda\mu'} \rangle = (\mu' - \mu)\hbar \langle \psi_{\lambda\mu} | \hat{J}_+ | \psi_{\lambda\mu'} \rangle , \tag{11.12}$$

which leads to

$$(\mu' - \mu + 1) \langle \psi_{\lambda\mu} | \hat{J}_+ | \psi_{\lambda\mu'} \rangle = 0 \implies \mu = \mu' + 1 , \tag{11.13}$$

since it cannot hold that $\langle \psi_{\lambda\mu} | \hat{J}_+ | \psi_{\lambda\mu'} \rangle = 0$ for all μ and μ', because that would imply that matrix elements of $\hat{J}_+$ vanish identically on the basis $|\psi_{\lambda\mu}\rangle$ and therefore that $\hat{J}_+$ is the null operator. Similarly, we find

$$(\mu' - \mu - 1) \langle \psi_{\lambda\mu} | \hat{J}_- | \psi_{\lambda\mu'} \rangle = 0 \implies \mu = \mu' - 1 . \tag{11.14}$$

From these last two relations we conclude that $\hat{J}_\pm |\psi_{\lambda\mu}\rangle$ must be proportional to the eigenstates $|\psi_{\lambda\mu\pm1}\rangle$, that is

$$\hat{J}_+ |\psi_{\lambda\mu}\rangle = a(\lambda, \mu) |\psi_{\lambda\mu+1}\rangle , \qquad \hat{J}_- |\psi_{\lambda\mu}\rangle = b(\lambda, \mu) |\psi_{\lambda\mu-1}\rangle . \tag{11.15}$$

The coefficients $a(\lambda, \mu)$ and $b(\lambda, \mu)$ are related to each other, since, using $\hat{J}_+^\dagger = \hat{J}_-$, we have

$$a^*(\lambda, \mu) = \langle\psi_{\lambda\mu+1}| \hat{J}_+ |\psi_{\lambda\mu}\rangle^* = \langle\psi_{\lambda\mu}| \hat{J}_- |\psi_{\lambda\mu+1}\rangle = b(\lambda, \mu + 1) , \tag{11.16}$$

and therefore

$$\hat{J}_+ |\psi_{\lambda\mu}\rangle = a(\lambda, \mu) |\psi_{\lambda\mu+1}\rangle , \qquad \hat{J}_- |\psi_{\lambda\mu}\rangle = a^*(\lambda, \mu - 1) |\psi_{\lambda\mu-1}\rangle . \tag{11.17}$$

3. Consider the second of equations (11.7) from right to left,

$$2\hbar \underbrace{\langle\psi_{\lambda\mu}| \hat{J}_3 |\psi_{\lambda\mu}\rangle}_{\mu\hbar} = \langle\psi_{\lambda\mu}| \underbrace{\hat{J}_+ \hat{J}_- - \hat{J}_- \hat{J}_+}_{[\hat{J}_+, \hat{J}_-]} |\psi_{\lambda\mu}\rangle = |a(\lambda, \mu - 1)|^2 - |a(\lambda, \mu)|^2 , \tag{11.18}$$

since, for example, from Eq. (11.17)

$$\hat{J}_+ \hat{J}_- |\psi_{\lambda\mu}\rangle = \hat{J}_+ \left[a^*(\lambda, \mu - 1)|\psi_{\lambda\mu-1}\rangle \right] = a^*(\lambda, \mu - 1)a(\lambda, \mu - 1) |\psi_{\lambda\mu}\rangle , \tag{11.19}$$

$$\hat{J}_- \hat{J}_+ |\psi_{\lambda\mu}\rangle = \hat{J}_- \left[a(\lambda, \mu)|\psi_{\lambda\mu+1}\rangle \right] = a(\lambda, \mu)a^*(\lambda, \mu) |\psi_{\lambda\mu}\rangle . \tag{11.20}$$

We arrive at the following recurrence relation:

$$|a(\lambda, \mu - 1)|^2 - |a(\lambda, \mu)|^2 = 2\mu\hbar^2 , \tag{11.21}$$

which has the solution[2]

$$|a(\lambda, \mu)|^2 = c(\lambda) - \mu(\mu + 1)\hbar^2 . \tag{11.22}$$

4. It turns out that $c(\lambda) = \lambda\hbar^2$ and that $\lambda \geq 0$. Since $\hat{\mathbf{J}}^2$ can be expressed in terms of raising and lowering operators,

$$\hat{\mathbf{J}}^2 = \frac{1}{2}(\hat{J}_+ \hat{J}_- + \hat{J}_- \hat{J}_+) + \hat{J}_3^2 , \tag{11.23}$$

we find that

$$\begin{aligned}
\lambda\hbar^2 &= \langle\psi_{\lambda\mu}| \hat{\mathbf{J}}^2 |\psi_{\lambda\mu}\rangle = \langle\psi_{\lambda\mu}| \frac{1}{2}(\hat{J}_+ \hat{J}_- + \hat{J}_- \hat{J}_+) + \hat{J}_3^2 |\psi_{\lambda\mu}\rangle \\
&= \frac{1}{2}\left[|a(\lambda, \mu - 1)|^2 + |a(\lambda, \mu)|^2 \right] + \mu^2\hbar^2 \\
&= \frac{1}{2}[\underbrace{c(\lambda) - \mu(\mu - 1)\hbar^2}_{|a(\lambda,\mu-1)|^2} + \underbrace{c(\lambda) - \mu(\mu + 1)\hbar^2}_{|a(\lambda,\mu)|^2}] + \mu^2\hbar^2 = c(\lambda) ,
\end{aligned} \tag{11.24}$$

where we have used Eqs. (11.19) and (11.20). Furthermore, since $\langle\psi_{\lambda\mu}| \hat{\mathbf{J}}^2 |\psi_{\lambda\mu}\rangle = \sum_i \langle\psi_{\lambda\mu}| \hat{J}_i^2 |\psi_{\lambda\mu}\rangle = \sum_i \|\hat{J}_i |\psi_{\lambda\mu}\rangle\|^2 \geq 0$, where $\|\hat{J}_i |\psi_{\lambda\mu}\rangle\|$ is the norm of the state $\hat{J}_i |\psi_{\lambda\mu}\rangle$, we conclude that $\lambda \geq 0$.

---

[2] To see how this comes about, define $f(\mu) = |a(\lambda, \mu)|^2$; the recurrence relation can then be written (taking $\mu \longrightarrow \mu + 1$) as

$$f(\mu) - f(\mu + 1) = 2(\mu + 1)\hbar^2 .$$

In the limit $\mu \gg 1$, we have $f'(\mu) \approx -2\mu\hbar^2$, since in this limit the left-hand side of the equation above is nothing other than $-f'(\mu)$; thus we conclude that, for $\mu \gg 1$, $f(\mu) = -\mu^2\hbar^2$. Assuming that $f(\mu)$ is a polynomial in μ, it must have the general form $f(\mu) = c + d\mu - \mu^2\hbar^2$. Inserting this into the equation above determines d:

$$\underbrace{c + d\mu - \mu^2\hbar^2}_{f(\mu)} - \underbrace{[c + d(\mu + 1) - (\mu + 1)^2\hbar^2]}_{f(\mu+1)} = -d + 2\mu\hbar^2 + \hbar^2 = 2(\mu + 1)\hbar^2 \implies d = -\hbar^2 .$$

Note that c can depend on λ.

Having established that

$$\lambda \geq 0, \qquad |a(\lambda, \mu)|^2 = \left[\lambda - \mu(\mu + 1)\right]\hbar^2, \tag{11.25}$$

we can now easily obtain the values taken on by the eigenvalues λ and μ. To begin with, if we apply $\hat{J}_\pm$ to $|\psi_{\lambda\mu}\rangle$ we change μ by ± 1. We can apply the raising or lowering operators to $|\psi_{\lambda\mu}\rangle$ repeatedly, but it is clear that such a procedure must terminate at some point otherwise the positivity condition above for $|a(\lambda, \mu)|^2$ would be violated. We also note that the norms squared (necessarily ≥ 0) of the two states $\hat{J}_- |\psi_{\lambda\mu}\rangle$ and $\hat{J}_+ |\psi_{\lambda\mu}\rangle$ are

$$\|\hat{J}_- |\psi_{\lambda\mu}\rangle\|^2 = \langle\psi_{\lambda\mu}| \underbrace{\hat{J}_-^\dagger \hat{J}_-}_{\hat{J}_+ \hat{J}_-} |\psi_{\lambda\mu}\rangle = |a(\lambda, \mu - 1)|^2 = \left[\lambda - \mu(\mu - 1)\right]\hbar^2 \geq 0, \tag{11.26}$$

$$\|\hat{J}_+ |\psi_{\lambda\mu}\rangle\|^2 = \langle\psi_{\lambda\mu}| \underbrace{\hat{J}_+^\dagger \hat{J}_+}_{\hat{J}_- \hat{J}_+} |\psi_{\lambda\mu}\rangle = |a(\lambda, \mu)|^2 = \left[\lambda - \mu(\mu + 1)\right]\hbar^2 \geq 0. \tag{11.27}$$

Since, for a given λ, these inequalities are satisfied if

$$\text{Eq. (11.26):} \qquad \frac{1 - \sqrt{1 + 4\lambda}}{2} \leq \mu \leq \frac{1 + \sqrt{1 + 4\lambda}}{2},$$

$$\text{Eq. (11.27):} \qquad \frac{-1 - \sqrt{1 + 4\lambda}}{2} \leq \mu \leq \frac{-1 + \sqrt{1 + 4\lambda}}{2},$$

then they will be satisfied simultaneously if μ is in the following range:

$$\frac{1 - \sqrt{1 + 4\lambda}}{2} \leq \mu \leq \frac{-1 + \sqrt{1 + 4\lambda}}{2}. \tag{11.28}$$

We now express λ as $j(j + 1)$ with $j \geq 0$, and, in terms of j, noting that $\sqrt{1 + 4j(j + 1)} = 2j + 1$, the inequality above simply reads

$$-j \leq \mu \leq j. \tag{11.29}$$

We still need to establish that j is an integer or a half integer. Let us denote by μ_0 the minimum allowed value for the eigenvalue of $\hat{J}_3$. The corresponding eigenstate must be such that $\hat{J}_- |\psi_{j\mu_0}\rangle = 0$, and, $\hat{J}_- |\psi_{j\mu_0}\rangle$ being the null state, its norm must vanish. Then, using Eq. (11.26) with $\lambda = j(j + 1)$, we find

$$0 = \|\hat{J}_- |\psi_{j\mu_0}\rangle\|^2 = \langle\psi_{j\mu_0}|\hat{J}_+ \hat{J}_- |\psi_{j\mu_0}\rangle = \left[j(j + 1) - \mu_0(\mu_0 - 1)\right]\hbar^2 \implies \mu_0 = -j. \tag{11.30}$$

Thus, the minimum allowed value has $\mu_0 = -j$. Now, we act p times with $\hat{J}_+$ on the state $|\psi_{j,-j}\rangle$. This process must also terminate, and so there must be an integer p such that $(\hat{J}_+)^p |\psi_{j,-j}\rangle$ is the null state with $\|(\hat{J}_+)^p|\psi_{j,-j}\rangle\|^2 = 0$. To calculate this norm, we observe that

$$\|(\hat{J}_+)^p|\psi_{j,-j}\rangle\|^2 = \langle\psi_{j,-j}| \underbrace{\hat{J}_- \cdots \hat{J}_-}_{p \text{ times}} \underbrace{\hat{J}_+ \cdots \hat{J}_+}_{p \text{ times}} |\psi_{j,-j}\rangle. \tag{11.31}$$

Next, we use p times the first of equations (11.17) along with its adjoint,

$$(\hat{J}_- |\psi_{j,-j}\rangle)^\dagger = (\langle\psi_{j,-j}|\hat{J}_-) = a^*(\lambda, -j)\langle\psi_{j,-j+1}|, \tag{11.32}$$

where λ stands for $j(j + 1)$, to arrive at

$$\|(\hat{J}_+)^p|\psi_{j,-j}\rangle\|^2 = |a(\lambda, -j)|^2 |a(\lambda, -j + 1)|^2 \cdots |a(\lambda, -j + p - 1)|^2. \tag{11.33}$$

This norm vanishes if

$$|a(\lambda, -j + p - 1)|^2 = [j(j + 1) - (-j + p)(-j + p - 1)]\hbar^2 = 0 \,, \tag{11.34}$$

from which we obtain either $p = 0$ or $p = 2j + 1$. The first solution is not acceptable since it would say that $|\psi_{j,-j}\rangle$ is the null state, thus contradicting that it is the state with the minimum eigenvalue of $\hat{J}_3$. The other solution says that, since p is an integer ≥ 1, j is either an integer or half integer, namely $j = 0, 1/2, 1, 3/2, \ldots$.

11.3 Basis Consisting of Eigenstates of $\hat{\mathbf{J}}^2$ and $\hat{J}_3$

We provide in this section a summary of the results obtained so far. The simultaneous eigenstates of $\hat{\mathbf{J}}^2$ and $\hat{J}_3$ are labeled as $|\psi_{j\mu}\rangle$, with

$$\hat{\mathbf{J}}^2 |\psi_{j\mu}\rangle = j(j + 1)\hbar^2 |\psi_{j\mu}\rangle \,, \qquad \hat{J}_3 |\psi_{j\mu}\rangle = \mu\hbar |\psi_{j\mu}\rangle \,, \tag{11.35}$$

where $j = 0, 1/2, 1, \ldots$ and $\mu = -j, -j + 1, \ldots, +j$. The raising and lowering operators have the following properties:

$$\hat{J}_+ |\psi_{j\mu}\rangle = \hbar \underbrace{\sqrt{j(j + 1) - \mu(\mu + 1)}}_{a(\lambda, \mu)} |\psi_{j\mu+1}\rangle \,, \tag{11.36}$$

$$\hat{J}_- |\psi_{j\mu}\rangle = \hbar \underbrace{\sqrt{j(j + 1) - \mu(\mu - 1)}}_{a^*(\lambda, \mu-1)} |\psi_{j\mu-1}\rangle \,, \tag{11.37}$$

where we have made a choice of phase, in the sense that the coefficient $a(\lambda, \mu) = |a(\lambda, \mu)| \, e^{i\phi(\lambda,\mu)}$ with $\lambda = j(j + 1)$ has been taken to be real (the arbitrary phase factor $\phi(\lambda, \mu)$ has been set to zero). In particular, note that

$$\hat{J}_+ |\psi_{j,j}\rangle = 0 \,, \qquad \hat{J}_- |\psi_{j,-j}\rangle = 0 \,. \tag{11.38}$$

We can also obtain the matrix representations of the $\hat{J}_i$ in this basis. We have already established that the matrix elements $\langle \psi_{j\mu} | \hat{J}_i | \psi_{j'\mu'} \rangle$ vanish unless $j = j'$; the matrices representing $\hat{J}_i$ in each subspace with a given j are square matrices of dimensions $(2j + 1) \times (2j + 1)$. In particular, the matrix representing $\hat{J}_3$ will be a diagonal matrix, with the $2j + 1$ eigenvalues along the diagonal. The matrices representing $\hat{J}_1$ and $\hat{J}_2$ can be obtained from those of $\hat{J}_+$ and $\hat{J}_-$, since

$$\hat{J}_1 = (\hat{J}_- + \hat{J}_+)/2 \,, \qquad \hat{J}_2 = i(\hat{J}_- - \hat{J}_+)/2 \,, \tag{11.39}$$

and

$$[\underline{J}_\pm^{(j)}]_{\mu'\mu} = \langle \psi_{j\mu'} | \hat{J}_\pm | \psi_{j\mu} \rangle = \hbar \sqrt{j(j + 1) - \mu(\mu \pm 1)} \, \delta_{\mu',\mu\pm1} \,, \tag{11.40}$$

where the superscript (j) specifies the subspace with fixed j. The matrices $\underline{J}_i$ are obviously hermitian, and they satisfy the same commutator algebra as the observables $\hat{J}_i$, in the sense that, in each subspace with fixed j,

$$\underline{J}_i^{(j)} \underline{J}_j^{(j)} - \underline{J}_j^{(j)} \underline{J}_i^{(j)} = i\hbar \sum_k \epsilon_{ijk} \underline{J}_k^{(j)} \,. \tag{11.41}$$

11.4 Problems

Problem 1   Commutation Relations of Matrices Representing the $\hat{J}_l$

Show that the matrices representing the angular momentum components $\hat{J}_l$ in subspace j satisfy the same commutation relations as the $\hat{J}_l$ themselves.

Solution

The commutation relations satisfied by the $\hat{J}_l$ imply that

$$\langle \psi_{j\mu'} | \hat{J}_l \hat{J}_m - \hat{J}_m \hat{J}_l | \psi_{j\mu} \rangle = i\hbar \sum_k \epsilon_{lmk} \underbrace{\langle \psi_{j\mu'} | \hat{J}_k | \psi_{j\mu} \rangle}_{[\underline{J}_k^{(j)}]_{\mu'\mu}} .$$

In subspace j we have

$$\sum_{\nu=-j}^{j} |\psi_{j\nu}\rangle\langle\psi_{j\nu}| = \hat{\mathbb{1}}^{(j)} ,$$

where the superscript (j) on the identity operator indicates that the completeness is relative to the subspace with fixed j, and therefore

$$\sum_\nu \Big[\underbrace{\langle \psi_{j\mu'} | \hat{J}_m | \psi_{j\nu} \rangle}_{[\underline{J}_m^{(j)}]_{\mu'\nu}} \underbrace{\langle \psi_{j\nu} | \hat{J}_l | \psi_{j\mu} \rangle}_{[\underline{J}_m^{(j)}]_{\nu\mu}} - \langle \psi_{j\mu'} | \hat{J}_m | \psi_{j\nu} \rangle\langle \psi_{j\nu} | \hat{J}_l | \psi_{j\mu} \rangle \Big] = i\hbar \sum_k \epsilon_{lmk} [\underline{J}_k^{(j)}]_{\mu'\mu} ,$$

or, in matrix notation,

$$\underline{J}_l^{(j)} \underline{J}_m^{(j)} - \underline{J}_m^{(j)} \underline{J}_l^{(j)} = i\hbar \sum_k \epsilon_{lmk} \underline{J}_k^{(j)} .$$

Problem 2   Angular Momentum $j = 1/2$ (or Spin 1/2)

Consider the case of angular momentum, or spin, $j = 1/2$. For brevity, denote the two basis states as $|\pm\rangle = |\psi_{1/2, \pm 1/2}\rangle$.

1. Obtain the matrices representing the angular momentum components in the subspace $j = 1/2$. Verify explicitly that $\hat{\mathbf{J}}^2$ is diagonal with eigenvalue $(3/4)\hbar^2$.

2. Introduce the Pauli matrices, defined as

$$\underline{J}_i^{(1/2)} = \frac{\hbar}{2} \underline{\sigma}_i ,$$

and show the following properties: (i) the determinant of each $\underline{\sigma}_i$ is equal to -1; (ii) the matrices $\underline{\sigma}_i$ and $\underline{\sigma}_j$ for $i \neq j$ anticommute; (iii) the identity

$$\underline{\sigma}_i \underline{\sigma}_j = \delta_{ij} \underline{\sigma}_0 + i \sum_k \epsilon_{ijk} \underline{\sigma}_k ,$$

where $\underline{\sigma}_0$ denotes the 2×2 identity matrix, holds; and (iv) the identity

$$\underline{\sigma} \cdot \hat{\mathbf{A}} \, \underline{\sigma} \cdot \hat{\mathbf{B}} = \hat{\mathbf{A}} \cdot \hat{\mathbf{B}} + i \underline{\sigma} \cdot (\hat{\mathbf{A}} \times \hat{\mathbf{B}}) ,$$

valid for any two generic vector operators $\hat{\mathbf{A}}$ and $\hat{\mathbf{B}}$ that do not necessarily commute with each other, holds.

Solution

Part 1

Since

$$\hat{J}_+ |+\rangle = 0 , \qquad \hat{J}_+ |-\rangle = \hbar \sqrt{1/2(1/2 + 1) - (-1/2)(-1/2 + 1)} \, |+\rangle = \hbar |+\rangle ,$$
$$\hat{J}_- |-\rangle = 0 , \qquad \hat{J}_- |+\rangle = \hbar \sqrt{1/2(1/2 + 1) - 1/2(1/2 - 1)} \, |-\rangle = \hbar |-\rangle ,$$

the 2×2 matrices representing $\hat{J}_\pm$ in the subspace with $j = 1/2$ are

$$\underline{J}_+^{(1/2)} = \begin{pmatrix} \langle +| \hat{J}_+ |+\rangle & \langle +| \hat{J}_+ |-\rangle \\ \langle -| \hat{J}_+ |+\rangle & \langle -| \hat{J}_+ |-\rangle \end{pmatrix} = \hbar \begin{pmatrix} 0 & 1 \\ 0 & 0 \end{pmatrix} , \qquad \underline{J}_-^{(1/2)} = \begin{pmatrix} \langle +| \hat{J}_- |+\rangle & \langle +| \hat{J}_- |-\rangle \\ \langle -| \hat{J}_- |+\rangle & \langle -| \hat{J}_- |-\rangle \end{pmatrix} = \hbar \begin{pmatrix} 0 & 0 \\ 1 & 0 \end{pmatrix} = (\underline{J}_+^{(1/2)})^\dagger ,$$

and therefore taking the appropriate linear combinations, namely, $\underline{J}_1^{(1/2)} = [\underline{J}_+^{(1/2)} + \underline{J}_-^{(1/2)}]/2$, etc., yields

$$\underline{J}_1^{(1/2)} = \frac{\hbar}{2} \begin{pmatrix} 0 & 1 \\ 1 & 0 \end{pmatrix} , \qquad \underline{J}_2^{(1/2)} = \frac{\hbar}{2} \begin{pmatrix} 0 & -i \\ i & 0 \end{pmatrix} , \qquad \underline{J}_3^{(1/2)} = \frac{\hbar}{2} \begin{pmatrix} 1 & 0 \\ 0 & -1 \end{pmatrix} .$$

The matrix representing $\hat{\mathbf{J}}^2 = \hat{J}_1^2 + \hat{J}_2^2 + \hat{J}_3^2$ is diagonal, having for diagonal elements the eigenvalue $(3/4)\hbar^2$, as it can be immediately verified by noting that, for each component i,

$$[\underline{J}_i^{(1/2)}]^2 = \frac{\hbar^2}{4} \begin{pmatrix} 1 & 0 \\ 0 & 1 \end{pmatrix} .$$

Part 2

It follows from the definition that

$$\underline{\sigma}_1 = \begin{pmatrix} 0 & 1 \\ 1 & 0 \end{pmatrix} , \qquad \underline{\sigma}_2 = \begin{pmatrix} 0 & -i \\ i & 0 \end{pmatrix} , \qquad \underline{\sigma}_3 = \begin{pmatrix} 1 & 0 \\ 0 & -1 \end{pmatrix} .$$

Clearly, the determinant of $\underline{\sigma}_i$ is -1 and $\underline{\sigma}_i^2 = \underline{\sigma}_0$. By direct computation we find

$$\underline{\sigma}_1 \underline{\sigma}_2 = \begin{pmatrix} 0 & 1 \\ 1 & 0 \end{pmatrix} \begin{pmatrix} 0 & -i \\ i & 0 \end{pmatrix} = \begin{pmatrix} i & 0 \\ 0 & -i \end{pmatrix} , \qquad \underline{\sigma}_2 \underline{\sigma}_1 = \begin{pmatrix} 0 & -i \\ i & 0 \end{pmatrix} \begin{pmatrix} 0 & 1 \\ 1 & 0 \end{pmatrix} = \begin{pmatrix} -i & 0 \\ 0 & i \end{pmatrix} = -\underline{\sigma}_1 \underline{\sigma}_2 ,$$

and $\underline{\sigma}_1$ and $\underline{\sigma}_2$ anticommute; similarly for the remaining cases. Recalling that $\underline{J}_i^{(1/2)} = (\hbar/2) \, \underline{\sigma}_i$, Problem 1 and the fact that the $\underline{\sigma}_i$ anticommute lead to

$$\underline{\sigma}_i \underline{\sigma}_j - \underline{\sigma}_j \underline{\sigma}_i = 2i \sum_k \epsilon_{ijk} \underline{\sigma}_k \implies \underline{\sigma}_i \underline{\sigma}_j = i \sum_k \epsilon_{ijk} \underline{\sigma}_k \qquad i \neq j .$$

The latter two properties can be combined into

$$\underline{\sigma}_i \underline{\sigma}_j = \delta_{ij} \underline{\sigma}_0 + i \sum_k \epsilon_{ijk} \underline{\sigma}_k ,$$

which is often written with the 2×2 identity matrix understood. Note that, by repeated application of this relation, a product of p Pauli matrices $\underline{\sigma}_{i_1} \underline{\sigma}_{i_2} \cdots \underline{\sigma}_{i_p}$ can always be reduced to a linear combination of the identity matrix and a single Pauli matrix. If $\hat{\mathbf{A}}$ and $\hat{\mathbf{B}}$ are any two vector operators, such as, for example, $\hat{\mathbf{r}}$ and $\hat{\mathbf{p}}$ or $\hat{\mathbf{p}}$ and $\hat{\mathbf{L}}$ (the orbital angular momentum), then

$$\underbrace{\sum_{ij} \underline{\sigma}_i \underline{\sigma}_j \hat{A}_i \hat{B}_j}_{\underline{\sigma}\cdot\hat{\mathbf{A}}\,\underline{\sigma}\cdot\hat{\mathbf{B}}} = \sum_{ij} \left(\delta_{ij} \underline{\sigma}_0 + i \sum_k \epsilon_{ijk} \underline{\sigma}_k \right) \hat{A}_i \hat{B}_j = \underline{\sigma}_0 \underbrace{\sum_i \hat{A}_i \hat{B}_i}_{\hat{\mathbf{A}}\cdot\hat{\mathbf{B}}} + i \sum_k \underline{\sigma}_k \underbrace{\left(\sum_{ij} \epsilon_{ijk} \hat{A}_i \hat{B}_j \right)}_{(\hat{\mathbf{A}}\times\hat{\mathbf{B}})_k}$$

$$= \underline{\sigma}_0 \hat{\mathbf{A}} \cdot \hat{\mathbf{B}} + i \underline{\sigma} \cdot (\hat{\mathbf{A}} \times \hat{\mathbf{B}}) .$$

Problem 3 Matrix Representation of the Angular Momentum Components

Construct the matrices representing $\hat{J}_x$, $\hat{J}_y$, and $\hat{J}_z$ in a basis consisting of the common eigenstates of $\hat{\mathbf{J}}^2$ and $\hat{J}_z$. Show that they are hermitian and traceless.

Solution

In terms of raising and lowering operators we have

$$\hat{J}_x = \frac{\hat{J}_+ + \hat{J}_-}{2} , \qquad \hat{J}_y = \frac{\hat{J}_+ - \hat{J}_-}{2i} ,$$

and hence the corresponding matrix elements follow as

$$(\underline{J}_x)_{\mu\mu'} = \frac{1}{2} \langle \psi_{j\mu} | \hat{J}_+ + \hat{J}_- | \psi_{j\mu'} \rangle = \frac{\hbar}{2} \left(\sqrt{j(j+1) - \mu'(\mu'+1)}\, \delta_{\mu,\mu'+1} + \sqrt{j(j+1) - \mu'(\mu'-1)}\, \delta_{\mu,\mu'-1} \right)$$

and

$$(\underline{J}_y)_{\mu\mu'} = \frac{\hbar}{2i} \left(\sqrt{j(j+1) - \mu'(\mu'+1)}\, \delta_{\mu,\mu'+1} - \sqrt{j(j+1) - \mu'(\mu'-1)}\, \delta_{\mu,\mu'-1} \right) ,$$

where $\mu, \mu' = -j, -j+1, \ldots, j$. We have also used the fact that these matrix elements vanish unless the eigenvalue j of $\hat{\mathbf{J}}^2$ is the same for the left-hand and right-hand states. Of course, J_z is diagonal in this basis, and

$$(\underline{J}_z)_{\mu\mu'} = \hbar \mu' \delta_{\mu,\mu'} .$$

These matrices are obviously hermitian (as they must be), since $(\underline{J}_i)_{\mu'\mu} = (\underline{J}_i)_{\mu\mu'}^*$; for example,

$$(\underline{J}_y)_{\mu'\mu} = \frac{\hbar}{2i} \left(\sqrt{j(j+1) - \mu(\mu+1)}\, \delta_{\mu',\mu+1} - \sqrt{j(j+1) - \mu(\mu-1)}\, \delta_{\mu',\mu-1} \right)$$

$$= \frac{\hbar}{2i} \left(\sqrt{j(j+1) - \mu'(\mu'-1)}\, \delta_{\mu,\mu'-1} - \sqrt{j(j+1) - \mu'(\mu'+1)}\, \delta_{\mu,\mu'+1} \right) = (\underline{J}_y)_{\mu\mu'}^* .$$

We also note that these matrices are traceless, a result we could have anticipated on the basis of the commutation relations since (see Problem 10 for a derivation) of this next relation)

$$\underline{J}_i \underline{J}_j - \underline{J}_j \underline{J}_i = i\hbar \sum_k \epsilon_{ijk} \underline{J}_k \implies \underline{J}_i = \frac{1}{i\hbar} \sum_{jk} \epsilon_{ijk} \underline{J}_j \underline{J}_k$$

and

$$\mathrm{tr}(\underline{J}_i) = \frac{1}{i\hbar} \sum_{jk} \epsilon_{ijk}\, \mathrm{tr}(\underline{J}_j \underline{J}_k) = \frac{1}{2i\hbar} \sum_{jk} \epsilon_{ijk} \underbrace{\left[\mathrm{tr}(\underline{J}_j \underline{J}_k) + \mathrm{tr}(\underline{J}_k \underline{J}_j) \right]}_{\text{symmetric in } j \rightleftharpoons k} = 0 ,$$

where we have used the cyclic property of the trace, so that $\mathrm{tr}(\underline{J}_j\underline{J}_k) = \mathrm{tr}(\underline{J}_k\underline{J}_j)$, and the fact that the contraction of ϵ_{ijk} with a symmetric tensor vanishes.

Problem 4   Averages of $\hat{J}_x$, $\hat{J}_z$ and $\hat{J}_x^2$, $\hat{J}_z^2$ on a State with $j = 1$

Consider a system, with angular momentum $j = 1$, whose state space is spanned by the basis $|+\rangle$, $|0\rangle$, and $|-\rangle$, eigenstates of $\hat{\mathbf{J}}^2$ with eigenvalue $2\hbar^2$ and of $\hat{J}_z$ with eigenvalues $+\hbar$, 0, $-\hbar$, respectively. The normalized state of the system is

$$|\psi\rangle = \alpha\,|+\rangle + \beta\,|0\rangle + \gamma\,|-\rangle\,,$$

where $|\alpha|^2 + |\beta|^2 + |\gamma|^2 = 1$.

1. Calculate the mean values of the components $\hat{J}_x$ and $\hat{J}_z$, that is, $\langle\psi|\hat{J}_x|\psi\rangle$ and $\langle\psi|\hat{J}_z|\psi\rangle$, in terms of α, β, and γ.
2. Calculate the mean values $\langle\psi|\hat{J}_x^2|\psi\rangle$ and $\langle\psi|\hat{J}_z^2|\psi\rangle$ in terms of the same quantities.

Solution

Part 1

It is convenient to express $\hat{J}_x$ in terms of raising and lowering operators as

$$\hat{J}_x = \frac{\hat{J}_+ + \hat{J}_-}{2}\,.$$

We label the eigenstates of $\hat{\mathbf{J}}^2$ and $\hat{J}_z$ as $|m\rangle$ with $m = \pm 1$ and 0, and

$$\hat{J}_\pm\,|m\rangle = \hbar\,\sqrt{2 - m(m\pm 1)}\,|m\pm 1\rangle \implies \langle m'|\hat{J}_\pm|m\rangle = \hbar\,\sqrt{2 - m(m\pm 1)}\,\delta_{m',m\pm 1}\,.$$

The matrix representations of $\hat{J}_\pm$ are

$$\underline{J}_+ = \sqrt{2}\hbar\begin{pmatrix} 0 & 1 & 0 \\ 0 & 0 & 1 \\ 0 & 0 & 0 \end{pmatrix}, \qquad \underline{J}_- = [\underline{J}_+]^\dagger = \sqrt{2}\hbar\begin{pmatrix} 0 & 0 & 0 \\ 1 & 0 & 0 \\ 0 & 1 & 0 \end{pmatrix} \implies \underline{J}_x = \frac{\hbar}{\sqrt{2}}\begin{pmatrix} 0 & 1 & 0 \\ 1 & 0 & 1 \\ 0 & 1 & 0 \end{pmatrix}.$$

On the other hand, the matrix representing $\hat{J}_z$ is diagonal,

$$\underline{J}_z = \hbar\begin{pmatrix} 1 & 0 & 0 \\ 0 & 0 & 0 \\ 0 & 0 & -1 \end{pmatrix}.$$

The expectation values are now easily evaluated,

$$\langle\psi|\hat{J}_x|\psi\rangle \longrightarrow \frac{\hbar}{\sqrt{2}}\begin{pmatrix} \alpha^* & \beta^* & \gamma^* \end{pmatrix}\begin{pmatrix} 0 & 1 & 0 \\ 1 & 0 & 1 \\ 0 & 1 & 0 \end{pmatrix}\begin{pmatrix} \alpha \\ \beta \\ \gamma \end{pmatrix} = \frac{\hbar}{\sqrt{2}}\left[\alpha^*\beta + \beta^*(\alpha + \gamma) + \gamma^*\beta\right] = \sqrt{2}\hbar\,\mathrm{Re}(\alpha\beta^* + \beta\gamma^*)$$

and

$$\langle\psi|\hat{J}_z|\psi\rangle \longrightarrow \hbar\begin{pmatrix} \alpha^* & \beta^* & \gamma^* \end{pmatrix}\begin{pmatrix} 1 & 0 & 0 \\ 0 & 0 & 0 \\ 0 & 0 & -1 \end{pmatrix}\begin{pmatrix} \alpha \\ \beta \\ \gamma \end{pmatrix} = \hbar\,(|\alpha|^2 - |\gamma|^2)\,.$$

Part 2

Note that

$$[\underline{J}_x]^2 = \frac{\hbar^2}{2}\begin{pmatrix} 0 & 1 & 0 \\ 1 & 0 & 1 \\ 0 & 1 & 0 \end{pmatrix}\begin{pmatrix} 0 & 1 & 0 \\ 1 & 0 & 1 \\ 0 & 1 & 0 \end{pmatrix} = \frac{\hbar^2}{2}\begin{pmatrix} 1 & 0 & 1 \\ 0 & 2 & 0 \\ 1 & 0 & 1 \end{pmatrix}, \qquad [\underline{J}_z]^2 = \hbar^2\begin{pmatrix} 1 & 0 & 0 \\ 0 & 0 & 0 \\ 0 & 0 & 1 \end{pmatrix}.$$

The expectation values are given by

$$\langle\psi|\hat{J}_x^2|\psi\rangle \longrightarrow \frac{\hbar^2}{2}\begin{pmatrix} \alpha^* & \beta^* & \gamma^* \end{pmatrix}\begin{pmatrix} 1 & 0 & 1 \\ 0 & 2 & 0 \\ 1 & 0 & 1 \end{pmatrix}\begin{pmatrix} \alpha \\ \beta \\ \gamma \end{pmatrix} = \frac{\hbar^2}{2}\left[\alpha^*(\alpha+\gamma) + 2|\beta|^2 + \gamma^*(\alpha+\gamma)\right]$$

$$= \frac{\hbar^2}{2}\left[|\alpha|^2 + 2|\beta|^2 + |\gamma|^2 + 2\,\mathrm{Re}(\alpha\gamma^*)\right]$$

and

$$\langle\psi|\hat{J}_z^2|\psi\rangle \longrightarrow \hbar^2\begin{pmatrix} \alpha^* & \beta^* & \gamma^* \end{pmatrix}\begin{pmatrix} 1 & 0 & 0 \\ 0 & 0 & 0 \\ 0 & 0 & 1 \end{pmatrix}\begin{pmatrix} \alpha \\ \beta \\ \gamma \end{pmatrix} = \hbar^2\left(|\alpha|^2 + |\gamma|^2\right).$$

Problem 5 Construction of the State of a Spin-1 Particle Polarized in a General Direction

The normalized state $|\psi\rangle$ is in the subspace spanned by the eigenstates of $\hat{\mathbf{J}}^2$ having eigenvalue $2\hbar^2$. Suppose $|\psi\rangle$ is also an eigenstate of $\mathbf{n}\cdot\hat{\mathbf{J}}$ with eigenvalue $+\hbar$; here, $\mathbf{n}$ is the unit vector with components $(\sin\theta\cos\phi, \sin\theta\sin\phi, \cos\theta)$. Obtain $|\psi\rangle$ as a linear combination of the eigenstates $|m\rangle$ of $\hat{J}_z$ with $m = \pm 1, 0$.

Solution

Since $|\psi\rangle$ is an eigenstate of $\mathbf{n}\cdot\hat{\mathbf{J}}$, we have

$$\mathbf{n}\cdot\hat{\mathbf{J}}\,|\psi\rangle = \hbar|\psi\rangle,$$

where

$$|\psi\rangle = \sum_{m=\pm 1,0} c_m|m\rangle,$$

and $|m\rangle$ are the eigenstates of $\hat{\mathbf{J}}^2$ and $\hat{J}_z$ with eigenvalues $2\hbar^2$ and $m\hbar$, with $m = \pm 1, 0$, respectively. It is convenient to express the projection of $\hat{\mathbf{J}}$ along $\mathbf{n}$ in terms of raising and lowering operators:

$$\mathbf{n}\cdot\hat{\mathbf{J}} = \sin\theta\cos\phi\,\hat{J}_x + \sin\theta\sin\phi\,\hat{J}_y + \cos\theta\,\hat{J}_z = \frac{\sin\theta}{2}\left(\hat{J}_+\,e^{-i\phi} + \hat{J}_-\,e^{i\phi}\right) + \cos\theta\,\hat{J}_z.$$

In the basis $|m\rangle$ this operator is therefore represented by the matrix

$$\mathbf{n}\cdot\hat{\mathbf{J}} \longrightarrow \hbar\begin{pmatrix} \cos\theta & \sin\theta\,e^{-i\phi}/\sqrt{2} & 0 \\ \sin\theta\,e^{i\phi}/\sqrt{2} & 0 & \sin\theta\,e^{-i\phi}/\sqrt{2} \\ 0 & \sin\theta\,e^{i\phi}/\sqrt{2} & -\cos\theta \end{pmatrix},$$

and the above eigenvalue problem reduces to

$$\begin{pmatrix} \cos\theta & \sin\theta\, e^{-i\phi}/\sqrt{2} & 0 \\ \sin\theta\, e^{i\phi}/\sqrt{2} & 0 & \sin\theta\, e^{-i\phi}/\sqrt{2} \\ 0 & \sin\theta\, e^{i\phi}/\sqrt{2} & -\cos\theta \end{pmatrix} \begin{pmatrix} c_{+1} \\ c_0 \\ c_{-1} \end{pmatrix} = \begin{pmatrix} c_{+1} \\ c_0 \\ c_{-1} \end{pmatrix},$$

which has the solutions

$$\cos\theta\, c_{+1} + \frac{1}{\sqrt{2}}\sin\theta\, e^{-i\phi}\, c_0 = c_{+1} \implies c_{+1} = \frac{1}{\sqrt{2}}\cot(\theta/2)\, e^{-i\phi}\, c_0$$

and

$$\frac{1}{\sqrt{2}}\sin\theta\, e^{i\phi}\, c_0 - \cos\theta\, c_{-1} = c_{-1} \implies c_{-1} = \frac{1}{\sqrt{2}}\tan(\theta/2)\, e^{i\phi}\, c_0\,,$$

with $\theta \neq 0, \pi$ (these values correspond to the state $|\psi\rangle$ being polarized along the $+z$ and $-z$ directions, respectively). Thus, we find the eigenstate is given by

$$|\psi\rangle \longrightarrow c_0 \begin{pmatrix} \cot(\theta/2)\, e^{-i\phi}/\sqrt{2} \\ 1 \\ \tan(\theta/2)\, e^{i\phi}/\sqrt{2} \end{pmatrix}.$$

We normalize it via

$$|c_0|^2 \left[\frac{\cos^2(\theta/2)}{2\sin^2(\theta/2)} + 1 + \frac{\sin^2(\theta/2)}{2\cos^2(\theta/2)} \right] = \frac{|c_0|^2}{2\sin^2(\theta/2)\cos^2(\theta/2)} = 1\,,$$

which yields (up to a phase factor)

$$|c_0| = \sqrt{2}\,\sin(\theta/2)\cos(\theta/2) \qquad \text{and} \qquad |\psi\rangle \longrightarrow \begin{pmatrix} \cos^2(\theta/2)\, e^{-i\phi} \\ \sqrt{2}\,\sin(\theta/2)\cos(\theta/2) \\ \sin^2(\theta/2)\, e^{i\phi} \end{pmatrix}.$$

Problem 6   Construction of Spherical Harmonics

The eigenstates of the orbital angular momentum satisfy the eigenvalue equations

$$\hat{L}^2|\psi_{l,m}\rangle = l(l+1)\hbar^2|\psi_{l,m}\rangle\,, \qquad \hat{L}_z|\psi_{l,m}\rangle = m\hbar\,|\psi_{l,m}\rangle\,,$$

where $l = 0, 1, 2, \ldots$ and $m = -l, \ldots, l$. In the coordinate representation, the corresponding wave functions are the spherical harmonics,

$$Y_{l,m}(\theta,\phi) = \langle\phi_{\mathbf{r}}|\psi_{l,m}\rangle\,.$$

1. Using the expression for L_z as a differential operator, solve the differential equation implied by the second eigenvalue equation above to show that the ϕ dependence of $Y_{l,m}(\theta,\phi)$ is proportional to $e^{im\phi}$.
2. Using the condition $L_+|\psi_{l,l}\rangle = 0$ and the fact that $Y_{l,l}(\theta,\phi) = F_l(\theta)\, e^{il\phi}$, show that

$$Y_{l,l}(\theta,\phi) = c_l \sin^l\theta\, e^{il\phi}\,,$$

where c_l is a normalization factor.

3. Assume that the orbital angular momentum can also take on half-integer values, say $l=1/2$. Construct the "spherical harmonic $Y_{1/2,-1/2}(\theta, \phi)$" by (i) applying the lowering operator L_- to $Y_{1/2,1/2}(\theta, \phi) \propto \sin^{1/2} \theta\, e^{i\phi/2}$ and (ii) by solving the differential equation resulting from $L_- Y_{1/2,-1/2}(\theta, \phi)=0$. Show that the two procedures are problematic and yield contradictory results. Thus, half-integer values cannot occur for the orbital angular momentum operator.

(Adapted from J. J. Sakurai and J. Napolitano (2020), *Modern Quantum Mechanics*, Cambridge University Press.)

Solution

Part 1

In the coordinate representation we have

$$L_z \longrightarrow -i\hbar \frac{\partial}{\partial \phi} \,,$$

and hence

$$L_z |\psi_{l,m}\rangle = m\hbar |\psi_{l,m}\rangle \implies -i\hbar \frac{\partial}{\partial \phi} Y_{l,m}(\theta, \phi) = m\hbar\, Y_{l,m}(\theta, \phi) \implies Y_{l,m}(\theta, \phi) = F_{l,m}(\theta)\, e^{im\phi} \,.$$

Part 2

The raising and lowering (orbital) angular momentum operators are given by the following differential operators:

$$L_\pm = L_x \pm i L_y = \pm \hbar\, e^{\pm i\phi} \left(\frac{\partial}{\partial \theta} \pm i \cot\theta\, \frac{\partial}{\partial \phi} \right) \,,$$

and hence

$$L_+ |\psi_{l,l}\rangle = 0 \implies \hbar\, e^{i\phi} \left(\frac{\partial}{\partial \theta} + i \cot\theta\, \frac{\partial}{\partial \phi} \right) F_{l,l}(\theta)\, e^{il\phi} = 0 \,,$$

yielding the differential equation

$$\frac{d}{d\theta} F_{l,l}(\theta) = l \cot\theta\, F_{l,l}(\theta) \,.$$

This latter equation is easily solved writing,

$$\frac{dF_{l,l}}{F_{l,l}} = l \cot\theta\, d\theta \implies \int^{F_{l,l}} \frac{dF'_{l,l}}{F'_{l,l}} = l \int^{\theta} \cot\theta'\, d\theta' \implies \ln|F_{l,l}| = l \ln|\sin\theta| + \text{constant} \,,$$

and so

$$F_{l,l}(\theta) = c_l \sin^l \theta \,,$$

with c_l constant.

Part 3

We first apply the lowering operator to $Y_{1/2,1/2}(\theta, \phi)$, that is

$$L_- \, Y_{1/2,1/2}(\theta, \phi) = \hbar \, \underbrace{\sqrt{1/2(1/2 + 1) - 1/2(1/2 - 1)}}_{\text{unity}} \, Y_{1/2,-1/2}(\theta, \phi) \, ,$$

yielding

$$Y_{1/2,-1/2}(\theta, \phi) = \frac{1}{\hbar} L_- \, Y_{1/2,1/2}(\theta, \phi) \, ,$$

where, using the result of part 2, we have

$$Y_{1/2,1/2}(\theta, \phi) = c_{1/2} \, \sin^{1/2} \theta \, e^{i\phi/2} \, .$$

We find

$$Y_{1/2,-1/2}(\theta, \phi) = -c_{1/2} \, e^{-i\phi} \left(\frac{\partial}{\partial \theta} - i \cot \theta \, \frac{\partial}{\partial \phi} \right) \sin^{1/2} \theta \, e^{i\phi/2} = -c_{1/2} \, e^{-i\phi/2} \cot \theta \, \sin^{1/2} \theta \, ,$$

and we observe that the solution is singular at the end points $\theta = 0$ and π.

We now obtain $Y_{1/2,-1/2}(\theta, \phi) = G_{1/2}(\theta) \, e^{-i\phi/2}$ (note that the ϕ dependence follows from part 1) by determining the function $G_{1/2}(\theta)$ via the requirement

$$L_- \, Y_{1/2,-1/2}(\theta, \phi) = 0 \implies -\hbar \, e^{-i\phi} \left(\frac{\partial}{\partial \theta} - i \cot \theta \, \frac{\partial}{\partial \phi} \right) G_{1/2}(\theta) \, e^{-i\phi/2} = 0 \, ,$$

leading to the differential equation

$$\frac{d}{d\theta} G_{1/2}(\theta) = \frac{1}{2} \cot \theta \, G_{1/2}(\theta) \implies G_{1/2}(\theta) = d_{1/2} \, \sin^{1/2} \theta \, ,$$

where $d_{1/2}$ is a normalization constant. Thus, we see that the two procedures lead to inconsistent results, namely

$$Y_{1/2,-1/2} = -c_{1/2} \, e^{-i\phi/2} \cot \theta \, \sin^{1/2} \theta \qquad \text{and} \qquad Y_{1/2,-1/2} = d_{1/2} \, e^{-i\phi/2} \sin^{1/2} \theta \, .$$

Problem 7   Energy Spectrum of an Asymmetric Rotator

The Hamiltonian of a rotator is given by

$$\hat{H} = \frac{\hat{L}_x^2 + \hat{L}_y^2}{2I_1} + \frac{\hat{L}_z^2}{2I_3} \, ,$$

where I_1 and I_3 are moments of inertia and $\hat{L}_x$, $\hat{L}_y$, and $\hat{L}_z$ are the components of the orbital angular momentum operator.

1. Determine the eigenvalues of the Hamiltonian and their degeneracy in the two limits $I_1 = I_3$ and $I_1 > I_3$. Sketch the energy spectrum in these two limits.
2. What are the eigenvalues and their degeneracy in the limit $I_1 \gg I_3$?

Solution

Part 1

Using $\hat{L}_x^2 + \hat{L}_y^2 = \hat{\mathbf{L}}^2 - \hat{L}_z^2$, we write the Hamiltonian as

$$\hat{H} = \frac{\hat{\mathbf{L}}^2}{2 I_1} + \frac{1}{2} \left(\frac{1}{I_3} - \frac{1}{I_1} \right) \hat{L}_z^2 \, .$$

The Hamiltonian commutes with $\hat{\mathbf{L}}^2$ and $\hat{L}_z$, and hence the eigenstates $|\psi_{lm}\rangle$ of $\hat{\mathbf{L}}^2$ and $\hat{L}_z$ are also eigenstates of $\hat{H}$,

$$\hat{H} |\psi_{lm}\rangle = \underbrace{\left[\frac{\hbar^2 l(l+1)}{2 I_1} + \frac{\hbar^2}{2} \left(\frac{1}{I_3} - \frac{1}{I_1} \right) m^2 \right]}_{E_{l|m|}} |\psi_{lm}\rangle \, .$$

We consider two cases.

1. $I_1 = I_3$: the eigenvalues are independent of the azimuthal quantum number m and are given by

$$E_l = \frac{\hbar^2 l(l+1)}{2 I_1}$$

with degeneracy $g_l = 2l + 1$.

2. $I_1 > I_3$: the eigenvalues are given by

$$E_{l|m|} = \frac{\hbar^2 l(l+1)}{2 I_1} + \frac{\hbar^2}{2} \left(\frac{1}{I_3} - \frac{1}{I_1} \right) m^2$$

and the states with positive or negative m are degenerate; specifically, the eigenvalues E_{l0} are non-degenerate while the eigenvalues $E_{l|m|}$ are two-fold degenerate.

Part 2

If $I_1 \gg I_3$ then we have

$$E_{|m|} = \frac{\hbar^2 m^2}{2 I_3} \, ,$$

and $E_{|m|}$ is infinitely degenerate, since for $m = 0$ the states with $l \geq 0$ all have vanishing energy, for $|m| = 1$ the states with $l \geq 1$ all have energy $\hbar^2/(2 I_3)$, and so on.

Problem 8 Angular Momentum Algebra and the Harmonic Oscillator in Two Dimensions

Let $\hat{a}_r$ and $\hat{a}_r^\dagger$ with $r = 1, 2$ be the annihilation and creation operators of a two-dimensional harmonic oscillator, satisfying

$$[a_r, a_s] = 0 = [a_r^\dagger, a_s^\dagger], \qquad [a_r, a_s^\dagger] = \delta_{rs} \, .$$

We define

$$\hat{S} = \frac{1}{2} \left(\hat{a}_1^\dagger \hat{a}_1 + \hat{a}_2^\dagger \hat{a}_2 \right)$$

and

$$\hat{J}_1 = \frac{1}{2} \left(\hat{a}_2^\dagger \hat{a}_1 + \hat{a}_1^\dagger \hat{a}_2 \right) , \qquad \hat{J}_2 = \frac{i}{2} \left(\hat{a}_2^\dagger \hat{a}_1 - \hat{a}_1^\dagger \hat{a}_2 \right) , \qquad \hat{J}_3 = \frac{1}{2} \left(\hat{a}_1^\dagger \hat{a}_1 - \hat{a}_2^\dagger \hat{a}_2 \right) ,$$

and the $\hat{J}_i$ may be considered as the cartesian components of a certain vector operator.

1. Show that the components of $\hat{\mathbf{J}}$ as defined above satisfy the commutation relations characteristic of an angular momentum, up to factors of $\hbar$, that is,

$$[\hat{J}_i, \hat{J}_j] = i \sum_k \epsilon_{ijk} \hat{J}_k \, ,$$

and show that

$$\hat{\mathbf{J}}^2 = \hat{S}\left(\hat{S} + \hat{\mathbb{1}}\right) \, .$$

2. Hereafter, $\hat{\mathbf{J}}$ will be considered to be the angular momentum of the system. We denote the eigenvalues of $\hat{\mathbf{J}}^2$ and $\hat{J}_3$ by $j(j + 1)$ and m, respectively. Show that $\hat{\mathbf{J}}^2$ and $\hat{J}_3$ form a complete set of commuting observables and that j may take all integral or half-integral values ≥ 0.

3. Show that the states

$$\frac{1}{\sqrt{(j + m)! \, (j - m)!}} \, (a_1^\dagger)^{j+m} \, (a_2^\dagger)^{j-m} \, |0; 0\rangle \, ,$$

form a basis of common eigenstates of $\hat{\mathbf{J}}^2$ and $\hat{J}_3$.

Solution

Part 1

It is convenient to define the number operator

$$\hat{N}_r = \hat{a}_r^\dagger \, \hat{a}_r \, ,$$

with the property that

$$[\hat{N}_r, \hat{a}_s] = -\delta_{rs} \, \hat{a}_s \, , \qquad [\hat{N}_r, \hat{a}_s^\dagger] = \delta_{rs} \, \hat{a}_s^\dagger \, .$$

In terms of the number operator we have

$$\hat{S} = \frac{1}{2}\left(\hat{N}_1 + \hat{N}_2\right) , \qquad \hat{J}_3 = \frac{1}{2}\left(\hat{N}_1 - \hat{N}_2\right) .$$

Consider first

$$[\hat{J}_1, \hat{J}_2] = \frac{i}{4}[\hat{a}_2^\dagger \hat{a}_1 + \hat{a}_1^\dagger \hat{a}_2, \hat{a}_2^\dagger \hat{a}_1 - \hat{a}_1^\dagger \hat{a}_2] = \frac{i}{4}\left([\hat{a}_2^\dagger \hat{a}_1, -\hat{a}_1^\dagger \hat{a}_2] + [\hat{a}_1^\dagger \hat{a}_2, \hat{a}_2^\dagger \hat{a}_1]\right) = \frac{i}{2}[\hat{a}_1^\dagger \hat{a}_2, \hat{a}_2^\dagger \hat{a}_1]$$

$$= \frac{i}{2}\left(\hat{a}_1^\dagger [\hat{a}_2, \hat{a}_2^\dagger \hat{a}_1] + [\hat{a}_1^\dagger, \hat{a}_2^\dagger \hat{a}_1] \hat{a}_2\right) = \frac{i}{2}\left(\hat{a}_1^\dagger \hat{a}_1 - \hat{a}_2^\dagger \hat{a}_2\right) = i\hat{J}_3 \, ,$$

then

$$[\hat{J}_2, \hat{J}_3] = \frac{i}{4}[\hat{a}_2^\dagger \hat{a}_1 - \hat{a}_1^\dagger \hat{a}_2, \hat{N}_1 - \hat{N}_2] = \frac{i}{4}\left([\hat{a}_2^\dagger \hat{a}_1, \hat{N}_1] - [\hat{a}_1^\dagger \hat{a}_2, \hat{N}_1] - [\hat{a}_2^\dagger \hat{a}_1, \hat{N}_2] + [\hat{a}_1^\dagger \hat{a}_2, \hat{N}_2]\right)$$

$$= \frac{i}{4}\left(\hat{a}_2^\dagger \hat{a}_1 + \hat{a}_1^\dagger \hat{a}_2 + \hat{a}_2^\dagger \hat{a}_1 + \hat{a}_1^\dagger \hat{a}_2\right) = \frac{i}{2}\left(\hat{a}_2^\dagger \hat{a}_1 + \hat{a}_1^\dagger \hat{a}_2\right) = i\hat{J}_1 \, ,$$

and lastly

$$[\hat{J}_3, \hat{J}_1] = \frac{1}{4}[\hat{N}_1 - \hat{N}_2, \hat{a}_2^\dagger \hat{a}_1 + \hat{a}_1^\dagger \hat{a}_2] = \frac{1}{4}\left([\hat{N}_1, \hat{a}_2^\dagger \hat{a}_1] + [\hat{N}_1, \hat{a}_1^\dagger \hat{a}_2] - [\hat{N}_2, \hat{a}_2^\dagger \hat{a}_1] - [\hat{N}_2, \hat{a}_1^\dagger \hat{a}_2]\right)$$

$$= \frac{1}{4}\left(-\hat{a}_2^\dagger \hat{a}_1 + \hat{a}_1^\dagger \hat{a}_2 - \hat{a}_2^\dagger \hat{a}_1 + \hat{a}_1^\dagger \hat{a}_2\right) = -\frac{1}{2}\left(\hat{a}_2^\dagger \hat{a}_1 - \hat{a}_1^\dagger \hat{a}_2\right) = i\hat{J}_2 \, .$$

We can express these commutation relations as

$$[\hat{J}_i, \hat{J}_j] = i \sum_k \epsilon_{ijk} \hat{J}_k \, .$$

We now consider the squares of these operators,

$$\hat{J}_1^2 = \frac{1}{4}\left[(\hat{a}_2^\dagger \hat{a}_1)^2 + (\hat{a}_1^\dagger \hat{a}_2)^2 + \hat{a}_2^\dagger \hat{a}_1 \hat{a}_1^\dagger \hat{a}_2 + \hat{a}_1^\dagger \hat{a}_2 \hat{a}_2^\dagger \hat{a}_1\right],$$

$$\hat{J}_2^2 = -\frac{1}{4}\left[(\hat{a}_2^\dagger \hat{a}_1)^2 + (\hat{a}_1^\dagger \hat{a}_2)^2 - \hat{a}_2^\dagger \hat{a}_1 \hat{a}_1^\dagger \hat{a}_2 - \hat{a}_1^\dagger \hat{a}_2 \hat{a}_2^\dagger \hat{a}_1\right],$$

and so

$$\hat{J}_1^2 + \hat{J}_2^2 = \frac{1}{2}\left(\hat{a}_2^\dagger \hat{a}_1 \hat{a}_1^\dagger \hat{a}_2 + \hat{a}_1^\dagger \hat{a}_2 \hat{a}_2^\dagger \hat{a}_1\right) = \frac{1}{2}\left(\hat{N}_1 \hat{N}_2 + \hat{N}_2 + \hat{N}_1 \hat{N}_2 + \hat{N}_1\right).$$

Thus, we find

$$\hat{J}_1^2 + \hat{J}_2^2 + \hat{J}_3^2 = \frac{1}{2}\left(\hat{N}_1 \hat{N}_2 + \hat{N}_2 + \hat{N}_1 \hat{N}_2 + \hat{N}_1\right) + \frac{1}{4}\left(\hat{N}_1^2 + \hat{N}_2^2 - 2\hat{N}_1 \hat{N}_2\right)$$

$$= \frac{1}{4}\left(\hat{N}_1^2 + \hat{N}_2^2 - 2\hat{N}_1 \hat{N}_2 + 4\hat{N}_1 \hat{N}_2\right) + \frac{1}{2}\left(\hat{N}_1 + \hat{N}_2\right) = \hat{S}^2 + \hat{S} \, .$$

Part 2

Since the $\hat{J}_i$ operators satisfy the same commutation relations as the angular momentum components (up to factors of $\hbar$), it follows that $\hat{\mathbf{J}}^2$ and $\hat{J}_3$ commute. It also follows that the possible eigenvalues of $\hat{\mathbf{J}}^2$ and $\hat{J}_3$ are $j = 0, 1/2, 1, \ldots$ and $m = -j, -j+1, \ldots, j$, respectively (using the same proof as for angular momentum operators).

Part 3

The states are the normalized eigenstates of the two oscillators, that is,

$$|j+m, j-m\rangle = \frac{1}{\sqrt{(j+m)! \, (j-m)!}} \, (a_1^\dagger)^{j+m} \, (a_2^\dagger)^{j-m} |0; 0\rangle \, ,$$

and therefore

$$\hat{J}_3 |j+m, j-m\rangle = \frac{1}{2}(\hat{N}_1 - \hat{N}_2)|j+m, j-m\rangle = \frac{1}{2}[j+m-(j-m)]\,|j+m, j-m\rangle = m\,|j+m, j-m\rangle$$

and

$$\hat{S} |j+m, j-m\rangle = \frac{1}{2}(\hat{N}_1 + \hat{N}_2)|j+m, j-m\rangle = \frac{1}{2}[j+m+(j-m)]\,|j+m, j-m\rangle = j\,|j+m, j-m\rangle \, .$$

Thus, we find

$$\hat{\mathbf{J}}^2 |j+m, j-m\rangle = \hat{S}(\hat{S} + \hat{\mathbb{1}})|j+m, j-m\rangle = j(j+1)|j+m, j-m\rangle \, .$$

Problem 9 Construction of Angular Momentum Eigenstates

Show that

$$|\psi_{j, \pm\mu}\rangle = \frac{1}{\hbar^{j-\mu}} \sqrt{\frac{(j+\mu)!}{(2j)! \, (j-\mu)!}} \, (\hat{J}_\mp)^{j-\mu} |\psi_{j, \pm j}\rangle \, ,$$

and

$$|\psi_{j,\pm j}\rangle = \frac{1}{\hbar^{j-\mu}} \sqrt{\frac{(j+\mu)!}{(2j)!\,(j-\mu)!}}\,(\hat{J}_{\pm})^{j-\mu}\,|\psi_{j,\pm\mu}\rangle\,,$$

where $|\psi_{j,\mu}\rangle$ are the eigenstates of $\hat{\mathbf{J}}^2$ and $\hat{J}_z$.

Solution

Recall the properties of the raising and lowering operators $\hat{J}_{\pm}$:

$$\hat{J}_{\pm}\,|\psi_{j,\mu}\rangle = \hbar\,\underbrace{\sqrt{(j\mp\mu)(j\pm\mu+1)}}_{\sqrt{j(j+1)-\mu(\mu\pm1)}}\,|\psi_{j,\mu\pm1}\rangle\,.$$

Start from the state $|\psi_{j,-j}\rangle$; repeated applications of $\hat{J}_+$ yield

$$\hat{J}_+\,|\psi_{j,-j}\rangle = \hbar\,\sqrt{2j}\,\sqrt{1}\,|\psi_{j,1-j}\rangle\,,$$
$$\hat{J}_+^2\,|\psi_{j,-j}\rangle = \hbar\,\sqrt{2j}\,\sqrt{1}\,\hat{J}_+\,|\psi_{j,1-j}\rangle = \hbar^2\,\sqrt{2j(2j-1)}\sqrt{1\times2}\,|\psi_{j,2-j}\rangle\,,$$

and so on; after k applications we have

$$\hat{J}_+^k\,|\psi_{j,-j}\rangle = \hbar\,\sqrt{2j}\,\sqrt{1}\,\hat{J}_+^{k-1}\,|\psi_{j,1-j}\rangle = \hbar^2\,\sqrt{2j(2j-1)}\sqrt{1\times2}\,\hat{J}_+^{k-2}\,|\psi_{j,2-j}\rangle = \cdots$$

$$= \hbar^k\,\sqrt{2j(2j-1)\cdots(2j-k+1)}\sqrt{1\times2\times\cdots\times k}\,|\psi_{j,k-j}\rangle\,,$$

which can be written, using $1\times2\times\cdots\times k = k!$ and $2j(2j-1)\cdots(2j-k+1) = (2j)!\,/(2j-k)!$, as

$$|\psi_{j,k-j}\rangle = \frac{1}{\hbar^k}\,\sqrt{\frac{(2j-k)!}{(2j)!\,k!}}\,\hat{J}_+^k\,|\psi_{j,-j}\rangle\,.$$

Setting $k = j - \mu$ leads to the required relation,

$$|\psi_{j,-\mu}\rangle = \frac{1}{\hbar^{j-\mu}}\,\sqrt{\frac{(j+\mu)!}{(2j)!\,(j-\mu)!}}\,\hat{J}_+^{j-\mu}\,|\psi_{j,-j}\rangle\,.$$

A similar argument for $\hat{J}_-$ gives the general relation

$$\hat{J}_-^k\,|\psi_{j,j}\rangle = \hbar\,\sqrt{2j}\,\sqrt{1}\,\hat{J}_-^{k-1}\,|\psi_{j,j-1}\rangle = \hbar^2\,\sqrt{2j(2j-1)}\sqrt{1\times2}\,\hat{J}_-^{k-2}\,|\psi_{j,j-2}\rangle = \cdots$$

$$= \hbar^k\,\sqrt{2j(2j-1)\cdots(2j-k+1)}\sqrt{1\times2\times\cdots\times k}\,|\psi_{j,j-k}\rangle\,,$$

or

$$|\psi_{j,j-k}\rangle = \frac{1}{\hbar^k}\,\sqrt{\frac{(2j-k)!}{(2j)!\,k!}}\,\hat{J}_-^k\,|\psi_{j,j}\rangle \implies |\psi_{j,\mu}\rangle = \frac{1}{\hbar^{j-\mu}}\,\sqrt{\frac{(j+\mu)!}{(2j)!\,(j-\mu)!}}\,\hat{J}_-^{j-\mu}\,|\psi_{j,j}\rangle\,,$$

after the replacement $k \longrightarrow j - \mu$.

The second set of relations can be proven similarly. For example, start from the state $|\psi_{j,j-k}\rangle$ and hence

$$
\begin{aligned}
\hat{J}_+^k \, |\psi_{j,j-k}\rangle &= \hbar \sqrt{2j-k+1}\, \sqrt{k}\, \hat{J}_+^{k-1} \, |\psi_{j,j-k+1}\rangle \\
&= \hbar^2 \sqrt{(2j-k+1)(2j-k+2)}\, \sqrt{k(k-1)}\, \hat{J}_+^{k-2} \, |\psi_{j,j-k+2}\rangle = \cdots \\[2mm]
&= \hbar^k \sqrt{(2j-k+1)(2j-k+2)\cdots 2j}\, \sqrt{k(k-1)\cdots 1}\, |\psi_{j,j}\rangle ,
\end{aligned}
$$

which after the replacement $k \longrightarrow j - \mu$ yields

$$
|\psi_{j,j}\rangle = \frac{1}{\hbar^{j-\mu}} \sqrt{\frac{(j+\mu)!}{(2j)!\,(j-\mu)!}}\, \hat{J}_+^{j-\mu} \, |\psi_{j,\mu}\rangle .
$$

Problem 10   Observables Commuting with Two Components of $\hat{\mathbf{J}}$

Show that if an operator commutes with any two components of the angular momentum operator $\hat{\mathbf{J}}$ then it commutes with the third.

Solution

We first express $\hat{J}_i$ in terms of the commutator of $\hat{J}_j$ and $\hat{J}_k$. To this end, start from

$$
[\hat{J}_j, \hat{J}_k] = i\hbar \sum_l \epsilon_{jkl}\, \hat{J}_l
$$

and consider

$$
\sum_{jk} \epsilon_{jki}\, [\hat{J}_j, \hat{J}_k] = i\hbar \sum_{jkl} \epsilon_{jki}\, \epsilon_{jkl}\, \hat{J}_l = i\hbar \sum_{kl} (\delta_{k,k}\, \delta_{i,l} - \delta_{k,l}\, \delta_{i,k})\, \hat{J}_l = 2i\hbar \hat{J}_i .
$$

Also note that

$$
\sum_{jk} \epsilon_{jki}\, [\hat{J}_j, \hat{J}_k] = \sum_{jk} \epsilon_{ijk}\, (\hat{J}_j \hat{J}_k - \hat{J}_k \hat{J}_j) = 2 \sum_{jk} \epsilon_{ijk}\, \hat{J}_j \hat{J}_k .
$$

We conclude that

$$
\sum_{jk} \epsilon_{ijk}\, \hat{J}_j \hat{J}_k = i\hbar \hat{J}_i .
$$

Now, if an operator $\hat{A}$ commutes with $\hat{J}_j$ and $\hat{J}_k$ then it will also commute with $\hat{J}_i$, since

$$
[\hat{J}_i, \hat{A}] = -\frac{i}{\hbar} \sum_{jk} \epsilon_{ijk}\, [\hat{J}_j \hat{J}_k, \hat{A}] = -\frac{i}{\hbar} \sum_{jk} \epsilon_{ijk}\, \left(\hat{J}_j [\hat{J}_k, \hat{A}] + [\hat{J}_j, \hat{A}] \hat{J}_k \right) = 0 .
$$

Problem 11   Orbital Angular Momentum and Parity

Let $\hat{\mathbf{L}}$ be the orbital angular momentum of a particle. The parity operator $\hat{P}$ acting on a generic state $|\psi\rangle$ leads in the coordinate representation to

$$
\hat{P}\,|\psi\rangle \longrightarrow P\psi(r,\theta,\phi) = \psi(r, \pi-\theta, \phi+\pi) ,
$$

where $\psi(r,\theta,\phi)$ is the wave function in spherical coordinates.

1. Show that $[\hat{P}, \hat{\mathbf{L}}]|\psi\rangle = 0$ on any state $|\psi\rangle$ by using the explicit expression for $\mathbf{L}$ in spherical coordinates. This implies that $[\hat{P}, \hat{\mathbf{L}}] = 0$.
2. Deduce from part 1 that the spherical harmonics have a well-defined parity depending only on the quantum number l of $\hat{\mathbf{L}}^2$. Determine this parity.

Solution

Part 1

Recall that, in the coordinate representation,

$$L_x = \hbar\left(\sin\phi\,\frac{\partial}{\partial\theta} + \cot\theta\,\cos\phi\,\frac{\partial}{\partial\phi}\right), \quad L_y = i\hbar\left(-\cos\phi\,\frac{\partial}{\partial\theta} + \cot\theta\,\sin\phi\,\frac{\partial}{\partial\phi}\right), \quad L_z = -i\hbar\,\frac{\partial}{\partial\phi}.$$

Under parity we have $\theta \longrightarrow \theta' = \pi - \theta$ and $\phi \longrightarrow \phi' = \phi + \pi$, and hence

$$\frac{\partial}{\partial\theta} = \frac{\partial\theta'}{\partial\theta}\frac{\partial}{\partial\theta'} = -\frac{\partial}{\partial\theta'}, \qquad \frac{\partial}{\partial\phi} = \frac{\partial}{\partial\phi'}.$$

We first consider

$$L_x P\,\psi(r,\theta,\phi) = L_x\,\psi(r,\theta',\phi')$$
$$= \hbar\left[-\sin(\phi'-\pi)\,\frac{\partial}{\partial\theta'} + \cot(\pi-\theta')\,\cos(\phi'-\pi)\,\frac{\partial}{\partial\phi'}\right]\psi(r,\theta',\phi')$$
$$= \hbar\left[\sin\phi'\,\frac{\partial}{\partial\theta'} + \cot\theta'\,\cos\phi'\,\frac{\partial}{\partial\phi'}\right]\psi(r,\theta',\phi').$$

Next, we examine

$$PL_x\,\psi(r,\theta,\phi) = \hbar P\left(\sin\phi\,\frac{\partial}{\partial\theta} + \cot\theta\,\cos\phi\,\frac{\partial}{\partial\phi}\right)\psi(r,\theta,\phi)$$
$$= \hbar\left[\sin\phi'\,\frac{\partial}{\partial\theta'} + \cot\theta'\,\cos\phi'\,\frac{\partial}{\partial\phi'}\right]\psi(r,\theta',\phi').$$

Thus, for any wave function $\psi(r,\theta,\phi)$, we have $L_x P\,\psi(r,\theta,\phi) = PL_x\,\psi(r,\theta,\phi)$, implying that the operators $\hat{L}_x$ and $\hat{P}$ commute with each other. A similar argument works for $\hat{L}_y$ and $\hat{L}_z$. An alternative proof consists of noting that under parity the position and momentum operators are odd, namely $\hat{\mathbf{r}} \longrightarrow \hat{\mathbf{r}}' = \hat{P}\hat{\mathbf{r}}\hat{P}^\dagger = -\hat{\mathbf{r}}$ and $\hat{\mathbf{p}} \longrightarrow \hat{\mathbf{p}}' = \hat{P}\hat{\mathbf{p}}\hat{P}^\dagger = -\hat{\mathbf{p}}$, and hence $\hat{\mathbf{L}}$ is invariant, since $\hat{\mathbf{L}} \longrightarrow \hat{\mathbf{L}}' = \hat{P}\hat{\mathbf{L}}\hat{P}^\dagger = \hat{P}\hat{\mathbf{r}}\hat{P}^\dagger \times \hat{P}\hat{\mathbf{p}}\hat{P}^\dagger = \hat{\mathbf{r}} \times \hat{\mathbf{p}} = \hat{\mathbf{L}}$.

Part 2

First note that, since $\hat{P}^2 = \hat{\mathbb{1}}$, the eigenvalues of $\hat{P}$ are ± 1. Furthermore, since it commutes with $\hat{\mathbf{L}}^2$ and $\hat{L}_z$, it follows that the spherical harmonics are also eigenfunctions of $\hat{P}$; indeed,

$$L_z P\,Y_{lm}(\Omega) = P L_z\,Y_{lm}(\Omega) = m\hbar\,P\,Y_{lm}(\Omega),$$

but the Y_{lm} are non-degenerate on the unit sphere (in other words, $\mathbf{L}^2$ and L_z form a complete set of commuting observables on the unit sphere), and hence $P\,Y_{lm}(\Omega)$ must be proportional to $Y_{lm}(\Omega)$, that is

$$P\,Y_{lm}(\Omega) = c_{lm}Y_{lm}(\Omega).$$

The eigenvalue c_{lm} must be independent of m, since for example

$$c_{lm} Y_{lm}(\Omega) = P Y_{lm}(\Omega) = \frac{P L_+ Y_{lm-1}(\Omega)}{\hbar\sqrt{l(l+1) - m(m-1)}} = \frac{L_+ P Y_{lm-1}(\Omega)}{\hbar\sqrt{l(l+1) - m(m-1)}}$$

$$= \frac{c_{lm-1}}{\hbar\sqrt{l(l+1) - m(m-1)}} L_+ Y_{lm-1}(\Omega) = c_{lm-1} Y_{lm}(\Omega) \,,$$

implying that $c_{lm} = c_{lm-1}$. In order to determine the parity of the $Y_{lm}(\Omega)$, we consider $Y_{l0}(\Omega)$, which is given by

$$Y_{l0}(\Omega) = \sqrt{\frac{2l+1}{4\pi}}\, P_l(\cos\theta) \,,$$

where ($x = \cos\theta$)

$$P_l(x) = \frac{1}{2^l\, l!} \frac{d^l}{dx^l} (x^2 - 1)^l \,.$$

Under parity $x \longrightarrow -x$, and hence

$$P_l(x) = \frac{1}{2^l\, l!} \frac{d^l}{d(-x)^l} [(-x)^2 - 1]^l = (-1)^l P_l(x) \,,$$

from which we deduce that

$$P Y_{lm}(\theta, \phi) = Y_{lm}(\pi - \theta, \phi + \pi) = (-1)^l Y_{lm}(\theta, \phi) \,.$$

Problem 12 Verification of the Properties Satisfied by $\hat{L}_\pm$ when Acting on $Y_{lm}(\Omega)$

Show by direct evaluation that, in the coordinate representation (so we omit carets from the operators),

$$L_\pm Y_{lm}(\Omega) = \hbar\sqrt{l(l+1) - m(m \pm 1)}\, Y_{l,m\pm 1}(\Omega) \,,$$

where $L_\pm$ are the raising and lowering orbital angular momentum operators.

Hint: Use the explicit expressions for $L_\pm$ in spherical coordinates and recall the following definitions:

$$Y_{lm}(\theta, \phi) = (-1)^m \sqrt{\frac{2l+1}{4\pi} \frac{(l-m)!}{(l+m)!}}\, P_l^m(\cos\theta)\, e^{im\phi} \qquad 0 \le m \le l$$

and

$$Y_{lm}(\theta, \phi) = (-1)^m\, Y_{l,-m}^*(\theta, \phi) \qquad -l \le m \le -1 \,,$$

as well as the alternative definitions of the associated Legendre functions of the first kind:

$$P_l^m(x) = \frac{1}{2^l\, l!} (1 - x^2)^{m/2} \frac{d^{l+m}}{dx^{l+m}} (x^2 - 1)^l$$

and

$$P_l^m(x) = \frac{(-1)^m}{2^l\, l!} \frac{(l+m)!}{(l-m)!} (1 - x^2)^{-m/2} \frac{d^{l-m}}{dx^{l-m}} (x^2 - 1)^l \,.$$

Solution

In spherical coordinates, the $L_\pm$ differential operators read

$$L_\pm = L_x \pm i L_y = \pm \hbar \, e^{\pm i\phi} \left(\frac{\partial}{\partial \theta} \pm i \cot \theta \, \frac{\partial}{\partial \phi} \right) .$$

Using explicit expressions for the $Y_{lm}(\Omega)$, we obtain

$$\frac{\partial}{\partial \phi} Y_{lm}(\theta, \phi) = im \, Y_{lm}(\theta, \phi) .$$

As for the derivative with respect to θ, note that ($x = \cos \theta$)

$$\frac{d}{d\theta} = \frac{dx}{d\theta} \frac{d}{dx} = -\sqrt{1 - x^2} \, \frac{d}{dx}$$

and

$$\frac{d}{d\theta} P_l^m(\theta) = -\sqrt{1 - x^2} \, \frac{d}{dx} P_l^m(x) .$$

We now use the first of the explicit expressions for $P_l^m(x)$ to obtain

$$P_l^{m\prime}(x) = \frac{1}{2^l \, l!} \left[-mx(1 - x^2)^{m/2 - 1} \frac{d^{l+m}(x^2 - 1)^l}{dx^{l+m}} + (1 - x^2)^{m/2} \frac{d^{l+m+1}(x^2 - 1)^l}{dx^{l+m+1}} \right]$$

$$= -\frac{mx}{1 - x^2} P_l^m(x) + \frac{P_l^{m+1}(x)}{\sqrt{1 - x^2}} ,$$

and the second to obtain

$$P_l^{m\prime}(x) = \frac{(-1)^m}{2^l \, l!} \frac{(l+m)!}{(l-m)!} \left[mx(1 - x^2)^{-m/2 - 1} \frac{d^{l-m}(x^2 - 1)^l}{dx^{l-m}} + (1 - x^2)^{-m/2} \frac{d^{l-m+1}(x^2 - 1)^l}{dx^{l-m+1}} \right]$$

$$= \frac{mx}{1 - x^2} P_l^m(x) - (l+m)(l-m+1) \frac{P_l^{m-1}(x)}{\sqrt{1 - x^2}} ,$$

for the derivatives of $P_l^m(x)$. Thus we find

$$L_+ Y_{lm}(\theta, \phi) = \hbar \, (-1)^m \sqrt{\frac{2l+1}{4\pi} \frac{(l-m)!}{(l+m)!}} \, e^{i\phi} \left(\frac{\partial}{\partial \theta} + i \cot \theta \, \frac{\partial}{\partial \phi} \right) P_l^m(\cos \theta) \, e^{im\phi}$$

$$= \hbar \, (-1)^m \sqrt{\frac{2l+1}{4\pi} \frac{(l-m)!}{(l+m)!}} \, e^{i\phi} \left[-e^{im\phi} \sqrt{1 - x^2} \, \frac{P_l^m(x)}{dx} - \frac{mx}{\sqrt{1 - x^2}} e^{im\phi} P_l^m(x) \right]$$

$$= \hbar \, (-1)^{m+1} \sqrt{\frac{2l+1}{4\pi} \frac{(l-m)!}{(l+m)!}} \, e^{i(m+1)\phi} P_l^{m+1}(x) ,$$

after inserting the first expression for the derivative of the associated Legendre functions and simplifying terms. Using

$$Y_{lm+1}(\theta, \phi) = (-1)^{m+1} \sqrt{\frac{2l+1}{4\pi} \frac{(l-m-1)!}{(l+m+1)!}} \, P_l^{m+1}(\cos \theta) \, e^{i(m+1)\phi} ,$$

we see that

$$L_+ Y_{lm}(\theta, \phi) = \hbar \sqrt{\frac{(l-m)!}{(l+m)!}} \sqrt{\frac{(l+m+1)!}{(l-m-1)!}} \, Y_{lm+1}(\theta, \phi) = \hbar \, \underbrace{\sqrt{(l-m)(l+m+1)}}_{\sqrt{l(l+1) - m(m+1)}} \, Y_{lm+1}(\theta, \phi) .$$

Similarly, we obtain

$$L_- Y_{lm}(\theta, \phi) = -\hbar (-1)^m \sqrt{\frac{2l+1}{4\pi} \frac{(l-m)!}{(l+m)!}} \, e^{-i\phi} \left(\frac{\partial}{\partial\theta} - i \cot\theta \, \frac{\partial}{\partial\phi}\right) P_l^m(\cos\theta) \, e^{im\phi}$$

$$= \hbar (-1)^{m+1} \sqrt{\frac{2l+1}{4\pi} \frac{(l-m)!}{(l+m)!}} \, e^{i(m-1)\phi} \left[-\sqrt{1-x^2} \, \frac{P_l^m(x)}{dx} + \frac{mx}{\sqrt{1-x^2}} P_l^m(x)\right]$$

$$= \hbar (-1)^{m-1} \sqrt{\frac{2l+1}{4\pi} \frac{(l-m)!}{(l+m)!}} \, (l+m)(l-m+1) \, e^{i(m-1)\phi} P_l^{m-1}(x) \, ,$$

after inserting the second expression for the derivative of the associated Legendre functions and simplifying terms. Using

$$Y_{lm-1}(\theta, \phi) = (-1)^{m-1} \sqrt{\frac{2l+1}{4\pi} \frac{(l-m+1)!}{(l+m-1)!}} \, P_l^{m-1}(\cos\theta) \, e^{i(m-1)\phi} \, ,$$

we find

$$L_- Y_{lm}(\theta, \phi) = \hbar (l+m)(l-m+1) \sqrt{\frac{(l-m)!}{(l+m)!}} \sqrt{\frac{(l+m-1)!}{(l-m+1)!}} \, Y_{lm-1}(\theta, \phi)$$

$$= \hbar \underbrace{\sqrt{(l+m)(l-m+1)}}_{\sqrt{l(l+1)-m(m-1)}} \, Y_{lm-1}(\theta, \phi) \, .$$

Problem 13 Transformation of States and Vector Operators Under Rotations

Consider a vector operator $\hat{\mathbf{V}}$. By definition it satisfies the following commutation relations with the components of the angular momentum $\hat{\mathbf{J}}$:

$$[\hat{J}_i, \hat{V}_j] = i\hbar \sum_k \epsilon_{ijk} \hat{V}_k \, .$$

1. Consider the operator

$$\hat{U}_y(\beta) = e^{-i\beta \hat{J}_y/\hbar},$$

   and prove that it represents a rotation by an angle β about the y-direction by showing that

$$\hat{U}_y(\beta) \, \hat{V}_i \, \hat{U}_y^\dagger(\beta) = \sum_j \hat{V}_j R_{y,ji}(\beta) = \sum_j R_{y,ij}^{-1}(\beta) \hat{V}_j \, ,$$

   where $\underline{R}_y(\beta)$ is the rotation matrix about the y-axis,

$$\underline{R}_y = \begin{pmatrix} \cos\beta & 0 & \sin\beta \\ 0 & 1 & 0 \\ -\sin\beta & 0 & \cos\beta \end{pmatrix} .$$

   Hint: Define $\hat{\mathbf{V}}(\beta) = \hat{U}_y(\beta) \, \hat{\mathbf{V}} \, \hat{U}_y^\dagger(\beta)$ with $\hat{\mathbf{V}}(\beta = 0) = \hat{\mathbf{V}}$ and consider the derivatives of the components $\hat{V}_i(\beta)$ with respect to β.

2. Assuming the above transformation properties for the components $\hat{V}_i$, show that in the limit $\beta \longrightarrow 0$ they imply the commutation relations of $\hat{J}_y$ with $\hat{V}_j$ given in the problem text.

3. Show that the state $\hat{U}_y(\pi)\,|\psi_{jm}\rangle$, obtained by applying a rotation by π about the y-axis to an eigenstate $|\psi_{jm}\rangle$ of $\hat{\mathbf{J}}^2$ and $\hat{J}_z$ with eigenvalues $j(j+1)\hbar^2$ and $m\hbar$, remains an eigenstate of these operators with eigenvalues $j(j+1)\hbar^2$ and $-m\hbar$, respectively.

4. Show the following identity:

$$e^{-i\beta\hat{J}_z/\hbar} = e^{-i(\pi/2)\hat{J}_x/\hbar}\,e^{-i\beta\hat{J}_y/\hbar}\,e^{i(\pi/2)\hat{J}_x/\hbar}\,,$$

which, as we will see in a later chapter, says that a rotation by β about the z-axis can be achieved by the following sequence of rotations: (i) a rotation by $\pi/2$ about the x-axis; then (ii) a rotation by β about the y'-axis (the new y-axis); and lastly (iii) a rotation by $-\pi/2$ about the x''-axis (the new x-axis).[3]

Solution

Part 1

Following the hint, we consider

$$\frac{d}{d\beta}\hat{V}_i(\beta) = -\frac{i}{\hbar}\,e^{-i\beta\hat{J}_y/\hbar}\left(\hat{J}_y\,\hat{V}_i - \hat{V}_i\hat{J}_y\right)e^{i\beta\hat{J}_y/\hbar} = \epsilon_{2ij}\,e^{-i\beta\hat{J}_y/\hbar}\,\hat{V}_j\,e^{i\beta\hat{J}_y/\hbar} = \epsilon_{2ij}\,\hat{V}_j(\beta)\,,$$

or, explicitly,

$$\frac{d}{d\beta}\hat{V}_x(\beta) = -\hat{V}_z(\beta)\,, \qquad \frac{d}{d\beta}\hat{V}_y(\beta) = 0\,, \qquad \frac{d}{d\beta}\hat{V}_z(\beta) = \hat{V}_x(\beta)\,.$$

We immediately obtain that $\hat{V}_y(\beta) = \hat{V}_y(0) = \hat{V}_y$ is independent of β. We also find

$$\frac{d^2}{d\beta^2}\hat{V}_x(\beta) = -\frac{d}{d\beta}\hat{V}_z(\beta) = -\hat{V}_x(\beta)\,,$$

with the initial conditions

$$\hat{V}_x(0) = \hat{V}_x\,, \qquad \left.\frac{d}{d\beta}\hat{V}_x(\beta)\right|_{\beta=0} = -\hat{V}_z(0) = -\hat{V}_z\,.$$

The above second-order differential equation has the solution

$$\hat{V}_x(\beta) = \hat{A}\,\cos\beta + \hat{B}\,\sin\beta\,.$$

Imposing the initial conditions yields

$$\hat{A} = \hat{V}_x\,, \qquad \hat{B} = -\hat{V}_z \implies V_x(\beta) = \hat{V}_x\,\cos\beta - \hat{V}_z\,\sin\beta\,,$$

which in turn implies that

$$\hat{V}_z(\beta) = -\frac{d}{d\beta}\hat{V}_x(\beta) = \hat{V}_x\,\sin\beta + \hat{V}_z\,\cos\beta\,.$$

---

[3] The attentive reader will have noticed that the sequence of operators in $e^{-i(\pi/2)\hat{J}_x/\hbar}e^{-i\beta\hat{J}_y/\hbar}e^{i(\pi/2)\hat{J}_x/\hbar}$ appears to be reversed with respect to the order in which the rotations are actually carried out (the rotation by $-\pi/2$ about the x-axis is carried out first rather than last, while that by $\pi/2$ about the same axis is carried out last rather than first. For an explanation, see Section 17.1.

In matrix notation, we have

$$\begin{pmatrix} \hat{V}_x(\beta) \\ \hat{V}_y(\beta) \\ \hat{V}_z(\beta) \end{pmatrix} = \underbrace{\begin{pmatrix} \cos\beta & 0 & -\sin\beta \\ 0 & 1 & 0 \\ \sin\beta & 0 & \cos\beta \end{pmatrix}}_{\underline{R}_y^T = \underline{R}_y^{-1}} \begin{pmatrix} \hat{V}_x \\ \hat{V}_y \\ \hat{V}_z \end{pmatrix},$$

where

$$\underline{R}_y = \begin{pmatrix} \cos\beta & 0 & \sin\beta \\ 0 & 1 & 0 \\ -\sin\beta & 0 & \cos\beta \end{pmatrix}.$$

Part 2

In the limit $\beta = \eta \longrightarrow 0$, up to linear terms in η we have

$$\hat{V}_i(\eta) = \left(\mathbb{1} - \frac{i}{\hbar}\eta\hat{J}_y + \cdots\right)\hat{V}_i\left(\mathbb{1} + \frac{i}{\hbar}\eta\hat{J}_y + \cdots\right) = \hat{V}_i - \frac{i}{\hbar}\eta[\hat{J}_y, \hat{V}_i] + \cdots$$

and therefore

$$\begin{pmatrix} \hat{V}_x \\ \hat{V}_y \\ \hat{V}_z \end{pmatrix} - \frac{i}{\hbar}\eta\begin{pmatrix} [\hat{J}_y, \hat{V}_x] \\ [\hat{J}_y, \hat{V}_y] \\ [\hat{J}_y, \hat{V}_z] \end{pmatrix} = \begin{pmatrix} 1 & 0 & -\eta \\ 0 & 1 & 0 \\ \eta & 0 & 1 \end{pmatrix}\begin{pmatrix} \hat{V}_x \\ \hat{V}_y \\ \hat{V}_z \end{pmatrix} = \begin{pmatrix} \hat{V}_x - \eta\hat{V}_z \\ \hat{V}_y \\ \hat{V}_z + \eta\hat{V}_x \end{pmatrix},$$

which leads to

$$[\hat{J}_y, \hat{V}_x] = -i\hbar\hat{V}_z, \qquad [\hat{J}_y, \hat{V}_y] = 0, \qquad [\hat{J}_y, \hat{V}_z] = i\hbar\hat{V}_x.$$

By considering infinitesimal rotations about the x- and z-axes, we can establish the commutation relations of $\hat{J}_x$ and $\hat{J}_z$ with the components of $\hat{\mathbf{V}}$. Note that the orthogonal matrices for finite rotations about these axes are given by

$$\underline{R}_x = \begin{pmatrix} 1 & 0 & 0 \\ 0 & \cos\beta & -\sin\beta \\ 0 & \sin\beta & \cos\beta \end{pmatrix}, \qquad \underline{R}_z = \begin{pmatrix} \cos\beta & -\sin\beta & 0 \\ \sin\beta & \cos\beta & 0 \\ 0 & 0 & 1 \end{pmatrix},$$

so that

$$\hat{U}_x(\beta)\,\hat{V}_i\,\hat{U}_x^\dagger(\beta) = \sum_j \hat{V}_j\,R_{x,ji}(\beta) = \sum_j R_{x,ij}^{-1}(\beta)\hat{V}_j$$

and

$$\hat{U}_z(\beta)\,\hat{V}_i\,\hat{U}_z^\dagger(\beta) = \sum_j \hat{V}_j\,R_{z,ji}(\beta) = \sum_j R_{z,ij}^{-1}(\beta)\hat{V}_j.$$

Part 3

We first note that, since $\hat{\mathbf{J}}^2$ commutes with the components $\hat{J}_i$, it commutes, in particular, with $U_y(\beta)$ and hence

$$\hat{\mathbf{J}}^2\,\hat{U}_y(\beta)\,|\psi_{jm}\rangle = \hat{U}_y(\beta)\,\hat{\mathbf{J}}^2\,|\psi_{jm}\rangle = j(j+1)\hbar^2\,\hat{U}_y(\beta)\,|\psi_{jm}\rangle,$$

showing that $\hat{U}_y(\beta)\,|\psi_{jm}\rangle$ is an eigenstate of $\hat{\mathbf{J}}^2$ for any β. Now, consider

$$\hat{J}_z\,\hat{U}_y(\beta)\,|\psi_{jm}\rangle = \hat{U}_y(\beta)\,\hat{U}_y^\dagger(\beta)\,\hat{J}_z\,\hat{U}_y(\beta)\,|\psi_{jm}\rangle = \hat{U}_y(\beta)\,\underbrace{\hat{U}_y(-\beta)\,\hat{J}_z\,\hat{U}_y^\dagger(-\beta)}_{\hat{J}_z(-\beta)}\,|\psi_{jm}\rangle$$

and

$$\hat{J}_z(-\beta) = -\hat{J}_x\,\sin\beta + \hat{J}_z\,\cos\beta \implies \hat{J}_z(-\pi) = -\hat{J}_z\,.$$

We conclude that, for $\beta = \pi$,

$$\hat{J}_z\,\hat{U}_y(\pi)\,|\psi_{jm}\rangle = \hat{U}_y(\pi)(-\hat{J}_z)|\psi_{jm}\rangle = -m\hbar\,\hat{U}_y(\pi)|\psi_{jm}\rangle\,,$$

showing that the state $\hat{U}_y(\pi)\,|\psi_{jm}\rangle$ is an eigenstate of $\hat{J}_z$ with eigenvalue $-m\hbar$; in fact, it can differ from $|\psi_{j,-m}\rangle$ by at most a phase factor.

Part 4

As noted in part 2, we have that, for a rotation about the x-axis by angle β, a vector operator, such as $\hat{\mathbf{J}}$, transforms as

$$\hat{J}_i(\beta) = \underbrace{e^{-i\beta\hat{J}_x/\hbar}}_{\hat{U}_x(\beta)}\,\hat{J}_i\,\underbrace{e^{i\beta\hat{J}_x/\hbar}}_{\hat{U}_x^\dagger(\beta)} = \sum_j R_{x,ij}^{-1}\,\hat{J}_j\,,$$

or, in matrix notation,

$$\begin{pmatrix}\hat{J}_x(\beta)\\ \hat{J}_y(\beta)\\ \hat{J}_z(\beta)\end{pmatrix} = \begin{pmatrix}1 & 0 & 0\\ 0 & \cos\beta & \sin\beta\\ 0 & -\sin\beta & \cos\beta\end{pmatrix}\begin{pmatrix}\hat{J}_x\\ \hat{J}_y\\ \hat{J}_z\end{pmatrix}\,,$$

where $\underline{R}_x$ is as given in part 2. In particular we find, for $\beta = \pi/2$,

$$\begin{pmatrix}\hat{J}_x(\pi/2)\\ \hat{J}_y(\pi/2)\\ \hat{J}_z(\pi/2)\end{pmatrix} = \begin{pmatrix}\hat{J}_x\\ \hat{J}_z\\ -\hat{J}_y\end{pmatrix}\,.$$

Now, consider

$$\begin{aligned}
\hat{U}_x(\pi/2)\,e^{-i\beta\hat{J}_y/\hbar}\,\hat{U}_x^\dagger(\pi/2) &= \sum_{n=0}^{\infty}\frac{(-i\beta/\hbar)^n}{n!}\,\hat{U}_x(\pi/2)\,(\hat{J}_y)^n\,\hat{U}_x^\dagger(\pi/2)\\
&= \sum_{n=0}^{\infty}\frac{(-i\beta/\hbar)^n}{n!}\,[\hat{U}_x(\pi/2)\,\hat{J}_y\,\hat{U}_x^\dagger(\pi/2)]^n\\
&= \sum_{n=0}^{\infty}\frac{(-i\beta/\hbar)^n}{n!}\,(\hat{J}_z)^n = e^{-i\beta\hat{J}_z/\hbar} = \hat{U}_z(\beta)\,,
\end{aligned}$$

as expected.

Problem 14　Algebraic Derivation of the Hydrogen Atom Spectrum

Consider the following operator (in this problem the caret notation for operators is understood):

$$\mathbf{K} = \frac{1}{2m}\,(\mathbf{p}\times\mathbf{L} - \mathbf{L}\times\mathbf{p}) - Ze^2\,\frac{\mathbf{r}}{r}\,.$$

Show that:

1. **K** is hermitian;
2. $\mathbf{K}^2 = Z^2 e^4 + (2H/m)(\mathbf{L}^2 + \hbar^2)$, where H is the hydrogen-like atom Hamiltonian

$$H = \frac{\mathbf{p}^2}{2m} - \frac{Ze^2}{r} \; ;$$

3. $[L_i, K_j] = i\hbar \sum_k \epsilon_{ijk} K_k$;
4. $[K_i, K_j] = -2i(\hbar/m) \sum_k \epsilon_{ijk} H L_k$.
5. **In the remainder of the problem, we restrict ourselves to the subspace of negative energies of the Hamiltonian**, and define the vector operator **A** as

$$\mathbf{A} = \sqrt{\frac{m}{2|E|}} \, \mathbf{K} \; .$$

Show that in this subspace

$$[A_i, A_j] = i\hbar \sum_k \epsilon_{ijk} L_k \; ,$$

and that

$$\mathbf{A}^2 + \mathbf{L}^2 = \frac{Z^2 e^4 m}{2|E|} - \hbar^2 \; .$$

6. Consider the operators

$$\mathbf{J}_\pm = \frac{1}{2}(\mathbf{L} \pm \mathbf{A}) \; ,$$

and show that

$$[J_{\pm,i}, J_{\pm,j}] = i\hbar \sum_k \epsilon_{ijk} J_{\pm,k} \; , \qquad [J_{\pm,i}, J_{\mp,j}] = 0 \; , \qquad \mathbf{J}_+^2 = \mathbf{J}_-^2 \; ,$$

and so $\mathbf{J}_+$ and $\mathbf{J}_-$ satisfy the commutation relations of an angular momentum, commute with each other, and their squares are the same.

7. Show that

$$2\left(\mathbf{J}_+^2 + \mathbf{J}_-^2\right) = \frac{Z^2 e^4 m}{2|E|} - \hbar^2 \; ,$$

and recalling that $\mathbf{J}_\pm$ are hermitian operators satisfying the commutation relations of an angular momentum, obtain the eigenvalues of the hydrogen-like atom Hamiltonian. What is their degeneracy?

Solution

Part 1

Consider

$$(\mathbf{p} \times \mathbf{L} - \mathbf{L} \times \mathbf{p})_i^\dagger = \sum_{jk} \epsilon_{ijk}(L_k p_j - p_k L_j) = -\sum_{jk} \epsilon_{ikj}(L_k p_j - p_k L_j) = -(\mathbf{L} \times \mathbf{p} - \mathbf{p} \times \mathbf{L})_i \; ,$$

where we have used $\epsilon_{ijk} = -\epsilon_{ikj}$. This shows that the combination $(\mathbf{p} \times \mathbf{L} - \mathbf{L} \times \mathbf{p})$ is hermitian, and hence

$$\mathbf{K} = \frac{1}{2m}(\mathbf{p} \times \mathbf{L} - \mathbf{L} \times \mathbf{p}) - Ze^2 \frac{\mathbf{r}}{r} \equiv \frac{1}{2m}\mathbf{O} - Ze^2 \frac{\mathbf{r}}{r}$$

is hermitian.

Part 2

We note that

$$O_i = (\mathbf{p} \times \mathbf{L})_i - \sum_{jk} \epsilon_{ijk} L_j p_k = (\mathbf{p} \times \mathbf{L})_i - \sum_{jk} \epsilon_{ijk} (p_k L_j + [L_j, p_k])$$

$$= 2(\mathbf{p} \times \mathbf{L})_i - i\hbar \sum_{jkl} \epsilon_{ijk} \epsilon_{jkl} p_l = 2(\mathbf{p} \times \mathbf{L})_i - i\hbar \sum_{jkl} (\delta_{kk}\delta_{il} - \delta_{kl}\delta_{ik}) p_l = 2(\mathbf{p} \times \mathbf{L})_i - 2i\hbar p_i,$$

and $\mathbf{K}$ can also be written as

$$\mathbf{K} = \underbrace{\frac{1}{m}\mathbf{p} \times \mathbf{L}}_{1} - \underbrace{i\frac{\hbar}{m}\mathbf{p}}_{2} - \underbrace{Ze^2 \frac{\mathbf{r}}{r}}_{3} = 1 - 2 - 3 \; .$$

In anticipation of squaring this operator, we consider

$$12 + 21 = i\frac{\hbar}{m^2}\left[(\mathbf{p} \times \mathbf{L}) \cdot \mathbf{p} + \mathbf{p} \cdot (\mathbf{p} \times \mathbf{L})\right],$$

and the second term vanishes since the components of the momentum operator commute with each other and

$$\mathbf{p} \cdot (\mathbf{p} \times \mathbf{L}) = \sum_{ijk} \epsilon_{ijk} p_i p_j L_k = \sum_k \overbrace{(\mathbf{p} \times \mathbf{p})_k}^{\text{vanishes}} L_k = 0 \; .$$

The first term can be written as

$$(\mathbf{p} \times \mathbf{L}) \cdot \mathbf{p} = \sum_{ijk} \epsilon_{ijk} p_i L_j p_k = \sum_{ijk} \epsilon_{ijk} p_i \left(p_k L_j + i\hbar \sum_l \epsilon_{jkl} p_l \right) = -\sum_{ijk} \epsilon_{ikj} p_i p_k L_j + i\hbar \sum_{ijkl} \epsilon_{ijk} \epsilon_{jkl} p_i p_l$$

$$= -\underbrace{\mathbf{p} \cdot (\mathbf{p} \times \mathbf{L})}_{\text{vanishes}} + i\hbar \sum_{ikl} (\delta_{kk}\delta_{il} - \delta_{kl}\delta_{ik}) p_i p_l = 2i\hbar\,\mathbf{p}^2 \; .$$

and hence

$$12 + 21 = -2\frac{\hbar^2}{m^2}\mathbf{p}^2 \; .$$

Next, we consider

$$13 + 31 = \frac{Ze^2}{m}\left[(\mathbf{p} \times \mathbf{L}) \cdot \frac{\mathbf{r}}{r} + \frac{\mathbf{r}}{r} \cdot (\mathbf{p} \times \mathbf{L})\right] \; .$$

The second term gives

$$\frac{\mathbf{r}}{r} \cdot (\mathbf{p} \times \mathbf{L}) = \sum_{ijk} \epsilon_{ijk} \frac{r_i}{r} p_j L_k = \frac{1}{r}\sum_k L_k L_k = \frac{1}{r}\mathbf{L}^2 \; ,$$

while the first can be written as

$$(\mathbf{p} \times \mathbf{L}) \cdot \frac{\mathbf{r}}{r} = \sum_{ijk} \epsilon_{ijk} p_i L_j \frac{r_k}{r} = \sum_{ijk} \epsilon_{ijk} \left(L_j p_i - i\hbar \sum_l \epsilon_{jil} p_l \right) \frac{r_k}{r}$$

$$= \sum_{ijk} \epsilon_{ijk} L_j r_k p_i \frac{1}{r} + i\hbar \sum_{ikl} (\delta_{ii}\delta_{kl} - \delta_{ik}\delta_{ik}) p_l \frac{r_k}{r}$$

$$= \mathbf{L}^2 \frac{1}{r} + 2i\hbar\,\mathbf{p} \cdot \mathbf{r} \frac{1}{r} \; ,$$

where in the first line we have expressed $p_i L_j = L_j p_i + [p_i, L_j]$ and have used $p_i r_k = r_k p_i$, valid for $i \neq k$ (as is the case here because of the ϵ_{ijk} tensor). Thus, we find (recall that $\mathbf{L}$ and $1/r$ commute)

$$13 + 31 = 2 \frac{Ze^2}{m} (\mathbf{L}^2 + i\hbar \, \mathbf{p} \cdot \mathbf{r}) \frac{1}{r} \, .$$

Now, we come to

$$23 + 32 = i \frac{Ze^2 \hbar}{m} \Big(\mathbf{p} \cdot \frac{\mathbf{r}}{r} + \frac{\mathbf{r}}{r} \cdot \mathbf{p} \Big) \, ,$$

where we write

$$\frac{\mathbf{r}}{r} \cdot \mathbf{p} = \mathbf{p} \cdot \frac{\mathbf{r}}{r} + \sum_i \Big[\frac{r_i}{r}, p_i \Big] = \mathbf{p} \cdot \frac{\mathbf{r}}{r} + \sum_i \Big(\frac{i\hbar}{r} \delta_{ii} + \Big[\frac{1}{r}, p_i \Big] r_i \Big) = \mathbf{p} \cdot \frac{\mathbf{r}}{r} + i\hbar \Big(3 \frac{1}{r} - \frac{1}{r} \Big) \, ,$$

and so

$$23 + 32 = i \frac{2 Ze^2 \hbar}{m} (\mathbf{p} \cdot \mathbf{r} + i\hbar) \frac{1}{r} \, .$$

Lastly, we consider

$$11 = \frac{1}{m^2} (\mathbf{p} \times \mathbf{L})^2 = \frac{1}{m^2} \sum_{ijklm} \epsilon_{ijk} \epsilon_{ilm} p_j L_k p_l L_m = \frac{1}{m^2} \sum_{jklm} (\delta_{jl} \delta_{km} - \delta_{jm} \delta_{kl}) p_j L_k p_l L_m$$

$$= \frac{1}{m^2} \sum_{jklm} (\delta_{jl} \delta_{km} - \delta_{jm} \delta_{kl}) p_j \Big(p_l L_k + i\hbar \sum_n \epsilon_{kln} p_n \Big) L_m$$

$$= \frac{1}{m^2} \Big(\mathbf{p}^2 \mathbf{L}^2 - \sum_j p_j \, \mathbf{p} \cdot \mathbf{L} \, L_j \Big) + i \frac{\hbar}{m^2} \sum_{jklmn} (\delta_{jl} \delta_{km} - \delta_{jm} \delta_{kl}) \epsilon_{kln} p_j p_n L_m$$

$$= \frac{1}{m^2} \Big(\mathbf{p}^2 \mathbf{L}^2 - \sum_j p_j \, \mathbf{p} \cdot \mathbf{L} \, L_j \Big) = \frac{1}{m^2} \mathbf{p}^2 \mathbf{L}^2 \, ,$$

where in the third line we used

$$\sum_{jklmn} (\delta_{jl} \delta_{km} - \delta_{jm} \delta_{kl}) \epsilon_{kln} p_j p_n L_m = \underbrace{(\mathbf{p} \times \mathbf{p})}_{\text{vanishes}} \cdot \mathbf{L} - \underbrace{\Big(\sum_{kn} \epsilon_{kkn} p_n \Big)}_{\text{vanishes}} \mathbf{p} \cdot \mathbf{L} = 0 \, ,$$

and in the last line we used

$$\mathbf{p} \cdot \mathbf{L} = \mathbf{L} \cdot \mathbf{p} = 0 \, ,$$

since (recall that r_j and p_k commute for $j \neq k$, as enforced by ϵ_{ijk})

$$\mathbf{p} \cdot \mathbf{L} = \sum_{ijk} \epsilon_{ijk} p_i r_j p_k = - \sum_{ijk} \epsilon_{ikj} p_i p_k r_j = -(\mathbf{p} \times \mathbf{p}) \cdot \mathbf{r} = 0 \, ,$$

and similarly for $\mathbf{L} \cdot \mathbf{p}$. Note that we also have

$$\mathbf{r} \cdot \mathbf{L} = \mathbf{L} \cdot \mathbf{r} = 0 \, ,$$

which can be shown similarly. Collecting results, we find for $\mathbf{K}^2$ (recall that $\mathbf{L}$ and hence $\mathbf{L}^2$ commute with $\mathbf{p}^2$ and $1/r$)

$$
\begin{aligned}
\mathbf{K}^2 &= 11 + 22 + 33 - (12 + 21) - (13 + 31) + (23 + 32) \\
&= \frac{1}{m^2}\mathbf{p}^2\mathbf{L}^2 - \frac{\hbar^2}{m^2}\mathbf{p}^2 + (Ze^2)^2 + 2\frac{\hbar^2}{m^2}\mathbf{p}^2 - 2\frac{Ze^2}{m}(\mathbf{L}^2 + i\hbar\,\mathbf{p}\cdot\mathbf{r})\frac{1}{r} + i\frac{2Ze^2\hbar}{m}(\mathbf{p}\cdot\mathbf{r} + i\hbar)\frac{1}{r} \\
&= \frac{1}{m^2}\mathbf{p}^2\mathbf{L}^2 + \frac{\hbar^2}{m^2}\mathbf{p}^2 + (Ze^2)^2 - 2\frac{Ze^2}{m}\mathbf{L}^2\frac{1}{r} - \frac{2Ze^2\hbar^2}{m}\frac{1}{r} \\
&= (Ze^2)^2 + \left(\frac{\mathbf{p}^2}{m^2} - \frac{2}{m}\frac{Ze^2}{r}\right)\mathbf{L}^2 + \left(\frac{\mathbf{p}^2}{m^2} - \frac{2}{m}\frac{Ze^2}{r}\right)\hbar^2 \\
&= (Ze^2)^2 + \frac{2}{m}H(\mathbf{L}^2 + \hbar^2)\,.
\end{aligned}
$$

Part 3

Since $\mathbf{K}$ is a linear combination of vector operators, it must satisfy the commutation relation

$$
[L_i,\, K_j] = i\hbar\sum_k \epsilon_{ijk}K_k\,.
$$

This can also be verified explicitly. Using the expression

$$
\mathbf{K} = \frac{1}{m}\mathbf{p}\times\mathbf{L} - i\frac{\hbar}{m}\mathbf{p} - Ze^2\frac{\mathbf{r}}{r}\,,
$$

it is obvious that the last two terms satisfy the commutation relations appropriate for vector operators. The first term gives

$$
\begin{aligned}
\left[L_i,\, \sum_{kl}\epsilon_{jkl}p_kL_l\right] &= i\hbar\sum_{kl}\epsilon_{jkl}\left(p_k\sum_m\epsilon_{ilm}L_m + \sum_m\epsilon_{ikm}p_mL_l\right) \\
&= i\hbar(p_iL_j - p_jL_i) = i\hbar\sum_{klm}\epsilon_{ijk}\epsilon_{klm}p_lL_m = i\hbar\sum_k\epsilon_{ijk}(\mathbf{p}\times\mathbf{L})_k\,.
\end{aligned}
$$

Part 4

Using the earlier notation for the terms in $\mathbf{K}$, we have

$$
[K_i,\, K_l] = [1_i - 2_i - 3_i,\, 1_l - 2_l - 3_l]\,.
$$

We begin by considering

$$
\begin{aligned}
[1_i,\, 1_l] &= \frac{1}{m^2}\sum_{jkmn}\epsilon_{ijk}\epsilon_{lmn}[p_jL_k,\, p_mL_n] \\
&= \frac{1}{m^2}\sum_{jkmn}\epsilon_{ijk}\epsilon_{lmn}(p_jp_m[L_k,\, L_n] + p_j[L_k,\, p_m]L_n + p_m[p_j,\, L_n]L_k + [p_j,\, p_m]L_nL_k) \\
&= i\frac{\hbar}{m^2}\sum_{jkmn}\epsilon_{ijk}\epsilon_{lmn}\sum_p(\epsilon_{knp}p_jp_mL_p + \epsilon_{kmp}p_jp_pL_n - \epsilon_{njp}p_mp_pL_k)\,.
\end{aligned}
$$

The first term vanishes since

$$
\sum_{jkmnp}\epsilon_{ijk}\epsilon_{lmn}\epsilon_{knp}p_jp_mL_p = \sum_{jmnp}\epsilon_{lmn}(\delta_{in}\delta_{jp} - \delta_{ip}\delta_{jn})p_jp_mL_p = \sum_m\epsilon_{lmi}p_m\mathbf{p}\cdot\mathbf{L} - (\mathbf{p}\times\mathbf{p})_lL_i = 0\,.
$$

The second and third terms give, respectively,

$$\sum_{jkmnp} \epsilon_{ijk}\,\epsilon_{lmn}\,\epsilon_{kmp}p_j p_p L_n = \sum_{jmnp} \epsilon_{lmn}(\delta_{im}\,\delta_{jp} - \delta_{ip}\,\delta_{jm})p_j p_p L_n = \sum_{n} \epsilon_{lin}\,\mathbf{p}^2\,L_n - p_i\,(\mathbf{p}\times\mathbf{L})_l$$

and

$$\sum_{jkmnp} \epsilon_{ijk}\,\epsilon_{lmn}\,\epsilon_{njp}\,p_m p_p L_k = \sum_{jkmp} \epsilon_{ijk}(\delta_{jl}\,\delta_{mp} - \delta_{lp}\,\delta_{jm})p_m p_p L_k = \sum_{k} \epsilon_{ilk}\,\mathbf{p}^2\,L_k - p_l\,(\mathbf{p}\times\mathbf{L})_i\;.$$

Combining terms, we find

$$[1_i\,,\,1_l] = -i\frac{2\hbar}{m^2}\,\mathbf{p}^2\sum_{k}\epsilon_{ilk}L_k - i\frac{\hbar}{m^2}\,[p_i\,(\mathbf{p}\times\mathbf{L})_l - p_l\,(\mathbf{p}\times\mathbf{L})_i]\;,$$

where the second term can be further reduced to

$$p_i\,(\mathbf{p}\times\mathbf{L})_l - p_l\,(\mathbf{p}\times\mathbf{L})_i = \sum_{kmn}\epsilon_{ilk}\,\epsilon_{kmn}\,p_m\,(\mathbf{p}\times\mathbf{L})_n = \sum_{k}\epsilon_{ilk}\,[\mathbf{p}\times(\mathbf{p}\times\mathbf{L})]_k$$

$$= \sum_{k}\epsilon_{ilk}(p_k\,\mathbf{p}\cdot\mathbf{L} - \mathbf{p}^2\,L_k) = -\mathbf{p}^2\sum_{k}\epsilon_{ilk}L_k\;.$$

Thus we obtain

$$[1_i\,,\,1_l] = -i\frac{\hbar}{m^2}\,\mathbf{p}^2\sum_{k}\epsilon_{ilk}L_k\;.$$

Next, we examine

$$[1_i\,,\,2_l] = i\frac{\hbar}{m^2}\sum_{jk}\epsilon_{ijk}\,[p_j L_k\,,\,p_l] = -\frac{\hbar^2}{m^2}\sum_{jkm}\epsilon_{ijk}\,p_j\,\epsilon_{klm}\,p_m = -\frac{\hbar^2}{m^2}(\delta_{il}\,\mathbf{p}^2 - p_i p_l)\;.$$

and

$$[1_i\,,\,3_l] = \frac{Ze^2}{m}\sum_{jk}\epsilon_{ijk}\left[p_j L_k\,,\,\frac{r_l}{r}\right] = \frac{Ze^2}{m}\sum_{jk}\epsilon_{ijk}\left(p_j\left[L_k\,,\,r_l\right]\frac{1}{r} + \left[p_j\,,\,\frac{r_l}{r}\right]L_k\right)$$

$$= i\frac{Ze^2\hbar}{m}\sum_{jk}\epsilon_{ijk}\left(\sum_{m}\epsilon_{klm}p_j r_m\frac{1}{r} - \delta_{jl}\frac{1}{r}L_k + \frac{1}{r^3}r_j r_l L_k\right)$$

$$= i\frac{Ze^2\hbar}{m}\left[\delta_{il}\,\mathbf{p}\cdot\mathbf{r}\,\frac{1}{r} - p_l r_i\frac{1}{r} - \sum_{k}\epsilon_{ilk}\frac{1}{r}L_k + \frac{1}{r^3}r_l\,(\mathbf{r}\times\mathbf{L})_i\right]\;.$$

We note that

$$[2_i\,,\,2_l] = [3_i\,,\,3_l] = 0\;,$$

while

$$[2_i\,,\,3_l] = i\frac{Ze^2\hbar}{m}\left[p_i\,,\,\frac{r_l}{r}\right] = \frac{Ze^2\hbar^2}{m}\left(\delta_{il}\frac{1}{r} - \frac{r_i r_l}{r^3}\right)\;.$$

Now, we consider

$$[K_i\,,\,K_l] = [1_i\,,\,1_l] - ([1_i\,,\,2_l] + [2_i\,,\,1_l]) - ([1_i\,,\,3_l] + [3_i\,,\,1_l]) + ([2_i\,,\,3_l] + [3_i\,,\,2_l])\;,$$

and observe that

$$[2_i\,,\,1_l] = -[1_l\,,\,2_i] = \frac{\hbar^2}{m^2}(\delta_{il}\,\mathbf{p}^2 - p_i p_l)\;,$$

where the last step follows from exchanging i with l in the commutator $[1_i, 2_l]$ (note that this commutator is symmetric under such an exchange). It follows that

$$[1_i, 2_l] + [2_i, 1_l] = 0 .$$

Similarly, we find

$$[2_i, 3_l] + [3_i, 2_l] = 0 .$$

Lastly, we have

$$[3_i, 1_l] = -[1_l, 3_i] = -i \frac{Ze^2\hbar}{m} \Big[\delta_{il} \, \mathbf{p} \cdot \mathbf{r} \, \frac{1}{r} - p_i r_l \frac{1}{r} - \sum_k \epsilon_{lik} \frac{1}{r} L_k + \frac{1}{r^3} r_i \, (\mathbf{r} \times \mathbf{L})_l \Big] ,$$

and

$$[1_i, 3_l] + [3_i, 1_l] = i \frac{Ze^2\hbar}{m} \Big[(p_i r_l - p_l r_i) \frac{1}{r} - 2 \sum_k \epsilon_{ilk} \frac{1}{r} L_k + \frac{1}{r^3} r_l \, (\mathbf{r} \times \mathbf{L})_i - \frac{1}{r^3} r_i \, (\mathbf{r} \times \mathbf{L})_l \Big] .$$

We observe that

$$p_i r_l - p_l r_i = \sum_{kmn} \epsilon_{ilk} \, \epsilon_{kmn} \, p_m r_n = - \sum_{kmn} \epsilon_{ilk} \, \epsilon_{knp} \, r_n p_m = - \sum_k \epsilon_{ilk} \, L_k ,$$

and

$$r_i \, (\mathbf{r} \times \mathbf{L})_l - r_l \, (\mathbf{r} \times \mathbf{L})_i = \sum_{kmn} \epsilon_{ilk} \, \epsilon_{kmn} \, r_m \, (\mathbf{r} \times \mathbf{L})_n = \sum_k \epsilon_{ilk} \, [\mathbf{r} \times (\mathbf{r} \times \mathbf{L})]_k$$

$$= \sum_k \epsilon_{ilk} (r_k \, \mathbf{r} \cdot \mathbf{L} - \mathbf{r}^2 \, L_k) = -\mathbf{r}^2 \sum_k \epsilon_{ilk} \, L_k .$$

Thus, we have

$$[1_i, 3_l] + [3_i, 1_l] = i \frac{Ze^2\hbar}{m} \sum_k \epsilon_{ilk} \Big[-\frac{1}{r} L_k - \frac{2}{r} L_k + \frac{1}{r} L_k \Big] = -i \frac{2Ze^2\hbar}{m} \sum_k \epsilon_{ilk} \frac{1}{r} L_k .$$

We finally arrive at

$$[K_i, K_l] = -i \frac{\hbar}{m^2} \mathbf{p}^2 \sum_k \epsilon_{ilk} \, L_k + i \frac{2Ze^2\hbar}{m} \frac{1}{r} \sum_k \epsilon_{ilk} \, L_k = -i \frac{2\hbar}{m} \underbrace{\Big(\frac{\mathbf{p}^2}{2m} - \frac{Ze^2}{r} \Big)}_{H} \sum_k \epsilon_{ilk} \, L_k .$$

Part 5

The commutator between the components A_i follows from that in part 2, since in the subspace of negative energies we have

$$\Big[\underbrace{\sqrt{\frac{2|E|}{m}} A_i}_{K_i} , \underbrace{\sqrt{\frac{2|E|}{m}} A_j}_{K_j} \Big] = \frac{2i\hbar}{m} |E| \sum_k \epsilon_{ijk} \, L_k \implies [A_i, A_j] = i\hbar \sum_k \epsilon_{ijk} \, L_k .$$

We also have

$$\frac{2|E|}{m} \mathbf{A}^2 = Z^2 e^4 - \frac{2}{m} |E| \, (\mathbf{L}^2 + \hbar^2) \implies \mathbf{A}^2 + \mathbf{L}^2 + \hbar^2 = \frac{Z^2 e^4 m}{2|E|} .$$

Part 6

We have

$$[J_{\pm,i}, J_{\pm,j}] = \frac{1}{4}[L_i \pm A_i, L_j \pm A_j] = \frac{i\hbar}{4}\sum_k \epsilon_{ijk}(L_k \pm A_k \pm A_k + L_k)$$

$$= \frac{i\hbar}{2}\sum_k \epsilon_{ijk}(L_k \pm A_k) = i\hbar \sum_k \epsilon_{ijk} J_{\pm,k},$$

where we have used the commutators obtained earlier, in particular

$$[A_i, L_j] = -[L_j, A_i] = -i\hbar \sum_k \epsilon_{jik} A_k = i\hbar \sum_k \epsilon_{ijk} A_k.$$

We also find that

$$[J_{\pm,i}, J_{\mp,j}] = \frac{1}{4}[L_i \pm A_i, L_j \mp A_j] = \frac{i\hbar}{4}\sum_k \epsilon_{ijk}(L_k \pm A_k \mp A_k - L_k) = 0$$

and lastly

$$\mathbf{J}_\pm^2 = \frac{1}{4}\left(\mathbf{L}^2 + \mathbf{A}^2 \pm \mathbf{L}\cdot\mathbf{A} \pm \mathbf{A}\cdot\mathbf{L}\right) = \frac{1}{4}\left(\mathbf{L}^2 + \mathbf{A}^2\right) \implies \mathbf{J}_+^2 = \mathbf{J}_-^2,$$

where we have used $\mathbf{L}\cdot\mathbf{A} = \mathbf{A}\cdot\mathbf{L} = 0$. To show this last result, first note that, using the identity obtained in part 2

$$(\mathbf{p}\times\mathbf{L})_i - (\mathbf{L}\times\mathbf{p})_i = 2(\mathbf{p}\times\mathbf{L})_i - 2i\hbar p_i,$$

the operator $\mathbf{A}$ can also be written as

$$\mathbf{A} = \sqrt{\frac{m}{2|E|}}\left(\frac{1}{m}\mathbf{p}\times\mathbf{L} - i\frac{\hbar}{m}\mathbf{p} - Ze^2\frac{\mathbf{r}}{r}\right).$$

We have already established that $\mathbf{r}\cdot\mathbf{L} = \mathbf{L}\cdot\mathbf{r} = 0$ and $\mathbf{p}\cdot\mathbf{L} = \mathbf{L}\cdot\mathbf{p} = 0$. So it is left to show that $(\mathbf{p}\times\mathbf{L})\cdot\mathbf{L} = \mathbf{L}\cdot(\mathbf{p}\times\mathbf{L}) = 0$, to conclude that $\mathbf{L}\cdot\mathbf{A} = \mathbf{A}\cdot\mathbf{L} = 0$. To this end, consider

$$(\mathbf{p}\times\mathbf{L})\cdot\mathbf{L} = \sum_{ijk}\epsilon_{ijk}p_i L_j L_k = \frac{1}{2}\sum_{ijk}\epsilon_{ijk}p_i\left(L_j L_k - L_k L_j\right) = \frac{i\hbar}{2}\sum_{ijkl}\epsilon_{ijk}\epsilon_{jkl}p_i L_l$$

$$= \frac{i\hbar}{2}\sum_{ikl}(\delta_{kk}\delta_{il} - \delta_{ik}\delta_{kl})p_i L_l = i\hbar\,\mathbf{p}\cdot\mathbf{L} = 0,$$

and similarly

$$\mathbf{L}\cdot(\mathbf{p}\times\mathbf{L}) = \sum_{ijk}\epsilon_{ijk}L_i p_j L_k = \sum_{ijk}\epsilon_{ijk}\left(p_j L_i + [L_i, p_j]\right)L_k$$

$$= -\underbrace{(\mathbf{p}\times\mathbf{L})\cdot\mathbf{L}}_{\text{vanishes}} + i\hbar\sum_{ijkl}\epsilon_{ijk}\epsilon_{ijl}p_l L_k = 2i\hbar\,\mathbf{p}\cdot\mathbf{L} = 0.$$

Part 7

Using the above result, we have

$$\mathbf{L}^2 + \mathbf{A}^2 = 2(\mathbf{J}_+^2 + \mathbf{J}_-^2).$$

Since the hermitian operators $\mathbf{J}_\pm$ satisfy the angular momentum commutation relations, their squares must have eigenvalues given by $\hbar^2 j_\pm(j_\pm + 1)$ with $j_\pm = 0, 1/2, 1, \ldots$ (that is, integer and half-integer values). Furthermore, the requirement that $\mathbf{J}_+^2 = \mathbf{J}_-^2$ dictates that $j_+ = j_- = j$. This yields

$$\mathbf{A}^2 + \mathbf{L}^2 + \hbar^2 = 2(\mathbf{J}_+^2 + \mathbf{J}_-^2) + \hbar^2 = \frac{Z^2 e^4 m}{2|E|} \implies 4j(j+1)\hbar^2 + \hbar^2 = (2j+1)^2\hbar^2 = \frac{Z^2 e^4 m}{2|E|} ,$$

and hence

$$E_n = -\frac{Z^2}{2n^2}\frac{e^4 m}{\hbar^2} = -\frac{(Z\alpha)^2}{2n^2} mc^2 \qquad n = 2j + 1 = 1, 2, 3, \ldots$$

The states

$$|j_+ = j, m_+\rangle \otimes |j_- = j, m_-\rangle \equiv |j, j; m_+, m_-\rangle, \qquad m_\pm = \underbrace{-j_\pm, -j_\pm + 1, \ldots, j_\pm}_{2j_\pm + 1 \text{ values}},$$

are simultaneous eigenstates of the commuting observables $\mathbf{J}_+^2$ and $J_{+,z}$, and $\mathbf{J}_-^2$ and $J_{-,z}$; they have the same energy. As a consequence, the degeneracy of the energy eigenvalues is $(2j_+ + 1)(2j_- + 1) = (2j + 1)^2 = n^2$, as expected.

12 Spin; Charged Particle in an Electromagnetic Field

We have established that the square of the angular momentum operator generally admits eigenvalues $j(j + 1)\hbar^2$ with j integer or semi-integer. On the other hand, in our treatment of a particle in a central potential we obtained the eigenvalues of the square of the *orbital* angular momentum operator as $l(l + 1)\hbar^2$ with l integer. The question arises then whether semi-integer values occur in nature. The answer is in the positive, as particles, in addition to *orbital* angular momentum, have an *intrinsic* angular momentum that we call spin. Spin can assume semi-integer or integer values – for example, the electron and photon have spin 1/2 and spin 1, respectively. In Chapter 13 we will study how to combine the orbital and spin angular momenta to form the total angular momentum of a particle (this is the general problem of the addition of angular momenta). Here, we present the treatment of spin degrees of freedom and discuss some of the physical consequences of spin, in particular the interaction of its magnetic dipole moment with a magnetic field. We also briefly review the classical and quantum mechanical treatments of a charged particle in an external electromagnetic field. Below, we specialize to the case of spin 1/2, but most of what we will say can be generalized to the case of a particle with spin ≥ 1.

12.1 Treatment of a Spin-1/2 Particle

For a spin-1/2 particle, let $|+\rangle$ and $|-\rangle$ represent the two eigenstates of $\hat{\mathbf{S}}^2$ and $\hat{S}_z$ (we use the symbol $\hat{\mathbf{S}}$ rather than $\hat{\mathbf{J}}$ to denote the spin angular momentum) with eigenvalues given by

$$\hat{\mathbf{S}}^2 |\pm\rangle = \frac{3}{4} \hbar^2 |\pm\rangle , \qquad \hat{S}_z |\pm\rangle = \pm \frac{1}{2} \hbar |\pm\rangle . \tag{12.1}$$

The spin operator commutes with the position and momentum operators of the particle, $[\hat{S}_i, \hat{r}_j] = 0 = [\hat{S}_i, \hat{p}_j]$, and therefore also commutes with any operator function of $\hat{\mathbf{r}}$ and $\hat{\mathbf{p}}$, in particular the orbital angular momentum $\hat{\mathbf{L}} = \hat{\mathbf{r}} \times \hat{\mathbf{p}}$. Orbital and spin operators act in different state spaces; the orbital state space is infinite dimensional, while the spin state space is two dimensional (or $2s + 1$ dimensional for a generic spin $s \geq 1$).

Since $\hat{\mathbf{r}}$, $\hat{\mathbf{S}}^2$, and $\hat{S}_z$ commute, we can form a continuous basis of common eigenstates of these observables. These eigenstates consist of tensor products of the eigenstates $|\phi_{\mathbf{r}}\rangle$ of $\hat{\mathbf{r}}$ and the eigenstates $|\sigma\rangle$ of $\hat{\mathbf{S}}^2$ and $\hat{S}_z$, with $\sigma = \pm$, and will be denoted as

$$|\phi_{\mathbf{r},\sigma}\rangle \equiv |\phi_{\mathbf{r}}\rangle \otimes |\sigma\rangle . \tag{12.2}$$

Since they form a basis, they satisfy the properties

$$\langle \phi_{\mathbf{r},\sigma} | \phi_{\mathbf{r}',\sigma'} \rangle = \delta(\mathbf{r} - \mathbf{r}') \, \delta_{\sigma,\sigma'} , \qquad \sum_{\sigma=\pm} \int d\mathbf{r} \, |\phi_{\mathbf{r},\sigma}\rangle\langle\phi_{\mathbf{r},\sigma}| = \hat{\mathbb{1}} . \tag{12.3}$$

Any state $|\psi\rangle$ of the particle in this combined orbital and spin space can be expanded in the basis above:

$$|\psi\rangle = \sum_{\sigma=\pm} \int d\mathbf{r}\, \psi_\sigma(\mathbf{r})\, |\phi_{\mathbf{r},\sigma}\rangle, \qquad \psi_\sigma(\mathbf{r}) = \langle\phi_{\mathbf{r},\sigma}|\psi\rangle, \qquad (12.4)$$

and the wave function $\psi_\sigma(\mathbf{r})$ now depends on the spin projection. We therefore interpret $|\psi_\sigma(\mathbf{r})|^2\, d\mathbf{r}$ as the probability for the particle to be in a volume $d\mathbf{r}$ centered at $\mathbf{r}$ with spin projection σ. In particular, $\sum_{\sigma=\pm} |\psi_\sigma(\mathbf{r})|^2\, d\mathbf{r}$ and $\int d\mathbf{r}\,|\psi_\sigma(\mathbf{r})|^2$ are interpreted as the probabilities, respectively, for the particle to be at $\mathbf{r}$ in $d\mathbf{r}$ regardless of its spin projection, and for the particle to be in spin projection σ regardless of its location in space.

In the basis $|\phi_{\mathbf{r},\sigma}\rangle$, a generic operator $\hat{A}$ acting on both space and spin degrees of freedom has matrix elements given by

$$\underline{A}_{\sigma\sigma'}(\mathbf{r}, \mathbf{r}') = \langle\phi_{\mathbf{r},\sigma}|\hat{A}|\phi_{\mathbf{r}',\sigma'}\rangle. \qquad (12.5)$$

Thus in this basis the state $|\psi\rangle$ can be thought of as a two-dimensional vector in spin space (a spinor wave function),

$$|\psi\rangle \longrightarrow \begin{pmatrix} \psi_+(\mathbf{r}) \\ \psi_-(\mathbf{r}) \end{pmatrix}, \qquad (12.6)$$

and the operator $\hat{A}$ as a 2×2 square matrix,

$$\hat{A} \longrightarrow \begin{pmatrix} A_{++}(\mathbf{r}, \mathbf{r}') & A_{+-}(\mathbf{r}, \mathbf{r}') \\ A_{-+}(\mathbf{r}, \mathbf{r}') & A_{--}(\mathbf{r}, \mathbf{r}') \end{pmatrix}. \qquad (12.7)$$

The action of $\hat{A}$ on $|\psi\rangle$ can be represented as

$$\langle\phi_{\mathbf{r},\sigma}|\hat{A}|\psi\rangle = \sum_{\sigma'=\pm} \int d\mathbf{r}'\langle\phi_{\mathbf{r},\sigma}|\hat{A}|\phi_{\mathbf{r}',\sigma'}\rangle\langle\phi_{\mathbf{r}',\sigma'}|\psi\rangle = \sum_{\sigma'=\pm} \int d\mathbf{r}' A_{\sigma\sigma'}(\mathbf{r}, \mathbf{r}')\psi_{\sigma'}(\mathbf{r}'), \qquad (12.8)$$

or, in matrix notation,

$$\hat{A}|\psi\rangle \longrightarrow \int d\mathbf{r}' \begin{pmatrix} A_{++}(\mathbf{r}, \mathbf{r}') & A_{+-}(\mathbf{r}, \mathbf{r}') \\ A_{-+}(\mathbf{r}, \mathbf{r}') & A_{--}(\mathbf{r}, \mathbf{r}') \end{pmatrix} \begin{pmatrix} \psi_+(\mathbf{r}') \\ \psi_-(\mathbf{r}') \end{pmatrix}. \qquad (12.9)$$

Of course, we could have chosen the continuous basis of common eigenstates $|\phi_{\mathbf{p},\sigma}\rangle$ of $\hat{\mathbf{p}}$, $\hat{\mathbf{S}}^2$, and $\hat{S}_z$ (or, in fact, any other basis of common eigenstates of a complete set of commuting observables that includes $\hat{\mathbf{S}}^2$ and $\hat{S}_z$).

12.2 Charged Particle in an Electromagnetic Field

In this section we review the treatment of a charged particle in an external electromagnetic (EM) field, first classically and then quantum mechanically. The EM field consists of an electric and a magnetic field, respectively $\mathbf{E}(\mathbf{r}, t)$ and $\mathbf{B}(\mathbf{r}, t)$, which satisfy Maxwell's equations (we use Gaussian units; in these units the Coulomb force between two charges q_1 and q_2 separated by a distance r is given by $q_1 q_2/r^2$)

$$1. \quad \mathbf{\nabla} \cdot \mathbf{E}(\mathbf{r}, t) = 4\pi\rho(\mathbf{r}, t) \,, \qquad 3. \quad \mathbf{\nabla} \times \mathbf{E}(\mathbf{r}, t) = -\frac{\partial \mathbf{B}(\mathbf{r}, t)}{c\,\partial t} \,, \qquad (12.10)$$

$$2. \quad \mathbf{\nabla} \cdot \mathbf{B}(\mathbf{r}, t) = 0 \,, \qquad 4. \quad \mathbf{\nabla} \times \mathbf{B}(\mathbf{r}, t) = \frac{4\pi}{c}\,\mathbf{j}(\mathbf{r}, t) + \frac{\partial \mathbf{E}(\mathbf{r}, t)}{c\,\partial t} \,. \qquad (12.11)$$

The fields can be described by a scalar and a vector potential, which we denote respectively as $U(\mathbf{r}, t)$ and $\mathbf{A}(\mathbf{r}, t)$. To see this, note that the first of Eqs. (12.11) implies that

$$\mathbf{B}(\mathbf{r}, t) = \mathbf{\nabla} \times \mathbf{A}(\mathbf{r}, t) \,, \qquad (12.12)$$

and equation 3 then gives

$$\mathbf{\nabla} \times \left[\mathbf{E}(\mathbf{r}, t) + \frac{\partial}{c\,\partial t}\mathbf{A}(\mathbf{r}, t) \right] = 0 \,, \qquad (12.13)$$

which says that the quantity in square brackets can be expressed as the gradient of a scalar function,

$$\mathbf{E}(\mathbf{r}, t) + \frac{\partial}{c\,\partial t}\mathbf{A}(\mathbf{r}, t) = -\mathbf{\nabla}U(\mathbf{r}, t) \,, \qquad (12.14)$$

where a minus sign is conventionally introduced. Note that the physical fields $\mathbf{E}$ and $\mathbf{B}$ can be described by different pairs of U and $\mathbf{A}$ fields (using different gauges). For, suppose that we have two pairs of vector and scalar potentials $\mathbf{A}, U$ and $\mathbf{A}', U'$ both of which give rise to the same EM fields. Then we have

$$\mathbf{\nabla} \times \mathbf{A}(\mathbf{r}, t) = \mathbf{B}(\mathbf{r}, t) = \mathbf{\nabla} \times \mathbf{A}'(\mathbf{r}, t) \implies \mathbf{\nabla} \times [\mathbf{A}'(\mathbf{r}, t) - \mathbf{A}(\mathbf{r}, t)] = 0 \,, \qquad (12.15)$$

and hence

$$\mathbf{A}'(\mathbf{r}, t) = \mathbf{A}(\mathbf{r}, t) + \mathbf{\nabla}\Lambda(\mathbf{r}, t) \,, \qquad (12.16)$$

where Λ is an arbitrary scalar function. Furthermore, we have

$$-\mathbf{\nabla}U(\mathbf{r}, t) - \frac{\partial}{c\,\partial t}\mathbf{A}(\mathbf{r}, t) = \mathbf{E}(\mathbf{r}, t) = -\mathbf{\nabla}U'(\mathbf{r}, t) - \frac{\partial}{c\,\partial t}\mathbf{A}'(\mathbf{r}, t) \,, \qquad (12.17)$$

from which we deduce that

$$\mathbf{\nabla}\left[U'(\mathbf{r}, t) - U(\mathbf{r}, t) + \frac{\partial}{c\,\partial t}\Lambda(\mathbf{r}, t) \right] = 0 \,. \qquad (12.18)$$

The scalar quantity in square brackets can only be a function of time, since its gradient with respect to the space variables vanishes identically. We conveniently choose this arbitrary function of time to be zero, so that

$$U'(\mathbf{r}, t) = U(\mathbf{r}, t) - \frac{\partial}{c\,\partial t}\Lambda(\mathbf{r}, t) \,. \qquad (12.19)$$

If $\mathbf{A}'$ and U' are inserted into Eqs. (12.12) and (12.14), we find the given EM fields. Thus different pairs of vector and scalar potentials are related to each other as in Eqs. (12.16) and (12.19). Of course, once a gauge $\mathbf{A}, U$ has been chosen, the remaining two of Maxwell's equations (1 and 4) can be expressed in terms of these potentials.

In an EM field a particle of charge q and mass m is subject to the Lorentz force,

$$\dot{\mathbf{p}} = q\left[\mathbf{E}(\mathbf{r}, t) + \frac{\dot{\mathbf{r}}}{c} \times \mathbf{B}(\mathbf{r}, t) \right] \,, \qquad (12.20)$$

where $\mathbf{r}$ and $\mathbf{p}$ are the position and momentum of the particle (in the non-relativistic limit $\mathbf{p} = m\dot{\mathbf{r}}$) and the dot on top of a symbol denotes the time derivative. The fields are to be evaluated at the position $\mathbf{r}$ of the particle, which changes in time. The Lorentz force can be expressed in terms of the scalar and vector potentials as (see Problem 4)

$$q\left[\mathbf{E}(\mathbf{r},t) + \frac{\dot{\mathbf{r}}}{c} \times \mathbf{B}(\mathbf{r},t)\right]_i = \left(-\frac{\partial}{\partial r_i} + \frac{d}{dt}\frac{\partial}{\partial \dot{r}_i}\right)\underbrace{\left[qU(\mathbf{r},t) - \frac{q}{c}\dot{\mathbf{r}}\cdot\mathbf{A}(\mathbf{r},t)\right]}_{\text{velocity-dependent potential}}, \tag{12.21}$$

where we note that the result of operating with $\partial/\partial\dot{r}_i$ on U vanishes since U depends only on $\mathbf{r}$.

The Lagrangian for a charged particle in an EM field now follows easily by taking $L = T - V$, where T is the kinetic energy and V is the velocity-dependent potential introduced above, namely

$$L = \frac{m\dot{\mathbf{r}}^2}{2} - qU + \frac{q}{c}\dot{\mathbf{r}}\cdot\mathbf{A}, \tag{12.22}$$

and the Euler–Lagrange equations yield

$$m\ddot{r}_i = -\frac{\partial}{\partial r_i}\left(qU - \frac{q}{c}\dot{\mathbf{r}}\cdot\mathbf{A}\right) + \frac{d}{dt}\frac{\partial}{\partial \dot{r}_i}\left(qU - \frac{q}{c}\dot{\mathbf{r}}\cdot\mathbf{A}\right) = q\left(\mathbf{E} + \frac{\dot{\mathbf{r}}}{c}\times\mathbf{B}\right)_i. \tag{12.23}$$

Note that under a different gauge the Lagrangian L' would involve the fields U' and $\mathbf{A}'$, and hence we would have

$$L' - L = \frac{q}{c}\left[\frac{\partial}{\partial t}\Lambda(\mathbf{r},t) + \dot{\mathbf{r}}\cdot\boldsymbol{\nabla}\Lambda(\mathbf{r},t)\right] = \frac{q}{c}\frac{d}{dt}\Lambda(\mathbf{r},t), \tag{12.24}$$

that is, L and L' would differ by the total time derivative of an arbitrary function. Consequently, when taking variations of the action integral relative to L', this total time derivative drops out, and the equations of motions derived from L and L' are the same. This is apparent from Eq. (12.23): it involves the $\mathbf{E}$ and $\mathbf{B}$ fields, which are independent of the gauge choice.

The Hamiltonian follows from

$$H(\mathbf{r},\mathbf{p}) = \mathbf{p}\cdot\dot{\mathbf{r}} - L(\mathbf{r},\dot{\mathbf{r}}) \qquad \text{with } \mathbf{p} = \frac{\partial L}{\partial \dot{\mathbf{r}}}, \tag{12.25}$$

where the conjugate momentum is

$$\mathbf{p} = m\dot{\mathbf{r}} + \frac{q}{c}\mathbf{A}, \tag{12.26}$$

and $\mathbf{p}$ differs from the mechanical momentum $m\dot{\mathbf{r}}$, associated with the velocity of the particle. Substituting $\dot{\mathbf{r}}$ by $[\mathbf{p} - (q/c)\mathbf{A}]/m$, we easily find

$$H = \frac{1}{2m}\left(\mathbf{p} - \frac{q}{c}\mathbf{A}\right)^2 + qU. \tag{12.27}$$

Hamilton's equations then leads to the equation of motion given in (12.23) (see Problem 5).

The transition to quantum mechanics is effected by interpreting $\mathbf{r}$ and $\mathbf{p}$ (the momentum conjugate to $\mathbf{r}$) as (hermitian) operators satisfying the canonical commutation relations $[\hat{r}_i, \hat{p}_j] = i\hbar\,\delta_{ij}$. The generally time-dependent quantum Hamiltonian for a charged particle then reads

$$\hat{H}(t) = \frac{1}{2m}\left[\hat{\mathbf{p}} - \frac{q}{c}\hat{\mathbf{A}}(\hat{\mathbf{r}},t)\right]^2 + q\,\hat{U}(\hat{\mathbf{r}},t). \tag{12.28}$$

This Hamiltonian governs the interactions of an atom with the EM field. We will see later how from it we can calculate, for example, the probabilities for photon absorption or emission in transitions between different atomic states. In the coordinate representation, the Schrödinger equation follows as

$$i\hbar\frac{\partial}{\partial t}\Psi(\mathbf{r},t) = \left[\frac{1}{2m}\left[-i\hbar\boldsymbol{\nabla} - \frac{q}{c}\mathbf{A}(\mathbf{r},t)\right]^2 + qU(\mathbf{r},t)\right]\Psi(\mathbf{r},t), \tag{12.29}$$

and we face the problem of gauge invariance. Under a gauge transformation $\mathbf{A} \longrightarrow \mathbf{A}' = \mathbf{A} + \nabla\Lambda$ and $U \longrightarrow U' = U - \partial\Lambda/(c\,\partial t)$ the physics cannot change, since the $\mathbf{E}$ and $\mathbf{B}$ fields (the physical fields) are left invariant under such a transformation. Therefore, the wave function must also change in order to leave the Schrödinger equation invariant in form; indeed, as shown in Problem 6, we must have

$$\Psi(\mathbf{r}, t) \longrightarrow \Psi'(\mathbf{r}, t) = e^{iq\Lambda(\mathbf{r},t)/(\hbar c)}\,\Psi(\mathbf{r}, t) \,. \tag{12.30}$$

Thus Ψ and Ψ' differ only by a phase factor (recall that Λ must be real since $\mathbf{E}$ and $\mathbf{B}$, and hence $\mathbf{A}$ and U, are real), and consequently represent the same physical state. Note that the phase factor is not a global one but, rather, depends on $\mathbf{r}$ and t.

12.3 Charged Particle with Spin in a Uniform Magnetic Field

Consider a particle of mass m, charge q, and spin $\mathbf{S}$ in a uniform and constant magnetic field $\mathbf{B}$. A vector potential associated with this magnetic field is given by

$$\mathbf{A}(\mathbf{r}) = \frac{1}{2}\,\mathbf{B} \times \mathbf{r} \,, \tag{12.31}$$

with $\nabla \times \mathbf{A} = \mathbf{B}$.[1] Ignoring the spin degrees of freedom for the time being, the quantum Hamiltonian relative to the orbital degrees of freedom (denoted by a the subscript o) reads in this case

$$\hat{H}_o = \frac{1}{2m}\left[\hat{\mathbf{p}} - \frac{q}{c}\hat{\mathbf{A}}(\hat{\mathbf{r}})\right]^2 = \frac{\hat{\mathbf{p}}^2}{2m} - \frac{q}{2mc}\left[\hat{\mathbf{p}}\cdot\hat{\mathbf{A}}(\hat{\mathbf{r}}) + \hat{\mathbf{A}}(\hat{\mathbf{r}})\cdot\hat{\mathbf{p}}\right] + \frac{q^2}{2mc^2}\hat{\mathbf{A}}^2(\hat{\mathbf{r}}) \,. \tag{12.32}$$

The term in square brackets on the right-hand side can be simplified by noting that

$$[\cdots] = \sum_i\left(\hat{p}_i\hat{A}_i + \hat{A}_i\hat{p}_i\right) = \sum_i\left([\hat{p}_i, \hat{A}_i] + 2\hat{A}_i\hat{p}_i\right) = \sum_i\left(-i\hbar\,\partial_i\hat{A}_i + 2\hat{A}_i\hat{p}_i\right) = -i\hbar\nabla\cdot\hat{\mathbf{A}} + 2\,\hat{\mathbf{A}}\cdot\hat{\mathbf{p}} = 2\,\hat{\mathbf{A}}\cdot\hat{\mathbf{p}} \,, \tag{12.33}$$

since the vector potential of Eq. (12.31) has vanishing divergence. The Hamiltonian H_o is then written as

$$\hat{H}_o = \frac{\hat{\mathbf{p}}^2}{2m} - \frac{q}{2mc}(\mathbf{B} \times \hat{\mathbf{r}})\cdot\hat{\mathbf{p}} + \frac{q^2}{8mc^2}(\mathbf{B} \times \hat{\mathbf{r}})^2 = \frac{\hat{\mathbf{p}}^2}{2m} - \frac{q}{2mc}\mathbf{B}\cdot\hat{\mathbf{L}} + \frac{q^2}{8mc^2}\mathbf{B}^2\,\hat{\mathbf{r}}_\perp^2 \,, \tag{12.34}$$

where $\hat{\mathbf{r}}_\perp$ is the position operator of the particle in the plane perpendicular to the magnetic field. The second term represents the interaction $-\mathbf{B}\cdot\hat{\boldsymbol{\mu}}_o$ of a magnetic dipole with a magnetic field, where the magnetic moment operator associated with the orbital motion of the particle is given (as in classical electrodynamics) by

$$\hat{\boldsymbol{\mu}}_o = \frac{q}{2mc}\hat{\mathbf{L}} \,. \tag{12.35}$$

In fact, experiments indicate that a charged particle with spin has, in addition to $\hat{\boldsymbol{\mu}}_o$, a magnetic dipole operator associated with the spin angular momentum. Consequently, the Hamiltonian describing such a system must include the interaction of the spin magnetic dipole with the $\mathbf{B}$ field,

$$\hat{H} = \hat{H}_o - \hat{\boldsymbol{\mu}}_s\cdot\mathbf{B} \,, \tag{12.36}$$

---

[1] Of course, because of gauge freedom the vector potential is not unique; for example, taking $\mathbf{B} = B\hat{\mathbf{z}}$, an equally valid choice for the vector potential is $\mathbf{A}' = (0, xB, 0)$, since $\nabla \times \mathbf{A}' = B\,\hat{\mathbf{z}}$. This vector potential and that given in Eq. (12.31) as $\mathbf{A} = (-yB/2, xB/2, 0)$ differ by the gradient of a scalar function Λ, with $\Lambda = xyB/2$ and $\mathbf{A}' = \mathbf{A} + \nabla\Lambda$.

where $\hat{\mu}_s$ is given by

$$\hat{\mu}_s = g \frac{q}{2mc} \hat{\mathbf{S}} . \tag{12.37}$$

Here, m is the particle mass and g is known as the gyromagnetic factor. The (non-dimensional) g-value depends on the particle: for an electron $q_e = -e$ and $g_e = 2 \times (1.001159652\ldots)$ while for a proton $q_p = e$ and $g_p = 2 \times (2.792847\ldots)$. Interestingly, even though the neutron is neutral and therefore has no orbital magnetic moment, it has nevertheless a spin magnetic moment, written as $[g_n e/(2m_p c)] \hat{\mathbf{S}}$, where $g_n = 2 \times (-1.913043\ldots)$ and m_p is the proton mass. The fact that both the proton and neutron have a quark substructure is the cause of the deviation of the gyromagnetic factor from the value 2 predicted by the Dirac equation for a structureless spin-1/2 particle, such as the electron (this is up to radiative corrections from quantum electrodynamics, the quantum theory of the EM field; for the electron, the leading radiative correction is given by $\alpha/(2\pi)$ with α being the fine structure constant). Since $m_p/m_e \approx 1836$, in atoms and molecules, effects associated with the proton and/or neutron magnetic moment ($\propto 1/m_p$) are much smaller than those due to the electron magnetic moment ($\propto 1/m_e$).

12.4 Problems

Problem 1   Coordinate Representation of the Operator $\hat{\mathbf{p}} \cdot \hat{\mathbf{S}}$

Obtain the coordinate representation of the operator $\hat{A} = \hat{\mathbf{p}} \cdot \hat{\mathbf{S}}$ (proportional to the helicity operator) and of $\hat{A} |\psi\rangle$, where $|\psi\rangle$ is a generic state.

Solution

In the basis $|\phi_{\mathbf{r},\sigma}\rangle$ the operator $\hat{A}$ has the representation

$$\underline{A}_{\sigma\sigma'}(\mathbf{r}, \mathbf{r}') = (\langle\phi_{\mathbf{r}}| \otimes \langle\sigma|) \hat{\mathbf{p}} \cdot \hat{\mathbf{S}} (|\phi_{\mathbf{r}'}\rangle \otimes |\sigma'\rangle) = \langle\phi_{\mathbf{r}}|\hat{\mathbf{p}}|\phi_{\mathbf{r}'}\rangle \cdot \langle\sigma|\hat{\mathbf{S}}|\sigma'\rangle = [-i\hbar\boldsymbol{\nabla}\delta(\mathbf{r}-\mathbf{r}')] \cdot \langle\sigma|\hat{\mathbf{S}}|\sigma'\rangle$$
$$= \sum_i \langle\sigma|\hat{S}_i|\sigma'\rangle \, [-i\hbar\partial_i \, \delta(\mathbf{r}-\mathbf{r}')] \; ,$$

which in terms of the Pauli matrices can be expressed as

$$\underline{A}_{\sigma\sigma'}(\mathbf{r}, \mathbf{r}') = \frac{\hbar}{2} \left[\begin{pmatrix} 0 & 1 \\ 1 & 0 \end{pmatrix}(-i\hbar\partial_x) + \begin{pmatrix} 0 & -i \\ i & 0 \end{pmatrix}(-i\hbar\partial_y) + \begin{pmatrix} 1 & 0 \\ 0 & -1 \end{pmatrix}(-i\hbar\partial_z) \right] \delta(\mathbf{r}-\mathbf{r}')$$
$$= \frac{\hbar}{2} \begin{pmatrix} -i\hbar\partial_z & -i\hbar(\partial_x - i\,\partial_y) \\ -i\hbar(\partial_x + i\,\partial_y) & i\hbar\partial_z \end{pmatrix} \delta(\mathbf{r}-\mathbf{r}') \; .$$

Given a state $|\psi\rangle$, the action of $\hat{A}$ on $|\psi\rangle$ can be represented as

$$\hat{A}|\psi\rangle \longrightarrow \int d\mathbf{r}' \frac{\hbar}{2} \begin{pmatrix} -i\hbar\partial_z & -i\hbar(\partial_x - i\partial_y) \\ -i\hbar(\partial_x + i\partial_y) & i\hbar\partial_z \end{pmatrix} \delta(\mathbf{r}-\mathbf{r}') \begin{pmatrix} \psi_+(\mathbf{r}') \\ \psi_-(\mathbf{r}') \end{pmatrix}$$
$$= \frac{\hbar}{2} \begin{pmatrix} -i\hbar\partial_z & -i\hbar(\partial_x - i\partial_y) \\ -i\hbar(\partial_x + i\partial_y) & i\hbar\partial_z \end{pmatrix} \begin{pmatrix} \psi_+(\mathbf{r}) \\ \psi_-(\mathbf{r}) \end{pmatrix}$$
$$= \frac{\hbar}{2} \begin{pmatrix} -i\hbar\partial_z\psi_+(\mathbf{r}) - i\hbar(\partial_x - i\partial_y)\psi_-(\mathbf{r}) \\ i\hbar\partial_z\psi_-(\mathbf{r}) - i\hbar(\partial_x + i\partial_y)\psi_+(\mathbf{r}) \end{pmatrix} \; .$$

Problem 2 Spinor Wave Function of the Hydrogen Atom: An Example

At time t, suppose that the electron in a hydrogen atom is in the state

$$|\psi\rangle = \frac{1}{\sqrt{5}}\,|\psi_{21-1,+}\rangle + \frac{2}{\sqrt{5}}\,|\psi_{100,-}\rangle\,.$$

Here, $|\psi_{nlm,\sigma}\rangle = |\psi_{nlm}\rangle \otimes |\sigma\rangle$, where $|\psi_{nlm}\rangle$ are the eigenstates of the (orbital) Hamiltonian H, the squared orbital angular momentum $\mathbf{L}^2$, and its projection L_z, with eigenvalues given by, respectively, ϵ_n, $l(l+1)\hbar^2$, and $m\hbar$, while $|\sigma\rangle = |\pm\rangle$ are the eigenstates of S_z with eigenvalues $\pm\hbar/2$.

1. What are the expectation values of the operators H, $\mathbf{L}^2$, L_y, L_z, $\mathbf{S}^2$, S_x, S_z, and parity P? List those operators for which $|\psi\rangle$ is an eigenstate.
2. Express the state $|\psi\rangle$ in the position representation as a spinor wave function,

$$\psi(\mathbf{r}) = \begin{pmatrix} \psi_+(\mathbf{r}) \\ \psi_-(\mathbf{r}) \end{pmatrix}.$$

Solution

Part 1

We obtain

$$\langle\psi|H|\psi\rangle = \langle\psi|\left(\frac{1}{\sqrt{5}}\,\epsilon_2\,|\psi_{21-1,+}\rangle + \frac{2}{\sqrt{5}}\,\epsilon_1\,|\psi_{100,-}\rangle\right) = \frac{1}{5}(\epsilon_2 + 4\,\epsilon_1)\,,$$

where we have used the fact that the states $|\psi_{100,-}\rangle$ and $|\psi_{21-1,+}\rangle$ are orthonormal and eigenstates of H with, respectively, eigenvalues ϵ_1 and ϵ_2. Similarly, we find

$$\langle\psi|\mathbf{L}^2|\psi\rangle = \langle\psi|\left(\frac{2}{\sqrt{5}}\,\hbar^2\,|\psi_{21-1,+}\rangle\right) = \frac{2}{5}\,\hbar^2\,,$$

$$\langle\psi|L_z|\psi\rangle = \langle\psi|\left(-\frac{1}{\sqrt{5}}\,\hbar\,|\psi_{21-1,+}\rangle\right) = -\frac{1}{5}\,\hbar\,,$$

$$\langle\psi|\mathbf{S}^2|\psi\rangle = \langle\psi|\,\frac{3}{4}\,\hbar^2\underbrace{\left(\frac{1}{\sqrt{5}}\,|\psi_{21-1,+}\rangle + \frac{2}{\sqrt{5}}\,|\psi_{100,-}\rangle\right)}_{|\psi\rangle} = \frac{3}{4}\,\hbar^2\,,$$

$$\langle\psi|S_z|\psi\rangle = \langle\psi|\,\frac{1}{2}\,\hbar\left(\frac{1}{\sqrt{5}}\,|\psi_{21-1,+}\rangle - \frac{2}{\sqrt{5}}\,|\psi_{100,-}\rangle\right) = -\frac{3}{10}\,\hbar\,,$$

$$\langle\psi|P|\psi\rangle = \langle\psi|\left(-\frac{1}{\sqrt{5}}\,|\psi_{21-1,+}\rangle + \frac{2}{\sqrt{5}}\,|\psi_{100,-}\rangle\right) = \frac{3}{5}\,.$$

The expectation values of L_y and S_x vanish see Problem 10 in Chapter 11. Explicitly, we have

$$\underbrace{\frac{L_+ - L_-}{2i}}_{L_y}\underbrace{\left(\frac{1}{\sqrt{5}}\,|\psi_{21-1,+}\rangle + \frac{2}{\sqrt{5}}\,|\psi_{100,-}\rangle\right)}_{|\psi\rangle} = \frac{1}{2i}\,L_+\left(\frac{1}{\sqrt{5}}\,|\psi_{21-1,+}\rangle\right) = \frac{\hbar}{2i}\,\sqrt{\frac{2}{5}}\,|\psi_{210,+}\rangle\,,$$

using the properties of the raising and lowering operators. The resulting state is orthogonal to $|\psi\rangle$, and hence $\langle\psi|L_y|\psi\rangle = 0$. A similar argument shows that $\langle\psi|S_x|\psi\rangle = 0$. The only operator for which $|\psi\rangle$ is an eigenstate is $\mathbf{S}^2$.

Part 2

In the coordinate representation we have

$$\langle\phi_{\mathbf{r}}|\psi\rangle = \frac{1}{\sqrt{5}}\,\psi_{21-1}(\mathbf{r})\otimes|+\rangle + \frac{2}{\sqrt{5}}\,\psi_{100}(\mathbf{r})\otimes|-\rangle \longrightarrow \frac{1}{\sqrt{5}}\begin{pmatrix}\psi_{21-1}(\mathbf{r})\\2\psi_{100}(\mathbf{r})\end{pmatrix} .$$

Problem 3   The Lorentz Force in Terms of Scalar and Vector Potentials

Express the combination of electric and magnetic fields in the Lorentz force in terms of scalar and vector potentials, that is,

$$\mathbf{E}(\mathbf{r},t) + \frac{\dot{\mathbf{r}}}{c}\times\mathbf{B}(\mathbf{r},t) = \left(-\boldsymbol{\nabla} + \frac{d}{dt}\frac{\partial}{\partial\dot{\mathbf{r}}}\right)\left[U(\mathbf{r},t) - \frac{1}{c}\dot{\mathbf{r}}\cdot\mathbf{A}(\mathbf{r},t)\right] .$$

Solution

Consider, say, the x component of the term $\dot{\mathbf{r}}\times\mathbf{B}$,

$$(\dot{\mathbf{r}}\times\mathbf{B})_x = \dot{r}_y(\boldsymbol{\nabla}\times\mathbf{A})_z - \dot{r}_z(\boldsymbol{\nabla}\times\mathbf{A})_y = \dot{r}_y(\partial_x A_y - \partial_y A_x) - \dot{r}_z(\partial_z A_x - \partial_x A_z)$$

$$= \underbrace{\dot{r}_x\partial_x A_x + \dot{r}_y\partial_x A_y + \dot{r}_z\partial_x A_z}_{\partial_x(\dot{\mathbf{r}}\cdot\mathbf{A})} - \dot{r}_x\partial_x A_x - \dot{r}_y\partial_y A_x - \dot{r}_z\partial_z A_x ,$$

where in the second line we have added and subtracted the term $\dot{r}_x\partial_x A_x$. Next, we observe that if we take the total time derivative of A_x, we have by the chain rule

$$\frac{dA_x}{dt} = \frac{\partial A_x}{\partial t} + \dot{r}_x\,\partial_x A_x + \dot{r}_y\partial_y A_x + \dot{r}_z\partial_z A_x ,$$

since A_x depends on the position $\mathbf{r}(t)$ of the particle, a function of time. Thus we find that

$$(\dot{\mathbf{r}}\times\mathbf{B})_x = \partial_x(\dot{\mathbf{r}}\cdot\mathbf{A}) - \frac{dA_x}{dt} + \frac{\partial A_x}{\partial t} ,$$

where ∂_x acts only on the $\mathbf{r}$ dependence contained in the vector potential. Similar relations are obtained for the y and z components, so that

$$\mathbf{E} + \frac{\dot{\mathbf{r}}}{c}\times\mathbf{B} = -\boldsymbol{\nabla}U - \frac{1}{c}\frac{\partial\mathbf{A}}{\partial t} + \frac{1}{c}\boldsymbol{\nabla}(\dot{\mathbf{r}}\cdot\mathbf{A}) - \frac{1}{c}\frac{d\mathbf{A}}{dt} + \frac{1}{c}\frac{\partial\mathbf{A}}{\partial t} = -\boldsymbol{\nabla}U + \frac{1}{c}\boldsymbol{\nabla}(\dot{\mathbf{r}}\cdot\mathbf{A}) - \frac{1}{c}\frac{d\mathbf{A}}{dt} .$$

Lastly, we can write

$$A_i = \frac{\partial}{\partial\dot{r}_i}(\dot{\mathbf{r}}\cdot\mathbf{A}) \implies \mathbf{A} = \frac{\partial}{\partial\dot{\mathbf{r}}}(\dot{\mathbf{r}}\cdot\mathbf{A}) ,$$

since $\mathbf{A}$ depends only on $\mathbf{r}$, and therefore

$$\mathbf{E} + \frac{\dot{\mathbf{r}}}{c}\times\mathbf{B} = -\boldsymbol{\nabla}U + \frac{1}{c}\boldsymbol{\nabla}(\dot{\mathbf{r}}\cdot\mathbf{A}) - \frac{1}{c}\frac{d}{dt}\frac{\partial}{\partial\dot{\mathbf{r}}}(\dot{\mathbf{r}}\cdot\mathbf{A}) = -\boldsymbol{\nabla}\left(U - \frac{1}{c}\dot{\mathbf{r}}\cdot\mathbf{A}\right) + \frac{d}{dt}\frac{\partial}{\partial\dot{\mathbf{r}}}\left(U - \frac{1}{c}\dot{\mathbf{r}}\cdot\mathbf{A}\right) ,$$

i.e., this is the Lorentz force (up to the charge q of the particle) in terms of a velocity-dependent potential. Note that in the last expression we have used the fact that $\partial U/\partial\dot{\mathbf{r}} = 0$.

Problem 4   Lagrange and Hamilton Equations of Motion for Charged Particle in EM Field

Obtain the (classical) equations of motion for a particle of charge q in an electromagnetic field, starting from either the Lagrangian or the Hamiltonian.

Solution

Lagrange's equations of motion are given by

$$\frac{d}{dt}\frac{\partial L}{\partial \dot{r}_i} - \frac{\partial L}{\partial r_i} = 0 \,,$$

where

$$\frac{\partial L}{\partial r_i} = -\frac{\partial}{\partial r_i}\left(q\,U - \frac{q}{c}\,\dot{\mathbf{r}}\cdot\mathbf{A}\right)\,, \qquad \frac{d}{dt}\frac{\partial L}{\partial \dot{r}_i} = m\ddot{r}_i + \frac{d}{dt}\frac{\partial}{\partial \dot{r}_i}\left(\frac{q}{c}\,\dot{\mathbf{r}}\cdot\mathbf{A}\right) = m\ddot{r}_i - \frac{d}{dt}\frac{\partial}{\partial \dot{r}_i}\left(q\,U - \frac{q}{c}\,\dot{\mathbf{r}}\cdot\mathbf{A}\right)\,,$$

and we have used the fact that $\partial U/\partial \dot{r}_i = 0$. We obtain (see Problem 3)

$$m\ddot{r}_i = q\left(\mathbf{E} + \frac{\dot{\mathbf{r}}}{c}\times\mathbf{B}\right)_i \,.$$

Hamilton's equations read

$$\dot{r}_i = \frac{\partial H}{\partial p_i}\,, \qquad \dot{p}_i = -\frac{\partial H}{\partial r_i}\,,$$

where

$$\dot{r}_i = \frac{p_i - (q/c)A_i}{m}\,, \qquad \dot{p}_i = \frac{q}{c}\frac{\partial}{\partial r_i}\dot{\mathbf{r}}\cdot\mathbf{A} - q\frac{\partial U}{\partial r_i}\,.$$

Differentiating both sides of the first equation with respect to time yields

$$m\ddot{r}_i = \dot{p}_i - \frac{q}{c}\left(\dot{\mathbf{r}}\cdot\boldsymbol{\nabla}A_i + \frac{\partial A_i}{\partial t}\right) = \underbrace{\frac{q}{c}\frac{\partial}{\partial r_i}\dot{\mathbf{r}}\cdot\mathbf{A} - q\frac{\partial U}{\partial r_i}}_{\dot{p}_i} - \frac{q}{c}\left(\dot{\mathbf{r}}\cdot\boldsymbol{\nabla}A_i + \frac{\partial A_i}{\partial t}\right)$$

$$= q\underbrace{\left(-\frac{\partial U}{\partial r_i} - \frac{\partial A_i}{c\,\partial t}\right)}_{E_i} + \frac{q}{c}\underbrace{\left(\frac{\partial}{\partial r_i}\dot{\mathbf{r}}\cdot\mathbf{A} - \dot{\mathbf{r}}\cdot\boldsymbol{\nabla}A_i\right)}_{[\dot{\mathbf{r}}\times(\boldsymbol{\nabla}\times\mathbf{A})]_i} = q\left(\mathbf{E} + \frac{\dot{\mathbf{r}}}{c}\times\mathbf{B}\right)_i \,,$$

where we have used the identity

$$[\mathbf{X}\times(\mathbf{Y}\times\mathbf{Z})]_i = \sum_j(X_jY_iZ_j - X_jY_jZ_i)\,,$$

with $\mathbf{X}=\dot{\mathbf{r}}$, $\mathbf{Y}=\boldsymbol{\nabla}$, and $\mathbf{Z}=\mathbf{A}$.

Problem 5   Schrödinger Equation for Charged Particle in EM Field and Gauge Invariance

Show that the Schrödinger equation for a particle of charge q in an electromagnetic field is invariant under a gauge transformation of the scalar and vector potentials

$$\Psi(\mathbf{r},t) \longrightarrow \Psi'(\mathbf{r},t) = e^{iq\Lambda(\mathbf{r},t)/(\hbar c)}\,\Psi(\mathbf{r},t)\,.$$

Solution

The following identities are useful:

$$
i\hbar \frac{\partial}{\partial t} \underbrace{e^{-iq\Lambda/(\hbar c)} \Psi'}_{\Psi} = e^{-iq\Lambda/(\hbar c)} \left(i\hbar \frac{\partial}{\partial t} + \frac{q}{c} \frac{\partial \Lambda}{\partial t} \right) \Psi',
$$

$$
- i\hbar \boldsymbol{\nabla} \underbrace{e^{-iq\Lambda/(\hbar c)} \Psi'}_{\Psi} = e^{-iq\Lambda/(\hbar c)} \left(-i\hbar \boldsymbol{\nabla} - \frac{q}{c} \boldsymbol{\nabla} \Lambda \right) \Psi',
$$

and

$$
\begin{aligned}
\left(-i\hbar \boldsymbol{\nabla} - \frac{q}{c} \mathbf{A} \right)^2 \underbrace{e^{-iq\Lambda/(\hbar c)} \Psi'}_{\Psi} &= \sum_i \left(-i\hbar \boldsymbol{\nabla} - \frac{q}{c} \mathbf{A} \right)_i \left(-i\hbar \boldsymbol{\nabla} - \frac{q}{c} \mathbf{A} \right)_i e^{-iq\Lambda/(\hbar c)} \Psi' \\
&= \sum_i \left(-i\hbar \boldsymbol{\nabla} - \frac{q}{c} \mathbf{A} \right)_i e^{-iq\Lambda/(\hbar c)} \underbrace{\left(-i\hbar \boldsymbol{\nabla} - \frac{q}{c} \boldsymbol{\nabla} \Lambda - \frac{q}{c} \mathbf{A} \right)_i \Psi'}_{\Psi_i''} \\
&= e^{-iq\Lambda/(\hbar c)} \sum_i \left(-i\hbar \boldsymbol{\nabla} - \frac{q}{c} \boldsymbol{\nabla} \Lambda - \frac{q}{c} \mathbf{A} \right)_i \Psi_i'' \\
&= e^{-iq\Lambda/(\hbar c)} \left(-i\hbar \boldsymbol{\nabla} \underbrace{- \frac{q}{c} \boldsymbol{\nabla} \Lambda - \frac{q}{c} \mathbf{A}}_{-(q/c)\mathbf{A}'} \right)^2 \Psi' = e^{-iq\Lambda/(\hbar c)} \left(-i\hbar \boldsymbol{\nabla} - \frac{q}{c} \mathbf{A}' \right)^2 \Psi'.
\end{aligned}
$$

Then, the Schrödinger equation

$$
i\hbar \frac{\partial}{\partial t} \Psi = \left[\frac{1}{2m} \left(-i\hbar \boldsymbol{\nabla} - \frac{q}{c} \mathbf{A} \right)^2 + qU \right] \Psi
$$

in terms of Ψ' reads

$$
e^{-iq\Lambda/(\hbar c)} \left(i\hbar \frac{\partial}{\partial t} + \frac{q}{c} \frac{\partial \Lambda}{\partial t} \right) \Psi' = e^{-iq\Lambda/(\hbar c)} \left[\frac{1}{2m} \left(-i\hbar \boldsymbol{\nabla} - \frac{q}{c} \mathbf{A}' \right)^2 + qU \right] \Psi',
$$

or, after removing the common phase factor on the left- and right-hand sides of the above equation,

$$
i\hbar \frac{\partial}{\partial t} \Psi' = \left[\frac{1}{2m} \left(-i\hbar \boldsymbol{\nabla} - \frac{q}{c} \mathbf{A}' \right)^2 + \underbrace{q U - \frac{q}{c} \frac{\partial \Lambda}{\partial t}}_{q U'} \right] \Psi',
$$

and so Ψ' satisfies the same Schrödinger equation as (12.29), but with $\mathbf{A}$ and U replaced by the corresponding primed quantities.

We could turn the argument around and *demand* gauge invariance in the presence of fields, which requires making the replacements $\mathbf{p} \longrightarrow \mathbf{p} - (q/c)\mathbf{A}$ and $H \longrightarrow H + qU$ in the Hamiltonian without fields; in this sense, then, gauge invariance dictates how particles interact with fields.

Problem 6 Spin-1/2 Particle in a Time-dependent Magnetic Field

Consider an electron at a fixed position in an oscillating magnetic field

$$
\mathbf{B}(t) = B_0 \cos(\omega t) \, \hat{\mathbf{z}},
$$

where B_0 and ω are constants and the caret denotes a unit vector

1. At time $t = 0$ the electron is in the spin state with respect to the x-axis having eigenvalue $\hbar/2$. Determine the spin state of the electron at later times.
2. Obtain the probability of obtaining $-\hbar/2$ if one measures S_x.
3. What is the minimum field B_0 required to force a complete flip in S_x?

Solution

Part 1

Ignoring orbital degrees of freedom, the Hamiltonian reads

$$H(t) = -\boldsymbol{\mu} \cdot \mathbf{B}(t) = g \frac{e B_0}{2mc} S_z \cos(\omega t) \, ,$$

where m and $-e$ are the mass and charge of the electron and g is its gyromagnetic factor. The time-dependent Schrödinger equation in the spin space of the electron is

$$i\hbar \frac{d}{dt} |\psi(t)\rangle = H(t) |\psi(t)\rangle \, , \qquad |\psi(t)\rangle \longrightarrow \begin{pmatrix} c_+(t) \\ c_-(t) \end{pmatrix} \, ,$$

where the second step gives the representation of the state $|\psi(t)\rangle$ in the basis consisting of eigenstates of S_z. Since $H(t)$ is diagonal in this basis, it easily follows that

$$i \begin{pmatrix} \dot{c}_+(t) \\ \dot{c}_-(t) \end{pmatrix} = g \frac{e B_0}{4mc} \cos(\omega t) \begin{pmatrix} c_+(t) \\ -c_-(t) \end{pmatrix} \, ,$$

with the initial condition

$$|\psi(0)\rangle = |+\rangle_x \implies \begin{pmatrix} c_+(0) \\ c_-(0) \end{pmatrix} = \frac{1}{\sqrt{2}} \begin{pmatrix} 1 \\ 1 \end{pmatrix} \, ,$$

that is, an eigenstate of S_x with eigenvalue $\hbar/2$. It is convenient to define the angular frequency

$$\omega_0 = g \frac{e B_0}{4mc} \, ,$$

so that

$$\dot{c}_+(t) = -i\omega_0 \cos(\omega t) \, c_+(t) \, , \qquad \dot{c}_-(t) = i\omega_0 \cos(\omega t) \, c_-(t) \, ,$$

which have the solutions

$$c_\pm(t) = c_\pm(0) \, e^{\mp i(\omega_0/\omega) \sin(\omega t)} = \frac{1}{\sqrt{2}} \, e^{\mp i(\omega_0/\omega) \sin(\omega t)} \, .$$

The state $|\psi(t)\rangle$ is thus uniquely determined (up to an overall phase factor, of course).

Part 2

The eigenstate of S_x with eigenvalue $-\hbar/2$ is given by

$$|-\rangle_x = \frac{1}{\sqrt{2}} \begin{pmatrix} 1 \\ -1 \end{pmatrix} \, ,$$

and the probability of measuring $-\hbar/2$ is

$$p_{x,-}(t) = |\,_x\langle -|\psi(t)\rangle|^2 \, ,$$

where

$$_x\langle-|\psi(t)\rangle = \frac{1}{\sqrt{2}} \begin{pmatrix} 1 & -1 \end{pmatrix} \begin{pmatrix} c_+(t) \\ c_-(t) \end{pmatrix} = \frac{1}{\sqrt{2}}[c_+(t) - c_-(t)] = \frac{1}{2} \left[e^{-i(\omega_0/\omega)\sin(\omega t)} - e^{i(\omega_0/\omega)\sin(\omega t)} \right] ,$$

or

$$_x\langle-|\psi(t)\rangle = -i\,\sin[(\omega_0/\omega)\sin(\omega t)] ,$$

yielding

$$p_{x,-}(t) = \sin^2\left[(\omega_0/\omega)\sin(\omega t)\right] .$$

Part 3

To achieve a complete flip of S_x, we must have $p_{x,-}(t^*) = 1$ at some time t^*, that is,

$$\frac{\omega_0}{\omega}\sin(\omega t^*) = (2n+1)\frac{\pi}{2} ,$$

with n some integer. The minimum ω_0 corresponds to $n = 0$ and is given by

$$\omega_0 = \frac{\pi}{2}\frac{\omega}{\sin(\omega t^*)} \implies B_0 = 2\pi\frac{mc}{ge}\frac{\omega}{\sin(\omega t^*)} .$$

Problem 7 Scattering of Spin-1/2 Particle in Spin-dependent δ-Function Potential

Consider a spin-1/2 particle of mass m in one dimension subject to a spin-dependent potential given by

$$v(x) = v_0\,\underline{\sigma}_x\,\delta(x) , \qquad\qquad v_0 > 0 .$$

Here $\underline{\sigma}_x$ is the Pauli matrix

$$\underline{\sigma}_x = \begin{pmatrix} 0 & 1 \\ 1 & 0 \end{pmatrix} .$$

The particle approaches the interaction region from the far left, $x \longrightarrow -\infty$, and has energy $E > 0$ and spin projection $+\hbar/2$ in the $\hat{z}$-direction. Calculate the probability for the particle to have spin projection $-\hbar/2$ relative to the $\hat{z}$-direction after it has traversed the interaction region and is at the far right, $x \longrightarrow +\infty$ (that is, calculate the transmission coefficient for the spin-flip transition).

Solution

It is convenient to work in the basis in which $\underline{\sigma}_x$ is diagonal, that is, a basis with states $|\pm\rangle_x$ having spin projections $\pm\hbar/2$ along the $\hat{x}$-direction, given by

$$|\pm\rangle_x = \frac{1}{\sqrt{2}}\left(|+\rangle_z \pm |-\rangle_z\right) ,$$

where $|\pm\rangle_z$ are the states with spin projections $\pm\hbar/2$ along the $\hat{z}$-direction. The spinor wave function of energy $E = (\hbar k)^2/(2m)$ is written as

$$\psi_k(x) = \psi_{k+}(x)\,|+\rangle_x + \psi_{k-}(x)\,|-\rangle_x ,$$

and, since $\underline{\sigma}_x |\pm\rangle_x = \pm |\pm\rangle_x$, the Schrödinger equation reads in this basis

$$\left[-\frac{d^2}{dx^2} + v_0 \underline{\sigma}_x \delta(x) \right] \psi_k(x) = k^2 \psi_k(x) \implies \psi''_{k\pm}(x) = \left(\pm v_0 \delta(x) - k^2 \right) \psi_{k\pm}(x) ,$$

that is, it decouples. The solutions with the particle incoming from the left with spin projections $\pm\hbar/2$ along the $\hat{x}$-direction are, respectively,

$$\begin{aligned}
\psi_{k\pm}(x) &= A_\pm e^{ikx} + B_\pm e^{-ikx} & x &< 0 \\
&= C_\pm e^{ikx} & x &> 0 .
\end{aligned}$$

Imposing the boundary conditions at the origin,

$$\psi_{k\pm}(0^-) = \psi_{k\pm}(0^+), \qquad \psi'_{k\pm}(0^+) - \psi'_{k\pm}(0^-) = \pm v_0 \psi_{k\pm}(0) ,$$

leads to

$$A_\pm + B_\pm = C_\pm , \qquad ikC_\pm - ik(A_\pm - B_\pm) = \pm v_0 C_\pm ,$$

which can be easily solved to yield

$$\frac{C_\pm}{A_\pm} = \frac{1}{1 \pm i\alpha} , \qquad \frac{B_\pm}{A_\pm} = \mp \frac{i\alpha}{1 \pm i\alpha} ,$$

where $\alpha = v_0/(2k)$. Expanding the spinor wave function on the basis of states with spin projections $\pm\hbar/2$ along the $\hat{z}$-direction yields

$$\psi_k(x) = \psi_{k+}(x) \frac{|+\rangle_z + |-\rangle_z}{\sqrt{2}} + \psi_{k-}(x) \frac{|+\rangle_z - |-\rangle_z}{\sqrt{2}} = \underbrace{\frac{\psi_{k+}(x) + \psi_{k-}(x)}{\sqrt{2}}}_{\overline{\psi}_{k+}(x)} |+\rangle_z + \underbrace{\frac{\psi_{k+}(x) - \psi_{k-}(x)}{\sqrt{2}}}_{\overline{\psi}_{k-}(x)} |-\rangle_z ,$$

where

$$\begin{aligned}
\overline{\psi}_{k+}(x) &= \frac{A_+ + A_-}{\sqrt{2}} e^{ikx} + \frac{B_+ + B_-}{\sqrt{2}} e^{-ikx} & x &< 0 \\
&= \frac{C_+ + C_-}{\sqrt{2}} e^{ikx} & x &> 0
\end{aligned}$$

and

$$\begin{aligned}
\overline{\psi}_{k-}(x) &= \frac{A_+ - A_-}{\sqrt{2}} e^{ikx} + \frac{B_+ - B_-}{\sqrt{2}} e^{-ikx} & x &< 0 \\
&= \frac{C_+ - C_-}{\sqrt{2}} e^{ikx} & x &> 0 .
\end{aligned}$$

According to the initial condition the particle is incident from the left, with spin up along $\hat{z}$, which corresponds to the choice $A_+ = A_-$,[2] so that

$$\begin{aligned}
\overline{\psi}_{k+}(x) &= \sqrt{2} A_+ e^{ikx} + \frac{B_+ + B_-}{\sqrt{2}} e^{-ikx} & x &< 0 \\
&= \frac{C_+ + C_-}{\sqrt{2}} e^{ikx} & x &> 0
\end{aligned}$$

---

[2] By contrast, if the initial condition had had the particle with spin down along $\hat{z}$ then the choice would have been $A_+ = -A_-$.

and

$$\overline{\psi}_{k-}(x) = \frac{B_+ - B_-}{\sqrt{2}} \, e^{-ikx} \qquad\qquad x < 0$$

$$= \frac{C_+ - C_-}{\sqrt{2}} \, e^{ikx} \qquad\qquad x > 0 \, .$$

There are reflected and transmitted waves with spin up as well as reflected and transmitted waves with spin down, but only a spin-up incident wave. Since the interaction does not commute with $\underline{\sigma}_z$, the spin of the particle can be flipped. This is in contrast with the situation in which the particle has spin up or down along the $\hat{x}$-direction, in which case no spin flip can occur.

The transmission coefficient for the spin-flip transition (from up to down) is obtained from the incident current for spin up and the transmitted current for spin down, which are respectively

$$j_{i+} = 2\,\frac{\hbar}{m}\,k\,|A_+|^2 \, , \qquad j_{t-} = \frac{\hbar}{2m}\,k\,|C_+ - C_-|^2 \, ,$$

as

$$T_{\text{flip}} = \frac{|j_{t-}|}{|j_{i+}|} = \frac{|C_+ - C_-|^2}{4|A_+|^2} = \frac{\alpha^2}{(1 + \alpha^2)^2} \, .$$

The reflection coefficient for the spin-flip transition follows similarly:

$$R_{\text{flip}} = \frac{|B_+ - B_-|^2}{4|A_+|^2} = \frac{\alpha^2}{(1 + \alpha^2)^2} \, ,$$

whereas the transmission and reflection coefficients for the non-spin-flip case are given by

$$T_{\text{no-flip}} = \frac{|C_+ + C_-|^2}{4|A_+|^2} = \frac{1}{(1 + \alpha^2)^2} \, , \qquad R_{\text{no-flip}} = \frac{|B_+ + B_-|^2}{4|A_+|^2} = \frac{\alpha^4}{(1 + \alpha^2)^2} \, .$$

Conservation of probability requires

$$T_{\text{flip}} + R_{\text{flip}} + T_{\text{no-flip}} + R_{\text{no-flip}} = 1 \, ,$$

which is easily seen to be satisfied.

Problem 8 Pauli Hamiltonian for Electron in EM Field

The Hamiltonian for an electron of mass m, charge $-e$, and spin $\hbar\sigma/2$, placed in an electromagnetic field described by the vector and scalar potentials $\mathbf{A}(\mathbf{r}, t)$ and $U(\mathbf{r}, t)$, is written as

$$H = \frac{1}{2m} \left[\mathbf{p} + \frac{e}{c}\mathbf{A}(\mathbf{r}, t) \right]^2 - eU(\mathbf{r}, t) + \frac{e\hbar}{2mc}\,\sigma \cdot \mathbf{B}(\mathbf{r}, t) \, ,$$

where the last term represents the interaction between the spin magnetic moment and the magnetic field (here, we have taken the gyromagnetic factor to be exactly 2, ignoring radiative corrections). Show, using the properties of the Pauli matrices, that this Hamiltonian can be written in the form

$$H = \frac{1}{2m} \left[\sigma \cdot \mathbf{p} + \frac{e}{c}\sigma \cdot \mathbf{A}(\mathbf{r}, t) \right]^2 - eU(\mathbf{r}, t) \, ,$$

known as the Pauli Hamiltonian.

Solution

Use the property

$$\sigma \cdot \mathbf{A}\, \sigma \cdot \mathbf{B} = \mathbf{A} \cdot \mathbf{B} + i\, \sigma \cdot (\mathbf{A} \times \mathbf{B})\,,$$

to obtain

$$\sigma \cdot \left(\mathbf{p} + \frac{e}{c}\mathbf{A}\right) \sigma \cdot \left(\mathbf{p} + \frac{e}{c}\mathbf{A}\right) = \left(\mathbf{p} + \frac{e}{c}\mathbf{A}\right)^2 + i\,\sigma \left[\left(\mathbf{p} + \frac{e}{c}\mathbf{A}\right) \times \left(\mathbf{p} + \frac{e}{c}\mathbf{A}\right)\right]\,.$$

The cross product can be written as follows:

$$\begin{aligned}
\left[\left(\mathbf{p} + \frac{e}{c}\mathbf{A}\right) \times \left(\mathbf{p} + \frac{e}{c}\mathbf{A}\right)\right]_i \\
= \frac{e}{c}(\mathbf{p} \times \mathbf{A} + \mathbf{A} \times \mathbf{p})_i = \frac{e}{c}\sum_{jk}(\epsilon_{ijk}\,p_j A_k + \epsilon_{ijk}\,A_j p_k) = \frac{e}{c}\sum_{jk}(\epsilon_{ijk}\,p_j A_k + \epsilon_{ikj}\,A_k p_j) \\
= \frac{e}{c}\sum_{jk}\epsilon_{ijk}(p_j A_k - A_k p_j) = \frac{e}{c}\sum_{jk}\epsilon_{ijk}\,[p_j\,,\,A_k] = -i\frac{e\hbar}{c}\sum_{jk}\epsilon_{ijk}\,\partial_j A_k = -i\,\frac{e\hbar}{c}\,B_i\,,
\end{aligned}$$

which then leads to

$$H = \frac{1}{2m}\left(\mathbf{p} + \frac{e}{c}\mathbf{A}\right)^2 + \frac{e\hbar}{2mc}\,\sigma \cdot \mathbf{B}\,.$$

Problem 9   A Simplified Analysis of the Stern–Gerlach Experiment

In the Stern–Gerlach experiment a beam of neutral atoms is sent into a region of space where a constant but non-uniform magnetic field $\mathbf{B}(\mathbf{r})$ is present, having the form

$$\mathbf{B}(\mathbf{r}) = \mathbf{B}_0 + \mathbf{B}_1(\mathbf{r})\,,$$

where the non-uniform term $\mathbf{B}_1(\mathbf{r})$ is much weaker than $\mathbf{B}_0 = B_0\,\hat{\mathbf{z}}$. The atom enters the region where the magnetic fields are present at time $t = 0$.

1. Suppose that the atom has mass M and total angular momentum $\mathbf{J}$; ignore its internal structure. The eigenvalues of $\mathbf{J}^2$ and J_z are, respectively, $j(j+1)\hbar^2$ and $m\hbar$. Associated with the total angular momentum is a magnetic moment operator, given by

$$\mu = \frac{\mu_0}{j\hbar}\,\mathbf{J}\,,$$

where μ_0 is a constant. Show that the expectation values of the position and momentum operators of the atom satisfy Ehrenfest's relations

$$\frac{d}{dt}\langle\psi(t)|\mathbf{r}|\psi(t)\rangle = \frac{\langle\psi(t)|\mathbf{p}|\psi(t)\rangle}{M}\,, \qquad \frac{d}{dt}\langle\psi(t)|\mathbf{p}|\psi(t)\rangle = \frac{\mu_0}{j\hbar}\,\langle\psi(t)|\mathbf{\nabla}[\mathbf{J}\cdot\mathbf{B}_1(\mathbf{r})]|\psi(t)\rangle\,,$$

where $|\psi(t)\rangle$ is the state of the atom at time t.

2. At time $t = 0$, the atom is in a state described by the following superposition of simultaneous eigenstates of the momentum operator and J_z,

$$|\psi(0)\rangle = \sum_{m=-j}^{j}\int d\mathbf{p}\, g_m(\mathbf{p})\,|\phi_{\mathbf{p}}\rangle \otimes |m\rangle\,,$$

where $g_m(\mathbf{p})$ is the momentum-space wave function. Suppose the subsequent time evolution of the state is governed by the Hamiltonian H_0, where

$$H_0 = \frac{\mathbf{p}^2}{2M} - \frac{\mu_0 B_0}{j\hbar} J_z \,,$$

that is, the contribution from the much weaker $\mathbf{B}_1(\mathbf{r})$ field is ignored. Evaluate the expectation values of the three components of the angular momentum, $\langle \psi(t)|J_i|\psi(t)\rangle$.

3. Ignoring rapidly oscillating time-dependent terms, show that

$$\frac{d}{dt}\langle \psi(t)|\mathbf{p}|\psi(t)\rangle = \frac{\mu_0}{j\hbar} \langle \psi(t)|J_z \boldsymbol{\nabla} B_{1,z}(\mathbf{r})|\psi(t)\rangle \,.$$

4. Now suppose that the non-uniform field depends linearly on position, namely, the components $B_{1,i}(\mathbf{r})$ can be expanded as

$$B_{1,i}(\mathbf{r}) = \sum_j b_{ij} r_j \,.$$

Maxwell's equations impose conditions on the coefficients b_{ij}. Establish the properties that the 3×3 matrix $\underline{b}$ must satisfy.

5. Assume that $|g_m(\mathbf{p})|^2$ is negligible except when $m = m^*$, that is, the atom is in a state with total angular momentum projection $m^*\hbar$. Show that

$$\frac{d^2}{dt^2} \langle \psi(t)|r_i|\psi(t)\rangle = \frac{\mu_0 m^*}{j} \frac{b_{zi}}{M} \,.$$

Since $m^* = -j, \ldots, j$, there are $2j + 1$ possible trajectories, and observation of the actual trajectory followed by the atom determines the value of m^*.

Solution

Part 1

We use the following relation for expectation values,

$$\frac{d}{dt} \langle A(t)\rangle = \frac{i}{\hbar} \langle \psi(t)|[H, A]|\psi(t)\rangle \,,$$

where

$$H = \frac{\mathbf{p}^2}{2M} - \frac{\mu_0}{j\hbar} [B_0 J_z + \mathbf{B}_1(\mathbf{r}) \cdot \mathbf{J}] \,,$$

to find that

$$\frac{d}{dt} \langle \mathbf{r}(t)\rangle = \frac{i}{2\hbar M} \langle \psi(t)|[\mathbf{p}^2, \mathbf{r}]|\psi(t)\rangle = \frac{\langle \psi(t)|\mathbf{p}|\psi(t)\rangle}{M}$$

and

$$\frac{d}{dt} \langle \mathbf{p}(t)\rangle = -i\frac{\mu_0}{j\hbar^2} \langle \psi(t)|[B_0 J_z + \mathbf{B}_1(\mathbf{r}) \cdot \mathbf{J}, \mathbf{p}]|\psi(t)\rangle = \frac{\mu_0}{j\hbar} \langle \psi(t)|\boldsymbol{\nabla}[\mathbf{B}_1(\mathbf{r}) \cdot \mathbf{J}]|\psi(t)\rangle \,.$$

Part 2

The state at time t is obtained from

$$|\psi(t)\rangle = e^{-iH_0 t/\hbar}|\psi(0)\rangle = \sum_{m=-j}^{j}\int d\mathbf{p}\, g_m(\mathbf{p})\, e^{-iH_0 t/\hbar}|\phi_{\mathbf{p},m}\rangle$$

$$= \sum_{m=-j}^{j}\int d\mathbf{p}\, g_m(\mathbf{p})\, \underbrace{e^{-iE_p t/\hbar}\, e^{i(\mu_0 B_0/j)mt/\hbar}}_{e^{-iE_p^m t/\hbar}}|\phi_{\mathbf{p},m}\rangle \,,$$

and the expectation values of J_i are given by

$$\langle J_z(t)\rangle = \sum_{m,m'=-j}^{j}\int d\mathbf{p}'\int d\mathbf{p}\, g_{m'}^*(\mathbf{p}')\, g_m(\mathbf{p})\, e^{iE_{p'}^{m'} t/\hbar}\, e^{-iE_p^m t/\hbar}\langle\phi_{\mathbf{p}',m'}|J_z|\phi_{\mathbf{p},m}\rangle$$

$$= \sum_{m,m'=-j}^{j}\int d\mathbf{p}'\int d\mathbf{p}\, g_{m'}^*(\mathbf{p}')\, g_m(\mathbf{p})\, e^{iE_{p'}^{m'} t/\hbar}\, e^{-iE_p^m t/\hbar}\, m\hbar\, \delta(\mathbf{p}-\mathbf{p}')\, \delta_{m,m'}$$

$$= \sum_{m=-j}^{j} m\hbar\int d\mathbf{p}\, |g_m(\mathbf{p})|^2 \,.$$

Evaluation of the J_x and J_y expectation values proceeds similarly. We need

$$\langle\phi_{\mathbf{p}',m'}|J_x|\phi_{\mathbf{p},m}\rangle = \frac{\hbar}{2}\,\delta(\mathbf{p}-\mathbf{p}')\Big[\sqrt{j(j+1)-m(m+1)}\,\delta_{m',m+1} + \sqrt{j(j+1)-m(m-1)}\,\delta_{m',m-1}\Big]$$

and

$$\langle\phi_{\mathbf{p}',m'}|J_y|\phi_{\mathbf{p},m}\rangle = \frac{\hbar}{2i}\,\delta(\mathbf{p}-\mathbf{p}')\Big[\sqrt{j(j+1)-m(m+1)}\,\delta_{m',m+1} - \sqrt{j(j+1)-m(m-1)}\,\delta_{m',m-1}\Big] \,.$$

Inserting these expressions, the expectation values are obtained as

$$\langle J_x(t)\rangle = \frac{\hbar}{2}\sum_{m=-j}^{j}\int d\mathbf{p}\,\Big[g_{m+1}^*(\mathbf{p})\, g_m(\mathbf{p})\, e^{i(\mu_0 B_0/j)t/\hbar}\,\sqrt{j(j+1)-m(m+1)}$$

$$+ g_{m-1}^*(\mathbf{p})\, g_m(\mathbf{p})\, e^{-i(\mu_0 B_0/j)t/\hbar}\,\sqrt{j(j+1)-m(m-1)}\Big] \,,$$

$$\langle J_y(t)\rangle = \frac{\hbar}{2i}\sum_{m=-j}^{j}\int d\mathbf{p}\,\Big[g_{m+1}^*(\mathbf{p})\, g_m(\mathbf{p})\, e^{i(\mu_0 B_0/j)t/\hbar}\,\sqrt{j(j+1)-m(m+1)}$$

$$- g_{m-1}^*(\mathbf{p})\, g_m(\mathbf{p})\, e^{-i(\mu_0 B_0/j)t/\hbar}\,\sqrt{j(j+1)-m(m-1)}\Big] \,.$$

Part 3

If the rapidly oscillating terms are ignored in the expectation value of $\mathbf{J}$, we find

$$\langle\mathbf{J}(t)\rangle \longrightarrow \langle J_z(t)\rangle$$

and hence

$$\frac{d}{dt}\langle\psi(t)|\mathbf{p}|\psi(t)\rangle = \frac{\mu_0}{j\hbar}\sum_i\langle\psi(t)|J_i\,\boldsymbol{\nabla}B_{1,i}(\mathbf{r})|\psi(t)\rangle \approx \frac{\mu_0}{j\hbar}\,\langle\psi(t)|J_z\,\boldsymbol{\nabla}B_{1,z}(\mathbf{r})|\psi(t)\rangle \,.$$

Part 4

Maxwell's equations require (in the absence of time-varying electric fields)

$$\nabla \cdot \mathbf{B}(\mathbf{r}) = 0 , \qquad \nabla \times \mathbf{B}(\mathbf{r}) = 0 .$$

The first condition gives

$$\sum_i \partial_i B_{1,i}(\mathbf{r}) = \sum_{ij} \partial_i (b_{ij} r_j) = \sum_{ij} b_{ij} \delta_{ij} = \sum_i b_{ii} = 0 ,$$

while the second yields

$$[\nabla \times \mathbf{B}(\mathbf{r})]_i = \sum_{jk} \epsilon_{ijk} \partial_j B_{1,k}(\mathbf{r}) = \sum_{jkl} \epsilon_{ijk} \partial_j (b_{kl} r_l) = \sum_{jk} \epsilon_{ijk} b_{kj} = 0 \implies b_{jk} = b_{kj} .$$

The first relation says that the matrix $\underline{b}$ must be traceless; the second that it must be symmetric.

Part 5

If $|g_m(\mathbf{p})|^2$ is negligible for $m \neq m^*$, then

$$|\psi(t)\rangle \approx \int d\mathbf{p} \, g_{m^*}(\mathbf{p}) \, e^{-iE_p^{m^*} t/\hbar} \, |\phi_{\mathbf{p},m^*}\rangle ,$$

and hence

$$\frac{d}{dt}\langle\psi(t)|p_i|\psi(t)\rangle = \frac{\mu_0}{j\hbar} \langle\psi(t)|J_z \partial_i B_{1,z}(\mathbf{r})|\psi(t)\rangle = \frac{\mu_0}{j\hbar} b_{zi}\langle\psi(t)|J_z|\psi(t)\rangle = \frac{\mu_0 m^*}{j} b_{zi} .$$

It follows that

$$\frac{d^2}{dt^2}\langle r_i(t)\rangle = \frac{d}{dt}\frac{\langle p_i(t)\rangle}{M} = \frac{\mu_0 m^*}{j}\frac{b_{zi}}{M} ,$$

and the "trajectories" followed by the atoms depend on m^*, thus providing a means of experimental verification for quantization of the angular momentum projection.

Problem 10   Polarizing a Beam of Spin-1/2 Particles by a Magnetic Interaction

A mono-energetic beam of neutral spin-1/2 particles is perpendicularly incident on a region where a uniform and constant magnetic field is present. The direction of the incident beam is along the x-axis, while the magnetic field fills the region with $x > 0$. The incident particles have energy E, mass m, and magnetic moment $\boldsymbol{\mu} = -(\mu_0/\hbar)\mathbf{S}$, where $\mu_0 > 0$ is a constant. Their potential energy $V(x)$ consists of two terms,

$$V(x) = \theta(x)\left(V_0 - \boldsymbol{\mu} \cdot \mathbf{B}_0\right) ,$$

where $\theta(x)$ is the step function: $\theta(x) = 1$ for $x > 0$ and 0 otherwise. The field $\mathbf{B}_0$ is along the z-axis, and is strong enough that $0 < V_0 < \mu_0 B_0/2$.

1. Suppose that the energy E of the incident particles is in the range

$$0 < E < V_0 + \mu_0 B_0/2 .$$

Determine the eigenfunctions of these particles (of energy E in the range above) corresponding to a positive incident momentum along the x-axis and spin either parallel or antiparallel to the z-axis. In particular, calculate the transmission and reflection coefficients for these two cases.

2. Assume the incident beam is unpolarized, that is, it consists of $N/2$ particles with spin up, and $N/2$ particles with spin down, along the z-axis. Denote by N_+ and N_- the numbers of reflected particles with spin up and spin down (along the z-axis), respectively. Calculate the polarization of the reflected beam, defined as

$$\mathcal{P} = \frac{N_+ - N_-}{N_+ + N_-} \, .$$

Solution

Part 1

Let

$$\psi(x) = \begin{pmatrix} \psi_+(x) \\ \psi_-(x) \end{pmatrix} \, ,$$

be the spinor wave function. The Schrödinger equation for $\psi(x)$ reads

$$-\frac{\hbar^2}{2m} \frac{d^2}{dx^2} \psi(x) + \Theta(x) \left(V_0 + \frac{\mu_0 B_0}{\hbar} S_z \right) \psi(x) = E\psi(x) \, .$$

Observing that

$$S_z \begin{pmatrix} \psi_+(x) \\ \psi_-(x) \end{pmatrix} = \frac{\hbar}{2} \begin{pmatrix} \psi_+(x) \\ -\psi_-(x) \end{pmatrix} \, ,$$

we find, for the $\pm$ components,

$$\psi''_\pm(x) + [\epsilon - \Theta(x)(v_0 \pm w_0)]\psi_\pm(x) = 0 \, ,$$

where we define

$$\epsilon = \frac{2mE}{\hbar^2} \, , \qquad v_0 = \frac{2mV_0}{\hbar^2} \qquad w_0 = \frac{m\mu_0 B_0}{\hbar^2} \, .$$

In the region $x < 0$ we have a free-particle solution,

$$\psi_\pm(x) = A_\pm \, e^{ikx} + B_\pm \, e^{-ikx} \qquad k = \sqrt{\epsilon} \, ,$$

while in the region $x > 0$ we have, for the $+$ component,

$$\psi_+(x) = C_+ e^{-\kappa_+ x} \, , \qquad \kappa_+ = \sqrt{v_0 + w_0 - \epsilon} \, ,$$

and for the $-$ component

$$\psi_-(x) = C_- e^{ik_- x} \, , \qquad k_- = \sqrt{\epsilon - v_0 + w_0} \, .$$

Imposing the boundary conditions at the origin yields the relations

$$A_+ + B_+ = C_+ \, , \qquad A_+ - B_+ = i\frac{\kappa_+}{k} C_+$$

and

$$A_- + B_- = C_- \, , \qquad A_- - B_- = \frac{k_-}{k} C_- \, ,$$

which we can solve in terms of the incoming wave coefficients $A_\pm$ to find

$$B_+ = \frac{1 - i\kappa_+/k}{1 + i\kappa_+/k} A_+ \, , \qquad C_+ = \frac{2}{1 + i\kappa_+/k} A_+$$

and

$$B_- = \frac{1 - k_-/k}{1 + k_-/k} A_- , \qquad C_- = \frac{2}{1 + k_-/k} A_- .$$

We obtain for the transmission coefficients

$$T_+ = \frac{|j_t^+|}{|j_i^+|} = 0 , \qquad T_- = \frac{|j_t^-|}{|j_i^-|} = \frac{k_- |C_-|^2}{k |A_-|^2} = \frac{4kk_-}{(k + k_-)^2}$$

and similarly for the reflection coefficients,

$$R_+ = 1 , \qquad R_- = \frac{(k - k_-)^2}{(k + k_-)^2} .$$

Part 2

The incident beam is unpolarized and can be thought of as consisting of $N/2$ particles having spin up and $N/2$ particles with spin down. The probabilities that a reflected particle has spin up or spin down are, respectively, $R_+ = 1$ and R_-. The fractions of reflected particles with spin up and with spin down are therefore

$$N_+ = R_+ \frac{N}{2} = \frac{N}{2} , \qquad N_- = R_- \frac{N}{2} ,$$

and the polarization of the reflected beam is then given by

$$\mathcal{P} = \frac{N_+ - N_-}{N_+ + N_-} = \frac{1 - R_-}{1 + R_-} .$$

Problem 11 Time Evolution of a Spin-1 State in a Time-Dependent Magnetic Field

Consider a spin-1 particle with magnetic moment $\boldsymbol{\mu} = -(\mu_0/\hbar)\,\mathbf{S}$, where μ_0 is a constant (ignore the orbital degrees of freedom). The particle is subject to the action of a time-dependent magnetic field given by $\mathbf{B}(t) = B(t)\,\hat{\mathbf{z}}$, where $B(t)$ is an assigned function of time. Denote as $|m\rangle$, with $m = \pm 1, 0$, the eigenstates of $\mathbf{S}^2$ and S_z with eigenvalues $2\hbar^2$ and $m\hbar$, respectively.

1. Consider the state

$$|\psi(t)\rangle = \sum_{m=\pm 1,0} c_m(t) |m\rangle$$

and show that the time-dependent coefficients satisfy the equation

$$i\hbar \dot{c}_m(t) = m\mu_0 B(t) c_m(t) .$$

Solve the equation above, expressing your solution in terms of an integral over time of $B(t)$.

2. Construct the states $|\pm 1\rangle_x$ with polarizations ± 1 along the x-axis as linear combinations of the eigenstates $|m\rangle$ defined above.

3. Suppose that at time $t = 0$ the particle is in the state $|+1\rangle_x$. Calculate the probability that at time t the spin-flip transition $|+1\rangle_x \longrightarrow |-1\rangle_x$ will occur.

Solution

Part 1

The time-dependent Schrödinger equation

$$i\hbar \frac{d}{dt} |\psi(t)\rangle = H(t) |\psi(t)\rangle ,$$

yields in the present case

$$i\hbar \sum_{m'} \dot{c}_{m'}(t) |m'\rangle = \frac{\mu_0 B(t)}{\hbar} \sum_{m'} c_{m'}(t) S_z |m'\rangle = \sum_{m'} m' \mu_0 B(t) c_{m'}(t) |m'\rangle ,$$

or, projecting onto the state $|m\rangle$ (that is, taking the inner product with $\langle m|$ of both sides of the equation above),

$$i\hbar \dot{c}_m(t) = m \mu_0 B(t) c_m(t) .$$

The above equation has the solution

$$c_m(t) = c_m(0) \exp\left[-\frac{i}{\hbar} m \mu_0 \int_0^t dt' \, B(t') \right] .$$

It is convenient to introduce

$$\lambda(t) = \frac{\mu_0}{\hbar} \int_0^t dt' \, B(t') ,$$

so that

$$c_m(t) = c_m(0) \, e^{-i m \lambda(t)} .$$

Part 2

In the basis $|m\rangle$ the matrix representing S_x is given by

$$S_x \longrightarrow \frac{\hbar}{\sqrt{2}} \begin{pmatrix} 0 & 1 & 0 \\ 1 & 0 & 1 \\ 0 & 1 & 0 \end{pmatrix}$$

and the eigenstates of S_x with eigenvalues $\pm\hbar$ are represented by the column vectors

$$|\pm 1\rangle_x \longrightarrow \begin{pmatrix} c_1^{\pm} \\ c_0^{\pm} \\ c_{-1}^{\pm} \end{pmatrix} .$$

They satisfy

$$\frac{\hbar}{\sqrt{2}} \begin{pmatrix} 0 & 1 & 0 \\ 1 & 0 & 1 \\ 0 & 1 & 0 \end{pmatrix} \begin{pmatrix} c_1^{\pm} \\ c_0^{\pm} \\ c_{-1}^{\pm} \end{pmatrix} = \pm\hbar \begin{pmatrix} c_1^{\pm} \\ c_0^{\pm} \\ c_{-1}^{\pm} \end{pmatrix} ,$$

or

$$\frac{c_0^{\pm}}{\sqrt{2}} = \pm c_1^{\pm} , \qquad \frac{1}{\sqrt{2}} \left(c_1^{\pm} + c_{-1}^{\pm} \right) = \pm c_0^{\pm} , \qquad \frac{c_0^{\pm}}{\sqrt{2}} = \pm c_{-1}^{\pm} .$$

Hence, we find

$$|\pm 1\rangle_x = c_0^\pm \begin{pmatrix} \pm 1/\sqrt{2} \\ 1 \\ \pm 1/\sqrt{2} \end{pmatrix}$$

with $|c_0^\pm|$ fixed by the normalization condition $|c_0^\pm| = 1/\sqrt{2}$.

Part 3

Given the initial condition $|\psi(0)\rangle = |+1\rangle_x$, we have

$$|\psi(t)\rangle = \frac{1}{2} e^{-i\lambda(t)} |+1\rangle + \frac{1}{\sqrt{2}} |0\rangle + \frac{1}{2} e^{i\lambda(t)} |-1\rangle ,$$

and the amplitude for the spin-flip transition is obtained as

$$_x\langle -1|\psi(t)\rangle = -\frac{1}{4} e^{-i\lambda(t)} + \frac{1}{2} - \frac{1}{4} e^{i\lambda(t)} = \sin^2[\lambda(t)/2] ,$$

yielding the spin-flip probability $\sin^4[\lambda(t)/2]$.

Problem 12　Neutron Interferometry and 4π Rotations of Spinor Wave Functions

Consider an experiment in which a neutron interferometer is arranged so that one of the paths (path A) available to the neutrons passes through a magnetic field $\mathbf{B}$, while the other (path B) goes through a magnetic-field-free region.[3] The length of the region where the magnetic field is present is L. The magnetic field is uniform in this region and directed along the z-axis. The neutron is an electrically neutral spin-1/2 particle with magnetic moment $\boldsymbol{\mu} = -(\mu_0/\hbar)\,\mathbf{S}$, where μ_0 is written as $|g_n|\, e\hbar/(2m_p c)$ where $g_n \approx -1.913$ and m_p are, respectively, the gyromagnetic factor and proton mass; the combination $e\hbar/(2m_p c)$ is known as the nuclear magneton. The neutrons at the source have spin projection $+\hbar/2$ along the z-axis (so, the beam is polarized), and are approximately mono-energetic with a very narrow spread in energy about the central value $E = p^2/(2m_n)$, m_n being the neutron mass.

1. Denote with $\phi_A(\mathbf{r}, t)$ and $\phi_B(\mathbf{r}, t)$ the orbital parts of the neutron spinor wave functions $\psi_A(\mathbf{r}, t)$ and $\psi_B(\mathbf{r}, t)$ for neutrons moving along paths A and B, respectively. Justify why $\phi_A(\mathbf{r}, t)$ and $\phi_B(\mathbf{r}, t)$ are free-particle wave packets. Then, obtain the spinor wave functions $\psi_A(\mathbf{r}, t)$ and $\psi_B(\mathbf{r}, t)$ by assuming that it takes a time interval τ for a neutron along path A to traverse the region where the magnetic field is present.

2. Assuming that the free-particle wave packets $\phi_A(\mathbf{r}, t)$ and $\phi_B(\mathbf{r}, t)$ differ by a phase factor, namely

$$\phi_A(\mathbf{r}, t) = e^{i\alpha}\, \phi_B(\mathbf{r}, t) ,$$

determine under what condition the counting rate in the interference region is largest. Show that the difference in magnetic fields that produces two successive maxima in the counting rates can be written as

$$\Delta B = \frac{8\pi\hbar c}{|g_n| e \lambda^* L} , \qquad \lambda^* = \frac{\lambda}{2\pi} ,$$

where λ is the de Broglie wavelength of the neutron with momentum p and energy $E = p^2/(2m_n)$ (ignore the neutron–proton mass difference).

---

[3] This experiment was actually carried out by Werner et al. (1975), *Phys. Rev. Lett.* **35**, 1053, using a beam of thermal neutrons.

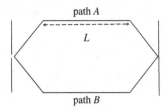

path A

L

path B

Schematics of the neutron interferometry experiment.

Solution

Part 1

The neutron along path B is free and is described by the spinor wave function

$$\psi_B(\mathbf{r}, t) = \phi_B(\mathbf{r}, t)|+\rangle ,$$

where $|+\rangle$ is the eigenstate of S_z with eigenvalue $+\hbar/2$ and the orbital part can be represented by the free-particle wave packet $\phi_B(\mathbf{r}, t)$. By contrast, the neutron along path A interacts, via its spin magnetic moment, with the magnetic field when traversing the region of length L. After it has exited this region, its spinor wave function reads

$$\psi_A(\mathbf{r}, t) = \phi_A(\mathbf{r}, t)\, e^{-i\mu_0 B\tau/(2\hbar)}\, |+\rangle ,$$

given that the interaction $H = -\boldsymbol{\mu} \cdot \mathbf{B} = \mu_0 B S_z/\hbar$ acts only on the spin degrees of freedom for a time interval τ, and

$$e^{-iH\tau/\hbar}\, |+\rangle = e^{-i\mu_0 B\tau/(2\hbar)}\, |+\rangle .$$

Here, $\phi_A(\mathbf{r}, t)$ is also a free-particle wave packet (the neutron has no electric charge and its orbital motion is unaffected by the magnetic field); however, it can differ from $\phi_B(\mathbf{r}, t)$ by a phase factor as follows:

$$\phi_A(\mathbf{r}, t) = e^{i\alpha}\, \phi_B(\mathbf{r}, t) .$$

Part 2

In the interference region, the amplitude is given by $\psi_A(\mathbf{r}, t) + \psi_B(\mathbf{r}, t)$, and the counting rate is therefore proportional to

$$\text{counting rate} \propto |\psi_A(\mathbf{r}, t) + \psi_B(\mathbf{r}, t)|^2 = |\phi_B(\mathbf{r}, t)|^2 \left| e^{i[\alpha - \mu_0 B\tau/(2\hbar)]} + 1 \right|^2 ,$$

yielding

$$\text{counting rate} \propto 1 + \cos\left(\alpha - \frac{\mu_0 B\tau}{2\hbar}\right) .$$

The counting rate is largest when

$$\alpha - \frac{\mu_0 B\tau}{2\hbar} = 2n\pi \qquad n = 0, \pm 1, \pm 2, \dots$$

Let B_0 and B_1 be the magnetic fields producing successive maxima, corresponding to n_0 and $n_0 + 1$, that is

$$B_0 = \frac{2\hbar}{\mu_0 \tau} (\alpha - 2n_0 \pi) , \qquad B_1 = \frac{2\hbar}{\mu_0 \tau} [\alpha - 2(n_0 + 1)\pi] .$$

The difference in magnetic fields is then obtained as

$$\Delta B = B_0 - B_1 = \frac{4\pi\hbar}{\mu_0 \tau} .$$

Assuming that the neutron beam at the source is approximately mono-energetic and of energy E, we have $\tau = L/(p/m_n)$, where m_n is the neutron mass and $p = h/\lambda$ (λ is the de Broglie wavelength) is the momentum corresponding to $E = p^2/(2m_n)$. Inserting the expression for μ_0 and ignoring the small neutron–proton mass difference leads to

$$\Delta B = \frac{8\pi\hbar c}{|g_n| e\lambda^* L} , \qquad \lambda^* = \frac{\lambda}{2\pi} .$$

A comment is in order here. The time evolution operator for a spin in a time-independent magnetic field is formally identical to the rotation operator, in the sense that

$$e^{-iH\tau/\hbar} = e^{-i(\mu_0/\hbar)B\hat{\mathbf{n}}\cdot\mathbf{S}\tau/\hbar} = U_R(\varphi\,\hat{\mathbf{n}}) , \qquad U_R(\varphi\,\hat{\mathbf{n}}) = e^{-i\varphi\hat{\mathbf{n}}\cdot\mathbf{S}/\hbar} \quad \text{with} \quad \varphi = \frac{\mu_0 B\tau}{\hbar} ,$$

where the unit vector $\hat{\mathbf{n}}$ specifies the direction of the magnetic field. For integer spins (or, in general, integer angular momenta), we have, as expected, $U_R(2\pi\,\hat{\mathbf{n}}) = \mathbb{1}$. By contrast, semi-integer spins (or, generally, semi-integer angular momenta) have the property that $U_R(2\pi\,\hat{\mathbf{n}}) = -\mathbb{1}$; it is only for $\varphi = 4\pi$ that the rotation operator reduces to the identity. This fact is reflected in the value of ΔB obtained above. If semi-integer spins were left invariant under a rotation of 2π (rather than 4π) then ΔB would have been one-half of this value. Experiments utilizing neutron beams have verified that the correct ΔB is indeed $4\pi\hbar/(\mu_0\tau)$.

Problem 13   Charged Spinless Particle Confined in Plane Perpendicular to Uniform Magnetic Field

Consider a spinless particle of mass μ and charge q ($q > 0$) in a time-independent magnetic field $\mathbf{B}(\mathbf{r})$. The velocity operator is given by

$$\mathbf{v} = \frac{1}{\mu}\left[-i\hbar\boldsymbol{\nabla} - \frac{q}{c}\,\mathbf{A}(\mathbf{r})\right] , \qquad\qquad \mathbf{B}(\mathbf{r}) = \boldsymbol{\nabla}\times\mathbf{A}(\mathbf{r}) ,$$

where $\mathbf{A}(\mathbf{r})$ is the vector potential. Show that

$$\left[v_i,\, v_j\right] = i\frac{q\hbar}{\mu^2 c}\sum_k \epsilon_{ijk}\, B_k(\mathbf{r}) .$$

Suppose the particle is constrained to move in the xy-plane under the influence of a uniform magnetic field directed along the $\hat{\mathbf{z}}$-axis. The Hamiltonian is given by

$$H = \frac{\mu}{2}\left(v_x^2 + v_y^2\right) .$$

1. Define the operators

$$\hat{a} = \frac{\alpha}{\sqrt{2}}(v_x + i\,v_y) , \qquad \hat{a}^\dagger = \frac{\alpha}{\sqrt{2}}(v_x - i\,v_y)$$

and determine the real parameter α such that $[\hat{a}, \hat{a}^\dagger] = 1$. Write the Hamiltonian in terms of $\hat{a}$ and $\hat{a}^\dagger$.

2. Obtain the eigenvalues and eigenstates of the Hamiltonian.

3. Determine the (infinitely degenerate) eigenfunctions corresponding to the ground-state energy from the condition $\hat{a}|0\rangle = 0$.

Solution

In a magnetic field we have

$$\mathbf{v} = \frac{1}{\mu}\left[\mathbf{p} - \frac{q}{c}\mathbf{A}(\mathbf{r})\right], \qquad \nabla \times \mathbf{A}(\mathbf{r}) = \mathbf{B}(\mathbf{r}),$$

and

$$[v_i, v_j] = -\frac{q}{\mu^2 c}\left([p_i, A_j] + [A_i, p_j]\right) = i\frac{\hbar q}{\mu^2 c}\left(\nabla_i A_j - \nabla_j A_i\right),$$

where we have used $[p_i, p_j] = 0 = [A_i, A_j]$. We now observe that, for $i \neq j$,

$$\nabla_i A_j - \nabla_j A_i = \sum_k \epsilon_{ijk} B_k,$$

which can be explicitly verified:

$$\sum_k \epsilon_{ijk} B_k = \sum_{klm} \epsilon_{ijk}\,\epsilon_{klm}\,\nabla_l A_m = \sum_{lm}(\delta_{il}\,\delta_{jm} - \delta_{im}\,\delta_{jl})\nabla_l A_m = \nabla_i A_j - \nabla_j A_i.$$

Thus, we obtain the required relation

$$[v_i, v_j] = i\frac{\hbar q}{\mu^2 c}\sum_k \epsilon_{ijk} B_k.$$

Part 1

Using the above relation, we find

$$[\hat{a}, \hat{a}^\dagger] = \frac{\alpha^2}{2}[v_x + iv_y, v_x - iv_y] = i\frac{\alpha^2}{2}([v_y, v_x] - [v_x, v_y]) = \alpha^2\frac{\hbar qB}{\mu^2 c},$$

where by assumption the magnetic field is uniform and directed along the z-axis, that is, it is given by $B\,\hat{\mathbf{z}}$ with B constant. We fix α:

$$\alpha = \left(\frac{\mu^2 c}{\hbar q B}\right)^{1/2},$$

so that $[\hat{a}, \hat{a}^\dagger] = 1$, and express v_x and v_y in terms of $\hat{a}$ and $\hat{a}^\dagger$ as

$$v_x = \frac{1}{\sqrt{2}\alpha}(\hat{a} + \hat{a}^\dagger), \qquad v_y = -\frac{i}{\sqrt{2}\alpha}(\hat{a} - \hat{a}^\dagger).$$

The Hamiltonian then reads

$$\hat{H} = \frac{\mu}{2}\frac{1}{2\alpha^2}\left[(\hat{a} + \hat{a}^\dagger)^2 - (\hat{a} - \hat{a}^\dagger)^2\right] = \frac{\mu}{2\alpha^2}(\hat{a}\hat{a}^\dagger + \hat{a}^\dagger\hat{a}) = \frac{\mu}{\alpha^2}(\hat{a}^\dagger\hat{a} + 1/2),$$

where in the last step we have made use of the identity $\hat{a}\,\hat{a}^\dagger = \hat{a}^\dagger\,\hat{a} + 1$, which follows from the commutator of $\hat{a}$ and $\hat{a}^\dagger$. Inserting the expression for α, we finally obtain for the Hamiltonian

$$\hat{H} = \frac{\hbar q B}{\mu c}(\hat{a}^\dagger\,\hat{a} + 1/2)\,.$$

Since $\hat{a}$ and $\hat{a}^\dagger$ satisfy the same commutation relations as in the harmonic oscillator problem, the energy eigenvalues are

$$E_n = \frac{\hbar q B}{\mu c}(n + 1/2)\qquad n = 0, 1, 2, \ldots ,$$

with corresponding eigenstates given by

$$|n\rangle = \frac{1}{\sqrt{n!}}\,(\hat{a}^\dagger)^n\,|0\rangle\,,$$

where the ground state $|0\rangle$ has the property that it is annihilated by the action of $\hat{a}$, that is, $\hat{a}\,|0\rangle = 0$. Note that, even though the Hamiltonian looks the same as that for a one-dimensional harmonic oscillator, this is not a one-dimensional problem. Indeed, the energy eigenvalues are infinitely degenerate (rather than being non-degenerate as in case of the one-dimensional harmonic oscillator).

Part 3

The condition $\hat{a}|0\rangle = 0$ implies $(v_x + iv_y)|0\rangle = 0$, which in the coordinate representation reduces to

$$\left[\left(-i\hbar\,\nabla_x - \frac{q}{c}\,A_x\right) + i\left(-i\hbar\,\nabla_y - \frac{q}{c}\,A_y\right)\right]\psi_0(x, y) = 0\,.$$

The vector potential for a uniform field $\mathbf{B} = B\,\hat{z}$ can be taken as

$$\mathbf{A} = \frac{1}{2}\,\mathbf{B}\times\mathbf{r}\implies A_x = -\frac{By}{2}\quad\text{and}\quad A_y = \frac{Bx}{2}\,.$$

Inserting these expressions into the equation above and rearranging terms, it follows that

$$\left[\left(\nabla_y + \frac{qB}{2\hbar c}\,y\right) - i\left(\nabla_x + \frac{qB}{2\hbar c}\,x\right)\right]\psi_0(x, y) = 0\,.$$

We now introduce polar coordinates

$$x = \rho\cos\phi\qquad y = \rho\sin\phi\,,\qquad \rho = \sqrt{x^2 + y^2}\qquad \phi = \tan^{-1}(y/x)\,,$$

with

$$\nabla_x = \frac{\partial}{\partial x} = \frac{\partial\rho}{\partial x}\frac{\partial}{\partial\rho} + \frac{\partial\phi}{\partial x}\frac{\partial}{\partial\phi} = \frac{x}{\sqrt{x^2 + y^2}}\frac{\partial}{\partial\rho} - \frac{y/x^2}{1 + (y/x)^2}\frac{\partial}{\partial\phi} = \cos\phi\,\frac{\partial}{\partial\rho} - \frac{\sin\phi}{\rho}\frac{\partial}{\partial\phi}\,,$$

and similarly

$$\nabla_y = \frac{\partial}{\partial y} = \sin\phi\,\frac{\partial}{\partial\rho} + \frac{\cos\phi}{\rho}\frac{\partial}{\partial\phi}\,.$$

The differential equation becomes

$$\left[\left(\sin\phi\,\frac{\partial}{\partial\rho} + \frac{\cos\phi}{\rho}\frac{\partial}{\partial\phi} + \frac{qB}{2\hbar c}\,\rho\sin\phi\right) - i\left(\cos\phi\,\frac{\partial}{\partial\rho} - \frac{\sin\phi}{\rho}\frac{\partial}{\partial\phi} + \frac{qB}{2\hbar c}\,\rho\cos\phi\right)\right]\psi_0(\rho, \phi) = 0\,,$$

or, combining terms,

$$-i\,e^{i\phi}\left(\frac{\partial}{\partial\rho} + \frac{i}{\rho}\frac{\partial}{\partial\phi} + \frac{qB}{2\hbar c}\,\rho\right)\psi_0(\rho,\phi) = 0\,.$$

After canceling the common factor $-i\,e^{i\phi}$, the resulting equation can be solved by separation of variables: setting $\psi_0(\rho,\phi) = R(\rho)\,F(\phi)$ yields

$$FR' + i\frac{R}{\rho}F' + \frac{qB}{2\hbar c}\,\rho RF = 0 \implies \rho\left(\frac{R'}{R} + \frac{qB}{2\hbar c}\,\rho\right) = -i\frac{F'}{F} = m\,,$$

where m is a constant. The equation for F gives

$$\frac{F'}{F} = im \implies F(\phi) = e^{im\phi}\,.$$

The periodic boundary condition $F(\phi + 2\pi) = F(\phi)$ requires that

$$e^{2\pi m i} = 1 \implies m = 0, \pm 1, \pm 2, \ldots\,.$$

The equation for R reduces to

$$\frac{1}{R}\frac{dR}{d\rho} + \frac{qB}{2\hbar c}\,\rho - \frac{m}{\rho} = 0 \implies \frac{dR}{R} = -\left(\frac{qB}{2\hbar c}\,\rho - \frac{m}{\rho}\right)d\rho$$

which has the solution

$$\int_{R_0}^{R}\frac{dR'}{R'} = -\int_{\rho_0}^{\rho}d\rho'\left(\frac{qB}{2\hbar c}\,\rho' - \frac{m}{\rho'}\right) \implies \ln\frac{R}{R_0} = m\ln\frac{\rho}{\rho_0} - \frac{qB}{4\hbar c}\,(\rho^2 - \rho_0^2)\,,$$

or

$$R(\rho) = C_m\rho^m\,e^{-qB\rho^2/(4\hbar c)}\,,$$

where C_m is a normalization constant. Note that we need $m \geq 0$, otherwise the solutions would be singular at the origin. The complete solutions are

$$\psi_0^{(m)}(\rho,\phi) = C_m\rho^m\,e^{-qB\rho^2/(4\hbar c)}\,e^{im\phi}\,,$$

and the ground state (as well as any excited state) is infinitely degenerate.

Problem 14   Particle Confined to a Cylindrical Region

Consider a particle of mass μ confined to the interior region of a hollow cylinder whose axis coincides with the z-axis. The wave function is required to vanish on the inner and outer walls, with radii, respectively, a and b ($0 < a < b$), and also at the bottom and top surfaces, located at $z = 0$ and $z = L$.

1. Knowing that the Laplacian in cylindrical coordinates (ρ, ϕ, z), where $x = \rho\,\cos\phi$ and $y = \rho\,\sin\phi$, is given by

$$\nabla^2 = \frac{\partial^2}{\partial\rho^2} + \frac{1}{\rho}\frac{\partial}{\partial\rho} + \frac{1}{\rho^2}\frac{\partial^2}{\partial\phi^2} + \frac{\partial^2}{\partial z^2}\,,$$

and expressing the electron wave function as

$$\psi(\rho,\phi,z) = R(\rho)\,\Phi(\phi)\,Z(z)\,,$$

show that the Schrödinger equation reduces to the following equations for $R(\rho)$, $\Phi(\phi)$, and $Z(z)$:

$$\frac{\partial^2 R(\rho)}{\partial \rho^2} + \frac{1}{\rho} \frac{\partial R(\rho)}{\partial \rho} + \left(k^2 - \lambda^2 - \frac{\nu^2}{\rho^2} \right) R(\rho) = 0 , \qquad \frac{\partial^2 \Phi(\phi)}{\partial \phi^2} = -\nu^2 \Phi(\phi) ,$$

and

$$\frac{\partial^2 Z(z)}{\partial z^2} = -\lambda^2 Z(z) ,$$

where $k^2 = 2\mu E/\hbar^2$ ($E > 0$ is the particle's energy), and ν^2 and λ^2 are (for the time being!) real parameters.

2. Solve the equations for $\Phi(\phi)$ and $Z(z)$, imposing the appropriate boundary conditions, namely periodic boundary conditions on $\Phi(\phi)$ and confinement in the region $0 \leq z \leq L$ on $Z(z)$.

3. Knowing that the two independent solutions of the equation

$$\frac{\partial^2 R(\rho)}{\partial \rho^2} + \frac{1}{\rho} \frac{\partial R(\rho)}{\partial \rho} + \left(\gamma^2 - \frac{n^2}{\rho^2} \right) R(\rho) = 0 ,$$

with $n = 0, \pm 1, \pm 2, \ldots$ are the regular and irregular Bessel functions $J_n(\gamma \rho)$ and $N_n(\gamma \rho)$ with $J_{-n}(\gamma \rho) = (-1)^n J_n(\gamma \rho)$ and $N_{-n}(\gamma \rho) = (-1)^n N_n(\gamma \rho)$, show that the electron eigenenergies are given by

$$E_{mnp} = \frac{\hbar^2}{2\mu} \left[\gamma_{mn}^2 + \left(\frac{p\pi}{L} \right)^2 \right] \quad \text{with} \quad m = 1, 2 \ldots \quad \text{and} \quad p = 1, 2, \ldots ,$$

where for a given n the parameters γ_{mn} are the solutions of a certain transcendental equation, which will be derived. Do not solve this equation. Write down the complete eigenfunctions.

Solution

Part 1

The Schrödinger equation in cylindrical coordinates reads

$$\left(\frac{\partial^2}{\partial \rho^2} + \frac{1}{\rho} \frac{\partial}{\partial \rho} + \frac{1}{\rho^2} \frac{\partial^2}{\partial \phi^2} + \frac{\partial^2}{\partial z^2} \right) \psi(r, \phi, z) = -k^2 \psi(r, \phi, z) , \qquad k^2 = \frac{2\mu E}{\hbar^2} .$$

We solve it by separation of variables, by assuming that

$$\psi(\rho, \phi, z) = R(\rho) \, \Phi(\phi) \, Z(z) ,$$

to obtain

$$\underbrace{\frac{1}{R(\rho)\Phi(\phi)} \left(\frac{\partial^2}{\partial \rho^2} + \frac{1}{\rho} \frac{\partial}{\partial \rho} + \frac{1}{\rho^2} \frac{\partial^2}{\partial \phi^2} + k^2 \right) R(\rho)\Phi(\phi)}_{\text{function of } \rho \text{ and } \phi} + \underbrace{\frac{1}{Z(z)} \frac{\partial^2 Z(z)}{\partial z^2}}_{\text{function of } z} = 0 ,$$

and hence

$$\frac{1}{R(\rho)\Phi(\phi)} \left(\frac{\partial^2}{\partial \rho^2} + \frac{1}{\rho} \frac{\partial}{\partial \rho} + \frac{1}{\rho^2} \frac{\partial^2}{\partial \phi^2} + k^2 \right) R(\rho)\Phi(\phi) = \lambda^2 , \qquad \frac{1}{Z(z)} \frac{\partial^2 Z(z)}{\partial z^2} = -\lambda^2 .$$

Similarly, we have

$$\frac{\rho^2}{R(\rho)} \left(\frac{\partial^2}{\partial \rho^2} + \frac{1}{\rho} \frac{\partial}{\partial \rho} + k^2 - \lambda^2 \right) R(\rho) + \underbrace{\frac{1}{\Phi(\phi)} \frac{\partial^2 \Phi(\phi)}{\partial \phi^2}}_{\text{function of } \phi} = 0 \, ,$$

$$\underbrace{\phantom{\frac{\rho^2}{R(\rho)} \left(\frac{\partial^2}{\partial \rho^2} + \frac{1}{\rho} \frac{\partial}{\partial \rho} + k^2 - \lambda^2 \right) R(\rho)}}_{\text{function of } \rho}$$

yielding

$$\frac{\rho^2}{R(\rho)} \left(\frac{\partial^2}{\partial \rho^2} + \frac{1}{\rho} \frac{\partial}{\partial \rho} + k^2 - \lambda^2 \right) R(\rho) = \nu^2 \, , \qquad \frac{1}{\Phi(\phi)} \frac{\partial^2 \Phi(\phi)}{\partial \phi^2} = -\nu^2 \, .$$

Rearranging terms, we arrive at the set of equations given in the text of the problem.

Part 2

The solution of the equation for $Z(z)$ with the boundary conditions $Z(0) = Z(L) = 0$ yields the (normalized) eigenfunctions

$$Z_p(z) = \sqrt{\frac{2}{L}} \, \sin(p\pi z/L) \, , \qquad \lambda_p^2 = \left(\frac{p\pi}{L} \right)^2 \qquad p = 1, 2, \ldots$$

The solution of the equation for $\Phi(\phi)$ with periodic boundary conditions $\Phi(\phi + 2\pi) = \Phi(\phi)$ leads to the eigenfunctions ($\nu^2 = n^2$)

$$\Phi_n(\phi) = \frac{1}{\sqrt{2\pi}} \, e^{in\phi} \, , \qquad n = 0, \pm 1, \pm 2, \ldots \, ,$$

which are of course also eigenfunction of L_z with eigenvalue $n\hbar$, that is,

$$-i\hbar \frac{\partial}{\partial \phi} \Phi_n(\phi) = n\hbar \, \Phi_n(\phi) \, .$$

Part 3

We set $\gamma^2 = k^2 - \lambda^2$ to obtain the Bessel equation

$$\frac{\partial^2 R(\rho)}{\partial \rho^2} + \frac{1}{\rho} \frac{\partial R(\rho)}{\partial \rho} + \left(\gamma^2 - \frac{n^2}{\rho^2} \right) R(\rho) = 0 \, ,$$

with the general solution in $\rho_a < \rho < \rho_b$

$$R(\rho) = \alpha J_n(\gamma \rho) + \beta N_n(\gamma \rho) \, .$$

Imposing the boundary conditions $R(a) = R(b) = 0$ yields

$$\alpha J_n(\gamma a) + \beta N_n(\gamma a) = 0 \, , \qquad \alpha J_n(\gamma b) + \beta N_n(\gamma b) = 0 \, ,$$

and the homogeneous linear system in α and β has a non-trivial solution iff

$$\det \begin{bmatrix} J_n(\gamma a) & N_n(\gamma a) \\ J_n(\gamma b) & N_n(\gamma b) \end{bmatrix} = 0 \implies J_n(\gamma a) \, N_n(\gamma b) - J_n(\gamma b) \, N_n(\gamma a) = 0 \, .$$

We label the parameters γ satisfying the above equation as γ_{mn}, where $m = 1, 2, \ldots$ indicates the solution for a given n. The eigenenergies result from

$$\gamma^2 = k^2 - \lambda_p^2 \implies k_{mnp}^2 = \gamma_{mn}^2 + \left(\frac{p\pi}{L} \right)^2 \qquad \text{or} \qquad E_{mnp} = \frac{\hbar^2}{2\mu} \left[\gamma_{mn}^2 + \left(\frac{p\pi}{L} \right)^2 \right] \, ,$$

with corresponding radial functions given by

$$R_{mn}(\rho) = C_{mn}\left[N_n(\gamma_{mn}a)\,J_n(\gamma_{mn}\,\rho) - J_n(\gamma_{mn}a)\,N_n(\gamma_{mn}\,\rho)\right]\,,$$

where the C_{mn} are normalization constants determined via

$$|C_{mn}|^2 \int_a^b d\rho\,\rho|N_n(\gamma_{mn}a)\,J_n(\gamma_{mn}\,\rho) - J_n(\gamma_{mn}a)\,N_n(\gamma_{mn}\,\rho)|^2 = 1\,.$$

The normalized eigenfunctions of the Hamiltonian are then given by

$$\psi_{mnp}(\mathbf{r}) = R_{mn}(\rho)\,\frac{e^{in\phi}}{\sqrt{2\pi}}\,Z_p(z)\,.$$

Problem 15   Ahronov–Bohm Effect for Charged Spinless Particle Confined in Cylindrical Shell

Consider a particle of mass m and charge q confined in a hollow cylindrical shell whose axis coincides with the z-axis. The wave function is required to vanish on the inner and outer walls, with radii, respectively, a and b ($0 < a < b$), and also at the bottom and top surfaces, located at $z = 0$ and $z = L$. In the inner cylindrical region with $0 < \rho < a$ there is a uniform and constant magnetic field $\mathbf{B} = B\,\hat{z}$; you may assume that this field is produced by a very long ($L \gg 0$) solenoid whose axis coincides with the z-axis and which is enclosed by the inner cylindrical surface.

1. Obtain the vector potential $\mathbf{A}$ in the regions $0 < \rho < a$ and $\rho > a$.
2. Obtain the eigenenergies of the particle; note that they depend on the magnetic field even though the particle is confined in a region where there is no magnetic field (but there is a vector potential!).
3. Suppose the flux of the magnetic field is quantized,

$$\pi a^2 B = 2\pi\,\frac{\hbar c}{q}\,l \qquad \text{with} \;\; l = \pm 1, \pm 2, \dots .$$

What then happens to the eigenenergies?

Solution

Part 1

Stokes' theorem states that

$$\int_S dS\,\underbrace{(\boldsymbol{\nabla} \times \mathbf{A})}_{\mathbf{B}} \cdot \hat{\mathbf{n}} = \oint_C d\mathbf{l} \cdot \mathbf{A}\,,$$

namely, the flux of $\mathbf{B}$ through the surface S equals the line integral of $\mathbf{A}$ along the curve C enclosing the surface; here, $\hat{\mathbf{n}}$ is perpendicular to the surface, and the curve C is traversed either counterclockwise or clockwise as specified by the right-hand rule relative to $\hat{\mathbf{n}}$. If S is a disk parallel to the xy-plane (that is, perpendicular to the axis of the solenoid), we find that the flux of $\mathbf{B}$ is given by

$$\int_S dS\,\mathbf{B} \cdot \hat{\mathbf{n}} = \pi\rho^2 B \qquad \text{for} \;\; \rho < a$$

$$= \pi a^2 B \qquad \text{for} \;\; \rho > a\,.$$

The symmetry of the problem is such that $\mathbf{A}$ is tangent at any point along the edge of the disk, and hence

$$\oint_C d\mathbf{l} \cdot \mathbf{A} = 2\pi\rho A ,$$

yielding

$$A = \frac{\rho B}{2} \qquad \text{for } \rho < a$$

$$= \frac{a^2 B}{2\rho} \qquad \text{for } \rho > a .$$

In cylindrical coordinates (ρ, ϕ, z) the vector potential is $\mathbf{A} = A\,\hat{\boldsymbol{\phi}}$, where the unit vector $\hat{\boldsymbol{\phi}}$ is given by

$$\hat{\boldsymbol{\phi}} = -\sin\phi\,\hat{\mathbf{x}} + \cos\phi\,\hat{\mathbf{y}} ,$$

and is orthogonal to both $\hat{\boldsymbol{\rho}} = \cos\phi\,\hat{\mathbf{x}} + \sin\phi\,\hat{\mathbf{y}}$ and $\hat{\mathbf{z}}$. Thus, we see that, in the region $\rho > a$, the vector potential is given by

$$\mathbf{A} = -\frac{a^2 B}{2\rho}\sin\phi\,\hat{\mathbf{x}} + \frac{a^2 B}{2\rho}\cos\phi\,\hat{\mathbf{y}} \qquad \text{for } \rho > a .$$

Part 2

The Hamiltonian is

$$H = \frac{\mathbf{p}^2}{2m} - \frac{q}{2mc}\,(\mathbf{A}\cdot\mathbf{p} + \mathbf{p}\cdot\mathbf{A}) + \frac{q^2}{2mc^2}\,\mathbf{A}^2 ,$$

The term linear in $\mathbf{A}$ can be written as

$$\mathbf{A}\cdot\mathbf{p} + \mathbf{p}\cdot\mathbf{A} = 2\,\mathbf{A}\cdot\mathbf{p} + \sum_i [p_i, A_i] = 2\,\mathbf{A}\cdot\mathbf{p} - i\hbar\,\boldsymbol{\nabla}\cdot\mathbf{A} = 2\mathbf{A}\cdot\mathbf{p} ,$$

since the divergence of $\mathbf{A}$ vanishes; explicitly,

$$\boldsymbol{\nabla}\cdot\mathbf{A} = \frac{\partial}{\partial x}\underbrace{\left[-\frac{a^2 By}{2(x^2 + y^2)}\right]}_{A_x} + \frac{\partial}{\partial y}\underbrace{\left[\frac{a^2 Bx}{2(x^2 + y^2)}\right]}_{A_y} = 0 .$$

Inserting the explicit expression for $\mathbf{A}$ into the Hamiltonian, we obtain

$$H = \frac{\mathbf{p}^2}{2m} - \frac{q}{mc}\frac{a^2 B}{2\rho^2}\underbrace{\left(-yp_x + xp_y\right)}_{L_z} + \frac{q^2}{2mc^2}\underbrace{\frac{a^4 B^2}{4\rho^2}}_{\mathbf{A}^2} .$$

In cylindrical coordinates, we have

$$L_z = -i\hbar\,\frac{\partial}{\partial\phi} ,$$

and, using

$$\boldsymbol{\nabla}^2 = \frac{\partial^2}{\partial\rho^2} + \frac{1}{\rho}\frac{\partial}{\partial\rho} + \frac{1}{\rho^2}\frac{\partial^2}{\partial\phi^2} + \frac{\partial^2}{\partial z^2} ,$$

the Schrödinger equation reads

$$\left[-\frac{\hbar^2}{2m}\left(\frac{\partial^2}{\partial\rho^2} + \frac{1}{\rho}\frac{\partial}{\partial\rho} + \frac{\partial^2}{\partial z^2} \right) + \frac{L_z^2}{2m\rho^2} - \frac{q\,a^2 B}{2mc\rho^2}L_z + \frac{q^2 a^4 B^2}{8mc^2\rho^2} \right]\psi(\rho,\phi,z) = E\psi(\rho,\phi,z)\,,$$

which can also be written as

$$\left[-\frac{\hbar^2}{2m}\left(\frac{\partial^2}{\partial\rho^2} + \frac{1}{\rho}\frac{\partial}{\partial\rho} + \frac{\partial^2}{\partial z^2} \right) + \frac{1}{2m\rho^2}\left(L_z - \frac{qa^2 B}{2c} \right)^2 \right]\psi(\rho,\phi,z) = E\psi(\rho,\phi,z)\,.$$

We now proceed by separation of variables, positing $\psi(\rho,\phi,z) = R(\rho)\,\Phi(\phi)\,Z(z)$ as in Problem 15, to obtain

$$\frac{\partial^2 R(\rho)}{\partial\rho^2} + \frac{1}{\rho}\frac{\partial R(\rho)}{\partial\rho} + \left(k^2 - \lambda^2 - \frac{\mu^2}{\rho^2} \right)R(\rho) = 0\,, \qquad \left(L_z - \frac{qa^2 B}{2c} \right)^2\Phi(\phi) = \mu^2\hbar^2\Phi(\phi)$$

and

$$\frac{\partial^2 Z(z)}{\partial z^2} = -\lambda^2 Z(z)\,,$$

where $k^2 = 2mE/\hbar^2$ ($E > 0$ is the particle's energy) and μ^2 and λ^2 are for the moment taken to be real parameters. The boundary conditions are

$$R(a) = R(b) = 0\,, \qquad \Phi(\phi + 2\pi) = \Phi(\phi)\,, \qquad Z(0) = Z(L) = 0\,.$$

We can solve the following eigenvalue problem in ϕ,

$$\left(L_z - \frac{qa^2 B}{2c} \right)\Phi(\phi) = \mu\hbar\,\Phi(\phi) \implies \frac{\partial\Phi(\phi)}{\partial\phi} = i\left(\mu + \frac{qa^2 B}{2\hbar c} \right)\Phi(\phi)\,,$$

since it implies the eigenvalue problem obtained above for the square of the operator. The solution is

$$\Phi(\phi) \propto e^{i(\mu+\eta)\phi} \qquad \eta = \frac{qa^2 B}{2\hbar c}\,,$$

with the boundary condition requiring

$$e^{2\pi i(\mu+\eta)} = 1 \implies \mu_n = n - \eta \qquad \text{with} \qquad n = 0, \pm 1, \pm 2, \ldots$$

The normalized eigenfunctions in ϕ and z are given by

$$\Phi_n(\phi) = \frac{1}{\sqrt{2\pi}}\,e^{in\phi}\,, \qquad Z_p(z) = \sqrt{\frac{2}{L}}\,\sin(p\pi z/L) \qquad \text{with} \quad p = 1, 2, \ldots\,,$$

while the equation in ρ now reads

$$\frac{\partial^2 R(\rho)}{\partial\rho^2} + \frac{1}{\rho}\frac{\partial R(\rho)}{\partial\rho} + \left[k^2 - \left(\frac{p\pi}{L} \right)^2 - \frac{\mu_n^2}{\rho^2} \right]R(\rho) = 0\,,$$

which has the solution

$$R(\rho) = \alpha J_{\mu_n}(\gamma\rho) + \beta N_{\mu_n}(\gamma\rho)\,,$$

where $J_{\mu_n}(x)$ and $N_{\mu_n}(x)$ are the regular and irregular Bessel functions of order $\mu_n = n - \eta$; here μ_n is (in general) a real number and $\gamma^2 = k^2 - (p\pi/L)^2$. The boundary conditions (see again Problem 15) then yield the requirement that

$$J_{\mu_n}(\gamma a)\,N_{\mu_n}(\gamma b) - J_{\mu_n}(\gamma b)\,N_{\mu_n}(\gamma a) = 0\,,$$

leading to the eigenenergies

$$k^2_{mnp} = \gamma^2_{mn} + \left(\frac{p\pi}{L}\right)^2 \quad \text{or} \quad E_{mnp} = \frac{\hbar^2}{2m}\left[\gamma^2_{mn} + \left(\frac{p\pi}{L}\right)^2\right],$$

where γ_{mn} are the solutions of the eigenvalue equation with $m = 1, 2, \ldots$ for fixed μ_n. It is important to note that these energies depend on the magnetic field through the order μ_n of the Bessel functions and hence the roots γ_{mn}. Therefore, even though in the region of space in which the particle is confined there is no magnetic field, nevertheless the energy eigenvalues do depend on the field in an adjacent region (this is just one version of the Aharonov–Bohm effect); this is so because the conjugate momentum of a particle in an EM field is not simply the kinetic momentum $m\dot{\mathbf{r}}$ but also includes a dependence on the vector potential.

Part 3

If the flux of the magnetic field through the surface is quantized then the parameter η defined above is simply an integer, that is,

$$\eta = \pi a^2 B \frac{q}{2\pi\hbar c} = l.$$

In such a case, the transcendental equation determining the solution for γ is given by

$$J_{n-l}(\gamma a)\, N_{n-l}(\gamma b) - J_{n-l}(\gamma b)\, N_{n-l}(\gamma a) = 0,$$

which can be seen to be the same as that obtained for the case in which there is no magnetic field, after making the replacement $n - l \longrightarrow n$. We see that eigenstates with quantum numbers m, n, p with no **B**-field and $m, n + l, p$ in the presence of the **B**-field have the same energies.

Problem 16 Electron in a Uniform Magnetic Field

Consider an electron (charge $-e$, mass m, and spin $\mathbf{S}$) in a uniform magnetic field $\mathbf{B}$.

1. Write down the Hamiltonian including the magnetic interaction of the spin with the magnetic field (here, the electron gyromagnetic factor is taken as $g_e \approx 2$). Show, in particular, that H can also be written as

$$H = \frac{m}{2}(\boldsymbol{\sigma} \cdot \mathbf{v})^2, \qquad \mathbf{v} = \frac{1}{m}\left(\mathbf{p} + \frac{e}{c}\mathbf{A}\right),$$

   where $\mathbf{S} = (\hbar/2)\,\boldsymbol{\sigma}$.
2. Take $\mathbf{A} = \mathbf{B} \times \mathbf{r}/2$ as the vector potential, with $\mathbf{B}$ along the z-axis, and show that the orbital part of the Hamiltonian consists of two commuting terms, which we denote as $H_\perp$ and $H_\parallel$ (here, the symbols $\perp$ and $\parallel$ mean perpendicular and parallel to $\mathbf{B}$, respectively).
3. Show that $H_\perp$ can be cast into the form of a one-dimensional harmonic oscillator Hamiltonian,

$$H_\perp = \frac{P^2}{2m} + \frac{m\omega^2}{2}Q^2 \quad \text{with} \quad \omega = \frac{eB}{mc},$$

   by a suitable definition of the operators Q and P, with $[Q, P] = i\hbar$. Introduce the annihilation and creation operators a and $a^\dagger$ and obtain the energy eigenvalues and eigenstates of $H_\perp$.
4. Show that $H_\perp$ commutes with the component of the orbital angular momentum along the z-axis, denoted as L_z, and that L_z satisfies the commutation relation $[L_z, a^{\dagger n}] = n\hbar\, a^{\dagger n}$.

5. Obtain the $H_\perp$ ground-state wave function in plane polar coordinates by utilizing the relation $a|0_\perp\rangle = 0$ (see Problem 14), and show that it is also an eigenstate of L_z, with eigenvalues $-k\hbar$ where k is equal to 0 or a positive integer.

6. Using the results of parts 4 and 5, show that the states $|n_\perp, k\rangle$ are eigenstates of L_z with eigenvalues $-(k - n_\perp)\hbar$. Further, by examining the behavior of the associated wave functions $\psi_{n_\perp,k}(\rho, \phi)$ near the origin, show that $k \geq n_\perp$ must hold.

7. Obtain the eigenvalues of $H_\parallel$ and write down the common eigenstates of $H_\perp$, $H_\parallel$, and L_z.

8. Show that the spin magnetic interaction H_S can be written as

$$H_S = \hbar\omega \left(b^\dagger b - \frac{1}{2} \right) \qquad \text{with } b = \frac{1}{\hbar}(S_y + i S_x) \,,$$

and that b and $b^\dagger$ satisfy the anticommutation relation $\{ b, b^\dagger \} = 1$ and the properties $b^2 = 0 = b^{\dagger 2}$. Show, further, that the number operator $N_S = b^\dagger b$ has eigenvalues 0 or 1 and that the normalized eigenstate $|0\rangle$ corresponding to eigenvalue 0 is such that $b|0\rangle = 0$ and the normalized eigenstate $|1\rangle$ corresponding to eigenvalue 1 is given by $b^\dagger|0\rangle$.

9. Define the operator $R = \sqrt{\hbar\omega}\, a\, b^\dagger$ (here, a is the annihilation operator defined earlier) and show that R satisfies the anticommutation relation

$$\{R, R^\dagger\} = H_\perp + H_S \,.$$

10. The eigenstates of $H_\perp + H_S$ are the same as those of the number operators $N_\perp = a^\dagger a$ and $N_S = b^\dagger b$. We write these eigenstates as $|n_\perp, n_S\rangle = |n_\perp\rangle \otimes |n_S\rangle$, where $n_S = 0$ or 1. Show that R and $R^\dagger$ commute with $H_\perp + H_S$ and that, as a consequence, the eigenstates $|n_\perp, 0\rangle$ and $|n_\perp - 1, 1\rangle$ are degenerate in energy.

Solution

Part 1

The Hamiltonian is given by

$$H = \frac{1}{2m} \left(\mathbf{p} + \frac{e}{c} \mathbf{A} \right)^2 + \frac{e}{mc} \mathbf{S} \cdot \mathbf{B} \,,$$

or, in terms of the velocity operator,

$$H = \frac{m}{2} \mathbf{v}^2 + \frac{e\hbar}{2mc} \boldsymbol{\sigma} \cdot \mathbf{B} \,.$$

First, we consider the following commutator

$$[v_i, v_{j.}] = \frac{1}{m^2} [p_i + (e/c)A_i, p_j + (e/c)A_j] = \frac{e}{m^2 c} \left([p_i, A_j] + [A_i, p_j]\right)$$

$$= -i \frac{e\hbar}{m^2 c} \left(\partial_i A_j - \partial_j A_i\right) = -i \frac{e\hbar}{m^2 c} \sum_k \epsilon_{ijk} B_k \,.$$

This relation can also be written as

$$\mathbf{v} \times \mathbf{v} = -i \frac{e\hbar}{m^2 c} \mathbf{B} \,,$$

by noting that

$$\sum_k \epsilon_{ijk} (\mathbf{v} \times \mathbf{v})_k = \sum_{klm} \epsilon_{ijk} \epsilon_{klm} v_l v_m = \sum_{lm} (\delta_{il}\delta_{jm} - \delta_{im}\delta_{jl}) v_l v_m = v_i v_j - v_j v_i$$

$$= [v_i, v_j] = -i \frac{e\hbar}{m^2 c} \sum_k \epsilon_{ijk} B_k \,.$$

Then, the Pauli identity yields

$$H = \frac{m}{2} (\sigma \cdot \mathbf{v})^2 = \frac{m}{2} \left[\mathbf{v}^2 + i\sigma \cdot (\mathbf{v} \times \mathbf{v}) \right] = \frac{m}{2} \mathbf{v}^2 + i \frac{m}{2} \sigma \cdot \left(-i \frac{e\hbar}{m^2 c} \mathbf{B} \right) = \frac{m}{2} \mathbf{v}^2 + \frac{e\hbar}{2mc} \sigma \cdot \mathbf{B} \,.$$

Part 2

Take $\mathbf{B}$ along the z-axis, so that $\mathbf{A} = (-By/2, Bx/2, 0)$. We find

$$H_O = \frac{m}{2} \mathbf{v}^2 = \underbrace{\frac{1}{2m} \left(p_x - \frac{eB}{2c} y \right)^2 + \frac{1}{2m} \left(p_y + \frac{eB}{2c} x \right)^2}_{H_\perp} + \underbrace{\frac{p_z^2}{2m}}_{H_\parallel} \,,$$

and manifestly $[H_\perp, H_\parallel] = 0$.

Part 3

Using the commutation relation for the velocity components (with $\mathbf{B}$ along the z-axis), we have

$$[v_x, v_y] = -i \frac{e\hbar B}{m^2 c} = -i\hbar \frac{\omega}{m} \,, \qquad \omega = \frac{eB}{mc} > 0 \,.$$

We set

$$Q = \alpha \sqrt{\frac{m}{\omega}} v_y \,, \qquad P = \frac{1}{\alpha} \sqrt{\frac{m}{\omega}} v_x \implies [Q, P] = i\hbar \,,$$

where the parameter α is determined below. We have

$$H_\perp = \frac{m}{2} (v_x^2 + v_y^2) = \frac{\omega}{2} \left(\alpha^2 P^2 + \frac{Q^2}{\alpha^2} \right) \,,$$

yielding

$$\alpha^2 \frac{\omega}{2} = \frac{1}{2m} \quad \text{or} \quad \alpha = \pm \frac{1}{\sqrt{m\,\omega}} \implies H_\perp = \frac{P^2}{2m} + \frac{m\omega^2}{2} Q^2$$

and

$$Q = \frac{v_y}{\omega} \,, \qquad P = m v_x \,,$$

where the choice of positive α has been made. We now introduce the annihilation and creation operators, respectively a and $a^\dagger$, as

$$a = \frac{1}{\sqrt{2m\hbar\omega}} (iP + m\omega Q) = \sqrt{\frac{m}{2\hbar\omega}} (v_y + i v_x) \,, \qquad a^\dagger = \sqrt{\frac{m}{2\hbar\omega}} (v_y - i v_x) \,,$$

in terms of which we have

$$H_\perp = \hbar\omega \left(a^\dagger a + 1/2 \right) \,.$$

The operators a and $a^\dagger$ satisfy the usual commutation relations $\left[a, a^\dagger \right] = 1$. The eigenvalues and eigenstates are

$$E_{n_\perp} = \hbar\omega(n_\perp + 1/2) , \qquad |n_\perp\rangle = \frac{a^{\dagger n_\perp}}{\sqrt{n_\perp!}} |0_\perp\rangle , \qquad a|0_\perp\rangle = 0 .$$

It is clear that these eigenvalues must be degenerate, since the dimensionality of the original Hamiltonian is 2.

Part 4

Following the treatment in Section 12.3, the orbital Hamiltonian can be written as

$$H_O = \frac{\mathbf{p}^2}{2m} + \frac{eB}{2mc} L_z + \frac{e^2 B^2}{8mc^2} (x^2 + y^2) ,$$

which makes it plain that L_z commutes with H_O (and, of course, it commutes with H_S too). This result also follows by noting that the components of $\mathbf{L}$ (the generators of rotations) commute with the scalar operator $\mathbf{v}^2$ and hence with H_O.

Since $\mathbf{v}$ is a vector operator, we obtain that

$$[L_z, v_j] = i\hbar \sum_k \epsilon_{zjk} v_k \implies [L_z, v_x] = i\hbar v_y \text{ and } [L_z, v_y] = -i\hbar v_x ,$$

and hence

$$[L_z, a] = \sqrt{\frac{m}{2\hbar\omega}} [L_z, v_y + i v_x] = \sqrt{\frac{m}{2\hbar\omega}} \left(-i\hbar v_x - \hbar v_y\right) = -\hbar a \implies [L_z, a^\dagger] = \hbar a^\dagger ,$$

the last relation following from $-\hbar a^\dagger = [L_z, a]^\dagger = [a^\dagger, L_z] = -[L_z, a^\dagger]$. We also find that

$$[L_z, a^{\dagger n}] = \hbar a^{\dagger n} ,$$

since by induction we have

$$[L_z, a^{\dagger p}] = p\hbar a^{\dagger p} \implies [L_z, a^{\dagger p+1}] = a^\dagger \underbrace{[L_z, a^{\dagger p}]}_{p\hbar a^{\dagger p}} + \underbrace{[L_z, a^\dagger]}_{\hbar a^\dagger} a^{\dagger p} = (p+1)\hbar a^{\dagger p+1} .$$

Part 5

The condition $a|0_\perp\rangle = 0$ implies $(v_y + i v_x)|0\rangle = 0$, which in the coordinate representation reduces to

$$\left[\left(-i\hbar\partial_y + \frac{m\omega}{2} x\right) + i\left(-i\hbar\partial_x - \frac{m\omega}{2} y\right) \right] \psi_{0_\perp}(x, y) = 0 .$$

Multiplying both sides by $i/\hbar$ and rearranging terms yields

$$\left[\left(\partial_y + \frac{m\omega}{2\hbar} y\right) + i\left(\partial_x + \frac{m\omega}{2\hbar} x\right) \right] \psi_{0_\perp}(x, y) = 0 .$$

In plane polar coordinates (see part 3 of Problem 14),

$$x = \rho\cos\phi \qquad y = \rho\sin\phi ,$$

with

$$\partial_x = \cos\phi \frac{\partial}{\partial\rho} - \frac{\sin\phi}{\rho} \frac{\partial}{\partial\phi} , \qquad \partial_y = \frac{\partial}{\partial y} = \sin\phi \frac{\partial}{\partial\rho} + \frac{\cos\phi}{\rho} \frac{\partial}{\partial\phi} ,$$

the differential equation becomes

$$\left[\left(\sin\phi\,\frac{\partial}{\partial\rho} + \frac{\cos\phi}{\rho}\,\frac{\partial}{\partial\phi} + \frac{m\omega}{2\hbar}\,\rho\sin\phi\right) + i\left(\cos\phi\,\frac{\partial}{\partial\rho} - \frac{\sin\phi}{\rho}\,\frac{\partial}{\partial\phi} + \frac{m\omega}{2\hbar}\,\rho\cos\phi\right)\right]\psi_{0_\perp}(\rho,\phi) = 0\,,$$

or combining terms (and canceling a common factor $i\,e^{-i\phi}$)

$$\left(\frac{\partial}{\partial\rho} - \frac{i}{\rho}\,\frac{\partial}{\partial\phi} + \frac{m\omega}{2\hbar}\,\rho\right)\psi_{0_\perp}(\rho,\phi) = 0\,.$$

This last equation has the solution (see part 3 of Problem 13)

$$\psi_{0_\perp,k}(\rho,\phi) = \underbrace{C_k\,\rho^k\,e^{-m\omega\rho^2/(4\hbar)}}_{R_{0_\perp,k}(\rho)}\,e^{-ik\phi}\,,\qquad k = 0,1,2,\dots\,,$$

where k is zero or a positive integer (in order to make the solution non-singular at the origin) and C_k is a normalization factor. The ground-state energy of $H_\perp$ is infinitely degenerate and

$$L_z|0_\perp,k\rangle = -k\hbar\,|0_\perp,k\rangle\,,$$

since

$$-i\hbar\frac{\partial}{\partial\phi}\,\psi_{0_\perp,k}(\rho,\phi) = -i\hbar\frac{\partial}{\partial\phi}\,R_{0_\perp,k}(\rho)\,e^{-ik\phi} = -\hbar k\psi_{0_\perp,k}(\rho,\phi)\,.$$

Part 6

Using the result of part 4, we have

$$L_z|n_\perp,k\rangle = \frac{1}{\sqrt{n_\perp!}}\,L_z\,a^{\dagger n_\perp}\,|0_\perp,k\rangle = \frac{1}{\sqrt{n_\perp!}}\,\underbrace{\left(a^{\dagger n_\perp}\,L_z + n_\perp\hbar\,a^{\dagger n_\perp}\right)}_{\text{from }[L_z,\,a^{\dagger n_\perp}]=n_\perp\hbar a^{\dagger n_\perp}}|0_\perp,k\rangle$$

$$= \frac{1}{\sqrt{n_\perp!}}\,(-k + n_\perp)\hbar\,a^{\dagger n_\perp}\,|0_\perp,k\rangle = -(k - n_\perp)\hbar\,|n_\perp,k\rangle\,,$$

from which it is clear that the states $|n_\perp,k\rangle$ are eigenstates of L_z with eigenvalues $-\hbar(k - n_\perp)$. However, not all positive k are allowed; indeed, we must have $k \geq n_\perp$, to make the eigenvalues of L_z equal to 0 or a negative multiple of $\hbar$. This can be seen by considering the behavior of the eigenfunctions $\psi_{n_\perp,k}(\rho,\phi)$ near the origin. To this end, we use the relation

$$|n_\perp,k\rangle = \frac{1}{\sqrt{n_\perp!}}\,a^{\dagger n_\perp}\,|0_\perp,k\rangle \longrightarrow \psi_{n_\perp,k}(\rho,\phi) = \frac{1}{\sqrt{n_\perp!}}\,\left[a^\dagger(\rho,\phi)\right]^{n_\perp}\psi_{0_\perp,k}(\rho,\phi)\,.$$

The representation of $a^\dagger$ in plane polar coordinates follows from

$$a^\dagger(\rho,\phi) = \sqrt{\frac{m}{2\hbar\omega}}\,(v_y - i\,v_x) = \frac{1}{\sqrt{2m\hbar\omega}}\,\left[p_y + m\omega\,x - i\,(p_x - m\omega\,y)\right]$$

$$= \sqrt{\frac{\hbar}{2m\omega}}\,\left[-\partial_x + \frac{m\omega}{\hbar}\,x - i\left(\partial_y - \frac{m\omega}{\hbar}\,y\right)\right] = -\sqrt{\frac{\hbar}{2m\omega}}\,e^{i\phi}\left(\frac{\partial}{\partial\rho} + \frac{i}{\rho}\,\frac{\partial}{\partial\phi} - \frac{m\omega}{\hbar}\,\rho\right)\,,$$

so that

$$\psi_{n_\perp,k}(\rho,\phi) = \frac{(-1)^{n_\perp}}{\sqrt{n_\perp!}}\left(\frac{\hbar}{2m\omega}\right)^{n_\perp/2}\left[e^{i\phi}\left(\frac{\partial}{\partial\rho} + \frac{i}{\rho}\,\frac{\partial}{\partial\phi} - \frac{m\omega}{\hbar}\,\rho\right)\right]^{n_\perp}R_{0_\perp,k}(\rho)\,e^{-ik\phi}\,.$$

Expanding the contents of the square brackets $[\cdots]^{n_\perp}$ leads to terms of the type (among several others)

$$\left(e^{i\phi}\,\frac{\partial}{\partial\rho}\right)^{n_\perp} R_{0_\perp,k}(\rho)\,e^{-ik\phi} = e^{-i(k-n_\perp)\phi}\,\frac{\partial^{n_\perp}}{\partial\rho^{n_\perp}}\,\rho^k\,e^{-m\omega\rho^2/(4\hbar)}$$

$$\propto e^{-i(k-n_\perp)\phi}\,\rho^{k-n_\perp}\,e^{-m\omega\rho^2/(4\hbar)} + \cdots ,$$

and, to avoid singularities at the origin, we must have $k \geq n_\perp$. We conclude that the states $|n_\perp,k\rangle$ are simultaneous eigenstates of $H_\perp$ and L_z with eigenvalues $E_{n_\perp}$ and $-(k-n_\perp)\hbar$, with $k = n_\perp, n_\perp+1, n_\perp+2, \dots$.

Part 7

The Hamiltonian $H_\parallel = p_z^2/(2m)$, a free-particle Hamiltonian. It has eigenvalues $p_z^2/(2m)$, namely

$$H_\parallel\,|p_z\rangle = \frac{p_z^2}{2m}\,|p_z\rangle ,$$

where $|p_z\rangle$ are the eigenstates of the momentum operator component p_z. The common eigenstates of $H_\perp$, $H_\parallel$, and L_z are then given by (recall that L_z and $H_\parallel$ commute)

$$|n_\perp,k,p_z\rangle = |n_\perp,k\rangle \otimes |p_z\rangle .$$

Part 8

Given the definition of b, we have

$$b^\dagger b = \frac{1}{\hbar^2}\left(S_y - iS_x\right)\left(S_y + iS_x\right) = \frac{1}{\hbar^2}\left[S_x^2 + S_y^2 - i\left(S_x S_y - S_y S_x\right)\right] = \frac{1}{\hbar^2}\left(\mathbf{S}^2 - S_z^2 + \hbar S_z\right) = \frac{1}{2} + \frac{S_z}{\hbar} ,$$

and hence

$$H_S = \hbar\omega\left(b^\dagger b - \frac{1}{2}\right) = \hbar\omega\left[\left(\frac{1}{2} + \frac{S_z}{\hbar}\right) - \frac{1}{2}\right] = \omega S_z .$$

We also note that

$$b\,b^\dagger = \frac{1}{\hbar^2}\left(S_y + iS_x\right)\left(S_y - iS_x\right) = \frac{1}{\hbar^2}\left[S_x^2 + S_y^2 + i\left(S_x S_y - S_y S_x\right)\right] = \frac{1}{\hbar^2}\left(\mathbf{S}^2 - S_z^2 - \hbar S_z\right) = \frac{1}{2} - \frac{S_z}{\hbar} ,$$

which yields the anticommutator

$$\underbrace{b^\dagger b + b\,b^\dagger}_{\{b,b^\dagger\}} = \frac{1}{2} + \frac{S_z}{\hbar} + \frac{1}{2} - \frac{S_z}{\hbar} = 1 .$$

Similarly, we have

$$b^2 = \frac{1}{\hbar^2}\left(S_y + iS_x\right)^2 = \frac{1}{\hbar^2}\left[S_y^2 - S_x^2 + i\left(S_x S_y + S_y S_x\right)\right] = \frac{1}{4}\left[\sigma_y^2 - \sigma_x^2 + i\left(\sigma_x \sigma_y + \sigma_y \sigma_x\right)\right] = 0 ,$$

using the fact that $\sigma_\alpha^2 = 1$ and that different components of the Pauli matrices anticommute, that is $\sigma_\alpha \sigma_\beta + \sigma_\beta \sigma_\alpha = 0$ with $\alpha \neq \beta$. Taking the adjoint of the above relation, it follows that $b^{\dagger\,2} = 0$. Finally, suppose $|\phi\rangle$ is an eigenstate of $N_S = b^\dagger b$ with $N_S\,|\phi\rangle = c\,|\phi\rangle$. Then, it follows that $N_S^2\,|\phi\rangle = c^2\,|\phi\rangle$. Now, consider

$$\underbrace{(b^\dagger b)(b^\dagger b)}_{N_S^2} = b^\dagger \underbrace{\left(1 - b^\dagger b\right)}_{\text{from } \{b,b^\dagger\}=1} b = b^\dagger b - b^{\dagger\,2}\,b^2 = N_S ,$$

and $N_S^2 = N_S$. Thus, N_S is a projection operator and its eigenvalues must be necessarily 0 or 1, since

$$c^2 |\phi\rangle = N_S^2 |\phi\rangle = N_S |\phi\rangle = c |\phi\rangle \implies c^2 = c ,$$

yielding $c = 0$ or 1. Let $|0\rangle$ be the normalized eigenstate of N_S corresponding to eigenvalue 0; we have

$$b^\dagger b |0\rangle = 0 \implies 0 = \langle 0|b^\dagger b|0\rangle = \| b |0\rangle\|^2 ,$$

and, since the norm vanishes, we must have $b |0\rangle = 0$, and so b annihilates the state $|0\rangle$. Also, the state $b^\dagger |0\rangle$ is the normalized eigenstate $|1\rangle$ of N_S with eigenvalue 1; indeed,

$$N_S |1\rangle = b^\dagger b \, b^\dagger |0\rangle = b^\dagger (1 - b^\dagger b)|0\rangle = b^\dagger |0\rangle = |1\rangle , \qquad \langle 1|1\rangle = \langle 0| b \, b^\dagger |0\rangle = \langle 0| 1 - b^\dagger b |0\rangle = 1 .$$

Part 9

Since a and b act in different state spaces (respectively, orbital space and spin space), they commute with each other. Thus, we have

$$\underbrace{R R^\dagger + R^\dagger R}_{\{R,R^\dagger\}} = \hbar\omega \left(a b^\dagger b a^\dagger + b a^\dagger a b^\dagger \right) = \hbar\omega \left[b^\dagger b a a^\dagger + (1 - b^\dagger b) a^\dagger a \right]$$

$$= \hbar\omega \left[a^\dagger a + b^\dagger b \left(a a^\dagger - a^\dagger a \right) \right] = \hbar\omega \left(a^\dagger a + b^\dagger b \right) = H_\perp + H_S .$$

Part 10

Using the expressions for $H_\perp$ and H_S, we find

$$H_\perp + H_S = \hbar\omega(a^\dagger a + 1/2) + \hbar\omega(b^\dagger b - 1/2) = \hbar\omega(N_\perp + N_S) ,$$

and hence

$$\hbar\omega(N_\perp + N_S)|n_\perp, n_S\rangle = \hbar\omega \underbrace{(n_\perp + n_S)}_{n} |n_\perp, n_S\rangle .$$

The eigenvalue $n\hbar\omega$ is obtained from $n_\perp = n$ and $n_S = 0$ or $n_\perp = n - 1$ and $n_S = 1$. The commutator of $H_\perp + H_S$ with R vanishes,

$$[R, H_\perp + H_S] = (\hbar\omega)^{3/2}[a b^\dagger , a^\dagger a + b^\dagger b] = (\hbar\omega)^{3/2} \left([a, a^\dagger a] b^\dagger + a [b^\dagger , b^\dagger b] \right)$$

$$= (\hbar\omega)^{3/2} \left[a b^\dagger + a (b^\dagger b^\dagger b - b^\dagger b b^\dagger) \right] = (\hbar\omega)^{3/2} \left[a b^\dagger - a b^\dagger (1 - b^\dagger b) \right] = 0 ,$$

where we have used $b^{\dagger 2} = 0$ and the fundamental commutator $[a, a^\dagger] = 1$ and anticommutator $\{b, b^\dagger\} = 1$. Note that $R^\dagger$ also commutes with $H_\perp + H_S$; to see this just take the adjoint of $[R, H_\perp + H_S]$.

We now consider the effects of R and $R^\dagger$ on the basis of eigenstates of $H_\perp$ and H_S,

$$\underbrace{\sqrt{\hbar\omega} \, a b^\dagger}_{R} |n_\perp, 0\rangle = \sqrt{\hbar\omega \, n_\perp} \, |n_\perp - 1, 1\rangle , \qquad \underbrace{\sqrt{\hbar\omega} \, a^\dagger b}_{R^\dagger} |n_\perp - 1, 1\rangle = \sqrt{\hbar\omega \, n_\perp} \, |n_\perp, 0\rangle ,$$

and

$$R |n_\perp - 1, 1\rangle = R^\dagger |n_\perp, 0\rangle = 0 .$$

We conclude that the states $|n_\perp, 0\rangle$ and $|n_\perp - 1, 1\rangle$ are degenerate in energy, since R commutes with $H_\perp + H_S$ and hence

$$E_{n_\perp - 1,1} = \langle n_\perp - 1, 1 | H_\perp + H_S | n_\perp - 1, 1\rangle = \frac{\langle n_\perp, 0 | R^\dagger (H_\perp + H_S) R | n_\perp, 0\rangle}{\hbar \omega n_\perp}$$

$$= \frac{\langle n_\perp, 0 | R^\dagger R (H_\perp + H_S) | n_\perp, 0\rangle}{\hbar \omega n_\perp} = E_{n_\perp, 0} \frac{\langle n_\perp, 0 | R^\dagger R | n_\perp, 0\rangle}{\hbar \omega n_\perp} = E_{n_\perp, 0} \,,$$

where in the last step we used $\langle n_\perp, 0 | R^\dagger R | n_\perp, 0\rangle = \|R |n_\perp, 0\rangle\|^2 = \hbar \omega n_\perp \| |n_\perp - 1, 1\rangle\|^2 = \hbar \omega n_\perp$.

Problem 17   Spin Precession in a Magnetic Field

Consider a spin-1/2 particle with magnetic moment $g[q/(2mc)]\hat{\mathbf{S}}$ (g is the gyromagnetic factor) in a uniform and constant magnetic field $\mathbf{B}$ directed along the positive z-axis. Ignore orbital degrees of freedom. Suppose the particle at time $t = 0$ is polarized along the positive x-axis (that is, it is in state $|+\rangle_x$, an eigenstate of $\hat{S}_x$). Obtain the state at time t and calculate the expectation values of $\hat{S}_x$ and $\hat{S}_y$ at this time. Interpret the results.

Solution

Neglecting the motion of the particle in orbital space, the Hamiltonian simply reads

$$\hat{H} = -g \frac{qB}{2mc} \hat{S}_z \,,$$

with eigenstates and eigenvalues given by

$$\hat{H}|\pm\rangle = \pm \frac{\hbar\omega}{2} |\pm\rangle \,, \qquad \omega = -g \frac{q}{2mc} B \,,$$

where $|\pm\rangle$ are the eigenstates of $\hat{S}_z$ with eigenvalues $\pm\hbar/2$, respectively. At time $t = 0$ the particle is in the eigenstate of $\hat{S}_x$ corresponding to the eigenvalue $+\hbar/2$,

$$|\psi(0)\rangle = |+\rangle_x = \frac{1}{\sqrt{2}}(|+\rangle + |-\rangle) \,,$$

and hence at time t the state is given by

$$|\psi(t)\rangle = e^{-i\hat{H}t/\hbar} |\psi(0)\rangle = \frac{1}{\sqrt{2}} \left(e^{-i\omega t/2} |+\rangle + e^{i\omega t/2} |-\rangle \right) \,.$$

Up to an irrelevant phase factor, the state $|\psi(t)\rangle$ can also be written as

$$|\psi(t)\rangle = \frac{1}{\sqrt{2}} \left(|+\rangle + e^{i\omega t} |-\rangle \right) \,.$$

Recalling that the eigenstates of $\hat{S}_x$ and $\hat{S}_y$ are given by

$$|\pm\rangle_x = \frac{1}{\sqrt{2}}(|+\rangle \pm |-\rangle) \,, \qquad |\pm\rangle_y = \frac{1}{\sqrt{2}}(|+\rangle \pm i |-\rangle) \,,$$

we see that, as time progresses, the state $|\psi(t)\rangle$ "rotates" (that is, precesses) in the xy-plane with period $2\pi/\omega$; for example, at times $t_k = k\pi/(2\omega)$ we have

$$|\psi(t_1)\rangle = |+\rangle_y \,, \qquad |\psi(t_2)\rangle = |-\rangle_x \,, \qquad |\psi(t_3)\rangle = |-\rangle_y \qquad |\psi(t_4)\rangle = |+\rangle_x \,.$$

Note that our definition of the angular frequency is such that ω is either positive or negative depending on whether the charge q is either negative or positive. When $\omega > 0$ the precession of the state about $\mathbf{B}$ is anticlockwise, while it is clockwise when $\omega < 0$. The expectation values of $\hat{S}_x$ and $\hat{S}_y$ on the state $|\psi(t)\rangle$ are obtained as

$$\langle\hat{S}_x(t)\rangle = \langle\psi(t)|\hat{S}_x|\psi(t)\rangle \longrightarrow \underbrace{\frac{1}{\sqrt{2}}\begin{pmatrix}1 & e^{-i\omega t}\end{pmatrix}}_{\langle\psi(t)|}\underbrace{\frac{\hbar}{2}\begin{pmatrix}0 & 1\\1 & 0\end{pmatrix}}_{\hat{S}_x}\underbrace{\frac{1}{\sqrt{2}}\begin{pmatrix}1\\e^{i\omega t}\end{pmatrix}}_{|\psi(t)\rangle},$$

or $\langle\hat{S}_x(t)\rangle = (\hbar/2)\cos(\omega t)$. As one would expect, at times t_2 and t_4 the expectation values are given by, respectively, $-\hbar/2$ and $\hbar/2$, since at t_2 (t_4) the particle is in the eigenstate $|-\rangle_x$ ($|+\rangle_x$) of $\hat{S}_x$. Similarly, we find $\langle\hat{S}_y(t)\rangle = (\hbar/2)\sin(\omega t)$ and $\langle\hat{S}_z(t)\rangle = 0$. We could have anticipated the last result: since $\hat{S}_z$ commutes with the Hamiltonian, its expectation value is a constant of motion. It vanishes at $t=0$, and hence will vanish at any other time.

Problem 18 Spin Precession in a Magnetic Field: Alternative Treatment

Consider a spin-1/2 particle with magnetic moment $g[q/(2mc)]\hat{\mathbf{S}}$ (g is the gyromagnetic factor) in a uniform and constant magnetic field $\mathbf{B}$. Ignore orbital degrees of freedom. Obtain an equation for the time evolution of the expectation value of $\hat{\mathbf{S}}$ at time time t. Assuming that the magnetic field is along the z-axis and that the particle is polarized along the positive x-axis at time $t=0$, solve explicitly the equation for $\langle\hat{\mathbf{S}}(t)\rangle$ (see also Problem 17).

Solution

Utilizing Eq. (7.21), we find

$$\frac{d}{dt}\langle\hat{S}_j(t)\rangle = \frac{i}{\hbar}\langle\psi(t)|[\hat{H}, \hat{S}_j]|\psi(t)\rangle .$$

For a $\mathbf{B}$ field in a generic direction, the commutator is given by

$$[\hat{H}, \hat{S}_j] = -g\frac{q}{2mc}\sum_i B_i\underbrace{[\hat{S}_i, \hat{S}_j]}_{i\hbar\sum_k\epsilon_{ijk}\hat{S}_k} = -ig\frac{q\hbar}{2mc}\sum_{ik}\epsilon_{ijk}B_i\hat{S}_k = -ig\frac{q\hbar}{2mc}(\hat{\mathbf{S}}\times\mathbf{B})_j ,$$

which leads to

$$\frac{d}{dt}\langle\hat{\mathbf{S}}(t)\rangle = g\frac{q}{2mc}\langle\hat{\mathbf{S}}(t)\rangle\times\mathbf{B} .$$

This is the quantum analog of the classical equation of motion describing the precession of the angular momentum $\mathbf{L}$ about the axis defined by $\mathbf{B}$ due to the torque $\boldsymbol{\mu}_o\times\mathbf{B}$, namely

$$\frac{d\mathbf{L}}{dt} = \frac{q}{2mc}\mathbf{L}\times\mathbf{B} \qquad \text{(classical mechanics)} .$$

By taking $\mathbf{B}$ along $\hat{z}$ and defining the angular frequency $\omega = -gqB/(2mc)$, we arrive at

$$\frac{d}{dt}\langle\hat{S}_x(t)\rangle = -\omega\langle\hat{S}_y(t)\rangle , \qquad \frac{d}{dt}\langle\hat{S}_y(t)\rangle = \omega\langle\hat{S}_x(t)\rangle , \qquad \frac{d}{dt}\langle\hat{S}_z(t)\rangle = 0 ,$$

namely a set of first-order coupled differential equations. In order to solve them, we introduce the auxiliary function $\xi(t) = \langle \hat{S}_x(t) \rangle + i \langle \hat{S}_y(t) \rangle$, which has real and imaginary parts given by

$$\mathrm{Re}[\xi(t)] = \langle \hat{S}_x(t) \rangle , \qquad \mathrm{Im}[\xi(t)] = \langle \hat{S}_y(t) \rangle$$

and satisfies

$$\dot{\xi}(t) = i\omega\, \xi(t) \implies \xi(t) = \xi(0)\, e^{i\omega t} .$$

Since at $t = 0$ the expectation values are $\langle \hat{S}_x(0) \rangle = \hbar/2$ and $\langle \hat{S}_y(0) \rangle = 0$ given the initial state $|\psi(0)\rangle = |+\rangle_x$, we immediately obtain

$$\xi(t) = \frac{\hbar}{2} e^{i\omega t} \implies \langle \hat{S}_x(t) \rangle = \frac{\hbar}{2} \cos(\omega t) \qquad \text{and} \qquad \langle \hat{S}_y(t) \rangle = \frac{\hbar}{2} \sin(\omega t) ,$$

The precessional effect just discussed is used to determine precisely the magnetic moment of the muon μ^- (one of the lepton family, a heavier version of the electron e^-). The muon is unstable, and decays via the weak interaction according to $\mu^- \longrightarrow e^- \overline{\nu}_e \nu_\mu$, where $\overline{\nu}_e$ and ν_μ are respectively the electron anti-neutrino and muon neutrino. This decay has the interesting property that the electron tends to come off, in the center of mass frame where the initial μ^- is at rest, in the direction opposite to the spin of the muon. One can measure the magnetic moment of the μ^- by preparing a beam of muons polarized in the direction opposite to their direction of motion, which is along the $\hat{x}$-axis, say. This beam is then passed through a uniform magnetic field directed along the $\hat{z}$-axis for a time t, and the direction in which the electrons come off in the xy-plane is measured (this direction is preferentially opposite, as noted above, to that of the precessing muon spin). Taking into account the Lorentz force, the angle with respect to the direction of motion of the muons provides a direct measurement of the gyromagnetic factor for the muon, g_μ.

Problem 19 Magnetic Resonance

Consider a spin-1/2 particle with magnetic moment $g[q/(2mc)]\,\hat{\mathbf{S}}$ (g is the gyromagnetic factor) in a magnetic field consisting of a constant term $\mathbf{B}_0$ along the $\hat{z}$-axis and weak oscillating (time-dependent) term $\mathbf{B}_1(t)$ along the $\hat{x}$-axis, namely

$$\mathbf{B}(t) = B_0\, \hat{\mathbf{z}} + B_1\, \cos(\omega t)\, \hat{\mathbf{x}} .$$

Assume $B_0 \gg B_1$ and that the angular frequency ω of the oscillating field is close to $2\omega_0$, where

$$\omega_0 = -g\, \frac{qB_0}{4mc} .$$

Neglecting rapidly oscillating transients, solve explicitly for the time evolution of the spin state $|\psi(t)\rangle$ (ignore orbital degrees of freedom). Next, suppose that at time $t = 0$ the particle is polarized along the positive z-axis, that is, it is in the state $|+\rangle$. Calculate the probability for the spin-flip transition $|+\rangle \longrightarrow |-\rangle$ at time t, and show that it can be expressed as

$$p_{\text{flip}}(t) = \frac{1}{4} \underbrace{\frac{\omega_1^2}{(\omega_0 - \omega/2)^2 + \omega_1^2/4}}_{f(\omega)} \sin^2\left[\sqrt{\left(\omega_0 - \frac{\omega}{2}\right)^2 + \frac{\omega_1^2}{4}}\; t \right] ,$$

where we define

$$\omega_1 = -g\, \frac{qB_1}{4mc} .$$

Provide a plot of $f(\omega)$ as function of the the oscillating-field angular frequency ω for the case in which $\omega_1/\omega_0 = 1/10$.

Solution

The time-dependent Hamiltonian describing the interaction of a spin-1/2 particle with the combined magnetic field reads

$$\hat{H}(t) = -g\frac{q}{2mc}\left[B_0\,\hat{S}_z + B_1\,\cos(\omega t)\,\hat{S}_x\right] = \underbrace{2\omega_0\,\hat{S}_z}_{\hat{H}_0} + \underbrace{2\omega_1\,\cos(\omega t)\,\hat{S}_x}_{\hat{H}_1(t)}\,.$$

The spin state $|\psi(t)\rangle$ satisfies the time-dependent Schrödinger equation

$$i\hbar\frac{d}{dt}\,|\psi(t)\rangle = \hat{H}(t)\,|\psi(t)\rangle\,.$$

When projected over the basis $|+\rangle, |-\rangle$ which diagonalizes $\hat{S}_z$, the state and the Schrödinger equation are given by

$$|\psi(t)\rangle \longrightarrow \begin{pmatrix}\psi_+(t)\\ \psi_-(t)\end{pmatrix}\,,\qquad \psi_\pm(t) = \langle\pm|\psi(t)\rangle\,,$$

and

$$i\hbar\begin{pmatrix}\dot{\psi}_+(t)\\ \dot{\psi}_-(t)\end{pmatrix} = \hbar\underbrace{\begin{pmatrix}\omega_0 & \omega_1\,\cos(\omega t)\\ \omega_1\,\cos(\omega t) & -\omega_0\end{pmatrix}}_{\omega_0\,\underline{\sigma}_z + \omega_1\,\underline{\sigma}_x\,\cos(\omega t)}\begin{pmatrix}\psi_+(t)\\ \psi_-(t)\end{pmatrix}\,;$$

the dot symbol indicates the time derivative. Since we have assumed $B_0 \gg B_1$, we have $|\omega_0| \gg |\omega_1|$ (the sign of these angular frequencies depends on the sign of the charge). As a consequence, the "fast" time dependence is determined by ω_0. For this reason it is convenient to extract this dependence by defining

$$\begin{pmatrix}\psi_+(t)\\ \psi_-(t)\end{pmatrix} = \begin{pmatrix}e^{-i\omega_0 t}\,\phi_+(t)\\ e^{+i\omega_0 t}\,\phi_-(t)\end{pmatrix}\,.$$

Note that, if the weak time-dependent field were to be absent ($\omega_1 = 0$), then the state

$$\begin{pmatrix}e^{-i\omega_0 t}\,\phi_+\\ e^{+i\omega_0 t}\,\phi_-\end{pmatrix}\,,$$

would satisfy the Schrödinger equation with (the time-independent) $\hat{H}_0$. The (generally complex) constants $\phi_\pm$ would specify the components of the initial state $|\psi(0)\rangle$ in the basis $|\pm\rangle$.

Substituting the above ansatz into the Schrödinger equation gives

$$\begin{pmatrix}i\dot{\phi}_+(t)\,e^{-i\omega_0 t} + \omega_0\phi_+(t)\,e^{-i\omega_0 t}\\ i\dot{\phi}_-(t)\,e^{+i\omega_0 t} - \omega_0\phi_-(t)\,e^{+i\omega_0 t}\end{pmatrix} = \begin{pmatrix}\omega_0\phi_+(t)\,e^{-i\omega_0 t} + \omega_1\,\cos(\omega t)\,\phi_-(t)\,e^{+i\omega_0 t}\\ \omega_1\,\cos(\omega t)\,\phi_+(t)\,e^{-i\omega_0 t} - \omega_0\,\phi_-(t)\,e^{+i\omega_0 t}\end{pmatrix}\,,$$

and simplifying the above expression yields

$$i\begin{pmatrix}\dot{\phi}_+(t)\\ \dot{\phi}_-(t)\end{pmatrix} = \omega_1\,\cos(\omega t)\begin{pmatrix}\phi_-(t)\,e^{+2i\omega_0 t}\\ \phi_+(t)\,e^{-2i\omega_0 t}\end{pmatrix}\,;$$

this equation is equivalent to the initial Schrödinger equation (no approximations have been made up until this point). However, we are interested in solutions of this equation for driving frequencies ω

in the oscillating magnetic field which are close to $2\omega_0$. This allows us to introduce approximations. Consider the factor

$$\cos(\omega t)\, e^{2i\omega_0 t} = \frac{1}{2}\left(e^{i\omega t} + e^{-i\omega t}\right) e^{2i\omega_0 t} = \frac{1}{2}\left[e^{i(2\omega_0+\omega)t} + e^{i(2\omega_0-\omega)t}\right].$$

When ω is approximately equal to $2\omega_0$, the term $e^{i(2\omega_0-\omega)t}$ will vary slowly with time, while $e^{i(2\omega_0+\omega)t}$ will oscillate rapidly with time. If the solution is to be averaged over some time interval where the second term is varying rapidly, its contribution will tend to average to zero. By contrast, the slowly varying term will not be noticeably changed by the averaging. As a result we can make the "rotating-field approximation"

$$\cos(\omega t)\, e^{2i\omega_0 t} \approx \frac{1}{2}\, e^{i(2\omega_0-\omega)t}, \qquad \cos(\omega t)\, e^{-2i\omega_0 t} \approx \frac{1}{2}\, e^{-i(2\omega_0-\omega)t}.$$

With these approximations the time-dependent Schrödinger equation reduces to the following pair of coupled first-order differential equations:

$$\dot\phi_+(t) = -\frac{i}{2}\,\omega_1\, e^{i(2\omega_0-\omega)t}\,\phi_-(t), \qquad \dot\phi_-(t) = -\frac{i}{2}\,\omega_1\, e^{-i(2\omega_0-\omega)t}\,\phi_+(t).$$

These equations can be decoupled by taking the time derivative of, for example, the first equation to obtain

$$\ddot\phi_+(t) = \frac{1}{2}\,\omega_1(2\omega_0-\omega)\, e^{i(2\omega_0-\omega)t}\,\phi_-(t) - \frac{i}{2}\,\omega_1\, e^{i(2\omega_0-\omega)t}\,\dot\phi_-(t).$$

By re-expressing the term with $\phi_-(t)$ via

$$\phi_-(t) = i\,\frac{2}{\omega_1}\, e^{-i(2\omega_0-\omega)t}\,\dot\phi_+(t),$$

and the term with $\dot\phi_-(t)$ via the second equation above, we find

$$\ddot\phi_+(t) = i(2\omega_0-\omega)\,\dot\phi_+(t) - \frac{\omega_1^2}{4}\,\phi_+(t),$$

namely a second-order differential equation with constant coefficients. To solve it, assume a solution of the type $\phi_+(t) \propto e^{i\Omega t}$ which, when inserted into the equation above, requires

$$\Omega^2 = (2\omega_0-\omega)\,\Omega + \frac{\omega_1^2}{4},$$

which gives

$$\Omega_\pm = \omega_0 - \frac{\omega}{2} \pm \sqrt{\left(\omega_0 - \frac{\omega}{2}\right)^2 + \frac{\omega_1^2}{4}},$$

and hence $\phi_+(t)$ can be written as a linear combination of the two solutions,

$$\phi_+(t) = A_+\, e^{i\Omega_+ t} + A_-\, e^{i\Omega_- t}.$$

Using the equation above relating $\phi_-(t)$ to $\dot\phi_+(t)$, we can write

$$\phi_-(t) = -\frac{2}{\omega_1}\, e^{-i(2\omega_0-\omega)t}\left(A_+\,\Omega_+\, e^{i\Omega_+ t} + A_-\,\Omega_-\, e^{i\Omega_- t}\right)$$

$$= -\frac{2}{\omega_1}\left[A_+\,\Omega_+\, e^{i(\Omega_+ - 2\omega_0+\omega)t} + A_-\,\Omega_-\, e^{i(\Omega_- - 2\omega_0+\omega)t}\right],$$

which can be simplified by noting that $\Omega_\pm - 2\omega_0 + \omega = -\Omega_\mp$, to finally arrive at

$$\phi_-(t) = -\frac{2}{\omega_1}\left(A_+\,\Omega_+\,e^{-i\Omega_- t} + A_-\,\Omega_-\,e^{-i\Omega_+ t}\right).$$

The components $\psi_\pm(t)$ of the time-dependent state $|\psi(t)\rangle$ are then given by

$$\begin{pmatrix}\psi_+(t)\\ \psi_-(t)\end{pmatrix} = \begin{pmatrix}e^{-i\omega_0 t}\,\phi_+(t)\\ e^{+i\omega_0 t}\,\phi_-(t)\end{pmatrix} = \begin{pmatrix} A_+\,e^{i(\Omega_+ - \omega_0)t} + A_-\,e^{i(\Omega_- - \omega_0)t}\\ -(2/\omega_1)\left[A_+\,\Omega_+\,e^{-i(\Omega_- - \omega_0)t} + A_-\,\Omega_-\,e^{-i(\Omega_+ - \omega_0)t}\right]\end{pmatrix},$$

this approximate solution being valid in the limit $\omega \approx 2\omega_0$.

Now, we consider the special case where the spin is initially aligned along the $\hat{z}$-axis, namely

$$|\psi(0)\rangle \longrightarrow \begin{pmatrix}\psi_+(0)\\ \psi_-(0)\end{pmatrix} = \begin{pmatrix}1\\ 0\end{pmatrix}.$$

Matching the general solution at $t=0$ with the initial condition above allows us to determine the constants $A_\pm$; we must have

$$A_+ + A_- = 1\,, \qquad A_+\,\Omega_+ + A_-\,\Omega_- = 0\,,$$

and consequently

$$A_+ = \frac{\Omega_-}{\Omega_- - \Omega_+}\,, \qquad A_- = -\frac{\Omega_+}{\Omega_- - \Omega_+}\,.$$

Using these coefficients, the state components at time t become

$$\psi_+(t) = \frac{\Omega_-\,e^{i(\Omega_+ - \omega_0)t} - \Omega_+\,e^{i(\Omega_- - \omega_0)t}}{\Omega_- - \Omega_+}$$

and

$$\psi_-(t) = -\frac{2}{\omega_1}\frac{\Omega_+\Omega_-}{\Omega_- - \Omega_+}\left[e^{-i(\Omega_- - \omega_0)t} - e^{-i(\Omega_+ - \omega_0)t}\right].$$

The state is normalized at $t=0$ and will remain so at time t, since the time evolution operator is unitary; hence, $|\psi_+(t)|^2 + |\psi_-(t)|^2 = 1$. The term $|\psi_-(t)|^2$ represents the spin-flip probability $p_{\text{flip}}(t) = |\langle -|\psi(t)\rangle|^2$. We have

$$p_{\text{flip}}(t) = \frac{4}{\omega_1^2}\left(\frac{\Omega_+\Omega_-}{\Omega_- - \Omega_+}\right)^2\left|e^{-i\Omega_- t} - e^{-i\Omega_+ t}\right|^2 = \frac{8}{\omega_1^2}\left(\frac{\Omega_+\Omega_-}{\Omega_- - \Omega_+}\right)^2[1 - \cos[(\Omega_- - \Omega_+)t]]\,,$$

which, by inserting the definitions for $\Omega_\pm$ given above, can be expressed as

$$p_{\text{flip}}(t) = \frac{1}{4}\underbrace{\frac{\omega_1^2}{(\omega_0 - \omega/2)^2 + \omega_1^2/4}}_{f(\omega)}\,\sin^2\left[\sqrt{\left(\omega_0 - \frac{\omega}{2}\right)^2 + \frac{\omega_1^2}{4}}\,t\right],$$

where we recall that ω_0 and ω_1 are proportional to the magnetic field strengths and ω is the angular frequency of the oscillating field. As time progresses, the probability $p_{\text{flip}}(t)$ oscillates between 0 and $f(\omega) > 0$,

$$f(\omega) = \frac{\omega_1^2}{(2\omega_0 - \omega)^2 + \omega_1^2}\,,$$

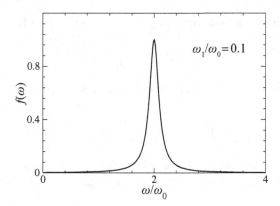

Fig. 12.2   The function $f(\omega)$.

where $f(\omega)$ determines the maximum probability for the given parameters. If the angular frequency of the oscillating field is tuned to match $2\omega_0$ (the precession frequency in the constant field $\mathbf{B}_0$) then we have $f(2\omega_0) = 1$ and

$$p_{\text{flip}}(t) = \sin^2 (\omega_1 t/2) \ ,$$

which shows that even a very weak oscillating field will eventually flip the initial spin-up state. The function $f(\omega)$ falls off rapidly for values of ω above or below $2\omega_0$, as can be seen in Fig. 12.2, which plots $f(\omega)$ as a function of ω/ω_0.

13

Addition of Angular Momenta

In this section, we consider the problem of combining two angular momenta, for example the orbital angular momentum and spin of a particle or the spins (or orbital angular momenta) of two particles. We denote these two angular momentum operators as $\hat{\mathbf{J}}_1$ and $\hat{\mathbf{J}}_2$. It is assumed that they commute with each other,

$$[\hat{J}_{1i}, \hat{J}_{2j}] = 0, \tag{13.1}$$

as is the case for the orbital angular momentum and spin of a particle or the the spins of two particles. The eigenstates and eigenvalues of each of these angular momenta are assumed to be known,

$$\hat{\mathbf{J}}_1^2 |\psi_{j_1 m_1}\rangle = j_1(j_1 + 1)\hbar^2 |\psi_{j_1 m_1}\rangle, \qquad \hat{J}_{1z} |\psi_{j_1 m_1}\rangle = m_1 \hbar |\psi_{j_1 m_1}\rangle, \tag{13.2}$$

$$\hat{\mathbf{J}}_2^2 |\psi_{j_2 m_2}\rangle = j_2(j_2 + 1)\hbar^2 |\psi_{j_2 m_2}\rangle, \qquad \hat{J}_{2z} |\psi_{j_2 m_2}\rangle = m_2 \hbar |\psi_{j_2 m_2}\rangle, \tag{13.3}$$

where j_1 and j_2 are either integers or semi-integers, $-j_1 \leq m_1 \leq j_1$, and $-j_2 \leq m_2 \leq j_2$. The set of commuting operators $\hat{\mathbf{J}}_1^2, \hat{\mathbf{J}}_2^2, \hat{J}_{1z}$, and $\hat{J}_{2z}$ have simultaneous eigenstates consisting of tensor products of the eigenstates of $\hat{\mathbf{J}}_1^2, \hat{J}_{1z}$ and those of $\hat{\mathbf{J}}_2^2, \hat{J}_{2z}$,

$$|\psi_{m_1 m_2}^{j_1 j_2}\rangle = |\psi_{j_1 m_1}\rangle \otimes |\psi_{j_2 m_2}\rangle. \tag{13.4}$$

For fixed j_1 and j_2, there are $(2j_1 + 1)(2j_2 + 1)$ of these eigenstates. Next, we define the operator

$$\hat{\mathbf{J}} = \hat{\mathbf{J}}_1 + \hat{\mathbf{J}}_2, \tag{13.5}$$

itself an angular momentum, since it satisfies the commutation relations

$$[\hat{J}_i, \hat{J}_j] = [\hat{J}_{1i} + \hat{J}_{2i}, \hat{J}_{1j} + \hat{J}_{2j},] = i\hbar \sum_k \epsilon_{ijk}(\hat{J}_{1k} + \hat{J}_{2k}) = i\hbar \sum_k \epsilon_{ijk} \hat{J}_k. \tag{13.6}$$

The operators $\hat{\mathbf{J}}_1^2, \hat{\mathbf{J}}_2^2, \hat{\mathbf{J}}^2$, and $\hat{J}_z$ form a commuting set, and the simultaneous eigenstates of these operators are denoted as $|\psi_{jm}^{j_1 j_2}\rangle$, where

$$\hat{\mathbf{J}}_1^2 |\psi_{jm}^{j_1 j_2}\rangle = j_1(j_1 + 1)\hbar^2 |\psi_{jm}^{j_1 j_2}\rangle, \qquad \hat{\mathbf{J}}_2^2 |\psi_{jm}^{j_1 j_2}\rangle = j_2(j_2 + 1)\hbar^2 |\psi_{jm}^{j_1 j_2}\rangle \tag{13.7}$$

and

$$\hat{\mathbf{J}}^2 |\psi_{jm}^{j_1 j_2}\rangle = j(j + 1)\hbar^2 |\psi_{jm}^{j_1 j_2}\rangle, \qquad \hat{J}_z |\psi_{jm}^{j_1 j_2}\rangle = m\hbar |\psi_{jm}^{j_1 j_2}\rangle; \tag{13.8}$$

j is either an integer or a semi-integer with $-j \leq m \leq j$. We ask: given j_1 and j_2, what are the allowed j-values? How are the eigenstates $|\psi_{jm}^{j_1 j_2}\rangle$ related to the eigenstates $|\psi_{m_1 m_2}^{j_1 j_2}\rangle$ (and vice versa)? We address these questions in the next two sections.

13.1 Eigenvalues of the Total Angular Momentum $\hat{\mathbf{J}}^2$

Below we establish that, for given j_1 and j_2, the allowed j-values are

$$j = |j_1 - j_2|, |j_1 - j_2| + 1, \ldots, j_1 + j_2 , \tag{13.9}$$

so that j runs from a minimum value of $|j_1 - j_2|$ to a maximum value of $j_1 + j_2$ in steps of one unit; for example, adding together an orbital angular momentum $l = 1$ and a spin $s = 1/2$, the allowed values for j are $1/2$ and $3/2$. To see how this comes about, we note the following.

- There are two "special" eigenstates of the set $\hat{\mathbf{J}}_1^2, \hat{\mathbf{J}}_2^2, \hat{J}_{1z}$, and $\hat{J}_{2z}$; these also happen to be eigenstates of the set $\hat{\mathbf{J}}_1^2, \hat{\mathbf{J}}_2^2, \hat{\mathbf{J}}^2$, and $\hat{J}_z$:

$$|\psi_{j_1 j_2}^{j_1 j_2}\rangle = |\psi_{j_1 j_1}\rangle \otimes |\psi_{j_2 j_2}\rangle , \qquad |\psi_{-j_1 - j_2}^{j_1 j_2}\rangle = |\psi_{j_1 - j_1}\rangle \otimes |\psi_{j_2 - j_2}\rangle , \tag{13.10}$$

where m_1 and m_2 assume their maximum values j_1 and j_2, or minimum values $-j_1$ and $-j_2$, respectively. It is easily seen that these are also eigenstates of $\hat{\mathbf{J}}^2$ and $\hat{J}_z$ (they are obviously eigenstates of $\hat{\mathbf{J}}_1^2$ and $\hat{\mathbf{J}}_2^2$); indeed,

$$\hat{J}_z |\psi_{j_1 j_2}^{j_1 j_2}\rangle = (\hat{J}_{1z} + \hat{J}_{2z})|\psi_{j_1 j_1}\rangle \otimes |\psi_{j_2 j_2}\rangle = (j_1 + j_2)\hbar |\psi_{j_1 j_1}\rangle \otimes |\psi_{j_2 j_2}\rangle , \tag{13.11}$$

and so the eigenvalue of $\hat{J}_z$ is $(j_1 + j_2)\hbar$. We find similarly that $|\psi_{-j_1 - j_2}^{j_1 j_2}\rangle$ is an eigenstate of $\hat{J}_z$ with eigenvalue $-(j_1 + j_2)\hbar$. From the known properties of the spectrum of an angular momentum operator, we expect that both these states are eigenstates of $\hat{\mathbf{J}}^2$ with eigenvalue $(j_1 + j_2)(j_1 + j_2 + 1)\hbar^2$. We verify this fact explicitly:

$$\hat{\mathbf{J}}^2 |\psi_{j_1 j_2}^{j_1 j_2}\rangle = [\hat{\mathbf{J}}_1^2 + \hat{\mathbf{J}}_2^2 + 2\hat{J}_{1z}\hat{J}_{2z} + \underbrace{\hat{J}_{1+}\hat{J}_{2-} + \hat{J}_{1-}\hat{J}_{2+}}_{2(\hat{J}_{1x}\hat{J}_{2x} + \hat{J}_{1y}\hat{J}_{2y})}]|\psi_{j_1 j_1}\rangle \otimes |\psi_{j_2 j_2}\rangle$$

$$= \hbar^2 [\underbrace{j_1(j_1 + 1) + j_2(j_2 + 1) + 2j_1 j_2}_{j_1(j_1 + j_2 + 1) + j_2(j_1 + j_2 + 1)}]|\psi_{j_1 j_1}\rangle \otimes |\psi_{j_2 j_2}\rangle = \hbar^2 (j_1 + j_2)(j_1 + j_2 + 1)|\psi_{j_1 j_2}^{j_1 j_2}\rangle ,$$

$$\tag{13.12}$$

where in the first line we introduced the raising and lowering operators

$$\hat{J}_{1\pm} = \hat{J}_{1x} \pm i\hat{J}_{1y} , \qquad \hat{J}_{2\pm} = \hat{J}_{2x} \pm i\hat{J}_{2y} , \tag{13.13}$$

and in obtaining the second line we used the property

$$\hat{J}_{1+}|\psi_{j_1 j_1}\rangle = 0 , \qquad \hat{J}_{2+}|\psi_{j_2 j_2}\rangle = 0 . \tag{13.14}$$

A similar reasoning, but using

$$\hat{J}_{1-}|\psi_{j_1 - j_1}\rangle = 0 , \qquad \hat{J}_{2-}|\psi_{j_2 - j_2}\rangle = 0 , \tag{13.15}$$

shows that $|\psi_{-j_1 - j_2}^{j_1 j_2}\rangle$ is an eigenstate of $\hat{\mathbf{J}}^2$ corresponding to the same eigenvalue as above. We conclude that $j^* = j_1 + j_2$ is one, in fact the largest, of the allowed j-values, and

$$\hat{\mathbf{J}}^2 |\psi_{j^* \pm j^*}^{j_1 j_2}\rangle = j^*(j^* + 1)\hbar^2 |\psi_{j^* \pm j^*}^{j_1 j_2}\rangle , \qquad \hat{J}_z |\psi_{j^* \pm j^*}^{j_1 j_2}\rangle = \pm j^*\hbar |\psi_{j^* \pm j^*}^{j_1 j_2}\rangle , \tag{13.16}$$

where the eigenstates of $|\psi_{j^* \pm j^*}^{j_1 j_2}\rangle$ of $\hat{\mathbf{J}}_1^2, \hat{\mathbf{J}}_2^2, \hat{\mathbf{J}}^2$, and $\hat{J}_z$ are simply related to those of $\hat{\mathbf{J}}_1^2, \hat{\mathbf{J}}_2^2, \hat{J}_{1z}$, and $\hat{J}_{2z}$,

$$|\psi_{j^* \pm j^*}^{j_1 j_2}\rangle = |\psi_{\pm j_1 \pm j_2}^{j_1 j_2}\rangle , \tag{13.17}$$

and it should be emphasized that we have made a choice of phase here.

- We can start from the state $|\psi_{j^*j^*}^{j_1j_2}\rangle$, and by repeated applications of the lowering operator

$$\hat{J}_- = \hat{J}_x - i\hat{J}_y = \hat{J}_{1-} + \hat{J}_{2-}\,, \tag{13.18}$$

generate the eigenstates of $\hat{J}_z$ having eigenvalues $(j^*-1)\hbar, (j^*-2)\hbar, \ldots, -j^*\hbar$, all with fixed eigenvalue $j^*(j^*+1)\hbar^2$ of $\hat{\mathbf{J}}^2$, namely the multiplet of $\hat{\mathbf{J}}^2$ having $j=j^*$; specifically,

$$|\psi_{j^*,j^*-1}^{j_1j_2}\rangle \propto \hat{J}_-|\psi_{j^*j^*}^{j_1j_2}\rangle\,, \qquad |\psi_{j^*,j^*-2}^{j_1j_2}\rangle \propto (\hat{J}_-)^2|\psi_{j^*j^*}^{j_1j_2}\rangle\,, \ldots \tag{13.19}$$

There is a total $2j^*+1$ of these states. Since $\hat{J}_-$ commutes with both $\hat{\mathbf{J}}_1^2$ and $\hat{\mathbf{J}}_2^2$, each is also an eigenstate of these operators, with eigenvalues, respectively, $j_1(j_1+1)\hbar^2$ and $j_2(j_2+1)\hbar^2$. We could have started equally well from the state $|\psi_{j^*,-j^*}^{j_1j_2}\rangle$ and moved up the ladder by repeatedly applying $\hat{J}_+$.

- We now consider the eigenstate of $\hat{J}_z$ with eigenvalue $(j^*-1)\hbar$, that is,

$$\begin{aligned}
|\psi_{j^*,j^*-1}^{j_1j_2}\rangle &= \frac{1}{\hbar\sqrt{j^*(j^*+1)-j^*(j^*-1)}}\hat{J}_-|\psi_{j^*j^*}^{j_1j_2}\rangle = \frac{1}{\hbar\sqrt{2j^*}}(\hat{J}_{1-}+\hat{J}_{2-})|\psi_{j_1j_1}\rangle \otimes |\psi_{j_2j_2}\rangle \\
&= \frac{1}{\sqrt{2j^*}}\Big[\sqrt{j_1(j_1+1)-j_1(j_1-1)}|\psi_{j_1j_1-1}\rangle \otimes |\psi_{j_2j_2}\rangle \\
&\quad + \sqrt{j_2(j_2+1)-j_2(j_2-1)}|\psi_{j_1j_1}\rangle \otimes |\psi_{j_2j_2-1}\rangle\Big] \\
&= \sqrt{\frac{j_1}{j^*}}|\psi_{j_1j_1-1}\rangle \otimes |\psi_{j_2j_2}\rangle + \sqrt{\frac{j_2}{j^*}}|\psi_{j_1j_1}\rangle \otimes |\psi_{j_2j_2-1}\rangle\,;
\end{aligned} \tag{13.20}$$

thus it is a (normalized) linear combination of the two states $|\psi_{j_1j_1-1}\rangle\otimes|\psi_{j_2j_2}\rangle$ and $|\psi_{j_1j_1}\rangle\otimes|\psi_{j_2j_2-1}\rangle$, each having eigenvalue $(j^*-1)\hbar$ of $\hat{J}_z$. Consider the (normalized) linear combination

$$|\psi_{j^*-1,j^*-1}^{j_1j_2}\rangle = -\sqrt{\frac{j_2}{j^*}}|\psi_{j_1j_1-1}\rangle \otimes |\psi_{j_2j_2}\rangle + \sqrt{\frac{j_1}{j^*}}|\psi_{j_1j_1}\rangle \otimes |\psi_{j_2j_2-1}\rangle\,, \tag{13.21}$$

orthogonal to that of Eq. (13.20) (another choice of phase has been made). It is clear that it is an eigenstate of $\hat{J}_z$ with eigenvalue $(j^*-1)\hbar$ as well as an eigenstate of $\hat{\mathbf{J}}_1^2$ and $\hat{\mathbf{J}}_2^2$ with eigenvalues, respectively $j_1(j_1+1)\hbar^2$ and $j_2(j_2+1)\hbar^2$. It is straightforward to verify that it is also an eigenstate of $\hat{\mathbf{J}}^2$ with eigenvalues $j^*(j^*-1)\hbar^2$, by expressing $\hat{\mathbf{J}}^2$ as in Eq. (13.12).

- By repeated applications of $\hat{J}_-$ we construct the eigenstates

$$|\psi_{j^*-1,j^*-2}^{j_1j_2}\rangle \propto \hat{J}_-|\psi_{j^*-1,j^*-1}^{j_1j_2}\rangle\,, \qquad |\psi_{j^*-1,j^*-3}^{j_1j_2}\rangle \propto (\hat{J}_-)^2|\psi_{j^*-1,j^*-1}^{j_1j_2}\rangle\,, \ldots \tag{13.22}$$

of $\hat{J}_z$ with eigenvalues $(j^*-2)\hbar, (j^*-3)\hbar, \ldots, -(j^*-1)\hbar$, all with the fixed eigenvalue $j^*(j^*-1)\hbar^2$ of $\hat{\mathbf{J}}^2$ (these form, the multiplet corresponding to $j=j^*-1$).

- Consider the two eigenstates of $\hat{J}_z$ with eigenvalue $(j^*-2)\hbar$ belonging to the multiplets of $\hat{\mathbf{J}}^2$ corresponding to $j=j^*$ and $j=j^*-1$. They are easily seen to be (orthogonal) linear combinations of the three states $|\psi_{j_1j_1-2}\rangle\otimes|\psi_{j_2j_2}\rangle$, $|\psi_{j_1j_1-1}\rangle\otimes|\psi_{j_2j_2-1}\rangle$, and $|\psi_{j_1j_1}\rangle\otimes|\psi_{j_2j_2-2}\rangle$. Out of these three states, we can construct a third linear combination orthogonal to the previous two (and yet another phase choice is made). Then by applying $\hat{J}_z$ and $\hat{\mathbf{J}}^2$ to it, we verify that it is an eigenstate of these operators with eigenvalues, respectively, $(j^*-2)\hbar$ and $(j^*-1)(j^*-2)\hbar^2$, corresponding to $j(j+1)\hbar^2$ with $j=j^*-2$. By repeated application of $\hat{J}_-$ we generate all remaining eigenstates in the multiplet having $j=j^*-2$.

- It should now be clear that the procedure can be iterated to generate the multiplets of $\hat{\mathbf{J}}^2$ having $j=j^*-3, j^*-4, \ldots$; clearly, however, the sequence of multiplets has to terminate at some point, for two reasons. First, since the operator $\hat{\mathbf{J}}$ is an angular momentum, the eigenvalues of $\hat{\mathbf{J}}^2$ are

$j(j+1)\hbar^2$ with $j \geq 0$. Second, as we have already noted, for fixed j_1 and j_2 the total number of states is $(2j_1 + 1)(2j_2 + 1)$. This number must be preserved in going from the eigenstates of the set $\hat{\mathbf{J}}_1^2$, $\hat{\mathbf{J}}_2^2$, $\hat{J}_{1z}$, and $\hat{J}_{2z}$ to those of the set $\hat{\mathbf{J}}_1^2$, $\hat{\mathbf{J}}_2^2$, $\hat{\mathbf{J}}^2$, and $\hat{J}_z$. Indeed, this fact allows us to determine that the minimum value of j is $|j_1 - j_2|$; see Problem 1

To summarize, we have outlined a procedure to construct the simultaneous eigenstates of the set $\hat{\mathbf{J}}_1^2$, $\hat{\mathbf{J}}_2^2$, $\hat{\mathbf{J}}^2$, and $\hat{J}_z$. For given j_1 and j_2, we have determined that the allowed j-values range from $|j_1 - j_2|$ to $j_1 + j_2$ in steps of one unit. The eigenvalues of $\hat{\mathbf{J}}^2$ are $j(j+1)\hbar^2$ and those of $\hat{J}_z$ are $m\hbar$ with $m = -j, \ldots, j$, namely, there are $2j + 1$ values. In the process, we have also determined, in the subspace with fixed j_1 and j_2, the coefficients of the transformation that maps the basis consisting of eigenstates $|\psi_{m_1 m_2}^{j_1 j_2}\rangle$ of $\hat{\mathbf{J}}_1^2$, $\hat{J}_{1z}$, $\hat{\mathbf{J}}_2^2$, and $\hat{J}_{2z}$ into the basis of eigenstates $|\psi_{jm}^{j_1 j_2}\rangle$ of $\hat{\mathbf{J}}_1^2$, $\hat{\mathbf{J}}_2^2$, $\hat{\mathbf{J}}^2$, and $\hat{J}_z$. We now turn our attention to discuss a few properties of these coefficients, known as Clebsch–Gordan coefficients.

13.2 Clebsch–Gordan Coefficients

In the subspace with fixed j_1 and j_2, we have established that the states in the two bases are related via

$$|\psi_{jm}^{j_1 j_2}\rangle = \sum_{m_1 m_2} C_{j_1 j_2}(m_1 m_2; jm) |\psi_{m_1 m_2}^{j_1 j_2}\rangle, \qquad C_{j_1 j_2}(m_1 m_2; jm) = \langle \psi_{m_1 m_2}^{j_1 j_2} | \psi_{jm}^{j_1 j_2} \rangle, \qquad (13.23)$$

and the procedure we outlined in the previous section, concurrently with the phase choices we made, shows that the Clebsch–Gordan coefficients $C_{j_1 j_2}(m_1 m_2; jm)$ are all real,[1]

$$C_{j_1 j_2}(m_1 m_2; jm) = C_{j_1 j_2}^*(m_1 m_2; jm) \implies \langle \psi_{m_1 m_2}^{j_1 j_2} | \psi_{jm}^{j_1 j_2} \rangle = \langle \psi_{jm}^{j_1 j_2} | \psi_{m_1 m_2}^{j_1 j_2} \rangle. \qquad (13.24)$$

Of course, they are non-vanishing only if $j = |j_1 - j_2|, |j_1 - j_2| + 1, \ldots, j_1 + j_2$ and $m = m_1 + m_2$, as is apparent from the explicit construction of the previous section. The last requirement also follows by considering that, since $\hat{J}_z = \hat{J}_{1z} + \hat{J}_{2z}$, we have

$$m\hbar |\psi_{jm}^{j_1 j_2}\rangle = \hat{J}_z |\psi_{jm}^{j_1 j_2}\rangle = \sum_{m_1 m_2} C_{j_1 j_2}(m_1 m_2; jm)(\hat{J}_{1z} + \hat{J}_{2z}) |\psi_{m_1 m_2}^{j_1 j_2}\rangle$$

$$= \hbar \sum_{m_1 m_2} (m_1 + m_2) C_{j_1 j_2}(m_1 m_2; jm) |\psi_{m_1 m_2}^{j_1 j_2}\rangle, \qquad (13.25)$$

which can be written as

$$\sum_{m_1 m_2} [m - (m_1 + m_2)] C_{j_1 j_2}(m_1 m_2; jm) |\psi_{m_1 m_2}^{j_1 j_2}\rangle = 0, \qquad (13.26)$$

yielding

$$C_{j_1 j_2}(m_1 m_2; jm) = 0 \qquad \text{if } m \neq m_1 + m_2. \qquad (13.27)$$

The orthonormality and completeness (in the subspace with fixed j_1 and j_2) of the states $|\psi_{jm}^{j_1 j_2}\rangle$ has two important implications for the Clebsch–Gordan coefficients:

---

[1] Several different notations are used for the Clebsch–Gordan coefficients. Apart from that adopted above, other common notations are $\langle j_1 m_1, j_2 m_2 | j_1 j_2 jm\rangle$ and $C_{m_1 m_2 m}^{j_1 j_2 j}$. More importantly, different authors may use different phase choices.

1. Since $\langle \psi_{jm}^{j_1 j_2} | \psi_{j'm'}^{j_1 j_2} \rangle = \delta_{jj'} \delta_{mm'}$, it follows that

$$
\delta_{jj'} \delta_{mm'} = \sum_{m_1 m_2} \sum_{m'_1 m'_2} \underbrace{C_{j_1 j_2}(m_1 m_2; jm)}_{\text{real hence c.c. unnecessary}} C_{j_1 j_2}(m'_1 m'_2; j'm') \underbrace{\langle \psi_{m_1 m_2}^{j_1 j_2} | \psi_{m'_1 m'_2}^{j_1 j_2} \rangle}_{\delta_{m_1 m'_1} \delta_{m_2 m'_2}}
$$

$$
= \sum_{m_1 m_2} C_{j_1 j_2}(m_1 m_2; jm) \, C_{j_1 j_2}(m_1 m_2; j'm') \,, \tag{13.28}
$$

where in the last line we have exploited the orthonormality of the basis $|\psi_{m_1 m_2}^{j_1 j_2}\rangle$.

2. Since both bases are complete in the subspace (j_1, j_2),

$$
\mathbb{1}_{(j_1, j_2)} = \sum_{j=|j_1-j_2|}^{j_1+j_2} \sum_{m=-j}^{j} |\psi_{jm}^{j_1 j_2}\rangle \langle \psi_{jm}^{j_1 j_2}| = \sum_{m_1=-j_1}^{j_1} \sum_{m_2=-j_2}^{j_2} |\psi_{m_1 m_2}^{j_1 j_2}\rangle \langle \psi_{m_1 m_2}^{j_1 j_2}| \,, \tag{13.29}
$$

it follows that

$$
\delta_{m_1 m'_1} \delta_{m_2 m'_2} = \underbrace{\langle \psi_{m_1 m_2}^{j_1 j_2} | \psi_{m'_1 m'_2}^{j_1 j_2} \rangle}_{\text{insert completeness}} = \sum_{jm} \underbrace{C_{j_1 j_2}(m_1 m_2; jm)}_{\langle \psi_{m_1 m_2}^{j_1 j_2} | \psi_{jm}^{j_1 j_2} \rangle} \underbrace{C_{j_1 j_2}(m'_1 m'_2; jm)}_{\langle \psi_{jm}^{j_1 j_2} | \psi_{m'_1 m'_2}^{j_1 j_2} \rangle} \,. \tag{13.30}
$$

The Clebsch–Gordan coefficients also satisfy a number of "symmetry" properties. As an illustration, we show below that

$$
C_{j_1 j_2}(m_1 m_2; jm) = (-1)^{j_1+j_2-j} \, C_{j_2 j_1}(m_2 m_1; jm) \,, \tag{13.31}
$$

that is, under the exchange $j_1 m_1 \rightleftharpoons j_2 m_2$ the Clebsch–Gordan coefficient is either even or odd depending on whether $j_1 + j_2 - j$ is even or odd. To see how this comes about, we note that the states $|\psi_{jm}^{j_1 j_2}\rangle$ and $|\psi_{jm}^{j_2 j_1}\rangle$ can differ only by a phase factor, since they are (normalized) eigenstates of the complete set of commuting observables $\hat{\mathbf{J}}_1^2$, $\hat{\mathbf{J}}_2^2$, $\hat{\mathbf{J}}^2$, and $\hat{J}_z$, corresponding to the *same* eigenvalues of these observables. Under the exchange $j_1 m_1 \rightleftharpoons j_2 m_2$, we have $|\psi_{jm}^{j_1 j_2}\rangle \longrightarrow e^{i\phi} |\psi_{jm}^{j_2 j_1}\rangle$; exchanging $j_1 m_1 \rightleftharpoons j_2 m_2$ twice, we then arrive at $|\psi_{jm}^{j_1 j_2}\rangle \longrightarrow e^{i\phi} |\psi_{jm}^{j_2 j_1}\rangle \longrightarrow e^{2i\phi} |\psi_{jm}^{j_1 j_2}\rangle$, which implies $e^{2i\phi} = 1$ or $\phi = 0$ or $\phi = \pi$. Thus, this phase factor is ± 1. It cannot depend on m, since states with different m are related to each other by application of the raising or lowering operators $\hat{J}_{\pm} = \hat{J}_{1\pm} + \hat{J}_{2\pm}$, which are symmetric under the exchange $1 \rightleftharpoons 2$. We conclude that these states all have phases $+1$ or all have phases -1,

$$
|\psi_{jm}^{j_1 j_2}\rangle = (\pm 1)_{j_1 j_2 j} \, |\psi_{jm}^{j_2 j_1}\rangle \,, \tag{13.32}
$$

where $(\pm 1)_{j_1 j_2 j}$ indicates that the sign depends only on the j_1-, j_2-, and j-values. This in turn leads to

$$
C_{j_2 j_1}(m_2 m_1; jm) = (\pm 1)_{j_1 j_2 j} \, C_{j_1 j_2}(m_1 m_2; jm) \,. \tag{13.33}
$$

From Eq. (13.10) we see that the state with maximum $m = j = j_1 + j_2$ is given by

$$
|\psi_{j_1+j_2, j_1+j_2}^{j_1 j_2}\rangle = |\psi_{j_1 j_1}\rangle \otimes |\psi_{j_2 j_2}\rangle \implies C_{j_1 j_2}(\underbrace{j_1 j_2}_{m_1 m_2}; \underbrace{j_1+j_2}_{j} \, \underbrace{j_1+j_2}_{m}) = 1 \,, \tag{13.34}
$$

and so the phase is $+1$ in this case as it is for all other states in the multiplet having $j = j_1 + j_2$ but different m; hence,

$$
C_{j_2 j_1}(m_2 m_1; \underbrace{j_1+j_2}_{j}, m) = C_{j_1 j_2}(m_1 m_2; \underbrace{j_1+j_2}_{j}, m) \,. \tag{13.35}
$$

Next, consider the state with $m = j = j_1 + j_2 - 1$ (the next state "down") in Eq. (13.21),

$$|\psi^{j_1 j_2}_{j_1+j_2-1,j_1+j_2-1}\rangle = -\sqrt{\frac{j_2}{j_1+j_2}}\,|\psi_{j_1 j_1-1}\rangle \otimes |\psi_{j_2 j_2}\rangle + \sqrt{\frac{j_1}{j_1+j_2}}\,|\psi_{j_1 j_1}\rangle \otimes |\psi_{j_2 j_2-1}\rangle\,, \tag{13.36}$$

which is antisymmetric under the exchange $j_1 \rightleftharpoons j_2$, and hence (in order to arrive at the condition in Eq. (13.37), denote the coefficients in the linear combination above and that obtained by exchanging $j_1 m_1 \rightleftharpoons j_2 m_2$ with the appropriate Clebsch–Gordan coefficients)

$$C_{j_2 j_1}(m_2 m_1; \underbrace{j_1 + j_2 - 1}_{j}, m) = -C_{j_1 j_2}(m_1 m_2; \underbrace{j_1 + j_2 - 1}_{j}, m)\,. \tag{13.37}$$

The argument can be repeated for the state with $m = j = j_1+j_2-2$, then for the state with $m = j = j_1+j_2-3$, and so on, to arrive at Eq. (13.31).

13.3 Problems

Problem 1   Determining the Minimum Value $|j_1 - j_2|$ in the Addition of Two Angular Momenta

Let j' be the lowest non-negative integer or semi-integer obtained in the addition of two angular momenta, following the procedure outlined in Section 13.1. Using the fact that

$$\sum_{j=j'}^{j_1+j_2}(2j+1) = (2j_1+1)(2j_2+1)\,,$$

that is, that the number of states in the subspace (j_1, j_2) is independent of whether the basis $|\psi^{j_1 j_2}_{m_1 m_2}\rangle$ or $|\psi^{j_1 j_2}_{jm}\rangle$ is adopted, show that $j' = |j_1 - j_2|$.

Solution

Two cases can occur. The first is when j_1 and j_2 are both integer or both semi-integer, which implies that $j^* = j_1 + j_2$ and j' are both integer. We obtain

$$\sum_{j=j'}^{j^*}(2j+1) = \sum_{j=0}^{j^*}(2j+1) - \sum_{j=0}^{j'-1}(2j+1) = 2\frac{j^*(j^*+1)}{2} + j^* + 1 - \left[2\frac{j'(j'-1)}{2} + j'\right]$$

$$= (j^*+1)^2 - j'^2 = (2j_1+1)(2j_2+1)\,,$$

and, solving with respect to j', we find

$$j'^2 = (j_1+j_2+1)^2 - (2j_1+1)(2j_2+1) = j_1^2 + j_2^2 - 2j_1 j_2 \implies j' = |j_1 - j_2|\,.$$

The other case occurs when j_1 is semi-integer and j_2 integer or vice versa; then j^* and j' are both semi-integer, and $2j^*$ and $2j'$ are odd, and it must hold that

$$\sum_{k\,\text{odd}=2j'}^{2j^*}(k+1) = (2j_1+1)(2j_2+1)\,,$$

where we have defined $k = 2j$ and the sum runs over odd integers k. Note that (n is assumed to be odd below)

$$\sum_{k \text{ odd}=1}^{n} k = \left(\frac{n+1}{2}\right)^2 , \qquad \sum_{k \text{ odd}=1}^{n} 1 = \frac{n+1}{2}$$

and

$$\sum_{k \text{ odd}=2j'}^{2j^*} (k+1) = \sum_{k \text{ odd}=1}^{2j^*} (k+1) - \sum_{k \text{ odd}=1}^{2j'-2} (k+1) = \frac{2j^* + 1}{2}\left(\frac{2j^* + 1}{2} + 1\right) - \frac{2j' - 1}{2}\left(\frac{2j' - 1}{2} + 1\right)$$

$$= \frac{(2j^* + 1)(2j^* + 3)}{4} - \frac{4j'^2 - 1}{4} = j^{*2} + 2j^* + 1 - j'^2 = (2j_1 + 1)(2j_2 + 1) ,$$

which leads to the same condition found earlier, $j' = |j_1 - j_2|$.

Problem 2   Angular Momentum in the Deuterium Atom

A deuterium atom consists of a deuteron nucleus – a proton and neutron bound together by the strong interaction – and an electron. The deuteron has total angular momentum $\mathbf{I}$, including the relative orbital angular momentum between the proton and neutron and their respective spins, equal to 1, while the total angular momentum of the electron is $\mathbf{J} = \mathbf{L} + \mathbf{S}$, where $\mathbf{L}$ and $\mathbf{S}$ are its orbital angular momentum and spin. The total angular momentum of the atom is $\mathbf{F} = \mathbf{J} + \mathbf{I}$, and the components of $\mathbf{I}$ commute with those of $\mathbf{J}$. What are the possible values of the quantum numbers j and f for a deuterium atom in either the 1s ground state or the 2p excited state?

Solution

In the 1s state the relative orbital angular momentum between the electron and deuteron is zero, and hence $j = 1/2$ and $f = 1/2$ or $3/2$. For the 2p excited state (having relative angular momentum quantum number $l = 1$), it follows that $j = 1/2$ or $3/2$ and $f = 1/2, 3/2$, or $5/2$. Note that $f = 1/2$ or $3/2$ come from combining $j = 1/2$ with $i = 1$ and $j = 3/2$ with $i = 1$, while $f = 5/2$ comes only from combining $j = 3/2$ with $i = 1$.

Problem 3   Combining Angular Momenta 1 and 1/2

Consider an angular momentum $j_1 = 1$ with eigenstates $|1\rangle$, $|0\rangle$, and $|-1\rangle$, such that

$$\mathbf{J}_1^2 |m\rangle = 2\hbar^2 |m\rangle , \qquad J_{1,z} |m\rangle = m\hbar |m\rangle , \qquad m = 0, \pm 1 .$$

Combine this angular momentum with an angular momentum $j_2 = 1/2$ with eigenstates $|+\rangle$ and $|-\rangle$,

$$\mathbf{J}_2^2 |\pm\rangle = \frac{3}{4}\hbar^2 |\pm\rangle , \qquad J_{2,z} |\pm\rangle = \pm\frac{\hbar}{2} |\pm\rangle .$$

What are the possible values for the total angular momentum $\mathbf{J} = \mathbf{J}_1 + \mathbf{J}_2$? Use a table of Clebsch–Gordan coefficients to express the states in the basis consisting of eigenstates of $\mathbf{J}_1^2, \mathbf{J}_2^2, \mathbf{J}^2$, and J_z, labeled simply as $|jm\rangle$ with

$$\mathbf{J}^2 |jm\rangle = j(j + 1)\hbar^2 |jm\rangle , \qquad J_z |jm\rangle = m\hbar |jm\rangle ,$$

as linear combinations of the states in the basis consisting of eigenstates of $\mathbf{J}_1^2, \mathbf{J}_2^2, J_{1,z}$, and $J_{2,z}$, labeled as $|m, \pm\rangle$ with $m = 0, \pm 1$.

Solution

The possible values of the total angular momentum are $j = 1/2$ and $3/2$. We can use the properties of the Clebsch–Gordan coefficients to write

$$|\psi_{j,m}^{1,1/2}\rangle = \sum_{m_1,m_2} C_{1,1/2}(m_1, m_2; j, m) |\psi_{m_1,m_2}^{1,1/2}\rangle \, ,$$

where $m_1 = \pm 1, 0$ and $m_2 = \pm 1/2$. We then find, for $j = 3/2$,

$$|\psi_{3/2,3/2}^{1,1/2}\rangle = |\psi_{1,1/2}^{1,1/2}\rangle \, , \qquad |\psi_{3/2,1/2}^{1,1/2}\rangle = \sqrt{\frac{1}{3}} |\psi_{1,-1/2}^{1,1/2}\rangle + \sqrt{\frac{2}{3}} |\psi_{0,1/2}^{1,1/2}\rangle \, ,$$

$$|\psi_{3/2,-3/2}^{1,1/2}\rangle = |\psi_{-1,-1/2}^{1,1/2}\rangle \, , \qquad |\psi_{3/2,-1/2}^{1,1/2}\rangle = \sqrt{\frac{2}{3}} |\psi_{0,-1/2}^{1,1/2}\rangle + \sqrt{\frac{1}{3}} |\psi_{-1,1/2}^{1,1/2}\rangle \, ,$$

and, for $j = 1/2$,

$$|\psi_{1/2,1/2}^{1,1/2}\rangle = \sqrt{\frac{2}{3}} |\psi_{1,-1/2}^{1,1/2}\rangle - \sqrt{\frac{1}{3}} |\psi_{0,1/2}^{1,1/2}\rangle \, , \qquad |\psi_{1/2,-1/2}^{1,1/2}\rangle = \sqrt{\frac{1}{3}} |\psi_{0,-1/2}^{1,1/2}\rangle - \sqrt{\frac{2}{3}} |\psi_{-1,1/2}^{1,1/2}\rangle \, .$$

Of course, if access to a tabulation of Clebsch–Gordan coefficients is not available, an alternative approach is sketched in Section 13.1.

Problem 4   Combining the Spins of Three Spin-1/2 Particle

Let $\mathbf{S} = \mathbf{S}_1 + \mathbf{S}_2 + \mathbf{S}_3$ be the total spin of three spin-1/2 particles (whose orbital degrees of freedom are ignored). Let $|\sigma_1, \sigma_2, \sigma_3\rangle$ be the eigenstates common to $S_{1,z}, S_{2,z}, S_{3,z}$, with respective eigenvalues $(\hbar/2)\sigma_1, (\hbar/2)\sigma_2, (\hbar/2)\sigma_3$ and $\sigma_i = \pm$. Give a basis of eigenstates common to $\mathbf{S}^2$ and S_z in terms of the basis states $|\sigma_1, \sigma_2, \sigma_3\rangle$. Do these two operators form a complete set of commuting observables?

Solution

We first combine the spins of particles 1 and 2 to obtain pair-12 spins 0 and 1 (of course, we could have chosen instead pair-13 or pair-23). We have, for the pair-12 state with spin 0,

$$|0, 0; 12\rangle = \frac{1}{\sqrt{2}} (|+, -\rangle - |-, +\rangle) \, ,$$

and, for the pair-12 state with spin 1,

$$|1, 0; 12\rangle = \frac{1}{\sqrt{2}} (|+, -\rangle + |-, +\rangle) \, , \qquad |1, \pm 1; 12\rangle = |\pm, \pm\rangle \, ,$$

where the notation on the right-hand sides of these relations is as follows: is

$$|\sigma, \sigma'\rangle = |\sigma; 1\rangle \otimes |\sigma'; 2\rangle \, .$$

Next, we combine pair-12 spin 0 with the spin of particle 3, to obtain a total spin 1/2. Using the properties of the Clebsch–Gordan coefficients,

$$|\psi_{j,m}^{j_1, j_2}\rangle = \sum_{m_1, m_2} C_{j_1, j_2}(m_1, m_2; j, m) |\psi_{j_1, m_1}\rangle \otimes |\psi_{j_2, m_2}\rangle \, ,$$

with $j_1 = 0$ and $j_2 = 1/2$, we find

$$|1/2, 1/2; 12(0)3\rangle = |0, 0; 12\rangle \otimes |+; 3\rangle \, , \qquad |1/2, -1/2; 12(0)3\rangle = |0, 0; 12\rangle \otimes |-; 3\rangle \, .$$

In a similar way, we combine pair-12 spin 1 with the spin of particle 3 to obtain the four total-spin-3/2 states:

$$|3/2, 3/2; 12(1)3\rangle = |1, 1; 12\rangle \otimes |+; 3\rangle,$$

$$|3/2, 1/2; 12(1)3\rangle = \sqrt{\frac{1}{3}} |1, 1; 12\rangle \otimes |-; 3\rangle + \sqrt{\frac{2}{3}} |1, 0; 12\rangle \otimes |+; 3\rangle,$$

$$|3/2, -1/2; 12(1)3\rangle = \sqrt{\frac{2}{3}} |1, 0; 12\rangle \otimes |-; 3\rangle + \sqrt{\frac{1}{3}} |1, -1; 12\rangle \otimes |+; 3\rangle,$$

$$|3/2, -3/2; 12(1)3\rangle = |1, -1; 12\rangle \otimes |-; 3\rangle.$$

We combine pair-12 spin 1 with the spin of particle 3 to obtain the two total-spin-1/2 states:

$$|1/2, 1/2; 12(1)3\rangle = \sqrt{\frac{2}{3}} |1, 1; 12\rangle \otimes |-; 3\rangle - \sqrt{\frac{1}{3}} |1, 0; 12\rangle \otimes |+; 3\rangle,$$

$$|1/2, -1/2; 12(1)3\rangle = \sqrt{\frac{1}{3}} |1, 0; 12\rangle \otimes |-; 3\rangle - \sqrt{\frac{2}{3}} |1, -1; 12\rangle \otimes |+; 3\rangle.$$

In terms of the states $|\sigma, \sigma', \sigma''\rangle$, the eight states above read

$$|1/2, 1/2; 12(0)3\rangle = \frac{1}{\sqrt{2}} (|+, -, +\rangle - |-, +, +\rangle),$$

$$|1/2, -1/2; 12(0)3\rangle = \frac{1}{\sqrt{2}} (|+, -, -\rangle - |-, +, -\rangle)$$

$$|1/2, 1/2; 12(1)3\rangle = \sqrt{\frac{1}{6}} (2 |+, +, -\rangle - |+, -, +\rangle - |-, +, +\rangle),$$

$$|1/2, -1/2; 12(1)3\rangle = \sqrt{\frac{1}{6}} (|+, -, -\rangle + |-, +, -\rangle - 2 |-, -, +\rangle),$$

and

$$|3/2, 3/2; 12(1)3\rangle = |+, +, +\rangle,$$

$$|3/2, 1/2; 12(1)3\rangle = \sqrt{\frac{1}{3}} (|+, +, -\rangle + |+, -, +\rangle + |-, +, +\rangle),$$

$$|1/2, -1/2; 12(1)3\rangle = \sqrt{\frac{1}{3}} (|+, -, -\rangle + |-, +, -\rangle + |-, -, +\rangle),$$

$$|3/2, -3/2; 12(1)3\rangle = |-, -, -\rangle.$$

Note that the total-spin-3/2 states are symmetric under the exchange of any pair of particles, while the total-spin-1/2 states are either symmetric or antisymmetric under the exchange of particles 1 and 2 depending on whether the pair-12 spin is either 1 or 0, respectively. The operators $\mathbf{S}^2$ and S_z alone do not form a complete set of commuting observables, since specifying their eigenvalues does not uniquely identify a single common eigenstate; in particular, there are two eigenstates corresponding to the pair of eigenvalues $s = 1/2$ and $s_z = 1/2$ (or $s = 1/2$ and $s_z = -1/2$).

Problem 5 Projection Operators onto $j = l \pm 1/2$ States

Consider a spin-1/2 particle. Show that, in the state space with given orbital angular momentum l, the operators

$$P_+ = \frac{l + 1 + \mathbf{L} \cdot \boldsymbol{\sigma}/\hbar}{2l + 1}, \qquad P_- = \frac{l - \mathbf{L} \cdot \boldsymbol{\sigma}/\hbar}{2l + 1}$$

are projection operators onto the states of total angular momentum $j = l + 1/2$ and $j = l - 1/2$, respectively.

Solution

It is obvious that $P_\pm$ are hermitian, since the components of the orbital angular momentum and spin commute with each other. In order to show that $P_\pm^2 = P_\pm$, we first note that

$$\sigma \cdot \mathbf{L}\, \sigma \cdot \mathbf{L} = \mathbf{L}^2 + i\,\sigma \cdot (\mathbf{L} \times \mathbf{L}) = \mathbf{L}^2 - \hbar\,\sigma \cdot \mathbf{L}\,,$$

where we have used $\mathbf{L} \times \mathbf{L} = i\hbar\,\mathbf{L}$, since

$$(\mathbf{L} \times \mathbf{L})_1 = \sum_{j,k} \epsilon_{1jk}\, L_j L_k = L_2 L_3 - L_3 L_2 = [L_2\,,\, L_3] = i\hbar\, L_1\,,$$

and similarly for the components 2 and 3. Considering P_+ (a similar argument holds for P_-), it follows that

$$P_+^2 = \frac{(l+1)^2 + 2(l+1)\mathbf{L} \cdot \sigma/\hbar + \mathbf{L}^2/\hbar^2 - \mathbf{L} \cdot \sigma/\hbar}{(2l+1)^2}\,.$$

In the subspace with fixed l, we have $\mathbf{L}^2 \longrightarrow l(l+1)\hbar^2$, so that

$$P_+^2 = \frac{(l+1)^2 + (2l+1)\mathbf{L} \cdot \sigma/\hbar + l(l+1)}{(2l+1)^2} = \frac{l+1+\mathbf{L} \cdot \sigma/\hbar}{2l+1} = P_+\,.$$

We also observe that

$$P_+ P_- = \frac{l(l+1) - \mathbf{L} \cdot \sigma/\hbar - \mathbf{L}^2/\hbar^2 + \mathbf{L} \cdot \sigma/\hbar}{(2l+1)^2} = 0\,.$$

Lastly, since

$$\mathbf{L} \cdot \sigma = 2\,\mathbf{L} \cdot \mathbf{S}/\hbar = [\mathbf{J}^2 - \mathbf{L}^2 - \mathbf{S}^2]/\hbar \longrightarrow \mathbf{J}^2/\hbar - [l(l+1) + 3/4]\hbar\,,$$

the projectors can also be written as

$$P_+ = \frac{l+1+\mathbf{J}^2/\hbar^2 - l(l+1) - 3/4}{2l+1} = \frac{\mathbf{J}^2/\hbar^2 - l^2 + 1/4}{2l+1}$$

and

$$P_- = \frac{l - \mathbf{J}^2/\hbar^2 + l(l+1) + 3/4}{2l+1} = \frac{-\mathbf{J}^2/\hbar^2 + l(l+2) + 3/4}{2l+1}\,.$$

In the subspace with $j = l + 1/2$, we have $\mathbf{J}^2/\hbar^2 \longrightarrow (l+1/2)(l+3/2) = l^2 + 2l + 3/4$ and we see that $P_+ = 1$ while $P_- = 0$. On the other hand, in the subspace with $j = l - 1/2$, we have $\mathbf{J}^2/\hbar^2 \longrightarrow (l-1/2)(l+1/2) = l^2 - 1/4$, so that $P_+ = 0$ and $P_- = 1$.

Problem 6   Angular Momentum and Parity Conservation in a Two-Particle Decay

An unstable particle – the $N(1520)$ – has spin $\mathbf{S}_i$ equal to 3/2 and is at rest. It decays into a nucleon (a proton or neutron) and a pion. The nucleon has spin $\mathbf{S}_N$ equal to 1/2, while the pion is spinless. Denote by $\mathbf{L}_f$ the relative orbital angular momentum between these final particles, and assume that total angular momentum is conserved in the decay process. What are the allowed values for $\mathbf{L}_f$? Suppose that parity is also conserved in the decay: what consequence would this fact have for the previous question?

Assume that initially the $N(1520)$ is at rest and in spin state $|3/2, m_i\rangle$, an eigenstate of $\mathbf{S}_i^2$ and $S_{i,z}$, with eigenvalues, respectively, $(15/4)\hbar^2$ and $m_i\hbar$, and that parity is conserved. Show that the relative orbital angular momentum (and hence parity) can be determined by measuring the probability for the final nucleon to be in the state $|1/2, +\rangle$.

Hint: The final state can be written as

$$|l_f m_f\rangle \otimes |1/2, m_N\rangle, \qquad m_f = -l_f, \ldots, l_f \text{ and } m_N = \pm 1/2.$$

Solution

The initial angular momentum $j_i = 3/2$. The final angular momentum consists of the relative orbital angular momentum $\mathbf{L}_f$ between the nucleon (N) and pion (π) and the nucleon spin $\mathbf{S}_N$, and therefore $j_f = |l_f - 1/2|, |l_f + 1/2|$. Since total angular momentum is conserved, the only possibilities are $l_f = 1$ and 2. The final state is the tensor product of the eigenstate $|l_f, m_f\rangle$ with $\mathbf{L}^2$ and L_z and the nucleon spin state $|1/2, m_N\rangle$, an eigenstate of $\mathbf{S}_N^2$ and $S_{N,z}$. Hence, the parity of the final state[2] is $(-1)^{l_f}$, while that of the initial state is $+$. Parity conservation requires $l_f = 2$.

The initial state is $|3/2, m_i\rangle$. The final state $|f\rangle$ must have $j_f = 3/2$ and angular momentum projection m_i, and therefore

$$|f\rangle = \sum_{m_f, m_N} C_{l_f\,1/2}(m_f, m_N; 3/2, m_i)\, |l_f, m_f\rangle\, |1/2, m_N\rangle$$
$$= \sum_{m_N} C_{l_f\,1/2}(m_i - m_N, m_N; 3/2, m_i)\, |l_f, m_i - m_N; 1/2, m_N\rangle.$$

The probability of finding the nucleon with spin up is

$$P_+(l_f) = \sum_{l_f' m_f'} |\langle l_f', m_f'; 1/2, 1/2|f\rangle|^2 ;$$

furthermore,

$$\langle l_f', m_f'; 1/2, 1/2|f\rangle = \sum_{m_N} C_{l_f\,1/2}(m_i - m_N, m_N; 3/2, m_i)\, |\langle l_f', m_f'; 1/2, 1/2|l_f, m_i - m_N; 1/2, m_N\rangle$$
$$= C_{l_f\,1/2}(m_i - 1/2, 1/2; 3/2, m_i)\, \delta_{l_f', l_f}\, \delta_{m_f', m_i - m_N}\, \delta_{m_N, 1/2},$$

which then yields

$$P_+(l_f) = [C_{l_f\,1/2}(m_i - 1/2, 1/2; 3/2, m_i)]^2.$$

The Clebsch–Gordan coefficients are real. We find

$$P_+(1) = [C_{1\,1/2}(m_i - 1/2, 1/2; 3/2, m_i)]^2 = \frac{3/2 + m_i}{3},$$

and

$$P_+(2) = [C_{2\,1/2}(m_i - 1/2, 1/2; 3/2, m_i)]^2 = \frac{5/2 - m_i}{5}.$$

For any m_i, $P_+(1) \neq P_+(2)$, and so a measurement of $P_+(l_f)$ leads to a determination of l_f and, hence, of parity.

---

[2] Here, we are ignoring the intrinsic parities of $N(1520)$, the nucleon, and the pion, respectively, $-$, $+$ and $-$. If we account for these, the parity of the initial state $\pi_i = -1$ and that of the final state $\pi_f = (-1)^{l_f+1}$. However, parity conservation leads to the same conclusion as that in the main text.

Problem 7 The Addition Formula for Two Spherical Harmonics

A state $|\psi\rangle$ is invariant under rotations if $\mathbf{J}|\psi\rangle = 0$. Now, consider two angular momenta $\mathbf{J}_1$ and $\mathbf{J}_2$, with $\mathbf{J}_1^2$ and J_{1z} and $\mathbf{J}_2^2$ and J_{2z} having eigenvalues $j_1(j_1 + 1)\hbar^2$ and $m_1\hbar$, and $j_2(j_2 + 1)\hbar^2$ and $m_2\hbar$, respectively.

1. What condition must j_1 and j_2 satisfy in order for a state of total angular momentum that is invariant under rotations to be constructed?
2. As implied by $\mathbf{J}|\psi\rangle = 0$, such a state is annihilated by the operator $J_+ = J_{1+} + J_{2+}$, that is, $J_+|\psi\rangle = 0$. Show that this requirement yields a relation between the Clebsch–Gordan coefficients.
3. Making use of the relation in part 2, show that the Clebsch–Gordan coefficients are given by

$$C_{jj}(m, -m; 00) = \frac{(-1)^{j-m}}{\sqrt{2j + 1}} \, .$$

4. Use the result in part 3 to obtain the formula for the addition of two spherical harmonics:

$$P_l(\hat{\mathbf{a}} \cdot \hat{\mathbf{b}}) = \frac{4\pi}{2l + 1} \sum_{m=-l}^{l} (-1)^m \, Y_{l,m}(\hat{\mathbf{a}}) \, Y_{l,-m}(\hat{\mathbf{b}}) \, ,$$

where $\hat{\mathbf{a}}$ and $\hat{\mathbf{b}}$ are two unit vectors. To this end, first note that the function

$$F_l(\hat{\mathbf{a}} \cdot \hat{\mathbf{b}}) = \sum_{m=-l}^{l} \frac{(-1)^{l-m}}{\sqrt{2l + 1}} \, Y_{l,m}(\hat{\mathbf{a}}) \, Y_{l,-m}(\hat{\mathbf{b}}) \, ,$$

can depend only on the quantity $\hat{\mathbf{a}} \cdot \hat{\mathbf{b}}$, a scalar and hence invariant under rotations. Then, by taking $\hat{\mathbf{a}} = (\sin\theta\cos\phi, \sin\theta\sin\phi, \cos\theta)$ and $\hat{\mathbf{b}} = (0, 0, 1)$, obtain the addition formula for spherical harmonics.

Solution

Part 1

A state that is invariant under rotation must have $j = 0$, that is, it must be the state $|\psi_{jm}\rangle = |\psi_{00}\rangle$. This state has the properties

$$J_{\pm}|\psi_{00}\rangle = 0 \qquad \text{and} \qquad J_z|\psi_{00}\rangle = 0 \implies \mathbf{J}|\psi_{00}\rangle = 0 \, .$$

Therefore a state invariant under rotations can result only from combining two angular momenta having $j_1 = j_2 = j$.

Part 2

We have

$$|\psi_{00}^{jj}\rangle = \sum_m C_{jj}(m, -m; 0, 0)|\psi_{j,m}\rangle|\psi_{j,-m}\rangle \, .$$

The condition $J_+|\psi_{00}^{jj}\rangle = 0$ yields

$$0 = \sum_m C_{jj}(m, -m; 0, 0)\left[(J_{1+}|\psi_{j,m}\rangle)|\psi_{j,-m}\rangle + |\psi_{j,m}\rangle(J_{2+}|\psi_{j,-m}\rangle)\right] \, ,$$

which implies

$$0 = \sum_m C_{jj}(m, -m; 0, 0)\left[\sqrt{j(j + 1) - m(m + 1)}\,|\psi_{j,m+1}\rangle|\psi_{j,-m}\rangle\right.$$
$$\left. + \sqrt{j(j + 1) + m(-m + 1)}\,|\psi_{j,m}\rangle|\psi_{j,-m+1}\rangle\right]\,.$$

In the second term, we change the summation from m to $m + 1$ to find that

$$0 = \sum_m C_{jj}(m, -m; 0, 0)\sqrt{j(j + 1) - m(m + 1)}\,|\psi_{j,m+1}\rangle|\psi_{j,-m}\rangle$$
$$+ \sum_m C_{jj}(m + 1, -m - 1; 0, 0)\sqrt{j(j + 1) - m(m + 1)}\,|\psi_{j,m+1}\rangle|\psi_{j,-m}\rangle$$
$$= \sum_m \left[C_{jj}(m, -m; 0, 0) + C_{jj}(m + 1, -m - 1; 0, 0)\right]\sqrt{j(j + 1) - m(m + 1)}\,|\psi_{j,m+1}\rangle|\psi_{j,-m}\rangle\,,$$

which implies that

$$C_{jj}(m + 1, -m - 1; 0, 0) = -C_{jj}(m, -m; 0, 0)\,,$$

that is, a recurrence relation between the Clebsch–Gordan coefficients.

Part 3

The above recurrence relation can be satisfied by taking

$$C_{jj}(m, -m; 0, 0) = (-1)^{j-m}\,N_j\,,$$

where N_j is a positive normalization constant to be determined shortly. Note that we have made the choice of phase such that

$$C_{jj}(j, -j, ; 0, 0) > 0\,,$$

that is, the Clebsch–Gordan coefficient with the maximum allowed m, that is, $m = j$, is positive. The normalization constant results from the orthonormality condition of the Clebsch–Gordan coefficients,

$$\sum_{m_1 m_2} C_{j_1 j_2}(m_1 m_2; jm)\, C_{j_1 j_2}(m_1 m_2; j'm') = \delta_{jj'}\,\delta_{mm'}\,,$$

which for $j_1 = j_2 = j, j = j' = 0$, and $m = m' = 0$ yields

$$\sum_m C_{jj}^2(m, -m; 0, 0) = 1\,.$$

It follows that (note that $j - m$ is always an integer, regardless of whether j is either integer or semi-integer)

$$\sum_m N_j^2\,(-1)^{2(j-m)} = N_j^2 \sum_m 1 = (2j + 1)\,N_j^2 = 1 \implies N_j = \frac{1}{\sqrt{2j + 1}}$$

and hence

$$C_{jj}(m, -m; 0, 0) = \frac{(-1)^{j-m}}{\sqrt{2j + 1}}\,.$$

Part 4

The combination below must be rotationally invariant, and therefore it can only depend on $\hat{\mathbf{a}} \cdot \hat{\mathbf{b}}$:

$$\sum_{m=-l}^{l} \frac{(-1)^{l-m}}{\sqrt{2l+1}} \, Y_{l,m}(\hat{\mathbf{a}}) \, Y_{l,-m}(\hat{\mathbf{b}}) \ .$$

Consider the special case $\hat{\mathbf{a}} = (\sin\theta\cos\phi, \sin\theta\sin\phi, \cos\theta)$ and $\hat{\mathbf{b}} = (0,0,1)$, which gives

$$Y_{lm}(\hat{\mathbf{b}}) \longrightarrow \sqrt{\frac{2l+1}{4\pi}} \, \delta_{m,0} \ ,$$

and the combination above then reduces to

$$\frac{(-1)^l}{\sqrt{4\pi}} \, Y_{l,0}(\hat{\mathbf{a}}) = \frac{(-1)^l}{4\pi} \, \sqrt{2l+1} \, P_l(\cos\theta) \ ,$$

where we have used

$$Y_{l,0}(\hat{\mathbf{a}}) = \sqrt{\frac{2l+1}{4\pi}} \, P_l(\cos\theta) \ .$$

Expressing $\cos\theta$ as $\hat{\mathbf{a}} \cdot \hat{\mathbf{b}}$, we deduce in general

$$\sum_{m=-l}^{l} \frac{(-1)^{l-m}}{\sqrt{2l+1}} \, Y_{l,m}(\hat{\mathbf{a}}) \, Y_{l,-m}(\hat{\mathbf{b}}) = (-1)^l \, \frac{\sqrt{2l+1}}{4\pi} \, P_l(\hat{\mathbf{a}} \cdot \hat{\mathbf{b}}) \ ,$$

or, since $(-1)^{-m} = (-1)^m$,

$$P_l(\hat{\mathbf{a}} \cdot \hat{\mathbf{b}}) = \frac{4\pi}{2l+1} \sum_{m=-l}^{l} (-1)^m \, Y_{l,m}(\hat{\mathbf{a}}) \, Y_{l,-m}(\hat{\mathbf{b}}) \ .$$

Problem 8   Positronium in a Magnetic Field

The Hamiltonian of the positronium atom (that is, the bound system made up of an electron of charge $-e$ and a positron, the electron's antiparticle, of charge e) in the 1s state (that is, the state with relative orbital angular momentum $l = 0$) in a magnetic field $\mathbf{B} = B\hat{\mathbf{z}}$ is given by

$$H = A\,\mathbf{S}_1 \cdot \mathbf{S}_2 + \frac{eB}{mc}\,(S_{1z} - S_{2z}) \ ,$$

where the electron and positron are labeled, respectively, as particles 1 and 2 and each has mass m. Using the coupled representation in which $\mathbf{S}^2 = (\mathbf{S}_1 + \mathbf{S}_2)^2$ and $S_z = S_{1z} + S_{2z}$ are diagonal, obtain the energy eigenvalues and eigenvectors and classify them according to the quantum numbers associated with constants of motion.

Solution

The electron and positron are spin-1/2 particles. Using the identity

$$\mathbf{S}_1 \cdot \mathbf{S}_2 = \frac{1}{2}\left(\mathbf{S}^2 - \mathbf{S}_1^2 - \mathbf{S}_2^2\right) = \frac{1}{2}\,\mathbf{S}^2 - \frac{3}{4}\,\hbar^2 \ ,$$

it can be seen that the first term in the Hamiltonian is diagonal in the basis $|sm_s\rangle$ of eigenstates of $\mathbf{S}^2$ and S_z (here $s = 0, 1$ and $m_s = -s, \ldots, s$). The second term, proportional to $S_{1z} - S_{2z}$, does not commute with $\mathbf{S}^2$ and is therefore not diagonal in the basis $|s m_s\rangle$. Note, however, that it commutes with $S_z = S_{1z} + S_{2z}$. In particular, we obtain

$$(S_{1z} - S_{2z}) |11\rangle = (S_{1z} - S_{2z}) |++\rangle = 0 , \qquad (S_{1z} - S_{2z}) |1-1\rangle = (S_{1z} - S_{2z}) |--\rangle = 0 ,$$

$$(S_{1z} - S_{2z}) |10\rangle = \frac{1}{\sqrt{2}} (S_{1z} - S_{2z}) (|+-\rangle + |-+\rangle) = \frac{\hbar}{\sqrt{2}} (|+-\rangle - |-+\rangle) = \hbar |00\rangle ,$$

and

$$(S_{1z} - S_{2z}) |00\rangle = \frac{1}{\sqrt{2}} (S_{1z} - S_{2z}) (|+-\rangle - |-+\rangle) = \frac{\hbar}{\sqrt{2}} (|+-\rangle + |-+\rangle) = \hbar |10\rangle ,$$

where we have used $S_{1z}|\pm\rangle = \pm (\hbar/2)|\pm\rangle$ and similarly for S_{2z}. In the basis $|s m_s\rangle$, the Hamiltonian of positronium is represented by the 4×4 matrix given by

$$\underline{H} = \begin{pmatrix} \langle 00|H|00\rangle & \langle 00|H|10\rangle & \langle 00|H|1-1\rangle & \langle 00|H|11\rangle \\ \langle 10|H|00\rangle & \langle 10|H|10\rangle & \langle 10|H|1-1\rangle & \langle 10|H|11\rangle \\ \langle 1-1|H|00\rangle & \langle 1-1|H|10\rangle & \langle 1-1|H|1-1\rangle & \langle 1-1|H|11\rangle \\ \langle 11|H|00\rangle & \langle 11|H|10\rangle & \langle 11|H|1-1\rangle & \langle 11|H|11\rangle \end{pmatrix}$$

$$= \begin{pmatrix} -(3/4)A\hbar^2 & \hbar\Omega & 0 & 0 \\ \hbar\Omega & (1/4)A\hbar^2 & 0 & 0 \\ 0 & 0 & (1/4)A\hbar^2 & 0 \\ 0 & 0 & 0 & (1/4)A\hbar^2 \end{pmatrix} ,$$

where we define

$$\Omega = \frac{eB}{mc} .$$

We obtain $E_0 = (1/4)A\hbar^2$ with twofold degeneracy and eigenstates $|1-1\rangle$ and $|11\rangle$. The remaining eigenvalues result from diagonalizing the 2×2 block,

$$\det (\underline{H} - E\,\underline{\mathbb{1}}) = 0 \implies E^2 + \frac{1}{2} A\hbar^2 E - \frac{3}{16} A^2\hbar^4 - \hbar^2\Omega^2 = 0 ,$$

yielding

$$E_\pm = -\frac{1}{4} A\hbar^2 \left(1 \mp 2 \sqrt{1 + \eta^2} \right) , \qquad \eta = 2 \frac{\Omega}{A\hbar} ,$$

with corresponding eigenstates

$$|0_+\rangle = \alpha_+ \left[\eta \, |00\rangle + \left(1 + \sqrt{1 + \eta^2} \right) |10\rangle \right] , \qquad |0_-\rangle = \alpha_- \left[\left(1 + \sqrt{1 + \eta^2} \right) |00\rangle - \eta \, |10\rangle \right] ,$$

where $\alpha_\pm$ are normalization factors determined from

$$\langle 0_\pm | 0_\pm \rangle = 1 \implies |\alpha_+|^2 = |\alpha_-|^2 = \frac{1}{2(1 + \eta^2)(1 + 1/\sqrt{1 + \eta^2})} .$$

Note that, in the limit of vanishing magnetic field, corresponding to $\eta \longrightarrow 0$, we recover the previous results, that is, $E_+ \longrightarrow (1/4)A\hbar^2$ with $|0_+\rangle \longrightarrow |10\rangle$ and $E_- \longrightarrow -(3/4)A\hbar^2$ with $|0_-\rangle \longrightarrow |00\rangle$. We see that the magnetic field mixes the states having total spin 0 and 1 with spin projection 0. Furthermore, since H and S_z commute with each other, matrix elements of H between states with different m_s projections vanish, as is apparent from the matrix representation obtained above.

Problem 9   Decay of a Spin-1/2 Particle into a Spin-0 Particle and a Spin-1/2 Particle

Consider a particle (a) of spin 1/2 (say, an $N(1440)$ baryon) which can disintegrate into two particles, one (b) of spin 1/2 (a nucleon, that is a proton or neutron) and the other (c) of spin 0 (a pion). Place yourself in the rest frame of particle a. Total angular momentum is conserved during the disintegration. Note that (i) the total angular momentum of the initial state is just $\mathbf{S}_a$ (the spin of particle a), while that of the final state is $\mathbf{L} + \mathbf{S}_b$, where $\mathbf{L}$ is the relative angular momentum and $\mathbf{S}_b$ is the spin of particle b; and (ii) the final state can be written as

$$|lm\rangle \otimes |m_b\rangle , \qquad m = -l, \ldots, l \text{ and } m_b = \pm 1/2 .$$

1. What values can be taken on by the relative orbital angular momentum of the two final particles? Explain why there is only one possible value if the parity of the relative orbital state is fixed.
2. Assuming that the parity of the relative orbital state is not fixed, show that the spinor wave function of the final state with total angular momentum 1/2 and projection $m_f = \pm 1/2$ can be written in momentum space as

$$\psi_{1/2 m_f}(\mathbf{p}) = A_S\, Y_{00}(\hat{\mathbf{p}})\, |m_f\rangle + A_P \sum_{m_b = \pm 1/2} C_{1,1/2}(m_f - m_b, m_b; 1/2, m_f)\, Y_{1(m_f - m_b)}(\hat{\mathbf{p}})\, |m_b\rangle ,$$

where the $|\mathbf{p}|$ dependence of the relative S- and P-wave amplitudes is ignored.
3. Show that the states $\psi_{1/2,\pm1/2}(\mathbf{p})$ are given explicitly by the following expressions:

$$\psi_{1/2,1/2}(\mathbf{p}) = \frac{1}{\sqrt{4\pi}}\left[A_S\,|+\rangle - A_P\left(\cos\theta\,|+\rangle + \sin\theta\,e^{i\phi}\,|-\rangle\right)\right]$$

and

$$\psi_{1/2,-1/2}(\mathbf{p}) = \frac{1}{\sqrt{4\pi}}\left[A_S\,|-\rangle + A_P\left(\cos\theta\,|-\rangle - \sin\theta\,e^{-i\phi}\,|+\rangle\right)\right] ,$$

where $|\pm\rangle$ are the spin states of particle b with $m_b = \pm 1/2$ and θ and ϕ are the polar and azimuthal angles specifying the direction $\hat{\mathbf{p}}$ relative to the spin quantization axis, the $\hat{\mathbf{z}}$-axis.
4. Assume that the initial particle a is in a state with spin projection along the direction

$$\hat{\mathbf{P}} = (\sin\theta^\star \cos\phi^\star, \sin\theta^\star \sin\phi^\star, \cos\theta^\star) ,$$

that is, it is polarized along $\hat{\mathbf{P}}$ relative to the spin quantization axis. Show that such a state can be written as

$$|\hat{\mathbf{P}}\rangle = \cos(\theta^*/2)\,|+\rangle + \sin(\theta^*/2)\,e^{i\phi^*}\,|-\rangle ,$$

where $|\pm\rangle$ are the spin states of particle a with $m_a = \pm 1/2$.
5. Show that if the initial particle a is polarized along $\hat{\mathbf{P}}$ then the angular distribution of the final-state particles b and c is proportional to $1 - \eta\,\hat{\mathbf{P}} \cdot \hat{\mathbf{p}}$. Determine η in terms of A_S and A_P.

Solution

Part 1

Since total angular momentum is conserved, the final relative orbital angular momentum l and spin 1/2 must combine to give total angular momentum 1/2. Hence, only $l = 0$ or 1 are possible. If it is further assumed that the parity of the relative orbital state is fixed, then even (odd) parity implies $l = 0$ (1), since the parity of the orbital states $|lm\rangle$ is $(-1)^l$.

Part 2

The final state with total angular momentum 1/2 and projection m_f can be written as

$$|\psi_{1/2,m_f}\rangle = \sum_{l=0,1} A_l \sum_{m,m_b} C_{l,1/2}(m, m_b; 1/2, m_f) \, |lm\rangle \otimes |m_b\rangle$$

$$= \sum_{l=0,1} A_l \sum_{m_b} C_{l,1/2}(m_f - m_b, m_b; 1/2, m_f) \, |l(m_f - m_b)\rangle \otimes |m_b\rangle \,,$$

where in the second line we have enforced the condition $m + m_b = m_f$ implicit in the Clebsch–Gordan coefficients. In the basis of eigenstates $|\phi_{\mathbf{p}}\rangle$ of the relative momentum operator $\mathbf{p}$ it follows that

$$\psi_{1/2,m_f}(\mathbf{p}) = \langle\phi_{\mathbf{p}}|\psi_{1/2,m_f}\rangle = A_S \, Y_{00}(\hat{\mathbf{p}}) \, |m_f\rangle + A_P \sum_{m_b=\pm1/2} C_{1,1/2}(m_f - m_b, m_b; 1/2, m_f) \, Y_{1(m_f-m_b)}(\hat{\mathbf{p}}) \, |m_b\rangle \,,$$

since for the S-wave term the Clebsch–Gordan coefficient is $C_{0,1/2}(0, m_b; 1/2, m_f) = \delta_{m_b,m_f}$.

Part 3

Inserting the explicit expressions for the spherical harmonics,

$$Y_{00}(\hat{\mathbf{p}}) = \frac{1}{\sqrt{4\pi}} \,, \qquad Y_{10}(\hat{\mathbf{p}}) = \sqrt{\frac{3}{4\pi}} \cos\theta \,, \qquad Y_{1\pm1}(\hat{\mathbf{p}}) = \mp\sqrt{\frac{3}{8\pi}} \sin\theta \, e^{\pm i\phi} \,,$$

and the following values for the Clebsch–Gordan coefficients,

$$C_{1,1/2}(0, 1/2; 1/2, 1/2) = -\frac{1}{\sqrt{3}} \,, \qquad C_{1,1/2}(1, -1/2; 1/2, 1/2) = \sqrt{\frac{2}{3}} \,,$$

and

$$C_{1,1/2}(0, -1/2; 1/2, -1/2) = \frac{1}{\sqrt{3}} \,, \qquad C_{1,1/2}(-1, 1/2; 1/2, -1/2) = -\sqrt{\frac{2}{3}} \,,$$

into $\psi_{1/2m_f}(\mathbf{p})$ immediately yields the required relations,

$$\psi_{1/2,1/2}(\mathbf{p}) = \frac{1}{\sqrt{4\pi}} \left[A_S \, |+\rangle - A_P \left(\cos\theta \, |+\rangle + \sin\theta \, e^{i\phi} \, |-\rangle \right) \right]$$

and

$$\psi_{1/2,-1/2}(\mathbf{p}) = \frac{1}{\sqrt{4\pi}} \left[A_S \, |-\rangle + A_P \left(\cos\theta \, |-\rangle - \sin\theta \, e^{-i\phi} \, |+\rangle \right) \right] \,.$$

Part 4

A particle polarized along $\hat{\mathbf{P}}$ is in an eigenstate of $\hat{\mathbf{P}} \cdot \mathbf{S}$ with eigenvalue $\hbar/2$, or, in terms of Pauli matrices,

$$\hat{\mathbf{P}} \cdot \sigma \, |\hat{\mathbf{P}}\rangle = |\hat{\mathbf{P}}\rangle \,,$$

which, using the explicit expressions for σ, can be written as

$$\begin{pmatrix} \cos\theta^\star & \sin\theta^\star \, e^{-i\phi^*} \\ \sin\theta^\star \, e^{i\phi^*} & -\cos\theta^\star \end{pmatrix} \begin{pmatrix} a \\ b \end{pmatrix} = \begin{pmatrix} a \\ b \end{pmatrix} \implies (1 - \cos\theta^*)a = \sin\theta^\star \, e^{-i\phi^*} b \,,$$

yielding

$$b = \frac{1 - \cos\theta^*}{\sin\theta^\star} e^{i\phi^*} a = \frac{\sin(\theta^\star/2)}{\cos(\theta^\star/2)} e^{i\phi^*} a .$$

Choosing $a = \cos(\theta^\star/2)$ leads to the normalized state

$$|\hat{\mathbf{P}}\rangle \longrightarrow \begin{bmatrix} \cos(\theta^\star/2) \\ \sin(\theta^\star/2)\, e^{i\phi^*} \end{bmatrix} \qquad \text{or} \qquad |\hat{\mathbf{P}}\rangle = \cos(\theta^\star/2)|+\rangle + \sin(\theta^\star/2)\, e^{i\phi^*}\, |-\rangle .$$

Part 5

The final state corresponding to the initial (decaying) state polarized along $\hat{\mathbf{P}}$ is given by

$$\psi_{\hat{\mathbf{P}}}(\mathbf{p}) = \cos(\theta^\star/2)\, \psi_{1/2,1/2}(\mathbf{p}) + \sin(\theta^\star/2)\, e^{i\phi^*}\, \psi_{1/2,-1/2}(\mathbf{p}) .$$

Inserting the expressions for $\psi_{1/2,\pm1/2}(\mathbf{p})$ obtained above and collecting terms proportional to the $|\pm\rangle$ states of particle b yields

$$\psi_{\hat{\mathbf{P}}}(\mathbf{p}) = \underbrace{\frac{1}{\sqrt{4\pi}} \left[A_S \cos(\theta^\star/2) - A_P \cos(\theta^\star/2)\, \cos\theta - A_P \sin(\theta^\star/2)\, \sin\theta\, e^{i(\phi^*-\phi)} \right] |+\rangle}_{\chi_+(\mathbf{p})}$$

$$+ \underbrace{\frac{e^{i\phi^*}}{\sqrt{4\pi}} \left[A_S \sin(\theta^\star/2) + A_P \sin(\theta^\star/2)\, \cos\theta - A_P \cos(\theta^\star/2)\, \sin\theta\, e^{-i(\phi^*-\phi)} \right] |-\rangle}_{\chi_-(\mathbf{p})} ,$$

and the angular distribution results from the probability density (below, we set $\Delta\phi = \phi^\star - \phi$)

$$\psi^\dagger_{\hat{\mathbf{P}}}(\mathbf{p})\, \psi_{\hat{\mathbf{P}}}(\mathbf{p}) = |\chi_+(\mathbf{p})|^2 + |\chi_-(\mathbf{p})|^2$$

$$= \frac{1}{4\pi}\Big[|A_S|^2 + |A_P|^2 - 2\,\mathrm{Re}(A_S A_P^*)\left[\cos^2(\theta^\star/2) - \sin^2(\theta^\star/2) \right] \cos\theta$$

$$- \sin(\theta^\star/2)\cos(\theta^\star/2)\sin\theta \left(A_S A_P^* + A_S^* A_P \right) \left(e^{-i\Delta\phi} + e^{i\Delta\phi} \right) \Big],$$

where in the last line the cross terms proportional to $|A_P|^2 \sin(\theta^\star/2)\cos(\theta^\star/2)\sin\theta\cos\theta$ cancel. Therefore, we have

$$\psi^\dagger_{\hat{\mathbf{P}}}(\mathbf{p})\, \psi_{\hat{\mathbf{P}}}(\mathbf{p}) = \frac{1}{4\pi}\Big[|A_S|^2 + |A_P|^2 - 2\,\mathrm{Re}(A_S A_P^*) \left[\cos\theta^\star \cos\theta + \sin\theta^\star \sin\theta\, \cos(\phi^\star - \phi) \right] \Big] ,$$

which can also be written as

$$\psi^\dagger_{\hat{\mathbf{P}}}(\mathbf{p})\, \psi_{\hat{\mathbf{P}}}(\mathbf{p}) = \frac{|A_S|^2 + |A_P|^2}{4\pi} \left[1 - 2\,\frac{\mathrm{Re}(A_S A_P^*)}{|A_S|^2 + |A_P|^2}\, \hat{\mathbf{P}} \cdot \hat{\mathbf{p}} \right] ,$$

since

$$\hat{\mathbf{P}} \cdot \hat{\mathbf{p}} = \sin\theta^\star \sin\theta\, \underbrace{\left(\cos\phi^\star \cos\phi + \sin\phi^\star \sin\phi \right)}_{\cos(\phi^\star - \phi)} + \cos\theta^\star \cos\theta .$$

We obtain

$$\psi^\dagger_{\hat{\mathbf{P}}}(\mathbf{p})\, \psi_{\hat{\mathbf{P}}}(\mathbf{p}) \propto 1 - \eta\, \hat{\mathbf{P}} \cdot \hat{\mathbf{p}} \qquad \text{with} \qquad \eta = 2\,\frac{\mathrm{Re}(A_S A_P^*)}{|A_S|^2 + |A_P|^2} .$$

Problem 10 Addition of Three Angular Momenta

Let $\mathbb{E}_a$, $\mathbb{E}_b$, and $\mathbb{E}_c$ be the state spaces of three systems a, b, and c with angular momenta $\mathbf{J}_a$, $\mathbf{J}_b$, and $\mathbf{J}_c$ (the components J_{ai} and J_{bj} commute if $a \neq b$). The eigenstates of $\mathbf{J}_a^2$ and J_{az} are denoted by $|\psi_{j_a m_a}\rangle$, and similarly for systems b and c. We want to add these three angular momenta to form eigenstates of $\mathbf{J}^2$ and J_z – here $\mathbf{J}$ is the total angular momentum – corresponding to the quantum numbers j_f and m_f. We denote these eigenstates as

$$|\psi_{j_f m_f}^{j_a,(j_b j_c)j_e}\rangle,$$

where j_b and j_c are added first to obtain j_e, and then j_a and j_e are combined to give j_f. Of course, we could also add j_a and j_b to form j_d and then j_d and j_c to form j_f, with the corresponding eigenstates denoted as

$$|\psi_{j_f m_f}^{(j_a j_b)j_d j_c}\rangle.$$

1. Express the eigenstates $|\psi_{j_f m_f}^{j_a,(j_b j_c)j_e}\rangle$ and $|\psi_{j_f m_f}^{(j_a j_b)j_d j_c}\rangle$ as linear combinations of those of $\mathbf{J}_a^2, J_{az}, \mathbf{J}_b^2, J_{bz}$, and $\mathbf{J}_c^2, J_{cz}$, that is,

$$|\psi_{m_a m_b m_c}^{j_a j_b j_c}\rangle = |\psi_{j_a m_a}\rangle \otimes |\psi_{j_b m_b}\rangle \otimes |\psi_{j_c m_c}\rangle.$$

2. Show that the sets of kets $|\psi_{j_f m_f}^{j_a,(j_b j_c)j_e}\rangle$ and $|\psi_{j_f m_f}^{(j_a j_b)j_d j_c}\rangle$ corresponding to the various possible values of, respectively, j_e, j_f and j_d, j_f form orthonormal bases.

3. Show, by using the operators $J_\pm$, that the scalar product $\langle\psi_{j_f m_f}^{(j_a j_b)j_d j_c}|\psi_{j_f m_f}^{j_a,(j_b j_c)j_e}\rangle$ does not depend on m_f. Hence, denote this scalar product as

$$\langle\psi_{j_f}^{(j_a j_b)j_d j_c}|\psi_{j_f}^{j_a,(j_b j_c)j_e}\rangle.$$

4. Show that

$$|\psi_{j_f,m_f}^{j_a,(j_b j_c)j_e}\rangle = \sum_{j_d}\langle\psi_{j_f}^{(j_a j_b)j_d j_c}|\psi_{j_f}^{j_a,(j_b j_c)j_e}\rangle\,|\psi_{j_f m_f}^{(j_a j_b)j_d j_c}\rangle.$$

5. Using the results of part 1, show that

$$\sum_{m_e} C_{j_b j_c}(m_b, m_c; j_e, m_e)\, C_{j_a j_e}(m_a, m_e; j_f, m_f)$$

$$= \sum_{j_d, m_d} C_{j_a j_b}(m_a, m_b; j_d, m_d) \times C_{j_d j_c}(m_d, m_c; j_f, m_f)\langle\psi_{j_f}^{(j_a j_b)j_d j_c}|\psi_{j_f}^{j_a,(j_b j_c)j_e}\rangle.$$

6. Starting from the result in part 5, prove, using the Clebsch–Gordan orthogonality relations, that the following identities hold:

$$\sum_{m_a m_b m_e} C_{j_b j_c}(m_b, m_c; j_e, m_e)\, C_{j_a j_e}(m_a, m_e; j_f, m_f)\, C_{j_a j_b}(m_a, m_b; j_d, m_d)$$

$$= C_{j_d j_c}(m_d, m_c; j_f, m_f)\langle\psi_{j_f}^{(j_a j_b)j_d j_c}|\psi_{j_f}^{j_a,(j_b j_c)j_e}\rangle$$

and

$$\langle\psi_{j_f}^{(j_a j_b)j_d j_c}|\psi_{j_f}^{j_a,(j_b j_c)j_e}\rangle = \frac{1}{2j_f + 1}\sum_{m_a m_b m_c m_d m_e m_f} C_{j_b j_c}(m_b, m_c; j_e, m_e)\, C_{j_a j_e}(m_a, m_e; j_f, m_f)$$

$$\times C_{j_a j_b}(m_a, m_b; j_d, m_d)\, C_{j_d j_c}(m_d, m_c; j_f, m_f).$$

(Adapted from C. Cohen-Tannoudji, B. Diu, and F. Laloë (1997), *Quantum Mechanics*, vol. 2, Wiley.)

Solution

Part 1

We obtain

$$|\psi_{j_e m_e}^{j_b j_c}\rangle = \sum_{m_b m_c} C_{j_b j_c}(m_b, m_c; j_e, m_e) \, |\psi_{j_b m_b}\rangle \otimes \psi_{j_c m_c}\rangle \, ,$$

and then combine j_a with j_e to give j_f, m_f:

$$|\psi_{j_f m_f}^{j_a,(j_b j_c)j_e}\rangle = \sum_{m_a m_e} C_{j_a j_e}(m_a, m_e; j_f, m_f) \, |\psi_{j_a m_a}\rangle \otimes |\psi_{j_e m_e}^{j_b j_c}\rangle$$

$$= \sum_{m_a m_b m_c m_e} C_{j_b j_c}(m_b, m_c; j_e, m_e) \, C_{j_a j_e}(m_a, m_e; j_f, m_f) \, \underbrace{|\psi_{j_a m_a}\rangle \otimes |\psi_{j_b m_b}\rangle \otimes |\psi_{j_c m_c}\rangle}_{|\psi_{m_a m_b m_c}^{j_a j_b j_c}\rangle} \, .$$

We obtain similarly

$$|\psi_{j_f m_f}^{(j_a j_b)j_d, j_c}\rangle = \sum_{m_a m_b m_c m_d} C_{j_a j_b}(m_a, m_b; j_d, m_d) \, C_{j_d j_c}(m_d, m_c; j_f, m_f) \, |\psi_{m_a m_b m_c}^{j_a j_b j_c}\rangle \, .$$

Part 2

In the basis $|\psi_{m_a m_b m_c}^{j_a j_b j_c}\rangle$ (for fixed j_a, j_b, j_c) there are $(2j_a + 1)(2j_b + 1)(2j_c + 1)$ independent states. The number of states $|\psi_{j_f m_f}^{j_a,(j_b j_c)j_e}\rangle$ is the same, since (see Problem 1 for an evaluation of these sums)

$$\sum_{j_e=|j_b-j_c|}^{j_b+j_c} \sum_{j_f=|j_a-j_e|}^{j_a+j_e} (2j_f + 1) = \sum_{j_e=|j_b-j_c|}^{j_b+j_c} (2j_a + 1)(2j_e + 1) = (2j_a + 1)(2j_b + 1)(2j_c + 1) \, .$$

The states are orthonormal and hence independent, that is, they themselves form a basis; that they are orthonormal follows from

$$\langle \psi_{j_f' m_f'}^{j_a,(j_b j_c)j_e'} | \psi_{j_f m_f}^{j_a,(j_b j_c)j_e}\rangle = \sum_{m_a' m_b' m_c' m_e'} \sum_{m_a m_b m_c m_e} C_{j_b j_c}(m_b', m_c'; j_e', m_e') \, C_{j_a j_e'}(m_a', m_e'; j_f', m_f')$$

$$\times C_{j_b j_c}(m_b, m_c; j_e, m_e) \, C_{j_a j_e}(m_a, m_e; j_f, m_f) \langle \psi_{m_a' m_b' m_c'}^{j_a j_b j_c} | \psi_{m_a m_b m_c}^{j_a j_b j_c}\rangle$$

$$= \sum_{m_e, m_e'} \sum_{m_a m_b m_c} C_{j_b j_c}(m_b, m_c; j_e', m_e') \, C_{j_a j_e'}(m_a, m_e'; j_f', m_f')$$

$$\times C_{j_b j_c}(m_b, m_c; j_e, m_e) \, C_{j_a j_e}(m_a, m_e; j_f, m_f)$$

$$= \sum_{m_e, m_e'} \sum_{m_a} \delta_{j_e j_e'} \, \delta_{m_e, m_e'} \, C_{j_a j_e'}(m_a, m_e'; j_f', m_f') \, C_{j_a j_e}(m_a, m_e; j_f, m_f)$$

$$= \delta_{j_e j_e'} \sum_{m_a m_e} C_{j_a j_e}(m_a, m_e; j_f', m_f') \, C_{j_a j_e}(m_a, m_e; j_f, m_f)$$

$$= \delta_{j_e j_e'} \, \delta_{j_f j_f'} \, \delta_{m_f, m_f'} \, ,$$

where we have repeatedly used the orthogonality relation for the Clebsch–Gordan coefficients.

Part 3

Using the property of the raising operator J_+, we find

$$|\psi_{j_f m_f}^{j_a,(j_b j_c)j_e}\rangle = \frac{1}{\hbar\sqrt{j_f(j_f+1)-m_f(m_f-1)}} J_+ |\psi_{j_f,m_f-1}^{j_a,(j_b j_c)j_e}\rangle\,,$$

and hence

$$\langle\psi_{j_f m_f}^{(j_a j_b)j_d j_c}|\psi_{j_f m_f}^{j_a,(j_b j_c)j_e}\rangle = \frac{1}{\hbar\sqrt{j_f(j_f+1)-m_f(m_f-1)}}\langle\psi_{j_f m_f}^{(j_a j_b)j_d j_c}|J_+|\psi_{j_f,m_f-1}^{j_a,(j_b j_c)j_e}\rangle$$

$$= \frac{1}{\hbar\sqrt{j_f(j_f+1)-m_f(m_f-1)}}(J_-|\psi_{j_f m_f}^{(j_a j_b)j_d j_c}\rangle)^\dagger|\psi_{j_f,m_f-1}^{j_a,(j_b j_c)j_e}\rangle$$

$$= \langle\psi_{j_f,m_f-1}^{(j_a j_b)j_d j_c}|\psi_{j_f,m_f-1}^{j_a,(j_b j_c)j_e}\rangle\,,$$

where we have used $J_- = (J_+)^\dagger$ and

$$J_-|\psi_{j_f m_f}^{j_a,(j_b j_c)j_e}\rangle = \hbar\sqrt{j_f(j_f+1)-m_f(m_f-1)}\,|\psi_{j_f,m_f-1}^{j_a,(j_b j_c)j_e}\rangle\,.$$

Thus, we see that the scalar product is independent of m_f,

$$\langle\psi_{j_f,m_f}^{(j_a j_b)j_d j_c}|\psi_{j_f,m_f}^{j_a,(j_b j_c)j_e}\rangle = \langle\psi_{j_f}^{(j_a j_b)j_d j_c}|\psi_{j_f}^{j_a,(j_b j_c)j_e}\rangle\,.$$

Part 4

The completeness relation for the basis $|\psi_{j_f,m_f}^{(j_a j_b)j_d j_c}\rangle$ reads

$$\sum_{j_d j_f m_f} |\psi_{j_f,m_f}^{(j_a j_b)j_d j_c}\rangle\langle\psi_{j_f,m_f}^{(j_a j_b)j_d j_c}| = (\mathbb{1})_{j_a j_b j_c}\,,$$

and therefore the state $|\psi_{j_f,m_f}^{j_a,(j_b j_c)j_e}\rangle$ can be expanded in the subspace with fixed j_a, j_b, j_c as

$$|\psi_{j_f,m_f}^{j_a,(j_b j_c)j_e}\rangle = \sum_{j_d j_f' m_f'}|\psi_{j_f',m_f'}^{(j_a j_b)j_d j_c}\rangle\underbrace{\langle\psi_{j_f',m_f'}^{(j_a j_b)j_d j_c}|\psi_{j_f,m_f}^{j_a,(j_b j_c)j_e}\rangle}_{\delta_{j_f j_f'}\,\delta_{m_f,m_f'}} = \sum_{j_d}|\psi_{j_f,m_f}^{(j_a j_b)j_d j_c}\rangle\langle\psi_{j_f}^{(j_a j_b)j_d j_c}|\psi_{j_f}^{j_a,(j_b j_c)j_e}\rangle\,,$$

where in the second step we have used the orthogonality of the states with total j_f, m_f (which holds regardless of how these j_f, m_f are obtained, i.e., whether by first coupling j_a, j_b to j_d or j_b, j_c to j_e) and in the last step the result obtained in part 3.

Part 5

In the expression derived in part 4 we expand the states $|\psi_{j_f m_f}^{j_a,(j_b j_c)j_e}\rangle$ and $|\psi_{j_f m_f}^{(j_a j_b)j_d j_c}\rangle$ in the basis $|\psi_{m_a m_b m_c}^{j_a j_b j_c}\rangle$, to obtain

$$\sum_{m_a' m_b' m_c' m_e} C_{j_b j_c}(m_b', m_c'; j_e, m_e)\, C_{j_a j_e}(m_a', m_e; j_f, m_f)|\psi_{m_a' m_b' m_c'}^{j_a j_b j_c}\rangle$$

$$= \sum_{j_d}\langle\psi_{j_f}^{(j_a j_b)j_d j_c}|\psi_{j_f}^{j_a,(j_b j_c)j_e}\rangle \times \sum_{m_a' m_b' m_c' m_d} C_{j_a j_b}(m_a', m_b'; j_d, m_d)\, C_{j_d j_c}(m_d, m_c'; j_f, m_f)|\psi_{m_a' m_b' m_c'}^{j_a j_b j_c}\rangle\,,$$

projecting out both sides onto $|\psi_{m_a m_b m_c}^{j_a j_b j_c}\rangle$. Using the orthonormality of the states $|\psi_{m_a m_b m_c}^{j_a j_b j_c}\rangle$, we deduce that

$$\sum_{m_e} C_{j_b j_c}(m_b, m_c; j_e, m_e)\, C_{j_a j_e}(m_a, m_e; j_f, m_f) = \sum_{j_d, m_d} C_{j_a j_b}(m_a, m_b; j_d, m_d)$$
$$\times C_{j_d j_c}(m_d, m_c; j_f, m_f)\langle \psi_{j_f}^{(j_a j_b) j_d j_c}|\psi_{j_f}^{j_a, (j_b j_c) j_e}\rangle\,.$$

Part 6

Multiply both sides of the above identity by the Clebsch–Gordan coefficient $C_{j_a j_b}(m_a, m_b; j_d, m_d)$ and sum over $m_a m_b$:

$$\sum_{m_a m_b m_e} C_{j_b j_c}(m_b, m_c; j_e, m_e)\, C_{j_a j_e}(m_a, m_e; j_f, m_f)\, C_{j_a j_b}(m_a, m_b; j_d, m_d)$$
$$= \sum_{m_a m_b j'_d, m'_d} C_{j_a j_b}(m_a, m_b; j'_d, m'_d)\, C_{j_a j_b}(m_a, m_b; j_d, m_d)$$
$$\times C_{j'_d j_c}(m'_d, m_c; j_f, m_f)\langle \psi_{j_f}^{(j_a j_b) j'_d j_c}|\psi_{j_f}^{j_a, (j_b j_c) j_e}\rangle\,.$$

Now, we use the orthogonality relation

$$\sum_{m_a m_b} C_{j_a j_b}(m_a, m_b; j'_d, m'_d)\, C_{j_a j_b}(m_a, m_b; j_d, m_d) = \delta_{j_d j'_d}\,\delta_{m_d, m'_d}$$

to find, after summing over j'_d, m'_d, the required relation:

$$\sum_{m_a m_b m_e} C_{j_b j_c}(m_b, m_c; j_e, m_e)\, C_{j_a j_e}(m_a, m_e; j_f, m_f)\, C_{j_a j_b}(m_a, m_b; j_d, m_d)$$
$$= C_{j_d j_c}(m_d, m_c; j_f, m_f)\langle \psi_{j_f}^{(j_a j_b) j_d j_c}|\psi_{j_f}^{j_a, (j_b j_c) j_e}\rangle\,.$$

Starting from this latter identity, we multiply both sides by $C_{j_d j_c}(m_d, m_c; j'_f, m'_f)$ and sum over m_c, m_d:

$$\sum_{m_a m_b m_c m_d m_e} C_{j_b j_c}(m_b, m_c; j_e, m_e) C_{j_a j_e}(m_a, m_e; j_f, m_f) C_{j_a j_b}(m_a, m_b; j_d, m_d) C_{j_d j_c}(m_d, m_c; j'_f, m'_f)$$
$$= \langle \psi_{j_f}^{(j_a j_b) j_d j_c}|\psi_{j_f}^{j_a, (j_b j_c) j_e}\rangle \underbrace{\sum_{m_c m_d} C_{j_d j_c}(m_d, m_c; j_f, m_f) C_{j_d j_c}(m_d, m_c; j'_f, m'_f)}_{\delta_{j_f j'_f}\,\delta_{m_f, m'_f}}\,.$$

We can sum both sides of the above identity over m_f, which yields

$$\sum_{m_a m_b m_c m_d m_e m_f} C_{j_b j_c}(m_b, m_c; j_e, m_e) C_{j_a j_e}(m_a, m_e; j_f, m_f) C_{j_a j_b}(m_a, m_b; j_d, m_d) C_{j_d j_c}(m_d, m_c; j_f, m_f)$$
$$= \sum_{m_f} \langle \psi_{j_f}^{(j_a j_b) j_d j_c}|\psi_{j_f}^{j_a, (j_b j_c) j_e}\rangle = (2 j_f + 1)\langle \psi_{j_f}^{(j_a j_b) j_d j_c}|\psi_{j_f}^{j_a, (j_b j_c) j_e}\rangle\,,$$

as indicated in the text of the problem.

14

Approximation Methods

There are only few problems for which we can determine the energy spectrum and eigenstates exactly; a particle in a Coulomb or harmonic oscillator potential are two important examples. For most other problems we must rely on approximation methods. In this section we introduce two of these methods, known as time-independent perturbation theory and the variational method, primarily for bound-state problems.

14.1 Non-Degenerate Perturbation Theory

We begin by writing the Hamiltonian as

$$\hat{H} = \hat{H}_0 + \hat{V}, \tag{14.1}$$

where it is assumed that the eigenvalues and (normalized) eigenstates of $\hat{H}_0$ are known:

$$\hat{H}_0 |\phi_m\rangle = \epsilon_m |\phi_m\rangle. \tag{14.2}$$

Some the eigenvalues of $\hat{H}_0$ may be degenerate; however, we assume that the eigenvalue ϵ_n to which we are interested in determining corrections is non-degenerate. We also assume that matrix elements of the perturbation $\hat{V}$ are small relative to those of $\hat{H}_0$; a more precise characterization of what "small" means in the present context is given below. We introduce a real parameter λ varying in the interval $[0,1]$, and define

$$\hat{H}(\lambda) = \hat{H}_0 + \lambda \hat{V}. \tag{14.3}$$

As λ is varied between 0 and 1, the non-degenerate eigenstate $|\phi_n\rangle$ of $\hat{H}_0$ will smoothly become a non-degenerate eigenstate $|\psi_n\rangle$ of $\hat{H}$ and the energy ϵ_n will smoothly become the energy E_n corresponding to $|\psi_n\rangle$. We look for expansions of the eigenstate $|\psi_n(\lambda)\rangle$ and eigenvalue $E_n(\lambda)$ of $\hat{H}$ in powers of λ (eventually, λ will be set to 1, but for now it is a convenient way to keep track of powers of $\hat{V}$):

$$|\psi_n(\lambda)\rangle = |\phi_n\rangle + \lambda |\psi_n^{(1)}\rangle + \lambda^2 |\psi_n^{(2)}\rangle + \cdots \tag{14.4}$$

$$E_n(\lambda) = \epsilon_n + \lambda E^{(1)} + \lambda^2 E^{(2)} + \cdots. \tag{14.5}$$

It is important to stress here that the existence of such expansions is a rather strong assumption. Expanding the states in a power series implies that the physics for, say, a small positive $V(\mathbf{r})$ is only slightly different from the physics for a small negative $V(\mathbf{r})$. But this is certainly not the case for a bound state. A weak attractive potential can have a bound state, but a weak repulsive one cannot – a very different situation.[1] In particular, bound states in a weak attractive potential cannot be obtained using perturbation theory based on the free-particle Hamiltonian.

With this proviso, we return to the expansions above. The states $|\phi_n\rangle$ are normalized ($\langle\phi_n|\phi_n\rangle = 1$). However, we *choose* to normalize the states $|\psi_n(\lambda)\rangle$ in such a way that

$$\langle\phi_n|\psi_n(\lambda)\rangle = 1 \implies \lambda\langle\phi_n|\psi_n^{(1)}\rangle + \lambda^2\langle\phi_n|\psi_n^{(2)}\rangle + \cdots = 0 , \tag{14.6}$$

and the coefficient of each power λ must vanish,

$$\langle\phi_n|\psi_n^{(p)}\rangle = 0 \qquad p = 1, 2, \ldots \tag{14.7}$$

We insert the expansions into the Schrödinger equation $(\hat{H}_0 + \lambda\hat{V})|\psi_n(\lambda)\rangle = E_n(\lambda)|\psi_n(\lambda)\rangle$ and match the powers of λ on the left- and right-hand sides to obtain

$$
\begin{aligned}
\lambda^0: &\quad \hat{H}_0|\phi_n\rangle = \epsilon_n|\phi_n\rangle , \\
\lambda^1: &\quad \hat{H}_0|\psi_n^{(1)}\rangle + \hat{V}|\phi_n\rangle = \epsilon_n|\psi_n^{(1)}\rangle + E_n^{(1)}|\phi_n\rangle , \\
&\vdots \\
\lambda^p: &\quad \hat{H}_0|\psi_n^{(p)}\rangle + \hat{V}|\psi_n^{(p-1)}\rangle = \epsilon_n|\psi_n^{(p)}\rangle + E_n^{(1)}|\psi_n^{(p-1)}\rangle + \cdots + E_n^{(p)}|\phi_n\rangle .
\end{aligned}
\tag{14.8}
$$

We project onto the state $|\phi_n\rangle$ both sides of each of the relations above (the first one, for λ^0, gives nothing new, and so we can ignore it),

$$
\begin{aligned}
\lambda^1: &\quad \langle\phi_n|(\hat{H}_0|\psi_n^{(1)}\rangle + \hat{V}|\phi_n\rangle) = \langle\phi_n|(\epsilon_n|\psi_n^{(1)}\rangle + E_n^{(1)}|\phi_n\rangle) , \\
\lambda^2: &\quad \langle\phi_n|(\hat{H}_0|\psi_n^{(2)}\rangle + \hat{V}|\psi_n^{(1)}\rangle) = \langle\phi_n|(\epsilon_n|\psi_n^{(2)}\rangle + E_n^{(1)}|\psi_n^{(1)}\rangle + E_n^{(2)}|\phi_n\rangle) , \\
&\vdots \\
\lambda^p: &\quad \langle\phi_n|(\hat{H}_0|\psi_n^{(p)}\rangle + \hat{V}|\psi_n^{(p-1)}\rangle) = \langle\phi_n|(\epsilon_n|\psi_n^{(p)}\rangle + E_n^{(1)}|\psi_n^{(p-1)}\rangle + \cdots + E_n^{(p)}|\phi_n\rangle) ,
\end{aligned}
\tag{14.9}
$$

and use the fact that $|\phi_n\rangle$ is an eigenstate of $\hat{H}_0$ along with the condition in Eq. (14.7),

$$
\begin{aligned}
\lambda^1: &\quad E_n^{(1)} = \langle\phi_n|\hat{V}|\phi_n\rangle , \\
\lambda^2: &\quad E_n^{(2)} = \langle\phi_n|\hat{V}|\psi_n^{(1)}\rangle , \\
&\vdots \\
\lambda^p: &\quad E_n^{(p)} = \langle\phi_n|\hat{V}|\psi_n^{(p-1)}\rangle .
\end{aligned}
\tag{14.10}
$$

We see that the correction to the energy at any particular order requires knowledge of the correction to the state at the next lower order.[2] Note, however, that the procedure has provided us with the

---

[1] As a specific example, consider a particle in one dimension under the influence of a potential $V(x) = -V_0$ for $|x| \leq a/2$ and $V(x) = 0$ for $|x| > a/2$. For *any* value of V_0 there is at least one bound state of energy $E_0 = -ma^2V_0^2/(2\hbar^2)$, however small V_0 is. By contrast, the potential with the sign of V_0 flipped does not have any bound states. Hence, the physics for small negative $V(x)$ is very different from that with small positive $V(x)$. In the former case, perturbation theory based on free states will fail.

[2] On multiplying both sides of the last equation by λ^p and then summing over p from 1 to ∞, we obtain

$$\underbrace{\sum_{p=1}^{\infty}\lambda^p E_n^{(p)}}_{E_n - \epsilon_n} = \underbrace{\sum_{p=1}^{\infty}\lambda^p\langle\phi_n|\hat{V}|\psi_n^{(p-1)}\rangle}_{\langle\phi_n|\lambda\hat{V}(\sum_{p=1}^{\infty}\lambda^{p-1}|\psi_n^{(p-1)}\rangle)} \implies E_n - \epsilon_n = \langle\phi_n|\hat{V}|\psi_n\rangle ,$$

where in the last step we have set $\lambda = 1$. This relation also follows directly from the Schrödinger equation $(\hat{H}_0 + \hat{V})|\psi_n\rangle = E_n|\psi\rangle$ by projecting onto the unperturbed state $|\phi_n\rangle$ and using the normalization condition $\langle\phi_n|\psi_n\rangle = 1$.

first-order correction to the energy – the first of the equations above: it is obtained as the expectation value of the perturbation $\hat{V}$ on the unperturbed state $|\phi_n\rangle$. We are left with the problem of determining the state $|\psi^{(p)}\rangle$, representing the pth-order correction to $|\phi_n\rangle$. Such a state can be expanded over the basis of eigenstates of $\hat{H}_0$ as

$$|\psi_n^{(p)}\rangle = \sum_{m\neq n} |\phi_m\rangle\langle\phi_m|\psi_n^{(p)}\rangle \,, \tag{14.11}$$

where we have omitted the term with $m = n$, since $\langle\phi_n|\psi_n^{(p)}\rangle = 0$. To obtain the component $\langle\phi_m|\psi^{(p)}\rangle$, project both sides of Eq. (14.8) onto $|\phi_m\rangle$:

$$\lambda^1: \qquad \langle\phi_m|(\hat{H}_0|\psi_n^{(1)}\rangle + \hat{V}|\phi_n\rangle) = \langle\phi_m|(\epsilon_n|\psi_n^{(1)}\rangle + E_n^{(1)}|\phi_n\rangle) \,,$$

$$\lambda^2: \qquad \langle\phi_m|(\hat{H}_0|\psi_n^{(2)}\rangle + \hat{V}|\psi_n^{(1)}\rangle) = \langle\phi_m|(\epsilon_n|\psi_n^{(2)}\rangle + E_n^{(1)}|\psi_n^{(1)}\rangle + E_n^{(2)}|\phi_n\rangle) \,,$$

$$\vdots \tag{14.12}$$

$$\lambda^p: \qquad \langle\phi_m|(\hat{H}_0|\psi_n^{(p)}\rangle + \hat{V}|\psi_n^{(p-1)}\rangle) = \langle\phi_m|(\epsilon_n|\psi_n^{(p)}\rangle + E_n^{(1)}|\psi_n^{(p-1)}\rangle + \cdots + E_n^{(p)}|\phi_n\rangle) \,,$$

or, letting $\hat{H}_0$ act to the left on the left-hand sides of the equations above and noting that $\langle\phi_m|\phi_n\rangle = 0$ for $m \neq n$, we obtain

$$\lambda^1: \qquad \langle\phi_m|\psi_n^{(1)}\rangle = \frac{1}{\epsilon_n - \epsilon_m}\langle\phi_m|\hat{V}|\phi_n\rangle \,,$$

$$\lambda^2: \qquad \langle\phi_m|\psi_n^{(2)}\rangle = \frac{1}{\epsilon_n - \epsilon_m}\left[\langle\phi_m|\hat{V}|\psi_n^{(1)}\rangle - E_n^{(1)}\langle\phi_m|\psi_n^{(1)}\rangle\right] \,,$$

$$\vdots \tag{14.13}$$

$$\lambda^p: \qquad \langle\phi_m|\psi_n^{(p)}\rangle = \frac{1}{\epsilon_n - \epsilon_m}\left[\langle\phi_m|\hat{V}|\psi_n^{(p-1)}\rangle - E_n^{(1)}\langle\phi_m|\psi_n^{(p-1)}\rangle - \cdots - E_n^{(p-1)}\langle\phi_m|\psi_n^{(1)}\rangle\right] \,,$$

which allows us to find the pth order correction. Note that the energy denominators never vanish.

We can work out explicitly the formulae above in the special cases of the first- and second-order corrections (by assumption, the energy ϵ_n being perturbed is non-degenerate). As already noted earlier, the first-order correction to the energy is given by

$$E_n^{(1)} = \langle\phi_n|\hat{V}|\phi_n\rangle \,, \tag{14.14}$$

while the first-order correction to the unperturbed state $|\phi_n\rangle$ reads

$$|\psi_n^{(1)}\rangle = \sum_{m\neq n} |\phi_m\rangle\langle\phi_m|\psi_n^{(1)}\rangle = \sum_{m\neq n} |\phi_m\rangle\frac{\langle\phi_m|\hat{V}|\phi_n\rangle}{\epsilon_n - \epsilon_m} \,. \tag{14.15}$$

Note that $|\psi_n^{(1)}\rangle$ is "small" if the absolute value of the matrix element $|\langle\phi_m|\hat{V}|\phi_n\rangle|$ is much less than the smallest energy difference $|\epsilon_n - \epsilon_m|$; in such an instance, we can expect the expansion to converge rapidly.

The second-order correction to the energy follows from the second of equations (14.10):

$$E_n^{(2)} = \langle\phi_n|\hat{V}|\psi_n^{(1)}\rangle = \sum_{m\neq n}\langle\phi_n|\hat{V}|\phi_m\rangle\frac{\langle\phi_m|\hat{V}|\phi_n\rangle}{\epsilon_n - \epsilon_m} = \sum_{m\neq n}\frac{|\langle\phi_m|\hat{V}|\phi_n\rangle|^2}{\epsilon_n - \epsilon_m} \,. \tag{14.16}$$

Note that if the state in which we are interested is the ground state $|\phi_0\rangle$ of $\hat{H}_0$ then the second-order correction $E_0^{(2)}$ is always negative ($\epsilon_0 - \epsilon_m < 0$). It is often the case that the first-order correction $E_0^{(1)}$ vanishes because of selection rules; in such a case the second-order correction $E_0^{(2)}$ always lowers the ground-state energy of $\hat{H}_0$. The second-order correction $|\psi_n^{(2)}\rangle$ to the state is obtained in Problem 1.

14.2 Degenerate Perturbation Theory

Perturbation theory is, in essence, an expansion in terms of ratios given by

$$\frac{\langle \phi_m | \hat{V} | \phi_n \rangle}{\epsilon_n - \epsilon_m}. \tag{14.17}$$

Therefore, one may expect the expansion to be more rapidly converging the smaller are the matrix elements of $\hat{V}$ compared with the level spacings of the unperturbed system. However, if there are any states for which $\epsilon_m = \epsilon_n$ but $\langle \phi_m | \hat{V} | \phi_n \rangle \neq 0$, the method as formulated so far breaks down. This is the case for the perturbation theory for a degenerate eigenvalue ϵ_n of $\hat{H}_0$,

$$\hat{H}_0 | \phi_n^{(k)} \rangle = \epsilon_n | \phi_n^{(k)} \rangle \qquad\qquad k = 1, 2, \dots, g_n, \tag{14.18}$$

where the states $| \phi_n^{(k)} \rangle$ span a subspace, which we denote as $\mathbb{E}_n$, of the state space $\mathbb{E}$ of the system. Note that any linear combination of the $| \phi_n^{(k)} \rangle$ is still an eigenstate of $\hat{H}_0$ with the same eigenvalue ϵ_n. We can exploit this freedom to remove (at least partially, depending on the specific perturbation $\hat{V}$; see below) the aforementioned problem. We choose a set of g_n orthonormal states

$$| \chi_l \rangle = \sum_{k=1}^{g_n} c_k^{(l)} | \phi_n^{(k)} \rangle \qquad\qquad l = 1, \dots, g_n, \tag{14.19}$$

such that

$$\langle \chi_l | \hat{V} | \chi_m \rangle = 0 \qquad\qquad \text{if } l \neq m, \tag{14.20}$$

namely, the matrix representing $\hat{V}$ in the basis $| \chi_l \rangle$ is diagonal. We can then use the perturbation methods developed earlier, since vanishing energy denominators are always accompanied by vanishing numerators, as will become apparent below. The correct choice of basis states to use in doing perturbation theory is therefore the basis that diagonalizes $\hat{V}$ within each group of degenerate states. Should we be interested in the perturbation of only one specific such group of energy ϵ_n, then it is only necessary to diagonalize $\hat{V}$ within that one group of states. The reason is that all the energy denominators that occur in the perturbative expansion are energy differences between ϵ_n and the energies of other states of $\hat{H}_0$, which are not in the subspace $\mathbb{E}_n$ spanned by the basis $| \chi_l \rangle$: that the energies of these other states may be degenerate is not an issue, since they do not give vanishing denominators.

The first task is to determine the coefficients $c_k^{(l)}$ of the linear combination in Eq. (14.19). To this end, we report below the first two equations of the set (14.8)

$$\begin{aligned} \lambda^0: &\qquad \hat{H}_0 | \chi_l \rangle = \epsilon_n | \chi_l \rangle, \\ \lambda^1: &\qquad \hat{H}_0 | \chi_l^{(1)} \rangle + \hat{V} | \chi_l \rangle = \epsilon_n | \chi_l^{(1)} \rangle + E_{nl}^{(1)} | \chi_l \rangle, \end{aligned} \tag{14.21}$$

where $| \chi_l \rangle$ is any of the g_n states in the subspace $\mathbb{E}_n$ of energy ϵ_n, and $| \chi_l^{(1)} \rangle$ and $E_{nl}^{(1)}$ are the first-order corrections to that state and the corresponding energy, respectively. The λ^0 equation does not allow us to determine the $c_k^{(l)}$; it just says that $| \chi_l \rangle$, being a linear combination of the $| \phi_n^{(k)} \rangle$, is an eigenstate of $\hat{H}_0$ with eigenvalue ϵ_n. Now, project both sides of the λ^1 equation onto the state $| \phi_n^{(m)} \rangle$ in the subspace $\mathbb{E}_n$,

$$\langle \phi_n^{(m)} | \hat{H}_0 | \chi_l^{(1)} \rangle + \langle \phi_n^{(m)} | \hat{V} | \chi_l \rangle = \epsilon_n \langle \phi_n^{(m)} | \chi_l^{(1)} \rangle + E_{nl}^{(1)} \langle \phi_n^{(m)} | \chi_l \rangle, \tag{14.22}$$

and note that letting $\hat{H}_0$ act to the left in the first term on the left-hand side leads to $\langle \phi_n^{(m)} | \hat{H}_0 | \chi_l^{(1)} \rangle = \epsilon_n \langle \phi_n^{(m)} | \chi_l^{(1)} \rangle$, which exactly cancels the first term on the right-hand side. We are left with

$$\sum_{k=1}^{g_n} \langle \phi_n^{(m)} | \hat{V} | \phi_n^{(k)} \rangle c_k^{(l)} = E_{nl}^{(1)} c_m^{(l)} , \tag{14.23}$$

and hence the first-order corrections $E_{nl}^{(1)}$ with $l = 1, \ldots, g_n$ are the eigenvalues of the $g_n \times g_n$ matrix representing the perturbation in the subspace $\mathbb{E}_n$ and the states $|\chi_l\rangle$ are the corresponding eigenvectors.

We can also use the λ^1 equation to determine the first-order corrections $|\chi_l^{(1)}\rangle$ to $|\chi_l\rangle$. We project both sides onto states $|\phi_m\rangle$ of energies $\epsilon_m \neq \epsilon_n$. These states are orthogonal to each of the g_n states $|\phi_n^{(k)}\rangle$ or $|\chi_l\rangle$, namely, they are in the subspace complement of $\mathbb{E}_n$. We find

$$\underbrace{\langle \phi_m | \hat{H}_0 | \chi_l^{(1)} \rangle}_{\epsilon_m \langle \phi_m |} + \langle \phi_m | \hat{V} | \chi_l \rangle = \epsilon_n \langle \phi_m | \chi_l^{(1)} \rangle + E_{nl}^{(1)} \underbrace{\langle \phi_m | \chi_l \rangle}_{\text{vanishes}} , \tag{14.24}$$

which we can solve for $\langle \phi_m | \chi_l^{(1)} \rangle$,

$$\langle \phi_m | \chi_l^{(1)} \rangle = \frac{\langle \phi_m | \hat{V} | \chi_l \rangle}{\epsilon_n - \epsilon_m} , \tag{14.25}$$

from which we obtain

$$|\chi_l^{(1)}\rangle = \sum_{m \neq \mathbb{E}_n} |\phi_m\rangle \langle \phi_m | \chi_l^{(1)} \rangle = \sum_{m \neq \mathbb{E}_n} |\phi_m\rangle \frac{\langle \phi_m | \hat{V} | \chi_l \rangle}{\epsilon_n - \epsilon_m} , \tag{14.26}$$

where the notation $\sum_{m \neq \mathbb{E}_n}$ means that the sum is over all states *except* the states $|\chi_1\rangle, \ldots, |\chi_{g_n}\rangle$ spanning the subspace $\mathbb{E}_n$.

Having determined the first-order corrections $E_{nl}^{(1)}$, two situations can occur: (i) the $E_{nl}^{(1)}$ are all distinct from each other, in which case $\hat{V}$ has completely removed the degeneracy of the level ϵ_n; (ii) some of the eigenvalues $E_{nl}^{(1)}$ are the same, in which case $\hat{V}$ has partially removed the degeneracy (it may happen that some degeneracy persists to all orders in perturbation theory; this is the case when the perturbation is invariant under some symmetry transformation – for example, when $\hat{V}$ commutes with the total angular momentum $\hat{\mathbf{J}}$).

Case (i)

Consider the λ^2 equation in Eq. (14.8), which we report below to assist the reader,

$$\lambda^2 : \qquad \hat{H}_0 |\chi_l^{(2)}\rangle + \hat{V} |\chi_l^{(1)}\rangle = \epsilon_n |\chi_l^{(2)}\rangle + E_{nl}^{(1)} |\chi_l^{(1)}\rangle + E_{nl}^{(2)} |\chi_l\rangle , \tag{14.27}$$

and project it onto the state $|\chi_l\rangle$:

$$\underbrace{\langle \chi_l | \hat{H}_0 | \chi_l^{(2)} \rangle}_{\epsilon_n \langle \chi_l |} + \langle \chi_l | \hat{V} | \chi_l^{(1)} \rangle = \epsilon_n \langle \chi_l | \chi_l^{(2)} \rangle + E_{nl}^{(1)} \underbrace{\langle \chi_l | \chi_l^{(1)} \rangle}_{\text{vanishes}} + E_{nl}^{(2)} \underbrace{\langle \chi_l | \chi_l \rangle}_{=1} , \tag{14.28}$$

where the second term on the right-hand side vanishes because the state $|\chi_l^{(1)}\rangle$ is in the subspace orthogonal to $\mathbb{E}_n$, and hence $\langle \chi_l | \chi_l^{(1)} \rangle = 0$. Also, the first terms on the left- and right-hand sides cancel out, which leaves

$$E_{nl}^{(2)} = \langle \chi_l | \hat{V} | \chi_l^{(1)} \rangle = \sum_{m \neq \mathbb{E}_n} \frac{|\langle \phi_m | \hat{V} | \chi_l \rangle|^2}{\epsilon_n - \epsilon_m} , \tag{14.29}$$

and thus the second-order correction $|\chi_l^{(2)}\rangle$ to the state is as was obtained in the non-degenerate case.

Case (ii)

Let the energy $E_{nl}^{(1)}$ have degeneracy $g_{nl} \leq g_n$, and label the corresponding eigenstates as $|\chi_1\rangle, \ldots, |\chi_{g_{nl}}\rangle$. We intend to find the "optimal" linear combinations of the g_{nl} eigenstates (in a sense that will become clear below) of the type

$$|\xi_p\rangle = \sum_{l=1}^{g_{nl}} c_l^{(p)} |\chi_l\rangle \qquad p = 1, \ldots, g_{nl} . \tag{14.30}$$

These states all have energy $\epsilon_n + E_{nl}^{(1)}$. We go back to the λ^2 equation in (14.27), now written in terms of the states $|\xi_p\rangle$:

$$\lambda^2 : \qquad \hat{H}_0 |\xi_p^{(2)}\rangle + \hat{V} |\xi_p^{(1)}\rangle = \epsilon_n |\xi_p^{(2)}\rangle + E_{nl}^{(1)} |\xi_p^{(1)}\rangle + E_{nl}^{(2)} |\xi_p\rangle , \tag{14.31}$$

where $|\xi_p^{(1)}\rangle$ is as in Eq. (14.26) but with $|\chi_l\rangle$ replaced by $|\xi_p\rangle$,

$$|\xi_p^{(1)}\rangle = \sum_{m \neq \mathbb{E}_n} |\phi_m\rangle \frac{\langle \phi_m | \hat{V} | \xi_p \rangle}{\epsilon_n - \epsilon_m} , \tag{14.32}$$

and where $|\xi_p^{(2)}\rangle$ is to be determined, although it is not needed if the second-order correction to the energy is all that concerns us. We now project Eq. (14.31) onto the subspace spanned by $|\chi_1\rangle, \ldots, |\chi_{g_{nl}}\rangle$ (which is contained in $\mathbb{E}_n$); that is, we take the scalar product

$$\underbrace{\langle \chi_m | \hat{H}_0}_{\epsilon_n \langle \chi_m |} |\xi_p^{(2)}\rangle + \langle \chi_m | \hat{V} | \xi_p^{(1)}\rangle = \epsilon_n \langle \chi_m | \xi_p^{(2)}\rangle + E_{nl}^{(1)} \underbrace{\langle \chi_m | \xi_p^{(1)}\rangle}_{\text{vanishes}} + E_{nl}^{(2)} \langle \chi_m | \xi_p\rangle , \tag{14.33}$$

and obtain

$$\langle \chi_m | \hat{V} | \xi_p^{(1)}\rangle = E_{nl}^{(2)} \langle \chi_m | \xi_p\rangle \implies \sum_{k \neq \mathbb{E}_n} \frac{\langle \chi_m | \hat{V} | \phi_k\rangle \langle \phi_k | \hat{V} | \xi_p\rangle}{\epsilon_n - \epsilon_k} = E_{nl}^{(2)} \langle \chi_m | \xi_p\rangle . \tag{14.34}$$

We define an operator $\hat{Z}$ by

$$\hat{Z} = \sum_{k \neq \mathbb{E}_n} \hat{V} \frac{|\phi_k\rangle \langle \phi_k|}{\epsilon_n - \epsilon_k} \hat{V} , \tag{14.35}$$

in terms of which Eq. (14.34) reads

$$\langle \chi_m | \hat{Z} | \xi_p\rangle = E_{nl}^{(2)} \langle \chi_m | \xi_p\rangle \implies \sum_{l=1}^{g_{nl}} \langle \chi_m | \hat{Z} | \chi_l\rangle c_l^{(p)} = E_{nl}^{(2)} c_m^{(p)} , \tag{14.36}$$

which is the eigenvalue equation for the $g_{nl} \times g_{nl}$ matrix representing $\hat{Z}$ in the basis $|\chi_l\rangle$. The g_{nl} eigenvalues are the second-order corrections $E_{nlp}^{(2)}$, with $p = 1, \ldots, g_{nl}$, that we are seeking. The calculation of these corrections is typically quite complex, since the construction of the matrix $\underline{Z}$ requires the evaluation of a sum, that is in general infinite, over intermediate states:

$$Z_{ml} = \sum_{k \neq \mathbb{E}_n} \frac{\langle \chi_m | \hat{V} | \phi_k\rangle \langle \phi_k | \hat{V} | \chi_l\rangle}{\epsilon_n - \epsilon_k} . \tag{14.37}$$

Of course, some second-order eigenvalues might be degenerate. One has then to look at the third-order corrections.

14.3 The Variational Method

For some problems perturbation theory may not be suitable, because the expansion converges poorly or because the Hamiltonian of interest is not close to a Hamiltonian with known eigenvalues and eigenstates. An example of the former is the perturbative treatment of dense neutron matter starting from the unperturbed Hamiltonian of a free Fermi gas of neutrons. An example of the latter occurs in molecular physics: there is no small parameter in which to expand eigenvalues and eigenstates of the Hamiltonian of a molecule with several nuclei. In such cases, the variational method provides a useful (and often very effective) alternative to perturbation theory.

The method is based on a general theorem that states that the expectation value of the Hamiltonian on any state $|\psi\rangle$ is greater than, or equal to, the exact ground-state energy E_0 of the system, that is,

$$\frac{\langle\psi|\hat{H}|\psi\rangle}{\langle\psi|\psi\rangle} \geq E_0 \qquad \text{for any } |\psi\rangle. \tag{14.38}$$

The proof follows easily by noting that, if E_n and $|\varphi_n^{(k)}\rangle$ are the (unknown) eigenvalues and eigenstates of $\hat{H}$,

$$\hat{H}|\varphi_n^{(k)}\rangle = E_n|\varphi_n^{(k)}\rangle \qquad k = 1, \ldots, g_n, \tag{14.39}$$

then the state $|\psi\rangle$ can be expanded in terms of the basis $|\varphi_n^{(k)}\rangle$ as

$$|\psi\rangle = \sum_n \sum_{k=1}^{g_n} c_n^{(k)}|\varphi_n^{(k)}\rangle, \qquad c_n^{(k)} = \langle\varphi_n^{(k)}|\psi\rangle. \tag{14.40}$$

Using the above expansion and the orthonormality of the basis states, $\langle\varphi_n^{(k)}|\varphi_p^{(l)}\rangle = \delta_{np}\,\delta_{kl}$, yields

$$\langle\psi|\hat{H}|\psi\rangle = \sum_n \sum_{k=1}^{g_n} E_n|c_n^{(k)}|^2 \geq E_0 \sum_n \sum_{k=1}^{g_n} |c_n^{(k)}|^2, \tag{14.41}$$

since $E_n > E_0$ for $n \geq 1$, and

$$\langle\psi|\psi\rangle = \sum_n \sum_{k=1}^{g_n} |c_n^{(k)}|^2. \tag{14.42}$$

Equations (14.41) and (14.42) imply Eq. (14.38), namely that the expectation value on any $|\psi\rangle$ is an upper bound to the true ground-state energy. Of course, if $|\psi\rangle$ happens to be a linear combination of the $|\varphi_0^{(k)}\rangle$ then the expectation value gives the exact ground-state energy E_0.

It is interesting to compare this result with that obtained for the ground-state energy in first-order perturbation theory,

$$\epsilon_0 + E_0^{(1)} = \epsilon_0 + \langle\phi_0|\hat{V}|\phi_0\rangle = \langle\phi_0|\underbrace{\hat{H}_0 + \hat{V}}_{\hat{H}}|\phi_0\rangle. \tag{14.43}$$

Note that this relation remains true also in degenerate perturbation theory, as long as $|\phi_0\rangle$ is understood, in that case, to correspond to the linear combination of degenerate unperturbed states $|\phi_0^{(k)}\rangle$ which diagonalizes the perturbation $\hat{V}$ (with $E_0^{(1)}$ its lowest eigenvalue). We can now see that the expectation value $\langle\phi_0|\hat{H}|\phi_0\rangle$ represents not only a first-order perturbative estimate but also an upper bound to the exact ground-state energy.

The simplest implementation of the variational method consists in choosing a trial state which depends on a set of parameters $\beta_1, \ldots, \beta_n$, and in minimizing the function

$$E(\beta_1, \ldots, \beta_n) = \frac{\langle \psi(\beta_1, \ldots, \beta_n) | \hat{H} | \psi(\beta_1, \ldots, \beta_n) \rangle}{\langle \psi(\beta_1, \ldots, \beta_n) | \psi(\beta_1, \ldots, \beta_n) \rangle} . \tag{14.44}$$

with respect to $\beta_1, \ldots, \beta_n$. Of course, the success of the method depends on the choice of trial state, which should include as many of the known features of the true ground state as possible. More sophisticated trial wave functions can be obtained by adopting more flexible parameterizations; for example, for a system of N helium atoms (which are bosons) strongly interacting with each other via a Lennard–Jones potential (a "quantum liquid"), a suitable and properly symmetrized trial wave function has the form $\psi(\mathbf{r}_1, \ldots, \mathbf{r}_N) = \prod_{i<j=1}^{N} f(r_{ij})$, where $f(r)$ is known as the correlation function. The function $f(r)$ can be obtained by minimizing the functional $E[\psi]$ with respect to variations in the correlation function $f(r)$.

There is a general result which applies to all discrete energy eigenstates. For an excited state the expectation value of the Hamiltonian is not a minimum, of course; however, we can show that the quantity

$$\langle \hat{H} \rangle_\psi = \frac{\langle \psi | \hat{H} | \psi \rangle}{\langle \psi | \psi \rangle} , \tag{14.45}$$

viewed as a functional of the state $|\psi\rangle$, is stationary with respect to infinitesimal variations $|\delta\psi\rangle$; indeed, we have

$$\begin{aligned}
\delta \langle \hat{H} \rangle_\psi &= \frac{\langle \delta\psi | \hat{H} | \psi \rangle}{\langle \psi | \psi \rangle} + \frac{\langle \psi | \hat{H} | \delta\psi \rangle}{\langle \psi | \psi \rangle} - \langle \psi | \hat{H} | \psi \rangle \left[\frac{\langle \delta\psi | \psi \rangle}{(\langle \psi | \psi \rangle)^2} + \frac{\langle \psi | \delta\psi \rangle}{(\langle \psi | \psi \rangle)^2} \right] \\
&= \frac{\langle \delta\psi | \hat{H} | \psi \rangle}{\langle \psi | \psi \rangle} + \frac{\langle \psi | \hat{H} | \delta\psi \rangle}{\langle \psi | \psi \rangle} - \langle \hat{H} \rangle_\psi \left[\frac{\langle \delta\psi | \psi \rangle}{\langle \psi | \psi \rangle} + \frac{\langle \psi | \delta\psi \rangle}{\langle \psi | \psi \rangle} \right] \\
&= \frac{\langle \delta\psi | \hat{H} - \langle \hat{H} \rangle_\psi | \psi \rangle}{\langle \psi | \psi \rangle} + \frac{\langle \psi | \hat{H} - \langle \hat{H} \rangle_\psi | \delta\psi \rangle}{\langle \psi | \psi \rangle} .
\end{aligned} \tag{14.46}$$

Since $\hat{H}$ is hermitian and the expectation value $\langle \hat{H} \rangle_\psi$ real, it follows that

$$\delta \langle \hat{H} \rangle_\psi = \frac{2 \operatorname{Re} [\langle \delta\psi | \hat{H} - \langle \hat{H} \rangle_\psi | \psi \rangle]}{\langle \psi | \psi \rangle} . \tag{14.47}$$

The condition for an extremum is that for arbitrary variations $|\delta\psi\rangle$ we have

$$\delta \langle \hat{H} \rangle_\psi = 0 \implies (\hat{H} - \langle \hat{H} \rangle_\psi) | \psi \rangle = 0 , \tag{14.48}$$

and the last relation is satisfied iff $|\psi\rangle$ is one of the eigenstates of $\hat{H}$ and $\langle \hat{H} \rangle_\psi$ is the corresponding eigenvalue. A by-product of this result is that, for example, a first-order change in the exact eigenstate $|\varphi_n\rangle$ to $|\psi\rangle = |\varphi_n\rangle + |\delta\psi\rangle$ produces only a second-order change in the exact eigenvalue E_n, namely

$$\frac{\langle \psi | \hat{H} | \psi \rangle}{\langle \psi | \psi \rangle} = E_n + O(|\delta\psi|^2) . \tag{14.49}$$

This is only true for the Hamiltonian; the expectation value of operators other than $\hat{H}$ will generally have errors linear in $|\delta\psi\rangle$.

14.4 Problems

Problem 1 Second-Order Correction to an Energy Eigenstate

Obtain the second-order correction $|\psi_n^{(2)}\rangle$ to an energy eigenstate.

Solution

The second-order correction to the state is derived from Eq. (14.13) as

$$\langle\phi_m|\psi_n^{(2)}\rangle = \frac{1}{\epsilon_n-\epsilon_m}\left[\langle\phi_m|\hat{V}|\psi_n^{(1)}\rangle - E_n^{(1)}\langle\phi_m|\psi_n^{(1)}\rangle\right] = \frac{1}{\epsilon_n-\epsilon_m}\langle\phi_m|\left(\hat{V}-E_n^{(1)}\right)|\psi_n^{(1)}\rangle$$

$$= \frac{1}{\epsilon_n-\epsilon_m}\langle\phi_m|\left(\hat{V}-E_n^{(1)}\right)\sum_{k\neq n}|\phi_k\rangle\frac{\langle\phi_k|\hat{V}|\phi_n\rangle}{\epsilon_n-\epsilon_k}$$

$$= \left[\sum_{k\neq n}\frac{\langle\phi_m|\hat{V}|\phi_k\rangle\langle\phi_k|\hat{V}|\phi_n\rangle}{(\epsilon_n-\epsilon_m)(\epsilon_n-\epsilon_k)}\right] - \frac{\langle\phi_n|\hat{V}|\phi_n\rangle\langle\phi_m|\hat{V}|\phi_n\rangle}{(\epsilon_n-\epsilon_m)^2},$$

which then yields

$$|\psi_n^{(2)}\rangle = \sum_{m\neq n}\sum_{k\neq n}|\phi_m\rangle\frac{\langle\phi_m|\hat{V}|\phi_k\rangle\langle\phi_k|\hat{V}|\phi_n\rangle}{(\epsilon_n-\epsilon_m)(\epsilon_n-\epsilon_k)} - \sum_{m\neq n}|\phi_m\rangle\frac{\langle\phi_m|\hat{V}|\phi_n\rangle\langle\phi_n|\hat{V}|\phi_n\rangle}{(\epsilon_n-\epsilon_m)^2}.$$

Problem 2 Wave-Function Renormalization Constant

The perturbed state $|\psi_n\rangle$ was constructed in Section 14.1 so as to have the normalization $\langle\phi_n|\psi_n\rangle = 1$. In order to correctly normalize $|\psi_n\rangle$, we must multiply it by the factor $Z_n^{1/2} = 1/\sqrt{\langle\psi_n|\psi_n\rangle}$, where Z_n is the so-called wave-function renormalization constant. The state $|\overline{\psi}_n\rangle = Z_n^{1/2}|\psi_n\rangle$ is such that $\langle\overline{\psi}_n|\overline{\psi}_n\rangle = 1$. Calculate Z_n to second order in $\hat{V}$, and show that it can be expressed as $Z_n = \partial E_n/\partial\epsilon_n$, where E_n is the correct energy up to second order, that is, $E_n = \epsilon_n + E_n^{(1)} + E_n^{(2)}$. What is the physical meaning of Z_n?

Solution

To second order, we have (recall that $\langle\phi_n|\psi_n^{(p)}\rangle = 0$ with $p \geq 1$)

$$\frac{1}{Z_n} = \langle\psi_n|\psi_n\rangle = \left(\langle\phi_n| + \lambda\langle\psi_n^{(1)}| + \lambda^2\langle\psi_n^{(2)}|\right)\left(|\phi_n\rangle + \lambda|\psi_n^{(1)}\rangle + \lambda^2|\psi_n^{(2)}\rangle\right) = 1 + \lambda^2\langle\psi_n^{(1)}|\psi_n^{(1)}\rangle.$$

On inserting the expression for $|\psi_n^{(1)}\rangle$, we find

$$\frac{1}{Z_n} = 1 + \lambda^2\left(\sum_{p\neq n}\frac{\langle\phi_p|\hat{V}|\phi_n\rangle^*}{\epsilon_n-\epsilon_p}\langle\phi_p|\right)\left(\sum_{m\neq n}|\phi_m\rangle\frac{\langle\phi_m|\hat{V}|\phi_n\rangle}{\epsilon_n-\epsilon_m}\right) = 1 + \lambda^2\sum_{m\neq n}\frac{|\langle\phi_m|\hat{V}|\phi_n\rangle|^2}{(\epsilon_n-\epsilon_m)^2},$$

and we see that the state $|\psi_n\rangle$ is normalized in first order, with the first correction occurring at second order. We can express the renormalization constant as

$$Z_n \approx 1 - \lambda^2 \sum_{m \neq n} \frac{|\langle \phi_m | \hat{V} | \phi_n \rangle|^2}{(\epsilon_n - \epsilon_m)^2} = \frac{\partial}{\partial \epsilon_n} \underbrace{\left(\epsilon_n + \lambda \langle \phi_n | \hat{V} | \phi_n \rangle + \lambda^2 \sum_{m \neq n} \frac{|\langle \phi_m | \hat{V} | \phi_n \rangle|^2}{\epsilon_n - \epsilon_m} \right)}_{\epsilon_n + E_n^{(1)} + E_n^{(2)}},$$

where the partial derivative with respect to ϵ_n does not act on the matrix elements of $\hat{V}$, or on the ϵ_m with $m \neq n$. It is conceivable that the relation above remains valid to all orders, namely $Z_n = \partial E_n / \partial \epsilon_n$ where E_n is the exact energy eigenvalue – not just its second-order expansion – but this of course needs to be proven. We note that

$$\langle \phi_n | \overline{\psi}_n \rangle = Z_n^{1/2} \langle \phi_n | \psi_n \rangle = Z_n^{1/2} \,.$$

Therefore, we can interpret Z_n as the probability of observing the interacting system (described by the Hamiltonian $\hat{H}$) in the unperturbed state $|\phi_n\rangle$.

Problem 3   One-Dimensional Harmonic Oscillator in a Uniform Electric Field

Consider a particle of charge q in one dimension in a harmonic-oscillator potential and under the influence of a uniform electric field. The Hamiltonian reads

$$\hat{H} = \hat{H}_0 + \hat{V} \,, \qquad \hat{H}_0 = \frac{\hat{p}^2}{2m} + \frac{m\omega^2}{2} \hat{x}^2 \,, \qquad \hat{V} = -qE\hat{x} \,.$$

1. Assume the electric field is weak, so that a perturbative calculation is permissible, and calculate the first- and second-order corrections to a generic eigenvalue ϵ_n of $\hat{H}_0$ (recall that the eigenvalues of the one-dimensional harmonic oscillator are non-degenerate).
2. Solve for the exact eigenvalues of $\hat{H}$ and compare the results with those obtained in the perturbative calculation. What do you conclude?

Solution

Part 1

The eigenenergies and eigenstates of the harmonic oscillator are well known: $\hat{H}_0 |n\rangle = \epsilon_n |n\rangle$, where $\epsilon_n = \hbar\omega(n + 1/2)$. We also recall that the position operator can be expressed in terms of lowering and raising operators as

$$\hat{x} = \sqrt{\frac{\hbar}{2m\omega}} (\hat{a} + \hat{a}^\dagger) \,,$$

with the properties

$$\hat{a} |n\rangle = \sqrt{n} |n - 1\rangle \,, \qquad \hat{a} |0\rangle = 0 \,, \qquad \hat{a}^\dagger |n\rangle = \sqrt{n + 1} |n + 1\rangle \,.$$

It is now easy to see that the first-order correction to any unperturbed level ϵ_n vanishes,

$$E_n^{(1)} = \langle n | \hat{V} | n \rangle = -qE \sqrt{\frac{\hbar}{2m\omega}} \langle n | \hat{a} + \hat{a}^\dagger | n \rangle = 0 \,,$$

since the operator $\hat{a}$ ($\hat{a}^\dagger$) can only connect the state $|n\rangle$ to the state $|n-1\rangle$ ($|n+1\rangle$); in other words, the only non-zero matrix elements of the perturbation $\hat{V}$ are $\langle n \pm 1|\hat{V}|n\rangle$. In view of this fact, the second-order correction is easy to evaluate,

$$E_n^{(2)} = \sum_{m \neq n} \frac{|\langle m|\hat{V}|n\rangle|^2}{\epsilon_n - \epsilon_m} = \frac{|\langle n+1|\hat{V}|n\rangle|^2}{\epsilon_n - \epsilon_{n+1}} + \frac{|\langle n-1|\hat{V}|n\rangle|^2}{\epsilon_n - \epsilon_{n-1}}.$$

The energy differences between the unperturbed energies are given by $\epsilon_n - \epsilon_{n\pm1} = \mp\hbar\omega$, while the matrix elements read

$$\langle n-1|\hat{V}|n\rangle = -qE\sqrt{\frac{\hbar}{2m\omega}}\sqrt{n}, \qquad \langle n+1|\hat{V}^\dagger|n\rangle = -qE\sqrt{\frac{\hbar}{2m\omega}}\sqrt{n+1},$$

so that

$$E_n^{(2)} = \frac{\hbar(qE)^2}{2m\omega}\left(-\frac{n+1}{\hbar\omega} + \frac{n}{\hbar\omega}\right) = -\frac{(qE)^2}{2m\omega^2}.$$

Part 2

We observe that $\hat{H}$ can be written as

$$\hat{H} = \frac{\hat{p}^2}{2m} + \frac{m\omega^2}{2}\left(\hat{x} - \frac{qE}{m\omega^2}\right)^2 - \frac{(qE)^2}{2m\omega^2},$$

that is, the Hamiltonian of a harmonic oscillator with shifted position and energy operators. Thus we perform the transformation (in fact, a canonical transformation, since it preserves the commutation relations)

$$\hat{x} \longmapsto \hat{X} = \hat{x} - \frac{qE}{m\omega^2}, \qquad \hat{P} \longmapsto \hat{P} = \hat{p}, \qquad [\hat{X}, \hat{P}] = i\hbar.$$

In terms of the new variables the Hamiltonian reads

$$\hat{H} = \frac{\hat{P}^2}{2m} + \frac{m\omega^2}{2}\hat{X}^2 - \frac{(qE)^2}{2m\omega^2},$$

with (exact) eigenvalues given by

$$E_n = \epsilon_n - \frac{(qE)^2}{2m\omega^2}.$$

We see that the correction is independent of the particular state $|n\rangle$ under consideration, and consists of an overall shift downward as found in part 1. We conclude that all corrections $E_n^{(p)}$ with $p \geq 3$ must vanish.

Problem 4 Particle in an Infinite Potential Well Subject to a Barrier Perturbation

A particle of mass m is placed in a potential

$$V(|x| > b) = \infty, \qquad V(b/2 < |x| < b) = 0, \qquad V(|x| < b/2) = V_0,$$

where the barrier height $V_0 > 0$. Note that $V(x)$ is invariant under space inversion, $V(x) = V(-x)$.

1. Assuming that the barrier can be treated as a perturbation, determine the first-order corrections to the unperturbed energies.

2. Assuming that the energy E of the particle is larger than V_0, solve the problem exactly and show that the eigenvalues of the even- and odd-parity solutions result from

$$\cot(kb/2) = \frac{K}{k} \tan(Kb/2), \qquad \text{even parity},$$

and

$$\tan(kb/2) = -\frac{k}{K} \tan(Kb/2), \qquad \text{odd parity},$$

where (in the notation of Section 4.1)

$$k = \sqrt{\epsilon}, \qquad K = \sqrt{\epsilon - v_0}.$$

Write down the corresponding eigenfunctions.

3. Show that, when the barrier can be treated as a perturbation, the eigenvalue equations yield energies in agreement with those obtained in part 1 (do this just for the even solutions).

Solution

Part 1

The unperturbed energies and eigenfunctions (for an infinite well of width $2b$) are

$$E_n = \frac{\hbar^2}{2m} \left(\frac{n\pi}{2b} \right)^2$$

and

$$\psi_n(x) = \frac{1}{\sqrt{b}} \cos\left(\frac{n\pi}{2b} x \right) \qquad n = 1, 3, \ldots$$

$$= \frac{1}{\sqrt{b}} \sin\left(\frac{n\pi}{2b} x \right) \qquad n = 2, 4, \ldots$$

Assuming that $V_0 \ll E_1$, we find, for the first-order corrections induced by the barrier perturbation,

$$E_n^{(1)} = \frac{V_0}{b} \int_{-b/2}^{b/2} dx \, \cos^2\left(\frac{n\pi}{2b} x \right) = V_0 \left[\frac{1}{2} + \frac{\sin(n\pi/2)}{n\pi} \right] \qquad n = 1, 3, \ldots$$

$$= \frac{V_0}{b} \int_{-b/2}^{b/2} dx \, \sin^2\left(\frac{n\pi}{2b} x \right) = V_0 \left[\frac{1}{2} - \frac{\sin(n\pi/2)}{n\pi} \right] \qquad n = 2, 4, \ldots$$

Part 2

Since the potential is invariant under space inversion, the states have alternating parities: even, odd, even, odd, and so on, the ground state and first excited state being even and odd, respectively. We have for energies $E > V_0$

$$\psi_I'' = -k^2 \psi_I(x), \qquad \psi_{II}'' = -K^2 \psi_{II}(x), \qquad \psi_{III}'' = -k^2 \psi_{III}(x),$$

where $k = \sqrt{\epsilon}$, $K = \sqrt{\epsilon - v_0}$, and the regions I, II, and III are $-b < x < -b/2$, $-b/2 < x < b/2$, and $b/2 < x < b$, respectively. We can solve for, say, $x > 0$, and then obtain the complete solution in $-b < x < b$ by exploiting the known parity of the state. In particular, the first derivatives of even-parity solutions $\psi^e(x)$, and the odd-parity solutions $\psi^o(x)$, vanish at $x = 0$, namely

$$\psi^{e\prime}(0) = 0, \qquad \psi^{o\prime}(0) = 0.$$

For $x > 0$, we have

$$\psi_{II}(x) = A \cos(Kx) , \qquad \psi_{III}(x) = B \sin[k(b - x)] , \qquad \text{even parity} ,$$
$$\psi_{II}(x) = C \sin(Kx) , \qquad \psi_{III}(x) = D \sin[k(b - x)] , \qquad \text{odd parity} ,$$

where the boundary condition $\psi_{III}(b) = 0$ has already been imposed. It should be noted that these solutions also satisfy the requirement $\psi^e{}'(0) = 0$ and $\psi^o(0) = 0$. The boundary conditions at $x = b/2$ yield

$$A \cos(Kb/2) = B \sin(kb/2) , \qquad -KA \sin(Kb/2) = -kB \cos(kb/2)$$
$$C \sin(Kb/2) = D \sin(kb/2) , \qquad KC \cos(Kb/2) = -kD \cos(kb/2)$$

and the eigenvalues result from

$$\cot(kb/2) = \frac{K}{k} \tan(Kb/2) , \quad \text{even parity} , \qquad \tan(kb/2) = -\frac{k}{K} \tan(Kb/2) , \quad \text{odd parity} .$$

The complete even-parity solutions are then

$$\psi_I^e(x) = B \sin[k_e(b + x)] , \qquad \psi_{II}^e(x) = A \cos(K_e x) , \qquad \psi_{III}^e(x) = B \sin[k_e(b - x)] ,$$

and the odd-parity solutions are

$$\psi_I^o(x) = -D \sin[k_o(b + x)] , \qquad \psi_{II}^o(x) = C \sin(K_o x) , \qquad \psi_{III}^o(x) = D \sin[k_o(b - x)] .$$

where, having determined the eigenvalues ϵ_e and ϵ_o, we can express A and C in terms of B and D as

$$A = \frac{\sin(k_e b/2)}{\cos(K_e b/2)} B , \qquad C = \frac{\sin(k_o b/2)}{\sin(K_o b/2)} D ,$$

and $|B|$ and $|D|$ can be fixed by normalizing the states. Here, we have defined

$$k_e = \sqrt{\epsilon_e} , \qquad K_e = \sqrt{\epsilon_e - v_0} ,$$

and similarly for k_o and κ_o.

Part 3

Introduce the variables

$$x = bk/2 , \qquad \epsilon = 4x^2/b^2 , \qquad x_0 = b\sqrt{v_0}/2 , \qquad v_0 = 4x_0^2/b^2 ,$$

so that the eigenvalue equations obtained in part 2 can be written for the even solutions as

$$\tan x \, \tan \sqrt{x^2 - x_0^2} = \frac{x}{\sqrt{x^2 - x_0^2}} ,$$

and for the odd solutions as

$$\cot x \, \tan \sqrt{x^2 - x_0^2} = -\frac{\sqrt{x^2 - x_0^2}}{x} ,$$

for $x > x_0$. Hereafter, we consider only the eigenvalue equation for the even solution, for the purpose of illustration. We define

$$f(x) = \tan x \, \tan \sqrt{x^2 - x_0^2} , \qquad g(x) = \frac{x}{\sqrt{x^2 - x_0^2}} ,$$

and in the asymptotic regime $x \gg x_0$, where $f(x \gg x_0) = \tan^2 x$ and $g(x \gg x_0) = 1$, the eigenvalue equation reduces to

$$\tan^2 x = 1 \implies \tan x = \pm 1 \implies x_n = \frac{\pi}{4} n, \qquad n = 1, 3, 5, \ldots$$

Note that these solutions correspond to the energies

$$E_n = \frac{\hbar^2}{2m} \epsilon_n = \frac{\hbar^2}{2m} \frac{4 x_n^2}{b^2} = \frac{\hbar^2}{2m} \left(\frac{n\pi}{2b} \right)^2,$$

that is, the unperturbed eigenenergies of the infinite well corresponding to the even eigenfunctions, as expected (for $x_0 = 0$ the potential is just that of an infinite well). Next, we want to determine the corrections linear in x_0^2 to these solutions x_n. To this end, we expand $f(x)$ and $g(x)$ around x_n,

$$f(x) = f(x_n) + (x - x_n) f'(x_n) + \cdots,$$

and similarly for $g(x)$, so that to linear terms in $x - x_n$ we have

$$f(x_n) + (x - x_n) f'(x_n) = g(x_n) + (x - x_n) g'(x_n) \implies x = x_n + \frac{g(x_n) - f(x_n)}{f'(x_n) - g'(x_n)}.$$

To leading order in x_0^2, we find

$$g(x_n) = \frac{1}{\sqrt{1 - x_0^2/x_n^2}} = 1 + \frac{x_0^2}{2 x_n^2} + \cdots, \qquad g'(x) = -\frac{x_0^2}{(x^2 - x_0^2)^{3/2}} \implies g'(x_n) = -\frac{x_0^2}{x_n^3} + \cdots.$$

Before evaluating $f(x_n)$, we note that

$$\sqrt{x_n^2 - x_0^2} = x_n \left(1 - \frac{x_0^2}{2 x_n^2} + \cdots \right) = x_n - \frac{x_0^2}{2 x_n} + \cdots$$

and, using the addition formula $\tan(\alpha + \beta) = (\tan \alpha + \tan \beta)/(1 - \tan \alpha \tan \beta)$, we find, to linear terms in x_0^2,

$$\begin{aligned}
\tan \sqrt{x_n^2 - x_0^2} &= \tan \left[x_n - x_0^2/(2 x_n) + \cdots \right] = \frac{\tan x_n - \tan[x_0^2/(2 x_n) - \cdots]}{1 + (\tan x_n) \, \tan[x_0^2/(2 x_n) - \cdots]} \\
&= \frac{\tan x_n - x_0^2/(2 x_n) + \cdots}{1 + (\tan x_n) \, x_0^2/(2 x_n) + \cdots} \\
&= \tan x_n - x_0^2/(2 x_n) - (\tan^2 x_n) \, x_0^2/(2 x_n) + \cdots = \tan x_n - x_0^2/x_n + \cdots,
\end{aligned}$$

since $\tan^2 x_n = 1$. We obtain

$$f(x_n) = \tan x_n \left(\tan x_n - x_0^2/x_n + \cdots \right) = 1 - (\tan x_n) \, x_0^2/x_n + \cdots.$$

The derivative of $f(x)$ is given by

$$f'(x) = \frac{\tan \sqrt{x^2 - x_0^2}}{\cos^2 x} + \frac{\tan x}{\cos^2 \sqrt{x^2 - x_0^2}} \frac{x}{\sqrt{x^2 - x_0^2}},$$

so that to linear terms in x_0^2 we have (using $\cos^2 x_n = 1/2$ and the identity $1/\cos^2 x = 1 + \tan^2 x$)

$$f'(x_n) = 2\left(\tan x_n - x_0^2/x_n + \cdots\right) + \tan x_n\left[1 + x_0^2/(2x_n^2) + \cdots\right]\underbrace{\left[1 + \left(\tan x_n - x_0^2/x_n + \cdots\right)^2\right]}_{1/\cos^2\sqrt{x_n^2 - x_0^2}}$$

$$= 4\tan x_n - 4x_0^2/x_n + (\tan x_n)\,x_0^2/x_n^2 + \cdots .$$

Inserting this expansion into the (approximate) solution for x finally yields

$$x = x_n + \frac{[1 + x_0^2/(2x_n^2) + \cdots] - (1 - (\tan x_n)\,x_0^2/x_n + \cdots)}{(4\tan x_n - 4x_0^2/x_n + (\tan x_n)\,x_0^2/x_n^2 + \cdots) - (-x_0^2/x_n^3 + \cdots)}$$

$$= x_n + \frac{x_0^2/(2x_n^2) + (\tan x_n)\,x_0^2/x_n}{4\tan x_n} + \cdots .$$

The energy follows from

$$\epsilon_n = \frac{4}{b^2}\left(x_n + \frac{x_0^2}{8x_n^2\tan x_n} + \frac{x_0^2}{4x_n} + \cdots\right)^2 = \frac{4x_n^2}{b^2} + \frac{4x_0^2}{b^2}\left(\frac{1}{2} + \frac{1}{4x_n\tan x_n} + \cdots\right),$$

or

$$E_n = \frac{\hbar^2}{2m}\,\epsilon_n = \frac{\hbar^2}{2m}\left(\frac{n\pi}{2b}\right)^2 + V_0\left[\frac{1}{2} + \frac{1}{n\pi\tan(n\pi/4)}\right] + \cdots ,$$

where the ellipses indicate terms involving higher powers of V_0. The term proportional to V_0 is exactly the first-order correction obtained in part 1, since $\sin(n\pi/2) = \tan(n\pi/4) = (\pm i)^{n-1}$ for $n = 1, 3, 5, \ldots$.

Problem 5   Alternative Derivation of First-Order Energy in Degenerate Perturbation Theory

Introduce a projection operator onto a degenerate subspace of energy ϵ_n,

$$\hat{P}_n = \sum_{k=1}^{g_n} |\phi_n^{(k)}\rangle\langle\phi_n^{(k)}| ,$$

and redefine the unperturbed Hamiltonian and the perturbation as

$$\hat{H}_0' = \hat{H}_0 + \hat{P}_n\hat{V}\hat{P}_n , \qquad \hat{V}' = \hat{V} - \hat{P}_n\hat{V}\hat{P}_n , \qquad \hat{H} = \hat{H}_0' + \hat{V}_0' .$$

Show that this redefinition is such that the matrix elements of $\hat{V}'$ vanish in the degenerate subspace, while they are left unchanged between eigenstates $|\phi_m\rangle$ of $\hat{H}_0$ corresponding to eigenvalues ϵ_m other than ϵ_n. Obtain the matrix elements of $\hat{H}_0'$ in the degenerate subspace, that is, the $\langle\phi_n^{(p)}|\hat{H}_0'|\phi_n^{(q)}\rangle$, and determine its eigenvalues. Do they agree with those found in Section 14.2?

Solution

In the degenerate subspace we find

$$\langle\phi_n^{(p)}|\hat{V}'|\phi_n^{(q)}\rangle = \langle\phi_n^{(p)}|\hat{V}|\phi_n^{(q)}\rangle - \underbrace{\langle\phi_n^{(p)}|\hat{P}_n}_{\langle\phi_n^{(p)}|}\hat{V}\underbrace{\hat{P}_n|\phi_n^{(q)}\rangle}_{|\phi_n^{(q)}\rangle} = 0 ,$$

that is, $\hat{V}'$ is the null operator in this subspace. By contrast, the matrix elements of $\hat{V}'$ between eigenstates $|\phi_m\rangle$ of $\hat{H}_0$ corresponding to eigenvalues ϵ_m other than ϵ_n are left unchanged,

$$\langle\phi_l|\,\hat{V}'\,|\phi_m\rangle = \langle\phi_l|\,\hat{V}\,|\phi_m\rangle - \underbrace{\langle\phi_l|\,\hat{P}_n}_{\text{vanishes}}\,\hat{V}\,\underbrace{\hat{P}_n\,|\phi_m\rangle}_{\text{vanishes}} = \langle\phi_l|\,\hat{V}\,|\phi_m\rangle \qquad \text{for } l, m \neq n\,.$$

In a similar fashion, it is easy to see that the matrix elements of $\hat{H}_0'$ are given by

$$\langle\phi_l|\,\hat{H}_0'\,|\phi_m\rangle = \epsilon_m\,\delta_{lm} \qquad \text{for } l, m \neq n\,, \qquad \langle\phi_n^{(p)}|\,\hat{H}_0'\,|\phi_n^{(q)}\rangle = \epsilon_n\,\delta_{pq} + \langle\phi_n^{(p)}|\,\hat{V}\,|\phi_n^{(q)}\rangle\,,$$

and so $\hat{H}_0'$ is no longer diagonal in the degenerate subspace. We now proceed to diagonalize $\hat{H}_0'$ – or equivalently $\hat{V}$ – in this subspace. We denote its g_n eigenvalues and eigenstates as

$$\hat{V}|\chi_l\rangle = E_{nl}^{(1)}\,|\chi_l\rangle \qquad l = 1, 2, \ldots, g_n\,,$$

These are obviously eigenstates of $\hat{H}_0'$ too (the $|\chi_l\rangle$ consist of linear combinations of the $|\phi_n^{(k)}\rangle$),

$$\hat{H}_0'\,|\chi_l\rangle = (\epsilon_n + E_{nl}^{(1)})|\chi_l\rangle\,,$$

and the g_n eigenvalues $E_{nl}^{(1)}$ represent the first-order corrections to ϵ_n determined in Section 14.2.

Problem 6 Particle Confined in a Box and Subject to a Perturbation

Consider a particle of mass m confined to a box of volume Ω, that is, the potential vanishes if $0 \leq x \leq \Omega^{1/3}$, $0 \leq y \leq \Omega^{1/3}$, and $0 \leq z \leq \Omega^{1/3}$ and is infinite otherwise. Inside the box, the particle is subject to the perturbation described by the potential

$$V(x, y, z) = V_0\,\frac{xyz}{\Omega}\,.$$

Calculate the perturbed energies of the ground and first excited states to linear terms in V_0. Obtain the first excited states to leading order in V_0.

Solution

The normalized eigenfunctions of the Hamiltonian describing a particle confined in the box are

$$\phi_{lmn}(x, y, z) = \sqrt{\frac{8}{\Omega}}\,\sin(l\pi x/L)\,\sin(m\pi y/L)\,\sin(n\pi z/L)\,, \qquad l, m, n = 1, 2, 3, \ldots\,,$$

with eigenenergies given by

$$E_{lmn} = \frac{\hbar^2}{2m}\left(\frac{\pi}{L}\right)^2\left(l^2 + m^2 + n^2\right)\,,$$

where $L = \Omega^{1/3}$. The ground state corresponds to $l = m = n = 1$ and is non-degenerate. The first-order correction to its energy follows from

$$E_{\text{gs}}^{(1)} = \langle\phi_{111}|\,V\,|\phi_{111}\rangle = \frac{8V_0}{\Omega}\left[\int_0^L dx\,\frac{x}{L}\,\sin^2(\pi x/L)\right]\left[\int_0^L dy\,\frac{y}{L}\,\sin^2(\pi y/L)\right]\left[\int_0^L dx\,\frac{z}{L}\,\sin^2(\pi z/L)\right]\,;$$

since the integral factorizes into the product of three identical terms. We find

$$\int_0^L dx\,(x/L)\,\sin^2(\pi x/L) = \frac{L}{2}\int_0^1 dt\,t\,[1 - \cos(2\pi t)] = \frac{L}{4}\,,$$

after carrying out a partial integration on the term $t \cos(2\pi t)$; hence, we obtain for the first-order correction

$$E_{\text{gs}}^{(1)} = \frac{V_0}{8} .$$

The first excited level is three-fold degenerate, with eigenfunctions $\phi_{211}(x, y, z)$, $\phi_{121}(x, y, z)$, and $\phi_{112}(x, y, z)$. In the subspace spanned by these three states, the 3×3 matrix has the form

$$\underline{V} = \begin{pmatrix} \langle \phi_{211}|V|\phi_{211}\rangle & \langle \phi_{211}|V|\phi_{121}\rangle & \langle \phi_{211}|V|\phi_{112}\rangle \\ \langle \phi_{121}|V|\phi_{211}\rangle & \langle \phi_{121}|V|\phi_{121}\rangle & \langle \phi_{121}|V|\phi_{112}\rangle \\ \langle \phi_{112}|V|\phi_{211}\rangle & \langle \phi_{112}|V|\phi_{121}\rangle & \langle \phi_{112}|V|\phi_{112}\rangle \end{pmatrix} = \begin{pmatrix} \alpha & \beta & \beta \\ \beta & \alpha & \beta \\ \beta & \beta & \alpha \end{pmatrix} ,$$

where in the present case the coefficients α and β are real. We have defined

$$\alpha = \langle \phi_{211}|V|\phi_{211}\rangle = \langle \phi_{121}|V|\phi_{121}\rangle = \langle \phi_{112}|V|\phi_{112}\rangle$$

$$= \frac{8 V_0}{\Omega} \left[\int_0^L dx \, (x/L) \sin^2(2\pi x/L) \right] \left[\int_0^L dy \, (y/L) \sin^2(\pi y/L) \right]^2 .$$

The second integral has already been evaluated above; the first gives

$$\int_0^L dx \, (x/L) \sin^2(2\pi x/L) = \frac{L}{2} \int_0^1 dt \, t \, [1 - \cos(4\pi t)] = \frac{L}{4} ,$$

so that $\alpha = V_0/8$. Next, we obtain

$$\beta = \langle \phi_{211}|V|\phi_{121}\rangle = \langle \phi_{211}|V|\phi_{112}\rangle = \langle \phi_{121}|V|\phi_{112}\rangle$$

$$= \frac{8 V_0}{\Omega} \left[\int_0^L dx \, (x/L) \sin(\pi x/L) \, \sin(2\pi x/L) \right]^2 \left[\int_0^L dy \, (y/L) \sin^2(\pi x/L) \right] ,$$

with

$$\int_0^L dx \, (x/L) \sin(\pi x/L) \, \sin(2\pi x/L) = \frac{L}{2} \int_0^1 dt \, t \, [\cos(\pi t) - \cos(3\pi t)] = -\frac{8L}{9\pi^2} ,$$

where use has been made of the identity $2 \sin \alpha \, \sin \beta = \cos(\alpha - \beta) - \cos(\alpha + \beta)$; so

$$\beta = \frac{128}{81\pi^4} V_0 .$$

The matrix representing the perturbation reads

$$\underline{V} = V_0 \begin{pmatrix} \lambda & \mu & \mu \\ \mu & \lambda & \mu \\ \mu & \mu & \lambda \end{pmatrix} , \qquad \lambda = \frac{1}{8} , \qquad \mu = \frac{128}{81\,\pi^4} .$$

The first-order corrections correspond to the eigenvalues of this matrix; we find

$$\det \begin{pmatrix} E - \lambda V_0 & -\mu V_0 & -\mu V_0 \\ -\mu V_0 & E - \lambda V_0 & -\mu V_0 \\ -\mu V_0 & -\mu V_0 & E - \lambda V_0 \end{pmatrix} = (E - \lambda V_0)^3 - 3(\mu V_0)^2 (E - \lambda V_0) - 2(\mu V_0)^3 = 0 ,$$

which, introducing $E = x V_0$, can be more conveniently written in terms of x as

$$(x - \lambda)^3 - 3\mu^2 (x - \lambda) - 2\mu^3 = 0 \implies t^3 - 3\mu^2 t - 2\mu^3 = (t + \mu)^2 (t - 2\mu) = 0 , \qquad t = x - \lambda .$$

The eigenvalues are finally obtained as

$$E_{\text{ex},1}^{(1)} = V_0(\lambda - \mu) \quad \text{(twice degenerate)} , \qquad\qquad E_{\text{ex},2}^{(1)} = V_0(\lambda + 2\mu) \quad \text{(non-degenerate)} .$$

The eigenvectors consist of linear combinations $a|\phi_{211}\rangle + b|\phi_{121}\rangle + c|\phi_{112}\rangle$; those corresponding to the twice degenerate eigenvalue $E_{\text{ex},1}^{(1)}$ follow from solving the (linearly dependent) system for the unknowns a, b, and c, obtained after substituting E with $V_0(\lambda - \mu)$ in the matrix above, that is,

$$
\begin{pmatrix} 1 & 1 & 1 \\ 1 & 1 & 1 \\ 1 & 1 & 1 \end{pmatrix} \begin{pmatrix} a \\ b \\ c \end{pmatrix} = 0 ,
$$

where the common factor $-\mu V_0$ has been removed. Two independent solutions are $a = 1$, $b = 0$, $c = -1$ and $a = 1$, $b = -1$, $c = 0$, with corresponding (unnormalized and non-orthogonal) eigenvectors $|\phi_{211}\rangle - |\phi_{112}\rangle$ and $|\phi_{211}\rangle - |\phi_{121}\rangle$. Gram–Schmidt orthogonalization of these yields

$$
|\phi_{\text{ex},1}^{(1)}\rangle = \frac{1}{\sqrt{2}} \left(|\phi_{211}\rangle - |\phi_{112}\rangle \right) , \qquad |\phi_{\text{ex},1}^{(2)}\rangle = \frac{1}{\sqrt{6}} \left(|\phi_{211}\rangle - 2|\phi_{121}\rangle + |\phi_{112}\rangle \right) .
$$

The (normalized) eigenvector corresponding to the non-degenerate eigenvalue $E_{\text{ex},2}^{(1)}$ can be obtained similarly:

$$
|\phi_{\text{ex},2}\rangle = \frac{1}{\sqrt{3}} \left(|\phi_{211}\rangle + |\phi_{121}\rangle + |\phi_{112}\rangle \right) .
$$

The three eigenstates above are the first excited states at leading order in V_0.

Problem 7 Effect of Finite Nuclear Size on Energy Spectrum of Hydrogen-Like Atom

In this problem the effect of the finite size of the nucleus on the spectrum of hydrogen-like atoms is considered. We assume for simplicity that the nuclear charge distribution has spherical symmetry (so that the nucleus has zero total angular momentum) and that it is described by a function $\rho_N(r)$ normalized as $\int d\mathbf{r}\, \rho_N(r) = 1$, where $\rho_N(r)$ vanishes for $r \gg R$, R being the nuclear radius, of the order of a few 10^{-13} cm (note that the proton in the hydrogen atom has a finite size too). The (attractive) Coulomb potential felt by the electron is given by

$$
\overline{V}(\mathbf{r}) = -Ze^2 \int d\mathbf{r}' \frac{\rho_N(r')}{|\mathbf{r} - \mathbf{r}'|} ,
$$

where Z is the atomic number (the number of protons in the nucleus).[3] Ignore effects associated with the electron's spin, and write the Hamiltonian as

$$
H = \frac{\mathbf{p}^2}{2\mu} + \overline{V}(\mathbf{r}) = \underbrace{\frac{\mathbf{p}^2}{2\mu} - \frac{Ze^2}{r}}_{H_0} + \underbrace{\overline{V}(\mathbf{r}) + \frac{Ze^2}{r}}_{V(\mathbf{r})} ,
$$

where $V(\mathbf{r})$ is identified as the perturbation,

$$
V(\mathbf{r}) = -Ze^2 \int d\mathbf{r}'\, \rho_N(r') \left(\frac{1}{|\mathbf{r} - \mathbf{r}'|} - \frac{1}{r} \right) .
$$

Obtain the first-order corrections to the ground- and excited-state energy levels of H_0, and discuss your results (the calculations are facilitated by observing that the nuclear radius is much smaller than the atomic radius). Compare the first-order correction to the ground-state energy level with its

---

[3] In our earlier treatment of hydrogen-like atoms, we assumed that the nucleus was point-like with its charge distribution given by $Ze\,\delta(\mathbf{r})$ (a δ-function centered at the origin).

unperturbed value in the hydrogen atom (the proton radius is approximately 0.8×10^{-13} cm). The following expansion is useful:

$$\frac{1}{|\mathbf{r} - \mathbf{r}'|} = \frac{1}{r_>} \sum_{l=0}^{\infty} \left(\frac{r_<}{r_>}\right)^l P_l(\cos\theta), \qquad r_< = \min(r, r'), \qquad r_> = \max(r, r'),$$

where $P_l(\cos\theta)$ are Legendre polynomials.

Solution

We have

$$\int d\mathbf{r}' \, \rho_N(r') \frac{1}{|\mathbf{r} - \mathbf{r}'|} = \int_0^\infty dr' r'^2 \rho_N(r') \frac{1}{r_>} \sum_{l=0}^{\infty} \left(\frac{r_<}{r_>}\right)^l \underbrace{\int_{-1}^{1} d(\cos\theta) P_l(\cos\theta)}_{\neq 0 \text{ only if } l=0} \int_0^{2\pi} d\phi$$

$$= \int_0^\infty dr' r'^2 \rho_N(r') \frac{1}{r_>} \int_{-1}^{1} d(\cos\theta) \int_0^{2\pi} d\phi = \int d\mathbf{r}' \rho_N(r') \frac{1}{r_>},$$

where we have used the orthogonality of Legendre polynomials and the fact that $P_0(\cos\theta) = 1$. Note that $r_>$ is equal to r if $r > r'$ and equal to r' if $r' > r$, and so it cannot be taken out of the integral sign. Thus, because of the spherical symmetry of the nuclear charge distribution, the perturbation simply reads

$$V(r) = -Ze^2 \int d\mathbf{r}' \, \rho_N(r') \left(\frac{1}{r_>} - \frac{1}{r}\right),$$

which shows that $V(r)$ is different from zero only if r is within the nuclear volume $r \lesssim R$ and vanishes when $r \gtrsim R$. The (bound) eigenstates of H_0 are given by $H_0 |\phi_{nlm}\rangle = \epsilon_n |\phi_{nlm}\rangle$, where $\epsilon_n = -(Z\alpha)^2 \mu c^2/(2n^2)$ with degeneracy $g_n = n^2$; the principal quantum number, the orbital angular momentum, and its projection take on the values $n = 1, 2, \ldots$, $l = 0, 1, \ldots, n-1$, and $m = -l, -l+1, \ldots, l$, respectively.

Since the ground-state is non-degenerate, the first-order correction requires calculation of the expectation value

$$E_{n=1}^{(1)} = \langle \phi_{100}| \hat{V} |\phi_{100}\rangle = \int d\mathbf{r} \, \phi_{100}^*(r) \, V(r) \, \phi_{100}(r).$$

The integral above gives

$$E_{n=1}^{(1)} = -Ze^2 \int d\mathbf{r} \, |\phi_{100}(r)|^2 \int d\mathbf{r}' \, \rho_N(r') \left(\frac{1}{r_>} - \frac{1}{r}\right) = -Ze^2 \int d\mathbf{r}' \, \rho_N(r') \int d\mathbf{r} \, |\phi_{100}(r)|^2 \left(\frac{1}{r_>} - \frac{1}{r}\right)$$

$$= -Ze^2 \int d\mathbf{r}' \, \rho_N(r') \left[\int_{r<r'} d\mathbf{r} \, |\phi_{100}(r)|^2 \left(\frac{1}{r'} - \frac{1}{r}\right) + \underbrace{\int_{r>r'} d\mathbf{r} \, |\phi_{100}(r)|^2 \left(\frac{1}{r} - \frac{1}{r}\right)}_{\text{vanishes}} \right]$$

$$= -Ze^2 |\phi_{100}(0)|^2 \int d\mathbf{r}' \, \rho_N(r') \int_{r<r'} d\mathbf{r} \left(\frac{1}{r'} - \frac{1}{r}\right),$$

where in the first line we have exchanged the order of integration and in the last line we have approximated $|\phi_{100}(r)|^2$ by its value at the origin, since the $\mathbf{r}$-integration is over a sphere of radius r', r' being of the order of the nuclear charge radius; this is much smaller than the atomic radius, which is of the order $a_0/Z \approx (0.53/Z) \times 10^{-8}$ cm. The final integrations can be easily done:

$$\int d\mathbf{r}' \rho_N(r') \int_{r<r'} d\mathbf{r} \left(\frac{1}{r'} - \frac{1}{r}\right) = 4\pi \int d\mathbf{r}' \rho_N(r') \int_0^{r'} dr\, r^2 \left(\frac{1}{r'} - \frac{1}{r}\right)$$

$$= 4\pi \int d\mathbf{r}' \rho_N(r') \left(\frac{r'^2}{3} - \frac{r'^2}{2}\right) = -\frac{2\pi}{3} \langle R_N^2 \rangle\,,$$

where we have defined the nuclear mean-square radius as

$$\langle R_N^2 \rangle = \int d\mathbf{r}'\, r'^2 \rho_N(r')\,.$$

Collecting results, we finally obtain

$$E_{n=1}^{(1)} = \frac{2\pi}{3} Ze^2 |\phi_{100}(0)|^2 \langle R_N^2 \rangle > 0\,,$$

and so the first-order correction makes the ground state less bound, as one might have expected (the Coulomb attraction felt by the electron is reduced when the electron penetrates the nuclear region).

An excited energy level ϵ_n has degeneracy $g_n = n^2$, and hence we need to use degenerate perturbation theory. The first-order corrections to the energy ϵ_n are the eigenvalues of the matrix of dimension g_n whose matrix elements are given by

$$\langle \phi_{nl'm'} | V | \phi_{nlm} \rangle\,,$$

in the subspace with fixed n. However, the perturbation $V(r)$ is spherically symmetric and is therefore invariant under rotations (for the case we are considering, that of a spherical nucleus), specifically

$$[\mathbf{L}, V] = 0\,.$$

In particular, $V(r)$ commutes with $\mathbf{L}^2$ and L_z, and as a consequence its matrix elements vanish between eigenstates of $\mathbf{L}^2$ and L_z belonging to different eigenvalues, which implies that

$$\langle \phi_{nl'm'} | V | \phi_{nlm} \rangle = \langle \phi_{nlm} | V | \phi_{nlm} \rangle\, \delta_{ll'} \delta_{mm'}\,,$$

and the matrix is diagonal in the basis $|\phi_{nlm}\rangle$. The first-order corrections are simply the diagonal matrix elements. Further, these diagonal matrix elements are independent of the azimuthal quantum number,[4] and are given by

$$E_{nl}^{(1)} = \int d\mathbf{r}\, |\phi_{nlm}(\mathbf{r})|^2\, V(r)\,.$$

This integral can be evaluated in exactly the same fashion as $E_{n=1}^{(1)}$ to obtain

$$E_{nl}^{(1)} = \frac{2\pi}{3} Ze^2 |\phi_{nlm}(0)|^2 \langle R_N^2 \rangle\,,$$

where we have ignored the spatial dependence of the atomic wave function by approximating it by its value at the origin. However, recalling that the wave function behaves as r^l for small r, we find that the correction above is non-vanishing only for s-wave states, and

$$|\phi_{nlm}(0)|^2 = \frac{1}{\pi} \left(\frac{Z\alpha}{n}\right)^3 \left(\frac{\mu c}{\hbar}\right)^3 \delta_{l0}\, \delta_{m0}\,.$$

---

[4] The perturbation V is a scalar operator and, as we will see later, the Wigner–Eckart theorem ensures that its matrix elements are independent of m. This can also be seen by using the property of the raising or lowering operator, $L_\pm |\phi_{nlm}\rangle = \hbar\sqrt{l(l+1) - m(m\pm 1)}\, |\phi_{nl,m\pm 1}\rangle$, and the fact that $L_\pm$ commute with V to show that

$$\langle \phi_{nlm} | V | \phi_{nlm} \rangle = \langle \phi_{nl,m\pm 1} | V | \phi_{nl,m\pm 1} \rangle\,,$$

thus proving its independence of m.

We conclude that at first order in the perturbative expansion only s-wave states are affected and the corresponding energies are shifted up. The upward shift is tiny; indeed,

$$\left| E_{n0}^{(1)}/\epsilon_n \right| = \frac{4}{3} \frac{(Z\alpha)^2}{n} \frac{\langle R_N^2 \rangle}{(\hbar/\mu c)^2} \, ,$$

which in the case of the hydrogen atom, using $\langle R^2 \rangle \approx 6.4 \times 10^{-27}$ cm^2 for the square of the proton radius and $\lambda_e = \hbar/m_e c \approx 3.9 \times 10^{-11}$ cm for the Compton wavelength of the electron (here $\mu \approx m_e$), gives for the 1s level

$$\left| E_{10}^{(1)}/\epsilon_n \right| \approx 3 \times 10^{-10} \, .$$

However, this correction becomes appreciable for μ-mesic atoms, in which one electron is replaced by a (negative) muon. The Compton wavelength of the muon is a factor ≈ 207 smaller than that of the electron and consequently the upward shift relative to the 1s level is of the order of 1.3×10^{-5}. By measuring this shift in μ-mesic atoms, information can be obtained on the nuclear mean square charge radius.

Problem 8 Hydrogen Atom in a Static External Electric Field: Stark Effect

We neglect spin degrees of freedom and take the unperturbed Hamiltonian as

$$H_0 = \frac{\mathbf{p}^2}{2\mu} - \frac{e^2}{r} \, .$$

Assume that the (static) electric field is produced by an external electrostatic potential $U(\mathbf{r})$. This leads to an additional term in the Hamiltonian, given by $-e\, U(\mathbf{r})$. Since atomic dimensions are much smaller than the length scale over which $U(\mathbf{r})$ varies appreciably, we can expand $U(\mathbf{r})$ about the origin to obtain

$$U(\mathbf{r}) = U(0) + \sum_i r_i \frac{\partial U(\mathbf{r})}{\partial r_i}\bigg|_{\mathbf{r}=0} + \cdots \, .$$

By setting the arbitrary value $U(0)$ to zero and noting that $\mathbf{E}(\mathbf{r}) = -\nabla U(\mathbf{r})$, we obtain the perturbation as[5]

$$V = e\mathbf{E} \cdot \mathbf{r} \qquad \text{with} \quad \mathbf{E} \equiv \mathbf{E}(0) \, ,$$

and arrange the coordinate system so that $\mathbf{E} = E\hat{\mathbf{z}}$, namely the external electric field is directed along the z-axis.

Calculate the first-order corrections to the unperturbed ground- and first excited-state energy levels, and compare them with the corresponding unperturbed energies (tabulations of the radial wave functions are readily available). Obtain the approximate eigenstates.

Solution

The perturbation $\hat{V}$ commutes with $\hat{L}_z$ (it is invariant under rotations about the z-axis) but not with $\hat{L}^2$. Under parity, the unperturbed eigenstates transform as $\hat{U}_P |\phi_{nlm}\rangle = (-1)^l |\phi_{nlm}\rangle$, and so they are even or odd depending on whether l is even or odd. As a consequence, except for the

---

[5] This is the interaction $-\mathbf{D} \cdot \mathbf{E}$ of an electric dipole $\mathbf{D} = -e\,\mathbf{r}$ with a uniform electric field $\mathbf{E}$. The dipole is formed by the proton (fixed at the origin) and electron.

non-degenerate ground state (corresponding to $n = 1$), which, being s-wave, is even under parity, the degenerate excited states with principal quantum number $n \geq 2$ have no definite parity. However, the perturbation $\hat{V}$ is odd, $\hat{U}_P \hat{V} \hat{U}_P^{\dagger} = -\hat{V}$, which implies that its matrix elements $\langle \phi_{n'l'm'} | \hat{V} \phi_{nlm} \rangle$ in the basis of eigenstates of $\hat{H}_0$ vanish unless $l + l'$ is odd; indeed, consider

$$\langle \phi_{n'l'm'} | \hat{V} | \phi_{nlm} \rangle = \langle \phi_{n'l'm'} | \underbrace{\hat{U}_P^{\dagger} \hat{U}_P}_{\hat{\mathbb{1}}} \hat{V} \underbrace{\hat{U}_P^{\dagger} \hat{U}_P}_{\hat{\mathbb{1}}} | \phi_{nlm} \rangle = (-1)^{l+l'} \langle \phi_{n'l'm'} | \underbrace{\hat{U}_P \hat{V} \hat{U}_P^{\dagger}}_{-\hat{V}} | \phi_{nlm} \rangle \,,$$

and so

$$\langle \phi_{n'l'm'} | \hat{V} | \phi_{nlm} \rangle = -(-1)^{l+l'} \langle \phi_{n'l'm'} | \hat{V} | \phi_{nlm} \rangle \,.$$

Parity puts no restrictions on the remaining quantum numbers. Further, since $\hat{V}$ commutes with $\hat{L}_z$, we have the selection rule $m' = m$ on the azimuthal quantum numbers, that is,

$$\langle \phi_{n'l'm'} | \hat{V} | \phi_{nlm} \rangle = 0 \qquad \text{if } m \neq m' \,.$$

After these preliminary considerations, we evaluate the first-order corrections to the ground-state and first-excited-state levels of hydrogen induced by an external electric field. For the (non-degenerate) ground-state level we find

$$E_{n=1}^{(1)} = \langle \phi_{100} | \hat{V} | \phi_{100} \rangle = 0 \,,$$

because of the selection rule on parity. For the first-excited-state level we must diagonalize the matrix of $\hat{V}$ in the degenerate subspace having principal quantum number $n = 2$,

$$\underline{V} = \begin{pmatrix} \langle \phi_{200} | \hat{V} | \phi_{200} \rangle & \langle \phi_{200} | \hat{V} | \phi_{210} \rangle & \langle \phi_{200} | \hat{V} | \phi_{211} \rangle & \langle \phi_{200} | \hat{V} | \phi_{21-1} \rangle \\ \langle \phi_{210} | \hat{V} | \phi_{200} \rangle & \langle \phi_{210} | \hat{V} | \phi_{210} \rangle & \langle \phi_{210} | \hat{V} | \phi_{211} \rangle & \langle \phi_{210} | \hat{V} | \phi_{21-1} \rangle \\ \langle \phi_{211} | \hat{V} | \phi_{200} \rangle & \langle \phi_{211} | \hat{V} | \phi_{210} \rangle & \langle \phi_{211} | \hat{V} | \phi_{211} \rangle & \langle \phi_{211} | \hat{V} | \phi_{21-1} \rangle \\ \langle \phi_{21-1} | \hat{V} | \phi_{200} \rangle & \langle \phi_{21-1} | \hat{V} | \phi_{210} \rangle & \langle \phi_{21-1} | \hat{V} | \phi_{211} \rangle & \langle \phi_{21-1} | \hat{V} | \phi_{21-1} \rangle \end{pmatrix} \,.$$

The selection rules above lead to

$$\underline{V} = \begin{pmatrix} 0 & a & 0 & 0 \\ a^* & 0 & 0 & 0 \\ 0 & 0 & 0 & 0 \\ 0 & 0 & 0 & 0 \end{pmatrix} \,, \qquad a = \langle \phi_{200} | \hat{V} | \phi_{210} \rangle \,,$$

and we are left with a single matrix element to evaluate. For this, we need explicit expressions for the (normalized) wave functions (with a_0 the Bohr radius)

$$\phi_{200}(r) = \frac{1}{(8\pi a_0^3)^{1/2}} \left(1 - \frac{r}{2a_0} \right) e^{-r/2a_0} \,, \qquad \phi_{210}(r) = \frac{1}{(32\pi a_0^3)^{1/2}} \frac{r}{a_0} e^{-r/2a_0} \cos \theta \,,$$

and therefore ($z = r \cos \theta$)

$$a = \frac{eE}{16\pi a_0^3} \int_0^{\infty} dr\, r^3 \frac{r}{a_0} \left(1 - \frac{r}{2a_0} \right) e^{-r/a_0} \underbrace{\int_{-1}^{1} d(\cos \theta) \cos^2 \theta \int_0^{2\pi} d\phi}_{4\pi/3} = \frac{eEa_0}{12} \int_0^{\infty} dx\, x^4 \left(1 - \frac{x}{2} \right) e^{-x} \,.$$

Using the following identity for $\alpha = 1$,

$$\int_0^{\infty} dx\, x^n\, e^{-\alpha x} = \left(-\frac{d}{d\alpha} \right)^n \int_0^{\infty} dx\, e^{-\alpha x} = \frac{n!}{\alpha^{n+1}} \,,$$

we finally obtain

$$a = -3\,eEa_0 \,.$$

The eigenvalues of the $\underline{V}$ matrix are easily found; they are 0 (twice) and $\pm 3eEa_0$, and so the perturbation partially removes the degeneracy. The eigenvalues and eigenstates up to and including first order are given by

$$-\frac{e^2}{8a_0}(1+\eta) \implies |\phi_-\rangle = \frac{1}{\sqrt{2}}\left(|\phi_{200}\rangle + |\phi_{210}\rangle\right) \,,$$

$$-\frac{e^2}{8a_0}(1-\eta) \implies |\phi_+\rangle = \frac{1}{\sqrt{2}}\left(|\phi_{200}\rangle - |\phi_{210}\rangle\right) \,,$$

where $\epsilon_{n=2} = -e^2/(8a_0)$ is the unperturbed energy, and the correction η is given by

$$\eta = 24\,\frac{E}{e/a_0^2} \approx 4.7 \times 10^{-9}\;E(\text{volts/cm}) \,.$$

Note that the linear combinations $|\phi_\pm\rangle$ are not eigenstates of the parity operator; as a matter of fact, $\hat{U}_P|\phi_\pm\rangle = |\phi_\mp\rangle$. The admixture of 2s and 2p states is independent of the strength of the electric field. Even a very weak field will maximally mix these two states. As we will see, the radiative decay of the 2s into the 1s state by single-photon emission is forbidden; the decay of the 2p into the 1s state is not. For this reason, the 2s state (in the absence of an external electric field) is metastable. However, in the presence of even a very weak electric field, it can decay rapidly through its mixing with the 2p state by emission of a single photon.

Problem 9   Spin-1/2 System in a Uniform Magnetic Field in a Generic Direction

Consider a spin-1/2 system described by the following Hamiltonian:

$$H_0 = 2\,\frac{v_x}{\hbar}\,S_x + 4\,\frac{v_z}{\hbar^2}\,S_z^2 \,,$$

where $\mathbf{S}$ is the spin operator and v_x and v_z $(v_z > v_x)$ are positive constants with dimensions of energy.

1. Obtain the eigenvalues $\epsilon_\pm$ and corresponding eigenstates $|\epsilon_\pm\rangle$ of H_0. Why are the $|\epsilon_\pm\rangle$ also eigenstates of S_x?
2. The system has magnetic moment $\boldsymbol{\mu} = -(\mu_0/\hbar)\,\mathbf{S}$, with μ_0 assumed to be positive. A static magnetic field $\mathbf{B}$ is applied with components

$$\mathbf{B} = B_0(\sin\theta\,\cos\phi, \sin\theta\,\sin\phi, \cos\theta) \,,$$

where θ and ϕ are the polar angles. Obtain the matrix representing the interaction V of the system with the magnetic field in the basis $|\pm\rangle$, consisting of eigenstates of S_z.
3. Assume that the magnetic field is weak enough that the effects of the interaction V can be treated in perturbation theory. Calculate the energies of the system to first and second order in B_0 and obtain the eigenstates to first order in B_0.
4. Denote by $|\psi_0\rangle$ the ground state of the system including the first-order correction calculated above, and evaluate the mean value $\langle\psi_0|\boldsymbol{\mu}|\psi_0\rangle$. Define the induced magnetic moment as $\langle\psi_0|\boldsymbol{\mu}|\psi_0\rangle - \langle\epsilon_-|\mu_x|\epsilon_-\rangle\,\hat{\mathbf{x}}$, where $|\epsilon_-\rangle$ turns out to be the (unperturbed) ground state corresponding to H_0. Is this induced magnetic moment parallel to $\mathbf{B}$? Show that

$$\langle \psi_0 | \mu_\alpha | \psi_0 \rangle - \delta_{\alpha,x} \langle \epsilon_- | \mu_x | \epsilon_- \rangle = \sum_\beta \chi_{\alpha\beta} B_\beta \,,$$

and calculate the coefficients $\chi_{\alpha\beta}$ (that is, the components of the magnetic susceptibility tensor).

Solution

Part 1

In the basis $|+\rangle$ and $|-\rangle$ that diagonalizes $\mathbf{S}^2$ and S_z, the matrix representing H_0 is given by

$$\underline{H}_0 = \begin{pmatrix} v_z & v_x \\ v_x & v_z \end{pmatrix} \,,$$

since $S_z^2 |\pm\rangle = (\hbar^2/4)|\pm\rangle$ and $S_x |\pm\rangle = (\hbar/2) |\mp\rangle$; this last relationship follows from

$$S_x = (S_+ + S_-)/2 \,, \qquad S_\pm |\mp\rangle = \hbar |\pm\rangle \,, \qquad S_\pm |\pm\rangle = 0 \,,$$

where S_+ and S_- are the spin raising and lowering operators, respectively. The (non-degenerate) eigenvalues $\epsilon_\pm$ and corresponding eigenstates $|\epsilon_\pm\rangle$ of H_0 are easily obtained,

$$\epsilon_\pm = v_z \pm v_x \,, \qquad |\epsilon_\pm\rangle = \frac{1}{\sqrt{2}} (|+\rangle \pm |-\rangle) \,,$$

and $|\epsilon_\pm\rangle$ are also eigenstates of S_x with eigenvalues $\pm\hbar/2$. This is so because S_x commutes with H_0, as is easily verified by noting that S_x anticommutes with S_z and hence commutes with S_z^2.

Part 2

The interaction V is given by

$$V = -\boldsymbol{\mu} \cdot \mathbf{B} = \frac{\mu_0 B_0}{\hbar} \left(\sin\theta \, \cos\phi \, S_x + \sin\theta \, \sin\phi \, S_y + \cos\theta \, S_z \right) \,.$$

In the basis of eigenstates $|\pm\rangle$, the components of $\mathbf{S}$ have the following matrix representations ($\mathbf{S} \longrightarrow (\hbar/2) \, \underline{\sigma}$, where $\underline{\sigma}$ are the Pauli matrices)

$$\underline{S}_x = \frac{\hbar}{2} \begin{pmatrix} 0 & 1 \\ 1 & 0 \end{pmatrix} \,, \qquad \underline{S}_y = \frac{\hbar}{2} \begin{pmatrix} 0 & -i \\ i & 0 \end{pmatrix} \,, \qquad \underline{S}_z = \frac{\hbar}{2} \begin{pmatrix} 1 & 0 \\ 0 & -1 \end{pmatrix} \,,$$

and hence

$$\underline{V} = \begin{pmatrix} V_{++} & V_{+-} \\ V_{-+} & V_{--} \end{pmatrix} = \frac{\mu_0 B_0}{2} \begin{pmatrix} \cos\theta & \sin\theta \, e^{-i\phi} \\ \sin\theta \, e^{i\phi} & -\cos\theta \end{pmatrix} \,.$$

Part 3

The first-order corrections to the unperturbed energies $\epsilon_\pm$ follow from

$$E_\pm^{(1)} = \langle \epsilon_\pm | V | \epsilon_\pm \rangle = \frac{1}{2} [V_{++} + V_{--} \pm (V_{+-} + V_{-+})] = \pm \frac{\mu_0 B_0}{2} \sin\theta \cos\phi \,,$$

while the second-order corrections follow from

$$E_\pm^{(2)} = \frac{|\langle \epsilon_\mp | V | \epsilon_\pm \rangle|^2}{\epsilon_\pm - \epsilon_\mp} \,,$$

where

$$\langle \epsilon_{\mp} | V | \epsilon_{\pm} \rangle = \frac{1}{2} [V_{++} - V_{--} \pm (V_{+-} - V_{-+})] = \frac{\mu_0 B_0}{2} \left(\cos\theta \mp i \sin\theta \sin\phi \right), \qquad \epsilon_{\pm} - \epsilon_{\mp} = \pm 2v_x.$$

Therefore, the second-order corrections are

$$E_{\pm}^{(2)} = \pm \frac{(\mu_0 B_0)^2}{8v_x} \left(\cos^2\theta + \sin^2\theta \sin^2\phi \right).$$

The first-order corrections to the eigenstates $|\epsilon_{\pm}\rangle$ induced by the perturbation V are given by

$$|\epsilon_{\pm}\rangle^{(1)} = \frac{\langle \epsilon_{\mp} | V | \epsilon_{\pm} \rangle}{\epsilon_{\pm} - \epsilon_{\mp}} |\epsilon_{\mp}\rangle = \pm \frac{\mu_0 B_0}{4v_x} \left(\cos\theta \mp i \sin\theta \sin\phi \right) |\epsilon_{\mp}\rangle.$$

Part 4

The ground state of the system, including the first-order correction, is

$$|\psi_0\rangle = |\epsilon_-\rangle + |\epsilon_-\rangle^{(1)} = \frac{1}{\sqrt{2}} [\, 1 - \underbrace{\lambda(\cos\theta + i \sin\theta \sin\phi)}_{\chi} \,]|+\rangle - \frac{1}{\sqrt{2}} [\, 1 + \underbrace{\lambda(\cos\theta + i \sin\theta \sin\phi)}_{\chi} \,]|-\rangle, \quad \lambda = \frac{\mu_0 B_0}{4v_x},$$

and the mean value of $\boldsymbol{\mu} = -(\mu_0/\hbar)\mathbf{S}$ is easily obtained from the matrix representation of $\mathbf{S}$. We have

$$\langle \psi_0 | \mu_z | \psi_0 \rangle = -\frac{\mu_0}{4} \left(1 - \chi^* \;\; -1 - \chi^* \right) \begin{pmatrix} 1 & 0 \\ 0 & -1 \end{pmatrix} \begin{pmatrix} 1 - \chi \\ -1 - \chi \end{pmatrix} = \mu_0 \lambda \cos\theta.$$

Similarly, we find

$$\langle \psi_0 | \mu_y | \psi_0 \rangle = -\frac{\mu_0}{4} \left(1 - \chi^* \;\; -1 - \chi^* \right) \begin{pmatrix} 0 & -i \\ i & 0 \end{pmatrix} \begin{pmatrix} 1 - \chi \\ -1 - \chi \end{pmatrix} = \mu_0 \lambda \sin\theta \sin\phi$$

and

$$\langle \psi_0 | \mu_x | \psi_0 \rangle = -\frac{\mu_0}{4} \left(1 - \chi^* \;\; -1 - \chi^* \right) \begin{pmatrix} 0 & 1 \\ 1 & 0 \end{pmatrix} \begin{pmatrix} 1 - \chi \\ -1 - \chi \end{pmatrix} = \frac{\mu_0}{2} [1 - \lambda^2(\cos^2\theta + \sin^2\theta \sin^2\phi)].$$

The induced magnetic moment $\langle \boldsymbol{\mu} \rangle - \langle \epsilon_- | \mu_x | \epsilon_- \rangle \, \hat{\mathbf{x}}$ and $\mathbf{B}$ are not parallel; while $\langle \mu_y \rangle$ and $\langle \mu_z \rangle$ are proportional to, respectively, B_y and B_z with the same coefficient, $\langle \mu_x \rangle - \langle \epsilon_- | \mu_x | \epsilon_- \rangle$ is not proportional to B_x but rather depends on $B_y^2 + B_z^2$. Indeed,

$$\langle \mu_\alpha \rangle - \delta_{\alpha x} \langle \epsilon_- | \mu_x | \epsilon_- \rangle = \sum_\beta \chi_{\alpha\beta} B_\beta \implies \underline{\chi} = \frac{\mu_0^2}{4v_x} \begin{pmatrix} 0 & -\mu_0 B_y/(8v_x) & -\mu_0 B_z/(8v_x) \\ 0 & 1 & 0 \\ 0 & 0 & 1 \end{pmatrix} \begin{pmatrix} B_x \\ B_y \\ B_z \end{pmatrix}.$$

Problem 10 Charged Particle Constrained on a Circle in a Uniform Electric Field

A particle of mass m and charge q is constrained to move in a circle of radius R. Its unperturbed Hamiltonian consists of the kinetic energy operator, which is written as

$$\hat{H}_0 = \frac{\hat{L}_z^2}{2I},$$

where $\hat{L}_z = -i\hbar \, d/d\phi$ is the z-component of the orbital angular momentum operator $\hat{\mathbf{L}}$, and $I = mR^2$ is the moment of inertia.

1. Obtain the eigenvalues, including their degeneracy, and eigenstates of $\hat{H}_0$.

2. Suppose that the particle is now under the influence of a uniform (time-independent) electric field directed along the x-direction and of magnitude $\mathcal{E}$. Assuming that this interaction can be treated in perturbation theory, obtain the corrections to the ground-state energy to first and second order in $\mathcal{E}$, and the perturbed ground state to first order in $\mathcal{E}$. Calculate the electric susceptibility

$$\chi = -\frac{\partial E_0}{\partial \mathcal{E}},$$

where E_0 is the ground-state energy including these first- and second-order corrections.

3. The ethane molecule has the chemical formula CH_3–CH_3. In the CH_3 group, the three H atoms form the basis of a regular three-sided pyramid with the C atom at its vertex. The ethane molecule therefore consists of two inverted pyramids joined by an axis going through the two C atoms. One group rotates relative to the other. To first approximation, this rotation is free, and the associated kinetic energy operator is just the Hamiltonian $\hat{H}_0$ above, where I is now the moment of inertia of the CH_3 group relative to the rotational axis. The electrostatic energy between the two groups is approximately modeled by adding to $\hat{H}_0$ a term having the form

$$\hat{V} = V_0 \cos(3\phi),$$

where V_0 is a real constant. Give a physical justification for the ϕ dependence of $\hat{V}$. Calculate the energy to second order and the wave function to first order of the new ground state. Give a physical interpretation of the result.

Solution

Part 1

The eigenfunctions of $\hat{L}_z$ are also eigenfunctions of $\hat{H}_0$. They satisfy

$$-i\hbar \psi'(\phi) = \overline{m}\psi(\phi) \implies \psi(\phi) = c\,e^{i\overline{m}\phi/\hbar}.$$

Periodic boundary conditions demand that

$$\psi(\phi) = \psi(\phi + 2\pi) \implies e^{2\pi i \overline{m}/\hbar} = 1,$$

requiring $\overline{m}/\hbar = m$ integer, with $m = 0, \pm 1, \pm 2, \ldots$; the normalized eigenfunctions are given by

$$\psi_m(\phi) = \frac{1}{\sqrt{2\pi}}\,e^{im\phi}.$$

The eigenvalues of $\hat{H}_0$ follow from

$$\hat{H}_0 |\psi_m\rangle = \epsilon_m |\psi_m\rangle,$$

and $\epsilon_m = \hbar^2 m^2/(2I)$ is two-fold degenerate for $m \neq 0$ and non-degenerate for $m = 0$.

Part 2

The perturbation reads $\hat{V} = -q\mathcal{E}x = -q\mathcal{E}R\cos\phi$. The first-order correction to the ground-state energy vanishes, since

$$E_0^{(1)} = \langle \psi_0 | \hat{V} | \psi_0 \rangle = -\frac{q\mathcal{E}R}{2\pi} \int_0^{2\pi} d\phi \, \cos\phi = 0,$$

while the second-order correction is

$$E_0^{(2)} = \sum_{m \neq 0} \frac{|\langle \psi_m | \hat{V} | \psi_0 \rangle|^2}{\epsilon_0 - \epsilon_m} ,$$

where

$$\langle \psi_m | \hat{V} | \psi_0 \rangle = -\frac{q\mathcal{E}R}{2} \int_0^{2\pi} d\phi \, \psi_m^*(\phi)(e^{i\phi} + e^{-i\phi})\psi_0(\phi) .$$

Recalling that $\psi_0(\phi) = 1/\sqrt{2\pi}$ and the form of the generic wave function $\psi_m(\phi)$, we find

$$\langle \psi_m | \hat{V} | \psi_0 \rangle = -\frac{q\mathcal{E}R}{2} \int_0^{2\pi} d\phi \, \psi_m^*(\phi) \left[\psi_1(\phi) + \psi_{-1}(\phi) \right] = -\frac{q\mathcal{E}R}{2} (\delta_{m,1} + \delta_{m,-1}) \implies E_0^{(2)} = -\frac{(q\mathcal{E}R)^2}{2\epsilon_1} .$$

The electric susceptibility follows:

$$\chi = -\frac{\partial}{\partial \mathcal{E}} \left(\epsilon_0 + E_0^{(1)} + E_0^{(2)} \right) = \frac{(qR)^2 \mathcal{E}}{\epsilon_1} .$$

The first-order correction to the ground state is given by

$$|\psi_0^{(1)}\rangle = \sum_{m \neq 0} \frac{\langle \psi_m | \hat{V} | \psi_0 \rangle}{\epsilon_0 - \epsilon_m} |\psi_m\rangle = \frac{q\mathcal{E}R}{2\epsilon_1} \left(|\psi_1\rangle + |\psi_{-1}\rangle \right) ,$$

since $\epsilon_0 = 0$ and the states with $m = \pm 1$ are degenerate.

Part 3

The form of the electrostatic energy $\hat{V}$ is invariant for rotations of one pyramid base by 120° ($\phi \longrightarrow \phi + 120°$), and reflects the symmetry of the CH_3–CH_3 molecule, discussed in the text of the problem. The calculation is similar to that above, the only difference being that the matrix element of $\hat{V}$ is now given by

$$\langle \psi_m | \hat{V} | \psi_0 \rangle = \frac{V_0}{2} \int_0^{2\pi} d\phi \, \psi_m^*(\phi) \left[\psi_3(\phi) + \psi_{-3}(\phi) \right] = \frac{V_0}{2} (\delta_{m,3} + \delta_{m,-3}) .$$

The first-order correction to the unperturbed ground state is obtained as

$$|\psi_0^{(1)}\rangle = -\frac{V_0}{2\epsilon_3} \left(|\psi_3\rangle + |\psi_{-3}\rangle \right) ,$$

and the second-order correction to the ground-state energy is given by

$$E_0^{(2)} = -\frac{V_0^2}{2\epsilon_3} .$$

The ground-state wave function up and including to first order is

$$\psi_0(\phi) - \frac{V_0}{2\epsilon_3} \left[\psi_3(\phi) + \psi_{-3}(\phi) \right] = \frac{1}{\sqrt{2\pi}} \left[1 - \frac{V_0}{\epsilon_3} \cos(3\phi) \right] ,$$

and reflects the 120° rotational symmetry of one CH_3 group relative to the other.

Problem 11   Electron Interacting with Two ^{3}He Nuclei in a Magnetic Field

An electron and two ^{3}He nuclei are in a uniform magnetic field $B\,\hat{z}$. The electron has a magnetic moment given by

$$\boldsymbol{\mu}_e = -g_e \frac{e}{2m_e c}\,\mathbf{S} \equiv -\gamma_e\,\mathbf{S}\,,$$

where m_e and g_e are the electron mass and gyromagnetic factor, respectively. The ^{3}He nucleus is a spin-1/2 particle consisting of two protons and a neutron, bound together by the strong interaction. It has a magnetic moment given by

$$\boldsymbol{\mu} = -g\frac{e}{2m_p c}\,\mathbf{I} \equiv -\gamma\,\mathbf{I}\,,$$

where m_p is the proton mass and $\mathbf{I}$ and g are the ^{3}He-nucleus spin-1/2 operator and gyromagnetic factor, respectively. Here, the parameters γ_e and γ are both positive, and $\gamma_e \gg \gamma$. The orbital degrees of freedom of the electron and ^{3}He nuclei are to be ignored throughout; further, the two ^{3}He nuclei (which are fermions) are to be considered as distinguishable particles.

1. Accounting only for the interactions of these particles with the external magnetic field, write down the Hamiltonian H_0 for this system and obtain the possible eigenvalues and their degree of degeneracy.
2. Now assume that, while the interaction between the nuclear spins is negligible, the hyperfine interaction between the electron and ^{3}He spins is given by

$$V = v\,\mathbf{S}\cdot(\mathbf{I}_1 + \mathbf{I}_2)\,,$$

   where the parameter $v > 0$ and $v\hbar^2$ has dimension of energy. Treat V as a small perturbation and obtain the first-order corrections to the energy levels, found in part 1.
3. Switch off the magnetic field, so that the Hamiltonian reduces to V. Let $\mathbf{I} = \mathbf{I}_1 + \mathbf{I}_2$ be the total nuclear spin. Show that V has no matrix elements between eigenstates of $\mathbf{I}^2$ belonging to different eigenvalues. Let $\mathbf{F} = \mathbf{S} + \mathbf{I}$ be the total angular momentum. What are the eigenvalues of $\mathbf{F}^2$ and their degeneracies? Determine the energy eigenvalues of the three-spin system and corresponding degeneracies. Do $\mathbf{F}^2$ and F_z form a complete set of commuting observables? And how about $\mathbf{I}^2, \mathbf{F}^2$, and F_z?

Solution

Part 1

The Hamiltonian H_0 of the system is given by

$$H_0 = -(\boldsymbol{\mu}_e + \boldsymbol{\mu}_1 + \boldsymbol{\mu}_2)\cdot\mathbf{B} = \Omega S_z + \omega(I_{1z} + I_{2z}) \qquad \text{with } \Omega = \gamma_e B \text{ and } \omega = \gamma B\,,$$

where the angular frequencies Ω and ω are both positive, and $\Omega \gg \omega$. The eigenstates of H_0 are the orthonormal kets $|m_s, m_1, m_2\rangle$, which are common eigenstates of S_z, I_{1z}, and I_{2z}, with corresponding eigenvalues given by

$$H_0\,|m_s, m_1, m_2\rangle = \hbar\,[\Omega m_s + \omega\,(m_1 + m_2)]\,|m_s, m_1, m_2\rangle\,, \qquad m_s, m_1, m_2 = \pm 1/2\,.$$

These eigenvalues are

$$\epsilon_{++} = \frac{\hbar}{2}(\Omega + 2\omega), \qquad \epsilon_{+0} = \frac{\hbar}{2}\Omega, \qquad \epsilon_{+-} = \frac{\hbar}{2}(\Omega - 2\omega)$$

and

$$\epsilon_{-+} = -\frac{\hbar}{2}(\Omega - 2\omega), \qquad \epsilon_{-0} = -\frac{\hbar}{2}\Omega, \qquad \epsilon_{--} = -\frac{\hbar}{2}(\Omega + 2\omega),$$

where the first subscript specifies whether the electron is spin up (+) or down (−), while the second subscript specifies whether the nuclear spins are both up (+), or the first up and the second down or vice versa (0), or both down (−). The eigenvalues ϵ_{+0} and ϵ_{-0} each have degeneracy 2.

Part 2

As a preliminary, we note that

$$\langle m_s', m_1', m_2'| \, \mathbf{S} \cdot \mathbf{I}_1 \, |m_s, m_1, m_2\rangle = \delta_{m_2', m_2} \langle m_s', m_1'| \, \mathbf{S} \cdot \mathbf{I}_1 \, |m_s, m_1\rangle \,.$$

Define $\mathbf{J}_1 = \mathbf{S} + \mathbf{I}_1$, so that

$$\mathbf{S} \cdot \mathbf{I}_1 = \frac{1}{2}\left(\mathbf{J}_1^2 - \frac{3}{2}\hbar^2\right) \,.$$

Since $J_{1z} = S_z + I_{1z}$ commutes with $\mathbf{S} \cdot \mathbf{I}_1$ and the states $|m_s, m_1\rangle$ are eigenstates of J_{1z}, $J_{1z}|m_s, m_1\rangle = \hbar(m_s + m_1)|m_s, m_1\rangle$, it follows that

$$\langle m_s', m_1'| \, \mathbf{S} \cdot \mathbf{I}_1 \, |m_s, m_1\rangle = 0 \qquad \text{if } m_s' + m_1' \neq m_s + m_1 \,.$$

Similarly, we have for $\mathbf{S} \cdot \mathbf{I}_2$ that its matrix elements vanish unless $m_1' = m_1$ and $m_s' + m_2' = m_s + m_2$.

We now proceed with the evaluation of the first-order corrections induced by V on the levels of H_0. We begin by considering the non-degenerate energy levels. We find

$$E_{++}^{(1)} = \langle +, +, +|V|+, +, +\rangle = v\left(\langle +, +|\mathbf{S} \cdot \mathbf{I}_1|+, +\rangle + \langle +, +|\mathbf{S} \cdot \mathbf{I}_2|+, +\rangle\right) \,.$$

Since the state $|++\rangle$ has $j_1 = 1$, we have, for the first matrix element within the parentheses,

$$\langle +, +|\mathbf{S} \cdot \mathbf{I}_1|+, +\rangle = \frac{1}{2}\langle +, +|\mathbf{J}_1^2 - 3\hbar^2/2|+, +\rangle = \frac{1}{4}\hbar^2$$

and similarly for the second matrix element. Thus, the first-order correction follows,

$$E_{++}^{(1)} = \frac{v\hbar^2}{2} \,,$$

and it is easy to see that $E_{--}^{(1)} = v\hbar^2/2$ also.

Next, we consider

$$E_{+-}^{(1)} = \langle +, -, -|V|+, -, -\rangle = v\left(\langle +, -|\mathbf{S} \cdot \mathbf{I}_1|+, -\rangle + \langle +, -|\mathbf{S} \cdot \mathbf{I}_2|+, -\rangle\right) \,.$$

Recalling that the state $|+, -\rangle$ is the following linear combination of $j_1 = 1$ and $j_1 = 0$ states with $m_{j_1} = 0$, namely

$$|+, -\rangle = \frac{1}{\sqrt{2}}\left(|1, 0\rangle + |0, 0\rangle\right) \,,$$

we obtain

$$\mathbf{S} \cdot \mathbf{I}_1 |+, -\rangle = \frac{1}{2}\left(\mathbf{J}_1^2 - \frac{3}{2}\hbar^2\right)\frac{1}{\sqrt{2}}\left(|1, 0\rangle + |0, 0\rangle\right) = \frac{1}{\sqrt{2}}\left(\frac{\hbar^2}{4}|1, 0\rangle - \frac{3\hbar^2}{4}|0, 0\rangle\right)$$

and

$$\langle+, -|\mathbf{S} \cdot \mathbf{I}_1|+, -\rangle = \frac{\hbar^2}{2}\left(\langle1, 0| + \langle0, 0|\right)\left(\frac{1}{4}|1, 0\rangle - \frac{3}{4}|0, 0\rangle\right) = -\frac{\hbar^2}{4} \,.$$

Of course, the second matrix element, involving $\mathbf{S} \cdot \mathbf{I}_2$, is the same. Thus we arrive at

$$E_{+-}^{(1)} = -\frac{v\hbar^2}{2} \,.$$

The correction $E_{-+}^{(1)}$ can be obtained similarly; the two-particle states are written as

$$|-, +\rangle = \frac{1}{\sqrt{2}}\left(|1, 0\rangle - |0, 0\rangle\right) \,,$$

which leads to

$$\mathbf{S} \cdot \mathbf{I}_1 |-, +\rangle = \frac{1}{\sqrt{2}}\left(\frac{\hbar^2}{4}|1, 0\rangle + \frac{3\hbar^2}{4}|0, 0\rangle\right)$$

and hence

$$\langle-, +|\mathbf{S} \cdot \mathbf{I}_1|-, +\rangle = \frac{\hbar^2}{2}\left(\langle1, 0| - \langle0, 0|\right)\left(\frac{1}{4}|1, 0\rangle + \frac{3}{4}|0, 0\rangle\right) = -\frac{\hbar^2}{4} \,,$$

so that $E_{-+}^{(1)} = E_{+-}^{(1)}$.

To obtain the corrections to the degenerate levels, we need to evaluate the matrix representing the perturbation in the degenerate subspace,

$$\underline{V} = \begin{pmatrix} \langle+, +, -|V|+, +, -\rangle & \langle+, +, -|V|+, -, +\rangle \\ \langle+, -, +|V|+, +, -\rangle & \langle+, -, +|V|+, -, +\rangle \end{pmatrix} = \begin{pmatrix} \langle+, +, -|V|+, +, -\rangle & 0 \\ 0 & \langle+, -, +|V|+, -, +\rangle \end{pmatrix} \,,$$

where the off-diagonal matrix elements vanish because either $m_1' \neq m_1$ or $m_2' \neq m_2$, and so

$$\langle+, +, -|V|+, +, -\rangle = v\left(\langle+, +|\mathbf{S} \cdot \mathbf{J}_1|+, +\rangle + \langle+, -|\mathbf{S} \cdot \mathbf{J}_2|+, -\rangle\right) = \frac{v\hbar^2}{4} - \frac{v\hbar^2}{4} = 0$$

and

$$\langle+, -, +|V|+, -, +\rangle = v\left(\langle+, -|\mathbf{S} \cdot \mathbf{J}_1|+, -\rangle + \langle+, +|\mathbf{S} \cdot \mathbf{J}_2|+, +\rangle\right) = -\frac{v\hbar^2}{4} + \frac{v\hbar^2}{4} = 0 \,.$$

We conclude that the states have vanishing first-order corrections and remain degenerate. The matrix for the remaining two degenerate states is given by

$$\underline{V} = \begin{pmatrix} \langle-, +, -|V|-, +, -\rangle & \langle-, +, -|V|-, -, +\rangle \\ \langle-, -, +|V|-, +, -\rangle & \langle-, -, +|V|-, -, +\rangle \end{pmatrix} = \begin{pmatrix} \langle-, +, -|V|-, +, -\rangle & 0 \\ 0 & \langle-, -, +|V|-, -, +\rangle \end{pmatrix} = \begin{pmatrix} 0 & 0 \\ 0 & 0 \end{pmatrix} \,,$$

yielding vanishing corrections.

Part 3

The Hamiltonian reduces to $V = v\, \mathbf{S} \cdot \mathbf{I}$ with $\mathbf{I} = \mathbf{I}_1 + \mathbf{I}_2$. Since V commutes with $\mathbf{I}^2$, it follows that matrix elements of V between eigenstates of $\mathbf{I}^2$ belonging to different eigenvalues vanish. We can combine the electron spin 1/2 with the total nuclear spin 0 (1) to obtain total angular momentum 1/2 (1/2, 3/2). The eigenvalues of $\mathbf{F}^2$ are $f(f+1)\hbar^2$, and so $3\hbar^2/4$ for $f = 1/2$ with degeneracy 4, and $15\hbar^2/4$ for $f = 3/2$, also with degeneracy 4 (recall the m_f degeneracy is $2f + 1$; however, $f = 1/2$ occurs twice). We write V as

$$V = \frac{v}{2} \left(\mathbf{F}^2 - \mathbf{I}^2 - \mathbf{S}^2 \right) = \frac{v}{2} \mathbf{F}^2 - \frac{v\hbar^2}{2} \left[i(i+1) + \frac{3}{4} \right] .$$

The eigenvalues of V are: 0 for $f = 1/2$ and $i = 0$ with degeneracy 2; $-v\hbar^2$ for $f = 1/2$ and $i = 1$ with degeneracy 2; and $v\hbar^2/2$ for $f = 3/2$ and $i = 1$ with degeneracy 4. Clearly, $\mathbf{F}^2$ and F_z do not form a complete set of commuting observables, since the eigenvalue 1/2 occurs twice; however, $\mathbf{I}^2$, $\mathbf{F}^2$, and F_z constitute a complete set of commuting observables, since specifying their respective eigenvalues identifies uniquely the corresponding common eigenstate.

Problem 12   Nucleus with Spin in Non-Uniform Electric Field and Weak Magnetic Field

Consider a nucleus of spin $I = 3/2$, whose state space is spanned by the four states $|m\rangle$ with $m = \pm 3/2$ and $\pm 1/2$, which are common eigenstates of $\mathbf{I}^2$ and I_z with eigenvalues $15\hbar^2/4$ and $m\hbar$. The nucleus is placed in a non-uniform electric field derived from the scalar potential $U(\mathbf{r})$, which satisfies the Laplace equation, $\nabla^2 U(\mathbf{r}) = 0$. The Hamiltonian describing the interaction between the electric field gradient and the nuclear electric quadrupole moment is given by

$$H_0 = \frac{eQ}{2I(2I-1)} \frac{1}{\hbar^2} \left(a_x I_x^2 + a_y I_y^2 + a_z I_z^2 \right) ,$$

where e is the unit charge, Q is a constant that has the dimensions of a surface and is proportional to the quadrupole moment of the nucleus, and

$$a_x = \left. \frac{\partial^2 U}{\partial x^2} \right|_0 , \qquad a_y = \left. \frac{\partial^2 U}{\partial y^2} \right|_0 , \qquad a_z = \left. \frac{\partial^2 U}{\partial z^2} \right|_0 ,$$

that is, the second partial derivatives are evaluated at the origin.

1. Show that if $U(\mathbf{r})$ is symmetrical with respect to revolutions about the z-axis then H_0 has the following form:

$$H_0 = A \left[3 I_z^2 - I(I+1) \right] ,$$

   where the constant A is to be determined. What are the eigenvalues of H_0, their degree of degeneracy, and the corresponding eigenstates?

2. Show that in the general case H_0 can be written as

$$H_0 = A \left[3 I_z^2 - I(I+1) \right] + B \left(I_+^2 + I_-^2 \right) ,$$

   where the constants A and B are to be determined. What is the matrix representing H_0 in the basis $|m\rangle$? Show that it can be broken down into two 2×2 sub-matrices. Calculate the eigenvalues of H_0, their degeneracy, and the corresponding eigenstates.

3. In addition to its quadrupole moment, the nucleus has magnetic moment $\boldsymbol{\mu} = \gamma\, \mathbf{I}$ (here γ is the gyromagnetic factor). A magnetic field $\mathbf{B}_0$ of arbitrary direction $\hat{\mathbf{n}}$ is superimposed on the

electrostatic field. Set $\omega_0 = -\gamma |\mathbf{B}_0|$. Include the interaction of the magnetic field with the nuclear magnetic moment and obtain the energies of the system to first order in B_0. The unperturbed Hamiltonian is that of part 2.

4. Assume that $\mathbf{B}_0$ is along the z-axis and is weak enough for the eigenstates in part 2 and the energies found in part 3 to be good approximations, to first order in ω_0. What are the Bohr frequencies that can appear in the expectation value $\langle I_x(t) \rangle$?

(Adapted from C. Cohen-Tannoudji, B. Diu, and F. Laloë (1997), *Quantum Mechanics*, vol. 2, Wiley.)

Solution

Part 1

Expand the scalar potential around the origin (the location of the nucleus) to obtain

$$U(\mathbf{r}) = U_0 + \sum_i r_i \left.\frac{\partial U}{\partial r_i}\right|_0 + \frac{1}{2} \sum_{i,j} r_i r_j \left.\frac{\partial^2 U}{\partial r_j \partial r_i}\right|_0 + \cdots .$$

Since $U(\mathbf{r})$ satisfies the Laplace equation $\nabla^2 U(\mathbf{r}) = 0$ everywhere in space and, in particular, at the origin, this puts a constraint on the parameters,

$$a_x + a_y + a_z = 0 ,$$

where we define

$$a_i = \left.\frac{\partial^2 U}{\partial r_i^2}\right|_0 .$$

If $U(\mathbf{r})$ is symmetrical with respect to revolutions about the z-axis, then it can only be a function of $\rho = \sqrt{x^2 + y^2}$ and z, that is, $U(\rho, z)$. This implies $a_x = a_y$ and, in turn, $a_z = -2a_x$. Under these conditions, the Hamiltonian becomes

$$H_0 = \frac{eQ}{2I(2I-1)} \frac{1}{\hbar^2} \left[a_x \underbrace{(I_x^2 + I_y^2)}_{\mathbf{I}^2 - I_z^2} - 2a_x I_z^2 \right] = -\frac{eQ}{2I(2I-1)} \frac{a_x}{\hbar^2} \left(3I_z^2 - \mathbf{I}^2 \right) ,$$

from which we read off

$$A = -\frac{eQ}{2I(2I-1)} \frac{a_x}{\hbar^2} .$$

We have

$$H_0 |m\rangle = \underbrace{A\hbar^2 (3m^2 - 15/4)}_{\epsilon_{|m|}} |m\rangle ,$$

and the eigenvalues $\epsilon_{|m|}$ with $m = \pm 1/2$ and $\pm 3/2$ are each two-fold degenerate, specifically $\epsilon_{1/2} = -3A\hbar^2$ and $\epsilon_{3/2} = 3A\hbar^2$.

Part 2

In the general case, it is convenient to express the parameters a_x and a_y as

$$b = (a_x + a_y)/2 \quad \text{and} \quad c = (a_x - a_y)/2 \implies a_x = b + c \quad \text{and} \quad a_y = b - c .$$

Consider

$$a_x I_x^2 + a_y I_y^2 + a_z I_z^2 = (b+c)I_x^2 + (b-c)I_y^2 - 2b I_z^2 = b(\mathbf{I}^2 - 3I_z^2) + c(I_x^2 - I_y^2) \,.$$

Recall that, in terms of raising and lowering operators we have

$$I_x = \frac{I_+ + I_-}{2} \,, \qquad I_y = \frac{I_+ - I_-}{2i} \,,$$

so that

$$I_x^2 - I_y^2 = \frac{1}{4} \left[I_+^2 + I_-^2 + (I_+ I_- + I_- I_+) + I_+^2 + I_-^2 - (I_+ I_- + I_- I_+) \right] = \frac{1}{2}(I_+^2 + I_-^2) \,.$$

The Hamiltonian is now given by

$$H_0 = \frac{eQ}{2I(2I-1)} \frac{1}{\hbar^2} \left[-b\,(3I_z^2 - \mathbf{I}^2) + \frac{c}{2}(I_+^2 + I_-^2) \right] = A\,(3I_z^2 - \mathbf{I}^2) + B\,(I_+^2 + I_-^2) \,,$$

where we define

$$A = -\frac{eQ}{2I(2I-1)} \frac{a_x + a_y}{2\,\hbar^2} \,, \qquad B = \frac{eQ}{2I(2I-1)} \frac{a_x - a_y}{4\hbar^2} \,.$$

In order to determine the matrix representation of H_0 in the basis $|m\rangle$, we note that

$$I_+^2 |m\rangle = \hbar\sqrt{15/4 - m(m+1)}\, I_+ |m+1\rangle = \hbar^2\,\sqrt{15/4 - m(m+1)}\,\sqrt{15/4 - (m+1)(m+2)}\,|m+2\rangle \,,$$

or, explicitly,

$$I_+^2\,|3/2\rangle = I_+^2\,|1/2\rangle = 0 \,, \qquad I_+^2\,|-1/2\rangle = 2\sqrt{3}\,\hbar^2\,|3/2\rangle \,, \qquad I_+^2\,|-3/2\rangle = 2\sqrt{3}\,\hbar^2|1/2\rangle \,.$$

The matrix representing I_+^2 is given by

$$\underline{I_+^2} = 2\sqrt{3}\,\hbar^2 \begin{pmatrix} 0 & 0 & 1 & 0 \\ 0 & 0 & 0 & 1 \\ 0 & 0 & 0 & 0 \\ 0 & 0 & 0 & 0 \end{pmatrix} \,.$$

Since $I_-^2 = (I_+^2)^\dagger$, the matrix representing I_-^2 is obtained by taking the adjoint of the above. The expression $3I_z^2 - \mathbf{I}^2$ is diagonal in the basis $|m\rangle$, and hence the matrix representing H_0 is obtained as

$$\underline{H_0} = 3A\hbar^2 \begin{pmatrix} 1 & 0 & 0 & 0 \\ 0 & -1 & 0 & 0 \\ 0 & 0 & -1 & 0 \\ 0 & 0 & 0 & 1 \end{pmatrix} + 2\sqrt{3}\,B\hbar^2 \begin{pmatrix} 0 & 0 & 1 & 0 \\ 0 & 0 & 0 & 1 \\ 1 & 0 & 0 & 0 \\ 0 & 1 & 0 & 0 \end{pmatrix}$$

It is convenient to change basis as (in practice, relabeling the basis states)

$$|1\rangle = |3/2\rangle \,, \qquad |2\rangle = |-1/2\rangle \,, \qquad |3\rangle = |-3/2\rangle \,, \qquad |4\rangle = |1/2\rangle \,.$$

The unitary (orthogonal in this case) matrix which induces the transformation from the $|m\rangle$ basis to the $|i\rangle$ basis is given by

$$S_{mi} = \langle m|i\rangle \implies \underline{S} = \begin{pmatrix} 1 & 0 & 0 & 0 \\ 0 & 0 & 0 & 1 \\ 0 & 1 & 0 & 0 \\ 0 & 0 & 1 & 0 \end{pmatrix} \,.$$

In the new basis, the matrix $\underline{H}'_0$ is given by

$$\underline{H}'_0 = \underline{S}^\dagger \, \underline{H}_0 \, \underline{S} \implies \underline{H}'_0 = 3A\hbar^2 \begin{pmatrix} 1 & 0 & 0 & 0 \\ 0 & -1 & 0 & 0 \\ 0 & 0 & 1 & 0 \\ 0 & 0 & 0 & -1 \end{pmatrix} + 2\sqrt{3}B\hbar^2 \begin{pmatrix} 0 & 1 & 0 & 0 \\ 1 & 0 & 0 & 0 \\ 0 & 0 & 0 & 1 \\ 0 & 0 & 1 & 0 \end{pmatrix} .$$

It is now clear that the matrix $\underline{H}'_0$ breaks down into two identical 2×2 submatrices, each given by

$$\underline{h}'_0 = 3A\hbar^2 \begin{pmatrix} 1 & 0 \\ 0 & -1 \end{pmatrix} + 2\sqrt{3}B\hbar^2 \begin{pmatrix} 0 & 1 \\ 1 & 0 \end{pmatrix} .$$

The eigenvalues of $\underline{h}'_0$ result from the characteristic equation

$$(\epsilon - 3A\hbar^2)(\epsilon + 3A\hbar^2) - (2\sqrt{3}B\hbar^2)^2 = 0 \implies \epsilon_\pm = \pm\hbar^2\sqrt{9A^2 + 12B^2} ,$$

and therefore the eigenvalues of $\underline{H}'_0$ are $\epsilon_\pm$ and each is two-fold degenerate. Before proceeding further, it is convenient to define

$$\cos\varphi = \frac{3A}{\sqrt{9A^2 + 12B^2}} , \qquad \sin\varphi = \frac{2\sqrt{3}B}{\sqrt{9A^2 + 12B^2}}$$

The eigenstates corresponding to $\epsilon_\pm$ are obtained from

$$\begin{pmatrix} \epsilon_\pm - 3A\hbar^2 & -2\sqrt{3}B\hbar^2 \\ -2\sqrt{3}B\hbar^2 & \epsilon_\pm + 3A\hbar^2 \end{pmatrix} \begin{pmatrix} a_\pm \\ b_\pm \end{pmatrix} = 0 \implies (\epsilon_\pm - 3A\hbar^2)a_\pm = 2\sqrt{3}B\hbar^2 \, b_\pm ,$$

which in terms of the definition above can be written simply as

$$(\pm 1 - \cos\varphi)a_\pm = \sin\varphi \, b_\pm \implies a_+ = \frac{\sin\varphi}{1 - \cos\varphi} b_+ \quad \text{and} \quad a_- = -\frac{\sin\varphi}{1 + \cos\varphi} b_- .$$

The degenerate (normalized) eigenstates of H'_0 and the original Hamiltonian H_0 corresponding to ϵ_+ are

$$|+; 1\rangle = \cos(\varphi/2)\,|1\rangle + \sin(\varphi/2)\,|2\rangle = \cos(\varphi/2)|3/2\rangle + \sin(\varphi/2)\,|-1/2\rangle ,$$

$$|+; 2\rangle = \cos(\varphi/2)\,|3\rangle + \sin(\varphi/2)\,|4\rangle = \cos(\varphi/2)|-3/2\rangle + \sin(\varphi/2)\,|1/2\rangle ,$$

while those corresponding to ϵ_- are

$$|-; 1\rangle = -\sin(\varphi/2)\,|1\rangle + \cos(\varphi/2)\,|2\rangle = -\sin(\varphi/2)|3/2\rangle + \cos(\varphi/2)\,|-1/2\rangle ,$$

$$|-; 2\rangle = -\sin(\varphi/2)\,|3\rangle + \cos(\varphi/2)\,|4\rangle = -\sin(\varphi/2)|-3/2\rangle + \cos(\varphi/2)\,|1/2\rangle .$$

Part 3

In the presence of the magnetic field $\mathbf{B}_0$, there is an additional interaction given by

$$V = -\boldsymbol{\mu} \cdot \mathbf{B}_0 = \omega_0 \, \hat{\mathbf{n}} \cdot \mathbf{I} \qquad \text{with} \quad \omega_0 = -\gamma \, |\mathbf{B}_0| ,$$

with unit vector $\hat{\mathbf{n}} = (\cos\theta \sin\phi, \sin\theta \sin\phi, \cos\phi)$. This interaction can be written as

$$V = \omega_0 \left[\frac{1}{2} \sin\theta \left(e^{-i\phi} I_+ + e^{i\phi} I_- \right) + \cos\theta \, I_z \right] .$$

In order to determine the first-order corrections induced by V, we need to diagonalize V in the subspaces $\pm$ of the degenerate eigenvalues,

$$\underline{V}|_{\pm} = \begin{pmatrix} \langle \pm; 1|V|\pm; 1\rangle & \langle \pm; 1|V|\pm; 2\rangle \\ \langle \pm; 2|V|\pm; 1\rangle & \langle \pm; 2|V|\pm; 2\rangle \end{pmatrix} .$$

We find for the diagonal matrix elements

$$\langle +; 1|V|+; 1\rangle = \frac{\hbar\omega_0}{2} \cos\theta \left[3\cos^2(\varphi/2) - \sin^2(\varphi/2) \right] \equiv \alpha ,$$

$$\langle +; 2|V|+; 2\rangle = -\frac{\hbar\omega_0}{2} \cos\theta \left[3\cos^2(\varphi/2) - \sin^2(\varphi/2) \right] \equiv -\alpha .$$

Before evaluating the off-diagonal matrix elements, we note that

$$I_+|+; 2\rangle = \sqrt{3}\,\hbar \left[\cos(\varphi/2)\,|-1/2\rangle + \sin(\varphi/2)\,|3/2\rangle \right] ,$$

$$I_-|+; 2\rangle = 2\hbar\,\sin(\varphi/2)\,|-1/2\rangle ,$$

and hence

$$\langle +; 1|V|+; 2\rangle = \hbar\omega_0 \sin\theta \left[\sqrt{3}\,\sin(\varphi/2)\,\cos(\varphi/2)\,e^{-i\phi} + \sin^2(\varphi/2)\,e^{i\phi} \right] \equiv \beta ,$$

and the matrix element $\langle +; 2|V|+; 1\rangle$ is obtained by taking the complex conjugate of the expression above. The matrix elements of V in the $(-)$ subspace can be obtained from those above by making the replacements

$$\cos(\varphi/2) \longrightarrow -\sin(\varphi/2) , \qquad \sin(\varphi/2) \longrightarrow \cos(\varphi/2)$$

in the expressions above. In the subspace $(+)$, the matrix V has the form

$$\underline{V}|_+ = \begin{pmatrix} \alpha & \beta \\ \beta^* & -\alpha \end{pmatrix} ,$$

which has eigenvalues $\pm\sqrt{\alpha^2 + |\beta|^2}$.

Part 4

Assuming the magnetic field to be along the z-axis, we find the perturbation is diagonal,

$$\underline{V}|_+ = \begin{pmatrix} \alpha & 0 \\ 0 & -\alpha \end{pmatrix} , \qquad \underline{V}|_- = \begin{pmatrix} \gamma & 0 \\ 0 & -\gamma \end{pmatrix} ,$$

where

$$\alpha = \frac{\hbar\omega_0}{2} \left[3\cos^2(\varphi/2) - \sin^2(\varphi/2) \right] , \qquad \gamma = \frac{\hbar\omega_0}{2} \left[3\sin^2(\varphi/2) - \cos^2(\varphi/2) \right] .$$

Therefore the energies of the states $|+; 1\rangle$ and $|+; 2\rangle$ are

$$E_{+,1} = \epsilon_+ + \alpha , \qquad E_{+,2} = \epsilon_+ - \alpha ,$$

and those of the states $|-; 1\rangle$ and $|-; 2\rangle$ are

$$E_{-,1} = \epsilon_- + \gamma , \qquad E_{-,2} = \epsilon_- - \gamma .$$

Suppose that, at time $t = 0$, the system is in the state

$$|\psi(0)\rangle = \sum_{p=\pm} \sum_{i=1,2} c_{p,i} |p;i\rangle, \qquad c_{p,i} = \langle p;i|\psi(0)\rangle.$$

It will evolve in time according to (here $H = H_0 + V$)

$$|\psi(t)\rangle = e^{-iHt/\hbar}|\psi(0)\rangle \approx \sum_{p=\pm} \sum_{i=1,2} c_{p,i} e^{-iE_{p,i}t/\hbar} |p;i\rangle.$$

The expectation value of I_x on this state is given by

$$\langle I_x(t)\rangle = \sum_{p,p'=\pm} \sum_{i,i'=1,2} c_{p',i'}^* c_{p,i} e^{iE_{p',i'}t/\hbar} e^{-iE_{p,i}t/\hbar} \langle p';i'|I_x|p;i\rangle.$$

Since I_x increases or decreases the m state by one unit, we see that the only non-vanishing matrix elements must be in general

$$\langle I_x(t)\rangle = \sum_{p,p'=\pm} \left(c_{p',1}^* c_{p,2} e^{iE_{p',1}t/\hbar} e^{-iE_{p,2}t/\hbar} \langle p';1|I_x|p;2\rangle + \text{c.c.} \right),$$

or, more explicitly

$$\langle I_x(t)\rangle = \left(c_{+,1}^* c_{+,2} e^{iE_{+,1}t/\hbar} e^{-iE_{+,2}t/\hbar} \langle +;1|I_x|+;2\rangle + c_{-,1}^* c_{+,2} e^{iE_{-,1}t/\hbar} e^{-iE_{+,2}t/\hbar} \langle -;1|I_x|+;2\rangle \right.$$
$$\left. + c_{+,1}^* c_{-,2} e^{iE_{+,1}t/\hbar} e^{-iE_{-,2}t/\hbar} \langle +;1|I_x|-;2\rangle + c_{-,1}^* c_{-,2} e^{iE_{-,1}t/\hbar} e^{-iE_{-,2}t/\hbar} \langle -;1|I_x|-;2\rangle + \text{c.c.} \right),$$

from which we can read off the Bohr frequencies

$$\omega_+ = \frac{E_{+,1} - E_{+,2}}{\hbar} = \frac{2\alpha}{\hbar}, \qquad \omega_- = \frac{E_{-,1} - E_{-,2}}{\hbar} = \frac{2\gamma}{\hbar},$$

and so

$$\omega_{+-} = -\omega_{-+} = \frac{E_{+,1} - E_{-,2}}{\hbar} \approx 2\hbar\sqrt{9A^2 + 12B^2}.$$

Thus, measuring transitions between these states yields information on the nuclear quadrupole moment.

Problem 13 Degenerate Perturbation Theory in Second Order: An Example

The orthonormal basis of a system consists of three states $|\phi_1\rangle$, $|\phi_2\rangle$, and $|\phi_3\rangle$. In this basis the unperturbed Hamiltonian H_0 is diagonal,

$$\underline{H_0} = \begin{pmatrix} \epsilon_1 & 0 & 0 \\ 0 & \epsilon_1 & 0 \\ 0 & 0 & \epsilon_2 \end{pmatrix}, \qquad \epsilon_2 > \epsilon_1.$$

A perturbation V is present, which in the basis $|\phi_1\rangle$, $|\phi_2\rangle$, $|\phi_3\rangle$ is represented by the following matrix:

$$\underline{V} = \begin{pmatrix} 0 & 0 & a \\ 0 & 0 & b \\ a^* & b^* & 0 \end{pmatrix}, \qquad |a| \approx |b| \ll \epsilon_2 - \epsilon_1.$$

1. Obtain the exact energy eigenvalues of $H_0 + V$.
2. Use perturbation theory to obtain the corrections up to second order to the energy levels ϵ_1 and ϵ_2. Compare your results with the exact results obtained above.

(Adapted from J. J. Sakurai and J. Napolitano (2020), *Modern Quantum Mechanics*, Cambridge University Press.)

Solution

Part 1

We find by direct diagonalization

$$\det(\underline{H}_0 + \underline{V} - E\underline{I}) = 0 \implies \det\begin{pmatrix} \epsilon_1 - E & 0 & a \\ 0 & \epsilon_1 - E & b \\ a^* & b^* & \epsilon_2 - E \end{pmatrix} = 0 ,$$

which has the solutions

$$E_0 = \epsilon_1 , \qquad E_\pm = \frac{\epsilon_2 + \epsilon_1}{2} \pm \sqrt{\frac{(\epsilon_2 - \epsilon_1)^2}{4} + |a|^2 + |b|^2} .$$

Note that, in the limit $|a|^2, |b|^2 \ll \epsilon_2 - \epsilon_1$, we have

$$E_\pm = \frac{\epsilon_2 + \epsilon_1}{2} \pm \frac{\epsilon_2 - \epsilon_1}{2} \sqrt{1 + 4\frac{|a|^2 + |b|^2}{(\epsilon_2 - \epsilon_1)^2}} \approx \frac{\epsilon_2 + \epsilon_1}{2} \pm \frac{\epsilon_2 - \epsilon_1}{2} \left[1 + 2\frac{|a|^2 + |b|^2}{(\epsilon_2 - \epsilon_1)^2} \right] ,$$

yielding

$$E_+ \approx \epsilon_2 + \frac{|a|^2 + |b|^2}{\epsilon_2 - \epsilon_1} , \qquad E_- \approx \epsilon_1 - \frac{|a|^2 + |b|^2}{\epsilon_2 - \epsilon_1} .$$

Part 2

We see that H_0 has an eigenvalue ϵ_1 with degeneracy $g_1 = 2$ and a non-degenerate eigenvalue ϵ_2. We first evaluate the correction up to second order for the non-degenerate eigenvalue:

$$E_2^{(1)} = \langle \phi_3 | V | \phi_3 \rangle = 0 , \qquad E_2^{(2)} = \sum_{m=1,2} \frac{|\langle \phi_m | V | \phi_3 \rangle|^2}{\epsilon_2 - \epsilon_m} = \frac{|a|^2 + |b|^2}{\epsilon_2 - \epsilon_1} ,$$

where in the last step we used

$$\langle \phi_1 | V | \phi_3 \rangle = a , \qquad \langle \phi_2 | V | \phi_3 \rangle = b ,$$

as it is obvious from the matrix representation of V. For the degenerate case, to obtain the first-order corrections we need to diagonalize V in the subspace spanned by the states $|\phi_1\rangle$ and $|\phi_2\rangle$. However, in this space,

$$\langle \phi_m | V | \phi_n \rangle = 0 \qquad m, n = 1, 2 ,$$

and hence the first-order corrections vanish in this case, and the degeneracy persists. In order to calculate the second-order corrections in the case of degeneracy, we construct a matrix corresponding to the operator Z, where

$$Z = \sum_{m \neq \mathbb{E}_1} \frac{V|\phi_m\rangle\langle\phi_m|V}{\epsilon_1 - \epsilon_m} = \frac{V|\phi_3\rangle\langle\phi_3|V}{\epsilon_1 - \epsilon_2} ,$$

in the subspace spanned by $|\phi_1\rangle$ and $|\phi_2\rangle$); the matrix elements of Z are given by

$$\langle \phi_1 | Z | \phi_1 \rangle = \frac{|a|^2}{\epsilon_1 - \epsilon_2} , \qquad \langle \phi_2 | Z | \phi_2 \rangle = \frac{|b|^2}{\epsilon_1 - \epsilon_2} , \qquad \langle \phi_1 | Z | \phi_2 \rangle = \frac{ab^*}{\epsilon_1 - \epsilon_2} , \qquad \langle \phi_2 | Z | \phi_1 \rangle = \frac{a^*b}{\epsilon_1 - \epsilon_2} ,$$

and the matrix is given by

$$Z = \frac{1}{\epsilon_1 - \epsilon_2} \begin{pmatrix} |a|^2 & ab^* \\ a^*b & |b|^2 \end{pmatrix} ,$$

which has eigenvalues

$$E_0^{(2)} = 0 , \qquad E_-^{(2)} = \frac{|a|^2 + |b|^2}{\epsilon_1 - \epsilon_2} .$$

Thus, the second-order corrections are in agreement with the second-order expansions obtained in part 1.

Problem 14   Charged Particle in Anisotropic Harmonic Oscillator Potential and Uniform Magnetic Field

Consider a charged particle (mass m and charge q) in an anisotropic three-dimensional oscillator potential with $\omega_x \neq \omega_y = \omega_z = \omega$. The particle is also exposed to a uniform magnetic field in the $\hat{\mathbf{x}}$-direction, $\mathbf{B} = B\,\hat{\mathbf{x}}$.

1. Write down the Hamiltonian for this system.
2. Assuming that the magnetic field is weak and the anisotropy of the oscillator potential is small (namely, $|\omega_x - \omega| \ll \omega$), split the Hamiltonian obtained in part 1 into an unperturbed term H_0 and a perturbative term V.
3. Assuming that the Zeeman splitting is comparable with the energy splitting produced by the anisotropy but small compared with $\hbar\omega$, calculate the first-order corrections to the energy of the first excited level. Does the perturbation remove the degeneracy completely?
4. Suppose that the oscillator potential is isotropic and that the term quadratic in the magnetic field is negligible. What are the exact eigenvalues and eigenstates of the resulting Hamiltonian?

Solution

Part 1

We take the vector potential $\mathbf{A} = \mathbf{B} \times \mathbf{r}/2$, and the Hamiltonian then reads

$$H = \frac{\mathbf{p}^2}{2m} - \frac{q}{2mc}\left(\mathbf{A}\cdot\mathbf{p} + \mathbf{p}\cdot\mathbf{A}\right) + \frac{q^2}{2mc^2}\mathbf{A}^2 + \frac{m\omega_x^2}{2}x^2 + \frac{m\omega^2}{2}(y^2 + z^2) ,$$

where

$$\mathbf{A}\cdot\mathbf{p} + \mathbf{p}\cdot\mathbf{A} = \mathbf{B}\cdot\mathbf{L} , \qquad \mathbf{A}^2 = \frac{1}{4}\left[\mathbf{B}^2\,\mathbf{r}^2 - (\mathbf{B}\cdot\mathbf{r})^2\right] .$$

Since the $\mathbf{B}$ field is along $\hat{\mathbf{x}}$, the Hamiltonian is given by

$$H = \frac{\mathbf{p}^2}{2m} - \frac{qB}{2mc}L_x + \frac{m\omega_x^2}{2}x^2 + \frac{m}{2}\left(\omega^2 + \frac{q^2 B^2}{4m^2 c^2}\right)(y^2 + z^2) .$$

Part 2

After defining the angular frequencies

$$\omega_L = \frac{qB}{2mc} , \qquad \Omega = \sqrt{\omega^2 + \omega_L^2} ,$$

we can express the Hamiltonian H_0 and perturbation V as

$$H_0 = \frac{\mathbf{p}^2}{2m} + \frac{m\Omega^2}{2}\mathbf{r}^2 , \qquad V = -\omega_L L_x + \frac{m}{2}(\omega_x^2 - \Omega^2)x^2 .$$

Part 3

A basis of H_0 eigenstates is given by the kets $|n_x, n_y, n_z\rangle$ with eigenvalues

$$\epsilon_n = \hbar\Omega\,(n + 3/2) , \qquad n = n_x + n_y + n_z = 0, 1, 2, \dots ,$$

and degeneracy $g_n = (n + 1)(n + 2)/2$. While the ground-state level $(n = 0)$ is non-degenerate, the first excited level $(n = 1)$ is three-fold degenerate. To construct the matrix $\underline{V}$ in the degenerate subspace spanned by the states $|1, 0, 0\rangle, |0, 1, 0\rangle, |0, 0, 1\rangle$, it is convenient to first express the perturbation in terms of raising and lowering operators. Recalling that

$$r_\alpha = \sqrt{\frac{\hbar}{2m\Omega}}\,(a_\alpha^\dagger + a_\alpha) , \qquad p_\alpha = i\sqrt{\frac{m\Omega\hbar}{2}}\,(a_\alpha^\dagger - a_\alpha) , \qquad \alpha = x, y, z ,$$

we find

$$x^2 = \frac{\hbar}{2m\Omega}(a_x^\dagger a_x^\dagger + a_x a_x + \underbrace{a_x^\dagger a_x + a_x a_x^\dagger}_{2N_x+1}) , \qquad L_x = i\hbar(a_z^\dagger a_y - a_y^\dagger a_z) ,$$

where $N_x = a_x^\dagger a_x$ is the number operator. The perturbation consists of the two terms V_x and V_{yz},

$$V = V_x + V_{yz} , \qquad V_x = \hbar\frac{\omega_x^2 - \Omega^2}{4\Omega}(a_x^\dagger a_x^\dagger + a_x a_x + 2N_x + 1) , \qquad V_{yz} = i\hbar\omega_L(a_y^\dagger a_z - a_z^\dagger a_y) .$$

In the degenerate subspace, the matrix $\underline{V}_x$ is given by

$$\langle n_x', n_y', n_z'|V_x|n_x, n_y, n_z\rangle = \delta_{n_y',n_y}\,\delta_{n_z',n_z}\,\langle n_x'|V_x|n_x\rangle \qquad \text{with} \qquad n_x, n_x' = 0, 1 ,$$

where

$$\langle 0|V_x|0\rangle = \frac{\hbar}{4}\frac{\omega_x^2 - \Omega^2}{\Omega} , \qquad \langle 1|V_x|1\rangle = \frac{3\hbar}{4}\frac{\omega_x^2 - \Omega^2}{\Omega} , \qquad \langle 0|V_x|1\rangle = \langle 1|V_x|0\rangle = 0 ,$$

and

$$\underline{V}_x = \frac{\hbar}{4}\frac{\omega_x^2 - \Omega^2}{\Omega}\begin{pmatrix} 3 & 0 & 0 \\ 0 & 1 & 0 \\ 0 & 0 & 1 \end{pmatrix} ,$$

which is a diagonal matrix. By contrast, the matrix $\underline{V}_{yz}$ has off-diagonal matrix element; we note that

$$\langle n_x', n_y', n_z'|V_{yz}|n_x, n_y, n_z\rangle = \delta_{n_x',n_x}\,\langle n_y', n_z'|V_{yz}|n_y, n_z\rangle \qquad \text{with} \qquad n_y, n_z, n_y', n_z' = 0, 1 ,$$

where

$$\langle n_y', n_z'|V_{yz}|n_y, n_z\rangle = i\hbar\omega_L\left(\sqrt{n_y + 1}\,\sqrt{n_z}\,\delta_{n_y',n_y+1}\,\delta_{n_z',n_z-1} - \sqrt{n_y}\,\sqrt{n_z + 1}\,\delta_{n_y',n_y-1}\,\delta_{n_z',n_z+1}\right) ,$$

from which we see that the only non-vanishing matrix elements (for the subspace) are

$$\langle 1, 0|V_{yz}|0, 1\rangle = i\hbar\omega_L , \qquad \langle 0, 1|V_{yz}|1, 0\rangle = -i\hbar\omega_L$$

where

$$\underline{V}_{yz} = i\hbar\omega_L \begin{pmatrix} 0 & 0 & 0 \\ 0 & 0 & 1 \\ 0 & -1 & 0 \end{pmatrix} .$$

The complete matrix can be written as

$$\underline{V} = \begin{pmatrix} 3a & 0 & 0 \\ 0 & a & ib \\ 0 & -ib & a \end{pmatrix} , \qquad a = \frac{\hbar}{4} \frac{\omega_x^2 - \Omega^2}{\Omega} , \qquad b = \hbar\omega_L ,$$

and the first-order corrections to the first excited levels are obtained as

$$E_0^{(1)} = 3a , \qquad E_\pm^{(1)} = a \pm |b| ;$$

the perturbation completely removes the degeneracy unless the magnetic field strength is tuned so as to make $3a$ equal to $a + |b|$ or $a - |b|$.

Part 4

If the term quadratic in the magnetic field is neglected then $\Omega = \omega$; under the further assumption $\omega_x \approx \omega$, we find

$$a = \frac{\hbar}{4} \frac{\omega_x^2 - \omega^2}{\omega} \approx \frac{\hbar}{2} (\omega_x - \omega) ,$$

to linear terms in the anisotropy. The linear term in the magnetic field and the anisotropy term give comparable corrections if

$$|a| \approx |b| \implies |\omega_x - \omega| \approx \left| \frac{qB}{mc} \right| ,$$

in essence, if the difference in angular frequencies is of the same order as the Larmor frequency ω_L. If the anisotropy term can be neglected relative to the Larmor frequency term then the Zeeman splittings of the level are given by

$$E_m^{(1)} = \hbar\omega_L m \qquad m = 0, \pm 1 .$$

Note that in such a limit the (full) Hamiltonian reads

$$H = \frac{\mathbf{p}^2}{2m} + \frac{m\omega^2 \mathbf{r}^2}{2} - \omega_L L_x ,$$

ignoring the term quadratic in B. This Hamiltonian commutes with $\mathbf{L}^2$ and L_x and can be diagonalized simultaneously along with these observables. We consider the states $|\psi_{nlm}\rangle$, which are eigenstates of H, $\mathbf{L}^2$, and L_x, to obtain

$$H|\psi_{nlm}\rangle = \underbrace{[\hbar\omega(n + 3/2) - \hbar\omega_L m]}_{E_{nm}} |\psi_{nlm}\rangle$$

and

$$\mathbf{L}^2|\psi_{nlm}\rangle = l(l+1)\hbar^2|\psi_{nlm}\rangle , \qquad L_x|\psi_{nlm}\rangle = m\hbar|\psi_{nlm}\rangle .$$

Here $n = 0, 1, 2, \ldots$ with $n = p + l$, and the allowed p are the even positive integers $p = 0, 2, 4, \ldots$, whereas $l = 0, 1, 2, \ldots$ (for an isotropic harmonic oscillator in three dimensions).

Problem 15 Hydrogen Atom in a Uniform Electric Field: Induced Electric Dipole Moment

Consider a hydrogen atom exposed to a uniform electric field $\mathcal{E}\,\hat{z}$. Ignore spin degrees of freedom.

1. Calculate the corrections to the ground-state energy level up to second order in perturbation theory. You may neglect the contribution from the continuum states in the second-order calculation.
2. Calculate the electric dipole moment $\mathbf{D} = -e\,\mathbf{r}$ induced by the electric field for the ground state of hydrogen, and verify that $\mathbf{D}$ has a non-vanishing component only along the $\hat{z}$-direction.
3. Prove the Hellmann–Feynman theorem: it states that, if the Hamiltonian depends on a parameter λ and $|\psi_n(\lambda)\rangle$ is a normalized eigenstate with eigenvalue $E_n(\lambda)$, it follows that

$$\frac{dE_n(\lambda)}{d\lambda} = \langle \psi_n(\lambda)|\frac{dH(\lambda)}{d\lambda}|\psi_n(\lambda)\rangle \,.$$

Apply the theorem to the case of the hydrogen atom in an electric field ($\lambda \longrightarrow \mathcal{E}$), and obtain the following relation for the ground state:

$$\langle \psi_{\mathrm{gs}}|D_z|\psi_{\mathrm{gs}}\rangle = -\frac{dE_{\mathrm{gs}}}{d\mathcal{E}} \,.$$

Verify that it is satisfied by the second-order calculation carried out in part 2.

Solution

Part 1

Up to and including second order, we have

$$E_{\mathrm{gs}} = \epsilon_{\mathrm{gs}} + E_{\mathrm{gs}}^{(1)} + E_{\mathrm{gs}}^{(2)} \,,$$

where

$$E_{\mathrm{gs}}^{(1)} = e\,\mathcal{E}\langle \phi_{100}|z|\phi_{100}\rangle = 0$$

and

$$E_{\mathrm{gs}}^{(2)} = (e\mathcal{E})^2 \left(\sum_{n=2}^{\infty}\sum_{l=0}^{n-1}\sum_{m=-l}^{l} \frac{|\langle \phi_{nlm}|z|\phi_{100}\rangle|^2}{\epsilon_{\mathrm{gs}} - \epsilon_n} + \underbrace{\sum_{l=0}^{\infty}\sum_{m=-l}^{l}\int_0^{\infty} d\epsilon\, \frac{|\langle \phi_{\epsilon lm}|z|\phi_{100}\rangle|^2}{\epsilon_{\mathrm{gs}} - \epsilon}}_{\text{continuum contribution}} \right) \,.$$

The first-order correction vanishes because of the parity selection rule. In the second-order correction we neglect the contribution of the continuum states; these are eigenstates of H_0 with continuum eigenvalues $\epsilon > 0$ and continuum normalization given by $\langle \phi_{\epsilon'l'm'}|\phi_{\epsilon lm}\rangle = \delta(\epsilon - \epsilon')\,\delta_{l,l'}\,\delta_{m,m'}$ (note that the sum over l is unrestricted for these states). Since the ground state is an s-wave, states contributing to $E_{\mathrm{gs}}^{(2)}$ must have $m = 0$ (the perturbation commutes with L_z). We are thus led to consider matrix elements of the type

$$\langle \phi_{nl0}|z|\phi_{100}\rangle = 0 \,, \qquad l \text{ even} \,,$$

and so l must be odd, again because of the parity selection rule. The (real) wave functions are

$$\phi_{100}(r) = R_{1s}(r) \underbrace{\frac{1}{\sqrt{4\pi}}}_{Y_{00}(\Omega)} , \qquad \phi_{nl0}(r) = R_{nl}(r) \underbrace{\sqrt{\frac{2l+1}{4\pi}} P_l(\cos\theta)}_{Y_{l0}(\Omega)} ,$$

and hence

$$\langle \phi_{nl0} | z | \phi_{100} \rangle = \frac{1}{\sqrt{4\pi}} \sqrt{\frac{2l+1}{4\pi}} \int_0^\infty dr\, r^3 R_{nl}(r) R_{1s}(r) \int_{-1}^1 d(\cos\theta) P_l(\cos\theta) \underbrace{\cos\theta}_{P_1(\cos\theta)} \int_0^{2\pi} d\phi$$

$$= \frac{1}{\sqrt{3}} \delta_{l,1} \underbrace{\int_0^\infty dr\, r^3 R_{np}(r) R_{1s}(r)}_{\gamma_n} ,$$

where in the first line we have used $P_1(\cos\theta) = \cos\theta$, and in the second the orthogonality of the Legendre polynomials. We see that only p-waves contribute. Ignoring continuum contributions, the second-order correction is then given by

$$E_{gs}^{(2)} = \frac{(e\mathcal{E})^2}{3} \sum_{n=2}^\infty \frac{|\gamma_n|^2}{\epsilon_{gs} - \epsilon_n} .$$

Part 2

The ground-state electric dipole is defined as the expectation value

$$\langle \mathbf{D} \rangle = -e \left(\langle \phi_{1s} | + \langle \psi_{1s}^{(1)} | \right) \mathbf{r} \left(|\phi_{1s}\rangle + |\psi_{1s}^{(1)}\rangle \right) ,$$

where $|\psi_1^{(1)}\rangle$ is the first-order correction to the unperturbed ground state $|\phi_{1s}\rangle$, and is given by

$$|\psi_1^{(1)}\rangle = e\mathcal{E} \sum_{n=2}^\infty |\phi_{np}\rangle \frac{\langle \phi_{np} | z | \phi_{1s} \rangle}{\epsilon_{gs} - \epsilon_n} .$$

The expectation value of $\mathbf{r}$ on the unperturbed state vanishes (owing to the parity selection rule), so that up to terms linear in the external electric field we have (noting that the term quadratic in $\mathcal{E}$ also vanishes, since it involves matrix elements of the type $\langle \phi_{n'p} | \mathbf{r} | \phi_{np} \rangle$ between states of the same parity)

$$\langle \mathbf{D} \rangle = -e^2 \mathcal{E} \sum_{n=2}^\infty \left(\langle \phi_{1s} | \mathbf{r} | \phi_{np} \rangle \frac{\langle \phi_{np} | z | \phi_{1s} \rangle}{\epsilon_{gs} - \epsilon_n} + \langle \phi_{np} | \mathbf{r} | \phi_{1s} \rangle \frac{\langle \phi_{1s} | z | \phi_{np} \rangle}{\epsilon_{gs} - \epsilon_n} \right) .$$

In the above matrix elements only the z component of $\mathbf{r}$ survives, since for $x = r \sin\theta \cos\phi$ we have

$$\langle \phi_{np} | x | \phi_{1s} \rangle = \frac{\sqrt{3}}{4\pi} \int_0^\infty dr\, r^3 R_{np}(r) R_{1s}(r) \int_{-1}^1 d(\cos\theta) P_1(\cos\theta) \sin\theta \int_0^{2\pi} d\phi \cos\phi = 0 ,$$

and similarly for the y component. Thus we arrive at

$$\langle \mathbf{D} \rangle = \alpha \mathcal{E} \hat{\mathbf{z}} , \qquad \alpha = -2e^2 \sum_{n=2}^\infty \frac{|\langle \phi_{np} | z | \phi_{1s} \rangle|^2}{\epsilon_{gs} - \epsilon_n} = -\frac{2}{3} e^2 \sum_{n=2}^\infty \frac{|\gamma_n|^2}{\epsilon_{gs} - \epsilon_n} ,$$

where α is the induced polarization of the hydrogen atom ground state.

Part 3

The theorem follows easily, since the normalization condition gives

$$\langle\psi_n(\lambda)|\psi_n(\lambda)\rangle = 1 \implies 0 = \frac{d}{d\lambda}\langle\psi_n(\lambda)|\psi_n(\lambda)\rangle = \langle\psi_n'(\lambda)|\psi_n(\lambda)\rangle + \langle\psi_n(\lambda)|\psi_n'(\lambda)\rangle \,,$$

where we have denoted as $|\psi_n'(\lambda)\rangle$ the state obtained by taking the derivative of $|\psi_n(\lambda)\rangle$ with respect to λ. This result in turn leads to

$$\frac{dE_n(\lambda)}{d\lambda} = \frac{d}{d\lambda}\langle\psi_n(\lambda)|H(\lambda)|\psi_n(\lambda)\rangle = \langle\psi_n'(\lambda)|H(\lambda)|\psi_n(\lambda)\rangle + \langle\psi_n(\lambda)|H(\lambda)|\psi_n'(\lambda)\rangle + \langle\psi_n(\lambda)|H'(\lambda)|\psi_n(\lambda)\rangle$$

$$= E_n(\lambda)\underbrace{\left[\langle\psi_n'(\lambda)|\psi_n(\lambda)\rangle + \langle\psi_n(\lambda)|\psi_n'(\lambda)\rangle\right]}_{\text{vanishes}} + \langle\psi_n(\lambda)|H'(\lambda)|\psi_n(\lambda)\rangle \,.$$

In our case we have, with $\lambda \longrightarrow \mathcal{E}$,

$$\frac{dE_{\mathrm{gs}}(\mathcal{E})}{d\mathcal{E}} = \langle\psi_{\mathrm{gs}}|\frac{dH(\mathcal{E})}{d\mathcal{E}}|\psi_{\mathrm{gs}}\rangle \,,$$

where we have denoted the exact ground state in the presence of the electric field simply by $|\psi_{\mathrm{gs}}\rangle$ and

$$H(\mathcal{E}) = \frac{\mathbf{p}^2}{2\mu} - \frac{e^2}{r} + e\mathcal{E}z \implies \frac{dH(\mathcal{E})}{d\mathcal{E}} = ez \implies e\langle\psi_{\mathrm{gs}}|z|\psi_{\mathrm{gs}}\rangle = \frac{dE_{\mathrm{gs}}(\mathcal{E})}{d\mathcal{E}} \,;$$

hence in perturbation theory

$$\langle D_z\rangle \approx -\frac{\partial}{\partial\mathcal{E}}\left(\epsilon_{\mathrm{gs}} + E_{\mathrm{gs}}^{(1)} + E_{\mathrm{gs}}^{(2)}\right) \,,$$

which is easily verified.

Problem 16   A Symmetric Rotator under the Influence of a Small Perturbation

A system with only angular degrees of freedom θ and ϕ is described by the following Hamiltonian:

$$H = \frac{\mathbf{L}^2}{2I} + \lambda\,\hbar^2\cos(2\phi) \,,$$

where I is the moment of inertia and λ is a (positive) parameter such that $\lambda I \ll 1$. Considering only the S- and P-wave levels, calculate their energies in first-order perturbation theory, and work out the corresponding zeroth-order energy eigenfunctions.

Solution

The unperturbed Hamiltonian is taken as $H_0 = \mathbf{L}^2/(2I)$. It has eigenvalues $\epsilon_l = l(l+1)\hbar^2/(2I)$ with degeneracy $g_l = 2l+1$ and corresponding eigenstates $|\psi_{lm}\rangle$, which, in the coordinate representation, are given by the spherical harmonics (here m is the azimuthal quantum number, the eigenvalue of L_z being $m\hbar$). Hence, the ground-state (S-wave) level is non-degenerate, whereas the first excited (P-wave) level is three-fold degenerate. The first-order correction to the ground state is

$$E_{\mathrm{S}}^{(1)} = \langle\psi_{00}|V|\psi_{00}\rangle = \lambda\,\hbar^2\int d\Omega\,Y_{00}^*(\Omega)\cos(2\phi)\,Y_{00}^*(\Omega) = \frac{\lambda\hbar^2}{4\pi}\int_{-1}^{1}dx\int_{0}^{2\pi}d\phi\cos(2\phi) = 0 \,,$$

where Ω denotes the solid angle and hereafter $x = \cos\theta$. For the P-wave level, we need to diagonalize $\underline{V}$ in the degenerate subspace. The matrix elements of $\underline{V}$ are given by

$$V_{m'm} = \langle\psi_{1m'}|V|\psi_{1m}\rangle = \frac{\lambda\hbar^2}{2}\int_{-1}^{1}dx\int_{0}^{2\pi}d\phi\, Y_{1m'}^*(x)\,\mathrm{e}^{-im'\phi}\left(\mathrm{e}^{2i\phi}+\mathrm{e}^{-2i\phi}\right)Y_{1m}(x)\,\mathrm{e}^{im\phi}\,,$$

where we have isolated the ϕ dependence of $Y_{lm}(\theta,\phi) \equiv Y_{lm}(x)\,\mathrm{e}^{im\phi}$. The ϕ integration yields

$$\int_{0}^{2\pi}d\phi\,\mathrm{e}^{-im'\phi}\left(\mathrm{e}^{2i\phi}+\mathrm{e}^{-2i\phi}\right)\mathrm{e}^{im\phi}=2\pi(\delta_{m',m+2}+\delta_{m',m-2})\,,$$

and hence the only non-vanishing matrix elements are those connecting $m'=1$ to $m=-1$ and vice versa ($\underline{V}$ is hermitian):

$$\underline{V}=\begin{pmatrix}0 & 0 & a\\ 0 & 0 & 0\\ a^* & 0 & 0\end{pmatrix},$$

where

$$a = \pi\lambda\hbar^2\int_{-1}^{1}dx\,Y_{11}^*(x)\,Y_{1-1}(x) = -\frac{3}{8}\lambda\hbar^2\int_{-1}^{1}dx\,(1-x^2) = -\frac{\lambda\hbar^2}{2}\,.$$

The above matrix has eigenvalues

$$E_{\mathrm{P0}}^{(1)} = 0\,,\qquad E_{\mathrm{P\pm}}^{(1)} = \pm|a|\,,$$

with corresponding eigenstates (the zeroth-order eigenstates of H) given by

$$|\psi_{\mathrm{P\pm}}\rangle = \frac{1}{\sqrt{2}}\left(|\psi_{11}\rangle \mp |\psi_{1-1}\rangle\right)\,,\qquad |\psi_{\mathrm{P0}}\rangle = |\psi_{10}\rangle\,,$$

and the perturbation mixes the states with $m=\pm1$.

Problem 17   Electron in a Harmonic Potential under the Action of a Uniform Electric Field

An electron (mass m and charge $-e$) is bound in a harmonic potential $V(r)=m\omega^2r^2/2$, with wave functions given by

$$\langle\phi_{\mathbf{r}}|\phi_{nlm}\rangle = \phi_{nlm}(\mathbf{r}) = R_{nl}(r)\,Y_{lm}(\Omega)\,.$$

It has an electric dipole moment along the z-axis given by $D_z = -ez$.

1. Show that

$$[[H_0\,,z]\,,z] = -\frac{\hbar^2}{m}\,,\qquad H_0 = \frac{\mathbf{p}^2}{2m}+V(r)\,,$$

and hence obtain the sum rule

$$\sum_{n'l'm'}(\epsilon_n-\epsilon_{n'})|\langle\phi_{n'l'm'}|D_z|\phi_{nlm}\rangle|^2 = -\frac{\hbar^2e^2}{2m}\,,$$

where $\epsilon_n=\hbar\omega(n+3/2)$ are the unperturbed energies.

2. Now suppose the electron also interacts with a uniform and weak electric field $\mathcal{E}\,\hat{z}$. Obtain the first- and second-order corrections to the ground-state energy, and the first-order correction to the ground-state wave function. Show that the expectation value of the harmonic potential in the perturbed ground state is the same as that in the unperturbed ground state, to linear terms in $\mathcal{E}$. Show explicitly that the sum rule is satisfied for the ground state.

3. Under the same conditions as above, obtain the first- and second-order corrections to the first-excited-state energy and show explicitly that the sum rule is satisfied for each $|\phi_{11m}\rangle$ state.

4. Obtain the exact energies for the Hamiltonian $H = H_0 + W$, where W is the interaction term with the electric field, and compare them with the energies obtained in parts 2 and 3.

Hint: You will find the following radial wave functions useful:

$$R_{00}(r) = \frac{2\alpha^{3/4}}{\pi^{1/4}}\,e^{-\alpha r^2/2}\,, \qquad R_{11}(r) = \frac{2\sqrt{2}\,\alpha^{5/4}}{\sqrt{3}\pi^{1/4}}\,r\,e^{-\alpha r^2/2}$$

and

$$R_{20}(r) = \frac{\sqrt{6}\alpha^{3/4}}{\pi^{1/4}}\left(1 - \frac{2\alpha r^2}{3}\right)e^{-\alpha r^2/2}\,, \qquad R_{22}(r) = \frac{4\alpha^{7/4}}{\sqrt{15}\pi^{1/4}}\,r^2\,e^{-\alpha r^2/2}\,,$$

where $\alpha = m\omega/\hbar$. Also, recall that the $R_{nl}(r)$ are orthonormal in the following sense (note the same l values)

$$\int_0^\infty dr\,r^2\,R_{nl}(r)\,R_{pl}(r) = \delta_{np}\,.$$

Further, note that $rR_{00}(r)$ is proportional to $R_{11}(r)$ and that $rR_{11}(r)$ can be written either as a linear combination of $R_{00}(r)$ and $R_{20}(r)$ or as proportional to $R_{22}(r)$. These observations allow one to carry out the radial integrations easily. Finally, we report for convenience the following expressions for the spherical harmonics:

$$Y_{00}(\Omega) = \frac{1}{\sqrt{4\pi}}\,, \qquad Y_{10}(\Omega) = \sqrt{\frac{3}{4\pi}}\,\cos\theta\,, \qquad Y_{1\pm1}(\Omega) = \mp\sqrt{\frac{3}{8\pi}}\,\sin\theta\,e^{\pm i\phi}$$

and

$$Y_{20}(\Omega) = \sqrt{\frac{5}{4\pi}}\,\frac{3\cos^2\theta - 1}{2}\,, \qquad Y_{2\pm1}(\Omega) = \mp\sqrt{\frac{15}{8\pi}}\,\sin\theta\,\cos\theta\,e^{\pm i\phi}\,;$$

the $Y_{2\pm2}(\Omega)$ are not needed. Again, the observation that $Y_{00}(\Omega)\cos\theta$ and $Y_{1m}(\Omega)\cos\theta$ can be written as linear combinations of spherical harmonics simplifies the angular integrations by making use of the orthonormality of the $Y_{lm}(\Omega)$.

Solution

Part 1

The potential energy operator obviously commutes with the z-component of the position operator, while the kinetic energy operator has a non-vanishing commutator with it,

$$[T, z] = \frac{1}{2m}\sum_i [p_i^2, z] = -\frac{i\hbar}{m}\sum_i \delta_{iz}p_i = -\frac{i\hbar}{m}p_z\,,$$

and hence

$$[[T + V(r), z], z] = -\frac{i\hbar}{m}[p_z, z] = -\frac{\hbar^2}{m}.$$

Now, we observe that

$$[[H_0, z], z] = [H_0, z]z - z[H_0, z] = H_0 z^2 - zH_0 z - zH_0 z + z^2 H_0,$$

yielding the expectation value on a generic eigenstate of H_0 as

$$\langle\phi_{nlm}|[[H_0, z], z]|\phi_{nlm}\rangle = 2\epsilon_n\langle\phi_{nlm}|z^2|\phi_{nlm}\rangle - 2\langle\phi_{nlm}|zH_0 z|\phi_{nlm}\rangle.$$

Inserting the expression for the completeness of H_0 eigenstates,

$$\underbrace{\sum_{n=0}^{\infty}\sum_{l=0\,\text{or}\,1}^{[n]}\sum_{m=-l}^{l}|\phi_{nlm}\rangle\langle\phi_{nlm}|}_{\Sigma_{nlm}\ \text{for short}} = \mathbb{1},$$

where

$$\sum_{l=0\,\text{or}\,1}^{[n]} \implies \text{either } l = 0, 2, \ldots, n \text{ if } n \text{ even, or } l = 1, 3, \ldots, n \text{ if } n \text{ odd},$$

into the last term in the expectation value leads to

$$\langle\phi_{nlm}|[[H_0, z], z]|\phi_{nlm}\rangle = 2\epsilon_n\langle\phi_{nlm}|z^2|\phi_{nlm}\rangle - 2\sum_{n'l'm'}\underbrace{\langle\phi_{nlm}|zH_0|\phi_{n'l'm'}\rangle}_{\epsilon_{n'}\langle\phi_{nlm}|z|\phi_{n'l'm'}\rangle}\langle\phi_{n'l'm'}|z|\phi_{nlm}\rangle$$

$$= 2\sum_{n'l'm'}(\epsilon_n - \epsilon_{n'})\langle\phi_{nlm}|z|\phi_{n'l'm'}\rangle\langle\phi_{n'l'm'}|z|\phi_{nlm}\rangle,$$

and hence

$$\sum_{n'l'm'}(\epsilon_n - \epsilon_{n'})|\langle\phi_{n'l'm'}|D_z|\phi_{nlm}\rangle|^2 = -\frac{\hbar^2 e^2}{2m}.$$

Part 2

The term giving the interaction of the particle with the electric field is

$$W = -\mathcal{E}D_z = e\mathcal{E}z.$$

The ground-state energy of the (three-dimensional) harmonic oscillator is non-degenerate. The first-order correction vanishes, since the perturbation W and the ground state are, respectively, odd and even under parity. The second-order correction is

$$E_0^{(2)} = (e\mathcal{E})^2\sum_{n=1}^{\infty}\sum_{l=0\,\text{or}\,1}^{[n]}\sum_{m=-l}^{l}\frac{|\langle\phi_{nlm}|z|\phi_{000}\rangle|^2}{\epsilon_0 - \epsilon_n}.$$

The angular momentum selection rules immediately yield that the matrix element above vanishes if $l, m \neq 1, 0$; indeed, we have

$$
\begin{aligned}
\langle \phi_{nlm} | z | \phi_{000} \rangle &= \int_0^\infty d\mathbf{r} \, \phi_{nlm}^*(\mathbf{r}) \, r \cos\theta \, \phi_{000}(r) \\
&= \int_0^\infty dr \, r^3 \, R_{nl}(r) \, R_{00}(r) \int d\Omega \, Y_{lm}^*(\Omega) \underbrace{\cos\theta \, Y_{00}(\Omega)}_{Y_{10}(\Omega)/\sqrt{3}} \\
&= \frac{1}{\sqrt{3}} \, \delta_{l,1} \, \delta_{m,0} \int_0^\infty dr \, r^3 \, R_{n1}(rr) \, R_{00}(r) \,.
\end{aligned}
$$

Further, the radial integral can be easily performed by noting that

$$
r R_{00}(r) = \sqrt{\frac{3}{2\alpha}} \, R_{11}(r)
$$

and, by using the orthonormality of the $R_{nl}(r)$ and $R_{n'l}(r)$ for $n \neq n'$ (but with the same quantum number l),

$$
\int_0^\infty dr \, r^2 \, R_{nl}(r) \, R_{pl}(r) = \delta_{np} \,,
$$

we finally obtain

$$
\langle \phi_{nlm} | z | \phi_{000} \rangle = \frac{1}{\sqrt{2\alpha}} \, \delta_{n,1} \, \delta_{l,1} \, \delta_{m,0} \,.
$$

In the sum over states only a single term survives and we are left with

$$
E_0^{(2)} = \frac{(e\mathcal{E})^2}{2\alpha} \frac{1}{\epsilon_0 - \epsilon_1} = -\frac{(e\mathcal{E})^2}{2m\omega^2} \,.
$$

The first-order correction to the ground state follows easily:

$$
|\psi_{000}\rangle^{(1)} = e\mathcal{E} \sum_{nlm} |\phi_{nlm}\rangle \frac{\langle \phi_{nlm} | z | \phi_{000} \rangle}{\epsilon_0 - \epsilon_n} = \frac{e\mathcal{E}}{\hbar\omega} \sqrt{\frac{\hbar}{2m\omega}} |\phi_{110}\rangle
$$

and so we see that the electric field induces an odd-parity component into the perturbed ground state. It is clear that to linear terms in $\mathcal{E}$ the expectation value of $V(r)$ is unchanged, since

$$
\left(\langle \phi_{000} | + {}^{(1)}\langle \psi_{000} | \right) V(r) \left(|\phi_{000}\rangle + |\psi_{000}\rangle^{(1)} \right) \approx \langle \phi_{000} | V(r) | \phi_{000} \rangle + \left(\langle \phi_{000} | V(r) | \psi_{000} \rangle^{(1)} + \text{c.c.} \right) \,,
$$

and the matrix element $\langle \phi_{000} | V(r) | \psi_{000} \rangle^{(1)}$ vanishes by the parity selection rule. By the virial theorem it follows that

$$
\langle \phi_{000} | V(r) | \phi_{000} \rangle = \frac{3}{4} \hbar\omega \,.
$$

Lastly, the sum rule is easily seen to be satisfied for the ground state:

$$
\sum_{nlm} (\epsilon_0 - \epsilon_n) \, |\langle \phi_{nlm} | D_z | \phi_{000} \rangle|^2 = (\epsilon_0 - \epsilon_1) e^2 \, |\langle \phi_{110} | z | \phi_{000} \rangle|^2 = -\hbar\omega e^2 \frac{1}{2\alpha} = -\frac{\hbar^2 e^2}{2m} \,.
$$

Part 3

The first excited state has $l=1$ and its energy is three-fold degenerate (corresponding to $m=\pm1$ and 0). In first order, we need to diagonalize the matrix representing the perturbation in the subspace spanned by the states $|\phi_{11m}\rangle$. However, $\langle\phi_{11m}|V(r)|\phi_{11m'}\rangle=0$ (by the parity selection rule) and the resulting matrix is identically zero. Thus, there are no first-order corrections. In second order, we need to construct the matrix

$$Z_{mm'} = \langle\phi_{11m}|Z|\phi_{11m'}\rangle\,, \qquad Z = (e\mathcal{E})^2 \sum_{nl\neq11} \sum_{m=-l}^{l} \frac{1}{\epsilon_1 - \epsilon_n}\, z|\phi_{nlm}\rangle\langle\phi_{nlm}|z\,.$$

Since the parity of $|\phi_{nlm}\rangle$ is $(-1)^l$ (and the perturbation is odd under parity),

$$\langle\phi_{11m}|z|\phi_{nlm'}\rangle = 0 \qquad \text{if } l \text{ is odd}\,.$$

Further, we must have $l=0$ or 2 and $m=m'$. To verify this, consider

$$\langle\phi_{11m}|z|\phi_{nlm'}\rangle = \int_0^\infty dr\, r^3\, R_{11}(r)\, R_{nl}(r) \underbrace{\int d\Omega\, Y_{1m}^*(\Omega)\, \cos\theta\, Y_{lm'}(\Omega)}_{I^l_{mm'}}\,,$$

and the angular integrations give

$$I^l_{mm'} = \int_{-1}^{1} d(\cos\theta)\, Y_{1m}^*(\theta,0)\, \cos\theta\, Y_{lm'}(\theta,0) \underbrace{\int_0^{2\pi} d\phi\, e^{-i(m-m')\phi}}_{2\pi\delta_{m,m'}}\,,$$

and the selection rule on the azimuthal quantum numbers follows (above, we have made explicit the ϕ dependence of $Y_{lm}(\theta,\phi)=Y_{lm}(\theta,0)\,e^{im\phi}$). Next, the following relations can be expressed as linear combinations of spherical harmonics of order 0 and 2:

$$Y_{10}(\Omega)\, \cos\theta = \sqrt{\frac{3}{4\pi}}\, \cos^2\theta = \frac{2}{\sqrt{15}}\, Y_{20}(\Omega) + \frac{1}{\sqrt{3}}\, Y_{00}(\Omega)\,,$$

and

$$Y_{1\pm1}(\Omega)\, \cos\theta = \mp\sqrt{\frac{3}{8\pi}}\, \sin\theta\, \cos\theta\, e^{\pm i\phi} = \frac{1}{\sqrt{5}}\, Y_{2\pm1}(\Omega)\,.$$

Therefore, we have

$$I^0_{m,0} = \delta_{m,0} \int d\Omega\, Y_{10}^*(\Omega)\, \cos\theta\, Y_{00} = \frac{1}{\sqrt{3}}\,, \qquad I^2_{m,m'} = \int d\Omega\, Y_{1m}^*(\Omega)\, \cos\theta\, Y_{2m'}(\Omega) = \delta_{m,m'}\, c_m\,,$$

where

$$c_{\pm1} = \frac{1}{\sqrt{5}}\,, \qquad c_0 = \frac{2}{\sqrt{15}}\,.$$

Incidentally, it is worthwhile pointing out here that the selection rules obtained above could have been derived more easily by means of the Wigner–Eckart theorem

$$\langle\phi_{11m}|z|\phi_{nlm'}\rangle = C_{1l}(0,m';1m)\,\langle\phi_{11}||z||\phi_{nl}\rangle \implies l = 0, 1, 2 \text{ and } m = m'\,,$$

where $l=1$ is excluded here by the parity selection rule.

We obtain

$$\langle \phi_{11m} | z | \phi_{nlm'} \rangle = \int_0^\infty dr\, r^3\, R_{11}(rr)\, R_{nl}(r)\, I^l_{mm'} = \frac{1}{\sqrt{3}}\, \delta_{l,0}\, \delta_{m,0}\, \delta_{m',0} \underbrace{\int_0^\infty dr\, r^3\, R_{11}(r)\, R_{n0}(r)}_{a_{n0}}$$

$$+ \delta_{l,2}\, \delta_{m,m'}\, c_m \underbrace{\int_0^\infty dr\, r^3\, R_{11}(r)\, R_{n2}(r)}_{a_{n2}}\ .$$

Now, consider the product $rR_{11}(r)$, which can be expressed either as

$$rR_{11}(r) = \frac{2\sqrt{2}\alpha^{5/4}}{\sqrt{3}\pi^{1/4}}\, r^2\, e^{-\alpha r^2/2} = \sqrt{\frac{3}{2\alpha}}\, R_{00}(r) - \frac{1}{\sqrt{\alpha}}\, R_{20}(r)$$

or as

$$rR_{11}(r) = \sqrt{\frac{5}{2\alpha}}\, R_{22}(r)\ .$$

We insert the first expression into a_{n0} to obtain

$$a_{n0} = \int_0^\infty dr\, r^2 \left[\sqrt{\frac{3}{2\alpha}}\, R_{00}(r) - \frac{1}{\sqrt{\alpha}}\, R_{20}(r) \right] R_{n0}(r) = \sqrt{\frac{3}{2\alpha}}\, \delta_{n,0} - \frac{1}{\sqrt{\alpha}}\, \delta_{n,2}\ ,$$

Similarly, by inserting the second expression and using orthonormality, we find

$$a_{n2} = \sqrt{\frac{5}{2\alpha}}\, \delta_{n,2}\ .$$

The matrix element finally follows:

$$\langle \phi_{11m} | z | \phi_{nlm'} \rangle = \frac{1}{\sqrt{3}}\, \delta_{l,0}\, \delta_{m,0}\, \delta_{m',0} \left(\sqrt{\frac{3}{2\alpha}}\, \delta_{n,0} - \frac{1}{\sqrt{\alpha}}\, \delta_{n,2} \right) + \delta_{l,2}\, \delta_{m,m'}\, c_m \left(\sqrt{\frac{5}{2\alpha}}\, \delta_{n,2} \right)\ .$$

We are now in position to construct the 3×3 matrix

$$Z_{mm'} = \sum_{nl \neq 11} \sum_{\overline{m}=-l}^{l} \frac{(e\mathcal{E})^2}{\epsilon_1 - \epsilon_n}\, \langle \phi_{11m} | z | \phi_{nl\overline{m}} \rangle \langle \phi_{nl\overline{m}} | z | \phi_{11m'} \rangle$$

$$= \sum_{nl \neq 11} \sum_{\overline{m}=-l}^{l} \frac{(e\mathcal{E})^2}{\epsilon_1 - \epsilon_n} \left[\frac{1}{\sqrt{3}}\, \delta_{l,0}\, \delta_{m,0}\, \delta_{\overline{m},0} \left(\sqrt{\frac{3}{2\alpha}}\, \delta_{n,0} - \frac{1}{\sqrt{\alpha}}\, \delta_{n,2} \right) + \delta_{l,2}\, \delta_{m,\overline{m}}\, c_m \left(\sqrt{\frac{5}{2\alpha}}\, \delta_{n,2} \right) \right]$$

$$\times \left[\frac{1}{\sqrt{3}}\, \delta_{l,0}\, \delta_{\overline{m},0}\, \delta_{m',0} \left(\sqrt{\frac{3}{2\alpha}}\, \delta_{n,0} - \frac{1}{\sqrt{\alpha}}\, \delta_{n,2} \right) + \delta_{l,2}\, \delta_{\overline{m},m'}\, c_{m'} \left(\sqrt{\frac{5}{2\alpha}}\, \delta_{n,2} \right) \right]\ ,$$

which can be simplified by exploiting the Kronecker deltas and noting that $\epsilon_1 - \epsilon_0 = \hbar\omega$ and $\epsilon_1 - \epsilon_2 = -\hbar\omega$:

$$Z_{mm'} = \frac{(e\mathcal{E})^2}{\epsilon_1 - \epsilon_0}\, \frac{1}{2\alpha}\, \delta_{m,0}\, \delta_{m',0} + \frac{(e\mathcal{E})^2}{\epsilon_1 - \epsilon_2} \left(\frac{1}{3\alpha}\, \delta_{m,0}\, \delta_{m',0} + \frac{5}{2\alpha}\, c_m^2\, \delta_{m,m'} \right)$$

$$= \frac{(e\mathcal{E})^2}{\hbar\omega} \left[\left(\frac{1}{2\alpha} - \frac{1}{3\alpha} \right) \delta_{m,0}\, \delta_{m',0} - \frac{5}{2\alpha}\, c_m^2\, \delta_{m,m'} \right]\ ,$$

or, in matrix form,

$$\underline{Z} = \frac{(e\mathcal{E})^2}{\hbar\omega}\begin{pmatrix} 0 & 0 & 0 \\ 0 & 1/(6\alpha) & 0 \\ 0 & 0 & 0 \end{pmatrix} - \frac{(e\mathcal{E})^2}{\hbar\omega}\begin{pmatrix} 1/(2\alpha) & 0 & 0 \\ 0 & 2/(3\alpha) & 0 \\ 0 & 0 & 1/(2\alpha) \end{pmatrix} = -\frac{(e\mathcal{E})^2}{\hbar\omega}\frac{1}{2\alpha}\begin{pmatrix} 1 & 0 & 0 \\ 0 & 1 & 0 \\ 0 & 0 & 1 \end{pmatrix}.$$

We find the eigenvalue of Z to be three-fold degenerate and to have the value

$$E_1^{(2)} = -\frac{(e\mathcal{E})^2}{2m\omega^2},$$

which is identical to the (second-order) ground-state shift $E_0^{(2)}$. It turns out this is true for any level, as will be shown in part 4 below. It is also easily verified that the sum rule relative to the states $|\phi_{11m}\rangle$ gives

$$\sum_{nlm'}(\epsilon_1 - \epsilon_n)\,|\langle\phi_{nlm'}|\,D_z\,|\phi_{11m}\rangle|^2 = -\frac{\hbar^2 e^2}{2m},$$

on noting that

$$|\langle\phi_{nlm'}|\,z\,|\phi_{11m}\rangle|^2 = \left(\frac{1}{2\alpha}\,\delta_{n,0} + \frac{1}{3\alpha}\,\delta_{n,2}\right)\delta_{l,0}\,\delta_{m,0}\,\delta_{m',0} + \frac{5c_m^2}{2\alpha}\,\delta_{n,2}\,\delta_{l,2}\,\delta_{m,m'}.$$

For the state $|\phi_{111}\rangle$ we find

$$|\langle\phi_{nlm}|\,z\,|\phi_{111}\rangle|^2 = \frac{5c_1^2}{2\alpha}\,\delta_{n,2}\,\delta_{l,2}\,\delta_{m,1} = \frac{\hbar}{2m\omega}\,\delta_{n,2}\,\delta_{l,2}\,\delta_{m,1},$$

and hence

$$\sum_{nlm}(\epsilon_1 - \epsilon_n)\,|\langle\phi_{nlm}|\,D_z\,|\phi_{111}\rangle|^2 = (\epsilon_1 - \epsilon_2)\frac{\hbar e^2}{2m\omega} = -\frac{\hbar^2 e^2}{2m},$$

and similarly for $|\phi_{110}\rangle$ and $|\phi_{11-1}\rangle$.

Part 4

The eigenvalues and eigenstates of H can be determined exactly by noting that

$$H = \underbrace{\frac{\mathbf{p}^2}{2m} + \frac{m\omega^2}{2}r^2}_{H_0} + \underbrace{e\,\mathbf{E}\cdot\mathbf{r}}_{W} = \frac{\mathbf{p}^2}{2m} + \frac{m\omega^2}{2}\left(\mathbf{r} + \frac{e\,\mathbf{E}}{m\omega^2}\right)^2 - \frac{(e\mathcal{E})^2}{2m\omega^2},$$

where $\mathbf{E} = \mathcal{E}\,\hat{\mathbf{z}}$. We now introduce new position and momentum operators as

$$\mathbf{R} = \mathbf{r} + \frac{e\,\mathbf{E}}{m\omega^2}, \qquad \mathbf{P} = \mathbf{p},$$

satisfying the canonical commutation relations $[R_i, P_j] = i\hbar\delta_{ij}$ (the electric field $\mathbf{E}$ is uniform and hence independent of $\mathbf{r}$). In terms of these, the Hamiltonian reads

$$H = \frac{\mathbf{P}^2}{2m} + \frac{m\omega^2}{2}\mathbf{R}^2 - \frac{(e\mathcal{E})^2}{2m\omega^2},$$

and the eigenvalues of H are simply given by

$$E_n = \hbar\omega(n + 3/2) - \frac{(e\mathcal{E})^2}{2m\omega^2},$$

with degeneracy $g_n = (n + 1)(n + 2)/2$.

Problem 18 Leading-Order Correction for the Hydrogen-Atom Ground-State Energy in an Electric Field

The second-order correction to an unperturbed energy level generally involves a summation over an infinite number of states, which is often difficult to carry out in closed form. In this problem, we investigate an alternative method based on the following relation:

$$E_n^{(2)} = \langle \phi_n | \hat{V} | \psi_n^{(1)} \rangle \,,$$

where $\hat{V}$ is the perturbation and $|\psi_n^{(1)}\rangle$ is the first-order correction to the unperturbed state $|\phi_n\rangle$ (an eigenstate of $\hat{H}_0$ with eigenvalue ϵ_n, the latter assumed to be non-degenerate).

1. Show that expanding $E_n(\lambda)$ and $|\psi_n(\lambda)\rangle$ – respectively, the eigenvalue and eigenstate of $\hat{H}(\lambda) = \hat{H}_0 + \lambda \hat{V}$ – in a power series in λ leads, to linear terms in λ, to the following relation satisfied by $|\psi_n^{(1)}\rangle$:

$$(\hat{H}_0 - \epsilon_n)|\psi_n^{(1)}\rangle = (\underbrace{\langle \phi_n | \hat{V} | \phi_n \rangle}_{E_n^{(1)}} - \hat{V})|\phi_n\rangle \,.$$

2. Assuming that both $\hat{H}_0$ and $\hat{V}$ are local in the coordinate representation,

$$\langle \phi_{\mathbf{r}} | \hat{H}_0 | \phi_{\mathbf{r}'} \rangle = H_0(\mathbf{r}, -i\hbar\boldsymbol{\nabla}) \, \delta(\mathbf{r} - \mathbf{r}') \,,$$

and similarly for $\hat{V}$ (here, $|\phi_{\mathbf{r}}\rangle$ are the eigenstates of the position operator $\mathbf{r}$), show that the relation in part 1 can be viewed as an inhomogeneous differential equation satisfied by the wave function $\psi_n^{(1)}(\mathbf{r})$,

$$[H_0(\mathbf{r}, -i\hbar\boldsymbol{\nabla}) - \epsilon_n] \, \psi_n^{(1)}(\mathbf{r}) = \left[E_n^{(1)} - V(\mathbf{r}) \right] \phi_n(\mathbf{r}) \,.$$

The solution of this equation is not unique, since an arbitrary multiple $\alpha|\phi_n\rangle$ can be added to it. However, it can be made unique by imposing the auxiliary condition $\langle \phi_n | \psi_n^{(1)} \rangle = 0$.

3. Consider a hydrogen atom under the influence of a uniform electric field directed along the z-axis, namely $\mathbf{E} = \mathcal{E}\,\hat{\mathbf{z}}$. Show that the first-order correction $\psi_{100}^{(1)}(\mathbf{r})$ to the ground-state wave function $\phi_{100}(r)$ satisfies

$$\left(-\frac{\hbar^2}{2\mu} \boldsymbol{\nabla}^2 - \frac{e^2}{r} + \frac{e^2}{2a_0} \right) \psi_{100}^{(1)}(\mathbf{r}) = -e\mathcal{E}r \cos\theta \, \phi_{100}(r) \,,$$

where

$$\phi_{100}(r) = \frac{1}{\sqrt{\pi a_0^3}} \, e^{-r/a_0} \,,$$

and a_0 is the Bohr radius. Explain why the wave function $\psi_{100}^{(1)}(\mathbf{r})$ must have the form

$$\psi_{100}^{(1)}(r, \theta) = g(r) \, \cos\theta \,.$$

Hence, show that such a wave function $\psi_{100}^{(1)}(\mathbf{r})$ is orthogonal to $\phi_{100}(r)$ (thus ensuring its uniqueness) and that $g(r)$ satisfies

$$-\frac{\hbar^2}{2\mu} \frac{1}{r^2} \frac{d}{dr} r^2 \frac{d}{dr} g(r) + \left(\frac{\hbar^2}{\mu r^2} - \frac{e^2}{r} + \frac{e^2}{2a_0} \right) g(r) = -\frac{e\mathcal{E}}{\sqrt{\pi a_0^3}} \, r\, e^{-r/a_0}$$

4. Posit the functional form

$$g(r) = p(r) \, e^{-r/a_0} \, ,$$

and show that $p(r)$ satisfies

$$x^2 p''(x) + 2x(1 - x)p'(x) - 2p(x) = \alpha x^3 \, , \qquad \alpha = \frac{2}{\sqrt{\pi}} \, (\mathcal{E}/e) \sqrt{a_0} \, ,$$

where $x = r/a_0$. Solve the equation for $p(x)$ by assuming that it is a polynomial.

5. Obtain the second-order correction by computing

$$E_{100}^{(2)} = \langle \phi_{100} | \hat{V} | \psi_{100}^{(1)} \rangle \, ,$$

and compare the result with that obtained in Problem 15.

Solution

Part 1

As in Section 14.1, we write

$$(\hat{H}_0 + \lambda \, \hat{V}) \, (|\phi_n\rangle + \lambda |\psi_n^{(1)}\rangle + \cdots) = (\epsilon_n + E_n^{(1)} + \cdots) \, (|\phi_n\rangle + \lambda |\psi_n^{(1)}\rangle + \cdots)$$

and the linear terms in λ lead to the following relation:

$$\hat{H}_0 \, |\psi_n^{(1)}\rangle + \hat{V} |\phi_n\rangle = \epsilon_n |\phi_n^{(1)}\rangle + E_n^{(1)} |\phi_n\rangle \, ,$$

or

$$(\hat{H}_0 - \epsilon_n) |\psi_n^{(1)}\rangle = (\langle \phi_n | \hat{V} \phi_n \rangle - \hat{V}) |\phi_n\rangle \, .$$

Part 2

In the coordinate representation, we have

$$\int dr' \, \langle \phi_{\mathbf{r}} | \hat{H}_0 - \epsilon_n | \phi_{\mathbf{r}'} \rangle \langle \phi_{\mathbf{r}'} | \psi_n^{(1)} \rangle = \int dr' \, \langle \phi_{\mathbf{r}} | \langle \phi_n | \hat{V} \phi_n \rangle - \hat{V} | \phi_{\mathbf{r}'} \rangle \langle \phi_{\mathbf{r}'} | \phi_n \rangle \, ,$$

and hence

$$[H_0(\mathbf{r}, -i\hbar\boldsymbol{\nabla}) - \epsilon_n] \underbrace{\int dr' \, \delta(\mathbf{r} - \mathbf{r}') \, \psi_n^{(1)}(\mathbf{r}')}_{\psi_n^{(1)}(\mathbf{r})} = \left[\langle \phi_n | \hat{V} \phi_n \rangle - \hat{V}(\mathbf{r}) \right] \underbrace{\int d\mathbf{r}' \, \delta(\mathbf{r} - \mathbf{r}') \, \phi_n(\mathbf{r}')}_{\phi_n(\mathbf{r})} \, ,$$

that is, an inhomogeneous partial differential equation. Clearly, the solution of the inhomogeneous equation is not unique, given that $\psi_n^{(1)}(\mathbf{r}) + \alpha \phi_n(\mathbf{r})$, with α an arbitrary constant, also satisfies the equation, since

$$[H_0(\mathbf{r}, -i\hbar\boldsymbol{\nabla}) - \epsilon_n]\phi_n(\mathbf{r}) = 0 \, .$$

It is made unique by imposing the auxiliary condition

$$\langle \phi_n | \psi_n^{(1)} \rangle = 0 \implies \langle \phi_n | (|\psi_n^{(1)}\rangle + \alpha |\phi_n\rangle) = \alpha = 0 \, ,$$

since $|\phi_n\rangle$ is normalized.

Part 3

The first-order correction $\langle\phi_{100}|\,e\mathcal{E}z\,|\phi_{100}\rangle$ to the ground-state energy vanishes (owing to the parity selection rule), and therefore the equation reads

$$\underbrace{\left(-\frac{\hbar^2}{2\mu}\nabla^2 - \frac{e^2}{r} + \underbrace{\frac{e^2}{2a_0}}_{-\epsilon_1}\right)}_{H_0(\mathbf{r},-i\hbar\nabla)}\psi_{100}^{(1)}(\mathbf{r}) = -\,e\mathcal{E}\underbrace{r\cos\theta}_{V}\,\phi_{100}(r)\,.$$

The unperturbed Hamiltonian on the left-hand side commutes with the components of the orbital angular momentum $\mathbf{L}$ (it is a scalar operator) and hence with $\mathbf{L}^2$, while the function $r\cos\theta\,\phi_{100}(r)$ on the right-hand side is proportional to the spherical harmonic $Y_{10}(\theta)$ and is an eigenfunction of $\mathbf{L}^2$ and L_z, with eigenvalues $2\hbar^2$ and 0, respectively. It follows that $\psi_{100}^{(1)}(\mathbf{r})$ must also be an eigenfunction of $\mathbf{L}^2$ and L_z with the same eigenvalues. This implies that $\psi_{100}(\mathbf{r})$ can be written as $g(r)\cos\theta$. Recalling that

$$-\frac{\hbar^2}{2\mu}\nabla^2 = -\frac{\hbar^2}{2\mu}\frac{1}{r^2}\frac{d}{dr}r^2\frac{d}{dr} + \frac{\mathbf{L}^2}{2\mu r^2}$$

and

$$\mathbf{L}^2\cos\theta = 2\hbar^2\cos\theta\,,$$

we find

$$\cos\theta\left[-\frac{\hbar^2}{2\mu}\frac{1}{r^2}\frac{d}{dr}r^2\frac{d}{dr} + \left(\frac{\hbar^2}{\mu r^2} - \frac{e^2}{r} + \frac{e^2}{2a_0}\right)\right]g(r) = -\frac{e\mathcal{E}}{\sqrt{\pi a_0^3}}\,r\,e^{-r/a_0}\cos\theta\,,$$

and the common $\cos\theta$ factor can be canceled, resulting in the radial equation given in the text of the problem. Note that $\psi_{100}^{(1)}(\mathbf{r}) = \cos\theta\,g(r)$ is obviously orthogonal to the (spherically symmetric) ground-state wave function, and so the requirement $\langle\phi_{100}|\psi_{100}^{(1)}\rangle = 0$ is automatically satisfied.

Part 4

Using

$$\frac{1}{r^2}\frac{d}{dr}r^2\frac{d}{dr} = \frac{d^2}{dr^2} + \frac{2}{r}\frac{d}{dr}\,,$$

and introducing the non-dimensional variable $x = r/a_0$, with $a_0 = \hbar^2/(\mu e^2)$, the differential equation for $g(x)$ is more conveniently written as

$$g''(x) + \frac{2}{x}g'(x) - \left(\frac{2}{x^2} - \frac{2}{x} + 1\right)g(x) = \underbrace{\frac{2(\mathcal{E}/e)\sqrt{a_0}}{\sqrt{\pi}}}_{\alpha}x\,e^{-x}$$

Inserting the ansatz $g(x) = p(x)\,e^{-x}$ and using

$$g'(x) = [p'(x) - p(x)]\,e^{-x}\,, \qquad g''(x) = [p''(x) - 2p'(x) + p(x)]\,e^{-x}\,,$$

we find (after canceling the common e^{-x} factor)

$$x^2 p''(x) + 2x(1 - x)p'(x) - 2p(x) = \alpha x^3\,.$$

Since $p(x)$ is a polynomial, it is clear from the structure of the equation that the highest power that can occur in $p(x)$ is x^2; indeed, if a power x^n with $n \geq 3$ were to be present then the term $x^2 p'(x)$ in the equation would yield a power x^{n+1} which could not be canceled. Thus, we take

$$p(x) = p_0 + p_1 x + p_2 x^2 \,,$$

and upon substitution in the equation we have

$$2p_2 x^2 + 2x(1-x)(p_1 + 2p_2 x) - 2p_0 - 2p_1 x - 2p_2 x^2 = \alpha x^3 \,.$$

Matching powers of x leads to

$$p_0 = 0 \,, \qquad p_1 = -\frac{\alpha}{2} \,, \qquad p_2 = -\frac{\alpha}{4} \,,$$

finally yielding

$$\psi_{100}^{(1)}(\mathbf{r}) = -\frac{\alpha}{2} \frac{r}{a_0} \left(1 + \frac{r}{2a_0}\right) e^{-r/a_0} \cos\theta \,.$$

A finer point: we should consider the homogeneous equation

$$x^2 p''(x) + 2x(1-x)p'(x) - 2p(x) = 0 \,,$$

since its solution can always be added to that obtained above. It turns out, however, that such a solution would lead to a wave function $\psi_{100}^{(1)}(\mathbf{r})$ that diverges exponentially in the asymptotic region, and therefore would not be acceptable. To see how this comes about, assume

$$p(x) = \sum_{n=0}^{\infty} a_n x^{n+s} \,, \qquad a_0 \neq 0 \,.$$

Insertion of the above expansion yields (after dropping a common factor x^s)

$$\sum_{n=0}^{\infty} \left[a_n(n+s)(n+s-1)x^n + 2a_n(n+s)(x^n - x^{n+1}) - 2a_n x^n \right] = 0 \,,$$

which can be written as

$$\sum_{n=0}^{\infty} [(n+s)(n+s-1) + 2(n+s) - 2] a_n x^n = \sum_{n=0}^{\infty} 2(n+s)a_n x^{n+1} \,.$$

Matching powers of x on the left- and right-hand sides, we see that the $n=0$ coefficient on the left-hand side must vanish

$$[s(s-1) + 2s - 2]a_0 = 0 \qquad \text{or} \qquad s^2 + s - 2 = 0 \implies s = 1, -2 \,,$$

and the only acceptable solution is $s = 1$ (the solution with $s = -2$ would lead to a wave function that was singular at the origin). Substituting $s = 1$ into the expansion, we find

$$\sum_{n=0}^{\infty} [n(n+1) + 2(n+1) - 2] a_n x^n = \sum_{n=0}^{\infty} 2(n+1)a_n x^{n+1}$$

or, setting $n+1 \longrightarrow n$ in the sum on the right (and noting that the term with $n=0$ does not contribute in the sum on the left),

$$\sum_{n=1}^{\infty} n(n+3)a_n x^n = \sum_{n=1}^{\infty} 2na_{n-1} x^n \implies a_n = \frac{2}{n+3} a_{n-1} \,,$$

which has the solution

$$a_n = 2^n \frac{3!}{(n+3)!} a_0 .$$

Note that the series cannot terminate (so the solution $p(x)$ of the homogeneous equation cannot be a polynomial). It can be identically zero if $a_0 = 0$, which in fact turns out to be the only possibility, the reason being that otherwise $p(x)$ diverges exponentially as $x \longrightarrow \infty$. Indeed, the series for $p(x)$ can be summed to give

$$p(x) = x \sum_{n=0}^{\infty} a_n x^n = 6a_0 x \sum_{n=0}^{\infty} \frac{(2x)^n}{(n+3)!} = 6a_0 x \sum_{m=3}^{\infty} \frac{1}{m!} (2x)^{m-3}$$

$$= \frac{3}{4} \frac{a_0}{x^2} \left[\sum_{m=0}^{\infty} \frac{(2x)^m}{m!} - 1 - 2x - 2x^2 \right] = a_0' \frac{e^{2x} - 1 - 2x - 2x^2}{x^2} ,$$

where $a_0' = (3/4)a_0$; hence $p(x)$ is asymptotically proportional to e^{2x}/x^2 and so, as anticipated, would lead to a wave function $\psi_{100}^{(1)}(\mathbf{r})$ that diverges exponentially as e^{r/a_0}.

Part 5

We obtain

$$E_{100}^{(2)} = -\frac{1}{\sqrt{\pi a_0^3}} e\mathcal{E} \frac{\alpha}{2} \int_0^{\infty} dr \, r^2 \int d\Omega \, e^{-r/a_0} \, r \cos\theta \, \frac{r}{a_0} \left(1 + \frac{r}{2a_0}\right) e^{-r/a_0} \cos\theta$$

$$= -\underbrace{\frac{a_0^{5/2}}{\sqrt{\pi}} e\mathcal{E} \frac{\alpha}{2} \frac{4\pi}{3}}_{(4/3)\,\mathcal{E}^2 a_0^3} \underbrace{\int_0^{\infty} dx \, x^4 \left(1 + \frac{x}{2}\right) e^{-2x}}_{4!/2^5 + 5!/2^7} = -\frac{9}{4} \mathcal{E}^2 a_0^3 .$$

In Problem 15 we found the result

$$E_{100}^{(2)} = \frac{(e\mathcal{E})^2}{3} \sum_{n=2}^{\infty} \frac{|\gamma_n|^2}{\epsilon_1 - \epsilon_n} = -\frac{2}{3} a_0 \mathcal{E}^2 \sum_{n=2}^{\infty} \frac{|\gamma_n|^2}{1 - 1/n^2} ,$$

where

$$\gamma_n = \int_0^{\infty} dr \, r^3 R_{np}(r) R_{1s}(r) .$$

On dimensional grounds, γ_n should scale as $a_0 x_n$ with x_n non-dimensional – the radial wave functions have the dimension of $(\text{length})^{-3/2}$ – and hence

$$E_{100}^{(2)} = -\frac{2}{3} a_0^3 \mathcal{E}^2 \sum_{n=2}^{\infty} \frac{|x_n|^2}{1 - 1/n^2} .$$

This estimate is, however, incomplete, since the contribution of the continuum states has been neglected. By contrast, the result obtained earlier does account for this contribution and is correct to order $\mathcal{E}^2$. The perturbation induces an electric dipole moment (since it deforms the spherically symmetric ground state), leading to an energy $-\lambda \mathcal{E}^2/2$ with $\lambda = (9/2) a_0^3$ the polarizability of the hydrogen atom in the ground state.

Problem 19 Perturbative Calculation of the Relativistic Kinetic Energy Term in Hydrogen-Like-Atom Levels

Consider a hydrogen-like atom described by the Hamiltonian

$$H = \left(c\sqrt{\mathbf{p}^2 + (mc)^2} - mc^2 \right) - \frac{Ze^2}{r} \,,$$

where m is the mass of the electron and Ze is the nuclear charge, with $Z \gg 1$. Since the typical velocity of the electron in such a system is of the order of $Z\alpha\,c$ (here, $\alpha \approx 1/137$ is the fine structure constant and c is the speed of light), it is physically sensible to represent the electron kinetic energy operator by its relativistic expression.

1. Expand the kinetic energy operator in powers of $[\mathbf{p}/(m\,c)]^2$, and show that the Hamiltonian can be written as

$$H = H_0 + V \qquad \text{with} \qquad H_0 = \frac{\mathbf{p}^2}{2m} - \frac{Ze^2}{r}$$

where V is the perturbation, consisting of the leading correction to the non-relativistic kinetic energy operator. Provide an expression for V.

2. Show that the operator V can also be expressed as

$$V = -\frac{1}{2mc^2}\left(H_0 + \frac{Ze^2}{r} \right)^2 \,.$$

Evaluate the first-order correction to the ground-state energy (corresponding to the unperturbed Hamiltonian H_0).

3. Obtain an expression for the second-order correction $E_{100}^{(2)}$ to the (unperturbed) ground-state energy, ignoring the contribution of the continuum states. Explain why the sum over (discrete) states includes only those having $l=0$ (s-waves). Do not carry out the integrations explicitly; however, reduce them to dimensional constants times non-dimensional numbers, by an appropriate change of the integration variable. Express your result in terms of Z, α, and mc^2.

4. Now consider the first excited level, having $n=2$, which is four-fold degenerate. Obtain the first-order corrections to the unperturbed energy ϵ_2. Does the perturbation lift the degeneracy of this level completely or only partially? Would you expect degeneracy to persist in higher orders of perturbation theory? Justify your answer. Compare the first-order correction in part 2 with the corrections obtained here and order them from smallest to largest. Explain why this ordering was to be expected.

Hints: The unperturbed bound-state energies are

$$\epsilon_n = -\frac{(Z\alpha)^2}{2n^2}\, mc^2 \,, \qquad n = 1, 2, \dots \,,$$

where $\alpha = e^2/(\hbar c)$ is the fine structure constant. The corresponding normalized eigenfunctions are written as

$$\phi_{nlm}(\mathbf{r}) = R_{nl}(x)\, Y_{lm}(\Omega) \,, \qquad l = 0, 1, \dots, n-1 \,, \qquad m = -l, -l+1, \dots, l \,,$$

with $x = Zr/a_0$ and

$$\int d\mathbf{r}\, \phi_{nlm}^*(\mathbf{r})\, \phi_{n'l'm'}(\mathbf{r}) = \delta_{n,n'}\, \delta_{l,l'}\, \delta_{m,m'} \,.$$

In particular, the radial wave functions with $n = 1$ and 2 are given by

$$R_{10}(x) = 2\left(\frac{Z}{a_0}\right)^{3/2} e^{-x},$$

$$R_{20}(x) = \frac{1}{\sqrt{2}}\left(\frac{Z}{a_0}\right)^{3/2}\left(1 - \frac{x}{2}\right)e^{-x/2}, \qquad R_{21}(x) = \frac{1}{2\sqrt{6}}\left(\frac{Z}{a_0}\right)^{3/2} x\, e^{-x/2}.$$

In the formulae above, a_0 is the Bohr radius,

$$a_0 = \frac{\hbar^2}{me^2} = \frac{1}{\alpha}\frac{\hbar}{mc} \qquad \text{and} \qquad \frac{e^2}{a_0} = \alpha^2 mc^2.$$

The following integral may be useful:

$$\int_0^\infty dx\, x^n\, e^{-\gamma x} = \frac{n!}{\gamma^{n+1}} \qquad n \geq 0.$$

Solution

Part 1

Using a Taylor expansion valid for $z \ll 1$,

$$\sqrt{1 + z^2} = 1 + \frac{z^2}{2} - \frac{z^4}{8} + \cdots,$$

the kinetic energy operator can be written as

$$T = mc^2\sqrt{1 + \frac{\mathbf{p}^2}{(mc)^2}} - mc^2 = mc^2\left[1 + \frac{\mathbf{p}^2}{2(mc)^2} - \frac{\mathbf{p}^4}{8(mc)^4} + \cdots\right] - mc^2 = \frac{\mathbf{p}^2}{2m} - \frac{\mathbf{p}^4}{8m^3c^2}.$$

It follows that

$$H = \frac{\mathbf{p}^2}{2m} - \frac{Ze^2}{r} - \frac{\mathbf{p}^4}{8m^3c^2} = H_0 + V,$$

with

$$V = -\frac{\mathbf{p}^4}{8m^3c^2}.$$

Part 2

Noting the operator identity

$$\frac{\mathbf{p}^2}{2m} = H_0 + \frac{Ze^2}{r} \implies \frac{\mathbf{p}^4}{4m^2} = \left(H_0 + \frac{Ze^2}{r}\right)^2,$$

we find

$$V = -\frac{1}{2mc^2}\left(H_0 + \frac{Ze^2}{r}\right)^2.$$

The first-order correction to the ground-state energy follows:

$$E_{100}^{(1)} = -\frac{1}{2mc^2} \langle \phi_{100} | \left(H_0 + \frac{Ze^2}{r} \right)^2 |\phi_{100}\rangle = -\frac{1}{2mc^2} \langle \phi_{100} | \left(\epsilon_1 + \frac{Ze^2}{r} \right)^2 |\phi_{100}\rangle$$

$$= -\frac{1}{2mc^2} \left(\epsilon_1^2 + 2\epsilon_1 \langle \phi_{100} | \frac{Ze^2}{r} |\phi_{100}\rangle + \langle \phi_{100} | \frac{Z^2 e^4}{r^2} |\phi_{100}\rangle \right)$$

where in the first line we have used the fact that $|\phi_{100}\rangle$ is the eigenstate of H_0 corresponding to the (non-degenerate) eigenvalue ϵ_1. We are left with the evaluation of the expectation values of $1/r$ and $1/r^2$:

$$\langle \phi_{100} | \frac{1}{r^n} |\phi_{100}\rangle = 4 \frac{Z^3}{a_0^3} \int_0^\infty dr\, r^{2-n}\, e^{-2Zr/a_0} = 4 \frac{Z^3}{a_0^3} \frac{(2-n)!}{(2Z/a_0)^{3-n}} = \frac{(2-n)!}{2} \left(\frac{2Z}{a_0} \right)^n ,$$

where $n = 1$ or 2. Thus we find

$$E_{100}^{(1)} = -\frac{1}{2mc^2} \left(\epsilon_1^2 + 2\,\epsilon_1 \frac{Z^2 e^2}{a_0} + \frac{2Z^4 e^4}{a_0^2} \right) .$$

Expressing e^2/a_0 as $\alpha\, e^2 mc/\hbar = \alpha^2 mc^2$ yields

$$E_{100}^{(1)} = -\frac{1}{2mc^2} \left(\frac{(Z\alpha)^4}{4} (mc^2)^2 - (Z\alpha)^4 (mc^2)^2 + 2(Z\alpha)^4 (mc^2)^2 \right) = -\frac{5}{8} (Z\alpha)^4 mc^2 ,$$

and the correction is suppressed relative to the ground-state energy by a factor $(5/4)(Z\alpha)^2$.

Part 3

Ignoring the contribution of the continuum states, the second-order correction is given by

$$E_{100}^{(2)} = \sum_{n=2}^\infty \sum_{l=0}^{n-1} \sum_{m=-l}^{l} \frac{|\langle \phi_{nlm} | V |\phi_{100}\rangle|^2}{\epsilon_1 - \epsilon_n} .$$

However, V commutes with $\mathbf{L}$, and hence the matrix representing V is diagonal in l and m, yielding

$$\langle \phi_{nlm} | V |\phi_{100}\rangle = -\frac{1}{2mc^2} \delta_{l,0}\, \delta_{m,0} \langle \phi_{n00} | \left(H_0 + \frac{Ze^2}{r} \right)^2 |\phi_{100}\rangle .$$

This result follows by considering the vanishing matrix element $\langle \phi_{nlm} | [\mathbf{L}^2, V] |\phi_{100}\rangle$ and similarly for L_z. By applying the operator to the right (but it could equally well be applied to the left, or "half" to the left and "half" to the right),

$$\langle \phi_{n00} | V |\phi_{100}\rangle = -\frac{1}{2mc^2} \left(2\,\epsilon_1 \langle \phi_{n00} | \frac{Ze^2}{r} |\phi_{100}\rangle + \langle \phi_{n00} | \frac{Z^2 e^4}{r^2} |\phi_{100}\rangle \right) ,$$

and the term proportional to ϵ_1^2 is absent, since the states $|\phi_{nlm}\rangle$ form an orthonormal basis. The relevant matrix elements are

$$\langle \phi_{n00} | \frac{1}{r^m} |\phi_{100}\rangle = \int_0^\infty dr\, r^{2-m} R_n(Zr/a_0) R_1(Zr/a_0) = \left(\frac{a_0}{Z} \right)^{3-m} \left(\frac{Z}{a_0} \right)^3 \gamma_{n1}^{(m)} ,$$

where the pure numbers $\gamma_{n1}^{(m)}$ are given by

$$\gamma_{n1}^{(m)} = \int_0^\infty dx\, x^{2-m} \overline{R}_n(x)\, \overline{R}_1(x) ,$$

and $R_{nlm}(x) = (Z/a_0)^{3/2} \overline{R}_{nlm}(x)$ with $\overline{R}_{nlm}(x)$ a non-dimensional function of x. Thus, we find

$$\langle \phi_{n00}|V|\phi_{100}\rangle = -\frac{1}{2mc^2}\left(2\,\epsilon_1\,\frac{Z^2 e^2}{a_0}\,\gamma_{n1}^{(1)} + \frac{Z^4 e^4}{a_0^2}\,\gamma_{n1}^{(2)}\right) = (Z\alpha)^4\left(\gamma_{n1}^{(1)} - \gamma_{n1}^{(2)}\right)\frac{mc^2}{2}\,,$$

and the second-order correction follows:

$$E_{100}^{(2)} = -\sum_{n=2}^{\infty}\frac{(Z\alpha)^8\left(\gamma_{n1}^{(1)} - \gamma_{n1}^{(2)}\right)^2 (mc^2)^2/4}{(1 - 1/n^2)(Z\alpha)^2 mc^2/2} = -\frac{(Z\alpha)^6}{2}\,mc^2\sum_{n=2}^{\infty}\frac{\left(\gamma_{n1}^{(1)} - \gamma_{n1}^{(2)}\right)^2}{1 - 1/n^2}\,,$$

and is smaller by a factor $(Z\alpha)^4$ relative to the unperturbed ground-state energy.

Part 4

The first-order corrections to the first excited level follow from diagonalizing the 4×4 matrix in the degenerate subspace spanned by the states $|\phi_{200}\rangle$ and $|\phi_{21m}\rangle$. Since V is a scalar operator, the matrix elements satisfy the selection rules

$$\langle \phi_{2l'm'}|V|\phi_{2lm}\rangle = \delta_{l,l'}\,\delta_{m,m'}\,\langle \phi_{2lm}|V|\phi_{2lm}\rangle\,,$$

and so the matrix is diagonal. Further, the p-wave matrix elements (having $l = 1$) are independent of m. This can be seen by noting that

$$\langle \phi_{21m}|V|\phi_{21m}\rangle = \frac{\langle \phi_{21m}|VL_+|\phi_{21m-1}\rangle}{\hbar\sqrt{2 - m(m-1)}} = \frac{\langle \phi_{21m}|L_+ V|\phi_{21m-1}\rangle}{\hbar\sqrt{2 - m(m-1)}} = \langle \phi_{21m-1}|V|\phi_{21m-1}\rangle\,,$$

where in the last step we applied $L_+ = L_-^\dagger$ to the left bra to obtain (here, $L_\pm$ are raising and lowering angular momentum operators)

$$\langle \phi_{21m}|L_+ = (L_-|\phi_{21m}\rangle)^\dagger = \hbar\sqrt{2 - m(m-1)}\,\langle \phi_{21m-1}|\,.$$

Thus, the matrix $\underline{V}$ is given by

$$\underline{V} = \begin{pmatrix} v_0 & 0 & 0 & 0 \\ 0 & v_1 & 0 & 0 \\ 0 & 0 & v_1 & 0 \\ 0 & 0 & 0 & v_1 \end{pmatrix}\,,$$

where

$$v_l = \langle \phi_{2l0}|V|\phi_{2l0}\rangle\,,\qquad l = 0,1\,.$$

The perturbation lifts the degeneracy on l. However, the degeneracy on m persists as expected, given that V is a scalar operator. The calculation of the corrections v_0 and v_1 is straightforward,

$$v_l = -\frac{1}{2mc^2}\left(\epsilon_2^2 + 2\epsilon_2\langle \phi_{2l0}|\frac{Ze^2}{r}|\phi_{2l0}\rangle + \langle \phi_{2l0}|\frac{Z^2 e^4}{r^2}|\phi_{2l0}\rangle\right)\,.$$

We find

$$\langle Ze^2/r\rangle_{20} = \frac{Ze^2}{2}\left(\frac{Z}{a_0}\right)^3\int_0^\infty dr\,r\left(1 - \frac{x}{2}\right)^2 e^{-x} = \frac{Z^2}{4}\frac{e^2}{a_0} = \frac{(Z\alpha)^2}{4}mc^2\,,$$

and

$$\langle Z^2 e^4/r^2 \rangle_{20} = \frac{Z^2 e^4}{2} \left(\frac{Z}{a_0} \right)^3 \int_0^\infty dr \left(1 - \frac{x}{2} \right)^2 e^{-x} = \frac{Z^4}{4} \frac{e^4}{a_0^2} = \frac{(Z\alpha)^4}{4} (mc^2)^2 \, ,$$

and similarly

$$\langle Ze^2/r \rangle_{10} = \frac{Ze^2}{24} \left(\frac{Z}{a_0} \right)^3 \int_0^\infty dr \, rx^2 \, e^{-x} = \frac{Z^2}{4} \frac{e^2}{a_0} = \frac{(Z\alpha)^2}{4} mc^2 \, ,$$

and

$$\langle Z^2 e^4/r^2 \rangle_{10} = \frac{Z^2 e^4}{24} \left(\frac{Z}{a_0} \right)^3 \int_0^\infty dr \, x^2 \, e^{-x} = \frac{Z^4}{12} \frac{e^4}{a_0^2} = \frac{(Z\alpha)^4}{12} (mc^2)^2 \, ,$$

so that

$$v_0 = -\frac{13}{128} (Z\alpha)^4 mc^2 \, , \qquad v_1 = -\frac{7}{384} (Z\alpha)^4 mc^2 \, ,$$

and $|v_0| > |v_1|$ as expected, given that the perturbation, being proportional to $(\epsilon_2 + Ze^2/r)^2$, receives most of the contribution from the small-r region, and that $R_{20}(x)$ is constant and $R_{21}(x) \propto x$ as $x \longrightarrow 0$. It is also clear that the first-order correction to the ground state is larger in magnitude than v_0. Since $R_{10}(x)$ and $R_{20}(x)$ must be orthogonal and $R_{10}(x)$ is nodeless, then $R_{20}(x)$ must have a node (which is obvious from the expression given in the text of the problem). As a consequence, the expectation value of the positive definite operator $(\epsilon_2 + Ze^2/r)^2$ is smaller for the first excited (s-wave) level than for the ground state.

Problem 20   Derivation of the Brillouin–Wigner Perturbation Theory

This problem deals with the derivation of the Brillouin–Wigner perturbation theory (as opposed to the Rayleigh–Schrödinger perturbation theory presented in Section 14.1). The starting point is the eigenvalue problem

$$(\hat{H}_0 + \hat{V})|\psi_n\rangle = E_n|\psi_n\rangle \, ,$$

where the unperturbed wave function and energy, $|\phi_n\rangle$ and ϵ_n, are assumed to be known,

$$\hat{H}_0|\phi_n\rangle = \epsilon_n|\phi_n\rangle \, .$$

The (unknown) state $|\psi_n\rangle$ is normalized as follows:

$$\langle \phi_n|\psi_n\rangle = 1 \, .$$

1. Define the projection operator

$$\hat{Q}_n = \sum_{m \neq n} |\phi_m\rangle\langle\phi_m| = \mathbb{1} - |\phi_n\rangle\langle\phi_n| \, .$$

   Show that

$$(E_n - \hat{H}_0) \, \hat{Q}_n|\psi_n\rangle = \hat{Q}_n \hat{V}|\psi_n\rangle$$

2. Noting that

$$|\psi_n\rangle = |\phi_n\rangle + \hat{Q}_n|\psi_n\rangle$$

   satisfies the normalization condition $\langle \phi_n|\psi_n\rangle = 1$, show that

$$|\psi_n\rangle = |\phi_n\rangle + \hat{R}_n \, \hat{V}|\psi_n\rangle \, ,$$

where the operator $\hat{R}_n$ is defined as follows:

$$\hat{R}_n = (E_n - \hat{H}_0)^{-1}\hat{Q}_n = \hat{Q}_n(E_n - \hat{H}_0)^{-1},$$

and $(\cdots)^{-1}$ indicates the inverse operator. Show further that the equation above can be solved by iteration to give

$$|\psi_n\rangle = |\phi_n\rangle + (\hat{R}_n\hat{V})|\phi_n\rangle + (\hat{R}_n\hat{V})^2|\phi_n\rangle + (\hat{R}_n\hat{V})^3|\phi_n\rangle + \cdots,$$

which can be formally summed to yield

$$|\psi_n\rangle = (\mathbb{1} - \hat{R}_n\hat{V})^{-1}|\phi_n\rangle,$$

even though this last form is not often used.

3. Show that

$$E_n = \epsilon_n + \langle\phi_n|\hat{V}|\psi_n\rangle,$$

and, hence, that

$$E_n = \epsilon_n + \langle\phi_n|\hat{V}|\phi_n\rangle + \sum_{m\neq n}\frac{\langle\phi_n|\hat{V}|\phi_m\rangle\langle\phi_m|\hat{V}|\phi_n\rangle}{E_n - \epsilon_m}$$

$$+ \sum_{k\neq n}\sum_{m\neq n}\frac{\langle\phi_n|\hat{V}|\phi_k\rangle\langle\phi_k|\hat{V}|\phi_m\rangle\langle\phi_m|\hat{V}|\phi_n\rangle}{(E_n - \epsilon_k)(E_n - \epsilon_m)} + \cdots.$$

Note that the expression above involves the unknown E_n on the right-hand side. Referring to this expression, suppose you were asked to calculate E_n at the third perturbative order. How would you proceed?

4. By formally expanding $(E_n - \epsilon_m)^{-1}$ as

$$\frac{1}{E_n - \epsilon_m} = \frac{1}{\epsilon_n - \epsilon_m} - \frac{1}{(\epsilon_n - \epsilon_m)^2}E_n^{(1)} + \cdots,$$

show that the expression above reduces to the Rayleigh–Schrödinger perturbation theory for E_n up to and including third order.

5. Consider a two-state system described by

$$\underline{H}_0 = \begin{pmatrix} \epsilon_1 & 0 \\ 0 & \epsilon_2 \end{pmatrix}, \qquad \underline{V} = \begin{pmatrix} 0 & a \\ a^* & 0 \end{pmatrix}.$$

Show that the Brillouin–Wigner formulation of perturbation theory yields the exact eigenvalues in both the non-degenerate ($\epsilon_1 \neq \epsilon_2$) and degenerate ($\epsilon_1 = \epsilon_2$) case. What happens in the Rayleigh–Schrödinger version?

Solution

Part 1

From the eigenvalue equation satisfied by $|\psi_n\rangle$ we have

$$(E_n - \hat{H}_0)|\psi_n\rangle = \hat{V}|\psi_n\rangle.$$

The projection operator $\hat{Q}_n$ commutes with $\hat{H}_0$, since

$$\hat{Q}_n \hat{H}_0 = \sum_{m \neq n} |\phi_m\rangle\langle\phi_m| \hat{H}_0 = \sum_{m \neq n} \epsilon_m |\phi_m\rangle\langle\phi_m| = \hat{H}_0 \sum_{m \neq n} |\phi_m\rangle\langle\phi_m| = H_0 \hat{Q}_n .$$

Applying $\hat{Q}_n$ from the left to both sides of the eigenvalue equation and using the above commutativity property, we obtain

$$\hat{Q}_n(E_n - \hat{H}_0)|\psi_n\rangle = (E_n - \hat{H}_0)\,\hat{Q}_n|\psi_n\rangle = \hat{Q}_n \hat{V}|\psi_n\rangle .$$

Part 2

Using the completeness of the H_0 eigenstates yields

$$|\psi_n\rangle = \sum_m |\phi_m\rangle\langle\phi_m|\psi_n\rangle = c_n|\phi_n\rangle + \left(\sum_{m \neq n} |\phi_m\rangle\langle\phi_m|\right)|\psi_n\rangle ,$$

and the normalization condition $\langle\phi_n|\psi_n\rangle = 1$ then gives $c_n = 1$, and so

$$|\psi_n\rangle = |\phi_n\rangle + \hat{Q}_n|\psi_n\rangle .$$

From part 1, after multiplying both sides by $(E_n - \hat{H}_0)^{-1}$, it follows that

$$\hat{Q}_n|\psi_n\rangle = (E_n - \hat{H}_0)^{-1}\hat{Q}_n \hat{V}|\psi_n\rangle \implies |\psi_n\rangle = |\phi_n\rangle + \underbrace{(E_n - \hat{H}_0)^{-1}\hat{Q}_n}_{R_n} \hat{V}|\psi_n\rangle .$$

Since $\hat{Q}_n$ commutes with $\hat{H}_0$, it also commutes with any function of $\hat{H}_0$ and so

$$\hat{R}_n = (E_n - \hat{H}_0)^{-1}\hat{Q}_n = \hat{Q}_n(E_n - \hat{H}_0)^{-1} .$$

The equation for $|\psi_n\rangle$ can be solved by iteration, by setting

$$
\begin{aligned}
|\psi_n\rangle^{(0)} &= |\phi_n\rangle , \\
|\psi_n\rangle^{(1)} &= |\phi_n\rangle + \hat{R}_n \hat{V}|\phi_n\rangle , \\
|\psi_n\rangle^{(2)} &= |\phi_n\rangle + \hat{R}_n \hat{V}|\phi_n\rangle + \hat{R}_n \hat{V}\hat{R}_n \hat{V}|\phi_n\rangle ,
\end{aligned}
$$

and so on. Formally, the resulting geometric series can be summed to yield

$$|\psi_n\rangle = \sum_{p=0}^{\infty} (\hat{R}_n \hat{V})^p |\phi_n\rangle = \left(\mathbb{1} - \hat{R}_n \hat{V}\right)^{-1} |\phi_n\rangle ,$$

This formal solution, even assuming convergence for the series, is not very useful, as the inverse $(\mathbb{1} - \hat{R}_n \hat{V})^{-1}$ is not easily calculable.

Part 3

Starting from the eigenvalue equation satisfied by $|\psi_n\rangle$, we immediately obtain by projecting both sides onto $\langle\phi_n|$

$$\langle\phi_n|\, E_n - \hat{H}_0 \,|\psi_n\rangle = \langle\phi_n|\, \hat{V}|\psi_n\rangle \implies (E_n - \epsilon_n)\underbrace{\langle\phi_n|\psi_n\rangle}_{=1} = \langle\phi_n|\, \hat{V}|\psi_n\rangle .$$

Now using the iterative solution obtained above, we find

$$E_n = \epsilon_n + \langle\phi_n|\, \hat{V}|\phi_n\rangle + \langle\phi_n|\, \hat{V}\hat{R}_n \hat{V}|\phi_n\rangle + \langle\phi_n|\, \hat{V}\hat{R}_n \hat{V}\hat{R}_n \hat{V}|\phi_n\rangle + \cdots ,$$

where $\hat{R}_n$ can expressed as follows:

$$\hat{R}_n = (E_n - \hat{H}_0)^{-1} \sum_{m \neq n} |\phi_m\rangle\langle\phi_m| = \sum_{m \neq n} \frac{|\phi_m\rangle\langle\phi_m|}{E_n - \epsilon_m} \, .$$

Insertion of the expression above into the perturbative expansion finally leads to

$$E_n = \epsilon_n + \langle\phi_n| \hat{V}|\phi_n\rangle + \sum_{m \neq n} \frac{\langle\phi_n| \hat{V}|\phi_m\rangle\langle\phi_m| \hat{V}|\phi_n\rangle}{E_n - \epsilon_m} + \sum_{k \neq n}\sum_{m \neq n} \frac{\langle\phi_n| \hat{V}|\phi_k\rangle\langle\phi_k| \hat{V}|\phi_m\rangle\langle\phi_m| \hat{V}|\phi_n\rangle}{(E_n - \epsilon_k)(E_n - \epsilon_m)} + \cdots \, .$$

Note that the exact (unknown) energy E_n appears in the denominators. As an example, a calculation up to and including third order would require setting $E_n \longrightarrow \epsilon_n$ in the double summation term (the last term above) and $E_n \longrightarrow \epsilon_n + \langle\phi_n| \hat{V}|\phi_n\rangle$ in the single summation term, that is,

$$E_n \text{ to third order} = \epsilon_n + \langle\phi_n| \hat{V}|\phi_n\rangle + \sum_{m \neq n} \frac{\langle\phi_n| \hat{V}|\phi_m\rangle\langle\phi_m| \hat{V}|\phi_n\rangle}{\epsilon_n + \langle\phi_n| \hat{V}|\phi_n\rangle - \epsilon_m}$$

$$+ \sum_{k \neq n}\sum_{m \neq n} \frac{\langle\phi_n| \hat{V}|\phi_k\rangle\langle\phi_k| \hat{V}|\phi_m\rangle\langle\phi_m| \hat{V}|\phi_n\rangle}{(\epsilon_n - \epsilon_k)(\epsilon_n - \epsilon_m)} \, .$$

The expression above includes all terms up to those cubic in $\hat{V}$.

Another method for obtaining E_n (in Brillouin–Wigner perturbation theory) is to guess an initial value for it (for instance, a variational estimate) and then to iterate this estimate by inserting it into the expansion for E_n up to a predetermined order until convergence within an assigned tolerance is achieved.

Part 4

First, note that, in Rayleigh–Schrödinger perturbation theory, the third-order correction reads

$$E_n^{(3)} = \langle\phi_n| \hat{V}|\psi_n^{(2)}\rangle = \sum_{k \neq n}\sum_{m \neq n} \frac{\langle\phi_n| \hat{V}|\phi_k\rangle\langle\phi_k| \hat{V}|\phi_m\rangle\langle\phi_m| \hat{V}|\phi_n\rangle}{(\epsilon_n - \epsilon_k)(\epsilon_n - \epsilon_m)} - \sum_{m \neq n} \frac{|\langle\phi_m|\hat{V}|\phi_n\rangle|^2 \langle\phi_n|\hat{V}|\phi_n\rangle}{(\epsilon_n - \epsilon_m)^2} \, ,$$

using the formulae derived in Section 14.1. On the other hand, consider the single summation term in the expression obtained in part 3, which we write as

$$\sum_{m \neq n} \frac{|\langle\phi_n| \hat{V}|\phi_m\rangle|^2}{\epsilon_n - \epsilon_m} \, \frac{1}{1 + \langle\phi_n|\hat{V}|\phi_n\rangle/(\epsilon_n - \epsilon_m)} \approx \underbrace{\sum_{m \neq n} \frac{|\langle\phi_n| \hat{V}|\phi_m\rangle|^2}{\epsilon_n - \epsilon_m}}_{E_n^{(2)}} - \sum_{m \neq n} \frac{|\langle\phi_n| \hat{V}|\phi_m\rangle|^2\langle\phi_n|\hat{V}|\phi_n\rangle}{(\epsilon_n - \epsilon_m)^2} \, ,$$

where the last term represents a third-order correction. It can be combined with the double-sum term to yield $E_n^{(3)}$ as obtained in Rayleigh–Schrödinger perturbation theory.

Part 5

In Brillouin–Wigner perturbation theory we have

$$E_1 = \epsilon_1 + \underbrace{\langle\phi_1|\hat{V}|\phi_1\rangle}_{\text{vanishes}} + \frac{\langle\phi_1|\hat{V}|\phi_2\rangle\langle\phi_2|\hat{V}|\phi_1\rangle}{E_1 - \epsilon_2} = \epsilon_1 + \frac{|a|^2}{E_1 - \epsilon_2} \, ,$$

and a similar equation for E_2. These equations can be solved with respect to E_1 (or E_2) to yield

$$E_\pm = \frac{\epsilon_1 + \epsilon_2}{2} \pm \sqrt{\frac{(\epsilon_1 - \epsilon_2)^2}{4} + |a|^2} \,,$$

which values are in agreement with the exact eigenvalues of $\underline{H}_0 + \underline{V}$. In the case of degeneracy, then $\epsilon_1 = \epsilon_2 = \epsilon$ and

$$E_\pm = \epsilon \pm |a| \,.$$

In Rayleigh–Schrödinger perturbation theory we find, for the non-degenerate case to second order,

$$E_1 = \epsilon_1 + \frac{|\langle \phi_2 | \hat{V} | \phi_1 \rangle|^2}{\epsilon_1 - \epsilon_2} = \epsilon_1 + \frac{|a|^2}{\epsilon_1 - \epsilon_2} \,, \qquad E_2 = \epsilon_2 + \frac{|a|^2}{\epsilon_2 - \epsilon_1} \,,$$

in agreement with the expansion of $E_\pm$ to leading order in $|a|^2$. In the degenerate case, we diagonalize V, to find $\pm|a|$ as eigenvalues, and hence energies equal to $\epsilon \pm |a|$ including these first-order corrections, in agreement with the exact results in this case. Note that higher-order terms in the expansion of E_n vanish as there are no additional states beyond the degenerate states (in other words, the subspace orthogonal to the degenerate subspace spanned by $|\phi_1\rangle$ and $|\phi_2\rangle$ is the null space).

Problem 21 Variational Calculation of the Ground-State Energy in a Screened Coulomb Potential

Consider an electron (mass m and charge $-e$) in a multi-electron atom and assume that the electrostatic potential felt by the electron can be adequately described by a screened Coulomb potential of the form

$$V(r) = -\frac{Ze^2}{r} e^{-r/R} \,,$$

where R has the dimension of length. Use as a (normalized) trial wave function

$$\psi_T(r) = \frac{1}{\sqrt{\pi a^3}} e^{-r/a} \,,$$

where a is a free parameter, and perform a variational calculation of the ground-state energy. Determine the condition that the screening radius must satisfy for a minimum of this energy to exist.

Solution

The variational estimate follows as (after an integration by parts)

$$E_T(a) = \langle \psi_T | H | \psi_T \rangle = \int d\mathbf{r} \left[\frac{\hbar^2}{2m} \left| \boldsymbol{\nabla} \psi_T(r) \right|^2 + V(r)\, \psi_T^2(r) \right] \,,$$

where

$$\boldsymbol{\nabla} \psi_T(r) = \hat{\mathbf{r}} \psi_T'(r) = -\frac{\hat{\mathbf{r}}}{\sqrt{\pi a^5}} e^{-r/a} \,,$$

so that

$$E_T(a) = \frac{1}{\pi a^3} \int d\mathbf{r} \left[\frac{\hbar^2}{2ma^2} - \frac{Ze^2}{r} e^{-r/R} \right] e^{-2r/a} = \frac{\hbar^2}{2ma^2} - \frac{Ze^2}{a} \frac{4R^2}{(2R + a)^2} \,.$$

The requirement $E'_T(a) = 0$, with

$$E'_T(a) = -\frac{\hbar^2}{ma^3} + \frac{Ze^2}{a^2}\frac{4R^2}{(2R+a)^2}\left(1 + \frac{2a}{2R+a}\right),$$

yields

$$\underbrace{\frac{x}{(1+x)^2}\left(1 + \frac{2x}{1+x}\right)}_{f(x)} = \frac{r_0}{2R}.$$

where $x = a/(2R)$ and $r_0 = a_0/Z$, and a_0 is the Bohr radius. The function $f(x)$ is such that $f(0) = 0$ and $f(x \longrightarrow \infty) = 0$, and it assumes a maximum at $x_0 \approx 1.55$ with $f(x_0) \approx 0.528$. If $r_0/(2R) > f(x_0)$, there are no solutions, suggesting that, if the screening radius is too small, there are no bound states. For a minimum (and a bound state) to exist, we must have $R > (r_0/2)/f(x_0)$.

Problem 22 Variational Calculation of the Hydrogen Atom Ground-State Energy

Perform a variational calculation of the hydrogen atom ground state using the (spherically symmetric) trial wave function $\psi_T(r) = u(r; \beta)/r$, with two possible choices for $u(r, \beta)$:

$$u_1(x; \beta) = \frac{x}{x^2 + \beta^2}, \qquad u_2(x; \beta) = x^2\,e^{-\beta x},$$

where $x = r/a_0$, a_0 is the Bohr radius, and β is a variational parameter. Which of these two trial wave functions provides the better variational estimate for the ground-state energy? Interpret the results. Also, calculate the quantity

$$\delta = 1 - |\langle \psi_0 | \psi_i \rangle|^2,$$

where $\psi_0(r)$ is the exact (and normalized) ground-state wave function of the hydrogen atom and $\psi_i(r) = u_i(r; \beta_0)/r$ is the (normalized) trial wave function corresponding to the optimal β_0 that minimizes the energy.

Solution

We compute the numerator N and denominator D of Eq. (14.44) to obtain

$$N = 4\pi \int_0^\infty dr\, u^*(r; \beta)\left(-\frac{\hbar^2}{2\mu}\frac{d^2}{dr^2} - \frac{e^2}{r}\right)u(r; \beta), \qquad D = 4\pi \int_0^\infty dr\, |u(r; \beta)|^2.$$

It is convenient to rescale the integration variable as $x = r/a_0$, where a_0 is the Bohr radius $a_0 = \hbar^2/(\mu\, e^2)$. We find

$$E(\beta) = \underbrace{-\frac{e^2}{2a_0}}_{E_0}\left[\int_0^\infty dx\, u^*(x; \beta)\left(\frac{d^2}{dx^2} + \frac{2}{x}\right)u(x; \beta)\right]\left[\int_0^\infty dx\, |u(x; \beta)|^2\right]^{-1},$$

and the pre-factor happens to be the exact ground state energy. Inserting the trial wave functions provided in the text and carrying out the integrations yields the results for the function $E(\beta)$, the value β_0 that minimizes $E(\beta)$, the variational estimate $E(\beta_0)$ to the true ground-state energy E_0, and the overlap δ, reported in Table 14.1. The parameter δ provides a rough estimate of the difference between the trial and true wave functions.

Table 14.1 The variational energies $E(\beta_0)$ corresponding to the two choices of trial wave functions; β_0 is the value that minimizes $E(\beta)$, and $\delta = 1 - |\langle \psi_0 | \psi_i \rangle|^2$.

	u_1	u_2
$E(\beta)/E_0$	$(8\beta - \pi)/(2\pi\beta^2)$	$\beta - \beta^2/3$
β_0	$\pi/4$	$3/2$
$E(\beta_0)/E_0$	$8/\pi^2$	$3/4$
δ	0.21	0.05

The wave function $u_1(r)$ has the correct behavior at the origin (namely, it is linear in r) but incorrect behavior at ∞ (it vanishes only as $1/r$ in this limit, not fast enough); nevertheless it gives the best variational estimate for the energy. By contrast, $u_2(r)$ falls off exponentially, and hence correctly, as $r \longrightarrow \infty$ but behaves incorrectly as $r \longrightarrow 0$. It gives a worse variational energy than $u_1(r)$ because it has the wrong behavior near the origin, where the potential is strongly attractive. Interestingly, $\delta_1 > \delta_2$ even though $E_1(\beta_0) < E_2(\beta_0)$.

Problem 23   Variational Calculation of the Helium Atom Ground-State Energy

Carry out a variational calculation of the helium atom ground-state energy using a (normalized) trial wave function of the form

$$\psi_T(r_1, r_2) = \frac{1}{\pi a^3}\, e^{-(r_1 + r_2)/a}, \qquad r_i = |\mathbf{r}_i| \,,$$

where a is the variational parameter. Ignore the spin degrees of freedom of the two electrons, and take the Hamiltonian as

$$H = -\frac{\hbar^2}{2m_e}\left(\mathbf{\nabla}_1^2 + \mathbf{\nabla}_2^2\right) - \frac{2e^2}{r_1} - \frac{2e^2}{r_2} + \frac{e^2}{|\mathbf{r}_1 - \mathbf{r}_2|}\,,$$

with the nucleus (of charge $2e$) fixed at the origin. The following expansion may be useful:

$$\frac{1}{|\mathbf{r}_1 - \mathbf{r}_2|} = \frac{1}{r_>}\sum_{l=0}^{\infty}\left(\frac{r_<}{r_>}\right)^l P_l(\cos\theta), \qquad r_< = \min(r_1, r_2), \qquad r_> = \max(r_1, r_2)\,,$$

where the $P_l(\cos\theta)$ are Legendre polynomials.

Solution

The trial wave function consists of the product of two (normalized) single-electron wave functions:

$$\psi_T(r_1, r_2) = \psi_1(r_1)\,\psi_2(r_2)\,, \qquad \psi_i(r_i) = \frac{1}{\sqrt{\pi a^3}}\, e^{-r_i/a}\,.$$

We define

$$H_i = T_i + V_i = -\frac{\hbar^2}{2m_e}\mathbf{\nabla}_i^2 - \frac{2e^2}{r_i}\,,$$

and calculate $\langle \psi_i | H_i | \psi_i \rangle$ to obtain

$$\langle \psi_i | T_i | \psi_i \rangle = \frac{\hbar^2}{2m_e} \int d\mathbf{r}_i \left| \boldsymbol{\nabla}_i \, \psi_i(r_i) \right|^2 = \frac{\hbar^2}{2m_e} \frac{1}{\pi a^3} \frac{4\pi}{a^2} \int_0^\infty dr_i \, r_i^2 \, e^{-2r_i/a} = \frac{\hbar^2}{2m_e a^2} \, ,$$

$$\langle \psi_i | V_i | \psi_i \rangle = -\frac{2e^2}{\pi a^3} \int d\mathbf{r}_i \frac{\left| \psi_i(r_i) \right|^2}{r_i} = -\frac{8e^2}{a^3} \int_0^\infty dr_i \, r_i \, e^{-2r_i/a} = -\frac{2e^2}{a} \, .$$

We introduce the Bohr radius

$$a_0 = \frac{\hbar^2}{m_e e^2} \, ,$$

in terms of which the kinetic energy expectation value is given by

$$\langle \psi_i | T_i | \psi_i \rangle = \frac{e^2}{2a} \frac{a_0}{a} \, ,$$

and therefore

$$\langle \psi_T | \sum_i H_i | \psi_T \rangle = \frac{e^2}{a} \frac{a_0}{a} - \frac{4e^2}{a} \, .$$

We are left with the evaluation of the Coulomb repulsive term

$$\langle \psi_T | V_{12} | \psi_T \rangle = \frac{e^2}{\pi^2 a^6} \int d\mathbf{r}_1 d\mathbf{r}_2 \, e^{-2r_1/a} \, e^{-2r_2/a} \frac{1}{|\mathbf{r}_1 - \mathbf{r}_2|} \, .$$

Using an expansion in terms of Legendre polynomials, the above expectation value can be written as

$$\langle \psi_T | V_{12} | \psi_T \rangle = \frac{e^2}{\pi^2 a^6} 4\pi \int_0^\infty dr_1 \, r_1^2 \, e^{-2r_1/a} \int_0^\infty dr_2 \, r_2^2 \, e^{-2r_2/a} \int d\Omega_{12} \frac{1}{r_>} \sum_l \left(\frac{r_<}{r_>} \right)^l P_l(\cos \theta_{12}) \, ,$$

where θ_{12} is the angle between $\mathbf{r}_1$ and $\mathbf{r}_2$. The solid-angle integration gives

$$\int d\Omega_{12} \frac{1}{r_>} \sum_l \left(\frac{r_<}{r_>} \right)^l P_l(\cos \theta_{12}) = 2\pi \frac{1}{r_>} \sum_l \left(\frac{r_<}{r_>} \right)^l \int_{-1}^1 d(\cos \theta_{12}) \, P_l(\cos \theta_{12}) = \frac{4\pi}{r_>} \, ,$$

using the orthogonality of the Legendre polynomials, that is,

$$\int_{-1}^1 dx \, P_l(x) = \int_{-1}^1 dx \, P_0(x) \, P_l(x) = 2 \, \delta_{l,0} \, .$$

Thus, we have

$$\langle \psi_T | V_{12} | \psi_T \rangle = 16 \frac{e^2}{a^6} \int_0^\infty dr_1 \, r_1^2 \, e^{-2r_1/a} \underbrace{\int_0^\infty dr_2 \, r_2^2 \, e^{-2r_2/a} \frac{1}{r_>}}_{I} \, ,$$

and the integration over r_2 is written as (with $\alpha = 2/a$)

$$I = \frac{1}{r_1} \int_0^{r_1} dr_2 \, r_2^2 \, e^{-\alpha r_2} + \int_{r_1}^\infty dr_2 \, r_2 \, e^{-\alpha r_2} = \frac{1}{r_1} \frac{d^2}{d\alpha^2} \int_0^{r_1} dr_2 \, e^{-\alpha r_2} - \frac{d}{d\alpha} \int_{r_1}^\infty dr_2 \, e^{-\alpha r_2}$$

$$= \frac{1}{r_1} \frac{d^2}{d\alpha^2} \frac{1}{\alpha} \left(1 - e^{-\alpha r_1} \right) - \frac{d}{d\alpha} \frac{e^{-\alpha r_1}}{\alpha} = \frac{2}{r_1 \alpha^3} - e^{-\alpha r_1} \left(\frac{2}{r_1 \alpha^3} + \frac{1}{\alpha^2} \right) \, .$$

Hence, it follows that

$$\langle \psi_T | V_{12} | \psi_T \rangle = 16 \frac{e^2}{a^6} \int_0^\infty dr_1 \, r_1^2 \, e^{-ar_1} \left(\frac{2}{r_1 \alpha^3} - e^{-ar_1} \left(\frac{2}{r_1 \alpha^3} + \frac{1}{\alpha^2} \right) \right) = \frac{5}{8} \frac{e^2}{a} .$$

We finally arrive at

$$E_T(a) = \frac{e^2}{a} \frac{a_0}{a} - \frac{4e^2}{a} + \frac{5}{8} \frac{e^2}{a} = \frac{e^2}{a} \left(\frac{a_0}{a} - \frac{27}{8} \right) ,$$

which has a minimum at

$$a^* = \frac{16}{27} a_0 \implies E_T(a^*) = - \left(\frac{27}{16} \right)^2 \frac{e^2}{a_0} .$$

We note that the experimental value for the helium atom ground-state energy is $-2.90 e^2/a_0$, while the variational estimate above gives $E_T(a^*) \approx -2.85 e^2/a_0$: not bad! In a more realistic treatment in which the spins of the electrons are accounted for, the present trial wave function would be written as

$$\psi_T(r_1, r_2; S = 0) = \frac{1}{\pi a^3} e^{-(r_1 + r_2)/a} \frac{1}{\sqrt{2}} (|+, -\rangle - |-, +\rangle) ,$$

so that it is antisymmetric under the exchange of electron 1 with electron 2, namely, the individual spins are coupled to total spin zero (a singlet state).

Problem 24   The Born–Oppenheimer Approximation

The purpose of this problem is to derive the Born–Oppenheimer approximation for a molecule. The molecule consists of M nuclei of charge $Z_i e$ and masses M_i with $i = 1, \ldots, M$ and N electrons of mass m with

$$\sum_{i=1}^M Z_i = N ,$$

so that the molecule is neutral.

1. Denote by $\hat{\mathbf{R}}_i$ and $\hat{\mathbf{P}}_i$ the position and momentum operators of nucleus i, and with $\hat{\mathbf{r}}_k$ and $\hat{\mathbf{p}}_k$ those of electron k. Recall that $[\hat{R}_{i\alpha}, \hat{P}_{j\beta}] = i\hbar \, \delta_{ij} \, \delta_{\alpha\beta}$ and $[\hat{r}_{k\alpha}, \hat{p}_{l\beta}] = i\hbar \, \delta_{kl} \, \delta_{\alpha\beta}$, with all other commutators vanishing. Ignoring spin degrees of freedom, write down the Hamiltonian of the molecule. It consists of three terms

$$\hat{H} = \hat{T}_{\text{nuc}} + \hat{T}_{\text{el}} + \hat{V}$$

where $\hat{T}_{\text{nuc}}$ and $\hat{T}_{\text{el}}$ are the kinetic energy operators for the nuclei and electrons, respectively, and $\hat{V}$ is the potential energy operator.

2. Ignore the kinetic energy of nuclei – why is it reasonable to do so? – and explain why the resulting Hamiltonian $\hat{H}' = \hat{T}_{\text{el}} + \hat{V}$ can be diagonalized simultaneously with the position operators $\hat{\mathbf{R}}_1, \ldots, \hat{\mathbf{R}}_M$ of the M nuclei. Denote the simultaneous eigenstates compactly as $|\psi_{a,\mathbf{R}}\rangle$, where

$$|\psi_{a,\mathbf{R}}\rangle \equiv |\chi_a\rangle \otimes |\phi_{\mathbf{R}_1}\rangle \otimes \cdots \otimes |\phi_{\mathbf{R}_M}\rangle ,$$

and $|\phi_{\mathbf{R}_i}\rangle$ are the position eigenstates of nucleus i and $|\chi_a\rangle$ are the eigenstates of $\hat{H}'$. The eigenstates $|\psi_{a,\mathbf{R}}\rangle$ form a basis. Write down the normalization condition that they satisfy.

3. Show that in the subspace of the M-nuclei with eigenvalues $\mathbf{R}_1, \ldots, \mathbf{R}_M$ the states $|\chi_a\rangle$ satisfy

$$\left[\sum_{k=1}^{N} \frac{\hat{\mathbf{p}}_k^2}{2m} + \hat{V}(\hat{\mathbf{r}}_1, \ldots, \hat{\mathbf{r}}_N; \mathbf{R})\right] |\chi_a\rangle = \mathcal{E}_a(\mathbf{R})|\chi_a\rangle ,$$

where hereafter $\mathbf{R}$ is a shorthand for $\mathbf{R}_1, \ldots, \mathbf{R}_M$. In the coordinate representation for the N electrons $|\phi_\mathbf{r}\rangle \equiv |\phi_{\mathbf{r}_1}\rangle \otimes \cdots \otimes |\phi_{\mathbf{r}_N}\rangle$, show that the Schrödinger equation reads

$$\left[-\sum_{k=1}^{N} \frac{\hbar^2 \mathbf{\nabla}_k^2}{2m} + V(\mathbf{r}_1, \ldots, \mathbf{r}_N; \mathbf{R})\right] \chi_a(\mathbf{r}_1, \ldots, \mathbf{r}_N; \mathbf{R}) = \mathcal{E}_a(\mathbf{R})\chi_a(\mathbf{r}_1, \ldots, \mathbf{r}_N; \mathbf{R}) ,$$

where $\mathbf{\nabla}_k^2$ is the Laplacian with respect to $\mathbf{r}_k$.

4. Exploiting the completeness of the basis $|\psi_{a,\mathbf{R}}\rangle$ introduced earlier, expand a generic eigenstate $|\Psi\rangle$ of the full Hamiltonian $\hat{H}$ corresponding to eigenvalue E as

$$|\Psi\rangle = \sum_{a'} \int d\mathbf{R}' \, |\psi_{a',\mathbf{R}'}\rangle \, \Psi_{a'}(\mathbf{R}') ,$$

and show that the Schrödinger equation for $|\Psi\rangle$ can be written as

$$\sum_{a'} \int d\mathbf{R}' \, \Psi_{a'}(\mathbf{R}') \left[\sum_{i=1}^{M} \frac{\hat{\mathbf{P}}_i^2}{2M_i} + \mathcal{E}_{a'}(\mathbf{R}') - E\right] |\psi_{a',\mathbf{R}'}\rangle = 0 .$$

This equation is exact; no approximations have been made up to until now.

5. Project the equation obtained above onto the basis of position eigenstates of the N electrons and M nuclei,

$$|\phi_{\mathbf{r},\mathbf{R}}\rangle \equiv |\phi_\mathbf{r}\rangle \otimes |\phi_\mathbf{R}\rangle \overset{i}{=} |\phi_{\mathbf{r}_1}\rangle \otimes \cdots \otimes |\phi_{\mathbf{r}_N}\rangle \otimes |\phi_{\mathbf{R}_1}\rangle \otimes \cdots \otimes |\phi_{\mathbf{R}_M}\rangle ,$$

and show that if derivatives of the wave functions $\psi_{a'}(\mathbf{r}; \mathbf{R})$ with respect to the $\mathbf{R}$ variables are neglected then $\Psi_a(\mathbf{R})$ satisfies the equation

$$\left[-\sum_{i=1}^{M} \frac{\hbar^2 \mathbf{\nabla}_i^2}{2M_i} + \mathcal{E}_a(\mathbf{R})\right] \Psi_a(\mathbf{R}) = E\Psi_a(\mathbf{R}) ,$$

that is, the wave function $\Psi_a(\mathbf{R})$ satisfies a Schrödinger equation in which the electronic energy $\mathcal{E}_a(\mathbf{R})$ obtained for fixed nuclear positions acts as the potential for the nuclei. This is the essence of the Born–Oppenheimer approximation. Define the Hamiltonian $\hat{H}_{\text{nuc}}$ as follows:

$$\hat{H}_{\text{nuc}} = \sum_{i=1}^{M} \frac{\hat{\mathbf{P}}_i^2}{2M_i} + \mathcal{E}_a(\hat{\mathbf{R}}) .$$

6. Consider the case in which $\mathcal{E}_a(\mathbf{R}) \longrightarrow \mathcal{E}_0(\mathbf{R})$ is the ground-state energy of the electronic configuration. Estimates for $\mathcal{E}_0(\mathbf{R})$ can be obtained by the variational method, by minimizing the Hamiltonian $\hat{H}'$. Suppose that the potential $\mathcal{E}_0(\mathbf{R})$ has a minimum when the nuclei are at positions $\mathbf{R}_i^0$. Show that H_{nuc} can be approximated (up to a constant) as

$$\hat{H}_{\text{nuc}} = \sum_{i=1}^{M} \frac{\hat{\mathbf{P}}_i^2}{2M_i} + \frac{1}{2} \sum_{i,j=1}^{M} \sum_{\alpha,\beta=1}^{3} K_{i\alpha,j\beta} \left(\hat{R}_{i\alpha} - R_{i\alpha}^0\right) \left(\hat{R}_{j\beta} - R_{j\beta}^0\right) ,$$

where the $K_{i\alpha,j\beta}$ are constants. How are the $K_{i\alpha,j\beta}$ related to $\mathcal{E}_0(\mathbf{R})$? What kind of system does the above Hamiltonian describe?

7. Using the Hellmann–Feynman theorem derived in Problem 15, show that partial derivatives of $\mathcal{E}_0(\mathbf{R})$ relative to the components $R_{i\alpha}$ can be expressed as expectation values of corresponding derivatives of the potential energy $V(\mathbf{r}, \mathbf{R})$, on the wave function $\chi_0(\mathbf{r}; \mathbf{R})$.

8. Show that the energies of the vibrational modes of the molecule are suppressed relative to those of the electrons by a factor $(m/M)^{1/2}$, where M is a typical nuclear mass.

Solution

Part 1

Ignoring spin couplings among electrons and between electrons and nuclei, the interactions consist of the Coulomb attraction between nuclei and electrons and the Coulomb repulsion among electrons and among nuclei, that is,

$$H = \underbrace{\sum_{i=1}^{M} \frac{\hat{\mathbf{P}}_i^2}{2M_i}}_{\hat{T}_{\text{nuc}}} + \underbrace{\sum_{k=1}^{N} \frac{\hat{\mathbf{p}}_k^2}{2m}}_{\hat{T}_{\text{el}}} \underbrace{- \sum_{i=1}^{M} \sum_{k=1}^{N} \frac{Z_i e^2}{|\hat{\mathbf{R}}_i - \hat{\mathbf{r}}_k|} + \sum_{i<j=1}^{M} \frac{Z_i Z_j e^2}{|\hat{\mathbf{R}}_i - \hat{\mathbf{R}}_j|} + \sum_{k<l=1}^{M} \frac{e^2}{|\hat{\mathbf{r}}_k - \hat{\mathbf{r}}_l|}}_{\hat{V}} .$$

Part 2

Since the nuclear masses are three orders of magnitude larger than the electron mass ($m_p/m \approx 1836$, m_p being the proton mass), it is reasonable to ignore the nuclear kinetic energies and approximate the full Hamiltonian by retaining only the kinetic energies of the electrons and the interaction terms, that is,

$$\hat{H}' = \hat{T}_{\text{el}} + \hat{V} ,$$

where it should be noted that $\hat{H}'$ depends not only on the position and momentum operators of the individual electrons but also on the position operators of the nuclei. Since $[\hat{R}_{i\alpha}, \hat{R}_{j\beta}] = 0$ and $[\hat{R}_{i\alpha}, \hat{H}'] = 0$, it is possible to construct a basis of common eigenstates of all these observables. The position eigenstates relative to $\hat{\mathbf{R}}_i$ are denoted as $|\phi_{\mathbf{R}_i}\rangle$, and

$$\hat{\mathbf{R}}_i |\phi_{\mathbf{R}_i}\rangle = \mathbf{R}_i |\phi_{\mathbf{R}_i}\rangle , \qquad \langle \phi_{\mathbf{R}_i'} | \phi_{\mathbf{R}_i}\rangle = \delta(\mathbf{R}_i - \mathbf{R}_i') ;$$

the corresponding eigenvalues $\mathbf{R}_i$ form a continuum. The basis of position eigenstates for nuclei $1, \ldots, M$ consists of the tensor product $|\phi_{\mathbf{R}_1}\rangle \otimes \cdots \otimes |\phi_{\mathbf{R}_M}\rangle$. In the state space of electrons and nuclei, we write the simultaneous eigenstates of $\hat{H}'$ and $\hat{\mathbf{R}}_1, \ldots, \hat{\mathbf{R}}_M$ as

$$|\psi_{a,\mathbf{R}}\rangle \equiv |\chi_a\rangle \otimes |\phi_{\mathbf{R}_1}\rangle \otimes \cdots \otimes |\phi_{\mathbf{R}_M}\rangle \equiv |\chi_a\rangle \otimes |\phi_{\mathbf{R}}\rangle .$$

They satisfy the continuum normalization condition

$$\langle \psi_{a,\mathbf{R}} | \psi_{b,\mathbf{R}'}\rangle = \delta_{a,b} \underbrace{\prod_{i=1}^{M} \delta(\mathbf{R}_i - \mathbf{R}_i')}_{\delta(\mathbf{R}-\mathbf{R}')} .$$

Part 3

Using the completeness relation for the nuclear position eigenstates,

$$\int d\mathbf{R}' \, |\phi_{\mathbf{R}'}\rangle\langle\phi_{\mathbf{R}'}| \equiv \int d\mathbf{R}'_1 \, |\phi_{\mathbf{R}'_1}\rangle\langle\phi_{\mathbf{R}'_1}| \otimes \cdots \otimes \int d\mathbf{R}'_M \, |\phi_{\mathbf{R}'_M}\rangle\langle\phi_{\mathbf{R}'_M}| = \hat{\mathbb{1}} \, ,$$

it follows (in a compact notation) that

$$
\begin{aligned}
\hat{H}' \, |\psi_{a,\mathbf{R}}\rangle &= \int d\mathbf{R}' \, |\phi_{\mathbf{R}'}\rangle\langle\phi_{\mathbf{R}'}|\hat{H}'|\chi_a\rangle \otimes |\phi_{\mathbf{R}}\rangle \\
&= \int d\mathbf{R}' \, |\phi_{\mathbf{R}'}\rangle \left[\hat{T}_{\mathrm{el}} + \hat{V}(\hat{\mathbf{r}}_1, \dots, \hat{\mathbf{r}}_N; \mathbf{R}')\right] |\chi_a\rangle \underbrace{\langle\phi_{\mathbf{R}'}|\phi_{\mathbf{R}}\rangle}_{\delta(\mathbf{R}-\mathbf{R}')} \\
&= |\phi_{\mathbf{R}}\rangle \left[\hat{T}_{\mathrm{el}} + \hat{V}(\hat{\mathbf{r}}_1, \dots, \hat{\mathbf{r}}_N; \mathbf{R})\right] |\chi_a\rangle \, ,
\end{aligned}
$$

where in the last step we have integrated out the δ-functions. Thus, we have

$$0 = (\hat{H}' - \mathcal{E}_a)|\psi_{a,\mathbf{R}}\rangle = |\phi_{\mathbf{R}}\rangle \left[\hat{T}_{\mathrm{el}} + \hat{V}(\hat{\mathbf{r}}_1, \dots, \hat{\mathbf{r}}_N; \mathbf{R}) - \mathcal{E}_a\right] |\chi_a\rangle \, ,$$

yielding the required relation in the subspace with eigenvalues $\mathbf{R}$,

$$\left[\hat{T}_{\mathrm{el}} + \hat{V}(\hat{\mathbf{r}}_1, \dots, \hat{\mathbf{r}}_N; \mathbf{R})\right] |\chi_a(\mathbf{R})\rangle = \mathcal{E}_a(\mathbf{R}) \, |\chi_a(\mathbf{R})\rangle \, ,$$

and we have made explicit the dependence on $\mathbf{R}$ of both the energy $\mathcal{E}_a$ and state $|\chi_a\rangle$. Projecting out the above relation onto the position eigenstates of the electrons, we find

$$\underbrace{\langle\phi_{\mathbf{r}}| \left[\hat{T}_{\mathrm{el}} + \hat{V}(\hat{\mathbf{r}}_1, \dots, \hat{\mathbf{r}}_N; \mathbf{R})\right] |\chi_a(\mathbf{R})\rangle}_{\left[-\sum_{k=1}^{N} \hbar^2 \boldsymbol{\nabla}_k^2/(2m) + V(\mathbf{r};\mathbf{R})\right]\chi_a(\mathbf{r};\mathbf{R})} = \mathcal{E}_a(\mathbf{R}) \, \underbrace{\langle\phi_{\mathbf{r}}|\chi_a(\mathbf{R})\rangle}_{\chi_a(\mathbf{r};\mathbf{R})} \, .$$

Part 4

The eigenvalue problem for the full Hamiltonian is given by

$$(\hat{T}_{\mathrm{nuc}} + \hat{H}')|\Psi\rangle = E|\Psi\rangle \, .$$

Inserting the expansion of $|\Psi\rangle$ into the basis $|\psi_{a,\mathbf{R}}\rangle$, we find

$$
\begin{aligned}
0 &= (\hat{T}_{\mathrm{nuc}} + \hat{H}' - E)|\Psi\rangle = (\hat{T}_{\mathrm{nuc}} + \hat{H}' - E) \sum_{a'} \int d\mathbf{R}' \, \Psi_{a'}(\mathbf{R}') \, |\psi_{a',\mathbf{R}'}\rangle \\
&= \sum_{a'} \int d\mathbf{R}' \, \Psi_{a'}(\mathbf{R}') \left[\hat{T}_{\mathrm{nuc}} + \mathcal{E}_{a'}(\mathbf{R}') - E\right] |\psi_{a',\mathbf{R}'}\rangle \, ,
\end{aligned}
$$

where in the last line we have used $\hat{H}' \, |\psi_{a',\mathbf{R}'}\rangle = \mathcal{E}_{a'}(\mathbf{R}')|\psi_{a',\mathbf{R}'}\rangle$.

Part 5

When projected onto position eigenstates $|\phi_{\mathbf{r},\mathbf{R}}\rangle$, the eigenvalue equation obtained above is written (again, in compact notation) as

$$0 = \sum_{a'} \int d\mathbf{R}' \, \Psi_{a'}(\mathbf{R}') \langle \phi_{\mathbf{r},\mathbf{R}} | \left[\hat{T}_{\text{nuc}} + \mathcal{E}_{a'}(\mathbf{R}') - E \right] |\psi_{a',\mathbf{R}'}\rangle$$

$$= \sum_{a'} \int d\mathbf{R}' \, \Psi_{a'}(\mathbf{R}') \left[-\sum_{i=1}^{M} \frac{\hbar^2}{2M_i} \boldsymbol{\nabla}_{R_i}^2 + \mathcal{E}_{a'}(\mathbf{R}') - E \right] \delta(\mathbf{R} - \mathbf{R}') \, \chi_{a'}(\mathbf{r}; \mathbf{R}') \,,$$

where the Laplacian acts on the unprimed $\mathbf{R}_i$ (and not on $\mathbf{R}'_i$). The last line follows from considering

$$\langle \phi_{\mathbf{r},\mathbf{R}} | \left[\hat{T}_{\text{nuc}} + \mathcal{E}_{a'}(\mathbf{R}') - E \right] |\psi_{a',\mathbf{R}'}\rangle = \int d\mathbf{r}'' \, d\mathbf{R}'' \, \langle \phi_{\mathbf{r},\mathbf{R}} | \left[\hat{T}_{\text{nuc}} + \mathcal{E}_{a'}(\mathbf{R}') - E \right] |\phi_{\mathbf{r}'',\mathbf{R}''}\rangle \langle \phi_{\mathbf{r}'',\mathbf{R}''} |\psi_{a',\mathbf{R}'}\rangle$$

$$= \int d\mathbf{r}'' \, d\mathbf{R}'' \, \delta(\mathbf{r} - \mathbf{r}'') \left[-\sum_{i=1}^{M} \frac{\hbar^2}{2M_i} \boldsymbol{\nabla}_{R_i}^2 + \mathcal{E}_{a'}(\mathbf{R}') - E \right] \delta(\mathbf{R} - \mathbf{R}'')$$

$$\times \delta(\mathbf{R}'' - \mathbf{R}') \chi_{a'}(\mathbf{r}''; \mathbf{R}')$$

$$= \left[-\sum_{i=1}^{M} \frac{\hbar^2}{2M_i} \boldsymbol{\nabla}_{R_i}^2 + \mathcal{E}_{a'}(\mathbf{R}') - E \right] \delta(\mathbf{R} - \mathbf{R}') \, \chi_{a'}(\mathbf{r}; \mathbf{R}') \,,$$

where we have used the completeness of the position eigenstates and in the last step we integrated out δ-functions. Because of the remaining δ-function, we can replace $\mathcal{E}_{a'}(\mathbf{R}')$ by $\mathcal{E}_{a'}(\mathbf{R})$ and hence arrive at

$$0 = \sum_{a'} \left[-\sum_{i=1}^{M} \frac{\hbar^2}{2M_i} \boldsymbol{\nabla}_{R_i}^2 + \mathcal{E}_{a'}(\mathbf{R}) - E \right] \int d\mathbf{R}' \, \Psi_{a'}(\mathbf{R}') \, \delta(\mathbf{R} - \mathbf{R}') \, \chi_{a'}(\mathbf{r}; \mathbf{R}')$$

$$= \sum_{a'} \left[-\sum_{i=1}^{M} \frac{\hbar^2}{2M_i} \boldsymbol{\nabla}_{R_i}^2 + \mathcal{E}_{a'}(\mathbf{R}) - E \right] \Psi_{a'}(\mathbf{R}) \, \chi_{a'}(\mathbf{r}; \mathbf{R}) \,,$$

and the Laplacian acts on both the "nuclear wave function" $\Psi_{a'}(\mathbf{R})$ as well as on the "electronic wave function" $\chi_{a'}(\mathbf{r}; \mathbf{R})$. At this stage no approximations have been made. The Born–Oppenheimer approximation consists in neglecting the derivatives acting on the electronic wave function:

$$\boldsymbol{\nabla}_{R_i}^2 \Psi_{a'}(\mathbf{R}) \chi_{a'}(\mathbf{r}; \mathbf{R}) = \left[\boldsymbol{\nabla}_{R_i}^2 \Psi_{a'} \right] \chi_{a'} + 2 \left[\boldsymbol{\nabla}_{R_i} \Psi_{a'} \right] \cdot \boldsymbol{\nabla}_{R_i} \chi_{a'} + \Psi_{a'} \boldsymbol{\nabla}_{R_i}^2 \chi_{a'} \approx \left[\boldsymbol{\nabla}_{R_i}^2 \Psi_{a'}(\mathbf{R}) \right] \chi_{a'}(\mathbf{r}; \mathbf{R}) \,.$$

Under this approximation, the eigenvalue problem reduces to

$$\sum_{a'} \chi_{a'}(\mathbf{r}; \mathbf{R}) \left[-\sum_{i=1}^{M} \frac{\hbar^2}{2M_i} \boldsymbol{\nabla}_{R_i}^2 + \mathcal{E}_{a'}(\mathbf{R}) - E \right] \Psi_{a'}(\mathbf{R}) = 0 \,.$$

The electronic wave functions are orthonormal,

$$\int d\mathbf{r} \, \chi_a^*(\mathbf{r}; \mathbf{R}) \, \chi_{a'}(\mathbf{r}; \mathbf{R}) = \delta_{a,a'} \,,$$

and therefore on multiplying the above relation by $\chi_a^*(\mathbf{r}; \mathbf{R})$ and integrating over $d\mathbf{r}$, we arrive at

$$\left[-\sum_{i=1}^{M} \frac{\hbar^2 \boldsymbol{\nabla}_i^2}{2M_i} + \mathcal{E}_a(\mathbf{R}) \right] \Psi_a(\mathbf{R}) = E \Psi_a(\mathbf{R}) \,,$$

and we see that the electronic degrees of freedom enter the nuclear eigenvalue problem only through the "potential" $\mathcal{E}_a(\mathbf{R})$. Of course, if derivatives of the electronic wave functions are retained then the problem becomes much more complex, as there are then couplings between $\Psi_a(\mathbf{R})$ and $\Psi_{a'}(\mathbf{R})$.

Part 6

Around the minimum $\mathbf{R}^0$, the "potential" $\mathcal{E}_0(\mathbf{R})$ can be expanded as

$$\mathcal{E}_0(\mathbf{R}) = \mathcal{E}_0(\mathbf{R}^0) + \frac{1}{2} \sum_{i,j=1}^{M} \sum_{\alpha,\beta=1}^{3} \left(R_{i\alpha} - R_{i\alpha}^0 \right) \left(R_{j\beta} - R_{j\beta}^0 \right) \left. \frac{\partial^2}{\partial R_{i\alpha} \partial R_{j\beta}} \mathcal{E}_0(\mathbf{R}) \right|_{\mathbf{R}=\mathbf{R}^0} + \cdots .$$

We define the constants

$$K_{i\alpha,j\beta} = \left. \frac{\partial^2}{\partial R_{i\alpha} \partial R_{j\beta}} \mathcal{E}_0(\mathbf{R}) \right|_{\mathbf{R}=\mathbf{R}^0} ,$$

and the $3M \times 3M$ matrix $\underline{K}$ has non-negative eigenvalues (it is positive definite). Up to an overall energy shift, by the constant $\mathcal{E}_0(\mathbf{R}^0)$, the nuclear Hamiltonian is reduced to that for a set of $3M$ coupled harmonic oscillators. On introducing normal coordinates the oscillators decouple, that is, we arrive at a system of $3M$ independent oscillators (see Problem 11 in Chapter 8). We can then easily obtain the energy spectrum of the vibrational modes of the molecule.

Part 7

We first observe that, since $\chi_a(\mathbf{r}; \mathbf{R})$ is a normalized eigenstate of H', the corresponding eigenvalue $\mathcal{E}_a(\mathbf{R})$ can be obtained from

$$\mathcal{E}_a(\mathbf{R}) = \int d\mathbf{r} \, \chi_a^*(\mathbf{r}; \mathbf{R}) \left[-\sum_{k=1}^{N} \frac{\hbar^2 \boldsymbol{\nabla}_k^2}{2m} + V(\mathbf{r}; \mathbf{R}) \right] \chi_a(\mathbf{r}; \mathbf{R}) ;$$

the nuclear positions $\mathbf{R}$ can be viewed as parameters. A direct application of the Hellmann–Feynman theorem yields

$$\frac{\partial}{\partial R_{i\alpha}} \mathcal{E}_a(\mathbf{R}) = \int d\mathbf{r} \, \chi_a^*(\mathbf{r}; \mathbf{R}) \left[\frac{\partial}{\partial R_{i\alpha}} V(\mathbf{r}; \mathbf{R}) \right] \chi_a(\mathbf{r}; \mathbf{R}) = \int d\mathbf{r} \, |\chi_a(\mathbf{r}; \mathbf{R})|^2 \frac{\partial}{\partial R_{i\alpha}} V(\mathbf{r}; \mathbf{R})$$

and similarly

$$\frac{\partial^2}{\partial R_{i\alpha} \partial R_{j\beta}} \mathcal{E}_a(\mathbf{R}) = \int d\mathbf{r} \, |\chi_a(\mathbf{r}; \mathbf{R})|^2 \frac{\partial^2}{\partial R_{i\alpha} \partial R_{j\beta}} V(\mathbf{r}; \mathbf{R}) ,$$

which simplifies the problem considerably, since in order to evaluate derivatives of $\mathcal{E}_a(\mathbf{R})$ we do not need to calculate derivatives of the electronic wave functions relative to the nuclear positions.

Part 8

The typical excitation energies of electrons in molecules are similar to those in atoms; they are of the order of e^2/a_0 where $a_0 = \hbar^2/(me^2)$ is the Bohr radius. By contrast, the vibrational energy of the nuclear modes is of the order $\hbar \omega = (\hbar^2 K/M)^{1/2}$, where M is a typical nuclear mass and K is obtained from the second derivatives of the electronic energy $\mathcal{E}_a(\mathbf{R})$. The parameter K has the dimension of energy over (length)^2, and as a consequence it is of order $(e^2/a_0)/a_0^2$, that is, the ratio of the typical electronic excitation energy divided by the square of the typical length – the Bohr radius – over which the electronic wave function $\chi_a(\mathbf{r}; \mathbf{R})$ changes appreciably. Thus, we obtain

$$\frac{\text{nuclear vibrational energy}}{\text{electronic excitation energy}} = \frac{[\hbar^2(e^2/a_0)/(Ma_0^2)]^{1/2}}{e^2/a_0} = \sqrt{\frac{m}{M}} \,.$$

Problem 25   Variational Calculation of the H_2^+ Molecular Ion Binding Energy

This problem concerns the variational ground-state energy calculation of the H_2^+ molecule, consisting of two protons, to be considered fixed at positions $\mathbf{r}_1$ and $\mathbf{r}_2$, and an electron "shared" by the two protons. Ignoring spin degrees of freedom (and hence spin–orbit interactions, magnetic interactions between the spins of the electron and protons, etc.), here the Hamiltonian for this system is taken to include only the Coulomb attraction between the electron and protons and the Coulomb repulsion between the two protons:

$$H = \frac{\mathbf{p}^2}{2m} - \frac{e^2}{|\mathbf{r} - \mathbf{r}_1|} - \frac{e^2}{|\mathbf{r} - \mathbf{r}_2|} + \frac{e^2}{|\mathbf{r}_1 - \mathbf{r}_2|} \,,$$

where m is the mass of the electron and $\mathbf{r}$ and $\mathbf{p}$ are its position and momentum operators. As a trial wave function, consider

$$\psi_T(\mathbf{r}) = N\left[\phi(|\mathbf{r} - \mathbf{r}_1|; \lambda) + \phi(|\mathbf{r} - \mathbf{r}_2|; \lambda)\right] \,,$$

where N is a normalization factor, λ is a variational parameter,

$$\phi(r; \lambda) = \sqrt{\frac{\lambda^3}{\pi a_0^3}} \, e^{-\lambda r/a_0} \,,$$

and a_0 is the Bohr radius. Note that for $\lambda \longrightarrow 1$ the trial wave function above reduces to a linear combination of two hydrogen atom ground-state wave functions, centered at $\mathbf{r}_1$ and $\mathbf{r}_2$. Define the binding energy of H_2^+ as

$$B(R; \lambda) = \frac{E_T(R; \lambda) - \epsilon_{\text{gs}}}{e^2/(2a_0)} \quad \text{with} \quad E_T(R; \lambda) = \frac{\langle \psi_T | H | \psi_T \rangle}{\langle \psi_T | \psi_T \rangle} \,,$$

where $\epsilon_{\text{gs}} = -e^2/(2a_0)$ is the hydrogen atom ground-state energy, and obtain a variational binding energy for it as a function of the distance $R = |\mathbf{r}_1 - \mathbf{r}_2|/a_0$ (in units of a_0) between the two protons. What do you conclude? Is the molecule bound? And if so, for what value of R is the binding energy largest?

Now consider the expression that was derived above for $B(R; \lambda)$ in the case $\lambda = 1$. It turns out that the corresponding electronic binding energy is not much less accurate than the variational estimate. In the spirit of the Born–Oppenheimer approximation, we can interpret $B(R; \lambda = 1)$ as the potential energy, in units of $e^2/(2a_0)$, between the two protons, that is,

$$V(R) = \frac{e^2}{2a_0} B(R; \lambda = 1) \,.$$

Write down the corresponding Hamiltonian (denote the proton mass by M), and discuss how to obtain the vibrational spectrum of the H_2^+ molecule.

Hint: You will find the following results useful:

$$\int d\Omega_x \, \frac{e^{-\gamma|\mathbf{x}-\mathbf{X}|}}{|\mathbf{x}-\mathbf{X}|} = \frac{2\pi}{\gamma \, x_< x_>} \left[e^{-\gamma(x_> - x_<)} - e^{-\gamma(x_> + x_<)} \right]$$

where $x_<$ and $x_>$ are, respectively, the smallest and largest values of x and X. Note also the following indefinite integral:

$$\int^x dy \, y^n \, e^{-\gamma y} = -\frac{e^{-\gamma x}}{\gamma^{n+1}} \left[(\gamma x)^n + n(\gamma x)^{n-1} + \cdots + n! \right] .$$

Solution

The following solid-angle integrations are useful:

$$\int d\Omega_x \, \frac{e^{-\gamma|\mathbf{x}\pm\mathbf{X}|}}{|\mathbf{x}\pm\mathbf{X}|} = \frac{2\pi}{\gamma \, x_< x_>} \left[e^{-\gamma(x_> - x_<)} - e^{-\gamma(x_> + x_<)} \right] ,$$

$$\int d\Omega_x \, \frac{1}{|\mathbf{x}\pm\mathbf{X}|} = \lim_{\gamma \to 0} \int d\Omega_x \, \frac{e^{-\gamma|\mathbf{x}\pm\mathbf{X}|}}{|\mathbf{x}\pm\mathbf{X}|} = \frac{4\pi}{x_>} ,$$

$$\int d\Omega_x \, e^{-\gamma|\mathbf{x}\pm\mathbf{X}|} = -\frac{d}{d\gamma} \int d\Omega_x \, \frac{e^{-\gamma|\mathbf{x}\pm\mathbf{X}|}}{|\mathbf{x}\pm\mathbf{X}|}$$

$$= \frac{2\pi}{\gamma^2 x_< x_>} \left[(1 + \gamma x_> - \gamma x_<) e^{-\gamma(x_> - x_<)} - (1 + \gamma x_> + \gamma x_<) e^{-\gamma(x_> + x_<)} \right] .$$

The first integral follows by noting that

$$I = \int d\Omega_x \, \frac{e^{-\gamma|\mathbf{x}-\mathbf{X}|}}{|\mathbf{x}\pm\mathbf{X}|} = 2\pi \int_{-1}^{1} d\alpha \, \frac{e^{-\gamma\sqrt{x^2+X^2-2xX\alpha}}}{\sqrt{x^2+X^2-2xX\alpha}} ,$$

where $\alpha = \cos\theta$. We change variable, setting

$$y = \sqrt{x^2+X^2-2xX\alpha} , \qquad dy = -\frac{xX}{\sqrt{x^2+X^2-2xX\alpha}} \, d\alpha \implies d\alpha = -\frac{y}{xX} \, dy ,$$

so that

$$I = -\frac{2\pi}{xX} \int_{x+X}^{|x-X|} dy \, y \, \frac{e^{-\gamma y}}{y} = \frac{2\pi}{xX} \int_{|x-X|}^{x+X} dy \, e^{-\gamma y} = \frac{2\pi}{\gamma xX} \left[e^{-\gamma|x-X|} - e^{-\gamma(x+X)} \right] .$$

We note that if $x < X$ then $|x - X| = X - x$ and, by contrast, if $x > X$ then $|x - X| = x - X$, yielding the expression as given above.

The following indefinite integrals are also useful:

$$\int^x dx \, e^{-\gamma x} = -\frac{e^{-\gamma x}}{\gamma} ,$$

$$\int^x dx \, x \, e^{-\gamma x} = -\frac{e^{-\gamma x}}{\gamma^2} (\gamma x + 1) ,$$

$$\int^x dx \, x^2 \, e^{-\gamma x} = -\frac{e^{-\gamma x}}{\gamma^3} [(\gamma x)^2 + 2\gamma x + 2] .$$

They are easily obtained by repeated partial integration.

It is convenient to introduce the shorthand notation

$$\phi_k = \phi(|\mathbf{r} - \mathbf{r}_k|; \lambda) .$$

The normalization of the trial wave function is given by

$$\int d\mathbf{r}\, |\psi_T|^2 = |N|^2 \left[\int d\mathbf{r}\, |\phi_1|^2 + \int d\mathbf{r}\, |\phi_2|^2 + 2 \int d\mathbf{r}\, \phi_1 \phi_2 \right] = 2|N|^2 \left[1 + \underbrace{\int d\mathbf{r}\, \phi_1 \phi_2}_{I} \right],$$

where in the first two terms, after shifting the integration variable acccording to $\mathbf{r} - \mathbf{r}_k \longrightarrow \mathbf{r}$, we have used the fact that each $\phi(|\mathbf{r} - \mathbf{r}_k|; \lambda)$ is normalized. The remaining integration reads

$$I = \frac{\lambda^3}{\pi a_0^3} \int d\mathbf{r}\, e^{-\lambda |\mathbf{r} - \mathbf{r}_1|/a_0}\, e^{-\lambda |\mathbf{r} - \mathbf{r}_2|/a_0} = \frac{1}{\pi} \int d\mathbf{x}\, e^{-x}\, e^{-|\mathbf{x} - \mathbf{X}|},$$

where in the last step we introduced the variable $\mathbf{x} = \lambda(\mathbf{r} - \mathbf{r}_2)/a_0$, and have defined

$$\mathbf{X} = \frac{\lambda}{a_0} (\mathbf{r}_1 - \mathbf{r}_2) = \lambda \mathbf{R};$$

$\mathbf{R}$ is the inter-proton distance in units of the Bohr radius. Using the result above, we have

$$\frac{1}{\pi} \int d\Omega_x\, e^{-|\mathbf{x} - \mathbf{X}|} = \frac{2}{xX} \left[(1 + X - x)\, e^{-(X-x)} - (1 + X + x)\, e^{-(X+x)} \right] \qquad x < X$$

$$= \frac{2}{xX} \left[(1 + x - X)\, e^{-(x-X)} - (1 + x + X)\, e^{-(x+X)} \right] \qquad x > X.$$

The integral I is then written as follows:

$$I = \frac{2}{X} \int_0^X dx\, x\, e^{-x} \left[1 + X - x)\, e^{-(X-x)} - (1 + X + x)\, e^{-(X+x)} \right]$$

$$+ \frac{2}{X} \int_X^\infty dx\, x\, e^{-x} \left[1 + x - X)\, e^{-(x-X)} - (1 + x + X)\, e^{-(x+X)} \right]$$

$$= \frac{2\, e^{-X}}{X} \underbrace{\int_0^X dx\, x\, (1 + X - x)}_{I_1} + \frac{2\, e^X}{X} \underbrace{\int_X^\infty dx\, x\, e^{-2x}\, (1 + x - X)}_{I_2} - \frac{2\, e^{-X}}{X} \underbrace{\int_0^\infty dx\, x\, e^{-2x}\, (1 + x + X)}_{I_3}.$$

Now, using the indefinite integrals listed above, we have

$$I_1 = (1 + X) \frac{X^2}{2} - \frac{X^3}{3},$$

$$I_2 = (1 - X) \frac{e^{-2X}}{4} (2X + 1) + \frac{e^{-2X}}{8} (4X^2 + 4X + 2),$$

$$I_3 = \frac{1 + X}{4} + \frac{2}{8},$$

and hence

$$I = \frac{2\, e^{-X}}{X} (I_1 - I_3) + \frac{2\, e^X}{X} I_2 = e^{-X} \left(1 + X + \frac{X^2}{3} \right).$$

The normalization is obtained as

$$\int d\mathbf{r}\, |\psi_T|^2 = 2|N|^2 \left[1 + e^{-X} \left(1 + X + \frac{X^2}{3} \right) \right].$$

We are left with the task of evaluating the matrix element $\langle \psi_T | H | \psi_T \rangle$:

$$\langle \psi_T | H | \psi_T \rangle = |N|^2 \int d\mathbf{r}\, (\phi_1 + \phi_2)\, (t + v_1 + v_2 + v_{12})\, (\phi_1 + \phi_2),$$

where

$$t = \frac{\mathbf{p}^2}{2m}, \qquad v_1 = -\frac{e^2}{|\mathbf{r} - \mathbf{r}_1|}, \qquad v_2 = -\frac{e^2}{|\mathbf{r} - \mathbf{r}_2|}, \qquad v_{12} = \frac{e^2}{|\mathbf{r}_1 - \mathbf{r}_2|}.$$

The above matrix element can also be written as

$$\langle \psi_T | H | \psi_T \rangle = |N|^2 \int d\mathbf{r} \, (\phi_1 + \phi_2) \, (t + v_1 + v_2) \, (\phi_1 + \phi_2) + v_{12} \int d\mathbf{r} \, |\psi_T|^2.$$

Now, we note that ϕ_k is an eigenfunction of $t - \lambda e^2 / |\mathbf{r} - \mathbf{r}_k|$ with eigenvalue $-(\lambda e)^2/(2a_0)$ (the hydrogen-like atom with $Z \longrightarrow \lambda$) and hence

$$t\phi_k = \left(-\frac{\lambda^2 e^2}{2\,a_0} + \frac{\lambda e^2}{|\mathbf{r} - \mathbf{r}_k|} \right) \phi_k = \left(\lambda^2 \epsilon_{gs} - \lambda v_k \right) \phi_k, \qquad \epsilon_{gs} = -\frac{e^2}{2a_0}.$$

Using this result, we have

$$(t + v_1 + v_2) \, (\phi_1 + \phi_2) = \left[\lambda^2 \epsilon_{gs} + (1 - \lambda) v_1 + v_2 \right] \phi_1 + \left[\lambda^2 \epsilon_{gs} + (1 - \lambda) v_2 + v_1 \right] \phi_2,$$

and consequently

$$\langle \psi_T | H | \psi_T \rangle = |N|^2 \int d\mathbf{r} \left[\phi_1 \left[\lambda^2 \epsilon_{gs} + (1 - \lambda) v_1 + v_2 \right] \phi_1 + \phi_1 \left[\lambda^2 \epsilon_{gs} + (1 - \lambda) v_2 + v_1 \right] \phi_2 \right]$$

$$+ |N|^2 \int d\mathbf{r} \left[1 \rightleftharpoons 2 \right] + v_{12} \int d\mathbf{r} \, |\psi_T|^2.$$

Since the matrix element is unchanged under the exchange $1 \rightleftharpoons 2$, it follows that the second term is identical to the first, and so

$$\langle \psi_T | H | \psi_T \rangle = 2|N|^2 \int d\mathbf{r} \left[\phi_1 \left[\lambda^2 \epsilon_{gs} + (1 - \lambda) v_1 + v_2 \right] \phi_1 + \phi_1 \left[\lambda^2 \epsilon_{gs} + (1 - \lambda) v_2 + v_1 \right] \phi_2 \right] + v_{12} \int d\mathbf{r} \, |\psi_T|^2$$

$$= 2|N|^2 (1 - \lambda) \left[\underbrace{\int d\mathbf{r} \, \phi_1 v_1 \phi_1}_{K_1} + \underbrace{\int d\mathbf{r} \, \phi_1 v_2 \phi_2}_{K_2} \right] + 2|N|^2 \left[\underbrace{\int d\mathbf{r} \, \phi_1 v_2 \phi_1}_{K_3} + \underbrace{\int d\mathbf{r} \, \phi_1 v_1 \phi_2}_{K_4} \right]$$

$$+ \left(\lambda^2 \epsilon_{gs} + v_{12} \right) \int d\mathbf{r} \, |\psi_T|^2.$$

Changing variable by setting $\mathbf{x} = (\lambda/a_0)(\mathbf{r} - \mathbf{r}_1)$, we have

$$K_1 = -e^2 \frac{\lambda^3}{\pi a_0^3} \int d\mathbf{r} \, \frac{e^{-2\lambda |\mathbf{r} - \mathbf{r}_1|/a_0}}{|\mathbf{r} - \mathbf{r}_1|} = -\frac{e^2}{a_0} \lambda \left(\frac{1}{\pi} \int d\mathbf{x} \, \frac{e^{-2x}}{x} \right) = -\frac{e^2}{a_0} \lambda.$$

Similarly, we obtain

$$K_2 = -e^2 \frac{\lambda^3}{\pi a_0^3} \int d\mathbf{r} \, e^{-\lambda |\mathbf{r} - \mathbf{r}_1|/a_0} \, \frac{e^{-\lambda |\mathbf{r} - \mathbf{r}_2|/a_0}}{|\mathbf{r} - \mathbf{r}_2|} = -\frac{e^2}{a_0} \lambda \left(\frac{1}{\pi} \int d\mathbf{x} \, e^{-x} \, \frac{e^{-|\mathbf{x} + \mathbf{X}|}}{|\mathbf{x} + \mathbf{X}|} \right)$$

$$= -2 \frac{e^2}{a_0} \lambda \int_0^\infty dx \, x^2 \, e^{-x} \frac{1}{x_< x_>} \left(e^{-x_> + x_<} - e^{-x_> - x_<} \right)$$

$$= -2 \frac{e^2}{a_0} \lambda \left[\frac{e^{-X}}{X} \int_0^X dx \, x - \frac{e^{-X}}{X} \int_0^X dx \, x \, e^{-2x} + \frac{e^X}{X} \int_X^\infty dx \, x \, e^{-2x} - \frac{e^{-X}}{X} \int_X^\infty dx \, x \, e^{-2x} \right]$$

$$= -2 \frac{e^2}{a_0} \lambda \left[\frac{e^{-X}}{X} \frac{X^2}{2} - \frac{e^{-X}}{X} \left(-e^{-2X} \frac{2X+1}{4} + \frac{1}{4} \right) + \left(\frac{e^X}{X} - \frac{e^{-X}}{X} \right) e^{-2X} \frac{2X+1}{4} \right] = -\frac{e^2}{a_0} \lambda \, e^{-X} (1 + X)$$

and

$$K_3 = -e^2 \frac{\lambda^3}{\pi a_0^3} \int d\mathbf{r} \, \frac{e^{-2\lambda|\mathbf{r}-\mathbf{r}_1|/a_0}}{|\mathbf{r}-\mathbf{r}_2|} = -\frac{e^2}{a_0} \lambda \left(\frac{1}{\pi} \int d\mathbf{x} \, \frac{e^{-2x}}{|\mathbf{x}+\mathbf{X}|} \right)$$

$$= -\frac{4\,e^2}{a_0} \lambda \left[\frac{1}{X} \int_0^X dx\, x^2\, e^{-2x} + \int_X^\infty dx\, x\, e^{-2x} \right]$$

$$= -\frac{4e^2}{a_0} \lambda \left[-\frac{e^{-2X}}{8X}(4X^2+4X+2) + \frac{1}{4X} + \frac{e^{-2X}}{4}(2X+1) \right] = -\frac{e^2}{a_0} \frac{\lambda}{X} \left[1 - e^{-2X}(1+X) \right].$$

In the integral K_4 we change variable by setting $\mathbf{x} = (\lambda/a_0)(\mathbf{r}-\mathbf{r}_2)$, to obtain

$$K_4 = -e^2 \frac{\lambda^3}{\pi a_0^3} \int d\mathbf{r} \, \frac{e^{-\lambda|\mathbf{r}-\mathbf{r}_1|/a_0}}{|\mathbf{r}-\mathbf{r}_1|} e^{-\lambda|\mathbf{r}-\mathbf{r}_2|/a_0} = -\frac{e^2}{a_0} \lambda \left(\frac{1}{\pi} \int d\mathbf{x} \, e^{-x} \frac{e^{-|\mathbf{x}-\mathbf{X}|}}{|\mathbf{x}-\mathbf{X}|} \right) = K_2.$$

Collecting results (recall that $X = \lambda R$), we find, for the expectation value,

$$E_T(R; \lambda) = \frac{\langle \psi_T | H | \psi_T \rangle}{\langle \psi_T | \psi_T \rangle} = -\lambda^2 \frac{e^2}{2a_0} + \lambda(\lambda-1) \frac{e^2}{a_0} \frac{1 + e^{-X}(1+X)}{1 + e^{-X}\,(1+X+X^2/3)}$$

$$+ \frac{e^2}{a_0 R} \left[1 - \frac{1 - e^{-2X}(1+X)}{1 + e^{-X}\,(1+X+X^2/3)} \right]$$

$$- \lambda \frac{e^2}{a_0} \frac{e^{-X}(1+X)}{1 + e^{-X}\,(1+X+X^2/3)}.$$

In the limit in which one proton is very far from the other proton, the H_2^+ molecule ground-state energy should reduce to that of an isolated hydrogen atom, that is, $\epsilon_{gs} = -e^2/(2a_0)$, so ϵ_{gs} represents the minimum energy required to disassociate the H_2^+ molecular ion. In units of $e^2/(2a_0)$, we conveniently define the molecular binding energy as

$$B(R; \lambda) = \frac{E_T(R; \lambda) - \epsilon_{gs}}{e^2/(2\,a_0)}.$$

Combining terms, this latter quantity can be written as

$$B(R; \lambda) = \frac{\lambda}{X} \frac{(6-4X^2)\,e^{-X} + 6(1+X)\,e^{-2X}}{3 + e^{-X}(3+3X+X^2)} + (\lambda-1) \frac{3(\lambda-1) + e^{-X}\left[3(\lambda-1)(1+X) - (\lambda+1)X^2 \right]}{3 + e^{-X}(3+3X+X^2)}.$$

Note that, for $R \gg 1$, the function $B(R \gg 1; \lambda) \longrightarrow (\lambda-1)^2$ and so in this limit we must have $\lambda = 1$, yielding a vanishing binding energy. For each R, let $\lambda^*(R)$ be the λ-value that minimizes $B(R; \lambda)$. We denote the corresponding energy by $B^*(R)$, which then represents the variational estimate. It is plotted in Fig. 14.1 along with $B(R; \lambda = 1)$. We find that the minimum of $B^*(R)$ occurs for $R_0 = 2.0a_0 = 1.06 \times 10^{-8}$ cm, which is the equilibrium separation between the two protons (the λ-value at R_0 is 1.239). The corresponding energy is $B_0^* = -0.173$ in units of $e^2/(2a_0)$, or approximately we have $B_0^* = -2.35$ eV. These values compare well with the experimental values, respectively, $R_0^{\exp} = 1.06 \times 10^{-8}$ cm and $B^{\exp} = -2.64$ eV.

The "nuclear" Hamiltonian reads

$$H_{\text{nuc}} = \frac{\mathbf{P}^2}{2\mu} + V(R),$$

where $\mathbf{P}$ is the relative momentum between the two protons and μ is their reduced mass ($\mu = M/2$). We expand $V(R)$ around the minimum occurring at R_0 to obtain

$$V(R) = V(R_0) + \frac{1}{2} V''(R_0)(\mathbf{R} - \mathbf{R}_0)^2 + \cdots,$$

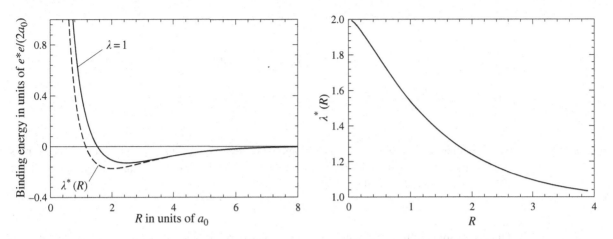

Fig. 14.1 Left panel: Binding energy of the H_2^+ molecular ion for $\lambda = 1$ and for $\lambda^*(R)$. Right panel: The function $\lambda^*(R)$ that minimizes $B(R; \lambda)$.

from which the angular frequency is obtained as $\omega = \sqrt{V''(R_0)/\mu}$. The vibrational spectrum of this system corresponds to the energy spectrum of a (shifted) harmonic oscillator, $E_n = \hbar\omega(n + 3/2)$.

Problem 26 Estimating Bound-State Energies of a Hamiltonian with the Variational Method

We are interested in estimating the bound-state energies of a Hamiltonian H with the variational method. Let $\phi_i(\mathbf{r})$ with $i = 1, \ldots, n$ be a set of linearly independent and square integrable functions. We will assume that these functions are also orthonormal without loss of generality; this can always be accomplished by Gram–Schmidt orthogonalization of the $\phi_i(\mathbf{r})$. Consider the trial wave function

$$\psi(\mathbf{r}) = \sum_{i=1}^{n} c_i \phi_i(\mathbf{r}) \,,$$

where the (generally complex) parameters c_i are to be determined by the variational method.

1. Since the c_i are generally complex, we minimize H by independently varying the c_i and c_i^* (which is equivalent to independently varying the real and imaginary parts of c_i). Show that this procedure leads to the following set of equations:

$$\sum_{j=1}^{n} H_{ij} c_j = \langle H \rangle c_i, \qquad i = 1, \ldots, n \,,$$

and a second set of equations that are equivalent to those above since H is hermitian. Here, we have defined

$$H_{ij} = \langle \phi_i | H | \phi_j \rangle = \int d\mathbf{r}\, \phi_i^*(\mathbf{r})\, H\, \phi_j(\mathbf{r}) \,, \qquad \langle H \rangle \equiv \frac{\langle \psi | H | \psi \rangle}{\langle \psi | \psi \rangle} \,.$$

2. Denote the (presently unknown) real expectation value $\langle H \rangle$ as E; then, the linear system above reduces to the following eigenvalue problem:

$$\sum_{j=1}^{n} H_{ij} c_j = E c_i, \qquad i = 1, \ldots, n \,.$$

Let $\psi_k(\mathbf{r})$ with $k = 1, \ldots, n$ be the optimal trial wave functions,

$$\psi_k(\mathbf{r}) = \sum_{i=1}^{n} c_i^{(k)}\, \phi_i(\mathbf{r})\,,$$

where in matrix notation

$$\underline{H}\,\underline{c}^{(k)} = E_k'\, \underline{c}^{(k)}\,.$$

Show that the matrix elements of H are diagonal if the set of optimal trial wave functions $\psi_k(\mathbf{r})$ is used, namely, that $\langle \psi_k | H | \psi_l \rangle = E_k'\, \delta_{kl}$. Hereafter, assume that the eigenvalues of the $n \times n$ matrix $\underline{H}$ have been ordered, so that

$$E_1' \le E_2' \le \cdots \le E_{n-1}' \le E_n'\,.$$

Of course, some of these E_k' may be the same (degenerate).

3. An additional function $\phi_{n+1}(\mathbf{r})$ is added to the set $\psi_1(\mathbf{r}), \ldots, \psi_n(\mathbf{r})$; it can be chosen to be normalized and orthogonal to the $\psi_k(\mathbf{r})$, that is, $\langle \psi_k | \phi_{n+1} \rangle = 0$ for $k = 1, \ldots, n$. Show that the eigenvalue problem has a non-trivial solution if

$$D_{n+1}(E) = \det \begin{pmatrix} E_1' - E & 0 & \cdots & \langle \psi_1 | H | \phi_{n+1} \rangle \\ 0 & E_2' - E & \cdots & \langle \psi_2 | H | \phi_{n+1} \rangle \\ \vdots & \vdots & \cdots & \vdots \\ \langle \phi_{n+1} | H | \psi_1 \rangle & \langle \phi_{n+1} | H | \psi_2 \rangle & \cdots & \langle \phi_{n+1} | H | \phi_{n+1} \rangle - E \end{pmatrix} = 0\,.$$

4. The characteristic equation $D_{n+1}(E) = 0$ in part 3 has $n+1$ solutions. The values of the determinants $D_{n+1}(E)$ for $E = E_k'$ and for $E = E_{k+1}'$ can be shown to have opposite signs, provided $E_k' < E_{k+1}'$. Hence, there must be a zero of $D_{n+1}(E) = 0$ between E_k' and E_{k+1}'. It can also be shown that there are two additional roots, one less than E_1' and the other greater than E_n'. Since there are $n+1$ roots of the characteristic equation, $n-1$ roots lie in the intervals $[E_1', E_2'], [E_2', E_3'], \ldots, [E_{n-1}', E_n']$, and the two additional roots lie respectively below and above the old spectrum. Prove these statements.

The procedure described above is known as the Rayleigh–Ritz variational method. It provides approximate upper bounds to the lowest n eigenvalues of the Hamiltonian. The bounds can be improved by choosing larger and larger values of n. The limit $n \longrightarrow \infty$ is equivalent to solving for the exact eigenvalues of H. However, given a set of initial functions $\phi_i(\mathbf{r})$, it is difficult to establish whether convergence to the exact eigenvalues will be achieved slowly or rapidly (of course, it will be rapid if the $\phi_i(\mathbf{r})$ are close to the eigenfunctions of H).

Solution

Part 1

We express the (real) expectation value of H as

$$\langle H \rangle = \frac{\langle \psi | H | \psi \rangle}{\langle \psi | \psi \rangle} = \frac{\sum_{l,m=1}^{n} c_l^* c_m H_{lm}}{\sum_{l=1}^{n} c_l^* c_l}\,,$$

where

$$H_{lm} = \langle \phi_l | H | \phi_m \rangle = \int d\mathbf{r}\, \phi_l^*(\mathbf{r})\, H(-i\hbar\boldsymbol{\nabla}, \mathbf{r})\, \phi_m(\mathbf{r})\,,$$

and in the denominator of $\langle H \rangle$ we have used the fact that the set of $\phi_l(\mathbf{r})$ is orthonormal. We minimize $\langle H \rangle$ with respect to independent variations of the c_l and c_l^* (which is equivalent to minimizing with respect to the real and imaginary parts of c_l, respectively $c_l + c_l^*$ and $c_l - c_l^*$). We find

$$\frac{\partial}{\partial c_i^*}\langle H \rangle = 0 \implies \frac{\sum_{m=1}^{n} H_{im}\, c_m}{\sum_{l=1}^{n} c_l^*\, c_l} - \frac{c_i \sum_{l,m=1}^{n} c_l^*\, c_m\, H_{lm}}{\left(\sum_{l=1}^{n} c_l^*\, c_l\right)^2} = \frac{\sum_{m=1}^{n} H_{im}\, c_m - c_i\,\langle H \rangle}{\sum_{l=1}^{n} c_l^*\, c_l} = 0\,,$$

which is equivalent to the eigenvalue problem

$$\sum_{m=1}^{n} H_{im}\, c_m = c_i\,\langle H \rangle$$

in the n-dimensional subspace spanned by the $\phi_i(\mathbf{r})$. Minimizing with respect to the c_i yields

$$\sum_{m=1}^{n} H_{mi}\, c_m^* = c_i^*\,\langle H \rangle\,,$$

which can be obtained from the previous eigenvalue problem by taking the complex conjugate of both sides and using the fact that H is hermitian and hence $H_{im}^* = H_{mi}$. Thus, it is enough to consider one of these eigenvalue problems.

Part 2

The problem is now reduced to the diagonalization of the $n \times n$ matrix $\underline{H}$. The eigenvalues result from the roots of the characteristic equation (here $E = \langle H \rangle$)

$$\det(\underline{H} - E\,\underline{I}) = 0\,.$$

This equation has n roots E_k' (eigenvalues), some of which may be the same (degenerate). The corresponding normalized eigenvectors $|\psi_k\rangle$ consist of the linear combinations

$$|\psi_k\rangle = \sum_{i=1}^{n} c_i^{(k)}\, |\phi_i\rangle\,, \qquad \sum_{m=1}^{n} H_{im}\, c_m^{(k)} = E_k'\, c_i^{(k)}\,, \qquad k = 1,\ldots,n\,.$$

It follows that

$$\langle \psi_l | H | \psi_m \rangle = \sum_{i,j=1}^{n} c_i^{(l)*}\, H_{ij}\, c_j^{(m)} = \sum_{i=1}^{n} c_i^{(l)*} \sum_{j=1}^{n} H_{ij}\, c_j^{(m)} = E_m' \sum_{i=1}^{n} c_i^{(l)*}\, c_i^{(m)} = \delta_{l,m}\, E_m'\,,$$

and the matrix is diagonal in this basis.

Part 3

We expand the set by adding a single state $|\phi_{n+1}\rangle$, which we choose to be orthogonal to all $|\psi_l\rangle$ and normalized, that is, $\langle \phi_{n+1} | \phi_{n+1} \rangle = 1$. Now, we define the projectors

$$P = \sum_{l=1}^{n} |\psi_l\rangle\langle\psi_l|\,, \qquad Q = \mathbb{1} - P\,,$$

where P projects states onto the subspace spanned by the $|\psi_1\rangle, \ldots, |\psi_n\rangle$ and Q projects states onto the complement of this subspace. Clearly, we have

$$P|\psi_l\rangle = |\psi_l\rangle\,, \qquad Q|\psi_l\rangle = 0\,, \qquad P|\phi_{n+1}\rangle = 0\,, \qquad Q|\phi_{n+1}\rangle = |\phi_{n+1}\rangle\,.$$

The full Hamiltonian can be written as

$$H = (P + Q)H(P + Q) = PHP + PHQ + QHP + QHQ \,.$$

In the P-subspace H is diagonal, since

$$\langle \psi_l | PHP + PHQ + QHP + QHQ | \psi_m \rangle = \langle \psi_l | PHP | \psi_m \rangle = \langle \psi_l | H | \psi_m \rangle = E_l' \, \delta_{l,m} \,.$$

Note that $H|\psi_l\rangle \neq E_l'|\psi_l\rangle$, since H is the Hamiltonian in the full space. However, we have for the Hamiltonian restricted to the P-space,

$$PHP = \sum_{l=1}^{n} E_l' \, |\psi_l\rangle\langle\psi_l| \,.$$

Similarly, we find

$$\langle \psi_l | PHP + PHQ + QHP + QHQ | \phi_{n+1} \rangle = \langle \psi_l | PHQ | \phi_{n+1} \rangle = \langle \psi_l | H | \phi_{n+1} \rangle = H_{l,n+1}$$

and

$$\langle \phi_{n+1} | H | \psi_l \rangle = H_{l,n+1}^* \,, \qquad \langle \phi_{n+1} | H | \phi_{n+1} \rangle = H_{n+1,n+1} \,,$$

yielding an $(n + 1)$-dimensional matrix given by

$$\begin{pmatrix} E_1' & 0 & \cdots & H_{1,n+1} \\ 0 & E_2' & \cdots & H_{2,n+1} \\ \vdots & \vdots & \cdots & \vdots \\ H_{1,n+1}^* & H_{2,n+1}^* & \cdots & H_{n+1,n+1} \end{pmatrix} \,,$$

with characteristic equation resulting from

$$D_{n+1}(E) = \det(\underline{H} - E\underline{I}) = \det \begin{pmatrix} E_1' - E & 0 & \cdots & H_{1,n+1} \\ 0 & E_2' - E & \cdots & H_{2,n+1} \\ \vdots & \vdots & \cdots & \vdots \\ H_{1,n+1}^* & H_{2,n+1}^* & \cdots & H_{n+1,n+1} - E \end{pmatrix} = 0 \,.$$

By expanding the determinant along the last column, we find

$$D_{n+1}(E) = (E_1' - E) \cdots (E_n' - E)(H_{n+1,n+1} - E) - |H_{1,n+1}|^2 (E_2' - E) \cdots (E_n' - E)$$
$$- |H_{2,n+1}|^2 (E_1' - E)(E_3' - E) \cdots (E_n' - E) - \cdots - |H_{n,n+1}|^2 (E_1' - E) \cdots (E_{n-1}' - E) \,,$$

which can be written compactly as

$$D_{n+1}(E) = \left[\prod_{l=1}^{n} (E_l' - E) \right] (H_{n+1,n+1} - E) - \sum_{l=1}^{n} |H_{l,n+1}|^2 \left[\prod_{m=1(m \neq l)}^{n} (E_m' - E) \right] \,.$$

It is now clear that for any $E = E_k'$ with $k = 1, \ldots, n$ the first term above vanishes and that in the sum over l only the term with $l = k$ survives, that is,

$$D_{n+1}(E_k') = -|H_{k,n+1}|^2 \left[\prod_{m=1(m \neq k)}^{n} (E_m' - E) \right] \,.$$

Having established this and since $E_1' \leq E_2' \leq \cdots \leq E_{n-1}' \leq E_n'$, it follows that

$$D_{n+1}(E_1') = -|H_{1,n+1}|^2 \prod_{m=2}^{n} (E_m' - E_1') < 0 \,,$$

while

$$D_{n+1}(E_2') = -|H_{2,n+1}|^2 (E_1' - E_2') \prod_{m=3}^{n} (E_m' - E_2') > 0 ,$$

and so on (namely, the signs alternate). Now, there are $n+1$ roots of the characteristic equation since $D_{n+1}(E)$ is a polynomial of order $n+1$ in E. Being a polynomial, $D_{n+1}(E)$ is also a continuous function, and therefore at least $n-1$ of these roots must lie in the intervals $[E_1', E_2'], [E_2', E_3'], \ldots, [E_{n-1}', E_n']$. There are two more roots: one must be less than E_1' and the other larger than E_n'. To show this, we note that the leading power of E in the polynomial $D_{n+1}(E)$ is $(-1)^{n+1} E^{n+1}$. It follows that, for E sufficiently large and negative, that is, $E \longrightarrow -\infty$, $D_{n+1}(E)$ is positive and hence there must be a root below E_1', since $D_{n+1}(E_1')$ is negative (there is a change of sign of the polynomial between $-\infty$ and E_1'). However, in the limit $E \longrightarrow \infty$, the sign of $D_{n+1}(E)$ is $(-1)^{n+1}$, that is, it is positive or negative depending on whether n is respectively odd or even. By contrast, the sign of $D_{n+1}(E_n')$ is $(-1)^n$, and therefore there is another change of sign between E_n' and ∞, requiring the presence of a root above E_n'.

In the above argument, we have tacitly assumed that the eigenvalues E_k' are all non-degenerate. In the presence of degeneracy, the conclusions above remain valid. As an extreme case, suppose that the E_k' are all degenerate, that is, $E_k' = E'$; the determinant $D_{n+1}(E)$ is then given by

$$D_{n+1}(E) = (E' - E)^n (H_{n+1,n+1} - E) - (E' - E)^{n-1} \sum_{l=1}^{n} |H_{l,n+1}|^2 .$$

It has $n-1$ roots with $E = E'$, and two additional roots that result from

$$(E' - E)(H_{n+1,n+1} - E) - \sum_{l=1}^{n} |H_{l,n+1}|^2 = 0 ,$$

yielding

$$E_\pm = \frac{E' + H_{n+1,n+1}}{2} \pm \frac{|E' - H_{n+1,n+1}|}{2} \left[1 + 4 \frac{\sum_{l=1}^{n} |H_{l,n+1}|^2}{(E' - H_{n+1,n+1})^2} \right]^{1/2}$$

Regardless of whether E' is greater or less than $H_{n+1,n+1}$, it is easily seen that $E_- < E'$ and $E_+ > E'$. Thus, as expected, we obtain the eigenvalue E' with degeneracy $n-1$, and two additional eigenvalues $E_\mp$ which bracket E'.

15

Scattering by a Potential

We have already analyzed scattering phenomena in one dimension. Here, we extend the treatment to three dimensions, introduce the concept of the scattering cross section, and relate the latter to the asymptotic behavior of the wave function.

In a scattering process, a beam consisting of particles of type 1 collides with a target consisting of particles of type 2. The products of the collisions between particles 1 and 2 – that is, the particles produced in the final state – are measured by detectors situated far from the region of space where the collisions have occurred. The detectors measure properties of the final-state particles (their energies, momenta, charges, etc.).

Scattering phenomena can be complex. For example, if particles 1 and 2 are not elementary but have a substructure (such as nuclei, which are made up of protons and neutrons, which in turn are made up of quarks and gluons), then in a collision their constituents can rearrange themselves into two or more composite particles that are different from 1 and 2. Similarly, if the energy of the beam particles is high enough then part of this energy can be converted into the creation of new particles, in fact a great many of them if the energy is very high.

The description of these complex scattering processes is outside the scope of the present textbook. In this section, we will concern ourselves with the simpler, but nevertheless physically relevant, case of elastic scattering, the process $1 + 2 \longrightarrow 1 + 2$, where the initial and final state are composed of the same particles and their substructure (if they have one) is left unchanged during the collision. We will assume that the interaction between particles 1 and 2 can be described (in the coordinate representation) by a potential energy operator $V(\mathbf{r}_1 - \mathbf{r}_2)$ that depends only on the relative positions of particles 1 and 2. We will further assume that (i) the target is thin enough to avoid multiple scattering, that is, processes in which particle 1 suffers multiple collisions with particles of type 2 before leaving the target and (ii) coherence effects,[1] that is, effects produced by destructive or constructive interference between waves scattered by different particles in the target (such as the Bragg reflection phenomenon mentioned at the end of the introduction to Chapter 2).

15.1 Cross Section and Scattering Wave Function

In a scattering experiment a beam of particles, all with the same momentum and energy, collides with a target. We define the incident flux F_i as the number of beam particles crossing a unit surface perpendicular to the beam direction per unit time,

$$F_i = n_i \, v, \tag{15.1}$$

---

[1] This requires that the wavelength associated with the scattered-particle wave is small compared with the distance between particles in the target.

where n_i is the number of beam particles per unit volume and v is the magnitude of the velocity relative to the target (its direction specifies the beam direction). We assume implicitly that the mutual interactions between beam particles are negligible, and that they undergo their collisions independently of each other. We place a detector far from the region where the collisions have occurred. The detector is positioned in a direction specified by the polar angles θ and ϕ relative to the beam direction, and subtends a solid angle $d\Omega$ (the "collision region" dimensions are negligible, since the detector is far from it). We then count the number dn_s of particles scattered per unit time into the solid angle $d\Omega$ about the direction specified by θ and ϕ. This number must be proportional to the incident flux and to $d\Omega$ (obviously, if F_i and/or $d\Omega$ increase then dn_s increases),

$$dn_s = F_i\, \sigma(\theta, \phi)\, d\Omega \,, \tag{15.2}$$

and the constant of proportionality, which will generally depend on θ and ϕ, is known as the differential scattering cross section. Since F_i has dimensions of $1/(\text{area}\times\text{time})$ and dn_s has dimensions of $1/\text{time}$, the differential cross section has dimensions of an area.

Having defined the cross section, we now need to connect it to solutions of the Schrödinger equation. We assume that the interaction of a beam particle with a particle in a target is described by a potential that depends on their relative positions. We can then use center-of-mass and relative variables to reduce the problem to that of the scattering of a particle of (reduced) mass μ from a center of force situated at the origin of the coordinate system. We will assume that the potential vanishes, namely as $r \longrightarrow \infty$ faster than $1/r$

$$\lim_{r\to\infty} rV(\mathbf{r}) = 0 \,. \tag{15.3}$$

This condition defines what we mean by a "short-range" potential. Scattering in a Coulomb potential must be treated separately. The Schrödinger equation is written as

$$\left(\boldsymbol{\nabla}^2 + k^2\right)\psi_{\mathbf{k}}(\mathbf{r}) = v(\mathbf{r})\,\psi_{\mathbf{k}}(\mathbf{r}) \,, \qquad v(\mathbf{r}) = \frac{2\mu}{\hbar^2}\,V(\mathbf{r}) \,, \qquad \mathbf{k}^2 = \frac{2\mu}{\hbar^2}\,E \,. \tag{15.4}$$

It might have bound-state solutions that vanish as $r \longrightarrow \infty$ and are therefore normalizable. We are not interested in those here. It will certainly have a continuum of solutions having $E > 0$ (or $k > 0$). As we will see below, we can relate the differential cross section to the continuum solutions that behave asymptotically:

$$\psi_{\mathbf{k}}(\mathbf{r}) = e^{i\mathbf{k}\cdot\mathbf{r}} + f_{\mathbf{k}}(\theta, \phi)\,\frac{e^{ikr}}{r} \qquad \text{as } r \longrightarrow \infty \,. \tag{15.5}$$

In the asymptotic region the wave function consists of a plane wave and a spherical wave e^{ikr}/r modulated by a function $f_{\mathbf{k}}(\Omega)$. In fact, this function is precisely the scattering amplitude, and the differential cross section is given by (see below)

$$\sigma_{\mathbf{k}}(\Omega) = |f_{\mathbf{k}}(\Omega)|^2 \,. \tag{15.6}$$

Before proceeding any further we need (i) to show that the asymptotic solution above does indeed satisfy the Schrödinger equation (in the asymptotic limit) and (ii) to derive the connection between the cross section and $f_k(\Omega)$. The first issue is settled in Problem 1, while to address the second issue we proceed in two steps. We first introduce the wave packet

$$\Psi(\mathbf{r}, t) = \int dk\, g(k)\, \underbrace{e^{-iEt/\hbar}\,\psi_{k\hat{z}}(\mathbf{r})}_{\text{solution of Eq. (15.4)}} \,, \tag{15.7}$$

where we have assumed that $\mathbf{k}$ is along the z-axis and that the profile function $g(k)$ is real, for simplicity, and strongly peaked at k_0. Since the wave packet is a superposition of solutions $\psi_{k\hat{z}}(\mathbf{r})$, it obviously satisfies the Schrödinger equation. In the asymptotic region, we have the sum of two terms, an incident and a scattered wave packet,

$$\Psi(|\mathbf{r}| \longrightarrow \infty, t) = \underbrace{\int dk\, g(k)\, e^{-iEt/\hbar}\, e^{ikz}}_{\Psi_i(\mathbf{r},t)} + \underbrace{\int dk\, g(k)\, e^{-iEt/\hbar}\, f_{k\hat{z}}(\Omega)\, \frac{e^{ikr}}{r}}_{\Psi_s(\mathbf{r},t)} . \tag{15.8}$$

These wave packets are largest when the phase factor in the integrand is stationary around k_0 (the stationary phase method, see Section 2.1). This requires that the incident wave packet satisfies (recall that $g(k)$ is real)

$$\frac{d}{dk}(kz - Et/\hbar)\Big|_{k=k_0} \approx 0 \implies z_i(t) \approx (\hbar k_0/\mu)t , \tag{15.9}$$

and so the center of the incident wave packet travels along the z-axis with a velocity given by $\hbar k_0/m = p_0/m$. On the other hand, for the scattered wave packet we have

$$\Psi_s(\mathbf{r}, t) = \int dk\, g(k)\, e^{-iEt/\hbar}\, |f_{k\hat{z}}(\Omega)|\, e^{i\chi_{k\hat{z}}(\Omega)}\, \frac{e^{ikr}}{r} , \qquad f_{k\hat{z}}(\Omega) = |f_{k\hat{z}}(\Omega)|\, e^{i\chi_{k\hat{z}}(\Omega)} , \tag{15.10}$$

since $f_{k\hat{z}}(\Omega)$ is a generally complex function of k and Ω. We therefore find that

$$\frac{d}{dk}[kr + \chi_{k\hat{z}}(\Omega) - Et/\hbar]\Big|_{k=k_0} \approx 0 \implies r_s(t) \approx (\hbar k_0/\mu)t - \frac{d}{dk}\chi_{k\hat{z}}(\Omega)\Big|_{k=k_0} , \tag{15.11}$$

and the center of the scattered wave particle in the direction Ω is located at a distance $r_s(t)$ from the center of force. Since $r_s(t)$ must be large and positive, there is no scattered wave packet as $t \longrightarrow -\infty$ ($r_s(t)$ would be large and negative!). Therefore, in this limit only the incident wave packet is present: it approaches the scattering region traveling from $z \longrightarrow -\infty$. On the other hand, for $t \longrightarrow \infty$ both wave packets are present; the incident wave packet continues to travel along the z-axis while the scattered wave packet propagates in all directions as a spherical wave emanating from the center of force.

Having clarified the interpretation of the two terms present in the asymptotic form of $\psi_{\mathbf{k}}(\mathbf{r})$, we calculate the probability current densities associated with the incident plane wave and scattered spherical wave.[2] We find, for the plane wave, $\mathbf{j}_i = \hbar\mathbf{k}/\mu$, a result independent of $\mathbf{r}$, while for the scattered wave the component along $\hat{\mathbf{r}}$ is given by

$$\hat{\mathbf{r}} \cdot \mathbf{j}_s(\mathbf{r}) = \frac{\hbar}{2\mu i}\left[f_{\mathbf{k}}^*(\Omega)\, \frac{e^{-ikr}}{r}\, \underbrace{\hat{\mathbf{r}} \cdot \boldsymbol{\nabla}}_{\partial/\partial r}\, f_{\mathbf{k}}(\Omega)\, \frac{e^{ikr}}{r} - \text{c.c.} \right] = \frac{\hbar}{2\mu i}\, |f_{\mathbf{k}}(\Omega)|^2 \left[i\frac{k}{r^2} - \text{c.c.} \right] = \frac{\hbar k}{\mu}\, \frac{|f_{\mathbf{k}}(\Omega)|^2}{r^2} . \tag{15.12}$$

If n_i is the number of incident particles per unit volume then the incident flux F_i and the number of scattered particle per unit time dn_s crossing a surface element dS at r are respectively

$$F_i = n_i\, |\mathbf{j}_i| = n_i\, \frac{\hbar k}{\mu} \qquad \text{and} \qquad dn_s = n_i\, \hat{\mathbf{r}} \cdot \mathbf{j}_s(\mathbf{r})\, \underbrace{r^2\, d\Omega}_{dS} = F_i\, |f_{\mathbf{k}}(\Omega)|^2\, d\Omega , \tag{15.13}$$

which leads to the desired relation, $\sigma_{\mathbf{k}}(\Omega) = |f_{\mathbf{k}}(\Omega)|^2$.

---

[2] The attentive reader will notice that in computing the probability current density we ignored the interference terms between the plane and spherical waves. As shown in Problem 6 in the context of the derivation of the optical theorem, these interference terms are important only in the forward direction with $\Omega = 0$.

15.2 Integral Equation for Scattering, Lippmann–Schwinger Equation, and Born Approximation

In this section, we derive a relation between the scattering amplitude $f_{\mathbf{k}}(\Omega)$ and the the wave function $\psi_{\mathbf{k}}(\mathbf{r})$, which turns out to be useful for developing a perturbative expansion for $f_{\mathbf{k}}(\Omega)$ (see Problem 9). The first step in the derivation consists in converting the Schrödinger equation into an integral equation. Suppose that we can find a function $G_k(\mathbf{r}, \mathbf{r}')$ – the Green's function – such that

$$\left(\nabla^2 + k^2\right) G_k(\mathbf{r}, \mathbf{r}') = -\delta(\mathbf{r} - \mathbf{r}') , \tag{15.14}$$

where the Laplacian ∇^2 acts on $\mathbf{r}$ (and not on $\mathbf{r}'$). We then write $\psi_{\mathbf{k}}(\mathbf{r})$ as follows:

$$\psi_{\mathbf{k}}(\mathbf{r}) = e^{i\mathbf{k}\cdot\mathbf{r}} - \int d\mathbf{r}' \, G_k(\mathbf{r}, \mathbf{r}') \, v(\mathbf{r}') \, \psi_{\mathbf{k}}(\mathbf{r}') . \tag{15.15}$$

It is now easily verified that $\psi_{\mathbf{k}}(\mathbf{r})$, as given above, satisfies the Schrödinger equation

$$
\begin{aligned}
\left(\nabla^2 + k^2\right) \psi_{\mathbf{k}}(\mathbf{r}) &= \left(\nabla^2 + k^2\right) \left[e^{i\mathbf{k}\cdot\mathbf{r}} - \int d\mathbf{r}' \, G_k(\mathbf{r}, \mathbf{r}') \, v(\mathbf{r}') \, \psi_{\mathbf{k}}(\mathbf{r}') \right] \\
&= - \int d\mathbf{r}' \, \underbrace{\left(\nabla^2 + k^2\right) G_k(\mathbf{r}, \mathbf{r}')}_{-\delta(\mathbf{r}-\mathbf{r}')} \, v(\mathbf{r}') \, \psi_{\mathbf{k}}(\mathbf{r}') \\
&= \int d\mathbf{r}' \, \delta(\mathbf{r} - \mathbf{r}') v(\mathbf{r}') \, \psi_{\mathbf{k}}(\mathbf{r}') = v(\mathbf{r}) \, \psi_{\mathbf{k}}(\mathbf{r}) ,
\end{aligned}
\tag{15.16}
$$

where we have used the fact that $\nabla^2 \, e^{i\mathbf{k}\cdot\mathbf{r}} = -\mathbf{k}^2 \, e^{i\mathbf{k}\cdot\mathbf{r}}$ and have passed $\left(\nabla^2 + k^2\right)$ under the integral sign, since it acts on the unprimed variables.

The second step in the derivation consists in showing that the solution of this integral equation satisfies the asymptotic boundary condition

$$\psi_{\mathbf{k}}(\mathbf{r}) = e^{i\mathbf{k}\cdot\mathbf{r}} + f_{\mathbf{k}}(\Omega) \, \frac{e^{ikr}}{r} \qquad \text{as } r \longrightarrow \infty . \tag{15.17}$$

For this we need the explicit form of the Green's function. It can be expressed as the following Fourier transform ($\eta \longrightarrow 0$ at the end), which can be evaluated by contour integration, as detailed in Problem 8; here, we only quote the result,

$$G_k(\mathbf{r}, \mathbf{r}') = \int \frac{d\mathbf{q}}{(2\pi)^3} \, e^{i\mathbf{q}\cdot(\mathbf{r}-\mathbf{r}')} \, \frac{1}{q^2 - k^2 - i\eta} = \frac{1}{4\pi} \, \frac{e^{ik|\mathbf{r}-\mathbf{r}'|}}{|\mathbf{r} - \mathbf{r}'|} . \tag{15.18}$$

Substituting into the scattering integral equation, we find

$$\psi_{\mathbf{k}}(\mathbf{r}) = e^{i\mathbf{k}\cdot\mathbf{r}} - \frac{1}{4\pi} \int d\mathbf{r}' \, \frac{e^{ik|\mathbf{r}-\mathbf{r}'|}}{|\mathbf{r} - \mathbf{r}'|} \, v(\mathbf{r}') \, \psi_{\mathbf{k}}(\mathbf{r}') , \tag{15.19}$$

which is fully equivalent to the Schrödinger equation with, as shown next, the additional advantage of automatically incorporating the correct asymptotic behavior of $\psi_{\mathbf{k}}(\mathbf{r})$. We note that, because of the presence of the short-range potential $v(\mathbf{r}')$, which is negligible for $|\mathbf{r}'| \gtrsim R$, the $\mathbf{r}'$ integration is in practice confined to a finite sphere of radius R. In the asymptotic region $|\mathbf{r}| \gg R$, we can therefore expand $|\mathbf{r} - \mathbf{r}'|$ as

$$|\mathbf{r} - \mathbf{r}'| = \left(r^2 + r'^2 - 2rr'\cos\theta\right)^{1/2} = r\left(1 - 2\frac{r'}{r}\cos\theta + \frac{r'^2}{r^2}\right)^{1/2} \approx r\left(1 - \frac{r'}{r}\cos\theta\right) = r - \hat{\mathbf{r}}\cdot\mathbf{r}',$$

$$(15.20)$$

where corrections of order $(r'/r)^n$ with $n \geq 2$ have been neglected. In this limit, the Green's function reads

$$G_k(\mathbf{r}, \mathbf{r}') \approx \frac{1}{4\pi}\frac{e^{ik(r-\hat{\mathbf{r}}\cdot\mathbf{r}')}}{r}\frac{1}{1 - \hat{\mathbf{r}}\cdot\mathbf{r}'/r} \approx \frac{1}{4\pi}\frac{e^{ik(r-\hat{\mathbf{r}}\cdot\mathbf{r}')}}{r}, \qquad (15.21)$$

where in the last step we have accounted only for the leading-order term in $1/r$. Substituting this result into Eq. (15.19) leads to

$$\psi_{\mathbf{k}}(\mathbf{r}) = e^{i\mathbf{k}\cdot\mathbf{r}} + \frac{e^{ikr}}{r}\left[-\frac{1}{4\pi}\int d\mathbf{r}'\, e^{-ik\hat{\mathbf{r}}\cdot\mathbf{r}'}\, v(\mathbf{r}')\,\psi_{\mathbf{k}}(\mathbf{r}')\right], \qquad (15.22)$$

which allows us to identify the scattering amplitude $f_{\mathbf{k}}(\Omega)$ as

$$f_{\mathbf{k}}(\Omega) = -\frac{1}{4\pi}\int d\mathbf{r}'\, e^{-ik\hat{\mathbf{r}}\cdot\mathbf{r}'}\, v(\mathbf{r}')\,\psi_{\mathbf{k}}(\mathbf{r}') = -\frac{1}{4\pi}\frac{2\mu}{\hbar^2}\langle\phi_{\mathbf{k}'}|\hat{V}|\psi_{\mathbf{k}}\rangle, \qquad \langle\mathbf{r}|\phi_{\mathbf{k}'}\rangle = e^{i\mathbf{k}'\cdot\mathbf{r}}, \quad (15.23)$$

where the θ and ϕ dependences of the scattering on the right-hand side amplitude are implicit the direction of the unit vector $\hat{\mathbf{r}} = (\sin\theta\cos\phi, \sin\theta\sin\phi, \cos\theta)$ on the right-hand side. The last expression makes it clear that $f_{\mathbf{k}}(\Omega)$ can be viewed as the matrix element of the potential operator – recall that $v(r) = (2\mu/\hbar^2)\,V(r)$ – between the initial scattering state $\psi_{\mathbf{k}}(\mathbf{r})$ and a final *free-particle* state of momentum $\mathbf{k}' = k\hat{\mathbf{r}}$. The energy of this final state is the same as the energy of the initial state, since

$$E' = \frac{\hbar^2\mathbf{k}'^2}{2\mu} = \frac{\hbar^2 k^2}{2\mu} = E, \qquad (15.24)$$

which is of course in line with the fact that we are dealing with elastic scattering. These relations are exact, albeit in their present forms they are not useful for the calculation of $f_{\mathbf{k}}(\Omega)$, since they require a knowledge of $\psi_{\mathbf{k}}(\mathbf{r})$ in the region $r \lesssim R$ where the potential is effective, and not just of its asymptotic behavior.

The above treatment is suitable for a potential that vanishes faster than $1/r$ as $r \longrightarrow \infty$. The Coulomb potential is therefore excluded and must be dealt with separately. We will not discuss it here (we refer the reader to A. Messiah (1961). *Quantum Mechanics*, North Holland), but will just note that Coulomb-scattering wave functions $\psi_{\mathbf{k}}^C(\mathbf{r})$ follow from solutions of the Schrödinger equation

$$\left(\nabla^2 \mp \underbrace{\frac{2\mu}{\hbar^2}\frac{Z_1 Z_2 e^2}{r}}_{2\gamma k/r} + k^2\right)\psi_{\mathbf{k}}^C(\mathbf{r}) = 0, \qquad \gamma = \frac{Z_1 Z_2 e^2}{\hbar v}, \qquad v = \frac{\hbar k}{\mu}, \qquad (15.25)$$

where $Z_1 e$ and $Z_2 e$ are the magnitudes of the charges of the incident and target particles respectively, and the negative sign applies for charges of the same sign and the + sign for charges of opposite sign. In this case, the scattered spherical wave is described by

$$f_{\mathbf{k}}(\Omega)\frac{e^{ikr}}{r} \longrightarrow f_{\mathbf{k}}^C(\Omega)\frac{e^{i[kr-\gamma\ln(2kr)]}}{r} \qquad \text{with } f_{\mathbf{k}}^C(\Omega) = -\frac{\gamma}{2k\sin^2\theta/2} \times \text{(phase factor)}, \quad (15.26)$$

and the differential cross section for Coulomb scattering is obtained as

$$\sigma_k^C(\Omega) = \left|f_k^C(\Omega)\right|^2 = \frac{\gamma^2}{4k^2\sin^4\theta/2} = \left(\frac{Z_1 Z_2 e^2}{4E\sin^2\theta/2}\right)^2. \qquad (15.27)$$

The differential cross section is independent of the signs of the colliding charges. It is also independent of the azimuthal angle ϕ and is therefore symmetric about the incident axis; it falls off as $1/E^2$ as the relative energy (see Eq. (15.24)). In particular, it diverges as $1/\theta^4$ in the forward direction, where $\theta \approx 0$. As a consequence the total cross section obtained by integrating over θ and ϕ is divergent. This divergence is characteristic of the pure $1/r$ Coulomb field. In nature such a field never occurs. For example, in the scattering of a charged particle by a nucleus, the Coulomb field due to the nucleus is screened by that of opposite sign generated by the orbiting atomic electrons, so that at distances large relative to the dimensions of the atom, the potential felt by the incoming charged particle effectively vanishes.

There is an interesting rewriting of the integral equation (15.15) known as the Lippmann–Schwinger equation. To obtain it, we first note that the Green's function (times $2\mu/\hbar^2$) can be expressed as

$$\frac{2\mu}{\hbar^2} G_k(\mathbf{r}, \mathbf{r}') = \int \frac{d\mathbf{q}}{(2\pi)^3} e^{i\mathbf{q}\cdot(\mathbf{r}-\mathbf{r}')} \frac{2\mu/\hbar^2}{q^2 - k^2 - i\eta} = \int \frac{d\mathbf{q}}{(2\pi)^3} e^{i\mathbf{q}\cdot(\mathbf{r}-\mathbf{r}')} \frac{1}{E_q - E_k - i\eta}, \tag{15.28}$$

where we have defined an energy $E_q = \hbar^2 q^2/(2\mu)$ and it is understood that $\eta \longrightarrow 0$. We can reinterpret Eq. (15.28) as the coordinate-space representation of an operator, since

$$\frac{2\mu}{\hbar^2} G_k(\mathbf{r}, \mathbf{r}') = \int \frac{d\mathbf{q}}{(2\pi)^3} \frac{\langle \phi_\mathbf{r} | \phi_\mathbf{q} \rangle \langle \phi_\mathbf{q} | \phi_{\mathbf{r}'} \rangle}{E_q - E_k - i\eta} = \int \frac{d\mathbf{q}}{(2\pi)^3} \left\langle \phi_\mathbf{r} \left| \frac{1}{\hat{H}_0 - E_k - i\eta} \right| \phi_\mathbf{q} \right\rangle \left\langle \phi_\mathbf{q} | \phi_{\mathbf{r}'} \right\rangle$$

$$= \left\langle \phi_\mathbf{r} \left| \frac{1}{\hat{H}_0 - E_k - i\eta} \right| \phi_{\mathbf{r}'} \right\rangle, \tag{15.29}$$

where we have used the completeness relation for the basis of momentum eigenstates $|\phi_\mathbf{q}\rangle$ and the fact that these states are also eigenstates of the free Hamiltonian, $\hat{H}_0 |\phi_\mathbf{q}\rangle = E_q |\phi_\mathbf{q}\rangle$. The integral equation (15.15) can now be written as

$$\langle \phi_\mathbf{r} | \psi_\mathbf{k} \rangle = \langle \phi_\mathbf{r} | \phi_\mathbf{k} \rangle + \int d\mathbf{r}' \left\langle \phi_\mathbf{r} \left| \frac{1}{E_k - \hat{H}_0 + i\eta} \right| \phi_{\mathbf{r}'} \right\rangle \langle \phi_{\mathbf{r}'} | \hat{V} \psi_\mathbf{k} \rangle, \tag{15.30}$$

and, using the completeness of the position eigenstates, reduces to

$$|\psi_\mathbf{k}\rangle = |\phi_\mathbf{k}\rangle + \frac{1}{E_k - \hat{H}_0 + i\eta} \hat{V} |\psi_\mathbf{k}\rangle, \tag{15.31}$$

the Lippmann–Schwinger equation precisely. It is equivalent to the Schrödinger equation ($\hat{H}_0 + \hat{V}) |\psi_\mathbf{k}\rangle = E_k |\psi_\mathbf{k}\rangle$, as can easily be verified by applying the operator $E_k - \hat{H}_0$ to both sides and using the fact that the free-particle state $|\phi_\mathbf{k}\rangle$ is an eigenstate of $\hat{H}_0$ with eigenvalue E_k. However, it has the advantage that it includes the correct asymptotic behavior for the wave function $\psi_\mathbf{k}(\mathbf{r})$. The Lippmann–Schwinger equation can be solved by iteration,

$$|\psi_\mathbf{k}\rangle = |\phi_\mathbf{k}\rangle + \frac{1}{E_k - \hat{H}_0 + i\eta} \hat{V} |\phi_\mathbf{k}\rangle + \frac{1}{E_k - \hat{H}_0 + i\eta} \hat{V} \frac{1}{E_k - \hat{H}_0 + i\eta} \hat{V} |\phi_\mathbf{k}\rangle + \cdots. \tag{15.32}$$

When such an expansion is inserted into the scattering amplitude (15.23), it leads to

$$f_\mathbf{k}(\mathbf{k}') = -\frac{\mu}{2\pi\hbar^2} \langle \phi_{\mathbf{k}'} | \underbrace{\hat{V} + \hat{V} \frac{1}{E_k - \hat{H}_0 + i\eta} \hat{V} + \hat{V} \frac{1}{E_k - \hat{H}_0 + i\eta} \hat{V} \frac{1}{E_k - \hat{H}_0 + i\eta} \hat{V} + \cdots}_{\hat{T}} | \phi_\mathbf{k} \rangle$$

$$= -\frac{\mu}{2\pi\hbar^2} \langle \phi_{\mathbf{k}'} | \hat{T} | \phi_\mathbf{k} \rangle, \tag{15.33}$$

and the scattering amplitude is seen to be proportional to the matrix element of the operator $\hat{T}$ (known as the T-matrix) between initial and final plane-wave states. The definition above makes it clear that $\hat{T}$ satisfies the following equation:

$$\hat{T} = \hat{V} + \hat{V} \frac{1}{E_k - \hat{H}_0 + i\eta} \hat{T} . \tag{15.34}$$

In the perturbative expansion of Eq. (15.33), the leading-order correction to the scattering amplitude is known as the Born approximation (BA),

$$f_{\mathbf{k}}^{\mathrm{BA}}(\Omega) = -\frac{\mu}{2\pi\hbar^2} \int d\mathbf{r}' e^{-i\mathbf{k}' \cdot \mathbf{r}'} V(\mathbf{r}') e^{i\mathbf{k} \cdot \mathbf{r}'} = -\frac{\mu}{2\pi\hbar^2} \widetilde{V}(\mathbf{q}) \qquad \text{with } \mathbf{q} = \mathbf{k}' - \mathbf{k} . \tag{15.35}$$

The BA is simply given by the Fourier transform of the potential. Here, $\hbar\mathbf{q}$ is the momentum transfer, where

$$q^2 = |\mathbf{k}' - \mathbf{k}|^2 = 2k^2 - 2k^2 \cos\theta = 4k^2 \sin^2 \theta/2 , \tag{15.36}$$

and the dependence of $f_{\mathbf{k}}^{\mathrm{BA}}(\Omega)$ on the scattering angles comes via the dependence on $\mathbf{q}$. If we now assume that $V(r)$ has rotational symmetry, that is, it only depends on the magnitude of the relative position vector, then the Born amplitude is real, since

$$\widetilde{V}^*(q) = \left[\int d\mathbf{r}\, e^{-i\mathbf{q} \cdot \mathbf{r}}\, V(r) \right]^* = \underbrace{\int d\mathbf{r}\, e^{i\mathbf{q} \cdot \mathbf{r}}\, V(r)}_{\text{change variable: } \mathbf{r} \longrightarrow -\mathbf{r}} = \int d\mathbf{r}\, e^{-i\mathbf{q} \cdot \mathbf{r}}\, V(r) = \widetilde{V}(q) , \tag{15.37}$$

implying that the BA violates the optical theorem; see Problem 6. The Fourier transform $\widetilde{V}(q)$ depends only on q, the magnitude of the "wave-number transfer,"

$$\widetilde{V}(q) = \frac{4\pi}{q} \int_0^\infty dr\, r V(r) \sin(qr) , \tag{15.38}$$

and the total cross section follows from

$$\sigma_{\mathbf{k}}^{\mathrm{BA}} = \left(\frac{\mu}{\sqrt{2\pi\hbar^2}} \right)^2 \int_0^\pi d\theta\, \sin\theta\, \widetilde{V}^2[2k\sin(\theta/2)] . \tag{15.39}$$

The conditions for the validity of the BA are examined in Problem 5.

15.3 Scattering by a Central Potential: Phase-Shift Method

We now turn our attention to the case in which the potential $V(r)$ has spherical symmetry. We can then expand the scattering wave function in a basis of simultaneous eigenfunctions of the Hamiltonian, the square of the angular momentum operator $\mathbf{L}$, and its z-component L_z,

$$\psi_{k\hat{z}}(\mathbf{r}) = \sum_{l=0}^\infty \sum_{m=-l}^l \frac{u_{kl}(r)}{r} Y_{lm}(\theta, \phi) , \tag{15.40}$$

where the radial function $u_{kl}(r)$ satisfies the radial equation

$$u_{kl}''(r) + \left[k^2 - \frac{l(l+1)}{r^2} - v(r) \right] u_{kl}(r) = 0 , \tag{15.41}$$

and $v(r) = 2\mu V(r)/\hbar^2 \longrightarrow 0$ as $r \longrightarrow \infty$. We are interested in the solutions for positive energy $\hbar^2 k^2/(2\mu) = E > 0$. They form a continuum; for fixed E, there is infinite degeneracy, since $l = 0, 1, 2, \ldots$ and $m = -l, \ldots, l$. The boundary conditions satisfied by $u_{kl}(r)$ are (assuming that $r^2 v(r) \longrightarrow 0$ as $r \longrightarrow 0$, see Section 9.2)

$$u_{kl}(r) \sim r^{l+1} \qquad \text{as } r \longrightarrow 0 \tag{15.42}$$

and

$$u_{kl}(r) = a_{kl} \sin(kr - l\pi/2 + \delta_{kl}) \qquad \text{as } r \longrightarrow \infty, \tag{15.43}$$

where the factor $l\pi/2$ is included for convenience. Equation (15.43) follows from the asymptotic form of the radial equation, which reads

$$u_{kl}''(r) + k^2 u_{kl}(r) = 0 \qquad \text{as } r \longrightarrow \infty. \tag{15.44}$$

The constant a_{kl} and so-called phase shift δ_{kl} (for fixed k and l) are obtained by matching the wave function $\psi_{k\hat{z}}(\mathbf{r})$ of Eq. (15.40) to the asymptotic form given in Eq. (15.17). As we will see, this allows us to relate the scattering amplitude to the phase shifts δ_{kl}. But, before establishing this connection, we need to briefly examine the free-particle problem.

The solution $u_{kl}(r)$ of the radial equation for a free particle – set $v(r) = 0$ in Eq. (15.41) – is proportional to the regular spherical Bessel function $r j_l(kr)$ (see Problem 8 in Chapter 10 and Problem 17 in the present chapter, which are relevant to the present discussion), and the wave functions

$$\phi_{klm}(\mathbf{r}) = b_{kl} j_l(kr) Y_{lm}(\Omega) \tag{15.45}$$

form a basis of simultaneous eigenfunctions of the free-particle Hamiltonian, $\mathbf{L}^2$, and L_z with eigenvalues given by $\hbar^2 k^2/(2\mu)$, $l(l+1)\hbar^2$, and $m\hbar$, respectively. The basis is continuous, and the "normalization constants" b_{kl} are determined by requiring that

$$\int d\mathbf{r}\, \phi_{klm}^*(\mathbf{r})\, \phi_{k'l'm'}(\mathbf{r}) = \delta(k - k')\, \delta_{ll'}\, \delta_{mm'}, \tag{15.46}$$

which, because of the orthonormality of the $Y_{lm}(\Omega)$ implies that, since the $j_l(kr)$ are real functions,

$$|b_{kl}|^2 \int_0^\infty dr\, r^2 j_l(kr) j_l(k'r) = \delta(k - k'). \tag{15.47}$$

Up to an irrelevant phase factor, b_{kl} is independent of l and is explicitly given by $b_{kl} = \sqrt{2/\pi}\, k$ (a clear derivation of this result and others relating to free-particle wave functions is provided in Chapter VIII of C. Cohen-Tannoudji, B. Diu, and F. Laloë (1997). *Quantum Mechanics*, Wiley).

Next, we want to relate the continuous basis of eigenfunctions $e^{i\mathbf{k}\cdot\mathbf{r}}$ of the momentum operator to the continuous basis consisting of the eigenfunctions $\phi_{klm}(\mathbf{r})$. Since the plane waves are also eigenfunctions of the free-particle Hamiltonian, we must have

$$e^{i\mathbf{k}\cdot\mathbf{r}} = \sum_{l=0}^\infty \sum_{m=-l}^l c_{lm}(k, \Omega_k)\, \phi_{klm}(\mathbf{r}) = \sum_{l=0}^\infty \sum_{m=-l}^l b_{kl}\, c_{lm}(k, \Omega_k) j_l(kr) Y_{lm}(\theta, \phi) \qquad \text{with } |\mathbf{k}| = k, \tag{15.48}$$

where Ω_k specifies the angles θ_k and ϕ_k of the direction $\hat{\mathbf{k}}$. Thus a given momentum eigenfunction is generally a linear superposition of all possible angular momentum eigenfunctions. In particular, if $\hat{\mathbf{k}}$ is taken to be along the z-axis, which we indicate with $\Omega_k = 0$ so that the expansion coefficient is $c_{lm}(k, 0)$ in this case, then only $m = 0$ terms can occur on the right-hand side of Eq. (15.48), since the left-hand side is independent of the angle ϕ; recall that $Y_{lm}(\theta, \phi)$ is proportional to $P_l^m(\cos\theta)\,e^{im\phi}$. Hence, the expansion simplifies to

$$e^{ikr\cos\theta} = \sum_{l=0}^{\infty} b_{kl}\,c_{l0}(k,0)\,j_l(kr)\,\underbrace{\sqrt{\frac{2l+1}{4\pi}}\,P_l(\cos\theta)}_{Y_{l0}(\theta,\phi)} = \sum_{l=0}^{\infty} i^l(2l+1)\,j_l(kr)\,P_l(\cos\theta)\,, \qquad (15.49)$$

where $P_l(x)$ is a Legendre polynomial and the combination $\sqrt{(2l+1)/(4\pi)}\,b_{kl}\,c_{l0}(k,0)$ is found to equal $i^l(2l+1)$. This expansion can be generalized to the case where $\mathbf{k}$, rather than being along the z-axis, is in a generic direction specified by Ω_k. In that case, θ is interpreted as the angle between the directions $\hat{\mathbf{k}}$ and $\hat{\mathbf{r}}$, that is, $\cos\theta \longrightarrow \cos\theta_{kr} = \hat{\mathbf{k}}\cdot\hat{\mathbf{r}}$, and we obtain

$$e^{i\mathbf{k}\cdot\mathbf{r}} = \sum_{l=0}^{\infty} i^l(2l+1)\,j_l(kr)\,P_l(\cos\theta_{kr}) = 4\pi\sum_{l=0}^{\infty}\sum_{m=-l}^{l} i^l\,j_l(kr)\,Y_{lm}^*(\theta_k,\phi_k)\,Y_{lm}(\theta,\phi)\,, \qquad (15.50)$$

where we have made use of the addition theorem for spherical harmonics (see Problem 7 in Chapter 13 for a derivation),

$$\frac{2l+1}{4\pi}\,P_l(\cos\theta_{kr}) = \sum_{m=-l}^{l} Y_{lm}^*(\theta_k,\phi_k)\,Y_{lm}(\theta,\phi)\,. \qquad (15.51)$$

We now return to the problem of matching the wave function of Eq. (15.40) to the asymptotic form given in Eq. (15.17). Because of the spherical symmetry of the problem, the scattering amplitude is independent of ϕ, $f_{k\hat{z}}(\Omega) = f_{k\hat{z}}(\theta)$, and therefore it can be expanded in Legendre polynomials:

$$f_{k\hat{z}}(\theta) = \sum_{l=0}^{\infty} f_{kl}\,P_l(\cos\theta)\,. \qquad (15.52)$$

In the asymptotic region the wave function can be written as

$$\lim_{r\to\infty}\psi_{k\hat{z}}(r,\theta) = e^{ikr\cos\theta} + f_{k\hat{z}}(\theta)\,\frac{e^{ikr}}{r} = \sum_{l=0}^{\infty}\left[i^l(2l+1)\,j_l(kr) + f_{kl}\,\frac{e^{ikr}}{r}\right]P_l(\cos\theta)\,. \qquad (15.53)$$

In order for this to match $\psi_{k\hat{z}}(\mathbf{r})$ of Eq. (15.40), only terms with $m=0$ must enter in the expansion; namely, the wave function must be independent of ϕ, so that

$$\psi_{k\hat{z}}(r,\theta) = \sum_{l=0}^{\infty}\frac{u_{kl}(r)}{r}\,P_l(\cos\theta)\,, \qquad (15.54)$$

where we have absorbed the factor $\sqrt{(2l+1)/(4\pi)}$ from $Y_{l0}(\theta)$ into $u_{kl}(r)$, which is permissible since $u_{kl}(r)$ satisfies a homogenous equation. Given $u_{kl}(r) = a_{kl}\sin(kr - l\pi/2 + \delta_{kl})$, this matching yields (see Problem 10)

$$a_{kl} = i^l\,\frac{2l+1}{k}\,e^{i\delta_{kl}}\,, \qquad f_{kl} = \frac{2l+1}{k}\,e^{i\delta_{kl}}\,\sin\delta_{kl}\,. \qquad (15.55)$$

Knowledge of the phase shift δ_{kl} determines the l-wave contribution to the scattering amplitude and also fixes the normalization a_{kl}. The full scattering amplitude reads

$$f_{k\hat{z}}(\theta) = \frac{1}{k}\sum_{l=0}^{\infty}(2l+1)\,e^{i\delta_{kl}}\,\sin\delta_{kl}\,P_l(\cos\theta)\,, \qquad (15.56)$$

and the differential cross section follows as

$$\sigma_{k\hat{z}}(\theta) = \frac{1}{k^2} \sum_{l,l'=0}^{\infty} \underbrace{(2l+1)(2l'+1) \, e^{-i(\delta_{kl}-\delta_{kl'})} \, \sin\delta_{kl} \, \sin\delta_{kl'}}_{g_{kl,kl'}} P_l(\cos\theta) \, P_{l'}(\cos\theta) \, ; \qquad (15.57)$$

the different l and l' waves generally interfere. The total cross section can be obtained either through direct integration over the solid angle or by the optical theorem. In the former case, we have (setting $x = \cos\theta$)

$$\sigma_k = 2\pi \int_{-1}^{1} dx \, \sigma_{k\hat{z}}(\cos^{-1} x) = \frac{2\pi}{k^2} \sum_{l,l'=0}^{\infty} g_{kl,kl'} \underbrace{\int_{-1}^{1} dx \, P_l(x) \, P_{l'}(x)}_{2/(2l+1)\,\delta_{ll'}} = \frac{4\pi}{k^2} \sum_{l=0}^{\infty} \frac{1}{2l+1} \, g_{kl,kl}$$

$$= \frac{4\pi}{k^2} \sum_{l=0}^{\infty} (2l+1) \, \sin^2\delta_{kl} \, , \qquad (15.58)$$

while in the latter case ($x = 1$ and $P_l(1) = 1$)

$$\sigma_k = \frac{4\pi}{k} \, \text{Im}[f_{k\hat{z}}(0)] = \frac{4\pi}{k^2} \sum_{l=0}^{\infty} (2l+1) \, \underbrace{\text{Im}(e^{i\delta_{kl}})}_{\sin\delta_{kl}} \, \sin\delta_{kl} \, , \qquad (15.59)$$

and the two expressions for σ_k coincide, of course. The contribution of partial wave l to the total cross section has the upper bound

$$\sigma_{kl} = \frac{4\pi}{k^2} \, (2l+1) \, \sin^2\delta_{kl} \le \frac{4\pi}{k^2} \, (2l+1) \, , \qquad (15.60)$$

which is reached when (and if) $\delta_{kl} = n\pi/2$ with n a (positive or negative) odd integer. Further, while in principle the sum in σ_k extends over all partial waves, in fact for a short-range potential we expect only the lowest-l waves to contribute appreciably at a given energy. To see how this comes about, suppose that $v(r)$ vanishes for $r \gtrsim R$. Reasoning classically for the time being, the magnitude of the (conserved) angular momentum is given by $L = |\mathbf{r} \times \mathbf{p}| = bp$, where $\mathbf{p}$ is the momentum of the particle when it is at $r \gg R$ and $b = r\sin\theta$ is the impact parameter, the distance of closest approach to the center of force. If $b \gtrsim R$ then the classical trajectory of the particle will be unaffected by the potential and will continue as a straight line. This condition can be expressed as $b/R = L/(Rp) \gtrsim 1$. Quantum mechanically, we have $L = \hbar\sqrt{l(l+1)}$ and $p = \hbar k$, and therefore the condition reads

$$\sqrt{l(l+1)} \gtrsim kR \, . \qquad (15.61)$$

Therefore, for a given energy $E = \hbar^2 k^2/(2\mu)$, only partial waves with $l \lesssim kR$ will be affected by the potential. As the energy increases, partial waves of progressively higher order will contribute to the cross section; conversely, as the energy decreases, the contribution of the $l = 0$ partial wave (the s-wave) will become dominant. This conclusion could also have been inferred from the behavior of the radial wave function for small r, $u_{kl}(r) \propto r^{l+1}$. Clearly, for large l the wave function will be suppressed at small r and will not feel the presence of the potential.

The procedure for obtaining the phase shifts and hence the differential cross section is straightforward. For a given energy $E = \hbar^2 k^2/(2\mu) > 0$, we solve the radial equation (15.41) with the boundary condition at the origin $u_{kl}(r) \propto r^{l+1}$. Since $v(r)$ is short range, there is an R such that $v(r \gtrsim R)$ is negligible. For $r \gtrsim R$ we have a free particle, and therefore the radial wave function $\psi_{kl}(r) = u_{kl}(r)/r$ reduces to a linear combination of the regular and irregular spherical Bessel functions, respectively $j_l(kr)$ and $n_l(kr)$, which we write as

$$A_{kl}\, j_l(kr) - B_{kl}\, n_l(kr) \qquad \text{for } r \gtrsim R. \tag{15.62}$$

In order to determine the constants A_{kl} and B_{kl} (for fixed k and l), we "match" $\psi_{kl}(r \gtrsim R)$ to the linear combination above, that is, we require that $\psi_{kl}(r)$ and its first derivative are continuous at some (arbitrary) $R_0 > R$,

$$\psi_{kl}(R_0) = A_{kl}\, j_l(x_0) - B_{kl}\, n_l(x_0)\,, \qquad \psi'_{kl}(R_0) = A_{kl}\, k j'_l(x_0) - B_{kl}\, k n'_l(x_0), \tag{15.63}$$

where $x_0 = kR_0$. This leads to an inhomogeneous linear system in the unknowns A_{kl} and B_{kl}, which is easily solved to give

$$A_{kl} = \frac{k n'_l(x_0)\, \psi_{kl}(R_0) - n_l(x_0)\, \psi'_{kl}(R_0)}{k n'_l(x_0)\, j_l(x_0) - k j'_l(x_0)\, n_l(x_0)}\,, \qquad B_{kl} = \frac{k j'_l(x_0)\, \psi_{kl}(R_0) - j_l(x_0)\, \psi'_{kl}(R_0)}{k n'_l(x_0)\, j_l(x_0) - k j'_l(x_0)\, n_l(x_0)}\,. \tag{15.64}$$

Having determined A_{kl} and B_{kl}, the phase shift is easily obtained as

$$\delta_{kl} = \tan^{-1}(B_{kl}/A_{kl})\,. \tag{15.65}$$

To see how this comes about, insert the asymptotic behavior of the spherical Bessel functions, to obtain

$$\lim_{r\to\infty}[A_{kl}\, j_l(kr) - B_{kl}\, n_l(kr)] = A_{kl}\, \frac{\sin(kr - l\pi/2)}{kr} + B_{kl}\, \frac{\cos(kr - l\pi/2)}{kr} = C_{kl}\, \frac{\sin(kr - l\pi/2 + \delta_{kl})}{kr}\,, \tag{15.66}$$

with

$$C_{kl}\, \cos\delta_{kl} = A_{kl}\,, \qquad C_{kl}\, \sin\delta_{kl} = B_{kl}\,, \tag{15.67}$$

and Eq. (15.66) shows that $\psi_{kl}(r)$ has the expected asymptotic behavior of Eq. (15.43), except for the normalization. By inserting the solutions for A_{kl} and B_{kl} found earlier, we obtain the phase shift as

$$\delta_{kl} = \tan^{-1}\left[\frac{k j'_l(x_0) - \Delta_{kl}\, j_l(x_0)}{k n'_l(x_0) - \Delta_{kl}\, n_l(x_0)}\right]\,, \tag{15.68}$$

where we have defined the logarithmic derivative

$$\Delta_{kl} = \frac{\psi'_{kl}(R_0)}{\psi_{kl}(R_0)}\,. \tag{15.69}$$

Now, all we need to determine the phase shift is the value of the logarithmic derivative of the radial wave function at a point $R > R_0$, where the potential is negligible. In order to match the normalization constant in Eq. (15.55), we need to rescale $\psi_{kl}(r)$ by the factor $i^l(2l + 1)e^{i\delta_{kl}}/C_{kl}$:

$$\psi_{kl}(r) \longrightarrow i^l(2l + 1)\, \frac{e^{i\delta_{kl}}}{C_{kl}}\, \psi_{kl}(r)\,, \tag{15.70}$$

which then leads to the asymptotic behavior

$$\lim_{r\to\infty}\psi_{kl}(r) = \underbrace{i^l(2l + 1)\frac{e^{i\delta_{kl}}}{k}}_{a_{kl}}\, \frac{\sin(kr - l\pi/2 + \delta_{kl})}{r}\,. \tag{15.71}$$

15.4 Problems

Problem 1 Verifying That the Asymptotic Wave Function Satisfies the Schrödinger Equation

Verify explicitly that the asymptotic wave function in Eq. (15.5) satisfies the Schrödinger equation.

Solution

In the asymptotic region ($v(r)$ vanishes in this limit), the Schrödinger equation reduces to the homogeneous partial differential equation

$$\left(\nabla^2 + k^2\right)\left[e^{i\mathbf{k}\cdot\mathbf{r}} + f_{\mathbf{k}}(\Omega)\,\frac{e^{ikr}}{r}\right] = 0\,.$$

The plane wave satisfies the equation $(\nabla^2 + k^2)\,e^{i\mathbf{k}\cdot\mathbf{r}} = (-\mathbf{k}^2 + k^2)\,e^{i\mathbf{k}\cdot\mathbf{r}} = 0$. Writing

$$\nabla^2 = \frac{1}{r^2}\frac{\partial}{\partial r}r^2\frac{\partial}{\partial r} - \frac{\mathbf{L}^2}{\hbar^2 r^2} = \frac{1}{r}\frac{\partial^2}{\partial r^2}r - \frac{1}{r^2}\left(\cdots\right),$$

where $(\cdots)$ involves only partial derivatives with respect to θ and ϕ, we find

$$\left(\nabla^2 + k^2\right)f_{\mathbf{k}}(\Omega)\,\frac{e^{ikr}}{r} = f_{\mathbf{k}}(\Omega)\left(\underbrace{\frac{1}{r}\frac{\partial^2}{\partial r^2}r + k^2}_{-k^2\,e^{ikr}/r}\right)\frac{e^{ikr}}{r} + \underbrace{\frac{e^{ikr}}{r^3}\left(\cdots\right)f_{\mathbf{k}}(\Omega)}_{\propto\,1/r^3\text{ so ignore}} = 0\,.$$

Problem 2 Born Approximation for Scattering in Yukawa and Coulomb Potentials

The central component of the nuclear potential between proton and neutron has the Yukawa form $V_Y(r) = V_0\,e^{-r/\lambda}/r$, where the parameter λ with the dimension of length is related to the pion mass via $\lambda = \hbar/(m_\pi c)$ (here c is the speed of light; λ is known as the Compton wavelength of the pion). Calculate the amplitude and differential cross section for scattering in the Born approximation (BA). Noting that the attractive or repulsive Coulomb potential between particles of charges $\pm Z_1 e$ and $\pm Z_2 e$ is proportional to $V_Y(r)$ in the limit $\lambda \longrightarrow \infty$, obtain the BA amplitude and differential cross section for Coulomb scattering.

Solution

The BA differential cross section is given by $\sigma_k^{\mathrm{BA}}(\Omega) = |f_{\mathbf{k}}^{\mathrm{BA}}(\Omega)|^2$, where the BA amplitude generally reads

$$f_{\mathbf{k}}^{\mathrm{BA}}(\Omega) = -\frac{\mu}{2\pi\hbar^2}\int d\mathbf{r}\,V(\mathbf{r})\,e^{-i\mathbf{q}\cdot\mathbf{r}}\,,$$

with $\mathbf{q} = \mathbf{k}' - \mathbf{k}$ and $k = k' = \sqrt{2\mu E/\hbar^2}$ (μ is the reduced mass). For a spherically symmetric potential, as is the case here, the Fourier transform depends only on the magnitude of the momentum transfer and is given by

$$\widetilde{V}_Y(q) = \frac{4\pi}{q} \int_0^\infty dr\, r V_Y(r) \sin(qr) = \frac{4\pi}{q} \frac{V_0}{2i} \int_0^\infty dr\, e^{-r/\lambda} \left(e^{iqr} - e^{-iqr} \right)$$

$$= \frac{4\pi V_0}{2iq} \left(-\frac{1}{iq - 1/\lambda} - \frac{1}{iq + 1/\lambda} \right) = \frac{4\pi V_0}{q^2 + 1/\lambda^2},$$

and the BA differential cross section follows as

$$\sigma_{Y,k}^{BA}(\Omega) = \left(\frac{\mu}{2\pi\hbar^2} \right)^2 \frac{(4\pi V_0)^2}{\left(4k^2 \sin^2 \theta/2 + 1/\lambda^2 \right)^2} = \left(\frac{2\mu V_0}{\hbar^2} \right)^2 \frac{1}{\left(4k^2 \sin^2 \theta/2 + 1/\lambda^2 \right)^2}.$$

It is independent of the azimuthal angle ϕ, as is always the case for a central potential, and decreases as θ varies from the forward to the backward direction, or equivalently as the momentum transfer $\hbar q$ varies from 0 at $\theta = 0$ to its maximum allowed value $2\hbar k$ at $\theta = \pi$.

The Coulomb potential between two charged particles is given by $V_C(r) = \pm Z_1 Z_2 e^2/r$, where the $+$ $(-)$ sign corresponds to the particles having charges of the same (opposite) sign. The BA amplitude and differential cross section can be obtained by setting $V_0 = \pm Z_1 Z_2 e^2$ and by taking the limit $\lambda \longrightarrow \infty$ in the Yukawa expressions. We find

$$\widetilde{V}_C^{BA}(q) = \pm 4\pi \frac{Z_1 Z_2 e^2}{q^2}, \qquad \sigma_{C,k}(\Omega) = \left(\frac{2\mu Z_1 Z_2 e^2}{\hbar^2} \right)^2 \frac{1}{16 k^4 \sin^4 \theta/2} = \left(\frac{Z_1 Z_2 e^2}{4E \sin^2 \theta/2} \right)^2.$$

It turns out that $\sigma_{C,k}(\Omega)$ is the *exact* result (the Rutherford differential cross section): an exact solution of the Coulomb scattering problem leads to an amplitude that differs from the BA solution simply by a phase factor; this phase factor becomes irrelevant when one is evaluating the differential cross section (see Section 15.2).

Problem 3   Born Approximation for Scattering in a Gaussian Potential

Consider a particle of mass μ in the central potential $V(r) = V_0\, e^{-r^2/a^2}$. Obtain the differential and total cross sections in the Born approximation.

Solution

The Born approximation (BA) amplitude is given by

$$f_{\mathbf{k}}^{BA}(\theta) = -\frac{\mu}{2\pi\hbar^2} \int d\mathbf{r}\, V(r)\, e^{-i\mathbf{q}\cdot\mathbf{r}} \qquad \text{with } \mathbf{q} = \mathbf{k}' - \mathbf{k}.$$

Because of the spherical symmetry of the potential, the Fourier transform depends only on the magnitude q of the momentum transfer. Thus, we take $\mathbf{q}$ to be along $\hat{\mathbf{z}}$ to find

$$f_{\mathbf{k}}^{BA}(\theta) = -\frac{\mu}{2\pi\hbar^2} V_0 \int d\mathbf{r}\, e^{-r^2/a^2}\, e^{-iqz} = -\frac{\mu}{2\pi\hbar^2} V_0 \pi a^2 \int_{-\infty}^{\infty} dz\, e^{-z^2/a^2}\, e^{-iqz}$$

$$= -\frac{\mu}{2\pi\hbar^2} V_0\, \pi^{3/2} a^3\, e^{-(qa)^2/4} = -\sqrt{\pi}\, \frac{\mu a^3 V_0}{2\hbar^2}\, e^{-(qa)^2/4},$$

where in the second step we carried out the Gaussian over x and y. The BA total cross section follows from

$$\sigma_k^{BA} = \int d\Omega\, |f_{\mathbf{k}}^{BA}(\theta)|^2 = \frac{1}{2} \left(\frac{\pi \mu a^3 V_0}{\hbar^2} \right)^2 \underbrace{\int_0^{\pi} d\theta\, \sin\theta\, e^{-(ka)^2(1-\cos\theta)/2}}_{I},$$

where we have inserted $q^2 = 2k^2(1 - \cos\theta)$. The θ-integration yields $I = 2[1 - e^{-(ka)^2}]/(ka)^2$, and we finally obtain

$$\sigma_k^{BA} = \left(\frac{\pi\mu a^2 V_0}{\hbar^2 k}\right)^2 \left[1 - e^{-(ka)^2}\right],$$

and thus for large energy σ_k^{BA} decreases as $1/E$, where $E = \hbar^2 k^2/(2\mu)$.

Problem 4 Alternative Derivation of the PDE Satisfied by the Scattering Green's Function

Consider the following derivation of the partial differential equation satisfied by the Green's functions $G_k^{(\pm)}(x) = e^{\pm ikx}/(4\pi x)$, where $x = r - r'$.

1. Show that, for $x \neq 0$, the functions $e^{\pm ikx}/(4\pi x)$ satisfy the wave equation

$$\left(\nabla^2 + k^2\right) \frac{e^{\pm ikx}}{4\pi x} = 0, \qquad x \neq 0.$$

2. Now consider a small sphere of radius R centered at the origin ($x = 0$), and show that

$$\lim_{R\to 0} \int_{x\leq R} dx\, \nabla^2 \frac{e^{\pm ikx}}{4\pi x} = -1.$$

What is then the equation satisfied by $G_k^{(\pm)}(x)$ over *all* space, including the origin?

Solution

Part 1

Recalling that

$$\nabla^2 = \frac{1}{x}\frac{\partial^2}{\partial x^2} x - \frac{L^2(\theta_x, \phi_x)/\hbar^2}{x^2},$$

where L^2, the square of the orbital angular momentum, is a differential operator that acts only on the θ_x and ϕ_x dependences, we have ($x \neq 0$)

$$(\nabla^2 + k^2)\frac{e^{\pm ikx}}{4\pi x} = \frac{1}{4\pi}\left(\frac{1}{x}\frac{\partial^2}{\partial x^2}e^{\pm ikx} + k^2\frac{e^{\pm ikx}}{x}\right) = \frac{e^{\pm ikx}}{4\pi x}\left(-k^2 + k^2\right) = 0.$$

Part 2

Having established that $G_k^{(\pm)}(x)$ satisfies the wave equation for $x \neq 0$, consider the following quantity:

$$\lim_{R\to 0}\int_{x\leq R} dx\,\left(\nabla^2 + k^2\right)\frac{e^{\pm ikx}}{4\pi x} = \lim_{R\to 0}\left(\int_{S_R} dS\,\hat{x}\cdot\nabla\frac{e^{\pm ikx}}{4\pi x} + k^2\underbrace{\int_0^R dx\, x\, e^{\pm ikx}}_{\text{vanishes as } R\to 0}\right),$$

where in the second step we have used Gauss's theorem to convert the volume integral into a surface integral. Now, the surface integral reduces to

$$\int_{S_R} dS\,\hat{x}\cdot\nabla\frac{e^{\pm ikx}}{4\pi x} = \frac{R^2}{4\pi}\int d\Omega_x \frac{\partial}{\partial x}\frac{e^{\pm ikx}}{x}\bigg|_R = \frac{R^2}{4\pi}\int d\Omega_x \left(\pm\frac{ik}{R} - \frac{1}{R^2}\right)e^{\pm ikR} = (\pm ikR - 1)\,e^{\pm ikR},$$

so that in the limit $R \longrightarrow 0$ we obtain

$$\lim_{R \to 0} \int_{x \le R} d\mathbf{x} \left(\nabla^2 + k^2 \right) \frac{e^{\pm ikx}}{4\pi x} = -1 \,.$$

The results for $x \ne 0$ and $x = 0$ can be combined by writing

$$\left(\nabla^2 + k^2 \right) \frac{e^{\pm ikx}}{4\pi x} = -\delta(\mathbf{x}) \,.$$

Problem 5   On the Validity of the Born Approximation

Determine under what conditions the Born approximation (BA) is valid in the low- and high-energy limits, by considering the ratio of the first-order correction $\psi_{\mathbf{k}}^{(1)}(\mathbf{r})$ and the leading-order term $\psi_{\mathbf{k}}^{(0)}(\mathbf{r}) = e^{i\mathbf{k}\cdot\mathbf{r}}$ in the scattering wave function. Assume that the potential $V(r)$ is largest when the particles are close to each, that is, when $r \approx 0$.

Solution

The first-order correction $\psi_{\mathbf{k}}^{(1)}(\mathbf{r})$ to the scattering wave function is given by

$$\psi_{\mathbf{k}}^{(1)}(\mathbf{r}) = -\frac{1}{4\pi} \int d\mathbf{r}' \frac{e^{ik|\mathbf{r}-\mathbf{r}'|}}{|\mathbf{r}-\mathbf{r}'|} v(r') e^{i\mathbf{k}\cdot\mathbf{r}'} \,, \qquad v(r) = \frac{2\mu}{\hbar^2} V(r) \,.$$

We expect the BA to be valid when $|\psi_{\mathbf{k}}^{(1)}(\mathbf{r})/\psi_{\mathbf{k}}^{(0)}(\mathbf{r})| \ll 1$. When r is very large and therefore the incident and target particles are far from each other, we have

$$\psi_{\mathbf{k}}^{(1)}(\mathbf{r}) \approx f^{\mathrm{BA}}(\Omega) \frac{e^{ikr}}{r} \,,$$

and the ratio above always vanishes as $|f^{\mathrm{BA}}(\Omega)|/r \longrightarrow 0$ in the limit $r \longrightarrow \infty$. By contrast, when r is in the region where the potential is most effective, namely, when the incident and target particles are close to each other, that is, $r \approx 0$, then $\psi_{\mathbf{k}}^{(0)}(0) = 1$ and

$$\psi_{\mathbf{k}}^{(1)}(0) = -\frac{1}{4\pi} \underbrace{\int d\mathbf{r}' \frac{e^{ikr'}}{r'} v(r') e^{i\mathbf{k}\cdot\mathbf{r}'}}_{\text{carry out angular integrations}} = -\int_0^\infty dr'\, r'^2 \frac{e^{ikr'}}{r'} v(r') \frac{\sin(kr')}{kr'}$$

$$= \frac{i}{2k} \int_0^\infty dr'\, v(r') e^{ikr'} \left(e^{ikr'} - e^{-ikr'} \right) \,,$$

so that the ratio of the first- and leading-order corrections is

$$\frac{1}{2k} \left| \int_0^\infty dr'\, v(r') \left(e^{2ikr'} - 1 \right) \right| \ll 1 \,.$$

It turns out this is a sufficient but not a necessary condition. For example, in the case of the Coulomb potential, the whole correction goes into a phase factor. After this proviso, we can now examine the validity of the BA under the two different regimes, of low and high energy.

At low energy, $k \longrightarrow 0$ and $e^{2ikr'} = 1 + 2ikr' + \cdots$ and so the above requirement reads

$$\left| \int_0^\infty dr'\, r'\, v(r') \right| \ll 1 \,,$$

and the integral converges as long as $v(r')$ decreases faster than $1/r'^2$. At high energy, $k \longrightarrow \infty$ and the rapidly oscillating term $v(r')\,e^{2ikr'}$ gives a vanishing contribution to the integral, so that

$$\frac{1}{2k} \left| \int_0^\infty dr'\, v(r') \right| \ll 1 ,$$

and, provided that the integral converges, we see that because of the factor $1/k$ we can always expect the BA to hold when the energy is sufficiently high (how high this is will depend on the potential being considered).

Problem 6 Optical Theorem

Exploiting the conservation of the probability current density $\mathbf{j}(\mathbf{r})$ associated with stationary solutions of the Schrödinger equation, which behave asymptotically as

$$\psi_{\mathbf{k}}(\mathbf{r}) = e^{i\mathbf{k}\cdot\mathbf{r}} + f_{\mathbf{k}}(\theta, \phi)\, \frac{e^{ikr}}{r} ,$$

show that

$$\sigma_{\mathbf{k}} = \frac{4\pi}{k}\, \mathrm{Im}[f_{\mathbf{k}}(0)] ,$$

where $f_{\mathbf{k}}(0)$ is the scattering amplitude in the forward direction $\theta = \phi = 0$, and $\sigma_{\mathbf{k}}$ is the total cross section.

Hint: The conservation of the probability current density, $\boldsymbol{\nabla} \cdot \mathbf{j}(\mathbf{r}) = 0$, implies that

$$0 = \int_V d\mathbf{r}\, \boldsymbol{\nabla} \cdot \mathbf{j}(\mathbf{r}) = \int_S dS\, \hat{\mathbf{r}} \cdot \mathbf{j}(\mathbf{r}) = R^2 \int d\Omega\, \hat{\mathbf{r}} \cdot \mathbf{j}(R, \Omega) ,$$

where V and S are, respectively, the volume and surface of a sphere of radius $R \longrightarrow \infty$. Evaluate the last integral using the asymptotic form of $\psi_{\mathbf{k}}(\mathbf{r})$.

Solution

Since $R \longrightarrow \infty$, we only need the asymptotic form of $\psi_{\mathbf{k}}(\mathbf{r})$ to evaluate $\mathbf{j}(\mathbf{r})$, and so (noting that $\mathbf{R} = R\hat{\mathbf{r}}$ on the surface of the sphere and $\hat{\mathbf{r}}$ depends only on Ω)

$$0 = R^2 \int d\Omega \underbrace{\left[e^{-iR\mathbf{k}\cdot\hat{\mathbf{r}}} + f_{\mathbf{k}}^*(\Omega)\, \frac{e^{-ikR}}{R} \right]}_{\psi_{\mathbf{k}}^*(R\to\infty,\Omega)} \underbrace{\hat{\mathbf{r}} \cdot \boldsymbol{\nabla}}_{\partial/\partial R} \underbrace{\left[e^{iR\mathbf{k}\cdot\hat{\mathbf{r}}} + f_{\mathbf{k}}(\Omega)\, \frac{e^{ikR}}{R} \right]}_{\psi_{\mathbf{k}}(R\to\infty,\Omega)} - \mathrm{c.c.} ,$$

where the prefactor $\hbar/(2\mu i)$ has been dropped, since it is irrelevant. There are three terms we need to concern ourselves with:

Plane-wave (pp) term: Since we have

$$\frac{\partial}{\partial R}\, e^{iR\mathbf{k}\cdot\hat{\mathbf{r}}} = i\mathbf{k}\cdot\hat{\mathbf{r}}\, e^{iR\mathbf{k}\cdot\hat{\mathbf{r}}} = ik\,\cos\theta\, e^{iR\mathbf{k}\cdot\hat{\mathbf{r}}} ,$$

the contribution from this term vanishes identically,

$$\mathrm{pp} = R^2 \underbrace{\int_0^{2\pi} d\phi \int_{-1}^1 d(\cos\theta)}_{d\Omega}\, e^{-iR\mathbf{k}\cdot\hat{\mathbf{r}}} ik\,\cos\theta\, e^{iR\mathbf{k}\cdot\hat{\mathbf{r}}} - \mathrm{c.c.} = ikR^2 \int_0^{2\pi} d\phi \int_{-1}^1 d(\cos\theta)\,\cos\theta - \mathrm{c.c.} = 0 .$$

Spherical-wave (ss) term: Since

$$\frac{\partial}{\partial R}\frac{e^{ikR}}{R} = \left(ik - \frac{1}{R}\right)\frac{e^{ikR}}{R} \,,$$

this contribution is

$$\text{ss} = R^2 \int d\Omega\, f_{\mathbf{k}}^*(\Omega)\,\frac{e^{-ikR}}{R}\,f_{\mathbf{k}}(\Omega)\left(ik - \frac{1}{R}\right)\frac{e^{ikR}}{R} - \text{c.c.} = \left(ik - \frac{1}{R}\right)\int d\Omega\,|f_{\mathbf{k}}(\Omega)|^2 - \text{c.c.} = 2ik\sigma_{\mathbf{k}} \,.$$

Plane–spherical-wave and spherical–plane-wave (ps) interference terms: We have

$$\text{ps} = R^2 \int d\Omega\Bigg[e^{-iR\mathbf{k}\cdot\hat{\mathbf{r}}}\,f_{\mathbf{k}}(\Omega)\left(ik - \frac{1}{R}\right)\frac{e^{ikR}}{R} - \underbrace{e^{iR\mathbf{k}\cdot\hat{\mathbf{r}}}\,f_{\mathbf{k}}^*(\Omega)\left(-ik - \frac{1}{R}\right)\frac{e^{-ikR}}{R}}_{\text{c.c.}}$$

$$+\, f_{\mathbf{k}}^*(\Omega)\,\frac{e^{-ikR}}{R}ik\cos\theta\,e^{iR\mathbf{k}\cdot\hat{\mathbf{r}}} - \underbrace{f_{\mathbf{k}}(\Omega)\,\frac{e^{ikR}}{R}(-ik\cos\theta)\,e^{-iR\mathbf{k}\cdot\hat{\mathbf{r}}}}_{\text{c.c.}}\Bigg]$$

$$= \int d\Omega\Bigg[ikR\left(1 + \cos\theta - \frac{1}{ikR}\right)f_{\mathbf{k}}(\Omega)e^{-iR\mathbf{k}\cdot\hat{\mathbf{r}}+ikR} + ikR\left(1 + \cos\theta + \frac{1}{ikR}\right)f_{\mathbf{k}}^*(\Omega)\,e^{iR\mathbf{k}\cdot\hat{\mathbf{r}}-ikR}\Bigg] \,,$$

where the last line is obtained by combining terms proportional to $f_{\mathbf{k}}(\Omega)$ and to $f_{\mathbf{k}}^*(\Omega)$. In the limit $R \longrightarrow \infty$ the terms $1/(ikR)$ in the parentheses can be dropped, and so we are left with

$$\text{ps} = \int d\Omega\Bigg[ikR\,(1 + \cos\theta)\,f_{\mathbf{k}}(\Omega)e^{-iR\mathbf{k}\cdot\hat{\mathbf{r}}+ikR} - \text{c.c.}\Bigg]$$

$$= \Bigg[ikR\,e^{ikR}\int_0^{2\pi} d\phi \int_{-1}^{1} d(\cos\theta)\,(1 + \cos\theta)\,f_{\mathbf{k}}(\theta, \phi)\,e^{-ikR\cos\theta}\Bigg] - \Big[\cdots\Big]^* \,.$$

We now examine the solid-angle integration, and set $x = \cos\theta$ and

$$g(x, \phi) = (1 + x)f_{\mathbf{k}}(\cos^{-1} x, \phi) \,.$$

Integrating by parts with respect to x, we find

$$\int_{-1}^{1} dx\, g(x, \phi)\, e^{-ikRx} = -\frac{e^{-ikRx}}{ikR}\, g(x, \phi)\Bigg|_{-1}^{1} + \frac{1}{ikR}\int_{-1}^{1} dx\, g'(x, \phi)\, e^{-ikRx} \,,$$

where $g'(x, \phi)$ denotes the derivative with respect to x. In the limit $R \longrightarrow \infty$, only the first term survives:[3]

$$\int_{-1}^{1} dx\, g(x, \phi)\, e^{-ikRx} = \frac{i}{kR}\Big[g(1, \phi)\, e^{-ikR} - \underbrace{g(-1, \phi)}_{=0}\, e^{ikR}\Big] = \frac{2i}{kR}\, f_{\mathbf{k}}(0, \phi)\, e^{-ikR} \,.$$

---

[3] To see why this is so, just keep integrating by parts:

$$\int_{-1}^{1} dx\, g(x, \phi)\, e^{-ikRx} = -\frac{e^{-ikRx}}{ikR}\, g(x, \phi)\Bigg|_{-1}^{1} - \frac{e^{-ikRx}}{(ikR)^2}\, g'(x, \phi)\Bigg|_{-1}^{1} + \frac{1}{(ikR)^2}\int_{-1}^{1} dx\, g''(x, \phi)\, e^{-ikRx} \,,$$

to obtain an expansion in powers of $1/(ikR)$. It is now obvious that in the limit $R \longrightarrow \infty$ only the first term is left, the other terms having been suppressed by powers of ikR.

Since $\theta = 0$ implies that $\hat{\mathbf{r}}$ is in the forward direction (along z), we have $\phi = 0$, which leads to $f_{\mathbf{k}}(0, \phi) \longrightarrow f_{\mathbf{k}}(0)$. Collecting results, we finally arrive at

$$ps = \underbrace{\left[ikR\, e^{ikR} \int_0^{2\pi} d\phi\, \frac{2i}{kR} f_{\mathbf{k}}(0) e^{-ikR} \right]}_{-2 f_{\mathbf{k}}(0) \int_0^{2\pi} d\phi} - \left[\cdots \right]^* = -4\pi \left[f_{\mathbf{k}}(0) - f_{\mathbf{k}}^*(0) \right] .$$

We see, then, that the conservation of probability leads to the following relation:

$$0 = \underbrace{0}_{pp} + \underbrace{2ik\,\sigma_{\mathbf{k}}}_{ss} - \underbrace{8\pi i\, \text{Im}[f_{\mathbf{k}}(0)]}_{ps} \implies \sigma_{\mathbf{k}} = \frac{4\pi}{k} \text{Im}[f_{\mathbf{k}}(0)] ,$$

and the scattering amplitude must be complex, otherwise the total cross section would vanish. Also note that the crucial interference between plane waves and spherical waves occurs only in the forward direction (the terms ps).

Problem 7 Integral Equation for Bound States

Consider a particle under the influence of a potential $V(\mathbf{r})$ that vanishes faster than $1/r$ as $r \longrightarrow \infty$ and admits a bound state of energy $E_b < 0$. Knowing that

$$G_b(\mathbf{r}, \mathbf{r}') = \frac{1}{4\pi} \frac{e^{-\kappa_b |\mathbf{r} - \mathbf{r}'|}}{|\mathbf{r} - \mathbf{r}'|} ,$$

where we have defined $\kappa_b = \sqrt{2\mu |E_b|/\hbar^2}$, satisfies the inhomogeneous partial differential equation

$$\left(\nabla^2 - \kappa_b^2 \right) G_b(\mathbf{r}, \mathbf{r}') = -\delta(\mathbf{r} - \mathbf{r}') ,$$

where the Laplacian acts on the unprimed variables, obtain an integral equation for the eigenfunction $\psi_b(\mathbf{r})$. Show that $\psi_b(\mathbf{r})$ satisfies the expected asymptotic behavior for a bound state.

Solution

The Schrödinger equation reads

$$\left[-\frac{\hbar^2}{2\mu} \nabla^2 + V(\mathbf{r}) \right] \psi_b(\mathbf{r}) = -|E_b| \psi_b(\mathbf{r}) \longrightarrow \left(\nabla^2 - \kappa_b^2 \right) \psi_b(\mathbf{r}) = v(\mathbf{r})\, \psi_b(\mathbf{r}) ,$$

which we can convert into a homogenous integral equation for $\psi_b(\mathbf{r})$:

$$\psi_b(\mathbf{r}) = -\int d\mathbf{r}'\, G_b(\mathbf{r}, \mathbf{r}')\, v(\mathbf{r}')\, \psi_b(\mathbf{r}') .$$

Indeed, we have

$$\left(\nabla^2 - \kappa_b^2 \right) \psi_b(\mathbf{r}) = -\int d\mathbf{r}'\, \left(\nabla^2 - \kappa_b^2 \right) G_b(\mathbf{r}, \mathbf{r}')\, v(\mathbf{r}')\, \psi_b(\mathbf{r}')$$

$$= \int d\mathbf{r}'\, \delta(\mathbf{r} - \mathbf{r}')\, v(\mathbf{r}')\, \psi_b(\mathbf{r}') = v(\mathbf{r})\, \psi_b(\mathbf{r}) .$$

Note that, in the case of scattering, the continuum eigenfunctions $\psi_{\mathbf{k}}(\mathbf{r})$ satisfy an inhomogeneous partial differential equation, see Eq. (15.19).

Because of the presence of the "short-range" potential $v(\mathbf{r}')$, which is negligible for $r' \gtrsim R$, the $\mathbf{r}'$ integration is in practice confined to a finite sphere of radius R. In asymptotic region $r \gg R$, we can therefore expand $|\mathbf{r} - \mathbf{r}'|$ as $|\mathbf{r} - \mathbf{r}'| = r - \hat{\mathbf{r}} \cdot \mathbf{r}'$ (see Section 15.2) and write for the Green's function

$$G_b(\mathbf{r}, \mathbf{r}') \approx \frac{1}{4\pi} \frac{e^{-\kappa_b(r - \hat{\mathbf{r}} \cdot \mathbf{r}')}}{r} \, .$$

Substituting this result into the equation for $\psi_b(\mathbf{r})$ leads to the asymptotic behavior

$$\psi_b(\mathbf{r}) = \frac{e^{-\kappa_b r}}{r} \underbrace{\left[-\frac{1}{4\pi} \int d\mathbf{r}' \, e^{\kappa_b \hat{\mathbf{r}} \cdot \mathbf{r}'} \, v(\mathbf{r}') \, \psi_b(\mathbf{r}') \right]}_{g_b(\Omega)} ,$$

where the function $g_b(\Omega)$ depends only on the direction $\hat{\mathbf{r}}$. Of course, this asymptotic behavior matches that obtained from the Schrödinger equation: in the limit $r \longrightarrow \infty$ and dropping a term proportional to $1/r^3$, we have

$$\underbrace{\left(\frac{1}{r} \frac{\partial^2}{\partial r^2} r - \frac{\mathbf{L}^2}{2\hbar^2 \mu r^2} - \kappa_b^2 \right)}_{\nabla^2} \underbrace{\frac{u_b(\mathbf{r})}{r}}_{\psi_b(\mathbf{r})} = v(\mathbf{r}) \underbrace{\frac{u_b(\mathbf{r})}{r}}_{\psi_b(\mathbf{r})} \longrightarrow \frac{\partial^2 u_b(\mathbf{r})}{\partial r^2} - \kappa_b^2 \, u_b(\mathbf{r}) = 0 \, ;$$

the solution is $\psi_b(\mathbf{r}) = g_b(\Omega) \, e^{-\kappa_b r}/r$.

Problem 8   Derivation of the Scattering and Bound-State Green's Functions

Obtain the scattering and bound-state Green's functions by solving the inhomogeneous partial differential equations

$$\left(\nabla^2 + k^2 \right) G_k(\mathbf{r}, \mathbf{r}') = -\delta(\mathbf{r} - \mathbf{r}') , \qquad \left(\nabla^2 - \kappa_b^2 \right) G_b(\mathbf{r}, \mathbf{r}') = -\delta(\mathbf{r} - \mathbf{r}') ,$$

where the Laplacians act on the unprimed variables.

Hint: By translational invariance the Green's functions can only be functions of $\mathbf{x} = \mathbf{r} - \mathbf{r}'$. Introduce the Fourier transforms of $G_k(\mathbf{x})$ and $\delta(\mathbf{x})$ and solve the equation in wave-number space. Fourier-transform back to configuration space by first shifting the singularity at k away from the real axis by the replacement $k \longrightarrow k + i\eta$ and then taking the limit $\eta \longrightarrow 0$.

Solution

Because of translational invariance, $G_k(\mathbf{r}, \mathbf{r}')$ depends only on the difference $\mathbf{r} - \mathbf{r}'$ rather than on $\mathbf{r}$ and $\mathbf{r}'$ separately. This can also be seen directly from the defining relation by introducing the variables $\mathbf{x} = \mathbf{r} - \mathbf{r}'$, so that $\nabla_x^2 = \nabla^2$ by the chain rule and

$$\left(\nabla_x^2 + k^2 \right) \underbrace{G_k(\mathbf{x} + \mathbf{r}', \mathbf{r}')}_{\text{independent of } \mathbf{r}'} = -\delta(\mathbf{x}) \implies \left(\nabla_x^2 + k^2 \right) G_k(\mathbf{x}) = -\delta(\mathbf{x}) ,$$

since the relation must hold for any $\mathbf{r}'$. We introduce the Fourier transforms

$$G_k(\mathbf{x}) = \int \frac{d\mathbf{q}}{(2\pi)^3} \, e^{i\mathbf{q} \cdot \mathbf{x}} \, \widetilde{G}_k(\mathbf{q}) , \qquad \delta(\mathbf{x}) = \int \frac{d\mathbf{q}}{(2\pi)^3} \, e^{i\mathbf{q} \cdot \mathbf{x}} ,$$

and insert them into the equation for $G_k(\mathbf{x})$ to obtain

$$\int \frac{d\mathbf{q}}{(2\pi)^3} \, e^{i\mathbf{q} \cdot \mathbf{x}} \left(-q^2 + k^2 \right) \widetilde{G}_k(\mathbf{q}) = -\int \frac{d\mathbf{q}}{(2\pi)^3} \, e^{i\mathbf{q} \cdot \mathbf{x}} \implies \widetilde{G}_k(\mathbf{q}) = \frac{1}{q^2 - k^2} ;$$

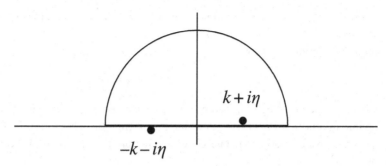

Fig. 15.1 Contour integration.

$\widetilde{G}_k(\mathbf{q})$ is singular at $q=k$, and, as such, its Fourier transform is not well defined. To remove the singularity, we shift the pole off the real axis by setting $k \longrightarrow k + i\eta$, so that

$$G_k(x) = \int \frac{d\mathbf{q}}{(2\pi)^3} e^{i\mathbf{q}\cdot\mathbf{x}} \frac{1}{q^2 - (k + i\eta)^2} ,$$

and take the limit $\eta \longrightarrow 0$ at the end.[4] We have

$$G_k(x) = \frac{1}{8\pi^3} \int_0^\infty dq\, q^2 \frac{1}{q^2 - (k + i\eta)^2} \int_{-1}^1 d(\cos\theta)\, e^{iqx\cos\theta} \int_0^{2\pi} d\phi$$

$$= \frac{1}{4\pi^2} \int_0^\infty dq\, q^2 \frac{1}{q^2 - (k + i\eta)^2} \frac{e^{iqx} - e^{-iqx}}{iqx}$$

$$= -\frac{i}{4\pi^2 x} \left[\int_0^\infty dq\, q \frac{e^{iqx}}{q^2 - (k + i\eta)^2} - \underbrace{\int_0^\infty dq\, q \frac{e^{-iqx}}{q^2 - (k + i\eta)^2}}_{\text{change variable: } q \longrightarrow -q} \right]$$

$$= -\frac{i}{4\pi^2 x} \left[\int_0^\infty dq\, q \frac{e^{iqx}}{q^2 - (k + i\eta)^2} + \underbrace{\int_{-\infty}^0 dq\, q \frac{e^{iqx}}{q^2 - (k + i\eta)^2}}_{\text{combine with first integral}} \right]$$

$$= -\frac{i}{4\pi^2 x} \int_{-\infty}^\infty dq\, q \frac{e^{iqx}}{q^2 - (k + i\eta)^2} ,$$

and the initial three-dimensional integral is now reduced to a one-dimensional integral over the whole real axis. Standard contour integration can be employed, see Fig. 15.1. The integrand has simple poles at $q = \pm(k + i\eta)$. Since $x = |\mathbf{x}| > 0$, we must close the contour in the upper half plane, and therefore we pick up the residue at the pole $q = k + i\eta$, that is

$$G_k(x) = -\frac{i}{4\pi^2 x} 2\pi i \underbrace{\left[q \frac{e^{iqx}}{q + (k + i\eta)} \right]_{q=k+i\eta}}_{\text{residue}} = \frac{1}{2\pi x}(k + i\eta) \frac{e^{i(k+i\eta)x}}{2(k + i\eta)} \longrightarrow \frac{e^{ikx}}{4\pi x} ,$$

the desired result.

---

[4] In this limit $(k + i\eta)^2 = k^2 - \eta^2 + 2ik\eta \longrightarrow k^2 - i\eta$ and the Fourier transform can be written as

$$G_k(x) = \int \frac{d\mathbf{q}}{(2\pi)^3} e^{i\mathbf{q}\cdot\mathbf{x}} \frac{1}{q^2 - k^2 - i\eta} .$$

Before moving on, we should note that we could have chosen to shift the pole off the real axis by setting $k \longrightarrow k - i\eta$ (rather than $k + i\eta$). Proceeding along the same lines as above, we would have found the Green's function

$$G_k^{(-)}(x) = \frac{e^{-ikx}}{4\pi x} \, ,$$

which has an incoming rather than an outgoing spherical wave. Thus we are forced to select the first option, in order for Eq. (15.15) to match the asymptotic behavior of Eq. (15.17).

In the case of the bound-state Green's function, the inversion of the Fourier transform poses no problem,

$$G_b(x) = \int \frac{d\mathbf{q}}{(2\pi)^3} \, e^{i\mathbf{q}\cdot\mathbf{x}} \, \frac{1}{q^2 + \kappa_b^2} \, ,$$

since the integrand has no singularity on the real axis. Following the manipulations above, we write $G_b(x)$ as

$$G_b(x) = -\frac{i}{4\pi^2 x} \int_{-\infty}^{\infty} dq \, q \, \frac{e^{iqx}}{q^2 + \kappa_b^2} \, .$$

By integrating over the upper half plane, we pick up the residue at the simple pole $q = i\kappa_b$, and hence

$$G_b(x) = \frac{e^{-\kappa_b x}}{4\pi x} \, .$$

We could also have obtained this result directly from $G_k(x)$ by the replacement $k \longrightarrow i\kappa_b$.

Problem 9 Perturbative Expansion of the Scattering Wave Function and Scattering Amplitude

Obtain a perturbative expansion for the scattering wave function $\psi_\mathbf{k}(\mathbf{r})$ and scattering amplitude $f_\mathbf{k}(\Omega)$, starting from the integral equation satisfied by $\psi_\mathbf{k}(\mathbf{r})$. Devise a diagrammatic representation of the terms in the expansion for $f_\mathbf{k}(\Omega)$.

Solution

The integral equation

$$\psi_\mathbf{k}(\mathbf{r}) = e^{i\mathbf{k}\cdot\mathbf{r}} - \int d\mathbf{r}' \, G_k(\mathbf{r}, \mathbf{r}') \, v(\mathbf{r}') \, \psi_\mathbf{k}(\mathbf{r}') \, , \qquad G_k(\mathbf{r}, \mathbf{r}') = \frac{e^{ik|\mathbf{r}-\mathbf{r}'|}}{4\pi|\mathbf{r} - \mathbf{r}'|} \, ,$$

can be solved by iteration to obtain a perturbative expansion for $\psi_\mathbf{k}(\mathbf{r})$,

$$\psi_\mathbf{k}(\mathbf{r}) = \psi_\mathbf{k}^{(0)}(\mathbf{r}) + \psi_\mathbf{k}^{(1)}(\mathbf{r}) + \psi_\mathbf{k}^{(2)}(\mathbf{r}) + \cdots \, ,$$

where

$$\psi_\mathbf{k}^{(0)}(\mathbf{r}) = e^{i\mathbf{k}\cdot\mathbf{r}} \, ,$$

$$\psi_\mathbf{k}^{(1)}(\mathbf{r}) = -\int d\mathbf{r}_1 \, G_k(\mathbf{r}, \mathbf{r}_1) \, v(\mathbf{r}_1) \, e^{i\mathbf{k}\cdot\mathbf{r}_1} \, ,$$

$$\psi_\mathbf{k}^{(2)}(\mathbf{r}) = \int d\mathbf{r}_1 \, G_k(\mathbf{r}, \mathbf{r}_1) \, v(\mathbf{r}_1) \int d\mathbf{r}_2 \, G_k(\mathbf{r}_1, \mathbf{r}_2) \, v(\mathbf{r}_2) \, e^{i\mathbf{k}\cdot\mathbf{r}_2}$$

$$= \int d\mathbf{r}_1 \, d\mathbf{r}_2 \, G_k(\mathbf{r}, \mathbf{r}_1) \, v(\mathbf{r}_1) \, G_k(\mathbf{r}_1, \mathbf{r}_2) \, v(\mathbf{r}_2) \, e^{i\mathbf{k}\cdot\mathbf{r}_2} \, ,$$

and so on. The general term of order n has the form

$$\psi_{\mathbf{k}}^{(n)}(\mathbf{r}) = (-1)^n \int d\mathbf{r}_1 \cdots d\mathbf{r}_n \underbrace{G_k(\mathbf{r}, \mathbf{r}_1)\, v(\mathbf{r}_1) \cdots G_k(\mathbf{r}_{n-1}, \mathbf{r}_n)\, v(\mathbf{r}_n)}_{\prod_i^n G_k(\mathbf{r}_{i-1}, \mathbf{r}_i)\, v(\mathbf{r}_i) \text{ with } \mathbf{r}_0 \equiv \mathbf{r}}\, e^{i\mathbf{k}\cdot\mathbf{r}_n} \, ,$$

and involves the nth power of the potential. If $v(\mathbf{r})$ is sufficiently weak, an accurate expression for $\psi_{\mathbf{k}}(\mathbf{r})$ can be obtained by only keeping the first few lowest-order corrections. The expansion above suggests that the total wave at $\mathbf{r}$ results from the superposition of the incident plane wave and an infinite number of waves coming from secondary sources and induced by the potential (this interpretation recalls Huygens' principle in wave optics). For example, the second-order term represents the wave at $\mathbf{r}$ induced by a secondary wave generated at $\mathbf{r}_1$ by $v(\mathbf{r}_1)$, which in turn was induced by a secondary wave generated at $\mathbf{r}_2$ by $v(\mathbf{r}_2)$.

Inserting the above expansion into the scattering amplitude

$$f_{\mathbf{k}}(\Omega) = -\frac{1}{4\pi} \int d\mathbf{r}'\, e^{-ik\hat{\mathbf{r}}\cdot\mathbf{r}'}\, v(\mathbf{r}')\, \psi_{\mathbf{k}}(\mathbf{r}')$$

yields a corresponding expansion for $f_{\mathbf{k}}(\Omega)$,

$$f_{\mathbf{k}}(\Omega) = f_{\mathbf{k}}^{(1)}(\Omega) + f_{\mathbf{k}}^{(2)}(\Omega) + f_{\mathbf{k}}^{(3)}(\Omega) + \cdots \, ,$$

where the general term has the form

$$f_{\mathbf{k}}^{(n)}(\Omega) = -\frac{1}{4\pi} \int d\mathbf{r}'\, e^{-i\mathbf{k}'\cdot\mathbf{r}'}\, v(\mathbf{r}')\, \psi_{\mathbf{k}}^{(n-1)}(\mathbf{r}') \qquad \text{with } \mathbf{k}' = k\hat{\mathbf{r}}.$$

A term of given order n in the expansion of $f_{\mathbf{k}}(\Omega)$ can be represented diagrammatically in the following way: (i) draw n vertices labeled $1, \ldots, n$ and connect them by $n-1$ line segments (say, $\mathbf{r}_1$ to $\mathbf{r}_2$, $\mathbf{r}_2$ to $\mathbf{r}_3$, $\ldots$, $\mathbf{r}_{n-1}$ to $\mathbf{r}_n$), each segment representing the Green's function (or propagator) $G_k(\mathbf{r}_i, \mathbf{r}_j)$; (ii) at each vertex i include the potential $v(\mathbf{r}_i)$; (iii) the external line segment coming into $\mathbf{r}_n$ represents the incoming plane wave $e^{i\mathbf{k}\cdot\mathbf{r}_n}$, while the external line segment going out of $\mathbf{r}_1$ represents the outgoing plane wave $e^{-i\mathbf{k}'\cdot\mathbf{r}_1}$; (iv) integrate over all vertices $\int d\mathbf{r}_1 \cdots \int d\mathbf{r}_n$ and multiply the resulting expression by the factor $(-1)^n/(4\pi)$.

Problem 10 Relating the Partial Wave Amplitude f_{kl} to the Phase Shift δ_{kl}

Consider the expansions

$$\psi_{k\hat{z}}(r, \theta) = \sum_{l=0}^{\infty} \frac{u_{kl}(r)}{r}\, P_l(\cos\theta)\, , \qquad \lim_{r\to\infty} \frac{u_{kl}(r)}{r} = a_{kl}\, \frac{\sin(kr - l\pi/2 + \delta_{kl})}{r}\, ,$$

and

$$\lim_{r\to\infty} \psi_{k\hat{z}}(r, \theta) = e^{ikr\cos\theta} + f_{k\hat{z}}(\theta)\, \frac{e^{ikr}}{r} = \sum_{l=0}^{\infty} \left[i^l (2l+1) j_l(kr) + f_{kl}\frac{e^{ikr}}{r} \right] P_l(\cos\theta)\, .$$

Match these two expressions, thus determining f_{kl} and a_{kl} in terms of the phase shift δ_{kl}.

Solution

The spherical Bessel functions have the asymptotic behavior

$$\lim_{r\to\infty} j_l(kr) = \frac{\sin(kr - l\pi/2)}{kr}\, ,$$

and so the matching requires

$$\underbrace{a_{kl} \frac{\sin(kr - l\pi/2 + \delta_{kl})}{r}}_{\text{l.h.s.}} = \underbrace{i^l (2l+1) \frac{\sin(kr - l\pi/2)}{kr} + f_{kl} \frac{e^{ikr}}{r}}_{\text{r.h.s.}} \, .$$

Using

$$\sin(kr - l\pi/2) = \frac{e^{i(kr - l\pi/2)}}{2i} - \frac{e^{-i(kr - l\pi/2)}}{2i} = (-i)^l \frac{e^{ikr}}{2i} - i^l \frac{e^{-ikr}}{2i} \, ,$$

we can write the above right-hand side as

$$\text{r.h.s.} = \frac{2l+1}{2ik} \frac{e^{ikr}}{r} - i^{2l} \frac{2l+1}{2ik} \frac{e^{-ikr}}{r} + f_{kl} \frac{e^{ikr}}{r} = \left(f_{kl} + \frac{2l+1}{2ik} \right) \frac{e^{ikr}}{r} - (-1)^l \frac{2l+1}{2ik} \frac{e^{-ikr}}{r} \, ,$$

and the left-hand side as

$$\text{l.h.s.} = a_{kl} (-i)^l \frac{e^{i\delta_{kl}}}{2i} \frac{e^{ikr}}{r} - a_{kl} i^l \frac{e^{-i\delta_{kl}}}{2i} \frac{e^{-ikr}}{r} \, .$$

We then obtain

$$a_{kl} (-i)^l \frac{e^{i\delta_{kl}}}{2i} = f_{kl} + \frac{2l+1}{2ik} \, , \qquad a_{kl} i^l \frac{e^{-i\delta_{kl}}}{2i} = (-1)^l \frac{2l+1}{2ik} \, ,$$

from which Eq. (15.55) follows.

Problem 11 Low-Energy Behavior of the Phase Shifts

Starting from Eq. (15.68), establish the general behavior of the phase shifts δ_{kl} in the limit of low energies. Note that this energy dependence enters via the spherical Bessel functions as well as the logarithmic derivative of the wave function. The small $-x$ behaviors of $j_l(x)$ and $n_l(x)$ are given by, respectively,

$$j_l(x) = \frac{x^l}{(2l+1)!!} \left[1 - \frac{x^2}{2(2l+3)} + \cdots \right] \, , \qquad n_l(x) = -\frac{(2l-1)!!}{x^{l+1}} \left[1 - \frac{x^2}{2(1-2l)} + \cdots \right] \, .$$

Solution

The energy dependence of the radial wave function, and hence of its logarithmic derivative, comes in via the term $k^2 u_{kl}(r)$ in the Schrödinger equation. However, as $k^2 \longrightarrow 0$, this term is suppressed, so that the logarithmic derivative becomes in practice independent of energy, $\Delta_{kl} \longrightarrow \Delta_l$. Thus we find to leading order

$$\begin{aligned}
\tan \delta_{kl} &= \frac{klx_0^{l-1}/(2l+1)!! - \Delta_l x_0^l/(2l+1)!!}{-k(-l-1)(2l-1)!! \, x_0^{-l-2} + \Delta_l (2l-1)!! \, x_0^{-l-1}} \\[2mm]
&= \frac{1}{(2l+1)!! \, (2l-1)!!} \frac{lx_0^l/R_0 - \Delta_l x_0^l}{(l+1)x_0^{-l-1}/R_0 + \Delta_l x_0^{-l-1}} \\[2mm]
&= \underbrace{\frac{l - R_0 \Delta_l}{l + 1 + R_0 \Delta_l} \frac{R_0^{2l+1}}{(2l+1)!! \, (2l-1)!!}}_{\gamma_0} k^{2l+1} \, ,
\end{aligned}$$

which shows that the leading-order term in the low-energy expansion of $\tan \delta_{kl} \approx \delta_{kl}$ is $\gamma_0 k^{2l+1}$. As a matter of fact, since Δ_{kl} depends on k^2 through the Schrödinger equation, it can be thought of as a function of k^2 and, as such, can be expanded in powers of k^2. Then, given that $x^{-l} j_l(x)$, $x^{l+1} n_l(x)$, $x^{-l+1} j_l'(x)$, and $x^{l+2} n_l'(x)$ have expansions in terms of x^2, as is evident from expressions given in the text of the problem, we have from Eq. (15.68) that

$$\frac{\tan \delta_{kl}}{k^{2l+1}} = \gamma_0 + \gamma_2 k^2 + \gamma_4 k^4 + \cdots .$$

Problem 12 Hard-Sphere Potential: Phase Shifts, Low- and High-Energy Limits of Total Cross Section

Consider the hard-sphere potential $V(r) = \infty$ for $r \leq a$ and $V(r) = 0$ for $r > a$. Determine the phase shift for a generic partial wave l. For the s-wave case, sketch the wave function obtained and compare it with the free-particle wave function. Calculate the scattering amplitude at low energies, and compare the corresponding total cross section with the classical result. Show that, in the high-energy limit $ka \longrightarrow \infty$, the total cross section is given by $2\pi a^2$.

Hints: To evaluate the high-energy limit, express the total cross section as follows,

$$\sigma_k = 4\pi a^2 \lim_{ka \to \infty} \frac{1}{(ka)^2} \sum_{l=0}^{ka} (2l + 1) \sin^2 \delta_l(k) ,$$

and recall that the regular and irregular spherical Bessel functions satisfy the recurrence relation

$$f_l(x) = \frac{2l - 1}{x} f_{l-1}(x) - f_{l-2}(x) \implies f_l(x) = -f_{l-2}(x) \qquad \text{as} \qquad x \longrightarrow \infty ,$$

where $f_l(x) = j_l(x)$ or $n_l(x)$.

Solution

The particle cannot penetrate the region $r \leq a$, and so the radial equation reads

$$u_{kl}''(r) + \left[k^2 - \frac{l(l+1)}{r^2} \right] u_{kl}(r) = 0 \qquad \text{with } r > a , \qquad u_{kl}(a) = 0 ,$$

corresponding to a free-particle equation having the general solution

$$\psi_{kl}(r) = \frac{u_{kl}(r)}{r} = A_{kl} j_l(kr) - B_{kl} n_l(kr) .$$

The boundary condition at $r = a$ requires

$$\psi_{kl}(a) = 0 = A_{kl} j_l(ka) - B_{kl} n_l(ka) \implies B_{kl}/A_{kl} = j_l(ka)/n_l(ka) ,$$

so that the solution for $r > a$ is simply given by

$$\psi_{kl}(r) = A_{kl} \left[j_l(kr) - \underbrace{\frac{j_l(ka)}{n_l(ka)}}_{\tan \delta_{kl}} n_l(kr) \right] ,$$

from which we can immediately read off the phase shift,

$$\delta_{kl} = \tan^{-1} \left[\frac{j_l(ka)}{n_l(ka)} \right] .$$

Using the small-x expansions for $j_l(x)$ and $n_l(x)$, we find that δ_{kl} in the limit of low energy ($k \longrightarrow 0$) is proportional to k^{2l+1}, as expected.

For the s-wave case, the phase shift is obtained as $\delta_{k0} = -ka$, since $j_0(ka)/n_0(ka) = -\tan(ka)$, and the wave function reads

$$u_{k0}(r) = A_{k0} \left[\frac{\sin(kr)}{k} - \tan(ka) \frac{\cos(kr)}{k} \right] = \frac{A_{k0}}{k} \cos(ka) \sin[k(r-a)].$$

The last expression follows directly from the Schrödinger equation above for $l=0$ and the boundary condition at $r=a$. When compared with the free solution $\sin(kr)$, we see that $u_{k0}(r)$ is "pushed out" (since the potential is repulsive); indeed, while the nodes of the free-particle solution occur at $n\pi/k$ with $n = 1, 2, \ldots$, those of the exact solution are shifted by an amount a, that is, they occur at $n\pi/k + a$.

The s-wave contribution to the total cross section can be read off from Eq. (15.60): $\sigma_{k0} = (4\pi/k^2) \sin^2(ka)$. Because of the behavior of the higher l-phase shifts when $k \longrightarrow 0$, this term dominates in the total cross section at low energy, $\sigma_k \approx 4\pi a^2$. In this limit, the differential cross section is isotropic, namely, independent of the scattering angle, since

$$f_k(\theta) = \frac{1}{k} \sum_{l=0}^{\infty} (2l+1) \, e^{i\delta_{kl}} \sin \delta_{kl} \, P_l(\cos \theta) \approx \frac{e^{i\delta_{k0}}}{k} \underbrace{\sin \delta_{k0}}_{\sin(-ka)} \underbrace{P_0(\cos \theta)}_{=1} \approx -a$$

and therefore $\sigma_k(\theta) = |f_k(\theta)|^2 \approx a^2$, in agreement with the result found above for the total cross section in the same limit. It is interesting to note that the total cross section obtained classically is πa^2 – a factor 4 smaller than that obtained quantum mechanically; it corresponds to the area of the disk seen by the incoming beam of particles (classically, only the particles with impact parameter $b \leq a$ will be scattered by the hard-sphere potential).

From $\tan \delta_l(k) = j_l(ka)/n_l(ka)$ and using the identity $\sin^2 x = \tan^2 x/(1 + \tan^2 x)$, we have

$$\sin^2 \delta_l(k) = \frac{j_l^2(ka)}{j_l^2(ka) + n_l^2(ka)} \, .$$

We are interested in the limit $x = ka \longrightarrow \infty$ (with a fixed). In such a limit, the recursion relation satisfied by the spherical Bessel functions gives

$$f_l(x) = -f_{l-2}(x) \implies f_l^2(x) = f_{l-2}^2(x) = f_{l-4}^2(x) = \cdots = f_0^2(x) \text{ if } l \text{ even or } f_1^2(x) \text{ if } l \text{ odd}.$$

Recalling that

$$j_0(x) = \frac{\sin x}{x} \, , \qquad n_0(x) = -\frac{\cos x}{x} \, ,$$

and

$$j_1(x) = \frac{\sin x}{x^2} - \frac{\cos x}{x} \, , \qquad n_1(x) = -\frac{\cos x}{x^2} - \frac{\sin x}{x} \, ,$$

we see that in the limit $x \longrightarrow \infty$ we have

$$j_0^2(x) = \frac{\sin^2 x}{x^2} \, , \qquad n_0^2(x) = \frac{\cos^2 x}{x^2} \qquad j_1^2(x) = \frac{\cos^2 x}{x^2} \, , \qquad n_1(x) = \frac{\sin^2 x}{x^2} \, .$$

Hence, we find in this limit that

$$\sin^2 \delta_l(k) = \sin^2 x \text{ if } l \text{ even}, \qquad \sin^2 \delta_l(k) = \cos^2 x \text{ if } l \text{ odd},$$

so that

$$\sigma_k = 4\pi a^2 \lim_{x \to \infty} \frac{1}{x^2} \left[\sin^2 x \sum_{l \text{ even}=0}^{[x]} (2l + 1) + \cos^2 x \sum_{l \text{ odd}=1}^{[x]} (2l + 1) \right],$$

where $[x]$ denotes the integer part of the real number $x = ka$. Now, we observe that

$$\sum_{l \text{ even}=0}^{[x]} = \frac{[x]}{2} + 1 \quad \text{and} \quad \sum_{l \text{ even}=0}^{[x]} l = \frac{[x]}{2}\left(\frac{[x]}{2} + 1\right) \qquad \text{for } [x] \text{ even},$$

$$\sum_{l \text{ even}=0}^{[x]} = \frac{[x] + 1}{2} \quad \text{and} \quad \sum_{l \text{ even}=0}^{[x]} l = \frac{[x]^2 - 1}{4} \qquad [x] \text{ odd},$$

and

$$\sum_{l \text{ odd}=1}^{[x]} = \frac{[x]}{2} \quad \text{and} \quad \sum_{l \text{ odd}=1}^{[x]} l = \frac{[x]^2}{4} \qquad [x] \text{ even},$$

$$\sum_{l \text{ odd}=1}^{[x]} = \frac{[x] + 1}{2} \quad \text{and} \quad \sum_{l \text{ odd}=1}^{[x]} l = \frac{([x] + 1)^2}{4} \qquad [x] \text{ odd},$$

from which we conclude that for $[x] \gg 1$, and regardless of whether $[x]$ is even or odd, we have

$$\sum_{l \text{ even}=0}^{[x]} (2l + 1) = \sum_{l \text{ odd}=0}^{[x]} (2l + 1) \longrightarrow \frac{[x]^2}{2} \qquad \text{for } [x] \gg 1.$$

The limits of the sums above could have been more easily obtained by observing that

$$\sum_{l=0}^{[x]} l = \frac{[x]([x] + 1)}{2} \implies \sum_{l=0}^{[x]} (2l + 1) \longrightarrow [x]^2 \qquad \text{for } [x] \gg 1,$$

and that the sums over even and odd l are the same in this limit. Inserting these relations into the expression for σ_k we arrive at

$$\sigma_k = 2\pi a^2 \lim_{x \to \infty} \frac{[x]^2}{x^2} = 2\pi a^2.$$

We conclude that a hard-sphere potential will induce scattering even in the limit of infinite incident energy.

Problem 13 Study of the Spherical Potential Well in S-Waves; Resonances

Consider a particle in a spherical potential well,

$$V(r) = -V_0 \qquad r \leq a, \qquad V(r) = 0 \qquad r > a,$$

where $V_0 > 0$. For simplicity, consider s-waves only. Define $k^2 = 2\mu E/\hbar^2$ and $k_0^2 = 2\mu V_0/\hbar^2$.

1. Determine under what conditions there are at least n bound states.
2. Obtain the s-wave phase shift δ_{k0} and its low- and high-energy limits.
3. For energies $E \geq 0$, the radial wave function can be written as

$$u_{k0}(r) = A_k \sin(Kr) \qquad r \leq a, \qquad u_{k0}(r) = B_k \sin(kr + \delta_{k0}) \qquad r > a,$$

where $K = \sqrt{k^2 + k_0^2}$. Obtain the ratio $|A_k/B_k|^2$; what is, roughly speaking, its physical meaning? Show that the ratio exhibits resonances (maxima) at certain energies. Determine these energies and corresponding (s-wave) phase shifts and associated contributions to the total cross section.

4. Suppose that the depth of the potential is close to an odd multiple of $\pi/2$, that is, $ak_0 = ak_0^* + \eta$ where $ak_0^* = (2n - 1)\pi/2$ and $\eta \ll 1$. Consider the case $\eta > 0$, and determine approximately the energy of the least (nth) bound state. Then consider $\eta < 0$, and show that this bound state disappears, becoming a resonance. Determine its energy, and show that, for energies close to the resonant energy E_R, the phase shift as a function of energy is given by

$$\tan \delta_{E0} = -\frac{\Gamma/2}{E - E_R} \quad \text{with } \Gamma = 2 \frac{\hbar^2 k_R}{a\mu} \;.$$

Obtain the s-wave contribution to the total cross section at these energies.

5. Show that the asymptotic wave function in s-waves can be written as

$$\psi_{k0}(r) = \frac{u_{k0}(r)}{r} = \frac{e^{i(kr+2\delta_{k0})}}{2ikr} - \frac{e^{-ikr}}{2ikr} \;.$$

Consider the wave packet

$$\Psi_0(r, t) = \int_0^\infty dk\, g(k)\, \psi_{k0}(r)\, e^{-iEt/\hbar} \;, \qquad E = \frac{\hbar^2 k^2}{2\mu} \;,$$

where the profile function $g(k)$ is strongly peaked at k_R and vanishes outside a narrow range centered at k_R. Calculate the time delay τ in the scattered wave packet. Interpret your result for τ.

Solution

Part 1

The Schrödinger equation for $l = 0$ is given by

$$u_0''(r) + \left(k^2 + k_0^2\right) u_0(r) = 0 \quad r \le a \;, \qquad u_0''(r) + k^2 u_0(r) = 0 \quad r > a \;.$$

The solution satisfies the boundary condition $u_0(r) \propto r$ as $r \longrightarrow 0$. Bound states (if they exist) have energies in the range $-V_0 < E < 0$ and satisfy the additional boundary condition $u_0(r \longrightarrow \infty) = 0$. We set

$$k^2 \longrightarrow -\kappa^2 = -\frac{2\mu|E|}{\hbar^2} \;, \qquad \kappa^2 < k_0^2 \;.$$

Thus the solution reads

$$u_0(r) = A \sin\left(r \sqrt{k_0^2 - \kappa^2}\right) \quad r \le a \;, \qquad u_0(r) = B\, e^{-\kappa r} \quad r > a \;.$$

Imposing continuity on $u_0(r)$ and its first derivative at $r = a$ leads to the homogeneous linear system

$$A \sin\left(a \sqrt{k_0^2 - \kappa^2}\right) = B\, e^{-\kappa a} \;, \qquad A \sqrt{k_0^2 - \kappa^2} \cos\left(a \sqrt{k^2 - \kappa^2}\right) = -B\, \kappa\, e^{-\kappa a} \;,$$

which has a non-trivial solution if the associated determinant vanishes, that is, if

$$\kappa\, e^{-\kappa a} \sin\left(a \sqrt{k_0^2 - \kappa^2}\right) + \sqrt{k_0^2 - \kappa^2}\, e^{-\kappa a} \cos\left(a \sqrt{k_0^2 - \kappa^2}\right) = 0 \;.$$

For $\sqrt{k_0^2 - \kappa^2}\, a \neq (2n + 1)\pi/2$ with $n = 0, 1, 2 \ldots$, the condition above can be expressed as

$$\tan\left(a\sqrt{k_0^2 - \kappa^2}\right) = -\frac{\sqrt{k_0^2 - \kappa^2}}{\kappa} \qquad \text{or} \qquad \tan\left(ak_0\sqrt{1 - x^2}\right) = -\frac{\sqrt{1 - x^2}}{x} \, ,$$

where $x = \kappa/k_0$ with x in the range $0 < x < 1$. The equation can be solved graphically. First, it is clear that, since the right-hand side is always negative, for a solution to exist the tangent of the left-hand side must be negative, which requires

$$ak_0 > \pi/2 \implies V_0 > \frac{\pi^2}{8}\frac{\hbar^2}{\mu a^2} \, .$$

Thus, if the well is not deep enough, there will be no bound states. On the other hand, it is easy to see that if $ak_0 > (2n - 1)\pi/2$ there will be n (s-wave) bound states.

Part 2

For energies $E \geq 0$ (and hence $k^2 \geq 0$), the radial wave function $u_{k0}(r)$ is given by

$$u_{k0}(r) = A_k \sin(Kr) \qquad r \leq a \, , \qquad u_{k0}(r) = B_k \sin(kr + \delta_{k0}) \qquad r > a \, ,$$

where δ_{k0} is the phase shift and $K = \sqrt{k_0^2 + k^2}$. Matching at $r = a$ leads to

$$A_k \sin(Ka) = B_k \sin(ka + \delta_{k0}) \, , \qquad A_k K \cos(Ka) = B_k k \cos(ka + \delta_{k0}) \, ,$$

from which by taking the ratio of the first and the second relation above we obtain

$$\delta_{k0} = -ka + \tan^{-1}\left[\frac{k}{K}\tan(Ka)\right] \, .$$

In the low-energy limit ($k \longrightarrow 0$), we have $K \longrightarrow k_0$ and

$$\delta_{k0} \approx -ka + \tan^{-1}\left[\frac{k}{k_0}\tan(k_0 a)\right] \implies \delta_{k0} \approx ka\left[\frac{\tan(k_0 a)}{k_0 a} - 1\right] \, .$$

The phase shift, as expected, behaves linearly with k (for s-waves). It is interesting to note that as $k_0 a$ is close to $(2n - 1)\pi/2$, recalling that for n bound states to exist we must have $k_0 a > (2n - 1)\pi/2$, it follows that $|\tan(k_0 a)|$ and hence $|\delta_{k0}|$ become very large. By contrast, in the high-energy limit, we have $K \longrightarrow k$ and hence $\delta_{k0} \approx 0$, which simply says that, in this limit, the particle does not undergo any scattering.

Part 3

We find for the ratio A_k/B_k

$$\frac{A_k}{B_k} = \frac{\sin(ka + \delta_{k0})}{\sin(Ka)} = \frac{1}{\sin(Ka)}\frac{\tan(ka + \delta_{k0})}{\sqrt{1 + \tan^2(ka + \delta_{k0})}}$$

$$= \frac{1}{\sin(Ka)}\frac{(k/K)\tan(Ka)}{\sqrt{1 + (k/K)^2\tan^2(Ka)}} = \frac{k}{\sqrt{k_0^2\cos^2(Ka) + k^2}} \, ,$$

and $|A_k/B_k|^2$ provides a rough estimate of how likely the particle is to be found inside the well rather than outside:

$$\frac{|A_k|^2}{|B_k|^2} = \frac{k^2}{k_0^2 \cos^2(Ka) + k^2} \ .$$

When plotted as a function of k, this ratio exhibits resonances when $Ka = (2n-1)\pi/2$, which corresponds to values of k^2 given by

$$k^2 = \underbrace{\frac{(2n-1)^2\pi^2}{4a^2}}_{k_0^{*2}} - k_0^2 = k_0^{*2} - k_0^2 \ .$$

At these values of k^2, $|A_k/B_k|^2 = 1$ and $\tan(Ka) \longrightarrow \infty$. It follows from the expression for δ_{k0} obtained in part 2 that $\delta_{k0} + ka$ must be an odd multiple of $\pi/2$. In particular, suppose that the depth of the well is such that $k_0^2 \approx k_0^{*2}$, and that the energy k^2 is very small, with $ka \ll 1$. Then we find that δ_{k0}, up to the negligible contribution $-ka$, is an odd multiple of $\pi/2$, and hence the contribution of s-wave scattering to the total cross section is essentially maximal,

$$\sigma_{k0} \approx \frac{4\pi}{k^2} \sin^2 \delta_{k0} \approx \frac{4\pi}{k^2} \ .$$

Part 4

It is interesting to explore more systematically what happens when the depth of the well is very close to an odd multiple of $\pi/2$,

$$k_0 a = k_0^* a + \eta \qquad \text{with } \eta \ll 1 \ .$$

We now consider the two possibilities, in which η is either positive or negative. If η is positive, as we have established above bound-state energies result from solving graphically

$$\tan\left(ak_0\sqrt{1-x^2}\right) = -\frac{\sqrt{1-x^2}}{x} \qquad \text{with } 0 < x < 1 \ ,$$

where $x = \kappa/k_0$ with $\kappa^2 = 2\mu|E|/\hbar^2$ and $E < 0$. Since for $\eta > 0$ we have $k_0 > k_0^*$, there are n solutions x_k with $k = 1, \ldots, n$, corresponding to energies $E_k = -\hbar^2(x_k k_0)^2/(2\mu)$. The least-bound state has energy E_n close to zero (that is, x_n is close to zero). We can determine x_n, and hence the energy E_n, by solving the above eigenvalue equation in this limit ($x \longrightarrow 0$), in which case we have[5]

$$\tan(ak_0) \approx -\frac{1}{x_n} \implies x_n = -\cot(ak_0^* + \eta) \approx \eta \ ,$$

and $E_n \approx -\hbar^2(\eta k_0^*)^2/(2\mu)$ to leading order.

If η is negative, the potential can only support $n-1$ bound states. However, the nth bound state becomes a scattering resonance with an energy from $k^2 = k_0^{*2} - k_0^2$ given by

$$k_R^2 = k_0^{*2} - (k_0^* - |\eta|/a)^2 \approx 2k_0^*|\eta|/a \implies E_R = \frac{\hbar^2|\eta|k_0^*}{\mu a} \ .$$

---

[5] We expand $\cot x$ around x_0 as

$$\cot x = \cot x_0 + (x - x_0)\frac{d}{dx}\cot x\Big|_{x_0} + \cdots = \cot x_0 - (x - x_0)\frac{1}{\sin^2 x_0} + \cdots \ .$$

For $x_0 = ak_0^* = (2n-1)\pi/2$ we have $\cot(ak_0^*) = 0$ and $\sin^2(ak_0^*) = 1$.

Thus, we see that if the depth of the well is gradually reduced (keeping a fixed), the bound states that disappear when $k_0 a$ passes through an odd multiple of $\pi/2$ produce low-energy scattering resonances.

To determine how the phase shift behaves for energies close to the resonant energy, we write the expression for the phase shift derived in part 2 as

$$\tan \delta_{k0} \approx \frac{k}{K} \tan(Ka) \,,$$

where on the left-hand side we have neglected the term $ak \longrightarrow ak_R \ll 1$. We expand the right-hand side around k_R^2; first, note that

$$K = \sqrt{k^2 + k_0^2} \approx \sqrt{k_R^2 + k_0^2} + \frac{1}{2\sqrt{k_R^2 + k_0^2}}(k^2 - k_R^2) + \cdots \,.$$

Since

$$k_R^2 + k_0^2 = 2\frac{k_0^* |\eta|}{a} + \left(k_0^* - \frac{|\eta|}{a}\right)^2 \approx k_0^{*2} \,,$$

we can rewrite the expansion for K as

$$K \approx k_0^* + \frac{1}{2k_0^*}(k^2 - k_R^2) + \cdots \,,$$

from which we deduce that

$$\tan(Ka) = \frac{1}{\cot(Ka)} = \frac{1}{\cot[ak_0^* + (a/2k_0^*)(k^2 - k_R^2)]} \approx -\frac{2k_0^*}{a}\frac{1}{k^2 - k_R^2} \,,$$

where we have expanded the cotangent around $ak_0^* = (2n - 1)\pi/2$. Collecting results, we find

$$\tan \delta_{k0} \approx \frac{k_R}{k_0^*}\left(1 - \frac{k^2 - k_R^2}{2k_0^{*2}}\right)\left(-\frac{2k_0^*}{a}\right)\frac{1}{k^2 - k_R^2} \approx -\frac{2k_R/a}{k^2 - k_R^2} \,,$$

and the singularity is consistent with our previous result that at $k = k_R$ the phase shift goes through an odd multiple of $\pi/2$. Rewriting the expression above in terms of the energy yields the expression given in the text of the problem. The constant Γ has dimensions of energy. We can now obtain the cross section in the resonance region (recall that we are dealing with s-wave scattering):

$$\sigma_{k0} \approx \frac{4\pi}{k^2} \sin^2 \delta_{E0} \approx \frac{4\pi}{k^2}\frac{\Gamma^2/4}{(E - E_R)^2 + \Gamma^2/4} \,,$$

where we have used the trigonometric identity $\sin^2 x = \tan^2 x/(1 + \tan^2 x)$. The equation above is known as the Breit–Wigner formula and describes the characteristic "bell" shape of the cross section in the resonance region; Γ gives the full width of the cross section at half maximum.

Part 5

We now investigate the behavior of the wave function in the resonance region. From Eqs. (15.53), (15.54), and (15.56), considering only s-wave terms we see that the asymptotic wave function reads

$$\psi_{k0}(r) = \underbrace{j_0(kr)}_{\sin(kr)/kr} + e^{i\delta_{k0}} \sin \delta_{k0} \frac{e^{ikr}}{kr} = \frac{e^{ikr} - e^{-ikr}}{2ikr} + \frac{e^{2i\delta_{k0}} - 1}{2i}\frac{e^{ikr}}{kr} = \frac{e^{i(kr+2\delta_{k0})}}{2ikr} - \frac{e^{-ikr}}{2ikr} \,,$$

that is, it consists of an incident spherical wave e^{-ikr} and a scattered spherical wave e^{ikr}. The effect of the potential is to introduce a phase shift in the latter. We now form the wave packet

$$\Psi_0(r, t) = \int_0^\infty dk\, g(k)\, \psi_{k0}(r)\, e^{-iEt/\hbar} \qquad E = \frac{\hbar^2 k^2}{2\mu}\ ,$$

with $g(k)$ real and strongly peaked at k_R.

Using the stationary phase method, we find that the scattered wave packet for $t \gg 0$ is given by

$$\Psi_{s,0}(r, t) = \frac{e^{i(kr+2\delta_{k0}-Et/\hbar)}}{2ikr}\bigg|_{k_R} \int_0^\infty dk\, g(k)\, e^{i(k-k_R)[r-r(t)]}\ ,$$

where we have defined

$$r(t) = \frac{\hbar k_R}{\mu} t - 2 \frac{d}{dk}\delta_{k0}\bigg|_{k_R} = \frac{\hbar k_R}{\mu}(t - \tau)\ ,$$

with the time delay τ given by

$$\tau = \frac{2\mu}{\hbar k_R} \frac{d}{dk}\delta_{k0}\bigg|_{k_R} = \frac{2\mu}{\hbar k_R} \frac{dE}{dk} \frac{d}{dE}\delta_{E0}\bigg|_{E_R} = 2\hbar \frac{d}{dE}\delta_{E0}\bigg|_{E_R} = \frac{4\hbar}{\Gamma}\ .$$

This time delay is the time during which the particle "hangs around" the region where the potential is present. Note that it is inversely proportional to Γ, so the narrower the resonance width, the longer the time delay (as a matter of fact, in the limit of an infinitely narrow width the particle never leaves the region of the potential). It is as if the particle is in an unstable bound state or, rather, the incident and target particles form an unstable bound state, which has a lifetime of order $\hbar/\Gamma$. In order to clarify this interpretation further, consider again the scattered wave packet but now more conveniently written as a superposition in energy:

$$\Psi_{s,0}(r, t) = \int_0^\infty dE\, g(E)\, \frac{e^{i(kr+2\delta_{E0})}}{2ikr} e^{-iEt/\hbar}\ .$$

We are interested in studying its behavior for large times at a fixed r in a region outside the potential. To this end, we first note that

$$e^{2i\delta_{E0}} = \frac{e^{i\delta_{E0}}}{e^{-i\delta_{E0}}} = \frac{\cos\delta_{E0} + i\sin\delta_{E0}}{\cos\delta_{E0} - i\sin\delta_{E0}} = \frac{1 + i\tan\delta_{E0}}{1 - i\tan\delta_{E0}} \approx \frac{E - E_R - i\Gamma/2}{E - E_R + i\Gamma/2}\ ,$$

which can also be written as

$$e^{2i\delta_{E0}} = 1 - \frac{i\Gamma}{E - E_R + i\Gamma/2}\ .$$

Thus we see that the scattered wave packet near the resonance has the form

$$\Psi_{s,0}(r, t) \approx \int_{-\infty}^\infty dE\, g(E)\, \frac{e^{ikr}}{2ikr}\left(1 - \frac{i\Gamma}{E - E_R + i\Gamma/2}\right)e^{-iEt/\hbar} \approx -g(E)\frac{e^{ikr}}{2ikr}\bigg|_{E_R} \int_{-\infty}^\infty dE\, \frac{i\Gamma}{E - E_R + i\Gamma/2} e^{-iEt/\hbar}$$

where in the first step we have taken the lower integration limit to $-\infty$ by formally extending the definition $g(E)$, setting $g(E) = 0$ for $E < 0$, and in the second step (i) we have assumed that the energy-dependent term $g(E)\, e^{ikr}/(kr)$ varies smoothly near the resonance and (ii) we have dropped a term proportional to a δ-function in time, that is, $\delta(t)$ (recall that we are in the regime $t \gg 0$). The time integral is easily found by contour integration, by closing the contour in the lower half of the complex E-plane, since $e^{-iEt/\hbar} = e^{-iE_r t/\hbar} e^{E_i t/\hbar}$, where E_r and E_i are the real and imaginary parts of E, and hence $E_i < 0$ for convergence. We find

$$\int_{-\infty}^{\infty} dE \, \frac{1}{E - E_R + i\Gamma/2} \, e^{-iEt/\hbar} = -2\pi i \underbrace{e^{-i(E_R - i\Gamma/2)t/\hbar}}_{\text{residue at } E_R - i\Gamma/2} .$$

Therefore, the wave packet develops a time dependence given by $e^{-iE_R t/\hbar} e^{-\Gamma t/(2\hbar)}$. This indeed supports the interpretation of the scattering process as proceeding through the formation of a quasi-stable state of energy E_R and with lifetime $\hbar/\Gamma$.

Problem 14 S-Wave Bound- and Scattering-State Problem in an Attractive δ-Shell Potential

Consider a particle of mass μ under the influence of a central potential $V(r) = -V_0 \, \delta(r - a)$ with $V_0 > 0$, the so-called δ-shell potential. Assume the particle is in an s-wave state.

1. Are there any bound states for $E < 0$? If so, under what conditions do these exist?
2. Take $E > 0$ and obtain an expression for the s-wave phase shift δ_k and determine its low- and high-energy limits.

Solution

Part 1

We refer to the region $0 < r < a$ as I and to the region $r > a$ as II. The reduced radial equation in s-wave reads

$$u''(r) + [\epsilon - v(r)]u(r) = 0 ,$$

where $v(r) = 2\mu V(r)/\hbar^2$ and $\epsilon = 2\mu E/\hbar^2$. Bound-state solutions possibly exist only for $-v_0 < \epsilon < 0$. Thus we have

$$u''(r) = \left[\kappa^2 - v_0 \, \delta(r - a) \right] u(r) \qquad \text{with } \kappa = \sqrt{|\epsilon|} ,$$

and, imposing the boundary conditions at the origin and ∞,

$$u_I(r) = A \, \sinh(\kappa r) , \qquad u_{II}(r) = B \, e^{-\kappa r} .$$

It remains to impose the conditions at $r = a$, which yields

$$A \, \sinh(\kappa a) = B \, e^{-\kappa a} , \qquad -B \, \kappa \, e^{-\kappa a} - A\kappa \, \cosh(\kappa a) = -v_0 B \, e^{-\kappa a} ,$$

leading to

$$\tanh(\kappa a) = \frac{\kappa}{v_0 - \kappa} \implies \tanh x = \frac{x}{x_0 - x} \qquad 0 < x < x_0 ,$$

where $x = \kappa a$ and $x_0 = v_0 a$. Graphically this equation is found to have a single solution if $x_0 > 1$ (just consider the slopes at the origin of the left- and right-hand side functions).

Part 2

Set $k^2 = \epsilon > 0$, and the radial equation now reads

$$u''(r) + \left[k^2 + v_0\, \delta(r-a)\right] u(r) = 0\,,$$

with solutions

$$u_I(r) = A\, \sin(kr)\,, \qquad u_{II}(r) = B\, \sin(kr + \delta_k)\,,$$

where δ_k is the (s-wave) phase shift. The boundary conditions at a require

$$A\, \sin(ka) = B\, \sin(ka + \delta_k)\,, \qquad Bk\, \cos(ka + \delta_k) - Ak\, \cos(ka) = -v_0 A\, \sin(ka)\,,$$

which yields

$$\tan(ka + \delta_k) = \frac{1}{\cot(ka) - v_0/k} \;\Longrightarrow\; \delta_k = -ka + \tan^{-1}\left[\frac{1}{\cot(ka) - v_0/k}\right].$$

In the low-energy limit, $\delta_k \approx -ka$ since in that limit the argument of the inverse tangent vanishes (recall that $\cot x \longrightarrow \infty$ as x goes to zero). In the high-energy limit, however, δ_k vanishes as it is obvious on neglecting the term v_0/k in the inverse tangent argument.

Problem 15 Scattering in a Repulsive δ-Shell Potential

Consider scattering by a repulsive δ-shell potential given by

$$V(r) = \frac{\hbar^2}{2\mu}\, v_0\, \delta(r-R), \qquad\qquad v_0 > 0\,.$$

1. Set up an equation that determines the s-wave phase shift $\delta(k)$ as function of k – the energy is $E = \hbar^2 k^2/(2\mu)$.
2. Assume hereafter that $v_0 \gg k$ and $v_0 \gg 1/R$. Show that if $\tan(kR)$ is *not* close to zero, the s-wave phase shift resembles the hard-sphere result obtained in Problem 12.
3. Show that if $\tan(kR)$ is close (but not exactly equal) to zero, resonance behavior is possible; that is, $\cot\delta_0(k)$ goes through zero from the positive side as k increases. Determine the energies of the resonances approximately, keeping terms of order $1/v_0$.
4. Obtain the s-wave bound-state energies for a particle confined in a spherical region of radius R; that is, a particle under the action of a potential $V(r) = 0$ if $r < R$ and $V(r) = \infty$ if $r > R$. Compare these energies with those obtained for the resonances in part 3.
5. Obtain an expression for the resonance width Γ, defined as

$$\frac{1}{\Gamma} = -\frac{1}{2} \left.\frac{d\cot\delta_0(E)}{dE}\right|_{E=E_R},$$

and note that the resonances become extremely sharp as v_0 becomes large.

(Adapted from J. J. Sakurai and J. Napolitano (2020), *Modern Quantum Mechanics*, Cambridge University Press.)

Solution

Part 1

The s-wave radial equation reads

$$u''(r) = \left[v_0\,\delta(r-R) - k^2\right]u(r)\,.$$

We define region I for $0 \le r < R$ and region II for $r > R$. The boundary conditions satisfied by $u(r)$ are given by

$$u_I(0) = 0\,,\qquad u_I(R) = u_{II}(R)\,,\qquad u'_{II}(R) - u'_I(R) = v_0 u_I(R)\,,$$

where the condition on the first derivative is obtained in the usual way by integrating the radial equation over a small interval containing the δ-function and then reducing this interval to zero. For $r \ne R$ the solutions in regions I and II are given by

$$u_I(r) = A_I\,\sin(kr)\,,\qquad u_{II}(r) = A_{II}\,\sin(kr + \delta)\,,$$

where we have already imposed the requirement that $u_I(r)$ vanishes at the origin. The remaining two boundary conditions yield

$$\sin(kR + \delta) = \frac{A_I}{A_{II}}\,\sin(kR)\,,\qquad \cos(kR+\delta) - \frac{A_I}{A_{II}}\,\cos(kR) = \frac{A_I}{A_{II}}\,\frac{v_0}{k}\,\sin(kR)\,,$$

which lead to the following relation for $\delta(k)$;

$$\tan[kR + \delta(k)] = \frac{k/v_0}{1 + (k/v_0)\,\cot(kR)}\,.$$

Part 2

In the limit $k/v_0 \ll 1$ and assuming that $kR \ne n\pi$ with $n = 1, 2, \ldots$ (at these points the cotangent is divergent and $(k/v_0)\cot(kR)$ may be finite), we have

$$\tan[kR + \delta(k)] \approx \frac{k}{v_0}\left(1 - \frac{k}{v_0}\,\cot(kR)\right)\,,$$

which implies, to terms linear in k/v_0 ($\tan^{-1} x \approx x$ for small x),

$$\delta(k) \approx -kR + \frac{k}{v_0}\,.$$

For the problem of hard-sphere scattering in an s-wave state, we found $\delta(k) = -kR$, which reproduces the leading-order term above.

Part 3

Using the addition formula for the tangent, we rewrite the relation obtained in part 1 as

$$\tan[kR + \delta(k)] = \frac{\tan(kR) + \tan\delta(k)}{1 - \tan(kR)\,\tan\delta(k)} = \frac{(k/v_0)\tan(kR)}{\tan(kR) + k/v_0}\,,$$

which we solve for $\tan\delta(k)$ to find

$$\tan\delta(k) = -\frac{\tan^2(kR)}{\tan(kR) + (k/v_0)(1 + \tan^2(kR))}\,.$$

We assume

$$k_n R = n\pi + \eta \qquad \text{with } |\eta| \ll 1 \,,$$

that is, $k_n R$ is close to a multiple of π. For these k_n, we obtain (recall that $\tan(n\pi + \eta) = \tan\eta \approx \eta$ for $\eta \ll 1$)

$$\tan\delta(k_n) = -\frac{\eta^2}{\eta + (k_n/v_0)(1 + \eta^2)} \implies \cot\delta(k_n) = -\frac{1}{\eta} - \frac{k_n}{v_0}\frac{1 + \eta^2}{\eta^2} \,.$$

Inserting the expression for k_n on the right-hand side and recalling that $Rv_0 \gg 1$, we find

$$\cot\delta(k_n) = -\frac{1}{\eta}\left(1 + \frac{1}{Rv_0}\right) - \frac{n\pi}{Rv_0}\left(1 + \frac{1}{\eta^2}\right) - \frac{\eta}{Rv_0} \approx -\frac{1}{\eta} - \frac{n\pi}{Rv_0}\frac{1}{\eta^2} \,,$$

or

$$\cot\delta(k_n) \approx -\frac{\eta + n\pi/(Rv_0)}{\eta^2} \,,$$

and we see that the cotangent vanishes, implying that the phase shift goes through $\pi/2$, if $\eta = \eta_n = -n\pi/(Rv_0)$; the wave number k_n^R and energy $E_n^R = \hbar^2 k_n^{R\,2}/(2\mu)$ of the corresponding resonance are then given by

$$k_n^R = \frac{n\pi}{R}\left(1 - \frac{1}{Rv_0}\right) \approx \frac{n\pi}{R} \,, \qquad E_n^R \approx \frac{(n\pi)^2\hbar^2}{2\mu R^2}\left(1 - \frac{2}{Rv_0}\right) \approx \frac{(n\pi)^2\hbar^2}{2\mu R^2} \,.$$

Part 4

The s-wave radial equation for a particle confined in the region $r \le R$ reads

$$u''(r) = -k^2 u(r) \qquad \text{for } 0 \le r \le R \,,$$

with boundary conditions $u(0) = u(R) = 0$. We have

$$u(r) = A \sin kr \qquad \sin kR = 0 \implies k_n = \frac{n\pi}{R} \,,$$

and the energies of the bound states are

$$E_n^B \approx \frac{(n\pi)^2\hbar^2}{2\mu R^2} \,,$$

which differ from the resonance energies found above by a term of order $1/(Rv_0) \ll 1$.

Part 5

In the energy region E close to a resonance energy E_n^R we have

$$\cot\delta(\eta) \approx -\frac{\eta - \eta_n}{\eta_n^2} \,, \qquad \eta_n = -\frac{n\pi}{Rv_0} \,.$$

Moreover, we have

$$E = \frac{\hbar^2}{2\mu R^2}(n\pi + \eta)^2 \,, \qquad E_n^R = \frac{\hbar^2}{2\mu R^2}(n\pi + \eta_n)^2 \approx \frac{(n\pi)^2\hbar^2}{2\mu R^2} \,.$$

We now use

$$\frac{d}{dE} \cot \delta(E) = \frac{d\eta}{dE} \frac{d}{d\eta} \cot \delta(\eta) \,,$$

where

$$dE = \frac{\hbar^2}{\mu R^2}(n\pi + \eta)d\eta \implies \frac{d\eta}{dE} = \frac{\mu R^2}{\hbar^2} \frac{1}{n\pi + \eta}$$

and

$$\frac{d}{d\eta} \cot \delta(\eta)\bigg|_{\eta_n} \approx -\frac{1}{\eta_n^2} \,.$$

Collecting results, we obtain at the resonant energies

$$\frac{d}{dE} \cot \delta(E)\bigg|_{E=E_n^R} = -\frac{\mu R^2}{\hbar^2} \frac{1}{\eta_n^2 (n\pi + \eta_n)} \,,$$

and hence the width of the nth resonance is given by

$$\Gamma_n = 2\eta_n^2 (n\pi + \eta_n) \frac{\hbar^2}{\mu R^2} = 2n\pi \left(\frac{n\pi}{Rv_0}\right)^2 \left(1 - \frac{1}{Rv_0}\right) \frac{\hbar^2}{\mu R^2} \approx 2n\pi \left(\frac{n\pi}{Rv_0}\right)^2 \frac{\hbar^2}{\mu R^2}.$$

We see that the width of the resonance increases as n^3 for fixed Rv_0. The sharpest resonance, namely the resonance with the smallest width, is that occurring for $n = 1$. As v_0 increases, the width of the resonances decreases as $1/v_0^2$.

In order to express the cotangent of the phase shift as function of energy, we note that, for $\eta \approx \eta_n$, keeping up to linear terms in $\eta - \eta_n$, we have

$$E = \frac{\hbar^2}{2\mu R^2}(n\pi + \eta)^2 = \frac{\hbar^2}{2\mu R^2}(n\pi + \eta_n + \eta - \eta_n)^2 = E_n^R + \frac{\hbar^2}{\mu R^2}(n\pi + \eta_n)(\eta - \eta_n) + \cdots \,,$$

which can also be written as

$$E - E_n^R = \frac{\hbar^2}{\mu R^2}(n\pi + \eta_n)^2 \frac{\eta - \eta_n}{n\pi + \eta_n} = 2E_n^R \frac{\eta - \eta_n}{n\pi + \eta_n} \,.$$

We then obtain

$$\cot \delta(E) = -\frac{E - E_n^R}{2E_n^R} \frac{n\pi + \eta_n}{\eta_n^2} = -\frac{E - E_n^R}{\Gamma_n/2} \,,$$

where in the last step we used

$$\frac{1}{2E_n^R} \frac{n\pi + \eta_n}{\eta_n^2} \approx \frac{1}{2E_n^R} \frac{1}{n\pi/(Rv_0)^2} = \frac{1}{\Gamma_n/2} \,.$$

We can now obtain the cross section in the resonance region (recall that we are dealing with s-wave scattering):

$$\sigma(E) \approx \frac{4\pi}{k^2} \sin^2 \delta(E) \approx \frac{4\pi}{k^2} \frac{\Gamma_n^2/4}{(E - E_n^R)^2 + \Gamma_n^2/4} \,,$$

where we have used the trigonometric identity $\sin^2 x = 1/(1 + \cot^2 x)$. The equation above is known as the Breit–Wigner formula, and describes the characteristic "bell" shape of the cross section in the resonance region; Γ_n gives the full width of the cross section at half maximum. For additional comments on the interpretation of these resonances we refer the reader to Problem 13.

Problem 16 Integral Equation for Scattering in One Dimension

The integral equation for scattering in one dimension for a particle of mass m and energy E in a potential $V(x)$ reads

$$\psi_k(x) = e^{ikx} - \int_{-\infty}^{\infty} dx' \, G_k(x, x') \, v(x') \, \psi_k(x') \,,$$

where the Green's function $G_k(x, x')$ satisfies

$$\left(\frac{d^2}{dx^2} + k^2 \right) G_k(x, x') = -\delta(x - x') \,,$$

$v(x) = 2mV(x)/\hbar^2$, and $k^2 = 2mE/\hbar^2$. The particle is assumed to be traveling initially from left to right.

1. Following the method illustrated in the construction of the scattering Green's function in Problem 8, that is, by Fourier-transforming both sides of the equation for $G_k(x, x')$ and removing the singularity in $\widetilde{G}_k(q)$ at $q = k$ by making the replacement $k \longrightarrow k + i\eta$ (with $\eta \longrightarrow 0$), show that

$$G_k(x, x') = \frac{i}{2k} e^{ik|x - x'|} \,.$$

2. Show that the integral equation for $\psi_k(x)$ can be written as

$$\psi_k(x) = e^{ikx} g_k(x; \mathrm{R}) + e^{-ikx} g_k(x; \mathrm{L}) \,,$$

and obtain expressions for $g_k(x; \mathrm{L})$ and $g_k(x; \mathrm{R})$.

3. Consider the wave function $\psi_k(x)$ in the asymptotic regions at $x \longrightarrow \pm\infty$. Show that

$$\lim_{x \to -\infty} \psi_k(x) = e^{ikx} + f_k(\mathrm{L}) \, e^{-ikx} \,, \qquad \lim_{x \to +\infty} \psi_k(x) = f_k(\mathrm{R}) \, e^{ikx} \,.$$

Obtain explicit expressions for the left and right scattering amplitudes $f_k(\mathrm{L})$ and $f_k(\mathrm{R})$.

4. In the rest of the problem, suppose that the particle is under the action of a δ-function potential centered at the origin,

$$V(x) = -V_0 \, \delta(x) \,, \qquad V_0 > 0 \,.$$

Solve the integral equation and obtain the wave function.

5. Obtain the left and right scattering scattering amplitudes and corresponding scattering cross sections $\sigma_k(\mathrm{L}) = |f_k(\mathrm{L})|^2$ and $\sigma_k(\mathrm{R}) = |f_k(\mathrm{R})|^2$. Compare these results with the reflection and transmission coefficients $R(k)$ and $T(k)$ obtained in Problem 5.1 with V_0 replaced by $-V_0$.

Solution

Part 1

Because of translational invariance, the Green's function depends only on the difference $z = x - x'$ rather than on x and x' separately. We introduce the Fourier transforms

$$G_k(z) = \int_{-\infty}^{\infty} \frac{dq}{2\pi} e^{iqz} \, \widetilde{G}_k(q) \,, \qquad \delta(z) = \int_{-\infty}^{\infty} \frac{dq}{2\pi} e^{iqz} \,,$$

and insert them into the equation for $G_k(x, x')$ to obtain

$$\int_{-\infty}^{\infty} \frac{dq}{2\pi} e^{iqz} \left(-q^2 + k^2 \right) \widetilde{G}_k(q) = - \int_{-\infty}^{\infty} \frac{dq}{2\pi} e^{iqz} \implies \widetilde{G}_k(q) = \frac{1}{q^2 - k^2} \,,$$

and so $\widetilde{G}_k(q)$ is singular at $q = k$, and therefore its Fourier transform is not well defined. To remove the singularity, we shift the pole off the real axis by setting $k \longrightarrow k + i\eta$, so that

$$G_k(z) = \int_{-\infty}^{\infty} \frac{dq}{2\pi} e^{iqz} \frac{1}{q^2 - (k + i\eta)^2} ,$$

and take the limit $\eta \longrightarrow 0$ at the end. This integral can be easily performed by contour integration in the complex-q plane. When $z > 0$, for convergence we must close the contour in the upper half plane, picking up the residue at the pole $q = k + i\eta$:

$$G_k(z) = 2\pi i \left[\frac{1}{2\pi} \frac{e^{iqz}}{q + (k + i\eta)} \right]_{q=k+i\eta} = \frac{i}{2} \frac{e^{i(k+i\eta)z}}{k + i\eta} \longrightarrow \frac{i}{2k} e^{ikz} \qquad z > 0 .$$

When $z < 0$, for convergence we must close the contour in the lower half plane, picking up the residue at the pole $q = -k - i\eta$:

$$G_k(z) = -2\pi i \left[\frac{1}{2\pi} \frac{e^{iqz}}{q - (k + i\eta)} \right]_{q=-k-i\eta} = -\frac{i}{2} \frac{e^{-i(k+i\eta)z}}{-k - i\eta} \longrightarrow \frac{i}{2k} e^{-ikz} \qquad z < 0 .$$

These two results can be combined to obtain

$$G_k(x, x') = \frac{i}{2k} e^{ik|x-x'|} .$$

Part 2

Insert the expression for $G_k(x, x')$ into the integral equation:

$$\begin{aligned} \psi_k(x) &= e^{ikx} - \frac{i}{2k} \int_{-\infty}^{\infty} dx' \, e^{ik|x-x'|} v(x') \, \psi_k(x') \\ &= e^{ikx} - \frac{i}{2k} \int_{-\infty}^{x} dx' \, e^{ik(x-x')} v(x') \, \psi_k(x') - \frac{i}{2k} \int_{x}^{\infty} dx' \, e^{-ik(x-x')} v(x') \, \psi_k(x') \\ &= e^{ikx} g_k(x; R) + e^{-ikx} g_k(x; L) , \end{aligned}$$

where

$$g_k(x; R) = 1 - \frac{i}{2k} \int_{-\infty}^{x} dx' \, e^{-ikx'} v(x') \, \psi_k(x') , \qquad g_k(x; L) = -\frac{i}{2k} \int_{x}^{\infty} dx' \, e^{ikx'} v(x') \, \psi_k(x') .$$

Part 3

In the limit $x \longrightarrow -\infty$, we have

$$\lim_{x \to -\infty} g_k(x; R) = 1 , \qquad \lim_{x \to -\infty} g_k(x; L) = \underbrace{-\frac{i}{2k} \int_{-\infty}^{\infty} dx' \, e^{ikx'} v(x') \, \psi_k(x')}_{f_k(L)} ,$$

whereas in the limit $x \longrightarrow \infty$ we have

$$\lim_{x \to \infty} g_k(x; R) = \underbrace{1 - \frac{i}{2k} \int_{-\infty}^{\infty} dx' \, e^{-ikx'} v(x') \, \psi_k(x')}_{f_k(R)} , \qquad \lim_{x \to \infty} g_k(x; L) = 0 .$$

Combining these results we obtain

$$\lim_{x \to -\infty} \psi_k(x) = e^{ikx} + f_k(L) e^{-ikx} , \qquad \lim_{x \to \infty} \psi_k(x) = f_k(R) e^{ikx} .$$

Part 4

In the case of an attractive δ-function potential, the integral equation reduces to

$$\psi_k(x) = e^{ikx} - \frac{i}{2k}\int_{-\infty}^{\infty} dx'\, e^{ik|x-x'|}\left[-\frac{2mV_0}{\hbar^2}\delta(x')\right]\psi_k(x') = e^{ikx} + i\frac{v_0}{2k}\, e^{ik|x|}\,\psi_k(0)\,,$$

and evaluating the left-hand side at $x = 0$, we find

$$\psi_k(0) = 1 + i\frac{v_0}{2k}\,\psi_k(0) \implies \psi_k(0) = \frac{1}{1 - iv_0/(2k)}\,,$$

and therefore

$$\psi_k(x) = e^{ikx} - \frac{e^{ik|x|}}{1 + 2ik/v_0}\,.$$

This wave function coincides with the wave function we found by solving the Schrödinger equation, and imposing the appropriate boundary conditions.

Part 5

The left and right scattering amplitudes found above are

$$f_k(L) = -\frac{i}{2k}\int_{-\infty}^{\infty} dx'\, e^{ikx'}\, v(x')\,\psi_k(x')\,, \qquad f_k(R) = 1 - \frac{i}{2k}\int_{-\infty}^{\infty} dx'\, e^{-ikx'}\, v(x')\,\psi_k(x')\,.$$

For the case of δ-function potentials, these expressions yield

$$f_k(L) = \frac{iv_0/(2k)}{1 - iv_0/(2k)}\,, \qquad f_k(R) = 1 + \frac{iv_0/(2k)}{1 - iv_0/(2k)} = \frac{1}{1 - iv_0/(2k)}\,,$$

and corresponding cross sections

$$\sigma_k(L) = |f_k(L)|^2 = \frac{v_0^2}{4k^2 + v_0^2}\,, \qquad \sigma_k(R) = |f_k(R)|^2 = \frac{4k^2}{4k^2 + v_0^2}\,,$$

which are in agreement with the reflection and transmission coefficients we obtained from solutions of the Schrödinger equation.

Problem 17 Partial Wave Expansion of the Free-Particle Wave Function

Consider a free particle of mass μ and energy E in three dimensions with Hamiltonian

$$H_0 = -\frac{\hbar^2}{2\mu}\,\boldsymbol{\nabla}^2\,.$$

As we have seen, the wave function can be expressed as

$$\psi_{klm}(\mathbf{r}) = R_{kl}(r)\, Y_{lm}(\Omega)\,, \qquad k^2 = \frac{2\mu E}{\hbar^2}\,,$$

where the $Y_{lm}(\Omega)$ are the spherical harmonics – the eigenfunctions of $\mathbf{L}^2$ and L_z.

1. By introducing the (non-adimensional) variable $\rho = kr$, show that $R_{kl}(\rho)$ satisfies the radial equation

$$R_{kl}''(\rho) + \frac{2}{\rho}R_{kl}'(\rho) + \left[1 - \frac{l(l+1)}{\rho^2}\right]R_{kl}(\rho) = 0\,.$$

2. Introduce the auxiliary function $R_{kl}(\rho) \equiv J_l(\rho)/\sqrt{\rho}$ and show that $J_l(\rho)$ satisfies

$$\rho^2 J_l''(\rho) + \rho J_l'(\rho) + [\rho^2 - (l + 1/2)^2]J_l(\rho) = 0 \qquad \text{with } l = 0, 1, 2, \ldots ,$$

which has the form of a Bessel equation. The general solution is given by

$$\frac{J_l(\rho)}{\sqrt{\rho}} = \alpha j_l(\rho) + \beta n_l(\rho) ,$$

that is, a linear combination of the regular and irregular spherical Bessel functions, respectively $j_l(\rho)$ and $n_l(\rho)$.

3. Explain why the only acceptable solution for $R_{kl}(r)$ is that proportional to $j_l(kr)$. The free-particle wave functions are written as

$$\psi_{klm}(\mathbf{r}) = \sqrt{\frac{2}{\pi}} \, k j_l(kr) \, Y_{lm}(\Omega) ,$$

and are simultaneous eigenfunctions of H, $\mathbf{L}^2$, and L_z. What is the degeneracy of the energy eigenvalue? Knowing that

$$\int_0^\infty dr \, r^2 \, j_l(k'r) j_l(kr) = \frac{\pi}{2k^2} \, \delta(k - k') ,$$

obtain the continuum normalization condition for $\psi_{klm}(\mathbf{r})$.

4. Knowing that

$$j_0(\rho) = \frac{\sin \rho}{\rho} ,$$

show by explicit calculation that

$$\int_0^\infty dr \, r^2 \, j_0(k'r) j_0(kr) = \frac{\pi}{2k^2} \, \delta(k - k') .$$

Solution

Part 1

The free-particle radial equation for $R_{kl}(r) = u_{kl}(r)/r$ is

$$\frac{1}{r^2} \frac{d}{dr} r^2 \frac{dR_{kl}(r)}{dr} + \left[k^2 - \frac{l(l + 1)}{r^2} \right] R_{kl}(r) = 0 .$$

Define the non-dimensional variable $\rho = kr$, so that in terms of ρ the above equation becomes

$$\frac{1}{\rho^2} \frac{d}{d\rho} \rho^2 \frac{dR_{kl}(\rho)}{d\rho} + \left[1 - \frac{l(l + 1)}{\rho^2} \right] R_{kl}(\rho) = 0 ,$$

or

$$R_{kl}''(\rho) + \frac{2}{\rho} R_{kl}'(\rho) + \left[1 - \frac{l(l + 1)}{\rho^2} \right] R_{kl}(\rho) = 0 .$$

Part 2

Now we introduce the auxiliary function $R_{kl}(\rho) \equiv J_l(\rho)/\sqrt{\rho}$, which leads to

$$R'_{kl}(\rho) = -\frac{J_l(\rho)}{2\rho^{3/2}} + \frac{J'_l(\rho)}{\rho^{1/2}}, \qquad R''_{kl}(\rho) = \frac{3J_l(\rho)}{4\rho^{5/2}} - \frac{J'_l(\rho)}{\rho^{3/2}} + \frac{J''_l(\rho)}{\rho^{1/2}},$$

and therefore

$$\frac{3J_l(\rho)}{4\rho^{5/2}} - \frac{J'_l(\rho)}{\rho^{3/2}} + \frac{J''_l(\rho)}{\rho^{1/2}} + \frac{2}{\rho}\left[-\frac{J_l(\rho)}{2\,\rho^{3/2}} + \frac{J'_l(\rho)}{\rho^{1/2}}\right] + \left[1 - \frac{(l+1/2)^2 - 1/4}{\rho^2}\right]\frac{J_l(\rho)}{\rho^{1/2}} = 0\,.$$

Canceling terms and multiplying both sides by $\rho^{5/2}$, we are finally left with

$$\rho^2 J''_l(\rho) + \rho J'_l(\rho) + [\rho^2 - \underbrace{(l+1/2)^2}_{\nu^2}]J_l(\rho) = 0\,,$$

which is a Bessel equation with $\nu = 1/2, 3/2, \ldots$, with solutions

$$\frac{J_l(\rho)}{\rho^{1/2}} = \alpha j_l(\rho) + \beta n_l(\rho)\,.$$

See M. Abramowitz and I. A. Stegun (1965). *Handbook of Mathematical Functions*, Dover, for a summary of properties satisfied by these functions.

Part 3

Recall that $R_{kl}(r) \propto r^l$ as $r \longrightarrow 0$. The irregular spherical Bessel function $n_l(\rho)$ behaves in this limit as $1/\rho^{l+1}$ and is therefore unacceptable; so $\beta = 0$ in the linear combination above. By contrast, the regular spherical Bessel function $j_l(\rho)$ behaves as ρ^l, in accord with the behavior of $R_{kl}(r)$. Since the $\psi_{klm}(\mathbf{r})$ are eigenfunctions of H for any l and m and since $l = 0, 1, 2, \ldots$ and $m = -l, \ldots, l$, we see that the degeneracy of the (continuous) energy eigenvalue $E = \hbar^2 k^2/(2m)$ is infinite. The wave functions are normalized as

$$\int d\mathbf{r}\, \psi^*_{k'l'm'}(\mathbf{r})\, \psi_{klm}(\mathbf{r}) = \frac{2}{\pi} k^2 \int_0^\infty dr\, r^2 j_{l'}(k'r) j_l(kr) \int d\Omega\, Y^*_{l'm'}(\Omega)\, Y_{lm}(\Omega)$$

$$= \delta_{l,l'}\, \delta_{m,m'}\, \frac{2}{\pi} k^2 \int_0^\infty dr\, r^2 j_l(k'r) j_l(kr) = \delta(k - k')\, \delta_{l,l'}\, \delta_{m,m'}\,.$$

Part 4

We have (note the integration limits)

$$\int_0^\infty dr\, r^2 j_0(kr) j_0(k'r) = \frac{1}{2kk'} \int_{-\infty}^\infty dr \sin(kr)\, \sin(k'r) = \frac{1}{4kk'} \int_{-\infty}^\infty dr\, [\cos(kr - k'r) - \cos(kr + k'r)]$$

$$= \frac{1}{4kk'} \operatorname{Re} \int_{-\infty}^\infty dr\, \left[e^{i(k-k')r} - e^{i(k+k')r}\right]$$

$$= \frac{\pi}{2kk'}\, [\delta(k - k') - \delta(k + k')] = \frac{\pi}{2k^2}\, \delta(k - k')\,,$$

since the second δ-function can never be satisfied for $k, k' > 0$.

Problem 18 Derivation of an Integral Relation for the Phase Shift

The scattering wave function in a central potential can be expanded in partial waves as

$$\psi_{\mathbf{k}}(\mathbf{r}) = \sum_{l=0}^{\infty} i^l (2l+1) P_l(\cos \theta_{kr}) \psi_l(r;k) \,,$$

where $\cos \theta_{kr} = \hat{\mathbf{k}} \cdot \hat{\mathbf{r}}$ and θ_{kr} denotes the angle between the directions $\hat{\mathbf{k}}$ and $\hat{\mathbf{r}}$, and $\psi_l(r;k)$ behaves asymptotically as

$$\lim_{r\to\infty} \psi_l(r;k) = \frac{e^{i\delta_l(k)}}{kr} \sin(kr - l\pi/2 + \delta_l) \,.$$

In the absence of interactions, $\psi_l(r;k)$ reduces to the (regular) spherical Bessel function $j_l(kr)$, and

$$\psi_{\mathbf{k}}(\mathbf{r}) \longrightarrow \phi_{\mathbf{k}}(\mathbf{r}) = \sum_{l=0}^{\infty} i^l (2l+1) P_l(\cos \theta_{kr}) j_l(kr) = \underbrace{\sum_{l=0}^{\infty} \sum_{m=-l}^{l} 4\pi i^l Y_{lm}^*(\Omega_{\hat{\mathbf{k}}}) Y_{lm}(\Omega_{\hat{\mathbf{r}}}) j_l(kr)}_{e^{i\mathbf{k}\cdot\mathbf{r}}} \,.$$

1. Starting from the scattering amplitude

$$f_{\mathbf{k}}(\hat{\mathbf{k}}') = -\frac{1}{4\pi} \int d\mathbf{r}\, e^{-i\mathbf{k}'\cdot\mathbf{r}} v(r) \psi_{\mathbf{k}}(\mathbf{r}) \qquad \text{with} \qquad \mathbf{k}' = k\hat{\mathbf{r}}' \,,$$

and inserting the above expansion for $\psi_{\mathbf{k}}(\mathbf{r})$, show that $f_{\mathbf{k}}(\hat{\mathbf{k}}')$ has the following partial wave expansion:

$$f_{\mathbf{k}}(\hat{\mathbf{k}}') = -\sum_{l=0}^{\infty} (2l+1) P_l(\hat{\mathbf{k}}' \cdot \hat{\mathbf{k}}) \int_0^{\infty} dr\, r^2 j_l(kr) v(r) \psi_l(r;k) \,.$$

2. Comparing the expression above with that given in Eq. (15.55) for the partial wave expansion of the scattering amplitude in terms of phase shifts, show that

$$e^{2i\delta_l(k)} = 1 - 2ik \int_0^{\infty} dr\, r^2 j_l(kr) v(r) \psi_l(r;k) \,.$$

Solution

Using the expansion

$$P_l(\cos \theta_{kr}) = \frac{4\pi}{2l+1} \sum_{m=-l}^{l} Y_{lm}^*(\hat{\mathbf{k}}) Y_{lm}(\hat{\mathbf{r}}) \,,$$

we express the wave function as

$$\psi_{\mathbf{k}}(\mathbf{r}) = \sum_{l=0}^{\infty} \sum_{m=-l}^{l} 4\pi i^l \psi_l(r;k) Y_{lm}^*(\hat{\mathbf{k}}) Y_{lm}(\hat{\mathbf{r}}) \,,$$

yielding, for the scattering amplitude,

$$f_{\mathbf{k}}(\hat{\mathbf{k}}') = -\frac{1}{4\pi} \sum_{l'm'} 4\pi (-i)^{l'} Y_{l'm'}(\hat{\mathbf{k}}') \sum_{lm} 4\pi i^l Y_{lm}^*(\hat{\mathbf{k}}) \int d\mathbf{r}\, j_{l'}(kr) Y_{l'm'}^*(\hat{\mathbf{r}}) v(r) \psi_l(r;k) Y_{lm}(\hat{\mathbf{r}})$$

$$= -4\pi \sum_{lm} Y_{lm}(\hat{\mathbf{k}}') Y_{lm}^*(\hat{\mathbf{k}}) \int_0^{\infty} dr\, r^2 j_l(kr) v(r) \psi_l(r;k)$$

$$= -\sum_l (2l+1) P_l(\hat{\mathbf{k}}' \cdot \hat{\mathbf{k}}) \int_0^{\infty} dr\, r^2 j_l(kr) v(r) \psi_l(r;k) \,,$$

where in the first line we have inserted the partial wave expansion of the plane wave $e^{-i\mathbf{k}'\cdot\mathbf{r}}$ (recall that $k' = k$), and in the second line we have used the orthonormality of the spherical harmonics.

Part 2

We derived (see Section 15.3)

$$f_{\mathbf{k}}(\hat{\mathbf{k}}') = \frac{1}{k}\sum_l(2l+1)\,P_l(\hat{\mathbf{k}}'\cdot\hat{\mathbf{k}})\,\sin\delta_l(k)\,e^{i\delta_l(k)} = \frac{1}{2ik}\sum_l(2l+1)\,P_l(\hat{\mathbf{k}}'\cdot\hat{\mathbf{k}})\left(e^{2i\delta_l(k)}-1\right),$$

and comparing this expression to that obtained in part 1, we deduce that

$$\frac{e^{2i\delta_l(k)}-1}{2ik} = -\int_0^\infty dr\,r^2\,j_l(kr)\,v(r)\,\psi_l(r;k),$$

which yields the desired relation.

Problem 19　Partial Wave Expansion of Integral Equation for Scattering in Central Potential

In this problem we will carry out the partial wave expansion of the integral equation for scattering in the case of a central potential $V(r)$. The following relations will be needed:

- Expansion of a plane wave in spherical harmonics,

$$e^{i\mathbf{k}\cdot\mathbf{r}} = 4\pi\sum_{lm}i^l j_l(kr)\,Y_{lm}^*(\hat{\mathbf{k}})\,Y_{lm}(\hat{\mathbf{r}}).$$

- Addition formula for spherical harmonics,

$$P_l(\hat{\mathbf{n}}\cdot\hat{\mathbf{n}}') = \frac{4\pi}{2l+1}\sum_m Y_{lm}^*(\hat{\mathbf{n}})\,Y_{lm}(\hat{\mathbf{n}}'),$$

where $\hat{\mathbf{n}}$ and $\hat{\mathbf{n}}'$ are directions.

- Definition of the spherical Hankel functions:

$$h_l^{(1)}(\rho) = j_l(\rho) + i\,n_l(\rho), \qquad h_l^{(2)}(\rho) = j_l(\rho) - i\,n_l(\rho) = h_l^{(1)*}(\rho).$$

- Symmetry properties of the regular and irregular spherical Bessel and Hankel functions,

$$j_l(-\rho) = (-1)^l j_l(\rho), \qquad n_l(-\rho) = (-1)^{l+1}\,n_l(\rho),$$

and

$$h_l^{(1)}(-\rho) = (-1)^l h_l^{(2)}(\rho), \qquad h_l^{(2)}(-\rho) = (-1)^l h_l^{(1)}(\rho).$$

- Asymptotic behavior for large ρ,

$$\lim_{\rho\to\infty} j_l(\rho) = \frac{\sin(\rho - l\pi/2)}{\rho}, \qquad \lim_{\rho\to\infty} n_l(\rho) = -\frac{\cos(\rho - l\pi/2)}{\rho},$$

and the corresponding asymptotic behavior for the spherical Hankel functions,

$$\lim_{\rho\to\infty} h_l^{(1)}(\rho) = \frac{e^{i[\rho-(l+1)\pi/2]}}{\rho}, \qquad \lim_{\rho\to\infty} h_l^{(2)}(\rho) = \frac{e^{-i[\rho-(l+1)\pi/2]}}{\rho}.$$

1. Using

$$G_k(\mathbf{r}, \mathbf{r}') = \int \frac{d\mathbf{q}}{(2\pi)^3} e^{i\mathbf{q}\cdot\mathbf{r}} \frac{1}{q^2 - (k + i\eta)^2} e^{-i\mathbf{q}\cdot\mathbf{r}'} ,$$

insert the above plane-wave expansion for $e^{i\mathbf{q}\cdot\mathbf{r}}$ and $e^{-i\mathbf{q}\cdot\mathbf{r}'}$ and carry out first the integration over $\hat{\mathbf{q}}$ and then the contour integration over q to obtain

$$G_k(\mathbf{r}, \mathbf{r}') = \sum_{lm} G_l(r, r'; k)\, Y_{lm}(\hat{\mathbf{r}})\, Y_{lm}^*(\hat{\mathbf{r}}') ,$$

where

$$G_l(r, r'; k) = ik\, j_l(kr_<)\, h_l^{(1)}(kr_>) ,$$

$r_< = \min\{r, r'\}$, and $r_> = \max\{r, r'\}$.

2. Since the scattering wave function $\psi_{\mathbf{k}}(\mathbf{r})$ can be thought of (for a spinless particle) as a scalar function, it can depend only on k, r, and $\hat{\mathbf{k}} \cdot \hat{\mathbf{r}}$, and therefore admits an expansion in Legendre polynomials,

$$\psi_{\mathbf{k}}(\mathbf{r}) = \sum_l i^l (2l + 1)\, \psi_l(r; k)\, P_l(\hat{\mathbf{k}} \cdot \hat{\mathbf{r}}) .$$

Using this fact and the integral equation satisfied by $\psi_{\mathbf{k}}(\mathbf{r})$, which reads

$$\psi_{\mathbf{k}}(\mathbf{r}) = e^{i\mathbf{k}\cdot\mathbf{r}} - \int d\mathbf{r}'\, G_k(\mathbf{r}, \mathbf{r}')\, v(r')\, \psi_{\mathbf{k}}(\mathbf{r}') ,$$

show that the partial wave component $\psi_l(r; k)$ satisfies the one-dimensional integral equation

$$\psi_l(r; k) = j_l(kr) - \int_0^\infty dr'\, r'^2\, G_l(r, r'; k)\, v(r')\, \psi_l(r'; k) .$$

3. Using the asymptotic expansions for the spherical Bessel and Hankel functions, show that

$$\lim_{r\to\infty} \psi_l(r; k) = \frac{1}{2kr} \left[e^{-i[kr - (l+1)\pi/2]} + S_l(k)\, e^{i[kr - (l+1)\pi/2]} \right] ,$$

where the complex number $S_l(k)$ (in fact it will turn out to be a phase factor) is given by

$$S_l(k) = 1 - 2ik \int_0^\infty dr\, r^2\, j_l(kr)\, v(r)\, \psi_l(r; k) .$$

4. By comparing the asymptotic expansion

$$\lim_{r\to\infty} \psi_{\mathbf{k}}(\mathbf{r}) = e^{i\mathbf{k}\cdot\mathbf{r}} + f_{\mathbf{k}}(\hat{\mathbf{k}} \cdot \hat{\mathbf{r}})\, \frac{e^{ikr}}{r}$$

with that following from part 3, show that $S_l(k)$ and the partial wave component $f_l(k)$ of the scattering amplitude, where

$$f_{\mathbf{k}}(\hat{\mathbf{k}} \cdot \hat{\mathbf{r}}) = \sum_l (2l + 1)\, f_l(k)\, P_l(\hat{\mathbf{k}} \cdot \hat{\mathbf{r}}) ,$$

are related via

$$f_l(k) = \frac{S_l(k) - 1}{2ik} , \qquad S_l(k) = 1 + 2ik\, f_l(k) .$$

5. Show that the optical theorem requires that $|S_l(k)| = 1$ and that $S_l(k) = e^{2i\delta_l(k)}$, where $\delta_l(k)$ is the phase shift.

6. Show that

$$\lim_{r \to \infty} \psi_l(r; k) = e^{i\delta_l(k)} \cos \delta_l(k) \left[j_l(kr) - \tan \delta_l(k) \, n_l(kr) \right] .$$

7. Solve by iteration the one-dimensional integral equation for $\psi_l(r; k)$ and obtain an expression for the phase shift in the Born approximation, $\delta_l^{BA}(k)$; assume that $\delta_l^{BA}(k) \ll 1$.

Solution

Part 1

Insert into the expression for the Green's function the expansions for the plane waves:

$$G_k(\mathbf{r}, \mathbf{r}') = \sum_{lm} \sum_{l'm'} \int_0^\infty \frac{dq}{8\pi^3} \frac{q^2}{q^2 - (k + i\eta)^2} \int d\Omega_q \left[4\pi i^l j_l(qr) \, Y_{lm}^*(\Omega_q) \, Y_{lm}(\Omega_r) \right] \left[4\pi (-i)^{l'} j_{l'}(qr') \, Y_{l'm'}^*(\Omega_q) \, Y_{l'm'}^*(\Omega_{r'}) \right]$$

$$= \sum_{lm} Y_{lm}(\hat{\mathbf{r}}) \, Y_{lm}^*(\hat{\mathbf{r}}') \underbrace{\left[\frac{2}{\pi} \int_0^\infty dq \, \frac{q^2}{q^2 - (k + i\eta)^2} \, j_l(qr) j_l(qr') \right]}_{G_l(r, r'; k)} .$$

In order to carry out the contour integration, we extend the lower integration limit to $-\infty$ using the symmetry property of the spherical Bessel function $j_l(-\rho) = (-1)^l j_l(\rho)$,

$$G_l(r, r'; k) = \frac{1}{\pi} \int_{-\infty}^\infty dq \, \frac{q^2}{q^2 - (k + i\eta)^2} \, j_l(qr) j_l(qr') .$$

Assume first that $r' > r$, and express $j_l(qr')$ as a linear combination of spherical Hankel functions; then

$$G_l(r, r'; k) = \frac{1}{2\pi} \int_{-\infty}^\infty dq \, \frac{q^2}{q^2 - (k + i\eta)^2} \, j_l(qr) \left[h_l^{(1)}(qr') + h_l^{(2)}(qr') \right] .$$

We note that, in the asymptotic region, j_l behaves as follows:

$$j_l(qr) \sim A_l(qr) \, e^{iqr} + B_l(qr) \, e^{-iqr} ,$$

whereas the Hankel functions behave as

$$h_l^{(1)}(qr') \sim C_l(qr') \, e^{iqr'} , \qquad h_l^{(2)}(qr') \sim C_l^*(qr') \, e^{-iqr'},$$

where A_l and B_l, and C_l, contain terms in $1/(qr)$ and $1/(qr')$, respectively. For $r' > r$ and q large we have

$$j_l(qr) \, h_l^{(1)}(qr') \sim A_l C_l \, e^{iq(r'+r)} + B_l C_l \, e^{iq(r'-r)} ,$$

and therefore for convergence we need to close the contour in the upper half plane (where the imaginary part of q is greater than 0), picking up the residue at $k + i\eta$. By contrast, in the same regime $r' > r$, we have

$$j_l(qr) \, h_l^{(2)}(qr') \sim A_l C_l^* \, e^{-iq(r'-r)} + B_l C_l^* \, e^{-iq(r'+r)} ,$$

and we must close the contour in the lower half plane (where the imaginary part of q is less than 0), picking up the residue at $-k - i\eta$. We obtain

$$
\begin{aligned}
G_l(r, r'; k) &= \frac{1}{2\pi} \left[2\pi i \left. \frac{q^2 j_l(qr) h_l^{(1)}(qr')}{q + (k + i\eta)} \right|_{k+i\eta} - 2\pi i \left. \frac{q^2 j_l(qr) h_l^{(2)}(qr')}{q - (k + i\eta)} \right|_{-k-i\eta} \right] \\
&= i \left[j_l(kr) h_l^{(1)}(kr') \frac{k^2}{2k} + j_l(-kr) h_l^{(2)}(-kr') \frac{k^2}{2k} \right] \\
&= i \frac{k}{2} \left[j_l(kr) h_l^{(1)}(kr') + (-1)^l j_l(kr) (-1)^l h_l^{(1)}(kr') \right] = ik j_l(kr) h_l^{(1)}(kr') ,
\end{aligned}
$$

where we have used the symmetry properties of the Bessel and Hankel functions. Repeating the above argument for $r > r'$, we arrive at the desired result,

$$
G_l(r, r'; k) = ik j_l(kr_<) h_l^{(1)}(kr_>) .
$$

Part 2

The wave function can be written as

$$
\psi_{\mathbf{k}}(\mathbf{r}) = \sum_l i^l (2l + 1) \psi_l(r; k) P_l(\hat{\mathbf{k}} \cdot \hat{\mathbf{r}}) = 4\pi \sum_{lm} i^l \psi_l(r; k) Y_{lm}^*(\hat{\mathbf{k}}) Y_{lm}(\hat{\mathbf{r}}) .
$$

Inserting the above expansion and that for the Green's function into the integral equation, we find

$$
\psi_{\mathbf{k}}(\mathbf{r}) = \underbrace{4\pi \sum_{lm} i^l j_l(r; k) Y_{lm}^*(\hat{\mathbf{k}}) Y_{lm}(\hat{\mathbf{r}})}_{e^{i\mathbf{k}\cdot\mathbf{r}}} - \int d\mathbf{r}' \underbrace{\left[\sum_{lm} G_l(r, r'; k) Y_{lm}(\hat{\mathbf{r}}) Y_{lm}^*(\hat{\mathbf{r}}') \right]}_{G_k(\mathbf{r},\mathbf{r}')}
$$

$$
v(r') \underbrace{\left[4\pi \sum_{l'm'} i^{l'} \psi_{l'}(r'; k) Y_{l'm'}^*(\hat{\mathbf{k}}) Y_{l'm'}(\hat{\mathbf{r}}') \right]}_{\psi_{\mathbf{k}}(\mathbf{r}')} .
$$

The integration over $d\Omega_{r'}$ is easily carried out using the orthonormality of the spherical harmonics, to obtain

$$
\psi_{\mathbf{k}}(\mathbf{r}) = 4\pi \sum_{lm} i^l \left[j_l(r; k) - \int_0^\infty dr' \, r'^2 \, G_l(r, r'; k) v(r') \psi_l(r'; k) \right] Y_{lm}^*(\hat{\mathbf{k}}) Y_{lm}(\hat{\mathbf{r}}) .
$$

Comparing with the expansion for $\psi_{\mathbf{k}}(\mathbf{r})$ given above, it follows that

$$
\psi_l(r; k) = j_l(r; k) - \int_0^\infty dr' \, r'^2 \, G_l(r, r'; k) v(r') \psi_l(r'; k) ,
$$

and we have reduced the three-dimensional integral equation to a one-dimensional integral equation.

Part 3

In the asymptotic region $r \longrightarrow \infty$, we have

$$
\lim_{r\to\infty} \psi_l(r; k) = \frac{\sin(kr - l\pi/2)}{kr} - ik \underbrace{\frac{e^{i[kr-(l+1)\pi/2]}}{kr}}_{h_l^{(1)}(kr)} \int_0^\infty dr' \, r'^2 j_l(kr') v(r') \psi_l(r'; k) ,
$$

since r' is confined to the region where the potential is non-vanishing, and

$$G_l(r, r'; k) = ik\, h_l^{(1)}(kr) j_l(kr') \qquad r \gg r' .$$

Now use

$$\frac{\sin(kr - l\pi/2)}{kr} = \frac{e^{i(kr-l\pi/2)}}{2ikr} - \frac{e^{-i(kr-l\pi/2)}}{2ikr} = \frac{e^{i[kr-(l+1)\pi/2]}}{2kr} + \frac{e^{-i[kr-(l+1)\pi/2]}}{2kr} ,$$

and therefore, hereafter setting $\varphi_l = kr - (l+1)\pi/2$, we find that

$$\lim_{r\to\infty} \psi_l(r; k) = \frac{e^{i\varphi_l}}{2kr} + \frac{e^{-i\varphi_l}}{2kr} - ik\frac{e^{i\varphi_l}}{kr} \int_0^\infty dr'\, r'^2\, j_l(kr')\, v(r')\, \psi_l(r'; k)$$

$$= \frac{e^{-i\varphi_l}}{2kr} + \frac{e^{i\varphi_l}}{2kr} \underbrace{\left[1 - 2ik \int_0^\infty dr\, r^2 j_l(kr)\, v(r)\, \psi_l(r; k) \right]}_{S_l(k)} .$$

Part 4

In the asymptotic region we have

$$\lim_{r\to\infty} \psi_{\mathbf{k}}(\mathbf{r}) = 4\pi \sum_{lm} \left[i^l \underbrace{\left(\frac{e^{i\varphi_l}}{2kr} + \frac{e^{-i\varphi_l}}{2kr} \right)}_{j_l(kr)} + f_l(k)\, \frac{e^{ikr}}{r} \right] Y_{lm}^*(\hat{\mathbf{k}})\, Y_{lm}(\hat{\mathbf{r}}) .$$

On the other hand, in part 3 we obtained

$$\lim_{r\to\infty} \psi_{\mathbf{k}}(\mathbf{r}) = 4\pi \sum_{lm} i^l \left(\frac{e^{-i\varphi_l}}{2kr} + S_l(k)\, \frac{e^{i\varphi_l}}{2kr} \right) Y_{lm}^*(\hat{\mathbf{k}})\, Y_{lm}(\hat{\mathbf{r}}) .$$

To facilitate a comparison, note that

$$f_l(k)\, \frac{e^{ikr}}{r} = i^{l+1}(-i)^{l+1} k\, f_l(k)\, \frac{e^{ikr}}{kr} = i^{l+1} k\, f_l(k)\, \frac{e^{i[kr-(l+1)\pi/2]}}{kr} = 2i^{l+1} k\, f_l(k)\, \frac{e^{i\varphi_l}}{2kr} ,$$

and therefore

$$\lim_{r\to\infty} \psi_{\mathbf{k}}(\mathbf{r}) = 4\pi \sum_{lm} i^l \left[\left(\frac{e^{i\varphi_l}}{2kr} + \frac{e^{-i\varphi_l}}{2kr} \right) + 2ik\, f_l(k)\, \frac{e^{i\varphi_l}}{2\,kr} \right] Y_{lm}^*(\hat{\mathbf{k}})\, Y_{lm}(\hat{\mathbf{r}})$$

$$= 4\pi \sum_{lm} i^l \left[\frac{e^{-i\varphi_l}}{2kr} + \frac{e^{i\varphi_l}}{2kr} [1 + 2ik\, f_l(k)] \right] Y_{lm}^*(\hat{\mathbf{k}})\, Y_{lm}(\hat{\mathbf{r}}) ,$$

which immediately leads to

$$S_l(k) = 1 + 2ik\, f_l(k) , \qquad f_l(k) = \frac{S_l(k) - 1}{2ik} .$$

Part 5

Write the expansion for the scattering amplitude in terms of $S_l(k)$ rather than $f_l(k)$:

$$f_{\mathbf{k}}(\hat{\mathbf{k}} \cdot \hat{\mathbf{r}}) = \sum_l (2l+1)\, \frac{S_l(k) - 1}{2ik}\, P_l(\hat{\mathbf{k}} \cdot \hat{\mathbf{r}}) \implies f_{\mathbf{k}} = \sum_l (2l+1)\, \frac{S_l(k) - 1}{2ik} ,$$

where $f_{\mathbf{k}}$ denotes the forward scattering amplitude corresponding to $\hat{\mathbf{k}} \cdot \hat{\mathbf{r}} = \cos\theta_{kr} = 1$, and

$$\text{Im}[f_{\mathbf{k}}] = \sum_l (2l+1)\,\text{Im}\left[\frac{S_l(k)-1}{2ik}\right] = \sum_l (2l+1)\,\text{Im}\left[\frac{i-iS_l(k)}{2k}\right] = \sum_l (2l+1)\,\text{Re}\left[\frac{1-S_l(k)}{2k}\right] .$$

On the other hand, the total cross section σ_k is given by

$$\sigma_{\mathbf{k}} = \sum_{ll'} (2l+1)\,(2l'+1)\,\frac{S_l(k)-1}{2ik}\,\frac{S_{l'}^*(k)-1}{-2ik}\int d\Omega_{kr}\,P_l(\theta_{kr})\,P_{l'}(\theta_{kr})$$

$$= 4\pi \sum_l (2l+1)\,\frac{[S_l(k)-1][S_l^*(k)-1]}{4k^2} ,$$

where we have used the orthogonality relation for Legendre polynomials,

$$\int d\Omega_{kr}\,P_l(\theta_{kr})\,P_{l'}(\theta_{kr}) = 2\pi \int_{-1}^{1} dx\,P_l(x)\,P_{l'}(x) = \frac{4\pi}{2l+1}\,\delta_{ll'} .$$

The optical theorem requires that

$$\text{Im}[f_{\mathbf{k}}] = \frac{k}{4\pi}\,\sigma_{\mathbf{k}} \implies \text{Re}\left[\frac{1-S_l(k)}{2k}\right] = \frac{[S_l(k)-1][S_l^*(k)-1]}{4k} ,$$

and hence

$$1 - \text{Re}[S_l(k)] = \frac{|S_l(k)|^2 + 1 - 2\,\text{Re}[S_l(k)]}{2} \implies |S_l(k)|^2 = 1 ;$$

the S-matrix element is just a phase factor. Indeed, since for a central potential

$$f_l(k) = \frac{e^{i\delta_l(k)}\,\sin\delta_l(k)}{k} = \frac{e^{2i\delta_l(k)}-1}{2ik} ,$$

we immediately obtain

$$S_l(k) = e^{2i\delta_l(k)} .$$

Part 6

Start from

$$\lim_{r\to\infty} \psi_l(r;k) = j_l(r;k) - ik\,h_l^{(1)}(kr)\int_0^\infty dr'\,r'^2\,j_l(kr')\,v(r')\,\psi_l(r';k) ,$$

without inserting the asymptotic expansions for the spherical Bessel and Hankel functions. Substituting $h_l^{(1)}(\rho) = j_l(\rho) + i\,n_l(\rho)$, we find

$$\lim_{r\to\infty}\psi_l(r;k) = \underbrace{\left[1 - ik\int_0^\infty dr'\,r'^2\,j_l(kr')\,v(r')\,\psi_l(r';k)\right]}_{[S_l(k)+1]/2} j_l(kr) + \underbrace{\left[k\int_0^\infty dr'\,r'^2\,j_l(kr')\,v(r')\,\psi_l(r';k)\right]}_{i[S_l(k)-1]/2} n_l(kr)$$

$$= \frac{e^{2i\delta_l(k)}+1}{2}\,j_l(kr) - \frac{e^{2i\delta_l(k)}-1}{2i}\,n_l(kr) = e^{i\delta_l(k)}\left[\cos\delta_l(k)\,j_l(kr) - \sin\delta_l(k)\,n_l(kr)\right] ,$$

or

$$\lim_{r\to\infty}\psi_l(r;k) = e^{i\delta_l(k)}\,\cos\delta_l(k)[j_l(kr) - \tan\delta_l(k)\,n_l(kr)] .$$

Part 7

Solve the integral equation

$$\psi_l(r;k) = j_l(r;k) - \int_0^\infty dr'\, r'^2\, G_l(r,r';k)\, v(r')\, \psi_l(r';k)$$

by iteration; thus

$$\psi_l^{(0)}(r;k) = j_l(kr)\,,$$

$$\psi_l^{(1)}(r;k) = -\int_0^\infty dr_1\, r_1^2\, G_l(r,r_1;k)\, v(r_1)\, j_l(kr_1)\,,$$

$$\psi_l^{(2)}(r;k) = \int_0^\infty dr_1\, r_1^2\, G_l(r,r_1;k)\, v(r_1) \int_0^\infty dr_2\, r_2^2\, G_l(r_1,r_2;k)\, v(r_2)\, j_l(kr_2)\,,$$

and so on. In generally, for $n \geq 1$,

$$\psi_l^{(n)}(r;k) = -\int_0^\infty dr'\, r'^2\, G_l(r,r';k)\, v(r')\, \psi_l^{(n-1)}(r';k)\,,$$

with the full solution given by

$$\psi_l(r;k) = \sum_{n=0}^\infty \psi_l^{(n)}(r;k)\,.$$

In the Born approximation, the full wave function is approximated by the leading-order term in the above expansion, and hence

$$S_l^{\text{BA}}(k) = 1 - 2ik \int_0^\infty dr\, r^2\, j_l(kr)\, v(r)\, j_l(kr)\,,$$

which can also be written as

$$e^{i\delta_l^{\text{BA}}(k)} \sin \delta_l^{\text{BA}}(k) = -k \int_0^\infty dr\, r^2\, [j_l(kr)]^2\, v(r)\,.$$

Assuming $\delta_l^{\text{BA}}(k)$ to be small, we have

$$e^{i\delta_l^{\text{BA}}(k)} \sin \delta_l^{\text{BA}}(k) \approx \delta_l^{\text{BA}}(k) = -k \int_0^\infty dr\, r^2\, [j_l(kr)]^2\, v(r)\,.$$

For small k, using $j_l(\rho) = \rho^l/(2l+1)!!$, we obtain a low-energy expansion for the phase shifts:

$$\delta_l^{\text{BA}}(k) = -\frac{k^{(2l+1)}}{[(2l+1)!!]^2} \int_0^\infty dr\, r^{2l+2}\, v(r)\,,$$

in agreement with the general result. Note that the contribution of high-l states is suppressed.

Problem 20   Phase Shift as Integral over Radial Scattering Solution of Schrödinger Equation

Consider two central potentials $V(r)$ and $\widetilde{V}(r)$, both vanishing as $r \longrightarrow \infty$ faster than $1/r$ and neither having a singularity as strong as $1/r^2$ at the origin. We are interested in the scattering solutions $u_{kl}(r)$ and $\widetilde{u}_{kl}(r)$ at the same energy $E = \hbar^2 k^2/(2\mu)$. The solutions $u_l(r;k)$ and $\widetilde{u}_l(r;k)$ satisfy

$$u_l''(r;k) = \left[\frac{l(l+1)}{r^2} + v(r) - k^2\right] u_l(r;k) ,$$

$$\widetilde{u}_l''(r;k) = \left[\frac{l(l+1)}{r^2} + \widetilde{v}(r) - k^2\right] \widetilde{u}_l(r;k) .$$

Take these solutions to be real and to have asymptotic normalizations given by

$$\lim_{r \to \infty} u_l(r;k) = \sin[kr - l\pi/2 + \delta_l(k)] , \qquad \lim_{r \to \infty} \widetilde{u}_l(r;k) = \sin[kr - l\pi/2 + \widetilde{\delta}_l(k)] ,$$

where $\delta_l(k)$ and $\widetilde{\delta}_l(k)$ are the phase shifts.

1. Multiply both sides of the first equation for $u_l(r;k)$ by $\widetilde{u}_l(r;k)$ and both sides of the second equation for $\widetilde{u}_l(r;k)$ by $u_l(r;k)$; subtract the second relation thus obtained from the first. By integrating both sides of this difference between 0 and ∞, show that

$$\sin[\delta_l(k) - \widetilde{\delta}_l(k)] = -\frac{1}{k} \int_0^\infty dr\, \widetilde{u}_l(r;k) [v(r) - \widetilde{v}(r)] u_l(r;k) .$$

2. Set $\widetilde{v}(r) = 0$ (corresponding to a free particle) and show that

$$\sin \delta_l(k) = - \int_0^\infty dr\, r j_l(kr)\, v(r)\, u_{kl}(r) .$$

Is this consistent with part 3 for $S_l(k) = e^{2i\delta_l(k)}$ and part 6 for $\psi_l(r;k) = u_l(r;k)/r$ of Problem 19?

Solution
Part 1

We find

$$\widetilde{u}_l(r;k)\, u_l''(r;k) = \widetilde{u}_l(r;k) \left[\frac{l(l+1)}{r^2} + v(r) - k^2\right] u_l(r;k) ,$$

$$u_l(r;k)\, \widetilde{u}_l''(r;k) = u_l(r;k) \left[\frac{l(l+1)}{r^2} + \widetilde{v}(r) - k^2\right] \widetilde{u}_l(r;k) ,$$

and therefore

$$\widetilde{u}_l(r;k)\, u_l''(r;k) - u_l(r;k)\, \widetilde{u}_l''(r;k) = [v(r) - \widetilde{v}(r)] \widetilde{u}_l(r;k)\, u_l(r;k) .$$

Integrating both sides between 0 and ∞ leads to

$$\int_0^\infty dr\, [\widetilde{u}_l(r;k)\, u_l''(r;k) - u_l(r;k)\, \widetilde{u}_l''(r;k)] = \int_0^\infty dr\, [v(r) - \widetilde{v}(r)] \widetilde{u}_l(r;k)\, u_l(r;k) .$$

Now note that

$$[\widetilde{u}_l(r;k)\, u_l''(r;k) - u_l(r;k)\, \widetilde{u}_l''(r;k)] = \frac{d}{dr} [\widetilde{u}_l(r;k)\, u_l'(r;k) - u_l(r;k)\, \widetilde{u}_l'(r;k)] ,$$

where the contents of the brackets on the right-hand side are known as the Wronskian. We therefore have

$$\underbrace{\widetilde{u}_l(r;k)\, u_l'(r;k) - u_l(r;k)\, \widetilde{u}_l'(r;k) \Big|_0^\infty}_{W} = \int_0^\infty dr\, [v(r) - \widetilde{v}(r)] \widetilde{u}_l(r;k)\, u_l(r;k) .$$

Since the solutions must vanish at the origin, we have

$$W = k \sin[kr - l\pi/2 + \widetilde{\delta}_l(k)] \cos[kr - l\pi/2 + \delta_l(k)]$$
$$-k \sin[kr - l\pi/2 + \delta_l(k)] \cos[kr - l\pi/2 + \widetilde{\delta}_l(k)] = -k \sin[\delta_l(k) - \widetilde{\delta}_l(k)] \,,$$

yielding the desired relation

$$\sin[\delta_l(k) - \widetilde{\delta}_l(k)] = -\frac{1}{k} \int_0^\infty dr\, \widetilde{u}_l(r;k) [v(r) - \widetilde{v}(r)] u_l(r;k) \,.$$

Part 2

Set $\widetilde{v}(r) = 0$ and recall that that $r j_l(kr)$ is the solution of the free-particle radial equation, with asymptotic behavior given by

$$\lim_{r \to \infty} r j_l(kr) = \frac{1}{k} \sin(kr - l\pi/2) \,.$$

Therefore, it is $\widetilde{u}_l(r;k) \longrightarrow kr j_l(kr)$ that has the asymptotic behavior of $\widetilde{u}_l(r;k)$ corresponding to $\widetilde{\delta}(k) = 0$. Thus, we arrive at

$$\sin \delta_l(k) = - \int_0^\infty dr\, r j_l(kr)\, v(r)\, u_l(r;k) \,.$$

This is consistent with what was found in Problem 19, as we will now show. In that problem we had for the radial wave function $\psi_l(r;k)$ the asymptotic behavior

$$\lim_{r \to \infty} \psi_l(r;k) = e^{i\delta_l(k)} [\cos \delta_l(k) j_l(kr) - \sin \delta_l(k) n_l(kr)] = \frac{e^{i\delta_l(k)}}{kr} \sin[kr - l\pi/2 + \delta_l(k)] \,,$$

so that $r \psi_l(r;k) = e^{i\delta_l(k)} u_l(r;k)/k$, and $u_l(r;k)$ has the asymptotic behavior adopted in the present problem. Using the expression for $S_l(k)$, we find

$$e^{2i\delta_l(k)} - 1 = -2ik \int_0^\infty dr\, r j_l(kr)\, v(r)\, \underbrace{\frac{e^{i\delta_l(k)}}{k} u_l(r;k)}_{r\psi_l(r;k)} \,,$$

or

$$e^{i\delta_l(k)} \underbrace{\left[e^{i\delta_l(k)} - e^{-i\delta_l(k)} \right]}_{2i \sin \delta_l(k)} = -2i\, e^{i\delta_l(k)} \int_0^\infty dr\, r j_l(kr)\, v(r)\, u_l(r;k) \,,$$

in agreement with the result found earlier.

Problem 21 Scattering in a Spin-Dependent Potential

Consider a proton and neutron (both spin-1/2 particles) interacting through a spin-dependent potential given by

$$V = V_c(r) + V_\sigma(r)\, \boldsymbol{\sigma}_1 \cdot \boldsymbol{\sigma}_2 \,,$$

where $V_c(r)$ is the central potential, r is the magnitude of the relative position operator and $\boldsymbol{\sigma}_1$ and $\boldsymbol{\sigma}_2$ are the Pauli operators for the proton and neutron, respectively; recall that the spin operator $\mathbf{S}_1$ is $(\hbar/2)\boldsymbol{\sigma}_1$ and similarly for the neutron.

1. Introduce the operators P_0 and P_1, defined as

$$P_0 = \frac{1 - \sigma_1 \cdot \sigma_2}{4} \,, \qquad P_1 = \frac{3 + \sigma_1 \cdot \sigma_2}{4} \,,$$

and show that they are projection operators onto states of pair-spin $S=0$ and $S=1$. Further, show that the potential can be written as

$$V = V_0(r)\, P_0 + V_1(r)\, P_1 \,.$$

2. Show that a spinor eigenfunction of energy E of the proton–neutron pair's (relative) Hamiltonian has the general expansion

$$\psi_E(\mathbf{r}) = c_{00}\, \phi_E^{(0)}(\mathbf{r})\, |00\rangle + \sum_{M=-1}^{1} c_{1M}\, \phi_E^{(1)}(\mathbf{r})\, |1M\rangle \,,$$

where $|SM\rangle$ are pair spin states with $S=0$ or 1 and $M=-S,\ldots,+S$, and $\phi_E^{(S)}(\mathbf{r})$ is the orbital wave function corresponding to pair-spin S. Further, show that the Schrödinger equation for $\psi_E(\mathbf{r})$ decouples into an equation for $\phi_E^{(0)}(\mathbf{r})$ and an equation for $\phi_E^{(1)}(\mathbf{r})$.

3. Now consider a collision between these two particles occurring at center-of-mass energy $E = \hbar^2 k^2/(2\mu)$, where μ is the proton–neutron reduced mass. Assume that before the collision the proton is in spin state $|\uparrow\rangle$ and the neutron in spin state $|\downarrow\rangle$. Show that the scattering (spinor) wave function is given by

$$\psi_k^{\uparrow\downarrow}(\mathbf{r}) = \frac{1}{\sqrt{2}}\, \phi_k^{(0)}(\mathbf{r})|00\rangle + \frac{1}{\sqrt{2}}\, \phi_k^{(1)}(\mathbf{r})|10\rangle \,,$$

where $\phi_k^{(0)}(\mathbf{r})$ and $\phi_k^{(1)}(\mathbf{r})$ are the scattering solutions in the central potentials $V_0(r)$ and $V_1(r)$, respectively. Denoting by $f_{k,0}(\theta)$ and $f_{k,1}(\theta)$ the scattering amplitudes corresponding to the potentials $V_0(r)$ and $V_1(r)$, obtain the differential cross section $\sigma_k(\theta)$ for the spin-exchange collision $\uparrow\downarrow \longrightarrow \downarrow\uparrow$; that is, a collision after which the proton spin state is $|\downarrow\rangle$ and the neutron spin state is $|\uparrow\rangle$.

4. By expanding the scattering amplitudes in partial waves, show that the total spin-exchange cross section is given by

$$\sigma_k = \frac{\pi}{k^2} \sum_{l=0}^{\infty} (2l+1)\, \sin^2(\delta_{kl}^1 - \delta_{kl}^0) \,,$$

where δ_{kl}^0 and δ_{kl}^1 are the phase shifts of the l partial waves corresponding to $f_{k,0}(\theta)$ and $f_{k,1}(\theta)$, respectively.

Solution

Part 1

The operators P_0 and P_1 are obviously hermitian. In order to show that they are projection operators, we need to verify that $P_S^2 = P_S$ with $S=0$ or 1. To this end, we note that

$$(\sigma_1 \cdot \sigma_2)^2 = \sigma_2 \cdot \sigma_2 + i\,\sigma_1 \cdot (\sigma_2 \times \sigma_2) = 3 - 2\,\sigma_1 \cdot \sigma_2 \,,$$

where we have used the Pauli identity

$$\sigma_1 \cdot \mathbf{A}\, \sigma_1 \cdot \mathbf{B} = \mathbf{A} \cdot \mathbf{B} + i\,\sigma_1 \cdot (\mathbf{A} \times \mathbf{B}) \,,$$

with $\mathbf{A} = \mathbf{B} = \sigma_2$, and the angular momentum property

$$\mathbf{S}_2 \times \mathbf{S}_2 = i\hbar\,\mathbf{S}_2 \implies \sigma_2 \times \sigma_2 = 2i\sigma_2 \;.$$

We then find

$$P_0^2 = \frac{1 - 2\,\sigma_1 \cdot \sigma_2 + (\sigma_1 \cdot \sigma_2)^2}{16} = \frac{4 - 4\,\sigma_1 \cdot \sigma_2}{16} = P_0 \;,$$

and similarly for P_1. We also have

$$P_0 + P_1 = 1 \;.$$

Expressing $\sigma_1 \cdot \sigma_2$ in terms of the pair-spin $\mathbf{S} = \mathbf{S}_1 + \mathbf{S}_2$ as

$$\sigma_1 \cdot \sigma_2 = \frac{4}{\hbar^2} \mathbf{S}_1 \cdot \mathbf{S}_2 = \frac{2}{\hbar^2}\left(\mathbf{S}^2 - \frac{3}{2}\hbar^2\right) = \frac{2}{\hbar^2}\mathbf{S}^2 - 3 \;,$$

we can also write

$$P_0 = 1 - \frac{1}{2\hbar^2}\mathbf{S}^2 \;, \qquad P_1 = \frac{1}{2\hbar^2}\mathbf{S}^2 \;,$$

which shows that for states $|SM\rangle$ with pair-spin $S = 0$ or 1 we have

$$\mathbf{S}^2\,|SM\rangle = S(S{+}1)\hbar^2\,|SM\rangle \implies P_0\,|00\rangle = |00\rangle \text{ and } P_1\,|00\rangle = 0 \;, \quad P_0\,|1M\rangle = 0 \text{ and } P_1\,|1M\rangle = |1M\rangle \;,$$

and hence P_0 and P_1 are projection operators onto states with pair-spin 0 and 1, respectively. Lastly, since $P_1 - 3P_0 = \sigma_1 \cdot \sigma_2$, the potential V can be written as

$$V = V_c(r)(P_0 + P_1) + V_\sigma(r)(P_1 - 3P_0) = V_0(r)\,P_0 + V_1(r)\,P_1 \;,$$

where we have defined

$$V_0(r) = V_c(r) - 3V_\sigma(r) \;, \qquad V_1(r) = V_c(r) + V_\sigma(r) \;.$$

Part 2

The center-of-mass Hamiltonian is given by

$$H = \frac{\mathbf{p}^2}{2\mu} + V_c(r) + V_\sigma(r)\,\sigma_1 \cdot \sigma_2 \;.$$

By making use of the properties of projection operators, H can also be written as

$$H = \frac{\mathbf{p}^2}{2\mu}(P_0 + P_1) + V_0(r)\,P_0 + V_1(r)\,P_1 = H_0 P_0 + H_1 P_1 \;,$$

where

$$H_S = \frac{\mathbf{p}^2}{2\mu} + V_S(r) \;, \qquad\qquad S = 0, 1 \;.$$

A general eigenstate of the Hamiltonian satisfies $H|\psi_E\rangle = E\,|\psi_E\rangle$. Such an eigenstate can be expanded in the combined orbital + spin space as

$$|\psi_E\rangle = \sum_{S=0}^{1} \sum_{M=-S}^{S} c_{SM}|SM\rangle \otimes |\phi_E^{(S)}\rangle \;,$$

where $|\phi_E^{(S)}\rangle$ is the orbital state. The latter is independent of M, since the Hamiltonian H_S commutes with $\mathbf{S}$, given that $\boldsymbol{\sigma}_1 \cdot \boldsymbol{\sigma}_2 = (2/\hbar^2)\mathbf{S}^2$. Inserting the above expansion into the eigenstate equation and using the property of the projection operator, we find

$$\sum_{S=0}^{1} \sum_{M=-S}^{S} c_{SM}(H_0 P_0 + H_1 P_1)|SM\rangle \otimes |\phi_E^{(S)}\rangle = c_{00}\,|00\rangle \otimes H_0\,|\phi_E^{(0)}\rangle + \sum_{M=-1}^{1} c_{1M}|1M\rangle \otimes H_1\,|\phi_E^{(1)}\rangle$$

$$= E \sum_{S=0}^{1} \sum_{M=-S}^{S} c_{SM}|SM\rangle \otimes |\phi_E^{(S)}\rangle\,,$$

where H_0 and H_1 act only in the orbital space. Projecting both sides onto $\langle S'M'|$ and dropping the irrelevant constant $c_{SM'}$ (which appears on both the left- and right-hand sides), it follows that, for $S' = 0$ or 1,

$$H_0\,|\phi_E^{(0)}\rangle = E\,|\phi_E^{(0)}\rangle\,, \qquad H_1\,|\phi_E^{(1)}\rangle = E\,|\phi_E^{(1)}\rangle$$

and we see that the eigenvalue equation decouples for the two pair-spin states. In the coordinate representation, the spinor wave function reads

$$\psi_E(\mathbf{r}) = \langle \phi_{\mathbf{r}}|\psi_E\rangle = \sum_{S=0}^{1} \sum_{M=-S}^{S} c_{SM}|SM\rangle\langle \phi_{\mathbf{r}}|\phi_E^{(S)}\rangle = \sum_{S=0}^{1} \sum_{M=-S}^{S} c_{SM}\phi_E^{(S)}(\mathbf{r})\,|SM\rangle\,.$$

Part 3

Since the problem decouples, we can solve the scattering problem separately for $S = 0$ and $S = 1$. For each case, we are dealing with scattering from a central potential. The asymptotic wave functions have the form (with $k = \sqrt{2\mu E/\hbar^2}$)

$$\lim_{r \to \infty} \phi_k^{(S)}(\mathbf{r}) = e^{i\mathbf{k}\cdot\mathbf{r}} + f_{k,S}(\theta)\,\frac{e^{ikr}}{r}\,,$$

where the first term (the plane wave) is the incident wave and the second term (the spherical wave) is the scattered wave. Consider now the spinor wave function $\psi_k^{\uparrow\downarrow}(\mathbf{r})$ given by

$$\psi_k^{\uparrow\downarrow}(\mathbf{r}) = \frac{1}{\sqrt{2}}\,\phi_k^{(0)}(\mathbf{r})\,|00\rangle + \frac{1}{\sqrt{2}}\,\phi_k^{(1)}(\mathbf{r})\,|10\rangle\,,$$

so that in the asymptotic region

$$\lim_{r \to \infty} \psi_k^{\uparrow\downarrow}(\mathbf{r}) = \frac{1}{\sqrt{2}}\left[e^{i\mathbf{k}\cdot\mathbf{r}} + f_{k,0}(\theta)\,\frac{e^{ikr}}{r}\right]\frac{|\uparrow\downarrow\rangle - |\downarrow\uparrow\rangle}{\sqrt{2}} + \frac{1}{\sqrt{2}}\left[e^{i\mathbf{k}\cdot\mathbf{r}} + f_{k,1}(\theta)\,\frac{e^{ikr}}{r}\right]\frac{|\uparrow\downarrow\rangle + |\downarrow\uparrow\rangle}{\sqrt{2}}$$

$$= e^{i\mathbf{k}\cdot\mathbf{r}}\,|\uparrow\downarrow\rangle + \frac{f_{k,1}(\theta) + f_{k,0}(\theta)}{2}\,\frac{e^{ikr}}{r}\,|\uparrow\downarrow\rangle + \frac{f_{k,1}(\theta) - f_{k,0}(\theta)}{2}\,\frac{e^{ikr}}{r}\,|\downarrow\uparrow\rangle\,,$$

and we see that this spinor wave function has an incident wave in which particle 1 (the proton) has spin up and particle 2 (the neutron) has spin down and scattered waves with particle 1 spin up and particle 2 spin down, and vice versa. Note that the interaction can flip the spins of the individual particles, since the Hamiltonian does not commute independently with $\mathbf{S}_1^2$ and $\mathbf{S}_2^2$. We can read off the amplitude for the spin-exchange transition as $[f_{k,1}(\theta) - f_{k,0}(\theta)]/2$, from which we obtain the spin-exchange differential cross section:

$$\sigma_k(\theta) = \frac{|f_{k,1}(\theta) - f_{k,0}(\theta)|^2}{4}\,.$$

Part 4

The amplitudes $f_{k,S}(\theta)$ have the partial wave expansions

$$f_{k,S}(\theta) = \frac{1}{k} \sum_{l=0}^{\infty} (2l+1)\, e^{i\delta_{kl}^S} \sin \delta_{kl}^S\, P_l(\cos\theta)\,,$$

so that (with $x = \cos\theta$)

$$f_{k,1}(\theta) - f_{k,0}(\theta) = \frac{1}{k} \sum_{l=0}^{\infty} (2l+1)\, \underbrace{\left(e^{i\delta_{kl}^1} \sin \delta_{kl}^1 - e^{i\delta_{kl}^0} \sin \delta_{kl}^0 \right)}_{g_{kl}} P_l(x)\,.$$

We can easily carry out the required integration by using the orthogonality of the Legendre polynomials, obtaining

$$\sigma_k = \frac{\pi}{2k^2} \sum_{l,l'} (2l+1)(2l'+1)\, g_{kl}^*\, g_{kl'} \underbrace{\int_{-1}^{1} dx\, P_l(x)\, P_{l'}(x)}_{2/(2l+1)\delta_{l,l'}} = \frac{\pi}{k^2} \sum_{l} (2l+1)|g_{kl}|^2\,.$$

We now consider $|g_{kl}|^2$. First note that

$$g_{kl} = \cos \delta_{kl}^1 \sin \delta_{kl}^1 - \cos \delta_{kl}^0 \sin \delta_{kl}^0 + i(\sin^2 \delta_{kl}^1 - \sin^2 \delta_{kl}^0)\,,$$

and hence

$$\begin{aligned}
|g_{kl}|^2 &= (\cos \delta_{kl}^1 \sin \delta_{kl}^1 - \cos \delta_{kl}^0 \sin \delta_{kl}^0)^2 + (\sin^2 \delta_{kl}^1 - \sin^2 \delta_{kl}^0)^2\\
&= \cos^2 \delta_{kl}^1 \sin^2 \delta_{kl}^1 + \cos^2 \delta_{kl}^0 \sin^2 \delta_{kl}^0 - 2 \cos \delta_{kl}^1 \sin \delta_{kl}^1 \cos \delta_{kl}^0 \sin \delta_{kl}^0 + \sin^4 \delta_{kl}^1 + \sin^4 \delta_{kl}^0\\
&\quad -2 \sin^2 \delta_{kl}^1 \sin^2 \delta_{kl}^0\\
&= \sin^2 \delta_{kl}^1 + \sin^2 \delta_{kl}^0 - 2 \cos \delta_{kl}^1 \sin \delta_{kl}^1 \cos \delta_{kl}^0 \sin \delta_{kl}^0 - 2 \sin^2 \delta_{kl}^1 \sin^2 \delta_{kl}^0\\
&= \sin^2 \delta_{kl}^1(1 - \sin^2 \delta_{kl}^0) + \sin^2 \delta_{kl}^0(1 - \sin^2 \delta_{kl}^1) - 2 \cos \delta_{kl}^1 \sin \delta_{kl}^1 \cos \delta_{kl}^0 \sin \delta_{kl}^0\\
&= \sin^2 \delta_{kl}^1 \cos^2 \delta_{kl}^0 + \sin^2 \delta_{kl}^0 \cos^2 \delta_{kl}^1 - 2 \cos \delta_{kl}^1 \sin \delta_{kl}^1 \cos \delta_{kl}^0 \sin \delta_{kl}^0 = \sin^2(\delta_{kl}^1 - \delta_{kl}^0)\,.
\end{aligned}$$

We obtain the spin-flip cross section as

$$\sigma_k = \frac{\pi}{k^2} \sum_{l} (2l+1)\, \sin^2(\delta_{kl}^1 - \delta_{kl}^0)\,.$$

Problem 22 Phase Shift in the Born Approximation

The goal of this problem is to derive an expression for the phase shift in the Born approximation.

1. By considering the reduced radial equation for $u_{kl}(r)$ in the presence of a central potential $V(r)$ and the reduced radial equation for a free particle $u_{kl}^f(r)$, show that

$$u_{kl}(r)\, u_{kl}^{f\prime}(r) - u_{kl}'(r)\, u_{kl}^f(r) = \frac{2\mu}{\hbar^2} \int_0^r dr'\, u_{kl}^f(r')\, V(r')\, u_{kl}(r')\,,$$

where μ is the mass of the particle and the $\prime$ indicates the derivative with respect to r.

2. Insert the asymptotic expressions for $u_{kl}(r)$ and $u_{kl}^f(r)$ and show that the phase shift in the Born approximation is given by

$$\sin \delta_{kl}^{\mathrm{BA}} \approx \delta_{kl}^{\mathrm{BA}} = -\frac{2\mu}{\hbar^2}\, k \int_0^{\infty} dr\, r^2\, V(r)\, [j_l(kr)]^2\,.$$

Hint: Take the solution $u_{kl}(r)$ to behave asymptotically as $u_{kl}(r) = \sin(kr - l\pi/2 + \delta_{kl})$ and the free solution $u_{kl}^f(r)$ as $kr j_l(kr)$, which matches the asymptotic behavior of $u_{kl}(r)$ in the absence of the potential (that is, for $\delta_{kl} = 0$).

Solution

Part 1

The reduced Schrödinger equations read

$$u_{kl}''(r) = \left[v(r) + \frac{l(l+1)}{r^2} - k^2 \right] u_{kl}(r) ,$$

for the interacting case, and

$$u_{kl}^{f''}(r) = \left[\frac{l(l+1)}{r^2} - k^2 \right] u_{kl}^f(r) ,$$

for the free-particle case. Multiplying the first equation by $u_{kl}^f(r)$ and the second equation by $u_{kl}(r)$, and subtracting one from the other, we find

$$u_{kl}^f(r) \, u_{kl}''(r) - u_{kl}(r) \, u_{kl}^{f''}(r) = v(r) \, u_{kl}(r) \, u_{kl}^f(r) .$$

The left-hand side of this relation can be expressed as

$$\text{l.h.s.} = \frac{dW(r)}{dr} = \frac{d}{dr} \left[u_{kl}^f(r) \, u_{kl}'(r) - u_{kl}(r) \, u_{kl}^{f'}(r) \right] ,$$

from which it follows that

$$\underbrace{u_{kl}^f(r) \, u_{kl}'(r) - u_{kl}(r) \, u_{kl}^{f'}(r)}_{W(r)} = \int_0^r dr' \, v(r') \, u_{kl}(r') \, u_{kl}^f(r') ,$$

after integrating both sides between 0 and r using the boundary condition at the origin, $u_{kl}(0) = u_{kl}^f(0) = 0$.

Part 2

The asymptotic behaviors of $u_{kl}(r)$ and $u_{kl}^f(r)$ are respectively,

$$u_{kl}(r) = \sin(kr - l\pi/2 + \delta_{kl}) , \qquad u_{kl}^f(r) = \sin(kr - l\pi/2) ,$$

and inserting these into the wronskian $W(r)$, we find

$$W = k \sin(kr - l\pi/2) \cos(kr - l\pi/2 + \delta_{lk}) - k \cos(kr - l\pi/2) \sin(kr - l\pi/2 + \delta_{lk})$$
$$= -k \sin \delta_{lk} .$$

Taking the limit $r \longrightarrow \infty$ yields

$$\sin \delta_{lk} = -\frac{1}{k} \int_0^\infty dr \, v(r) \, u_{kl}(r) \, u_{kl}^f(r) .$$

In the Born approximation we have $u_{kl}(r) \approx u_{kl}^f(r)$, and hence substituting $u_{kl}^f(r) = kr j_l(kr)$ we arrive at

$$\sin \delta_{kl}^{\text{BA}} \approx \delta_{kl}^{\text{BA}} = -\frac{2\mu}{\hbar^2} k \int_0^\infty dr \, r^2 \, V(r) \, [j_l(kr)]^2 .$$

Problem 23 Effective Range Theory

The low-energy expansion for the s-wave phase shift can be written as (see Problem 11 for the derivation of the low-energy behavior of phase shifts)

$$k \cot \delta_0(k) = -\frac{1}{a} + \frac{r_s}{2} k^2 + \cdots ,$$

where a is the scattering length and r_s is the effective range. The goal of the present problem is to derive an expression for r_s in terms of zero-energy radial wave functions. It is assumed that the potential is short range (it vanishes for $r \gtrsim R$). Denote by $u_0(r; k)$ the solution of the (radial) Schrödinger equation for the s-wave state,

$$u_0''(r; k) + \left[k^2 - v(r) \right] u_0(r; k) = 0 ,$$

where $v(r) = 2\mu V(r)/\hbar^2$ and $u_0(r; k)$ vanishes at the origin. In the asymptotic region, we normalize it to behave as

$$u_0(r \geq R; k) = \frac{\sin[kr + \delta_0(k)]}{\sin \delta_0(k)} = \cot \delta_0(k) \sin(kr) + \cos(kr) .$$

1. Define the radial function $\phi_0(r; k)$ by

$$\phi_0(r; k) \equiv \frac{\sin[kr + \delta_0(k)]}{\sin \delta_0(k)} ,$$

and write down the equation satisfied by $\phi_0(r; k)$ for $r \geq 0$. Using the expansion for $\cot \delta_0(k)$ given above, show in particular that, at $k = 0$,

$$\phi_0(r; 0) = 1 - \frac{r}{a} .$$

2. Show that

$$\underbrace{u_0(r; k_1) u_0'(r; k_2) - u_0'(r; k_1) u_0(r; k_2) \Big|_0^R}_{W_0} = (k_1^2 - k_2^2) \int_0^R dr \, u_0(r; k_1) u_0(r; k_2)$$

and

$$\underbrace{\phi_0(r; k_1) \phi_0'(r; k_2) - \phi_0'(r; k_1) \phi_0(r; k_2) \Big|_0^R}_{W_0^f} = (k_1^2 - k_2^2) \int_0^R dr \, \phi_0(r; k_1) \phi_0(r; k_2) .$$

3. Show that

$$W_0 - W_0^f = k_2 \cot \delta_0(k_2) - k_1 \cot \delta_0(k_1) ,$$

and hence that

$$k_1 \cot \delta_0(k_1) - k_2 \cot \delta_0(k_2) = (k_1^2 - k_2^2) \int_0^\infty dr \, \left[\phi_0(r; k_1) \phi_0(r; k_2) - u_0(r; k_1) u_0(r; k_2) \right] .$$

Why can the integral be extended to ∞?

4. From the relation obtained above, deduce that

$$k \cot \delta_0(k) = -\frac{1}{a} + k^2 \int_0^\infty dr \, \left[\phi_0(r; k) \phi_0(r; 0) - u_0(r; k) u_0(r; 0) \right] ,$$

and finally that

$$r_s = 2 \int_0^\infty dr \left[\phi_0^2(r;0) - u_0^2(r;0) \right] .$$

Solution
Part 1

Clearly, $\phi_0(r;k)$ is the free-particle solution of the Schrödinger equation; it satisfies, for $r \geq 0$,

$$\phi_0''(r;k) + k^2 \phi_0(r;k) = 0 ,$$

This solution can also be written as

$$\phi_0(r;k) = k \cot \delta_0(k) \frac{\sin(kr)}{k} + \cos(kr) ,$$

so that

$$\lim_{k \to 0} \phi_0(r;k) = \lim_{k \to 0} \left(-\frac{1}{a} + \frac{r_s}{2} k^2 + \cdots \right) \left(r - \frac{k^2 r^3}{6} + \cdots \right) + \left(1 - \frac{k^2 r^2}{2} + \cdots \right) = 1 - \frac{r}{a} .$$

Also note that the free-particle wave function, as defined above, is such that $\phi_0(0;k) = 1$.

Part 2

Multiplying the equation for $u_0(r;k_1)$ by $u_0(r;k_2)$ and vice versa, we obtain

$$u_0(r;k_2) u_0''(r;k_1) = u_0(r;k_2) \left[v(r) - k_1^2 \right] u_0(r;k_1) ,$$

and

$$u_0(r;k_1) u_0''(r;k_2) = u_0(r;k_1) \left[v(r) - k_2^2 \right] u_0(r;k_2) .$$

Subtracting the first relation from the second,

$$\underbrace{u_0(r;k_1) u_0''(r;k_2) - u_0(r;k_2) u_0''(r;k_1)}_{[u_0(r;k_1) u_0'(r;k_2) - u_0(r;k_2) u_0'(r;k_1)]'} = \left(k_1^2 - k_2^2 \right) u_0(r;k_1) u_0(r;k_2) ,$$

and then integrating both sides from 0 to R yields

$$\underbrace{u_0(r;k_1) u_0'(r;k_2) - u_0(r;k_2) u_0'(r;k_1) \Big|_0^R}_{W_0} = \left(k_1^2 - k_2^2 \right) \int_0^R dr\, u_0(r;k_1) u_0(r;k_2) .$$

Identical manipulations for the free-particle Schrödinger equation lead to a similar relation but with $u_0(r;k)$ replaced by $\phi_0(r;k)$ and the Wronskian W_0 replaced by the corresponding W_0^f.

Part 3

The boundary conditions satisfied by $u(r;k)$ at the origin and in the asymptotic region, respectively, $u_0(0;k) = 0$ and $u_0(r \geq R;k) = \phi_0(r;k)$, give for the Wronskian W_0

$$W_0 = \phi_0(R;k_1) \phi_0'(R;k_2) - \phi_0'(R;k_1) \phi_0(R;k_2) ,$$

so that the difference $W_0 - W_0^f$ follows:

$$W_0 - W_0^f = \phi_0(0; k_1)\, \phi_0'(0; k_2) - \phi_0'(0; k_1)\, \phi_0(0; k_2) = \phi_0'(0; k_2) - \phi_0'(0; k_1)\,,$$

where we have used $\phi_0(0; k) = 1$. Now, we observe that

$$\phi_0'(r; k) = k\, \frac{\cos[kr + \delta_0(k)]}{\sin \delta_0(k)}\,,$$

and hence

$$W_0 - W_0^f = k_2 \cot \delta_0(k_2) - k_1 \cot \delta_0(k_1)\,.$$

We arrive at

$$k_2 \cot \delta_0(k_2) - k_1 \cot \delta_0(k_1) = \left(k_1^2 - k_2^2\right) \int_0^R dr \left[u_0(r; k_1)\, u_0(r; k_2) - \phi_0(r; k_1)\, \phi_0(r; k_2) \right]\,,$$

and the upper integration limit can be extended to ∞, since the integrand vanishes for $r \geq R$. This is the desired relation.

Part 4

We take the limit $k_2 \longrightarrow 0$, and find (with $k_1 \longrightarrow k$)

$$-\frac{1}{a} - k \cot \delta_0(k) = k^2 \int_0^R dr \left[u_0(r; k)\, u_0(r; 0) - \phi_0(r; k)\, \phi_0(r; 0) \right]\,,$$

or

$$k \cot \delta_0(k) = -\frac{1}{a} + k^2 \int_0^\infty dr \left[\phi_0(r; k)\, \phi_0(r; 0) - u_0(r; k)\, u_0(r; 0) \right]\,.$$

We have made no approximations in obtaining the relation above; it is exact. In the limit $k \longrightarrow 0$, we obtain

$$-\frac{1}{a} + \frac{r_s}{2} k^2 + \cdots = -\frac{1}{a} + k^2 \int_0^\infty dr \left[\phi_0^2(r; 0) - u_0^2(r; 0) \right] + \cdots$$

from which we deduce that

$$r_s = 2 \int_0^\infty dr \left[\phi_0^2(r; 0) - u_0^2(r; 0) \right]\,,$$

which is the desired relation. Note that the effective range parameter r_s provides a measure of the difference between the free-particle and full wave function in the interaction region $r < R$.

Problem 24 Phase Shift in the High-Energy Approximation

The goal of this problem is to derive an approximation for the phase shift at high energy.

1. The asymptotic behavior of the solution of the reduced radial equation is given in Eqs. (15.43) and (15.55); factoring out $i^l(2l + 1)$, it can be written as namely

$$u(r) = \frac{e^{i\delta_{kl}}}{k}\, \sin(kr - l\pi/2 + \delta_{kl})\,.$$

It can also be written as

$$u(r) = \frac{1}{2k} \left[e^{2i\delta_{kl}} e^{i[kr - \pi(l+1)/2]} + e^{-i[kr - \pi(l+1)/2]} \right] .$$

Here we have dropped the subscript kl from $u(r)$ for brevity.

2. Following the WKB method, write the two independent solutions of the reduced radial equation as $u = e^{\pm ik\phi(r)}$, and show that $\phi(r)$ satisfies the following differential equation:

$$\pm ik\phi''(r) - k^2[\phi'(r)]^2 + k^2 - v(r) - \frac{l(l+1)}{r^2} = 0 ,$$

where the $\prime$ indicates a derivative with respect to r.

3. Show that in the limit $k \longrightarrow \infty$ (the high-energy limit), the term proportional to $\phi''(r)$ can be neglected if

$$\left| \frac{\phi''(r)}{k[\phi'(r)]^2} \right| = \frac{1}{2k^3} \frac{|v'_{\text{eff}}(r)|}{[1 - v_{\text{eff}}(r)/k^2]^{3/2}} \ll 1 ,$$

where we define

$$v_{\text{eff}}(r) = v(r) + \frac{l(l+1)}{r^2} .$$

The requirement above is satisfied when

$$\frac{|v'_{\text{eff}}(r)|}{k^3} \ll 1 , \qquad \frac{|v_{\text{eff}}(r)|}{k^2} \ll 1 .$$

Interpret these conditions.

4. Assume that the requirement above is satisfied. Show that $\phi(r)$ is then obtained as

$$\phi(r) = \phi(r_0) + \int_{r_0}^r dr' \left[1 - \frac{v_{\text{eff}}(r')}{k^2} \right]^{1/2} = \phi(r_0) + r - r_0 + \int_{r_0}^r dr' \left\{ \left[1 - \frac{v_{\text{eff}}(r')}{k^2} \right]^{1/2} - 1 \right\} .$$

5. Take r_0 to be the classical inversion point at which $V_{\text{eff}}(r_0) = E$, or equivalently $v_{\text{eff}}(r_0) = k^2$. In the high-energy limit that we are considering, this inversion point will be close to the origin. With the requirement $u(r_0) = 0$, show that

$$u(r) = a \left[e^{ik[\phi(r) - \phi(r_0)]} - e^{-ik[\phi(r) - \phi(r_0)]} \right] ,$$

where a is an overall normalization constant.

6. By comparing the solution above with that obtained in part 1 in terms of outgoing and ingoing spherical waves, obtain the following expression for the phase shift:

$$\delta_{kl} = l\frac{\pi}{2} - kr_0 + k \int_{r_0}^\infty dr \left\{ \left[1 - \frac{v_{\text{eff}}(r)}{k^2} \right]^{1/2} - 1 \right\} .$$

7. In the absence of the potential, the phase shift should vanish. Show that the expression found above **does not** satisfy this requirement (you may use tables of integrals). Show, however, that the latter requirement can indeed be satisfied by the following *ad hoc* replacements: $l(l+1) \longrightarrow (l + 1/2)^2$ in $v_{\text{eff}}(r)$ and $l\pi/2 \longrightarrow (l + 1/2)\pi/2$, namely

$$\delta_{kl}^{\text{WKB}} = \left(l + \frac{1}{2} \right) \frac{\pi}{2} - kr_0 + k \int_{r_0}^\infty dr \left\{ \left[1 - \frac{v(r) + (l + 1/2)^2/r^2}{k^2} \right]^{1/2} - 1 \right\} .$$

The above WKB approximation turns out to be very good at high k.

Solution

Part 1

We easily obtain

$$u(r) = \frac{e^{i\delta_{kl}}}{2ik}\left[e^{i(kr-l\pi/2+\delta_{kl})} - e^{-i(kr-l\pi/2+\delta_{kl})}\right] = \frac{e^{i(\delta_{kl}-\pi/2)}}{2k}\left[e^{i(kr-l\pi/2+\delta_{kl})} - e^{-i(kr-l\pi/2+\delta_{kl})}\right]$$

$$= \frac{1}{2k}\left[e^{2i\delta_{kl}}\, e^{i[kr-(l+1)\pi/2]} - e^{-i\pi/2}\, e^{-i(kr-l\pi/2)}\right] = \frac{1}{2k}\left[e^{2i\delta_{kl}}\, e^{i[kr-(l+1)\pi/2]} + e^{-i[kr-(l+1)\pi/2]}\right] \, .$$

Part 2

We find

$$u' = \pm\, ik\phi'\, e^{\pm ik\phi} \, , \qquad u'' = \left(\pm\, ik\phi'' - k^2\phi'^2\right) e^{\pm ik\phi} \, ,$$

and substituting into the reduced radial equation it follows that

$$\left(\pm\, ik\phi'' - k^2\phi'^2\right) e^{\pm ik\phi} + \left(k^2 - v_{\text{eff}}\right) e^{\pm ik\phi} = 0 \implies \pm\, ik\phi'' - k^2\phi'^2 + k^2 - v_{\text{eff}} = 0 \, ,$$

where hereafter we define

$$v_{\text{eff}}(r) = v(r) + \frac{l(l+1)}{r^2} \, .$$

Part 3

Ignoring the term proportional to the second derivative gives

$$\phi' = \pm\left(1 - v_{\text{eff}}/k^2\right)^{1/2} \implies \phi'' = \mp\frac{1}{2k^2}\, v_{\text{eff}}'\left(1 - v_{\text{eff}}/k^2\right)^{-1/2} \, ,$$

and hence

$$\left|\frac{\phi''}{k\phi'^2}\right| = \frac{1}{2k^2}\, \frac{|v_{\text{eff}}'|\left(1 - v_{\text{eff}}/k^2\right)^{-1/2}}{k(1 - v_{\text{eff}}/k^2)} = \frac{1}{2k^3}\, \frac{|v_{\text{eff}}'|}{(1 - v_{\text{eff}}/k^2)^{3/2}} \ll 1 \, .$$

This requires (i) $|v_{\text{eff}}|/k^2 \ll 1$, which says that the incident energy must be much larger than the effective potential energy and (ii) $|v_{\text{eff}}'|/k^3 \ll 1$, implying that the effective-potential change is small over a wavelength $\lambda = 2\pi/k$ of the incident wave.

Part 4

After neglecting the second derivative term, a straightforward integration of the differential equation leads to

$$\phi' = \pm\left(1 - v_{\text{eff}}/k^2\right)^{1/2} \implies \phi(r) = \phi(r_0) + \int_{r_0}^{r} dr'\left[1 - \frac{v_{\text{eff}}(r')}{k^2}\right]^{1/2} \, ,$$

where we have dropped the $\pm$ symbol and included these signs into the solution for $u(r)$, which is thus written as $e^{\pm ik\phi(r)}$. Adding and subtracting 1 in the integral, namely

$$\int_{r_0}^{r} dr'\left[1 + \left(1 - \frac{v_{\text{eff}}}{k^2}\right)^{1/2} - 1\right] = r - r_0 + \int_{r_0}^{r} dr'\left[\left(1 - \frac{v_{\text{eff}}}{k^2}\right)^{1/2} - 1\right]$$

yields the required result.

Part 5

The general solution reads

$$u(r) = A\,e^{ik\phi(r)} + B\,e^{-ik\phi(r)} \,,$$

and imposing the boundary condition at the classical inversion point r_0 yields

$$u(r_0) = 0 \implies B/A = -e^{2ik\phi(r_0)} \,,$$

and hence

$$u(r) = A\left[e^{ik\phi(r)} - e^{-ik\phi(r)+2ik\phi(r_0)}\right] = \underbrace{A\,e^{ik\phi(r_0)}}_{a}\left[e^{ik[\phi(r)-\phi(r_0)]} - e^{-ik[\phi(r)-\phi(r_0)]}\right].$$

Part 6

In the asymptotic region $r \longrightarrow \infty$ we set

$$\phi(r) - \phi(r_0) = r - r_0 + \int_{r_0}^{\infty} dr\left[\left(1 - \frac{v_{\text{eff}}}{k^2}\right)^{1/2} - 1\right] \equiv r + R$$

and

$$\begin{aligned}
u(r) &= a\left[e^{ik(r+R)} - e^{-ik(r+R)}\right] = a\,e^{-ikR}\left[e^{ikr}\,e^{2ikR} - e^{-ikr}\right] \\
&= a\,e^{-ikR}\left[e^{i[kr-(l+1)\pi/2]}\,e^{i(l+1)\pi/2}\,e^{2ikR} - e^{-i[kr-(l+1)\pi/2]}\,e^{-i(l+1)\pi/2}\right] \\
&= a\,e^{-ikR}\,e^{-i(l+1)\pi/2}\left[e^{i[kr-(l+1)\pi/2]}\,e^{i(l+1)\pi}\,e^{2ikR} - e^{-i[kr-(l+1)\pi/2]}\right] \\
&= -a\,e^{-ikR}\,e^{-i(l+1)\pi/2}\left[e^{i[kr-(l+1)\pi/2]}\,e^{i(2kR+l\pi)} + e^{-i[kr-(l+1)\pi/2]}\right] \implies 2\delta = 2kR + l\pi \,,
\end{aligned}$$

which yields (reinstating subscripts and arguments)

$$\delta_{kl} = l\,\frac{\pi}{2} - kr_0 + k\int_{r_0}^{\infty} dr\left\{\left[1 - \frac{v_{\text{eff}}(r)}{k^2}\right]^{1/2} - 1\right\} \,.$$

Part 7

In the absence of the potential, the integral in the expression for δ_{kl} reduces to

$$I = k\int_{r_0}^{\infty} dr\left\{\left[1 - \frac{l(l+1)}{(kr)^2}\right]^{1/2} - 1\right\} \,,$$

where r_0 is the inversion point given by

$$\frac{l(l+1)}{r_0^2} = k^2 \implies r_0 = \frac{\sqrt{l(l+1)}}{k}.$$

Introduce the variable $\rho = kr$ to obtain

$$I = \int_{\rho_0}^{\infty} d\rho\left[\left(1 - \frac{\rho_0^2}{\rho^2}\right)^{1/2} - 1\right] \,,$$

and then make the substitution

$$\sin\theta = \frac{\rho_0}{\rho}, \qquad d\theta\cos\theta = -d\rho\,\frac{\rho_0}{\rho^2},$$

to find

$$I = -\int_{\pi/2}^{0} d\theta\cos\theta\,\frac{\rho^2}{\rho_0}(\cos\theta - 1) = \rho_0\int_0^{\pi/2} d\theta\,\frac{\cos\theta}{\sin^2\theta}(\cos\theta - 1) = -\rho_0\int_0^{\pi/2} d\theta\,\frac{\cos\theta}{1 + \cos\theta}$$

$$= -\rho_0\int_0^{\pi/2} d\theta\,\frac{2\cos^2\theta/2 - 1}{2\cos^2\theta/2} = -\rho_0\left(\frac{\pi}{2} - \frac{1}{2}\int_0^{\pi/2} d\theta\,\frac{1}{\cos^2\theta/2}\right) = \rho_0\left(1 - \frac{\pi}{2}\right),$$

where in the last step we used

$$2\frac{d}{d\theta}\tan\theta/2 = \frac{1}{\cos^2\theta/2}.$$

In this limit, the phase shift becomes, with $r_0 = \sqrt{l(l+1)}/k$,

$$\delta_{kl} = l\frac{\pi}{2} - kr_0 + kr_0\left(1 - \frac{\pi}{2}\right) = \left[l - \sqrt{l(l+1)}\right]\frac{\pi}{2} \neq 0.$$

However, under the replacements $l(l+1) \longrightarrow (l+1/2)^2$ in the centrifugal term and $l\pi/2 \longrightarrow (l+1/2)\pi/2$ in δ_{kl}, the phase shift does vanish in the free-particle case, and the expression for δ_{kl}^{WKB} given in the text of the problem is obtained.

Problem 25 Eikonal Approximation for the Scattering Amplitude

The goal of this problem is to derive the high-energy eikonal (or Glauber) approximation to scattering. The starting point is the expression for the phase shift obtained in the WKB approximation (see Problem 24),

$$\delta_{kl}^{\text{WKB}} = \left(l + \frac{1}{2}\right)\frac{\pi}{2} - kr_0 + k\int_{r_0}^{\infty} dr\left\{\left[1 - \frac{v(r) + (l+1/2)^2/r^2}{k^2}\right]^{1/2} - 1\right\},$$

where r_0 is the classical inversion point given by

$$v(r_0) + \frac{(l+1/2)^2}{r_0^2} = k^2.$$

Following the semiclassical treatment of scattering by a short-range potential briefly discussed in Section 15.3, after Eq. (15.60), define the impact parameter b as

$$|\mathbf{L}| = pr\sin\theta = pb \implies (l+1/2)\hbar = \hbar kb \qquad \text{or} \qquad b = \frac{l+1/2}{k}.$$

1. In the limit $|v(r)|/k^2 \ll 1$ show that the classical inversion point can be approximated as

$$r_0 = b\left[1 + \frac{v(b)}{2k^2} + \cdots\right],$$

and that

$$\delta_{kl}^{\text{WKB}} = -\frac{1}{2k}\int_b^{\infty} dr\,\frac{r}{\sqrt{r^2 - b^2}}\,v(r).$$

The following Taylor expansion may be useful:

$$\int_{a+\eta}^{b} dx \, f(x) = \int_{a}^{b} dx \, f(x) - \eta f(a) + \cdots ,$$

where it is assumed that $\eta \ll |a|$.

2. Express r^2 as $b^2 + z^2$ and from δ_{kl}^{WKB} obtain the expression, denoted as δ_{kl}^{G} for the Glauber (or eikonal) approximation,

$$\delta_{kl}^{\text{G}} = -\frac{1}{4k} \int_{-\infty}^{\infty} dz \, v[(b^2 + z^2)^{1/2}] .$$

Thus the phase shift in the high-energy limit, where $|v(r)|/k^2 \ll 1$, follows simply by integrating $v(r)$ over a straight-line eikonal trajectory at impact parameter $b = (l + 1/2)/k$. Note that the l dependence in δ_{kl}^{G} comes in via b.

3. Show that the scattering amplitude, generally given by

$$f_{k\hat{z}}(\theta) = \frac{1}{k} \sum_{l=0}^{\infty} (2l + 1) \, e^{i\delta_{kl}} \sin \delta_{kl} \, P_l(\cos \theta) ,$$

can be expressed in this high-energy limit as

$$f_{k\hat{z}}(\theta) = -ik \int_0^{\infty} db \, b J_0(qb) \left[e^{i\chi_k(b)} - 1 \right] , \qquad \chi_k(b) = -\frac{1}{2k} \int_{-\infty}^{\infty} dz \, v[(b^2 + z^2)^{1/2}] ,$$

where $J_0(x)$ is the (cylindrical) Bessel function of order zero and $q^2 = 2k^2(1 - \cos \theta)$ is the momentum transfer. To obtain this result, use the following hint,

$$\sum_{l=0}^{\infty} \longrightarrow \int_0^{\infty} dl \longrightarrow k \int_0^{\infty} db ,$$

and Heine's relation

$$\lim_{l\to\infty} P_l[1 - x^2/(2l^2)] = J_0(x) .$$

Solution

Part 1

From the classical inversion-point relationship, we find

$$r_0 = \frac{l + 1/2}{k} \frac{1}{\sqrt{1 - v(r_0)/k^2}} = b \left[1 + \frac{v(r_0)}{2k^2} + \cdots \right] = b \left[1 + \frac{v(b)}{2k^2} + \cdots \right] ,$$

where in the last step the argument of the potential has been approximated by b, which is correct up to terms proportional to $1/k^2$. Replacing $l + 1/2$ by kb and using the above expansion in the expression for δ_{kl}^{WKB}, we obtain

$$\delta_{kl}^{\text{WKB}} = \frac{\pi}{2} kb - kb \left[1 + \frac{v(b)}{2k^2} \right] + k \int_{b+b\,v(b)/(2k^2)}^{\infty} dr \left\{ \left[1 - \frac{b^2}{r^2} - \frac{v(r)}{k^2} \right]^{1/2} - 1 \right\} .$$

Expanding the integrand in $|v(r)|/k^2 \ll 1$ yields

$$\left[1 - \frac{b^2}{r^2} - \frac{v(r)}{k^2}\right]^{1/2} = \left(1 - \frac{b^2}{r^2}\right)^{1/2}\left[1 - \frac{r^2}{r^2 - b^2}\frac{v(r)}{k^2}\right]^{1/2}$$

$$\approx \left(1 - \frac{b^2}{r^2}\right)^{1/2}\left[1 - \frac{r^2}{r^2 - b^2}\frac{v(r)}{2k^2}\right] = \left(1 - \frac{b^2}{r^2}\right)^{1/2} - \frac{r}{\sqrt{r^2 - b^2}}\frac{v(r)}{2k^2}\,.$$

The integral becomes

$$\int_{b+bv(b)/(2k^2)}^{\infty} dr \cdots = \int_{b+bv(b)/(2k^2)}^{\infty} dr\left[\left(1 - \frac{b^2}{r^2}\right)^{1/2} - 1\right] - \int_{b+bv(b)/(2k^2)}^{\infty} dr\,\frac{r}{\sqrt{r^2 - b^2}}\frac{v(r)}{2k^2}$$

$$\approx \int_{b}^{\infty} dr\,\underbrace{\left[\left(1 - \frac{b^2}{r^2}\right)^{1/2} - 1\right]}_{f(r)} - b\frac{v(b)}{2k^2}f(b) - \int_{b}^{\infty} dr\,\frac{r}{\sqrt{r^2 - b^2}}\frac{v(r)}{2k^2}$$

$$= b\left(1 - \frac{\pi}{2}\right) + b\frac{v(b)}{2k^2} - \int_{b}^{\infty} dr\,\frac{r}{\sqrt{r^2 - b^2}}\frac{v(r)}{2k^2}\,,$$

where in the second line we made use of the hint and the property

$$f(r) = \left(1 - \frac{b^2}{r^2}\right)^{1/2} - 1 \implies f(b) = -1\,.$$

Further, since we are keeping only terms proportional to $1/k^2$, and the last integral in the second line is already of order $1/k^2$, it is correct to make the approximation

$$\int_{b+bv(b)/(2k^2)}^{\infty} dr \cdots \approx \int_{b}^{\infty} dr \cdots\,.$$

In the third line we have carried out the integral (see the previous problem),

$$\int_{b}^{\infty} dr\left[\left(1 - \frac{b^2}{r^2}\right)^{1/2} - 1\right] = b\left(1 - \frac{\pi}{2}\right)\,.$$

Inserting these results into the expression for $\delta_{kl}^{\mathrm{WKB}}$ leads to

$$\delta_{kl}^{\mathrm{WKB}} = \underbrace{\frac{\pi}{2}kb - kb\left[1 + \frac{v(b)}{2k^2}\right] + kb\left(1 - \frac{\pi}{2}\right) + kb\frac{v(b)}{2k^2}}_{\text{cancel}} - \frac{1}{2k}\int_{b}^{\infty} dr\,\frac{r}{\sqrt{r^2 - b^2}}v(r)$$

and finally

$$\delta_{kl}^{\mathrm{WKB}} = -\frac{1}{2k}\int_{b}^{\infty} dr\,\frac{r}{\sqrt{r^2 - b^2}}v(r)\,.$$

Part 2

Change variables by setting $r^2 = b^2 + z^2$, so that $r\,dr = z\,dz$ and

$$\delta_{kl}^{\mathrm{G}} = -\frac{1}{2k}\int_{0}^{\infty} dz\,z\,\frac{1}{z}v(\sqrt{b^2 + z^2}) = -\frac{1}{4k}\int_{-\infty}^{\infty} dz\,v(\sqrt{b^2 + z^2}) \qquad \text{with } b = \frac{l + 1/2}{k}\,.$$

We see that the phase shift in the Glauber approximation results, up to an overall factor $-1/(4k)$, from integrating the potential along the straight-line trajectory at impact parameter b.

Part 3

Note

$$e^{i\delta_{kl}^G} \sin \delta_{kl}^G = \frac{1}{2i} \left(e^{2i\delta_{kl}^G} - 1 \right) = \frac{1}{2i} \left[e^{i\chi_k(b)} - 1 \right] ,$$

where we have defined

$$\chi_k(b) \equiv 2\,\delta_{kl}^G = -\frac{1}{2k} \int_{-\infty}^{\infty} dz\, v(\sqrt{b^2 + z^2}) .$$

The scattering amplitude in the Glauber approximation then reads

$$f_{k\hat{z}}(\theta) = \sum_{l=0}^{\infty} (2l + 1) \frac{e^{i\chi_k(b)} - 1}{2ik} P_l(\cos\theta) = -i \sum_{l=0}^{\infty} \frac{l + 1/2}{\underbrace{k}_{b}} \left[e^{i\chi_k(b)} - 1 \right] P_l(\cos\theta) .$$

At high energy, many partial waves contribute, and the sum over l can be replaced by an integral, as suggested in the hint, to obtain

$$f_{k\hat{z}}(\theta) = -ik \int_0^{\infty} db\, b \left[e^{i\chi_k(b)} - 1 \right] P_{kb-1/2}(\cos\theta) ,$$

where the order l of the Legendre polynomial is expressed as $kb - 1/2$. The Legendre polynomials depend on $\cos\theta$, which can be expressed as

$$\cos\theta = 1 - \frac{q^2}{2k^2} = 1 - \frac{(qb)^2}{2(kb)^2} .$$

In the limit $kb \longrightarrow \infty$ (the high-energy limit), we have

$$\lim_{kb\to\infty} P_{kb-1/2}(\cos\theta) = \lim_{kb\to\infty} P_{kb}(\cos\theta) = \lim_{kb\to\infty} P_{kb}\left[1 - \frac{(qb)^2}{2(kb)^2} \right] = J_0(qb) ,$$

and hence the desired result,

$$f_{k\hat{z}}(\theta) = -ik \int_0^{\infty} db\, b \left[e^{i\chi_k(b)} - 1 \right] J_0(qb) .$$

Note that it is exact in the high-energy limit.

In this chapter we consider symmetry transformations. A symmetry transformation on a system is generally a transformation that does not change the laws of nature as they apply to that system. In the specific context of quantum mechanics, the previous statement does not mean that the states and operators of a physical system are invariant under a symmetry transformation, but rather that the new states and new operators obey the same laws as the old states and operators. As an example, suppose that $\hat{A}$ is an observable. Then, a rotation of the whole system will map $\hat{A}$ into a new observable $\hat{A}'$. The statement that this is a symmetry transformation entails that the observables $\hat{A}$ and $\hat{A}'$ have the same spectrum, namely, the same eigenvalues since these are the only results that can be measured experimentally and that the eigenstates of each eigenvalue of $\hat{A}'$ are transforms of the eigenstates belonging to the same eigenvalue of $\hat{A}$. In particular, the probability of measuring a given eigenvalue a with the system in state $|\psi\rangle$ is invariant, so that $|\langle\varphi_a|\psi\rangle|^2 = |\langle\varphi_a'|\psi'\rangle|^2$, where $|\varphi_a'\rangle$ and $|\psi'\rangle$ are the transforms of the eigenstate $|\varphi_a\rangle$ and the state $|\psi\rangle$.

Symmetry transformations in quantum mechanics can be characterized generally as those transformations that leave transition probabilities unchanged, specifically

$$|\langle\phi|\psi\rangle|^2 = |\langle\phi'|\psi'\rangle|^2 \,, \tag{16.1}$$

for any two states $|\phi\rangle$ and $|\psi\rangle$. The mapping $|\psi\rangle \longmapsto |\psi'\rangle$ can be thought of as being effected by an operator $\hat{T}$:

$$|\psi'\rangle = \hat{T}|\psi\rangle \,, \tag{16.2}$$

and obviously $\hat{T}$ must be invertible. If $\hat{O}$ is a generic operator and $|\varphi\rangle = \hat{O}|\psi\rangle$, then after a symmetry transformation we must have $|\varphi'\rangle = \hat{O}'|\psi'\rangle$, which leads to the following transformation law for operators:

$$\hat{O} = \hat{T}^{-1}\hat{O}'\hat{T} \quad\text{or}\quad \hat{O}' = \hat{T}\hat{O}\hat{T}^{-1} \,. \tag{16.3}$$

Recalling that a unitary operator $\hat{U}$ is defined by the property that it preserves inner products, that is, $(U\phi, U\psi) = (\phi, \psi)$ for any two states $|\phi\rangle$ and $|\psi\rangle$, one would be tempted to conclude, erroneously, that $\hat{T}$ must be a unitary operator, since this would ensure that the requirement of Eq. (16.1) is automatically satisfied.[1] In fact, a theorem originally due to Wigner shows that $\hat{T}$ is either unitary or anti-unitary – an anti-unitary operator $\hat{\Omega}$ is an invertible operator that satisfies $(\Omega\phi, \Omega\psi) = (\phi, \psi)^*$ (anti-unitary operators are discussed in Section 16.3.1); in either case, however, we have $\hat{T}^{-1} = \hat{T}^\dagger$.

Symmetry transformations that can be implemented starting from the identity transformation and by varying continuously a parameter (or set of parameters) – that is, by a sequence of infinitesimal transformations – are known as continuous transformations. Important examples are space and time translations and rotations. In a space translation, the parameters are the components of the displacement **a** by which we translate the whole system. In a rotation, the parameters are the angle φ

---

[1] Incidentally, when the ket–bra notation is ambiguous, we will denote inner products as (ϕ, ψ), matrix elements as $(\phi, A\psi)$, etc.

and direction $\hat{\mathbf{n}}$ of the axis about which we rotate the system by φ. These transformations are induced by unitary operators. However, there are also transformations, known as discrete transformations, that cannot be implemented via the continuous variation of a parameter (or set of parameters). Examples of these are space inversion, or parity, and time reversal: the former is induced by a unitary operator and the latter by an anti-unitary operator.

We begin by discussing continuous transformations; they form a *group*.[2] Let $\hat{U}_T(\lambda)$ be the operator representing a continuous symmetry transformation characterized by a (real) parameter λ. In the limit in which λ is an infinitesimal, η, we expand $\hat{U}_T(\eta)$ up to linear terms in η as follows:

$$\hat{U}_T(\eta) = \hat{\mathbb{1}} - i\eta\,\hat{T} + \cdots , \qquad (16.4)$$

where the imaginary unit is introduced for later convenience and the operator $\hat{T}$ is independent of η. Continuous transformations must necessarily be represented by unitary rather than anti-unitary operators, since the identity is obviously a unitary operator. Therefore, it follows that

$$\hat{U}_T^\dagger(\eta)\,\hat{U}_T(\eta) = \hat{\mathbb{1}} , \qquad (16.5)$$

implying, up and including terms linear in η,

$$(\hat{\mathbb{1}} + i\eta\,\hat{T}^\dagger + \cdots)(\hat{\mathbb{1}} - i\eta\,\hat{T} + \cdots) = \hat{\mathbb{1}} - i\eta(\hat{T} - \hat{T}^\dagger) + \cdots = \hat{\mathbb{1}} , \qquad (16.6)$$

and so $\hat{T}$ must be hermitian (had the factor i not been included then $\hat{T}$ would have been found to be anti-hermitian). Any finite (continuous) transformation can be implemented via a sequence of infinitesimal transformations,

$$\hat{U}_T(\lambda) = \lim_{n\to\infty}\,\hat{U}_T(\eta)\,\hat{U}_T(\eta)\cdots\hat{U}_T(\eta) = \lim_{n\to\infty}\left[\hat{\mathbb{1}} - i(\lambda/n)\hat{T}\right]^n = e^{-i\lambda\hat{T}} ; \qquad (16.7)$$

see Problem 6 on Chapter 6 for a proof of the unitarity of $e^{-i\lambda\hat{T}}$. The hermitian operator $\hat{T}$ is aptly known as the generator of the symmetry. In principle, any hermitian operator is the generator of some symmetry. However, among hermitian operators, the momentum, angular momentum, and Hamiltonian are of particular relevance since they are the generators of the symmetries under, respectively, space translations, rotations, and time translations. Under an infinitesimal transformation, a generic operator $\hat{O}$ transforms as

$$\hat{O}' = \underbrace{(\hat{\mathbb{1}} - i\eta\,\hat{T} + \cdots)}_{\hat{U}_T(\eta)}\,\hat{O}\,\underbrace{(\hat{\mathbb{1}} + i\eta\,\hat{T} + \cdots)}_{\hat{U}_T^\dagger(\eta)} = \hat{O} - i\eta\,[\hat{T},\hat{O}] + \cdots . \qquad (16.8)$$

For a *finite* continuous transformation, $\hat{O}'$ is given by

$$\hat{O}' = \underbrace{e^{-i\lambda\hat{T}}}_{\hat{U}_T(\lambda)}\,\hat{O}\,\underbrace{e^{i\lambda\hat{T}}}_{\hat{U}_T^\dagger(\lambda)} = \hat{O} - i\lambda\,[\hat{T},\hat{O}] + \frac{(-i\lambda)^2}{2!}\,[\hat{T},[\hat{T},\hat{O}]] + \dots$$

$$+ \frac{(-i\lambda)^n}{n!}\,\underbrace{[\hat{T},[\hat{T},\dots[\hat{T},\hat{O}]\dots]]}_{n\text{ nested commutators}} + \cdots , \qquad (16.9)$$

where the right-hand side expansion in terms of nested commutators is derived in Problem VI.25 of Chapter 6.

---

[2] As a reminder, a group is defined by the following properties: (i) if $\hat{U}_1$ and $\hat{U}_2$ belong to the group then so does the product $\hat{U}_2\,\hat{U}_1$; (ii) the identity $\hat{\mathbb{1}}$ belongs to the group; and (iii) given a group element $\hat{U}$ there exists an inverse element $\hat{U}^{-1}$ in the group, such that $\hat{U}\,\hat{U}^{-1} = \hat{\mathbb{1}} = \hat{U}^{-1}\,\hat{U}$.

16.1 Space and Time Translations

We consider a single particle and the basis consisting of eigenstates $|\phi_\mathbf{r}\rangle$ of the position operator $\hat{\mathbf{r}}$. We suppose the particle to be localized at $\mathbf{r}_0$, that is, it is in the eigenstate $|\phi_{\mathbf{r}_0}\rangle$ with wave function given by $\langle\phi_\mathbf{r}|\phi_{\mathbf{r}_0}\rangle = \delta(\mathbf{r} - \mathbf{r}_0)$. After a translation by $\mathbf{a}$, the particle will be localized at $\mathbf{r}_0 + \mathbf{a}$, namely[3]

$$|\phi_{\mathbf{r}_0}\rangle \longmapsto |\phi'_{\mathbf{r}_0}\rangle = \hat{U}_T(\mathbf{a})\,|\phi_{\mathbf{r}_0}\rangle = |\phi_{\mathbf{r}_0+\mathbf{a}}\rangle\,, \tag{16.10}$$

and $\hat{U}_T(\mathbf{a})$ is the unitary operator inducing the translation with $\hat{U}_T^\dagger(\mathbf{a}) = \hat{U}_T^{-1}(\mathbf{a}) = \hat{U}_T(-\mathbf{a})$. Since we know the effect of $\hat{U}_T(\mathbf{a})$ on a basis, we can deduce its effect on any state $|\psi\rangle$; indeed, we have

$$|\psi\rangle \longmapsto |\psi'\rangle = \hat{U}_T(\mathbf{a})|\psi\rangle \implies \psi'(\mathbf{r}) = (\phi_\mathbf{r},\,\psi') = (\phi_\mathbf{r},\,U_T(\mathbf{a})\,\psi) = (U_T^\dagger(\mathbf{a})\,\phi_\mathbf{r},\,\psi) \tag{16.11}$$

$$= (\phi_{\mathbf{r}-\mathbf{a}},\,\psi) = \psi(\mathbf{r} - \mathbf{a})\,.$$

Note that the wave function on the right-hand side is evaluated, correctly, at $\mathbf{r} - \mathbf{a}$ rather than at $\mathbf{r} + \mathbf{a}$; for example, if the wave function is strongly peaked at the origin before the translation, after the translation it will peak at $\mathbf{a}$.

In order to determine $U_T(\mathbf{a})$ explicitly, we consider an infinitesimal translation along some direction specified by the unit vector $\hat{\mathbf{n}}$:

$$\hat{U}_T(\eta\,\hat{\mathbf{n}}) = \hat{\mathbb{1}} - i\,\eta\,\hat{T}\,, \tag{16.12}$$

which implies that

$$\psi'(\mathbf{r}) = \langle\phi_\mathbf{r}|\,\hat{\mathbb{1}} - i\,\eta\,\hat{T}\,|\psi\rangle = \psi(\mathbf{r} - \eta\,\hat{\mathbf{n}}) = \psi(\mathbf{r}) - \eta\,\hat{\mathbf{n}}\cdot\boldsymbol{\nabla}\,\psi(\mathbf{r})\,, \tag{16.13}$$

where in the last step we expanded the wave function in a Taylor series and kept terms linear in η. From the relation above, we infer that

$$\langle\phi_\mathbf{r}|\,\hat{T}\,|\psi\rangle = -i\,\hat{\mathbf{n}}\cdot\boldsymbol{\nabla}\,\psi(\mathbf{r}) \implies \hat{T} = \hat{\mathbf{n}}\cdot\hat{\mathbf{p}}/\hbar\,, \tag{16.14}$$

and the generator of the translation along $\hat{\mathbf{n}}$ is proportional to the component of the momentum operator along this direction. For a finite translation, we simply exponentiate the generator to obtain

$$U_T(\mathbf{a}) = \mathrm{e}^{-i\mathbf{a}\cdot\hat{\mathbf{p}}/\hbar}\,. \tag{16.15}$$

Any translation can be decomposed into independent translations along the unit vectors $\hat{\mathbf{e}}_1$, $\hat{\mathbf{e}}_2$, and $\hat{\mathbf{e}}_3$ – an orthonormal set specifying a right-handed coordinate system – and therefore

$$U_T(\mathbf{a}) = \mathrm{e}^{-i\mathbf{a}\cdot\hat{\mathbf{p}}/\hbar} = \mathrm{e}^{-i(a_1\hat{p}_1 + a_2\hat{p}_2 + a_3\hat{p}_3)/\hbar} = \hat{U}_T(a_1)\,\hat{U}_T(a_2)\,\hat{U}_T(a_3)\,, \tag{16.16}$$

and, since the generators commute among themselves, the order in which two (or more) translations are executed is immaterial (as we will see, this commutativity property does not generally hold for rotations, except for rotations about the same axis). Under a translation, the momentum and spin operators of a particle are invariant,

$$\hat{\mathbf{p}} \longmapsto \hat{\mathbf{p}}' = \hat{U}_T(\mathbf{a})\,\hat{\mathbf{p}}\,\hat{U}_T^\dagger(\mathbf{a}) = \hat{\mathbf{p}}\,, \qquad \hat{\mathbf{S}} \longmapsto \hat{\mathbf{S}}' = \hat{U}_T(\mathbf{a})\,\hat{\mathbf{S}}\,\hat{U}_T^\dagger(\mathbf{a}) = \hat{\mathbf{S}}\,, \tag{16.17}$$

---

[3] The states $|\phi'_{\mathbf{r}_0}\rangle$ and $|\phi_{\mathbf{r}_0+\mathbf{a}}\rangle$ can differ by a phase factor; here and in the sections that follow, unless otherwise specified we set it equal to unity.

since both $\hat{\mathbf{p}}$ and $\hat{\mathbf{S}}$ commute with $\hat{U}_T(\mathbf{a})$ or, equivalently, with the generators $\hat{\mathbf{p}}/\hbar$ of translations. On the other hand, for the position operator we obtain (see Problem 1)

$$\hat{\mathbf{r}} \longmapsto \hat{\mathbf{r}}' = \hat{U}_T(\mathbf{a})\,\hat{\mathbf{r}}\,\hat{U}_T^\dagger(\mathbf{a}) = \hat{\mathbf{r}} - \mathbf{a}\,. \tag{16.18}$$

If more than one particle is present then the generator of translations is proportional to the total momentum operator $\hat{\mathbf{P}}$,

$$\hat{\mathbf{P}} = \sum_n \hat{\mathbf{p}}_n\,, \qquad \hat{U}_T(\mathbf{a}) = e^{-i\mathbf{a}\cdot\hat{\mathbf{P}}/\hbar}\,, \tag{16.19}$$

and

$$\hat{\mathbf{r}}_n \longmapsto \hat{\mathbf{r}}'_n = \hat{U}_T(\mathbf{a})\,\hat{\mathbf{r}}_n\,\hat{U}_T^\dagger(\mathbf{a}) = \hat{\mathbf{r}}_n - \mathbf{a}\,, \tag{16.20}$$

with the $\hat{\mathbf{p}}_n$ and $\hat{\mathbf{S}}_n$ left unchanged (the fundamental operators $\hat{\mathbf{r}}$, $\hat{\mathbf{p}}$, and $\hat{\mathbf{S}}$ of different particles commute and so, for example, $[\,\hat{r}_{m,i}\,,\,\hat{p}_{n,j}\,] = i\hbar\,\delta_{mn}\,\delta_{ij}$). In particular, it follows that if a Hamiltonian is invariant under translations then the total momentum is conserved. This is the case for two particles interacting with each other through a potential which depends only on the relative position operator $\hat{\mathbf{r}}_1 - \hat{\mathbf{r}}_2$; see Section 10.3.

In a time translation $t \longmapsto t' = t + \tau$, we have

$$|\psi(t)\rangle \longmapsto |\psi'(t)\rangle = \hat{U}_{\mathcal{T}}(\tau)\,|\psi(t)\rangle = |\psi(t-\tau)\rangle\,, \tag{16.21}$$

and $|\psi'(t)\rangle$ corresponds to the state $|\psi(t-\tau)\rangle$ after a translation by τ. Taking τ to be infinitesimal, the relation above implies

$$(\hat{\mathbb{1}} - i\eta\,\hat{T})|\psi(t)\rangle = |\psi(t)\rangle - \eta\frac{d}{dt}|\psi(t)\rangle = |\psi(t)\rangle + i\frac{\eta}{\hbar}\hat{H}|\psi(t)\rangle\,, \tag{16.22}$$

where in the last step we have used the time-dependent Schrödinger equation to express the time derivative of the state, and have excluded the presence of time-dependent external fields, so that the Hamiltonian is time independent (if such fields are present, the system will not generally be invariant under time translations). We see that $\hat{T} = -\hat{H}/\hbar$ is the generator of time translations, and a finite transformation results from

$$\hat{U}_{\mathcal{T}}(\tau) = e^{i\hat{H}\tau/\hbar}\,. \tag{16.23}$$

If the time interval τ is taken as $t - t_0$, we have

$$|\psi'(t)\rangle = \hat{U}_{\mathcal{T}}(t-t_0)\,|\psi(t)\rangle = e^{i\hat{H}(t-t_0)/\hbar}\,\underbrace{e^{-i\hat{H}(t-t_0)/\hbar}\,|\psi(t_0)\rangle}_{|\psi(t)\rangle} = |\psi(t_0)\rangle\,. \tag{16.24}$$

and $|\psi'(t)\rangle$ is constant in time. Referring to Eq. (7.26), we conclude that the Heisenberg picture (H-picture) can be viewed as resulting from a translation in time. Operators transform as $\hat{A} \longmapsto \hat{A}' = \hat{U}_{\mathcal{T}}(\tau)\,\hat{A}\,\hat{U}_{\mathcal{T}}^\dagger(\tau)$, which for $\tau = t - t_0$ reduces to the transformation law for operators in the H-picture, given in Eq. (7.27).

16.2 Rotations

Before discussing the transformation of quantum states and operators, it is useful to remind the reader of a few facts about rotations. A rotation by φ about an axis specified by the unit vector $\hat{\mathbf{n}}$ can be represented by a 3×3 orthogonal matrix $\underline{R}(\varphi \hat{\mathbf{n}})$. In the special case of rotations about the unit vectors $\hat{\mathbf{e}}_1$, $\hat{\mathbf{e}}_2$, and $\hat{\mathbf{e}}_3$ (the orthonormal set of a right-handed coordinate system), these matrices are given by

$$\underline{R}(\varphi \hat{\mathbf{e}}_1) = \begin{pmatrix} 1 & 0 & 0 \\ 0 & \cos \varphi & -\sin \varphi \\ 0 & \sin \varphi & \cos \varphi \end{pmatrix}, \qquad \underline{R}(\varphi \hat{\mathbf{e}}_2) = \begin{pmatrix} \cos \varphi & 0 & \sin \varphi \\ 0 & 1 & 0 \\ -\sin \varphi & 0 & \cos \varphi \end{pmatrix}, \qquad \underline{R}(\varphi \hat{\mathbf{e}}_3) = \begin{pmatrix} \cos \varphi & -\sin \varphi & 0 \\ \sin \varphi & \cos \varphi & 0 \\ 0 & 0 & 1 \end{pmatrix}.$$

$$(16.25)$$

In the general case of a rotation about a direction $\hat{\mathbf{n}} = (\hat{n}_1, \hat{n}_2, \hat{n}_3)$, where $\hat{n}_i$ is the component of $\hat{\mathbf{n}}$ along $\hat{\mathbf{e}}_i$, the matrix $\underline{R}(\varphi \hat{\mathbf{n}})$ is obtained by exponentiation,

$$\underline{R}(\varphi \hat{\mathbf{n}}) = e^{\varphi \hat{\mathbf{n}} \cdot \underline{\mathbf{E}}}, \qquad \hat{\mathbf{n}} \cdot \underline{\mathbf{E}} = \hat{n}_1 \underline{E}_1 + \hat{n}_2 \underline{E}_2 + \hat{n}_3 \underline{E}_3, \tag{16.26}$$

where the antisymmetric matrix $\underline{E}_i$ is the infinitesimal generator for a rotation about the axis $\hat{\mathbf{e}}_i$, with matrix elements given by $(\underline{E}_i)_{jk} = -\epsilon_{ijk}$, specifically

$$\underline{E}_1 = \begin{pmatrix} 0 & 0 & 0 \\ 0 & 0 & -1 \\ 0 & 1 & 0 \end{pmatrix}, \qquad \underline{E}_2 = \begin{pmatrix} 0 & 0 & 1 \\ 0 & 0 & 0 \\ -1 & 0 & 0 \end{pmatrix}, \qquad \underline{E}_3 = \begin{pmatrix} 0 & -1 & 0 \\ 1 & 0 & 0 \\ 0 & 0 & 0 \end{pmatrix}. \tag{16.27}$$

These generators can be derived from Eq. (16.25) by noting that $\underline{R}(\eta \hat{\mathbf{e}}_i) = \underline{1} + \eta \underline{E}_i$ in the limit $\eta \longrightarrow 0$. They satisfy the commutation relations

$$\underline{E}_i \underline{E}_j - \underline{E}_j \underline{E}_i = \sum_k \epsilon_{ijk} \underline{E}_k,$$

that is, the same commutation relations as the angular momentum up to a factor $i\hbar$. Under an infinitesimal rotation, a generic vector $\mathbf{v}$ changes as follows:

$$v_i \longmapsto v_i' = \sum_k R_{ik}(\eta \hat{\mathbf{n}}) v_k = \sum_k \underbrace{\left(\delta_{ik} + \eta \sum_j \hat{n}_j E_{jik} \right)}_{(\underline{1} + \eta \hat{\mathbf{n}} \cdot \underline{\mathbf{E}})_{ik}} v_k$$

$$= v_i - \eta \sum_{jk} \epsilon_{jik} \hat{n}_j v_k = v_i - \eta (\mathbf{v} \times \hat{\mathbf{n}})_i = \hat{\mathbf{e}}_i \cdot (\mathbf{v} + \eta \hat{\mathbf{n}} \times \mathbf{v}), \tag{16.28}$$

and $\delta \mathbf{v} = \eta \hat{\mathbf{n}} \times \mathbf{v}$ represents the infinitesimal change in $\mathbf{v}$.

In order to construct the unitary operator $\hat{U}_R(\varphi \hat{\mathbf{n}})$ that induces rotations of the states describing a particle (spinless, for the time being), we consider, as in the case of translations, the particle to be initially localized at $\mathbf{r}_0$, which is represented by the eigenstate $|\phi_{\mathbf{r}_0}\rangle$ of the position operator. After the rotation, the particle will be localized at $\underline{R}(\varphi \hat{\mathbf{n}}) \mathbf{r}_0$, $R \mathbf{r}_0$ for short or, in terms of components, $(R \mathbf{r}_0)_i = \sum_j R_{ij} r_{0,j}$,

$$|\phi_{\mathbf{r}_0}\rangle \longmapsto |\phi_{\mathbf{r}_0}'\rangle = \hat{U}_R(\varphi \hat{\mathbf{n}}) |\phi_{\mathbf{r}_0}\rangle = |\phi_{R\mathbf{r}_0}\rangle. \tag{16.29}$$

We can evaluate the effect of $\hat{U}_R(\varphi \hat{\mathbf{n}})$ on a generic state $|\psi\rangle$ by noting that

$$|\psi\rangle \longmapsto |\psi'\rangle = \hat{U}_R(\varphi \hat{\mathbf{n}}) |\psi\rangle \implies \psi'(\mathbf{r}) = (\phi_{\mathbf{r}}, \psi') = (\phi_{\mathbf{r}}, U_R(\varphi \hat{\mathbf{n}}) \psi) = (U_R^\dagger(\varphi \hat{\mathbf{n}}) \phi_{\mathbf{r}}, \psi)$$

$$= (\phi_{R^{-1}\mathbf{r}}, \psi) = \psi(R^{-1} \mathbf{r}), \tag{16.30}$$

where we have used $\hat{U}_R^\dagger(\varphi \hat{\mathbf{n}}) |\phi_{\mathbf{r}}\rangle = \hat{U}_R^{-1}(\varphi \hat{\mathbf{n}}) |\phi_{\mathbf{r}}\rangle = |\phi_{R^{-1}\mathbf{r}}\rangle$. Note that the wave function on the final right-hand side is evaluated, correctly, at $R^{-1}\mathbf{r}$ rather than at $R \mathbf{r}$, since it is the position $R^{-1}\mathbf{r}$ that corresponds to $\mathbf{r}$ after the rotation. In an infinitesimal rotation we have

$$\psi'(\mathbf{r}) = \langle \phi_{\mathbf{r}} | \hat{\mathbb{1}} - i\eta \hat{T} | \psi \rangle = \psi(\underbrace{\mathbf{r} - \eta \hat{\mathbf{n}} \times \mathbf{r}}_{R^{-1}\mathbf{r}}) = \psi(\mathbf{r}) - \eta (\hat{\mathbf{n}} \times \mathbf{r}) \cdot \boldsymbol{\nabla} \psi(\mathbf{r}), \tag{16.31}$$

where in the next-to-last step we used $\underline{R}^{-1}(\eta\hat{\mathbf{n}})\mathbf{r} = \underline{R}(-\eta\hat{\mathbf{n}})\mathbf{r} = \mathbf{r} - \eta\,\hat{\mathbf{n}} \times \mathbf{r}$, and in the last step we expanded the wave function in a Taylor series and kept terms linear in η; hence,

$$\langle\phi_\mathbf{r}|\,\hat{T}|\psi\rangle = -i(\hat{\mathbf{n}} \times \mathbf{r}) \cdot \boldsymbol{\nabla}\,\psi(\mathbf{r}) = -i\hat{\mathbf{n}} \cdot (\mathbf{r} \times \boldsymbol{\nabla})\psi(\mathbf{r}) \implies \hat{T} = \hat{\mathbf{n}} \cdot \underbrace{(\hat{\mathbf{r}} \times \hat{\mathbf{p}})}_{\hat{\mathbf{L}}}/\hbar\,, \tag{16.32}$$

and so the generator of rotations about $\hat{\mathbf{n}}$ is proportional to the component of the orbital angular momentum operator along this direction; for a finite rotation, we obtain

$$\hat{U}_R(\varphi\hat{\mathbf{n}}) = \mathrm{e}^{-i\varphi\hat{\mathbf{n}}\cdot\hat{\mathbf{L}}/\hbar}\,. \tag{16.33}$$

For the general case of a particle with spin, the generator of rotations is the total angular momentum $\hat{\mathbf{J}} = \hat{\mathbf{L}} + \hat{\mathbf{S}}$, and the unitary operator that effects the transformation is then given by

$$\hat{U}_R(\varphi\hat{\mathbf{n}}) = \mathrm{e}^{-i\varphi\hat{\mathbf{n}}\cdot\hat{\mathbf{J}}/\hbar} = \underbrace{\mathrm{e}^{-i\varphi\hat{\mathbf{n}}\cdot\hat{\mathbf{S}}/\hbar}}_{\hat{U}_R^s(\varphi\hat{\mathbf{n}})}\underbrace{\mathrm{e}^{-i\varphi\hat{\mathbf{n}}\cdot\hat{\mathbf{L}}/\hbar}}_{\hat{U}_R^o(\varphi\hat{\mathbf{n}})} = \underbrace{\mathrm{e}^{-i\varphi\hat{\mathbf{n}}\cdot\hat{\mathbf{L}}/\hbar}}_{\hat{U}_R^o(\varphi\hat{\mathbf{n}})}\underbrace{\mathrm{e}^{-i\varphi\hat{\mathbf{n}}\cdot\hat{\mathbf{S}}/\hbar}}_{\hat{U}_R^s(\varphi\hat{\mathbf{n}})}\,, \tag{16.34}$$

since the orbital angular momentum and spin commute with each other. Thus rotations in orbital and spin space can be carried out independently (this is not the case in relativistic quantum mechanics). To see how states transform under rotations in the orbital + spin space, we first consider the effect of a rotation on the basis of eigenstates $|\phi_{\mathbf{r}\sigma}\rangle = |\phi_\mathbf{r}\rangle \otimes |\sigma\rangle$ of the position and spin operator (not necessarily spin-1/2); we have

$$\hat{U}_R(\varphi\hat{\mathbf{n}})\,|\phi_{\mathbf{r}\sigma}\rangle = |\phi_{R\mathbf{r}}\rangle \otimes \hat{U}_R^s(\varphi\hat{\mathbf{n}})\,|\sigma\rangle\,. \tag{16.35}$$

The spinor wave function of a generic state then transforms as

$$\psi_\sigma'(\mathbf{r}) = \langle\phi_{\mathbf{r}\sigma}|\hat{U}_R(\varphi\hat{\mathbf{n}})\,|\psi\rangle = \langle\phi_{R^{-1}\mathbf{r}}| \otimes \langle\sigma|\hat{U}_R^s(\varphi\hat{\mathbf{n}})|\psi\rangle = \sum_\tau \langle\sigma|\hat{U}_R^s(\varphi\hat{\mathbf{n}})|\tau\rangle\,\psi_\tau(R^{-1}\mathbf{r})\,, \tag{16.36}$$

where in the last step we have inserted the completeness operator over the spin states, $\sum_\tau |\tau\rangle\langle\tau| = \hat{\mathbb{1}}^{(s)}$. It should be clear that if, say, two successive rotations are performed, then the order in which these are carried out is important, since $\hat{U}_R(\varphi\hat{\mathbf{n}})$ and $\hat{U}_R(\varphi'\hat{\mathbf{n}}')$ do not commute (unless, of course, $\hat{\mathbf{n}} = \hat{\mathbf{n}}'$). Finally, in Problem 8 it is shown that for a particle with half-integer spin (or a system containing an odd number of particles with half-integer spins) the operator inducing a rotation by 2π is not equal to the identity; rather, it is $\hat{U}_R(2\pi\hat{\mathbf{n}}) = -\hat{\mathbb{1}}$. Note, however, that for such a particle (or system) we have $\hat{U}_R(4\pi\hat{\mathbf{n}}) = \hat{\mathbb{1}}$ (an experimental verification of this property by a neutron interferometry experiment is discussed in Problem 13 of Chapter 12).

Scalar operators commute with $\hat{\mathbf{J}}$ and hence with the rotation operator: they are invariant under rotations. By contrast, vector operators transform under rotations according to

$$\hat{\mathbf{v}} \longmapsto \hat{\mathbf{v}}' = \hat{U}_R(\varphi\hat{\mathbf{n}})\,\hat{\mathbf{v}}\,\hat{U}_R^\dagger(\varphi\hat{\mathbf{n}}) = \underline{R}^{-1}(\varphi\hat{\mathbf{n}})\,\hat{\mathbf{v}}\,, \tag{16.37}$$

indeed, this is what we mean when we say that a triplet of operators $\hat{v}_1$, $\hat{v}_2$ and $\hat{v}_3$ forms a *vector* operator. Here, the notation $[\underline{R}^{-1}(\varphi\hat{\mathbf{n}})\,\hat{\mathbf{v}}]_i$ means $\sum_j R_{ij}^{-1}(\varphi\hat{\mathbf{n}})\,\hat{v}_j = \sum_j R_{ji}(\varphi\hat{\mathbf{n}})\,\hat{v}_j$, since $\underline{R}(\varphi\hat{\mathbf{n}})$ is an orthogonal matrix (see Problem 12 for an explicit derivation of this transformation law). For an infinitesimal rotation, Eq. (16.37) dictates that

$$\underbrace{(\hat{\mathbb{1}} - i\eta\hat{\mathbf{n}}\cdot\hat{\mathbf{J}}/\hbar)\hat{\mathbf{v}}(\hat{\mathbb{1}} + i\eta\hat{\mathbf{n}}\cdot\hat{\mathbf{J}}/\hbar)}_{\hat{U}_R(\varphi\mathbf{n})\,\hat{\mathbf{v}}\,\hat{U}_R^\dagger(\varphi\hat{\mathbf{n}})} = \underbrace{\hat{\mathbf{v}} - \eta\hat{\mathbf{n}} \times \hat{\mathbf{v}}}_{\underline{R}^{-1}(\varphi\hat{\mathbf{n}})\,\hat{\mathbf{v}}} \implies [\hat{\mathbf{n}} \cdot \hat{\mathbf{J}}, \hat{\mathbf{v}}] = -i\hbar\,\hat{\mathbf{n}} \times \hat{\mathbf{v}}\,, \tag{16.38}$$

using Eq. (16.28). Therefore, a vector operator can be defined as that triplet of operators that satisfy the commutation relations above; in terms of components, these commutation relations read $[\hat{J}_i, \hat{v}_j] = i\hbar \sum_k \epsilon_{ijk}\,\hat{v}_k$.

16.3 Discrete Symmetries: Space Inversion and Time Reversal

In this section we consider the discrete symmetries of space inversion and time reversal. As noted earlier, time reversal is the only symmetry transformation that is induced by an anti-unitary operator, and the properties of anti-unitary operators are briefly reviewed below.

Space inversion transformations were introduced in Section 4.2. Here, we explore their properties further. For a spinless particle, space inversion, when acting on the basis consisting of position eigenstates, gives

$$|\phi_{\mathbf{r}}\rangle \longmapsto |\phi_{\mathbf{r}}'\rangle = \hat{U}_P|\phi_{\mathbf{r}}\rangle = |\phi_{-\mathbf{r}}\rangle . \tag{16.39}$$

Since $\hat{U}_P^2|\phi_{\mathbf{r}}\rangle = |\phi_{\mathbf{r}}\rangle$, we see that $\hat{U}_P^2 = \hat{\mathbb{1}}$, yielding $\hat{U}_P = \hat{U}_P^{-1} = \hat{U}_P^\dagger$, where in the last step we used the fact that $\hat{U}_P$ is unitary (its unitarity can be verified directly from the definition of $\hat{U}_P$). Consequently $\hat{U}_P$ is also hermitian, and hence an observable with eigenvalues ± 1. The action on a generic state $|\psi\rangle$ is obtained as follows:

$$|\psi\rangle \longmapsto |\psi'\rangle = \hat{U}_P|\psi\rangle \implies \psi'(\mathbf{r}) = (\phi_{\mathbf{r}}, \psi') = (\phi_{\mathbf{r}}, U_P\psi) = (U_P^\dagger \phi_{\mathbf{r}}, \psi) = (\phi_{-\mathbf{r}}, \psi) = \psi(-\mathbf{r}) , \tag{16.40}$$

in accordance with the result of Section 4.2. A particle in a central potential (for example, the electron in the hydrogen atom) has a wave function of the form $\psi_{lm}(\mathbf{r}) = R(r)\, Y_{lm}(\theta, \phi)$, and under space inversion the transformation properties of $\psi_{lm}(\mathbf{r})$ follow from those of the spherical harmonics, since

$$\psi_{lm}(-\mathbf{r}) = R(r)\, Y_{lm}(\pi - \theta, \phi + \pi) = (-1)^l\, R(r)\, Y_{lm}(\theta, \phi) = (-1)^l \psi_{lm}(\mathbf{r}) , \tag{16.41}$$

and in spherical coordinates $\mathbf{r} = (r, \theta, \phi) \longrightarrow (r, \pi - \theta, \phi + \pi) = -\mathbf{r}$ (see Problem 11 in Chapter 11 for a derivation of the transformation properties of spherical harmonics under parity).

The position and momentum operators transform as follows (since $\hat{U}_P = \hat{U}_P^\dagger$):

$$\hat{\mathbf{r}} \longmapsto \hat{\mathbf{r}}' = \hat{U}_P\, \hat{\mathbf{r}}\, \hat{U}_P = -\hat{\mathbf{r}} , \qquad \hat{\mathbf{p}} \longmapsto \hat{\mathbf{p}}' = \hat{U}_P\, \hat{\mathbf{p}}\, \hat{U}_P = -\hat{\mathbf{p}} , \tag{16.42}$$

as can be easily verified; for the basis $|\phi_{\mathbf{r}}\rangle$, for example, we have that the matrix elements of $\hat{\mathbf{p}}$ are $(\phi_{\mathbf{r}}, \hat{\mathbf{p}}\, \phi_{\mathbf{r}'}) = -i\hbar\, \boldsymbol{\nabla}\, \delta(\mathbf{r} - \mathbf{r}')$, where the gradient acts on the $\mathbf{r}$ dependence, and hence

$$(\phi_{\mathbf{r}}, \hat{\mathbf{p}}'\, \phi_{\mathbf{r}'}) = (\phi_{\mathbf{r}}, U_P\, \hat{\mathbf{p}}\, U_P\, \phi_{\mathbf{r}'}) = (U_P\, \phi_{\mathbf{r}}, \hat{\mathbf{p}}\, U_P\, \phi_{\mathbf{r}'}) = (\phi_{-\mathbf{r}}, \hat{\mathbf{p}}\, \phi_{-\mathbf{r}'})$$
$$= +i\hbar\, \boldsymbol{\nabla}\, \delta(-\mathbf{r} + \mathbf{r}') = -(\phi_{\mathbf{r}}, \hat{\mathbf{p}}\, \phi_{\mathbf{r}'}) . \tag{16.43}$$

As a consequence of Eq. (16.42), the orbital angular momentum $\hat{\mathbf{L}}$ is invariant under space inversion and hence commutes with $\hat{U}_P$. Consistency then requires that $\hat{U}_P$ also commutes with $\hat{\mathbf{J}}$ and, in particular, $\hat{\mathbf{S}}$, the spin operator. If a particle has spin then the basis of eigenstates $|\phi_{\mathbf{r}\sigma}\rangle$ of $\mathbf{r}$, $\hat{S}^2$, and $\hat{S}_z$ (a complete set of commuting observables for a single particle) transforms under space inversion as $\hat{U}_P|\phi_{\mathbf{r}\sigma}\rangle = |\phi_{-\mathbf{r}\sigma}\rangle$, and $\hat{U}_P|\phi_{\mathbf{r}\sigma}\rangle$ remains an eigenstate of $\hat{S}^2$ and $\hat{S}_z$ with the same eigenvalues as $|\phi_{\mathbf{r}\sigma}\rangle$.

16.3.1 Time Reversal; Properties of Anti-Unitary Operators

Classically, the equations of motion for a particle moving under the influence of a potential $v(\mathbf{r})$ are invariant under the time-reversal transformation $\mathbf{r}(t) \longmapsto \mathbf{r}(-t)$ and $\mathbf{p}(t) \longmapsto -\mathbf{p}(-t)$. As a consequence, for any solution there corresponds one that retraces the original trajectory backwards. More explicitly, suppose that the particle, moving along a given trajectory, reaches at time t_0 the position $\mathbf{r}_0 = \mathbf{r}(t_0)$ with momentum $\mathbf{p}_0 = \mathbf{p}(t_0)$. If the particle were to suddenly reverse its momentum, that is, if $\mathbf{p}_0 \longrightarrow -\mathbf{p}_0$ when it reaches the position $\mathbf{r}_0$, it would retrace the trajectory it had followed up to time t_0.

To discuss time reversal in quantum mechanics, we make explicit the time dependence of the position and momentum operators by adopting the Heisenberg picture (H-picture), that is,

$$\hat{\mathbf{r}}_{\mathrm{H}}(t) = e^{i\hat{H}t/\hbar}\,\hat{\mathbf{r}}\,e^{-i\hat{H}t/\hbar}\,, \qquad \hat{\mathbf{p}}_{\mathrm{H}}(t) = e^{i\hat{H}t/\hbar}\,\hat{\mathbf{p}}\,e^{-i\hat{H}t/\hbar}\,, \tag{16.44}$$

where $\hat{\mathbf{r}} = \hat{\mathbf{r}}_{\mathrm{H}}(0) = \hat{\mathbf{r}}_{\mathrm{S}}$ and similarly for $\hat{\mathbf{p}}$. Note that at time $t = 0$ the Schrödinger-picture (S-picture) and H-picture representations of states and operators coincide. Under time reversal we require that

$$\hat{\mathbf{r}}(t) \longmapsto \hat{\mathbf{r}}'_{\mathrm{H}}(t) = \hat{\Omega}_{\mathcal{T}}\,\hat{\mathbf{r}}_{\mathrm{H}}(t)\,\hat{\Omega}_{\mathcal{T}}^{\dagger} = \hat{\mathbf{r}}_{\mathrm{H}}(-t)\,, \qquad \hat{\mathbf{p}}(t) \longmapsto \hat{\mathbf{p}}'_{\mathrm{H}}(t) = \hat{\Omega}_{\mathcal{T}}\,\hat{\mathbf{p}}_{\mathrm{H}}(t)\,\hat{\Omega}_{\mathcal{T}}^{\dagger} = -\hat{\mathbf{p}}_{\mathrm{H}}(-t)\,, \tag{16.45}$$

where $\hat{\Omega}_{\mathcal{T}}$ is the operator that induces the transformation. This then leads to the following canonical (equal-time) commutation relations for the transformed operators $\left[\hat{r}'_{\mathrm{H},\alpha}(t)\,, \hat{p}'_{\mathrm{H},\beta}(t)\right] = -\left[\hat{r}_{\mathrm{H},\alpha}(-t)\,, \hat{p}_{\mathrm{H},\beta}(-t)\right] = -i\hbar\,\delta_{\alpha\beta}$, while, on the other hand, we also have

$$\left[\hat{r}'_{\mathrm{H},\alpha}(t)\,, \hat{p}'_{\mathrm{H},\beta}(t)\right] = \hat{\Omega}_{\mathcal{T}}\,\left[\hat{r}_{\mathrm{H},\alpha}(t)\,, \hat{p}_{\mathrm{H},\beta}(t)\right]\hat{\Omega}_{\mathcal{T}}^{\dagger} = \hat{\Omega}_{\mathcal{T}}\,(i\hbar\delta_{\alpha\beta})\,\hat{\Omega}_{\mathcal{T}}^{\dagger} \implies \hat{\Omega}_{\mathcal{T}}\,\left(i\hbar\,\delta_{\alpha\beta}\right)\hat{\Omega}_{\mathcal{T}}^{\dagger} = -i\hbar\,\delta_{\alpha\beta}\,, \tag{16.46}$$

and $\hat{\Omega}_{\mathcal{T}}$ must be anti-unitary, so that $\hat{\Omega}_{\mathcal{T}}\,c = c^*\,\hat{\Omega}_{\mathcal{T}}$, where c is a complex number. Generally, an anti-unitary operator $\hat{\Omega}$ is an invertible operator, such that

$$(\hat{\Omega}\,\phi, \hat{\Omega}\,\psi) = (\phi, \psi)^*\,. \tag{16.47}$$

The product of a unitary and an anti-unitary operator is anti-unitary, while the product of two anti-unitary operators is unitary; for example, let $\hat{\Omega}_1$ and $\hat{\Omega}_2$ be two anti-unitary operators; then $\hat{\Omega}_1\,\hat{\Omega}_2$ is unitary, since

$$(\hat{\Omega}_1\,\hat{\Omega}_2\,\phi, \hat{\Omega}_1\,\hat{\Omega}_2\,\psi) = (\hat{\Omega}_2\,\phi, \hat{\Omega}_2\,\psi)^* = \left[(\phi, \psi)^*\right]^* = (\phi, \psi)\,. \tag{16.48}$$

An anti-unitary operator is necessarily antilinear (see Problem 17), namely

$$\hat{\Omega}(\alpha|\phi\rangle + \beta|\psi\rangle) = \alpha^*\,\hat{\Omega}\,|\phi\rangle + \beta^*\,\hat{\Omega}\,|\psi\rangle\,, \tag{16.49}$$

and the adjoint of an antilinear operator is defined as

$$(\phi, \hat{\Omega}\,\psi) = (\hat{\Omega}^{\dagger}\,\phi, \psi)^*\,. \tag{16.50}$$

If each of the operators $\hat{\Omega}_1$, $\hat{\Omega}_2$, ..., $\hat{\Omega}_n$ is either linear or antilinear then the adjoint of the product $(\hat{\Omega}_1\,\hat{\Omega}_2\cdots\hat{\Omega}_n)^{\dagger}$ is given by the product of the individual adjoint operators in reverse order. For an anti-unitary operator it still holds true that $\hat{\Omega}^{\dagger}\,\hat{\Omega} = \hat{\mathbb{1}}$, since for any two states we have $(\phi, \psi)^* = (\hat{\Omega}\,\phi, \hat{\Omega}\,\psi) = (\hat{\Omega}^{\dagger}\,\hat{\Omega}\,\phi, \psi)^*$, yielding $\hat{\Omega}^{\dagger}\,\hat{\Omega} = \hat{\mathbb{1}}$. Note that the transformation $|\phi\rangle \longmapsto \hat{\Omega}\,|\phi\rangle$ is *not* generally equivalent to a transformation of the operators that act on these states, in the sense that

$$(\phi, A\,\psi) \longmapsto (\phi', A'\,\psi') = (\hat{\Omega}\,\phi, \hat{\Omega}A\,\hat{\Omega}^{\dagger}\,\hat{\Omega}\,\psi) = (\hat{\Omega}\,\phi, \hat{\Omega}\,A\,\psi) = (\phi, A\,\psi)^*\,. \tag{16.51}$$

In the case of a *hermitian operator*, Eq. (16.51) implies that $(\phi', A'\,\psi') = (\psi, A\,\phi)$.

An example of an anti-unitary operator is the complex conjugation operator $\hat{K}$. In the Hilbert space of a single (spinless) particle, we have

$$(\hat{K}\,\phi, \hat{K}\,\psi) = \int d\mathbf{r}\,\left[\hat{K}\,\phi(\mathbf{r})\right]^*\left[\hat{K}\,\psi(\mathbf{r})\right] = \int d\mathbf{r}\,\left[\phi^*(\mathbf{r})\right]^*\psi^*(\mathbf{r}) = \int d\mathbf{r}\,\phi(\mathbf{r})\,\psi^*(\mathbf{r}) = (\phi, \psi)^*\,. \tag{16.52}$$

Note that $\hat{K}^2 = \hat{\mathbb{1}}$, implying that $\hat{K} = \hat{K}^{-1} = \hat{K}^\dagger$. As a matter of fact, any anti-unitary operator $\hat{\Omega}$ can be written as the product of $\hat{K}$ and a unitary operator $\hat{U}$, namely $\hat{\Omega} = \hat{U}\hat{K}$: given an $\hat{\Omega}$, the $\hat{U}$ that effects this is simply $\hat{\Omega}\hat{K}$, since $\hat{U}\hat{K} = \hat{\Omega}\hat{K}\hat{K} = \hat{\Omega}$. In the specific case of the anti-unitary operator $\hat{\Omega}_{\mathcal{T}}$ associated with time reversal, we express it as

$$\hat{\Omega}_{\mathcal{T}} = \hat{U}_{\mathcal{T}}\hat{K}, \tag{16.53}$$

where $\hat{U}_{\mathcal{T}}$ is not to be confused with the time translation operator defined in Section 16.1. It is useful to dwell a little longer on the properties of these operators. Consider a basis $|\phi_m\rangle$ in the state space of a system. The action of $\hat{K}$ on a member $|\phi_p\rangle$ of this basis is $\hat{K}|\phi_p\rangle = |\phi_p\rangle$, the reason being that in such a basis the "wave function" corresponding to $|\phi_p\rangle$ is given by the (real) column vector with zeros in all its entries, except for the entry p which is a 1, and the complex conjugation of such a vector yields the same vector. Now, suppose we choose a different basis $|\chi_m\rangle$ in the state space. In such a basis, the state $|\phi_p\rangle$ is represented as

$$|\phi_p\rangle = \sum_m c_m^{(p)} |\chi_m\rangle, \qquad c_m^{(p)} = \langle \chi_m | \phi_p \rangle, \tag{16.54}$$

and the corresponding "wave function" is the generally complex column vector $|\phi_p\rangle \longrightarrow [c_1^{(p)}, c_2^{(p)}, \dots, c_m^{(p)}, \dots]^T$, where the superscript T indicates the transpose. The action of $\hat{K}$ on $|\phi_p\rangle$ is now given by

$$\hat{K}|\phi_p\rangle = \sum_m c_m^{(p)*} \hat{K}|\chi_m\rangle = \sum_m c_m^{(p)*} |\chi_m\rangle. \tag{16.55}$$

We conclude that the effect of $\hat{K}$ depends on the representation, that is, on the basis in which we choose to represent states and operators. For a discussion of this point, see Problem 18.

16.3.2 Transformation of States and Operators under Time Reversal

Spinless particle: Let $|\phi_{\mathbf{r}}\rangle$ be a basis of position eigenstates. Under time reversal, we posit that

$$\hat{\Omega}_{\mathcal{T}} |\phi_{\mathbf{r}}\rangle = \eta |\phi_{\mathbf{r}}\rangle. \tag{16.56}$$

In other words, up to a phase factor, the state of a particle localized at $\mathbf{r}$ is left unchanged by time reversal. The relation (16.56) completely defines $\hat{\Omega}_{\mathcal{T}}$; in particular

$$\hat{\Omega}_{\mathcal{T}}\left(\hat{\Omega}_{\mathcal{T}}|\phi_{\mathbf{r}}\rangle\right) = \hat{\Omega}_{\mathcal{T}}\left(\eta|\phi_{\mathbf{r}}\rangle\right) = \eta^* \hat{\Omega}_{\mathcal{T}}|\phi_{\mathbf{r}}\rangle = \eta^*\eta|\phi_{\mathbf{r}}\rangle = |\phi_{\mathbf{r}}\rangle, \tag{16.57}$$

which yields

$$\hat{\Omega}_{\mathcal{T}}^2 = \hat{\mathbb{1}} \qquad \text{for a spinless particle}, \tag{16.58}$$

and hence $\hat{\Omega}_{\mathcal{T}} = \hat{\Omega}_{\mathcal{T}}^{-1} = \hat{\Omega}_{\mathcal{T}}^\dagger$. The property Eq. (16.58) is independent of the phase factor η and only holds for a spinless particle; as we will see below, for a spin-1/2 particle (or, more generally, a particle with half-integer angular momentum) we have instead $\hat{\Omega}_{\mathcal{T}}^2 = -\hat{\mathbb{1}}$. Here, we set the arbitrary phase equal to 1, so that

$$\hat{\Omega}_{\mathcal{T}}|\phi_{\mathbf{r}}\rangle = |\phi_{\mathbf{r}}\rangle = \hat{\Omega}_{\mathcal{T}}^\dagger|\phi_{\mathbf{r}}\rangle. \tag{16.59}$$

A generic state transforms as $|\psi\rangle \longmapsto |\psi'\rangle = \hat{\Omega}_{\mathcal{T}}|\psi\rangle$, which implies for the associated wave function

$$\psi'(\mathbf{r}) = (\phi_{\mathbf{r}}, \psi') = (\phi_{\mathbf{r}}, \hat{\Omega}_{\mathcal{T}}\psi) = (\hat{\Omega}_{\mathcal{T}}\hat{\Omega}_{\mathcal{T}}\phi_{\mathbf{r}}, \hat{\Omega}_{\mathcal{T}}\psi) = (\hat{\Omega}_{\mathcal{T}}\phi_{\mathbf{r}}, \psi)^* = \psi^*(\mathbf{r}). \tag{16.60}$$

In particular, for the wave function of a momentum eigenstate $|\phi_{\mathbf{p}}\rangle$ we obtain

$$(\phi_{\mathbf{r}}, \phi'_{\mathbf{p}}) = \phi'_{\mathbf{p}}(\mathbf{r}) = \phi^*_{\mathbf{p}}(\mathbf{r}) = \frac{1}{(2\pi\hbar)^{3/2}}\, e^{-i\mathbf{p}\cdot\mathbf{r}/\hbar} = (\phi_{\mathbf{r}}, \phi_{-\mathbf{p}})\,, \qquad (16.61)$$

from which we deduce that under time reversal the momentum eigenstates transform as

$$|\phi_{\mathbf{p}}\rangle \longmapsto |\phi'_{\mathbf{p}}\rangle = \hat{\Omega}_{\mathcal{T}}\,|\phi_{\mathbf{p}}\rangle = |\phi_{-\mathbf{p}}\rangle\,, \qquad (16.62)$$

which is what one would naively have expected. As a consequence, the momentum-space wave function of a time-reversed state is not simply the complex conjugate of the original (momentum-space) wave function, but rather $\widetilde{\psi}'(\mathbf{p}) = \widetilde{\psi}^*(-\mathbf{p})$ holds; indeed, we have

$$\underbrace{(\phi_{\mathbf{p}}, \psi')}_{\widetilde{\psi}'(\mathbf{p})} = (\phi_{\mathbf{p}}, \Omega_{\mathcal{T}}\psi) = (\Omega_{\mathcal{T}}\,\Omega_{\mathcal{T}}\phi_{\mathbf{p}}, \Omega_{\mathcal{T}}\psi) = (\Omega_{\mathcal{T}}\phi_{\mathbf{p}}, \psi)^* = \underbrace{(\phi_{-\mathbf{p}}, \psi)^*}_{\widetilde{\psi}^*(-\mathbf{p})}\,. \qquad (16.63)$$

Thus, the "time-reversed wave function" depends on the adopted representation: $\psi'(\mathbf{r}) = \psi^*(\mathbf{r})$ in the position representation and $\widetilde{\psi}'(\mathbf{p}) = \widetilde{\psi}^*(-\mathbf{p})$ in the momentum representation.

The position and momentum operators transform according to

$$\hat{\mathbf{r}} \longmapsto \hat{\mathbf{r}}' = \hat{\Omega}_{\mathcal{T}}\,\hat{\mathbf{r}}\,\hat{\Omega}_{\mathcal{T}}^{\dagger} = \hat{\mathbf{r}}\,, \qquad \hat{\mathbf{p}} \longmapsto \hat{\mathbf{p}}' = \hat{\Omega}_{\mathcal{T}}\,\hat{\mathbf{p}}\,\hat{\Omega}_{\mathcal{T}}^{\dagger} = -\hat{\mathbf{p}}\,. \qquad (16.64)$$

In the basis $|\phi_{\mathbf{r}}\rangle$, we have, for example,

$$(\phi_{\mathbf{r}}, \mathbf{p}'\phi_{\mathbf{r}'}) = (\phi_{\mathbf{r}}, \Omega_{\mathcal{T}}\,\mathbf{p}\,\Omega_{\mathcal{T}}^{\dagger}\phi_{\mathbf{r}'}) = (\Omega_{\mathcal{T}}\,\Omega_{\mathcal{T}}\phi_{\mathbf{r}}, \Omega_{\mathcal{T}}\,\mathbf{p}\,\Omega_{\mathcal{T}}\phi_{\mathbf{r}'}) = (\Omega_{\mathcal{T}}\phi_{\mathbf{r}}, \mathbf{p}\,\Omega_{\mathcal{T}}\phi_{\mathbf{r}'})^*$$

$$= (\phi_{\mathbf{r}}, \mathbf{p}\,\phi_{\mathbf{r}'})^* = i\hbar\boldsymbol{\nabla}\,\delta(\mathbf{r} - \mathbf{r}') = -(\phi_{\mathbf{r}}, \mathbf{p}\,\phi_{\mathbf{r}'})\,, \qquad (16.65)$$

where we have used Eq. (16.58), the anti-unitarity of $\hat{\Omega}_{\mathcal{T}}$, and Eq. (16.59). These transformation laws are the same as those we had inferred in Eq. (16.45) from classical mechanics, and are independent of the representation. As a consequence of Eq. (16.64), the orbital angular momentum $\hat{\mathbf{L}}$ is odd under $\hat{\Omega}_{\mathcal{T}}$:

$$\hat{\mathbf{L}} \longmapsto \hat{\mathbf{L}}' = \hat{\Omega}_{\mathcal{T}}\,\hat{\mathbf{L}}\,\hat{\Omega}_{\mathcal{T}}^{\dagger} = -\hat{\mathbf{L}}\,. \qquad (16.66)$$

Since according to Eq. (16.60) the eigenfunctions of $\hat{\mathbf{L}}^2$ and $\hat{L}_z$ – the spherical harmonics – satisfy

$$(\phi_{\hat{\mathbf{r}}}, \psi'_{l,m}) = Y'_{l,m}(\hat{\mathbf{r}}) = Y^*_{l,m}(\hat{\mathbf{r}}) = (-1)^m\, Y_{l,-m}(\hat{\mathbf{r}}) = (-1)^m\,(\phi_{\hat{\mathbf{r}}}, \psi_{l,-m})\,, \qquad (16.67)$$

it follows that the eigenstates $|\psi_{lm}\rangle$ transform as follows:

$$|\psi_{l,m}\rangle \longmapsto |\psi'_{l,m}\rangle = \hat{\Omega}_{\mathcal{T}}|\psi_{l,m}\rangle = (-1)^m\,|\psi_{l,-m}\rangle = i^{2m}\,|\psi_{l,-m}\rangle\,. \qquad (16.68)$$

It is easily verified that the transformation law above for the states $|\psi_{lm}\rangle$ is consistent with that of the angular momentum $\hat{\mathbf{L}}$, in the sense that

$$(\Omega_{\mathcal{T}}\psi_{l',m'}, L_i\,\Omega_{\mathcal{T}}\psi_{l,m})^* = (\psi_{l',m'}, L'_i\psi_{l,m}) = -(\psi_{l',m'}, L_i\psi_{l,m})\,, \qquad L'_i = \hat{\Omega}_{\mathcal{T}}\,L_i\,\hat{\Omega}_{\mathcal{T}}^{\dagger}\,. \qquad (16.69)$$

Spin-1/2 particle: Under time reversal we require, for consistency with the transformation property of the orbital angular momentum $\hat{\mathbf{L}}$, that the spin operator is odd,

$$\hat{\mathbf{S}} \longmapsto \hat{\mathbf{S}}' = \hat{\Omega}_{\mathcal{T}}\,\hat{\mathbf{S}}\,\hat{\Omega}_{\mathcal{T}}^{\dagger} = -\hat{\mathbf{S}}\,, \qquad (16.70)$$

so that the total angular momentum $\hat{\mathbf{J}} = \hat{\mathbf{L}} + \hat{\mathbf{S}}$ is odd overall. In order to construct $\hat{\Omega}_{\mathcal{T}} = \hat{U}_{\mathcal{T}}\,\hat{K}$, we select the basis of eigenstates $|\psi_\sigma\rangle$ of $\hat{S}^2$ and $\hat{S}_z$ with $\hat{S}_z\,|\psi_\sigma\rangle = \hbar\sigma|\psi_s\rangle$ and $\sigma = \pm 1/2$. In such a basis, the matrices $(\psi_\sigma, \hat{S}_i\psi_{\sigma'})$ representing $\hat{S}_i$ are $(\hbar/2)\,\underline{\sigma}_i$, where $\underline{\sigma}_i$ are the standard Pauli matrices (see

Problem 2 of Chapter 11 for their explicit expressions). Recalling that $\hat{K}^\dagger = \hat{K}^{-1} = \hat{K}$, and further that $\hat{K}|\psi_\sigma\rangle = |\psi_\sigma\rangle$ and $(a, Kb) = (Ka, b)^*$, we find

$$(\psi_\sigma, KS_i K^\dagger \psi_{\sigma'}) = (K\psi_\sigma, S_i K^\dagger \psi_{\sigma'})^* = (\psi_\sigma, S_i \psi_{\sigma'})^* . \tag{16.71}$$

Since the matrices $\underline{S}_1$ and $\underline{S}_3$ are real and the matrix $\underline{S}_2$ is imaginary, the relation above implies that under $\hat{K}$ the operators $\hat{S}_i$ transform as follows:

$$\hat{K}\hat{S}_1\hat{K}^\dagger = \hat{S}_1 , \qquad \hat{K}\hat{S}_2\hat{K}^\dagger = -\hat{S}_2 , \qquad \hat{K}\hat{S}_3\hat{K}^\dagger = \hat{S}_3 . \tag{16.72}$$

We are left with the task of determining the unitary operator $\hat{U}_{\mathcal{T}}$ such that

$$\hat{U}_{\mathcal{T}} \hat{S}_1 \hat{U}_{\mathcal{T}}^\dagger = -\hat{S}_1 , \qquad \hat{U}_{\mathcal{T}} \hat{S}_2 \hat{U}_{\mathcal{T}}^\dagger = \hat{S}_2 , \qquad \hat{U}_{\mathcal{T}} \hat{S}_3 \hat{U}_{\mathcal{T}}^\dagger = -\hat{S}_3 . \tag{16.73}$$

Thus $\hat{U}_{\mathcal{T}}$ must commute with $\hat{S}_2$ and anticommute with $\hat{S}_1$ and $\hat{S}_3$. Recalling that the Pauli matrices anticommute with each other and that their square is the identity matrix, an operator $\hat{U}_{\mathcal{T}}$ that satisfies these requirements is simply proportional to σ_2; we choose

$$\hat{U}_{\mathcal{T}} = -i\eta\sigma_2 , \tag{16.74}$$

where η is a phase factor. In fact, up to the phase factor η, $\hat{U}_{\mathcal{T}}$ is just the unitary operator inducing a rotation by π about the axis $\hat{\mathbf{e}}_2$. The time-reversal operator can then be written as[4]

$$\hat{\Omega}_{\mathcal{T}} = \eta \, e^{-i\pi\hat{S}_2/\hbar} \, \hat{K} . \tag{16.75}$$

On the basis $|\psi_\sigma\rangle$ that diagonalizes S_z, $\Omega_{\mathcal{T}}$ has the following effect:

$$\hat{\Omega}_{\mathcal{T}}|\psi_{1/2}\rangle \longrightarrow \eta \underbrace{\begin{pmatrix} 0 & -1 \\ 1 & 0 \end{pmatrix}}_{-i\,\sigma_y} \begin{pmatrix} 1 \\ 0 \end{pmatrix}^* = \eta \begin{pmatrix} 0 \\ 1 \end{pmatrix} , \qquad \hat{\Omega}_{\mathcal{T}}|\psi_{-1/2}\rangle \longrightarrow \eta \underbrace{\begin{pmatrix} 0 & -1 \\ 1 & 0 \end{pmatrix}}_{-i\,\sigma_y} \begin{pmatrix} 0 \\ 1 \end{pmatrix}^* = -\eta \begin{pmatrix} 1 \\ 0 \end{pmatrix} , \tag{16.76}$$

where the complex conjugation in the intermediate steps (although it has no effect here) comes from the action of $\hat{K}$. We conclude that $\hat{\Omega}_{\mathcal{T}}|\psi_\sigma\rangle = \eta\,(-1)^{1/2-\sigma}|\psi_{-\sigma}\rangle$, that is, the time-reversal operator flips the spin projection (up to a phase factor). The general state $|\psi\rangle$ of a spin-1/2 particle can be represented by the spinor wave function $\psi(\mathbf{r}) = \psi_+(\mathbf{r})|\psi_{1/2}\rangle + \psi_-(\mathbf{r})|\psi_{-1/2}\rangle$, and under time reversal we have

$$|\psi'\rangle = \hat{\Omega}_{\mathcal{T}}|\psi\rangle \longrightarrow \begin{pmatrix} \psi'_+(\mathbf{r}) \\ \psi'_-(\mathbf{r}) \end{pmatrix} = \eta \begin{pmatrix} -\psi_-^*(\mathbf{r}) \\ \psi_+^*(\mathbf{r}) \end{pmatrix} \longrightarrow \psi'(\mathbf{r}) = \eta[-\psi_-^*(\mathbf{r})|\psi_{1/2}\rangle + \psi_+^*(\mathbf{r})|\psi_{-1/2}\rangle] , \tag{16.77}$$

and

$$\hat{\Omega}_{\mathcal{T}}^2 \, \psi(\mathbf{r}) = -\psi(\mathbf{r}) \implies \hat{\Omega}_{\mathcal{T}}^2 = -\hat{\mathbb{1}} \qquad \text{for a spin -1/2 particle} , \tag{16.78}$$

regardless of the phase convention η that we adopt. This is in contrast with the case of a spinless particle, for which, as we found earlier, $\hat{\Omega}_{\mathcal{T}}^2 = \hat{\mathbb{1}}$.

We conclude this section with a comment on the phase convention of the angular momentum eigenstates under time reversal. In our discussion of the orbital angular momentum eigenstates, we found that with the usual phase convention for the spherical harmonics – that is, $Y_{l,m}^*(\hat{\mathbf{r}}) = (-1)^m Y_{l,-m}(\hat{\mathbf{r}})$ – it was convenient to fix the phase of these states under time reversal as

---

[4] It is worthwhile pointing out that in this form the expression for $\hat{\Omega}_{\mathcal{T}}$ holds for any spin $\mathbf{S}$, and not just for spin 1/2.

$\hat{\Omega}_{\mathcal{T}}|\psi_{l,m}\rangle = i^{2m}|\psi_{l,-m}\rangle$. We will adopt this choice of phase also for the spin-1/2 states, namely (note that different authors adopt different phase conventions)

$$\hat{\Omega}_{\mathcal{T}}|\psi_{j,m}\rangle = i^{2m}|\psi_{j,-m}\rangle , \qquad j \text{ integer or half-integer} . \tag{16.79}$$

This is accomplished by setting $\eta = i$, that is, $\hat{\Omega}_{\mathcal{T}} = i\,e^{-i\pi\hat{S}_2/\hbar}\,\hat{K} = \sigma_2\,\hat{K}$, so that

$$\hat{\Omega}_{\mathcal{T}}|\psi_\sigma\rangle = i\,(-1)^{1/2-\sigma}|\psi_{-\sigma}\rangle = i^{2\sigma}|\psi_{-\sigma}\rangle . \tag{16.80}$$

This all works out consistently; for example, we can combine two spin-1/2 states to form states of total spin $S = 1$ to find

$$\hat{\Omega}_{\mathcal{T}}|\psi_{1-1}\rangle = [\hat{\Omega}_{\mathcal{T}}|\psi_{-1/2}(1)\rangle] \otimes [\hat{\Omega}_{\mathcal{T}}|\psi_{-1/2}(2)\rangle] = i^{-1}\,i^{-1}|\psi_{1/2}(1)\rangle \otimes |\psi_{1/2}(2)\rangle = i^{-2}|\psi_{11}\rangle , \tag{16.81}$$

in agreement with the convention of Eq. (16.79).

16.4 Problems

Problem 1 Transformation of the Position Operator under a Translation

Show that, under a translation by a displacement $\mathbf{a}$, the position operator of a particle transforms as $\hat{\mathbf{r}} \longmapsto \hat{\mathbf{r}}' = \hat{\mathbf{r}} - \mathbf{a}$.

Solution

We set $\mathbf{a} = a\,\hat{\mathbf{n}}$, where $\hat{\mathbf{n}}$ is a unit vector, and use Eq. (16.9) to find

$$\hat{\mathbf{r}}' = e^{-ia\hat{\mathbf{n}}\cdot\hat{\mathbf{p}}/\hbar}\,\hat{\mathbf{r}}\,e^{ia\hat{\mathbf{n}}\cdot\hat{\mathbf{p}}/\hbar} = \hat{\mathbf{r}} - (ia/\hbar)[\hat{\mathbf{n}}\cdot\hat{\mathbf{p}},\,\hat{\mathbf{r}}] + \frac{(-ia/\hbar)^2}{2!}\,[\hat{\mathbf{n}}\cdot\hat{\mathbf{p}},\,[\hat{\mathbf{n}}\cdot\hat{\mathbf{p}},\,\hat{\mathbf{r}}]] + \cdots .$$

The only non-vanishing commutator is the first, since

$$[\hat{\mathbf{n}}\cdot\hat{\mathbf{p}},\,\hat{r}_i] = \sum_j \hat{n}_j[\hat{p}_j,\,\hat{r}_i] = -i\hbar\sum_j \delta_{ij}\,\hat{n}_j = -i\hbar\,\hat{n}_i \implies [\hat{\mathbf{n}}\cdot\hat{\mathbf{p}},\,\hat{\mathbf{r}}] = -i\hbar\,\hat{\mathbf{n}}$$

is a c-number. Hence, we have $\hat{\mathbf{r}}' = \mathbf{r} - i(a/\hbar)(-i\hbar\,\hat{\mathbf{n}}) = \mathbf{r} - a\,\hat{\mathbf{n}}$.

Problem 2 Charged Particle in a Harmonic Oscillator Potential and a Uniform Electric Field

Consider a particle of charge q in a harmonic potential $(m\omega^2/2)\,\hat{x}^2$ and under the influence of a uniform electric field $\mathcal{E}$ in the x-direction. Obtain the eigenvalues and corresponding eigenfunctions for this system.

Solution

The Hamiltonian governing this system reads

$$\hat{H} = \frac{\hat{p}^2}{2m} + \frac{m\omega^2}{2}\,\hat{x}^2 - q\mathcal{E}\,\hat{x} = \frac{\hat{p}^2}{2m} + \frac{m\omega^2}{2}\Big(\hat{x} - \underbrace{\frac{q\mathcal{E}}{m\omega^2}}_{a}\Big)^2 - \underbrace{\frac{(q\mathcal{E})^2}{2m\omega^2}}_{b} .$$

Up to the constant term b, it can be obtained from $\hat{H}_0$, where

$$\hat{H}_0 = \frac{\hat{p}^2}{2m} + \frac{m\omega^2}{2}\hat{x}^2 \, ,$$

by carrying out a translation of the system by a, namely

$$\underbrace{\frac{\hat{p}^2}{2m} + \frac{m\omega^2}{2}(\hat{x} - a)^2}_{\hat{H}+b} = \hat{U}_T(a)\, H_0 \,\hat{U}_T^\dagger(a) \, ,$$

since the momentum operator is invariant and $\hat{U}_T(a)\,\hat{x}\,\hat{U}_T^\dagger(a) = \hat{x} - a$. In a symmetry transformation, such as the translation above, the spectra of the original and transformed observables are the same, and the eigenvalues of $\hat{H}$ are $\epsilon_n = \hbar\omega(n+1/2) - b$, where n is a non-negative integer. The eigenstates $|n'\rangle$ of $\hat{H}$ are related to the eigenstates $|n\rangle$ of $\hat{H}_0$ via

$$|n'\rangle = \hat{U}_T(a)|n\rangle = \mathrm{e}^{-ia\hat{p}/\hbar}\,|n\rangle \, ,$$

or, in terms of wave functions, $\psi_n'(x) = \psi_n(x-a)$, where $\psi_n(x)$ are the wave functions of the unshifted harmonic oscillator given in Eq. (8.26).

Problem 3 Periodic Potential and Bloch Waves

Consider a particle in one dimension in a periodic potential such that $V(x - L) = V(x)$, where L is a fixed displacement. Using only symmetry arguments, show that the eigenfunctions of the Hamiltonian have the form of Bloch waves, given by

$$\varphi_m^{(q)}(x) = \mathrm{e}^{iqx/\hbar}\, u_m^{(q)}(x) \, , \qquad -\pi\hbar/L < q \le \pi\hbar/L \, ,$$

where $u_m^{(q)}(x)$ is a periodic function of period L.

Solution

The Hamiltonian describing this system,

$$\hat{H} = \frac{\hat{p}^2}{2m} + \hat{V}(\hat{x}) \, , \qquad \hat{V}(\hat{x} - L) = \hat{V}(\hat{x}) \, ,$$

is not invariant under a generic translation. However, it is invariant under a subgroup of translations, namely translations by a multiple of L (an example of a discrete transformation), since

$$\hat{U}_T(L)\,\hat{V}(\hat{x})\,\hat{U}_T^\dagger(L) = \hat{V}(\hat{x} - L) = \hat{V}(\hat{x}) \implies [\,\hat{U}_T(L)\,,\hat{H}\,] = 0 \, .$$

The last relation says that $\hat{U}_T(L)$ and $\hat{H}$ can be diagonalized simultaneously. The eigenstates of the unitary operator $\hat{U}_T(L)$ are the same as those of the momentum operator $\hat{p}$; indeed, we have

$$\hat{U}_T(L)|\psi_q\rangle = \mathrm{e}^{-iqL/\hbar}\,|\psi_q\rangle \, .$$

However, each eigenvalue $\mathrm{e}^{-iqL/\hbar}$ (which is a phase factor, since $\hat{U}_T(L)$ is unitary) is infinitely degenerate: the eigenstates $|\psi_{q+(2\pi\hbar/L)n}\rangle$ with $n = 0, \pm 1, \pm 2, \ldots$ all have the same eigenvalue $\mathrm{e}^{-iqL/\hbar}$,

$$\hat{U}_T(L)|\psi_{q+(2\pi\hbar/L)n}\rangle = \mathrm{e}^{-i[q+(2\pi\hbar/L)n]L/\hbar}\,|\psi_{q+(2\pi\hbar/L)n}\rangle = \mathrm{e}^{-iqL/\hbar}\,\underbrace{\mathrm{e}^{-i2\pi n}}_{=1}\,|\psi_{q+(2\pi\hbar/L)n}\rangle \, .$$

As a consequence, we can restrict q to values in an interval of length $2\pi\hbar/L$, say $-\pi\hbar/L < q \leq \pi\hbar/L$. Then, each eigenvalue $e^{-iqL/\hbar}$ with q restricted to this range is infinitely degenerate, with corresponding eigenstates given by

$$|\psi_q\rangle, \quad |\psi_{q\pm2\pi\hbar/L}\rangle, \quad |\psi_{q\pm4\pi\hbar/L}\rangle, \quad \ldots = |\psi_{q+2\pi n\hbar/L}\rangle\Big|_{n=-\infty}^{\infty}.$$

Since the Hamiltonian commutes with $\hat{U}_T(L)$, its matrix elements between eigenstates of $\hat{U}_T(L)$ belonging to different eigenvalues vanish,

$$\langle\psi_{k+2\pi m\hbar/L}|H|\psi_{q+2\pi n\hbar/L}\rangle = 0 \qquad \text{if } k \neq q.$$

However, provided that $k=q$, there is no restriction on the m and n values, and the Hamiltonian will generally have non-vanishing matrix elements in the degenerate subspace spanned by the eigenstates belonging to a given eigenvalue $e^{-iqL/\hbar}$ of $\hat{U}_T(L)$. In other words, in the basis of eigenstates of $\hat{U}_T(L)$, $\hat{H}$ will be represented by a block matrix:

$$\underline{H} = \begin{pmatrix} \underline{H}^{(q_1)} & 0 & 0 & 0 & \ldots 0 \\ 0 & \underline{H}^{(q_2)} & 0 & 0 & \ldots 0 \\ 0 & 0 & \underline{H}^{(q_3)} & 0 & \ldots 0 \\ 0 & 0 & 0 & \underline{H}^{(q_4)} & \ldots \\ \cdot & \cdot & \cdot & \cdot & \ldots \\ \cdot & \cdot & \cdot & \cdot & \ldots \\ \cdot & \cdot & \cdot & \cdot & \ldots \\ \cdot & \cdot & \cdot & \cdot & \ldots \\ 0 & 0 & 0 & 0 & \ldots \end{pmatrix}.$$

Each block $\underline{H}^{(q_i)}$ is characterized by a value q_i in the range $-\pi\hbar/L < q_i \leq \pi\hbar/L$, is of infinite dimension, and has matrix elements

$$H_{mn}^{(q_i)} = \underbrace{\langle\psi_{q_i+2\pi m\hbar/L}|H|\psi_{q_i+2\pi n\hbar/L}\rangle}_{\text{same } q_i}.$$

In order to diagonalize $\underline{H}$, so that

$$0 = \det(\underline{H} - E) = \det(\underline{H}^{(q_1)} - E)\det(\underline{H}^{(q_2)} - E)\cdots = \prod_i \det(\underline{H}^{(q_i)} - E),$$

it is sufficient to diagonalize each block independently, setting $\det(\underline{H}^{(q_i)} - E) = 0$, in order to determine the eigenvalues $E_m^{(q_i)}$ with $m = 1, 2, \ldots$ in that block; some of these eigenvalues may be degenerate. The corresponding eigenstates are $|\varphi_m^{(q_i)}\rangle$ with the superscript q_i specifying the block under consideration. Each of these eigenstates has the following expansion in the basis $|\psi_{q_i+2\pi n\hbar/L}\rangle$:

$$|\varphi_m^{(q_i)}\rangle = \sum_{n=-\infty}^{\infty} c_n^{(q_i,m)} |\psi_{q_i+2\pi n\hbar/L}\rangle \qquad \text{with } c_n^{(q_i,m)} = \langle\psi_{q_i+2\pi n\hbar/L}|\varphi_m^{(q_i)}\rangle.$$

To gain some insight, it is useful to consider the wave functions $\varphi_m^{(q_i)}(x) = \langle\phi_x|\varphi_m^{(q_i)}\rangle$, where

$$\varphi_m^{(q_i)}(x) = \sum_{n=-\infty}^{\infty} c_n^{(q_i,m)} \langle\phi_x|\psi_{q_i+2\pi n\hbar/L}\rangle = \sum_{n=-\infty}^{\infty} c_n^{(q_i,m)} e^{i(q_i/\hbar+2\pi n/L)x} = e^{iq_ix/\hbar} u_m^{(q_i)}(x),$$

where we have defined

$$u_m^{(q_i)}(x) = \underbrace{\sum_{n=-\infty}^{\infty} c_n^{(q_i,m)} e^{i(2\pi n/L)x}}_{\text{Fourier series}}, \qquad\qquad u_m^{(q_i)}(x) = u_m^{(q_i)}(x-L),$$

and this function is periodic in L. Thus we have derived the interesting result that in a periodic potential the eigenvalues are of the form $E_m^{(q)}$, where q varies continuously in the range $-\pi\hbar/L < q \leq \pi\hbar/L$ and m varies over discrete values. For a fixed value of m, these eigenvalues occupy a band. The corresponding eigenfunctions are given in the text of the problem, and are known as Bloch waves. We emphasize that this result was derived solely from the invariance of $\hat{H}$ under translations by L.

The result above is relevant in solid-state physics. A crystal consists of a regular lattice of atoms or molecules. Such a lattice is generally invariant under translations $\mathbf{r} \longmapsto \mathbf{r} - \mathbf{L}_r$, where $\mathbf{L}_1$, $\mathbf{L}_2$, $\mathbf{L}_3$ are three independent displacement vectors. For example, a cubic lattice of side L is invariant under translations by $L\,\hat{\mathbf{x}}$ and/or $L\,\hat{\mathbf{y}}$ and/or $L\,\hat{\mathbf{z}}$. The Hamiltonian that describes the interactions of an electron with the crystal lattice is therefore invariant under the translations induced by the three unitary operators $U_T(\mathbf{L}_r)$. As a consequence, it has eigenvalues $E_m^{(\mathbf{q})}$ and eigenfunctions $\varphi_m^{(\mathbf{q})}(\mathbf{r}) = e^{i\mathbf{q}\cdot\mathbf{r}/\hbar}\, u_m^{(\mathbf{q})}(\mathbf{r})$, with $u_m^{(\mathbf{q})}(\mathbf{r}) = u_m^{(\mathbf{q})}(\mathbf{r} - \mathbf{L}_r)$ and $\mathbf{q}$ varying continuously in the ranges $|\mathbf{q}\cdot\mathbf{L}_r| \leq \pi\hbar$ for $r = 1, 2$, and 3; in particular, for a cubic lattice, we have $-\pi\hbar/L < q_x \leq \pi\hbar/L$ and similarly for q_y and q_z.

Problem 4   Periodic Potential and Bloch Waves: An Alternative Treatment

Consider a particle in a one-dimensional periodic potential $V(x)$ with $V(x + a) = V(x)$. In order to preserve the physics under a translation by a, the eigenfunctions of the Hamiltonian must be such that

$$\rho(x + a) = \rho(x) \qquad \text{and} \qquad j(x + a) = j(x)\,,$$

where $\rho(x)$ and $j(x)$ are, respectively, the probability density and probability current density.

1. Show that the two conditions above require

$$\psi(x + a) = e^{i\phi}\,\psi(x)\,, \qquad \phi = \text{constant}\,.$$

2. Set the constant $\phi = qa$ and write the solution $\psi_q(x) \equiv e^{iqx}\,u_q(x)$. By substituting into the relation between $\psi_q(x + a)$ and $\psi_q(x)$ obtained above, show that the function $u_q(x)$ must be periodic in a, namely $u_q(x + a) = u_q(x)$.

3. Rather than assuming the crystal lattice to be of infinite length, let its length be L, that is, $L = Na$ with N, say, an even integer much greater than 2. Use periodic boundary conditions,

$$\psi_q(x + L) = \psi_q(x)\,,$$

to show that the wave number q is restricted to the range $-\pi \leq qa < \pi$ with

$$q_p = \frac{2\pi p}{Na}\,, \qquad p = 0, \pm 1, \ldots, \pm\frac{N-1}{2}, -\frac{N}{2}\,.$$

Solution

Part 1

The condition on the probability density $|\psi(x + a)|^2 = |\psi(x)|^2$ implies that

$$\psi(x + a) = e^{i\phi(x)}\,\psi(x)\,, \qquad \phi(x) \text{ a real function of } x\,.$$

The probability current density at $x + a$ then follows:

$$j(x + a) = \frac{\hbar}{2im}\left[\psi^*\psi' - \text{c.c.}\right]_{x+a} = \frac{\hbar}{2im}\left[e^{-i\phi}\psi^*\,e^{i\phi}(\psi' + i\phi'\,\psi) - \text{c.c.}\right]_x = j(x) + \frac{\hbar}{m}\,\phi'(x)|\psi(x)|^2\,,$$

and the last term must vanish, requiring $\phi(x)$ to be constant. This yields the required result, $\psi(x+a) = e^{i\phi}\psi(x)$.

Part 2

Defining $\psi_q(x) \equiv e^{iqx}u_q(x)$ and using the relation $\psi_q(x + a) = e^{iqa}\psi_q(x)$ obtained above, we find

$$\psi_q(x + a) = e^{iq(x+a)}u_q(x + a) = e^{iqa}\psi_q(x) = e^{iqa}e^{iqx}u_q(x) \implies u_q(x + a) = u_q(x) \,,$$

and so the function $u_q(x)$ is periodic with period a.

Part 3

Using the result in part 1 N times, we obtain

$$\psi_q(x + Na) = e^{iqa}\psi_q[x + (N - 1)a] = \cdots = e^{iqNa}\psi_q(x) \,.$$

Imposing the periodic boundary condition $\psi_q(x + Na) = \psi_q(x)$ yields

$$e^{iqNa} = 1 \implies q_k = \frac{2\pi k}{Na} \,, \qquad k = 0, \pm1, \ldots, \pm\frac{N-1}{2}, -\frac{N}{2}$$

and there are N distinct q_k-values in the interval $[-\pi/2, \pi/2[$. Note that the shift $k \longrightarrow k + N$ or $aq_k \longrightarrow aq_k + 2\pi$ does not lead to a new, linearly independent, solution.

Problem 5 The Kronig–Penney Model

Consider the one-dimensional potential

$$V(x) = \frac{\hbar^2}{2m}\frac{\lambda}{a}\sum_{n=1}^{N}\delta(x - na) \,,$$

where $L = Na$ is the length of the crystal. Assume periodic boundary conditions, so that $V(x + L) = V(x)$.

1. Denote by R_n the region $(n - 1)a < x < na$, namely the region between the repulsive δ-function potentials at $(n - 1)a$ and na, respectively. Show that the solution $\psi_n(x)$ of the Schrödinger equation in R_n can be taken to be of the form

$$\psi_n(x) = A_n e^{ik(x-na)} + B_n e^{-ik(x-na)} \,, \qquad k^2 = \frac{2mE}{\hbar^2} \,.$$

Next, consider the neighboring region R_{n+1}, with corresponding solution

$$\psi_{n+1}(x) = A_{n+1} e^{ik[x-(n+1)a]} + B_{n+1} e^{-ik[x-(n+1)a]} \,.$$

Impose the appropriate boundary conditions at $x = na$ and make use of the Bloch-wave condition,

$$\psi_{n+1}(x + a) = e^{iqa}\psi_n(x) \qquad \text{with} \qquad qa = \frac{2\pi p}{N} \,,$$

where p is a positive or negative integer in a range to be determined, to obtain the following eigenvalue equation:

$$\cos(qa) = \cos(ka) + \frac{\lambda}{2}\frac{\sin(ka)}{ka} \,.$$

2. Set $\xi = ka$ and obtain a graphical solution of the eigenvalue equation

$$\cos(qa) = F(\xi) , \qquad F(\xi) = \cos \xi + \frac{\lambda}{2} \frac{\sin \xi}{\xi} ,$$

where the energy eigenvalue E follows as

$$E = \frac{\hbar^2}{2ma^2} \xi^2 .$$

Discuss the two limiting cases $\lambda = 0$ and $\lambda \longrightarrow \infty$.

Solution

Part 1

In region R_n the Schrödinger equation is given by

$$\psi_n''(x) + k^2 \psi_n(x) = 0 ,$$

which has the solution

$$\psi_n(x) = A_n' \, e^{ikx} + B_n' \, e^{-ikx} = A_n \, e^{ik(x-na)} + B_n \, e^{-ik(x-na)} .$$

Similarly, in region R_{n+1} we have

$$\psi_{n+1}(x) = A_{n+1} \, e^{ik[x-(n+1)a]} + B_{n+1} \, e^{-ik[x-(n+1)a]} .$$

We now impose the boundary conditions at $x = na$, namely

$$\psi_n(na) = \psi_{n+1}(na) , \qquad \psi_{n+1}'(na) - \psi_n'(na) = \frac{\lambda}{a} \, \psi_n(na) ,$$

where the discontinuity of the derivative is due to the presence of the δ-function potential at na. These boundary conditions lead to

$$A_n + B_n = A_{n+1} \, e^{-ika} + B_{n+1} \, e^{ika}$$

and

$$ik \left(A_{n+1} \, e^{-ika} - B_{n+1} \, e^{ika} \right) - ik(A_n - B_n) = \frac{\lambda}{a} \, (A_n + B_n) .$$

Solving with respect to A_{n+1} and B_{n+1} gives

$$A_{n+1} = e^{ika} \left[\left(1 - i \frac{\lambda}{2ka} \right) A_n - i \frac{\lambda}{2ka} B_n \right] , \qquad B_{n+1} = e^{-ika} \left[\left(1 + i \frac{\lambda}{2ka} \right) B_n + i \frac{\lambda}{2ka} A_n \right] .$$

In regions R_n and R_{n+1} the wave functions at, respectively, x and $x + a$ read

$$\psi_n(x) = A_n \, e^{ik(x-na)} + B_n \, e^{-ik(x-na)}$$

and

$$\psi_{n+1}(x + a) = A_{n+1} \, e^{ik[x+a-(n+1)a]} + B_{n+1} \, e^{-ik[x+a-(n+1)a]} = A_{n+1} \, e^{ik(x-na)} + B_{n+1} \, e^{-ik(x-na)} .$$

Use of the Bloch-wave condition between these wave functions in R_n and R_{n+1} (if x is in region R_n then $x + a$ is in region R_{n+1}), namely

$$\psi_{n+1}(x + a) = e^{iqa} \, \psi_n(x) ,$$

leads to

$$A_{n+1}\, e^{ik(x-na)} + B_{n+1}\, e^{-ik(x-na)} = e^{iqa}\left[A_n\, e^{ik(x-na)} + B_n\, e^{-ik(x-na)}\right],$$

requiring

$$A_{n+1} = e^{iqa}\, A_n\,, \qquad B_{n+1} = e^{iqa}\, B_n\,.$$

Because of the periodic boundary condition, the wave number q is restricted to have the values (assuming that N is large and even)

$$q = \frac{2\pi p}{Na}\,, \qquad p = 0, \pm 1, \ldots, \pm\frac{N-1}{2}, -\frac{N}{2}\,.$$

Substituting the expressions for A_{n+1} and B_{n+1} into the earlier relations derived from imposing the boundary conditions at na, we arrive at the homogenous linear system in A_n and B_n given by

$$\underbrace{\left[e^{iqa} - e^{ika}\left(1 - i\frac{\lambda}{2ka}\right)\right]}_{\alpha} A_n + \underbrace{i\frac{\lambda}{2ka}\, e^{ika}}_{\beta} B_n = 0\,, \qquad \underbrace{\left[e^{iqa} - e^{-ika}\left(1 + i\frac{\lambda}{2ka}\right)\right]}_{\gamma} B_n \underbrace{-i\frac{\lambda}{2ka}\, e^{-ika}}_{\beta^*} A_n = 0\,.$$

The condition for non-trivial solutions is $\alpha\gamma - |\beta|^2 = 0$ or, equivalently,

$$e^{2iqa} - e^{iqa}\left[e^{-ika}\left(1 + i\frac{\lambda}{2ka}\right) + e^{ika}\left(1 - i\frac{\lambda}{2ka}\right)\right] + 1 = 0\,.$$

Multiplying both sides of the above relation by e^{-iqa}, we arrive at

$$e^{iqa} + e^{-iqa} = e^{-ika}\left(1 + i\frac{\lambda}{2ka}\right) + e^{ika}\left(1 - i\frac{\lambda}{2ka}\right),$$

which can be written as

$$2\cos(qa) = 2\cos(ka) + \frac{\lambda}{ka}\sin(ka) \implies \cos(qa) = \cos(ka) + \frac{\lambda}{2}\frac{\sin(ka)}{ka}\,.$$

Part 2

Setting $\xi = ka$, the eigenvalue equation is simply

$$F(\xi) = \cos\xi + \frac{\lambda}{2}\frac{\sin\xi}{\xi} = \cos(2\pi p/N)\,, \qquad p = 0, \pm 1, \ldots, \pm\frac{N-1}{2}, -\frac{N}{2}\,,$$

and can be solved graphically; see Fig. 16.1. Note that, since $|\cos(2\pi p/N)| \leq 1$, the solutions for ξ must lie in the intervals delimited by the vertical lines and labeled 1, 2, … in the figure. Thus, there are gaps between the ξs belonging to different intervals. These gaps become smaller and smaller as ξ increases. For a fixed $\cos(2\pi p/N)$, there is an infinite number of solutions ξ_{np} with $n = 1, 2, \ldots$ and corresponding energies

$$E_{np} = \frac{\hbar^2}{2ma^2}\,\xi_{np}^2\,.$$

In the limiting case $\lambda = 0$, the problem becomes that of a free particle in $0 \leq x \leq L = Na$ with periodic boundary conditions, that is

$$\psi_q(x) = \frac{1}{\sqrt{L}}\, e^{iqx} \qquad \text{with} \qquad \psi_q(x) = \psi_q(x + L) \implies q = \frac{2\pi p}{Na}\,,$$

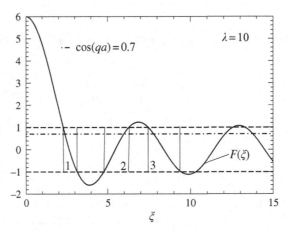

Fig. 16.1 Graphical solution of the eigenvalue equation. The ξ-solutions must lie in the intervals labeled 1, 2, ... delimited by the vertical lines, in which $|F(\xi)| \leq 1$.

and energies $E = (\hbar q)^2/(2m)$. Note that for $\lambda = 0$ the eigenvalue equation reduces to $\cos(qa) = \cos \xi$, which has the solution $\xi = qa = 2\pi p/N$, in agreement with the result above.

By contrast, in the limit $\lambda \longrightarrow \infty$, the solutions of the eigenvalue equation are $\xi_n = n\pi$ and are independent of p. In this limit, the intervals 1, 2, ... shrink to points, which are the ξ_n exactly (a glance at Fig. 16.1 above makes this obvious). At these values $F(\xi_n) = (-1)^n$, independently of λ. Thus, we see that in this "tight-binding limit" the energy eigenvalues reduce to those of a particle confined in a segment of length a; namely, the particle is confined in one of the regions R_n.

Problem 6   Construction of the Rotation Matrices from the Infinitesimal Generators

Starting from the expression for a rotation matrix given by $\underline{R}(\varphi\,\hat{\mathbf{n}}) = e^{\varphi\hat{\mathbf{n}}\cdot\underline{\mathbf{E}}}$, where $\underline{\mathbf{E}} = (\underline{E}_1, \underline{E}_2, \underline{E}_3)$ and the three antisymmetric matrices $\underline{E}_1$, $\underline{E}_2$, and $\underline{E}_3$ are defined in Eq. (16.27), verify explicitly that

$$\underline{R}(\varphi\,\hat{\mathbf{e}}_3) = \begin{pmatrix} \cos\varphi & -\sin\varphi & 0 \\ \sin\varphi & \cos\varphi & 0 \\ 0 & 0 & 1 \end{pmatrix}.$$

Solution

For a rotation about $\hat{\mathbf{e}}_3$, the rotation matrix is given by

$$\underline{R}(\varphi\,\hat{\mathbf{e}}_3) = e^{\varphi\underline{E}_3} = \sum_{p=0}^{\infty} \frac{\varphi^p}{p!} \left(\underline{E}_3\right)^p .$$

By direct multiplication of the matrices, we find

$$(\underline{E}_3)^2 = \begin{pmatrix} -1 & 0 & 0 \\ 0 & -1 & 0 \\ 0 & 0 & 0 \end{pmatrix} \equiv -\underline{E}_0 , \qquad (\underline{E}_3)^3 = \begin{pmatrix} 0 & 1 & 0 \\ -1 & 0 & 0 \\ 0 & 0 & 0 \end{pmatrix} = -\underline{E}_3 , \qquad (\underline{E}_3)^4 = \underline{E}_0 , \qquad (\underline{E}_3)^5 = \underline{E}_3\,\underline{E}_0 = \underline{E}_3 ,$$

from which it follows that

$$\underline{R}(\varphi\,\hat{e}_3) = \mathbb{1} + \frac{\varphi}{1!}\,\underline{E}_3 - \frac{\varphi^2}{2!}\,\underline{E}_0 - \frac{\varphi^3}{3!}\,\underline{E}_3 + \frac{\varphi^4}{4!}\,\underline{E}_0 + \frac{\varphi^5}{5!}\,\underline{E}_3 + \cdots$$

$$= \mathbb{1} + (\cos\varphi - 1)\underline{E}_0 + \sin\varphi\,\underline{E}_3 = \begin{pmatrix} \cos\varphi & -\sin\varphi & 0 \\ \sin\varphi & \cos\varphi & 0 \\ 0 & 0 & 1 \end{pmatrix}.$$

Problem 7 Explicit Expression for the Rotation Operator in Spin Space

Show that the rotation operator in spin space is given by

$$\hat{U}_R^s(\varphi\hat{n}) = \cos(\varphi/2) - i\,\sin(\varphi/2)\,\hat{n}\cdot\underline{\sigma}\,.$$

Consider a particle in a state $|+\rangle$ along the z-direction. Suppose that it is subject to a (counterclockwise) rotation by φ about the x-direction. Show that the resulting state is an eigenstate of the $\hat{S}$ component along the direction $(0, -\sin\varphi, \cos\varphi)$ with eigenvalue $\hbar/2$.

Solution

In spin space, we have

$$(\hat{n}\cdot\hat{S})^2 \longrightarrow (\hbar^2/4)\,(\hat{n}\cdot\underline{\sigma})\,(\hat{n}\cdot\underline{\sigma}) = (\hbar^2/4)\,\hat{n}\cdot\hat{n} = \hbar^2/4 \implies (\hat{n}\cdot\hat{S})^{2p} \longrightarrow (\hbar/2)^{2p} \quad \text{and}$$

$$(\hat{n}\cdot\hat{S})^{2p+1} \longrightarrow (\hbar/2)^{2p+1}\,\hat{n}\cdot\underline{\sigma}\,,$$

where p is a non-negative integer, and therefore

$$\hat{U}_R^s(\varphi\hat{n}) = \sum_{p=0}^{\infty} \frac{(-i\varphi/\hbar)^{2p}}{(2p)!}\,(\hat{n}\cdot\hat{S})^{2p} + \sum_{p=0}^{\infty} \frac{(-i\varphi/\hbar)^{2p+1}}{(2p+1)!}\,(\hat{n}\cdot\hat{S})^{2p+1}$$

$$= \underbrace{\sum_{p=0}^{\infty}(-1)^p\frac{(\varphi/2)^{2p}}{(2p)!}}_{\cos(\varphi/2)} - i\,\hat{n}\cdot\underline{\sigma}\underbrace{\sum_{p=0}^{\infty}(-1)^p\frac{(\varphi/2)^{2p+1}}{(2p+1)!}}_{\sin(\varphi/2)},$$

resulting in the expression given in the text of the problem. In the specific case of rotation by φ about the x-direction, namely $\hat{n} = (1, 0, 0)$, we have

$$\hat{U}_R^s(\varphi\,\hat{e}_x) = \cos(\varphi/2) - i\,\sin(\varphi/2)\,\underline{\sigma}_x = \begin{pmatrix} \cos(\varphi/2) & -i\,\sin(\varphi/2) \\ -i\,\sin(\varphi/2) & \cos(\varphi/2) \end{pmatrix},$$

so that

$$|+\rangle' = \hat{U}_R^s(\varphi\,\hat{e}_x)|+\rangle = \cos(\varphi/2)\,|+\rangle - i\,\sin(\varphi/2)\,|-\rangle\,.$$

The state $|+\rangle'$ represents a particle polarized in the direction $(0, -\sin\varphi, \cos\varphi)$ in the yz-plane, and hence it must be an eigenstate of the spin operator along this direction with eigenvalue $+\hbar/2$,

$$(-\sin\varphi\,\hat{S}_y + \cos\varphi\,\hat{S}_z)|+\rangle' = \frac{\hbar}{2}\,|+\rangle',$$

which can be verified directly:

$$\frac{\hbar}{2}\underbrace{\begin{pmatrix} \cos\varphi & i\sin\varphi \\ -i\sin\varphi & -\cos\varphi \end{pmatrix}}_{-\sin\varphi\,\hat{S}_y+\cos\varphi\,\hat{S}_z}\begin{pmatrix} \cos(\varphi/2) \\ -i\sin(\varphi/2) \end{pmatrix} = \frac{\hbar}{2}\begin{pmatrix} \cos\varphi\,\cos(\varphi/2)+\sin\varphi\,\sin(\varphi/2) \\ -i\sin\varphi\,\cos(\varphi/2)+i\cos\varphi\,\sin(\varphi/2) \end{pmatrix} = \frac{\hbar}{2}\begin{pmatrix} \cos(\varphi/2) \\ -i\sin(\varphi/2) \end{pmatrix},$$

where we have made use of the identities $\cos\varphi = 2\cos^2(\varphi/2) - 1$ and $\sin\varphi = 2\sin(\varphi/2)\cos(\varphi/2)$. We also see that under a rotation by 2π the state $|+\rangle'$ differs from $|+\rangle$ by the phase factor -1. This is obvious from the relation $\hat{U}_R^s(2\pi\,\hat{\mathbf{n}}) = -\hat{\mathbb{1}}$. Finally, we note that the spin operator $-\sin\varphi\,\hat{S}_y + \cos\varphi\,\hat{S}_z$ in the direction $(0, -\sin\varphi, \cos\varphi)$ can be obtained by rotating $\hat{S}_z$ in this direction; explicitly,

$$\hat{U}_R^s(\varphi\,\hat{\mathbf{e}}_x)\,\hat{S}_z\,\hat{U}_R^{s\dagger}(\varphi\,\hat{\mathbf{e}}_x) = -\sin\varphi\,\hat{S}_y + \cos\varphi\,\hat{S}_z\,.$$

Indeed, using $\underline{\sigma}_x\underline{\sigma}_z = -i\underline{\sigma}_y$ and $\underline{\sigma}_y\underline{\sigma}_x = -i\underline{\sigma}_z$, we find

$$[\cos(\varphi/2) - i\sin(\varphi/2)\,\underline{\sigma}_x]\underline{\sigma}_z[\cos(\varphi/2) + i\sin(\varphi/2)\,\underline{\sigma}_x] = \underbrace{[\cos^2(\varphi/2) - \sin^2(\varphi/2)]}_{\cos\varphi}\underline{\sigma}_z$$

$$-\underbrace{2\sin(\varphi/2)\cos(\varphi/2)}_{\sin\varphi}\underline{\sigma}_y\,.$$

Problem 8   Rotation by 2π

Consider a single particle and show that the rotation operator $U_R(2\pi\hat{\mathbf{n}})$ is $\pm\,\hat{\mathbb{1}}$, depending on whether the particle has integer (+ sign) or half-integer (− sign) spin. Show that a 2π rotation of a system containing an even or odd number of particles with half-integer spins (and any number of particles with integer spins) is equal to plus or minus the identity operator.

Solution

Consider a generic state $|\psi\rangle$ of a single particle and expand it in a basis of eigenstates $|\phi_{\alpha jm}\rangle$ of $\hat{\mathbf{J}}^2$ and $\hat{\mathbf{n}}\cdot\hat{\mathbf{J}}$ and other observables commuting with them (in other words, $\hat{\mathbf{J}}^2$ and $\hat{\mathbf{n}}\cdot\hat{\mathbf{J}}$ along with the other observables form a complete set of commuting observables); here α denotes collectively the eigenvalues corresponding to these other observables. We have

$$|\psi\rangle = \sum_{\alpha jm} c_{\alpha jm}|\phi_{\alpha jm}\rangle\,, \qquad c_{\alpha jm} = \langle\phi_{\alpha jm}|\psi\rangle\,,$$

and

$$\hat{U}_R(2\pi\hat{\mathbf{n}})|\psi\rangle = \sum_{\alpha jm} c_{\alpha jm}\,\hat{U}_R(2\pi\hat{\mathbf{n}})|\phi_{\alpha jm}\rangle = \sum_{\alpha jm} c_{\alpha jm}\,e^{-2\pi im}|\phi_{\alpha jm}\rangle = \sum_{\alpha jm}(-1)^{2m}\,c_{\alpha jm}|\phi_{\alpha jm}\rangle = \pm|\psi\rangle\,,$$

where we have used the eigenstate property $\hat{\mathbf{n}}\cdot\hat{\mathbf{J}}|\phi_{\alpha jm}\rangle = m\hbar\,|\phi_{\alpha jm}\rangle$. The phase $(-1)^{2m}$ is either $+1$ or -1 depending on whether j is either integer or half-integer. The last step in the equation above follows because if the particle has integer spin, only integer j can occur in the sum over j and $2m$ is always even whereas if the particle has half-integer spin, only half-integer j enter the sum and $2m$ is always odd (if the particle has integer spin, the total angular momentum $\mathbf{J} = \mathbf{L} + \mathbf{S}$ assumes integer values; conversely, if the particle has half-integer spin, the total angular momentum assumes half-integer values). These conclusions also apply to the case in which $|\psi\rangle$ represents the state of a system consisting of any number of particles having integer spins and either an even or an odd number of particles with half-integer spins. In the former case, the total angular momentum can only assume integer values, while in the latter it can only assume half-integer values.

Problem 9 Consequences of Rotational Invariance

Show that a state $|\psi\rangle$ is rotationally invariant iff $\hat{\mathbf{J}}|\psi\rangle = 0$. Let $\hat{A}$ be a scalar operator and $|\psi_{\alpha jm}\rangle$ a basis of simultaneous eigenstates of a complete set of commuting observables, including $\hat{\mathbf{J}}^2$ and $\hat{J}_z$ (α denotes collectively the eigenvalues of the remaining observables). Derive selection rules for the matrix elements $\langle \psi_{\alpha jm}|\hat{A}|\psi_{\alpha'j'm'}\rangle$ and show, in particular, that these matrix elements are independent of the azimuthal quantum numbers.

Solution

A state $|\psi\rangle$ is invariant under rotations when $|\psi\rangle \longmapsto |\psi'\rangle = \hat{U}_R(\varphi\hat{\mathbf{n}})|\psi\rangle = |\psi\rangle$ for *any choice* of the rotation axis $\hat{\mathbf{n}}$. For an infinitesimal rotation, this condition leads to

$$|\psi'\rangle = (\hat{\mathbb{1}} - i\eta\,\hat{\mathbf{n}}\cdot\hat{\mathbf{J}}/\hbar)|\psi\rangle = |\psi\rangle \implies \hat{\mathbf{n}}\cdot\hat{\mathbf{J}}|\psi\rangle = 0 \qquad \text{for any } \hat{\mathbf{n}}\,,$$

which implies $\hat{\mathbf{J}}|\psi\rangle = 0$. The converse statement is obviously true: if $\hat{\mathbf{J}}|\psi\rangle = 0$ then the state is rotationally invariant. For example, in the case of a spinless particle the state $|\psi\rangle$ is rotationally invariant when the corresponding wave function $\psi(\mathbf{r})$ is a function only of the magnitude $|\mathbf{r}|$, since then $\psi'(|\mathbf{r}|) = \psi(|R^{-1}\mathbf{r}|) = \psi(|\mathbf{r}|)$.

A necessary and sufficient condition for an operator $\hat{A}$ to be invariant under rotations – that is, to be a scalar operator – is that it commutes with $\hat{\mathbf{J}}$ (the generator of rotations), in which case $\hat{A} \longmapsto \hat{A}' = \hat{U}_R(\varphi\hat{\mathbf{n}})\,\hat{A}\,\hat{U}_R^\dagger(\varphi\hat{\mathbf{n}}) = \hat{A}$. Now, an operator that commutes with $\hat{\mathbf{J}}$ also commutes with $\hat{\mathbf{J}}^2$, and hence matrix elements of $\hat{A}$ between eigenstates of $\hat{\mathbf{J}}^2$ and $\hat{J}_z$ belonging to different eigenvalues vanish,

$$\langle \psi_{\alpha jm}|\hat{A}|\psi_{\alpha'j'm'}\rangle = 0 \qquad \text{if } j \neq j' \text{ and/or } m \neq m'\,;$$

generally, there are no restrictions on α and α'. Further, using

$$|\psi_{\alpha jm}\rangle = \frac{1}{\hbar\sqrt{j(j+1)-m(m-1)}}\,\hat{J}_+|\psi_{\alpha j,m-1}\rangle\,, \qquad |\psi_{\alpha j,m-1}\rangle = \frac{1}{\hbar\sqrt{j(j+1)-m(m-1)}}\,\hat{J}_-|\psi_{\alpha jm}\rangle\,,$$

and $\langle \psi_{\alpha jm}|\hat{J}_+ = (\hat{J}_-|\psi_{\alpha jm}\rangle)^\dagger$, we conclude that the matrix elements are independent of m,

$$\langle \psi_{\alpha jm}|\hat{A}|\psi_{\alpha'jm}\rangle = \frac{\langle \psi_{\alpha jm}|\hat{A}\hat{J}_+|\psi_{\alpha'j,m-1}\rangle}{\hbar\sqrt{j(j+1)-m(m-1)}} = \frac{((\langle \psi_{\alpha jm}|\hat{J}_+)\,\hat{A}|\psi_{\alpha'j,m-1}\rangle)}{\hbar\sqrt{j(j+1)-m(m-1)}} = \langle \psi_{\alpha j,m-1}|\hat{A}|\psi_{\alpha'j,m-1}\rangle\,.$$

Lastly, if the state $|\psi_{\alpha jm}\rangle$ happens to be also an eigenstate of $\hat{A}$ (because, for example, $\hat{A}$ is one of the observables in the complete set) and $\hat{A}|\psi_{\alpha jm}\rangle = a|\psi_{\alpha jm}\rangle$, then the eigenvalue a is $(2j+1)$-fold degenerate, since

$$\hat{A}\,\underbrace{\hat{J}_\pm|\psi_{\alpha jm}\rangle}_{\propto\,|\psi_{\alpha j(m\pm1)}\rangle} = \hat{J}_\pm\hat{A}|\psi_{\alpha jm}\rangle = a\hat{J}_\pm|\psi_{\alpha jm}\rangle\,,$$

and the $2j+1$ states $|\psi_{\alpha j,-j}\rangle, |\psi_{\alpha j,-(j-1)}\rangle, \ldots, |\psi_{\alpha j,+j}\rangle$ each have the same eigenvalue a. Therefore, the rotational invariance of the Hamiltonian implies, at least, the $(2j+1)$-degeneracy of each of its eigenvalues. Of course, if additional symmetries are present, as in the case of the Coulomb or harmonic oscillator potentials, the degeneracy can be larger.

Problem 10   Matrix Elements of Vector Operators and Rotations

Let $|lm\rangle$ be the basis of eigenstates of $\mathbf{L}^2$ and L_z, with eigenvalues $l(l+1)\hbar^2$ and $m\hbar$, respectively, and let $U_R(\varphi\hat{\mathbf{z}})$ be the unitary operator inducing a rotation by φ about the z-axis,

$$U_R(\varphi\hat{\mathbf{z}}) = e^{-i\varphi L_z/\hbar} \ .$$

1. Show that under such a rotation, the raising and lowering operators transform as

$$L_\pm \longmapsto L'_\pm = e^{\mp i\varphi}\, L_\pm \ .$$

   Hint: Consider the action of $L'_\pm$ on the basis $|lm\rangle$.

2. Express the transformed operators L'_x, L'_y, and L'_z in terms of L_x, L_y, and L_z. Are these expressions consistent with the transformation law under rotations of vector operators discussed in Section 16.2?

3. Let $\mathbf{V}$ be a generic vector operator. Calculate the commutators $[V_x \pm iV_y, L_z]$ and $[V_z, L_z]$. Show that the states $(V_x \pm iV_y)|lm\rangle$ and $V_z|lm\rangle$ are eigenstates of L_z and calculate their eigenvalues. What relation must exist between m and m' for the matrix element $\langle l'm'|V_x \pm iV_y|lm\rangle$ to be non-zero? Consider the same question for $\langle l'm'|V_z|lm\rangle$.

4. By comparing the matrix elements of $V'_x \pm i V'_y$ and V'_z with those of $V_x \pm iV_y$ and V_z, calculate V'_x, V'_y, and V'_z in terms of V_x, V_y, and V_z.

Solution
Part 1

Consider

$$
\begin{aligned}
L'_+|lm\rangle &= e^{-i\varphi L_z/\hbar}\, L_+\, e^{i\varphi L_z/\hbar}|lm\rangle = e^{im\varphi}\, e^{-i\varphi L_z/\hbar}\, L_+\,|lm\rangle \\
&= e^{im\varphi}\,\hbar\sqrt{l(l+1)-m(m+1)}\, e^{-i\varphi L_z/\hbar}\,|l(m+1)\rangle = e^{-i\varphi}\, L_+|lm\rangle \ ,
\end{aligned}
$$

and similarly $L'_-|lm\rangle = e^{i\varphi} L_-|lm\rangle$), and, since the states $|lm\rangle$ form a basis, it follows that $L'_\pm = e^{\mp i\varphi} L_\pm$.

Part 2

Take the linear combinations

$$
L'_x = \frac{L'_+ + L'_-}{2} = \frac{e^{-i\varphi} L_+ + e^{i\varphi} L_-}{2} = \frac{e^{-i\varphi} + e^{i\varphi}}{2} L_x + \frac{e^{i\varphi} - e^{-i\varphi}}{2i} L_y = \cos\varphi\, L_x + \sin\varphi\, L_y \ ,
$$

$$
L'_y = \frac{L'_+ - L'_-}{2i} = \frac{e^{-i\varphi} L_+ - e^{i\varphi} L_-}{2i} = -\frac{e^{i\varphi} - e^{-i\varphi}}{2i} L_x + \frac{e^{i\varphi} + e^{-i\varphi}}{2} L_y = -\sin\varphi\, L_x + \cos\varphi\, L_y \ ,
$$

$$
L'_z = e^{-i\varphi L_z/\hbar}\, L_z\, e^{i\varphi L_z/\hbar} = L_z \ ,
$$

where we have used $L_\pm = L_x \pm iL_y$ and similarly for the primed quantities. This is consistent with the transformation law of vector operators under rotations. Indeed,

$$
L'_i = U_R(\varphi\hat{\mathbf{z}})\, L_i\, U_R^\dagger(\varphi\hat{\mathbf{z}}) = \sum_j [\underline{R}^{-1}(\varphi\hat{\mathbf{z}})]_{ij}\, L_j \ ,
$$

or

$$\begin{pmatrix} L'_x \\ L'_y \\ L'_z \end{pmatrix} = \underbrace{\begin{pmatrix} \cos\varphi & \sin\varphi & 0 \\ -\sin\varphi & \cos\varphi & 0 \\ 0 & 0 & 1 \end{pmatrix}}_{\underline{R}^{-1}(\varphi\hat{z})} \begin{pmatrix} L_x \\ L_y \\ L_z \end{pmatrix} .$$

Part 3

A vector operator is defined by the following commutation relations with the angular momentum operator:

$$\left[L_i , V_j \right] = i\hbar \sum_k \epsilon_{ijk} V_k ,$$

We define $V_\pm = V_x \pm iV_y$, and obtain

$$[L_z , V_\pm] = [L_z , V_x] \pm i[L_z , V_y] = i\hbar V_y \pm \hbar V_x = \pm\hbar V_\pm , \qquad [L_z , V_z] = 0 ,$$

and hence (from the commutators above it follows that $L_z V_\pm = V_\pm L_z \pm \hbar V_\pm$)

$$L_z V_\pm |lm\rangle = (V_\pm L_z \pm \hbar V_\pm)|lm\rangle = (m \pm 1)\hbar V_\pm |lm\rangle , \qquad L_z V_z |lm\rangle = m\hbar V_z |lm\rangle .$$

Multiply the relations above on the left by the bra $\langle l'm'|$ to obtain

$$\underbrace{\langle l'm'|L_z V_\pm |lm\rangle}_{m'\hbar\langle l'm'|V_\pm |lm\rangle} = (m \pm 1)\hbar \langle l'm'|V_\pm |lm\rangle \implies m' = m \pm 1 ,$$

and

$$\underbrace{\langle l'm'|L_z V_z |lm\rangle}_{m'\hbar\langle l'm'|V_z |lm\rangle} = m\hbar \langle l'm'|V_z |lm\rangle \implies m' = m .$$

Part 4

Using the result above, we obtain

$$\langle l'm'|V'_\pm |lm\rangle = \langle l'm'|U_R V_\pm U_R^\dagger |lm\rangle = e^{-im'\varphi} \langle l'm'|V_\pm |lm\rangle e^{im\varphi} = e^{\mp i\varphi} \langle l'm'|V_\pm |lm\rangle ,$$

where in the last step we have used the selection rule $m' = m \pm 1$. Similarly we obtain

$$\langle l'm'|V'_z |lm\rangle = \langle l'm'|U_R V_z U_R^\dagger |lm\rangle = \langle l'm'| V_z |lm\rangle ,$$

since V_z and U_R commute. Therefore, we have

$$V'_\pm = e^{\mp i\varphi} V_\pm , \qquad V'_z = V_z .$$

Using the result of part 2, it follows that

$$V'_x = V_x \cos\varphi + V_y \sin\varphi , \qquad V'_y = -V_x \sin\varphi + V_y \cos\varphi , \qquad V'_z = V_z ,$$

in accordance with expectations, namely

$$\begin{pmatrix} V'_x \\ V'_y \\ V'_z \end{pmatrix} = \begin{pmatrix} \cos\varphi & \sin\varphi & 0 \\ -\sin\varphi & \cos\varphi & 0 \\ 0 & 0 & 1 \end{pmatrix} \begin{pmatrix} V_x \\ V_y \\ V_z \end{pmatrix} .$$

Problem 11 Transformation of a Spin-1/2 Angular Momentum Operator Under a General Rotation

Consider the transformation properties of the spin-1/2 angular momentum $\hat{\mathbf{S}}$ under a rotation. Show that, under a (finite) rotation by an angle φ about an axis $\hat{\mathbf{n}}$,

$$\hat{U}_R(\varphi\hat{\mathbf{n}}) \, \hat{\mathbf{S}} \, \hat{U}_R^\dagger(\varphi\hat{\mathbf{n}}) = \hat{\mathbf{n}}(\hat{\mathbf{n}} \cdot \hat{\mathbf{S}}) - \hat{\mathbf{n}} \times (\hat{\mathbf{n}} \times \hat{\mathbf{S}}) \, \cos\varphi - \hat{\mathbf{n}} \times \hat{\mathbf{S}} \, \sin\varphi \, ,$$

where the unitary operator $\hat{U}_R(\varphi\hat{\mathbf{n}})$ is $e^{-i\varphi\hat{\mathbf{n}}\cdot\hat{\mathbf{S}}/\hbar}$. Verify that in the limit $\varphi = \eta$, with η infinitesimal, the formula above reduces to the familiar result.

Solution

Using $\hat{\mathbf{S}} = (\hbar/2)\sigma$, the identity to be proven reduces to

$$\hat{U}_R(\varphi\hat{\mathbf{n}}) \, \sigma \, \hat{U}_R^\dagger(\varphi\hat{\mathbf{n}}) = \hat{\mathbf{n}} \, (\hat{\mathbf{n}} \cdot \sigma) - \hat{\mathbf{n}} \times (\hat{\mathbf{n}} \times \sigma) \, \cos\varphi - \hat{\mathbf{n}} \times \sigma \, \sin\varphi$$

$$= \sigma \cos\varphi + \hat{\mathbf{n}} \cdot (\hat{\mathbf{n}} \cdot \sigma) \, (1 - \cos\varphi) - \hat{\mathbf{n}} \times \sigma \, \sin\varphi \, ,$$

where in the last expression we have used the well-known formula for the double cross product. Recalling that

$$\hat{U}_R(\varphi\hat{\mathbf{n}}) = \cos(\varphi/2) - i \, \sigma \cdot \hat{\mathbf{n}} \, \sin(\varphi/2) \, ,$$

we first work out

$$\underbrace{\sigma \cdot \hat{\mathbf{e}}_i}_{\sigma_i} \underbrace{(\cos\varphi/2 + i \, \sigma \cdot \hat{\mathbf{n}} \, \sin\varphi/2)}_{\hat{U}_R^\dagger(\varphi\hat{\mathbf{n}})} = \sigma \cdot \hat{\mathbf{e}}_i \cos\varphi/2 + i \hat{\mathbf{e}}_i \cdot \hat{\mathbf{n}} \, \sin\varphi/2 - \sigma \cdot (\hat{\mathbf{e}}_i \times \hat{\mathbf{n}}) \sin\varphi/2 \, ,$$

where use has been made of the Pauli identity. Next, we consider

$$
\begin{aligned}
\hat{U}_R(\varphi\hat{\mathbf{n}}) \, \sigma_i \, \hat{U}_R^\dagger(\varphi\hat{\mathbf{n}}) = {}& = (\cos\varphi/2 - i \, \sigma \cdot \hat{\mathbf{n}} \, \sin\varphi/2)(\sigma \cdot \hat{\mathbf{e}}_i \cos\varphi/2 + i \hat{\mathbf{e}}_i \cdot \hat{\mathbf{n}} \, \sin\varphi/2 - \sigma \cdot (\hat{\mathbf{e}}_i \times \hat{\mathbf{n}}) \sin\varphi/2) \\
& = \sigma \cdot \hat{\mathbf{e}}_i \cos^2\varphi/2 + i \hat{\mathbf{e}}_i \cdot \hat{\mathbf{n}} \, \sin\varphi/2 \, \cos\varphi/2 - \sigma \cdot (\hat{\mathbf{e}}_i \times \hat{\mathbf{n}}) \sin\varphi/2 \, \cos\varphi/2 \\
& \quad - i \hat{\mathbf{n}} \cdot \hat{\mathbf{e}}_i \, \sin\varphi/2 \, \cos\varphi/2 + \sigma \cdot (\hat{\mathbf{n}} \times \hat{\mathbf{e}}_i) \sin\varphi/2 \, \cos\varphi/2 + \sigma \cdot \hat{\mathbf{n}} \, (\hat{\mathbf{e}}_i \cdot \hat{\mathbf{n}}) \sin^2\varphi/2 \\
& \quad + i \underbrace{\hat{\mathbf{n}} \cdot (\hat{\mathbf{e}}_i \times \hat{\mathbf{n}})}_{\text{vanishes}} \sin^2\varphi/2 - \sigma \cdot [\hat{\mathbf{n}} \times (\hat{\mathbf{e}}_i \times \hat{\mathbf{n}})] \sin^2\varphi/2 \\
& = \sigma \cdot \hat{\mathbf{e}}_i \cos^2\varphi/2 - 2 \hat{\mathbf{e}}_i \cdot (\hat{\mathbf{n}} \times \sigma) \sin\varphi/2 \, \cos\varphi/2 + \sigma \cdot \hat{\mathbf{n}} \, (\hat{\mathbf{e}}_i \cdot \hat{\mathbf{n}}) \sin^2\varphi/2 \\
& \quad - \sigma \cdot [\hat{\mathbf{e}}_i - \hat{\mathbf{n}}(\hat{\mathbf{e}}_i \cdot \hat{\mathbf{n}})] \sin^2\varphi/2 \\
& = \sigma \cdot \hat{\mathbf{e}}_i \cos\varphi - \hat{\mathbf{e}}_i \cdot (\hat{\mathbf{n}} \times \sigma) \sin\varphi + \sigma \cdot \hat{\mathbf{n}} \, (\hat{\mathbf{e}}_i \cdot \hat{\mathbf{n}})(1 - \cos\varphi) \\
& = \sigma \cdot \hat{\mathbf{n}} \, (\hat{\mathbf{e}}_i \cdot \hat{\mathbf{n}}) + \underbrace{[\sigma \cdot \hat{\mathbf{e}}_i - \sigma \cdot \hat{\mathbf{n}} \, (\hat{\mathbf{e}}_i \cdot \hat{\mathbf{n}})]}_{-\hat{\mathbf{e}}_i \cdot [\hat{\mathbf{n}}\times(\hat{\mathbf{n}}\times\sigma)]} \cos\varphi - \hat{\mathbf{e}}_i \cdot (\hat{\mathbf{n}} \times \sigma) \sin\varphi \\
& = \hat{\mathbf{e}}_i \cdot \underbrace{[\hat{\mathbf{n}} \, (\hat{\mathbf{n}} \cdot \sigma) - \hat{\mathbf{n}} \times (\hat{\mathbf{n}} \times \sigma) \, \cos\varphi - \hat{\mathbf{n}} \times \sigma \, \sin\varphi]}_{\sigma \cos\varphi + \hat{\mathbf{n}}\cdot(\hat{\mathbf{n}}\cdot\sigma)(1-\cos\varphi) - \hat{\mathbf{n}}\times\sigma \, \sin\varphi} \, ,
\end{aligned}
$$

which yields the desired relation. In the limit $\varphi \longrightarrow$ infinitesimal η, the above relation becomes, for the spin vector operator $\hat{\mathbf{S}}$ (to terms linear in η),

$$\hat{U}_R(\eta\hat{\mathbf{n}}) \, \hat{\mathbf{S}} \, \hat{U}_R^\dagger(\eta\hat{\mathbf{n}}) = \hat{\mathbf{S}} - \eta \, \hat{\mathbf{n}} \times \hat{\mathbf{S}} \, .$$

We have established that under a rotation a vector operator transforms as $\hat{U}_R(\varphi\mathbf{n})\,\hat{\mathbf{v}}\,\hat{U}_R^\dagger(\varphi\hat{\mathbf{n}}) = \underline{R}^{-1}(\varphi\hat{\mathbf{n}})\,\hat{\mathbf{v}}$, where

$$\underline{R}^{-1}(\varphi\hat{\mathbf{n}}) = e^{-\varphi\hat{\mathbf{n}}\cdot\underline{\mathbf{E}}} .$$

In our case we have

$$\hat{U}_R(\varphi\hat{\mathbf{n}})\,\hat{S}_j\,\hat{U}_R^\dagger(\eta\hat{\mathbf{n}}) = \sum_k \Big(\underline{\mathbb{1}} - \eta\sum_i \hat{n}_i\,\underline{E}_i\Big)_{jk}\,\hat{S}_k .$$

Using $(\underline{E}_i)_{jk} = -\epsilon_{ijk}$, we arrive at

$$\hat{U}_R(\varphi\hat{\mathbf{n}})\,\hat{S}_j\,\hat{U}_R^\dagger(\eta\hat{\mathbf{n}}) = \sum_k \Big(\delta_{jk} + \eta\sum_i \hat{n}_i\,\epsilon_{ijk}\Big)\,\hat{S}_k = \hat{S}_j - \eta(\hat{\mathbf{n}}\times\hat{\mathbf{S}})_j ,$$

in agreement with the relation found above.

Problem 12   Explicit Derivation of the Transformation Properties of $\hat{\mathbf{r}}$, $\hat{\mathbf{p}}$, and $\hat{\mathbf{L}}$ under Rotations

Show explicitly that under a rotation the components of the position operator transform as follows:

$$\hat{r}'_i = \hat{U}_R(\varphi\hat{\mathbf{n}})\,\hat{r}_i\,\hat{U}_R^\dagger(\varphi\hat{\mathbf{n}}) = \sum_j \big[R^{-1}(\varphi\hat{\mathbf{n}})\big]_{ij}\,\hat{r}_j ,$$

and similarly for the components of $\hat{\mathbf{p}}$. Deduce the transformation law of the orbital angular momentum $\hat{\mathbf{L}}$ from those for $\hat{\mathbf{r}}$ and $\hat{\mathbf{p}}$, obtained above.

Solution

Consider the matrix elements of $\hat{\mathbf{r}}'$ in the basis consisting of eigenstates of the position operator $\hat{\mathbf{r}}$. Dropping, for brevity, the argument $\varphi\hat{\mathbf{n}}$ from the rotation operator and using R as a short-hand notation for $\underline{R}(\varphi\hat{\mathbf{n}})$, we find

$$(\phi_{\mathbf{r}},\,\hat{\mathbf{r}}'\,\phi_{\mathbf{r}'}) = (\phi_{\mathbf{r}},\,\hat{U}_R\,\hat{\mathbf{r}}\,\hat{U}_R^\dagger\,\phi_{\mathbf{r}'}) = (\hat{U}_R^\dagger\,\phi_{\mathbf{r}},\,\hat{\mathbf{r}}\,\hat{U}_R^\dagger\,\phi_{\mathbf{r}'}) = (\phi_{R^{-1}\mathbf{r}},\,\hat{\mathbf{r}}\,\phi_{R^{-1}\mathbf{r}'}) = (R^{-1}\mathbf{r}')(\phi_{R^{-1}\mathbf{r}},\,\phi_{R^{-1}\mathbf{r}'})$$
$$= (R^{-1}\mathbf{r}')\,\delta(R^{-1}\mathbf{r}-R^{-1}\mathbf{r}') = (R^{-1}\mathbf{r}')\,\delta(\mathbf{r}-\mathbf{r}') = (\phi_{\mathbf{r}},\,R^{-1}\hat{\mathbf{r}}\,\phi_{\mathbf{r}'}) ,$$

in accordance with Eq. (16.37). In the first line we used the transformation properties of the eigenstates $|\phi_{\mathbf{r}}\rangle$ under rotations and the fact that $|\phi_{R^{-1}\mathbf{r}'}\rangle$ is an eigenstate of $\hat{\mathbf{r}}$ with eigenvalue $R^{-1}\mathbf{r}'$, and in the second line we used

$$\delta(R^{-1}\mathbf{r}-R^{-1}\mathbf{r}') = \int\frac{d\mathbf{k}}{(2\pi)^3}\,e^{i\mathbf{k}\cdot R^{-1}(\mathbf{r}-\mathbf{r}')} = \int\frac{d\mathbf{k}}{(2\pi)^3}\,e^{i(R\mathbf{k})\cdot(\mathbf{r}-\mathbf{r}')} = \int\frac{d\mathbf{k}'}{(2\pi)^3}\,e^{i\mathbf{k}'\cdot(\mathbf{r}-\mathbf{r}')} = \delta(\mathbf{r}-\mathbf{r}') ,$$

where

$$\mathbf{k}\cdot R^{-1}(\mathbf{r}-\mathbf{r}') = R^{-1}(R\mathbf{k})\cdot R^{-1}(\mathbf{r}-\mathbf{r}') = (R\mathbf{k})\cdot(\mathbf{r}-\mathbf{r}') ,$$

since rotations preserve scalar products, and the Jacobian of the transformation $R\mathbf{k} \longrightarrow \mathbf{k}'$ is $\det(\underline{R}^{-1}) = 1$.

A similar argument can be used for the momentum operator $\hat{\mathbf{p}}$. We have

$$
\begin{aligned}
\langle \phi_{\mathbf{r}} | \hat{\mathbf{p}}' | \phi_{\mathbf{r}'} \rangle &= \langle \phi_{\mathbf{r}} | \hat{U}_R \hat{\mathbf{p}} \, \hat{U}_R^\dagger | \phi_{\mathbf{r}'} \rangle = \langle \phi_{R^{-1}\mathbf{r}} | \hat{\mathbf{p}} | \phi_{R^{-1}\mathbf{r}'} \rangle = \int d\mathbf{p} \, \mathbf{p} \, \langle \phi_{R^{-1}\mathbf{r}} | \phi_{\mathbf{p}} \rangle \langle \phi_{\mathbf{p}} | \phi_{R^{-1}\mathbf{r}'} \rangle \\
&= \int d\mathbf{p} \, \mathbf{p} \, \frac{e^{i\mathbf{p} \cdot R^{-1}(\mathbf{r}-\mathbf{r}')/\hbar}}{(2\pi\hbar)^3} = \int d\mathbf{p} \, \mathbf{p} \, \frac{e^{i(R\mathbf{p}) \cdot (\mathbf{r}-\mathbf{r}')/\hbar}}{(2\pi\hbar)^3} = \int d\mathbf{p} \, (R^{-1}\mathbf{p}) \, \frac{e^{i\mathbf{p} \cdot (\mathbf{r}-\mathbf{r}')/\hbar}}{(2\pi\hbar)^3} \\
&= \int d\mathbf{p} \, (R^{-1}\mathbf{p}) \, \langle \phi_{\mathbf{r}} | \phi_{\mathbf{p}} \rangle \langle \phi_{\mathbf{p}} | \phi_{\mathbf{r}'} \rangle = \int d\mathbf{p} \, \langle \phi_{\mathbf{r}} | R^{-1} \hat{\mathbf{p}} | \phi_{\mathbf{p}} \rangle \langle \phi_{\mathbf{p}} | \phi_{\mathbf{r}'} \rangle = \langle \phi_{\mathbf{r}} | R^{-1} \hat{\mathbf{p}} | \phi_{\mathbf{r}'} \rangle \,,
\end{aligned}
$$

where in the first line we have inserted the completeness of momentum eigenstates and used the previous identity with $\mathbf{k}$ replaced by $\mathbf{p}$, and in the second line we made the change of variables $R\mathbf{p} \longrightarrow \mathbf{p}$, noting that the Jacobian of the transformation is 1, and used the eigenstate property of $|\phi_{\mathbf{p}}\rangle$, yielding $R^{-1}\hat{\mathbf{p}} |\phi_{\mathbf{p}}\rangle = R^{-1}\mathbf{p} |\phi_{\mathbf{p}}\rangle$.

Using the transformation law of $\hat{\mathbf{r}}$ and $\hat{\mathbf{p}}$ yields

$$
\begin{aligned}
L_i' &= \sum_{jk} \epsilon_{ijk} \, \hat{r}_j' \hat{p}_k' = \sum_{jk} \epsilon_{ijk} \, (R^{-1}\hat{\mathbf{r}})_j \, (R^{-1}\hat{\mathbf{p}})_k = \sum_{jkmn} \epsilon_{ijk} R_{mj} R_{nk} \, \hat{r}_m \hat{p}_n = \sum_{jkmnp} \delta_{ip} \, \epsilon_{pjk} R_{mj} R_{nk} \, \hat{r}_m \hat{p}_n \\
&= \sum_{jkmnp} (R^{-1}R)_{ip} \epsilon_{pjk} R_{mj} R_{nk} \, \hat{r}_m \hat{p}_n = \sum_{jkmnpq} \epsilon_{pjk} R_{qi} R_{qp} R_{mj} R_{nk} \, \hat{r}_m \hat{p}_n = \sum_{qmn} R_{qi} \, \hat{r}_m \hat{p}_n \sum_{pjk} \epsilon_{pjk} R_{qp} R_{mj} R_{nk} \\
&= \sum_{qmn} R_{qi} \, \hat{r}_m \hat{p}_n \, \epsilon_{qmn} \det(\underline{R}) = \sum_q R_{qi} \sum_{mn} \epsilon_{qmn} \, \hat{r}_m \hat{p}_n = \sum_q R_{qi} L_q = (R^{-1}\mathbf{L})_i \,,
\end{aligned}
$$

as expected. We used the property $\sum_{pjk} \epsilon_{pjk} R_{qp} R_{mj} R_{nk} = \det(\underline{R}) \, \epsilon_{qmn}$ and the fact that rotation matrices have unit determinant.

Problem 13 Construction of a Spin State Polarized Along a Generic Direction $\hat{\mathbf{n}}$

Given the spin-1/2 state $|+\rangle$ with $\hat{S}_z|+\rangle = (\hbar/2)|+\rangle$ (that is, the state is polarized along the $\hat{\mathbf{z}}$-axis), we want to construct the spin state $|+\rangle'$ polarized along a direction

$$
\hat{\mathbf{n}} = (\sin\theta \cos\phi, \sin\theta \sin\phi, \cos\theta) \,,
$$

such that

$$
\hat{\mathbf{n}} \cdot \hat{\mathbf{S}} |+\rangle' = (\hbar/2) |+\rangle' \,.
$$

1. Show that this can be accomplished by a sequence of two rotations, the first by θ about the $\hat{\mathbf{y}}$-axis and the second by ϕ about the $\hat{\mathbf{z}}$-axis. Using this insight, construct the state $|+\rangle'$ explicitly.
2. The state $|+\rangle'$ can also be obtained by solving the above eigenvalue problem explicitly, namely $\hat{\mathbf{n}} \cdot \hat{\mathbf{S}} |+\rangle' = (\hbar/2) |+\rangle'$. Do so and compare the result with that obtained in part 1.

Solution

Part 1

The spin in a given direction $\hat{\mathbf{n}}$ specified by angles θ and ϕ can be obtained by the following sequence of rotations (draw a figure): first by θ about $\hat{\mathbf{y}}$ and then by ϕ about $\hat{\mathbf{z}}$, namely

$$
|+\rangle' = e^{-i\phi\hat{S}_z/\hbar} \, e^{-i\theta\hat{S}_y/\hbar} \, |+\rangle \,.
$$

In the basis of S_z eigenstates, in which $|+\rangle' = a'_+ |+\rangle + a'_- |-\rangle$, the above equation says that

$$\begin{pmatrix} a'_+ \\ a'_- \end{pmatrix} = \underbrace{\begin{pmatrix} \cos(\phi/2) - i\sin(\phi/2) & 0 \\ 0 & \cos(\phi/2) + i\sin(\phi/2) \end{pmatrix}}_{e^{-i\phi\hat{S}_z/\hbar}} \underbrace{\begin{pmatrix} \cos(\theta/2) & -\sin(\theta/2) \\ \sin(\theta/2) & \cos(\theta/2) \end{pmatrix}}_{e^{-i\theta\hat{S}_y/\hbar}} \begin{pmatrix} 1 \\ 0 \end{pmatrix} .$$

We have used $e^{-i\theta\hat{S}_y/\hbar} = \cos(\theta/2) - i\sigma_y \sin(\theta/2)$ and similarly for the other rotation. Thus we find

$$\begin{pmatrix} a'_+ \\ a'_- \end{pmatrix} = \begin{pmatrix} e^{-i\phi/2} & 0 \\ 0 & e^{i\phi/2} \end{pmatrix} \begin{pmatrix} \cos(\theta/2) \\ \sin(\theta/2) \end{pmatrix} \implies |+\rangle' = e^{-i\phi/2} \cos(\theta/2) |+\rangle + e^{i\phi/2} \sin(\theta/2) |-\rangle .$$

In order to verify that this is correct, note that

$$\hat{\mathbf{n}} \cdot \hat{\mathbf{S}} = \frac{\hbar}{2} \begin{pmatrix} \cos\theta & \sin\theta\, e^{-i\phi} \\ \sin\theta\, e^{i\phi} & -\cos\theta \end{pmatrix} ,$$

and hence

$$\hat{\mathbf{n}} \cdot \hat{\mathbf{S}} |+\rangle' \longrightarrow \frac{\hbar}{2} \begin{pmatrix} \cos\theta & \sin\theta\, e^{-i\phi} \\ \sin\theta\, e^{i\phi} & -\cos\theta \end{pmatrix} \begin{pmatrix} e^{-i\phi/2} \cos(\theta/2) \\ e^{i\phi/2} \sin(\theta/2) \end{pmatrix}$$

$$= \frac{\hbar}{2} \begin{bmatrix} e^{-i\phi/2} (\cos\theta \cos(\theta/2) + \sin\theta \sin(\theta/2)) \\ e^{i\phi/2} (\sin\theta \cos(\theta/2) - \cos\theta \sin(\theta/2)) \end{bmatrix} = \frac{\hbar}{2} \begin{pmatrix} e^{-i\phi/2} \cos(\theta/2) \\ e^{i\phi/2} \sin(\theta/2) \end{pmatrix} \longrightarrow \frac{\hbar}{2} |+\rangle' ,$$

where in the last step we have made use of the identities $\cos\theta = 2\cos^2(\theta/2) - 1$ and $\sin\theta = 2\sin(\theta/2)\cos(\theta/2)$.

Part 2

The eigenvalue equation yields the linear system

$$\begin{pmatrix} \cos\theta & \sin\theta\, e^{-i\phi} \\ \sin\theta\, e^{i\phi} & -\cos\theta \end{pmatrix} \begin{pmatrix} a'_+ \\ a'_- \end{pmatrix} = \begin{pmatrix} a'_+ \\ a'_- \end{pmatrix} ,$$

implying (considering the upper equation)

$$(1 - \cos\theta)a'_+ = e^{-i\phi} \sin\theta\, a'_- \implies \sin(\theta/2)\, a'_+ = e^{-i\phi} \cos(\theta/2)\, a'_- \implies a'_- = a'_+ \frac{\sin(\theta/2)}{\cos(\theta/2)} e^{i\phi} ,$$

and the choice $a'_+ = e^{-i\phi/2} \cos(\theta/2)$ leads to the normalized eigenstate given by

$$\begin{pmatrix} a'_+ \\ a'_- \end{pmatrix} = \begin{pmatrix} e^{-i\phi/2} \cos(\theta/2) \\ e^{i\phi/2} \sin(\theta/2) \end{pmatrix} ,$$

as expected.

Problem 14 Construction of the Unitary Operator Inducing Galilean Transformations

In a Galilean transformation of the coordinate system, the position and momentum observables $\hat{x}$ and $\hat{p}$ transform as follows:

$$\hat{x}' = \hat{x} - vt , \qquad \hat{p}' = \hat{p} - mv ,$$

where v is the relative velocity between the two coordinate systems, m is the mass of the particle, and t is the time.

1. Construct the unitary operator $\hat{U}_G = e^{i\hat{K}}$ that induces Galilean transformations, knowing that the hermitian operator $\hat{K}$ is some linear combination of $\hat{x}$ and $\hat{p}$,

$$\hat{K} = a\hat{x} + b\hat{p} .$$

2. Construct the wave function corresponding to the state

$$|\Psi'_q(t)\rangle = \hat{U}_G |\Psi_q(t)\rangle ,$$

where $|\Psi_q(t)\rangle$ is a free-particle state of momentum q with wave function

$$\Psi_q(x, t) = \frac{1}{\sqrt{2\pi\hbar}} e^{i(qx - Et)/\hbar} , \qquad E = \frac{q^2}{2m} .$$

Give a physical interpretation of the result.

Hint: The following formulae may be useful:

$$e^{\hat{A}} \hat{B} e^{-\hat{A}} = \hat{B} + \frac{1}{1!}[\hat{A}, \hat{B}] + \frac{1}{2!}[\hat{A}, [\hat{A}, \hat{B}]] + \cdots$$

and

$$e^{\hat{A}+\hat{B}} = e^{\hat{A}} e^{\hat{B}} e^{-[\hat{A}, \hat{B}]/2} .$$

The last formula is valid when $\hat{A}$ and $\hat{B}$ each commute with their commutator, $[\hat{A}, \hat{B}]$.

Solution

Part 1

The hermitian operator $\hat{K}$ is assumed to be a linear combination,

$$\hat{K} = a\hat{x} + b\hat{p} ,$$

where a and b are real parameters. Use the expansion

$$\hat{x} - vt = \underbrace{e^{i\hat{K}}}_{\hat{U}_G} \hat{x} \underbrace{e^{-i\hat{K}}}_{\hat{U}_G^\dagger} = \hat{x} + \frac{i}{1!}[\hat{K}, \hat{x}] + \frac{i^2}{2!}[\hat{K}, [\hat{K}, \hat{x}]] + \cdots ,$$

and note that the following commutator is a c-number,

$$[\hat{K}, \hat{x}] = [a\hat{x} + b\hat{p}, \hat{x}] = -i\hbar b ,$$

which implies that all higher-order commutators vanish, since $[\hat{K}, [\hat{K}, \hat{x}]] = [\hat{K}, -i\hbar b] = 0$. We arrive at

$$\hat{x} - vt = \hat{x} + \hbar b \implies b = -vt/\hbar .$$

Similarly, we find

$$\hat{p} - mv = \underbrace{e^{i\hat{K}}}_{\hat{U}_G} \hat{p} \underbrace{e^{-i\hat{K}}}_{\hat{U}_G^\dagger} = \hat{p} + \frac{i}{1!}[\hat{K}, \hat{p}] + \frac{i^2}{2!}[\hat{K}, [\hat{K}, \hat{p}]] + \cdots ,$$

where $[\hat{K}, \hat{p}] = [a\hat{x} + b\hat{p}, \hat{p}] = i\hbar a$. Comparing the left- and right-hand sides, we obtain

$$\hat{p} - mv = \hat{p} - \hbar a \implies a = mv/\hbar ,$$

and the hermitian operator $\hat{K}$ is

$$\hat{K} = \frac{v}{\hbar}\,(m\hat{x} - \hat{p}\,t)\ .$$

Part 2

The wave function corresponding to the state $|\Psi'_q(t)\rangle$ is given by

$$\Psi'_q(x,t) = \langle\phi_x|\Psi'_q(t)\rangle = \langle\phi_x|\,e^{iv(m\hat{x}-\hat{p}t)/\hbar}\,|\Psi'_q(t)\rangle\ ,$$

where $|\phi_x\rangle$ are position eigenstates. Use

$$e^{iv(m\hat{x}-\hat{p}t)/\hbar} = e^{imv\hat{x}/\hbar}\,e^{-ivt\hat{p}/\hbar}\,e^{-[imv\hat{x}/\hbar,\,-ivt\hat{p}/\hbar]/2} = e^{imv\hat{x}/\hbar}\,e^{-ivt\hat{p}/\hbar}\,\underbrace{e^{-imv^2t/(2\hbar)}}_{c-number}\ ,$$

which yields

$$\Psi'_q(x,t) = e^{-imv^2t/(2\hbar)}\,\langle\phi_x|e^{imv\hat{x}/\hbar}\,e^{-ivt\hat{p}/\hbar}|\Psi_q(t)\rangle = e^{-imv^2t/(2\hbar)}\,e^{imvx/\hbar}\,\langle\phi_x|\,e^{-ivt\hat{p}/\hbar}|\Psi_q(t)\rangle$$

$$= e^{-imv^2t/(2\hbar)}\,e^{imvx/\hbar}\,e^{-ivtq/\hbar}\,\langle\phi_x|\Psi_q(t)\rangle = \frac{1}{\sqrt{2\pi\hbar}}\,e^{-imv^2t/(2\hbar)}\,e^{imvx/\hbar}\,e^{-ivtq/\hbar}\,e^{i(qx-Et)/\hbar}$$

$$= \frac{1}{\sqrt{2\pi\hbar}}\,e^{i(q+mv)x/\hbar}\,e^{-i(q+mv)^2t/(2m\hbar)}\ .$$

This last result shows that $\Psi'_q(x,t)$ is the wave function corresponding to a state of a particle with momentum q boosted to momentum $q + mv$.

Problem 15   Consequences of Space Inversion Symmetry

Obtain the selection rules for an operator $\hat{A}$ that is either odd or even under space inversion.

Solution

Since $\hat{U}_P$ is hermitian, it can be thought of as an observable, and, since its square is the identity operator, its eigenvalues are ± 1. Let $|\psi_a\rangle$ and $|\psi_b\rangle$ be eigenstates of $\hat{U}_P$ with eigenvalues π_a and π_b, respectively, and let $\hat{A}$ be an operator that is odd under space inversion, so that, $\hat{U}_P\,\hat{A}\,\hat{U}_P = -\hat{A}$. The matrix element $\langle\psi_a|\hat{A}|\psi_b\rangle$ vanishes unless $\pi_a\pi_b = -1$, namely the states have opposite parity, since

$$(\psi_a,\,\hat{A}\,\psi_b) = -(\psi_a,\,U_P\,\hat{A}\,U_P\,\psi_b) = -(U_P\,\psi_a,\,\hat{A}\,U_P\,\psi_b) = -\pi_a\,\pi_b\,(\psi_a,\,\hat{A}\,\psi_b)\ .$$

Similarly, an even operator, $\hat{U}_P\,\hat{A}\,\hat{U}_P = \hat{A}$, has non-vanishing matrix elements only between states of the same parity.

Space inversion symmetry leads to selection rules in the most common radiative transitions occurring in atoms. As we will see, in the approximation in which the wavelength of the emitted (or absorbed) photon is much larger than the atomic radius, the transition rate is proportional to the magnitude squared of the electric-dipole-operator matrix element; in hydrogen, for example, this operator is simply given by $\hat{\mathbf{D}} = -e\,\hat{\mathbf{r}}$ and is odd under space inversion. It follows that the transition must involve states of opposite parities. The parity of the states (in hydrogen) is simply $(-1)^l$ and therefore we must have $(-1)^{l_a+l_b} = -1$; for instance, the transitions $2p \rightleftharpoons 1s$ are allowed, while $3p \rightleftharpoons 2p$ are not.

Problem 16   A Model for the Ammonia Molecule and Broken Parity Symmetry

In the ammonia molecule NH_3, the three hydrogen atoms lie at the corners of an equilateral triangle and the nitrogen atom moves along a line transverse to the plane of the triangle passing through its center. Let x be the position of the N atom along this line. The N atom can be on either side of the triangle. A simple model for the potential felt by the N atom is given by

$$V(|x| > b) = \infty , \qquad V(a < |x| < b) = 0 , \qquad V(|x| < a) = V_0 ,$$

where the barrier $V_0 > 0$ represents the Coulomb repulsion between the positive charges of the hydrogen and nitrogen nuclei. Note that $V(x)$ is invariant under space inversion, $V(x) = V(-x)$.

1. Discuss the nature of the energy spectrum. What is the parity of the ground state? And that of the first-excited state? Define, respectively, as I, II, and III the regions $-b < x < -a$, $-a < x < a$, and $a < x < b$. Assume $E < V_0$ and show that the eigenvalues of the even- and odd-parity solutions result from

$$\cot[k(b - a)] = -\frac{\kappa}{k} \tanh(\kappa a) , \qquad \text{even parity} ,$$

and

$$\tan[k(b - a)] = -\frac{k}{\kappa} \tanh(\kappa a) , \qquad \text{odd parity} ,$$

where (in the notation of Section 4.1)

$$k = \sqrt{\epsilon} , \qquad \kappa = \sqrt{v_0 - \epsilon} .$$

2. Hereafter, assume that the energies of the ground state (of even parity) and first excited state (of odd parity) are much less than V_0, and show that these energies are given by

$$E_e \approx \frac{(\pi\hbar)^2}{2m(b - a)^2} (1 - 2\eta_e) , \qquad E_o \approx \frac{(\pi\hbar)^2}{2m(b - a)^2} (1 - 2\eta_o) ,$$

where $\eta_e, \eta_o (\ll 1)$ are given by

$$\eta_e = \frac{\coth(a\sqrt{v_0})}{(b - a)\sqrt{v_0}} , \qquad \eta_o = \frac{\tanh(a\sqrt{v_0})}{(b - a)\sqrt{v_0}} .$$

Obtain the corresponding wave functions (up to overall normalization factors), and briefly comment on them.

3. Consider the states

$$\psi_R(x) = \frac{1}{\sqrt{2}} \left[\psi^e(x) + \psi^o(x) \right] , \qquad \psi_L(x) = \frac{1}{\sqrt{2}} \left[\psi^e(x) - \psi^o(x) \right] .$$

Are $\psi_R(x)$ and $\psi_L(x)$ eigenfunctions of the Hamiltonian? Are they eigenfunctions of the parity operator U_P? Where are $\psi_R(x)$ and $\psi_L(x)$ mostly non-vanishing? At time $t = 0$ the N atom is in the state represented by $\psi_L(x)$, namely

$$\Psi(x, 0) = \psi_L(x) .$$

Obtain $\Psi(x, t)$. Show that the N atom oscillates between the left- and right-hand potential well with period $T = h/\Delta E$, where the energy difference $\Delta E = E_o - E_e$ is given by

$$\Delta E \approx \frac{(\pi\hbar)^2}{m(b - a)^2} (\eta_e - \eta_o) \approx \frac{(2\pi\hbar)^2}{m(b - a)^3 \sqrt{v_0}} e^{-2a\sqrt{v_0}} .$$

The time $T/2$ necessary for the N atom to tunnel through the barrier and materialize in the right-hand well will be very long. So, it happens that the broken symmetry wave function $\psi_L(x)$, which is not an eigenfunction of the parity operator but is very nearly an eigenfunction of the Hamiltonian (for a very high barrier), represents the physical state realized in nature. This phenomenon explains why certain molecules only exist in L or R states. These L and R states are separated by very thick barriers. For such molecules the transition from one broken symmetry state $\psi_L(x)$ to the other $\psi_R(x)$ takes so long as to be unobservable.

4. Now consider the case in which the barrier V_0 is infinitely high. What happens to the energies of the even and odd solutions $\psi^e(x)$ and $\psi^o(x)$? Sketch these wave functions. Are $\psi_L(x)$ and $\psi_R(x)$ eigenfunctions of the Hamiltonian? Sketch $\psi_L(x)$ and $\psi_R(x)$. Are they eigenfunctions of the parity operator? If the N atom is in the left-hand well can it ever be found in the right-hand well?

Solution

Part 1

Only bound states are possible. The states have alternating parities: even, odd, even, odd, and so on, the ground and first excited states being even and odd, respectively. We have, for energies $E < V_0$,

$$\psi_I'' = -k^2 \psi_I(x) , \qquad \psi_{II}'' = \kappa^2 \psi_{II}(x) , \qquad \psi_{III}'' = -k^2 \psi_{III}(x) ,$$

where $k = \sqrt{\epsilon}$ and $\kappa = \sqrt{v_0 - \epsilon}$. We can solve for, say, $x > 0$, and then obtain the complete solution in $-b < x < b$ by exploiting the known parity of the state. In particular, the first derivatives of even-parity solutions $\psi^e(x)$ and the odd-parity solutions $\psi^o(x)$ vanish at $x = 0$, namely

$$\psi^{e\,\prime}(0) = 0 , \qquad \psi^{o\,\prime}(0) = 0 .$$

For $x > 0$, we have

$$\psi_{II}(x) = A \cosh(\kappa x) , \qquad \psi_{III}(x) = B \sin[k(b - x)] , \qquad \text{even parity} ,$$
$$\psi_{II}(x) = C \sinh(\kappa x) , \qquad \psi_{III}(x) = D \sin[k(b - x)] , \qquad \text{odd parity} ,$$

where the boundary condition $\psi_{III}(b) = 0$ has already been imposed. It should be noted that these solutions also satisfy the requirements $\psi^{e\,\prime}(0) = 0$ and $\psi^{o\,\prime}(0) = 0$. The boundary conditions at $x = a$ yield

$$A \cosh(\kappa a) = B \sin[k(b - a)] , \qquad \kappa A \sinh(\kappa a) = -kB \cos[k(b - a)] ,$$
$$C \sinh(\kappa a) = D \sin[k(b - a)] , \qquad \kappa C \cosh(\kappa a) = -kD \cos[k(b - a)] ,$$

and the eigenvalues result from

$$\tanh(\kappa a) = -\frac{k}{\kappa} \cot[k(b - a)] , \quad \text{even parity} , \qquad \tanh(\kappa a) = -\frac{\kappa}{k} \tan[k(b - a)] , \quad \text{odd parity} .$$

The complete even-parity solutions are then

$$\psi_I^e(x) = B \sin[k_e(b + x)] , \qquad \psi_{II}^e(x) = A \cosh(\kappa_e x) , \qquad \psi_{III}^e(x) = B \sin[k_e(b - x)] ,$$

and the odd-parity solutions are

$$\psi_I^o(x) = -D \sin[k_o(b + x)] , \qquad \psi_{II}^o(x) = C \sinh(\kappa_o x) , \qquad \psi_{III}^o(x) = D \sin[k_o(b - x)] ,$$

where, having determined the eigenvalues ϵ_e and ϵ_o, we can express A and C in terms of B and D:

$$A = \frac{\sin[k_e(b-a)]}{\cosh(\kappa_e a)}\, B\,, \qquad C = \frac{\sin[k_o(b-a)]}{\sinh(\kappa_o a)}\, D\,,$$

and $|B|$ and $|D|$ can be fixed by normalizing the states. Here, we have defined

$$k_e = \sqrt{\epsilon_e}\,, \qquad \kappa_e = \sqrt{v_0 - \epsilon_e}\,,$$

and similarly for k_o and κ_o.

Part 2

In the limit $v_0 \gg \epsilon_e$ and $v_0 \gg \epsilon_o$ the eigenvalue equations for ϵ_e and ϵ_o read

$$\cot[k_e(b-a)] = -\frac{\sqrt{v_0}}{k_e}\,\tanh(\sqrt{v_0}\,a)\,, \qquad \tan[k_o(b-a)] = -\frac{k_o}{\sqrt{v_0}}\,\tanh(\sqrt{v_0}\,a)\,.$$

Note that the cotangent and tangent vanish for, respectively,

$$k_n = \frac{\pi}{b-a}\frac{2n-1}{2}\,, \qquad \bar{k}_n = \frac{\pi}{b-a}\,n \qquad n = 1, 2, \dots\,.$$

A plot of the cotangent shows that the first intersection of $\cot[k_e(b-a)]$ with the function $-(\sqrt{v_0}/k_e)\tanh(\sqrt{v_0}\,a) \longrightarrow -\infty$ occurs when the argument $k(b-a) \longrightarrow \pi$ from below. The Taylor expansion of $\tan x$ about π gives

$$\tan x \approx \tan \pi + \frac{1}{\cos^2 \pi}\,(x - \pi) + \cdots = x - \pi + \cdots\,,$$

and therefore $\cot x$ can be expanded as

$$\cot x = \frac{1}{\tan x} \approx \frac{1}{x - \pi}\,,$$

and

$$\frac{1}{k_e(b-a) - \pi} \approx -\frac{\sqrt{v_0}}{k_e}\,\tanh(\sqrt{v_0}\,a)\,,$$

which can be solved for k_e to give

$$k_e \approx \frac{\pi}{b - a + \coth(\sqrt{v_0}\,a)/\sqrt{v_0}} \approx \frac{\pi}{b-a}\frac{1}{1+\eta_e} \approx \frac{\pi}{b-a}\,(1 - \eta_e)\,,$$

where we have defined $\eta \ll 1$ as

$$\eta_e = \frac{\coth(a\sqrt{v_0})}{(b-a)\sqrt{v_0}}\,.$$

For the odd solutions we note that the first intersection of the tangent with the function $-(k_o/\sqrt{v_0})\tanh(\sqrt{v_0}\,a)$ also occurs when the argument $k_o(b-a) \longrightarrow \pi$ from below, and hence

$$k_o(b-a) - \pi \approx -\frac{k_o}{\sqrt{v_0}}\,\tanh(\sqrt{v_0}\,a)\,,$$

yielding

$$k_o \approx \frac{\pi}{b - a + \tanh(\sqrt{v_0}\,a)/\sqrt{v_0}} \approx \frac{\pi}{b-a}\frac{1}{1+\eta_o} \approx \frac{\pi}{b-a}\,(1 - \eta_o)\,,$$

where

$$\eta_o = \frac{\tanh(a\sqrt{v_0})}{(b-a)\sqrt{v_0}} \, .$$

The energies are given by

$$E_e \approx \frac{(\pi\hbar)^2}{2m(b-a)^2}(1-2\eta_e)\,, \qquad E_o \approx \frac{(\pi\hbar)^2}{2m(b-a)^2}(1-2\eta_o)\,,$$

and for the energy difference $\Delta E = E_o - E_e > 0$ we have

$$\Delta E \approx \frac{(\pi\hbar)^2}{m(b-a)^2}(\eta_e - \eta_o)\,,$$

where

$$\eta_e - \eta_o = \frac{\coth(\sqrt{v_0}\,a) - \tanh(\sqrt{v_0}\,a)}{(b-a)\sqrt{v_0}} \approx \frac{4}{(b-a)\sqrt{v_0}}\,e^{-2a\sqrt{v_0}} > 0\,.$$

We have used the expansion

$$\coth z - \tanh z = \frac{e^z + e^{-z}}{e^z - e^{-z}} - \frac{e^z - e^{-z}}{e^z + e^{-z}} = \frac{4}{e^{2z} - e^{-2z}} \longrightarrow 4\,e^{-2z} \qquad \text{for } z \longrightarrow \infty\,.$$

In the same limit $v_0 \gg \epsilon$, we find that, for the even solution,

$$\frac{A}{B} = \frac{\sin[k_e(b-a)]}{\cosh(\kappa_e a)} \approx 2\,e^{-a\sqrt{v_0}}\sin(\pi - \pi\eta_e) \approx 2\pi\,e^{-a\sqrt{v_0}}\,\eta_e \approx 2\pi\,\frac{e^{-a\sqrt{v_0}}}{(b-a)\sqrt{v_0}}\,,$$

and similarly, for the odd solution,

$$\frac{C}{D} \approx 2\pi\,\frac{e^{-a\sqrt{v_0}}}{(b-a)\sqrt{v_0}}\,.$$

The complete ground-state (even-parity) solution then reads (up to the normalization constant B)

$$\psi_I^e(x) = \sin[k_e(b+x)]\,, \quad \psi_{II}^e(x) = 2\pi\,\frac{e^{-a\sqrt{v_0}}}{(b-a)\sqrt{v_0}}\cosh(\sqrt{v_0}\,x)\,, \quad \psi_{III}^e(x) = \sin[k_e(b-x)]\,,$$

and the odd-parity solutions are (up to the normalization constant D)

$$\psi_I^o(x) = -\sin[k_o(b+x)]\,, \quad \psi_{II}^o(x) = 2\pi\,\frac{e^{-a\sqrt{v_0}}}{(b-a)\sqrt{v_0}}\sinh(\sqrt{v_0}\,x)\,, \quad \psi_{III}^o(x) = \sin[k_o(b-x)]\,;$$

the solutions in the classically forbidden region $|x| < a$ are strongly suppressed. In the limit that we are considering, we have $k_e = k_0$ up to exponentially small corrections, and the normalization constants $|B|$ and $|D|$ are approximately the same.

Part 3

The wave functions $\psi_R(x)$ and $\psi_L(x)$ are defined as

$$\psi_R(x) = \frac{1}{\sqrt{2}}\left[\psi^e(x) + \psi^o(x)\right]\,, \qquad \psi_L(x) = \frac{1}{\sqrt{2}}\left[\psi^e(x) - \psi^o(x)\right]\,,$$

and are not eigenfunctions of either the Hamiltonian or the parity operator U_P. As a matter of fact, we find that

$$U_P\psi_R(x) = \frac{1}{\sqrt{2}}\left[\psi^e(-x) + \psi^o(-x)\right] = \frac{1}{\sqrt{2}}\left[\psi^e(x) - \psi^o(x)\right] = \psi_L(x)$$

and $U_P \psi_L(x) = \psi_R(x)$, and so the parity operator transforms one wave function into the other. In particular, we have that $\psi_R(x)$ is non-vanishing primarily in the right-hand well $a < x < b$ (region III) and $\psi_L(x)$ is non-vanishing primarily in the left-hand well $-b < x < -a$ (region I); for example, the combination $\psi^e(x) + \psi^o(x)$ in the left-hand well is

$$-B \sin[k_e(b+x)] - D \sin[k_o(b+x)] \approx 0 \qquad \text{up to terms} \propto e^{-2a\sqrt{v_0}} .$$

The particle's wave function is $\psi_L(x)$ at time $t = 0$, and its wave function at time t is given by

$$\Psi(x,t) = \frac{1}{\sqrt{2}} \left[\psi^e(x)\, e^{-iE_e t/\hbar} - \psi^o(x)\, e^{-iE_o t/\hbar} \right] = \frac{e^{-iE_e t/\hbar}}{\sqrt{2}} \left[\psi^e(x) - \psi^o(x)\, e^{-i\Delta E t/\hbar} \right] .$$

As time increases the particle will oscillate between the left- and right-hand wells with angular frequency $\omega = \Delta E/\hbar$: at time π/ω it will be in the right-hand well, at time $2\pi/\omega$ in the left-hand well, at time $3\pi/\omega$ back in the right-hand well, and so on. In essence, if the particle is initially in the left-hand well then it penetrates the classically forbidden region and materializes on the other side of the barrier, in the right-hand well, and then goes back to the left-hand well, and so on. The leaking from one well to the other will occur in a time $\pi/\omega \propto e^{2a\sqrt{v_0}}$; this time will be very long if the barrier is very high, of course.

Part 4

In the limit $v_0 \longrightarrow \infty$, we have left-hand and right-hand infinite wells, located symmetrically on each side of the origin. The Hamiltonian H is still invariant under space inversion. The even and odd wave functions now have the same energy; they are degenerate eigenfunctions of H. So are the linear combinations $\psi_L(x)$ and $\psi_R(x)$. However, the latter are not eigenfunctions of the parity operator. In particular, if the system is found in state $\psi_L(x)$ then it will remain in that state forever, since the tunneling probability vanishes in this case (the time required for the particle to materialize on the other side becomes infinite). In the presence of degeneracy, the physically realized state is not necessarily an eigenstate of the parity operator, even though H is invariant under parity. The double-well problem provides a simple illustration of broken symmetry: we have an asymmetrical ground state in spite of the fact that H is symmetric under space inversion.

Problem 17 Unitarity Implies Linearity and Anti-Unitarity Implies Antilinearity

Show that a unitary operator is necessarily linear and, conversely, that an anti-unitary operator is necessarily antilinear.

Solution

A unitary operator $\hat{U}$ is defined by the property $(U\phi, U\psi) = (\phi, \psi)$ for any two states $|\phi\rangle$ and $|\psi\rangle$. To show that it is also necessarily linear, consider the state

$$|\varphi\rangle = \hat{U}(\alpha|\phi\rangle + \beta|\psi\rangle) - \alpha\, \hat{U}|\phi\rangle - \beta\, \hat{U}|\psi\rangle ,$$

where α and β are generally complex constants. Using only the unitarity condition, we obtain, for the squared norm of $|\varphi\rangle$,

$$(\varphi, \varphi) = (\hat{U}(\alpha\phi + \beta\psi) - \alpha\,\hat{U}\phi - \beta\,\hat{U}\psi, \; \hat{U}(\alpha\phi + \beta\psi) - \alpha\,\hat{U}\phi - \beta\,\hat{U}\psi)$$
$$= (\alpha\phi + \beta\psi, \; \alpha\phi + \beta\psi) - \alpha(\alpha\phi + \beta\psi, \; \phi) - \beta(\alpha\phi + \beta\psi, \; \psi) - \alpha^*(\phi, \; \alpha\phi + \beta\psi)$$
$$+ |\alpha|^2(\phi, \phi) + \alpha^*\beta(\phi, \psi) - \beta^*(\psi, \; \alpha\phi + \beta\psi) + \alpha\beta^*(\psi, \phi) + |\beta|^2(\psi, \psi) = 0,$$

and so $|\varphi\rangle$ is the null state, which in turn implies the linearity of $\hat{U}$. A similar proof holds for an anti-unitary operator $\hat{\Omega}$.

Problem 18 The Unitary Operator $\hat{U}_\mathcal{T}$ in $\hat{\Omega}_\mathcal{T} = \hat{U}_\mathcal{T}\,\hat{K}$ Depends on the Representation

In Section 16.3.1 we noted that the anti-unitary time-reversal operator $\hat{\Omega}_\mathcal{T}$ can be written as the product of a unitary operator $\hat{U}_\mathcal{T}$ and the anti-unitary operator $\hat{K}$ for hermitian conjugation, namely $\hat{\Omega}_\mathcal{T} = \hat{U}_\mathcal{T}\,\hat{K}$. We also noted that the effect of $\hat{K}$ (and hence $\hat{U}_\mathcal{T}$) depends on the basis that we have chosen to represent states and operators. Specifically, show that, in the basis of position eigenstates, $\hat{U}_\mathcal{T}$ is the identity operator whereas in the basis of momentum eigenstates it is the parity operator $\hat{U}_P$.

Solution

Under time reversal, we have $\hat{\mathbf{r}} \longmapsto \mathbf{r}' = \mathbf{r}$ and $\hat{\mathbf{p}} \longmapsto \mathbf{p}' = -\mathbf{p}$. Now, consider first the basis of position eigenstates. Under $\hat{K}$, these eigenstates transform as $\hat{K}|\phi_\mathbf{r}\rangle = |\phi_\mathbf{r}\rangle$; furthermore, recalling that $\hat{K}^2 = \hat{\mathbb{1}}$ and $\hat{K} = \hat{K}^\dagger$, we find that the momentum operator transforms as

$$(\phi_\mathbf{r}, \; \hat{K}\hat{\mathbf{p}}\hat{K}^\dagger\,\phi_{\mathbf{r}'}) = (\hat{K}^2\,\phi_\mathbf{r}, \; \hat{K}\hat{\mathbf{p}}\hat{K}\,\phi_{\mathbf{r}'}) = (\hat{K}\phi_\mathbf{r}, \; \hat{\mathbf{p}}\hat{K}\phi_{\mathbf{r}'})^* = (\phi_\mathbf{r}, \; \hat{\mathbf{p}}\phi_{\mathbf{r}'})^*$$
$$= [-i\hbar\boldsymbol{\nabla}_\mathbf{r}\,\delta(\mathbf{r} - \mathbf{r}')]^* = (\phi_\mathbf{r}, \; -\hat{\mathbf{p}}\,\phi_{\mathbf{r}'}),$$

or $\hat{K}\hat{\mathbf{p}}\hat{K}^\dagger = -\hat{\mathbf{p}}$; a similar proof shows that $\hat{K}\hat{\mathbf{r}}\hat{K}^\dagger = \hat{\mathbf{r}}$. We conclude that in the basis of position eigenstates $\hat{U}_\mathcal{T} = \hat{\mathbb{1}}$, up to an arbitrary phase factor, which has been set to 1 as in Eq. (16.59), and hence $\hat{\Omega}_\mathcal{T} = \hat{K}$.

Consider next the basis of momentum eigenstates $|\phi_\mathbf{p}\rangle$. Since $\hat{K}|\phi_\mathbf{p}\rangle = |\phi_\mathbf{p}\rangle$, it follows that

$$(\phi_\mathbf{p}, \; \hat{K}\hat{\mathbf{r}}\hat{K}^\dagger\,\phi_{\mathbf{p}'}) = (\hat{K}^2\,\phi_\mathbf{p}, \; \hat{K}\hat{\mathbf{r}}\hat{K}\,\phi_{\mathbf{p}'}) = (\hat{K}\phi_\mathbf{p}, \; \hat{\mathbf{r}}\hat{K}\phi_{\mathbf{p}'})^* = (\phi_\mathbf{p}, \; \hat{\mathbf{r}}\phi_{\mathbf{p}'})^*$$
$$= [i\hbar\,\boldsymbol{\nabla}_\mathbf{p}\,\delta(\mathbf{p} - \mathbf{p}')]^* = (\phi_\mathbf{p}, \; -\hat{\mathbf{r}}\,\phi_{\mathbf{p}'}).$$

In a similar fashion, we find that $(\phi_\mathbf{p}, \; \hat{K}\hat{\mathbf{p}}\hat{K}^\dagger\,\phi_{\mathbf{p}'}) = (\phi_\mathbf{p}, \; \hat{\mathbf{p}}\phi_{\mathbf{p}'})$, and in the basis of momentum eigenstates $\hat{K}\hat{\mathbf{r}}\hat{K}^\dagger = -\hat{\mathbf{r}}$ and $\hat{K}\hat{\mathbf{p}}\hat{K}^\dagger = \hat{\mathbf{p}}$. Therefore, in such a basis the unitary operator $\hat{U}_\mathcal{T}$ must satisfy

$$\hat{U}_\mathcal{T}\,\hat{\mathbf{r}}\,\hat{U}_\mathcal{T}^\dagger = -\hat{\mathbf{r}}, \qquad \hat{U}_\mathcal{T}\,\hat{\mathbf{p}}\,\hat{U}_\mathcal{T}^\dagger = -\hat{\mathbf{p}},$$

in order for $\hat{\Omega}_\mathcal{T}$ to induce the correct transformations of $\hat{\mathbf{r}}$ and $\hat{\mathbf{p}}$. Since $\hat{U}_\mathcal{T}$ induces the same transformations as the space inversion (unitary) operator $\hat{U}_P$, we conclude that, in the basis of momentum eigenstates, $\hat{\Omega}_\mathcal{T} = \hat{U}_P\,\hat{K}$.

Problem 19 Time Evolution of a State and of Its Time-Reversed Partner

Let $|\psi(0)\rangle$ be the state of a system at time $t = 0$ and $|\psi'(0)\rangle = \hat{\Omega}_\mathcal{T}\,|\psi(0)\rangle$ the corresponding time-reversed state. Assume that the Hamiltonian describing this system is time-reversal invariant. Establish the relation between the time-evolved states $|\psi(t)\rangle$ and $|\psi'(t)\rangle$.

Solution

The states $|\psi(0)\rangle$ and $|\psi'(0)\rangle$ evolve in time independently according to

$$|\psi(t)\rangle = e^{-i\hat{H}t/\hbar} |\psi(0)\rangle , \qquad |\psi'(t)\rangle = e^{-i\hat{H}t/\hbar} \hat{\Omega}_{\mathcal{T}} |\psi(0)\rangle .$$

Since the Hamiltonian is time-reversal invariant, the states $|\psi(t)\rangle$ and $|\psi'(t)\rangle$ are related to each other via

$$|\psi'(t)\rangle = e^{-i\hat{H}t/\hbar} \hat{\Omega}_{\mathcal{T}} |\psi(0)\rangle = \hat{\Omega}_{\mathcal{T}} e^{+i\hat{H}t/\hbar} |\psi(0)\rangle = \hat{\Omega}_{\mathcal{T}} |\psi(-t)\rangle ,$$

which follows from

$$\underbrace{\left[\sum_{n=0}^{\infty} \frac{(-i t/\hbar)^n}{n!} \hat{H}^n \right]}_{e^{-i\hat{H}t/\hbar}} \hat{\Omega}_{\mathcal{T}} = \sum_{n=0}^{\infty} \frac{(-it/\hbar)^n}{n!} \hat{\Omega}_{\mathcal{T}} \hat{H}^n = \hat{\Omega}_{\mathcal{T}} \underbrace{\left[\sum_{n=0}^{\infty} \frac{(+it/\hbar)^n}{n!} \hat{H}^n \right]}_{e^{+i\hat{H}t/\hbar}} ,$$

where we have used the property that $\hat{\Omega}_{\mathcal{T}}$ is antilinear (and so the imaginary unit changes sign when it passes through $\hat{\Omega}_{\mathcal{T}}$).

Problem 20   On Eigenfunctions of Non-Degenerate Eigenvalues of a Time-Reversal-Invariant Observable

Show that the eigenfunctions corresponding to non-degenerate eigenvalues of a time-reversal-invariant observable $\hat{A}$ can be chosen to be real up to an overall phase factor.

Solution

Since the observable $\hat{A}$ is taken to be invariant under time reversal, it commutes with $\hat{\Omega}_{\mathcal{T}}$. Hence, if $|\psi\rangle$ is an eigenstate of $\hat{A}$, so is the state $|\psi'\rangle = \hat{\Omega}_{\mathcal{T}}|\psi\rangle$. By assumption, the eigenvalue is non-degenerate and so $|\psi'\rangle = e^{i\alpha}|\psi\rangle$ (α real), that is, these two states can differ by at most a phase factor. On the other hand, the time-reversed wave function $\psi'(\mathbf{r})$ is simply $\psi^*(\mathbf{r})$. We conclude that $\psi^*(\mathbf{r}) = e^{i\alpha}\psi(\mathbf{r})$. Note that the assumption of non-degeneracy is crucial. For example, the Hamiltonian of a free particle is time-reversal invariant; however, each eigenvalue $E_p = \mathbf{p}^2/(2m)$ is infinitely degenerate, and the corresponding eigenfunctions $e^{i p \hat{\mathbf{n}} \cdot \mathbf{r}/\hbar}/(2\pi\hbar)^{3/2}$ are complex; here, $\hat{\mathbf{n}}$ is the arbitrary direction of the momentum $\mathbf{p}$.

Problem 21   Transformation of States under $\hat{\Omega}_{\mathcal{T}}^2$; Kramers' Degeneracy

Consider the action of $\hat{\Omega}_{\mathcal{T}}^2$ on the spinor wave function of a single spin-1/2 particle and then on a spinor wave function representing the state of N spin-1/2 particles; show that $\hat{\Omega}_{\mathcal{T}}^2 = (-1)^N \hat{\mathbb{1}}$. Show further that this implies that the eigenstates $|\psi_{JM}\rangle$ of the total angular momentum $\hat{\mathbf{J}}^2$ and component $\hat{J}_z$ of an N-spin-1/2-particle system must also be eigenstates of $\hat{\Omega}_{\mathcal{T}}^2$, with eigenvalues $(-1)^{2J}$. Next, suppose that the Hamiltonian of such a system consisting of an *odd* number of particles is time-reversal invariant: deduce that its eigenvalues are at least two-fold degenerate (Kramers' degeneracy).

Solution

Consider a single spin-1/2 particle. Using Eq. (16.77), we find

$$\hat{\Omega}_{\mathcal{T}}^2 \psi(\mathbf{r}) = \hat{\Omega}_{\mathcal{T}}^2 \sum_s \psi_s(\mathbf{r})|\psi_s\rangle = \hat{\Omega}_{\mathcal{T}} \sum_s \eta\,(-1)^{1/2-s}\,\psi_s^*(\mathbf{r})|\psi_{-s}\rangle$$

$$= \sum_s \eta^*(-1)^{1/2-s}\,\psi_s(\mathbf{r})\,\hat{\Omega}_{\mathcal{T}}|\psi_{-s}\rangle = \sum_s \underbrace{\eta^*\eta}_{=1}\,(-1)^{1/2-s}\,(-1)^{1/2+s}\,\psi_s(\mathbf{r})|\psi_s\rangle = -\psi(\mathbf{r})\,,$$

and therefore $\hat{\Omega}_{\mathcal{T}}^2 = -\hat{\mathbb{1}}$. For a system of N spin-1/2 particles (not necessarily identical), it follows similarly that under the action of $\hat{\Omega}_{\mathcal{T}}^2$ the spinor wave function $\psi(\mathbf{r}_1, \ldots, \mathbf{r}_N)$, given by

$$\psi(\mathbf{r}_1, \ldots, \mathbf{r}_N) = \sum_{s_1, \ldots, s_N} \psi_{s_1, \ldots, s_N}(\mathbf{r}_1, \ldots \mathbf{r}_N)|s_1, \ldots, s_N\rangle\,,$$

transforms as follows ($\hat{\Omega}_{\mathcal{T}}^2$ acts on each individual particle, yielding a factor -1):

$$\hat{\Omega}_{\mathcal{T}}^2 \psi(\mathbf{r}_1, \ldots, \mathbf{r}_N) = (-1)^N \psi(\mathbf{r}_1, \ldots, \mathbf{r}_n) \implies \hat{\Omega}_{\mathcal{T}}^2 = (-1)^N \hat{\mathbb{1}}\,,$$

that is, $\hat{\Omega}_{\mathcal{T}}^2$ is either plus or minus the identity, depending on whether N is even or odd. Suppose we now expand this N-particle state in a basis $|JM\rangle$ of eigenstates of $\hat{\mathbf{J}}^2$ and $\hat{J}_z$, where $\mathbf{J}$ is the total angular momentum. The latter is obtained by combining the angular momenta $\mathbf{J}_i = \mathbf{L}_i + \mathbf{S}_i$ of the N particles, and J will be integer (half-integer) depending on whether N is even (odd), that is,

$$|\psi\rangle = \sum_{J=0,1,\ldots} \sum_{M=-J}^{J} c_{JM}|\psi_{JM}\rangle \text{ if } N \text{ even}\,, \qquad |\psi\rangle = \sum_{J=1/2,3/2,\ldots} \sum_{M=-J}^{J} c_{JM}|\psi_{JM}\rangle \text{ if } N \text{ odd}\,.$$

Consistency with the equation derived above, $\hat{\Omega}_{\mathcal{T}}^2 = (-1)^N \hat{\mathbb{1}}$, requires that

$$\hat{\Omega}_{\mathcal{T}}^2 |\psi_{JM}\rangle = |\psi_{JM}\rangle \text{ if } J \text{ integer}\,, \qquad \hat{\Omega}_{\mathcal{T}}^2 |\psi_{JM}\rangle = -|\psi_{JM}\rangle \text{ if } J \text{ half-integer}\,,$$

which says that $|\psi_{JM}\rangle$ are eigenstates of $\hat{\Omega}_{\mathcal{T}}^2$ with eigenvalues $(-1)^{2J}$.

A consequence of invariance under time reversal is Kramers' degeneracy. If $|\psi\rangle$ is an eigenstate of a time-reversal-invariant Hamiltonian $\hat{H}$ with eigenvalue E then $\hat{\Omega}_{\mathcal{T}}|\psi\rangle$ is also an eigenstate of $\hat{H}$ corresponding to the same eigenvalue E. Suppose that the eigenvalue E is non-degenerate; then $|\psi\rangle$ and $\hat{\Omega}_{\mathcal{T}}|\psi\rangle$ can differ only by a phase factor, namely

$$\hat{\Omega}_{\mathcal{T}}|\psi\rangle = e^{i\alpha}|\psi\rangle \qquad \text{if } E \text{ non-degenerate}\,.$$

On the other hand, applying $\hat{\Omega}_{\mathcal{T}}$ to both sides of the relation above yields

$$\hat{\Omega}_{\mathcal{T}}^2|\psi\rangle = \hat{\Omega}_{\mathcal{T}}[\hat{\Omega}_{\mathcal{T}}|\psi\rangle] = e^{-i\alpha}\,\hat{\Omega}_{\mathcal{T}}|\psi\rangle = e^{-i\alpha}\,e^{i\alpha}|\psi\rangle = |\psi\rangle\,.$$

But this property is inconsistent with the property $\hat{\Omega}_{\mathcal{T}}^2 = -\hat{\mathbb{1}}$ that we have established for a system containing an *odd number* of spin-1/2 particles. Therefore we conclude that for such a system $|\psi\rangle$ and $\hat{\Omega}_{\mathcal{T}}|\psi\rangle$ must be independent states, namely, there must be at least two-fold degeneracy.

Problem 22 Time-Reversal Invariance and Scattering of Spinless Particles

We have shown that the scattering amplitude can be written as

$$f_{\mathbf{k}}(\mathbf{k}') = -\frac{\mu}{2\pi\hbar^2}\,\langle\varphi_{\mathbf{k}'}|\,\hat{T}\,|\varphi_{\mathbf{k}}\rangle\,,$$

where $|\varphi_{\mathbf{k}}\rangle$ and $|\varphi_{\mathbf{k}'}\rangle$ are free-particle states, and the $\hat{T}$ operator is defined as follows:

$$\hat{T} = \hat{V} + \hat{V}\frac{1}{E_k - \hat{H}_0 + i\eta}\hat{V} + \hat{V}\frac{1}{E_k - \hat{H}_0 + i\eta}\hat{V}\frac{1}{E_k - \hat{H}_0 + i\eta}\hat{V} + \cdots .$$

Assume the interaction $\hat{V}$ is invariant under time reversal, $\hat{\Omega}_{\mathcal{T}}\,\hat{V}\hat{\Omega}_{\mathcal{T}}^{\dagger} = \hat{V}$.

1. Show that

$$(\varphi_{\mathbf{k}'}, \hat{T}\varphi_{\mathbf{k}}) = (\varphi_{-\mathbf{k}}, \hat{T}\varphi_{-\mathbf{k}'})$$

and hence, in terms of scattering amplitudes, the implication of time-reversal invariance is that

$$f_{\mathbf{k}}(\mathbf{k}') = f_{-\mathbf{k}'}(-\mathbf{k}) ,$$

that is, the amplitude for the transition $\mathbf{k} \longrightarrow \mathbf{k}'$ is the same as that for the transition $-\mathbf{k}' \longrightarrow -\mathbf{k}$, in which the initial and final states and associated momenta are inverted.

2. Suppose now that $\hat{V}$ is also invariant under space inversion,

$$\hat{U}_P\,\hat{V}\hat{U}_P^{\dagger} = \hat{V} ,$$

and deduce that the combination of invariance under time reversal and under parity leads to

$$f_{\mathbf{k}}(\mathbf{k}') = f_{\mathbf{k}'}(\mathbf{k}) ,$$

that is, the amplitude is symmetric under the exchange $\mathbf{k} \rightleftharpoons \mathbf{k}'$, which in turn leads to the result that the cross section for $\mathbf{k} \longrightarrow \mathbf{k}'$ is the same for $\mathbf{k}' \longrightarrow \mathbf{k}$.

Solution

Part 1

Since

$$\hat{\Omega}_{\mathcal{T}}\,\hat{V}\hat{\Omega}_{\mathcal{T}}^{\dagger} = \hat{V}, \qquad \hat{\Omega}_{\mathcal{T}}\,\hat{H}_0\,\hat{\Omega}_{\mathcal{T}}^{\dagger} = \hat{H}_0 ,$$

and $\hat{\Omega}_{\mathcal{T}}$ is anti-unitary, it follows that

$$\hat{\Omega}_{\mathcal{T}}\,\hat{T}\hat{\Omega}_{\mathcal{T}}^{\dagger} = \hat{V} + \hat{V}\frac{1}{E_k - \hat{H}_0 - i\eta}\hat{V} + \hat{V}\frac{1}{E_k - \hat{H}_0 - i\eta}\hat{V}\frac{1}{E_k - \hat{H}_0 + i\eta}\hat{V} + \cdots = \hat{T}^{\dagger} ,$$

and therefore

$$(\varphi_{\mathbf{k}'}, \hat{T}\varphi_{\mathbf{k}}) = (\hat{\Omega}_{\mathcal{T}}\,\varphi_{\mathbf{k}'}, \hat{\Omega}_{\mathcal{T}}\,\hat{T}\varphi_{\mathbf{k}})^* = (\hat{\Omega}_{\mathcal{T}}\,\varphi_{\mathbf{k}'}, \underbrace{\hat{\Omega}_{\mathcal{T}}\,\hat{T}\hat{\Omega}_{\mathcal{T}}^{\dagger}}_{T^{\dagger}}\,\hat{\Omega}_{\mathcal{T}}\,\varphi_{\mathbf{k}})^* = (\varphi_{-\mathbf{k}'}, \hat{T}^{\dagger}\varphi_{-\mathbf{k}})^* = (\varphi_{-\mathbf{k}}, \hat{T}\varphi_{-\mathbf{k}'}).$$

In terms of scattering amplitudes, the implication of time-reversal invariance is that

$$f_{\mathbf{k}}(\mathbf{k}') = f_{-\mathbf{k}'}(-\mathbf{k}) .$$

Part 2

Since $\hat{V}$ is also invariant under space inversion,

$$\hat{U}_P\,\hat{V}\hat{U}_P^{\dagger} = \hat{V} \implies \hat{U}_P\,\hat{T}\hat{U}_P^{\dagger} = \hat{T} ,$$

we deduce (using the unitarity of $\hat{U}_P$) that

$$(\varphi_{-\mathbf{k}}, \hat{T}\varphi_{-\mathbf{k}'}) = (\hat{U}_P\,\varphi_{-\mathbf{k}}, \hat{U}_P\,\hat{T}\varphi_{-\mathbf{k}'}) = (\hat{U}_P\,\varphi_{-\mathbf{k}}, \hat{T}\hat{U}_P\,\varphi_{-\mathbf{k}'}) = (\varphi_{\mathbf{k}}, \hat{T}\varphi_{\mathbf{k}'})\;.$$

Thus the combination of invariance under time reversal and parity leads to

$$f_{\mathbf{k}}(\mathbf{k}') = f_{\mathbf{k}'}(\mathbf{k})\;,$$

that is, the amplitude is symmetric under the exchange $\mathbf{k} \rightleftharpoons \mathbf{k}'$.

Rotation Matrices and the Wigner–Eckart Theorem; Fine and Hyperfine Structure of Energy Levels in Hydrogen-Like Atoms

In this chapter we discuss the matrix representation of the unitary operator that induces rotations, introduce a general classification of operators in relation to their transformation properties under rotations (that is, irreducible tensor operators), and obtain an important result (the Wigner–Eckart theorem) regarding the matrix elements of irreducible tensor operators between given eigenstates of $\hat{\mathbf{J}}^2$ and $\hat{J}_z$. Applications of these results to the physically relevant cases of the fine and hyperfine structure of the hydrogen atom, including the splitting of its energy levels in weak magnetic fields (the Zeeman effect) and in strong magnetic fields (the Paschen-Back effect), are presented in a number of problems.

One way to characterize a rotation is to specify the axis of rotation $\hat{\mathbf{n}}$ and the angle φ by which we rotate the system counter-clockwise about this axis. The representation of the associated unitary operator $\hat{U}_R(\varphi\hat{\mathbf{n}}) = \exp(-i\varphi\hat{\mathbf{n}}\cdot\hat{\mathbf{J}}/\hbar)$ in terms of rotation matrices is introduced in Problem 17.1, where derivations of some properties satisfied by these matrices are also provided. An alternative and useful way to characterize a rotation, though, is in terms of the three Euler angles. An arbitrary rotation can be performed as a sequence of three rotations: the first about the $\hat{\mathbf{z}}$-axis by an angle α, the second about the $\hat{\mathbf{y}}'$-axis (that is, the $\hat{\mathbf{y}}$-axis rotated by α about $\hat{\mathbf{z}}$) by an angle β, and the third about the axis $\hat{\mathbf{z}}''$ (that is, the $\hat{\mathbf{z}}$-axis rotated by β about $\hat{\mathbf{y}}'$) by an angle γ. The angles α, β, and γ are taken to vary over the ranges $[0, 2\pi]$, $[0, \pi]$, and $[-\pi, \pi]$, respectively. The unitary operator inducing these three rotations is given by the product

$$\hat{U}_R(\alpha, \beta, \gamma) = \mathrm{e}^{-i\gamma\hat{\mathbf{z}}''\cdot\hat{\mathbf{J}}/\hbar} \, \mathrm{e}^{-i\beta\hat{\mathbf{y}}'\cdot\hat{\mathbf{J}}/\hbar} \, \mathrm{e}^{-i\alpha\hat{\mathbf{z}}\cdot\hat{\mathbf{J}}/\hbar} \,, \tag{17.1}$$

where it should be noted that the β and γ rotations involve the components of $\hat{\mathbf{J}}$ along the directions $\hat{\mathbf{y}}'$ and $\hat{\mathbf{z}}''$, respectively. It is convenient to express these components in terms of components relative to the fixed axes $\hat{\mathbf{x}}$, $\hat{\mathbf{y}}$, and $\hat{\mathbf{z}}$. To this end, using the general transformation properties of operators under rotations, we have

$$\hat{\mathbf{y}}' \cdot \hat{\mathbf{J}} = \hat{U}_R(\alpha\hat{\mathbf{z}}) \, \hat{\mathbf{y}} \cdot \hat{\mathbf{J}} \, \hat{U}_R^\dagger(\alpha\hat{\mathbf{z}}) = \mathrm{e}^{-i\alpha\hat{\mathbf{z}}\cdot\hat{\mathbf{J}}/\hbar} \, \hat{\mathbf{y}} \cdot \hat{\mathbf{J}} \, \mathrm{e}^{i\alpha\hat{\mathbf{z}}\cdot\hat{\mathbf{J}}/\hbar} \,, \tag{17.2}$$

yielding (see Problem 3)

$$\mathrm{e}^{-i\beta\hat{\mathbf{y}}'\cdot\hat{\mathbf{J}}/\hbar} = \mathrm{e}^{-i\alpha\hat{\mathbf{z}}\cdot\hat{\mathbf{J}}/\hbar} \, \mathrm{e}^{-i\beta\hat{\mathbf{y}}\cdot\hat{\mathbf{J}}/\hbar} \, \mathrm{e}^{i\alpha\hat{\mathbf{z}}\cdot\hat{\mathbf{J}}/\hbar} \,, \tag{17.3}$$

where the right-hand side involves only components of $\hat{\mathbf{J}}$ relative to the fixed axes. Similarly, we find

$$\mathrm{e}^{-i\gamma\hat{\mathbf{z}}''\cdot\hat{\mathbf{J}}/\hbar} = \mathrm{e}^{-i\beta\hat{\mathbf{y}}'\cdot\hat{\mathbf{J}}/\hbar} \, \mathrm{e}^{-i\gamma\hat{\mathbf{z}}\cdot\hat{\mathbf{J}}/\hbar} \, \mathrm{e}^{i\beta\hat{\mathbf{y}}'\cdot\hat{\mathbf{J}}/\hbar} \,, \tag{17.4}$$

and, using the above result,

$$\mathrm{e}^{-i\gamma\hat{\mathbf{z}}''\cdot\hat{\mathbf{J}}/\hbar} = \underbrace{\mathrm{e}^{-i\alpha\hat{\mathbf{z}}\cdot\hat{\mathbf{J}}/\hbar} \, \mathrm{e}^{-i\beta\hat{\mathbf{y}}\cdot\hat{\mathbf{J}}/\hbar} \, \mathrm{e}^{i\alpha\hat{\mathbf{z}}\cdot\hat{\mathbf{J}}/\hbar}}_{\mathrm{e}^{-i\beta\hat{\mathbf{y}}'\cdot\hat{\mathbf{J}}/\hbar}} \, \mathrm{e}^{-i\gamma\hat{\mathbf{z}}\cdot\hat{\mathbf{J}}/\hbar} \, \underbrace{\mathrm{e}^{-i\alpha\hat{\mathbf{z}}\cdot\hat{\mathbf{J}}/\hbar} \, \mathrm{e}^{i\beta\hat{\mathbf{y}}\cdot\hat{\mathbf{J}}/\hbar} \, \mathrm{e}^{i\alpha\hat{\mathbf{z}}\cdot\hat{\mathbf{J}}/\hbar}}_{\mathrm{e}^{i\beta\hat{\mathbf{y}}'\cdot\hat{\mathbf{J}}/\hbar}}$$
$$= \mathrm{e}^{-i\alpha\hat{\mathbf{z}}\cdot\hat{\mathbf{J}}/\hbar} \, \mathrm{e}^{-i\beta\hat{\mathbf{y}}\cdot\hat{\mathbf{J}}/\hbar} \, \mathrm{e}^{-i\gamma\hat{\mathbf{z}}\cdot\hat{\mathbf{J}}/\hbar} \, \mathrm{e}^{i\beta\hat{\mathbf{y}}\cdot\hat{\mathbf{J}}/\hbar} \, \mathrm{e}^{i\alpha\hat{\mathbf{z}}\cdot\hat{\mathbf{J}}/\hbar} \,, \tag{17.5}$$

where we have used the fact that $e^{-i\gamma\hat{z}\cdot\hat{\mathbf{J}}/\hbar}$ and $e^{-i\alpha\hat{z}\cdot\hat{\mathbf{J}}/\hbar}$ commute with each other (since they are rotations about the same axis), and that $e^{i\alpha\hat{z}\cdot\hat{\mathbf{J}}/\hbar}\,e^{-i\alpha\hat{z}\cdot\hat{\mathbf{J}}/\hbar} = \hat{\mathbb{1}}$. Collecting results, we arrive at

$$\hat{U}_R(\alpha,\beta,\gamma) = \underbrace{e^{-i\alpha\hat{z}\cdot\hat{\mathbf{J}}/\hbar}\,e^{-i\beta\hat{y}\cdot\hat{\mathbf{J}}/\hbar}\,e^{-i\gamma\hat{z}\cdot\hat{\mathbf{J}}/\hbar}\,e^{i\beta\hat{y}\cdot\hat{\mathbf{J}}/\hbar}\,e^{i\alpha\hat{z}\cdot\hat{\mathbf{J}}/\hbar}}_{e^{-i\gamma\hat{z}''\cdot\hat{\mathbf{J}}/\hbar}}\,\underbrace{e^{-i\alpha\hat{z}\cdot\hat{\mathbf{J}}/\hbar}\,e^{-i\beta\hat{y}\cdot\hat{\mathbf{J}}/\hbar}\,e^{i\alpha\hat{z}\cdot\hat{\mathbf{J}}/\hbar}}_{e^{-i\beta\hat{y}'\cdot\hat{\mathbf{J}}/\hbar}}\,e^{-i\alpha\hat{z}\cdot\hat{\mathbf{J}}/\hbar}\,,\qquad(17.6)$$

and after simplifying terms we obtain

$$\hat{U}_R(\alpha,\beta,\gamma) = e^{-i\alpha\hat{z}\cdot\hat{\mathbf{J}}/\hbar}\,e^{-i\beta\hat{y}\cdot\hat{\mathbf{J}}/\hbar}\,e^{-i\gamma\hat{z}\cdot\hat{\mathbf{J}}/\hbar}\,,\qquad(17.7)$$

which looks like Eq. (17.1) with the order of the operators reversed, but with one crucial difference: the components of angular momentum are now relative to the fixed axes. This facilitates the evaluation of the $\underline{D}^{(j)}$ matrices,

$$D^{(j)}_{m,m'}(\alpha,\beta,\gamma) = \langle\psi_{jm}|\,e^{-i\alpha\hat{J}_z/\hbar}\,e^{-i\beta\hat{J}_y/\hbar}\,e^{-i\gamma\hat{J}_z/\hbar}\,|\psi_{jm'}\rangle = e^{-im\alpha}\underbrace{\langle\psi_{jm}|\,e^{-i\beta\hat{J}_y/\hbar}\,|\psi_{jm'}\rangle}_{d^{(j)}_{m,m'}(\beta)}\,e^{-im'\gamma}\,,\qquad(17.8)$$

where in the basis of eigenstates $|\psi_{jm}\rangle$ of $\hat{\mathbf{J}}^2$ and $\hat{J}_z$ it should be stressed that $\hat{J}_z$ refers to the component of $\hat{\mathbf{J}}$ along the fixed $\hat{z}$-axis. The problem is reduced to the evaluation of the matrices $\underline{d}^{(j)}(\beta)$. For the case $j=1/2$, this is especially simple since, for $\hat{J}_y = \hbar\,\hat{\sigma}_y/2$,

$$e^{-i\beta\hat{\sigma}_y/2} = \cos(\beta/2)\,\hat{\mathbb{1}} - i\sin(\beta/2)\,\hat{\sigma}_y \implies \underline{d}^{(1/2)} = \begin{pmatrix}\cos(\beta/2) & -\sin(\beta/2)\\ \sin(\beta/2) & \cos(\beta/2)\end{pmatrix}\,.\qquad(17.9)$$

The $\underline{D}^{(j)}(\alpha,\beta,\gamma)$ are unitary matrices of dimensions $(2j+1)\times(2j+1)$ with determinant equal to 1, and whose inverses are simply $[\underline{D}^{(j)}(\alpha,\beta,\gamma)]^{-1} = \underline{D}^{(j)}(-\gamma,-\beta,-\alpha)$. Additional properties of these matrices are discussed in Problems 12 and 13.

17.1 Irreducible Tensor Operators and the Wigner–Eckart Theorem

It is useful to classify operators according to the their commutation relations with the angular momentum operator, the generator of rotations. As we will see, this allows us to deduce which matrix elements of these operators are non-vanishing and derive useful selection rules. Suppose we have a complete set of commuting observables consisting of $\hat{\mathbf{J}}^2$ and $\hat{J}_z$ and, possibly, additional observables which we collectively indicate by the symbol $\hat{\Gamma}$ (for example, in the case of the hydrogen atom a complete set of commuting observables consists of $\hat{\mathbf{L}}^2$, $\hat{L}_z$, and $\hat{H}$, if we ignore the electron and proton spins). The common eigenstates are $|\psi_{\gamma jm}\rangle$, where γ denotes the eigenvalues corresponding to $\hat{\Gamma}$.

We begin by considering scalar operators, which we generically denote as $\hat{S}$ (not to be confused with the spin angular momentum $\hat{\mathbf{S}}$). A scalar operator is characterized by the fact that it commutes with *all* components of $\hat{\mathbf{J}}$,

$$\text{scalar operator:}\qquad [\hat{J}_i,\hat{S}] = 0\,.\qquad(17.10)$$

Examples of scalar operators are the kinetic energy and central potential operators and the spin–orbit operator, defined as $\hat{V}(\hat{r})\,\hat{\mathbf{L}}\cdot\hat{\mathbf{S}}$ (note that this latter operator commutes with the total angular momentum $\hat{\mathbf{J}}$ but not with the orbital or spin angular momenta separately; it is a scalar in the combined orbital + spin space). It immediately follows that the matrix elements of $\hat{S}$ in the basis

$|\psi_{\gamma jm}\rangle$ vanish unless the left and right states have the same eigenvalues j and m; namely, we obtain the selection rule

$$\langle \psi_{\gamma jm}| \hat{S} |\psi_{\gamma' j' m'}\rangle = S_{\gamma, \gamma'}\, \delta_{j,j'}\, \delta_{m,m'}\,. \tag{17.11}$$

These selection rules follow easily by observing that

$$0 = \langle \psi_{\gamma jm}|[\, \hat{\mathbf{J}}^2,\, \hat{S}\,]|\psi_{\gamma' j' m'}\rangle = \hbar^2\,[j(j+1) - j'(j'+1)]\,\langle \psi_{\gamma jm}|\hat{S}|\psi_{\gamma' j' m'}\rangle\,, \tag{17.12}$$

and similarly for $m \neq m'$ by considering the vanishing commutator of $\hat{J}_z$ with $\hat{S}$ (the matrix element is also independent of the azimuthal quantum number, see Problem 9 in Chapter 16).

We next consider vector operators, generically denoted as $\hat{\mathbf{V}}$ (examples are the position and momentum operators, the orbital, spin, and total angular momenta, $\hat{\mathbf{L}} \times \hat{\mathbf{S}}$, etc.). A vector operator satisfies the commutation relations

$$\text{vector operator:} \qquad [\,\hat{J}_i,\, \hat{V}_j\,] = i\hbar \sum_k \epsilon_{ijk}\, \hat{V}_k\,. \tag{17.13}$$

Before examining matrix elements, we introduce the spherical components of $\hat{\mathbf{V}}$, defined as

$$\hat{V}_{+1} = -\frac{\hat{V}_x + i\,\hat{V}_y}{\sqrt{2}}\,, \qquad \hat{V}_{-1} = \frac{\hat{V}_x - i\,\hat{V}_y}{\sqrt{2}}\,, \qquad \hat{V}_0 = \hat{V}_z\,, \tag{17.14}$$

which, if $\hat{\mathbf{V}}$ is a hermitian operator ($\hat{\mathbf{V}}^\dagger = \hat{\mathbf{V}}$), imply that

$$\hat{V}_\mu^\dagger = (-1)^\mu\,\hat{V}_{-\mu}\,, \tag{17.15}$$

where $\mu = \pm 1, 0$. The commutation relations of these components with $\hat{J}_z$ and the angular momentum lowering and raising operators $\hat{J}_-$ and $\hat{J}_+$ are easily obtained from those above:

$$\text{vector operator:} \qquad [\,\hat{J}_z,\, \hat{V}_\mu\,] = \mu\hbar \hat{V}_\mu\,, \qquad [\,\hat{J}_\pm,\, \hat{V}_\mu\,] = \sqrt{2 - \mu(\mu \pm 1)}\,\hbar\, \hat{V}_{\mu \pm 1}\,. \tag{17.16}$$

The definition in Eq. (17.16) can be extended to the general case of tensor operators. Generally, a set of $2k+1$ operators $\hat{T}_q^{(k)}$ with $q = -k, -k+1, \ldots, k$ is said to form an irreducible tensor operator (ITO) of rank k if its components satisfy the following commutation relations with the angular momentum:

$$\text{ITO of rank } k\colon \qquad [\,\hat{J}_z,\, \hat{T}_q^{(k)}\,] = q\hbar\, \hat{T}_q^{(k)}\,, \qquad [\,\hat{J}_\pm,\, \hat{T}_q^{(k)}\,] = \sqrt{k(k+1) - q(q \pm 1)}\,\hbar\, \hat{T}_{q \pm 1}^{(k)}\,. \tag{17.17}$$

Obviously this definition reproduces Eqs. (17.10) and (17.16) when $k=0$ and 1, and so a scalar and vector operator are ITOs of rank 0 and 1, respectively. The attentive reader will observe that these relations are reminiscent of those of $\hat{J}_z$ and $\hat{J}_\pm$ acting on eigenstates $|\psi_{jm}\rangle$ of $\hat{\mathbf{J}}^2$ and $\hat{J}_z$.

The Wigner–Eckart theorem concerns matrix elements of ITOs between eigenstates of the angular momentum. It states that

$$\langle \psi_{\gamma' j' m'}| \hat{T}_q^{(k)} |\psi_{\gamma jm}\rangle = \underbrace{C_{kj}\,(qm; j'm')}_{\text{Clebsch–Gordan coefficient}}\, \overbrace{\langle \psi_{\gamma' j'}\|\hat{T}^{(k)}\|\psi_{\gamma j}\rangle}^{\text{reduced matrix element}}\,, \tag{17.18}$$

where $C_{kj}\,(qm; j'm')$ is the Clebsch–Gordan coefficient corresponding to addition of the two angular momenta kq and jm to obtain the total angular momentum $j'm'$, and $\langle \psi_{\gamma' j'}\|\hat{T}^{(k)}\|\psi_{\gamma j}\rangle$ is a coefficient known as the reduced matrix element (RME). The RME is independent of the azimuthal quantum numbers q, m, and m' but depends on everything else, including the quantum numbers k, j, and j'

(the conventional notation with the double vertical lines is meant to emphasize the independence of q, m, and m').[1] Two consequences of this important result are: (i) the dependence of the matrix element $\langle \psi_{\gamma'j'm'} | \hat{T}_q^{(k)} | \psi_{\gamma jm} \rangle$ on the azimuthal quantum numbers is made explicit – it is given by the Clebsch–Gordan coefficient and (ii) the matrix element $\langle \psi_{\gamma'j'm'} | \hat{T}_q^{(k)} | \psi_{\gamma jm} \rangle$ vanishes if the following two selection rules are not satisfied simultaneously:

$$|k - j| \le j' \le k + j \qquad \text{and} \qquad m' = q + m \,. \tag{17.19}$$

These selection rules are accounted for by the Clebsch–Gordan coefficients. To prove the Wigner–Eckart theorem, we consider the state obtained by applying the ITO $\hat{T}_q^{(k)}$ to the state $|\psi_{\gamma jm}\rangle$,

$$|\Psi_{kq;jm}\rangle = \hat{T}_q^{(k)} |\psi_{\gamma jm}\rangle \,. \tag{17.20}$$

When acting with the angular momentum operators $\hat{J}_z$ and $\hat{J}_\pm$ on $|\Psi_{kq;jm}\rangle$, we find

$$\hat{J}_z |\Psi_{kq;jm}\rangle = \underbrace{[\hat{J}_z, \hat{T}_q^{(k)}]}_{q\hbar \hat{T}_q^{(k)}} |\psi_{\gamma jm}\rangle + \hat{T}_q^{(k)} \underbrace{\hat{J}_z |\psi_{\gamma jm}\rangle}_{m\hbar |\psi_{\gamma jm}\rangle} = (q + m)\hbar |\Psi_{kq;jm}\rangle \tag{17.21}$$

and similarly

$$\hat{J}_\pm |\Psi_{kq;jm}\rangle = \sqrt{k(k+1) - q(q \pm 1)}\,\hbar\,|\Psi_{k,q\pm 1;jm}\rangle + \sqrt{j(j+1) - m(m \pm 1)}\,\hbar\,|\Psi_{kq;j,m\pm 1}\rangle \,. \tag{17.22}$$

Thus the action of $\hat{\mathbf{J}}$ on $|\Psi_{kq;jm}\rangle$ is the same as if $|\Psi_{kq;jm}\rangle$ corresponded to a state of two particles, one with angular momentum kq and the other with angular momentum jm; in other words, acting with $(\hat{\mathbf{J}}_1 + \hat{\mathbf{J}}_2)_{z,\pm}$ on the state $|\Psi_{kq;jm}\rangle$, viewed as a state of two particles with angular momentum kq and jm leads to the same states on the right-hand side of the relations derived above. We can therefore add these two angular momenta to form the total angular momentum and expand $|\Psi_{kq;jm}\rangle$ in the basis of eigenstates of this total angular momentum, see Eqs. (13.29), (13.30), to obtain

$$|\Psi_{kq;jm}\rangle = \sum_{j''=|k-j|}^{k+j} \sum_{m''=-j''}^{j''} C_{kj}(qm; j''m'') |\Psi_{j''m''}^{kj}\rangle \,. \tag{17.23}$$

We now take the inner product of the state $|\psi_{\gamma'j'm'}\rangle$ with the state $|\Psi_{kq;jm}\rangle$,

$$\langle \psi_{\gamma'j'm'} | \Psi_{kq;jm}\rangle = \sum_{j''=|k-j|}^{k+j} \sum_{m''=-j''}^{j''} C_{kj}(qm; j''m'') \underbrace{\langle \psi_{\gamma'j'm'} | \Psi_{j''m''}^{kj}\rangle}_{\text{vanishes if } j'' \ne j' \text{ and } m'' \ne m'} = C_{kj}(qm; j'm') \langle \psi_{\gamma'j'm'} | \Psi_{j'm'}^{kj}\rangle \,, \tag{17.24}$$

where we have used the fact that $|\psi_{\gamma'j'm'}\rangle$ and $|\Psi_{j'm'}^{kj}\rangle$ are both eigenstates of $\hat{\mathbf{J}}^2$ and $\hat{J}_z$. The inner product $\langle \psi_{\gamma'j'm'} | \Psi_{j'm'}^{kj}\rangle$ is independent of the azimuthal quantum number m', and hence we obtain

$$\langle \psi_{\gamma'j'm'} | \Psi_{kq;jm}\rangle = \langle \psi_{\gamma'j'm'} | \hat{T}_q^{(k)} | \psi_{\gamma jm}\rangle = C_{kj}(qm; j'm') \underbrace{\langle \psi_{\gamma'j'} || \Psi_{j'}^{kj}\rangle}_{\langle \psi_{\gamma'j'} || \hat{T}^{(k)} || \psi_{\gamma j}\rangle} \,. \tag{17.25}$$

---

[1] In the literature different conventions are adopted in the definition of the RME; for example,

$$\langle \psi_{\gamma'j'm'} | \hat{T}_q^{(k)} | \psi_{\gamma jm}\rangle = \frac{(-1)^{k-j+j'}}{\sqrt{2j'+1}} C_{kj}(qm; j'm') \langle \psi_{\gamma'j'} || \hat{T}^{(k)} || \psi_{\gamma j}\rangle \,.$$

17.2 Matrix Elements of Scalar and Vector Operators

Application of the Wigner–Eckart theorem to the matrix elements of a scalar operator $\hat{S}$ (an ITO of rank 0) immediately leads, using $C_{0j}(0m; jm) = 1$, to Eq. (17.11), derived earlier. In the case of a vector operator $\hat{V}_\mu$ (an ITO of rank 1, we have

$$\langle \psi_{\gamma'j'm'} | \hat{V}_\mu | \psi_{\gamma jm} \rangle = C_{1j}(\mu m; j'm') \langle \psi_{\gamma'j'} || \hat{V} || \psi_{\gamma j} \rangle, \tag{17.26}$$

where the selection rules require

$$j' = |j - 1|, j, j + 1 \qquad \text{and} \qquad m' = \mu + m. \tag{17.27}$$

A straightforward consequence is that the matrix elements of vector operators between eigenstates of angular momentum are proportional to each other. Given two vector operators $\hat{\mathbf{V}}$ and $\hat{\mathbf{W}}$, we have (assuming the RMEs to be both non-vanishing)

$$\frac{\langle \psi_{\gamma'j'm'} | \hat{V}_\mu | \psi_{\gamma jm} \rangle}{\langle \psi_{\gamma'j'} || \hat{V} || \psi_{\gamma j} \rangle} = C_{1j}(\mu m; j'm') = \frac{\langle \psi_{\gamma'j'm'} | \hat{W}_\mu | \psi_{\gamma jm} \rangle}{\langle \psi_{\gamma'j'} || \hat{W} || \psi_{\gamma j} \rangle}, \tag{17.28}$$

which can also be written, for the cartesian components of $\hat{\mathbf{V}}$ and $\hat{\mathbf{W}}$,

$$\langle \psi_{\gamma'j'm'} | \hat{V}_i | \psi_{\gamma jm} \rangle = \frac{\langle \psi_{\gamma'j'} || \hat{V} || \psi_{\gamma j} \rangle}{\langle \psi_{\gamma'j'} || \hat{W} || \psi_{\gamma j} \rangle} \langle \psi_{\gamma'j'm'} | \hat{W}_i | \psi_{\gamma jm} \rangle \qquad \text{and} \qquad \langle \psi_{\gamma'j'} || \hat{W} || \psi_{\gamma j} \rangle \neq 0. \tag{17.29}$$

The relation above has useful implications when $\hat{\mathbf{W}}$ is the angular momentum itself. Firstly, matrix elements of $\hat{J}_i$ are diagonal in γ and j, namely

$$\langle \psi_{\gamma'j'm'} | \hat{J}_i | \psi_{\gamma jm} \rangle = 0 \qquad \text{if } \gamma \neq \gamma' \text{ and/or } j \neq j', \tag{17.30}$$

the reason being that $\hat{\Gamma}$ and $\hat{\mathbf{J}}^2$ both commute with $\hat{\mathbf{J}}$, and matrix elements of $\hat{J}_i$ between eigenstates belonging to different eigenvalues of $\hat{\Gamma}$ and $\hat{\mathbf{J}}^2$ vanish. However, when $\gamma = \gamma'$ and $j = j'$ we have

$$\langle \psi_{\gamma jm'} | \hat{V}_i | \psi_{\gamma jm} \rangle = \underbrace{\frac{\langle \psi_{\gamma j} || \hat{V} || \psi_{\gamma j} \rangle}{\langle \psi_{\gamma j} || \hat{J} || \psi_{\gamma j} \rangle}}_{\alpha_{\gamma j}} \langle \psi_{\gamma jm'} | \hat{J}_i | \psi_{\gamma jm} \rangle. \tag{17.31}$$

We emphasize that this does not mean that in general a vector operator has vanishing matrix elements when $\gamma \neq \gamma'$ and/or $j \neq j'$ – in fact, the selection rules for such an operator are those given in Eq. (17.27). Equation (17.30) only holds for the angular momentum operator. It is not in contradiction with Eq. (17.29); that equation is only applicable when $\langle \psi_{\gamma'j'} || \hat{J} || \psi_{\gamma j} \rangle \neq 0$, which is not the case if $\gamma \neq \gamma'$ and/or $j \neq j'$. After this cautionary note, consider the scalar operator $\hat{\mathbf{J}} \cdot \hat{\mathbf{V}}$ and its expectation value

$$
\begin{aligned}
\langle \psi_{\gamma jm} | \hat{\mathbf{J}} \cdot \hat{\mathbf{V}} | \psi_{\gamma jm} \rangle &= \sum_i \sum_{m'} \langle \psi_{\gamma jm} | \hat{J}_i | \psi_{\gamma jm'} \rangle \langle \psi_{\gamma jm'} | \hat{V}_i | \psi_{\gamma jm} \rangle \\
&= \sum_i \sum_{m'} \langle \psi_{\gamma jm} | \hat{J}_i | \psi_{\gamma jm'} \rangle \, \alpha_{\gamma j} \, \langle \psi_{\gamma jm'} | \hat{J}_i | \psi_{\gamma jm} \rangle \\
&= \alpha_{\gamma j} \langle \psi_{\gamma jm} | \hat{\mathbf{J}}^2 | \psi_{\gamma jm} \rangle = j(j + 1)\hbar^2 \alpha_{\gamma j},
\end{aligned} \tag{17.32}
$$

which allows us to write Eq. (17.31) as

$$\langle \psi_{\gamma jm'} | \hat{V}_i | \psi_{\gamma jm} \rangle = \frac{\langle \psi_{\gamma jm} | \hat{\mathbf{J}} \cdot \hat{\mathbf{V}} | \psi_{\gamma jm} \rangle}{j(j+1)\hbar^2} \langle \psi_{\gamma jm'} | \hat{J}_i | \psi_{\gamma jm} \rangle \, . \tag{17.33}$$

This relation is used in Problem 18 to obtain the splitting of atomic levels in a magnetic field.

17.3 Relativistic Corrections to Hydrogen-Like-Atom Hamiltonian

In this section, we consider the effects of relativistic corrections on the energy spectrum of hydrogen-like atoms. These corrections emerge naturally in a non-relativistic expansion of the Dirac equation, which describes the relativistic behavior of a charged spin-1/2 particle in an external electromagnetic field. Here, we will provide a derivation that is not based on the Dirac equation but rather on a physical interpretation of these corrections. Evaluation of their effect on the energy levels is carried out using perturbation theory in a number of problems.

A non-relativistic expansion of the Dirac equation for an electron in the Coulomb field of a nucleus of charge Ze fixed at the origin leads to the following Hamiltonian (we will simplify the notation here and not include a caret on operators)

$$H = \underbrace{\frac{\mathbf{p}^2}{2m_e} + V_C(r)}_{H_0} - \underbrace{\frac{\mathbf{p}^4}{8m_e^3c^2}}_{V_K} + \underbrace{\frac{g_e - 1}{2m_e^2c^2} \frac{1}{r} \frac{dV_C(r)}{dr} \mathbf{L} \cdot \mathbf{S}}_{V_{SO}} + \underbrace{\frac{\hbar^2}{8m_e^2c^2} \boldsymbol{\nabla}^2 V_C(r)}_{V_{DF}} \, , \tag{17.34}$$

where $V_C(r)$ is the Coulomb potential, $\mathbf{L}$ and $\mathbf{S}$ are the orbital and spin angular momentum operators of the electron, $g_e \approx 2$ is its gyromagnetic factor, and c is the speed of light. The terms labeled V_K, V_{SO}, V_{DF} represent relativistic corrections, known respectively as the kinetic energy, spin–orbit, and Darwin–Foldy terms. We now proceed to provide a physical interpretation of these terms.

Kinetic energy term V_K: This term has a straightforward interpretation. The relativistic expression for the electron kinetic energy operator is given by $K = c\sqrt{\mathbf{p}^2 + m_e^2c^2} - m_ec^2$, where m_ec^2 is the electron rest-mass energy. In the limit $x = p/(m_ec) \ll 1$, we expand K as follows:

$$K = m_ec^2 \left(\sqrt{1 + x^2} - 1 \right) = m_ec^2 \left(\frac{x^2}{2} - \frac{x^4}{8} + \cdots \right) = \frac{\mathbf{p}^2}{2m_e} - \frac{\mathbf{p}^4}{8m_e^3c^2} + \cdots \, , \tag{17.35}$$

where the second term is precisely V_K. Note that the typical velocity of the electron is of order $v/c \approx Z\alpha$. To show this, recall the virial theorem for a particle in a $1/r$ potential[2]

$$\left\langle \frac{\mathbf{p}^2}{2m_e} \right\rangle_{nl} = -\frac{1}{2} \langle V(r) \rangle_{nl} \, . \tag{17.36}$$

---

[2] Consider a particle of mass m in a generic central potential $V(r) \propto r^n$ ($n = -1$ for the case of the Coulomb potential) and the commutator $[H, \mathbf{r} \cdot \mathbf{p}]$, where $H = \mathbf{p}^2/(2m) + V(r)$. The latter can be evaluated to give

$$[H, \mathbf{r} \cdot \mathbf{p}] = \mathbf{r} \cdot [H, \mathbf{p}] + [H, \mathbf{r}] \cdot \mathbf{p} = i\hbar \, \mathbf{r} \cdot \boldsymbol{\nabla} V(r) - i\hbar \frac{\mathbf{p}}{m} \cdot \mathbf{p} = i\hbar \left[nV(r) - \frac{\mathbf{p}^2}{m} \right] .$$

By evaluating the expectation value of the commutator on an eigenstate $|\psi_{nlm}\rangle$ of the Hamiltonian, we find $\langle \psi_{nlm} | [H, \mathbf{r} \cdot \mathbf{p}] | \psi_{nlm} \rangle = 0$, which in turn implies $\langle V \rangle_{nl} = (2/n) \langle \mathbf{p}^2/(2m) \rangle_{nl}$.

Since for a hydrogen-like atom in level nl the expectation values of the (non-relativistic) kinetic and potential energy must add up to the eigenenergy of this level, we find for the ground state

$$\left\langle \frac{\mathbf{p}^2}{2m_e} \right\rangle_{1s} + \langle V(r) \rangle_{1s} = -\frac{(Z\alpha)^2}{2} m_e c^2 . \tag{17.37}$$

By combining the two results above, we obtain

$$\sqrt{\langle \mathbf{p}^2 \rangle_{1s}}/(m_e c) = v/c = Z\alpha . \tag{17.38}$$

Hence, at least for moderate atomic numbers Z, the ratio v/c is much less than 1. We can roughly estimate the order of magnitude of the correction V_K in hydrogen by making the approximation

$$|V_K| \approx \frac{1}{2m_e c^2} \underbrace{\left(\frac{\alpha^2}{2} m_e c^2 \right)^2}_{[\mathbf{p}^4/4m_e^2]} \approx 1.8 \times 10^{-4}\, \text{eV} . \tag{17.39}$$

Spin-orbit term V_{SO}: This term has an interesting physical origin and comes about because of two distinct relativistic effects. We analyze them in turn. We introduce two frames, the external fixed frame (the unprimed frame O) and the rest frame of the electron (the primed frame O'). In the fixed frame the electron is subject only to the electric field generated by the nucleus, $\mathbf{E} = Ze\, \mathbf{r}/r^3$. However, in its rest frame the electron feels a magnetic field, which, to leading order in v/c, is given by[3]

$$\mathbf{B}' \approx -\frac{\mathbf{v}}{c} \times \mathbf{E} , \tag{17.40}$$

where $\mathbf{v}$ is the electron velocity in the fixed frame (note that to terms linear in v/c the electric field in this frame is given by $\mathbf{E}' \approx \mathbf{E}$). This magnetic field couples to the spin magnetic moment of the electron and leads to an interaction energy given by

$$U' = -\boldsymbol{\mu}_s \cdot \mathbf{B}' , \tag{17.41}$$

where from Eq. (12.37) the spin magnetic moment is

$$\boldsymbol{\mu}_s = -g_e \frac{e}{2m_e c} \mathbf{S} . \tag{17.42}$$

Viewed as a classical angular momentum, the spin $\mathbf{S}$ is subject to a torque originating from the interaction U'. The equation of motion for $\mathbf{S}$ is then

$$\left. \frac{d\mathbf{S}}{dt} \right|_{O'} = \boldsymbol{\mu}_s \times \mathbf{B}' , \tag{17.43}$$

where we emphasize that the time derivative is in frame O'. From classical mechanics we know that, if the rest frame O' were to be rotating relative to the fixed frame O with an

---

[3] In a general Lorentz transformation from a frame O to a frame O' moving with velocity $\mathbf{v}$ relative to O, the electric and magnetic fields transform as

$$\mathbf{E}' = \gamma\left(\mathbf{E} + \frac{\mathbf{v}}{c} \times \mathbf{B} \right) - \frac{\gamma^2}{\gamma + 1} \frac{\mathbf{v}}{c}\left(\frac{\mathbf{v}}{c} \cdot \mathbf{E} \right), \qquad \mathbf{B}' = \gamma\left(\mathbf{B} - \frac{\mathbf{v}}{c} \times \mathbf{E} \right) - \frac{\gamma^2}{\gamma + 1} \frac{\mathbf{v}}{c}\left(\frac{\mathbf{v}}{c} \cdot \mathbf{B} \right),$$

where $\gamma = [1 - (v/c)^2]^{-1/2}$. To leading order in v/c and in consideration of the fact that in the frame O there is only an electric field, we obtain the relation given in the text.

angular velocity ω, the time derivative of a generic vector $\mathbf{V}$ in O would be related to that in O' via[4]

$$\left.\frac{d\mathbf{V}}{dt}\right|_O = \left.\frac{d\mathbf{V}}{dt}\right|_{O'} + \omega \times \mathbf{V} . \tag{17.44}$$

Indeed, O' is rotating relative to O. The origin of this rotation is due to the fact that the motion of the electron is not along a straight line. The effect was discovered by Thomas and is known as the Thomas precession (see J. D. Jackson, *Classical Electrodynamics*, 3rd edition, Wiley). A direct calculation shows that the angular velocity associated with this rotation is given by

$$\omega = \frac{\gamma^2}{\gamma+1}\frac{\mathbf{a}\times\mathbf{v}}{c^2} \approx \frac{\mathbf{a}\times\mathbf{v}}{2c^2} , \tag{17.45}$$

where $\mathbf{a}$ is the acceleration of O' relative to O, and in the last step we have taken into account only the leading term. Note that when $\mathbf{a}$ and $\mathbf{v}$ are collinear, ω vanishes. Because of the Thomas precession, the equation of motion for $\mathbf{S}$ in the fixed frame is given by

$$\left.\frac{d\mathbf{S}}{dt}\right|_O = \underbrace{\mu_s \times \mathbf{B}' + \omega \times \mathbf{S}}_{d\mathbf{S}/dt|_{O'}} = \mu_s \times (\mathbf{B}' + \mathbf{B}_T) , \qquad \mathbf{B}_T = \frac{2m_e c}{g_e e}\omega , \tag{17.46}$$

where we have used Eq. (17.42) to express $\mathbf{S}$ in terms of μ_s and have introduced an effective magnetic field $\mathbf{B}_T$ proportional to ω. This torque results from the interaction

$$U = -\mu_s \cdot (\mathbf{B}' + \mathbf{B}_T) , \tag{17.47}$$

which can also be written as

$$U = \underbrace{g_e \frac{e}{2mc}\mathbf{S}}_{-\mu_s} \cdot \underbrace{\left(-\frac{\mathbf{v}}{c}\times\mathbf{E}\right)}_{\mathbf{B}'} + \underbrace{\frac{1}{2c^2}\mathbf{S}\cdot(\mathbf{a}\times\mathbf{v})}_{-\mu_s \cdot \mathbf{B}_T = \mathbf{S}\cdot\omega} , \tag{17.48}$$

where for a point-like nucleus we have

$$\mathbf{E} = \frac{Ze}{r^3}\mathbf{r} , \qquad \mathbf{a} = -\frac{e}{m_e}\mathbf{E} , \qquad \mathbf{v} = \frac{\mathbf{p}}{m_e} . \tag{17.49}$$

Inserting these expressions into Eq. (17.48) and introducing the orbital angular momentum $\mathbf{L} = \mathbf{r}\times\mathbf{p}$, we find

$$U = \frac{g_e - 1}{2m_e^2 c^2}\frac{Ze^2}{r^3}\mathbf{L}\cdot\mathbf{S} = \frac{g_e - 1}{2m_e^2 c^2}\frac{1}{r}\frac{dV_C(r)}{dr}\mathbf{L}\cdot\mathbf{S} , \tag{17.50}$$

where the last expression is valid in the general case of a finite, rather than just point-like, nuclear charge distribution (for a point-like distribution $V_C(r) = -Ze^2/r$).

---

[4] This relation establishes how the acceleration in the fixed (inertial) frame is related to that in the rotating frame, namely

$$\left.\frac{d\mathbf{r}}{dt}\right|_O = \left.\frac{d\mathbf{r}}{dt}\right|_{O'} + \omega \times \mathbf{r} , \qquad \left.\frac{d^2\mathbf{r}}{dt^2}\right|_O = \left.\frac{d^2\mathbf{r}}{dt^2}\right|_{O'} + \omega \times \left.\frac{d\mathbf{r}}{dt}\right|_{O'} + \frac{d\omega}{dt}\times\mathbf{r} + \omega \times \left(\left.\frac{d\mathbf{r}}{dt}\right|_{O'} + \omega \times \mathbf{r}\right) ,$$

where the time derivative of ω is the same in the fixed and rotating frames, since $\omega \times \omega = 0$. We see that

$$\left.\frac{d^2\mathbf{r}}{dt^2}\right|_O = \left.\frac{d^2\mathbf{r}}{dt^2}\right|_{O'} + \underbrace{2\omega \times \left.\frac{d\mathbf{r}}{dt}\right|_{O'}}_{\text{Coriolis term}} + \underbrace{\omega \times (\omega \times \mathbf{r})}_{\text{centrifugal term}} + \frac{d\omega}{dt}\times\mathbf{r} ;$$

the last term vanishes if ω is constant in time.

The previous derivation was based on classical physics. In quantum mechanics, $\mathbf{r}$, $\mathbf{L}$, and $\mathbf{S}$ are interpreted as vector operators. Because of the spherical symmetry of the Coulomb potential, the orbital angular momentum commutes with $dV_C(r)/dr$, and so the ordering of the various terms in V_{SO} is irrelevant. We can roughly estimate the order of magnitude of the spin–orbit correction in hydrogen by setting $1/r^3 \approx 1/a_0^3$ (a_0 is the Bohr radius) and $|\mathbf{L} \cdot \mathbf{S}| \approx \hbar^2$, so that

$$|V_{SO}| \approx \frac{e^2}{2a_0} \frac{(\hbar/m_e c)^2}{a_0^2} \approx 7.2 \times 10^{-4} \text{ eV}, \tag{17.51}$$

where $e^2/(2a_0) \approx 13.6$ eV (the binding energy of the hydrogen atom).

Darwin–Foldy term V_{DF}: As a consequence of the expansion of the Dirac equation in powers of v/c, the interaction of the electron with the Coulomb potential of the nucleus, rather than being local, turns out to be non-local. It can be modeled as

$$\widetilde{V}_C(r) = \int d\boldsymbol{\lambda}\, f(\lambda)\, V_C(|\mathbf{r} + \boldsymbol{\lambda}|), \qquad f(\lambda) = \overline{f}(\lambda) \bigg/ \int d\boldsymbol{\lambda}\, \overline{f}(\lambda), \tag{17.52}$$

where the normalized function $f(\lambda)$ vanishes when λ is of the order of, or larger than, the Compton wavelength of the electron, $\hbar/(m_e c)$. It is as if the electron felt a Coulomb interaction averaged over a sphere of radius $\hbar/(m_e c)$ centered at $\mathbf{r}$ rather than just $V_C(r)$. By expanding $V_C(|\mathbf{r} + \boldsymbol{\lambda}|)$ in a Taylor series in powers of λ_i, we obtain

$$V_C(|\mathbf{r} + \boldsymbol{\lambda}|) = V_C(r) + \sum_i \lambda_i \frac{\partial V_C(r)}{\partial r_i} + \frac{1}{2} \sum_{i,j} \lambda_i \lambda_j \frac{\partial^2 V_C(r)}{\partial r_i \partial r_j} + \cdots. \tag{17.53}$$

Inserting this expansion into the expression for $\widetilde{V}_C(r)$, we find

$$\begin{aligned}
\widetilde{V}_C(r) &= V_C(r) + \sum_i \frac{\partial V_C(r)}{\partial r_i} \underbrace{\int d\boldsymbol{\lambda}\, \lambda_i f(\lambda)}_{\text{vanishes}} + \frac{1}{2} \sum_{i,j} \frac{\partial^2 V_C(r)}{\partial r_i \partial r_j} \underbrace{\int d\boldsymbol{\lambda}\, \lambda_i \lambda_j f(\lambda)}_{\delta_{ij} \int d\boldsymbol{\lambda}\, f(\lambda)\, \lambda^2/3} + \cdots \\
&= V_C(r) + \frac{1}{6} \sum_i \frac{\partial^2 V_C(r)}{\partial r_i^2} \int d\boldsymbol{\lambda}\, \lambda^2 f(\lambda) + \cdots \\
&= V_C(r) + \frac{\hbar^2}{6m_e^2 c^2} \boldsymbol{\nabla}^2 V_C(r) \underbrace{\int d\mathbf{x}\, x^2 f(x)}_{\text{non--dimensional constant}} + \cdots,
\end{aligned} \tag{17.54}$$

where in the first line the terms linear in λ_i vanish (the integrand is odd under space inversion) and the terms quadratic in $\lambda_i \lambda_j$ are proportional to $\delta_{ij} \lambda^2/3$ when integrated over $\boldsymbol{\lambda}$ because of the spherical symmetry of $f(\lambda)$; in the last line we have rescaled the integration variable $\boldsymbol{\lambda} = (\hbar/m_e c)\, \mathbf{x}$. The first term in $\widetilde{V}_C(r)$ is the leading-order Coulomb potential included in H_0 and the second term with $\int d\mathbf{x}\, x^2 f(x) = 3/4$ is the required V_{DF} correction.

Note that use of the Poisson equation of electrostatics – recall that generally $V_C(\mathbf{r}) = -e\, U(\mathbf{r})$ and the scalar potential $U(\mathbf{r})$ is such that $\boldsymbol{\nabla}^2 U(\mathbf{r}) = -4\pi \rho_N(\mathbf{r})$ – allows us to write

$$\boldsymbol{\nabla}^2 V_C(r) = 4\pi e \rho_N(r), \tag{17.55}$$

where $\rho_N(r)$ is the nuclear charge distribution. In hydrogen and for a point-like proton, in which case $\boldsymbol{\nabla}^2 V_C(r) = 4\pi e^2 \delta(\mathbf{r})$, we can easily estimate the magnitude of V_{DF}:

$$\langle V_{DF} \rangle_{1s} = \frac{\pi}{2} \frac{e^2 \hbar^2}{m_e^2 c^2} |\phi_{100}(0)|^2 = \frac{\alpha^4}{2} m_e c^2 \approx 7.2 \times 10^{-4} \text{ eV}. \tag{17.56}$$

17.4 Problems

Problem 1 Some Properties of Rotation Matrices

Let $|\psi_{jm}\rangle$ be the basis consisting of the common eigenstates of $\hat{\mathbf{J}}^2$ and $\hat{J}_z$. Consider the matrix elements

$$\langle\psi_{jm}|\,\hat{U}_R(\varphi\hat{\mathbf{n}})\,|\psi_{j'm'}\rangle = \langle\psi_{jm}|\,e^{-i\varphi\hat{\mathbf{n}}\cdot\hat{\mathbf{J}}/\hbar}\,|\psi_{j'm'}\rangle = \delta_{j,j'}\,D^{(j)}_{m,m'}(\varphi\hat{\mathbf{n}})\,.$$

The $(2j+1)\times(2j+1)$ matrices $\underline{D}^{(j)}(\varphi\hat{\mathbf{n}})$ are known as rotation matrices. Explain why they are diagonal in j; show that $[\underline{D}^{(j)}(\varphi\hat{\mathbf{n}})]^{\dagger}$ is equal to $\underline{D}^{(j)}(-\varphi\hat{\mathbf{n}})$ (the matrix representing the inverse rotation) and that $\underline{D}^{(j)}(\varphi\hat{\mathbf{n}})$ is therefore unitary.

Next, consider the basis $|\psi_{lm}\rangle$ of eigenstates of $\hat{\mathbf{L}}^2$ and $\hat{L}_z$, and the spherical harmonics $Y_{lm}(\hat{\mathbf{r}})=\langle\phi_{\mathbf{r}}|\psi_{lm}\rangle$, where here $\hat{\mathbf{r}}$ denotes the unit vector specified by the angle θ and ϕ ($\hat{\mathbf{r}}$ is *not* the position operator). Show that under a rotation by φ about the $\hat{\mathbf{z}}$-axis we have $Y'_{lm}(\hat{\mathbf{r}})=\langle\phi_{\mathbf{r}}|\psi'_{lm}\rangle=e^{-im\varphi}\,Y_{lm}(\mathbf{r})$, and verify explicitly that $Y'_{lm}(\hat{\mathbf{r}})=Y_{lm}(R^{-1}\hat{\mathbf{r}})$.

Solution

Since $\hat{U}_R(\varphi\hat{\mathbf{n}})$ and $\hat{\mathbf{J}}^2$ commute with each other, the matrix representation of $\hat{U}_R(\varphi\hat{\mathbf{n}})$ in the basis $|\psi_{jm}\rangle$ is diagonal in j. As a consequence, a state $|\psi_{jm}\rangle$ under a rotation will transform into a linear combination of states having the same j but different m-values,

$$\hat{U}_R(\varphi\hat{\mathbf{n}})\,|\psi_{jm}\rangle = \sum_{j'}\sum_{m'=-j'}^{j'}|\psi_{j'm'}\rangle\langle\psi_{j'm'}|\hat{U}_R(\varphi\hat{\mathbf{n}})\,|\psi_{jm}\rangle = \sum_{m'=-j}^{j}|\psi_{jm'}\rangle D^{(j)}_{m',m}(\varphi\hat{\mathbf{n}})\,,$$

and the matrix elements are generally complex numbers. The matrix $\underline{D}^{(j)}$ has the property

$$\big[\underline{D}^{(j)\dagger}(\varphi\hat{\mathbf{n}})\big]_{m,m'}=D^{(j)*}_{m',m}(\varphi\hat{\mathbf{n}})=\langle\psi_{jm'}|\,e^{-i\varphi\hat{\mathbf{n}}\cdot\hat{\mathbf{J}}/\hbar}\,|\psi_{jm}\rangle^*=\langle\psi_{jm}|\,e^{i\varphi\hat{\mathbf{n}}\cdot\hat{\mathbf{J}}/\hbar}\,|\psi_{jm'}\rangle=D^{(j)}_{m,m'}(-\varphi\hat{\mathbf{n}})\,,$$

and $\underline{D}^{(j)\dagger}$ is the matrix representing the inverse rotation; namely, the matrix $\underline{D}^{(j)}$ is unitary,

$$\big[\underline{D}^{(j)\dagger}(\varphi\hat{\mathbf{n}})\,\underline{D}^{(j)}(\varphi\hat{\mathbf{n}})\big]_{m,m'}=\delta_{m,m'}\,.$$

Under a rotation the spherical harmonics transform as follows:

$$Y'_{lm}(\hat{\mathbf{r}}) = \langle\phi_{\mathbf{r}}|\psi'_{lm}\rangle = \langle\phi_{\mathbf{r}}|U_R(\varphi\hat{\mathbf{n}})|\psi_{lm}\rangle = \sum_{m'}\langle\phi_{\mathbf{r}}|\psi_{lm'}\rangle D^{(l)}_{m',m}(\varphi\hat{\mathbf{n}})$$

$$= \sum_{m'}Y_{lm'}(\hat{\mathbf{r}})\,D^{(l)}_{m',m}(\varphi\hat{\mathbf{n}}) = Y_{lm}(R^{-1}\hat{\mathbf{r}})\,,$$

where in the last step we used the transformation law of wave functions under rotations that we established earlier. The matrix representing a rotation about $\hat{\mathbf{z}}$ is especially simple, since in this case $\underline{D}^{(j)}(\varphi\hat{\mathbf{z}})$ is diagonal with respect to m and is given by

$$D^{(j)}_{m,m'}(\varphi\hat{\mathbf{z}}) = \langle\psi_{jm}|\,e^{-i\varphi\hat{J}_z/\hbar}\,|\psi_{jm'}\rangle = e^{-im\varphi}\,\delta_{m,m'} \implies Y'_{lm}(\hat{\mathbf{r}}) = e^{-im\varphi}\,Y_{lm}(\hat{\mathbf{r}}) = Y_{lm}(R^{-1}\hat{\mathbf{r}})\,.$$

This can be verified explicitly by noting that

$$R^{-1}\hat{\mathbf{r}} \longrightarrow \underline{R}^{-1}(\varphi\hat{\mathbf{z}})\begin{pmatrix}\hat{r}_x\\\hat{r}_y\\\hat{r}_z\end{pmatrix} = \begin{pmatrix}\cos\varphi & \sin\varphi & 0\\-\sin\varphi & \cos\varphi & 0\\0 & 0 & 1\end{pmatrix}\begin{pmatrix}\hat{r}_x\\\hat{r}_y\\\hat{r}_z\end{pmatrix} = \begin{pmatrix}\cos\varphi\,\hat{r}_x + \sin\varphi\,\hat{r}_y\\-\sin\varphi\,\hat{r}_x + \cos\varphi\,\hat{r}_y\\\hat{r}_z\end{pmatrix},$$

which in terms of the components $\hat{r}_\pm = \hat{r}_x \pm i\hat{r}_y$ and $\hat{r}_0 = \hat{r}_z$ yields

$$\begin{aligned}\left(R^{-1}\hat{\mathbf{r}}\right)_\pm &= (\cos\varphi\,\hat{r}_x + \sin\varphi\,\hat{r}_y) \pm i(-\sin\varphi\,\hat{r}_x + \cos\varphi\,\hat{r}_y)\\&= (\cos\varphi \mp i\sin\varphi)\,\hat{r}_x \pm i(\cos\varphi \mp i\sin\varphi)\,\hat{r}_y = e^{\mp i\varphi}\,\hat{r}_\pm\,,\end{aligned}$$

and $\hat{r}_0$ is left unchanged. Recalling that (for fixed l) the $Y_{lm}(\hat{\mathbf{r}})$ are homogenous polynomials of order l in $\hat{r}_\pm$ and $\hat{r}_0$ and using results obtained in Problem 5 of Chapter 9, we see that $Y_{lm}(R^{-1}\hat{\mathbf{r}}) = e^{-im\varphi}Y_{lm}(\hat{\mathbf{r}})$; as an example, for $l=1$ we have $Y_{1,\pm 1}(\hat{\mathbf{r}}) \propto \hat{r}_\pm$ and $Y_{10}(\hat{\mathbf{r}}) \propto \hat{r}_0$, and so $Y_{1,\pm 1}(R^{-1}\hat{\mathbf{r}}) \propto e^{\mp i\varphi}\hat{r}_\pm \propto e^{\mp i\varphi}Y_{1,\pm 1}(\hat{\mathbf{r}})$ and $Y_{10}(R^{-1}\hat{\mathbf{r}}) \propto \hat{r}_0 \propto Y_{10}(\hat{\mathbf{r}})$.

Problem 2   Physical Interpretation of the Matrix Elements $D^{(j)}_{m,m'}(\alpha,\beta,\gamma)$

Provide a physical interpretation of the matrix elements of the rotation matrix.

Solution

Consider the state $|\psi'_{jm}\rangle$, an eigenstate of $\hat{J}'_z$ corresponding to the eigenvalue $m\hbar$, where

$$|\psi'_{jm}\rangle = \hat{U}_R(\alpha,\beta,\gamma)|\psi_{jm}\rangle\,, \qquad \hat{J}'_z = \hat{U}_R(\alpha,\beta,\gamma)\,\hat{J}_z\,\hat{U}^\dagger_R(\alpha,\beta,\gamma)\,.$$

Then, by expanding $|\psi'_{jm}\rangle$ in the basis of eigenstates $|\psi_{jm}\rangle$ of $\hat{\mathbf{J}}^2$ and $\hat{J}_z$, we see that

$$|\psi'_{jm}\rangle = \sum_{m'=-j}^{j}|\psi_{jm'}\rangle D^{(j)}_{m',m}(\alpha,\beta,\gamma)\,,$$

and the probability that the state $|\psi'_{jm}\rangle$ has angular momentum with projection $m'\hbar$ relative to the fixed $\hat{\mathbf{z}}$-axis is given by

$$|\langle\psi_{jm'}|\psi'_{jm}\rangle|^2 = |D^{(j)}_{m',m}(\alpha,\beta,\gamma)|^2\,.$$

In other words, $D^{(j)}_{m',m}(\alpha,\beta,\gamma)$ represents the probability amplitude that a state with polarization $m\hbar$ along the axis whose direction is specified by the Euler angles α,β,γ is found in a state with polarization $m'\hbar$ along the fixed $\hat{\mathbf{z}}$-axis.

Problem 3   Explicit Calculation of the Transform of an Angular Momentum Component under Rotation

Obtain by explicit calculation $\hat{J}'_y$, the transform of $\hat{J}_y$ under a rotation by α about the $\hat{\mathbf{z}}$-axis. Show that the transformation law is consistent with that given in Eq. (16.37). Further, show that

$$e^{-i\beta\hat{J}'_y/\hbar} = e^{-i\alpha\hat{J}_z/\hbar}\,e^{-i\beta\hat{J}_y/\hbar}\,e^{i\alpha\hat{J}_z/\hbar}\,.$$

Solution

The relation in Eq. (16.9) gives

$$\hat{J}'_y = e^{-i\lambda\hat{J}_z}\,\hat{J}_y\,e^{i\lambda\hat{J}_z} = \hat{J}_y - i\lambda\,[\hat{J}_z,\hat{J}_y] + \frac{(-i\lambda)^2}{2!}\,[\hat{J}_z,[\hat{J}_z,\hat{J}_y]] + \frac{(-i\lambda)^3}{3!}\,[\hat{J}_z,[\hat{J}_z,[\hat{J}_z,\hat{J}_y]]] + \cdots,$$

where $\lambda = \alpha/\hbar$. The relevant commutators are $[\hat{J}_z,\hat{J}_y] = -i\hbar\hat{J}_x$ and $[\hat{J}_z,\hat{J}_x] = i\hbar\hat{J}_y$, and

$$e^{-i\lambda\hat{J}_z}\,\hat{J}_y\,e^{i\lambda\hat{J}_z} = \left[1 - \frac{(\hbar\lambda)^2}{2!} + \cdots\right]\hat{J}_y - \left[\hbar\lambda - \frac{(\hbar\lambda)^3}{3!} + \cdots\right]\hat{J}_x = \cos\alpha\,\hat{J}_y - \sin\alpha\,\hat{J}_x.$$

This relation is consistent with Eq. (16.37), since $\hat{\mathbf{J}}$ as a vector operator must transform as follows:

$$\hat{\mathbf{J}}' = \hat{U}_R(\alpha\hat{\mathbf{z}})\,\hat{\mathbf{J}}\,\hat{U}_R^\dagger(\alpha\hat{\mathbf{z}}) = \underline{R}^{-1}(\alpha\hat{\mathbf{z}})\,\hat{\mathbf{J}} = \begin{pmatrix} \cos\alpha & \sin\alpha & 0 \\ -\sin\alpha & \cos\alpha & 0 \\ 0 & 0 & 1 \end{pmatrix}\begin{pmatrix} \hat{J}_x \\ \hat{J}_y \\ \hat{J}_z \end{pmatrix},$$

It also follows that

$$e^{-i\beta\hat{J}'_y/\hbar} = \sum_{n=0}^{\infty}\frac{(-i\beta/\hbar)^n}{n!}\,(\hat{J}'_y)^n = \sum_{n=0}^{\infty}\frac{(-i\beta/\hbar)^n}{n!}\left(e^{-i\alpha\hat{J}_z/\hbar}\,\hat{J}_y\,e^{i\alpha\hat{J}_z/\hbar}\right)^n$$

$$= e^{-i\alpha\hat{J}_z/\hbar}\left[\sum_{n=0}^{\infty}\frac{(-i\beta/\hbar)^n}{n!}\,(\hat{J}_y)^n\right]e^{i\alpha\hat{J}_z/\hbar} = e^{-i\alpha\hat{J}_z/\hbar}\,e^{-i\beta\hat{J}_y/\hbar}\,e^{i\alpha\hat{J}_z/\hbar},$$

where in the second line we have made use of the identity $(U_R J_y U_R^\dagger)(U_R J_y U_R^\dagger)\cdots(U_R J_y U_R^\dagger) = U_R(J_y J_y \cdots J_y)U_R^\dagger$.

Problem 4 Construction of the Rotation Matrix $\underline{D}^{(1)}(\alpha,\beta,\gamma)$

Let $\mathbf{J}$ be an angular momentum with $j = 1$ (the eigenvalue of $\mathbf{J}^2$ is $2\hbar^2$). Define $S_u = \mathbf{J}\cdot\hat{\mathbf{u}}/\hbar$, where $\hat{\mathbf{u}}$ is a unit vector, and show that

$$S_u^3 = S_u,\qquad e^{-i\varphi S_u} = 1 - i\sin\varphi\,S_u - (1-\cos\varphi)S_u^2.$$

Give an explicit expression for the rotation matrix $\underline{D}^{(1)}(\alpha,\beta,\gamma)$.

Solution

In the basis of J_z eigenstates, the components of $\mathbf{J}$ have the representation

$$\underline{J}_x = \frac{\hbar}{\sqrt{2}}\begin{pmatrix} 0 & 1 & 0 \\ 1 & 0 & 1 \\ 0 & 1 & 0 \end{pmatrix},\qquad \underline{J}_y = \frac{\hbar}{\sqrt{2}}\begin{pmatrix} 0 & -i & 0 \\ i & 0 & -i \\ 0 & i & 0 \end{pmatrix},\qquad \underline{J}_z = \hbar\begin{pmatrix} 1 & 0 & 0 \\ 0 & 0 & 0 \\ 0 & 0 & -1 \end{pmatrix}.$$

These matrices for J_x and J_y can be easily obtained by expressing the operators J_x and J_y in terms of raising and lowering operators, respectively J_+ and J_-, and by making use of the properties of $J_\pm$. Thus, the matrix representing S_u follows:

$$\underline{S}_u = \begin{pmatrix} u_0 & u_{-1} & 0 \\ u_{+1} & 0 & u_{-1} \\ 0 & u_{+1} & -u_0 \end{pmatrix},\qquad u_{\pm 1} = \frac{\hat{u}_x \pm i\hat{u}_y}{\sqrt{2}},\qquad u_0 = \hat{u}_z.$$

We find

$$\underline{S}_u^2 = \begin{pmatrix} u_0 & u_{-1} & 0 \\ u_{+1} & 0 & u_{-1} \\ 0 & u_{+1} & -u_0 \end{pmatrix} \begin{pmatrix} u_0 & u_{-1} & 0 \\ u_{+1} & 0 & u_{-1} \\ 0 & u_{+1} & -u_0 \end{pmatrix} = \begin{pmatrix} u_0^2 + u_{+1}u_{-1} & u_0 u_{-1} & u_{-1}u_{-1} \\ u_0 u_{+1} & 2u_{+1}u_{-1} & -u_0 u_{-1} \\ u_{+1}u_{+1} & -u_0 u_{+1} & u_0^2 + u_{+1}u_{-1} \end{pmatrix}$$

and

$$\underline{S}_u^3 = \begin{pmatrix} u_0^2 + u_{+1}u_{-1} & u_0 u_{-1} & u_{-1}u_{-1} \\ u_0 u_{+1} & 2u_{+1}u_{-1} & -u_0 u_{-1} \\ u_{+1}u_{+1} & -u_0 u_{+1} & u_0^2 + u_{+1}u_{-1} \end{pmatrix} \begin{pmatrix} u_0 & u_{-1} & 0 \\ u_{+1} & 0 & u_{-1} \\ 0 & u_{+1} & -u_0 \end{pmatrix}$$

$$= \begin{pmatrix} u_0(u_0 u_0 + 2u_{+1}u_{-1}) & u_{-1}(u_0 u_0 + 2u_{+1}u_{-1}) & 0 \\ u_{+1}(u_0 u_0 + 2u_{+1}u_{-1}) & 0 & u_{-1}(u_0 u_0 + 2u_{+1}u_{-1}) \\ 0 & u_{+1}(u_0 u_0 + 2u_{+1}u_{-1}) & -u_0(u_0 u_0 + 2u_{+1}u_{-1}) \end{pmatrix} = \underline{S}_u \ ,$$

since

$$u_0 u_0 + 2u_{+1}u_{-1} = \hat{u}_x^2 + \hat{u}_y^2 + \hat{u}_z^2 = 1 \ .$$

Next, we obtain

$$e^{-i\varphi S_u} = 1 + \frac{(-i\varphi)}{1!} S_u + \frac{(-i\varphi)^2}{2!} S_u^2 + \frac{(-i\varphi)^3}{3!} S_u^3 + \frac{(-i\varphi)^4}{4!} S_u^4 + \frac{(-i\varphi)^5}{5!} S_u^5 + \frac{(-i\varphi)^6}{6!} S_u^6 + \cdots$$

$$= 1 + \frac{(-i\varphi)}{1!} S_u + \frac{(-i\varphi)^2}{2!} S_u^2 + \frac{(-i\varphi)^3}{3!} S_u + \frac{(-i\varphi)^4}{4!} S_u^2 + \frac{(-i\varphi)^5}{5!} S_u + \frac{(-i\varphi)^6}{6!} S_u^2 + \cdots$$

$$= 1 - i \underbrace{\left[\frac{\varphi}{1!} - \frac{\varphi^3}{3!} + \frac{\varphi^5}{5!} - \cdots \right]}_{\sin\varphi} S_u - \underbrace{\left[\frac{\varphi^2}{2!} - \frac{\varphi^4}{4!} + \frac{\varphi^6}{6!} - \cdots \right]}_{1 - \cos\varphi} S_u^2$$

$$= 1 - i \sin\varphi \, S_u - (1 - \cos\varphi) \, S_u^2 \ ,$$

where we have used the property

$$S_u^n = S_u \quad \text{if } n \text{ odd} \ , \qquad S_u^n = S_u^2 \quad \text{if } n \text{ even} \ .$$

The D-matrices are generally defined as

$$D_{m,m'}^{(j)}(\alpha, \beta, \gamma) = \langle \psi_{jm} | \, e^{-i\alpha J_z/\hbar} \, e^{-i\beta J_y/\hbar} \, e^{-i\gamma J_z/\hbar} \, | \psi_{jm'} \rangle = e^{-im\alpha} \underbrace{\langle \psi_{jm} | \, e^{-i\beta J_y/\hbar} \, | \psi_{jm'} \rangle}_{d_{m,m'}^{(j)}(\beta)} e^{-im'\gamma} \ .$$

In the present case, $j = 1$ and $J_y/\hbar \longrightarrow S_y$ (here, $\hat{\mathbf{u}} = \hat{\mathbf{y}}$) to obtain

$$d_{m,m'}^{(1)}(\beta) = \langle \psi_{1,m} | \, e^{-i\beta S_y} \, | \psi_{1,m'} \rangle = \delta_{m,m'} - i \sin\beta \, (S_y)_{m,m'} - (1 - \cos\beta) \, (S_y^2)_{m,m'} \ ,$$

and the matrices representing S_y and S_y^2 follow from the expressions given above as

$$\underline{S}_y = \frac{1}{\sqrt{2}} \begin{pmatrix} 0 & -i & 0 \\ i & 0 & -i \\ 0 & i & 0 \end{pmatrix} \ , \qquad \underline{S}_y^2 = \frac{1}{2} \begin{pmatrix} 1 & 0 & -1 \\ 0 & 2 & 0 \\ -1 & 0 & 1 \end{pmatrix} \ .$$

Collecting results, we find

$$\underline{d}^{(1)} - \mathbb{1} = \begin{pmatrix} 0 & -(\sin\beta)/\sqrt{2} & 0 \\ (\sin\beta)/\sqrt{2} & 0 & -(\sin\beta)/\sqrt{2} \\ 0 & (\sin\beta)/\sqrt{2} & 0 \end{pmatrix} + \begin{pmatrix} -\sin^2(\beta/2) & 0 & \sin^2(\beta/2) \\ 0 & -2\sin^2(\beta/2) & 0 \\ \sin^2(\beta/2) & 0 & -\sin^2(\beta/2) \end{pmatrix} \ ,$$

where $\underline{\mathbb{1}}$ is the identity matrix, and hence

$$
\underline{d}^{(1)} = \begin{pmatrix} \cos^2(\beta/2) & -(\sin\beta)/\sqrt{2} & \sin^2(\beta/2) \\ (\sin\beta)/\sqrt{2} & \cos\beta & -(\sin\beta)/\sqrt{2} \\ \sin^2(\beta/2) & (\sin\beta)/\sqrt{2} & \cos^2(\beta/2) \end{pmatrix}.
$$

Problem 5 Commutation Relations of $\hat{\mathbf{J}}$ with the Spherical Components of a Vector Operator

Obtain the commutation relations of $\hat{J}_\pm$ and $\hat{J}_z$ with the spherical components $\hat{V}_\pm = \mp(\hat{V}_x \pm i\,\hat{V}_y)/\sqrt{2}$ and $\hat{V}_0 = \hat{V}_z$, given in in Eq. (17.16) .

Solution

Considering first $\hat{J}_+$, we obtain

$$
[\hat{J}_+, \hat{V}_{+1}] = -\frac{1}{\sqrt{2}}[\hat{J}_x + i\hat{J}_y, \hat{V}_x + i\hat{V}_y] = -\frac{i}{\sqrt{2}}\underbrace{[\hat{J}_x, \hat{V}_y]}_{i\hbar\hat{V}_z} - \frac{i}{\sqrt{2}}\underbrace{[\hat{J}_y, \hat{V}_x]}_{-i\hbar\hat{V}_z} = 0 ,
$$

$$
[\hat{J}_+, \hat{V}_0] = \underbrace{[\hat{J}_x, \hat{V}_z]}_{-i\hbar\hat{V}_y} + i\underbrace{[\hat{J}_y, \hat{V}_z]}_{i\hbar\hat{V}_x} = -i\hbar\,\hat{V}_y - \hbar\,\hat{V}_x = \sqrt{2}\,\hbar\,\hat{V}_{+1} ,
$$

$$
[\hat{J}_+, \hat{V}_{-1}] = \frac{1}{\sqrt{2}}[\hat{J}_x + i\hat{J}_y, \hat{V}_x - i\hat{V}_y] = -\frac{i}{\sqrt{2}}\underbrace{[\hat{J}_x, \hat{V}_y]}_{i\hbar\hat{V}_z} + \frac{i}{\sqrt{2}}\underbrace{[\hat{J}_y, \hat{V}_x]}_{-i\hbar\hat{V}_z} = \sqrt{2}\,\hbar\,\hat{V}_0 .
$$

We also obtain

$$
[\hat{J}_0, \hat{V}_{+1}] = -\frac{1}{\sqrt{2}}[\hat{J}_z, \hat{V}_x + i\hat{V}_y] = -\frac{1}{\sqrt{2}}\underbrace{[\hat{J}_z, \hat{V}_x]}_{i\hbar\hat{V}_y} - \frac{i}{\sqrt{2}}\underbrace{[\hat{J}_z, \hat{V}_y]}_{-i\hbar\hat{V}_x} = \hbar\,\hat{V}_{+1} ,
$$

$$
[\hat{J}_0, \hat{V}_{-1}] = \frac{1}{\sqrt{2}}[\hat{J}_z, \hat{V}_x - i\hat{V}_y] = \frac{1}{\sqrt{2}}\underbrace{[\hat{J}_z, \hat{V}_x]}_{i\hbar\hat{V}_y} - \frac{i}{\sqrt{2}}\underbrace{[\hat{J}_z, \hat{V}_y]}_{-i\hbar\hat{V}_x} = -\hbar\,\hat{V}_{-1} ,
$$

and $[\hat{J}_0, \hat{V}_0] = [\hat{J}_z, \hat{V}_z] = 0$. The remaining set of commutation relations, involving $\hat{J}_-$, can be worked out in a similar fashion.

Problem 6 Combining Two ITOs

Two irreducible tensor operators of ranks k_1 and k_2 can be combined in the following way:

$$
\hat{T}^{(k_1)}_{q_1}\,\hat{T}^{(k_2)}_{q_2} = \sum_{k=|k_1-k_2|}^{k_1+k_2} C_{k_1 k_2}(q_1 q_2; kq)\,\hat{T}^{(k)}_q ,
$$

where $C_{k_1 k_2}(q_1 q_2; kq)$ are Clebsch–Gordan coefficients. Without doing any detailed calculations, justify the expression above. Consider two vector operators $\hat{\mathbf{A}}$ and $\hat{\mathbf{B}}$, and combine them into ITOs of ranks 0, 1, and 2. To what do the ITOs of rank 0 and 1 correspond?

Solution

The formula is made plausible by noting that the state

$$|\Psi_{k_1 q_1; k_2 q_2; jm}\rangle = \hat{T}_{q_1}^{(k_1)} \hat{T}_{q_2}^{(k_2)} |\psi_{\gamma jm}\rangle$$

behaves under the action of the angular momentum $\hat{\mathbf{J}}$ as if it were a state of three particles with angular momenta $k_1 q_1$, $k_2 q_2$, and jm (this assertion can be verified following the approach used in proving the Wigner–Eckart theorem in Section 17.1). Thus we can combine the first two angular momenta ($k_1 q_1$ and $k_2 q_2$) in the usual way to obtain the relation above. The formula above can be "inverted" in the sense that

$$\hat{T}_q^{(k)} = \sum_{q_1, q_2} C_{k_1 k_2}(q_1 q_2; kq) \, \hat{T}_{q_1}^{(k_1)} \hat{T}_{q_2}^{(k_2)} \, .$$

In the case of two vector operators $\hat{\mathbf{A}}$ and $\hat{\mathbf{B}}$, we first construct the relative ITOs of rank 1, consisting of the spherical components $\hat{A}_\mu$ and $\hat{B}_\mu$. Using the formula above, we have

$$\hat{T}_q^{(k)} = \sum_{\mu, \nu = \pm 1, 0} C_{11}(\mu\nu; kq) \, \hat{A}_\mu \, \hat{B}_\nu \, ,$$

and inserting the relevant Clebsch–Gordan coefficients, we obtain for the ITO of rank 2

$$\hat{T}_{\pm 2}^{(2)} = \hat{A}_{\pm 1} \hat{B}_{\pm 1} \, , \qquad \hat{T}_{\pm 1}^{(2)} = \frac{\hat{A}_{\pm 1} \hat{B}_0 + \hat{A}_0 \hat{B}_{\pm 1}}{\sqrt{2}} \, , \qquad \hat{T}_0^{(2)} = \frac{2 \hat{A}_0 \hat{B}_0 + \hat{A}_{+1} \hat{B}_{-1} + \hat{A}_{-1} \hat{B}_{+1}}{\sqrt{6}} \, ,$$

for the ITO's of rank 1 we obtain

$$\hat{T}_{\pm 1}^{(1)} = \pm \frac{\hat{A}_{\pm 1} \hat{B}_0 - \hat{A}_0 \hat{B}_{\pm 1}}{\sqrt{2}} \, , \qquad \hat{T}_0^{(1)} = \frac{\hat{A}_{+1} \hat{B}_{-1} - \hat{A}_{-1} \hat{B}_{+1}}{\sqrt{2}} \, ,$$

and for the ITO of rank 0 we obtain

$$\hat{T}_0^{(0)} = -\frac{1}{\sqrt{3}} \sum_{\mu = \pm 1, 0} (-1)^\mu \hat{A}_\mu \hat{B}_{-\mu} \, .$$

The rank-0 ITO is a scalar and hence proportional to $\hat{\mathbf{A}} \cdot \hat{\mathbf{B}}$, while the rank-1 ITO is a vector and hence proportional to $(\hat{\mathbf{A}} \times \hat{\mathbf{B}})_\mu$, given that $\hat{\mathbf{A}} \times \hat{\mathbf{B}}$ is the only vector operator that can be constructed from $\hat{\mathbf{A}}$ and $\hat{\mathbf{B}}$.

Problem 7   ITOs of Ranks 0, 1, and 2 from Two Vector Operators

Consider two vector operators $\mathbf{V}$ and $\mathbf{W}$ acting in the orbital state space of a system, and introduce the corresponding spherical components V_μ and W_μ with $\mu = \pm 1, 0$. Form the irreducible tensor operators resulting from the coupling of V_μ and W_μ,

$$X_q^{(k)} = \sum_{\mu_1 \mu_2} C_{11}(\mu_1, \mu_2; kq) \, V_{\mu_1} W_{\mu_2} \, .$$

Show that $X_0^{(0)}$ and $X_q^{(1)}$ are proportional to the scalar product $\mathbf{V} \cdot \mathbf{W}$ and to the three spherical components of $\mathbf{V} \times \mathbf{W}$, respectively. Express the five components $X_q^{(2)}$ in terms of $V_\pm = V_x \pm i V_y$ and V_z and $W_\pm = W_x \pm i W_y$ and W_z (these are not exactly the spherical components). In the notation adopted in this chapter, under what conditions is the matrix element $\langle \psi_{\gamma' l'm'} | X_q^{(2)} | \psi_{\gamma lm} \rangle$ non-vanishing?

Solution

Inserting the values for the Clebsch–Gordan coefficients, we have

$$X_0^{(0)} = -\frac{1}{\sqrt{3}} \sum_{\mu=\pm1,0} (-1)^\mu V_\mu W_{-\mu} = \frac{1}{\sqrt{3}} (V_{+1} W_{-1} + V_{-1} W_{+1} - V_0 W_0)$$

and

$$V_{+1} W_{-1} + V_{-1} W_{+1} = -\frac{1}{2}(V_x + i V_y)(W_x - i W_y) - \frac{1}{2}(V_x - i V_y)(W_x + i W_y) = -V_x W_x - V_y W_y ,$$

so that

$$X_0^{(0)} = \frac{1}{\sqrt{3}} \left(-V_x W_x - V_y W_y - V_z W_z \right) = -\frac{1}{\sqrt{3}} \mathbf{V} \cdot \mathbf{W} .$$

Similarly, we have

$$X_0^{(1)} = \frac{1}{\sqrt{2}} (V_{+1} W_{-1} - V_{+1} W_{-1}) , \qquad X_{\pm1}^{(1)} = \pm \frac{1}{\sqrt{2}} (V_{\pm1} W_0 - V_0 W_{\pm1}) ,$$

where

$$V_{+1} W_{-1} - V_{-1} W_{+1} = -\frac{1}{2}(V_x + i V_y)(W_x - i W_y) + \frac{1}{2}(V_x - i V_y)(W_x + i W_y)$$

$$= i \left(V_x W_y - V_y W_x \right) = i (\mathbf{V} \times \mathbf{W})_0$$

and

$$V_{\pm1} W_0 - V_0 W_{\pm1} = \mp \frac{1}{\sqrt{2}}(V_x \pm i V_y) W_z \pm \frac{1}{\sqrt{2}} V_z (W_x \pm i W_y)$$

$$= \mp \frac{1}{\sqrt{2}} (V_x W_z - V_z W_x) - \frac{i}{\sqrt{2}} \left(V_y W_z - V_z W_z \right)$$

$$= \pm \frac{1}{\sqrt{2}} (\mathbf{V} \times \mathbf{W})_y - \frac{i}{\sqrt{2}} (\mathbf{V} \times \mathbf{W})_x = -\frac{i}{\sqrt{2}} \left[(\mathbf{V} \times \mathbf{W})_x \pm i (\mathbf{V} \times \mathbf{W})_y \right] .$$

Therefore, it follows that

$$X_0^{(1)} = \frac{i}{\sqrt{2}} (\mathbf{V} \times \mathbf{W})_0 , \qquad X_{\pm1}^{(1)} = \mp \frac{i}{2} \left[(\mathbf{V} \times \mathbf{W})_x \pm i (\mathbf{V} \times \mathbf{W})_y \right] = \frac{i}{\sqrt{2}} (\mathbf{V} \times \mathbf{W})_{\pm1} .$$

Lastly, we have

$$X_{\pm2}^{(2)} = V_{\pm1} W_{\pm1} , \qquad X_{\pm1}^{(2)} = \frac{1}{\sqrt{2}} (V_{\pm1} W_0 + V_0 W_{\pm1}) , \qquad X_0^{(2)} = \frac{1}{\sqrt{6}} (2 V_0 W_0 + V_{+1} W_{-1} + V_{-1} W_{+1}) ,$$

where

$$2 V_0 W_0 + V_{+1} W_{-1} + V_{-1} W_{+1} = 2 V_z W_z - V_x W_x - V_y W_y = 3 V_z W_z - \mathbf{V} \cdot \mathbf{W} ,$$

$$V_{\pm1} W_0 + V_0 W_{\pm1} = \mp \frac{1}{\sqrt{2}} \left(V_x \pm i V_y \right) W_z \mp \frac{1}{\sqrt{2}} V_z \left(W_x \pm i W_y \right)$$

$$= \mp \frac{1}{\sqrt{2}} (V_x W_z + V_z W_x) - \frac{i}{\sqrt{2}} \left(V_y W_z + V_z W_y \right)$$

and

$$V_{\pm1} W_{\pm1} = \frac{1}{2}(V_x \pm i V_y)(W_x \pm i W_y) = \frac{1}{2} \left(V_x W_x - V_y W_y \right) \pm \frac{i}{2} \left(V_x W_y + V_y W_x \right) ,$$

yielding

$$X_0^{(2)} = \frac{1}{\sqrt{6}} \left(3V_z W_z - \mathbf{V} \cdot \mathbf{W}\right), \qquad X_{\pm 1}^{(2)} = \mp \frac{1}{2} \left(V_x W_z + V_z W_x\right) - \frac{i}{2}\left(V_y W_z + V_z W_y\right)$$

and

$$X_{\pm 2}^{(2)} = \frac{1}{2}\left(V_x W_x - V_y W_y\right) \pm \frac{i}{2}\left(V_x W_y + V_y W_x\right).$$

The Wigner–Eckart theorem gives

$$\langle \psi_{\gamma' l' m'} | X_q^{(2)} | \psi_{\gamma l m} \rangle = C_{2,l}(q, m; l', m') \langle \psi_{\gamma' l'} || X^{(2)} || \psi_{\gamma l} \rangle,$$

and the Clebsch–Gordan coefficient requires $m' = m + q$ and $l' = |l - 2|, \ldots, l + 2$.

Problem 8   Product of Two Rotation Matrices

Show that the rotation matrices satisfy the relation

$$D_{m_1,m_1'}^{(j_1)}(\varphi \hat{\mathbf{n}})\, D_{m_2,m_2'}^{(j_2)}(\varphi \hat{\mathbf{n}}) = \sum_{j=|j_1-j_2|}^{j_1+j_2} \sum_{m,m'=-j}^{j} C_{j_1 j_2}(m_1 m_2; jm)\, C_{j_1 j_2}(m_1' m_2'; jm')\, D_{m,m'}^{(j)}(\varphi \hat{\mathbf{n}}).$$

Hint: The left-hand side represents independent rotations in the state space of angular momentum j_1 and that of angular momentum j_2. However, the left-hand side can also be thought of as representing a rotation in the combined space $j_1 \otimes j_2$.

Solution

We have

$$\begin{aligned}
D_{m_1,m_1'}^{(j_1)}(\varphi \hat{\mathbf{n}})\, D_{m_2,m_2'}^{(j_2)}(\varphi \hat{\mathbf{n}}) &= \langle j_1, m_1 | e^{-i\varphi \hat{\mathbf{n}} \cdot \mathbf{J}_1/\hbar} | j_1, m_1' \rangle \langle j_2, m_2 | e^{-i\varphi \hat{\mathbf{n}} \cdot \mathbf{J}_2/\hbar} | j_2, m_2' \rangle \\
&= \left(\langle j_1, m_1 | \otimes \langle j_2, m_2 | \right) e^{-i\varphi \hat{\mathbf{n}} \cdot (\mathbf{J}_1 + \mathbf{J}_2)/\hbar} \left(| j_1, m_1' \rangle \otimes | j_2, m_2' \rangle \right) \\
&= \langle j_1, m_1, j_2, m_2 | e^{-i\varphi \hat{\mathbf{n}} \cdot \mathbf{J}/\hbar} | j_1, m_1', j_2, m_2' \rangle,
\end{aligned}$$

where we have defined $\mathbf{J} = \mathbf{J}_1 + \mathbf{J}_2$. We now write

$$|j_1, m_1, j_2, m_2\rangle = \sum_{j=|j_1-j_2|}^{j_1+j_2} \sum_{m=-j}^{j} C_{j_1 j_2}(m_1, m_2; jm) |jm\rangle$$

and similarly for the primed state. We obtain

$$D_{m_1,m_1'}^{(j_1)}(\varphi \hat{\mathbf{n}})\, D_{m_2,m_2'}^{(j_2)}(\varphi \hat{\mathbf{n}}) = \sum_{j,j'} \sum_{m,m'} C_{j_1 j_2}(m_1, m_2; jm) C_{j_1 j_2}(m_1', m_2'; j'm') \langle jm | e^{-i\varphi \hat{\mathbf{n}} \cdot \mathbf{J}/\hbar} | j'm' \rangle.$$

Since $\mathbf{J}^2$ commutes with $e^{-i\varphi \hat{\mathbf{n}} \cdot \mathbf{J}/\hbar}$, the matrix element above is diagonal in j; hence,

$$D_{m_1,m_1'}^{(j_1)}(\varphi \hat{\mathbf{n}})\, D_{m_2,m_2'}^{(j_2)}(\varphi \hat{\mathbf{n}}) = \sum_{j} \sum_{m,m'} C_{j_1 j_2}(m_1, m_2; jm) C_{j_1 j_2}(m_1', m_2'; jm') \underbrace{\langle jm | e^{-i\varphi \hat{\mathbf{n}} \cdot \mathbf{J}/\hbar} | jm' \rangle}_{D_{m,m'}^{(j)}(\varphi \hat{\mathbf{n}})}.$$

Problem 9 Transformation Law of ITOs under Rotations

An irreducible tensor operator can be defined as an operator that transforms under a rotation about an axis $\hat{\mathbf{n}}$ by an angle φ as[5]

$$U_R(\varphi\hat{\mathbf{n}})\, T_q^{(k)}\, U_R^\dagger(\varphi\hat{\mathbf{n}}) = \sum_{q'=-k}^{k} T_{q'}^{(k)}\, D_{q'q}^{(k)}(\varphi\hat{\mathbf{n}})\,.$$

Show that the above definition of an ITO is consistent with that given in Eq. (17.17).

Solution

Under an infinitesimal rotation, we have

$$U_R(\varphi\hat{\mathbf{n}}) = \mathbb{1} - i\varphi\,\hat{\mathbf{n}}\cdot\mathbf{J}/\hbar + \cdots\,,$$

and, keeping terms up to linear in φ, the left-hand side is given by

$$U_R(\varphi\hat{\mathbf{n}})\, T_q^{(k)}\, U_R^\dagger(\varphi\hat{\mathbf{n}}) = (\mathbb{1} - i\varphi\,\hat{\mathbf{n}}\cdot\mathbf{J}/\hbar)\, T_q^{(k)}\, (\mathbb{1} + i\varphi\,\hat{\mathbf{n}}\cdot\mathbf{J}/\hbar) = T_q^{(k)} - i(\varphi/\hbar)\,\hat{\mathbf{n}}\cdot\left[\mathbf{J},\, T_q^{(k)}\right]\,,$$

while the right-hand side yields

$$\sum_{q'=-k}^{k} T_{q'}^{(k)}\,\langle k,q'|\mathbb{1} - i\varphi\,\hat{\mathbf{n}}\cdot\mathbf{J}/\hbar|k,q\rangle = T_q^{(k)} - i(\varphi/\hbar)\,\hat{\mathbf{n}}\cdot\sum_{q'} T_{q'}^{(k)}\,\langle k,q'|\mathbf{J}|k,q\rangle\,,$$

using the orthonormality of the angular momentum eigenstates $|k,q\rangle$. Comparing these two relations, we arrive at

$$\left[\mathbf{J},\, T_q^{(k)}\right] = \sum_{q'} T_{q'}^{(k)}\,\langle k,q'|\,\mathbf{J}\,|k,q\rangle\,,$$

which implies

$$\left[J_z,\, T_q^{(k)}\right] = \sum_{q'} T_{q'}^{(k)}\,\underbrace{\langle k,q'|J_z|k,q\rangle}_{\hbar q\,\delta_{q,q'}} = \hbar q\, T_q^{(k)}$$

and

$$\left[J_\pm,\, T_q^{(k)}\right] = \sum_{q'} T_{q'}^{(k)}\,\underbrace{\langle k,q'|J_\pm|k,q\rangle}_{\hbar\sqrt{k(k+1)-q(q\pm1)}\,\delta_{q',q\pm1}} = \hbar\sqrt{k(k+1)-q(q\pm1)}\, T_{q\pm1}^{(k)}\,.$$

These are the commutation relations satisfied by ITOs.

Problem 10 Transformation Law under Rotations of ITOs of Rank 1 (Vector Operators)

Consider a spherical tensor of rank 1 (that is, a vector operator) V_μ. Show that, for a rotation by an angle β about the $\hat{\mathbf{y}}$-axis, the formula

$$U_R(\varphi\hat{\mathbf{n}})\, V_\mu\, U_R^\dagger(\varphi\hat{\mathbf{n}}) = \sum_{\mu'=\pm1,0} V_{\mu'}\, D_{\mu'\mu}^{(1)}(\varphi\hat{\mathbf{n}})\,,$$

leads to a result which is just what one would expect from the transformation properties of V_x, V_y, and V_z under rotations about $\hat{\mathbf{y}}$.

---

[5] These definitions are equivalent to the definition we adopted in Section 17.2 in terms of commutators of $T_q^{(k)}$ with the components of the angular momentum operator.

Hint: The matrix $d^{(1)}_{\mu\mu'}(\beta) = \langle\psi_{1\mu}|e^{-i\beta J_y/\hbar}|\psi_{1\mu'}\rangle$ is given explicitly by

$$\underline{d}^{(1)}(\beta) = \begin{pmatrix} \cos^2(\beta/2) & (-\sin\beta)/\sqrt{2} & \sin^2(\beta/2) \\ (\sin\beta)/\sqrt{2} & \cos\beta & (-\sin\beta)/\sqrt{2} \\ \sin^2(\beta/2) & (\sin\beta)/\sqrt{2} & \cos^2(\beta/2) \end{pmatrix}.$$

Solution

We have

$$U_R(0\beta0)\, V_\mu\, U_R^\dagger(0\beta0) = e^{-i\beta J_y/\hbar}\, V_\mu\, e^{i\beta J_y/\hbar} = \sum_{\mu'=\pm1,0} V_{\mu'} d^{(1)}_{\mu'\mu}(\beta),$$

yielding

$$V'_{+1} = e^{-i\beta J_y/\hbar} V_{+1} e^{i\beta J_y/\hbar} = V_{+1}\cos^2(\beta/2) + \frac{V_0}{\sqrt{2}}\sin\beta + V_{-1}\sin^2(\beta/2),$$

$$V'_{-1} = e^{-i\beta J_y/\hbar} V_{-1} e^{i\beta J_y/\hbar} = V_{+1}\sin^2(\beta/2) - \frac{V_0}{\sqrt{2}}\sin\beta + V_{-1}\cos^2(\beta/2),$$

$$V'_0 = e^{-i\beta J_y/\hbar} V_0 e^{i\beta J_y/\hbar} = -\frac{V_{+1}}{\sqrt{2}}\sin\beta + V_0\cos\beta + \frac{V_{-1}}{\sqrt{2}}\sin\beta.$$

Using

$$V_x = \frac{V_{-1} - V_{+1}}{\sqrt{2}}, \qquad V_y = i\frac{V_{-1} + V_{+1}}{\sqrt{2}}, \qquad V_z = V_0,$$

it follows that

$$V'_x = \frac{1}{\sqrt{2}}\Big[V_{+1}\underbrace{[\sin^2(\beta/2) - \cos^2(\beta/2)]}_{-\cos\beta} - \sqrt{2}\,V_0\sin\beta + V_{-1}\underbrace{[\cos^2(\beta/2) - \sin^2(\beta/2)]}_{\cos\beta}\Big]$$

$$= V_x\cos\beta - V_z\sin\beta,$$

$$V'_y = \frac{i}{\sqrt{2}}\Big[V_{+1}\underbrace{[\sin^2(\beta/2) + \cos^2(\beta/2)]}_{1} + V_{-1}\underbrace{[\cos^2(\beta/2) + \sin^2(\beta/2)]}_{1}\Big] = V_y,$$

$$V'_z = V_x\sin\beta + V_z\cos\beta.$$

We have also established that under a rotation a vector operator transforms as

$$V'_i = \hat{U}_R(\beta\hat{y})\,\hat{V}_i\,\hat{U}_R^\dagger(\beta\hat{y}) = \sum_j [\underline{R}^{-1}(\beta\hat{y})]_{ij}\,\hat{V}_j,$$

where

$$\underline{R}(\beta\hat{y}) = \begin{pmatrix} \cos\beta & 0 & \sin\beta \\ 0 & 1 & 0 \\ -\sin\beta & 0 & \cos\beta \end{pmatrix} \implies \underline{R}^{-1}(\beta\hat{y}) = \begin{pmatrix} \cos\beta & 0 & -\sin\beta \\ 0 & 1 & 0 \\ \sin\beta & 0 & \cos\beta \end{pmatrix}.$$

Thus, we have

$$\begin{pmatrix} V'_x \\ V'_y \\ V'_z \end{pmatrix} = \begin{pmatrix} \cos\beta & 0 & -\sin\beta \\ 0 & 1 & 0 \\ \sin\beta & 0 & \cos\beta \end{pmatrix}\begin{pmatrix} V_x \\ V_y \\ V_z \end{pmatrix} = \begin{pmatrix} V_x\cos\beta - V_z\sin\beta \\ V_y \\ V_x\sin\beta + V_z\cos\beta \end{pmatrix},$$

in agreement with the result found earlier via the rotation matrices.

Problem 11   The Rotation Matrices as Polynomials of Order $2j$ in $\sin(\beta/2)$ and $\cos(\beta/2)$

Show that

$$e^{-i\beta J_y/\hbar} = e^{i(\pi/2)J_x/\hbar}\, e^{-i\beta J_z/\hbar}\, e^{-i(\pi/2)J_x/\hbar} \,,$$

and deduce from this result that the matrix elements

$$d^{(j)}_{m'm}(\beta) = \langle \psi_{jm'} |\, e^{-i\beta J_y/\hbar}\, |\psi_{jm}\rangle \,,$$

are polynomials of degree $2j$ with respect to the variables $\sin(\beta/2)$ and $\cos(\beta/2)$.

Hint: For the second part, insert the completeness of the angular momentum eigenstates, note that twice the azimuthal quantum number is always an integer regardless of whether j is integer or half-integer, and use the binomial expansion.

Solution

We note that a *clockwise* rotation by $\pi/2$ about $\hat{\mathbf{x}}$ will transform J_z into J_y. Such a rotation is effected by $e^{i(\pi/2)J_x}$ (note the sign of the exponent) and hence

$$J_y = e^{i(\pi/2)J_x/\hbar}\, J_z\, e^{-i(\pi/2)J_x/\hbar} \,.$$

This can be verified explicitly via the formula

$$e^{-\lambda B} A\, e^{\lambda B} = A - \lambda[B\,,A] + \frac{\lambda^2}{2!}\,[B\,,[B\,,A]] + \cdots \,,$$

which yields, for $\lambda = -i(\pi/2)/\hbar$,

$$e^{i(\pi/2)J_x/\hbar}\, J_z\, e^{-i(\pi/2)J_x/\hbar} = J_z + i\frac{\pi}{2\hbar}\,[J_x\,,J_z] + \frac{1}{2!}\left(-i\frac{\pi}{2\hbar}\right)^2 [J_x\,,[J_x\,,J_z]] + \cdots$$

$$= J_z + \frac{\pi}{2}J_y - \frac{1}{2!}\left(\frac{\pi}{2}\right)^2 J_z - \frac{1}{3!}\left(\frac{\pi}{2}\right)^3 J_y + \cdots = J_z\cos\frac{\pi}{2} + J_y\sin\frac{\pi}{2} = J_y \,.$$

It then follows that

$$e^{-i\beta J_y/\hbar} = \sum_{n=0}^{\infty} \frac{(-i\beta/\hbar)^n}{n!}\, J_y^n = \sum_{n=0}^{\infty} \frac{(-i\beta/\hbar)^n}{n!}\left[e^{i(\pi/2)J_x/\hbar}\, J_z\, e^{-i(\pi/2)J_x/\hbar}\right]^n$$

$$= e^{i(\pi/2)J_x/\hbar}\left[\sum_{n=0}^{\infty} \frac{(-i\beta/\hbar)^n}{n!}\, J_z^n\right] e^{-i(\pi/2)J_x/\hbar} = e^{i(\pi/2)J_x/\hbar}\, e^{-i\beta J_z/\hbar}\, e^{-i(\pi/2)J_x/\hbar} \,.$$

The $\underline{d}^{(j)}$ matrices are given by

$$d^{(j)}_{m'm}(\beta) = \langle \psi_{jm'} |\, e^{-i\beta J_y/\hbar}\, |\psi_{jm}\rangle = \langle \psi_{jm'} |\, e^{i(\pi/2)J_x/\hbar}\, e^{-i\beta J_z/\hbar}\, e^{-i(\pi/2)J_x/\hbar}\, |\psi_{jm}\rangle$$

$$= \sum_{m''=-j}^{j} \langle \psi_{jm'} |\, e^{i(\pi/2)J_x/\hbar}\, |\psi_{jm''}\rangle\, e^{-i\beta m''}\, \langle \psi_{jm''} |\, e^{-i(\pi/2)J_x/\hbar}\, |\psi_{jm}\rangle \,,$$

where in the second line we have inserted the completeness and used the eigenstate property $J_z|\psi_{jm''}\rangle = m''\hbar |\psi_{jm''}\rangle$. Note that j is either integer or half-integer and therefore $2m''$ is necessarily integer. We also introduce the constants (which are independent of β)

$$c^{(j)}_{m'',m} \equiv \langle \psi_{jm''} |\, e^{-i(\pi/2)J_x/\hbar}\, |\psi_{jm}\rangle \,,$$

and, assuming j integer, express the d-matrix element as

$$
\begin{aligned}
d^{(j)}_{m'm}(\beta) &= \sum_{m''=-j}^{j} c^{(j)*}_{m'',m'} c^{(j)}_{m'',m} \left(e^{-i\beta/2}\right)^{2m''} \\
&= c^{(j)*}_{0,m'} c^{(j)}_{0,m} + \sum_{m''=1}^{j} \left[c^{(j)*}_{m'',m'} c^{(j)}_{m'',m} \left(e^{-i\beta/2}\right)^{2m''} + c^{(j)*}_{-m'',m'} c^{(j)}_{-m'',m} \left(e^{i\beta/2}\right)^{2m''} \right] \\
&= c^{(j)*}_{0,m'} c^{(j)}_{0,m} + \sum_{m''=1}^{j} c^{(j)*}_{m'',m'} c^{(j)}_{m'',m} [\cos(\beta/2) - i\,\sin(\beta/2)]^{2m''} \\
&\quad + \sum_{m''=1}^{j} c^{(j)*}_{-m'',m'} c^{(j)}_{-m'',m} [\cos(\beta/2) + i\,\sin(\beta/2)]^{2m''} .
\end{aligned}
$$

Note that, when j is half-integer, a relation similar to that above is obtained, except that the constant term is absent and the sums over m'' start from $1/2$. Using the binomial expansion, we obtain

$$
\begin{aligned}
d^{(j)}_{m'm}(\beta) &= c^{(j)*}_{0,m'} c^{(j)}_{0,m} + \sum_{m''=1}^{j} c^{(j)*}_{m'',m'} c^{(j)}_{m'',m} \sum_{k=0}^{2m''} \frac{(2m'')!}{k!\,(2m''-k)!} [\cos(\beta/2)]^{2m''-k} [-i\,\sin(\beta/2)]^{k} \\
&\quad + \sum_{m''=1}^{j} c^{(j)*}_{-m'',m'} c^{(j)}_{-m'',m} \sum_{k=0}^{2m''} \frac{(2m'')!}{k!\,(2m''-k)!} [\cos(\beta/2)]^{2m''-k} [i\,\sin(\beta/2)]^{k} \\
&= c^{(j)*}_{0,m'} c^{(j)}_{0,m} + \sum_{m''=1}^{j} \sum_{k=0}^{2m''} \frac{(2m'')!}{k!\,(2m''-k)!}\, i^{k} [\cos(\beta/2)]^{2m''-k} [\sin(\beta/2)]^{k} \\
&\quad \left[(-1)^{k} c^{(j)*}_{m'',m'} c^{(j)}_{m'',m} + c^{(j)*}_{-m'',m'} c^{(j)}_{-m'',m} \right] ,
\end{aligned}
$$

which shows that these are polynomials of degree $2j$ with respect to the variables $\sin(\beta/2)$ and $\cos(\beta/2)$.

Problem 12   Some Properties of the Rotation Matrices

This problem deals with the derivation of properties satisfied by the $\underline{d}^{(j)}$ matrices.

1. Knowing that $\underline{d}^{(j)}(\beta)$ is an orthogonal matrix, show that

$$
d^{(j)}_{m',m}(-\beta) = d^{(j)}_{m,m'}(\beta) .
$$

2. Use the transformation properties of vector operators under rotations to show that

$$
e^{-i\pi J_y/\hbar} J_x e^{i\pi J_y/\hbar} = -J_x , \qquad e^{-i\pi J_y/\hbar} J_z e^{i\pi J_y/\hbar} = -J_z , \qquad e^{-i\pi J_y/\hbar} J_\pm e^{i\pi J_y/\hbar} = -J_\mp .
$$

Using these results, show that

$$
|\psi'_{jm}\rangle = e^{-i\pi J_y/\hbar} |\psi_{j,m}\rangle = e^{i\phi_{j,m}} |\psi_{j,-m}\rangle ,
$$

namely, it is an eigenstate of $\mathbf{J}^2$ and J_z with eigenvalues $j(j+1)\hbar^2$ and $-m\hbar$ (here, $\phi_{j,m}$ is a real phase factor).

3. To determine the phase factor, first note that

$$
\left(e^{-i\pi J_y/\hbar}\right)^2 = e^{-2\pi i J_y/\hbar} = (-1)^{2j} .
$$

Next, consider

$$e^{-i\pi J_y/\hbar} |\psi_{j,-m}\rangle = \frac{1}{\hbar^{2m} c_m c_{m-1} \cdots c_{-m+1}} e^{-i\pi J_y/\hbar} (J_-)^{2m} |\psi_{j,m}\rangle,$$

with known c_m. Use these two relations to show that

$$\phi_{j,m} = \pi(j-m) \implies e^{-i\pi J_y/\hbar} |\psi_{j,m}\rangle = (-1)^{j-m} |\psi_{j,-m}\rangle.$$

4. Using the result in part 3, we obtain

$$d^{(j)}_{m',m}(\beta) = (-1)^{m-m'} d^{(j)}_{-m',-m}(\beta) = (-1)^{m'-m} d^{(j)}_{-m',-m}(\beta).$$

Solution

Part 1

Since the $\underline{d}^{(j)}$ matrix is orthogonal, we have

$$[\underline{d}^{(j)}(\beta)]^T = [\underline{d}^{(j)}(\beta)]^{-1} = \underline{d}^{(j)}(-\beta),$$

and hence

$$[\underline{d}^{(j)}(\beta)^T]_{m,m'} = [\underline{d}^{(j)}(-\beta)]_{m,m'} \implies d^{(j)}_{m',m}(\beta) = d^{(j)}_{m,m'}(-\beta).$$

Part 2

Consider the operator corresponding to a rotation by π about the $\hat{\mathbf{y}}$-axis, $e^{-i\pi J_y/\hbar}$. Direct evaluation, or recalling that for a generic vector operator $\mathbf{V}$ we have

$$e^{-i\pi J_y/\hbar} V_i e^{i\pi J_y/\hbar} = \sum_j [\underline{R}^{-1}(\pi\hat{\mathbf{y}})]_{ij} V_j, \qquad \underline{R}^{-1}(\pi\hat{\mathbf{y}}) = \begin{pmatrix} -1 & 0 & 0 \\ 0 & 1 & 0 \\ 0 & 0 & -1 \end{pmatrix},$$

leads to the transformation laws given in the text of the problem. We then find that

$$J_z e^{-i\pi J_y/\hbar} |\psi_{j,m}\rangle = -e^{-i\pi J_y/\hbar} J_z |\psi_{j,m}\rangle = -m\hbar e^{-i\pi J_y/\hbar} |\psi_{j,m}\rangle,$$

where in the second step we have used the fact that J_z anticommutes with $e^{-i\pi J_y/\hbar}$; we also have

$$\mathbf{J}^2 e^{-i\pi J_y/\hbar} |\psi_{j,m}\rangle = e^{-i\pi J_y/\hbar} \mathbf{J}^2 |\psi_{j,m}\rangle = j(j+1)\hbar^2 e^{-i\pi J_y/\hbar} |\psi_{j,m}\rangle.$$

We conclude that

$$e^{-i\pi J_y/\hbar} |\psi_{j,m}\rangle = e^{i\phi_{j,m}} |\psi_{j,-m}\rangle,$$

where the phase factor can depend on j and m; that is, the state $e^{-i\pi J_y/\hbar} |\psi_{j,m}\rangle$ is, up to a phase factor, equal to the eigenstate $|\psi_{j,-m}\rangle$ of J_z.

Part 3

In order to determine $\phi_{j,m}$, we note that by applying twice a rotation by π about $\hat{\mathbf{y}}$, we have (recalling that for half-integer angular momentum a rotation by 2π is equal to minus the identity operator)

$$\left(e^{-i\pi J_y/\hbar}\right)^2 = e^{-2\pi i J_y/\hbar} = (-1)^{2j} = \left(e^{\pi i}\right)^{2j} = e^{2\pi i j}$$

and

$$\underbrace{e^{-i\pi J_y/\hbar}\, e^{-i\pi J_y/\hbar}}_{(-)^{2j}=e^{2\pi i j}} |\psi_{j,m}\rangle = e^{i\phi_{j,m}}\, e^{-i\pi J_y/\hbar} |\psi_{j,-m}\rangle = e^{i(\phi_{j,m}+\phi_{j,-m})} |\psi_{j,m}\rangle\,,$$

and hence

$$\phi_{j,m} + \phi_{j,-m} = 2\pi j\,.$$

Next, we write the state $|\psi_{j,-m}\rangle$ as

$$|\psi_{j,-m}\rangle = \frac{1}{\hbar^{2m} c_m\, c_{m-1} \cdots c_{-m+1}}\, (J_-)^{2m}\, |\psi_{j,m}\rangle\,,$$

where we define

$$c_m = \sqrt{j(j+1) - m(m-1)}\,, \qquad c_{m-1} = \sqrt{j(j+1) - (m-1)(m-2)}$$

and so on, the last coefficient being given by

$$c_{-m+1} = \sqrt{j(j+1) - m(m-1)}\,.$$

Thus we obtain

$$\underbrace{e^{-i\pi J_y/\hbar} |\psi_{j,-m}\rangle}_{e^{i\phi_{j,-m}} |\psi_{j,m}\rangle} = \frac{1}{\hbar^{2m} c_m\, c_{m-1} \cdots c_{-m+1}} \underbrace{\left(e^{-i\pi J_y/\hbar}\, J_-\, e^{i\pi J_y/\hbar}\right)^{2m}}_{(-J_+)^{2m}} e^{-i\pi J_y/\hbar} |\psi_{j,m}\rangle$$

$$= (-1)^{2m}\, \frac{1}{\hbar^{2m} c_m\, c_{m-1} \cdots c_{-m+1}}\, J_+^{2m}\, e^{i\phi_{j,m}} |\psi_{j,-m}\rangle = (-1)^{2m}\, e^{i\phi_{j,m}} |\psi_{j,m}\rangle\,,$$

yielding

$$\phi_{j,-m} = \phi_{j,m} + 2\pi m\,.$$

Note that

$$J_+|\psi_{j,-m}\rangle = \hbar c_{-m+1} |\psi_{j,-m+1}\rangle\,, \qquad (J_+)^2 |\psi_{j,-m}\rangle = \hbar^2 c_{-m+1}\, c_{-m+2} |\psi_{j,-m+2}\rangle\,,$$

and so on, and the coefficients cancel out. We conclude that

$$\phi_{j,m} + \phi_{j,-m} = 2\pi j \qquad \text{and} \qquad \phi_{j,-m} = \phi_{j,m} + 2\pi m \implies \phi_{j,m} = \pi(j - m)\,,$$

where it should be noted that $j - m$ is always an integer (regardless of whether j is integer or half-integer). We finally arrive at

$$e^{-i\pi J_y/\hbar} |\psi_{j,m}\rangle = (-1)^{j-m} |\psi_{j,-m}\rangle\,,$$

up to a further overall phase factor, which can depend only on j.

Part 4

From these relations we derive the following symmetry property of the d-matrices:

$$d^{(j)}_{m',m}(\beta) = \langle \psi_{j,m'} | e^{-i\beta J_y/\hbar} | \psi_{j,m} \rangle = \langle \psi_{j,m'} | e^{i\pi J_y/\hbar} \underbrace{e^{-i\pi J_y/\hbar}\, e^{-i\beta J_y/\hbar}\, e^{i\pi J_y/\hbar}}_{e^{-i\beta J_y/\hbar}} e^{-i\pi J_y/\hbar} | \psi_{j,m} \rangle$$

$$= (-1)^{j-m}\, (-1)^{j-m'} \langle \psi_{j,-m'} | e^{-i\beta J_y/\hbar} | \psi_{j,-m} \rangle = (-1)^{2j-m-m'}\, d^{(j)}_{-m',-m}(\beta)\,,$$

and since $(-1)^{2j} = (-1)^{2m} = (-1)^{2m'}$ (if j is half integer then m and m' are also half integers, and hence $2m$ and $2m'$ are always odd; if j is integer, then m and m' are also integers, and hence $2m$ and $2m'$ are always even), we can also write

$$d^{(j)}_{m',m}(\beta) = (-1)^{m-m'} d^{(j)}_{-m',-m}(\beta) = (-1)^{m'-m} d^{(j)}_{-m',-m}(\beta) \ .$$

Problem 13 An Explicit Derivation of the Rotation Matrices

Let $\hat{a}_r$ and $\hat{a}_r^\dagger$ with $r = 1, 2$ be the annihilation and creation operators for a two-dimensional harmonic oscillator, satisfying

$$[a_r, a_s] = 0 = [a_r^\dagger, a_s^\dagger], \qquad [a_r, a_s^\dagger] = \delta_{r,s} \ .$$

We define

$$S = \frac{1}{2} \left(a_1^\dagger a_1 + a_2^\dagger a_2 \right) \ ,$$

and

$$J_1 = \frac{1}{2} \left(a_2^\dagger a_1 + a_1^\dagger a_2 \right) \ , \qquad J_2 = \frac{i}{2} \left(a_2^\dagger a_1 - a_1^\dagger a_2 \right) \ , \qquad J_3 = \frac{1}{2} \left(a_1^\dagger a_1 - a_2^\dagger a_2 \right) \ ,$$

and the J_i may be considered as the cartesian components of a certain vector operator. These components satisfy the commutation relations characteristic of an angular momentum up to factors of $\hbar$, that is,

$$[J_i, J_j] = i \sum_k \epsilon_{ijk} J_k$$

and

$$\mathbf{J}^2 = S(S + \mathbb{1}) \ .$$

Hereafter, $\mathbf{J}$ is in fact taken as the angular momentum of the system. We denote the eigenvalues of $\mathbf{J}^2$ and J_3 by $j(j + 1)$ and m, respectively. The states

$$|\psi_{jm}\rangle = \frac{1}{\sqrt{(j + m)! \ (j - m)!}} \ (a_1^\dagger)^{j+m} \ (a_2^\dagger)^{j-m} |0; 0\rangle \qquad \text{with} \qquad m = -j, \dots, j \ ,$$

form a basis of common eigenstates of $\mathbf{J}^2$ and J_3. See Problem 11.4.8 in Chapter 11 for a proof of these various statements. This approach can be used to derive an explicit formula for the d-matrices (the non-trivial ones).

1. Show that

$$e^{-i\beta J_2} |\psi_{jm}\rangle = \frac{(e^{-i\beta J_2} a_1^\dagger e^{i\beta J_2})^{j+m} \ (e^{-i\beta J_2} a_2^\dagger e^{i\beta J_2})^{j-m} \ e^{-i\beta J_2} |0; 0\rangle}{\sqrt{(j + m)! \ (j - m)!}} \ .$$

2. Show that

$$e^{-i\beta J_2} |0; 0\rangle = |0; 0\rangle$$

and

$$e^{-i\beta J_2} a_1^\dagger e^{i\beta J_2} = a_1^\dagger \cos(\beta/2) + a_2^\dagger \sin(\beta/2) \ ,$$
$$e^{-i\beta J_2} a_2^\dagger e^{i\beta J_2} = -a_1^\dagger \sin(\beta/2) + a_2^\dagger \cos(\beta/2) \ .$$

3. Show that

$$
e^{-i\beta J_2} |\psi_{jm}\rangle = \sum_{k=0}^{j+m} \frac{(j+m)!}{k!\,(j+m-k)!} \frac{[a_1^\dagger \cos(\beta/2)]^{j+m-k}\,[a_2^\dagger \sin(\beta/2)]^k}{\sqrt{(j+m)!}}
$$

$$
\times \sum_{l=0}^{j-m} \frac{(j-m)!}{l!\,(j-m-l)!} \frac{[-a_1^\dagger \sin(\beta/2)]^{j-m-l}\,[a_2^\dagger \cos(\beta/2)]^l}{\sqrt{(j-m)!}} |0;0\rangle .
$$

4. Show that

$$
e^{-i\beta J_2} |\psi_{jm}\rangle = \sum_{m'} |\psi_{jm'}\rangle d^{(j)}_{m'm}(\beta) = \sum_{m'} \frac{(a_1^\dagger)^{j+m'}\,(a_2^\dagger)^{j-m'}\,|0;0\rangle}{\sqrt{(j+m')!\,(j-m')!}}\, d^{(j)}_{m'm}(\beta) .
$$

Then, comparing this relation with that found in part 3, obtain (by matching powers of $a_1^\dagger$) the following explicit expression for the d-matrices:

$$
d^{(j)}_{m'm}(\beta) = \sum_{k} (-1)^{k-m+m'} \frac{\sqrt{(j+m)!\,(j-m)!\,(j+m')!\,(j-m')!}}{k!\,(j+m-k)!\,(k-m+m')!\,(j-k-m')!}
$$

$$
\times [\cos(\beta/2)]^{2j-2k+m-m'}\,[\sin(\beta/2)]^{2k-m+m'}
$$

and the sum over k is over all integers such that the arguments of the factorials in the denominator are either zero or positive integers.

Solution

Part 1

Denote the unitary operator $e^{-i\beta J_2}$ simply as U_R. Inserting the expansion for the state $|\psi_{jm}\rangle$, we find

$$
U_R |\psi_{jm}\rangle = \frac{U_R\,(a_1^\dagger)^{j+m}\,U_R^\dagger\,U_R\,(a_2^\dagger)^{j-m}\,U_R^\dagger\,U_R\,|0;0\rangle}{\sqrt{(j+m)!\,(j-m)!}}
$$

$$
= \frac{\left(U_R\,a_1^\dagger\,U_R^\dagger\right)^{j+m}\left(U_R\,a_2^\dagger\,U_R^\dagger\right)^{j-m}\,U_R\,|0;0\rangle}{\sqrt{(j+m)!\,(j-m)!}} ,
$$

where we have used the unitary property of U_R, in particular

$$
U_R\,(a^\dagger)^p\,U_R^\dagger = U_R\,\underbrace{a^\dagger\,a^\dagger \cdots a^\dagger}_{p\ \text{times}}\,U_R = U_R\,a^\dagger\,U_R^\dagger\,U_R\,a^\dagger\,U_R^\dagger\,U_R \cdots U_R\,a^\dagger\,U_R^\dagger = \left(U_R\,a^\dagger\,U_R^\dagger\right)^p .
$$

Part 2

The operator U_R applied to the vacuum state gives

$$
U_R |0;0\rangle = \sum_{p=0}^{\infty} \frac{(-i\beta)^p}{p!} \underbrace{\left[\frac{i}{2}\left(a_2^\dagger a_1 - a_1^\dagger a_2\right)\right]^p}_{J_2^p} |0;0\rangle = \sum_{p=0}^{\infty} \frac{(\beta/2)^p}{p!}\left(a_2^\dagger a_1 - a_1^\dagger a_2\right)^p |0;0\rangle = |0;0\rangle ,
$$

since a_1 and a_2 annihilate the vacuum, and hence

$$
a_2^\dagger a_1 |0;0\rangle - a_1^\dagger a_2 |0;0\rangle = 0 \implies \left(a_2^\dagger a_1 - a_1^\dagger a_2\right)^{p \geq 1} |0;0\rangle = 0 .
$$

We use the formula (derived in Problem 25 of Chapter 6)

$$e^{-i\beta J_2} a_i^\dagger e^{i\beta J_2} = a_i^\dagger - i\beta \, [J_2 , a_i^\dagger] + \frac{(-i\beta)^2}{2!} \, [J_2 , [J_2 , a_i^\dagger]] + \cdots ,$$

where

$$[J_2 , a_1^\dagger] = \frac{i}{2} [a_2^\dagger a_1 - a_1^\dagger a_2 , a_1^\dagger] = \frac{i}{2} a_2^\dagger ,$$

and similarly

$$[J_2 , a_2^\dagger] = \frac{i}{2} [a_2^\dagger a_1 - a_1^\dagger a_2 , a_2^\dagger] = -\frac{i}{2} a_1^\dagger ,$$

hence

$$[J_2 , [J_2 , a_1^\dagger]] = \frac{i}{2} [J_2 , a_2^\dagger] = \frac{1}{4} a_1^\dagger , \qquad [J_2 , [J_2 , a_2^\dagger]] = \frac{1}{4} a_2^\dagger ,$$

and so on. Thus, we find

$$e^{-i\beta J_2} a_1^\dagger e^{i\beta J_2} = a_1^\dagger + (\beta/2) a_2^\dagger - \frac{(\beta/2)^2}{2!} a_1^\dagger - \frac{(\beta/2)^3}{3!} a_2^\dagger + \cdots = a_1^\dagger \cos(\beta/2) + a_2^\dagger \sin(\beta/2)$$

and

$$e^{-i\beta J_2} a_2^\dagger e^{i\beta J_2} = a_2^\dagger - (\beta/2) a_1^\dagger - \frac{(\beta/2)^2}{2!} a_2^\dagger + \frac{(\beta/2)^3}{3!} a_1^\dagger + \cdots = -a_1^\dagger \sin(\beta/2) + a_2^\dagger \cos(\beta/2) .$$

Part 3

Using the binomial expansion

$$(x + y)^n = \sum_{k=0}^{n} \frac{n!}{k! \, (n-k)!} x^k y^{n-k} ,$$

we have

$$\left(U_R a_1^\dagger U_R^\dagger \right)^{j+m} = [a_1^\dagger \cos(\beta/2) + a_2^\dagger \sin(\beta/2)]^{j+m} = \sum_{k=0}^{j+m} \frac{(j+m)!}{k! \, (j+m-k)!} [a_1^\dagger \cos(\beta/2)]^{j+m-k} [a_2^\dagger \sin(\beta/2)]^k ,$$

$$\left(U_R a_2^\dagger U_R^\dagger \right)^{j-m} = [-a_1^\dagger \sin(\beta/2) + a_2^\dagger \cos(\beta/2)]^{j-m} = \sum_{l=0}^{j-m} \frac{(j-m)!}{l! \, (j-m-l)!} [-a_1^\dagger \sin(\beta/2)]^{j-m-l} [a_2^\dagger \cos(\beta/2)]^l .$$

Part 4

Using the property of the $\underline{d}^{(j)}$ matrices, we obtain

$$U_R |\psi_{jm}\rangle = \sum_{m'} |\psi_{jm'}\rangle d_{m'm}^{(j)}(\beta) ,$$

which, after inserting the expansion for the state $|\psi_{jm'}\rangle$, reduces to

$$U_R |\psi_{jm}\rangle = \sum_{m'} \frac{(a_1^\dagger)^{j+m'} (a_2^\dagger)^{j-m'} |0; 0\rangle}{\sqrt{(j+m')! \, (j-m')!}} d_{m'm}^{(j)}(\beta) .$$

This relation can be compared with that obtained above,

$$
e^{-i\beta J_2} |\psi_{jm}\rangle = \sum_{k=0}^{j+m} \frac{(j+m)!}{k!\,(j+m-k)!} \frac{[a_1^\dagger \cos(\beta/2)]^{j+m-k}\,[a_2^\dagger \sin(\beta/2)]^k}{\sqrt{(j+m)!}}
$$
$$
\times \sum_{l=0}^{l-k} \frac{(j-m)!}{l!\,(j-m-l)!} \frac{[-a_1^\dagger \sin(\beta/2)]^{j-m-l}\,[a_2^\dagger \cos(\beta/2)]^l}{\sqrt{(j-m)!}} |0;0\rangle,
$$

matching powers of $a_1^\dagger$. Compare the coefficient of $a_1^\dagger$ raised to the power $j + m'$ in the first formula (with the rotation matrix) together with the coefficient of, $a_1^\dagger$ raised to the power $2j - k - l$ in the second formula. For these powers to be the same, we must have

$$
j + m' = 2j - k - l \implies l = j - k - m'.
$$

Since we want to determine $d_{m'm}^{(j)}$ with m' fixed, it follows that the sums over k and l in the second formula are no longer independent of each other. Also note that the power of $a_2^\dagger$ in the second formula is $k + l = j - m'$ with l fixed as above, and therefore it automatically matches the same power in the first formula. By substituting for l as above, we find that

$$
\frac{d_{m'm}^{(j)}(\beta)}{\sqrt{(j+m')!\,(j-m')!}} = \sum_{k=0}^{j+m} \frac{\sqrt{(j+m)!}}{k!\,(j+m-k)!} \frac{\sqrt{(j-m)!}}{(j-k-m')!\,(k-m+m')!} (-1)^{k-m+m'}
$$
$$
\times [\cos(\beta/2)]^{2j-2k+m-m'}\,[\sin(\beta/2)]^{2k-m+m'},
$$

finally yielding

$$
d_{m'm}^{(j)}(\beta) = \sum_{k=0}^{j+m} (-1)^{k-m+m'} \frac{\sqrt{(j+m)!\,(j-m)!\,(j+m')!\,(j-m')!}}{k!\,(j+m-k)!\,(k-m+m')!\,(j-k-m')!}
$$
$$
\times [\cos(\beta/2)]^{2j-2k+m-m'}\,[\sin(\beta/2)]^{2k-m+m'},
$$

with the understanding that the arguments of the factorials cannot be negative (recall that $m, m' = -j, -j+1, \ldots, j$).

Problem 14 Rotating the States or Rotating the Operator: An Example

Consider the hydrogen atom and the vector operator (proportional to the electric dipole operator) with spherical components given by

$$
p_\mu = r\,Y_{1\mu}(\Omega), \quad p_{\pm 1} = \mp\sqrt{\frac{3}{4\pi}}\,\frac{1}{\sqrt{2}}(r_x \pm i r_y), \quad p_0 = \sqrt{\frac{3}{4\pi}}\,r_0.
$$

Note that the components of the position operator are relative to a reference system defined by the orthonormal set $\hat{\mathbf{x}}$, $\hat{\mathbf{y}}$, and $\hat{\mathbf{z}}$. We are interested in the matrix elements $P_{\mu m}$ connecting the initial 2p, m excited state to the final 1s, 0 ground state, namely

$$
P_{\mu m} = \langle \phi_{1s,0} | p_\mu | \phi_{2p,m} \rangle,
$$

where the wave functions corresponding to these states are

$$
\phi_{2p,m}(\mathbf{r}) = R_{2p}(r)\,Y_{1m}(\theta, \phi), \qquad \phi_{1s,0}(\mathbf{r}) = \frac{1}{\sqrt{4\pi}}\,R_{1s}(r).
$$

Throughout this problem you may assume that (recalling that the radial functions are taken to be real)

$$\frac{1}{\sqrt{4\pi}} \int_0^\infty dr\, r^3 \, R_{1s}(r)\, R_{2p}(r) = \lambda \, .$$

The constant λ has the dimension of length, but its precise value is irrelevant.

1. Using the Wigner–Eckart theorem, calculate the matrix elements $P_{\mu m}$. In particular, establish which matrix elements $P_{\mu m}$ vanish and which do not, and show that the reduced matrix element of p_μ is given by

$$\langle \phi_{1s} || p || \phi_{2p} \rangle = -\sqrt{3}\, \lambda \, .$$

2. Consider the basis consisting of the following unit vectors:

$$\hat{\mathbf{e}}_z = \hat{\mathbf{n}} \, , \qquad \hat{\mathbf{e}}_y = \frac{\hat{\mathbf{z}} \times \hat{\mathbf{n}}}{|\hat{\mathbf{z}} \times \hat{\mathbf{n}}|} \, , \qquad \hat{\mathbf{e}}_x = \hat{\mathbf{e}}_y \times \hat{\mathbf{e}}_z \, ,$$

where $\hat{\mathbf{n}}$ is the unit vector with components given by $(\sin\beta \cos\gamma, \sin\beta \sin\gamma, \cos\beta)$ in the frame defined by the unit vectors $\hat{\mathbf{x}}$, $\hat{\mathbf{y}}$, and $\hat{\mathbf{z}}$. Show that the set $\hat{\mathbf{e}}_x$, $\hat{\mathbf{e}}_y$, and $\hat{\mathbf{e}}_z$ can be obtained from the set $\hat{\mathbf{x}}$, $\hat{\mathbf{y}}$, and $\hat{\mathbf{z}}$ via the following sequence of two rotations: (i) by β about $\hat{\mathbf{y}}$ and (ii) by γ about $\hat{\mathbf{z}}'$ (note that $\hat{\mathbf{z}}' \neq \hat{\mathbf{z}}$).

3. The initial and final hydrogen-atom states are rotated via

$$U_R(\beta, \gamma) = e^{-i\beta L_y/\hbar}\, e^{-i\gamma L_z/\hbar} \, ,$$

where we note that the components of the (orbital) angular momentum are given relative to the fixed frame specified by $\hat{\mathbf{x}}$, $\hat{\mathbf{y}}$, and $\hat{\mathbf{z}}$. Calculate the matrix elements

$$P'_{\mu m} = \langle \phi'_{1s,0} | p_\mu | \phi'_{2p,m} \rangle \, ,$$

where p_μ are the the spherical components defined in part 1 and

$$|\phi'_{2p,m}\rangle = U_R(\beta, \gamma) |\phi_{2p,m}\rangle \, , \qquad |\phi'_{1s,0}\rangle = U_R(\beta, \gamma) |\phi_{1s,0}\rangle \, ,$$

knowing that the rotation matrix $d^{(1)}_{m',m}(\beta) = \langle Y_{1m'} | e^{-i\beta L_y/\hbar} | Y_{1m} \rangle$ is given explicitly by

$$\underline{d}^{(1)}(\beta) = \begin{pmatrix} \cos^2(\beta/2) & (-\sin\beta)/\sqrt{2} & \sin^2(\beta/2) \\ (\sin\beta)/\sqrt{2} & \cos\beta & (-\sin\beta)/\sqrt{2} \\ \sin^2(\beta/2) & (\sin\beta)/\sqrt{2} & \cos^2(\beta/2) \end{pmatrix} \, .$$

4. Using the transformation property of irreducible tensor operators under rotations, obtain $P'_{\mu m}$ and show that it is the same as that of part 3.

Solution

Part 1

Since p_μ is an irreducible tensor operator of rank 1, we find

$$P_{\mu m} = C_{11}(\mu, m; 0, 0)\, \langle \phi_{1s} || p || \phi_{2p} \rangle \, ,$$

where the Clebsch-Gordan coefficient for the coupling of two angular momenta each with $l = 1$ to form a total angular momentum 0 is given by

$$C_{11}(\mu, m; 0, 0) = -\frac{(-1)^\mu}{\sqrt{3}}\, \delta_{\mu+m,0} \, ,$$

and the matrix elements $P_{\mu m}$ vanish unless $\mu + m = 0$. In order to obtain the reduced matrix element, we first evaluate, for example, the matrix element (recall that $Y_{10}(\Omega)$ is real)

$$P_{00} = \langle \phi_{1s,0} | p_0 | \phi_{2p,0} \rangle = \frac{1}{\sqrt{4\pi}} \int_0^\infty dr\, r^3 R_{1s}(r)\, R_{2p}(r) \underbrace{\int d\Omega\, Y_{10}(\Omega)\, Y_{10}(\Omega)}_{\text{normalization}} = \lambda \;,$$

and by the Wigner–Eckart theorem, we obtain

$$P_{00} = -\frac{1}{\sqrt{3}} \langle \phi_{1s} \| p \| \phi_{2p} \rangle \implies \langle \phi_{1s} \| p \| \phi_{2p} \rangle = -\sqrt{3}\, \lambda \;.$$

In matrix notation, we have

$$\underline{P} = \lambda \begin{pmatrix} 0 & 0 & -1 \\ 0 & 1 & 0 \\ -1 & 0 & 0 \end{pmatrix}.$$

Part 2

Using the given definitions of $\hat{\mathbf{e}}_x$, $\hat{\mathbf{e}}_y$, and $\hat{\mathbf{e}}_z$, we obtain

$$\hat{\mathbf{e}}_z \longrightarrow \begin{pmatrix} \sin\beta \cos\gamma \\ \sin\beta \sin\gamma \\ \cos\beta \end{pmatrix}, \qquad \hat{\mathbf{e}}_y \longrightarrow \begin{pmatrix} -\sin\gamma \\ \cos\gamma \\ 0 \end{pmatrix}, \qquad \hat{\mathbf{e}}_x \longrightarrow \begin{pmatrix} \cos\beta \cos\gamma \\ \cos\beta \sin\gamma \\ -\sin\beta \end{pmatrix}.$$

These unit vectors can also be obtained by the following sequence of two (anticlockwise) rotations $\underline{R}(\gamma \hat{\mathbf{z}}')\underline{R}(\beta \hat{\mathbf{y}})$, where

$$\underline{R}(\beta \hat{\mathbf{y}}) = \begin{pmatrix} \cos\beta & 0 & \sin\beta \\ 0 & 1 & 0 \\ -\sin\beta & 0 & \cos\beta \end{pmatrix}, \qquad \underline{R}(\gamma \hat{\mathbf{z}}') = \begin{pmatrix} \cos\gamma & -\sin\gamma & 0 \\ \sin\gamma & \cos\gamma & 0 \\ 0 & 0 & 1 \end{pmatrix}.$$

We find

$$\underline{R}(\gamma \hat{\mathbf{z}}')\, R(\beta \hat{\mathbf{y}})\, \hat{\mathbf{x}} \longrightarrow \begin{pmatrix} \cos\gamma & -\sin\gamma & 0 \\ \sin\gamma & \cos\gamma & 0 \\ 0 & 0 & 1 \end{pmatrix} \begin{pmatrix} \cos\beta & 0 & \sin\beta \\ 0 & 1 & 0 \\ -\sin\beta & 0 & \cos\beta \end{pmatrix} \begin{pmatrix} 1 \\ 0 \\ 0 \end{pmatrix} = \begin{pmatrix} \cos\beta \cos\gamma \\ \cos\beta \sin\gamma \\ -\sin\beta \end{pmatrix},$$

$$\underline{R}(\gamma \hat{\mathbf{z}}')\, R(\beta \hat{\mathbf{y}})\, \hat{\mathbf{y}} \longrightarrow \begin{pmatrix} \cos\gamma & -\sin\gamma & 0 \\ \sin\gamma & \cos\gamma & 0 \\ 0 & 0 & 1 \end{pmatrix} \begin{pmatrix} \cos\beta & 0 & \sin\beta \\ 0 & 1 & 0 \\ -\sin\beta & 0 & \cos\beta \end{pmatrix} \begin{pmatrix} 0 \\ 1 \\ 0 \end{pmatrix} = \begin{pmatrix} -\sin\gamma \\ \cos\gamma \\ 0 \end{pmatrix},$$

$$\underline{R}(\gamma \hat{\mathbf{z}}')\, R(\beta \hat{\mathbf{y}})\, \hat{\mathbf{z}} \longrightarrow \begin{pmatrix} \cos\gamma & -\sin\gamma & 0 \\ \sin\gamma & \cos\gamma & 0 \\ 0 & 0 & 1 \end{pmatrix} \begin{pmatrix} \cos\beta & 0 & \sin\beta \\ 0 & 0 & 1 \\ -\sin\beta & 0 & \cos\beta \end{pmatrix} \begin{pmatrix} 0 \\ 0 \\ 1 \end{pmatrix} = \begin{pmatrix} \sin\beta \cos\gamma \\ \sin\beta \sin\gamma \\ \cos\beta \end{pmatrix},$$

in agreement with the direct evaluation of the components.

Part 3

We note that the 1s state is invariant under rotations, while the 2p state transforms as

$$|\phi'_{2p,m}\rangle = U_R(\beta, \gamma)\, |\phi_{2p,m}\rangle = \sum_{m'} |\phi_{2p,m'}\rangle \langle Y_{1m'}|\, U_R(\beta, \gamma)\, |Y_{1m}\rangle = \sum_{m'} |\phi_{2p,m'}\rangle\, d^{(1)}_{m',m}(\beta)\, e^{-im\gamma}\;,$$

from which it follows that

$$P'_{\mu m} = \sum_{m'} \langle \phi_{1s,0} | p_\mu | \phi_{2p,m'} \rangle \, d^{(1)}_{m',m}(\beta) \, e^{-im\gamma} = -\frac{(-)^\mu}{\sqrt{3}} \underbrace{\langle \phi_{1s} || p || \phi_{2p} \rangle}_{-\sqrt{3}\,\lambda} d^{(1)}_{-\mu,m}(\beta) \, e^{-im\gamma}$$

or simply

$$P'_{\mu m} = \lambda \left[(-1)^\mu \, d^{(1)}_{-\mu,m}(\beta) \, e^{-im\gamma} \right] .$$

Using the explicit expression for the rotation matrix, we find

$$\underline{P'} = \lambda \begin{pmatrix} -e^{-i\gamma} \sin^2(\beta/2) & (-\sin\beta)/\sqrt{2} & -e^{i\gamma} \cos^2(\beta/2) \\ e^{-i\gamma} (\sin\beta)/\sqrt{2} & \cos\beta & -e^{i\gamma} (\sin\beta)/\sqrt{2} \\ -e^{-i\gamma} \cos^2(\beta/2) & (\sin\beta)/\sqrt{2} & -e^{i\gamma} \sin^2(\beta/2) \end{pmatrix} .$$

Part 4

We have

$$P'_{\mu m} = \langle \phi_{1s,0} | U_R^\dagger(\beta,\gamma) \, p_\mu \, U_R(\beta,\gamma) | \phi_{2p,m} \rangle ,$$

and, using the transformation property of an irreducible tensor operator,

$$U_R^\dagger(\beta,\gamma) \, p_\mu \, U_R(\beta,\gamma) = U_R(-\gamma,-\beta) \, p_\mu \, U_R^\dagger(-\gamma,-\beta)$$
$$= \sum_{\mu'} p_{\mu'} \langle Y_{1\mu'} | U_R(-\gamma,-\beta) | Y_{1\mu} \rangle = \sum_{\mu'} p_{\mu'} \, e^{i\mu'\gamma} \, d^{(1)}_{\mu',\mu}(-\beta) ,$$

where the inverse of $U_R(\beta,\gamma)$ is

$$U_R(-\gamma,-\beta) = e^{i\gamma J_z/\hbar} \, e^{i\beta J_y/\hbar} .$$

We obtain

$$P'_{\mu m} = \sum_{\mu'} \langle \phi_{1s,0} | p_{\mu'} | \phi_{2p,m} \rangle \, e^{i\mu'\gamma} \, d^{(1)}_{\mu',\mu}(-\beta) = \lambda \left[(-1)^{-m} d^{(1)}_{-m,\mu}(-\beta) \, e^{-im\gamma} \right] .$$

We now use the properties of the rotation matrices derived in Problem 12 to find

$$P'_{\mu m} = \lambda \left[(-1)^{-m} d^{(1)}_{\mu,-m}(\beta) \, e^{-im\gamma} \right] = \lambda \left[(-1)^{-m} (-1)^{-\mu+m} d^{(1)}_{-\mu,m}(\beta) \, e^{-im\gamma} \right]$$
$$= \lambda \left[(-1)^\mu \, d^{(1)}_{-\mu,m}(\beta) \, e^{-im\gamma} \right] ,$$

in agreement with the result derived in part 3.

Problem 15 Fine Structure of the Hydrogen Atom

As we saw in Section 17.3, the relativistic terms in the Hamiltonian are expected to produce corrections that are suppressed by α^2 (where α is the fine-structure constant) relative to the unperturbed energy levels in hydrogen-like atoms. In this problem, we will evaluate their effect on the spectrum of these systems in first-order perturbation theory and determine their so-called fine structure. Note that, since we must now include the spin of the electron, a complete set of commuting observables consists of H_0, $\mathbf{L}^2$, L_z, $\mathbf{S}^2$, and S_z with common eigenstates $|\phi_{nlm,m_s}\rangle$, where $|\phi_{nlm,m_s}\rangle = |\phi_{nlm}\rangle \otimes |m_s\rangle$ and the principal quantum number is $n = 1, 2, \ldots, \infty$, the orbital angular momentum number is $l = 0, 1, \ldots, n - 1$, the orbital angular momentum projection is

$m = -l, -l+1, \ldots, +l$, and the spin projection is $m_s = \pm 1/2$. The spin quantum number $s = 1/2$ is fixed, of course. The eigenvalues of H_0 – referring to Eq. (17.34) – are known,

$$H_0 |\phi_{nlm,m_s}\rangle = \epsilon_n |\phi_{nlm,m_s}\rangle , \qquad \epsilon_n = -\frac{(Z\alpha)^2}{2n^2} \, m_e c^2 , \qquad g_n = 2n^2 ,$$

where, however, the degeneracy g_n of a generic level n is now doubled because of the two possible values for the spin azimuthal quantum number m_s. As a consequence, for example, the ground state is two-fold degenerate rather than being non-degenerate as when the electronic spin degrees of freedom are not considered.

1. Calculate the first-order correction to the unperturbed energies ϵ_n induced by the Darwin–Foldy term (assume the nuclear charge distribution to be point-like; in other words, ignore nuclear finite-size effects). Recall that the (squared) value of the wave function at the origin is given by

$$|\phi_{nlm}(0)|^2 = \frac{1}{\pi} \left(\frac{Z\alpha}{n}\right)^3 \left(\frac{m_e c}{\hbar}\right)^3 \delta_{l,0} \, \delta_{m,0} .$$

2. Calculate the first-order correction to the unperturbed energies ϵ_n induced by V_K. Note that the calculation is facilitated by observing that

$$V_K = -\frac{1}{2m_e c^2} \underbrace{(H_0 - V_C)^2}_{(\mathbf{p}^2/2m_e)^2} = -\frac{1}{2 \, m_e c^2} (H_0^2 - V_C H_0 - H_0 V_C + V_C^2);$$

The following result will be useful:

$$\langle 1/r^2 \rangle_{nl} = \int_0^\infty dr \, |R_{nl}(r)|^2 = \frac{2Z^2}{(2l+1)n^3 a_0^2} ,$$

where a_0 is the Bohr radius.

3. Write the spin–orbit term as

$$V_{SO} = \xi(r) \, \mathbf{L} \cdot \mathbf{S} , \qquad \xi(r) = \frac{g_e - 1}{2m_e^2 \, c^2} \frac{Ze^2}{r^3} .$$

Note that V_{SO} is a scalar in the combined orbital + spin space and, as such, it commutes with the total angular momentum $\mathbf{J} = \mathbf{L} + \mathbf{S}$. Thus it is convenient to work with a basis of eigenstates of H_0, $\mathbf{L}^2$, $\mathbf{S}^2$, $\mathbf{J}^2$, and J_z (a complete set of commuting observables); the transformation from the basis $|\phi_{nlm}\rangle \otimes |m_s\rangle$ of eigenstates of $\mathbf{L}^2, L_z$ and of $\mathbf{S}^2, S_z$ to the new basis $|\phi_{nl,jm_j}\rangle$ is implemented via the linear combination

$$|\phi_{nl,jm_j}\rangle = \sum_{m=-l}^{l} \sum_{m_s = \pm 1/2} C_{l,1/2}(m m_s; j m_j) \underbrace{|\phi_{nlm}\rangle \otimes |m_s\rangle}_{|\phi_{nlm,m_s}\rangle} ,$$

where $C_{l,1/2}(m m_s; j m_j)$ are the (real) Clebsch–Gordan coefficients corresponding to the addition of orbital angular momentum l and spin $1/2$ (this latter quantum number, being fixed, is not specified in the states). Exploiting these observations, calculate the first-order correction to the unperturbed energies ϵ_n induced by V_{SO}. The following result is helpful:

$$\langle 1/r^3 \rangle_{nl} = \int_0^\infty dr \, \frac{1}{r} \, |R_{nl}(r)|^2 = \frac{8Z^3}{(2l+1)[(2l+1)^2 - 1] \, n^3 a_0^3} .$$

Solution

Part 1

Since the finite size of the nuclear charge distribution is neglected, that is, $V_C(r) = -Ze^2/r$, the Darwin–Foldy term is simply given by

$$V_{DF} = \frac{\hbar^2}{8m_e^2 c^2}\, \boldsymbol{\nabla}^2 V_C(r) = \frac{\hbar^2}{8m_e^2 c^2} 4\pi Z e^2 \delta(\mathbf{r}) \,,$$

where we have used $\boldsymbol{\nabla}^2(1/r) = -4\pi\delta(\mathbf{r})$. This term has spherical symmetry in orbital and spin space, since it commutes with $\mathbf{L}$ and $\mathbf{S}$ separately. Thus V_{DF} is a scalar operators (an ITO of rank 0), and therefore the Wigner–Eckart theorem ensures that, in the degenerate subspace of dimension g_n, the matrix $\underline{V}_{DF}$ is in fact diagonal, that is,

$$\langle \phi_{nl'm',m'_s} | V_{DF} | \phi_{nlm,m_s} \rangle = \delta_{l,l'}\, \delta_{m,m'}\, \delta_{m_s,m'_s}\, \langle \phi_{nl} \| V_{DF} \| \phi_{nl} \rangle \,.$$

Since the reduced matrix element (RME) is independent of m and m_s, we can compute it by simply evaluating the expectation value for some m and m_s, to find

$$\langle \phi_{nl} \| V_{DF} \| \phi_{nl} \rangle = \langle \phi_{nlm,m_s} | V_{DF} | \phi_{nlm,m_s} \rangle = \langle \phi_{nlm} | V_{DF} | \phi_{nlm} \rangle \underbrace{\langle m_s | m_s \rangle}_{\text{unity}} \,,$$

where in the last step we have used the fact that V_{DF} does not depend on the spin operator. The presence of the δ-function makes the evaluation of this RME straightforward,

$$\langle \phi_{nl} \| V_{DF} \| \phi_{nl} \rangle = \int d\mathbf{r}\, V_{DF}(r)\, |\phi_{nlm}(\mathbf{r})|^2 = 4\pi Z e^2 \frac{\hbar^2}{8m_e^2 c^2} |\phi_{nlm}(0)|^2 = \frac{(Z\alpha)^4}{2n^3}\, m_e c^2 \delta_{l,0}\, \delta_{m,0} \,,$$

and the correction is repulsive (positive) and only affects s-states.

Part 2

The term V_K is also a scalar in the orbital and spin spaces separately, and hence

$$\langle \phi_{nl'm',m'_s} | V_K | \phi_{nlm,m_s} \rangle = \delta_{l,l'}\, \delta_{m,m'}\, \delta_{m_s,m'_s}\, \langle \phi_{nl} \| V_K \| \phi_{nl} \rangle \implies \langle \phi_{nl} \| V_K \| \phi_{nl} \rangle = \langle \phi_{nlm} | V_K | \phi_{nlm} \rangle \,.$$

Inserting the expression for V_K given in the text of the problem leads to

$$\langle \phi_{nl} \| V_K \| \phi_{nl} \rangle = -\frac{1}{2m_e c^2} \langle \phi_{nlm} | H_0^2 - V_C H_0 - H_0 V_C + V_C^2 | \phi_{nlm} \rangle$$

$$= -\frac{1}{2m_e c^2} \left(\epsilon_n^2 - 2\epsilon_n \langle \phi_{nlm} | V_C | \phi_{nlm} \rangle + \langle \phi_{nlm} | V_C^2 | \phi_{nlm} \rangle \right) \,.$$

Use of the virial theorem as in Eq. (17.36) gives, for the expectation value of V_C,

$$\langle \phi_{nlm} | V_C | \phi_{nlm} \rangle = 2\epsilon_n \,.$$

The expectation value of V_C^2 follows from

$$\langle \phi_{nlm} | V_C^2 | \phi_{nlm} \rangle = \int_0^\infty dr\, r^2\, |R_{nl}(r)|^2\, V_C^2(r) \int d\Omega\, Y_{lm}^*(\Omega) Y_{lm}(\Omega) = (Ze^2)^2 \underbrace{\int_0^\infty dr\, |R_{nl}(r)|^2}_{\langle 1/r^2 \rangle_{nl}} \,,$$

which, using the result for $\langle 1/r^2 \rangle_{nl}$ given in the text of the problem, can be written as

$$\langle \phi_{nlm} | V_C^2 | \phi_{nlm} \rangle = \frac{8n}{2l+1} \, \epsilon_n^2 \,,$$

in terms of the unperturbed energy ϵ_n. Combining terms, we arrive at

$$\langle \phi_{nl} || V_K || \phi_{nl} \rangle = -\frac{\epsilon_n^2}{2m_e c^2} \left(\frac{8n}{2l+1} - 3 \right) = - \left(\frac{8n}{2l+1} - 3 \right) \frac{(Z\alpha)^4}{8n^4} \, m_e c^2 \,.$$

The correction due to V_K is attractive (negative) for any level nl.

Part 3

Making use of the identity $\mathbf{J}^2 = \mathbf{L}^2 + \mathbf{S}^2 + 2\,\mathbf{L} \cdot \mathbf{S}$ allows one to write

$$V_{SO} = \frac{1}{2} \, \xi(r) \left(\mathbf{J}^2 - \mathbf{L}^2 - \mathbf{S}^2 \right) \,,$$

which makes it plain that V_{SO} commutes with $\mathbf{J}$. In the basis $|\phi_{nl,jm_j}\rangle$ the allowed values of j are

$$j = 1/2 \quad \text{for} \ \ l = 0 \,, \qquad\qquad j = l - 1/2 \,, l + 1/2 \quad \text{for} \ \ l \geq 1 \,,$$

with $m_j = -j, -j + 1, \ldots, +j$. The degenerate subspace of energy ϵ_n is spanned by the states $|\phi_{nl,jm_j}\rangle$ with $l = 0, \ldots, n-1$, $j = |l - 1/2|$ and $l + 1/2$, and $m_j = -j, -j + 1, \ldots, j$. For example, the subspace with $n = 1$ has $j = 1/2$ and dimension 2; the subspace with $n = 2$ has $j = 1/2$ twice and $j = 3/2$ once, and hence has dimension 8; the subspace with $n = 3$ has $j = 1/2$ twice, $j = 3/2$ twice, and $j = 5/2$ once, and hence has dimension 18; and so on. The dimension of the degenerate subspace is $g_n = 2n^2$ (the same as before, of course), as can be easily verified by noting that

$$g_n = \sum_{l=0}^{n-1} \sum_{j=|l-1/2|}^{l+1/2} (2j+1) = 2 + \sum_{l=1}^{n-1} (2l + 2l + 2) = 2 + 4\sum_{l=1}^{n-1} l + 2(n-1) = 2 + 2n(n-1) + 2n - 2 = 2n^2 \,.$$

The Wigner–Eckart theorem ensures that, in the degenerate subspace, the matrix $\underline{V}_{SO}$ is diagonal in j and m_j, since V_{SO} is a scalar in the combined orbital + spin space,

$$\langle \phi_{nl'j'm_j'} | V_{SO} | \phi_{nl,jm_j} \rangle = \delta_{j,j'} \, \delta_{m_j,m_j'} \, \langle \phi_{nl'j} || V_{SO} || \phi_{nl,j} \rangle \,.$$

Further, V_{SO} commutes with $\mathbf{L}^2$ (as well as with $\mathbf{S}^2$), and hence the matrix $\underline{V}_{SO}$ is also diagonal in l. We are left with the task of evaluating the reduced matrix elements,

$$\langle \phi_{nl,j} || V_{SO} || \phi_{nl,j} \rangle = \langle \phi_{nl,jm_j} | V_{SO} | \phi_{nl,jm_j} \rangle = \frac{\hbar^2}{2} \left[j(j+1) - l(l+1) - \frac{3}{4} \right] \langle \phi_{nl,jm_j} | \, \xi(r) \, | \phi_{nl,jm_j} \rangle \,,$$

with $j = |l - 1/2|$ or $j = l + 1/2$ and corresponding prefactor given by

$$\frac{1}{2} \left[j(j+1) - l(l+1) - \frac{3}{4} \right] = \frac{l}{2} \qquad\qquad \text{for} \ \ j = l + \frac{1}{2} \,,$$

$$= -\frac{l+1}{2} \qquad\qquad \text{for} \ \ j = l - \frac{1}{2} \,,$$

where we have used

$$\frac{1}{2} \left(\mathbf{J}^2 - \mathbf{L}^2 - \mathbf{S}^2 \right) \xi(r) \, | \phi_{nl,jm_j} \rangle = \frac{\hbar^2}{2} \left[j(j+1) - l(l+1) - \frac{3}{4} \right] \xi(r) \, | \phi_{nl,jm_j} \rangle \,.$$

Note that the prefactor is zero for $l = 0$ (and $j = 1/2$). By inserting the expansion for the states into the matrix element of $\xi(r)$, we find

$$\langle \phi_{nl,jm_j} | \xi(r) | \phi_{nl,jm_j} \rangle = \sum_{m,m_s} \sum_{m',m_s'} C_{l,1/2}(mm_s; jm_j) C_{l,1/2}(m'm_s'; jm_j) \langle \phi_{nlm'} | \xi(r) | \phi_{nlm} \rangle \langle m_s' | m_s \rangle$$

$$= \sum_{m,m_s} C_{l,1/2}(mm_s; jm_j) C_{l,1/2}(mm_s; jm_j) \langle \phi_{nlm} | \xi(r) | \phi_{nlm} \rangle \,,$$

where in the second line we have used the orthonormality of the spin states, $\langle m_s' | m_s \rangle = \delta_{m_s, m_s'}$, and the fact the matrix element of $\xi(r)$, which commutes with $\mathbf{L}$, vanishes for $m \neq m'$; that is, $\xi(r)$ is an ITO of rank 0 in orbital space. Further, by applying the Wigner–Eckart theorem to the matrix element $\langle \phi_{nlm} | \xi(r) | \phi_{nlm} \rangle$, we see that this matrix element is independent of m, and can therefore be pulled out of the summation symbol. We arrive at

$$\langle \phi_{nl,jm_j} | \xi(r) | \phi_{nl,jm_j} \rangle = \langle \phi_{nl} || \xi(r) || \phi_{nl} \rangle \sum_{m,m_s} C_{l,1/2}(mm_s; jm_j) C_{l,1/2}(mm_s; jm_j) = \langle \phi_{nl} || \xi(r) || \phi_{nl} \rangle \,,$$

where use has been made of the orthogonality relation (13.28) satisfied by the Clebsch–Gordan coefficients. We report it here for clarity:

$$\delta_{j,j'} \, \delta_{m_j, m_j'} = \sum_{m_1 m_2} C_{j_1 j_2} (m_1 m_2; j m_j) \, C_{j_1 j_2} (m_1 m_2; j' m_j') \,.$$

The calculation is reduced to evaluating the matrix element

$$\langle \phi_{nl} || \xi(r) || \phi_{nl} \rangle = \int d\mathbf{r} \, \xi(r) \, |\phi_{nlm}(\mathbf{r})|^2 = \frac{(g_e - 1)Ze^2}{2m_e^2 c^2} \underbrace{\int_0^\infty dr \, \frac{1}{r} |R_{nl}|^2}_{\langle 1/r^3 \rangle_{nl}}$$

$$= \frac{g_e - 1}{\hbar^2} \frac{4}{(2l + 1)[(2l + 1)^2 - 1]} \frac{(Z\alpha)^4}{n^3} m_e c^2 \,,$$

and l can never be 0 since $\mathbf{L} \cdot \mathbf{S}$ vanishes on s-wave states. We obtain, for the diagonal matrix elements of the spin–orbit interaction,

$$\langle \phi_{nl,jm_j} | V_{SO} | \phi_{nl,jm_j} \rangle = (g_e - 1) \frac{2l}{(2l + 1)[(2l + 1)^2 - 1]} \frac{(Z\alpha)^4}{n^3} m_e c^2 \qquad \text{for } j = l + 1/2 \,,$$

$$= -(g_e - 1) \frac{2(l + 1)}{(2l + 1)[(2l + 1)^2 - 1]} \frac{(Z\alpha)^4}{n^3} m_e c^2 \qquad \text{for } j = l - 1/2 \,,$$

with $l \geq 1$, and the matrix elements with $l = 0$ vanish.

Problem 16 Spin–Orbit Corrections to the Ground and First-Excited Levels of the Hydrogen Atom

Consider the spin-orbit interaction

$$V_{SO} = \xi(r) \, \mathbf{L} \cdot \mathbf{S} \,, \qquad \xi(r) = \frac{g_e - 1}{2m_e^2 c^2} \frac{Ze^2}{r^3} \,.$$

Using the basis of common eigenstates $|\phi_{nlm,m_s}\rangle$ of H_0, $\mathbf{L}^2$, L_z, $\mathbf{S}^2$, and S_z, evaluate the first-order corrections induced by V_{SO} on the ground and first-excited levels of hydrogen-like atoms.

Solution

We need to calculate the matrix $\underline{V}_{SO}$ in the degenerate subspace of level ϵ_n (here, $n = 1$ or 2), where

$$(\underline{V}_{SO})_{n\alpha',n\alpha} = \langle \phi_{nl'm',m_s'} | \xi(r) \, \mathbf{L} \cdot \mathbf{S} | \phi_{nlm,m_s} \rangle \,,$$

$l = 0, 1, \ldots, n - 1$, $m = -l, \ldots, l$, and $m_s = \pm 1/2$, and similarly for the primed quantities (we denote these quantum numbers collectively by α and α', respectively). It is convenient to express $\mathbf{L} \cdot \mathbf{S}$ in terms of spherical components as

$$\mathbf{L} \cdot \mathbf{S} = \sum_{\mu = \pm 1, 0} (-1)^{\mu} L_{\mu} \, S_{-\mu} \,,$$

where

$$L_{\pm 1} = \mp \frac{L_x \pm i L_y}{\sqrt{2}} = \mp \frac{1}{\sqrt{2}} L_{\pm} \,, \qquad L_0 = L_z \,,$$

and similarly for S_{μ}; note that the spherical components $L_{\pm 1}$ and $S_{\pm 1}$ are proportional to the corresponding raising and lowering operators $L_{\pm}$ and $S_{\pm}$. Since $\mathbf{L}$ and $\mathbf{S}$ act in different spaces, we have

$$
\begin{aligned}
(\underline{V}_{SO})_{n\alpha',n\alpha} &= \sum_{\mu = \pm 1, 0} (-1)^{\mu} \langle \phi_{nl'm'} | \xi(r) L_{\mu} | \phi_{nlm} \rangle \langle 1/2, m_s' | S_{-\mu} | 1/2, m_s \rangle \\
&= \sum_{\mu = \pm 1, 0} (-1)^{\mu} C_{1,l}(\mu m; l'm') \langle \phi_{nl'} || \xi(r) L || \phi_{nl} \rangle \, C_{1,1/2}(-\mu, m_s; 1/2, m_s') \langle 1/2 || S || 1/2 \rangle \,,
\end{aligned}
$$

where we have made use of the Wigner–Eckart theorem in orbital space and spin space separately. In the first line, we see that the states with $l = l' = 0$ and those with either $l = 0$ or $l' = 0$ cannot contribute, since, for any μ,

$$L_{\mu} | \phi_{n00} \rangle = 0 \,, \qquad \langle \phi_{n00} | L_{\mu} = 0 \,,$$

because of the spherical symmetry of s-states. In the second line, the second Clebsch–Gordan coefficient requires $\mu = m_s - m_s'$, which yields

$$(\underline{V}_{SO})_{n\alpha',n\alpha} = (-)^{m_s - m_s'} C_{1,l}(m_s - m_s', m; l', m') \, C_{1,1/2}(m_s' - m_s, m_s; 1/2, m_s') \langle \phi_{nl'} || \xi(r) L || \phi_{nl} \rangle \langle 1/2 || S || 1/2 \rangle \,.$$

The first Clebsch–Gordan coefficient requires

$$m_s - m_s' + m = m' \implies m + m_s = m' + m_s' \,.$$

We recall that the ground-state energy is two-fold degenerate with eigenstates $| \phi_{100,m_s} \rangle$, where $m_s = \pm 1/2$. In this subspace, the 2×2 matrix representing V_{SO} is the null matrix, owing to the selection rule on l. Similarly, the first excited state is eight-fold degenerate, with eigenstates $| \phi_{200,m_s} \rangle$ and $| \phi_{21m,m_s} \rangle$, and the 8×8 matrix representing V_{SO} is given by

$$\underline{V}_{SO} = \begin{pmatrix}
0 & 0 & 0 & 0 & 0 & 0 & 0 & 0 \\
0 & 0 & 0 & 0 & 0 & 0 & 0 & 0 \\
0 & 0 & v/\sqrt{6} & 0 & 0 & 0 & 0 & 0 \\
0 & 0 & 0 & -v/\sqrt{6} & v/\sqrt{3} & 0 & 0 & 0 \\
0 & 0 & 0 & v/\sqrt{3} & 0 & 0 & 0 & 0 \\
0 & 0 & 0 & 0 & 0 & 0 & v/\sqrt{3} & 0 \\
0 & 0 & 0 & 0 & 0 & v/\sqrt{3} & -v/\sqrt{6} & 0 \\
0 & 0 & 0 & 0 & 0 & 0 & 0 & v/\sqrt{6}
\end{pmatrix} \,,$$

where have defined the parameter

$$v = \langle \phi_{21} || \xi(r) L || \phi_{21} \rangle \langle 1/2 || S || 1/2 \rangle \,,$$

and the constants multiplying v result from the combination

$$(-1)^{m_s - m_s'} \, C_{1,1}(m_s - m_s', m; 1, m') \, C_{1,1/2}(m_s' - m_s, m_s; 1/2, m_s').$$

In the above matrix the rows and columns correspond to the quantum numbers l', m', m_s' and l, m, m_s, where, using labels 1–8 as follows,

$$1 \longrightarrow 0, 0, 1/2 \qquad 2 \longrightarrow 0, 0, -1/2 \,, \qquad 3 \longrightarrow 1, 1, 1/2 \,, \qquad 4 \longrightarrow 1, 1, -1/2 \,,$$
$$5 \longrightarrow 1, 0, 1/2 \qquad 6 \longrightarrow 1, 0, -1/2 \,, \qquad 7 \longrightarrow 1, -1, 1/2 \,, \qquad 8 \longrightarrow 1, -1, -1/2 \,,$$

we have

$$(\underline{V}_{SO})_{mn} = \langle m | V_{SO} | n \rangle \qquad \text{with} \qquad m, n = 1, \ldots, 8 \,.$$

The above matrix is of block form, and we only need to diagonalize the 2×2 matrices

$$\begin{pmatrix} -v/\sqrt{6} & v/\sqrt{3} \\ v/\sqrt{3} & 0 \end{pmatrix}, \qquad \begin{pmatrix} 0 & v/\sqrt{3} \\ v/\sqrt{3} & -v/\sqrt{6} \end{pmatrix} \,.$$

These two matrices have the same eigenvalues, $v/\sqrt{6}$ and $-2v/\sqrt{6}$, and we see that the matrix $\underline{V}_{SO}$ has eigenvalues 0 that are two-fold degenerate, $-2v/\sqrt{6}$ that are two-fold degenerate, and $v/\sqrt{6}$ that are four-fold degenerate. Clearly, the eigenvalues $v/\sqrt{6}$ and $-2v/\sqrt{6}$ correspond to those of the states having, respectively, $j = l + 1/2 = 3/2$ and $j = l - 1/2 = 1/2$ for p-waves ($l = 1$). We are left with the task of evaluating the reduced matrix elements. Using the Wigner–Eckart theorem, we have

$$\langle 1/2 || S || 1/2 \rangle = \frac{\langle 1/2, 1/2 | S_z | 1/2, 1/2 \rangle}{C_{1,1/2}(0, 1/2; 1/2, 1/2)} = -\frac{\hbar/2}{1/\sqrt{3}} = -\hbar \sqrt{3}/2$$

and

$$\langle \phi_{21} || \xi(r) L || \phi_{21} \rangle = \frac{\langle \phi_{211} | \xi(r) L_z | \phi_{211} \rangle}{C_{1,1}(0, 1; 1, 1)} = -\hbar \frac{\langle \phi_{211} | \xi(r) | \phi_{211} \rangle}{1/\sqrt{2}} = -\hbar \sqrt{2} \, \langle \phi_{21} || \xi(r) || \phi_{21} \rangle \,,$$

which yields

$$v = \sqrt{3/2} \, \hbar^2 \, \langle \phi_{211} | \xi(r) | \phi_{211} \rangle \,.$$

The first-order corrections due to the spin–orbit interaction and relative degeneracies are therefore

$$E_1^{(1)} = 0 \qquad\qquad\qquad g_1 = 2 \,,$$
$$E_2^{(1)} = -\hbar^2 \langle \phi_{n11} | \xi(r) | \phi_{n11} \rangle \qquad g_2 = 2 \,,$$
$$E_3^{(1)} = \frac{\hbar^2}{2} \langle \phi_{n11} | \xi(r) | \phi_{n11} \rangle \qquad g_3 = 4 \,.$$

and the expectation value of $\xi(r)$ can be evaluated using the result given in the text of Problem 15,

$$\langle \phi_{211} | \xi(r) | \phi_{211} \rangle = \frac{(g_e - 1) Z e^2}{2 m_e^2 c^2} \int_0^\infty dr \, \frac{1}{r} \, |R_{21}|^2 = \frac{g_e - 1}{\hbar^2} \frac{(Z\alpha)^4}{48} m_e c^2 \,.$$

The eigenstates corresponding to these eigenvalues are

$$E_1^{(1)}: \qquad |\phi_{200,1/2}\rangle$$

$$|\phi_{200,-1/2}\rangle$$

$$E_2^{(1)}: \qquad \sqrt{\frac{2}{3}}\,|\phi_{211,-1/2}\rangle - \sqrt{\frac{1}{3}}\,|\phi_{210,1/2}\rangle$$

$$\sqrt{\frac{1}{3}}\,|\phi_{210,-1/2}\rangle - \sqrt{\frac{2}{3}}\,|\phi_{21-1,1/2}\rangle$$

$$E_3^{(1)}: \qquad |\phi_{211,1/2}\rangle$$

$$\sqrt{\frac{1}{3}}\,|\phi_{211,-1/2}\rangle + \sqrt{\frac{2}{3}}\,|\phi_{210,1/2}\rangle$$

$$\sqrt{\frac{2}{3}}\,|\phi_{210,-1/2}\rangle + \sqrt{\frac{1}{3}}\,|\phi_{21-1,1/2}\rangle$$

$$|\phi_{21-1,-1/2}\rangle\,.$$

In the basis $|\phi_{nl,jm_j}\rangle$ in which the spin–orbit interaction is diagonal, these states correspond to $|\phi_{n0,1/2m_j}\rangle$, $|\phi_{n1,1/2m_j}\rangle$, and $|\phi_{n1,3/2m_j}\rangle$, respectively.

Problem 17   Effects of Relativistic Corrections on 1s, 2s, and 2p Levels of the Hydrogen Atom

Calculate explicitly the kinetic energy (V_K), spin-orbit (V_{SO}), and Darwin–Foldy (V_{DF}) relativistic corrections for the 1s, 2s, and 2p levels of hydrogen-like atoms in the basis $|\phi_{nl,jm_j}\rangle$ of eigenstates of H_0, $\mathbf{L}^2$, $\mathbf{J}^2$, and J_z (see Problem 15). Briefly comment on the results.

Solution

In Problem 15 the corrections due to V_K and V_{DF} were calculated using the basis $|\phi_{nlm,m_s}\rangle$ of eigenstates of H_0, $\mathbf{L}^2$, L_z, and S_z. However, they can also be calculated straightforwardly in the basis $|\phi_{nl,jm_j}\rangle$ that diagonalizes V_{SO}, since

$$\langle\phi_{nl,jm_j}|V_K|\phi_{nl,jm_j}\rangle = \sum_{m,m_s}\sum_{m',m_s'} C_{l,1/2}(mm_s;jm_j)\,C_{l,1/2}(m'm_s';jm_j)\underbrace{\langle\phi_{nlm'}|V_K|\phi_{nlm}\rangle}_{\text{vanishes if } m\neq m'}\langle m_s'|m_s\rangle$$

$$= \underbrace{\langle\phi_{nlm}|V_K|\phi_{nlm}\rangle}_{\text{independent of } m}\sum_{m,m_s} C_{l,1/2}(mm_s;jm_j)C_{l,1/2}(mm_s;jm_j) = \langle\phi_{nlm}|V_K|\phi_{nlm}\rangle\,,$$

and similarly for V_{DF}. Hence, the kinetic energy and Darwin–Foldy corrections are independent of j and also of m_j since they commute with $\mathbf{J}$ (in addition to commuting with $\mathbf{L}$). We finally arrive at

$$\langle\phi_{nl,jm_j}|V_K|\phi_{nl,jm_j}\rangle = -\left(\frac{8n}{2l+1}-3\right)\frac{(Z\alpha)^4}{8n^4}\,m_ec^2\,,$$

$$\langle\phi_{nl,jm_j}|V_{SO}|\phi_{nl,jm_j}\rangle = (g_e-1)\frac{2l}{(2l+1)[(2l+1)^2-1]}\frac{(Z\alpha)^4}{n^3}\,m_ec^2 \qquad \text{for } j=l+1/2\,,$$

$$= -(g_e-1)\frac{2(l+1)}{(2l+1)[(2l+1)^2-1]}\frac{(Z\alpha)^4}{n^3}\,m_ec^2 \qquad \text{for } j=l-1/2\,,$$

$$\langle\phi_{nl,jm_j}|V_{DF}|\phi_{nl,jm_j}\rangle = \frac{(Z\alpha)^4}{2n^3}\,m_ec^2\,\delta_{l,0}\,\delta_{m,0}\,,$$

where the spin-orbit term vanishes for s-states, and hence $l \geq 1$. The overall correction is obtained by summing the individual corrections. We find (the notation is nl_j):

- $1s_{1/2}$ level: the spin–orbit term vanishes, and the first-order correction to the energy is

$$E^{(1)}_{1s_{1/2}} = \left(-\frac{5}{8} + \frac{1}{2}\right)(Z\alpha)^4 \, m_e c^2 = -\frac{1}{8}(Z\alpha)^4 \, m_e c^2 \; ;$$

- $2s_{1/2}$ level: again there is no spin–orbit term, and

$$E^{(1)}_{2s_{1/2}} = -\frac{1}{16}\left(\frac{13}{8} - 1\right)(Z\alpha)^4 \, m_e c^2 = -\frac{5}{128}(Z\alpha)^4 \, m_e c^2 \; ;$$

- $2p_{1/2}$ level: there is no Darwin–Foldy term, and, using $g_e = 2$ in the spin–orbit term,

$$E^{(1)}_{2p_{1/2}} = -\frac{1}{16}\left(\frac{7}{24} + \frac{1}{3}\right)(Z\alpha)^4 \, m_e c^2 = -\frac{5}{128}(Z\alpha)^4 \, m_e c^2 \; ;$$

- $2p_{3/2}$ level: again there is no Darwin–Foldy term,

$$E^{(1)}_{2p_{3/2}} = -\frac{1}{16}\left(\frac{7}{24} - \frac{1}{6}\right)(Z\alpha)^4 \, m_e c^2 = -\frac{1}{128}(Z\alpha)^4 \, m_e c^2 \; .$$

The corrections are all negative and the energy levels are lowered by these relativistic corrections. Each level has degeneracy $2j + 1$ from spherical symmetry. More interesting is the fact that $2s_{1/2}$ and $2p_{1/2}$ remain degenerate to order $(Z\alpha)^4$. It turns out that they stay so to all orders of perturbation theory, in the sense that the exact solution of the Dirac equation gives the same energy for these two levels.

The degeneracy of the $2s_{1/2}$ and $2p_{1/2}$ levels is lifted by the interaction of the atom with the fluctuating (and quantized) electromagnetic field. As a consequence of this interaction, the $2s_{1/2}$ level is raised relative to the $2p_{1/2}$ level by 1060 MHz in frequency units (the famous Lamb shift), corresponding to an energy of 4.4×10^{-6} eV, about 10 times smaller than the fine-structure splitting of 4.5×10^{-5} eV between the $2p_{1/2}$ and $2p_{3/2}$ levels.

Problem 18 Hydrogen Atom in a Magnetic Field (Zeeman and Paschen–Back Effects)

In this problem we consider the effect of a uniform magnetic field on the levels of hydrogen-like atoms. The Hamiltonian, including relativistic corrections from the kinetic energy, spin–orbit, and Darwin–Foldy terms (collectively indicated as V_{RC} below), reads

$$H = \underbrace{\overbrace{\frac{\mathbf{p}^2}{2m_e} - \frac{Ze^2}{r}}^{H_0} + V_{RC}}_{H_0'} + \underbrace{\frac{e}{2m_e c}\mathbf{B}\cdot(\mathbf{L} + g_e\,\mathbf{S})}_{V_{MM}} + \underbrace{\frac{e^2}{8m_e c^2}\mathbf{B}^2\,\mathbf{r}_\perp^2}_{V_{DM}} \, ,$$

where the two terms denoted as V_{MM} represent the interaction of the orbital and spin magnetic moments of the electron with the magnetic field, and the last term, quadratic in the magnetic field, is known as the diamagnetic term.

1. Provide a rough estimate of the terms V_{MM} and V_{DM} as functions of the magnetic field, expressed in teslas (1 tesla is 10^4 gauss). Show that, even for relatively strong fields, the magnitude of the V_{DM} term is much smaller than that of the V_{MM} term. Hereafter, we ignore the diamagnetic term.

2. Consider the regime of a weak magnetic field (the Zeeman effect), in which $|V_{MM}| \ll |V_{RC}|$. Take the unperturbed Hamiltonian to be H_0', with (approximate) eigenvalues and eigenstates given by

$$H_0' |\phi_{nl,jm_j}\rangle = E_{nlj} |\phi_{nl,jm_j}\rangle , \qquad E_{nlj} = \epsilon_n + E_{nlj}^{(1)} ,$$

where $E_{nlj}^{(1)}$ represent the corrections induced by V_{RC}, which were determined in Problem 15. Making use of Eq. (17.33), obtain the Zeeman splittings of the energy levels E_{nlj} to first order in V_{MM}. Show that these splittings are independent of m_j and are given by $g_{lj}\mu_0 B$, where $\mu_0 = e\hbar/(2m_e c)$ and g_{lj} is a factor depending on l and j, to be determined. We set $g_e \approx 2$ here and in part 3.

3. Now, consider the regime of a strong magnetic field (the Paschen–Back effect), in which $|V_{RC}| \ll |V_{MM}|$. Take the unperturbed Hamiltonian to be H_0 with (exact) eigenvalues and eigenstates given by

$$H_0 |\phi_{nlm,m_s}\rangle = \epsilon_n |\phi_{nlm,m_s}\rangle ,$$

with degeneracy $g_n = 2n^2$. Obtain the splitting of these levels due to V_{MM}. Is the degeneracy completely resolved?

Solution

Part 1

The order of magnitude of the V_{MM} corrections is

$$|V_{MM}| \approx \mu_0 B \approx B(\mathrm{T}) \times 5.8 \times 10^{-5} \, \mathrm{eV} ,$$

where $\mu_0 = e\hbar/(2m_e c)$ and $B(\mathrm{T})$ is the magnetic field strength expressed in teslas (T). The order of magnitude of the V_{DM} term, proportional to $\mathbf{B}^2$, is

$$|V_{DM}| \approx \frac{e^2}{12 m_e c^2} \left(\frac{a_0}{Z}\right)^2 B^2 \approx B^2(\mathrm{T}) \frac{4.1}{Z^2} \times 10^{-11} \, \mathrm{eV} ,$$

where we have approximated $\mathbf{r}_\perp^2$ as $(2/3)(a_0/Z)^2$ and a_0/Z is the "Bohr radius" of hydrogen-like atoms. We see that V_{DM} is typically much smaller than V_{MM} and V_{RC}.

Part 2

When the magnetic field is weak, V_{MM} can be viewed as a perturbation relative to H_0'. In order to evaluate its first-order correction, we need to diagonalize the matrix representing V_{MM} in the degenerate subspace of nlj,

$$\left(\underline{V}_{MM}\right)_{m_j', m_j} = \frac{\mu_0 B}{\hbar} \langle \phi_{nl,jm_j'} | L_z + g_e S_z | \phi_{nl,jm_j}\rangle ,$$

where we have taken the magnetic field to be along the $\hat{z}$-axis. The evaluation of these matrix elements is facilitated by using the relation in Eq. (17.33). In our case, we have $\mathbf{V} = \mathbf{L} + g_e \mathbf{S}$ and, by considering the z-component, we find

$$\langle \phi_{nl,jm_j'} | L_z + g_e S_z | \phi_{nl,jm_j}\rangle = \frac{\langle \phi_{nl,jm_j} | \mathbf{J} \cdot (\mathbf{L} + g_e \mathbf{S}) | \phi_{nl,jm_j}\rangle}{j(j+1)\hbar^2} \underbrace{\langle \psi_{nl,jm_j'} | J_z | \psi_{nl,jm_j}\rangle}_{m_j \hbar \, \delta_{m_j, m_j'}} ,$$

and so the matrix $\underline{V}_{MM}$ is diagonal. Further, we have

$$\mathbf{J} \cdot (\mathbf{L} + g_e\mathbf{S}) = (\mathbf{L} + \mathbf{S}) \cdot (\mathbf{L} + g_e\mathbf{S}) = \mathbf{L}^2 + g_e\mathbf{S}^2 + (g_e + 1)\mathbf{L} \cdot \mathbf{S} = \mathbf{L}^2 + g_e\mathbf{S}^2 + \frac{g_e + 1}{2}(\mathbf{J}^2 - \mathbf{L}^2 - \mathbf{S}^2)$$

$$= \frac{g_e + 1}{2}\mathbf{J}^2 - \frac{g_e - 1}{2}\mathbf{L}^2 + \frac{g_e - 1}{2}\mathbf{S}^2$$

and the expectation value is simply given by (setting $g_e \approx 2$)

$$\langle \phi_{nl,jm_j}| \mathbf{J} \cdot (\mathbf{L} + 2\mathbf{S}) |\phi_{nl,jm_j}\rangle = \frac{\hbar^2}{2}\left[3j(j+1) - l(l+1) + 3/4\right] .$$

We finally arrive at

$$(\underline{V}_{MM})_{m'_j, m_j} = \frac{\mu_0 B}{\hbar}\left[\frac{\hbar^2}{2}\frac{3j(j+1) - l(l+1) + 3/4}{j(j+1)\hbar^2}\right] m_j\hbar \, \delta_{m_j, m'_j} = g_{lj}m_j\mu_0 B \, \delta_{m_j, m'_j} ,$$

where we have defined the so-called Landé factor,

$$g_{lj} = \frac{1}{2}\frac{3j(j+1) - l(l+1) + 3/4}{j(j+1)} = 1 + \frac{j(j+1) - l(l+1) + 3/4}{2j(j+1)} .$$

Thus we see that the magnetic field has completely removed the degeneracy in m_j. The splitting between Zeeman levels is given by $g_{lj}\mu_0 B$, and is independent of m_j.

Part 3

In this regime V_{MM} can be considered as a small perturbation relative to H_0, rather than to H'_0 as in the treatment of the Zeeman effect above. With $\mathbf{B}$ along the $\hat{z}$-axis, the perturbation commutes with $\mathbf{L}^2$, L_z and S_z; indeed, the states $|\phi_{nlm,m_s}\rangle$ are exact eigenstates of $H_0 + V_{MM}$, that is, $(H_0 + V_{MM})|\phi_{nlm,m_s}\rangle = E_{n,mm_s}|\phi_{nlm,m_s}\rangle$, where

$$E_{n,mm_s} = \epsilon_n + \mu_0 B(m + 2m_s) ,$$

with $l = 0, 1, \ldots, n - 1$ (and again $g_e \approx 2$). We see, for example, that the degeneracy of the ground-state level is completely removed, $E_{1,m=0,m_s=\pm 1/2} = \epsilon_1 \pm \mu_0 B$, while that of the first-excited-state level is only partially removed since $E_{2,m=0,m_s=\pm 1/2} = \epsilon_2 \pm \mu_0 B$ with $l = 0$ or 1, and $E_{21,m\pm 1,m_s=\mp 1/2} = \epsilon_2$; however, the eigenenergies $E_{2,m=\pm 1,m_s=\pm 1/2} = \epsilon_2 \pm 2\mu_0 B$ are non-degenerate.

Problem 19   Hyperfine Structure of the Hydrogen Atom

The proton, and generally any nucleus with a non-vanishing total angular momentum, has a magnetic moment and induces a magnetic field that interacts with the electron's orbital and spin degrees of freedom. This interaction leads to the so-called hyperfine structure of the atom. The nuclear magnetic moment is denoted by $\boldsymbol{\mu}_N$, and is written as

$$\boldsymbol{\mu}_N = g_N \frac{e\hbar}{2m_p c}\frac{\mathbf{S}_N}{\hbar} ,$$

where m_p is the proton mass and $\mathbf{S}_N$ and g_N are the nuclear spin (total angular momentum) and g-factor, respectively; in particular, for a proton the spin is $1/2$ and $g_p = 2 \times (2.792847\cdots)$. The combination $e\hbar/(2m_p c)$ is known as the nuclear magneton, as opposed to the Bohr (atomic) magneton $e\hbar/(2m_e c)$; the magnetic moments of nucleons and nuclei are expressed in units of nuclear magnetons. Because of the proton's large mass, the associated magnetic moment is much smaller than for an electron. For this reason hyperfine-structure effects are expected to be much smaller than fine-structure effects.

The current distribution responsible for producing $\boldsymbol{\mu}_N$ in the nucleus is confined to a volume whose radius is of the order of the nuclear charge's radius, which is much smaller than the atomic radius. A confined current distribution gives rise to a vector potential

$$\mathbf{A}_N(\mathbf{r}) = \frac{\boldsymbol{\mu}_N \times \mathbf{r}}{r^3} ,$$

and the resulting magnetic field can be obtained by evaluating the curl of $\mathbf{A}(\mathbf{r})$. It reads

$$\mathbf{B}_N(\mathbf{r}) = \frac{3\hat{\mathbf{r}}\left(\boldsymbol{\mu}_N \cdot \hat{\mathbf{r}}\right) - \boldsymbol{\mu}_N}{r^3} + \frac{8\pi}{3}\,\boldsymbol{\mu}_N\,\delta(\mathbf{r}) ,$$

where $\hat{\mathbf{r}}$ is the unit vector ($\hat{\mathbf{r}} = \mathbf{r}/r$). Due to the presence of these terms the Hamiltonian of the electron (charge $-e$), including relativistic corrections but in the absence of external magnetic fields, is now given by

$$H = \frac{1}{2m_e}\left[\mathbf{p} + \frac{e}{c}\mathbf{A}_N(\mathbf{r})\right]^2 + V_C(r) + V_{RC}(\mathbf{r}) - \boldsymbol{\mu}_e \cdot \mathbf{B}_N(\mathbf{r}) .$$

By ignoring terms quadratic in $\mathbf{A}_N(\mathbf{r})$, the above Hamiltonian can be more conveniently written as

$$H = \underbrace{H_0 + V_{RC}}_{H_0'} + \underbrace{\frac{e}{m_e c}\frac{\mathbf{L} \cdot \boldsymbol{\mu}_N}{r^3} - \frac{3(\boldsymbol{\mu}_e \cdot \hat{\mathbf{r}})(\boldsymbol{\mu}_N \cdot \hat{\mathbf{r}}) - \boldsymbol{\mu}_e \cdot \boldsymbol{\mu}_N}{r^3} - \frac{8\pi}{3}\boldsymbol{\mu}_e \cdot \boldsymbol{\mu}_N\,\delta(\mathbf{r})}_{V_{HF}} ,$$

where $\mathbf{L}$ is the electron's orbital angular momentum.

Treat V_{HF} as a small perturbation relative to H_0' and evaluate its effect on the ground-state level of a hydrogen atom to first order. Note that, since V_{HF} couples the electron's orbital angular momentum and spin to the proton's spin, we must account explicitly for the proton's spin degrees of freedom. Therefore, the states in the combined electron orbital + spin and proton spin spaces are written as $|\phi_{nl,jm_j,m_p}\rangle = |\phi_{nl,jm_j}\rangle \otimes |m_p\rangle$, where $|m_p\rangle$ are the states corresponding to the two possible spin projections for the proton, $m_p = \pm 1/2$; they are simultaneous (approximate) eigenstates of the commuting observables H_0', $\mathbf{L}^2$, $\mathbf{S}^2$, $\mathbf{J}^2$, and J_z, which act only on the electron degrees of freedom, and of $\mathbf{S}_p^2$ and $S_{p,z}$, which only act on the proton spin degrees of freedom; for example, we have $H_0'|\phi_{nl,jm_j,m_p}\rangle \approx E_{nlj}|\phi_{nl,jm_j,m_p}\rangle$. The eigenvalues of H_0' are E_{nlj} and their degeneracy is now doubled, since the E_{nlj} are the same regardless of which spin state the proton is in. In particular, the ground-state level $E_{1s_{1/2}}$,

$$E_{1s_{1/2}} = \epsilon_{1s} - \frac{\alpha^4}{8}\,m_e c^2 ,$$

is four-fold degenerate and its states can be written as (using spectroscopic notation for brevity)

$$|1s_{1/2}, m, m_p\rangle \equiv \underbrace{|\phi_{10,1/2\,m,m_p}\rangle}_{|\phi_{nl,jm,m_p}\rangle} = \underbrace{|\phi_{10,1/2\,m}\rangle}_{|\phi_{nl,jm}\rangle} \otimes |m_p\rangle = \underbrace{|\phi_{100}\rangle}_{|\phi_{nlm}\rangle} \otimes |m\rangle \otimes |m_p\rangle ,$$

where $|m\rangle = |\pm 1/2\rangle$ are the electron spin states and we have expressed the state $|\phi_{10,1/2\,m}\rangle$ simply as $|\phi_{100}\rangle \otimes |m\rangle$ (here, since $l = 0$, we have $m_j = m_s \equiv m$).

Solution

In first-order perturbation theory, we need to diagonalize the matrix

$$(\underline{V}_{HF})_{m'm'_p,mm_p} = \langle 1s_{1/2}, m', m'_p | V_{HF} | 1s_{1/2}, m, m_p \rangle .$$

Of the three terms entering V_{HF}, only the term proportional to the δ-function is non-vanishing. To see this, consider first the **L**-term,

$$\left\langle 1s_{1/2}, m', m'_p \left| \frac{\mathbf{L} \cdot \boldsymbol{\mu}_p}{r^3} \right| 1s_{1/2}, m, m_p \right\rangle = \underbrace{\left\langle 1s \left| \frac{\mathbf{L}}{r^3} \right| 1s \right\rangle}_{\text{vanishes}} \cdot \langle m'_p | \boldsymbol{\mu}_p | m_p \rangle \underbrace{\langle m' | m \rangle}_{\delta_{m,m'}} = 0 ,$$

since $\mathbf{L} | 1s \rangle = 0$ (the $| 1s \rangle$ state is spherically symmetric and depends only on r). This conclusion also follows from the Wigner–Eckart theorem, since a vector operator such as $\mathbf{L}$ cannot connect s-wave states (the associated Clebsch–Gordan coefficient vanishes). The spin-dependent term is written as

$$\frac{3(\boldsymbol{\mu}_e \cdot \hat{\mathbf{r}})(\boldsymbol{\mu}_p \cdot \hat{\mathbf{r}}) - \boldsymbol{\mu}_e \cdot \boldsymbol{\mu}_p}{r^3} = \sum_{ij} Q_{ij} \, \mu_{e,i} \, \mu_{p,j} , \qquad Q_{ij} = \frac{3\hat{r}_i \hat{r}_j - \delta_{ij}}{r^3} ,$$

so that its matrix elements read

$$\langle 1s_{1/2}, m', m'_p | [\cdots] | 1s_{1/2}, m, m_p \rangle = \sum_{ij} \underbrace{\langle 1s | Q_{ij} | 1s \rangle}_{\text{vanishes}} \langle m' | \mu_{e,i} | m \rangle \langle m'_p | \mu_{p,j} | m_p \rangle .$$

The matrix element of Q_{ij} vanishes by spherical symmetry,

$$\langle 1s | Q_{ij} | 1s \rangle = \underbrace{\frac{1}{4\pi}}_{|Y_{00}(\Omega)|^2} \int dr \, \frac{1}{r} |R_{1s}(r)|^2 \underbrace{\int d\Omega \, (3\hat{r}_i \hat{r}_j - \delta_{ij})}_{\text{vanishes for any } i \text{ and } j} = 0 ,$$

since $\int d\Omega \hat{r}_i \hat{r}_j = (4\pi/3) \delta_{ij}$. We are therefore left with the task of evaluating only the δ-function term, which we write as

$$
\begin{aligned}
(\underline{V}_{HF})_{m'm'_p,mm_p} &= \frac{8\pi}{3} \frac{g_e e}{2m_e c} \frac{g_p e}{2m_p c} \langle 1s_{1/2}, m', m'_p | \mathbf{S} \cdot \mathbf{S}_p \, \delta(\mathbf{r}) | 1s_{1/2}, m, m_p \rangle \\
&= \frac{8\pi}{3} \frac{g_e e}{2m_e c} \frac{g_p e}{2m_p c} |\phi_{1s}(0)|^2 \underbrace{\langle m', m'_p | \mathbf{S} \cdot \mathbf{S}_p | m, m_p \rangle}_{\text{m.e.}}
\end{aligned}
$$

where we have accounted for the fact that the charge of the electron is negative (hence the overall sign flip). On general grounds, we see that this term affects only s-wave states – recall that $\phi_{nlm}(0)$ vanishes for $l > 0$. Here $\mathbf{S}$ and $\mathbf{S}_p$ are the electron and proton spin operators. The evaluation of the matrix element in spin space is facilitated by introducing the total spin $\mathbf{F} = \mathbf{S} + \mathbf{S}_p$ and noting that

$$\mathbf{S} \cdot \mathbf{S}_p = \frac{1}{2} (\mathbf{F}^2 - \mathbf{S}^2 - \mathbf{S}_p^2) = \frac{1}{2} \left(\mathbf{F}^2 - \frac{3}{2} \hbar^2 \right) .$$

The eigenstates $| f m_f \rangle$ of $\mathbf{F}^2$ and F_z with $f = 0, 1$ are obviously also eigenstates of $\mathbf{S} \cdot \mathbf{S}_p$, and in such a basis the perturbation is diagonal;

$$(\underline{V}_{HF})_{f'm'_f, fm_f} = \frac{4\pi}{3} g_e g_p \left[f(f+1) - \frac{3}{2} \right] \frac{e\hbar}{2m_e c} \frac{e\hbar}{2m_p c} |\phi_{1s}(0)|^2 \, \delta_{ff'} \, \delta_{m_f m'_f} .$$

By using

$$|\phi_{1s}(0)|^2 = \frac{\alpha^3}{\pi}\left(\frac{m_e c}{\hbar}\right)^3 ,$$

we find that the hyperfine states with $f=0$ and 1 are, respectively, lowered and raised by an amount given by (setting $g_e \approx 2$)

$$f = 0: \quad -g_p \frac{m_e}{m_p} \alpha^4 m_e c^2 , \qquad f = 1: \quad \frac{g_p}{3} \frac{m_e}{m_p} \alpha^4 m_e c^2 .$$

When the hyperfine splitting between the $f=0$ and $f=1$ states is compared with the shift of the ground-state energy induced by relativistic corrections, we see that it is suppressed by a factor $(32/3)g_p(m_e/m_p) \approx 3.2 \times 10^{-2}$. The radiative transition between these hyperfine states gives rise to the famous 21-cm line in the radio spectrum of hydrogen.

Problem 20   The Tensor Term in the Hyperfine Interaction

Consider, for the hydrogen atom, the following term – known as the tensor coupling – in the hyperfine interaction;

$$V_T = -\frac{3(\boldsymbol{\mu}_e \cdot \hat{\mathbf{r}})(\boldsymbol{\mu}_p \cdot \hat{\mathbf{r}}) - \boldsymbol{\mu}_e \cdot \boldsymbol{\mu}_p}{r^3} ,$$

where

$$\boldsymbol{\mu}_e = -g_e \frac{e}{2m_e c} \mathbf{S}_e , \qquad \boldsymbol{\mu}_p = g_p \frac{e}{2m_p c} \mathbf{S}_p ,$$

and $\mathbf{S}_e$ and $\mathbf{S}_p$ are the spins of the electron and proton, respectively, and g_e and g_p are the corresponding gyromagnetic factors.

1. Define

$$T_{ep}(\hat{\mathbf{r}}) = 3(\mathbf{S}_e \cdot \hat{\mathbf{r}})(\mathbf{S}_p \cdot \hat{\mathbf{r}}) - \mathbf{S}_e \cdot \mathbf{S}_p , \qquad V_T = g_e \frac{e}{2m_e c} g_p \frac{e}{2m_p c} \frac{T_{ep}(\hat{\mathbf{r}})}{r^3} .$$

   Show that the tensor operator $T_{ep}(\hat{\mathbf{r}})$ can be written as follows:

$$T_{ep}(\hat{\mathbf{r}}) = \frac{3}{2}(\mathbf{F} \cdot \hat{\mathbf{r}})^2 - \frac{1}{2}\mathbf{F}^2 , \qquad \mathbf{F} = \mathbf{S}_e + \mathbf{S}_p .$$

2. Define the total angular momentum $\mathbf{J} = \mathbf{L} + \mathbf{F}$. Show that $T_{ep}(\hat{\mathbf{r}})$ is a scalar with respect to rotations induced by $\mathbf{J}$.

3. It turns out that $T_{ep}(\hat{\mathbf{r}})$ can be written in terms of a rank-2 irreducible tensor operator (ITO), constructed from the rank-1 ITO $\hat{r}_\mu$, and a rank-2 ITO constructed from the rank-1 ITO F_μ (see Problem 6 for how to construct these rank-2 ITOs from $\hat{r}_\mu$ and F_μ), so that

$$T_{ep}(\hat{\mathbf{r}}) = \frac{3}{2}\sum_{\mu=-2}^{2}(-1)^\mu R_\mu^{(2)} F_{-\mu}^{(2)} ,$$

   where $R_\mu^{(2)}$ and $F_\mu^{(2)}$ are the rank-2 ITOs in the orbital and spin spaces, respectively. Show this result and, without doing any detailed calculation, explain why the tensor term does not contribute in the ground state of the hydrogen atom, that is,

$$\langle \phi_{1s,f'm'_f} | V_T | \phi_{1s,fm_f} \rangle = 0 ,$$

   where $|\phi_{1s,fm_f}\rangle = |\phi_{1s}\rangle \otimes |fm_f\rangle$ with $f=0$ or 1 and $m_f = -f, \ldots, f$.

Solution

Part 1

Recall that $\mathbf{S} = (\hbar/2)\,\boldsymbol{\sigma}$ and that the electron and proton spin operators (and hence their associated Pauli matrices) commute with each other, and so we find

$$
\begin{aligned}
\frac{3}{2}\,(\mathbf{F}\cdot\hat{\mathbf{r}})^2 - \frac{1}{2}\,\mathbf{F}^2 &= \frac{3\hbar^2}{8}\left(\boldsymbol{\sigma}_e\cdot\hat{\mathbf{r}} + \boldsymbol{\sigma}_p\cdot\hat{\mathbf{r}}\right)^2 - \frac{\hbar^2}{8}\left(\boldsymbol{\sigma}_e + \boldsymbol{\sigma}_p\right)^2 \\
&= \frac{3\hbar^2}{8}\left[2\hat{\mathbf{r}}\cdot\hat{\mathbf{r}} + 2(\boldsymbol{\sigma}_e\cdot\hat{\mathbf{r}})(\boldsymbol{\sigma}_p\cdot\hat{\mathbf{r}})\right] - \frac{\hbar^2}{8}\left(6 + 2\boldsymbol{\sigma}_e\cdot\boldsymbol{\sigma}_p\right) \\
&= \frac{3\hbar^2}{4}\,(\boldsymbol{\sigma}_e\cdot\hat{\mathbf{r}})(\boldsymbol{\sigma}_p\cdot\hat{\mathbf{r}}) - \frac{\hbar^2}{4}\,\boldsymbol{\sigma}_e\cdot\boldsymbol{\sigma}_p = T_{ep}(\hat{\mathbf{r}})\,,
\end{aligned}
$$

where we have used the Pauli identity

$$
(\boldsymbol{\sigma}_e\cdot\hat{\mathbf{r}})(\boldsymbol{\sigma}_e\cdot\hat{\mathbf{r}}) = \hat{\mathbf{r}}\cdot\hat{\mathbf{r}} + i\boldsymbol{\sigma}_e\cdot(\hat{\mathbf{r}}\times\hat{\mathbf{r}}) = \hat{\mathbf{r}}\cdot\hat{\mathbf{r}} = 1\,, \qquad \boldsymbol{\sigma}_e\cdot\boldsymbol{\sigma}_e = 3\,,
$$

(a unit matrix is understood here) and similarly for $\boldsymbol{\sigma}_p$.

Part 2

We have

$$
\left[J_i,\, T_{ep}(\hat{\mathbf{r}})\right] = \frac{1}{2}\left[J_i,\, 3(\mathbf{F}\cdot\hat{\mathbf{r}})^2 - \mathbf{F}^2\right] = \frac{3}{2r^2}\left[J_i,\, (\mathbf{F}\cdot\mathbf{r})^2\right]\,,
$$

since J_i commutes with $\mathbf{F}^2$ and with any function of r. We consider

$$
\begin{aligned}
\left[J_i,\, (\mathbf{F}\cdot\mathbf{r})^2\right] &= \sum_{lm}[L_i + F_i,\, F_l F_m r_l r_m] = \sum_{lm}\left(F_l F_m\,[L_i,\, r_l r_m] + r_l r_m\,[F_i,\, F_l F_m]\right) \\
&= \sum_{lm}\left(F_l F_m r_l\,[L_i,\, r_m] + F_l F_m\,[L_i,\, r_l]\,r_m + r_l r_m F_l\,[F_i,\, F_m] + r_l r_m\,[F_i,\, F_l]\,F_m\right) \\
&= i\hbar \sum_{lmk}\left(\epsilon_{imk} F_l F_m r_l r_k + \epsilon_{ilk} F_l F_m r_k r_m + \epsilon_{imk} r_l r_m F_l F_k + \epsilon_{ilk} r_l r_m F_k F_m\right) \\
&= i\hbar\left[\mathbf{F}\cdot\mathbf{r}\,(\mathbf{F}\times\mathbf{r})_i + (\mathbf{F}\times\mathbf{r})_i\,\mathbf{F}\cdot\mathbf{r} + \mathbf{F}\cdot\mathbf{r}\,(\mathbf{r}\times\mathbf{F})_i + (\mathbf{r}\times\mathbf{F})_i\,\mathbf{F}\cdot\mathbf{r}\right] = 0\,,
\end{aligned}
$$

since any component of $\mathbf{F}$ commutes with any component of $\mathbf{r}$, and also $\mathbf{r}\times\mathbf{F} = -\mathbf{F}\times\mathbf{r}$. Thus, $T_{ep}(\hat{\mathbf{r}})$ and hence the tensor interaction are scalar operators in the combined orbital space and spin space (that is, the electron and proton spin space).

Part 3

Given two irreducible tensor operators (ITOs) of rank k_1 and k_2, we can combine them to form irreducible tensor operators of rank $k = |k_1 - k_2|, \ldots, k_1 + k_2$, according to

$$
T_q^{(k)} = \sum_{q_1, q_2} C_{k_1, k_2}(q_1 q_2; kq)\, T_{q_1}^{(k_1)}\, T_{q_2}^{(k_2)}.
$$

In the case of two generic ITOs of rank 1 (that is, two vector operators), A_μ and B_μ, after inserting the relevant Clebsch–Gordan coefficients this formula gives the following expressions for the rank-2 tensor operators:

$$T^{(2)}_{\pm 2} = A_{\pm 1}B_{\pm 1} \,, \qquad T^{(2)}_{\pm 1} = \frac{A_{\pm 1}B_0 + A_0 B_{\pm 1}}{\sqrt{2}} \,, \qquad T^{(2)}_0 = \frac{2A_0 B_0 + A_{+1}B_{-1} + A_{-1}B_{+1}}{\sqrt{6}} \,.$$

Now, consider $A_\mu = B_\mu = r_\mu$; we find, for $R^{(2)}_\mu$,

$$R^{(2)}_{\pm 2} = \frac{1}{2}(\hat{r}_x \pm i\hat{r}_y)^2 = \frac{1}{2}\sin^2\theta\,(\cos^2\phi - \sin^2\phi \pm 2i\sin\phi\cos\phi) = \frac{1}{2}\sin^2\theta\,e^{\pm 2i\phi} = \sqrt{\frac{8\pi}{15}}\,Y_{2,\pm 2}\,,$$

$$R^{(2)}_{\pm 1} = \mp \hat{r}_z(\hat{r}_x \pm i\hat{r}_y) = \mp\cos\theta\sin\theta(\cos\phi \pm i\sin\phi) = \mp\cos\theta\,\sin\theta\,e^{\pm i\phi} = \sqrt{\frac{8\pi}{15}}\,Y_{2,\pm 1}\,,$$

and

$$R^{(2)}_0 = \sqrt{\frac{2}{3}}\left[\hat{r}_z\hat{r}_z - \frac{1}{2}\left(\hat{r}_x + i\hat{r}_y\right)\left(\hat{r}_x - i\hat{r}_y\right)\right] = \sqrt{\frac{1}{6}}\left(2\cos^2\theta - \sin^2\theta\right) = \sqrt{\frac{8\pi}{15}}\,Y_{2,0}\,,$$

so that

$$R^{(2)}_\mu(\hat{\mathbf{r}}) = \sqrt{\frac{8\pi}{15}}\,Y_{2,\mu}(\hat{\mathbf{r}})\,.$$

Similarly, we find, for $F^{(2)}_\mu$,

$$F^{(2)}_{\pm 2} = \frac{1}{2}(F_x \pm iF_y)^2 = \frac{1}{2}\left[F_x^2 - F_y^2 \pm i(F_xF_y + F_yF_x)\right]\,,$$

$$F^{(2)}_{\pm 1} = \mp\frac{1}{2}\left[(F_x \pm iF_y)F_z + F_z(F_x \pm iF_y)\right] = \mp\frac{1}{2}\left[F_xF_z + F_zF_x \pm i(F_yF_z + F_zF_y)\right]\,,$$

$$F^{(2)}_0 = \frac{1}{\sqrt{6}}\left[2F_z^2 - \frac{1}{2}\left(F_x + iF_y\right)\left(F_x - iF_y\right) - \frac{1}{2}\left(F_x - iF_y\right)\left(F_x + iF_y\right)\right]$$

$$= \frac{1}{\sqrt{6}}\left(2F_z^2 - F_x^2 - F_y^2\right) = \frac{1}{\sqrt{6}}\left(3F_z^2 - \mathbf{F}^2\right)\,.$$

Now, out of these rank-2 ITOs we can form a scalar by combining them via

$$S^{(0)} = \sum_{\mu,\mu'} C_{2,2}(\mu,\mu';0,0)\,R^{(2)}_\mu F^{(2)}_{\mu'} = \sum_\mu C_{2,2}(\mu,-\mu;0,0)\,R^{(2)}_\mu F^{(2)}_{-\mu} = \frac{1}{\sqrt{5}}\sum_\mu (-1)^\mu R^{(2)}_\mu F^{(2)}_{-\mu}\,.$$

Since $T_{ep}(\hat{\mathbf{r}})$ is also a scalar operator, it must be proportional to $S^{(0)}$, that is,

$$T_{ep}(\hat{\mathbf{r}}) = c\,S^{(0)}\,,$$

and the proportionality constant can be determined by considering, for example, the case in which $\hat{\mathbf{r}} = \hat{\mathbf{z}}$, implying $\theta = 0$. In such a case, we find that the only non-vanishing component of $R^{(2)}_\mu$ is that with $\mu = 0$, giving

$$S^{(0)} = \frac{1}{\sqrt{5}}R^{(2)}_0 F^{(2)}_0 = \frac{1}{\sqrt{5}}\sqrt{\frac{2}{3}}\,\frac{3F_z^2 - \mathbf{F}^2}{\sqrt{6}} = \frac{1}{3\sqrt{5}}(3F_z^2 - \mathbf{F}^2)\,.$$

On the other hand, $T_{ep}(\hat{\mathbf{z}}) = (3F_z^2 - \mathbf{F}^2)/2$, which gives finally

$$T_{ep}(\hat{\mathbf{z}}) = \frac{1}{2}\sqrt{45}\,S^{(0)} \implies T_{ep}(\hat{\mathbf{r}}) = \frac{3}{2}\sum_\mu (-1)^\mu R^{(2)}_\mu F^{(2)}_{-\mu}\,.$$

Having established this result, we have

$$\langle \phi_{1s,f'm'_f}|\,V_T\,|\phi_{1s,fm_f}\rangle \propto \sum_\mu (-)^\mu \langle \phi_{1s}|\,R^{(2)}_\mu\,|\phi_{1s}\rangle\langle f'm'_f|\,F^{(2)}_\mu\,|fm_f\rangle\,,$$

and by the Wigner–Eckart theorem we immediately see that $\langle \phi_{1s}|\,R^{(2)}_\mu\,|\phi_{1s}\rangle = 0$.

Problem 21   ITOs and Time-Reversal Invariance

Let $\hat{A}$ be a linear hermitian operator that is either even or odd under time reversal, $\hat{\Omega}_{\mathcal{T}} \hat{A} \hat{\Omega}_{\mathcal{T}}^{\dagger} = \pm \hat{A}$; that is, $\hat{A}$ commutes or anticommutes with $\hat{\Omega}_{\mathcal{T}}$.

1. Consider the expectation value of $\hat{A}$ on the state $|\psi_{j,m}\rangle$, an eigenstate of the total angular momentum transforming under $\hat{\Omega}_{\mathcal{T}}$ according to $\hat{\Omega}_{\mathcal{T}} |\psi_{jm}\rangle = i^{2m} |\psi_{j,-m}\rangle$. What can you conclude about this expectation value?
2. Next, consider the irreducible tensor operator (ITO) $\hat{T}_0^{(k)}$, that is, the $q = 0$ component, and assume that it is hermitian and either even or odd under time reversal. Show that

$$(\psi_{j,m}, \hat{T}_0^{(k)} \psi_{j,m}) = \pm (-1)^k (\psi_{j,m}, \hat{T}_0^{(k)} \psi_{j,m}),$$

which implies the selection rule that k must be even or odd for an even or odd (under time reversal) $\hat{T}_0^{(k)}$, otherwise the matrix element vanishes. Explain why the above selection rule remains valid for any of the q components of $\hat{T}_q^{(k)}$. How do these components transform under time reversal?
3. Use the results obtained in part 2 to show that the expectation value of $\hat{r}$ on the hydrogen atom spinor state (in spectroscopic notation) $c_{ns} |\psi_{ns;1/2,m}\rangle + c_{np} |\psi_{np;1/2,m}\rangle$, having $j = 1/2$, azimuthal quantum number m, and real c_{ns} and c_{np}, vanishes.

Solution

Part 1

Time-reversal symmetry implies

$$(\psi_{j,m}, \hat{A} \psi_{j,m}) = (\hat{\Omega}_{\mathcal{T}} \psi_{j,m}, \hat{\Omega}_{\mathcal{T}} \hat{A} \psi_{j,m})^* = (\hat{\Omega}_{\mathcal{T}} \psi_{j,m}, \hat{\Omega}_{\mathcal{T}} \hat{A} \hat{\Omega}_{\mathcal{T}}^{\dagger} \hat{\Omega}_{\mathcal{T}} \psi_{j,m})^*$$
$$= (i^{2m} \psi_{j,-m}, \pm i^{2m} \hat{A} \psi_{j,-m})^* = \pm (\psi_{j,-m}, \hat{A} \psi_{j,-m}),$$

where we have used the fact that $\hat{A}$ is linear and hermitian along with the transformation properties of angular momentum eigenstates.

Part 2

Using the result of part 1, we have

$$(\psi_{j,m}, \hat{T}_0^{(k)} \psi_{j,m}) = \pm (\psi_{j,-m}, \hat{T}_0^{(k)} \psi_{j,-m}).$$

However, the state $\psi_{j,-m}$ can be obtained (possibly up to a phase factor) by executing a rotation by π about the y-axis, namely $|\psi_{j,-m}\rangle = \hat{U}_R(0, \pi, 0) |\psi_{j,m}\rangle$, and hence

$$(\psi_{j,-m}, \hat{T}_0^{(k)} \psi_{j,-m}) = (\hat{U}_R \psi_{j,m}, \hat{T}_0^{(k)} \hat{U}_R \psi_{j,m}) = (\psi_{j,m}, \hat{U}_R^{\dagger} \hat{T}_0^{(k)} \hat{U}_R \psi_{j,m}).$$

We can now use the transformation properties under rotation of ITOs to obtain

$$\hat{U}_R^{\dagger}(0, \pi, 0) \hat{T}_0^{(k)} \hat{U}_R(0, \pi, 0) = \hat{U}_R(0, -\pi, 0) \hat{T}_0^{(k)} \hat{U}_R^{\dagger}(0, -\pi, 0) = \sum_{q=-k}^{k} \hat{T}_q^{(k)} D_{q,0}(0, -\pi, 0).$$

When the expansion above is inserted into $(\psi_{j,m}, \hat{U}_R^{\dagger} \hat{T}_0^{(k)} \hat{U}_R \psi_{j,m})$, because of the selection rule on the azimuthal quantum numbers ($q + m = m \implies q = 0$) from the Wigner–Eckart theorem, only the term with $q = 0$ contributes to the matrix element; using $D_{0,0}^{(k)}(0, -\pi, 0) = P_k(\cos \pi) = (-1)^k$ then leads to

$$(\psi_{j,-m}, \hat{T}_0^{(k)} \psi_{j,-m}) = D_{0,0}^{(k)}(0, -\pi, 0) (\psi_{j,m}, \hat{T}_0^{(k)} \psi_{j,m}) = (-1)^k (\psi_{j,m}, \hat{T}_0^{(k)} \psi_{j,m}),$$

and hence to the relation provided in the text of the problem. This relation must hold for any component of the ITO, since the Wigner–Eckart theorem gives

$$(\psi_{j,m'}, \hat{T}_q^{(k)} \psi_{j,m}) = C_{kj}(qm; jm') \langle \psi_j || \hat{T}^{(k)} || \psi_j \rangle = \frac{C_{kj}(qm; jm')}{C_{kj}(0m; jm)} (\psi_{j,m}, \hat{T}_0^{(k)} \psi_{j,m}),$$

and the order k of the ITO is in the range $0 \le k \le 2j$; otherwise, $(\psi_{j,m'}, T_q^{(k)} \psi_{j,m})$ would vanish. Finally, note that since $\hat{T}_0^{(k)}$ is even or odd under time reversal, this implies that

$$\hat{\Omega}_{\mathcal{T}} \hat{T}_{\pm 1}^{(k)} \hat{\Omega}_{\mathcal{T}}^{\dagger} = \hat{\Omega}_{\mathcal{T}} \frac{1}{\hbar\sqrt{k(k+1)}} [\hat{J}_\pm, \hat{T}_0^{(k)}] \hat{\Omega}_{\mathcal{T}}^{\dagger} = -\frac{1}{\hbar\sqrt{k(k+1)}} [\hat{J}_\mp, \hat{\Omega}_{\mathcal{T}} \hat{T}_0^{(k)} \hat{\Omega}_{\mathcal{T}}^{\dagger}]$$

$$= \mp \frac{1}{\hbar\sqrt{k(k+1)}} [\hat{J}_\mp, \hat{T}_0^{(k)}] = \mp \hat{T}_{\mp 1}^{(k)},$$

where we have used the property of ITOs $[\hat{J}_\pm, \hat{T}_q^{(k)}] = \sqrt{k(k+1) - q(q \pm 1)} \hbar \hat{T}_{q \pm 1}^{(k)}$ and $\hat{\Omega}_{\mathcal{T}} \hat{J}_\pm \hat{\Omega}_{\mathcal{T}}^{\dagger} = -\hat{J}_\mp$. We can proceed similarly to establish the transformation properties of all components q.

Part 3

The z-component of the position operator is an ITO of rank $k = 1$. Since $\hat{z}$ is even under time reversal, it follows that its expectation value, and hence that of $\hat{\mathbf{r}}$, vanishes, namely $\langle \psi_{jm} | \hat{\mathbf{r}} | \psi_{jm} \rangle = 0$. Of course, if $|\psi_{jm}\rangle$ is a parity eigenstate, this result also follows from symmetry under space inversion. However, it remains valid even when $|\psi_{jm}\rangle$ is *not* a parity eigenstate, as is the case under consideration here. Note that since c_{ns} and c_{np} are real, this ensures that under time reversal the liner combination with j, m transforms into $j, -m$ up to an overall phase factor i^{2m}.

So far, we have primarily dealt with isolated systems, which are described by time-independent Hamiltonians. A major concern has been the (approximate) determination of the eigenvalues and eigenstates of the Hamiltonian of such a system, by time-independent perturbation theory or by variational methods. In this chapter, we consider systems that are not isolated, but, rather, interact with external time-dependent fields (an example is an atom under the influence of an electromagnetic field). Since the Hamiltonian describing these systems is necessarily time dependent, there are no longer stationary states – states of definite energy. Hence, our interest here is, typically, to calculate the probability for a transition from one state to another. Only in the simplest cases can this probability be calculated exactly. We therefore find it necessary to develop approximate methods for dealing with this type of problem. One such method is time-dependent perturbation theory.

18.1 Perturbative Expansion for the Time Evolution Operator

We consider the Hamiltonian

$$\hat{H}(t) = \hat{H}_0 + \hat{V}(t) , \tag{18.1}$$

where the eigenvalues and eigenstates of (the time-independent) $\hat{H}_0$ are assumed to be known (if some of the eigenvalues are degenerate, then the corresponding eigenstates carry an additional superscript; this superscript is to be understood below):

$$\hat{H}_0|\phi_n\rangle = E_n|\phi_n\rangle . \tag{18.2}$$

Suppose that the perturbation $\hat{V}(t)$ is turned on at time $t = t_0$ and that the system is in eigenstate $|\phi_i\rangle$ of $\hat{H}_0$ at this time,

$$|\phi_i(t_0)\rangle = |\phi_i\rangle . \tag{18.3}$$

The probability that at a later time $t > t_0$ the system is in eigenstate $|\phi_f\rangle$ of $\hat{H}_0$ is given by

$$P_{fi}(t) = |\langle\phi_f|\phi_i(t)\rangle|^2 , \qquad |\phi_i(t)\rangle = \hat{U}(t, t_0)|\phi_i(t_0)\rangle , \tag{18.4}$$

where $\hat{U}(t, t_0)$ is the time evolution operator corresponding to $\hat{H}(t)$,

$$i\hbar\frac{d}{dt}\hat{U}(t, t_0) = \hat{H}(t)\,\hat{U}(t, t_0) , \qquad \hat{U}(t_0, t_0) = \hat{\mathbb{1}} . \tag{18.5}$$

If the perturbation were absent, the evolution operator would simply be

$$\hat{U}_0(t, t_0) = e^{-i\hat{H}_0(t-t_0)/\hbar} . \tag{18.6}$$

With the goal of developing a perturbative expansion in powers of $\hat{V}(t)$, we make the ansatz

$$\hat{U}(t, t_0) = \hat{U}_0(t, t_0)\,\hat{U}_I(t, t_0) , \tag{18.7}$$

and, if the perturbation is weak, $\hat{U}_I(t, t_0)$ is close to the identity $\hat{\mathbb{1}}$. Inserting this ansatz into Eq. (18.5) for $\hat{U}(t, t_0)$, we obtain

$$i\hbar\frac{d}{dt}\left[\hat{U}_0(t, t_0)\,\hat{U}_I(t, t_0)\right] = \underbrace{\left[i\hbar\frac{d}{dt}\hat{U}_0(t, t_0)\right]}_{\hat{H}_0\,\hat{U}_0(t, t_0)}\hat{U}_I(t, t_0) + \hat{U}_0(t, t_0)\left[i\hbar\frac{d}{dt}\hat{U}_I(t, t_0)\right]$$

$$= [\hat{H}_0 + \hat{V}(t)]\,\hat{U}_0(t, t_0)\,\hat{U}_I(t, t_0)\,. \qquad (18.8)$$

By simplifying terms with $\hat{H}_0$ and multiplying both sides on the left by $\hat{U}_0^\dagger(t, t_0)$, we are left with

$$i\hbar\frac{d}{dt}\hat{U}_I(t, t_0) = \hat{U}_0^\dagger(t, t_0)\,\hat{V}(t)\,\hat{U}_0(t, t_0)\,\hat{U}_I(t, t_0)\,, \qquad \hat{U}_I(t_0, t_0) = \hat{\mathbb{1}}\,. \qquad (18.9)$$

It is convenient to define

$$\hat{V}_I(t) = \hat{U}_0^\dagger(t, t_0)\,\hat{V}(t)\,\hat{U}_0(t, t_0)\,, \qquad (18.10)$$

which is known as the interaction-picture expression for the perturbation, and to convert the first-order differential equation into an integral equation,

$$\hat{U}_I(t, t_0) = \hat{\mathbb{1}} - \frac{i}{\hbar}\int_{t_0}^t dt'\,\hat{V}_I(t')\,\hat{U}_I(t', t_0)\,. \qquad (18.11)$$

This can be solved by iteration in the usual way to obtain

$$\hat{U}_I(t, t_0) = \hat{\mathbb{1}} - \frac{i}{\hbar}\int_{t_0}^t dt_1\,\hat{V}_I(t_1) + \left(-\frac{i}{\hbar}\right)^2\int_{t_0}^t dt_1\int_{t_0}^{t_1} dt_2\,\hat{V}_I(t_1)\,\hat{V}_I(t_2) + \cdots \qquad (18.12)$$

$$\cdots + \left(-\frac{i}{\hbar}\right)^n\int_{t_0}^t dt_1\int_{t_0}^{t_1} dt_2\cdots\int_{t_0}^{t_{n-1}} dt_n\,\hat{V}_I(t_1)\hat{V}_I(t_2)\cdots\hat{V}_I(t_n) + \cdots\,.$$

Inserting this expansion into Eq. (18.7) for $\hat{U}(t, t_0)$, we finally arrive at

$$\hat{U}(t, t_0) = \hat{U}_0(t, t_0)\left[\hat{\mathbb{1}} - \frac{i}{\hbar}\int_{t_0}^t dt_1\,\hat{V}_I(t_1) + \cdots\right] = \hat{U}_0(t, t_0) + \sum_{n=1}^\infty \hat{U}^{(n)}(t, t_0)\,, \qquad (18.13)$$

where the generic term reads

$$\hat{U}^{(n)}(t, t_0) = \left(-\frac{i}{\hbar}\right)^n\int_{t_0}^t dt_1\int_{t_0}^{t_1} dt_2\cdots\int_{t_0}^{t_{n-1}} dt_n\,\underbrace{\hat{U}_0(t, t_0)\,\hat{V}_I(t_1)\hat{V}_I(t_2)\cdots\hat{V}_I(t_n)}_{\text{product}}\,. \qquad (18.14)$$

Recalling the definition of interaction-picture operators, the operator product above can be written as

$$\text{product} = \underbrace{e^{-i\hat{H}_0(t-t_0)/\hbar}\,e^{i\hat{H}_0(t_1-t_0)/\hbar}}_{e^{-i\hat{H}_0(t-t_1)/\hbar}}\,\hat{V}(t_1)\underbrace{e^{-i\hat{H}_0(t_1-t_0)/\hbar}\,e^{i\hat{H}_0(t_2-t_0)/\hbar}}_{e^{-i\hat{H}_0(t_1-t_2)/\hbar}}\,\hat{V}(t_2)\,e^{-i\hat{H}_0(t_2-t_0)/\hbar}\cdots$$

$$\times\,e^{i\hat{H}_0(t_{n-1}-t_0)/\hbar}\,\hat{V}(t_{n-1})\underbrace{e^{-i\hat{H}_0(t_{n-1}-t_0)/\hbar}\,e^{i\hat{H}_0(t_n-t_0)/\hbar}}_{e^{-i\hat{H}_0(t_{n-1}-t_n)/\hbar}}\,\hat{V}(t_n)\,e^{-i\hat{H}_0(t_n-t_0)/\hbar}\,, \qquad (18.15)$$

which then leads to

$$\hat{U}^{(n)}(t, t_0) = \left(-\frac{i}{\hbar}\right)^n\int_{t_0}^t dt_1\int_{t_0}^{t_1} dt_2\cdots\int_{t_0}^{t_{n-1}} dt_n\,\hat{U}_0(t, t_1)\,\hat{V}(t_1)\,\hat{U}_0(t_1, t_2)\,\hat{V}(t_2)\cdots\hat{U}_0(t_{n-1}, t_n)\,\hat{V}(t_n)\,\hat{U}_0(t_n, t_0)\,.$$

$$(18.16)$$

We see that the nth term above corresponds (rightmost to leftmost) free propagation from t_0 to t_n, interaction at t_n, free propagation from t_n to t_{n-1}, interaction at t_{n-1}, and so on, followed by integration over the intermediate times $t_1, \ldots, t_n$ (note the integration limits). This formal expansion can be evaluated in practice by inserting complete sets of $\hat{H}_0$ eigenstates between subsequent interactions $\hat{V}(t_i)$.

The first-order approximation corresponds to keeping the first term in the expansion for $\hat{U}(t, t_0)$. The amplitude for the transition $i \longrightarrow f$ reads

$$
\begin{aligned}
\langle \phi_f | \phi_i(t) \rangle &= \langle \phi_f | \left[\hat{U}_0(t, t_0) - \frac{i}{\hbar} \int_{t_0}^{t} dt_1 \, \hat{U}_0(t, t_1) \, \hat{V}(t_1) \, \hat{U}_0(t_1, t_0) \right] | \phi_i(t_0) \rangle \\
&= e^{-iE_f(t-t_0)/\hbar} \, \delta_{if} - \frac{i}{\hbar} \int_{t_0}^{t} dt_1 \, e^{-iE_f(t-t_1)/\hbar} \underbrace{\langle \phi_f | \hat{V}(t_1) | \phi_i \rangle}_{V_{fi}(t_1)} e^{-iE_i(t_1-t_0)/\hbar} \\
&= e^{-iE_f(t-t_0)/\hbar} \left[\delta_{if} - \frac{i}{\hbar} \int_{t_0}^{t} dt_1 \, e^{i\omega_{fi}(t_1-t_0)} \, V_{fi}(t_1) \right].
\end{aligned}
\tag{18.17}
$$

Setting aside the case in which the initial and final states are the same ($i = f$), the probability (18.4) in first order is given by

$$
P_{fi}(t) = \frac{1}{\hbar^2} \left| \int_{t_0}^{t} dt_1 \, e^{i\omega_{fi}(t_1-t_0)} \, V_{fi}(t_1) \right|^2, \qquad \omega_{fi} = \frac{E_f - E_i}{\hbar},
\tag{18.18}
$$

that is, it is proportional to the square of the time integral of $V_{fi}(t)$.

18.2 Special Cases: Constant and Periodic Perturbations

For the case of a constant perturbation the amplitude for the transition $i \longrightarrow f$ is given by

$$
\int_{t_0}^{t} dt_1 \, e^{i\omega_{fi}(t_1-t_0)} \, V_{fi} = V_{fi} \left. \frac{e^{i\omega_{fi}(t_1-t_0)}}{i\omega_{fi}} \right|_{t_0}^{t} = V_{fi} \frac{e^{i\omega_{fi}(t-t_0)/2}}{\omega_{fi}/2} \sin\left[\frac{\omega_{fi}(t-t_0)}{2} \right],
\tag{18.19}
$$

and the probability is then obtained as

$$
P_{fi}(t) = \frac{|V_{fi}|^2}{\hbar^2} \frac{\sin^2[\omega_{fi}(t-t_0)/2]}{(\omega_{fi}/2)^2}.
\tag{18.20}
$$

The time dependence is contained in the function

$$
f(\tau, \omega) = \left[\frac{\sin(\omega\tau/2)}{\omega/2} \right]^2,
\tag{18.21}
$$

where we have defined $\tau = t - t_0$ and $\omega = \omega_{fi}$. This function is such that

$$
\lim_{\omega \to 0} f(\tau, \omega) = \tau^2, \qquad \int_{-\infty}^{\infty} d\omega \, f(\tau, \omega) = 2\pi\tau.
\tag{18.22}
$$

For large τ, the function is sharply peaked at $\omega = 0$ and has width $\approx 2\pi/\tau$; the area under the curve $f(\tau, \omega)$ as a function of ω is $2\pi\tau$; indeed, in the limit $\tau \longrightarrow \infty$ this function is a representation of the δ-function,

$$
\lim_{\tau \to \infty} f(\tau, \omega) = 2\pi\tau \, \delta(\omega).
\tag{18.23}
$$

Thus, for a given τ, the probability $P_{fi}(t)$ is largest for states $|\phi_f\rangle$ whose energies are within the interval $E_i - \pi\hbar/\tau$ and $E_i + \pi\hbar/\tau$, that is, the transition $i \longrightarrow f$ occurs preferentially toward states

$|\phi_f\rangle$ with energies in a band of width $2\pi\hbar/\tau$. Consequently, the unperturbed energy E_i is conserved to within $\pm\,\pi\hbar/\tau$. The transition probability has a maximal value of $(\tau|V_{fi}|/\hbar)^2$, and therefore, for the present perturbative treatment to be valid, we must have

$$\tau \ll \frac{\hbar}{|V_{fi}|} \, . \tag{18.24}$$

There are situations in which this condition is not satisfied (for example, in the case of the time evolution of a spin state in the presence of a strong constant magnetic field and a weak time-dependent magnetic field, discussed in Problem 20 in Chapter 12), and in such situations a different approximation must be used. However, assuming that the condition $0 \ll \tau \ll \hbar/|V_{fi}|$ is satisfied, we have

$$P_{fi}(t) = 2\pi(t - t_0)\,\delta(\omega_{fi})\,\frac{|V_{fi}|^2}{\hbar^2} \implies R_{fi} = \frac{d}{dt}P_{fi}(t) = \frac{2\pi}{\hbar}\,\delta(E_f - E_i)\,|V_{fi}|^2 \, , \tag{18.25}$$

and we have defined the (constant) rate R_{fi} as the probability per unit time; note that $\delta[(E_f - E_i)/\hbar]= \hbar\,\delta(E_i - E_f)$. The formula for the transition rate R_{fi} is known as Fermi's golden rule.

The case of a (hermitian) perturbation that oscillates in time according to

$$\hat{V}(t) = \hat{W}e^{-i\omega t} + \hat{W}^\dagger e^{i\omega t} \, , \tag{18.26}$$

where $\hat{W}$ is a time-independent operator and $\omega > 0$, is considered next. From Eq. (18.17) the first-order amplitude reads

$$\langle \phi_f | \phi_i(t) \rangle = e^{-iE_f(t-t_0)/\hbar}\left[\delta_{if} - \frac{i}{\hbar}\int_{t_0}^t dt_1\, e^{i\omega_{fi}(t_1 - t_0)}\, V_{fi}(t_1)\right] , \tag{18.27}$$

where

$$V_{fi}(t) = e^{-i\omega t}\langle \phi_f|\hat{W}|\phi_i\rangle + e^{i\omega t}\langle \phi_f|\hat{W}^\dagger|\phi_i\rangle = e^{-i\omega t}\,W_{fi} + e^{i\omega t}\,W_{fi}^\dagger \, . \tag{18.28}$$

We find for the transition probability ($i \neq f$)

$$P_{fi}(t) = \frac{1}{\hbar^2}\left|e^{-i\omega t_0}\,W_{fi}\int_{t_0}^t dt_1\, e^{i(\omega_{fi}-\omega)(t_1-t_0)} + e^{i\omega t_0}\,W_{fi}^\dagger\int_{t_0}^t dt_1\, e^{i(\omega_{fi}+\omega)(t_1-t_0)}\right|^2 . \tag{18.29}$$

Each of the two time integrals above is identical to that obtained in the case of a constant perturbation, except for the replacement $\omega_- = \omega_{fi} - \omega$ in the first integral and $\omega_+ = \omega_{fi} + \omega$ in the second integral. The probability can be expressed as

$$P_{fi}(t) = \frac{|W_{fi}|^2}{\hbar^2}\,f(\tau, \omega_-) + \frac{|W_{fi}^\dagger|^2}{\hbar^2}\,f(\tau, \omega_+) + \text{interference term} \, , \tag{18.30}$$

where $\tau = t - t_0$ and the interference term is given by

$$\text{interference term} = \frac{W_{fi}(W_{fi}^\dagger)^*}{\hbar^2}\,e^{-2i\omega t_0}\,\frac{e^{i\,\omega_-\tau/2}}{\omega_-/2}\,\frac{e^{-i\,\omega_+\tau/2}}{\omega_+/2}\,\sin\left(\frac{\omega_-\tau}{2}\right)\sin\left(\frac{\omega_+\tau}{2}\right) + \text{c.c.}$$

$$= \frac{W_{fi}(W_{fi}^\dagger)^*}{\hbar^2}\,e^{-i\omega(t+t_0)}\,g(\tau, \omega_-)\,g(\tau, \omega_+) + \text{c.c.} \, , \tag{18.31}$$

with

$$g(\tau, \omega_\pm) = \frac{\sin(\omega_\pm\tau/2)}{\omega_\pm/2} \, . \tag{18.32}$$

This function is such that $g(\tau, 0) = \tau$ with an integral over $\omega_\pm$ given by $\int_{-\infty}^{\infty} d\omega_\pm\, g(\tau, \omega_\pm) = 2\pi$; indeed, we have

$$\lim_{\tau \to \infty} g(\tau, \omega_\pm) = 2\pi\, \delta(\omega_\pm) . \tag{18.33}$$

We then see that in this limit the interference term vanishes, since it is proportional to the product $\delta(\omega_-)\,\delta(\omega_+)$, which cannot be simultaneously satisfied, and the transition probability reduces to the first two terms in Eq. (18.30). Using Eq. (18.23), we obtain for the transition rate

$$R_{fi} = \frac{2\pi}{\hbar}\left[|W_{fi}|^2\, \delta(E_f - E_i - \hbar\omega) + |W_{fi}^\dagger|^2\, \delta(E_f - E_i + \hbar\omega)\right] . \tag{18.34}$$

The positive-frequency ($\propto e^{-i\omega t}$) and negative-frequency ($\propto e^{+i\omega t}$) parts of the perturbation act independently, the first increasing and the second decreasing the initial energy of the system by $\hbar\omega$. However, it should be kept in mind that the treatment above is valid only in the regime of Eq. (18.24).

18.3 Problems

Problem 1 Time Evolution in a Two-State System

A box containing a particle is divided into right and left compartments by a thin partition. If the particle is known to be on the right side with certainty, the state is represented by the ket $|R\rangle$; similarly, if it is known to be on the left side with certainty, the state is represented by the ket $|L\rangle$. The most general state can then be written as

$$|\phi\rangle = |R\rangle\langle R|\phi\rangle + |L\rangle\langle L|\phi\rangle .$$

The particle can tunnel through the partition. The Hamiltonian describing this system is given by

$$\hat{H} = \Delta\left(|L\rangle\langle R| + |R\rangle\langle L|\right) ,$$

where the parameter Δ is a real number with dimensions of energy.

1. Find the energy eigenvalues and corresponding normalized eigenstates.
2. Suppose that the system at time $t = 0$ is in state $|\phi\rangle$, that is, $|\psi(0)\rangle = |\phi\rangle$. Obtain the state $|\psi(t)\rangle$ by applying the time evolution operator to $|\psi(0)\rangle$.
   Hint: it is convenient to expand $|\phi\rangle$ in the basis of eigenstates of $\hat{H}$.
3. Suppose that at $t = 0$ the particle is on the right side with certainty. What is the probability for observing the particle on the left side as a function of time?
4. Expand the state $|\psi(t)\rangle$ in the basis consisting of the states $|R\rangle$ and $|L\rangle$ and, using the time-dependent Schrödinger equation satisfied by $|\psi(t)\rangle$, obtain a set of coupled differential equations satisfied by the (time-dependent) expansion coefficients. Show that the solutions of this set of equations are just what would be expected from part 2.

Solution

Part 1

In the basis consisting of the states $|R\rangle$ and $|L\rangle$ the matrix representing the Hamiltonian is given by

$$\underline{H} = \begin{pmatrix} \langle R|\hat{H}|R\rangle & \langle R|\hat{H}|L\rangle \\ \langle L|\hat{H}|R\rangle & \langle L|\hat{H}|L\rangle \end{pmatrix} = \begin{pmatrix} 0 & \Delta \\ \Delta & 0 \end{pmatrix} ,$$

and the eigenvalues follow from

$$\det(\underline{H} - E\,\underline{I}) = E^2 - \Delta^2 = 0 \implies E_\pm = \pm\Delta .$$

We denote the corresponding eigenstates as

$$|\varphi_\pm\rangle = c_R^{(\pm)} |R\rangle + c_L^{(\pm)} |L\rangle ,$$

and the coefficients satisfy

$$\begin{pmatrix} -E_\pm & \Delta \\ \Delta & -E_\pm \end{pmatrix} \begin{pmatrix} c_R^{(\pm)} \\ c_L^{(\pm)} \end{pmatrix} = \Delta \begin{pmatrix} \mp 1 & 1 \\ 1 & \mp 1 \end{pmatrix} \begin{pmatrix} c_R^{(\pm)} \\ c_L^{(\pm)} \end{pmatrix} = 0 \implies c_L^{(\pm)} = \pm c_R^{(\pm)} .$$

The normalized eigenstates read (up to irrelevant phase factors)

$$|\varphi_\pm\rangle = \frac{1}{\sqrt{2}} |R\rangle \pm \frac{1}{\sqrt{2}} |L\rangle .$$

Part 2

It is convenient to express the state $|\phi\rangle$ as a linear combination of the eigenstates $|\varphi_\pm\rangle$, since $\hat{H}$ is diagonal in this basis; from

$$|R\rangle = \frac{1}{\sqrt{2}} |\varphi_+\rangle + \frac{1}{\sqrt{2}} |\varphi_-\rangle , \qquad |L\rangle = \frac{1}{\sqrt{2}} |\varphi_+\rangle - \frac{1}{\sqrt{2}} |\varphi_-\rangle ,$$

we obtain

$$|\psi(0)\rangle = c_R(0)|R\rangle + c_L(0)|L\rangle = \underbrace{\frac{c_R(0) + c_L(0)}{\sqrt{2}}}_{c_+(0)} |\varphi_+\rangle + \underbrace{\frac{c_R(0) - c_L(0)}{\sqrt{2}}}_{c_-(0)} |\varphi_-\rangle ,$$

where we have defined

$$c_R(0) = \langle R|\psi(0)\rangle = \langle R|\phi\rangle , \qquad c_L(0) = \langle L|\psi(0)\rangle = \langle L|\phi\rangle .$$

The time-evolved state follows from

$$|\psi(t)\rangle = e^{-i\hat{H}t/\hbar} |\psi(0)\rangle = c_+(0)\, e^{-iE_+ t/\hbar} |\varphi_+\rangle + c_-(0)\, e^{-iE_- t/\hbar} |\varphi_-\rangle ,$$

which in the original basis reads

$$|\psi(t)\rangle = \frac{1}{\sqrt{2}} \left[c_+(0)\, e^{-it\Delta/\hbar} + c_-(0)\, e^{it\Delta/\hbar} \right] |R\rangle + \frac{1}{\sqrt{2}} \left[c_+(0)\, e^{-it\Delta/\hbar} - c_-(0)\, e^{it\Delta/\hbar} \right] |L\rangle .$$

Inserting the expressions for $c_\pm(0)$ also yields

$$|\psi(t)\rangle = \underbrace{[c_R(0)\, \cos(t\Delta/\hbar) - i\, c_L(0)\, \sin(t\Delta/\hbar)]}_{c_R(t)} |R\rangle + \underbrace{[c_L(0)\, \cos(t\Delta/\hbar) - ic_R(0)\, \sin(t\Delta/\hbar)]}_{c_L(t)} |L\rangle.$$

Part 3

If the particle is on the right side at $t=0$ then $c_R(0)=1$ and $c_L(0)=0$, which leads to

$$|\psi(t)\rangle = \cos(t\Delta/\hbar)\,|R\rangle - i\,\sin(t\Delta/\hbar)\,|L\rangle\,,$$

and the probability that the particle will be on the left side at time t follows from

$$\mathcal{P}_L(t) = |\langle L|\psi(t)\rangle|^2 = \sin^2(t\Delta/\hbar)\,;$$

it is unity at times such that $t\Delta_n/\hbar = (2n+1)\pi/2$ with $n = 0, 1, 2, \ldots$.

Part 4

Starting from the time-dependent Schrödinger equation for the state $|\psi(t)\rangle$,

$$i\hbar\frac{d}{dt}\,|\psi(t)\rangle = \hat{H}\,|\psi(t)\rangle\,,$$

we expand $|\psi(t)\rangle$ in the basis $|R\rangle$ and $|L\rangle$ as

$$|\psi(t)\rangle = c_R(t)|R\rangle + c_L(t)|L\rangle\,,$$

so that

$$i\hbar\,\dot{c}_R(t)|R\rangle + i\hbar\,\dot{c}_L(t)|L\rangle = c_R(t)\,\hat{H}\,|R\rangle + c_L(t)\,\hat{H}\,|L\rangle\,.$$

We now project out the above on the $|R\rangle$ and $|L\rangle$ states, namely $\langle R|\text{l.h.s.}\rangle = \langle R|\text{r.h.s.}\rangle$ and similarly for $|L\rangle$, to obtain a linear system of first-order differential equations

$$i\hbar\,\dot{c}_R(t) = \Delta\,c_L(t)\,, \qquad i\hbar\,\dot{c}_L(t) = \Delta\,c_R(t)\,,$$

with initial conditions $c_R(0)$ and $c_L(0)$. To solve them, take the time derivative of, say, the first equation:

$$\ddot{c}_R(t) = -i\frac{\Delta}{\hbar}\,\dot{c}_L(t) \implies \ddot{c}_R(t) = -\frac{\Delta^2}{\hbar^2}\,c_R(t) \qquad \text{with} \qquad \dot{c}_R(0) = -i\frac{\Delta}{\hbar}\,c_L(0)\,.$$

We find

$$c_R(t) = A\,\cos(t\Delta/\hbar) + B\,\sin(t\Delta/\hbar) \implies c_R(t) = c_R(0)\,\cos(t\Delta/\hbar) - i c_L(0)\,\sin(t\Delta/\hbar)\,,$$

where the last expression follows from imposing the initial conditions $A = c_R(0)$ and $B(\Delta/\hbar) = -i(\Delta/\hbar)c_L(0)$. We obtain $c_L(t)$ by noting that

$$c_L(t) = i\frac{\hbar}{\Delta}\,\dot{c}_R(t) = i\frac{\hbar}{\Delta}\left[-\frac{\Delta}{\hbar}\,c_R(0)\,\sin(t\Delta/\hbar) - i\frac{\Delta}{\hbar}\,c_L(0)\,\cos(t\Delta/\hbar)\right]\,,$$

or

$$c_L(t) = c_L(0)\,\cos(t\Delta/\hbar) - i c_R(0)\,\sin(t\Delta/\hbar)\,.$$

These relations are of course in agreement with those derived in part 2.

Problem 2   Spin-1 System Perturbed by an Oscillating Field

Consider a spin-1 system and denote by $|m\rangle$ the eigenstates of $\mathbf{S}^2$ and S_z, with $\mathbf{S}^2|m\rangle = 2\hbar^2|m\rangle$ and $S_z|m\rangle = m\hbar|m\rangle$. In the basis $|m\rangle$, the unperturbed Hamiltonian of the system is represented by the following diagonal matrix:

$$\underline{H}_0 = \begin{pmatrix} E_1 & 0 & 0 \\ 0 & E_0 & 0 \\ 0 & 0 & E_{-1} \end{pmatrix},$$

where $E_1 > E_0 > E_{-1}$. The system is subjected to a rotating magnetic field in the xy-plane and the perturbation describing the interaction with this field is given by

$$V(t) = \gamma \left[\cos(\omega t)\, S_x + \sin(\omega t)\, S_y \right],$$

where ω is the angular frequency of the rotating field and γ is a constant proportional to the product of the field amplitude and the system's magnetic moment.

1. Expand a generic state $|\psi(t)\rangle$ of the system as follows:

$$|\psi(t)\rangle = \sum_{m=0,\pm 1} c_m(t)\, e^{-iE_m t/\hbar} |m\rangle,$$

and obtain a set of differential equations for the coefficients $c_m(t)$ describing the exact time evolution of the state.

2. Obtain a perturbative expansion for $c_m(t)$ in powers of γ by positing

$$c_m(t) = \sum_{p=0}^{\infty} c_m^{(p)}(t)\, \gamma^p.$$

Assume that at time $t = 0$ the system is in the state $|-1\rangle$. Can the transition to the state $|+1\rangle$ proceed in first order? Justify your answer.

Solution

Part 1

The time-dependent Schrödinger equation is given by

$$i\hbar \frac{d}{dt}|\psi(t)\rangle = [H_0 + V(t)]\,|\psi(t)\rangle,$$

where

$$i\hbar \frac{d}{dt}|\psi(t)\rangle = \sum_m \left[i\hbar\,\dot{c}_m(t)\, e^{-iE_m t/\hbar} + E_m c_m(t)\, e^{-iE_m t/\hbar} \right] |m\rangle.$$

Inserting the latter expression into the left-hand side of the Schrödinger equation, we find

$$\sum_m \left[i\hbar\,\dot{c}_m(t)\, e^{-iE_m t/\hbar} + E_m c_m(t)\, e^{-iE_m t/\hbar} \right] |m\rangle = \sum_m c_m\, e^{-iE_m t/\hbar} [H_0 + V(t)] |m\rangle$$

$$= \sum_m c_m\, e^{-iE_m t/\hbar} [E_m + V(t)] |m\rangle,$$

where we have used $H_0|m\rangle = E_m|m\rangle$. Simplifying terms, we obtain

$$\sum_m i\hbar\,\dot{c}_m(t)\, e^{-iE_m t/\hbar} |m\rangle = \sum_m c_m(t)\, e^{-iE_m t/\hbar}\, V(t)|m\rangle.$$

We project both sides of the above relation onto state $\langle n|$ to find

$$i\hbar\, \dot{c}_n(t) = \sum_m c_m(t)\, e^{i(E_n - E_m)t/\hbar}\, \langle n|\, V(t)\, |m\rangle\, .$$

In terms of the spin raising and lowering operators $S_\pm$, the perturbation is more conveniently expressed as

$$V(t) = \gamma \left(\frac{e^{i\omega t} + e^{-i\omega t}}{2}\, \frac{S_+ + S_-}{2} + \frac{e^{i\omega t} - e^{-i\omega t}}{2i}\, \frac{S_+ - S_-}{2i} \right) = \frac{\gamma}{2}\left(e^{-i\omega t} S_+ + e^{i\omega t} S_- \right)\, .$$

Evaluation of the matrix element of the perturbation is now straightforward,

$$\langle n|\, V(t)\, |m\rangle = \frac{\gamma}{2}\left(e^{-i\omega t}\, \langle n|\, J_+\, |m\rangle + e^{i\omega t}\, \langle n|\, J_-\, |m\rangle \right)$$

$$= \frac{\hbar\gamma}{2}\left(e^{-i\omega t}\, \sqrt{2 - m(m+1)}\, \delta_{n,m+1} + e^{i\omega t}\, \sqrt{2 - m(m-1)}\, \delta_{n,m-1} \right)\, ,$$

which yields

$$i\dot{c}_n(t) = \frac{\gamma}{2}\left(\sqrt{2 - n(n-1)}\, e^{i(E_n - E_{n-1})t/\hbar}\, e^{-i\omega t}\, c_{n-1}(t) + \sqrt{2 - n(n+1)}\, e^{i(E_n - E_{n+1})t/\hbar}\, e^{i\omega t}\, c_{n+1}(t) \right)\, ,$$

or

$$i\dot{c}_1(t) = \frac{\gamma}{\sqrt{2}}\, e^{i(\omega_0 - \omega)t}\, c_0(t)\, , \qquad i\dot{c}_0(t) = \frac{\gamma}{\sqrt{2}}\left[e^{i(\omega_0' - \omega)t}\, c_{-1}(t) + e^{i(-\omega_0 + \omega)t}\, c_1(t) \right]\, , \qquad i\dot{c}_{-1}(t) = \frac{\gamma}{\sqrt{2}}\, e^{i(-\omega_0' + \omega)t}\, c_0(t)\, ,$$

where we have defined the Bohr frequencies

$$\omega_0 = \frac{E_1 - E_0}{\hbar}\, , \qquad \omega_0' = \frac{E_0 - E_{-1}}{\hbar}\, .$$

By introducing the parameters

$$\Delta\omega = \omega_0 - \omega\, , \qquad \Delta\omega' = \omega_0' - \omega\, ,$$

the linear set of differential equations above can be expressed as

$$i\dot{c}_1(t) = \frac{\gamma}{\sqrt{2}}\, e^{i\Delta\omega t}\, c_0(t)\, , \qquad i\dot{c}_0(t) = \frac{\gamma}{\sqrt{2}}\left[e^{i\Delta\omega' t}\, c_{-1}(t) + e^{-i\Delta\omega t}\, c_1(t) \right]\, , \qquad i\dot{c}_{-1}(t) = \frac{\gamma}{\sqrt{2}}\, e^{-i\Delta\omega' t}\, c_0(t)\, .$$

Part 2

We develop a perturbative expansion for the coefficients $c_n(t)$ by expanding them in powers of the coupling constant γ, that is,

$$c_n(t) = \sum_{p=0}^{\infty} c_n^{(p)}(t)\gamma^p\, .$$

Insert the expansion above in the equation for the $\dot{c}_n(t)$ and equate powers of γ to find, for $p \geq 1$,

$$i\dot{c}_1^{(p)}(t) = \frac{e^{i\Delta\omega t}}{\sqrt{2}}\, c_0^{(p-1)}(t)\, , \qquad i\dot{c}_0^{(p)}(t) = \frac{e^{i\Delta\omega' t}}{\sqrt{2}}\, c_{-1}^{(p-1)}(t) + \frac{e^{-i\Delta\omega t}}{\sqrt{2}}\, c_1^{(p-1)}(t)\, , \qquad i\dot{c}_{-1}^{(p)}(t) = \frac{e^{-i\Delta\omega' t}}{\sqrt{2}}\, c_0^{(p-1)}(t)\, .$$

The initial condition that at time $t=0$ the system is in state $|-1\rangle$ requires that the leading-order coefficients are given by

$$c_{-1}^{(0)}(t) = 1\, , \qquad c_0^{(0)}(t) = c_1^{(0)}(t) = 0\, .$$

In first order ($p = 1$), we obtain

$$c_1^{(1)}(t) = c_{-1}^{(1)}(t) = 0 \,, \qquad i\dot{c}_0^{(1)}(t) = \frac{e^{i\Delta\omega' t}}{\sqrt{2}} \implies c_0^{(1)}(t) = \frac{1 - e^{i\Delta\omega' t}}{\sqrt{2}\,\Delta\omega'} \,,$$

and the state $|\psi(t)\rangle$ reads

$$|\psi(t)\rangle = e^{-iE_{-1}t/\hbar}\,|-1\rangle + \frac{\gamma}{\sqrt{2}}\,\frac{1 - e^{i\Delta\omega' t}}{\Delta\omega'}\,e^{-iE_0 t/\hbar}\,|0\rangle + \cdots \,;$$

$\langle 1|\psi(t)\rangle = 0$ and no transitions are possible (in first order) to $|1\rangle$. This is obvious from the form of the perturbation, which can connect the state $|m\rangle$ only to states $|m \pm 1\rangle$ (in first order).

Problem 3   Alternative Derivation of Time-Dependent Perturbation Theory

Let $\hat{V}(t)$ be a time-dependent perturbation and $|\psi(t)\rangle$ the state satisfying the time-dependent Schrödinger equation corresponding to the Hamiltonian $\hat{H}_0 + \hat{V}(t)$ with initial condition $|\psi(t_0)\rangle = |\phi_k\rangle$, where $|\phi_k\rangle$ is the eigenstate of $\hat{H}_0$ of energy E_k. Expanding $|\psi(t)\rangle$ (at time t) in the basis of $\hat{H}_0$ eigenstates as $|\psi(t)\rangle = \sum_n c_n(t)\, e^{-iE_n t/\hbar}\,|\phi_n\rangle$ (the factor $e^{-iE_n t/\hbar}$ is included for convenience), obtain a perturbative expansion for the $c_n(t)$. Calculate the corrections up to second order in $\hat{V}(t)$, and show that they are in accordance with those obtained in Section 18.1.

Solution

We have

$$i\hbar\frac{\partial}{\partial t}|\psi(t)\rangle = \sum_n e^{-iE_n t/\hbar}\,[i\hbar\dot{c}_n(t) + E_n c_n(t)]\,|\phi_n\rangle$$

$$= \sum_n e^{-iE_n t/\hbar}\,c_n(t)\,\left[\hat{H}_0 + \lambda\hat{V}(t)\right]|\phi_n\rangle = \sum_n e^{-iE_n t/\hbar}\,c_n(t)\,\left[E_n + \lambda\hat{V}(t)\right]|\phi_n\rangle \,,$$

where we have multiplied the perturbation by the real parameter λ (with the understanding that $\lambda \longrightarrow 1$ at the end). After projecting onto the unperturbed eigenstate $|\phi_m\rangle$, the equation above reduces to

$$\dot{c}_m(t) = -\frac{i}{\hbar}\sum_n e^{i(E_m - E_n)t/\hbar}\,c_n(t)\,\underbrace{\langle\phi_m|\lambda\hat{V}(t)|\phi_n\rangle}_{\lambda V_{mn}(t)} \,,$$

and (for $\lambda = 1$) this is an exact rewriting of the time-dependent Schrödinger equation. Now, assume that the $c_n(t)$ can be expanded in a power series in λ as $\sum_{p=0}^{\infty}\lambda^p c_n^{(p)}(t)$. Inserting this expansion into the coupled differential equations for the $c_n(t)$, we find

$$\sum_{p=0}^{\infty}\dot{c}_m^{(p)}(t)\lambda^p = -\frac{i}{\hbar}\sum_n e^{i(E_m - E_n)t/\hbar}\,V_{mn}(t)\sum_{p=0}^{\infty}\lambda^{p+1} c_n^{(p)}(t) \,,$$

and matching powers of λ on the left- and right-hand sides yields

$$\dot{c}_m^{(0)}(t) = 0 \,, \qquad \dot{c}_m^{(p)}(t) = -\frac{i}{\hbar}\sum_n e^{i(E_m - E_n)t/\hbar}\,V_{mn}(t)\,c_n^{(p-1)}(t) \qquad \text{for } p \geq 1 \,,$$

which can be rewritten, by integrating both sides with respect to time, as

$$c_m^{(0)}(t) = c_m^{(0)}(t_0) \,, \qquad c_m^{(p)}(t) = c_m^{(p)}(t_0) - \frac{i}{\hbar} \sum_n \int_{t_0}^t dt' \, e^{i(E_m - E_n)t'/\hbar} \, V_{mn}(t') \, c_n^{(p-1)}(t') \,.$$

At time t_0, when the perturbation $\hat{V}(t)$ is turned on, the system is assumed to be in the eigenstate $|\phi_k\rangle$ of $\hat{H}_0$, so that

$$|\psi(t_0)\rangle = \sum_n e^{-iE_n t_0/\hbar} c_n(t_0)|\phi_n\rangle = |\phi_k\rangle \implies c_n(t_0) = \delta_{nk} e^{iE_k t_0/\hbar} \implies c_n^{(0)}(t_0) = \delta_{nk} e^{iE_k t_0/\hbar} \text{ and } c_n^{(p\geq 1)}(t_0) = 0 \,.$$

Hence we find the solutions (order-by-order)

$$c_m^{(0)}(t) = \delta_{mk} e^{iE_k t_0/\hbar}$$

$$c_m^{(1)}(t) = (-i/\hbar) \sum_n \int_{t_0}^t dt_1 \, e^{i(E_m - E_n)t_1/\hbar} \, V_{mn}(t_1) \, c_n^{(0)}(t_1) = \left[(-i/\hbar) \int_{t_0}^t dt_1 \, e^{i(E_m - E_k)t_1/\hbar} \, V_{mk}(t_1) \right] e^{iE_k t_0/\hbar} \,,$$

$$c_m^{(2)}(t) = (-i/\hbar) \sum_n \int_{t_0}^t dt_1 \, e^{i(E_m - E_n)t_1/\hbar} \, V_{mn}(t_1) \, c_n^{(1)}(t_1)$$

$$= \left[(-i/\hbar)^2 \sum_n \int_{t_0}^t dt_1 \, e^{i(E_m - E_n)t_1/\hbar} \, V_{mn}(t_1) \int_{t_0}^{t_1} dt_2 \, e^{i(E_n - E_k)t_2/\hbar} \, V_{nk}(t_2) \right] e^{iE_k t_0/\hbar} \,,$$

and so on. The perturbative expansion for the state $|\psi(t)\rangle$ is then given by

$$|\psi(t)\rangle = \sum_m e^{-iE_m t/\hbar} c_m(t)|\phi_m\rangle = \sum_m e^{-iE_m t/\hbar} e^{iE_k t_0/\hbar} \left[\delta_{mk} - \frac{i}{\hbar} \int_{t_0}^t dt_1 \, e^{i(E_m - E_k)t_1/\hbar} \, V_{mk}(t_1) \right.$$

$$\left. + \left(-\frac{i}{\hbar} \right)^2 \sum_n \int_{t_0}^t dt_1 \, e^{i(E_m - E_n)t_1/\hbar} \, V_{mn}(t_1) \int_{t_0}^{t_1} dt_2 \, e^{i(E_n - E_k)t_2/\hbar} \, V_{nk}(t_2) + \cdots \right] |\phi_m\rangle \,.$$

Given the initial condition $|\psi(t_0)\rangle = |\phi_k\rangle$ and using Eq. (18.13) we have (up to second order)

$$|\psi(t)\rangle = \sum_m |\phi_m\rangle\langle\phi_m|\hat{U}(t, t_0)|\phi_k\rangle = \sum_m |\phi_m\rangle\langle\phi_m|\hat{U}_0(t, t_0) + \hat{U}^{(1)}(t, t_0) + \hat{U}^{(2)}(t, t_0) + \cdots |\phi_k\rangle$$

which after insertion of Eq. (18.16) leads to the expression obtained above by noting that, for example,

$$\langle\phi_m|\hat{U}_0(t, t_0)|\phi_k\rangle = e^{-iE_k(t - t_0)/\hbar} \, \delta_{mk}$$

and

$$\langle\phi_m|\hat{U}^{(1)}(t, t_0)|\phi_k\rangle = -\frac{i}{\hbar} \int_{t_0}^t dt_1 \, \langle\phi_m|\hat{U}_0(t, t_1)\hat{V}(t_1)\hat{U}_0(t_1, t_0)|\phi_k\rangle$$

$$= -\frac{i}{\hbar} \int_{t_0}^t dt_1 \, \underbrace{e^{-iE_m(t - t_1)/\hbar} \, V_{mk}(t_1) \, e^{-iE_k(t_1 - t_0)/\hbar}}_{e^{-iE_m t/\hbar} \, V_{mk}(t_1) \, e^{i(E_m - E_k)t_1/\hbar} \, e^{iE_k t_0/\hbar}} \,.$$

The matrix element of $\langle\phi_m|\hat{U}^{(2)}(t, t_0)|\phi_k\rangle$ can be worked out similarly by inserting the completeness relation $\sum_n |\phi_n\rangle\langle\phi_n|$ between $\hat{V}(t_1)$ and $\hat{V}(t_2)$.

Problem 4   Elastic Scattering Cross Section in Born Approximation
from Fermi's Golden Rule

Use Fermi's golden rule to derive the cross section for elastic scattering in the Born approximation.

Solution

The unperturbed Hamiltonian consists of the free-particle Hamiltonian $\hat{H}_0 = \hat{\mathbf{p}}^2/(2m)$. We are interested in the transition induced by a (time-independent) interaction $\hat{V}$ from an initial free-particle state $|\phi_{\mathbf{p}_i}\rangle$ of energy $E_i = \mathbf{p}_i^2/(2m)$ to a final free-particle state $|\phi_{\mathbf{p}_f}\rangle$ of energy $E_f = \mathbf{p}_f^2/(2m)$; both are eigenstates of $\hat{H}_0$ and their energies are infinitely degenerate. We assume here that these states have the continuum normalization $\langle \phi_{\mathbf{p}'} | \phi_{\mathbf{p}} \rangle = \delta(\mathbf{p} - \mathbf{p}')$. The transition rate according to Fermi's golden rule is given by

$$R_{\mathbf{p}_f, \mathbf{p}_i} = \frac{2\pi}{\hbar} \delta(E_f - E_i) \, |\langle \phi_{\mathbf{p}_f} | \hat{V} | \phi_{\mathbf{p}_i} \rangle|^2 \qquad \text{with} \qquad \mathbf{p}_i \neq \mathbf{p}_f \, ,$$

and we are excluding forward scattering (recall that in the derivation of Fermi's golden rule we assumed $i \neq f$). We are dealing here with continuum states and therefore, rather than the rate into a specific final state $|\phi_{\mathbf{p}_f}\rangle$, we are interested in the rate into a set of states with momenta in a small shell of volume $\Delta \mathbf{p}_f$ centered at $\mathbf{p}_f$, namely

$$
\begin{aligned}
\int_{\Delta \mathbf{p}_f} d\mathbf{p}_f \, R_{\mathbf{p}_f, \mathbf{p}_i} &= \frac{2\pi}{\hbar} \int_{\Delta \mathbf{p}_f} d\Omega_f \, dp_f \, p_f^2 \, \delta(E_f - E_i) \, |\langle \phi_{\mathbf{p}_f} | \hat{V} | \phi_{\mathbf{p}_i} \rangle|^2 \\
&= \frac{2\pi}{\hbar} \int_{\Delta \mathbf{p}_f} d\Omega_f \, dE_f \, m \sqrt{2mE_f} \, \delta(E_f - E_i) \, |\langle \phi_{\mathbf{p}_f} | \hat{V} | \phi_{\mathbf{p}_i} \rangle|^2 \\
&= \frac{2\pi}{\hbar} \int_{\Delta \Omega_f} d\Omega_f \, m \sqrt{2mE_i} \, |\langle \phi_{\mathbf{p}_f} | \hat{V} | \phi_{\mathbf{p}_i} \rangle|^2 \, ,
\end{aligned}
$$

where we have used $p_f \, dp_f = m \, dE_f$ and have integrated out the δ-function enforcing $|\mathbf{p}_f| = |\mathbf{p}_i|$ for the momenta of the final states. The (differential) cross section $\sigma(\Omega_f)$ follows from

$$\int_{\Delta \Omega_f} d\Omega_f \, \sigma(\Omega_f) = \frac{\int_{\Delta \mathbf{p}_f} d\mathbf{p}_f \, R_{\mathbf{p}_f, \mathbf{p}_i}}{|\mathbf{j}_i|} \, ,$$

where the incident flux $\mathbf{j}_i$ reads (recall that with the continuum normalization we have adopted here, the free-particle wave function of momentum $\mathbf{p}$ is given by $\phi_{\mathbf{p}}(\mathbf{r}) = (2\pi\hbar)^{-3/2} \, e^{i\mathbf{p}\cdot\mathbf{r}}$)

$$\mathbf{j}_i = \frac{\hbar}{2mi} \left[\psi_{\mathbf{p}_i}^*(\mathbf{r}) \boldsymbol{\nabla} \psi_{\mathbf{p}_i}(\mathbf{r}) - \text{c.c.} \right] = \frac{1}{(2\pi\hbar)^3} \frac{\mathbf{p}_i}{m} \, .$$

By comparing the integrands with respect to $d\Omega_f$ of both sides in the relation above, we finally obtain

$$
\begin{aligned}
\sigma(\Omega_f) &= \frac{(2\pi\hbar)^3 \, m}{p_i} \frac{2\pi}{\hbar} \, m \, \underbrace{\sqrt{2mE_i}}_{p_i} \, |\langle \phi_{\mathbf{p}_f} | \hat{V} | \phi_{\mathbf{p}_i} \rangle|^2 \\
&= (2\pi)^4 \, (m\hbar)^2 \left| \int d\mathbf{r} \, \frac{1}{(2\pi\hbar)^{3/2}} \, e^{-i\mathbf{p}_f \cdot \mathbf{r}/\hbar} \, V(\mathbf{r}) \, \frac{1}{(2\pi\hbar)^{3/2}} \, e^{i\mathbf{p}_i \cdot \mathbf{r}/\hbar} \right|^2 \\
&= \left(\frac{m}{2\pi\hbar^2} \right)^2 \left| \int d\mathbf{r} \, e^{-i\mathbf{q}\cdot\mathbf{r}} \, V(\mathbf{r}) \right|^2 \, , \qquad \mathbf{q} = (\mathbf{p}_f - \mathbf{p}_i)/\hbar \, ,
\end{aligned}
$$

which, as expected, reproduces the scattering cross section we obtained previously using the Born approximation.

Problem 5 Positronium in Static and Oscillating Magnetic Fields

Consider positronium, a bound system consisting of an electron and a positron (the electron's antiparticle having the same mass and spin, but opposite charge). In the presence of a uniform and static magnetic field $\mathbf{B} = B\,\hat{\mathbf{z}}$ and ignoring orbital degrees of freedom, the Hamiltonian describing this system reads

$$H = \underbrace{\frac{\alpha}{\hbar^2}\,\mathbf{S}_1 \cdot \mathbf{S}_2}_{H_0} + \underbrace{\frac{\mu B}{\hbar}\,(S_{1z} - S_{2z})}_{V}\,,$$

where $\mathbf{S}_1$ and $\mathbf{S}_2$ are the electron and positron spin-1/2 operators, respectively, α is a constant with dimensions of energy, and

$$\mu = \frac{e\hbar}{mc}\,;$$

$-e$ $(+e)$ and m are the charge and mass of the electron (positron). Of course, any component $\mathbf{S}_1$ commutes with any component of $\mathbf{S}_2$.

1. Define the total spin operator $\mathbf{S} = \mathbf{S}_1 + \mathbf{S}_2$. Do $\mathbf{S}^2$ and S_z commute with H? Justify your answer by evaluating the relevant commutators.

2. Work in the basis of eigenstates $|SM_S\rangle$ of $\mathbf{S}^2$ and S_z and obtain the exact energy eigenvalues of the full Hamiltonian H by diagonalizing

$$\underline{H} = \begin{pmatrix} \langle 00|H|00\rangle & \langle 00|H|10\rangle & \langle 00|H|11\rangle & \langle 00|H|1-1\rangle \\ \langle 10|H|00\rangle & \langle 10|H|10\rangle & \langle 10|H|11\rangle & \langle 10|H|1-1\rangle \\ \langle 11|H|00\rangle & \langle 11|H|10\rangle & \langle 11|H|11\rangle & \langle 11|H|1-1\rangle \\ \langle 1-1|H|00\rangle & \langle 1-1|H|10\rangle & \langle 1-1|H|11\rangle & \langle 1-1|H|1-1\rangle \end{pmatrix}.$$

3. Regard H_0 as the unperturbed Hamiltonian and V as the perturbation, and use time-independent perturbation theory to obtain the **leading** (i.e., non-vanishing) corrections to the energies.

   Hint: To evaluate second-order corrections in degenerate perturbation theory, one needs to diagonalize the operator

$$Z = \sum_{k \neq \mathbb{E}_n} V \frac{|\phi_k\rangle\langle\phi_k|}{\epsilon_n - \epsilon_k}\,V\,,$$

   where the sum over the eigenstates $|\phi_k\rangle$ of H_0 (with eigenvalues ϵ_k) excludes those eigenstates in the degenerate subspace $\mathbb{E}_n$.

4. Expand the results in part 2 in powers of $\mu_0 B/\alpha$, and show that they agree with those of part 3. Therefore, up to and including first-order corrections, show that the approximate eigenstates of the Hamiltonian H are the states $|1 \pm 1\rangle$ with eigenenergy $E_0 = \alpha/4$, and also

$$|1\rangle = |10\rangle + \frac{\mu B}{\alpha}\,|00\rangle\,, \qquad |2\rangle = |00\rangle - \frac{\mu B}{\alpha}\,|10\rangle\,,$$

with eigenenergies given by, respectively,

$$E_1 = \frac{\alpha}{4} + \frac{(\mu B)^2}{\alpha}\,, \qquad E_2 = -\frac{3\alpha}{4} - \frac{(\mu B)^2}{\alpha}\,.$$

5. Suppose we introduce an oscillating magnetic field $\mathbf{B}_0 \cos(\omega t)$ (in addition to the static field $B\,\hat{\mathbf{z}}$) with the intention of inducing a transition between the states $|1\rangle$ and $|2\rangle$ via stimulated emission (so that $|1\rangle \longrightarrow |2\rangle$) or absorption (so that $|2\rangle \longrightarrow |1\rangle$). The electron and positron have magnetic moments given by, respectively,

$$\boldsymbol{\mu}_- = -\frac{\mu}{\hbar}\,\mathbf{S}_1\,, \qquad \boldsymbol{\mu}_+ = \frac{\mu}{\hbar}\,\mathbf{S}_2\,,$$

so that the interaction with the oscillating field reads

$$V(t) = \frac{\mu}{\hbar}\,\mathbf{B}_0 \cdot (\mathbf{S}_1 - \mathbf{S}_2)\,\cos(\omega t)\,.$$

To answer the following questions, use first-order time-dependent perturbation theory.

a. Should the oscillating field be directed along the $\hat{\mathbf{x}}$-(or $\hat{\mathbf{y}}$-)axis or along the $\hat{\mathbf{z}}$-axis? Justify your answer.

b. Assume $\mathbf{B}_0 = B_0\hat{\mathbf{z}}$. Obtain the transition rate for the absorption $|2\rangle \longrightarrow |1\rangle$.

Solution

Part 1

We note that

$$\mathbf{S}_1 \cdot \mathbf{S}_2 = \frac{1}{2}\left(\mathbf{S}^2 - \mathbf{S}_1^2 - \mathbf{S}_2^2\right) = \frac{1}{2}\mathbf{S}^2 - \frac{3}{4}\hbar^2\,,$$

and, hence, $\mathbf{S}^2$ commutes with itself and S_z. By contrast, we have

$$
\begin{aligned}
\left[\mathbf{S}^2, S_{1z} - S_{2z}\right] &= \sum_i \left(S_i\underbrace{[S_{1i} + S_{2i}}_{S_i}, S_{1z} - S_{2z}] + [\underbrace{S_{1i} + S_{2i}}_{S_i}, S_{1z} - S_{2z}]S_i\right)\\
&= i\hbar\sum_{ik}\epsilon_{i3k}\left[S_i(S_{1k} - S_{2k}) + (S_{1k} - S_{2k})S_i\right]\\
&= -i\hbar\left[(\mathbf{S}_1 + \mathbf{S}_2) \times (\mathbf{S}_1 - \mathbf{S}_2)\right]_z + i\hbar\left[(\mathbf{S}_1 - \mathbf{S}_2) \times (\mathbf{S}_1 + \mathbf{S}_2)\right]_z\\
&= -i\hbar\left[\underbrace{\mathbf{S}_1 \times \mathbf{S}_1}_{i\hbar\mathbf{S}_1} -2\mathbf{S}_1 \times \mathbf{S}_2 - \underbrace{\mathbf{S}_2 \times \mathbf{S}_2}_{i\hbar\mathbf{S}_2}\right]_z + i\hbar\left[\underbrace{\mathbf{S}_1 \times \mathbf{S}_1}_{i\hbar\mathbf{S}_1} +2\mathbf{S}_1 \times \mathbf{S}_2 - \underbrace{\mathbf{S}_2 \times \mathbf{S}_2}_{i\hbar\mathbf{S}_2}\right]_z\\
&= 4i\hbar\,(\mathbf{S}_1 \times \mathbf{S}_2)_z\,,
\end{aligned}
$$

while $S_z = S_{1z} + S_{2z}$ obviously commutes with $S_{1z} - S_{2z}$. We have used the property, valid for a generic angular momentum $\mathbf{J}$, that

$$(\mathbf{J} \times \mathbf{J})_i = \sum_{jk}\epsilon_{ijk}J_jJ_k = \frac{1}{2}\sum_{jk}\epsilon_{ijk}(J_jJ_k - J_kJ_j) = \frac{i\hbar}{2}\sum_{jkl}\epsilon_{ijk}\epsilon_{jkl}J_l = \frac{i\hbar}{2}\sum_{jl}(\delta_{il}\delta_{jj} - \delta_{ij}\delta_{jl})J_l = i\hbar J_i\,.$$

Another, simpler, approach is that suggested in the text, namely

$$
\begin{aligned}
\left[\mathbf{S}^2, S_{1z} - S_{2z}\right] &= \left[\mathbf{S}_1^2 + \mathbf{S}_2^2 + 2\mathbf{S}_1 \cdot \mathbf{S}_2, S_{1z} - S_{2z}\right] = 2\sum_i\left([S_{1i}, S_{1z}]S_{2i} - S_{1i}[S_{2i}, S_{2z}]\right)\\
&= 2i\hbar\sum_{ik}\epsilon_{i3k}(S_{1k}S_{2i} - S_{1i}S_{2k}) = 4i\hbar(\mathbf{S}_1 \times \mathbf{S}_2)_z\,.
\end{aligned}
$$

Part 2

It is convenient to use the basis of eigenstates $|SM_s\rangle$ of $\mathbf{S}^2$ and S_z, namely

$$|11\rangle = |++\rangle, \qquad |10\rangle = \frac{1}{\sqrt{2}}(|+-\rangle + |-+\rangle), \qquad |1-1\rangle = |--\rangle,$$

and

$$|10\rangle = \frac{1}{\sqrt{2}}(|+-\rangle - |-+\rangle),$$

where $|\sigma_1\sigma_2\rangle$ are the eigenstates of S_{1z} and S_{2z}. The Hamiltonian H_0 is diagonal in this basis,

$$\langle S'M'_S|H_0|SM_S\rangle = \frac{\alpha}{2}\left[S(S+1) - \frac{3}{2}\right]\delta_{S,S'}\,\delta_{M_S,M'_S},$$

and hence

$$\underline{H}_0 = \begin{pmatrix} \langle 00|H_0|00\rangle & \langle 00|H_0|10\rangle & \langle 00|H_0|11\rangle & \langle 00|H_0|1-1\rangle \\ \langle 10|H_0|00\rangle & \langle 10|H_0|10\rangle & \langle 10|H_0|11\rangle & \langle 10|H_0|1-1\rangle \\ \langle 11|H_0|00\rangle & \langle 11|H_0|10\rangle & \langle 11|H_0|11\rangle & \langle 11|H_0|1-1\rangle \\ \langle 1-1|H_0|00\rangle & \langle 1-1|H_0|10\rangle & \langle 1-1|H_0|11\rangle & \langle 1-1|H_0|1-1\rangle \end{pmatrix} = \frac{\alpha}{4}\begin{pmatrix} -3 & 0 & 0 & 0 \\ 0 & 1 & 0 & 0 \\ 0 & 0 & 1 & 0 \\ 0 & 0 & 0 & 1 \end{pmatrix}.$$

Since V commutes with S_z, its matrix elements between eigenstates of S_z belonging to different eigenvalues must vanish, and therefore the matrix $\underline{V}$ must have the form

$$\underline{V} = \begin{pmatrix} \langle 00|V|00\rangle & \langle 00|V|10\rangle & 0 & 0 \\ \langle 10|V|00\rangle & \langle 10|V|10\rangle & 0 & 0 \\ 0 & 0 & \langle 11|V|11\rangle & 0 \\ 0 & 0 & 0 & \langle 1-1|V|1-1\rangle \end{pmatrix}.$$

Furthermore, we have

$$V|11\rangle = \frac{\mu B}{\hbar}(S_{1z} - S_{2z})|++\rangle = \frac{\mu B}{2}(|++\rangle - |++\rangle) = 0,$$

and similarly $V|1-1\rangle = 0$. On the other hand, we find

$$V|00\rangle = \frac{1}{\sqrt{2}}\frac{\mu B}{\hbar}(S_{1z} - S_{2z})(|+-\rangle - |-+\rangle) = \frac{1}{\sqrt{2}}\frac{\mu B}{2}(|+-\rangle - (-)|+-\rangle - (-)|-+\rangle + |-+\rangle) = \mu B|10\rangle,$$

$$V|10\rangle = \frac{1}{\sqrt{2}}\frac{\mu B}{\hbar}(S_{1z} - S_{2z})(|+-\rangle + |-+\rangle) = \frac{1}{\sqrt{2}}\frac{\mu B}{2}(|+-\rangle - (-)|+-\rangle + (-)|-+\rangle - |-+\rangle) = \mu B|00\rangle.$$

We conclude that

$$\underline{V} = \mu B\begin{pmatrix} 0 & 1 & 0 & 0 \\ 1 & 0 & 0 & 0 \\ 0 & 0 & 0 & 0 \\ 0 & 0 & 0 & 0 \end{pmatrix}.$$

The matrix $\underline{H}$ has the following block form:

$$\underline{H} = \begin{pmatrix} -3\alpha/4 & \mu B & 0 & 0 \\ \mu B & \alpha/4 & 0 & 0 \\ 0 & 0 & \alpha/4 & 0 \\ 0 & 0 & 0 & \alpha/4 \end{pmatrix},$$

with eigenvalues resulting from

$$\det(\underline{H} - E\underline{I}) = 0 \implies \left(\frac{\alpha}{4} - E\right)^2 \left[\left(-\frac{3\alpha}{4} - E\right)\left(\frac{\alpha}{4} - E\right) - (\mu B)^2\right] = 0,$$

which yields

$$E_0 = \frac{\alpha}{4}, \qquad \text{twice degenerate},$$

and

$$E^2 + \frac{\alpha}{2}E - \frac{3}{16}\alpha^2 - (\mu B)^2 = 0 \implies E_\pm = -\frac{\alpha}{4} \pm \sqrt{\frac{\alpha^2}{4} + (\mu B)^2},$$

or

$$E_\pm = -\frac{\alpha}{4}\left[1 \mp 2\sqrt{1 + (2\mu B/\alpha)^2}\right], \qquad \text{non-degenerate}.$$

Part 3

The Hamiltonian H_0 has a non-degenerate eigenvalue $E_0 = -(3/4)\alpha$ with eigenstate $|00\rangle$ and a three-fold degenerate eigenvalue $E_1 = \alpha/4$ with eigenstates $|1M_S\rangle$. For the non-degenerate eigenvalue, the first-order correction vanishes since $\langle 00|V|00\rangle = 0$. The second-order correction is given by

$$E_0^{(2)} = \sum_{M_S = \pm 1, 0} \frac{|\langle 1M_S|V|00\rangle|^2}{E_0 - E_1} = \frac{|\langle 10|V|00\rangle|^2}{E_0 - E_1} = -\frac{(\mu B)^2}{\alpha},$$

so that the total energy to second order is

$$\overline{E}_0 = -\frac{3}{4}\alpha - \frac{(\mu B)^2}{\alpha},$$

with corresponding eigenstate up to and including first-order corrections

$$|\overline{00}\rangle = |00\rangle + \sum_{M_S = \pm 1, 0} |1M_S\rangle \frac{\langle 1M_S|V|00\rangle}{E_0 - E_1} = |00\rangle - \frac{\mu B}{\alpha}|10\rangle.$$

In the degenerate subspace corresponding to the eigenvalue E_1, the 3×3 matrix representing V vanishes identically, and so these states remain degenerate in first order. However, in second order we consider the 3×3 matrix $\underline{Z}$ with matrix elements

$$Z_{M_S', M_S} = \frac{\langle 1M_S'|V|00\rangle\langle 00|V|1M_S\rangle}{E_1 - E_0} = \frac{(\mu B)^2}{\alpha}\delta_{M_S', 0}\,\delta_{M_s, 0},$$

since the sum over the intermediate states consists only of a single state, the state $|00\rangle$. We obtain immediately

$$\underline{Z} = \begin{pmatrix} \langle 10|Z|10\rangle & \langle 10|Z|11\rangle & \langle 10|Z|1-1\rangle \\ \langle 11|Z|10\rangle & \langle 11|Z|11\rangle & \langle 11|Z|1-1\rangle \\ \langle 1-1|Z|10\rangle & \langle 1-1|Z|11\rangle & \langle 1-1|Z|1-1\rangle \end{pmatrix} = \frac{(\mu B)^2}{\alpha}\begin{pmatrix} 1 & 0 & 0 \\ 0 & 0 & 0 \\ 0 & 0 & 0 \end{pmatrix},$$

and so the states $|11\rangle$ and $|1-1\rangle$ remain degenerate, while the energy of the state $|10\rangle$ becomes (including this second-order correction)

$$\overline{E}_1 = \frac{\alpha}{4} + \frac{(\mu B)^2}{\alpha},$$

with corresponding eigenstate, up to and including first-order corrections (again the sum over intermediate states consists of the single state $|00\rangle$),

$$\overline{|10\rangle} = |10\rangle + |00\rangle \frac{\langle 00|V|10\rangle}{E_1 - E_0} = |10\rangle + \frac{\mu B}{\alpha} |00\rangle \ .$$

Note that the states $\overline{|00\rangle}$ and $\overline{|10\rangle}$ are orthogonal and normalized to 1, up to and including linear terms in $\mu B/\alpha$. The perturbation induces small admixtures into the leading-order states $|10\rangle$ and $|00\rangle$.

Part 4

Expanding the energies $E_\pm$ obtained in part 2 in powers of $\mu B/\alpha$ yields

$$E_\pm \approx -\frac{\alpha}{4} \left[1 \mp 2 \left(1 + \frac{2\mu^2 B^2}{\alpha^2} \right) \right] \ ,$$

or

$$E_+ = \frac{\alpha}{4} \left(1 + \frac{4\mu^2 B^2}{\alpha^2} \right) = \overline{E}_1 \ , \qquad E_- = -\frac{\alpha}{4} \left(3 + \frac{4\mu^2 B^2}{\alpha^2} \right) = \overline{E}_0 \ ,$$

in agreement with the perturbative results.

Part 5a

Suppose that the oscillating field is taken along the $\hat{\mathbf{x}}$-axis. The time-dependent perturbation is then given by

$$V(t) = \frac{\mu}{2\hbar} B_0 (S_{1x} - S_{2x}) \left(e^{-i\omega t} + e^{i\omega t} \right) = W e^{-i\omega t} + W e^{i\omega t} \ ,$$

since W is hermitian. The transition $|1\rangle \longrightarrow |2\rangle$ is driven by the amplitude

$$A_{21}(t) = -\frac{i}{\hbar} \int_0^t dt' \, e^{i(E_2 - E_1)t'/\hbar} \frac{\mu}{2\hbar} B_0 \langle 2|S_{1x} - S_{2x}|1\rangle \left(e^{-i\omega t'} + e^{i\omega t'} \right) \ .$$

Recall that

$$S_{ix} = \frac{1}{2} (S_{i+} + S_{i-}) \ ,$$

where $S_{i\pm}$ are the raising and lowering operators for particle i. When acting on the spin states of particle i, they yield

$$S_\pm |\mp\rangle = \hbar \, |\pm\rangle \qquad \text{and} \qquad S_\pm |\pm\rangle = 0 \ ,$$

and so, apart from the factor of $\hbar/2$, S_{ix} just flips the spin of particle i, that is,

$$S_{ix} |\pm\rangle = \frac{\hbar}{2} |\mp\rangle \ .$$

Now consider

$$\begin{aligned} (S_{1x} - S_{2x})|00\rangle &= \frac{1}{\sqrt{2}} (S_{1x} - S_{2x})| + -\rangle - \frac{1}{\sqrt{2}} (S_{1x} - S_{2x})| - +\rangle \\ &= \frac{\hbar}{2\sqrt{2}} (| - -\rangle - | + +\rangle) - \frac{\hbar}{2\sqrt{2}} (| + +\rangle - | - -\rangle) = \frac{\hbar}{\sqrt{2}} (|1{-}1\rangle - |11\rangle) \ . \end{aligned}$$

Similarly we find

$$(S_{1x} - S_{2x})\,|10\rangle = 0\,,$$

and therefore

$$(S_{1x} - S_{2x})|1\rangle = (S_{1x} - S_{2x})\left(|10\rangle + \frac{\mu B}{\alpha}\,|00\rangle\right) = \frac{\hbar \mu B}{\sqrt{2}\,\alpha}\,(|1\,{-}1\rangle - |11\rangle)\,.$$

We conclude that, since the state $|2\rangle$ is a linear combination of $|10\rangle$ and $|00\rangle$, it is orthogonal to both $|11\rangle$ and $1{-}1\rangle$, so that

$$\langle 2|S_{1x} - S_{2x}|1\rangle = 0\,.$$

Therefore, the transition cannot proceed if the $\mathbf{B}_0$ field is taken along the $\hat{\mathbf{x}}$-axis (or in fact along the $\hat{\mathbf{y}}$-axis; by a similar argument, the operator S_{iy} also flips the spin state of particle i, except that a phase factor is included that is different depending on whether, before the flip, the spin is up or down).

By contrast, if $\mathbf{B}_0$ is taken along the $\hat{\mathbf{z}}$-axis, we have (see the previous problem)

$$(S_{1z} - S_{2z})|10\rangle = \hbar\,|00\rangle \qquad \text{and} \qquad (S_{1z} - S_{2z})|00\rangle = \hbar\,|10\rangle\,,$$

and hence

$$(S_{1z} - S_{2z})|1\rangle = \hbar|00\rangle + \frac{\hbar \mu B}{\alpha}\,|10\rangle\,, \qquad (S_{1z} - S_{2z})|2\rangle = \hbar|10\rangle - \frac{\hbar \mu B}{\alpha}\,|00\rangle\,,$$

yielding

$$\langle 2|S_{1z} - S_{2z}|1\rangle = \langle 1|S_{1z} - S_{2z}|2\rangle^* = \hbar\left(\langle 00| - \frac{\mu B}{\alpha}\,\langle 10|\right)\left(|00\rangle + \frac{\mu B}{\alpha}\,|10\rangle\right) = \hbar + \cdots\,,$$

where we have neglected corrections proportional to $(\mu B/\alpha)^2$.

Part 5b

The transition $|2\rangle \longrightarrow |1\rangle$ involves the matrix element (see above)

$$\langle 1|W|2\rangle = \frac{\mu B_0}{2\hbar}\,\underbrace{\langle 1|S_{1z} - S_{2z}|2\rangle}_{\approx \hbar} = \frac{\mu B_0}{2}\,,$$

yielding the rate

$$R_{21} = \frac{2\pi}{\hbar}\,\frac{(\mu B_0)^2}{4}\,\delta(E_1 - E_2 - \hbar\,\omega)\,,$$

where it should be noted that

$$E_1 - E_2 = \alpha + 2\,\frac{(\mu B)^2}{\alpha} > 0\,.$$

The transition will proceed provided that the angular frequency of the oscillating field matches the Bohr angular frequency $(E_1 - E_2)/\hbar$.

Problem 6 The Transition $i \longrightarrow i$

Including terms up to second order in the perturbative expansion, calculate the probability for the transition $i \longrightarrow i$ at time t induced by a (time-independent) perturbation $\hat{V}$. Interpret the results.

Solution

We assume the perturbation is switched on at $t = 0$ and stays constant for $t > 0$. To second order, the amplitude for the transition $i \longrightarrow i$ is given by

$$\langle \phi_i | \phi_i(t) \rangle = \langle \phi_i | \left[\hat{U}_0(t, 0) - \frac{i}{\hbar} \int_0^t dt_1 \, \hat{U}_0(t, t_1) \, \hat{V} \hat{U}_0(t_1, 0) \right.$$
$$\left. - \frac{1}{\hbar^2} \int_0^t dt_1 \int_0^{t_1} dt_2 \, \hat{U}_0(t, t_1) \, \hat{V} \hat{U}_0(t_1, t_2) \, \hat{V} \hat{U}_0(t_2, 0) \right] | \phi_i(0) \rangle$$
$$= e^{-iE_i t/\hbar} - \frac{i}{\hbar} \int_0^t dt_1 \, e^{-iE_i(t-t_1)/\hbar} \, \langle \phi_i | \hat{V} | \phi_i \rangle \, e^{-iE_i t_1/\hbar}$$
$$+ \left(-\frac{i}{\hbar} \right)^2 \int_0^t dt_1 \int_0^{t_1} dt_2 \, e^{-iE_i(t-t_1)/\hbar} \, \langle \phi_i | \hat{V} e^{-iH(t_1-t_2)/\hbar} \, \hat{V} | \phi_i \rangle \, e^{-iE_i t_2/\hbar} \, .$$

To proceed further we insert the complete set of eigenstates of the unperturbed Hamiltonian between the two subsequent $\hat{V}$s, that is,

$$\langle \phi_i | \hat{V} e^{-iH(t_1-t_2)/\hbar} \, \hat{V} | \phi_i \rangle = \sum_p \langle \phi_i | \hat{V} | \phi_p \rangle \langle \phi_p | e^{-iH(t_1-t_2)/\hbar} \, \hat{V} | \phi_i \rangle = \sum_p e^{-iE_p(t_1-t_2)/\hbar} \, \langle \phi_i | \hat{V} | \phi_p \rangle \langle \phi_p | \hat{V} | \phi_i \rangle \, ,$$

to arrive at (after rearranging the time-dependent exponential factors)

$$\langle \phi_i | \phi_i(t) \rangle = e^{-iE_i t/\hbar} \left[1 - \frac{i}{\hbar} \langle \phi_i | \hat{V} | \phi_i \rangle \int_0^t dt_1 + \left(\frac{i}{\hbar} \right)^2 \sum_p |\langle \phi_i | \hat{V} | \phi_p \rangle|^2 \int_0^t dt_1 \, e^{i(E_i-E_p)t_1/\hbar} \int_0^{t_1} dt_2 \, e^{-i(E_i-E_p)t_2/\hbar} \right] .$$

We split the sum over intermediate states in the following way: $\sum_p |\phi_p\rangle\langle\phi_p| = |\phi_i\rangle\langle\phi_i| + \sum_{p \neq i} |\phi_p\rangle\langle\phi_p|$, so that

$$\langle \phi_i | \phi_i(t) \rangle = e^{-iE_i t/\hbar} \left[1 - \frac{i}{\hbar} \langle \phi_i | \hat{V} | \phi_i \rangle \int_0^t dt_1 + \left(-\frac{i}{\hbar} \right)^2 \langle \phi_i | \hat{V} | \phi_i \rangle^2 \int_0^t dt_1 \int_0^{t_1} dt_2 \right.$$
$$\left. + \left(-\frac{i}{\hbar} \right)^2 \sum_{p \neq i} |\langle \phi_i | \hat{V} | \phi_p \rangle|^2 \int_0^t dt_1 \, e^{i(E_i-E_p)t_1/\hbar} \int_0^{t_1} dt_2 \, e^{-i(E_i-E_p)t_2/\hbar} \right] ,$$

and carry out the time integrations,

$$\int_0^t dt_1 = t \, , \qquad \int_0^t dt_1 \int_0^{t_1} dt_2 = \frac{t^2}{2!} \, ,$$

and

$$\int_0^t dt_1 \, e^{i(E_i-E_p)t_1/\hbar} \int_0^{t_1} dt_2 \, e^{-i(E_i-E_p)t_2/\hbar} = \int_0^t dt_1 \, \frac{e^{i(E_i-E_p)t_1/\hbar}}{-i(E_i-E_p)/\hbar} \left[e^{-i(E_i-E_p)t_1/\hbar} - 1 \right]$$
$$= \frac{t}{-i(E_i-E_p)/\hbar} + \frac{1}{i(E_i-E_p)/\hbar} \frac{e^{i(E_i-E_p)t/\hbar} - 1}{i(E_i-E_p)/\hbar} \, .$$

Collecting results and introducing the notation $V_{pq} \equiv \langle \phi_p | \hat{V} | \phi_q \rangle$, we obtain for the second-order amplitude

$$\langle \phi_i | \phi_i(t) \rangle = e^{-iE_i t/\hbar} \left[1 + \left(-\frac{i}{\hbar} \right) V_{ii} \, t + \frac{1}{2!} \left(-\frac{i}{\hbar} \right)^2 V_{ii}^2 \, t^2 + \left(-\frac{i}{\hbar} \right) \left(\sum_{p \neq i} \frac{|V_{ip}|^2}{E_i - E_p} \right) t + \sum_{p \neq i} |V_{ip}|^2 \frac{e^{i(E_i-E_p)t/\hbar} - 1}{(E_i - E_p)^2} \right].$$

This is an interesting relation; the first four terms in the square brackets result from the expansion of

$$e^{-i[V_{ii}+\sum_{p\neq i}|V_{ip}|^2/(E_i-E_p)]t/\hbar} = 1 + \left(-\frac{i}{\hbar}\right)\left[V_{ii}+\sum_{p\neq i}\frac{|V_{ip}|^2}{E_i-E_p}\right]t + \frac{1}{2!}\left(-\frac{i}{\hbar}\right)^2\underbrace{\left[V_{ii}+\sum_{p\neq i}\frac{|V_{ip}|^2}{E_i-E_p}\right]^2}_{\approx V_{ii}^2}t^2 + \cdots,$$

where we have kept terms only up to quadratic in the perturbation. We recognize the first- and second-order corrections to the unperturbed energy E_i, namely

$$E_i^{(1)} = V_{ii}, \qquad E_i^{(2)} = \sum_{p\neq i}\frac{|V_{ip}|^2}{E_i-E_p},$$

which state that, after the perturbation has been turned on at $t=0$, the energy of the state $|\phi_i\rangle$ is no longer just E_i, it has been shifted. Thus, we can view stationary-state perturbation theory as a special case of time-dependent perturbation theory. To second order in the perturbation, the amplitude can be written as

$$\langle\phi_i|\phi_i(t)\rangle = e^{-i\overline{E}_i t/\hbar}\left[1 + \sum_{p\neq i}|V_{ip}|^2\frac{e^{i(E_i-E_p)t/\hbar}-1}{(E_i-E_p)^2}\right],$$

where we define the energy $\overline{E}_i$ (correct to second order)

$$\overline{E}_i = E_i + E_i^{(1)} + E_i^{(2)}.$$

We are now in position to calculate the probability for the transition $i \longrightarrow i$, which (to second order in the perturbation) is given by

$$|\langle\phi_i|\phi_i(t)\rangle|^2 = 1 + \frac{1}{\hbar^2}\sum_{p\neq i}|V_{ip}|^2\left[\frac{e^{i\omega_{ip}t}-1}{\omega_{ip}^2} + \text{c.c.}\right] = 1 - \frac{1}{\hbar^2}\sum_{p\neq i}|V_{ip}|^2 f(t,\omega_{ip}),$$

where we have defined $\omega_{ip} = (E_i - E_p)/\hbar$ and we have introduced the function $f(t,\omega)$ of Eq. (18.21), namely

$$\frac{e^{i\omega_{ip}t}-1}{\omega_{ip}^2} + \text{c.c.} = \frac{e^{i\omega_{ip}t}+e^{-i\omega_{ip}t}-2}{\omega_{ip}^2} = -2\frac{1-\cos(\omega_{ip}t)}{\omega_{ip}^2} = -\left[\frac{\sin(\omega_{ip}t/2)}{\omega_{ip}/2}\right]^2 = -f(t,\omega_{ip}).$$

In the limit of large t, we have $f(t,\omega_{ip}) \longrightarrow 2\pi t\,\delta(\omega_{ip})$, and the transition probability becomes

$$|\langle\phi_i|\phi_i(t)\rangle|^2 = 1 - \frac{2\pi}{\hbar}t\sum_{p\neq i}|V_{ip}|^2\,\delta(E_i-E_p),$$

after the replacement $\delta(\omega_{ip}) = \hbar\,\delta(E_i-E_p)$. The quantity

$$\Gamma_i = \frac{2\pi}{\hbar}\sum_{p\neq i}|V_{ip}|^2\,\delta(E_i-E_p)$$

is the total transition rate, and we see that

$$|\langle\phi_i|\phi_i(t)\rangle|^2 = 1 - \Gamma_i t \approx e^{-\Gamma_i t},$$

namely, the probability that the system will remain in the initial state decays exponentially when there is a constant rate of transitions to other states. We can say that the state $|\phi_i(t)\rangle = \hat{U}(t,0)\,|\phi_i\rangle$ behaves as if it had a complex energy $\overline{E}_i - i\hbar\Gamma_i/2$, that is,

$$|\phi_i(t)\rangle = e^{-i(\overline{E}_i-i\hbar\Gamma_i/2)t/\hbar}\,|\phi_i\rangle.$$

Problem 7 Hydrogen Atom in Time-Dependent EM Fields: Doppler Effect and Recoil Energy

The Hamiltonian of a free hydrogen atom reads

$$H_0 = \frac{\mathbf{P}^2}{2M} + \frac{\mathbf{p}^2}{2\mu} - \frac{e^2}{r} = H_{\text{cm}} + H_{\text{rel}} ,$$

where $\mathbf{P}$ is the center-of-mass momentum operator, $\mathbf{r}$ and $\mathbf{p}$ are the relative position and momentum operators, and M and μ are the total and reduced masses of the e–p system, respectively. Spin-dependent effects are neglected throughout. We are concerned here with the interaction of this system with the electromagnetic fields induced by the vector potential

$$\mathbf{A}(\mathbf{r}, t) = A_0\, \hat{e}\, e^{i(\mathbf{k}\cdot\mathbf{r}-\omega t)} + \text{c.c.} ,$$

where $\mathbf{k}$ and $\omega = ck$ are, respectively, the wave number and angular frequency, $\hat{e}$ is the polarization unit vector (perpendicular to $\mathbf{k}$), and A_0 is a constant.

1. Express the electromagnetic perturbation $V(t)$, ignoring terms quadratic in $\mathbf{A}(\mathbf{r}, t)$, in terms of the center-of-mass and relative position and momentum operators. Show that, in the long-wavelength approximation, in which the wavelength of the electromagnetic wave is large relative to the atomic dimensions, $V(t)$ can be written as

$$V(t) = W e^{-i\omega t} + W^\dagger e^{-i\omega t} , \qquad W = -\frac{qA_0}{\mu c} e^{i\mathbf{k}\cdot\mathbf{R}}\, \hat{e}\cdot\mathbf{p} ,$$

where $\mathbf{R}$ is the center-of-mass position operator.

2. Introduce the basis $|\phi_{\mathbf{K}}; \psi_\gamma\rangle$, consisting of the tensor product of eigenstates $|\phi_{\mathbf{K}}\rangle$ of the center-of-mass momentum operator and eigenstates $|\psi_\gamma\rangle$ of the relative Hamiltonian (the quantum numbers identifying the eigenstates of H_{rel} are denoted collectively by γ). Consider a transition from the state $|\phi_{\mathbf{K}}; \psi_\gamma\rangle$ of relative energy E_n to a state $|\phi_{\mathbf{K}'}; \psi_{\gamma'}\rangle$ of relative energy $E_{n'}$. Show that the matrix element of W between the states $|\phi_{\mathbf{K}}; \psi_\gamma\rangle$ and $|\phi_{\mathbf{K}'}; \psi_{\gamma'}\rangle$ is different from zero only if there exists a certain relation between $\mathbf{K}$, $\mathbf{k}$, and $\mathbf{K}'$. Give a physical interpretation of this relation.

3. Show that, for an absorption process ($E_{n'} > E_n$), resonance occurs when the photon energy $\hbar\omega$ differs from the energy $\hbar\omega_0 = E_{n'} - E_n$ by a quantity δE consisting of the sum of two terms δE_1 and δE_2, where δE_1 depends on the angle between $\mathbf{K}$ and $\mathbf{k}$ (the Doppler effect) and δE_2 is independent of $\mathbf{K}$. What is the physical meaning of δE_2? Show that for a typical atomic transition δE_2 is negligible relative to δE_1.

Solution

Part 1

The Hamiltonian of the e–p system in an electromagnetic wave reads

$$H(t) = \frac{1}{2m_p}\left[\mathbf{p}_p - \frac{e}{c}\mathbf{A}(\mathbf{r}_p, t)\right]^2 + \frac{1}{2m_e}\left[\mathbf{p}_e + \frac{e}{c}\mathbf{A}(\mathbf{r}_e, t)\right]^2 - \frac{e^2}{|\mathbf{r}_p - \mathbf{r}_e|} ,$$

where $\mathbf{A}(\mathbf{r}, t)$ is the vector potential associated with the electromagnetic wave; in our case

$$\mathbf{A}(\mathbf{r}, t) = A_0\, \epsilon_{\mathbf{k}}\left[e^{i(\mathbf{k}\cdot\mathbf{r}-\omega t)} + e^{-i(\mathbf{k}\cdot\mathbf{r}-\omega t)}\right] \qquad \text{with} \qquad \omega = c|\mathbf{k}| \text{ and } \mathbf{k}\cdot\epsilon_{\mathbf{k}} = 0 .$$

Note that the condition $\boldsymbol{\nabla} \cdot \mathbf{A}(\mathbf{r}, t) = 0$ (the Coulomb gauge) ensures that $\mathbf{p}_p$ and $\mathbf{A}(\mathbf{r}_p, t)$, and $\mathbf{p}_e$ and $\mathbf{A}(\mathbf{r}_e, t)$, respectively commute with each other. Therefore, we define

$$H_0 = \frac{\mathbf{p}_p^2}{2m_p} + \frac{\mathbf{p}_e^2}{2m_e} - \frac{e^2}{|\mathbf{r}_p - \mathbf{r}_e|} \,,$$

and the time-dependent perturbation $V(t)$ as

$$V(t) = -\frac{e}{m_p c}\, \mathbf{p}_p \cdot \mathbf{A}(\mathbf{r}_p, t) + \frac{e}{m_e c}\, \mathbf{p}_e \cdot \mathbf{A}(\mathbf{r}_e, t) \;;$$

terms quadratic in the vector potential are ignored. The position and momentum operators of the proton and electron in terms of the center-of-mass and relative position and momentum operators are given by the following expressions:

$$\mathbf{r}_p = \mathbf{R} + \frac{\mu}{m_p}\, \mathbf{r} \,, \qquad \mathbf{p}_p = \frac{m_p}{M}\, \mathbf{P} + \mathbf{p} \,,$$

$$\mathbf{r}_e = \mathbf{R} - \frac{\mu}{m_e}\, \mathbf{r} \,, \qquad \mathbf{p}_e = \frac{m_e}{M}\, \mathbf{P} - \mathbf{p} \,,$$

where $M = m_p + m_e$ and $\mu = m_p m_e/(m_p + m_e)$. Substituting these expression into H_0 and $V(t)$, we find

$$H_0 = \underbrace{\frac{\mathbf{P}^2}{2M}}_{H_{\mathrm{cm}}} + \underbrace{\frac{\mathbf{p}^2}{2\mu} + V(r)}_{H_{\mathrm{rel}}}$$

and

$$V(t) = -\frac{e}{m_p c}\left(\frac{m_p}{M}\, \mathbf{P} + \mathbf{p}\right) \cdot \mathbf{A}[\mathbf{R} + (\mu/m_p)\mathbf{r}, t] + \frac{e}{m_e c}\left(\frac{m_e}{M}\, \mathbf{P} - \mathbf{p}\right) \cdot \mathbf{A}[\mathbf{R} - (\mu/m_e)\mathbf{r}, t] \,.$$

Under the assumption that the wavelength of the electromagnetic wave is large compared with the dimension of the bound two-particle system – the long-wavelength approximation (LWA) $e^{i\mathbf{k}\cdot\mathbf{r}} \approx 1$ – we can ignore the dependence on the relative position in the vector potential, that is,

$$\mathbf{A}[\mathbf{R} + (\mu/m_p)\mathbf{r}, t] \longrightarrow \mathbf{A}(\mathbf{R}, t) \,, \qquad \mathbf{A}[\mathbf{R} - (\mu/m_e)\mathbf{r}, t] \longrightarrow \mathbf{A}(\mathbf{R}, t) \,.$$

We now see that the terms proportional to the total momentum cancel out, and we are left with

$$V(t) = -\frac{e}{c}\left(\frac{1}{m_p} + \frac{1}{m_e}\right) \mathbf{p} \cdot \mathbf{A}(\mathbf{R}, t) \equiv W e^{-i\omega t} + W^\dagger\, e^{i\omega t} \,,$$

where W is defined as

$$W = -\frac{eA_0}{\mu c}\, \mathbf{p} \cdot \boldsymbol{\epsilon}_\mathbf{k}\, e^{i\mathbf{k}\cdot\mathbf{R}} \,.$$

Part 2

We first evaluate the matrix element

$$W_{fi} = \langle \phi_{\mathbf{K}'}; \psi_{\gamma'} | W | \phi_\mathbf{K}; \psi_\gamma \rangle = -\frac{eA_0}{\mu c}\, \boldsymbol{\epsilon}_\mathbf{k} \cdot \int d\mathbf{R}\, d\mathbf{r}\, \frac{e^{-i\mathbf{K}'\cdot\mathbf{R}}}{(2\pi)^{3/2}}\, \psi_{\gamma'}^*(\mathbf{r})\, \underbrace{(-i\hbar\boldsymbol{\nabla}_\mathbf{r})}_{\mathbf{p}}\, e^{i\mathbf{k}\cdot\mathbf{R}}\, \frac{e^{i\mathbf{K}\cdot\mathbf{R}}}{(2\pi)^{3/2}}\, \psi_\gamma(\mathbf{r}) \,,$$

where the subscript of the gradient operator indicates that it acts only on $\mathbf{r}$. The matrix element above factorizes, and hence

$$W_{fi} = -\frac{eA_0}{\mu c} \underbrace{\left[\int d\mathbf{r} \; \psi^*_{\gamma'}(\mathbf{r}) \, \boldsymbol{\epsilon}_\mathbf{k} \cdot \mathbf{p} \, \psi_\gamma(\mathbf{r}) \right]}_{w_{fi}} \underbrace{\left[\int d\mathbf{R} \; \frac{e^{-i\mathbf{K}'\cdot\mathbf{R}}}{(2\pi)^{3/2}} \, e^{i\mathbf{k}\cdot\mathbf{R}} \, \frac{e^{i\mathbf{K}\cdot\mathbf{R}}}{(2\pi)^{3/2}} \right]}_{\delta(\mathbf{K}'-\mathbf{k}-\mathbf{K})} .$$

For the perturbation $V(t)$ we obtain

$$V_{fi}(t) = \delta(\mathbf{K}' - \mathbf{k} - \mathbf{K}) \, w_{fi} \, e^{-i\omega t} + \delta(\mathbf{K}' + \mathbf{k} - \mathbf{K}) \, w_{fi} \, e^{i\omega t} ,$$

and therefore the matrix element vanishes unless $\mathbf{K}' = \mathbf{K} + \mathbf{k}$ for absorption (negative-frequency term) or $\mathbf{K}' = \mathbf{K} - \mathbf{k}$ for emission (positive-frequency term). The δ-function enforces momentum conservation (the photon has momentum $\hbar\mathbf{k}$).

Part 3

The probability amplitude is given by

$$A_{fi}(t) = -\frac{i}{\hbar} \int_0^t dt' \, e^{i\omega_{fi}t'} \, V_{fi}(t') , \qquad \omega_{fi} = \frac{E_f - E_i}{\hbar} ,$$

and therefore

$$A_{fi}(t) = \delta(\mathbf{K}' - \mathbf{k} - \mathbf{K}) \left(-\frac{i}{\hbar} \int_0^t dt' \, e^{i\omega_{fi}t'} \, e^{-i\omega t} \right) w_{fi} + \delta(\mathbf{K}' + \mathbf{k} - \mathbf{K}) \left(-\frac{i}{\hbar} \int_0^t dt' \, e^{i\omega_{fi}t'} \, e^{i\omega t} \right) w_{fi} .$$

Here E_i and E_f are the eigenvalues of H_0, that is, $H_0|\phi_\mathbf{k}; \psi_\gamma\rangle = E_i|\phi_\mathbf{k}; \psi_\gamma\rangle$ and similarly for the final state; they are given by

$$E_i = \frac{\hbar^2 \mathbf{K}^2}{2M} + E_n , \qquad E_i = \frac{\hbar^2 \mathbf{K}'^2}{2M} + E_{n'} .$$

In order to obtain the transition probability, we need to interpret the square of the δ-function. Recall that

$$(2\pi)^3 \, \delta(\mathbf{q}) = \lim_{V\to\infty} \int_V d\mathbf{r} \, e^{-i\mathbf{q}\cdot\mathbf{r}} \implies \delta(0) = \frac{1}{(2\pi)^3} \left(\lim_{V\to\infty} \int_V d\mathbf{r} \right) = \lim_{V\to\infty} \frac{V}{(2\pi)^3}$$

and therefore

$$[\delta(\mathbf{q})]^2 = \delta(0) \, \delta(\mathbf{q}) = \lim_{V\to\infty} \frac{V}{(2\pi)^3} \, \delta(\mathbf{q}) .$$

Ignoring the interference-term contribution, which vanishes in the limit of large t, the transition rate per unit volume is then obtained as

$$R_{fi}(t) = \frac{1}{V} \frac{dP_{fi}(t)}{dt} = \frac{1}{(2\pi)^2} \frac{|w_{fi}|^2}{\hbar} \left[\delta(\mathbf{K}' - \mathbf{k} - \mathbf{K}) \, \delta(E_f - E_i - \hbar\omega) + \delta(\mathbf{K}' + \mathbf{k} - \mathbf{K}) \, \delta(E_f - E_i + \hbar\omega) \right] .$$

As expected, it is proportional to δ-functions enforcing energy and momentum conservation in the process. In the case of absorption, only the first term contributes and

$$E_f = E_i + \hbar\omega \implies \omega = \omega_0 + \frac{\hbar}{2M}[(\mathbf{K} + \mathbf{k})^2 - \mathbf{K}^2] ,$$

where $\omega_0 = (E_{n'} - E_n)/\hbar$ and we have used momentum conservation to write $\mathbf{K}' = \mathbf{K} + \mathbf{k}$. The angular frequency of the emitted photon differs from its resonance value ω_0 by two terms,

$$\delta\omega_1 = \frac{\delta E_1}{\hbar} = \frac{\hbar K \omega}{Mc} \cos\theta , \qquad \delta\omega_2 = \frac{\delta E_2}{\hbar} = \frac{\hbar \omega^2}{2Mc^2} ,$$

where θ is the angle between the photon momentum and the bound system (total) momentum. To leading order these corrections can be expressed as

$$\delta\omega_1 \approx \frac{\hbar K \omega_0}{Mc} \cos\theta , \qquad \delta\omega_2 \approx \frac{\hbar \omega_0^2}{2Mc^2} ,$$

by replacing ω with ω_0. We see that the sign of $\delta\omega_1$ – the Doppler shift – depends on the relative orientation of the atom and photon momenta, while $\delta\omega_2$ depends only on the resonance frequency. Since $\omega_0 = k_0/c$, δE_2 represents the kinetic energy of the atom as it recoils after absorption of the photon. The ratio of these corrections (assuming $2|\cos\theta| \approx 1$) is

$$|\delta\omega_2/\delta\omega_1| \sim \frac{\hbar \omega_0}{Pc} ,$$

where $P = \hbar K$ is the atom's momentum. For a typical atomic transition, we take $\hbar\omega_0 \approx 10$ eV; at a temperature of 300 K the average kinetic energy of the atom is (by the equipartition theorem)

$$\frac{\langle \mathbf{P}^2 \rangle}{2M} = \frac{3}{2} k_B T \implies P \approx \sqrt{M k_B T} \quad \text{and} \quad |\delta\omega_2/\delta\omega_1| \approx \frac{\hbar \omega_0}{\sqrt{k_B T M c^2}} \approx 2 \times 10^{-3} ,$$

where in the last step we took $Mc^2 \approx 10^9$ eV and $k_B T \approx 2.5 \times 10^{-2}$ eV.

Problem 8 Two Spin-1/2 Particles of Opposite Charge in a Time-Dependent Magnetic Field

Consider two spin-1/2 particles with opposite charges q and $-q$ coupled to an external time-dependent magnetic field $\hat{\mathbf{z}} B(t)$, which is adiabatically turned on and off as t goes from $-\infty$ to ∞. The Hamiltonian describing this system reads

$$H(t) = \frac{\mu B(t)}{\hbar} (S_{1z} - S_{2z}) ,$$

where $\mu = gq\hbar/(2mc)$ is the magnetic moment (m, c, and g are the mass, speed of light, and gyromagnetic factor, respectively). At time $t = -\infty$, the system is in state $|10\rangle$, that is, in an eigenstate of $\mathbf{S}^2$ and S_z, where $\mathbf{S} = \mathbf{S}_1 + \mathbf{S}_2$, with eigenvalues $2\hbar^2$ and 0, respectively.

1. Calculate exactly the state of the system at $t = \infty$. Without doing any detailed calculation, explain why the probabilities for the transitions $|10\rangle \longrightarrow |1 \pm 1\rangle$ vanish identically. Calculate the probability for the transition $|10\rangle \longrightarrow |00\rangle$, showing that it depends only on the integral

$$\int_{-\infty}^{\infty} dt\, B(t) ,$$

assumed to be convergent.

2. Calculate the probability for the transition $|10\rangle \longrightarrow |00\rangle$ in first-order perturbation theory. By comparing with the exact result above, justify under what conditions this leading-order calculation is valid.

Solution

Part 1

A generic state $|\psi(t)\rangle$ can be expanded in the of basis of eigenstates $|SM\rangle$ of $\mathbf{S}^2$ and S_z as

$$|\psi(t)\rangle = \sum_{S=0}^{1} \sum_{M=-S}^{S} c_{SM}(t)|SM\rangle .$$

The time-dependent Schrödinger equation,

$$i\hbar \frac{d}{dt} |\psi(t)\rangle = H(t) |\psi(t)\rangle ,$$

reads in this basis

$$\sum_{S'M'} i\hbar \, \dot{c}_{S'M'}(t)|S'M'\rangle = \frac{\mu B(t)}{\hbar} \sum_{S'M'} c_{S'M'}(t)(S_{1z} - S_{2z})|S'M'\rangle .$$

Projecting onto the state $|SM\rangle$, we arrive at

$$\dot{c}_{SM}(t) = -i\frac{\mu B(t)}{\hbar^2} \sum_{S'M'} \langle SM|S_{1z} - S_{2z}|S'M'\rangle \, c_{S'M'}(t) .$$

The operator $S_{1z} - S_{2z}$ commutes with S_z (but not with $\mathbf{S}^2$, see Problem 5); as a consequence, we have

$$\langle SM|S_{1z} - S_{2z}|S'M'\rangle = 0 \qquad \text{if } M \neq M' ,$$

and hence transitions between states with different M and M' cannot occur. Further, we note that

$$(S_{1z} - S_{2z})|1 \pm 1\rangle = 0 , \qquad (S_{1z} - S_{2z})|10\rangle = \hbar |00\rangle , \qquad (S_{1z} - S_{2z})|00\rangle = \hbar |10\rangle ,$$

and the equations for $c_{SM}(t)$ decouple for $SM = 11$ or $1{-}1$, yielding $\dot{c}_{1\pm1}(t) = 0$ or $c_{1\pm1}(t) = c_{1\pm1}(t_0)$, while for $SM = 00$ or 10 they give

$$\dot{c}_{00}(t) = -i\frac{\mu B(t)}{\hbar} c_{10}(t) , \qquad \dot{c}_{10}(t) = -i\frac{\mu B(t)}{\hbar} c_{00}(t)$$

Introducing $c_{\pm}(t) = c_{10}(t) \pm c_{00}(t)$, these equations are easily solved:

$$\dot{c}_{\pm}(t) = \mp i\frac{\mu B(t)}{\hbar} c_{\pm}(t) \implies c_{\pm}(t) = c_{\pm}(t_0) \, \mathrm{e}^{\mp i(\mu/\hbar) \int_{t_0}^{t} dt' B(t')} = c_{\pm}(t_0) \, \mathrm{e}^{\mp i\alpha(t)} .$$

We then find that

$$c_{10}(t) = \frac{c_{+}(t) + c_{-}(t)}{2} = \frac{c_{+}(t_0) \, \mathrm{e}^{-i\alpha(t)} + c_{-}(t_0) \, \mathrm{e}^{i\alpha(t)}}{2} , \qquad c_{00}(t) = \frac{c_{+}(t) - c_{-}(t)}{2} = \frac{c_{+}(t_0) \, \mathrm{e}^{-i\alpha(t)} - c_{-}(t_0) \, \mathrm{e}^{i\alpha(t)}}{2} .$$

In the present case, the initial state is $|10\rangle$ at $t_0 = -\infty$ and so $c_{10}(-\infty) = 1$ with all remaining coefficients vanishing; therefore $c_{+}(-\infty) = 1$ and $c_{-}(-\infty) = 1$. The state $|\psi(t)\rangle$ evolving from $|10\rangle$ at $t = -\infty$ is given by

$$|\psi(t)\rangle = c_{10}(t)|10\rangle + c_{00}(t)|00\rangle = \frac{1}{2} \left[\mathrm{e}^{-i\alpha(t)} + \mathrm{e}^{i\alpha(t)} \right] |10\rangle + \frac{1}{2} \left[\mathrm{e}^{-i\alpha(t)} - \mathrm{e}^{i\alpha(t)} \right] |00\rangle = \cos \alpha(t) \, |10\rangle - i \sin \alpha(t) \, |00\rangle .$$

The probability for the transition $|10\rangle \longrightarrow |00\rangle$ at $t = \infty$ is obtained from

$$P = |\langle 00|\psi(\infty)\rangle|^2 = \sin^2 \alpha(t) = \sin^2 \left[\frac{\mu}{\hbar} \int_{-\infty}^{\infty} dt' B(t') \right] .$$

Part 2

In first-order perturbation theory, the probability amplitude for the transition from the initial state $|10\rangle$ at $t = -\infty$ to the final state $|00\rangle$ at $t = +\infty$ is given by

$$A^{(1)} = -\frac{i}{\hbar} \int_{-\infty}^{\infty} dt \, \langle 00| \, V(t) \, |10\rangle \,,$$

which follows from the standard formula by setting the unperturbed energy of the initial and final states to zero. The matrix element is given by

$$\langle 00| \, V(t) \, |10\rangle = \frac{\mu B(t)}{\hbar} \langle 00|S_{1z} - S_{2z}|10\rangle = \mu B(t)$$

from which the probability in first order follows:

$$P^{(1)} = \frac{\mu^2}{\hbar^2} \left| \int_{-\infty}^{\infty} dt \, B(t) \right|^2 \,.$$

This result is seen to agree with the exact result by expanding the sine in the expression for P to leading order in the time-integral of $B(t)$; of course, such an expansion is valid if

$$\left| \frac{\mu}{\hbar} \int_{-\infty}^{\infty} dt \, B(t) \right| \ll 1 \,.$$

Problem 9 Hydrogen Atom in a Time-Dependent Electric Field

A hydrogen atom in the 1s ground state (we will ignore spin degrees of freedom) is placed in a uniform and weak electric field in the z-direction, $\hat{z} \mathcal{E}_0 \, e^{-t/\tau}$, which is turned on at time $t = 0$. The associated perturbation reads

$$V(t) = e\mathcal{E}_0 z \, e^{-t/\tau} \,,$$

where $-e$ and z are, respectively, the charge and z-component of the position operator of the electron.

1. State which transition (or transitions) to the first excited level of the hydrogen atom are allowed by this perturbation, in first order.
2. Calculate the probability in first order that the hydrogen atom is excited to the 2p level at $t \gg \tau$.
3. Can the transition 1s $\longrightarrow$ 2s occur in second order? Justify your answer.
4. Carry out the calculation of the 1s $\longrightarrow$ 2s transition to second order as far as you can. Ignore the contribution of the eigenfunctions in the continuum.

Hint: The normalized wave functions of the hydrogen atom are written as

$$\psi_{nlm}(\mathbf{r}) = a_0^{-3/2} \, g_{nl}(r/a_0) \, Y_{lm}(\hat{\mathbf{r}}) \,,$$

where the Y_{lm} are spherical harmonics, and the 1s, 2s, and 2p radial wave functions are given explicitly by

$$g_{10}(x) = 2 \, e^{-x} \,, \qquad g_{20}(x) = \frac{1}{\sqrt{2}} \left(1 - \frac{x}{2} \right) e^{-x/2} \,, \qquad g_{21}(x) = \frac{\sqrt{6}}{12} \, x \, e^{-x/2} \,,$$

with $x = r/a_0$ and a_0 the Bohr radius. The relevant spherical harmonics are

$$Y_{00}(\hat{\mathbf{r}}) = \frac{1}{\sqrt{4\pi}} \qquad Y_{11}(\hat{\mathbf{r}}) = -\sqrt{\frac{3}{8\pi}} \sin\theta \, e^{i\phi} \,, \qquad Y_{10}(\hat{\mathbf{r}}) = -\sqrt{\frac{3}{4\pi}} \cos\theta \,,$$

and $Y_{1-1}(\hat{\mathbf{r}}) = -Y_{11}^*(\hat{\mathbf{r}})$.

Solution

Part 1

In principle, the possible transitions are $1s \longrightarrow 2s$ and $1s \longrightarrow 2p_m$ with $m = 0$ and ± 1, although, as we will see, the first transition cannot occur in first-order perturbation theory. The first-order amplitude reads

$$A_{fi} = -\frac{i}{\hbar} \int_0^t dt'\, e^{i(\epsilon_f - \epsilon_i)t'/\hbar} \langle f | V(t') | i \rangle = -\frac{i}{\hbar} e\mathcal{E}_0 \int_0^t dt'\, e^{i(\epsilon_f - \epsilon_i)t'/\hbar}\, e^{-t'/\tau} \langle \phi_{2lm} | z | \phi_{100} \rangle \,.$$

Recalling that the eigenfunctions of the hydrogen atom Hamiltonian have parity $(-1)^l$ and the position operator is odd under parity, we see that the matrix element between the 1s and 2s state vanishes identically,

$$\langle \phi_{2s} | z | \phi_{1s} \rangle = 0 \,.$$

Alternatively, the Wigner–Eckart theorem states that $\langle \phi_{200} | z | \phi_{100} \rangle = C_{10}(00; 00)\, \langle \phi_{20} || r || \phi_{10} \rangle = 0$, since $C_{10}(00; 00) = 0$. By contrast, the matrix element between the 1s and 2p$_m$ states is given by

$$\langle \phi_{21m} | z | \phi_{100} \rangle = C_{10}(00; 1m)\, \langle \phi_{21} || r || \phi_{10} \rangle = \delta_{m,0}\, \langle \phi_{21} || r || \phi_{10} \rangle \,.$$

Note that parity does not yield any additional selection rule in this case (the initial and final states have opposite parity). Thus, the only allowed transition is $1s \longrightarrow 2p_0$.

Part 2

The probability in first order is given by

$$P_{fi}(t) = \left(\frac{e\mathcal{E}_0}{\hbar} \right)^2 \left| \int_0^t dt'\, e^{(i\omega_{21} - 1/\tau)t'} \langle \phi_{210} | z | \phi_{100} \rangle \right|^2 \,,$$

where $\omega_{21} = (\epsilon_2 - \epsilon_1)/\hbar$. The matrix element reads

$$\langle \phi_{210} | z | \phi_{100} \rangle = \int_0^\infty dr\, r^2 \int d\Omega\, \phi_{210}^*(\mathbf{r})\, r \cos\theta\, \phi_{100}(\mathbf{r}) = \frac{1}{a_0^3} \int_0^\infty dr\, r^3 g_{21}(r/a_0)\, g_{10}(r/a_0) \frac{1}{\sqrt{3}} \int d\Omega\, Y_{10}^*\, Y_{10} \,,$$

where in the last step we used

$$\cos\theta\, Y_{00} = \frac{1}{\sqrt{4\pi}}\, \cos\theta = \frac{1}{\sqrt{3}}\, Y_{10} \,.$$

The matrix element reduces to

$$\langle \phi_{210} | z | \phi_{100} \rangle = \frac{a_0}{\sqrt{3}} \underbrace{\int_0^\infty dx\, x^3 g_{21}(x)\, g_{10}(x)}_{I} \,.$$

The numerical factor I is obtained as

$$I = \frac{1}{\sqrt{6}} \int_0^\infty dx\, x^4\, e^{-3x/2} = \frac{1}{\sqrt{6}} \frac{4!}{(3/2)^5} = \frac{1}{\sqrt{6}} \frac{2^8}{3^4} \,.$$

resulting in the matrix element

$$\langle \phi_{210} | z | \phi_{100} \rangle = \frac{2^8}{3^5} \frac{a_0}{\sqrt{2}} \,.$$

The time integral gives

$$\int_0^t dt'\, e^{\beta t'} = \frac{1}{\beta}\left(e^{\beta t} - 1\right) , \qquad \beta = \frac{i\omega_{21}\tau - 1}{\tau} ,$$

which in the limit $t \gg \tau$ yields

$$\lim_{t \gg \tau} \int_0^t dt'\, e^{\beta t'} = -\frac{\tau}{i\omega_{21}\tau - 1} .$$

Collecting results, we find that the probability in this limit is

$$P_{fi}(t) = \frac{2^{15}}{3^{10}}\left(\frac{e\mathcal{E}_0 a_0}{\hbar}\right)^2 \frac{\tau^2}{1 + (\tau\omega_{21})^2} .$$

Part 3

The second-order probability amplitude is given by

$$A_{fi}^{(2)} = \left(-\frac{i}{\hbar}\right)^2 \int_0^t dt_1 \int_0^{t_1} dt_2\, \langle f| e^{-iH_0(t-t_1)/\hbar}\, \hat{V}(t_1)\, e^{-iH_0(t_1-t_2)/\hbar}\, \hat{V}(t_2)\, e^{-iH_0 t_2/\hbar} |i\rangle .$$

In our case, we have $V(t) = W e^{-t/\tau}$ with $W = e\mathcal{E}_0 z$, and hence for the 2s $\longrightarrow$ 1s transition

$$A_{fi}^{(2)} = \left(-\frac{i}{\hbar}\right)^2 \int_0^t dt_1 \int_0^{t_1} dt_2\, e^{-i\epsilon_2(t-t_1)/\hbar - t_1/\tau}\, \langle 2s| W e^{-iH_0(t_1-t_2)/\hbar}\, W |1s\rangle\, e^{-i\epsilon_1 t_2/\hbar - t_2/\tau} .$$

The matrix element can be evaluated by using the completeness of the eigenfunction of the hydrogen atom. However, below for simplicity we ignore the contribution of the continuum eigenfunctions, with energy $\epsilon > 0$, so that

$$\langle 2s| W e^{-iH_0(t_1-t_2)/\hbar}\, W |1s\rangle = \sum_{n=0}^{\infty}\sum_{l=0}^{n-1}\sum_{m=-l}^{l} e^{-i\epsilon_n(t_1-t_2)/\hbar}\, \langle 2s|W|\phi_{nlm}\rangle\langle\phi_{nlm}|W|1s\rangle + \text{continuum} .$$

Using the Wigner–Eckart theorem, we see that the operator W can connect the 1s and 2s states only to the states np_0, since

$$\langle 2s|z|\phi_{nlm}\rangle = C_{1l}(0m; 00)\, \langle 2s||r||\phi_{nl}\rangle = -\frac{1}{\sqrt{3}}\,\delta_{l,1}\,\delta_{m,0}\,\langle 2s||r||np_0\rangle ,$$

and

$$\langle\phi_{nlm}|z|1s\rangle = C_{10}(00; lm)\,\langle\phi_{nl}||r||1s\rangle = \delta_{l,1}\,\delta_{m,0}\,\langle np_0||r||1s\rangle .$$

We arrive at

$$\langle 2s| W e^{-iH_0(t_1-t_2)/\hbar}\, W |1s\rangle = -\frac{(e\mathcal{E}_0)^2}{\sqrt{3}}\sum_{n=2}^{\infty} e^{-i\epsilon_n(t_1-t_2)/\hbar}\,\alpha_n ,$$

where we have defined the product of reduced matrix elements

$$\alpha_n = \langle 2s||r||np_0\rangle\langle np_0||r||1s\rangle .$$

We conclude that the transition 1s to 2s can occur as a two-step process 1s $\longrightarrow np_0 \longrightarrow$ 2s.

Part 4

We can carry out the time integrations to obtain

$$
A_{fi}^{(2)} = \frac{1}{\sqrt{3}} \left(\frac{e\mathcal{E}_0}{\hbar} \right)^2 \sum_{n=2}^{\infty} \alpha_n \int_0^t dt_1 \int_0^{t_1} dt_2 \; e^{-i\epsilon_2(t-t_1)/\hbar - t_1/\tau} \; e^{-i\epsilon_n(t_1-t_2)/\hbar} \; e^{-i\epsilon_1 t_2/\hbar - t_2/\tau}
$$

$$
= \frac{e^{-i\epsilon_2 t/\hbar}}{\sqrt{3}} \left(\frac{e\mathcal{E}_0}{\hbar} \right)^2 \sum_{n=2}^{\infty} \alpha_n \underbrace{\int_0^t dt_1 \; e^{\beta_{2n} t_1} \int_0^{t_1} dt_2 \; e^{\beta_{n1} t_2}}_{I} \;,
$$

where in the last line we have defined

$$
\beta_{mn} = i\omega_{mn} - 1/\tau \;, \qquad \omega_{mn} = (\epsilon_m - \epsilon_n)/\hbar \;.
$$

The time integrals yield

$$
I = \int_0^t dt_1 \; e^{\beta_{2n} t_1} \frac{1}{\beta_{n1}} \left(e^{\beta_{n1} t_1} - 1 \right) = \frac{1}{\beta_{n1}} \left(\frac{1}{\beta_{2n} + \beta_{n1}} \left[e^{(\beta_{2n} + \beta_{1n})t} - 1 \right] - \frac{1}{\beta_{2n}} \left(e^{\beta_{2n} t} - 1 \right) \right) \;,
$$

which in the limit $t \gg \tau$ reduces to

$$
\lim_{t \gg \tau} I = -\frac{1}{\beta_{n1}} \left(\frac{1}{\beta_{2n} + \beta_{n1}} - \frac{1}{\beta_{2n}} \right) = \frac{1}{\beta_{2n}(\beta_{n1} + \beta_{2n})} \;, \qquad \beta_{2n} + \beta_{n1} = i\omega_{21} - 2/\tau \;,
$$

and hence

$$
A_{fi}^{(2)} = \frac{e^{-i\epsilon_2 t/\hbar}}{\sqrt{3}} \left(\frac{e\mathcal{E}_0}{\hbar} \right)^2 \frac{1}{i\omega_{21} - 2/\tau} \sum_{n=2}^{\infty} \frac{\alpha_n}{i\omega_{2n} - 1/\tau} \;.
$$

The transition probability for $t \gg \tau$ reads

$$
P_{fi}^{(2)} = \frac{1}{3} \left(\frac{e\mathcal{E}_0 \tau}{\hbar} \right)^4 \frac{1}{(\tau\omega_{21})^2 + 4} \left| \sum_{n=2}^{\infty} \frac{\alpha_n}{i\tau\omega_{2n} - 1} \right|^2 \;.
$$

Problem 10 Charged Harmonic Oscillator Subjected to an Electric Field Pulse

Suppose an electron (mass m and charge $-e$) is in the ground state of a one-dimensional Hamiltonian with a harmonic oscillator potential of angular frequency ω_0. An electric field of amplitude $\mathcal{E}_0$ is suddenly turned on at time $t = 0$ and then off at time $t = \tau$, that is, $\mathcal{E}(t) = \mathcal{E}_0 \left[\theta(t) - \theta(t - \tau) \right]$, where $\theta(u)$ is the step function, equal to 1 for $u > 0$ and equal to 0 for $u < 0$.

1. What is the leading-order probability $P_{10}(t \geq \tau)$ that the electron will be found in the first-excited state of the harmonic oscillator Hamiltonian at time $t \geq \tau$? Same question, but for $P_{20}(t \geq \tau)$, that is, the leading-order probability for a transition to the second excited state.
2. By making use of the properties of the translation operator, the probabilities $P_{10}(t > \tau)$ and $P_{20}(t > \tau)$ can be calculated exactly. Show that the exact expressions reproduce the results obtained above.

Solution

Part 1

In first-order perturbation theory, we have

$$P_{10}^{(1)}(t \geq \tau) = \frac{1}{\hbar^2} \left| \int_0^\tau dt\, e^{i\omega_{10}t} \langle 1| V(t) |0\rangle \right|^2 , \qquad V(t) = e\mathcal{E}_0 x[\theta(t) - \theta(t - \tau)] ,$$

where

$$\omega_{10} = \frac{E_1 - E_0}{\hbar} = \omega_0 .$$

We find

$$P_{10}^{(1)}(t \geq \tau) = \left(\frac{e\mathcal{E}_0}{\hbar} \right)^2 |\langle 1|x|0\rangle|^2 \left| \int_0^\tau dt\, e^{i\omega_0 t} \right|^2 = \left(\frac{e\mathcal{E}_0}{\hbar} \right)^2 |\langle 1|x|0\rangle|^2 \frac{\sin^2(\omega_0\tau/2)}{(\omega_0/2)^2} .$$

and, using

$$x = \sqrt{\frac{\hbar}{2m\omega_0}}(a + a^\dagger) \implies \langle 1|x|0\rangle = \sqrt{\frac{\hbar}{2m\omega_0}} \langle 1| a^\dagger |0\rangle = \sqrt{\frac{\hbar}{2m\omega_0}} ,$$

we arrive at

$$P_{10}^{(1)}(t \geq \tau) = \left(\frac{e\mathcal{E}_0}{\hbar} \right)^2 \frac{\hbar}{2m\omega_0} \frac{\sin^2(\omega_0\tau/2)}{(\omega_0/2)^2} .$$

We see that for fixed ω_0 the first-order probability is proportional to $\sin^2(\omega_0\tau/2)$, thus being largest for τ equal to $n\pi/\omega_0$ with n odd, and vanishing for τ equal to $n\pi/\omega_0$ with n even.

We now turn to calculating the probability for a transition to the second excited state. Since x has a vanishing matrix element between $|0\rangle$ and $|2\rangle$ (by explicit calculation or by parity selection rule), the transition probability between these two states cannot occur in first-order perturbation theory. In second order, we have

$$P_{20}^{(2)}(t \geq \tau) = |A_{20}^{(2)}(t \geq \tau)|^2 ,$$

where

$$A_{20}^{(2)}(t \geq \tau) = \left(\frac{-i}{\hbar} \right)^2 \int_0^\tau dt_1 \int_0^{t_1} dt_2\, e^{-iE_2(t-t_1)/\hbar} \langle 2| V(t_1) \, e^{-iH_0(t_1-t_2)/\hbar} \, V(t_2) |0\rangle\, e^{-iE_0 t_2/\hbar}$$

$$= \sum_{n=0}^{\infty} \left(\frac{-i}{\hbar} \right)^2 \int_0^\tau dt_1 \int_0^{t_1} dt_2\, e^{-iE_2(t-t_1)/\hbar} \langle 2| V(t_1) |n\rangle\, e^{-iE_n(t_1-t_2)/\hbar} \langle n| V(t_2) |0\rangle\, e^{-iE_0 t_2/\hbar} ;$$

in the second line we have inserted the completeness relation. In the sum over states, only the state $|1\rangle$ contributes, since $\langle m|x|n\rangle \neq 0$ iff $m = n \pm 1$, and hence

$$A_{20}^{(2)}(t \geq \tau) = \left(\frac{ie\mathcal{E}_0}{\hbar} \right)^2 \langle 2|x|1\rangle \langle 1|x|0\rangle\, e^{-iE_2 t/\hbar}\, F(\omega_0, \tau) ,$$

where we define

$$F(\omega_0, \tau) = \int_0^\tau dt_1\, e^{i\omega_0 t_1} [\theta(t_1) - \theta(t_1 - \tau)] \int_0^{t_1} dt_2\, e^{i\omega_0 t_2} [\theta(t_2) - \theta(t_2 - \tau)] = \int_0^\tau dt_1\, e^{i\omega_0 t_1} \frac{e^{i\omega_0 t_1} - 1}{i\omega_0}$$

$$= \frac{1}{i\omega_0} \frac{e^{2i\omega_0\tau} - 1}{2i\omega_0} - \frac{e^{i\omega_0\tau} - 1}{(i\omega_0)^2} = -\frac{i}{\omega_0^2} \left[e^{i\omega_0\tau} \sin(\omega_0\tau) - 2\, e^{i\omega_0\tau/2} \sin(\omega_0\tau/2) \right]$$

$$= -\frac{2i}{\omega_0^2}\, e^{i\omega_0\tau/2} \sin(\omega_0\tau/2) \left[e^{i\omega_0\tau/2} \cos(\omega_0\tau/2) - 1 \right] .$$

The magnitude squared of $F(\omega_0, \tau)$ is then given by

$$|F(\omega_0, \tau)|^2 = \frac{4}{\omega_0^4}\, \sin^2(\omega_0\tau/2)\, \left| e^{i\omega_0\tau/2}\, \cos(\omega_0\tau/2) - 1 \right|^2 = \frac{4}{\omega_0^4}\, \sin^4(\omega_0\tau/2)\,,$$

from which we obtain the transition probability

$$P_{20}^{(2)}(t \geq \tau) = 4\left(\frac{e\mathcal{E}_0}{\hbar\omega_0}\right)^4\, |\langle 2|x|1\rangle \langle 1|x|0\rangle|^2\, \sin^4(\omega_0\tau/2)\,,$$

where

$$\langle 2|x|1\rangle \langle 1|x|0\rangle = \frac{\hbar}{2m\omega_0}\, \langle 2|a^\dagger|1\rangle \langle 1|a^\dagger|0\rangle = \sqrt{2}\, \frac{\hbar}{2m\omega_0}\,,$$

and finally

$$P_{20}^{(2)}(t \geq \tau) = 2\left(\frac{e\mathcal{E}_0}{\hbar\omega_0}\right)^4 \left(\frac{\hbar}{m\omega_0}\right)^2\, \sin^4(\omega_0\tau/2)\,.$$

Part 2

In $0 \leq t \leq \tau$ the Hamiltonian reads

$$H = \frac{p^2}{2m} + \frac{m\omega_0^2}{2}\left(x + \frac{e\mathcal{E}_0}{m\omega_0^2}\right)^2 - \frac{(e\mathcal{E}_0)^2}{2m\omega_0^2}\,,$$

that is, the Hamiltonian of a shifted harmonic oscillator. This Hamiltonian, up to the constant term, can be obtained from H_0 by carrying out a translation of the system by $a_0 = -e\mathcal{E}_0/(m\omega_0^2)$, namely

$$\underbrace{\frac{p^2}{2m} + \frac{m\omega_0^2}{2}\left(x + \frac{e\mathcal{E}_0}{m\omega_0^2}\right)^2}_{H + (e\mathcal{E}_0)^2/(2m\omega_0^2)} = U_T(a_0)\, H_0\, U_T^\dagger(a_0)\,,$$

since

$$U_T(a_0)\, x\, U_T^\dagger(a_0) = x - a_0\,, \qquad U_T(a_0)\, p\, U_T^\dagger(a_0) = p\,.$$

A symmetry transformation, such as the translation above, preserves the eigenvalues and therefore H and H_0 have the same eigenvalues, $E_n = \hbar\omega(n + 1/2)$, up to a constant shift $(e\mathcal{E}_0)^2/(2m\omega_0^2)$. By contrast, the eigenstates $|n'\rangle$ of H are related to the eigenstates $|n\rangle$ of H_0 via

$$|n'\rangle = U_T(a_0)|n\rangle = e^{-ia_0\hat{p}/\hbar}|n\rangle\,.$$

Now, at time $t = 0^-$ the system is in the state $|0\rangle$ of H_0, and at time $t = 0$, when the electric field is switched on, we can expand the state $|0\rangle$ in the basis of H eigenstates and then evolve it in time with $e^{-iHt/\hbar}$, namely,

$$|0\rangle = \sum_{n'=0}^{\infty} |n'\rangle\langle n'|0\rangle\,, \qquad |0(t)\rangle = e^{-iHt/\hbar}|0\rangle \qquad \text{in} \qquad 0 \leq t \leq \tau\,.$$

At time $t = \tau$, the state of the system is

$$|0(\tau)\rangle = \sum_{n'=0}^{\infty} e^{-iE_n'\tau/\hbar}|n'\rangle\langle n'|0\rangle\,, \qquad E_n' = E_n - \frac{(e\mathcal{E}_0)^2}{2m\omega_0^2}\,.$$

For $t > \tau$, this state can be expanded in the basis of eigenstates of H_0, and it evolves in time with $e^{-iH_0(t-\tau)/\hbar}$, so that

$$|0(t > \tau)\rangle = \sum_{n=0}^{\infty} e^{-iE_n(t-\tau)/\hbar}|n\rangle\langle n|0(\tau)\rangle .$$

The (exact) probability that the system is found in a state $|m\rangle$ at time $t > \tau$ is given by

$$P_{m0}(t \geq \tau) = |\langle m|0(t > \tau)\rangle|^2 = \left|e^{-iE_m(t-\tau)/\hbar}\langle m|0(\tau)\rangle\right|^2 = |\langle m|0(\tau)\rangle|^2 ,$$

where

$$\langle m|0(\tau)\rangle = \sum_{n'=0}^{\infty} e^{-iE'_n\tau/\hbar}\langle m|n'\rangle\langle n'|0\rangle = \sum_{n=0}^{\infty} e^{-iE'_n\tau/\hbar}\langle m|U_T(a_0)|n\rangle\langle n|U_T^{\dagger}(a_0)|0\rangle .$$

To terms linear in $\mathcal{E}$, we have

$$|n'\rangle = \left(1 - \frac{ia_0}{\hbar}p\right)|n\rangle ,$$

and hence the corresponding probability amplitude for the transition $|0\rangle \longrightarrow |1\rangle$ is given by

$$A_{10}^{(1)}(t \geq \tau) = \sum_{n=0}^{\infty} e^{-iE'_n\tau/\hbar}\langle 1|\left[1 - i(a_0/\hbar)p\right]|n\rangle\langle n|\left[1 + i(a_0/\hbar)p\right]|0\rangle .$$

Since $p \propto a - a^{\dagger}$, only the terms with $n = 0$ and $n = 1$ can contribute in the sum over the states; hence,

$$\langle 1|\left[1 - i(a_0/\hbar)p\right]|0\rangle\langle 0|\left[1 + i(a_0/\hbar)p\right]|0\rangle = -i(a_0/\hbar)\langle 1|p|0\rangle$$

and

$$\langle 1|\left[1 - i(a_0/\hbar)p\right]|1\rangle\langle 1|\left[1 + i(a_0/\hbar)p\right]|0\rangle = i(a_0/\hbar)\langle 1|p|0\rangle ,$$

yielding

$$A_{10}^{(1)}(t \geq \tau) = -i\frac{a_0}{\hbar} e^{-iE'_0\tau/\hbar}\langle 1|p|0\rangle + i\frac{a_0}{\hbar} e^{-iE'_1\tau/\hbar}\langle 1|p|0\rangle = \frac{a_0}{\hbar}\langle 1|p|0\rangle e^{-iE'_0\tau/\hbar}\left(-i + i e^{-i\omega_0\tau}\right)$$

$$= \frac{a_0}{\hbar}\langle 1|p|0\rangle \underbrace{e^{-iE'_0\tau/\hbar} e^{-i\omega_0\tau/2}}_{\text{phase factor}} \underbrace{\left(-i e^{i\omega_0\tau/2} + i e^{-i\omega_0\tau/2}\right)}_{2\sin(\omega_0\tau/2)} .$$

Using the relationship $p = i(m/\hbar)[H_0, x]$, we can express the matrix element of p as

$$\langle 1|p|0\rangle = i\frac{m}{\hbar}\langle 1|H_0x - xH_0|0\rangle = i\frac{m}{\hbar}\underbrace{(E_1 - E_0)}_{\hbar\omega_0}\langle 1|x|0\rangle ,$$

and therefore

$$A_{10}^{(1)}(t \geq \tau) = -2\underbrace{\frac{e\mathcal{E}_0}{m\hbar\omega_0^2}}_{a/\hbar}\underbrace{im\omega_0\langle 1|x|0\rangle}_{\langle 1|p|0\rangle} e^{-iE'_0\tau/\hbar} e^{-i\omega_0\tau/2}\sin(\omega_0\tau/2) ,$$

yielding the transition probability in first order as

$$P_{10}^{(1)}(t \geq \tau) = \left(\frac{e\mathcal{E}_0}{\hbar}\right)^2 |\langle 1|x|0\rangle|^2 \frac{\sin^2(\omega_0\tau/2)}{(\omega_0/2)^2} ,$$

in agreement with the result obtained earlier.

To calculate $P_{20}^{(2)}(t \geq \tau)$, we proceed similarly, except that we need to consider terms quadratic in $\mathcal{E}$, that is,

$$A_{20}^{(2)}(t \geq \tau) = \sum_{n=0}^{\infty} e^{-iE_n'\tau/\hbar} \langle 2| \underbrace{\left[1 - i\frac{a_0}{\hbar}p - \frac{a_0^2}{2\hbar^2}p^2\right]}_{U_T(a_0)} |n\rangle\langle n| \underbrace{\left[1 + i\frac{a_0}{\hbar}p - \frac{a_0^2}{2\hbar^2}p^2\right]}_{U_T^\dagger(a_0)} |0\rangle \,.$$

To terms quadratic in $\mathcal{E}$, we have

$$\langle 2|U_T(a_0)|n\rangle\langle n|U_T^\dagger(a_0)|0\rangle = \underbrace{\langle 2|n\rangle\langle n|0\rangle}_{\text{vanishes}} - i\frac{a_0}{\hbar}\underbrace{\langle 2|p|n\rangle\langle n|0\rangle}_{\text{vanishes}} - \frac{a_0^2}{2\hbar^2}\langle 2|p^2|n\rangle\langle n|0\rangle$$

$$+ i\frac{a_0}{\hbar}\underbrace{\langle 2|n\rangle\langle n|p|0\rangle}_{\text{vanishes}} + \frac{a_0^2}{\hbar^2}\langle 2|p|n\rangle\langle n|p|0\rangle - \frac{a_0^2}{2\hbar^2}\langle 2|n\rangle\langle n|p^2|0\rangle + \cdots \,,$$

since $\langle 2|p|n\rangle$ and $\langle 2|p^2|n\rangle$ are non-vanishing if $n = 1, 3$ and $n = 0, 2, 4$, respectively (recall that $p^2 \propto a^2 + a^{\dagger 2} - aa^\dagger - a^\dagger a$). We conclude that, in the sum over states, the only contributing terms are those with $n = 0, 1, 2$, namely,

$$A_{20}^{(2)}(t \geq \tau) = -\frac{a_0^2}{2\hbar^2} e^{-iE_0'\tau/\hbar} \langle 2|p^2|0\rangle + \frac{a_0^2}{\hbar^2} e^{-iE_1'\tau/\hbar} \langle 2|p|1\rangle\langle 1|p|0\rangle - \frac{a_0^2}{2\hbar^2} e^{-iE_2'\tau/\hbar} \langle 2|p^2|0\rangle \,.$$

Furthermore, the matrix elements of p^2 can be evaluated by inserting the completeness relation for the H_0 basis to yield

$$\langle 2|p^2|0\rangle = \langle 2|p|1\rangle\langle 1|p|0\rangle \,,$$

and therefore

$$A_{20}^{(2)}(t \geq \tau) = -\frac{a_0^2}{2\hbar^2} e^{-iE_0'\tau/\hbar} \langle 2|p|1\rangle\langle 1|p|0\rangle \left[1 - 2 e^{-i\omega_0\tau} + e^{-2i\omega_0\tau}\right] \,.$$

Now, the contents of the square brackets can be written as

$$1 - 2 e^{-i\omega_0\tau} + e^{-2i\omega_0\tau} = \left(1 - e^{-i\omega_0\tau}\right)^2 = \left[e^{-i\omega_0\tau/2}\left(e^{i\omega_0\tau/2} - e^{-i\omega_0\tau/2}\right)\right]^2 = -4 e^{-i\omega_0\tau} \sin^2(\omega_0\tau/2) \,,$$

which yields the following expression for the amplitude:

$$A_{20}^{(2)}(t \geq \tau) = 2\frac{a_0^2}{\hbar^2} (im\,\omega_0)^2 \langle 2|x|1\rangle\langle 1|x|0\rangle e^{-iE_0'\tau/\hbar} e^{-i\omega_0\tau} \sin^2(\omega_0\tau/2) \,,$$

and hence the transition probability

$$P_{20}^{(2)}(t \geq \tau) = 4\left(\frac{e\mathcal{E}_0}{\hbar\omega_0}\right)^4 \sin^4(\omega_0\tau/2) \, |\langle 2|x|1\rangle\langle 1|x|0\rangle|^2 \,,$$

again in agreement with the previous result.

Problem 11 A One-Dimensional Model for the Photoelectric Effect

Consider a particle of mass m and charge q in one dimension, subject to the potential $V(x) = -\alpha\,\delta(x)$ with $\alpha > 0$. This potential admits a single bound state of energy E_0 and corresponding wave function given by

$$E_0 = -\frac{m\alpha^2}{2\hbar^2} \,, \qquad \varphi_0(x) = \sqrt{m\alpha/\hbar^2} \; e^{-m\alpha|x|/\hbar^2} \,.$$

For each positive energy $E = \hbar^2 k^2/(2m)$, there are two independent stationary wave functions $\varphi_{k,+}(x)$ and $\varphi_{k,-}(x)$, corresponding, respectively, to an incident particle coming from the left or from the right. They can be shown to satisfy the continuum normalization (see Problem 9 in Chapter 7),

$$\int_{-\infty}^{\infty} dx\, \varphi_{k,\alpha}^*(x)\, \varphi_{k',\alpha'}(x) = \delta(k - k')\, \delta_{\alpha,\alpha'}\,, \qquad \alpha, \alpha' = \pm\,.$$

The expression for the first wave function, for example, is

$$\varphi_{k,+}(x) = \frac{1}{\sqrt{2\pi}} \left[e^{ikx} - \frac{1}{1 + i\hbar^2 k/(m\alpha)}\, e^{-ikx} \right] \qquad x < 0$$

$$= \frac{1}{\sqrt{2\pi}} \frac{i\hbar^2 k/(m\alpha)}{1 + i\hbar^2 k/(m\alpha)}\, e^{ikx} \qquad x > 0\,.$$

The particle also interacts with an oscillating electric field of angular frequency ω; the perturbation is given by

$$V(t) = -q\mathcal{E}x\, \cos(\omega t)\,,$$

where $\mathcal{E}$ is the amplitude of the electric field. Assuming $\hbar\omega > -E_0$, calculate the transition probability per unit time Γ_{k0} to an arbitrary positive energy state, expressing it as a function of $\lambda = \hbar\omega/|E_0|$. How does Γ_{k0} depend on λ? This problem provides a simple one-dimensional model for photo-ionization (the photoelectric) effect.

Solution

The particle is subject to a periodic perturbation of the type

$$V(t) = -\frac{q\mathcal{E}x}{2}\, e^{i\omega t} - \frac{q\mathcal{E}x}{2}\, e^{-i\omega t} = W^\dagger\, e^{i\omega t} + W e^{-i\omega t}\,, \qquad W = -\frac{q\mathcal{E}x}{2}\,,$$

and the rate for the transition[1] $|\varphi_0\rangle \longrightarrow |\varphi_{k,+}\rangle$ is

$$R_{k,0} = \frac{2\pi}{\hbar} \big[\underbrace{|\langle\varphi_{k,+}| W |\varphi_0\rangle|^2\, \delta(E_k - E_0 - \hbar\omega)}_{\text{absorption}} + \underbrace{|\langle\varphi_{k,+}| W^\dagger |\varphi_0\rangle|^2\, \delta(E_k - E_0 + \hbar\omega)}_{\text{emission}} \big]\,.$$

We note that the energy-conserving δ-function in the second term requires $E_k = E_0 - \hbar\omega$, and this cannot be satisfied ($E_0 < 0$). It is convenient to introduce the parameter $a = \hbar^2/(m\alpha)$, and first to evaluate the matrix element of the position operator between the initial bound state and final continuum state,

$$\langle\varphi_{k,+}| x |\varphi_0\rangle = \int_{-\infty}^{\infty} dx\, \varphi_{k,+}^*(x)\, x\, \varphi_0(x) = \int_{-\infty}^{0} dx\, \frac{1}{\sqrt{2\pi}} \left(e^{-ikx} - \frac{1}{1 - iak}\, e^{ikx} \right) x\, \frac{e^{x/a}}{\sqrt{a}} - \int_{0}^{\infty} dx\, \frac{1}{\sqrt{2\pi}} \frac{iak}{1 - iak}\, e^{-ikx}\, x\, \frac{e^{-x/a}}{\sqrt{a}}$$

$$= \frac{1}{\sqrt{2\pi a}} \left[\int_{-\infty}^{0} dx\, x\, e^{x(1-iak)/a} - \frac{1}{1 - iak} \int_{-\infty}^{0} dx\, x\, e^{x(1+iak)/a} - \frac{iak}{1 - iak} \int_{0}^{\infty} dx\, x\, e^{-x(1+iak)/a} \right]\,.$$

We use

$$\int_{-\infty}^{0} dx\, x\, e^{\lambda x} = \frac{d}{d\lambda} \int_{-\infty}^{0} dx\, e^{\lambda x} = -\frac{1}{\lambda^2}\,, \qquad \int_{0}^{\infty} dx\, x\, e^{-\lambda x} = \frac{1}{\lambda^2}\,,$$

---

[1] One expects the rate for the transition $|\varphi_0\rangle \longrightarrow |\varphi_{k,-}\rangle$ to be the same as that for $|\varphi_0\rangle \longrightarrow |\varphi_{k,+}\rangle$, since it cannot depend on whether the final continuum state has an incoming wave from the left or from the right.

to obtain

$$\langle \varphi_{k,+} | x | \varphi_0 \rangle = \frac{1}{\sqrt{2\pi}\,a} \left[-\frac{a^2}{(1-iak)^2} + \frac{1}{1-iak} \frac{a^2}{(1+iak)^2} - \frac{iak}{1-iak} \frac{a^2}{(1+iak)^2} \right] = -4i \frac{a^{5/2}}{\sqrt{2\pi}} \frac{k}{(1+a^2k^2)^2} .$$

Thus, we find for the transition rate

$$R_{k,0} = \frac{2\pi}{\hbar} \frac{(q\mathcal{E})^2}{4} |\langle \varphi_{k,+} | x | \varphi_0 \rangle|^2 \, \delta(E_k - E_0 - \hbar\omega) = 4 \frac{(q\mathcal{E})^2 a^5}{\hbar} \frac{k^2}{(1+a^2k^2)^4} \delta(E_k - E_0 - \hbar\omega) .$$

For each $k > 0$, there are two independent continuum states, and the density of states as a function of the energy is therefore given by

$$\rho(E_k) = \frac{2}{dE_k/dk} = 2 \frac{m}{\hbar^2 k} ,$$

where $k = \sqrt{2mE_k/\hbar^2}$. Integrating out the δ-function, we arrive at

$$\Gamma_{k0} = \int dE_k \, \rho(E_k) \, R_{k0} = 8 \frac{m\,(q\mathcal{E})^2 a^5}{\hbar^3} \frac{k}{(1+a^2k^2)^4} \bigg|_{E_k = E_0 + \hbar\omega} .$$

and Γ_{k0} has the dimension (time)$^{-1}$. This rate can be expressed as a function of the energy absorbed from the oscillating field, by noting that

$$E_k = E_0 + \hbar\omega \implies k^2 = \frac{2m}{\hbar^2}(E_0 + \hbar\omega) \qquad \text{and} \qquad a^2k^2 = \frac{E_0 + \hbar\omega}{\hbar^2/(2ma^2)} = \frac{\hbar\omega}{|E_0|} - 1 ,$$

which yields

$$\Gamma_{k0} = 4 \frac{(q\mathcal{E}a)^2}{\hbar|E_0|} \frac{\sqrt{\hbar\omega - |E_0|}}{\sqrt{|E_0|}} \frac{|E_0|^4}{(\hbar\omega)^4} , \qquad\qquad \hbar\omega \geq |E_0| .$$

We define $\lambda = \hbar\omega/|E_0|$, and finally write the rate as

$$\Gamma_{k0} = 4 \frac{(q\mathcal{E}a)^2}{\hbar|E_0|} \frac{\sqrt{\lambda - 1}}{\lambda^4} , \qquad\qquad \lambda \geq 1 .$$

As a function of λ, we see that Γ_{k0} has a maximum (for a fixed field amplitude $\mathcal{E}$) at $\lambda = 8/7$, and decreases rapidly as $\lambda \longrightarrow \infty$.

Problem 12 Inelastic Scattering of a Projectile at a Target: A Simple Model

A heavy projectile of charge Ze travels by a heavy target in a straight (undeviated) trajectory (along the $\hat{\mathbf{z}}$-axis) described by $\mathbf{r}_p(t) = b\,\hat{\mathbf{x}} + vt\,\hat{\mathbf{z}}$. The projectile interacts with a particle of charge e in the target via the Coulomb potential

$$V(t) = \frac{Ze^2}{|\mathbf{r}_p(t) - \mathbf{r}|} ,$$

where $\mathbf{r}$ is the position of the target particle. Assume that the Hamiltonian describing this target particle can be modeled by the three-dimensional harmonic oscillator Hamiltonian,

$$H_0 = \frac{\mathbf{p}^2}{2m} + \frac{m\omega_0^2}{2} \mathbf{r}^2 .$$

The target particle is initially in the ground state represented by the wave function $\phi_i(\mathbf{r}) = \phi_0(x)\,\phi_0(y)\,\phi_0(z)$. Show that the probability for a transition to the first excited state, represented by the wave function $\phi_f(\mathbf{r}) = \phi_1(x)\,\phi_0(y)\,\phi_0(z)$ with one quantum of excitation in the transverse direction $\hat{\mathbf{x}}$, is given by

$$P_{fi}(b,v) = 2\left(\frac{Ze^2}{\hbar v}\right)^2\left(\frac{b_0\omega_0}{v}\right)^2 |K_1(\omega_0 b/v)|^2 \,, \qquad K_1(\gamma) = \int_0^\infty d\lambda\,\lambda\,\frac{\sin(\gamma\lambda)}{\sqrt{1+\lambda^2}}\,,$$

where $b_0 = \sqrt{\hbar/(m\omega_0)}$ and $K_1(\gamma)$ is the modified Bessel function of order 1. Assume that $|\mathbf{r}| \ll |\mathbf{r}_p(t)|$, working in the dipole approximation to simplify the calculation.

Let F be the number of projectiles per unit time and per unit area incident on the target. The number of projectiles per unit time traversing a ring of radius b and thickness db surrounding the target is $(2\pi b\,db)F$ and the number of transitions per unit time is then obtained as $F(2\pi b\,db)P_{fi}(b,v) \equiv F d\sigma_{fi}$, yielding the (inelastic) differential cross section $d\sigma_{fi}/db = 2\pi b P_{fi}(b,v)$. Calculate the total cross section under the assumption that $b \gtrsim b_0$ (that is, the projectile is "outside the target"). Note that most of the contribution to the cross section comes from $\omega_0 b/v \lesssim 1$, where $K_1(\omega_0 b/v) \approx v/(b\omega_0)$.

Solution

The perturbation can be expanded as (suppressing the time dependence of $\mathbf{r}_p$ for brevity)

$$V(t) = \frac{Ze^2}{|\mathbf{r}_p - \mathbf{r}|} = \frac{Ze^2}{\sqrt{r_p^2 + r^2 - 2\,\mathbf{r}_p\cdot\mathbf{r}}} = \frac{Ze^2}{r_p}\frac{1}{\sqrt{1 - 2\,\mathbf{r}_p\cdot\mathbf{r}/r_p^2 + (r/r_p)^2}} = \frac{Ze^2}{r_p} + \frac{Ze^2}{r_p^3}\,\mathbf{r}_p\cdot\mathbf{r} + \cdots\,.$$

In this limit the transition matrix element reduces to

$$V_{fi}(t) = \langle\phi_f|\,Ze^2/r_p + (Ze^2/r_p^3)\,\mathbf{r}_p\cdot\mathbf{r}\,|\phi_i\rangle = Ze^2/r_p\,\langle\phi_f|\phi_i\rangle + (Ze^2/r_p^3)\,\mathbf{r}_p\cdot\langle\phi_f|\,\mathbf{r}\,|\phi_i\rangle = (Ze^2/r_p^3)\,\mathbf{r}_p\cdot\langle\phi_f|\,\mathbf{r}\,|\phi_i\rangle\,,$$

and the first term vanishes because of the orthogonality of the initial and final states. Further, the y and z components of the position operator $\mathbf{r} = x\,\hat{\mathbf{x}} + y\,\hat{\mathbf{y}} + z\,\hat{\mathbf{z}}$ have vanishing matrix elements, and

$$\langle\phi_f|\mathbf{r}|\phi_i\rangle = \hat{\mathbf{x}}\,\langle\phi_1|x|\phi_0\rangle\langle\phi_0|\phi_0\rangle\langle\phi_0|\phi_0\rangle = \hat{\mathbf{x}}\,\sqrt{\hbar/(2m\omega_0)}\,\langle 1|a_x + a_x^\dagger|0\rangle = \hat{\mathbf{x}}\,\frac{b_0}{\sqrt{2}}\,,$$

where in the last step we introduced the constant $b_0 = \sqrt{\hbar/(m\omega_0)}$, recall that the ground-state wave function of the harmonic oscillator is $\phi_0(x) \propto \exp[-x^2/(2b_0^2)]$. The matrix element of the perturbation is given by

$$V_{fi}(t) = \frac{Ze^2}{r_p^3}\,\mathbf{r}_p\cdot\hat{\mathbf{x}}\,\frac{b_0}{\sqrt{2}} = Ze^2\,\frac{b}{[b^2 + (vt)^2]^{3/2}}\,\frac{b_0}{\sqrt{2}}\,.$$

From Eq. (18.18) the transition probability follows as (recalling that $E_f - E_i = \hbar\omega_0$)

$$P_{fi}(b,v) = \frac{1}{\hbar^2}\left|\int_{-\infty}^\infty dt\,e^{i\omega_0 t}\,V_{fi}(t)\right|^2 = \frac{1}{2}\left(\frac{Ze^2 b_0}{\hbar}\right)^2 |I|^2\,,$$

where

$$I = \int_{-\infty}^\infty dt\,e^{i\omega_0 t}\,\frac{b}{[b^2 + (vt)^2]^{3/2}} = -\frac{d}{db}\int_{-\infty}^\infty dt\,e^{i\omega_0 t}\,\frac{1}{\sqrt{b^2 + (vt)^2}} = -\frac{1}{v}\frac{d}{db}\int_{-\infty}^\infty d\lambda\,e^{i(\omega_0 b/v)\lambda}\,\frac{1}{\sqrt{1+\lambda^2}}$$

$$= -\frac{\omega_0}{v^2}\frac{d}{d\gamma}\int_{-\infty}^\infty d\lambda\,e^{i\gamma\lambda}\,\frac{1}{\sqrt{1+\lambda^2}} = -\frac{\omega_0}{v^2}\frac{d}{d\gamma}\int_{-\infty}^\infty d\lambda\,\frac{\cos(\gamma\lambda)}{\sqrt{1+\lambda^2}}$$

$$= -\frac{2\omega_0}{v^2}\frac{d}{d\gamma}\int_0^\infty d\lambda\,\frac{\cos(\gamma\lambda)}{\sqrt{1+\lambda^2}} = \frac{2\omega_0}{v^2}\underbrace{\int_0^\infty d\lambda\,\lambda\,\frac{\sin(\gamma\lambda)}{\sqrt{1+\lambda^2}}}_{K_1(\gamma)}\,,$$

and in the first line we changed the integration variable to $\lambda = vt/b$ and in the second line defined $\gamma = \omega_0 b/v$.[2] The transition probability is then obtained as given in the text of the problem.

The total cross section is given by

$$\sigma_{fi} \approx 2\pi \int_{b_0}^{v/\omega_0} db\, b P_{fi}(b,v) \approx 4\pi \left(\frac{Ze^2}{\hbar v}\right)^2 \left(\frac{b_0\omega_0}{v}\right)^2 \int_{b_0}^{v/\omega_0} db\, b \left(\frac{v}{b\,\omega_0}\right)^2 \approx 4\pi b_0^2 \left(\frac{Ze^2}{\hbar v}\right)^2 \ln\left(\frac{v}{b_0\omega_0}\right),$$

where the lower integration limit has been taken as b_0 and the upper one as v/ω_0, the latter limit ensuring that $b\omega_0/v \leq 1$, where most of the contribution to the integral comes. In such a region $K_1(b\omega_0/v) \approx v/(b\omega_0)$, hence the result above.

Problem 13   Ionization of a Hydrogen Atom by an External Electromagnetic Field

In this problem, we will calculate the cross section for the ionization of a hydrogen atom by an electromagnetic wave. The (real) vector potential $\mathbf{A}(\mathbf{r}, t)$ inducing the electric and magnetic fields associated with this wave is written as

$$\mathbf{A}(\mathbf{r}, t) = \mathbf{A_k} \left[e^{i(\mathbf{k}\cdot\mathbf{r} - \omega t)} + e^{-i(\mathbf{k}\cdot\mathbf{r} - \omega t)} \right], \qquad \mathbf{A_k} = A\,\hat{e}_\mathbf{k},$$

where $\hat{e}_\mathbf{k}$ is the unit vector specifying the polarization of the electromagnetic wave, and in the Coulomb gauge (which we assume here) we have $\nabla \cdot \mathbf{A}(\mathbf{r}, t) = 0$, implying $\mathbf{k} \cdot \hat{e}_\mathbf{k} = 0$.

1. Write down the Hamiltonian, including the coupling of the magnetic field $\mathbf{B}(\mathbf{r}, t) = \nabla \times \mathbf{A}(\mathbf{r}, t)$ to the electron spin magnetic moment. The proton is assumed to be fixed at the origin. Justify why in the long-wavelength approximation (LWA), where the wavelength λ associated with the electromagnetic wave is much larger than the Bohr radius a_0, the electron spin magnetic-moment term can be neglected.

2. Hereafter, ignore the quadratic term in the vector potential, and show that the Hamiltonian consists of the unperturbed term $\hat{H}_0$ (the hydrogen atom Hamiltonian) and a perturbation $\hat{V}(t)$ given by

$$\hat{V}(t) = \frac{e}{m_e c}\left(e^{-i\omega t}\,\hat{W} + e^{i\omega t}\,\hat{W}^\dagger\right).$$

Obtain the operator $\hat{W}$.

3. Consider the transition, induced by $\hat{V}(t)$, from the initial ground state $|\phi_{1s}\rangle = |\phi_{100}\rangle$ of energy $E_i = E_{1s} = -e^2/(2a_0)$ to a final unbound state (where the atom is ionized) of energy $E_f = E_p = \mathbf{p}^2/(2m_e)$. This final state should in principle be represented by a continuum solution of $\hat{H}_0$ (a Coulomb wave function). However, if the energy with which the electron is emitted is much larger than its binding energy $e^2/(2a_0)$, then it is reasonable simply to approximate the final state by a free-particle state $|\phi_\mathbf{p}\rangle$, an eigenstate of the momentum operator (but not of H_0!). Under these assumptions, namely

$$ka_0 \ll 1 \qquad \text{and} \qquad E_p \gg \frac{e^2}{2a_0},$$

calculate the rate $R_{\mathbf{p},1s}$ for the transition $1s \longrightarrow \mathbf{p}$, and hence the rate $\Delta\Gamma(1s \longrightarrow \mathbf{p})$ summed over a set of final electron states with momenta in a shell $\Delta\mathbf{p}$, that is,

---

[2] Note that (see M. Abramowitz and I. A. Stegun (1965), *Handbook of Mathematical Functions*, Dover)

$$K_1(\gamma) = \frac{1}{\gamma} \text{ as } \gamma \longrightarrow 0 \qquad \text{and} \qquad K_1(\gamma) = \sqrt{\frac{\pi}{2\gamma}}\, e^{-\gamma} \text{ as } \gamma \longrightarrow \infty.$$

$$\Delta\Gamma(1s \longrightarrow \mathbf{p}) = \int_{\Delta\mathbf{p}} d\mathbf{p}\, R_{\mathbf{p},1s} \equiv \int_{\Delta\Omega} d\Omega\, \underbrace{\frac{d}{d\Omega}\Gamma(1s \longrightarrow \mathbf{p})}_{\text{differential rate}} ,$$

showing that the differential rate can be expressed as

$$\frac{d}{d\Omega}\Gamma(1s \rightarrow \mathbf{p}) = \frac{16\alpha}{\pi} \frac{A^2}{m_e c} \frac{(\hat{\mathbf{e}}_{\mathbf{k}} \cdot \hat{\mathbf{p}})^2}{(pa_0/\hbar)^5} ,$$

where $\alpha = e^2/(\hbar c)$ is the fine-structure constant and $\hat{\mathbf{p}}$ is the unit vector specifying the the electron momentum direction (note that here it is *not* the electron momentum operator); this direction is relative to the incoming electromagnetic wave direction specified by $\hat{\mathbf{k}}$. Lastly, show that the conditions above for the validity of this result are not incompatible with each other.

4. Show that the energy density and energy flux (the Poynting vector), associated with the electromagnetic wave and averaged over the period $T = 2\pi/\omega$, are given by

$$\langle u \rangle_T = \frac{1}{2\pi} \frac{\omega^2}{c^2} A^2 , \qquad \langle \mathbf{S} \rangle_T = \frac{\omega}{2\pi} A^2\, \mathbf{k} .$$

Since the Poynting vector $\langle \mathbf{S} \rangle_T$ has dimension of energy per unit area and per unit time, a photon" flux (that is, the number of photons per unit area and per unit time) can be defined:[3]

$$\text{photon flux} = \frac{\langle \mathbf{S} \rangle_T}{\hbar\omega} = \frac{A^2}{2\pi\hbar}\, \mathbf{k} .$$

Having determined the photon flux, obtain the photo-ionization cross section in hydrogen corresponding to a given polarization $\boldsymbol{\epsilon}_{\mathbf{k}}$ of the incident wave.

5. There are two independent (mutually orthogonal) polarizations, $\hat{\mathbf{e}}_{\mathbf{k},1}$ and $\hat{\mathbf{e}}_{\mathbf{k},2}$, each orthogonal to $\mathbf{k}$, such that $\hat{\mathbf{e}}_{\mathbf{k},1}$, $\hat{\mathbf{e}}_{\mathbf{k},2}$, and $\hat{\mathbf{k}}$ constitute a basis in three-dimensional space (that is, these three unit vectors specify a right-handed coordinate system). The cross section obtained in part 4 corresponds to one of these polarizations $\hat{\mathbf{e}}_{\mathbf{k},i}$. Obtain the differential cross section averaged over the polarizations of the electromagnetic wave and the corresponding total cross section by integrating over the directions of the emitted electron.

Solution

Part 1

The Hamiltonian describing the coupled system of hydrogen atom and electromagnetic wave reads (with $-e$ the electron charge)

$$\hat{H}(t) = \frac{1}{2m_e}\left[\hat{\mathbf{p}} + \frac{e}{c}\mathbf{A}(\hat{\mathbf{r}}, t)\right]^2 - \frac{e^2}{r} + \frac{g_e e}{2m_e c}\hat{\mathbf{S}} \cdot \mathbf{B}(\hat{\mathbf{r}}, t) .$$

The ratio of the spin term and the orbital term is of the order (for this estimate, we take $g_e \approx 2$ and simply replace the spin operator with $\hbar/2$)

$$\left|\frac{\text{spin}}{\text{orbital}}\right| \approx \left|\underbrace{\frac{e\hbar B}{2m_e c}}_{\text{spin}} \underbrace{\frac{1}{eAp/(m_e c)}}_{\text{1/orbital}}\right| \approx \frac{ka_0}{2} ,$$

---

[3] At this stage, this definition is simply *ad hoc*, since the concept of photons has not emerged as a consequence of the quantization of the electromagnetic field. However, the expressions for the cross section and, in fact, radiative transition amplitudes that we will derive below and in Problems 13–15, when properly interpreted, remain correct, even when the fields are quantized. This treatment is known as the semiclassical approach.

where we have made the replacement $\hbar/p \approx a_0$ (using Heisenberg's uncertainty relation) and, noting that

$$\mathbf{B}(\mathbf{r}, t) = i\,\mathbf{k} \times \mathbf{A}_\mathbf{k} \left[e^{i(\mathbf{k}\cdot\mathbf{r}-\omega t)} - e^{-i(\mathbf{k}\cdot\mathbf{r}-\omega t)} \right] ,$$

have expressed the ratio B/A of the magnetic field magnitude and the vector potential magnitude as approximately given by the wave number k. In the LWA we have (recall that $k = 2\pi/\lambda$)

$$\frac{ka_0}{2} = \pi a_0/\lambda \ll 1 ,$$

and hence the spin term can be neglected in this limit.

Part 2

In the LWA the Hamiltonian can be simply taken as

$$\hat{H}(t) = \underbrace{\frac{\hat{\mathbf{p}}^2}{2m_e} - \frac{e^2}{r}}_{\hat{H}_0} + \underbrace{\frac{e}{m_e c}\,\mathbf{A}(\hat{\mathbf{r}}, t) \cdot \hat{\mathbf{p}}}_{\hat{V}(t)} ,$$

since in Coulomb gauge the linear terms in the vector potential reduce to

$$\hat{\mathbf{p}} \cdot \mathbf{A} + \mathbf{A} \cdot \hat{\mathbf{p}} = \sum_i \left(\underbrace{[\hat{p}_i, A_i]}_{-i\hbar\,\partial_i A_i} + 2A_i\hat{p}_i \right) = -i\hbar\,\underbrace{\boldsymbol{\nabla}\cdot\mathbf{A}}_{\text{vanishes}} + 2\mathbf{A}\cdot\hat{\mathbf{p}} .$$

The perturbation is of the periodic type and manifestly hermitian,

$$\hat{V}(t) = \frac{e}{m_e c}\left(e^{-i\omega t}\,\underbrace{e^{i\mathbf{k}\cdot\hat{\mathbf{r}}}\mathbf{A}_\mathbf{k}\cdot\hat{\mathbf{p}}}_{\hat{W}} + e^{i\omega t}\,\underbrace{e^{-i\mathbf{k}\cdot\hat{\mathbf{r}}}\mathbf{A}_\mathbf{k}\cdot\hat{\mathbf{p}}}_{\hat{W}^\dagger} \right) = \frac{e}{m_e c}\left(e^{-i\omega t}\,\hat{W} + e^{i\omega t}\,\hat{W}^\dagger \right) ,$$

given that the operator $\hat{W}$ satisfies

$$\hat{W}^\dagger = (e^{i\mathbf{k}\cdot\hat{\mathbf{r}}}\mathbf{A}_\mathbf{k}\cdot\hat{\mathbf{p}})^\dagger = \hat{\mathbf{p}}\cdot\mathbf{A}_\mathbf{k}\,e^{-i\mathbf{k}\cdot\hat{\mathbf{r}}} = e^{-i\mathbf{k}\cdot\hat{\mathbf{r}}}\mathbf{A}_\mathbf{k}\cdot\hat{\mathbf{p}} - \hbar\,\underbrace{\mathbf{k}\cdot\mathbf{A}_\mathbf{k}}_{\text{vanishes}}\,e^{-i\mathbf{k}\cdot\hat{\mathbf{r}}} ,$$

where we have used the condition $\mathbf{k}\cdot\mathbf{A}_\mathbf{k} = 0$.

Part 3

The transition rate follows from Eq. (18.34):

$$R_{\mathbf{p},1s} = \frac{2\pi}{\hbar}\,\frac{e^2}{m_e^2 c^2}\,|\langle\phi_\mathbf{p}|\,\hat{W}\,|\phi_{1s}\rangle|^2\,\delta(E_p - E_{1s} - \hbar\omega) ,$$

where only the first term contributes (the second δ-function can never be satisfied, since $E_p > 0$ and $E_{1s} < 0$). This term corresponds to the absorption of energy $\hbar\omega$ from the incident electromagnetic wave (i.e., the absorption of a "photon" of energy $\hbar\omega$), resulting in the emission of the electron. As already noted, treating it as a free particle is justified if the energy $E_p \gg |E_{1s}|$, and in this regime the rate reduces to

$$R_{\mathbf{p},1s} = \frac{2\pi}{\hbar}\,\frac{e^2}{m_e^2 c^2}\,|\langle\phi_\mathbf{p}|\,\hat{W}\,|\phi_{1s}\rangle|^2\,\delta(E_p - \hbar\omega) ,$$

with the matrix element given by

$$\langle\phi_{\mathbf{p}}|\,\hat{W}\,|\phi_{1s}\rangle = \langle\phi_{\mathbf{p}}|\,e^{i\mathbf{k}\cdot\hat{\mathbf{r}}}\mathbf{A_k}\cdot\hat{\mathbf{p}}\,|\phi_{1s}\rangle = -i\hbar\int d\mathbf{r}\,\frac{e^{-i\mathbf{p}\cdot\mathbf{r}/\hbar}}{(2\pi\hbar)^{3/2}}\,e^{i\mathbf{k}\cdot\mathbf{r}}\,\mathbf{A_k}\cdot\left[\boldsymbol{\nabla}\phi_{1s}(r)\right]\,.$$

In the LWA we can set the plane wave factor $e^{i\mathbf{k}\cdot\mathbf{r}}$ as approximately equal to unity.[4] This is so because the bound-state wave function (or, rather, its gradient) is exponentially small for $r\gg a_0$, and therefore the contribution to the integral is mostly coming from a spherical region of radius of order a_0, where $|\mathbf{k}\cdot\mathbf{r}|\ll 1$. Under these conditions the matrix element can be expressed as

$$\langle\phi_{\mathbf{p}}|\,\hat{W}\,|\phi_{1s}\rangle = \mathbf{A_k}\cdot\langle\phi_{\mathbf{p}}|\,\hat{\mathbf{p}}\,|\phi_{1s}\rangle = \mathbf{A_k}\cdot\mathbf{p}\,\langle\phi_{\mathbf{p}}|\phi_{1s}\rangle\,,$$

where we have taken advantage of the fact that $|\phi_{\mathbf{p}}\rangle$ is an eigenstate of the momentum operator.[5] Note that the inner product $\langle\phi_{\mathbf{p}}|\phi_{1s}\rangle$ does not vanish here. In contrast with $|\phi_{1s}\rangle$, $|\phi_{\mathbf{p}}\rangle$ is *not* in the basis of $\hat{H}_0$ eigenstates; as a matter of fact, it is an exact eigenstate of the kinetic energy operator alone, but not of $\hat{H}_0$. With the normalized ground-state wave function

$$\phi_{1s}(r) = \frac{1}{\sqrt{\pi a_0^3}}\,e^{-r/a_0}\,,$$

the overlap integral is given by

$$\langle\phi_{\mathbf{p}}|\phi_{1s}\rangle = \frac{1}{\sqrt{\pi a_0^3}}\,\frac{1}{\sqrt{(2\pi\hbar)^3}}\int d\mathbf{r}\,e^{-i\mathbf{p}\cdot\mathbf{r}/\hbar}\,e^{-r/a_0} = \frac{1}{\sqrt{\pi a_0^3}}\,\frac{1}{\sqrt{(2\pi\hbar)^3}}\,\frac{8\pi\,a_0^3}{[(pa_0/\hbar)^2+1]^2}\,,$$

where

$$\frac{p^2 a_0^2}{\hbar^2} = \frac{2m_e E_p a_0}{\hbar^2}\,\underbrace{\frac{\hbar^2}{m_e e^2}}_{a_0} = \frac{2\,\overbrace{\hbar c k}^{\hbar\omega}\,a_0}{e^2} = \frac{k a_0}{\alpha/2}\gg 1\,.$$

Therefore we have

$$R_{\mathbf{p},1s} = \frac{2\pi}{\hbar}\,\frac{e^2}{m_e^2 c^2}\,\frac{1}{\pi a_0^3}\,\frac{1}{(2\pi\hbar)^3}\,\frac{64\pi^2 a_0^6}{(pa_0/\hbar)^8}\,p^2 A^2\,(\hat{\mathbf{e}}_{\mathbf{k}}\cdot\hat{\mathbf{p}})^2\,\delta(E_p-\hbar\omega)\equiv\overline{R}_{\mathbf{p},1s}\,\delta(E_p-\hbar\omega)\,,$$

$$\overline{R}_{\mathbf{p},1s} = \frac{16\alpha}{\pi}\,\frac{A^2}{pm_e^2 c}\,\frac{(\hat{\mathbf{e}}_{\mathbf{k}}\cdot\hat{\mathbf{p}})^2}{(pa_0/\hbar)^5}\,,$$

from which the rate $\Delta\Gamma(1s\to\mathbf{p})$ follows:

$$\Delta\Gamma(1s\to\mathbf{p}) = \int_{\Delta\mathbf{p}}d\mathbf{p}\,R_{\mathbf{p},1s} = \int_{\Delta\mathbf{p}}d\Omega\,dE_p\,m_e p\,\delta(E_p-\hbar\omega)\,\overline{R}_{\mathbf{p},1s}\,.$$

---

[4] Of course, we can evaluate the transition rate without relying on the LWA; however, in that case for a realistic treatment we also need to account for spin-dependent interaction effects.

[5] Had we used the exact continuum state $|\phi_{E_p}\rangle$ to describe the final electron then this property would not be satisfied, since such a state would not be an eigenstate of the momentum operator. However, evaluation of the matrix element $\mathbf{A_k}\cdot\langle\phi_{E_p}|\,\hat{\mathbf{p}}\,|\phi_{1s}\rangle$ in that case would be facilitated by the identity $\hat{\mathbf{p}} = -i(m_e/\hbar)[\,\hat{\mathbf{r}},\hat{H}_0\,]$, yielding

$$\langle\phi_{E_p}|\,\hat{W}\,|\phi_{1s}\rangle = -i(m_e/\hbar)(E_{1s}-E_p)\,\mathbf{A_k}\cdot\langle\phi_{E_p}|\,\hat{\mathbf{r}}\,|\phi_{1s}\rangle = im_e\omega\,\mathbf{A_k}\cdot\langle\phi_{E_p}|\,\hat{\mathbf{r}}\,|\phi_{1s}\rangle\,,$$

where in the last line we have used the conservation of energy as enforced by the δ-function.

Integrating out the δ-function yields the differential ionization rate given in the text of the problem. This result is valid in the regime specified there: the conditions $ka_0 \ll 1$ and $E_p = \hbar\omega = \hbar ck \gg e^2/(2a_0)$ must be simultaneously satisfied,

$$\underbrace{\frac{e^2}{2\hbar c}}_{\alpha/2} \ll ka_0 \ll 1 \,,$$

where $\alpha \approx 1/137$ is the fine structure constant; hence, $1/274 \ll ka_0 \ll 1$. The two conditions are not incompatible with each other.

Part 4

The energy density and energy flux associated with the monochromatic electromagnetic wave follow from

$$u(\mathbf{r}, t) = \frac{1}{8\pi} \left[\mathbf{E}^2(\mathbf{r}, t) + \mathbf{B}^2(\mathbf{r}, t) \right] \,, \qquad \mathbf{S}(\mathbf{r}, t) = \frac{c}{4\pi} \, \mathbf{E}(\mathbf{r}, t) \times \mathbf{B}(\mathbf{r}, t) \,,$$

where

$$\mathbf{E}(\mathbf{r}, t) = -\frac{\partial}{c \, \partial t} \mathbf{A}(\mathbf{r}, t) \,, \qquad \mathbf{B}(\mathbf{r}, t) = \boldsymbol{\nabla} \times \mathbf{A}(\mathbf{r}, t) \,.$$

In the absence of charge and current densities, $u(\mathbf{r}, t)$ and $\mathbf{S}(\mathbf{r}, t)$ satisfy the local conservation law $\partial u(\mathbf{r}, t)/\partial t + \boldsymbol{\nabla} \cdot \mathbf{S}(\mathbf{r}, t) = 0$. Using the fact that the fields are real ($\mathbf{E} = \mathbf{E}^*$ and $\mathbf{B} = \mathbf{B}^*$), we find

$$\mathbf{E} \cdot \mathbf{E}^* = \frac{\omega^2}{c^2} \mathbf{A}_{\mathbf{k}}^2 \left(e^{i\phi} - e^{-i\phi} \right) \left(e^{-i\phi} - e^{i\phi} \right) = 2 \frac{\omega^2}{c^2} A^2 \left[1 - \cos(2\phi) \right] \,,$$

$$\mathbf{B} \cdot \mathbf{B}^* = |\mathbf{k} \times \mathbf{A}_{\mathbf{k}}|^2 \left(e^{i\phi} - e^{-i\phi} \right) \left(e^{-i\phi} - e^{i\phi} \right) = 2 \, k^2 A^2 \left[1 - \cos(2\phi) \right] \,,$$

$$\mathbf{E} \times \mathbf{B}^* = \frac{\omega}{c} \mathbf{A}_{\mathbf{k}} \times (\mathbf{k} \times \mathbf{A}_{\mathbf{k}}) \left(e^{i\phi} - e^{-i\phi} \right) \left(e^{-i\phi} - e^{i\phi} \right) = 2 \frac{\omega}{c} A^2 \mathbf{k} \left[1 - \cos(2\phi) \right] \,,$$

where $\phi = \mathbf{k} \cdot \mathbf{r} - \omega t$ and the condition $\mathbf{k} \cdot \mathbf{A}_{\mathbf{k}} = 0$ has been used. Averaging these quantities over the period $T = 2\pi/\omega$,

$$\frac{1}{T} \int_0^T dt \underbrace{[1 - \cos(2\mathbf{k} \cdot \mathbf{r} - 2\,\omega t)]}_{1-\cos(2\phi)} = 1 + \frac{1}{2\omega T} \left. \sin(2\mathbf{k} \cdot \mathbf{r} - 2\omega t) \right|_0^T = 1 \,,$$

removes the explicit time dependence, yielding the expressions for u and $\mathbf{S}$ given in the text of the problem.

The differential cross section for photo-ionization is obtained by dividing the differential transition rate by the incident "photon flux":

$$\frac{d}{d\Omega} \sigma(1s \to \mathbf{p}) = \frac{1}{kA^2/(2\pi\hbar)} \frac{d}{d\Omega} \Gamma(1s \to \mathbf{p}) = \frac{32\alpha}{k} \frac{\hbar}{m_e c} \frac{(\hat{\boldsymbol{\epsilon}}_{\mathbf{k}} \cdot \hat{\mathbf{p}})^2}{(pa_0/\hbar)^5} \,,$$

and the A^2 factors cancel out. Since the photon wavelength k has dimension 1/length and $\hbar/(m_e c)$ is the Compton wavelength of the electron, the cross section has dimension (length)^2, as expected.

Part 5

The differential cross section averaged over the polarization directions of the incoming electromagnetic wave is given by

$$\left\langle \frac{d}{d\Omega}\sigma(1s \to \mathbf{p})\right\rangle = \frac{1}{2}\sum_{i=1,2}\frac{32\alpha}{k}\frac{\hbar}{m_e c}\frac{(\hat{e}_{\mathbf{k},i}\cdot\hat{\mathbf{p}})^2}{(pa_0/\hbar)^5}\ .$$

Since $\hat{e}_{\mathbf{k},1}$, $\hat{e}_{\mathbf{k},2}$, and $\hat{\mathbf{k}}$ form an orthonormal basis, any vector $\mathbf{V}$ can be expanded as

$$\mathbf{V} = \hat{e}_{\mathbf{k},1}(\hat{e}_{\mathbf{k},1}\cdot\mathbf{V}) + \hat{e}_{\mathbf{k},2}(\hat{e}_{\mathbf{k},2}\cdot\mathbf{V}) + \hat{\mathbf{k}}(\hat{\mathbf{k}}\cdot\mathbf{V})\ ,$$

or, in terms of components,

$$V_\alpha = \sum_\beta (\hat{e}_{\mathbf{k},1}^\alpha \hat{e}_{\mathbf{k},1}^\beta + \hat{e}_{\mathbf{k},2}^\alpha \hat{e}_{\mathbf{k},2}^\beta + \hat{k}^\alpha \hat{k}^\beta)V_\beta \implies \hat{e}_{\mathbf{k},1}^\alpha \hat{e}_{\mathbf{k},1}^\beta + \hat{e}_{\mathbf{k},2}^\alpha \hat{e}_{\mathbf{k},2}^\beta + \hat{k}^\alpha \hat{k}^\beta = \delta_{\alpha\beta}\ .$$

It follows that

$$\sum_{i=1,2}(\mathbf{V}\cdot\hat{e}_{\mathbf{k},i})^2 = \sum_{\alpha\beta}\underbrace{(\hat{e}_{\mathbf{k},1}^\alpha \hat{e}_{\mathbf{k},1}^\beta + \hat{e}_{\mathbf{k},2}^\alpha \hat{e}_{\mathbf{k},2}^\beta)}_{\delta_{\alpha\beta}-\hat{k}^\alpha \hat{k}^\beta} V_\alpha V_\beta = \mathbf{V}\cdot\mathbf{V} - (\mathbf{V}\cdot\hat{\mathbf{k}})^2\ .$$

In the present case $\mathbf{V}$ is the unit vector $\hat{\mathbf{p}}$ along the momentum direction of the emitted electron, and therefore

$$\left\langle \frac{d}{d\Omega}\sigma(1s \to \mathbf{p})\right\rangle = \frac{16\alpha}{k}\frac{\hbar}{m_e c}\frac{\sin^2\theta}{(pa_0/\hbar)^5}\ .$$

When averaged over the initial polarizations, the differential cross section is independent of ϕ and is largest when $\theta = \pi/2$. The total cross section is then

$$\langle\sigma(1s \to \mathbf{p})\rangle = \frac{16\alpha}{k}\frac{\hbar}{m_e c}\frac{1}{(pa_0/\hbar)^5}\int_0^{2\pi}d\phi\int_{-1}^1 dx\,(1-x^2) = \frac{128\pi\alpha}{3k}\frac{\hbar}{m_e c}\frac{1}{(pa_0/\hbar)^5}\ ,$$

in the regime $\alpha/2 \ll ka_0 \ll 1$.

Problem 14   Cross Sections for Stimulated Absorption and Emission in Hydrogen

Calculate the cross sections for stimulated absorption and emission in hydrogen induced by a monochromatic electromagnetic wave of wave number $\mathbf{k}$, resulting in a transition from an initial bound state $|\phi_i\rangle$ to a final bound state $|\phi_f\rangle$ (for absorption, $E_i < E_f$ while for emission $E_i > E_f$). Treat the cases in which the electromagnetic wave has polarization $\hat{e}_{\mathbf{k}}$ or is unpolarized, and work in the long-wavelength approximation (LWA), utilizing the semiclassical approach developed in Problem 13, a prerequisite for the present problem. Comment the results.

Solution

As shown in Problem 13, in the LWA the coupling of the spin magnetic moment of the electron to the magnetic field induced by the electromagnetic wave can be neglected, and the Hamiltonian is simply taken as

$$\hat{H}(t) = \hat{H}_0 + \hat{V}(t)\ , \qquad \hat{V}(t) = \frac{e}{m_e c}\left(e^{-i\omega t}\hat{W} + e^{i\omega t}\hat{W}^\dagger\right),$$

where $\hat{H}_0$ is the hydrogen atom Hamiltonian (relativistic corrections in $\hat{H}_0$ due to kinetic energy, Darwin–Foldy, and spin–orbit terms are ignored here for simplicity but could be accounted for without incurring conceptual difficulties). We have also introduced the definitions of Problem 13,

$$\hat{W} = A_\mathbf{k}\, e^{i\mathbf{k}\cdot\hat{\mathbf{r}}}\, \hat{e}_\mathbf{k} \cdot \hat{\mathbf{p}}\,, \qquad \hat{W}^\dagger = A_\mathbf{k}\, e^{-i\mathbf{k}\cdot\hat{\mathbf{r}}}\, \hat{e}_\mathbf{k} \cdot \hat{\mathbf{p}}\,,$$

where $\hat{e}_\mathbf{k}$ is the polarization vector of the electromagnetic wave. From Eq. (18.34), the transition rate from an initial (bound) state $|\phi_i\rangle$ of energy E_i to a final (bound) state $|\phi_f\rangle$ of energy E_f is given by

$$R_{f,i} = \frac{2\pi}{\hbar}\,\frac{e^2}{m_e^2 c^2}\, \big[\underbrace{|\langle\phi_f|\,\hat{W}\,|\phi_i\rangle|^2\ \delta(E_f - E_i - \hbar\omega)}_{\text{absorption}} + \underbrace{|\langle\phi_f|\,\hat{W}^\dagger\,|\phi_i\rangle|^2\ \delta(E_f - E_i + \hbar\omega)}_{\text{emission}} \big]\,.$$

In the LWA (which amounts to setting $e^{i\mathbf{k}\cdot\mathbf{r}} \approx 1$), the matrix element is given by $\langle\phi_f|\,\hat{W}\,|\phi_i\rangle = A\,\hat{e}_\mathbf{k}\cdot \langle\phi_f|\,\hat{\mathbf{p}}\,|\phi_i\rangle$. Using the identity[6]

$$\hat{\mathbf{p}} = -i\,\frac{m_e}{\hbar}\,[\,\hat{\mathbf{r}}, \hat{H}_0\,]\,,$$

we find

$$\langle\phi_f|\,\hat{W}\,|\phi_i\rangle = -i\frac{m_e}{\hbar}(E_i - E_f)\, A\,\langle\phi_f|\,\hat{e}_\mathbf{k}\cdot\hat{\mathbf{r}}\,|\phi_i\rangle = \pm i m_e \omega A\langle\phi_f|\,\hat{e}_\mathbf{k}\cdot\hat{\mathbf{r}}\,|\phi_i\rangle\,,$$

where the $\pm$ sign depends on whether we are dealing with absorption $(+)$ or emission $(-)$, and the transition rate is then obtained as

$$R_{fi}^{\text{ab/em}}(\omega) = \frac{2\pi}{\hbar}\,\frac{e^2\omega^2}{c^2}\, A^2\,|\langle\phi_f|\,\hat{e}_\mathbf{k}\cdot\hat{\mathbf{r}}\,|\phi_i\rangle|^2\ \delta(E_f - E_i \mp \hbar\omega)$$

$$= 2\pi\alpha\,\frac{\omega^2}{c}\, A^2\,|\langle\phi_f|\,\hat{e}_\mathbf{k}\cdot\hat{\mathbf{r}}\,|\phi_i\rangle|^2\ \delta(E_f - E_i \mp \hbar\omega)\,,$$

where in the δ-function the $-$ sign is for absorption and the $+$ sign is for emission (α is the fine structure constant). The cross section for the discrete transition $i \longrightarrow f$ induced by the electromagnetic wave follows by dividing the rate by the incident "photon flux" (see Problem 13):

$$\sigma_{fi}^{\text{ab/em}}(\omega) = \frac{R_{fi}^{\text{ab/em}}(\omega)}{kA^2/(2\pi\hbar)} = 4\pi^2\alpha\,\hbar\omega\,|\langle\phi_f|\,\hat{e}_\mathbf{k}\cdot\hat{\mathbf{r}}\,|\phi_i\rangle|^2\ \delta(E_f - E_i \mp \hbar\omega)\,,$$

which removes the A^2 factor. The cross section is proportional to the matrix element squared of the electric dipole operator $\hat{\mathbf{D}} = -e\,\hat{\mathbf{r}}$ between the initial and final states, and it vanishes unless the energy difference between these states matches the "photon energy" $\hbar\omega$. Note that the calculation above ignores the energy of the recoiling atom after the absorption or emission process (see Problem 7 for a treatment of this effect). Lastly, in reality sharp δ-function peaks are not observed, but rather peaks that are spread out by the natural line width (see Problem 17). As a consequence, what is actually measured is the area under the curve, namely

$$\int_{\text{over peak}} d\omega\, \sigma_{fi}^{\text{ab/em}}(\omega) = 4\pi^2\alpha\,\omega_{fi}\,|\hat{e}_\mathbf{k}\cdot \underbrace{\langle\phi_f|\,\hat{\mathbf{r}}\,|\phi_i\rangle}_{\mathbf{r}_{fi}}|^2\,, \qquad \omega_{fi} = \frac{|E_f - E_i|}{\hbar}\,.$$

Of course, if the incoming electromagnetic wave is unpolarized, an average over polarizations must be taken, resulting in the replacement $|\hat{e}_\mathbf{k}\cdot\mathbf{r}_{fi}|^2 \longrightarrow \mathbf{r}_{fi}^2 - (\hat{\mathbf{k}}\cdot\mathbf{r}_{fi})^2$ (see Problem 13).

[6] We note that if relativistic corrections are retained in $\hat{H}_0$, this identity is no longer valid since the spin–orbit term does not commute with $\hat{\mathbf{r}}$ and the relativistic kinetic energy term leads to a commutator $[\,\hat{\mathbf{r}}, \hat{\mathbf{p}}^4\,]$ that is proportional to $\hat{\mathbf{p}}^3$. In such a case, one needs to evaluate directly the matrix element of $\hat{\mathbf{p}}$.

Problem 15 Spontaneous Emission: Selection Rules

The semiclassical approach introduced in Problem 13 allows one to describe not only stimulated absorption and emission but also spontaneous emission, that is, the process $i \longrightarrow f + \gamma$. Consider an isolated atom in a volume V and reinterpret the rate obtained in Problem 14,

$$R^{em}_{f\mathbf{k},i} = 2\pi\alpha\, \frac{\omega^2}{c}\, A^2\, |\langle\phi_f|\, \hat{\boldsymbol{\epsilon}}_\mathbf{k} \cdot \hat{\mathbf{r}}\, |\phi_i\rangle|^2\, \delta(E_f - E_i + \hbar\omega)\,,$$

as giving the probability per unit time for the emission of a photon of wave number $\mathbf{k}$ (and hence momentum $\mathbf{p} = \hbar\mathbf{k}$), energy $E = \hbar\omega = \hbar c|\mathbf{k}|$, and polarization $\hat{\boldsymbol{\epsilon}}_\mathbf{k}$, in the long-wavelength approximation (LWA).

1. Assuming that the energy density of the electromagnetic field is due to the emitted photon enclosed in the volume V, show that the factor A^2 in $R^{em}_{f\mathbf{k},i}$ is given by

$$A^2 = \frac{2\pi c^2 \hbar}{V\omega}\,.$$

2. Consider the rate $\Delta\Gamma^{em}_{i\to f}$ summed over a set of photon states with wave numbers in the shell $\Delta\mathbf{k}$ with solid angle $\Delta\Omega$, namely

$$\Delta\Gamma^{em}_{i\to f} = \frac{V}{(2\pi)^3} \int_{\Delta\mathbf{k}} d\mathbf{k}\, R^{em}_{f\mathbf{k},i}\,,$$

and, by integrating out the energy-conserving δ-function, obtain the differential rate $d\Gamma^{em}_{i\to f}/d\Omega$, defined by

$$\Delta\Gamma^{em}_{i\to f} = \int_{\Delta\Omega} d\Omega\, \frac{d}{d\Omega}\Gamma^{em}_{i\to f}\,.$$

3. What selection rules have to be satisfied for the spontaneous emission rate in hydrogen to be non-vanishing?

Solution

Part 1

As shown in Problem 13, the energy density of the electromagnetic field averaged over the period $T = 2\pi/\omega$ is given by

$$\langle u\rangle_T = \frac{1}{2\pi}\, \frac{\omega^2}{c^2}\, A^2\,,$$

and, if this energy is assumed to be due to the photon, then it follows that

$$\langle u\rangle_T = \frac{\hbar\omega}{V} \implies A^2 = \frac{2\pi c^2 \hbar}{V\omega}\,.$$

Part 2

Substituting the expression for A^2 obtained above, the rate for the emission of a photon of wave number $\mathbf{k}$ and polarization $\hat{\boldsymbol{\epsilon}}_\mathbf{k}$ can be expressed as

$$R^{em}_{f\mathbf{k},i} = \frac{(2\pi)^2\alpha c\hbar\,\omega}{V}\, |\langle\phi_f|\, \hat{\boldsymbol{\epsilon}}_\mathbf{k} \cdot \hat{\mathbf{r}}\, |\phi_i\rangle|^2\, \delta(E_f - E_i + \hbar\omega)\,,$$

and the rate $\Delta\Gamma^{\text{em}}_{i\to f}$ summed over a set of photon states with wave numbers in $\Delta\mathbf{k}$ is

$$\Delta\Gamma^{\text{em}}_{i\to f} = \frac{V}{(2\pi)^3}\int_{\Delta\mathbf{k}} d\mathbf{k}\, R^{\text{em}}_{f\mathbf{k},i} = \frac{V}{(2\pi)^3}\int_{\Delta\Omega} d\Omega \int_{\Delta k} dk\, k^2\, \delta(E_f - E_i - \hbar\omega)\frac{(2\pi)^2 \alpha c\hbar\,\omega}{V}|\hat{\mathbf{e}}_\mathbf{k}\cdot\mathbf{r}_{fi}|^2$$

$$= \int_{\Delta\Omega} d\Omega\, \frac{\alpha}{2\pi}\frac{\omega^3_{if}}{c^2}|\hat{\mathbf{e}}_\mathbf{k}\cdot\mathbf{r}_{fi}|^2\,,$$

where we have defined $\mathbf{r}_{fi}=\langle\phi_f|\hat{\mathbf{r}}|\phi_i\rangle$ and $\omega_{if}=(E_i-E_f)/\hbar$, and have used (recall $\omega=c\,k$)

$$\int_{\Delta\Omega} d\Omega \int_{\Delta k} dk\, k^2\,\delta(E_f - E_i - \hbar\omega)[\cdots] = \int_{\Delta\Omega} d\Omega \int_{\Delta\omega} d\omega\,\omega^2\,\frac{1}{\hbar c^3}\delta(\omega_{if}-\omega)[\cdots]\,,$$

yielding the result in the last step above. The differential rate for emission of a photon into the solid angle $d\Omega$ is then found as

$$\frac{d}{d\Omega}\Gamma^{\text{em}}_{i\to f} = \frac{\alpha}{2\pi}\frac{\omega^3_{if}}{c^2}|\hat{\mathbf{e}}_\mathbf{k}\cdot\mathbf{r}_{fi}|^2\,.$$

It is worthwhile pointing out here that the correct quantum treatment of the system consisting of an atom + electromagnetic fields leads to this same expression (at this order).

Part 3

The matrix element entering the transition rate can be written as

$$\hat{\mathbf{e}}_\mathbf{k}\cdot\mathbf{r}_{fi} = \hat{\mathbf{e}}_\mathbf{k}\cdot\langle\phi_f|\hat{\mathbf{r}}|\phi_i\rangle = \sum_{\mu=\pm1,0}(-1)^\mu\,\hat{e}_{\mathbf{k},-\mu}\langle\phi_{n_f l_f m_f}|\hat{r}_\mu|\phi_{n_i l_i m_i}\rangle\,,$$

where we have introduced the spherical components

$$\hat{r}_{\pm1} = \mp\frac{1}{\sqrt{2}}(\hat{r}_x \pm i\hat{r}_y)\,,\qquad \hat{r}_0 = \hat{r}_z\,,$$

with similar relations for the polarization vector. Using the fact that $\hat{r}_\mu$ is an irreducible tensor operator of rank 1, we have from the Wigner–Eckart theorem that

$$\langle\phi_{n_f l_f m_f}|\hat{r}_\mu|\phi_{n_i l_i m_i}\rangle = C_{1l_i}(\mu m_i; l_f m_f)\langle\phi_{n_f l_f}||\hat{r}||\phi_{n_i l_i}\rangle\,,$$

and hence the matrix element vanishes unless $l_f = 1$, if $l_i = 0$, or unless $l_f = l_i - 1, l_i, l_i + 1$ if $l_i \geq 1$; in either case, $m_f = \mu + m_i$. Further, $\hat{\mathbf{r}}$ is an odd operator under parity and hence the matrix element vanishes unless the initial and final states have opposite parity. Since the parity of the state $|\phi_{nlm}\rangle$ is $(-1)^l$, we must have $(-1)^{l_i+l_f} = -1$; hence l_i and l_f are, respectively, even and odd, or vice versa. From these selection rules, we conclude that the matrix element vanishes if $l_f \neq |l_i - 1|, l_i + 1$ and/or $m_f \neq \mu + m_i$.

Problem 16 The $2p \longrightarrow 1s$ Transition in Hydrogen

Calculate, in the long-wavelength approximation (LWA), the differential rate for the transition $2p \longrightarrow 1s$ in hydrogen, averaged over the azimuthal quantum numbers of the initial 2p state and summed over the final polarizations of the emitted photon. Also calculate the total rate integrated over the solid angle corresponding to all directions of emission of the photon, and provide a numerical value for the 2p-level lifetime. Problem 15 is a prerequisite to the present problem.

Solution

In the LWA, the differential rate is given by

$$\frac{d}{d\Omega}\Gamma^{\text{em}}_{2p_m \to 1s} = \frac{\alpha}{2\pi}\frac{\omega_{21}^3}{c^2}|\hat{\boldsymbol{\epsilon}}_{\mathbf{k}} \cdot \langle\phi_{1s}|\hat{\mathbf{r}}|\phi_{2p_m}\rangle|^2 , \qquad \omega_{21} = \frac{E_2 - E_1}{\hbar} = \frac{3}{8}\frac{e^2}{\hbar a_0} ,$$

where α is the fine structure constant, c is the speed of light, and a_0 is the Bohr radius. Since $l_i = 1$ and $l_f = 0$, such a transition is not forbidden by the selection rules (see Problem 15); indeed, we have

$$\langle\phi_{100}|\hat{r}_\mu|\phi_{21m}\rangle = C_{11}(\mu m; 00)\langle\phi_{10}||\hat{r}||\phi_{21}\rangle ,$$

and the reduced matrix element, being independent of the azimuthal quantum numbers, can be calculated from, for example,

$$\langle\phi_{10}||\hat{r}||\phi_{21}\rangle = \frac{\langle\phi_{100}|\hat{r}_0|\phi_{210}\rangle}{C_{11}(00;00)} = -\sqrt{3}\int d\mathbf{r}\, R_{1s}^*(r)\, Y_{00}^*(\Omega)\, \underbrace{r\cos\theta}_{r_0=z}\, R_{2p}(r)\, Y_{10}(\Omega) ,$$

where we have made a specific choice of μ and m that is compatible with the selection rules. Recalling that $Y_{00} = 1/\sqrt{4\pi}$ and $Y_{10} = \sqrt{3/(4\pi)}\cos\theta$, we easily find for the angle integrations

$$\int d\Omega\, Y_{00}(\Omega)\cos\theta\, Y_{10}(\Omega) = \frac{1}{\sqrt{4\pi}}\int d\Omega\, \underbrace{\sqrt{\frac{4\pi}{3}}Y_{10}^*}_{\cos\theta}\, Y_{10} = \frac{1}{\sqrt{3}} ,$$

where in the last step we have used the orthonormality of the spherical harmonics, and therefore (noting that the radial functions are real)

$$\langle\phi_{10}||\hat{r}||\phi_{21}\rangle = -\int_0^\infty dr\, r^3\, R_{1s}(r)\, R_{2p}(r) .$$

Inserting the expressions for the radial functions,

$$R_{1s}(r) = \frac{2}{\sqrt{a_0^3}}e^{-r/a_0} , \qquad R_{2p}(r) = \frac{1}{\sqrt{24a_0^3}}\frac{r}{a_0}e^{-r/(2a_0)} ,$$

we have

$$\langle\phi_{10}||\hat{r}||\phi_{21}\rangle = -\frac{2}{\sqrt{a_0^3}}\frac{1}{\sqrt{24a_0^3}}\int_0^\infty dr\, r^3\, \frac{r}{a_0}e^{-3r/(2a_0)}$$

$$= -\frac{2}{\sqrt{a_0^3}}\frac{a_0^4}{\sqrt{24a_0^3}}\left(\frac{2}{3}\right)^5 \underbrace{\int_0^\infty dx\, x^4\, e^{-x}}_{4!} = \underbrace{-\frac{24}{\sqrt{6}}\left(\frac{2}{3}\right)^5 a_0}_{\gamma} ,$$

from which we obtain

$$\hat{\boldsymbol{\epsilon}}_{\mathbf{k}} \cdot \langle\phi_{100}|\hat{\mathbf{r}}|\phi_{21m}\rangle = \sum_{\mu=\pm1,0}(-1)^\mu\hat{\epsilon}_{k,-\mu}C_{11}(\mu m; 00)(-\gamma a_0) = (-1)^{1-m}\gamma C_{11}(-m, m; 0, 0)a_0\hat{\epsilon}_{k,m} ,$$

where in the last step we have enforced the selection rule $\mu = -m$. Considering the three possibilities $m = \pm1, 0$ and inserting the corresponding Clebsch–Gordan coefficients,

$$C_{11}(\mp1, \pm1; 0, 0) = \frac{1}{\sqrt{3}} , \qquad C_{11}(0, 0; 0, 0) = -\frac{1}{\sqrt{3}} ,$$

we arrive at

$$\hat{\mathbf{e}}_{\mathbf{k}} \cdot \langle \phi_{100}| \, \hat{\mathbf{r}} \, |\phi_{21m}\rangle = \frac{\gamma}{\sqrt{3}} \, a_0 \, (\hat{\mathbf{e}}_{\mathbf{k}})_m \, ,$$

and the matrix element involves the spherical component m of the emitted photon's polarization. These components are relative to the coordinate system in which the z-axis defines the quantization axis of the angular momentum component L_z. Since $\mathbf{k} \cdot \hat{\mathbf{e}}_{\mathbf{k}} = 0$, the polarization vectors are in the plane perpendicular to the photon wave number $\mathbf{k}$. This means, for example, that if the photon is emitted in the z-direction then its polarization vector is in the xy-plane and therefore the only non-vanishing matrix elements are those involving transitions from the $m = \pm 1$ states, but not from the $m = 0$ state.

The $2\mathrm{p}_m \longrightarrow 1\mathrm{s}$ differential transition rate is (here m specifies the initial m state)

$$\frac{d}{d\Omega} \Gamma^{\mathrm{em}}_{2\mathrm{p}_m \to 1\mathrm{s}} = \frac{\alpha}{2\pi} \frac{\omega_{21}^3}{c^2} \frac{\gamma^2 a_0^2}{3} (\hat{\mathbf{e}}_{\mathbf{k}})_m \, [(\hat{\mathbf{e}}_{\mathbf{k}})_m]^* = \frac{\alpha^4}{9\pi} \left(\frac{2}{3}\right)^5 \frac{c}{a_0} (\hat{\mathbf{e}}_{\mathbf{k}})_m \, [(\hat{\mathbf{e}}_{\mathbf{k}})_m]^* \, ,$$

where we have made the replacement $\omega_{21} = (3/8)\alpha c/a_0$. The transition rate averaged over the initial m states and summed over the final polarizations is given by

$$\left\langle \frac{d}{d\Omega} \Gamma^{\mathrm{em}}_{2\mathrm{p} \to 1\mathrm{s}} \right\rangle = \frac{1}{3} \sum_{m=\pm 1,0} \sum_{\lambda=1,2} \frac{\alpha^4}{9\pi} \left(\frac{2}{3}\right)^5 \frac{c}{a_0} (\hat{\mathbf{e}}_{\mathbf{k},\lambda})_m \, [(\hat{\mathbf{e}}_{\mathbf{k},\lambda})_m]^* = \frac{\alpha^4}{9\pi} \left(\frac{2}{3}\right)^6 \frac{c}{a_0} \, ,$$

where in the last step we have used $[(\hat{\mathbf{e}}_{\mathbf{k},\lambda})_m]^* = (-1)^m (\hat{\mathbf{e}}_{\mathbf{k},\lambda})_{-m}$ and

$$\sum_{m=\pm 1,0} (\hat{\mathbf{e}}_{\mathbf{k},\lambda})_m [(\hat{\mathbf{e}}_{\mathbf{k},\lambda})_m]^* = \sum_{m=\pm 1,0} (-1)^m (\hat{\mathbf{e}}_{\mathbf{k},\lambda})_m (\hat{\mathbf{e}}_{\mathbf{k},\lambda})_{-m} = \hat{\mathbf{e}}_{\mathbf{k},\lambda} \cdot \hat{\mathbf{e}}_{\mathbf{k},\lambda} = 1 \, .$$

We see that the average over m is independent of the photon polarization, thus resulting in an overall factor of 2 when summing over the photon polarizations. The rate is isotropic, that is, independent of the direction in which a photon is emitted. When integrated over all directions, this rate reads

$$\Gamma_{2\mathrm{p} \to 1\mathrm{s}} = \int d\Omega \, \frac{\alpha^4}{9\pi} \left(\frac{2}{3}\right)^6 \frac{c}{a_0} = \alpha^4 \left(\frac{2}{3}\right)^8 \frac{c}{a_0} \approx 6.27 \times 10^8 \; \mathrm{sec}^{-1} \, ,$$

and the lifetime of the 2p level in hydrogen is $\tau_{2\mathrm{p} \to 1\mathrm{s}} = 1/\Gamma_{2\mathrm{p} \to 1\mathrm{s}} \approx 1.59 \times 10^{-9}$ sec. Since the state has a finite lifetime, the corresponding spectral line will have a width in energy of order $\hbar/\tau_{2\mathrm{p} \to 1\mathrm{s}} = \hbar \Gamma_{2\mathrm{p} \to 1\mathrm{s}} \approx 4.13 \times 10^{-7}$ eV, much smaller than the (unperturbed) 2p binding energy, which equals 3.4 eV.

Problem 17 Theory of the Line Width

The derivation of the line width illustrated here is due to Wigner and Weisskopf (Problems 14 and 15 are prerequisite to the present problem). It proceeds through the following three steps.

1. Let $|\phi_n\rangle$ be a basis of H_0 eigenstates with energies E_n, and let $\hat{V}$ be a perturbation. Consider the time-dependent Schrödinger equation

$$i\hbar \frac{d}{dt} |\psi(t)\rangle = (\hat{H}_0 + \hat{V})|\psi(t)\rangle \, ,$$

and, by expanding $|\psi(t)\rangle$ as

$$|\psi(t)\rangle = \sum_n c_n(t) \, \mathrm{e}^{-iE_n t/\hbar} \, |\phi_n\rangle \, ,$$

show that it is equivalent to the following set of coupled first-order differential equations for the expansion coefficients;

$$\dot{c}_m(t) = -\frac{i}{\hbar} \sum_n e^{i(E_m - E_n)t/\hbar} V_{mn} c_n(t), \qquad V_{mn} = \langle \phi_m | \hat{V} | \phi_n \rangle.$$

2. A hydrogen-like atom in an excited state $|\phi_1\rangle$ decays to a lower energy state $|\phi_0\rangle$ by emission of a single photon. Denote the initial state of the system atom + electromagnetic field as $|\phi_1; 0\rangle$ and the final state as $|\phi_0; \mathbf{k}\lambda\rangle$, where $\mathbf{k}$ and $\hat{e}_\lambda$ are the wave number and polarization of the photon (here $\lambda = 1, 2$ specify the polarization in the plane transverse to $\mathbf{k}$). In the state space of atom + electromagnetic field, expand $|\psi(t)\rangle$ as

$$|\psi(t)\rangle = c_{1,0}(t) e^{-iE_1 t/\hbar} |\phi_1; 0\rangle + \sum_{\mathbf{k}} \sum_\lambda c_{0,\mathbf{k}\lambda} e^{-i(E_0 + \hbar\omega_k)t/\hbar} |\phi_0; \mathbf{k}\lambda\rangle,$$

ignoring additional components (such as, for example, those having two photon states or involving atomic states other than $|\phi_0\rangle$ and $|\phi_1\rangle$). The initial conditions are $c_{1,0}(0) = 1$ and $c_{0,\mathbf{k}\lambda}(0) = 0$, that is, the atom is initially in an excited state with no photons present. Denoting by $\hat{V}$ the perturbation responsible for the transition and using the results of part 1, obtain a set of coupled equations for $c_{1,0}(t)$ and $c_{0,\mathbf{k}\lambda}(t)$. Insert the ansatz $c_{1,0}(t) = e^{-\gamma t/2} c_{1,0}(0) = e^{-\gamma t/2}$, and show that it leads to the following condition[7]

$$\gamma = \int_0^\infty d\omega_k \left[\frac{2}{\hbar^2} \sum_\lambda \frac{V}{(2\pi c)^3} \omega_k^2 \int d\Omega_k \, |\langle \phi_0; \mathbf{k}\lambda | \hat{V} | \phi_1; 0 \rangle|^2 \right] I(\omega_k - \omega_{10}, \gamma; t),$$

$$I(\omega_k - \omega_{10}, \gamma; t) = \frac{e^{-i(\omega_k - \omega_{10} + i\gamma/2)t} - 1}{-i(\omega_k - \omega_{10} + i\gamma/2)},$$

after converting the sum over $\mathbf{k}$ to an integral over $V/(2\pi)^3 \, dk \, k^2 d\Omega_k$ (V is the box volume) and replacing k by ω_k/c. We have defined $\omega_{10} = (E_1 - E_0)/\hbar$. For the equation above to make sense, the right-hand side must be independent of time. Assume that in the limits $\gamma \longrightarrow 0$ and $t \longrightarrow \infty$ the function I is strongly peaked at $\omega_k = \omega_{10}$ and that

$$\int_{-\infty}^\infty d\omega_k I(\omega_k - \omega_{10}, \gamma; t) = \int_{-\infty}^\infty d\omega_k \frac{\sin[(\omega_k - \omega_{10})t]}{\omega_k - \omega_{10}} = \pi,$$

where the integration has been extended from $-\infty$ to ∞, since I is significant only for ω_k close to ω_{10}. Under these assumptions, show that γ reduces the transition rate obtained by summing over the possible directions and polarizations of the emitted photon, calculated in Problems 14 and 15.

3. Show that in the limit $t \gg 1/\gamma$ the probability that a photon has been emitted is given by

$$\lim_{t \gg 1/\gamma} \sum_{\mathbf{k}} \sum_\lambda |c_{0,\mathbf{k}\lambda}(t)|^2 = \int_{-\infty}^\infty d\omega_k D(\omega_k), \qquad D(\omega_k) = \frac{\gamma/2}{(\omega_{10} - \omega_k)^2 + (\gamma/2)^2},$$

where $D(\omega_k)$ is the distribution in angular frequency (that is, the line width) of the emitted photon. Is probability conserved?

---

[7] The matrix element of the perturbation is precisely the same as given in Problems 14 and 15, that is,

$$\langle \phi_0; \mathbf{k}\lambda | \hat{V} | \phi_1; 0 \rangle = \frac{e}{mc} \langle \phi_0 | A e^{-i\mathbf{k}\cdot\hat{\mathbf{r}}} \hat{e}_\lambda \cdot \hat{\mathbf{p}} | \phi_1 \rangle, \qquad A = \sqrt{\frac{2\pi c^2 \hbar}{V\omega_k}},$$

where $\hat{\mathbf{r}}$ and $\hat{\mathbf{p}}$ are the position and momentum operators of the electron.

Solution

Part 1

Inserting the expansion into the time-dependent Schrödinger equation yields

$$
i\hbar \frac{\partial}{\partial t} |\psi(t)\rangle = \sum_n e^{-iE_n t/\hbar} \left[i\hbar \dot{c}_n(t) + E_n c_n(t) \right] |\phi_n\rangle
$$

$$
= \sum_n e^{-iE_n t/\hbar} c_n(t) \left(\hat{H}_0 + \hat{V} \right) |\phi_n\rangle = \sum_n e^{-iE_n t/\hbar} c_n(t) \left(E_n + \hat{V} \right) |\phi_n\rangle .
$$

After projecting onto the unperturbed eigenstate $|\phi_m\rangle$, the equation above reduces to

$$
\dot{c}_m(t) = -\frac{i}{\hbar} \sum_n e^{i(E_m - E_n)t/\hbar} c_n(t) \langle \phi_m | \hat{V} | \phi_n \rangle = -\frac{i}{\hbar} \sum_n e^{i(E_m - E_n)t/\hbar} V_{mn} c_n(t) .
$$

The equation above is the starting point for the perturbative calculation of the $c_n(t)$ discussed in Problem 3. However, below we do not expand in powers of the perturbation, but rather truncate the state space by retaining only the states $|\phi_1; 0\rangle$ and $|\phi_0; \mathbf{k}\lambda\rangle$.

Part 2

From part 1 it follows that

$$
\dot{c}_{1,0}(t) = -\frac{i}{\hbar} \sum_{\mathbf{k}} \sum_{\lambda} e^{i(\omega_{10} - \omega_k)t} \langle \phi_1; 0 | \hat{V} | \phi_0; \mathbf{k}\lambda \rangle c_{0,\mathbf{k}\lambda}(t) , \qquad \dot{c}_{0,\mathbf{k}\lambda}(t) = -\frac{i}{\hbar} e^{-i(\omega_{10} - \omega_k)t} \langle \phi_0; \mathbf{k}\lambda | \hat{V} | \phi_1; 0 \rangle c_{1,0}(t) .
$$

Inserting the ansatz $c_{1,0}(t) = e^{-\gamma t/2} c_{1,0}(0) = e^{-\gamma t/2}$ into the equation for $\dot{c}_{0,\mathbf{k}\lambda}(t)$, we find

$$
\dot{c}_{0,\mathbf{k}\lambda}(t) = -\frac{i}{\hbar} e^{-[\gamma/2 + i(\omega_{10} - \omega_k)]t} \langle \phi_0; \mathbf{k}\lambda | \hat{V} | \phi_1; 0 \rangle \implies c_{0,\mathbf{k}\lambda}(t) = \frac{i}{\hbar} \langle \phi_0; \mathbf{k}\lambda | \hat{V} | \phi_1; 0 \rangle \frac{e^{-[\gamma/2 + i(\omega_{10} - \omega_k)]t} - 1}{\gamma/2 + i(\omega_{10} - \omega_k)} .
$$

Substituting the latter back into the equation for $c_{1,0}(t)$ leads to the consistency condition

$$
-\frac{\gamma}{2} e^{-\gamma t/2} = \frac{1}{\hbar^2} \sum_{\mathbf{k}} \sum_{\lambda} \frac{|\langle \phi_0; \mathbf{k}\lambda | \hat{V} | \phi_1; 0 \rangle|^2}{\gamma/2 + i(\omega_{10} - \omega_k)} \left[e^{-\gamma t/2} - e^{i(\omega_{10} - \omega_k)t} \right] ,
$$

which, after multiplying both sides by $e^{\gamma t/2}$ and rearranging terms within the square brackets, can be written as

$$
\frac{\gamma}{2} = \frac{1}{\hbar^2} \sum_{\mathbf{k}} \sum_{\lambda} \frac{|\langle \phi_0; \mathbf{k}\lambda | \hat{V} | \phi_1; 0 \rangle|^2}{\gamma/2 + i(\omega_{10} - \omega_k)} \left[e^{\gamma t/2 + i(\omega_{10} - \omega_k)t} - 1 \right]
$$

$$
\implies \gamma = \frac{2}{\hbar^2} \sum_{\mathbf{k}} \sum_{\lambda} |\langle \phi_0; \mathbf{k}\lambda | \hat{V} | \phi_1; 0 \rangle|^2 \underbrace{\frac{e^{-i(\omega_k - \omega_{10} + i\gamma/2)t} - 1}{-i(\omega_k - \omega_{10} + i\gamma/2)}}_{I(\omega_k - \omega_{10}, \gamma; t)} ,
$$

or, converting sums to integrals over $\mathbf{k}$ (below we have replaced k by ω_k/c), as

$$
\gamma = \int_0^\infty d\omega_k \left[\frac{2}{\hbar^2} \sum_{\lambda} \frac{V}{(2\pi c)^3} \omega_k^2 \int d\Omega_k |\langle \phi_0; \mathbf{k}\lambda | \hat{V} | \phi_1; 0 \rangle|^2 \right] I(\omega_k - \omega_{10}, \gamma; t) .
$$

Since for $\gamma \longrightarrow 0$ and $t \longrightarrow \infty$ the function I is assumed to be strongly peaked at $\omega_k = \omega_{10}$, we can write

$$\gamma = \left[\frac{2}{\hbar^2} \sum_\lambda \frac{V}{(2\pi c)^3} \, \omega_k^2 \int d\Omega_k \, |\langle \phi_0; \mathbf{k}\lambda| \, \hat{V} \, |\phi_1; 0\rangle|^2 \right]_{\omega_k = \omega_{10}} \int_{-\infty}^{\infty} d\omega_k \, I(\omega_k - \omega_{10}, \gamma; t)$$

$$= \frac{2\pi}{\hbar^2} \sum_\lambda \frac{V}{(2\pi c)^3} \, \omega_{10}^2 \left[\int d\Omega_k \, |\langle \phi_0; \mathbf{k}\lambda| \, \hat{V} \, |\phi_1; 0\rangle|^2 \right]_{\omega_k = \omega_{10}},$$

where we have made use of the results provided in the text of the problem. This shows that γ is the rate (the probability per unit time) for the transition $i \longrightarrow f$ summed over the possible directions and final polarizations of the emitted photon, as obtained in Problems 14 and 15, that is

$$\gamma = \frac{2\pi}{\hbar} \sum_{\mathbf{k}} \sum_\lambda |\langle \phi_0; \mathbf{k}\lambda| \, \hat{V} \, |\phi_1; 0\rangle|^2 \, \delta(E_1 - E_0 - \hbar\omega_k),$$

where

$$\langle \phi_0; \mathbf{k}\lambda| \, \hat{V} \, |\phi_1; 0\rangle = \frac{e}{mc} \langle \phi_0| A \, e^{-i\mathbf{k}\cdot\hat{\mathbf{r}}} \, \hat{\boldsymbol{e}}_\lambda \cdot \hat{\mathbf{p}} |\phi_1\rangle, \qquad A = \sqrt{\frac{2\pi c^2 \hbar}{V\omega_k}},$$

and $\hat{\mathbf{r}}$ and $\hat{\mathbf{p}}$ are the position and momentum operators of the electron.

Part 3

The probability that the atom remains in state $|\phi_1; 0\rangle$ is given by $|c_{1,0}(t)|^2 = e^{-\gamma t}$ with γ as obtained in part 2. The probability that it decays to the state $|\phi_0\rangle$ and that a photon of wave number $\mathbf{k}$ and polarization $\hat{\boldsymbol{e}}_\lambda$ is emitted is given by $|c_{0,\mathbf{k}\lambda}(t)|^2$, where

$$|c_{0,\mathbf{k}\lambda}(t)|^2 = \frac{1}{\hbar^2} |\langle \phi_0; \mathbf{k}\lambda| \, \hat{V} \, |\phi_1; 0\rangle|^2 \left| \frac{e^{-(\gamma/2 + i\omega)t} - 1}{\gamma/2 + i\omega} \right|^2$$

$$= \frac{1}{\hbar^2} |\langle \phi_0; \mathbf{k}\lambda| \, \hat{V} \, |\phi_1; 0\rangle|^2 \, \frac{e^{-\gamma t} + 1 - 2e^{-\gamma t/2} \cos(\omega t)}{\omega^2 + (\gamma/2)^2}$$

$$= \frac{1}{\hbar^2} \frac{|\langle \phi_0; \mathbf{k}\lambda| \, \hat{V} \, |\phi_1; 0\rangle|^2}{\omega^2 + (\gamma/2)^2} \qquad \text{as } t \gg 1/\gamma \, ;$$

we have defined $\omega = \omega_{10} - \omega_k$. The probability that a photon is emitted (regardless of its wave number and polarization) follows:

$$\lim_{t \gg 1/\gamma} \sum_{\mathbf{k}} \sum_\lambda |c_{0,\mathbf{k}\lambda}(t)|^2$$

$$= \frac{1}{\hbar^2} \sum_\lambda \frac{V}{(2\pi c)^3} \int_0^\infty d\omega_k \, \omega_k^2 \int d\Omega_k \, \frac{|\langle \phi_0; \mathbf{k}\lambda| \, \hat{V} \, |\phi_1; 0\rangle|^2}{(\omega_{10} - \omega_k)^2 + (\gamma/2)^2}$$

$$\approx \underbrace{\frac{1}{\hbar^2} \sum_\lambda \frac{V}{(2\pi c)^3} \, \omega_{10}^2 \int d\Omega_k \left[|\langle \phi_0; \mathbf{k}\lambda| \, \hat{V} \, |\phi_1; 0\rangle|^2 \right]_{\omega_k = \omega_{10}}}_{\gamma/(2\pi)} \int_0^\infty d\omega_k \, \frac{1}{(\omega_{10} - \omega_k)^2 + (\gamma/2)^2},$$

where in the second line we have assumed that γ is small and that, again, $\omega_k^2 |\langle \cdots \rangle|^2$ varies slowly when ω_k is close to the resonant angular frequency ω_{10}. Under these conditions, we have

$$\lim_{t \gg 1/\gamma} \sum_{\mathbf{k}} \sum_{\lambda} |c_{0,\mathbf{k}\lambda}(t)|^2 = \frac{1}{\pi} \int_0^\infty d\omega_k \underbrace{\frac{\gamma/2}{(\omega_{10} - \omega_k)^2 + (\gamma/2)^2}}_{D(\omega_k)} ,$$

where $D(\omega_k)$ is the distribution in angular frequency for the emitted photon – the line width. Note that, since $D(\omega_k)$ is sharply peaked at ω_{10} (for γ "small"), we can extend the integration in ω_k from $-\infty$ to ∞, to obtain

$$\int_{-\infty}^\infty d\omega_k D(\omega_k) = \frac{1}{\pi} \int_{-\infty}^\infty d\omega_k \frac{\gamma/2}{(\omega_{10} - \omega_k)^2 + (\gamma/2)^2} = 1 .$$

This result is consistent with the conservation of probability, since $|c_{1,0}(t)|^2 \approx 0$ and $\sum_{\mathbf{k}} \sum_{\lambda} |c_{0,\mathbf{k}\lambda}(t)|^2 \approx 1$ in the limit $t \gg 1/\gamma$.

Comment: We now justify the assumptions for the function $I(\omega, \gamma; t)$ with $\omega = \omega_k - \omega_{10}$, made in the text of the problem. To this end, we consider the integral in the complex plane

$$\oint_C dz \underbrace{\frac{e^{-izt} - 1}{-iz}}_{f(z)} = 0 .$$

Here, the closed contour C is taken to consist of the following pieces: C_1 is along the real axis and runs from $-\omega^\star$ to $-\eta$, then around the semicircle of radius η centered at the origin (to avoid the singularity of $f(z)$ at the origin), and back on the real axis from η to $\omega^\star$ (with the understanding that $\eta \longrightarrow 0$ and $\omega^\star \longrightarrow \infty$ at the end of the calcuation); C_2 is parallel to the imaginary axis and runs from $\omega^\star$ to $\omega^\star + i\gamma/2$; C_3 is parallel to the real axis and runs from $\omega^\star + i\gamma/2$ to $-\omega^\star + i\gamma/2$; and C_4 is parallel to the imaginary axis and runs from $-\omega^\star + i\gamma/2$ back to the real axis at $-\omega^\star$. Apart from the semicircle around the origin, the closed contour $C = C_1 + C_2 + C_3 + C_4$ is a rectangle. The integrand is analytic inside C and hence $\oint_C dz f(z) = 0$, so that (schematically)

$$\int_{C_1} + \int_{C_2} + \int_{C_3} + \int_{C_4} = 0$$

We first examine the integral along C_2 ; we set $z = \omega^\star + iy$ with $0 \le y \le \gamma/2$, so that

$$\int_{C_2} dz f(z) = \int_0^{\gamma/2} (i\, dy) \frac{e^{-i(\omega^\star + iy)t} - 1}{-i(\omega^\star + iy)} = -\int_0^{\gamma/2} dy \frac{e^{(y - i\omega^\star)t} - 1}{\omega^\star + iy}$$

$$= -\int_0^{\gamma/2} dy \frac{e^{yt}[\cos(\omega^* t) - i\sin(\omega^* t)] - 1}{\omega^\star + iy} ,$$

and this integral vanishes as $\omega^\star \longrightarrow \infty$. A similar argument shows that C_4 also vanishes in this limit. The integral along C_1 is given by

$$\int_{C_1} dz f(z) = \int_{-\omega^\star}^{-\eta} d\omega \frac{e^{-i\omega t} - 1}{-i\omega} + \int_\pi^0 \underbrace{(i\eta\, e^{i\theta}\, d\theta)}_{dz} \frac{e^{-i\eta(\cos\theta + i\sin\theta)t} - 1}{-i\eta\, e^{i\theta}} + \int_\eta^{\omega^\star} d\omega \frac{e^{-i\omega t} - 1}{-i\omega} ,$$

where, along this path, $z = \omega$ with $-\omega^\star \le \omega \le -\eta$ and $\eta \le \omega \le \omega^\star$ on the two segments, and $z = \eta\, e^{i\theta}$ on the semicircle of radius η with θ varying from π down to 0. In the limit $\eta \longrightarrow 0$ the integral around the semicircle vanishes. We obtain

$$\int_{C_1} dz\, f(z) = \int_{-\omega^\star}^{\omega^\star} d\omega\, \frac{e^{-i\omega t} - 1}{-i\omega} = \int_{-\omega^\star}^{\omega^\star} d\omega\, \frac{\sin(\omega t)}{\omega} + \underbrace{\int_{-\omega^\star}^{\omega^\star} d\omega\, \frac{\cos(\omega t) - 1}{-i\omega}}_{\text{vanishes}},$$

and the second integral vanishes, since the integrand is an odd function. The integral along C_3, along which $z = \omega + i\gamma/2$ with ω varying from $\omega^\star$ to $-\omega^\star$, is given by

$$\int_{C_1} dz\, f(z) = \int_{\omega^\star}^{-\omega^\star} d\omega\, \frac{e^{-i(\omega + i\gamma/2)t} - 1}{-i(\omega + i\gamma/2)}.$$

We conclude that in the limit $\omega^\star \longrightarrow \infty$ we have

$$\int_{-\infty}^{\infty} d\omega\, \frac{\sin(\omega t)}{\omega} + \int_{\infty}^{-\infty} d\omega\, \frac{e^{-i(\omega + i\gamma/2)t} - 1}{-i(\omega + i\gamma/2)} = 0 \implies \int_{-\infty}^{\infty} d\omega\, I(\omega, \gamma; t) = \int_{-\infty}^{\infty} d\omega\, \frac{\sin(\omega t)}{\omega} = \pi.$$

For large enough times, $\sin(\omega t)/\omega$ (as a function of ω) is peaked at the origin, where it assumes the value t; further, in this limit it is rapidly oscillating, with zeros occurring at $n\pi/t$ for $n = \pm 1, \pm 2, \ldots$

Problem 18   Formal Scattering Theory

This problem deals with formal developments in scattering theory for a system described by a generally time-dependent Hamiltonian $\hat{H}(t) = \hat{H}_0 + \hat{V}(t)$.

1. Introduce the interaction representation state $|\psi_I(t)\rangle = e^{i\hat{H}_0 t/\hbar}|\psi(t)\rangle$ and obtain an integral equation for the time evolution operator $\hat{U}_I(t, t_0)$ such that $|\psi_I(t)\rangle = \hat{U}_I(t, t_0)|\psi_I(t_0)\rangle$ (see Problem 12 in Chapter 7 – a prerequisite to the present problem – and also Section 18.1). Next, consider the limits $t_0 \longrightarrow -\infty$ and $t \longrightarrow \infty$. In order to make the time integrals mathematically meaningful in the expansion for $\hat{U}_I(t, t_0)$, multiply each $\hat{V}_I(t)$ by a damping factor $e^{-\eta|t|}$ with the understanding that $\eta \longrightarrow 0$ at the end of the calculation (that is, the interaction is slowly turned on and then off as t changes from $-\infty$ to ∞), and define the $\hat{S}$ operator (the so-called S-matrix) as

$$\hat{S} = \lim_{\eta \to 0} \lim_{t \to +\infty} \lim_{t_0 \to -\infty} \hat{U}_I^\eta(t, t_0), \qquad |\psi_I(\infty)\rangle = \hat{S}|\psi_I(-\infty)\rangle.$$

Note that $|\psi_I(-\infty)\rangle$ is a free state and denote it by $|\phi_i\rangle$, so that $\hat{H}_0|\phi_i\rangle = E_i|\phi_i\rangle$ (namely, $|\phi_i\rangle$ is the initial state of the system before scattering has taken place). Obtain the probability that, after scattering has occurred, the system is in a free state $|\phi_f\rangle$ of energy E_f in terms of the S-matrix.

2. Hereafter, assume that the perturbation is time independent (apart from the damping factor). Carry out the time integrations, and show that the amplitude $\langle \phi_f|\hat{S}|\phi_i\rangle$ can expressed as

$$\langle \phi_f|\hat{S}|\phi_i\rangle = \delta_{fi} - 2\pi i\, \delta(E_f - E_i)\, \langle \phi_f|\hat{T}|\phi_i\rangle, \qquad \hat{T} = \lim_{\eta \to 0} \sum_{n=1}^{\infty} \hat{V}\left(\frac{1}{E_i - \hat{H}_0 + i\eta}\, \hat{V}\right)^{n-1},$$

where we have defined the $\hat{T}$ operator (the T-matrix). Show that

$$\langle \phi_f|\hat{T}|\phi_i\rangle = \langle \phi_f|\hat{V}|\psi_i^{(+)}\rangle,$$

where $|\psi_i^{(+)}\rangle$ satisfies the Lippmann–Schwinger equation introduced in Eq. (15.31). Verify that $|\psi_i^{(+)}\rangle$ is an eigenstate of the full Hamiltonian $\hat{H} = \hat{H}_0 + \hat{V}$ with energy E_i.

3. Obtain an expression for the probability amplitude $A_{fi}(t) = \langle \phi_f(t) | \psi_i(t) \rangle$ at time t, where $|\psi_i(t)\rangle$ is the state $e^{-i\hat{H}t/\hbar} |\psi_i(0)\rangle$, evolving in time according to the full Hamiltonian and $|\phi_f(t)\rangle$ is the free state at time t, namely $|\phi_f(t)\rangle = e^{-iE_f t/\hbar} |\phi_f\rangle$. To this end, first note that

$$|\psi_i(t)\rangle = e^{-i\hat{H}t/\hbar} |\psi_i(0)\rangle = e^{-i\hat{H}t/\hbar} |\psi_{I,i}(0)\rangle = e^{-i\hat{H}t/\hbar} \lim_{\eta \to 0} \hat{U}_I^\eta(0, -\infty) |\psi_{I,i}(-\infty)\rangle$$

$$= e^{-i\hat{H}t/\hbar} \lim_{\eta \to 0} \hat{U}_I^\eta(0, -\infty) |\phi_i\rangle \, ,$$

and then carry out the time integrations in $\hat{U}_I^\eta(0, -\infty)$ to arrive at

$$A_{fi}(t) = \langle \phi_f | e^{i(E_f - \hat{H})t/\hbar} |\psi_i^{(+)}\rangle \, .$$

Show that the transition rate $R_{fi} = d|A_{fi}(t)|^2/dt$ can be expressed as

$$R_{fi} = \frac{2}{\hbar} \delta_{fi} \, \text{Im} \, (T_{ii}) + \frac{2\pi}{\hbar} |T_{fi}|^2 \delta(E_i - E_f) \, .$$

Hint: In calculating R_{fi}, the time derivative of $A_{fi}(t)$ is needed; first take the time derivative of $e^{i(E_f - \hat{H})t/\hbar}$ and then let $e^{i(E_f - \hat{H})t/\hbar}$ act on $|\psi_i^{(+)}\rangle$.

Solution

Part 1

In the interaction picture, the state vector $|\psi_I(t)\rangle$ of the system is obtained from the Schrödinger-picture state vector $|\psi(t)\rangle$ via the unitary transformation

$$|\psi_I(t)\rangle = e^{i\hat{H}_0 t/\hbar} |\psi(t)\rangle \, .$$

It satisfies the equation

$$i\hbar \frac{d}{dt} |\psi_I(t)\rangle = \hat{V}_I(t) |\psi_I(t)\rangle \, , \qquad \hat{V}_I(t) = e^{i\hat{H}_0 t/\hbar} \hat{V}(t) \, e^{-i\hat{H}_0 t/\hbar} \, .$$

The state $|\psi_I(t)\rangle$ can be written as $|\psi_I(t)\rangle = \hat{U}_I(t, t_0) |\psi_I(t_0)\rangle$, where the (unitary) time-evolution operator for the state in the interaction representation obeys

$$i\hbar \frac{d}{dt} \hat{U}_I(t, t_0) = \hat{V}_I(t) \, \hat{U}_I(t, t_0) \qquad \text{or} \qquad \hat{U}_I(t, t_0) = \hat{\mathbb{1}} - \frac{i}{\hbar} \int_{t_0}^{t} dt' \, \hat{V}_I(t') \, \hat{U}_I(t', t_0) \, ,$$

and $\hat{U}_I(t_0, t_0) = \hat{\mathbb{1}}$ in order to satisfy the initial condition at t_0. The perturbative solution for the time evolution operator is obtained by iteration in the usual way (see Section 18.1),

$$\hat{U}_I^\eta(t, t_0) = \hat{\mathbb{1}} - \frac{i}{\hbar} \int_{t_0}^{t} dt_1 \, e^{-\eta|t_1|} \hat{V}_I(t_1) + \left(-\frac{i}{\hbar}\right)^2 \int_{t_0}^{t} dt_1 \, e^{-\eta|t_1|} \int_{t_0}^{t_1} dt_2 \, e^{-\eta|t_2|} \hat{V}_I(t_1) \, \hat{V}_I(t_2) + \cdots$$

$$+ \left(-\frac{i}{\hbar}\right)^n \int_{t_0}^{t} dt_1 \, e^{-\eta|t_1|} \int_{t_0}^{t_1} dt_2 \, e^{-\eta|t_2|} \cdots \int_{t_0}^{t_{n-1}} dt_n \, e^{-\eta|t_n|} \hat{V}_I(t_1) \hat{V}_I(t_2) \cdots \hat{V}_I(t_n) + \cdots \, ,$$

where we have introduced the converging factor $e^{-\eta|t|}$, namely we have defined

$$\hat{V}_I^\eta(t) = e^{-\eta|t|} \hat{V}_I(t) \, ,$$

with the understanding that $\eta \longrightarrow 0$ at the end of the calculation; that is, the perturbation is very slowly turned on and then off as t varies between $-\infty$ and ∞. The trick is to make the integrals converging (for example, in the case of an oscillating perturbation). We consider the limits $t \longrightarrow \infty$ and $t_0 \longrightarrow -\infty$ to obtain

$$|\psi_I(\infty)\rangle = \lim_{t \to +\infty} |\psi_I(t)\rangle = \lim_{t \to +\infty} \hat{U}_I^\eta(t, -\infty) |\psi_I(-\infty)\rangle = \left[\lim_{t \to +\infty} \lim_{t_0 \to -\infty} \hat{U}_I^\eta(t, t_0) \right] |\psi_I(-\infty)\rangle ,$$

which leads to the $\hat{S}$ operator defined in the text of the problem. Thus the probability for the transition from the initial free state $|\phi_i\rangle = |\psi_I(-\infty)\rangle$ to the final free state $|\phi_f\rangle$, after scattering has taken place, is given by

$$P_{fi} = |\langle \phi_f | \hat{S} | \phi_i \rangle|^2 .$$

Part 2

When the perturbation is time independent, we have, for a generic nth-order term contributing to the $\hat{S}$-operator,

$$\langle \phi_f | \hat{S}^{(n)} | \phi_i \rangle = \lim_{\eta \to 0} \left(-\frac{i}{\hbar} \right)^n \int_{-\infty}^{\infty} dt_1 \, e^{-\eta|t_1|} \int_{-\infty}^{t_1} dt_2 \, e^{-\eta|t_2|} \cdots \int_{-\infty}^{t_{n-1}} dt_n \, e^{-\eta|t_n|}$$
$$\times \langle \phi_f | e^{i\hat{H}_0 t_1/\hbar} \hat{V} e^{-i\hat{H}_0 t_1/\hbar} \, e^{i\hat{H}_0 t_2/\hbar} \hat{V} e^{-i\hat{H}_0 t_2/\hbar} \cdots e^{i\hat{H}_0 t_n/\hbar} \hat{V} e^{-i\hat{H}_0 t_n/\hbar} | \phi_i \rangle ,$$

which, using the fact that the initial and final states are eigenstates of the free Hamiltonian $\hat{H}_0$, can be rewritten as

$$\langle \phi_f | \hat{S}^{(n)} | \phi_i \rangle = \lim_{\eta \to 0} \left(-\frac{i}{\hbar} \right)^n \int_{-\infty}^{\infty} dt_1 \, e^{-\eta|t_1|} \int_{-\infty}^{t_1} dt_2 \, e^{-\eta|t_2|} \cdots \int_{-\infty}^{t_{n-1}} dt_n \, e^{-\eta|t_n|}$$
$$\times \langle \phi_f | e^{i(E_f - E_i)t_1/\hbar} \, \hat{V} e^{-i\hat{H}_0(t_1 - t_2)/\hbar} e^{iE_i(t_1 - t_2)/\hbar} \, \hat{V} e^{-i\hat{H}_0(t_2 - t_3)/\hbar} e^{iE_i(t_2 - t_3)/\hbar} \cdots e^{-i\hat{H}_0(t_{n-1} - t_n)/\hbar} e^{iE_i(t_{n-1} - t_n)/\hbar} \hat{V} | \phi_i \rangle ,$$

where the product of c-numbers $e^{i(E_f - E_i)t_1/\hbar} e^{iE_i(t_1 - t_2)/\hbar} e^{iE_i(t_2 - t_3)/\hbar} \cdots e^{iE_i(t_{n-1} - t_n)/\hbar}$ reduces to $e^{iE_f t_1/\hbar} e^{-iE_i t_n/\hbar}$ as it must (these are the factors that are obtained by letting the free evolution operators act on the free initial and final states). We introduce the change of variables

$$x_1 = t_1/\hbar , \qquad x_2 = (t_2 - t_1)/\hbar , \qquad x_3 = (t_3 - t_2)/\hbar , \qquad \ldots , \qquad x_n = (t_n - t_{n-1})/\hbar ,$$

which can be inverted to give

$$t_1 = \hbar x_1 , \qquad t_2 = \hbar(x_1 + x_2) , \qquad t_3 = \hbar(x_1 + x_2 + x_3) , \qquad \ldots , \qquad t_n = \hbar(x_1 + x_2 + \cdots + x_n) ,$$

so that the nested integrals reduce to

$$\langle \phi_f | \hat{S}^{(n)} | \phi_i \rangle = \lim_{\eta \to 0} (-i)^n \int_{-\infty}^{\infty} dx_1 \, e^{-\eta|x_1|} \, e^{i(E_f - E_i)x_1} \, \langle \phi_f | \hat{V} \left[\int_{-\infty}^{0} dx_2 \, e^{-\eta|x_1 + x_2|} e^{-i(E_i - \hat{H}_0)x_2} \right]$$
$$\times \hat{V} \left[\int_{-\infty}^{0} dx_2 \, e^{-\eta|x_1 + x_2 + x_3|} e^{-i(E_i - \hat{H}_0)x_3} \right] \cdots \hat{V} \left[\int_{-\infty}^{0} dx_n \, e^{-\eta|x_1 + x_2 + \cdots + x_n|} e^{-i(E_i - \hat{H}_0)x_n} \right] \hat{V} | \phi_i \rangle ,$$

where the $1/\hbar^n$ factor is canceled by the Jacobian of the transformation, and we have redefined $\hbar\eta$ as η (this is legitimate, since η will be taken to zero at the end of the calculation anyway). Consider the last integral over x_n; defining $X = x_1 + \cdots + x_{n-1}$, we have

$$I_n = \lim_{\eta \to 0} \int_{-\infty}^{0} dx_n \, e^{-\eta|X + x_n|} \, e^{-i(E_i - \hat{H}_0)x_n} = \lim_{\eta \to 0} \int_{-\infty}^{0} dx_n \, e^{\eta x_n} \, e^{-i(E_i - \hat{H}_0)x_n} = \lim_{\eta \to 0} \frac{i}{E_i - \hat{H}_0 + i\eta} ,$$

where in the intermediate step we have used the fact that for fixed X the converging factor is only important for very large and negative x_n (in the limit $\eta \longrightarrow 0$). The integrals over $x_{n-1}, \ldots, x_2$ can be performed similarly, while the last integral over x_1 gives

$$\lim_{\eta \to 0} \int_{-\infty}^{\infty} dx_1 \, e^{-\eta|x_1|} \, e^{i(E_f - E_i)x_1} = \lim_{\eta \to 0} \frac{2\eta}{\eta^2 + (E_f - E_i)^2} = 2\pi \, \delta(E_f - E_i) \, .$$

Having carried out the time integrations, the nth term in the S-operator follows:

$$\langle \phi_f | \hat{S}^{(n)} | \phi_i \rangle = -2\pi i \, \delta(E_f - E_i) \langle \phi_f | \hat{V} \frac{1}{E_i - \hat{H}_0 + i\eta} \, \hat{V} \frac{1}{E_i - \hat{H}_0 + i\eta} \cdots \hat{V} \frac{1}{E_i - \hat{H}_0 + i\eta} \, \hat{V} | \phi_i \rangle \, ,$$

and it contains n factors of $\hat{V}$ and $n - 1$ factors of $1/(E_i - \hat{H}_0 + i\eta)$. The complete S-operator reads

$$\langle \phi_f | \hat{S} | \phi_i \rangle = \delta_{fi} + \lim_{\eta \to 0} \sum_{n=1}^{\infty} \langle \phi_f | \hat{S}^{(n)} | \phi_i \rangle = \delta_{fi} - 2\pi i \, \delta(E_f - E_i) \lim_{\eta \to 0} \sum_{n=1}^{\infty} \langle \phi_f | \hat{V} \left(\frac{1}{E_i - \hat{H}_0 + i\eta} \, \hat{V} \right)^{n-1} | \phi_i \rangle \, ,$$

which can be written as $S_{fi} = \delta_{fi} - 2\pi i \, \delta(E_f - E_i) \, T_{fi}$ by introducing the T-operator,

$$\hat{T} = \lim_{\eta \to 0} \sum_{n=1}^{\infty} \hat{V} \left(\frac{1}{E_i - \hat{H}_0 + i\eta} \, \hat{V} \right)^{n-1} \, .$$

The operator relations above can be simplified by inserting complete sets of $\hat{H}_0$ eigenstates between subsequent terms; for example,

$$\hat{V} \frac{1}{E_i - \hat{H}_0 + i\eta} \, \hat{V} = \sum_k \frac{\hat{V} | \phi_k \rangle \langle \phi_k | \hat{V}}{E_i - E_k + i\eta} \, .$$

Note that the $\hat{T}$ operator matrix element can also be expressed as (the limit $\eta \longrightarrow 0$ is understood)

$$T_{fi} = \langle \phi_f | \hat{V} | \psi_i^{(+)} \rangle \, , \qquad | \psi_i^{(+)} \rangle = | \phi_i \rangle + \sum_{n=1}^{\infty} \left(\frac{1}{E_i - \hat{H}_0 + i\eta} \, \hat{V} \right)^n | \phi_i \rangle \, ,$$

and the state $| \psi_i^{(+)} \rangle$ satisfies the Lippmann–Schwinger equation:

$$| \psi_i^{(+)} \rangle = | \phi_i \rangle + \frac{1}{E_i - \hat{H}_0 + i\eta} \, \hat{V} | \psi_i^{(+)} \rangle \, .$$

Indeed, the iterative solution of this equation reproduces the expansion given above in powers of $\hat{V}$. On multiplying both sides by $E_i - \hat{H}_0 + i\eta$ and then taking the limit $\eta \longrightarrow 0$, the state $| \psi_i^{(+)} \rangle$ is easily seen to be an eigenstate of the full Hamiltonian $\hat{H}_0 + \hat{V}$, having energy E_i.

Part 3

We start from

$$| \psi_i(t) \rangle = e^{-i\hat{H}t/\hbar} | \psi_i(0) \rangle = e^{-i\hat{H}t/\hbar} | \psi_{I,i}(0) \rangle = e^{-i\hat{H}t/\hbar} \lim_{\eta \to 0} \hat{U}_I^{\eta}(0, -\infty) | \psi_{I,i}(-\infty) \rangle$$

$$= e^{-i\hat{H}t/\hbar} \lim_{\eta \to 0} \hat{U}_I^{\eta}(0, -\infty) | \phi_i \rangle \, ,$$

where $| \psi_i(0) \rangle$ is the state in the Schrödinger picture at time $t = 0$, which evolves in time with the full Hamiltonian, and $| \psi_{I,i}(t) \rangle$ is the state in the interaction picture, which evolves according to $\hat{U}_I^{\eta}(t, t_0)$; note that, at the reference time $t = 0$, we have $| \psi_i(0) \rangle = | \psi_{I,i}(0) \rangle$ (see part 1). Of course, in the limit

$t_0 \longrightarrow -\infty$, $|\psi_{I,i}(-\infty)\rangle$ reduces to the free state $|\phi_i\rangle$. Inserting the expansion for $U_I^\eta(0,-\infty)$, we find that the generic term of order n is given by

$$
n\text{th term} = \lim_{\eta \to 0} \left(-\frac{i}{\hbar}\right)^n \int_{-\infty}^0 dt_1\, e^{\eta t_1} \int_{-\infty}^{t_1} dt_2\, e^{\eta t_2} \cdots \int_{-\infty}^{t_{n-1}} dt_n\, e^{\eta t_n}\, e^{i\hat{H}_0 t_1/\hbar}\, \hat{V} e^{-i\hat{H}_0 t_1/\hbar}
$$
$$
\times\, e^{i\hat{H}_0 t_2/\hbar}\, \hat{V} e^{-i\hat{H}_0 t_2/\hbar} \cdots e^{i\hat{H}_0 t_n/\hbar}\, \hat{V} \underbrace{e^{-i\hat{H}_0 t_n/\hbar}|\phi_i\rangle}_{e^{-iE_i t_n/\hbar}|\phi_i\rangle}\,,
$$

where the time integrations are for $t \le 0$ and hence the converging factors are given by $e^{\eta t_k}$. The same change of variables as introduced in part 2 yields for this term (with the redefinition $\hbar\eta \longrightarrow \eta$)

$$
n\text{th term} = \lim_{\eta \to 0} (-i)^n \int_{-\infty}^0 dx_1\, e^{\eta x_1} \int_{-\infty}^0 dx_2\, e^{\eta(x_1+x_2)} \cdots \int_{-\infty}^0 dx_n\, e^{\eta(x_1+\cdots+x_n)}\, e^{i\hat{H}_0 x_1}\, \hat{V} e^{i\hat{H}_0 x_2}\, \hat{V} \cdots e^{i\hat{H}_0 x_n}\, \hat{V} e^{-iE_i(x_1+\cdots+x_n)}|\phi_i\rangle
$$
$$
= \lim_{\eta \to 0} (-i)^n \int_{-\infty}^0 dx_1\, e^{n\eta x_1} \int_{-\infty}^0 dx_2\, e^{(n-1)\eta x_2} \cdots \int_{-\infty}^0 dx_n\, e^{\eta x_n}\, e^{i(\hat{H}_0-E_i)x_1}\, \hat{V} e^{i(\hat{H}_0-E_i)x_2}\, \hat{V} \cdots e^{i(\hat{H}_0-E_n)x_n}\, \hat{V}|\phi_i\rangle\,,
$$

and the integrations can be carried out independently, for example

$$
\int_{-\infty}^0 dx_1\, e^{n\eta x_1}\, e^{i(\hat{H}_0-E_i)x_1} = \frac{1}{n\eta + i(\hat{H}_0 - E_i)} = \frac{i}{E_i - \hat{H}_0 + i\eta}\,, \qquad n\eta \longrightarrow \eta\,.
$$

We obtain

$$
n\text{th term} = \lim_{\eta \to 0} \frac{1}{E_i - \hat{H}_0 + i\eta}\, \hat{V} \frac{1}{E_i - \hat{H}_0 + i\eta}\, \hat{V} \cdots \frac{1}{E_i - \hat{H}_0 + i\eta}\, \hat{V} = \lim_{\eta \to 0} \left(\frac{1}{E_i - \hat{H}_0 + i\eta}\, \hat{V}\right)^n\,,
$$

so that

$$
\lim_{\eta \to 0} \hat{U}_I^\eta(0,-\infty)|\phi_i\rangle = \lim_{\eta \to 0} \sum_{n=0}^\infty \left(\frac{1}{E_i - \hat{H}_0 + i\eta}\, \hat{V}\right)^n |\phi_i\rangle = |\psi_i^{(+)}\rangle\,,
$$

where $|\psi_i^{(+)}\rangle$ is the exact eigenstate of the full Hamiltonian $\hat{H} = \hat{H}_0 + \hat{V}$ with energy E_i, introduced in part 2, and the state in the Schrödinger picture follows: $|\psi_i(t)\rangle = e^{-i\hat{H}t/\hbar}|\psi_i^{(+)}\rangle$. We can calculate the probability amplitude for $|\psi_i(t)\rangle$ to be in some free-particle state $|\phi_f(t)\rangle = e^{-i\hat{H}_0 t/\hbar}|\phi_f\rangle = e^{-iE_f t/\hbar}|\phi_f\rangle$ at time t as[8]

$$
A_{fi}(t) = \langle \phi_f | e^{i(E_f-\hat{H})t/\hbar} |\psi_i^{(+)}\rangle\,, \qquad \frac{d}{dt}A_{fi}(t) = \frac{i}{\hbar} \langle \phi_f |(E_f - \hat{H}) e^{i(E_f-\hat{H})t/\hbar} |\psi_i^{(+)}\rangle = -\frac{i}{\hbar} \langle \phi_f | \hat{V} e^{i(E_f-\hat{H})t/\hbar} |\psi_i^{(+)}\rangle
$$
$$
= -\frac{i}{\hbar} e^{i(E_f-E_i)t/\hbar} \langle \phi_f | \hat{V} |\psi_i^{(+)}\rangle\,,
$$

where we have used $\langle \phi_f |(E_f - \hat{H}) = \langle \phi_f |(\hat{H}_0 - \hat{H}) = \langle \phi_f |(-\hat{V})$ and the fact that $|\psi_i^{(+)}\rangle$ is an eigenstate of $\hat{H}$. The probability is given by $P_{fi}(t) = |A_{fi}(t)|^2$, and the transition rate is given by

$$
R_{fi} = \frac{d}{dt}P_{fi}(t) = A_{fi}^*(t) \frac{d}{dt}A_{fi}(t) + \text{c.c.} = -\frac{i}{\hbar}\left[e^{i(E_f-E_i)t/\hbar}\langle \phi_f|\psi_i^{(+)}\rangle\right]^* \left[e^{i(E_f-E_i)t/\hbar}\langle \phi_f| \hat{V}|\psi_i^{(+)}\rangle\right] + \text{c.c.}
$$
$$
= -\frac{i}{\hbar}\langle \phi_f|\psi_i^{(+)}\rangle^* \langle \phi_f| \hat{V}|\psi_i^{(+)}\rangle + \text{c.c.} = \frac{2}{\hbar}\,\text{Im}\left[\langle \phi_f|\psi_i^{(+)}\rangle^*\langle \phi_f| \hat{V}|\psi_i^{(+)}\rangle\right] = \frac{2}{\hbar}\,\text{Im}\left[\langle \phi_f|\psi_i^{(+)}\rangle^*\, T_{fi}\right]\,,
$$

and is time independent. The expression for the rate can be further manipulated by noting that the Lippmann–Schwinger equation allows one to write (the limit $\eta \longrightarrow 0$ is understood)

---

[8] There is a subtle point here: the time derivative has to be taken before letting $e^{i(E_f-\hat{H})t/\hbar}$ act on $|\psi_i^{(+)}\rangle$.

$$\langle \phi_f | \psi^{(+)} \rangle = \langle \phi_f | \left(|\phi_i\rangle + \frac{1}{E_i - \hat{H}_0 + i\eta} \hat{V} | \psi^{(+)} \rangle \right) = \delta_{fi} + \frac{1}{E_i - E_f + i\eta} \underbrace{\langle \phi_f | \hat{V} | \psi_i^{(+)} \rangle}_{T_{fi}} .$$

and hence

$$R_{fi} = \frac{2}{\hbar} \operatorname{Im} \left[\left(\delta_{fi} + \frac{1}{E_i - E_f - i\eta} T_{fi}^* \right) T_{fi} \right] = \frac{2}{\hbar} \left[\delta_{fi} \operatorname{Im}\left(T_{fi}\right) + |T_{fi}|^2 \frac{\eta}{(E_i - E_f)^2 + \eta^2} \right]$$

$$= \frac{2}{\hbar} \delta_{fi} \operatorname{Im}\left(T_{ii}\right) + \frac{2\pi}{\hbar} |T_{fi}|^2 \delta(E_i - E_f) .$$

The second term in R_{fi} looks like Fermi's golden rule, but it is in fact exact (the golden rule is obtained by approximating the exact state $|\psi_i^{(+)}\rangle$ by the free state $|\phi_i\rangle$ in T_{fi}). Since conservation of probability implies that $\sum_f P_{fi}(t) = 1$, it follows that (the optical theorem)

$$\sum_f R_{fi} = 0 \implies \sum_f |T_{fi}|^2 \delta(E_i - E_f) = -\frac{1}{\pi} \operatorname{Im}(T_{ii}) .$$

In this chapter we consider systems made up of identical particles. Identical particles have the same properties (mass, spin, charge, etc.) and, as far as we know, cannot be distinguished from each other by experiment. So every electron (or every proton, or He atom, or photon, ...) in the universe is identical to every other electron (or proton, or He atom, or photon, ...). This fact leads to a difficulty – known as exchange degeneracy – which must be resolved in order to avoid ambiguities in applying the quantum mechanical postulates. To see how this difficulty comes about, consider two identical spin-1/2 particles (we ignore other degrees of freedom, such as position or momentum, to keep things simple). We number the particles arbitrarily as 1 and 2, and the corresponding spin operators as $\hat{S}_1$ and $\hat{S}_2$. Suppose one particle has spin up and the other has spin down along the $\hat{z}$-axis. Since the particles are indistinguishable, the states $|1, +; 2, -\rangle$ and $|1, -; 2, +\rangle$, where in the first ket particle 1 has spin up and particle 2 has spin down and in the second ket vice versa, represent the same physical state. By the superposition principle, so does any linear combination of the form

$$\alpha |1, +; 2, -\rangle + \beta |1, -; 2, +\rangle \qquad \text{with} \qquad |\alpha|^2 + |\beta|^2 = 1 \ .$$

We now ask what is the probability that particles 1 and 2 both have spin up along the $\hat{x}$-axis. This state can be written as

$$\underbrace{\frac{1}{\sqrt{2}} \big(|1, +\rangle + |1, -\rangle \big)}_{|+\rangle_x \text{ eigenstate of } S_{1x}} \otimes \underbrace{\frac{1}{\sqrt{2}} \big(|2, +\rangle + |2, -\rangle \big)}_{|+\rangle_x \text{ eigenstate of } S_{2x}}$$

or equivalently

$$\frac{1}{2} \big(|1, +; 2, +\rangle + |1, +; 2, -\rangle + |1, -; 2, +\rangle + |1, -; 2, -\rangle \big) \ .$$

This probability is given by $|\alpha + \beta|^2/4$; it depends on the coefficients α and β, and consequently cannot be determined unambiguously.

The exchange degeneracy is removed by the **symmetrization postulate**, which states that the physical states of a system consisting of identical particles must be either symmetric or antisymmetric with respect to the exchange of any two (identical) particles, that is, under such an exchange the state is multiplied by +1 (symmetric case) or −1 (antisymmetric case). It turns out that the physical states of particles whose spin is an integer, including 0 (for example, the pion and photon, which have, respectively, spins 0 and 1) are symmetric, while those of particles with half-integer spin (for example, the electron and quark, which have both spin 1/2) are antisymmetric. The former particles are known as bosons, and the latter as fermions.[1]

The symmetrization postulate also holds for composite particles, such as, for example, the He atom which consists of a nucleus – two protons and two neutrons bound together by the strong force

---

[1] The symmetrization "postulate" can actually be derived in the context of relativistic quantum field theory on the basis of very general assumptions. In this sense, it can be viewed as a consequence in quantum mechanics of special relativity.

(the proton and neutron, which are themselves composite particles, have spin 1/2 and are therefore fermions) – and two electrons. When we exchange two He atoms in a state describing a system of He atoms, we exchange all their constituents, so that is, the two protons, the two neutrons, and the two electrons. Therefore, the state describing this system acquires an overall sign factor resulting from the product of the sign factors associated with the exchanges of the individual constituents. Since in the specific case of the He atom, the exchange involves six fermions and hence the product of six minus signs, the overall sign factor is plus and the He atom is therefore a boson. We conclude that a composite particle is either a boson or a fermion, depending only on whether the number of fermion constituents is either even or odd[2].

In the following, we consider the implications of the symmetrization postulate on the description of systems consisting of identical fermions or bosons.

19.1 States of Identical Particles

To begin, let $|\psi_\gamma\rangle$ be a basis in the state space of a single particle (this basis could be made up of the simultaneous eigenstates of a complete set of commuting observables, collectively denoted by $\hat\Gamma$, and so $\hat\Gamma|\psi_\gamma\rangle = \gamma\,|\psi_\gamma\rangle$). Next consider two particles, which we label as 1 and 2. A basis in the combined space of these two particles is provided by the tensor product $|\psi_{\gamma_1}(1)\rangle \otimes |\psi_{\gamma_2}(2)\rangle$, which we denote as $|\psi_{\gamma_1,\gamma_2}(1,2)\rangle$ (particle 1 is in state $|\psi_{\gamma_1}\rangle$ and particle 2 in state $|\psi_{\gamma_2}\rangle$). Note that

$$|\psi_{\gamma_1,\gamma_2}(1,2)\rangle = |\psi_{\gamma_2,\gamma_1}(2,1)\rangle\,, \tag{19.1}$$

since the order of the states – that is, $|\psi_{\gamma_1}(1)\rangle \otimes |\psi_{\gamma_2}(2)\rangle$ or $|\psi_{\gamma_2}(2)\rangle \otimes |\psi_{\gamma_1}(1)\rangle$ – is irrelevant in the tensor product. The symmetrization postulate requires that the physical states (those realized in nature) are either symmetric or antisymmetric with respect to the exchange of particles 1 and 2, depending on whether the two particles are bosons or fermions expressed as,

$$|\psi^S_{\gamma_1,\gamma_2}\rangle = \frac{1}{\sqrt{2}}\left[|\psi_{\gamma_1,\gamma_2}(1,2)\rangle + |\psi_{\gamma_1,\gamma_2}(2,1)\rangle\right]\,,\quad |\psi^A_{\gamma_1,\gamma_2}\rangle = \frac{1}{\sqrt{2}}\left[|\psi_{\gamma_1,\gamma_2}(1,2)\rangle - |\psi_{\gamma_1,\gamma_2}(2,1)\rangle\right]\,, \tag{19.2}$$

or as

$$|\psi^S_{\gamma_1,\gamma_2}\rangle = \frac{1}{\sqrt{2}}\left[|\psi_{\gamma_1,\gamma_2}(1,2)\rangle + |\psi_{\gamma_2,\gamma_1}(1,2)\rangle\right]\,,\quad |\psi^A_{\gamma_1,\gamma_2}\rangle = \frac{1}{\sqrt{2}}\left[|\psi_{\gamma_1,\gamma_2}(1,2)\rangle - |\psi_{\gamma_2,\gamma_1}(1,2)\rangle\right]\,, \tag{19.3}$$

where in the kets on the left-hand sides of the equations $|\psi^{S/A}_{\gamma_1,\gamma_2}\rangle$, we have removed the particle labels, since all we need to know is that there are two particles, one of which is in the (single-particle) state $|\psi_{\gamma_1}\rangle$ and the other of which is in the (single-particle) state $|\psi_{\gamma_2}\rangle$; in other words, we need to know only which single-particle states are occupied. Comparing Eqs. (19.2) and (19.3) we see that we can symmetrize or antisymmetrize the states $|\psi_{\gamma_1,\gamma_2}(1,2)\rangle$ with respect to either the particles or the quantum numbers.

An immediate consequence of the symmetrization postulate is Pauli's exclusion principle, that is, two fermions cannot occupy the same single-particle state, for then $|\psi^A_{\gamma,\gamma}\rangle$ would vanish identically. This restriction does not hold in the case of two bosons, indeed, for $\gamma_1 = \gamma_2$, we have

---

[2] The rule we have established for the addition of angular momenta is in line with the spin-statistics rule of the symmetrization postulate, in the sense that the addition of an even or odd number of half-integer spins results in a total spin which is either integer or half-integer – a boson in the former case and a fermion in the latter.

$$|\psi_{\gamma_1,\gamma_1}^{S}\rangle = |\psi_{\gamma_1,\gamma_1}(1,2)\rangle \,. \tag{19.4}$$

The states $|\psi_{\gamma_1,\gamma_2}^{S/A}\rangle$ obviously transform in accordance with the symmetrization postulate. The factor $1/\sqrt{2}$ ensures that they are correctly normalized (here $\gamma_1 \neq \gamma_2$):

$$\langle \psi_{\gamma_1',\gamma_2'}^{S/A} | \psi_{\gamma_1,\gamma_2}^{S/A} \rangle = \frac{1}{2} \left[\langle \psi_{\gamma_1',\gamma_2'}(1,2)| \pm \langle \psi_{\gamma_1',\gamma_2'}(2,1)| \right] \left[|\psi_{\gamma_1,\gamma_2}(1,2)\rangle \pm |\langle \psi_{\gamma_1,\gamma_2}(2,1)\rangle \right]$$

$$= \underbrace{\delta_{\gamma_1',\gamma_1} \, \delta_{\gamma_2',\gamma_2}}_{\text{direct}} \pm \underbrace{\delta_{\gamma_1',\gamma_2} \, \delta_{\gamma_2',\gamma_1}}_{\text{exchange}} \,, \tag{19.5}$$

where we have used $|\psi_{\gamma_1,\gamma_2}(1,2)\rangle = |\psi_{\gamma_1}(1)\rangle \otimes |\psi_{\gamma_2}(2)\rangle$ and $|\psi_{\gamma_1,\gamma_2}(2,1)\rangle = |\psi_{\gamma_2}(1)\rangle \otimes |\psi_{\gamma_1}(2)\rangle$ and similarly for the bras. For two bosons occupying the same state γ, the normalized state is given by $|\psi_{\gamma,\gamma}^{S}\rangle = |\psi_{\gamma,\gamma}(1,2)\rangle$. The states $|\psi_{\gamma_1,\gamma_2}^{S/A}\rangle$ form a basis in the state space of symmetric and antisymmetric two-particle states, respectively. The completeness relation in this space reads

$$\frac{1}{2!} \sum_{\gamma_1,\gamma_2} |\psi_{\gamma_1,\gamma_2}^{S/A}\rangle \langle \psi_{\gamma_1,\gamma_2}^{S/A}| = \hat{\mathbb{1}} \,, \tag{19.6}$$

where the factor $1/2!$ is needed to avoid double counting,

$$|\psi_{\alpha_1,\alpha_2}^{S/A}\rangle = \frac{1}{2!} \sum_{\gamma_1,\gamma_2} \cdot |\psi_{\gamma_1,\gamma_2}^{S/A}\rangle \langle \psi_{\gamma_1,\gamma_2}^{S/A} | \psi_{\alpha_1,\alpha_2}^{S/A}\rangle = \frac{1}{2!} \sum_{\gamma_1,\gamma_2} |\psi_{\gamma_1,\gamma_2}^{S/A}\rangle \left[\delta_{\gamma_1,\alpha_1} \, \delta_{\gamma_2,\alpha_2} \pm \delta_{\gamma_1,\alpha_2} \, \delta_{\gamma_2,\alpha_1} \right]$$

$$= \frac{1}{2!} \left[|\psi_{\alpha_1,\alpha_2}^{S/A}\rangle \pm \underbrace{|\psi_{\alpha_2,\alpha_1}^{S/A}\rangle}_{\pm |\psi_{\alpha_1,\alpha_2}^{S/A}\rangle} \right] = |\psi_{\alpha_1,\alpha_2}^{S/A}\rangle \,. \tag{19.7}$$

To make this more concrete, consider a system of two identical and free particles of spin $\mathbf{S}$; the eigenvalue of $\mathbf{S}^2$ is $\hbar^2 S(S+1)$. The wave function of a single particle of momentum $\mathbf{p}$ and spin projection s with $s = -S, -S+1, \ldots, S$ reads

$$\psi_{\mathbf{p}s}(\mathbf{r}) = \frac{1}{(2\pi\hbar)^{3/2}} \, e^{i\mathbf{p}\cdot\mathbf{r}/\hbar} \, |s\rangle \,, \tag{19.8}$$

where the (fixed) quantum number S is not explicitly indicated. Depending on whether S is integer or half-integer, the wave function of the two particles is either symmetric or antisymmetric,

$$\psi_{\mathbf{p}_1 s_1, \mathbf{p}_2 s_2}^{S/A}(\mathbf{r}_1, \mathbf{r}_2) = \frac{1}{\sqrt{2}} \left[\psi_{\mathbf{p}_1 s_1}(\mathbf{r}_1) \, \psi_{\mathbf{p}_2 s_2}(\mathbf{r}_2) \pm \psi_{\mathbf{p}_2 s_2}(\mathbf{r}_1) \, \psi_{\mathbf{p}_1 s_1}(\mathbf{r}_2) \right] \,, \tag{19.9}$$

which can also be written as a "permanent" (S) or determinant (A),

$$\psi_{\mathbf{p}_1 s_1, \mathbf{p}_2 s_2}^{S/A}(\mathbf{r}_1, \mathbf{r}_2) = \frac{1}{\sqrt{2}} \text{perm/det} \begin{pmatrix} \psi_{\mathbf{p}_1 s_1}(\mathbf{r}_1) & \psi_{\mathbf{p}_2 s_2}(\mathbf{r}_1) \\ \psi_{\mathbf{p}_1 s_1}(\mathbf{r}_2) & \psi_{\mathbf{p}_2 s_2}(\mathbf{r}_2) \end{pmatrix} \,, \tag{19.10}$$

where a permanent is defined as the determinant except that all minus signs are changed to plus signs. Note that when $\mathbf{p}_1 = \mathbf{p}_2 = \mathbf{p}$ and $s_1 = s_2 = s$, that is, the particles occupy the same (single-particle) state, the determinant vanishes (corresponding to Pauli's exclusion principle), while the permanent gives $\psi_{\mathbf{p}s,\mathbf{p}s}^{S}(\mathbf{r}_1, \mathbf{r}_2) = \sqrt{2} \, \psi_{\mathbf{p}s}(\mathbf{r}_1) \, \psi_{\mathbf{p}s}(\mathbf{r}_2)$, and so the wave function of Eq. (19.10) for two bosons is incorrectly normalized and needs to be divided by a further $\sqrt{2}$.

In the case of N identical particles, in the abstract state space of the system the physical states can be written as

$$|\psi^{S/A}_{\gamma_1,\ldots,\gamma_N}\rangle = C \,\mathrm{perm/det}\begin{pmatrix} |\psi_{\gamma_1}(1)\rangle & |\psi_{\gamma_2}(1)\rangle & \ldots & |\psi_{\gamma_N}(1)\rangle \\ |\psi_{\gamma_1}(2)\rangle & |\psi_{\gamma_2}(2)\rangle & \ldots & |\psi_{\gamma_N}(2)\rangle \\ \vdots & \vdots & & \vdots \\ |\psi_{\gamma_1}(N)\rangle & |\psi_{\gamma_2}(N)\rangle & \ldots & |\psi_{\gamma_N}(N)\rangle \end{pmatrix}, \tag{19.11}$$

and in expanding the permanent or determinant the product of single-particle states is to be understood as a tensor product in the N-particle state space. The normalization constant C is equal to $1/\sqrt{N!}$ for the fermion state, while for the boson state is $1/[\sqrt{N!}\,\sqrt{n_{\gamma_i}!\,n_{\gamma_j}!\cdots}]$, where the extra factors $\sqrt{n_{\gamma_i}!}, \sqrt{n_{\gamma_j}!}, \ldots$ account for the possibility that there might be $n_{\gamma_i} > 1$, $n_{\gamma_j} > 1$, ... bosons in the corresponding single-particle states (see the previous case of two bosons). It is clear that the state (19.11) obeys the symmetrization postulate: the determinant changes sign if either any two rows are interchanged (corresponding to the exchange of two particles) or any two columns are interchanged (corresponding to the exchange of two sets of quantum numbers). The determinant is known as a Slater determinant. In the permanent case, no minus sign occurs under either interchange. The completeness relation for the fully symmetrized basis now reads

$$\frac{1}{N!} \sum_{\gamma_1,\gamma_2,\ldots,\gamma_N} |\psi^{S/A}_{\gamma_1,\gamma_2,\ldots,\gamma_N}\rangle \langle \psi^{S/A}_{\gamma_1,\gamma_2,\ldots,\gamma_N}| = \hat{\mathbb{1}}, \tag{19.12}$$

and the factor $N!$ is needed to avoid overcounting.

These multi-particle states are adequate to describe systems for which the mutual interactions among particles vanish or are weak. Note that the individual particles need not be free. For example, the state of N bound electrons in the attractive Coulomb field of a nucleus can be written as in Eq. (19.11), where γ, in the single-particle state $|\psi_\gamma\rangle$, denotes the quantum numbers n, l, m, s. However, such an N-electron state would not be very realistic, since the Coulomb repulsion between electrons cannot be ignored.

19.2 Problems

Problem 1 Energy Levels of Three Identical Fermions or Three Identical Bosons

Let the Hamiltonian of a single spin-$1/2$ fermion be given by $h = (a/\hbar)\,L_z$, where L_z is the z-component of its orbital angular momentum and the constant $a > 0$ has the dimensions of energy. What are the energies and degeneracies for a system of three such identical and non-interacting fermions? Assume the fermions are all in p-wave orbitals (that is, they all have angular momentum $L = 1$). What are the energies and degeneracies if the system consists, instead, of three identical and non-interacting spin-0 bosons?

Solution

In a p-wave state the single-particle energy levels are $-a$, 0, a (the eigenvalues of L_z are $m\hbar$ with $m = \pm 1, 0$). We label the corresponding three orbital states as $|p_{-1}\rangle$, $|p_0\rangle$, and $|p_1\rangle$. The Hamiltonian describing the three (non-interacting) fermions is given by

Table 19.1 Energy levels E_n and degeneracies g_n for three spin-1/2 fermions in p-wave orbitals.

$(p_{-1})^2, p_0$	$(p_{-1})^2, p_1$ and $p_{-1}, (p_0)^2$	p_{-1}, p_0, p_1	$p_{-1}, (p_1)^2$ and $(p_0)^2, p_1$	$p_0, (p_1)^2$
$E_0 = -2a$	$E_1 = -a$	$E_2 = 0$	$E_3 = a$	$E_4 = 2a$
$g_0 = 2$	$g_1 = 4$	$g_2 = 8$	$g_3 = 4$	$g_4 = 2$

Table 19.2 Energy levels E_n and degeneracies g_n for three spin-0 bosons in p-wave orbitals.

$(p_{-1})^3$	$(p_{-1})^2, p_0$	$(p_{-1})^2, p_1$ and $p_{-1}, (p_0)^2$	p_{-1}, p_0, p_1 and $(p_0)^3$	$p_{-1}, (p_1)^2$ and $(p_0)^2, p_1$	$p_0, (p_1)^2$	$(p_1)^3$
$E_0 = -3a$	$E_1 = -2a$	$E_2 = -a$	$E_3 = 0$	$E_4 = a$	$E_5 = 2a$	$E_6 = 3a$
$g_0 = 1$	$g_1 = 1$	$g_2 = 2$	$g_3 = 2$	$g_4 = 2$	$g_5 = 1$	$g_6 = 1$

$$H = h(1) + h(2) + h(3) .$$

Each of the three single-particle levels can accommodate at the most two fermions (one with spin up and the other one with spin down). The possibilities are in Table 19.1, where, for example, $(p_{-1})^2, p_0$ means two of the fermions are in orbital state $|p_{-1}\rangle$ and the remaining fermion is in orbital state $|p_0\rangle$. In the bosonic case, each level can accommodate one, two, or three bosons. The resulting energies and degeneracies are given in Table 19.2.

Problem 2   When Can the Symmetrization Postulate Be Ignored?

The requirement that the wave functions of electrons (or, for that matter, any identical fermions) be antisymmetric raises the question whether it is really necessary to antisymmetrize the wave functions of all the electrons in the universe at the same time as we are trying to solve for the wave functions of, say, the helium atom. We expect that particles at large distances from one another should be independent and that it should only be necessary to consider the symmetry of particles which are located within small distances from each other. A reasonable criterion can be established for whether it is necessary to antisymmetrize, by considering the following example.

A system consists of two one-dimensional potential wells $V_l(x)$ and $V_r(x)$ with their centers located a distance a apart. Assume initially that a is large. One particle is in the left-hand well, while the other is in the right-hand well. The normalized states describing these particles are, respectively, $|\varphi_l(1)\rangle$ and $|\varphi_r(2)\rangle$, where we assume that the single-particle states $|\varphi_l\rangle$ and $|\varphi_r\rangle$ are eigenstates of, respectively, the left- and right-hand Hamiltonians, defined as

$$\hat{H}_l = \frac{\hat{p}^2}{2m} + \hat{V}_l , \qquad \hat{H}_r = \frac{\hat{p}^2}{2m} + \hat{V}_r .$$

1. Suppose the particles are distinguishable. Let $|\phi_x(i)\rangle\langle\phi_x(i)|$ be the projection operator onto the state in which particle i is at position x. Calculate the number density $n(x)$ that *either* particle is at position x; that is, the expectation value of $\sum_i |\phi_x(i)\rangle\langle\phi_x(i)|$ on the two-particle state. What is the normalization of $n(x)$?

2. Now, suppose the particles are fermions. Calculate the number density $n^A(x)$ in this case and the difference $n^A(x) - n(x)$. Interpret the results in light of the comments made in the introduction to the problem.

3. Assume that the potentials are harmonic oscillator wells centered at $x = \pm a/2$ and, further, that each particle is in the ground state of its respective well. Calculate the overlap (inner product) $\langle \varphi_l | \varphi_r \rangle$, thus justifying the interpretation above.

Solution

Part 1

The unsymmetrized state and corresponding wave function for the pair of particles are given by

$$|\psi\rangle = |\varphi_l(1)\rangle \otimes |\varphi_r(2)\rangle, \qquad \psi(x_1, x_2) = \varphi_l(x_1)\, \varphi_r(x_2) \,.$$

The number density that either particle is at x follows:

$$n(x) = \langle \psi | \phi_x(1)\rangle\langle \phi_x(1) | \psi \rangle + \langle \psi | \phi_x(2)\rangle\langle \phi_x(2) | \psi \rangle = |\varphi_l(x)|^2 + |\varphi_r(x)|^2 \,,$$

and its normalization is just given by the total number of particles

$$\int_{-\infty}^{\infty} dx\, n(x) = 2 \,.$$

Part 2

The state and corresponding wave function must be antisymmetric, so that

$$|\psi^A\rangle = \frac{1}{\sqrt{2}} \left[|\varphi_l(1)\rangle \otimes |\varphi_r(2)\rangle - |\varphi_l(2)\rangle \otimes |\varphi_r(1)\rangle \right] \,,$$

$$\psi^A(x_1, x_2) = \frac{1}{\sqrt{2}} \left[\varphi_l(x_1)\, \varphi_r(x_2) - \varphi_l(x_2)\, \varphi_r(x_1) \right] \,.$$

The number density is now given by

$$n^A(x) = \langle \psi^A | \phi_x(1)\rangle\langle \phi_x(1) | \psi^A \rangle + \langle \psi^A | \phi_x(2)\rangle\langle \phi_x(2) | \psi^A \rangle$$

$$= |\varphi_l(x)|^2 + |\varphi_r(x)|^2 - 2\,\mathrm{Re} \left[\varphi_l^*(x)\varphi_r(x) \int_{-\infty}^{\infty} dy\, \varphi_r^*(y)\, \varphi_l(y) \right] \,,$$

and hence

$$n^A(x) - n(x) = -2\,\mathrm{Re} \left[\varphi_l^*(x)\varphi_r(x) \int_{-\infty}^{\infty} dy\, \varphi_r^*(y)\, \varphi_l(y) \right] \,,$$

where it should be noted that the single-particle states $|\varphi_l\rangle$ and $|\varphi_r\rangle$ are not orthogonal to each other, since they are eigenstates of different Hamiltonians. Hence, the overlap $\langle \varphi_l | \varphi_r \rangle$ is strictly non-zero. However, since the wave functions $\varphi_l(x)$ and $\varphi_r(x)$ fall off exponentially outside their respective (left- and right-hand) potential wells, this overlap will be small if the two wells are separated by a distance which is large compared with the "sizes" of the two wave functions. Of course, while this argument holds strictly for bound states, it remains valid in the case of scattering states as long as the associated wave packets do not have a significant overlap.

Part 3

The ground-state wave functions are

$$\varphi_l(x) = (b/\pi)^{1/4} e^{-b(x+a/2)^2/2} \, , \qquad \varphi_r(x) = (b/\pi)^{1/4} e^{-b(x-a/2)^2/2} \, ,$$

where $b = m\omega/\hbar$. The overlap is given by

$$\int_{-\infty}^{\infty} dy \, \varphi_l^*(y) \, \varphi_r(y) = \sqrt{\frac{b}{\pi}} \int_{-\infty}^{\infty} dy \, e^{-b(y+a/2)^2/2} \, e^{-b(y-a/2)^2/2} = \sqrt{\frac{b}{\pi}} \int_{-\infty}^{\infty} dy \, e^{-by^2} \, e^{-ba^2/4} = e^{-ba^2/4} \, .$$

Clearly, for large a it will be very small and the effect of antisymmetrization will also be very small. In other words, antisymmetrization should be considered only in those cases where the overlap between wave functions is significant.

Problem 3 Properties of the Exchange Operator

Define the exchange operator $\hat{P}_{21}$ as follows:

$$\hat{P}_{21} |\psi_{\gamma_1,\gamma_2}(1,2)\rangle = |\psi_{\gamma_1,\gamma_2}(2,1)\rangle = |\psi_{\gamma_2,\gamma_1}(1,2)\rangle \, ,$$

so that $\hat{P}_{21}$ exchanges particle 1 with particle 2 or the corresponding quantum numbers. Since these states form a basis, the operator $\hat{P}_{21}$ is uniquely defined in the full two-particle state space. Show that the exchange operator is such that $\hat{P}_{21}^2 = \hat{\mathbb{1}}$, and that it is hermitian and unitary. Further, show that under the exchange operator a generic operator $\hat{O}(1,2)$ acting in the two-particle state space transforms according to

$$\hat{O}(1,2) \longmapsto \hat{O}'(1,2) = \hat{P}_{21} \, \hat{O}(1,2) \, \hat{P}_{21}^\dagger = \hat{O}(2,1) \, .$$

Solution

We find that

$$\hat{P}_{21}[\hat{P}_{21} |\psi_{\gamma_1,\gamma_2}(1,2)\rangle] = \hat{P}_{21} |\psi_{\gamma_2,\gamma_1}(1,2)\rangle = |\psi_{\gamma_1,\gamma_2}(1,2)\rangle \, ,$$

and therefore the eigenvalues of $\hat{P}_{21}$ are ± 1. That $\hat{P}_{21}$ is hermitian and unitary follows from

$$\langle \psi_{\gamma_1',\gamma_2'}(1,2)| \underbrace{\hat{P}_{21} |\psi_{\gamma_1,\gamma_2}(1,2)\rangle}_{|\psi_{\gamma_2,\gamma_1}(1,2)\rangle} = \delta_{\gamma_1',\gamma_2} \, \delta_{\gamma_2',\gamma_1} = \langle \psi_{\gamma_1,\gamma_2}(1,2)| \underbrace{\hat{P}_{21} |\psi_{\gamma_1',\gamma_2'}(1,2)\rangle^*}_{|\psi_{\gamma_2',\gamma_1'}(1,2)\rangle} \, ,$$

and

$$\hat{P}_{21}^2 = \hat{\mathbb{1}} \implies \hat{P}_{21}^{-1} = \hat{P}_{21} = \hat{P}_{21}^\dagger \, .$$

A generic operator $\hat{O}(1,2)$ transforms as

$$\langle \psi_{\gamma_1',\gamma_2'}(1,2)| \underbrace{\hat{P}_{21} \, O(1,2) \, \hat{P}_{21}^\dagger}_{O'(1,2)} |\psi_{\gamma_1,\gamma_2}(1,2)\rangle = \langle \psi_{\gamma_1',\gamma_2'}(2,1)| \, O(1,2) \, |\psi_{\gamma_1,\gamma_2}(2,1)\rangle$$

$$= \langle \psi_{\gamma_1',\gamma_2'}(1,2)| \, O(2,1) \, |\psi_{\gamma_1,\gamma_2}(1,2)\rangle \, ,$$

and $\hat{O}(1,2) \longmapsto \hat{O}'(1,2) = \hat{O}(2,1)$. An operator is said to be symmetric if $\hat{O}^S(1,2) = \hat{O}^S(2,1)$ or, equivalently, if $\hat{O}^S(1,2)$ commutes with $\hat{P}_{21}$.

Problem 4   Two Bosons or Two Fermions in a Central Potential

Consider a system of two identical particles interacting by a central potential depending only on the magnitude of their relative position. Its Hamiltonian reads

$$H = \frac{\mathbf{p}_1^2}{2m} + \frac{\mathbf{p}_2^2}{2m} + V(|\mathbf{r}_1 - \mathbf{r}_2|) .$$

1. Show that, in terms of the center-of-mass and relative variables, the above Hamiltonian simplifies to

$$H = \underbrace{\frac{\mathbf{P}^2}{4m}}_{H_{cm}} + \underbrace{\frac{\mathbf{p}^2}{m} + V(r)}_{H_{rel}} ,$$

where $\mathbf{P}$ and $\mathbf{p}$ are the total and relative moment a operators and $\mathbf{r}$ is the relative position operator.

2. In the two-particle state space, a basis is provided by the position eigenstates $|\phi_{\mathbf{r}_1,\mathbf{r}_2}\rangle = |\phi_{\mathbf{r}_1}(1)\rangle \otimes |\phi_{\mathbf{r}_2}(2)\rangle$ of particles 1 and 2. Define the exchange operator P_{21} in such a way that its action on this basis is as follows:

$$P_{21}|\phi_{\mathbf{r}_1,\mathbf{r}_2}\rangle = P_{21}|\phi_{\mathbf{r}_1}(1)\rangle \otimes |\phi_{\mathbf{r}_2}(2)\rangle = |\phi_{\mathbf{r}_1}(2)\rangle \otimes |\phi_{\mathbf{r}_2}(1)\rangle = |\phi_{\mathbf{r}_2,\mathbf{r}_1}\rangle .$$

Now, use the basis of position eigenstates $|\phi_{\mathbf{R},\mathbf{r}}\rangle = |\phi_{\mathbf{R}}\rangle \otimes |\phi_{\mathbf{r}}\rangle$ to show that in this basis $P_{21}|\phi_{\mathbf{R},\mathbf{r}}\rangle = |\phi_{\mathbf{R},-\mathbf{r}}\rangle$. Further show that, if $|\psi_{\mathbf{P},E_n lm}\rangle$ denotes the basis of common eigenstates of the observables $\mathbf{P}$, H_{rel}, $\mathbf{L}^2$, and L_z, these states are represented in the basis $|\phi_{\mathbf{R},\mathbf{r}}\rangle$ as follows:

$$|\psi_{\mathbf{P},E_n lm}\rangle = \frac{1}{(2\pi\hbar)^{3/2}} \int d\mathbf{R}\, e^{i\mathbf{P}\cdot\mathbf{R}/\hbar} \int d\mathbf{r}\, R_{nl}(r)\, Y_{lm}(\hat{\mathbf{r}})\, |\phi_{\mathbf{R},\mathbf{r}}\rangle .$$

Hence

$$P_{21}|\psi_{\mathbf{P},E_n lm}\rangle = (-1)^l |\psi_{\mathbf{P},E_n lm}\rangle .$$

3. Assume the two identical particles are bosons of spin 0. Are there any restrictions on the allowed values of l? Next, assume that the two identical particles are fermions of spin 1/2, and denote by $|\phi_{\mathbf{R},\mathbf{r}SM}\rangle$ the basis of common eigenstates of $\mathbf{R}$, $\mathbf{r}$, $\mathbf{S}^2$, and S_z, where $\mathbf{S}$ is the total spin. Show that, under the exchange operator, we have

$$P_{21}|\phi_{\mathbf{R},\mathbf{r}SM}\rangle = (-1)^{S+1} |\phi_{\mathbf{R},-\mathbf{r}SM}\rangle .$$

Show further that

$$P_{21}|\psi_{\mathbf{P},E_n lm,SM}\rangle = (-1)^{l+S+1} |\psi_{\mathbf{P},E_n lm,SM}\rangle ,$$

where $|\psi_{\mathbf{P},E_n lm,SM}\rangle$ is the basis of common eigenstates of $\mathbf{P}$, H_{rel}, $\mathbf{L}^2$, L_z, $\mathbf{S}^2$, and S_z. What values of l are allowed in, respectively, spin-singlet ($S = 0$) and spin-triplet ($S = 1$) states?

Solution

Part 1

In terms of center-of-mass and relative position and momentum operators ($\mathbf{R}$, $\mathbf{P}$ and $\mathbf{r}$, $\mathbf{p}$, respectively) we have

$$\mathbf{r}_1 = \mathbf{R} + \frac{\mathbf{r}}{2} , \qquad \mathbf{p}_1 = \frac{\mathbf{P}}{2} + \mathbf{p} , \qquad \mathbf{r}_2 = \mathbf{R} - \frac{\mathbf{r}}{2} , \qquad \mathbf{p}_2 = \frac{\mathbf{P}}{2} - \mathbf{p} ,$$

and hence

$$H = \frac{(\mathbf{P}/2 + \mathbf{p})^2}{2m} + \frac{(\mathbf{P}/2 - \mathbf{p})^2}{2m} + V(|\mathbf{r}_1 - \mathbf{r}_2|) = \frac{\mathbf{P}^2}{4m} + \frac{\mathbf{p}^2}{m} + V(r) \; .$$

Part 2

Consider the state $|\phi_{\mathbf{R},\mathbf{r}}\rangle$; it can be expanded as

$$|\phi_{\mathbf{R},\mathbf{r}}\rangle = \int d\mathbf{r}_1 \, d\mathbf{r}_2 \, |\phi_{\mathbf{r}_1,\mathbf{r}_2}\rangle\langle\phi_{\mathbf{r}_1,\mathbf{r}_2}|\phi_{\mathbf{R},\mathbf{r}}\rangle = \int d\mathbf{r}_1 \, d\mathbf{r}_2 \, \delta[(\mathbf{r}_1 + \mathbf{r}_2)/2 - \mathbf{R}] \, \delta(\mathbf{r}_1 - \mathbf{r}_2 - \mathbf{r}) \, |\phi_{\mathbf{r}_1,\mathbf{r}_2}\rangle \, ,$$

and hence

$$P_{21} |\phi_{\mathbf{R},\mathbf{r}}\rangle = \int d\mathbf{r}_1 \, d\mathbf{r}_2 \, \delta[(\mathbf{r}_1 + \mathbf{r}_2)/2 - \mathbf{R}] \, \delta(\mathbf{r}_1 - \mathbf{r}_2 - \mathbf{r}) \, \underbrace{P_{21} |\phi_{\mathbf{r}_1,\mathbf{r}_2}\rangle}_{|\phi_{\mathbf{r}_2,\mathbf{r}_1}\rangle}$$

$$= \int d\mathbf{r}_1 \, d\mathbf{r}_2 \, \overbrace{\delta[(\mathbf{r}_2 + \mathbf{r}_1)/2 - \mathbf{R}]}^{\delta[(\mathbf{r}_1+\mathbf{r}_2)/2-\mathbf{R}]} \, \underbrace{\delta(\mathbf{r}_2 - \mathbf{r}_1 - \mathbf{r})}_{\delta[\mathbf{r}_1-\mathbf{r}_2-(-\mathbf{r})]} \, |\phi_{\mathbf{r}_1,\mathbf{r}_2}\rangle = |\phi_{\mathbf{R},-\mathbf{r}}\rangle \, ,$$

where in the second line we have made the exchange of integration variables $\mathbf{r}_1 \rightleftharpoons \mathbf{r}_2$ (the Jacobian of the transformation is unity), which then yields the desired result.

The basis of eigenstates $|\psi_{\mathbf{P},E_n lm}\rangle$ consists of the tensor product

$$|\psi_{\mathbf{P},E_n lm}\rangle = |\psi_{\mathbf{P}}\rangle \otimes |\psi_{E_n lm}\rangle \, ,$$

where $|\psi_{\mathbf{P}}\rangle$ and $|\psi_{E_n lm}\rangle$ are, respectively, in the center-of-mass and relative state spaces, so that, for example,

$$H|\psi_{\mathbf{P}}\rangle \otimes |\psi_{E_n lm}\rangle = \left[H_{cm}|\psi_{\mathbf{P}}\rangle\right] \otimes |\psi_{E_n lm}\rangle + |\psi_{\mathbf{P}}\rangle \otimes H_{rel} |\psi_{E_n lm}\rangle = \left(\frac{\mathbf{P}^2}{4m} + E_n\right) |\psi_{\mathbf{P}}\rangle \otimes |\psi_{E_n lm}\rangle \, .$$

Thus, using the completeness relation for the basis $|\phi_{\mathbf{R},\mathbf{r}}\rangle$, we obtain

$$|\psi_{\mathbf{P},E_n lm}\rangle = \int d\mathbf{R} \, d\mathbf{r} \, |\phi_{\mathbf{R},\mathbf{r}}\rangle\langle\phi_{\mathbf{R},\mathbf{r}}|\psi_{\mathbf{P},E_n lm}\rangle = \left[\int d\mathbf{R} \, |\phi_{\mathbf{R}}\rangle\langle\phi_{\mathbf{R}}|\psi_{\mathbf{P}}\rangle\right] \otimes \left[\int d\mathbf{r} \, |\phi_{\mathbf{r}}\rangle\langle\phi_{\mathbf{r}}|\psi_{E_n lm}\rangle\right]$$

$$= \left[\frac{1}{(2\pi\hbar)^{3/2}} \int d\mathbf{R} \, e^{i\mathbf{P}\cdot\mathbf{R}/\hbar} |\psi_{\mathbf{P}}\rangle\right] \otimes \left[\int d\mathbf{r} \, R_{nl}(r) \, Y_{lm}(\hat{\mathbf{r}}) \, |\phi_{\mathbf{r}}\rangle\right] = \int d\mathbf{R} \, d\mathbf{r} \, \psi_{\mathbf{P},E_n lm}(\mathbf{R}, \mathbf{r}) \, |\phi_{\mathbf{R},\mathbf{r}}\rangle \, ,$$

where we have defined the wave function in the center-of-mass and relative orbital space as

$$\psi_{\mathbf{P},E_n lm}(\mathbf{R}, \mathbf{r}) = \frac{1}{(2\pi\hbar)^{3/2}} \, e^{i\mathbf{P}\cdot\mathbf{R}/\hbar} \, R_{nl}(r) \, Y_{lm}(\hat{\mathbf{r}}) \; .$$

Applying the exchange operator leads to

$$P_{21} \underbrace{\int d\mathbf{R} \, d\mathbf{r} \, \psi_{\mathbf{P},E_n lm}(\mathbf{R}, \mathbf{r}) \, |\phi_{\mathbf{R},\mathbf{r}}\rangle}_{|\psi_{\mathbf{P},E_n lm}\rangle} = \int d\mathbf{R} \, d\mathbf{r} \, \psi_{\mathbf{P},E_n lm}(\mathbf{R}, \mathbf{r}) \, |\phi_{\mathbf{R},-\mathbf{r}}\rangle = \int d\mathbf{R} \, d\mathbf{r} \, \psi_{\mathbf{P},E_n lm}(\mathbf{R}, -\mathbf{r}) |\phi_{\mathbf{R},\mathbf{r}}\rangle$$

$$= (-1)^l \int d\mathbf{R} \, d\mathbf{r} \, \psi_{\mathbf{P},E_n lm}(\mathbf{R}, \mathbf{r}) |\phi_{\mathbf{R},\mathbf{r}}\rangle = (-1)^l |\psi_{\mathbf{P},E_n lm}\rangle \, ,$$

as expected.

Part 3

If the particles are bosons, the state must be even under their exchange, and hence l must be even. If, instead, the particles are spin-1/2 fermions, the state $|\phi_{\mathbf{R},\mathbf{r}SM}\rangle$ consists of the tensor product $|\phi_{\mathbf{R},\mathbf{r}}\rangle \otimes |SM\rangle$, where these states are in the orbital and spin state spaces, respectively. Now, recall that the spin states can be written as

$$|11\rangle = |1,+;2,+\rangle, \qquad |10\rangle = \frac{1}{\sqrt{2}}(|1,+;2,-\rangle + |1,-;2,+\rangle), \qquad |1-1\rangle = |1,-;2,-\rangle,$$

and

$$|10\rangle = \frac{1}{\sqrt{2}}(|1,+;2,-\rangle - |1,-;2,+\rangle),$$

yielding

$$P_{21}|SM\rangle = (-1)^{S+1}|SM\rangle, \qquad S = 0, 1;$$

the spin-singlet and spin-triplet states are, respectively, odd and even under the exchange. It follows that

$$P_{21}|\phi_{\mathbf{R},\mathbf{r}SM}\rangle = (-1)^{S+1}|\phi_{\mathbf{R},-\mathbf{r}SM}\rangle,$$

which then leads, following the same steps as in part 2, to

$$P_{21}|\psi_{\mathbf{P},E_n lm,SM}\rangle = (-1)^{l+S+1}|\psi_{\mathbf{P},E_n lm,SM}\rangle.$$

Since the two-fermion state must be odd under exchange, we must have $(-1)^{l+S+1} = -1$, so that l is even for a spin singlet ($S=0$) and odd for a spin triplet ($S=1$).

Problem 5 Two-Particle Transition Amplitudes in a Central Potential

Two particles (1 and 2) with the same mass interact via a potential $V(r)$ that depends only on the distance r between them.

1. Show that the time evolution operator for such a system factorizes as follows:

$$U(t, t_0) = U_{\mathrm{cm}}(t, t_0)\, U_{\mathrm{rel}}(t, t_0),$$

   where $U_{\mathrm{cm}}(t, t_0)$ and $U_{\mathrm{rel}}(t, t_0)$ are the time evolution operators corresponding to the center-of-mass and relative Hamiltonians, respectively. Show that these operators are scalars with respect to independent rotations in the center-of-mass and relative state spaces.

2. Consider the two-particle free state $|\psi_{\mathbf{p}_1,\mathbf{p}_2}\rangle = |\psi_{\mathbf{p}_1}\rangle \otimes |\psi_{\mathbf{p}_2}\rangle$, where $|\psi_{\mathbf{p}_i}\rangle$ is the eigenstate of the momentum operator of particle i with eigenvalue $\mathbf{p}_i$. Show that under a rotation in the combined center-of-mass and relative orbital spaces it transforms as

$$|\psi_{\mathbf{P},\mathbf{p}}\rangle \longrightarrow |\psi_{\mathbf{P},\mathbf{p}}\rangle' = |\psi_{R\mathbf{P},R\mathbf{p}}\rangle,$$

   where $\mathbf{P} = \mathbf{p}_1 + \mathbf{p}_2$ and $\mathbf{p} = (\mathbf{p}_1 - \mathbf{p}_2)/2$ – the center-of-mass and relative momentum eigenvalues, respectively – and $R\mathbf{P}$ and $R\mathbf{p}$ are shorthand notations for $P_i' = \sum_j R_{ij}P_j$ and $p_i' = \sum_j R_{ij}p_j$, where $\underline{R}$ is the 3×3 (orthogonal) rotation matrix.

3. Assume that the two particles do not have spin and are distinguishable. At time t_0 the system is in a state $|\psi_{p\hat{\mathbf{z}},-p\hat{\mathbf{z}}}\rangle$, where particles 1 and 2 are in momentum eigenstates with eigenvalues $p\hat{\mathbf{z}}$ and $-p\hat{\mathbf{z}}$,

respectively. Let $U(t, t_0)$ be the time evolution operator of the system. Show that the probability amplitude of finding this system in a state $|\psi_{p\hat{\mathbf{n}}, -p\hat{\mathbf{n}}}\rangle$ at time $t_1 > t_0$ is given by

$$A(\hat{\mathbf{n}}) = \langle \psi_{p\hat{\mathbf{n}}} | U_{\text{rel}}(t_1, t_0) | \psi_{p\hat{\mathbf{z}}} \rangle \,,$$

where $|\psi_{p\hat{\mathbf{n}}}\rangle$ is the eigenstate of the relative momentum operator with eigenvalue $p\hat{\mathbf{n}}$ and where the direction $\hat{\mathbf{n}}$ is specified by the polar and azimuthal angles θ and ϕ. Show that the amplitude $A(\hat{\mathbf{n}})$ is independent of ϕ, that is, it is invariant under rotation about the $\hat{\mathbf{z}}$-axis. Calculate the probability of finding either of the two particles (without specifying which one) with momentum $p\hat{\mathbf{n}}$ and the other one with momentum $-p\hat{\mathbf{n}}$. What happens to this probability if θ is changed to $\pi - \theta$?

4. In the same spin-independent potential, now consider two identical particles, one of which is initially in the state $|\psi_{p\hat{\mathbf{z}},s}\rangle$ and the other in the state $|\psi_{-p\hat{\mathbf{z}},s'}\rangle$ with $s \neq s'$. Here, s and s' refer to the eigenvalues $s\hbar$ and $s'\hbar$ of the spin component along the $\hat{\mathbf{z}}$-axis. Express in terms of $A(\hat{\mathbf{n}})$ the probability of finding, at time t_1, one particle with momentum $p\hat{\mathbf{n}}$ and spin s and the other one with momentum $-p\hat{\mathbf{n}}$ and spin s'.

5. Now, treat the case $s = s'$. In particular, examine the $\theta = \pi/2$ direction, distinguishing between two possibilities: either the particles are bosons (integer spin) or fermions (half-integer spin). Show that the scattering probability is the same in the θ and the $\pi - \theta$ directions.

Solution

Part 1

The time evolution operator is given by

$$U(t, t_0) = e^{-iH(t-t_0)/\hbar} = \underbrace{e^{-iH_{\text{cm}}(t-t_0)/\hbar}}_{U_{\text{cm}}(t,t_0)} \underbrace{e^{-iH_{\text{rel}}(t-t_0)/\hbar}}_{U_{\text{rel}}(t,t_0)} \,, \qquad H = \underbrace{\frac{\mathbf{P}^2}{4m}}_{H_{\text{cm}}} + \underbrace{\frac{\mathbf{p}^2}{m} + V(r)}_{H_{\text{rel}}} \,,$$

where $\mathbf{P}$ and $\mathbf{p}$ are the center-of-mass and relative momentum operators, and $\mathbf{r}$ is the relative position operator (see Problem 4 for the definitions of these in terms of $\mathbf{r}_i$ and $\mathbf{p}_i$). The time evolution operator factorizes, since H_{cm} and H_{rel} commute with each other. The Hamiltonian, being a scalar, commutes with the components of the total orbital angular momentum $\mathbf{L} = \mathbf{L}_1 + \mathbf{L}_2 = \mathbf{L}_{\text{cm}} + \mathbf{L}_{\text{rel}}$ with $\mathbf{L}_{\text{cm}} = \mathbf{R} \times \mathbf{P}$ and $\mathbf{L}_{\text{rel}} = \mathbf{r} \times \mathbf{p}$, and hence it also commutes with the rotation operator given by

$$U_R(\varphi\hat{\mathbf{n}}) = e^{-i\varphi \mathbf{L} \cdot \hat{\mathbf{n}}/\hbar} = e^{-i\varphi \mathbf{L}_1 \cdot \hat{\mathbf{n}}/\hbar} e^{-i\varphi \mathbf{L}_2 \cdot \hat{\mathbf{n}}/\hbar} = e^{-i\varphi \mathbf{L}_{\text{cm}} \cdot \hat{\mathbf{n}}/\hbar} e^{-i\varphi \mathbf{L}_{\text{rel}} \cdot \hat{\mathbf{n}}/\hbar} \,,$$

where the intermediate steps hold since the components of $\mathbf{L}_1$ and $\mathbf{L}_2$ ($\mathbf{L}_{\text{cm}}$ and $\mathbf{L}_{\text{rel}}$) commute with each other. Thus, we have

$$[U_R(\varphi\hat{\mathbf{n}}), H] = 0 \,,$$

which implies that

$$H_{\text{cm}} = U_R^{\text{cm}\dagger}(\varphi\hat{\mathbf{n}}) H_{\text{cm}} U_R^{\text{cm}}(\varphi\hat{\mathbf{n}}) \qquad \text{and} \qquad H_{\text{rel}} = U_R^{\text{rel}\dagger}(\varphi\hat{\mathbf{n}}) H_{\text{rel}} U_R^{\text{rel}}(\varphi\hat{\mathbf{n}}) \,,$$

and, for the time evolution operators,

$$U_{\text{cm}}(t, t_0) = U_R^{\text{cm}\dagger}(\varphi\hat{\mathbf{n}}) U_{\text{cm}}(t, t_0) U_R^{\text{cm}}(\varphi\hat{\mathbf{n}}) \qquad \text{and} \qquad U_{\text{rel}}(t, t_0) = U_R^{\text{rel}\dagger}(\varphi\hat{\mathbf{n}}) U_{\text{rel}}(t, t_0) U_R^{\text{rel}}(\varphi\hat{\mathbf{n}}) \,.$$

Part 2

Under a rotation the eigenstates $|\psi_{\mathbf{p}_1,\mathbf{p}_2}\rangle = |\psi_{\mathbf{p}_1}\rangle \otimes |\psi_{\mathbf{p}_2}\rangle$ of the momentum operators $\mathbf{p}_1$ and $\mathbf{p}_2$ transform as

$$|\psi_{\mathbf{p}_1,\mathbf{p}_2}\rangle \longrightarrow |\psi_{\mathbf{p}_1,\mathbf{p}_2}\rangle' = U_R(\varphi\hat{\mathbf{n}})|\psi_{\mathbf{p}_1,\mathbf{p}_2}\rangle = |\psi_{R\mathbf{p}_1,R\mathbf{p}_2}\rangle \,,$$

or, in terms of center-of-mass and relative variables,

$$|\psi_{\mathbf{P},\mathbf{p}}\rangle \longrightarrow |\psi_{\mathbf{P},\mathbf{p}}\rangle' = \underbrace{[U_R^{\text{cm}}(\varphi\hat{\mathbf{n}})|\psi_{\mathbf{P}}\rangle]}_{|\psi_{R\mathbf{P}}\rangle} \otimes \underbrace{[U_R^{\text{rel}}(\varphi\hat{\mathbf{n}})|\psi_{\mathbf{p}}\rangle]}_{|\psi_{R\mathbf{p}}\rangle} = |\psi_{R\mathbf{P},R\mathbf{p}}\rangle \,,$$

where, for example, $|\psi_{R\mathbf{p}_1}\rangle$ is the eigenstate of the momentum operator $\mathbf{p}_1$ with eigenvalue $\mathbf{p}'_1 = R\,\mathbf{p}_1$. This follows by noting that (see Section 16.2)

$$\langle\phi_{\mathbf{r}}|\psi_{\mathbf{p}}\rangle' = \langle\phi_{\mathbf{r}}|U_R(\varphi\hat{\mathbf{n}})|\psi_{\mathbf{p}}\rangle = \langle\phi_{R^{-1}\mathbf{r}}|\psi_{\mathbf{p}}\rangle = \frac{e^{i\mathbf{p}\cdot(R^{-1}\mathbf{r})}}{(2\pi\hbar)^{3/2}} = \frac{e^{i(R\mathbf{p})\cdot\mathbf{r}}}{(2\pi\hbar)^{3/2}} = \langle\phi_{\mathbf{r}}|\psi_{R\mathbf{p}}\rangle \,,$$

where we have used $\mathbf{p}\cdot(R^{-1}\mathbf{r}) = (R^{-1}R\,\mathbf{p})\cdot(R^{-1}\mathbf{r}) = (R\,\mathbf{p})\cdot\mathbf{r}$.

Part 3

Since the total momentum operator $\mathbf{P}$ commutes with H and hence with $U(t,t_0)$, it is conserved. In the specific case under consideration and in terms of center-of-mass and relative variables, we have the state $|\psi_{0,p\hat{\mathbf{z}}}\rangle$ at t_0 and the state $|\psi_{0,p\hat{\mathbf{n}}}\rangle$ at t_1 (both with vanishing total momentum eigenvalue), and the transition amplitude reads

$$A(\hat{\mathbf{n}}) = \langle\psi_{0,p\hat{\mathbf{n}}}|U(t_1,t_0)|\psi_{0,p\hat{\mathbf{z}}}\rangle = \langle\psi_{p\hat{\mathbf{n}}}|U_{\text{rel}}(t_1,t_0)|\psi_{p\hat{\mathbf{z}}}\rangle = \langle\psi_{p\hat{\mathbf{n}}}|U_R^{\text{rel}\dagger}(\varphi\hat{\mathbf{z}})U_{\text{rel}}(t_1,t_0)U_R^{\text{rel}}(\varphi\hat{\mathbf{z}})|\psi_{p\hat{\mathbf{z}}}\rangle$$
$$= \langle\psi_{Rp\hat{\mathbf{n}}}|U_{\text{rel}}(t_1,t_0)|\psi_{Rp\hat{\mathbf{z}}}\rangle \,,$$

where we are considering here a rotation by φ about the $\hat{\mathbf{z}}$-axis. Clearly, the initial state is invariant under such a rotation and $|\psi_{Rp\hat{\mathbf{z}}}\rangle = |\psi_{p\hat{\mathbf{z}}}\rangle$, while for the final state we have ($\mathbf{p} = p\hat{\mathbf{n}}$)

$$R\,\mathbf{p} \implies \begin{pmatrix} \cos\varphi & -\sin\varphi & 0 \\ \sin\varphi & \cos\varphi & 0 \\ 0 & 0 & 1 \end{pmatrix}\begin{pmatrix} p\sin\theta\cos\phi \\ p\sin\theta\sin\phi \\ p\cos\theta \end{pmatrix} = p\begin{pmatrix} \sin\theta\cos(\phi+\varphi) \\ \sin\theta\sin(\phi+\varphi) \\ \cos\theta \end{pmatrix} \,.$$

We conclude that $A(\theta,\phi) = A(\theta,\phi+\varphi)$, that is, the transition amplitude is the same for any arbitrary angle φ, and hence it must be independent of the azimuthal angle ϕ, so that $\mathbf{p} = p\hat{\mathbf{n}}$ can be taken to be in the xz-plane, that is, $\hat{\mathbf{n}} = (\sin\theta, 0, \cos\theta)$.

Since the particles are distinguishable, the probability of observing one of them with momentum $p\hat{\mathbf{n}}$ and the other one with momentum $-p\hat{\mathbf{n}}$ (without specifying which is which) must be given by

$$P_{\text{dis}} = |A(\hat{\mathbf{n}})|^2 + |A(-\hat{\mathbf{n}})|^2 \,,$$

where $A(\hat{\mathbf{n}})$ and $A(-\hat{\mathbf{n}})$ are the amplitudes for a transition to a final state in which particle 1 has momentum $p\hat{\mathbf{n}}$ and particle 2 has momentum $-p\hat{\mathbf{n}}$ and vice versa, specifically

$$A(\hat{\mathbf{n}}) = \langle\psi_{p\hat{\mathbf{n}},-p\hat{\mathbf{n}}}|U(t_1,t_0)|\psi_{p\hat{\mathbf{z}},-p\hat{\mathbf{z}}}\rangle = \langle\psi_{p\hat{\mathbf{n}}}|U_{\text{rel}}(t_1,t_0)|\psi_{p\hat{\mathbf{z}}}\rangle \,,$$
$$A(-\hat{\mathbf{n}}) = \langle\psi_{-p\hat{\mathbf{n}},p\hat{\mathbf{n}}}|U(t_1,t_0)|\psi_{p\hat{\mathbf{z}},-p\hat{\mathbf{z}}}\rangle = \langle\psi_{-p\hat{\mathbf{n}}}|U_{\text{rel}}(t_1,t_0)|\psi_{p\hat{\mathbf{z}}}\rangle \,.$$

The relative momentum $-p\hat{\mathbf{n}}$ has components $p(-\sin\theta, 0, -\cos\theta)$, and under a rotation by π about the $\hat{\mathbf{z}}$-axis transforms as follows:

$$-p\hat{\mathbf{n}} \longrightarrow p\hat{\mathbf{n}}', \qquad \hat{\mathbf{n}}' = (\sin\theta, 0, -\cos\theta),$$

using the expression for $\mathbf{Rp}$ obtained above.

Since the amplitude is invariant under rotations about the $\hat{\mathbf{z}}$-axis it follows that $A(-\hat{\mathbf{n}}) = A(\hat{\mathbf{n}}')$, so that

$$P_{\text{dis}} = |A(\hat{\mathbf{n}})|^2 + |A(\hat{\mathbf{n}}')|^2,$$

which shows that this probability is invariant if $\theta \longrightarrow \pi - \theta$, since under such a transformation (a reflection relative to the xy-plane) we have $\hat{\mathbf{n}} \rightleftharpoons \hat{\mathbf{n}}'$.

Part 4

The initial and final states need to be symmetrized or antisymmetrized depending on whether the spin of the particles is integer or half-integer. Thus we have

$$|\psi_{p\hat{\mathbf{z}}s,-p\hat{\mathbf{z}}s'}\rangle^{S/A} = \frac{1}{\sqrt{2}} \left(|\psi_{p\hat{\mathbf{z}}s,-p\hat{\mathbf{z}}s'}\rangle \pm |\psi_{-p\hat{\mathbf{z}}s',p\hat{\mathbf{z}}s}\rangle \right),$$

or, in terms of center-of-mass and relative variables,

$$|\psi_{p\hat{\mathbf{z}},ss'}\rangle^{S/A} = \frac{1}{\sqrt{2}} \left(|\psi_{p\hat{\mathbf{z}},ss'}\rangle \pm |\psi_{-p\hat{\mathbf{z}},s's}\rangle \right),$$

and similarly for the final state. Note that here $|\psi_{p\hat{\mathbf{z}},ss'}\rangle$ and $|\psi_{-p\hat{\mathbf{z}},s's}\rangle$ are defined as follows:

$$|\psi_{p\hat{\mathbf{z}},ss'}\rangle = |\psi_{p\hat{\mathbf{z}}}(1)\rangle \otimes |s(1)\rangle \otimes |\psi_{-p\hat{\mathbf{z}}}(2)\rangle \otimes |s'(2)\rangle, \qquad |\psi_{-p\hat{\mathbf{z}},s's}\rangle = |\psi_{-p\hat{\mathbf{z}}}(1)\rangle \otimes |s'(1)\rangle \otimes |\psi_{p\hat{\mathbf{z}}}(2)\rangle \otimes |s(2)\rangle,$$

where we have made explicit the particle labeling. The amplitude for the transition reads

$$^{S/A}\langle\psi_{p\hat{\mathbf{n}},ss'}|U_{\text{rel}}(t_1,t_0)|\psi_{p\hat{\mathbf{z}},ss'}\rangle^{S/A} = \frac{1}{2}\big[\langle\psi_{p\hat{\mathbf{n}},ss'}|U_{\text{rel}}(t_1,t_0)|\psi_{p\hat{\mathbf{z}},ss'}\rangle + \langle\psi_{-p\hat{\mathbf{n}},s's}|U_{\text{rel}}(t_1,t_0)|\psi_{-p\hat{\mathbf{z}},s's}\rangle$$
$$\pm \langle\psi_{-p\hat{\mathbf{n}},s's}|U_{\text{rel}}(t_1,t_0)|\psi_{p\hat{\mathbf{z}},ss'}\rangle \pm \langle\psi_{p\hat{\mathbf{n}},ss'}|U_{\text{rel}}(t_1,t_0)|\psi_{-p\hat{\mathbf{z}},s's}\rangle\big].$$

Since the Hamiltonian is independent of spin, the time evolution operator acts only on the orbital component of the state, and hence the exchange terms vanish, given that $s \neq s'$,

$$\langle\psi_{-p\hat{\mathbf{n}},s's}|U_{\text{rel}}(t_1,t_0)|\psi_{p\hat{\mathbf{z}},ss'}\rangle = \langle\psi_{-p\hat{\mathbf{n}}}|U_{\text{rel}}(t_1,t_0)|\psi_{p\hat{\mathbf{z}}}\rangle \, \delta_{s',s}\, \delta_{s,s'} = 0, \qquad s \neq s'.$$

By contrast the direct terms give, for example,

$$\langle\psi_{-p\hat{\mathbf{n}},s's}|U_{\text{rel}}(t_1,t_0)|\psi_{-p\hat{\mathbf{z}},s's}\rangle = \underbrace{\langle\psi_{-p\hat{\mathbf{n}}}|U_{\text{rel}}(t_1,t_0)|\psi_{-p\hat{\mathbf{z}}}\rangle}_{\overline{A}(\hat{\mathbf{n}})} \, \delta_{s',s'}\, \delta_{s,s} = \overline{A}(\hat{\mathbf{n}}),$$

and therefore

$$^{S/A}\langle\psi_{p\hat{\mathbf{n}},ss'}|U_{\text{rel}}(t_1,t_0)|\psi_{p\hat{\mathbf{z}},ss'}\rangle^{S/A} = \frac{1}{2}\left[A(\hat{\mathbf{n}}) + \overline{A}(\hat{\mathbf{n}})\right] = A(\hat{\mathbf{n}}),$$

since these amplitudes are the same. To see this, without loss of generality, take $p\hat{\mathbf{n}}$ to be in the xz-plane.[3] Now, rotate about the $\hat{\mathbf{y}}$-axis by π. Under such a rotation, we have

$$|\psi_{-p\hat{\mathbf{z}}}\rangle \longrightarrow U_R(\pi\hat{\mathbf{y}})|\psi_{-p\hat{\mathbf{z}}}\rangle = |\psi_{p\hat{\mathbf{z}}}\rangle, \qquad |\psi_{-p\hat{\mathbf{n}}}\rangle \longrightarrow U_R(\pi\hat{\mathbf{y}})|\psi_{-p\hat{\mathbf{n}}}\rangle = |\psi_{p\hat{\mathbf{n}}}\rangle,$$

---

[3] If $p\hat{\mathbf{n}}$ is not in such a plane, we can rotate it about the $\hat{\mathbf{z}}$-axis by an angle $-\phi$, so that it is in this plane; recall that the amplitude is independent of ϕ.

yielding

$$\underbrace{\langle\psi_{-p\hat{\mathbf{n}}}|U_{\text{rel}}(t_1,t_0)|\psi_{-p\hat{\mathbf{z}}}\rangle}_{\overline{A}(\hat{\mathbf{n}})} = \langle\psi_{p\hat{\mathbf{n}}}|U_R(\pi\hat{\mathbf{y}})\,U_{\text{rel}}(t_1,t_0)\,U_R^{\dagger}(\pi\hat{\mathbf{y}})|\psi_{p\hat{\mathbf{z}}}\rangle = \underbrace{\langle\psi_{p\hat{\mathbf{n}}}|U_{\text{rel}}(t_1,t_0)|\psi_{p\hat{\mathbf{z}}}\rangle}_{A(\hat{\mathbf{n}})},$$

an obvious result (to see this, just draw a picture of the two configurations). The probability for the transition $ss' \longrightarrow ss'$ with $s \neq s'$ is given by

$$P_{s\neq s'}^{S/A} = |A(\hat{\mathbf{n}})|^2,$$

and is independent of the statistics of the two particles (whether they are bosons or fermions).

Part 5

If the spin states of the two particles are the same ($s = s'$), we obtain, for the amplitude,

$$^{S/A}\langle\psi_{p\hat{\mathbf{n}},ss}|U_{\text{rel}}(t_1,t_0)|\psi_{p\hat{\mathbf{z}},ss}\rangle^{S/A} = \frac{1}{2}\left[A(\hat{\mathbf{n}}) + \overline{A}(\hat{\mathbf{n}}) \pm A(-\hat{\mathbf{n}}) \pm \overline{A}(-\hat{\mathbf{n}})\right],$$

where we have used for the exchange term (with $s = s'$)

$$\langle\psi_{-p\hat{\mathbf{n}},ss}|U_{\text{rel}}(t_1,t_0)|\psi_{p\hat{\mathbf{z}},ss}\rangle = \langle\psi_{-p\hat{\mathbf{n}}}|U_{\text{rel}}(t_1,t_0)|\psi_{p\hat{\mathbf{z}}}\rangle\,\delta_{s,s}\,\delta_{s,s} = A(-\hat{\mathbf{n}}),$$

and similarly

$$\langle\psi_{p\hat{\mathbf{n}},ss}|U_{\text{rel}}(t_1,t_0)|\psi_{-p\hat{\mathbf{z}},ss}\rangle = \overline{A}(-\hat{\mathbf{n}}).$$

On the basis of the argument in parts 3 and 4, this can be further simplified as

$$^{S/A}\langle\psi_{p\hat{\mathbf{n}},ss}|U_{\text{rel}}(t_1,t_0)|\psi_{p\hat{\mathbf{z}},ss}\rangle^{S/A} = A(\hat{\mathbf{n}}) \pm A(-\hat{\mathbf{n}}) = A(\hat{\mathbf{n}}) \pm A(\hat{\mathbf{n}}'),$$

and the resulting probability in this case is given by

$$P_{s=s'}^{S/A} = |A(\hat{\mathbf{n}}) \pm A(\hat{\mathbf{n}}')|^2,$$

which shows that these probabilities are invariant under a reflection with respect to the xy-plane, since $\hat{\mathbf{n}} \rightleftharpoons \hat{\mathbf{n}}'$ under such a transformation. We also note that at $\theta = \pi/2$ we have $\hat{\mathbf{n}} = \hat{\mathbf{n}}' = \hat{\mathbf{x}}$, with $\hat{\mathbf{x}} = (1, 0, 0)$, and hence the probability above vanishes for fermions and is equal to $4|A(\hat{\mathbf{x}})|^2$ for bosons, which is twice as large as the value P_{dis} obtained earlier.

Problem 6 Properties of Permutation Operators for Three Particles

A basis in the three-particle state space is obtained as

$$|\psi_{\gamma_1,\gamma_2,\gamma_3}(1,2,3)\rangle = |\psi_{\gamma_1}(1)\rangle \otimes |\psi_{\gamma_2}(2)\rangle \otimes |\psi_{\gamma_2}(3)\rangle,$$

and the particular ordering of the single-particle states $|\psi_{\gamma_i}(i)\rangle$ within the tensor product is unimportant. There are 3! possible permutations of the particle labels $1, 2, 3$ (or γ_1, γ_2, γ_3):

$$\begin{pmatrix} 1\ 2\ 3 \\ 1\ 2\ 3 \end{pmatrix},\ \begin{pmatrix} 1\ 2\ 3 \\ 2\ 3\ 1 \end{pmatrix},\ \begin{pmatrix} 1\ 2\ 3 \\ 3\ 1\ 2 \end{pmatrix},\ \begin{pmatrix} 1\ 2\ 3 \\ 2\ 1\ 3 \end{pmatrix},\ \begin{pmatrix} 1\ 2\ 3 \\ 3\ 2\ 1 \end{pmatrix},\ \begin{pmatrix} 1\ 2\ 3 \\ 1\ 3\ 2 \end{pmatrix},$$

where the first row gives the reference ordering (which we have arbitrarily chosen as 123; it could be any permutation of 123) and the second row specifies the permutation actually being considered; for example,

$$\begin{pmatrix} 1 & 2 & 3 \\ 2 & 3 & 1 \end{pmatrix} \implies 1 \longrightarrow 2, \ 2 \longrightarrow 3, \ 3 \longrightarrow 1.$$

The ordering of the columns within the brackets is irrelevant, in the sense that

$$\begin{pmatrix} 1 & 2 & 3 \\ 2 & 3 & 1 \end{pmatrix}, \qquad \begin{pmatrix} 3 & 1 & 2 \\ 1 & 2 & 3 \end{pmatrix}, \qquad \cdots$$

all represent the same permutation. The corresponding 3! exchange operators are denoted as $\hat{P}_{123}$, $\hat{P}_{231}$, $\hat{P}_{312}$, $\hat{P}_{213}$, $\hat{P}_{321}$, and $\hat{P}_{132}$, and

$$\hat{P}_{mnp}|\psi_{\gamma_1,\gamma_2,\gamma_3}(1,2,3)\rangle = |\psi_{\gamma_1,\gamma_2,\gamma_3}(m,n,p)\rangle,$$

so that, for example,

$$\hat{P}_{231}|\psi_{\gamma_1,\gamma_2,\gamma_3}(1,2,3)\rangle = \underbrace{|\psi_{\gamma_1,\gamma_2,\gamma_3}(2,3,1)\rangle}_{|\psi_{\gamma_1}(2)\rangle \otimes |\psi_{\gamma_2}(3)\rangle \otimes |\psi_{\gamma_3}(1)\rangle} = |\psi_{\gamma_3,\gamma_1,\gamma_2}(1,2,3)\rangle.$$

The set of 3! permutation operators forms a finite group, defined by the following properties: (i) the product of two permutation operators is also a permutation operator; (ii) the permutation operator $\hat{P}_{123}$ is the identity operator; and (iii) for every permutation operator there exists an inverse permutation operator. Verify these facts by considering, as an example, the two permutation operators $\hat{P}_{231}$ and $\hat{P}_{213}$: obtain the permutation operator corresponding to $\hat{P}_{231}\hat{P}_{213}$ and determine the inverse $\hat{P}_{231}^{-1}$. Permutation operators generally do not commute: verify this fact by showing that $\hat{P}_{231}\hat{P}_{213} \neq \hat{P}_{213}\hat{P}_{231}$. Lastly, they are unitary but not hermitian (except in the case of two particles, for which the exchange operator is both hermitian and unitary, see Problem 3). Note that all these properties generalize to the case of N particles, and the set of $N!$ permutation operators forms a group.

Solution

The product of permutation operators $\hat{P}_{231}\hat{P}_{213}$ is seen to correspond to the permutation operator $\hat{P}_{132}$. One way to verify this consists in applying $\hat{P}_{231}\hat{P}_{213}$ to the basis $|\psi_{\gamma_1,\gamma_2,\gamma_3}(1,2,3)\rangle$ and showing that the resulting state is $\hat{P}_{132}|\psi_{\gamma_1,\gamma_2,\gamma_3}(1,2,3)\rangle = |\psi_{\gamma_1,\gamma_2,\gamma_3}(1,3,2)\rangle = |\psi_{\gamma_1,\gamma_3,\gamma_2}(1,2,3)\rangle$. Another way is to write

$$\underbrace{\begin{pmatrix} 1 & 2 & 3 \\ 2 & 3 & 1 \end{pmatrix}}_{\hat{P}_{231}} \underbrace{\begin{pmatrix} 1 & 2 & 3 \\ 2 & 1 & 3 \end{pmatrix}}_{\hat{P}_{213}} = \underbrace{\begin{pmatrix} 1 & 2 & 3 \\ 2 & 3 & 1 \end{pmatrix}}_{\hat{P}_{231}} \underbrace{\begin{pmatrix} 2 & 3 & 1 \\ 1 & 3 & 2 \end{pmatrix}}_{\hat{P}_{213}} = \underbrace{\begin{pmatrix} 1 & 2 & 3 \\ 1 & 3 & 2 \end{pmatrix}}_{\hat{P}_{132}},$$

where in the second step we have rearranged the columns of $\hat{P}_{213}$ so that the top line matches the lower line of $\hat{P}_{231}$. To determine the inverse of a permutation operator we can proceed similarly. The inverse $\hat{P}_{231}^{-1}$ is easily obtained by the rule above:

$$\underbrace{\begin{pmatrix} 1 & 2 & 3 \\ 1 & 2 & 3 \end{pmatrix}}_{\text{identity } \hat{P}_{123}} = \underbrace{\begin{pmatrix} 1 & 2 & 3 \\ 2 & 3 & 1 \end{pmatrix}}_{\hat{P}_{231}} \underbrace{\begin{pmatrix} 2 & 3 & 1 \\ 1 & 2 & 3 \end{pmatrix}}_{\hat{P}_{231}^{-1}} = \underbrace{\begin{pmatrix} 1 & 2 & 3 \\ 2 & 3 & 1 \end{pmatrix}}_{\hat{P}_{231}} \underbrace{\begin{pmatrix} 1 & 2 & 3 \\ 3 & 1 & 2 \end{pmatrix}}_{\hat{P}_{231}^{-1}},$$

which shows that $\hat{P}_{231}^{-1} = \hat{P}_{312}$; it can also be verified that $\hat{P}_{312}\hat{P}_{231} = \hat{P}_{231}^{-1}\hat{P}_{231} = \hat{P}_{123}$, that is, $\hat{P}_{231}^{-1} = \hat{P}_{312}$ and $\hat{P}_{231}$ commute, as they must. As noted in the text of the problem, generally permutation operators do not commute with each other; indeed, the product $\hat{P}_{213}\hat{P}_{231}$ is given by

$$\underbrace{\begin{pmatrix} 1 & 2 & 3 \\ 2 & 1 & 3 \end{pmatrix}}_{\hat{P}_{213}} \underbrace{\begin{pmatrix} 1 & 2 & 3 \\ 2 & 3 & 1 \end{pmatrix}}_{\hat{P}_{231}} = \underbrace{\begin{pmatrix} 1 & 2 & 3 \\ 2 & 1 & 3 \end{pmatrix}}_{\hat{P}_{213}} \underbrace{\begin{pmatrix} 2 & 1 & 3 \\ 3 & 2 & 1 \end{pmatrix}}_{\hat{P}_{231}} = \underbrace{\begin{pmatrix} 1 & 2 & 3 \\ 3 & 2 & 1 \end{pmatrix}}_{\hat{P}_{321}},$$

which is different from the permutation operator $\hat{P}_{132}$ corresponding to $\hat{P}_{231}\hat{P}_{213}$, obtained above. To show that permutation operators are unitary, consider, for example, $\hat{P}_{231}$:

$$\hat{P}_{231}|\psi_{\gamma_1,\gamma_2,\gamma_3}(1,2,3)\rangle = |\psi_{\gamma_3,\gamma_1,\gamma_2}(1,2,3)\rangle,$$

and therefore

$$\underbrace{[\langle\psi_{\gamma_1',\gamma_2',\gamma_3'}(1,2,3)|\hat{P}_{231}^{\dagger}]}_{\langle\psi_{\gamma_3',\gamma_1',\gamma_2'}(1,2,3)|} \underbrace{\hat{P}_{231}|\psi_{\gamma_1,\gamma_2,\gamma_3}(1,2,3)\rangle}_{|\psi_{\gamma_3,\gamma_1,\gamma_2}(1,2,3)\rangle} = \delta_{\gamma_3,\gamma_3'}\,\delta_{\gamma_1,\gamma_1'}\,\delta_{\gamma_2,\gamma_2'}$$

$$= \langle\psi_{\gamma_1',\gamma_2',\gamma_3'}(1,2,3)|\psi_{\gamma_1,\gamma_2,\gamma_3}(1,2,3)\rangle,$$

and hence $\hat{P}_{231}$ is unitary. It is now easy to understand that any N-particle permutation operator is unitary: it leads to the same ordering of quantum numbers on the left- and right-hand side states. Lastly, to verify that it is *not* hermitian, note that

$$\langle\psi_{\gamma_1',\gamma_2',\gamma_3'}(1,2,3)|\underbrace{\hat{P}_{231}|\psi_{\gamma_1,\gamma_2,\gamma_3}(1,2,3)\rangle}_{|\psi_{\gamma_3,\gamma_1,\gamma_2}(1,2,3)\rangle} = \delta_{\gamma_1,\gamma_2'}\,\delta_{\gamma_2,\gamma_3'}\,\delta_{\gamma_3,\gamma_1'}$$

and

$$\left[\langle\psi_{\gamma_1,\gamma_2,\gamma_3}(1,2,3)|\underbrace{\hat{P}_{231}|\psi_{\gamma_1',\gamma_2',\gamma_3'}(1,2,3)\rangle}_{|\psi_{\gamma_3',\gamma_1',\gamma_2'}(1,2,3)\rangle}\right]^* = \delta_{\gamma_1,\gamma_3'}\,\delta_{\gamma_2,\gamma_1'}\,\delta_{\gamma_3,\gamma_2'},$$

so that $\hat{P}_{231}^{\dagger} \neq \hat{P}_{231}$.

Problem 7 Symmetrizer and Antisymmetrizer and N-Particle States

Here, we consider permutation operators in an N-particle system. A permutation operator that only involves two particles and leaves the remaining $N-2$ particles unaffected is known as a pair-exchange or transposition operator. For example, in the three-particle case (see Problem 6), $\hat{P}_{213}$, $\hat{P}_{321}$, and $\hat{P}_{132}$ are transposition operators: the first involves the pair 12, the second involves the pair 13, and the third involves the pair 23. Transpositions are hermitian and equal to their inverse, which also makes them unitary (the proofs are similar to those for $\hat{P}_{21}$).

It turns out that any N-particle permutation operator can be decomposed into a product of pair-exchange operators. This decomposition is not unique. For example, for the three-particle case and $\hat{P}_{231}$, we have

$$\underbrace{\begin{pmatrix} 1 & 2 & 3 \\ 2 & 3 & 1 \end{pmatrix}}_{\hat{P}_{231}} = \underbrace{\begin{pmatrix} 1 & 2 & 3 \\ 2 & 1 & 3 \end{pmatrix}}_{\hat{P}_{213}} \underbrace{\begin{pmatrix} 2 & 1 & 3 \\ 2 & 3 & 1 \end{pmatrix}}_{\hat{P}_{321}}$$

$$= \underbrace{\begin{pmatrix} 1 & 2 & 3 \\ 1 & 3 & 2 \end{pmatrix}}_{\hat{P}_{132}} \underbrace{\begin{pmatrix} 1 & 3 & 2 \\ 2 & 3 & 1 \end{pmatrix}}_{\hat{P}_{213}}$$

$$= \underbrace{\begin{pmatrix} 1 & 2 & 3 \\ 3 & 2 & 1 \end{pmatrix}}_{\hat{P}_{321}} \underbrace{\begin{pmatrix} 3 & 2 & 1 \\ 3 & 1 & 2 \end{pmatrix}}_{\hat{P}_{213}} \underbrace{\begin{pmatrix} 3 & 1 & 2 \\ 2 & 1 & 3 \end{pmatrix}}_{\hat{P}_{132}} \underbrace{\begin{pmatrix} 2 & 1 & 3 \\ 2 & 3 & 1 \end{pmatrix}}_{\hat{P}_{321}}$$

However, the number p of pair-exchange operators into which a given N-particle permutation operator can be factored is always either even or odd. So, the parity $(-1)^p$ of any N-particle permutation operator is unique. Among the $N!$ (necessarily an even number for $N \geq 2$) permutation operators, half of them have parity plus and the remaining half have parity minus, so in the three-particle case, $\hat{P}_{123}, \hat{P}_{231}, \hat{P}_{312}$ have even parity and $\hat{P}_{213}, \hat{P}_{321}$, and $\hat{P}_{132}$ have odd parity. In particular, the identity permutation operator has even parity, while the pair-exchange operators all have odd parity.

1. Given a permutation operator $\hat{P}_\alpha$, where α specifies a permutation of $12 \cdots N$, let ϵ_α be its parity, either $+$ or $-$. Define the operators $\hat{S}$ and $\hat{A}$, known respectively as a symmetrizer and antisymmetrizer,

$$\hat{S} = \frac{1}{N!} \sum_\alpha \hat{P}_\alpha , \qquad \hat{A} = \frac{1}{N!} \sum_\alpha \epsilon_\alpha \hat{P}_\alpha ,$$

where the sum is over all $N!$ permutations, and show that they are projection operators and that $\hat{S}\hat{A} = 0$.

2. Properly symmetrized multi-boson or multi-fermion states can be written in terms of a symmetrizer or antisymmetrizer as

$$|\psi_{\gamma_1,\gamma_2,\ldots,\gamma_N}^{S/A}\rangle = c\,\hat{S}/\hat{A}\,|\psi_{\gamma_1,\gamma_2,\ldots,\gamma_N}(1,2,\ldots,N)\rangle ,$$

where $|\psi_{\gamma_1,\gamma_2,\ldots,\gamma_N}(1,2,\ldots,N)\rangle = |\psi_{\gamma_1}(1)\rangle \otimes |\psi_{\gamma_2}(2)\rangle \otimes \cdots \otimes |\psi_{\gamma_N}(N)\rangle$, and c is a normalization constant. How do these states change under a pair exchange? And under a generic permutation P_α? Determine the normalization constant.

Solution

Part 1

The symmetrizer is a projection operator if $\hat{S}^\dagger = \hat{S}$ and $\hat{S}^2 = \hat{S}$; similarly, for the antisymmetrizer. To verify the hermiticity (recall that the individual $\hat{P}_\alpha$ are not generally hermitian), note that $\hat{P}_\alpha$ is unitary and hence $\hat{P}_\alpha^\dagger = \hat{P}_\alpha^{-1}$ and, since $\hat{P}_\alpha \hat{P}_\alpha^{-1}$ is the identity permutation, which has even parity, it follows that $\hat{P}_\alpha^{-1}$ has the same parity as $\hat{P}_\alpha$. From the group property we know that $\hat{P}_\alpha^{-1}$ corresponds to one (and only one) of the $N!$ permutation operators, and therefore $\sum_\alpha \hat{P}_\alpha^{-1}$ is the same operator as $\sum_\alpha \hat{P}_\alpha$, the only difference being that the $N!$ terms are summed up in a different order.[4] A similar argument can be used to verify that $\hat{A}$ is hermitian.

To show that $\hat{S}^2 = \hat{S}$ and $\hat{A}^2 = \hat{A}$, first note that

$$\hat{P}_\beta \hat{S} = \frac{1}{N!} \sum_\alpha \hat{P}_\beta \hat{P}_\alpha = \frac{1}{N!} \sum_\gamma \hat{P}_\gamma = \hat{S}, \qquad \hat{S}\hat{P}_\beta = \frac{1}{N!} \sum_\alpha \hat{P}_\alpha \hat{P}_\beta = \frac{1}{N!} \sum_{\gamma'} \hat{P}_{\gamma'} = \hat{S},$$

where we have used the group property that $\hat{P}_\beta \hat{P}_\alpha$ corresponds to one (and only one) of the $N!$ permutation operators and that, as α cycles through all of them, the product $\hat{P}_\gamma = \hat{P}_\beta \hat{P}_\alpha$ runs over all $N!$ individual permutation operators. For the antisymmetrizer we have

$$\epsilon_\beta \hat{P}_\beta \hat{A} = \frac{1}{N!} \sum_\alpha \epsilon_\beta \epsilon_\alpha \hat{P}_\beta \hat{P}_\alpha = \frac{1}{N!} \sum_\gamma \epsilon_\gamma \hat{P}_\gamma = \hat{A},$$

since the parity of the product is equal to the product of the parities. Because $\epsilon_\beta^2 = 1$, by multiplying both side of the relation above by ϵ_β we can write

$$\hat{P}_\beta \hat{A} = \epsilon_\beta \hat{A},$$

We also have

$$\epsilon_\beta \hat{A} \hat{P}_\beta = \hat{A} \implies \hat{A} \hat{P}_\beta = \epsilon_\beta \hat{A}.$$

With these properties in hand, we now see that

$$\hat{S}^2 = \frac{1}{N!} \sum_\beta \hat{P}_\beta \hat{S} = \hat{S} \frac{1}{N!} \sum_\beta = \hat{S}, \qquad \hat{A}^2 = \frac{1}{N!} \sum_\beta \epsilon_\beta \hat{P}_\beta \hat{A} = \hat{A} \frac{1}{N!} \sum_\beta = \hat{A},$$

and

$$\hat{A}\hat{S} = \frac{1}{N!} \sum_\beta \epsilon_\beta \hat{P}_\beta \hat{S} = \hat{S} \frac{1}{N!} \sum_\beta \epsilon_\beta = 0,$$

since half the permutation operators have even parity and the remaining half have odd parity; further, the fact that $\hat{S}$ and $\hat{A}$ are hermitian ensures that $\hat{S}\hat{A} = 0$.

Part 2

It is easily verified that these states are either symmetric (S) or antisymmetric (A) under the action of a generic pair-exchange operator $\hat{P}_e$ (which has always odd parity), since $\hat{P}_e \hat{S} = \hat{S}$ and $\hat{P}_e \hat{A} = -\hat{A}$. Incidentally, this implies that if two or more fermions occupy the same single-particle state then

---

[4] Again for three-particle case, we have

$$\hat{S} = \frac{1}{3!}(\hat{P}_{123} + \hat{P}_{231} + \hat{P}_{312} + \hat{P}_{213} + \hat{P}_{321} + \hat{P}_{132}), \qquad \hat{A} = \frac{1}{3!}(\hat{P}_{123} + \hat{P}_{231} + \hat{P}_{312} - \hat{P}_{213} - \hat{P}_{321} - \hat{P}_{132}),$$

and since the pair-exchange operators are hermitian and

$$\hat{P}_{123}^\dagger = \hat{P}_{123}, \qquad \hat{P}_{231}^\dagger = \hat{P}_{312}, \qquad \hat{P}_{312}^\dagger = \hat{P}_{231},$$

we find that

$$\hat{S}^\dagger = \frac{1}{3!}(\hat{P}_{123}^\dagger + \hat{P}_{231}^\dagger + \hat{P}_{312}^\dagger + \hat{P}_{213}^\dagger + \hat{P}_{321}^\dagger + \hat{P}_{132}^\dagger) = \frac{1}{3!}(\hat{P}_{123} + \overset{\text{note order}}{\hat{P}_{312} + \hat{P}_{231}} + \hat{P}_{213} + \hat{P}_{321} + \hat{P}_{132}) = \hat{S},$$

and similarly for $\hat{A}$.

$|\psi^A_{\gamma_1,\dots,\gamma_N}\rangle$ vanishes identically, in accordance with Pauli's exclusion principle. However, under the action of a generic permutation operator $\hat{P}_\alpha$ (not necessarily a pair-exchange operator) we have

$$\hat{P}_\alpha|\psi^S_{\gamma_1,\dots,\gamma_n}\rangle = |\psi^S_{\gamma_1,\dots,\gamma_n}\rangle\,, \qquad \hat{P}_\alpha|\psi^A_{\gamma_1,\dots,\gamma_n}\rangle = \epsilon_\alpha|\psi^A_{\gamma_1,\dots,\gamma_n}\rangle\,,$$

where ϵ_α is the parity of $\hat{P}_\alpha$.

The orthogonality relation reads (recalling that $\hat{S}$ and $\hat{A}$ are projectors)

$$\langle\psi^S_{\gamma'_1,\dots,\gamma'_N}|\psi^S_{\gamma_1,\dots,\gamma_N}\rangle = \frac{|c|^2}{N!}\sum_\alpha\langle\psi_{\gamma'_1,\dots,\gamma'_N}(1,\dots,N)|\,\hat{P}_\alpha\,|\psi_{\gamma_1,\dots,\gamma_N}(1,\dots,N)\rangle = \frac{|c|^2}{N!}\sum_\alpha\delta_{\gamma',P_\alpha\gamma}\,,$$

where $\delta_{\gamma',P_\alpha\gamma}$ indicates a product of Kronecker δs:

$$\delta_{\gamma',P_\alpha\gamma} = \underbrace{\delta_{\gamma'_1,P_\alpha\gamma_1}\,\delta_{\gamma'_2,P_\alpha\gamma_2}\,\cdots\,\delta_{\gamma'_N,P_\alpha\gamma_N}}_{\text{product of }N\text{ Kronecker }\delta\text{'s}}\,,$$

and $P_\alpha\gamma_1,\dots,P_\alpha\gamma_N$ denotes a permutation of $\gamma_1,\dots,\gamma_N$; for example, the identity permutation gives

$$\delta_{\gamma',P_\alpha\gamma}\longrightarrow\delta_{\gamma'_1,\gamma_1}\,\delta_{\gamma'_2,\gamma_2}\,\cdots\,\delta_{\gamma'_N,\gamma_N}\,.$$

For the fermion states, we obtain a similar relation,

$$\langle\psi^A_{\gamma'_1,\dots,\gamma'_N}|\psi^A_{\gamma_1,\dots,\gamma_N}\rangle = \frac{|c|^2}{N!}\sum_\alpha\epsilon_\alpha\,\delta_{\gamma',P_\alpha\gamma}\,,$$

except for the presence of the factor ϵ_α, the parity of the permutation. Since in the case of the fermion state the γ_i must all be different from each other, we see that $|c| = \sqrt{N!}$ correctly normalizes $|\psi^A_{\gamma_1,\dots,\gamma_N}\rangle$. This is not so for the boson state, since more than one boson can occupy the same single-particle state, and hence $|c| = [N!/(n_{\gamma_i}!\,n_{\gamma_j}!\cdots)]^{1/2}$, where n_{γ_i}, n_{γ_j}, ... are the numbers of bosons in the single-particle states $|\psi_{\gamma_i}\rangle$, $|\psi_{\gamma_j}\rangle$, ...

Problem 8 Expectation Value of a Totally Symmetric Operator on an *N*-Particle Antisymmetric State

Consider a system of N identical fermions. Show that, in order to evaluate the expectation value of a totally symmetric operator $\hat{O}(1,\dots,N)$ (that is, $\hat{O}$ commutes with any pair-exchange operator), it suffices to antisymmetrize only the state immediately to the left or right of $\hat{O}$. Consider the cases in which $\hat{O}$ consists of the sum of one-body operators $\hat{o}(i)$ and two-body operators $\hat{o}(ij)$, specifically

$$\hat{O}(1,\dots,N) = \sum_{i=1}^N\hat{o}(i)\,, \qquad \hat{O}(1,\dots,N) = \frac{1}{2}\sum_{i,j=1}^N\hat{o}(ij)\,.$$

Solution

In the notation of Problem 7, the normalized antisymmetric state reads $|\psi^A_{\gamma_1\cdots\gamma_N}\rangle = \sqrt{N!}\,\hat{A}\,|\psi_{\gamma_1}(1)\cdots\psi_{\gamma_N}(N)\rangle$, where $\hat{A}$ is the antisymmetrizer. The expectation value can then be expressed as

$$\langle O\rangle = \langle\psi^A_{\gamma_1\cdots\gamma_N}|\hat{O}(1,\dots,N)|\psi^A_{\gamma_1\cdots\gamma_N}\rangle = N!\,(\langle\psi_{\gamma_1}(1)\cdots\psi_{\gamma_N}(N)|\hat{A}^\dagger)\hat{O}(1,\dots,N)\,\hat{A}|\psi_{\gamma_1}(1)\cdots\psi_{\gamma_N}(N)\rangle\,.$$

Since $\hat{A}$ is a projection operator ($\hat{A}^\dagger = \hat{A}$ and $\hat{A}^2 = \hat{A}$) and since it commutes with the totally symmetric operator $\hat{O}$, it follows that

$$\langle O \rangle = N! \, \langle \psi_{\gamma_1}(1) \cdots \psi_{\gamma_N}(N)| \, \hat{O}(1,\ldots,N) \, \hat{A} |\psi_{\gamma_1}(1) \cdots \psi_{\gamma_N}(N)\rangle$$
$$= \sqrt{N!} \, \langle \psi_{\gamma_1}(1) \cdots \psi_{\gamma_N}(N)| \, \hat{O}(1,\ldots,N) \, |\psi^A_{\gamma_1 \cdots \gamma_N}\rangle \, .$$

If $\hat{O}$ consists of the sum of one-body (1b) operators, we have

$$\langle O \rangle_{1b} = \sum_{i=1}^{N} \sum_\alpha \epsilon_\alpha \langle \psi_{\gamma_1}(1) \cdots \psi_{\gamma_N}(N)| \, \hat{o}(i) \, |\psi_{P_\alpha\gamma_1}(1) \cdots \psi_{P_\alpha\gamma_N}(N)\rangle$$

$$= \sum_{i=1}^{N} \langle \psi_{\gamma_1}(1) \cdots \psi_{\gamma_N}(N)| \, \hat{o}(i) \, |\psi_{\gamma_1}(1) \cdots \psi_{\gamma_N}(N)\rangle$$

$$= \sum_{i=1}^{N} \langle \psi_{\gamma_1}(1)|\psi_{\gamma_1}(1)\rangle \cdots \langle \psi_{\gamma_i}(i)| \, \hat{o}(i) \, |\psi_{\gamma_i}(i)\rangle \cdots \langle \psi_{\gamma_N}(N)|\psi_{\gamma_N}(N)\rangle = \sum_{i=1}^{N} \langle \psi_{\gamma_i}(i)| \, \hat{o}(i) \, |\psi_{\gamma_i}(i)\rangle,$$

and the only term that contributes in the sum over permutations is $|\psi_{P_\alpha\gamma_1}(1) \ldots \psi_{P_\alpha\gamma_N}(N)\rangle = |\psi_{\gamma_1}(1) \ldots \psi_{\gamma_N}(N)\rangle$ (that is, $\hat{P}_\alpha$ is the identity permutation, of even parity), since any other permutation involves the exchange of at least two γs and therefore vanishes by orthogonality. By contrast, if $\hat{O}$ consists of the sum of two-body (2b) operators, the expectation value reads

$$\langle O \rangle_{2b} = \frac{1}{2} \sum_{i,j=1}^{N} \sum_\alpha \epsilon_\alpha \langle \psi_{\gamma_1}(1) \cdots \psi_{\gamma_N}(N)| \, \hat{o}(ij) \, |\psi_{P_\alpha\gamma_1}(1) \cdots \psi_{P_\alpha\gamma_N}(N)\rangle \, ,$$

and in the sum over permutations only two terms contribute: the identity (of even parity),

$$|\psi_{P_\alpha\gamma_1}(1) \cdots \psi_{P_\alpha\gamma_i}(i) \cdots \psi_{P\gamma_j}(j) \cdots \psi_{P\gamma_N}(N)\rangle = |\psi_{\gamma_1}(1) \cdots \psi_{\gamma_i}(i) \cdots \psi_{\gamma_j}(j) \cdots \psi_{\gamma_N}(N)\rangle \, ,$$

and the permutation involving the exchange $\gamma_i \rightleftarrows \gamma_j$ (of odd parity),

$$|\psi_{P_\alpha\gamma_1}(1) \cdots \psi_{P_\alpha\gamma_i}(i) \cdots \psi_{P\gamma_j}(j) \cdots \psi_{P\gamma_N}(N)\rangle = |\psi_{\gamma_1}(1) \cdots \psi_{\gamma_j}(i) \cdots \psi_{\gamma_i}(j) \cdots \psi_{\gamma_N}(N)\rangle \, ,$$

yielding

$$\langle O \rangle_{2b} = \frac{1}{2} \sum_{i,j=1}^{N} \Big[\underbrace{\langle \psi_{\gamma_i}(i), \psi_{\gamma_j}(j)| \, \hat{o}(ij) \, |\psi_{\gamma_i}(i), \psi_{\gamma_j}(j)\rangle}_{\text{direct}} - \underbrace{\langle \psi_{\gamma_i}(i), \psi_{\gamma_j}(j)| \, \hat{o}(ij) \, |\psi_{\gamma_j}(i), \psi_{\gamma_i}(j)\rangle}_{\text{exchange}} \Big] \, ,$$

the sum of a direct and an exchange term.

Problem 9 The Fermi Gas

A simple example of a many-body system is a non-interacting gas of identical fermions, called a Fermi gas. The Fermi gas has been used as a simple model to describe a variety of different systems, including electrons in metals or protons and neutrons (nucleons) in nuclei.

1. Consider a single fermion of spin $\mathbf{S}$ ($S_z|m\rangle = m\hbar|m\rangle$ with $m = -S, \ldots, S$), confined in a box of volume V (and side $V^{1/3}$). Imposing periodic boundary conditions, obtain the eigenvalues and corresponding normalized eigenfunctions of the Hamiltonian.

2. Now, suppose that there are N non-interacting fermions confined in the box. Calculate the ground-state energy E_0 of such a system, expressing it in terms of N and the Fermi energy – the energy of the highest occupied single-particle level. Work in the limit $V \longrightarrow \infty$. Knowing that the pressure P_0 is given by $P_0 = -\partial E_0/\partial V$ (at zero temperature), obtain P_0.

Solution

Part 1

The Hamiltonian is $H = \mathbf{p}^2/(2m)$ and commutes with the momentum operator $\mathbf{p}$. The simultaneous eigenfunctions of H and $\mathbf{p}$ can be written as

$$\psi_{\mathbf{k},m}(\mathbf{r}) = \frac{1}{V^{1/2}} e^{i\mathbf{k}\cdot\mathbf{r}}|m\rangle , \qquad m = -S, -S+1, \ldots, S ,$$

where $\mathbf{k}$ is the wave number (the corresponding momentum $\mathbf{p}$ is $\hbar\mathbf{k}$) and m labels eigenstates of S_z. The wave numbers $\mathbf{k}$ are quantized, with

$$\mathbf{k} = \frac{2\pi}{V^{1/3}}\,\mathbf{n} , \qquad \mathbf{n} = (n_x, n_y, n_z) , \qquad n_i = 0, \pm 1, \pm 2, \ldots ,$$

as required by the periodic boundary condition; indeed,

$$\psi_{\mathbf{k},m}(x + V^{1/3}, y, z) = \frac{1}{V^{1/2}} e^{ik_x(x+V^{1/3})}\, e^{i(k_y y + k_z z)}|m\rangle = e^{i2\pi n_x}\,\psi_{\mathbf{k},m}(\mathbf{r}) = \psi_{\mathbf{k},m}(\mathbf{r}) ,$$

and similarly for y and z. These states form an orthonormal basis,

$$\langle \psi_{\mathbf{k}',m'}|\psi_{\mathbf{k},m}\rangle = \left[\frac{1}{V}\int d\mathbf{r}\, e^{-i(\mathbf{k}'-\mathbf{k})\cdot\mathbf{r}}\right]\langle m'|m\rangle = \delta_{\mathbf{k},\mathbf{k}'}\,\delta_{m,m'} ,$$

as can be easily verified by noting that the integral factorizes into three terms,

$$\int d\mathbf{r}\, e^{-i(\mathbf{k}'-\mathbf{k})\cdot\mathbf{r}} = \int_0^{V^{1/3}} dx\, e^{-i(k_x'-k_x)x} \int_0^{V^{1/3}} dy\, e^{-i(k_y'-k_y)y} \int_0^{V^{1/3}} dz\, e^{-i(k_z'-k_z)z} .$$

Assuming $k_x' \neq k_x$ or $n_x' \neq n_x$, we have for the integral in x (and obviously the same for the integrals in y and z)

$$\int_0^{V^{1/3}} dx\, e^{-i(k_x'-k_x)x} = \frac{i}{2\pi}\frac{V^{1/3}}{n_x'-n_x} e^{-2i\pi(n_x'-n_x)x/V^{1/3}}\Big|_0^{V^{1/3}} = \frac{i}{2\pi}\frac{V^{1/3}}{n_x'-n_x}\left[e^{-2i\pi(n_x'-n_x)} - 1\right] = 0 .$$

On the other hand, if $k_x' = k_x$ the integral is simply $V^{1/3}$. We conclude that

$$\int d\mathbf{r}\, e^{-i(\mathbf{k}'-\mathbf{k})\cdot\mathbf{r}} = V\delta_{n_x,n_x'}\,\delta_{n_y,n_y'}\,\delta_{n_z,n_z'} = V\delta_{\mathbf{k},\mathbf{k}'} .$$

The eigenenergies are obtained as

$$H\psi_{\mathbf{k},m}(\mathbf{r}) = -\frac{\hbar^2}{2m}\boldsymbol{\nabla}^2\,\psi_{\mathbf{k},m}(\mathbf{r}) = E_{\mathbf{k}}\psi_{\mathbf{k},m}(\mathbf{r}) \qquad \text{with} \qquad E_{\mathbf{k}} = \frac{\hbar^2\,\mathbf{k}^2}{2m} = \frac{\hbar^2}{2m}\frac{4\pi^2}{V^{2/3}}\left(n_x^2 + n_y^2 + n_z^2\right),$$

and are quantized. The state of the particle is uniquely determined by its wave number $\mathbf{k}$ and spin projection m. The energy above is degenerate: the ground-state energy is 0 and is $(2S + 1)$-fold degenerate, the first-excited-state energy is $2\pi^2\hbar^2/(mL^2)$ and is $6(2S + 1)$-fold degenerate, and so on.

Part 2

In accordance with Pauli's exclusion principle, the ground state is the state in which the N lowest single-particle states are occupied. Let us denote by E_F the energy of the highest occupied single-particle level,

$$E_F = \frac{\hbar^2 k_F^2}{2m},$$

and E_F is the Fermi energy. Since each state with wave number $\mathbf{k}$ can accommodate $2S + 1$ fermions, the total number N of fermions must equate

$$N = (2S + 1) \sum_{n_x} \sum_{n_y} \sum_{n_z} \theta(n_F^2 - n_x^2 - n_y^2 - n_z^2), \qquad E_F = \frac{2\pi^2 \hbar^2}{m V^{2/3}} n_F^2,$$

where $\theta(u)$ is the step function, which is equal to 1 if $u \geq 0$ and 0 if $u < 0$. This condition can also be written as

$$N = (2S + 1) \sum_{\mathbf{k}} \theta(k_F - |\mathbf{k}|) = (2S + 1) \sum_{\mathbf{k} \leq k_F},$$

and the role of the step function is to enforce the constraint $|\mathbf{k}| \leq k_F$. The ground-state energy E_0 of the fermion gas then follows from

$$E_0 = (2S + 1) \sum_{\mathbf{k}} \frac{\hbar^2 \mathbf{k}^2}{2m} \theta(k_F - |\mathbf{k}|).$$

In the limit $V \longrightarrow \infty$ (the thermodynamic limit) the wave numbers $\mathbf{k}$ are densely distributed, and we can describe their distribution with a function $\rho(\mathbf{k})$. Since there is a single wave number in each cell centered at $\mathbf{k}$ and volume $(2\pi)^3/V$, the function $\rho(\mathbf{k})$ must satisfy the condition

$$\rho(\mathbf{k}) \frac{(2\pi)^3}{V} = 1 \implies \rho(\mathbf{k}) = \frac{V}{(2\pi)^3}.$$

As a result, we can convert a sum over $\mathbf{k}$ into an integral over $\rho(\mathbf{k}) \, d\mathbf{k}$ via

$$\sum_{\mathbf{k}} \longrightarrow \int d\mathbf{k} \, \rho(\mathbf{k}) = \frac{V}{(2\pi)^3} \int d\mathbf{k}.$$

In the large-volume limit, we then have

$$N = (2S + 1) \frac{V}{(2\pi)^3} \int d\mathbf{k} \, \theta(k_F - k), \qquad E_0 = (2S + 1) \frac{V}{(2\pi)^3} \int d\mathbf{k} \, \frac{\hbar^2 k^2}{2m} \theta(k_F - k).$$

These integrations can easily be carried out to obtain

$$N = (2S+1) \frac{V}{2\pi^2} \int_0^{k_F} dk \, k^2 = (2S+1) \frac{k_F^3}{6\pi^2} V, \quad E_0 = (2S+1) \frac{V}{2\pi^2} \int_0^{k_F} dk \, k^2 \frac{\hbar^2 k^2}{2m} = (2S+1) \frac{\hbar^2 k_F^5}{20\pi^2 m} V.$$

Note that as the volume of the box goes to infinity, both the number of particles and the ground-state energy E_0 become infinite, as should be expected since they are both extensive quantities. If we are going to use the Fermi gas model to approximate physical systems, we need to take the infinite-volume limit into account in such a way that some intensive quantity remains fixed. A reasonable quantity to fix is the number density

$$\frac{N}{V} = (2S + 1) \frac{k_F^3}{6\pi^2} \implies k_F = \left(\frac{6\pi^2}{2S+1} \frac{N}{V} \right)^{1/3}.$$

The ground-state energy per particle can now be rewritten as

$$\frac{E_0}{N} = \frac{3}{5} E_F \qquad \text{with} \qquad E_F = \frac{\hbar^2}{2m} \left(\frac{6\pi^2}{2S+1} \frac{N}{V} \right)^{2/3} ,$$

where E_F is the Fermi energy, defined earlier. Note that the energy per particle increases as the volume is reduced while keeping N fixed (that is, the density is increased). This means that the Fermi gas is exerting a pressure on the walls of the box, the so-called degeneracy pressure given by

$$P_0 = -\frac{\partial E_0}{\partial V} = \left(\frac{6\pi^2}{2S+1} \right)^{2/3} \frac{\hbar^2}{5m} \left(\frac{N}{V} \right)^{5/3} = \frac{2}{5} \frac{N}{V} E_F .$$

This result also follows from the thermodynamic relationship $P = -\partial F(V, T)/\partial V$, where $F(V, T)$ is the free energy and T is the temperature, by observing that at zero temperature the free energy reduces to the system's ground-state energy.

Problem 10   White Dwarf Stars

In stars like our sun, energy is generated by the "burning" of hydrogen into helium, namely by converting, through a network of reactions known as the *pp* cycle, four protons into a helium nucleus (^{4}He), a bound system of two protons and two neutrons (the binding is provided by the strong interaction).[5] White dwarfs are (small) stars that have exhausted their hydrogen supply, and consist predominantly of ^{4}He nuclei and electrons. Their brightness is due to the release of gravitational energy by slow contraction. This contraction is counteracted by the Pauli pressure due to the electron gas. Typical parameters of a white dwarf are

$$\text{density } \rho \approx 10^7 \text{ g/cm}^3 , \qquad \text{mass } M \approx 10^{33} \text{ g} , \qquad \text{central temperature } T \approx 10^7 \text{ K} .$$

The thermal energy $k_B T$ corresponding to this temperature (k_B is the Boltzmann constant) is $\approx 10^3$ eV, and therefore the helium atoms are ionized. The electron number density N/V can be estimated to be of the order 10^{30} cm^{-3} by noting that the mass M of the star is given by

$$M \approx Nm_e + \frac{N}{2} \left(2m_p + 2m_n \right) \approx 2Nm_p ,$$

where we have ignored the binding energy of the ^{4}He nucleus, and have taken its mass to be $2m_p + 2m_n$ with m_p and m_n the proton and neutron masses, respectively (in the last step we have also ignored the p–n mass difference). We make the further simplifying assumption of neglecting interactions among the electrons and between these and the ^{4}He nuclei, and regard the electrons as a Fermi gas of free spin-1/2 particles. We have for the Fermi energy (the energy of the highest occupied level)

$$E_F = \frac{\hbar^2}{2m_e} \left(3\pi^2 \frac{N}{V} \right)^{2/3} \approx 20 \text{ MeV} , \qquad T_F = \frac{E_F}{k_B} \approx 10^{11} \text{ K} .$$

Since the central temperature $T \ll T_F$, we can consider the Fermi gas in the limit of zero temperature. The zero-point pressure P_0 is equilibrated by the gravitational attraction, due almost entirely to the nuclei. The white dwarf model then consists of N electrons and $N/2$ motionless ^{4}He nuclei (the star is electrically neutral). It turns out there is a critical mass M_c such that for masses larger than M_c

---

[5] More precisely, the *pp* sequence of reactions coverts $4p + 2e^-$ into ^{4}He $+ 2\nu_e + \gamma$ and releases energy $E_\gamma = (4m_p + 2m_e - M)c^2 - 2\langle E_{\nu_e} \rangle \approx 26.7$ MeV, where M is the mass of the ^{4}He nucleus.

no white dwarf can exist, that is, the zero-point pressure from the electron gas is insufficient to counteract the gravitational attraction. The aim of the problem is to estimate M_c.

1. The calculation of the ground-state energy E_0 of the Fermi gas needs to be amended, since the Fermi energy is much larger than the rest mass energy of the electron. Calculate E_0, using the relativistic expression for the single-particle energy, namely $E_{\mathbf{k}} = c\sqrt{(\hbar \mathbf{k})^2 + (m_e c)^2}$ where c is the speed of light, and show that it is can be expressed as

$$E_0 = \frac{1}{\pi^2} \frac{m_e^4 c^5}{\hbar^3} V f(x_F), \qquad f(x_F) = \int_0^{x_F} dx\, x^2 \sqrt{1+x^2} \qquad \text{with} \qquad x_F = \frac{\hbar k_F}{m_e c}.$$

Obtain the zero-point pressure P_0 and show that, in the two limits $x_F \ll 1$ (the non-relativistic regime) and $x_F \gg 1$ (the ultra-relativistic regime, in which the electron rest mass can be neglected),

$$P_0 = \frac{1}{15\pi^2} \frac{m_e^4 c^5}{\hbar^3} x_F^5 \qquad\qquad x_F \ll 1$$

$$= \frac{1}{12\pi^2} \frac{m_e^4 c^5}{\hbar^3} \left(x_F^4 - x_F^2\right) \qquad x_F \gg 1.$$

2. Express the zero-point pressure in the two regimes above as a function of the physical parameters of the star, its mass M and radius R rather than as a function of the electron number density N/V (the star can be assumed to be spherical).

3. Assume that the gravitational self-energy of the star is $-\gamma GM^2/R$, where G is the gravitational constant and γ is a parameter of the order of unity. From the condition of equilibrium between the (outward) zero-point pressure and (inward) gravitational attraction, establish a relationship between the mass and radius of the star in the non-relativistic and ultra-relativistic regimes. Interpret the results and deduce the critical mass M_c.

Solution

Part 1

The ground-state energy is given by

$$E_0 = 2\sum_{\mathbf{k}} \theta(k_F - |\mathbf{k}|) E_{\mathbf{k}} = \frac{V}{\pi^2} \int_0^{k_F} dk\, k^2 \sqrt{(\hbar k c)^2 + (m_e c^2)^2},$$

and the factor 2 accounts for the spin degeneracy. We set $x = \hbar k/(m_e c)$ and find the expression for E_0 given in text of the problem. The zero-point pressure follows from

$$P_0 = -\frac{\partial E_0}{\partial V} = -\frac{1}{\pi^2} \frac{m_e^4 c^5}{\hbar^3} \left[f(x_F) + V \frac{\partial f(x_F)}{\partial x_F} \frac{\partial x_F}{\partial V} \right],$$

where

$$\frac{\partial f(x_F)}{\partial x_F} = x_F^2 \sqrt{1+x_F^2}, \qquad \frac{\partial x_F}{\partial V} = \frac{\hbar}{m_e c} \frac{\partial k_F}{\partial V} = -\frac{1}{3V} \frac{\hbar k_F}{m_e c},$$

and we have used $k_F = (3\pi^2 N/V)^{1/3}$. Collecting results, we arrive at the following expression for the zero-point pressure:

$$P_0 = \frac{1}{\pi^2} \frac{m_e^4 c^5}{\hbar^3} \left[-f(x_F) + \frac{x_F^3}{3} \sqrt{1+x_F^2} \right].$$

In the two regimes of interest, we have

$$x_F \ll 1: \quad f(x_F) \approx \int_0^{x_F} dx\, x^2 \left(1 + \frac{x^2}{2} + \cdots\right) = \frac{x_F^3}{3} + \frac{x_F^5}{10} + \cdots,$$

$$x_F \gg 1: \quad f(x_F) = \int_0^{x_F} dx\, x^3 \sqrt{1 + \frac{1}{x^2}} \approx \int_0^{x_F} dx\, x^3 \left(1 + \frac{1}{2x^2} + \cdots\right) = \frac{x_F^4}{4} + \frac{x_F^2}{4} + \cdots,$$

which, in combination with the expansion $\sqrt{1 + x_F^2} = 1 + x_F^2/2 + \cdots$, lead to the expressions for P_0 given in the text.

Part 2

We have

$$N = \frac{M}{2m_p}, \qquad \frac{V}{N} = \frac{4\pi R^3}{3}\frac{2m_p}{M} = \frac{8\pi}{3}\frac{m_p}{M} R^3,$$

and hence

$$x_F = \frac{\hbar k_F}{m_e c} = \frac{\hbar/(m_e c)}{R}\left(\frac{9\pi}{8}\frac{M}{m_p}\right)^{1/3},$$

where $\hbar/(m_e c)$ is the electron Compton wavelength. It is convenient to define the non-dimensional variables

$$\overline{R} = \frac{R}{\hbar/(m_e c)}, \qquad \overline{M} = \frac{9\pi}{8}\frac{M}{m_p},$$

in terms of which the zero-point pressure reads

$$P_0(\overline{R}) = \frac{4}{5}\lambda\frac{\overline{M}^{5/3}}{\overline{R}^5} \qquad\qquad x_F \ll 1$$

$$= \lambda\left(\frac{\overline{M}^{4/3}}{\overline{R}^4} - \frac{\overline{M}^{2/3}}{\overline{R}^2}\right) \qquad x_F \gg 1,$$

where the constant λ is given by

$$\lambda = \frac{1}{12\pi^2}\frac{(m_e c^2)^4}{(\hbar c)^3}.$$

It can be seen that the behavior of $P_0(\overline{R})$ is quite different in the two regimes.

Part 3

The equilibrium condition is given by

$$4\pi R^2 P_0(R) = \gamma\frac{GM^2}{R^2},$$

which can be expressed in terms of the non-dimensional quantities $\overline{M}$ and $\overline{R}$ defined earlier as

$$P_0(\overline{R}) = \lambda'\frac{\overline{M}^2}{\overline{R}^4}, \qquad \lambda' = \gamma\frac{G}{4\pi}\left(\frac{8m_p}{9\pi}\right)^2\left(\frac{m_e c}{\hbar}\right)^4.$$

The regimes $x_F \ll 1$ and $x_F \gg 1$ correspond, respectively, to low mass (or large radius) and high mass (or small radius). The equilibrium conditions in these regimes read

$$x_F \ll 1: \qquad \frac{4}{5} \lambda \frac{\overline{M}^{5/3}}{\overline{R}^5} = \lambda' \frac{\overline{M}^2}{\overline{R}^4} \implies \overline{R} = \frac{4}{5} \frac{\lambda}{\lambda'} \overline{M}^{-1/3},$$

$$x_F \gg 1: \qquad \lambda \left(\frac{\overline{M}^{4/3}}{\overline{R}^4} - \frac{\overline{M}^{2/3}}{\overline{R}^2} \right) = \lambda' \frac{\overline{M}^2}{\overline{R}^4} \implies \overline{R} = \overline{M}^{1/3} \left[1 - \left(\frac{\overline{M}}{\overline{M}_c} \right)^{2/3} \right]^{1/2},$$

where we have defined

$$\overline{M}_c = (\lambda/\lambda')^{3/2} .$$

A plot of $\overline{R}$ as a function of $\overline{M}$ shows that for $x_F \ll 1$ (low $\overline{M}$) the radius diverges as $\overline{M}^{-1/3}$, while for $x_F \gg 1$ it vanishes when $\overline{M}$ equals $\overline{M}_c$, i.e., the critical mass. In physical units the latter is given by

$$M_c = \frac{8}{9\pi} m_p \left(\frac{\lambda}{\lambda'} \right)^{3/2} = \frac{8}{9\pi} m_p \left(\frac{27\pi}{64\gamma} \right)^{3/2} \left(\frac{\hbar c}{G m_p^2} \right)^{3/2} \approx M_\odot ,$$

where $M_\odot$ is the mass of the sun. A more accurate calculation leads to $M_c \approx 1.4 M_\odot$, known as the Chandrasekhar limit.

Problem 11 The Thomas–Fermi Approximation for Many-Electron Atoms

Consider a many-electron atom and denote by $n(r)$ the *local* electron number density at r (spherical symmetry is assumed throughout). We assume that the "electron gas" is in hydrostatic equilibrium, so that

$$\nabla P(r) = -en(r)\, \mathbf{E}(r\hat{\mathbf{r}}) ,$$

where $P(r)$ is the pressure, $-en(r)$ is the electron charge density (here $e > 0$), and $\mathbf{E}(r\hat{\mathbf{r}})$ is the radially directed electric field. This field includes the contribution $(Ze/r^2)\,\hat{\mathbf{r}}$ due to the nucleus as well as the contribution due to the electrons.

1. Use the Fermi gas model to express the local pressure $P(r)$ as function of the electron number density $n(r)$ (the so-called local density approximation) and obtain a relationship between the electrostatic potential $\phi(r)$, recalling that $\mathbf{E} = -\nabla\phi$, and $n(r)$, where $n(r \longrightarrow \infty) = 0$. Use this relationship to derive the following non-linear equation satisfied by $\phi(r)$,

$$\nabla^2 \phi(r) = \kappa\, \phi^{3/2}(r) ,$$

where κ is a constant to be determined in terms of e, the Bohr radius $a_0 = \hbar^2/(m\,e^2)$, and a numerical coefficient. Note that in the limits $r \longrightarrow 0$ and $r \longrightarrow \infty$, the potential satisfies the boundary conditions $\phi(r \longrightarrow 0) = Ze/r$ and $\phi(r \longrightarrow \infty) = 0$, where Ze is the nuclear charge.

2. Introduce the non-dimensional variable $\rho = r/a_0$ and show that the partial differential equation above reduces to

$$\frac{d^2}{d\rho^2} \chi(\rho) = \sqrt{\frac{Z}{\rho}} \left[\frac{\chi(\rho)}{c} \right]^{3/2} \quad \text{with} \quad c = \frac{1}{2} \left(\frac{3\pi}{4} \right)^{2/3} ,$$

where $\chi(\rho) = (\rho/Z)\, \phi(\rho)/(e/a_0)$. Lastly, define the variable $x = (Z^{1/3}/c)\rho$, and obtain the Thomas–Fermi equation for $\chi(x)$. What boundary conditions must $\chi(x)$ satisfy at $x = 0$ and $x = \infty$? Express the electrostatic potential $\phi(r)$ and electron number density $n(r)$ in terms of x and $\chi(x)$.

3. Evaluate the kinetic energy of the atomic electrons by assuming that the average kinetic energy per electron is given by the Fermi-gas result $(3/5)E_F(r)$, where $E_F(r)$ is the Fermi energy in the local density approximation. Evaluate the potential energy of the atomic electrons by observing that it consists of an attractive contribution (electron–nucleus interactions) and a repulsive contribution (electron–electron interactions). Obtain the total energy of the atom and show that it scales as $Z^{7/3}$.

Solution

Part 1

In the local-density approximation we have (m is the electron mass)

$$P(r) = \frac{\hbar^2}{5m} (3\pi^2)^{2/3} \, n^{5/3}(r) \,,$$

which in combination with the condition for hydrostatic equilibrium yields

$$\frac{\partial}{\partial r} P(r) = en(r) \frac{\partial}{\partial r} \phi(r) \implies \frac{\partial}{\partial r} \phi(r) = \frac{\hbar^2}{3me} (3\pi^2)^{2/3} \frac{1}{n^{1/3}(r)} \frac{\partial}{\partial r} n(r) = \frac{\hbar^2}{2me} \frac{\partial}{\partial r} [3\pi^2 n(r)]^{2/3} \,.$$

We obtain

$$\phi(r) = \frac{\hbar^2}{2me} [3\pi^2 n(r)]^{2/3} + \text{constant} = \frac{ea_0}{2} [3\pi^2 n(r)]^{2/3} + \text{constant} \implies n(r) = \frac{1}{3\pi^2} \left[\frac{2}{ea_0} \phi(r) \right]^{3/2} \,,$$

where we have introduced the Bohr radius a_0. Note that the constant must be zero, since $\phi(r)$ vanishes at infinity. The Poisson equation states that $\nabla^2 \phi(r) = 4\pi en(r)$, and this equation can be converted into one involving only $\phi(r)$ by replacing $n(r)$ as indicated above,

$$\nabla^2 \phi(r) = \kappa \phi^{3/2}(r) \,, \qquad \kappa = \frac{8}{3\pi} \sqrt{\frac{2}{ea_0^3}} \,.$$

Part 2

Expressing the Laplacian in spherical coordinates, the partial differential equation for $\phi(r)$ reduces to

$$\frac{1}{r} \frac{d^2}{dr^2} r \, \phi(r) = \kappa \phi^{3/2}(r) \,,$$

which, when rewritten in terms of the non-dimensional variable $\rho = r/a_0$ and non-dimensional potential $\overline{\phi}(r) = \phi(r)/(e/a_0)$, reads

$$\frac{1}{\rho} \frac{d^2}{d\rho^2} \rho \left[\frac{e}{a_0} \overline{\phi}(\rho) \right] = \frac{8}{3\pi} \sqrt{\frac{2a_0}{e}} \left[\frac{e}{a_0} \overline{\phi}(\rho) \right]^{3/2} \implies \frac{1}{\rho} \frac{d^2}{d\rho^2} \rho \, \overline{\phi}(\rho) = \frac{8\sqrt{2}}{3\pi} \overline{\phi}^{3/2}(\rho) \,,$$

or

$$\frac{1}{\rho} \frac{d^2}{d\rho^2} \rho \, \overline{\phi}(\rho) = \left[\frac{\overline{\phi}(\rho)}{c} \right]^{3/2} \quad \text{with} \quad c = \frac{1}{2} \left(\frac{3\pi}{4} \right)^{2/3} \,.$$

The boundary conditions on $\overline{\phi}(\rho)$ follow from

$$r\phi(r \longrightarrow 0) = a_0 \rho \left[\frac{e}{a_0} \overline{\phi}(\rho \longrightarrow 0) \right] = Ze \implies \rho\overline{\phi}(\rho \longrightarrow 0) = Z \,, \quad \phi(r \longrightarrow \infty) = 0 \implies \overline{\phi}(\rho \longrightarrow \infty) = 0 \,.$$

The equation for $\overline{\phi}(\rho)$ can be further simplified by defining the auxiliary function $\chi(\rho) = (\rho/Z)\,\overline{\phi}(\rho)$, such that $\chi(\rho \longrightarrow 0) = 1$; we find

$$\frac{1}{\rho}\frac{d^2}{d\rho^2}\left[\rho\,\frac{Z\chi(\rho)}{\rho}\right] = \left[\frac{1}{c}\frac{Z\chi(\rho)}{\rho}\right]^{3/2} \implies \frac{d^2}{d\rho^2}\chi(\rho) = \sqrt{\frac{Z}{\rho}}\left[\frac{\chi(\rho)}{c}\right]^{3/2},$$

and rescaling ρ as

$$\rho = \frac{c}{Z^{1/3}}x, \qquad \frac{d}{d\rho} = \frac{dx}{d\rho}\frac{d}{dx} = \frac{Z^{1/3}}{c}\frac{d}{dx},$$

we finally arrive at the Thomas–Fermi equation,

$$\frac{Z^{2/3}}{c^2}\frac{d^2}{dx^2}\chi(x) = \sqrt{\frac{Z^{4/3}}{c}\frac{1}{x}}\left[\frac{\chi(x)}{c}\right]^{3/2} \implies \sqrt{x}\,\frac{d^2}{dx^2}\chi(x) = \chi^{3/2}(x),$$

supplemented by the boundary conditions $\chi(0) = 1$ and $\chi(x \longrightarrow \infty) = 0$. The electrostatic potential and electron number density in terms of the above quantities are given by

$$\phi(r) = \frac{e}{a_0}\overline{\phi}(\rho) = \frac{Ze}{a_0}\frac{\chi(\rho)}{\rho} = \frac{Z^{4/3}}{c}\frac{e}{a_0}\frac{\chi(x)}{x}, \qquad n(r) = \frac{1}{3\pi^2}\frac{2\sqrt{2}}{c^{3/2}}\frac{Z^2}{a_0^3}\left[\frac{\chi(x)}{x}\right]^{3/2} = \frac{1}{4\pi}\frac{Z^2}{(ca_0)^3}\left[\frac{\chi(x)}{x}\right]^{3/2},$$

where

$$x = \frac{Z^{1/3}}{c}\rho = \frac{Z^{1/3}}{c}\frac{r}{a_0}.$$

Part 3

The kinetic energy per electron is $(3/5)E_F(r)$, where $E_F(r)$ is the Fermi energy, which in the local-density approximation is given by

$$E_F(r) = \frac{\hbar^2}{2m}\left[3\pi^2 n(r)\right]^{2/3}.$$

The total kinetic energy of the electrons in the atom then follows:

$$T = \frac{3}{5}\frac{\hbar^2}{2m}4\pi\int_0^\infty dr\,r^2 n(r)\left[3\pi^2 n(r)\right]^{2/3},$$

where $n(r)\,4\pi r^2\,dr$ is the number of electrons in a shell centered at r and of thickness dr. In terms of the variable x and the Thomas–Fermi function $\chi(x)$, we find

$$4\pi n(r)r^2\,dr = 4\pi\frac{1}{4\pi}\frac{Z^2}{(ca_0)^3}\left[\frac{\chi(x)}{x}\right]^{3/2}\frac{(ca_0)^3}{Z}x^2\,dx = Z\sqrt{x}\chi^{3/2}(x)\,dx,$$

and hence

$$T = \frac{3}{5}\frac{\hbar^2}{2m}\int_0^\infty dx\,Z\sqrt{x}\chi^{3/2}(x)\left\{3\pi^2\frac{1}{4\pi}\frac{Z^2}{(ca_0)^3}\left[\frac{\chi(x)}{x}\right]^{3/2}\right\}^{2/3} = Z^{7/3}\frac{e^2}{2a_0}\left[\frac{6}{5c}\int_0^\infty dx\,\frac{1}{\sqrt{x}}\chi^{5/2}(x)\right],$$

where $e^2/(2a_0)$ is the binding energy in hydrogen.

There are two terms contributing to the total potential energy: the attraction between the nucleus and electrons (labeled V_{eN}), and the repulsion among the electrons (labeled V_{ee}). The former is simply given by

$$V_{eN} = 4\pi \int_0^\infty dr\, r^2 n(r)\left(-\frac{Ze^2}{r}\right) = -\int_0^\infty dx\, Z\sqrt{x}\chi^{3/2}(x)\frac{Z^{4/3}}{c a_0}\frac{e^2}{x}\frac{1}{x} = -Z^{7/3}\frac{e^2}{2a_0}\left[\frac{2}{c}\int_0^\infty dx\frac{1}{\sqrt{x}}\chi^{3/2}(x)\right].$$

The latter results from[6]

$$V_{ee} = \frac{1}{2}\int_0^\infty 4\pi\, dr\, r^2 n(r)(-e)\left[\phi(r) - \frac{Ze}{r}\right] = -\frac{1}{2}\int_0^\infty 4\pi\, dr\, r^2 n(r)\, e\,\phi(r) - \frac{1}{2}V_{eN},$$

where $-en(r)4\pi r^2$ is the electron charge density, and the contribution Ze/r due to the nucleus, included in $\phi(r)$, needs to be removed from it in order to obtain the contribution due only to the electrons. We find

$$V_{ee} + \frac{1}{2}V_{eN} = -\frac{1}{2}\int_0^\infty dx\, Z\sqrt{x}\chi^{3/2}(x)\frac{Z^{4/3}}{c}\frac{e^2}{a_0}\frac{\chi(x)}{x} = -Z^{7/3}\frac{e^2}{2a_0}\left[\frac{1}{c}\int_0^\infty dx\frac{1}{\sqrt{x}}\chi^{5/2}(x)\right].$$

The total energy of the atom is then given by

$$E = T + V_{eN} + V_{ee} = -\frac{Z^{7/3}}{c}\frac{e^2}{2a_0}\int_0^\infty dx\frac{1}{\sqrt{x}}\left[\chi^{3/2}(x) - \frac{1}{5}\chi^{5/2}(x)\right] = -Z^{7/3}E_{TF},$$

and scales as $Z^{7/3}$. The constant E_{TF} is independent of the atom and has the value 20.9 eV. The Thomas–Fermi approximation predicts ground-state energies for high-Z atoms that are reasonably close to the experimental values; for example, in lead $_{82}$Pb the experimental energy/$Z^{7/3}$ is approximately -19.0 eV as compared with the -20.9 eV of the Thomas–Fermi approximation.

---

[6] The electrostatic potential energy is generally given by

$$\frac{1}{2}\int d\mathbf{r}\,\rho(\mathbf{r})\,\phi(\mathbf{r}),$$

where $\rho(\mathbf{r})$ is the charge density and $\phi(\mathbf{r})$ is the scalar potential induced by this charge distribution.

Bibliography

The following is a list of, primarily, textbooks in quantum mechanics at the upper-undergraduate and graduate level, a few of which are cited in the present book. The list also includes textbooks in mathematical methods, classical mechanics, and classical electrodynamics, as well as a few advanced texts covering more specialized topics. The list is far from being exhaustive, but students and instructors may find it useful for a more in-depth treatment of the various topics presented in this book and for additional problems.

Abramowitz, M., and I. A. Stegun (1965). *Handbook of Mathematical Functions*. Dover.

Arfken, G. B., H. J. Weber, and F. E. Harris (2013). *Mathematical Methods for Physicists*. Academic Press.

Baym, G. (1969). *Lectures on Quantum Mechanics*. Benjamin/Cummings.

Bethe, H. A., and R. W. Jackiw (1968). *Intermediate Quantum Mechanics*. Benjamin.

Blatt, J. M., and V. W. Weisskopf (1952). *Theoretical Nuclear Physics*. John Wiley & Sons.

Cohen-Tannoudji, C., B. Diu, and F. Laloë (1997). *Quantum Mechanics*, vols. I and II. Wiley.

Dirac, P. A. M. (1958). *The Principles of Quantum Mechanics*. Oxford University Press.

Edmonds, A. R. (1974). *Angular Momentum in Quantum Mechanics*. Princeton University Press.

Fetter, A. L., and J. D. Walecka (1971). *Quantum Theory of Many-Particle Systems*. McGraw-Hill.

Fetter, A. L., and J. D. Walecka (1980). *Theoretical Mechanics of Particles and Continua*. McGraw-Hill.

Goldberger, M. L., and K. M. Watson (1964). *Collision Theory*. Wiley.

Goldstein, H., C. Poole, and J. Safko (2002). *Classical Mechanics*. Addison-Wesley.

Gottfried, K., and T.-G. Yan (2004). *Quantum Mechanics: Fundamentals*. Springer-Verlag.

Griffiths, D. J. (2005). *Introduction to Quantum Mechanics*. Pearson.

Jackson, J. D. (1998). *Classical Electrodynamics*. Wiley.

Landau, L. D., and E. M. Lifshitz (1965). *Quantum Mechanics, Nonrelativistic Theory*. Pergamon Press.

Lipkin, H. J. (1973). *Quantum Mechanics, New Approaches to Selected Topics*. North Holland.

Merzbacher, E. (1970) *Quantum Mechanics*. Wiley.

Messiah, A. (1961). *Quantum Mechanics*. vols. I and II. North Holland.

Morse, P. M., and H. Feshbach (1953). *Methods of Theoretical Physics*. McGraw-Hill.

Sakurai, J. J., and J. Napolitano (2020). *Modern Quantum Mechanics*. Cambridge University Press.

Schiff, L. (1968). *Quantum Mechanics*. McGraw-Hill.

Walecka, J. D. (2013). *Topics in Modern Physics*. World Scientific.

Weinberg, S. (2012). *Lectures on Quantum Mechanics*. Cambridge University Press.

Index